Nondestructive Assay of Nuclear Materials
for Safeguards and Security

William H. Geist • Peter A. Santi
Martyn T. Swinhoe
Editors

Nondestructive Assay of Nuclear Materials for Safeguards and Security

Second Edition

Editors
William H. Geist
Los Alamos National Laboratory
Los Alamos, NM, USA

Peter A. Santi
Los Alamos National Laboratory
Los Alamos, NM, USA

Martyn T. Swinhoe
Los Alamos National Laboratory
Los Alamos, NM, USA

ISBN 978-3-031-58276-9   ISBN 978-3-031-58277-6   (eBook)
https://doi.org/10.1007/978-3-031-58277-6

This work was supported by Los Alamos National Laboratory

© The Editor(s) (if applicable) and The Author(s) 2024, Corrected Publication 2024. This book is an open access publication.
1st edition: © United States Regulatory Commission 1991
Open Access This book is licensed under the terms of the Creative Commons Attribution 4.0 International License (http://creativecommons.org/licenses/by/4.0/), which permits use, sharing, adaptation, distribution and reproduction in any medium or format, as long as you give appropriate credit to the original author(s) and the source, provide a link to the Creative Commons license and indicate if changes were made.
The images or other third party material in this book are included in the book's Creative Commons license, unless indicated otherwise in a credit line to the material. If material is not included in the book's Creative Commons license and your intended use is not permitted by statutory regulation or exceeds the permitted use, you will need to obtain permission directly from the copyright holder.
The use of general descriptive names, registered names, trademarks, service marks, etc. in this publication does not imply, even in the absence of a specific statement, that such names are exempt from the relevant protective laws and regulations and therefore free for general use.
The publisher, the authors and the editors are safe to assume that the advice and information in this book are believed to be true and accurate at the date of publication. Neither the publisher nor the authors or the editors give a warranty, expressed or implied, with respect to the material contained herein or for any errors or omissions that may have been made. The publisher remains neutral with regard to jurisdictional claims in published maps and institutional affiliations.

This Springer imprint is published by the registered company Springer Nature Switzerland AG
The registered company address is: Gewerbestrasse 11, 6330 Cham, Switzerland

If disposing of this product, please recycle the paper.

The editors would like to dedicate this version of the PANDA book to Howard Menlove of Los Alamos National Laboratory, who has made an outstanding contribution to the field of nondestructive assay. Howard pioneered most of the neutron methods and instruments described in this book. He has also passed on his knowledge to several generations of NDA practitioners, including many of the authors who contributed to this work.

Preface

This book is a general reference on the theory and application of nondestructive assay (NDA) techniques applied to the measurement of nuclear material. The intent of this book is to serve as an introduction for newcomers to NDA as well as a reference for experienced NDA practitioners.

This book is a revision to and an extension of *Passive Nondestructive Assay of Nuclear Materials* (PANDA) published in 1991, which is available both online and in print versions. This updated version also includes content and revisions from the 2007 PANDA Addendum, which is available only online. Although the basic physics has not changed, scientists have made many advances in analysis methods, instrumentation, and applications during the last 30 years. In this new version, we have updated basic descriptions of the origin and interactions of radiation and included some newer references. Extensive revisions include the description of gamma detection methods, attenuation correction procedures, and analysis methods, including for the measurement of uranium enrichment and the determination of plutonium isotopic composition. We have also revised and broadened the information regarding neutron detectors and the explanation of neutron coincidence techniques. Chapter 18, "Principles of Neutron Multiplicity," is a new addition to this version. To remove obsolete systems and to include many current applications, we have completely overhauled the information regarding the application of gamma and neutron techniques, and we have updated the values of and references to nuclear data.

Like the first version of PANDA, most of the material was prepared by staff of Los Alamos National Laboratory. The book also benefitted from contributions from other national laboratories and the International Atomic Energy Agency. You can find the complete list of contributors on Page xvii.

The production of this new version of PANDA has taken a large amount of effort by many people over several years. This work has been supported by multiple organizations and programs detailed in the acknowledgments that follow.

Los Alamos, NM, USA
August 2024

William H. Geist
Peter A. Santi
Martyn T. Swinhoe

Acknowledgments

The editors would like to express their appreciation to the US Department of Energy National Nuclear Security Administration Office of International Nuclear Security and DOE National Nuclear Security Administration Office of International Nuclear Safeguards for supporting the update to the PANDA book.

The authors would like to thank Tamara Hawman, Lead Production Editor, and Jen Patureau, both of Los Alamos National Laboratory, for technical editing and formatting support, and Eva Ciabattoni from the International Atomic Energy Agency (IAEA) who provided technical editing of all the IAEA contributions. They would also extend their thanks to Allen Hopkins, Graphic Designer, for his detailed expertise on the many diagrams, and to Elektra Caffrey and Joe Longo who provided support to create the plots in consistent format.

Disclaimer

Los Alamos National Laboratory strongly supports academic freedom and a researcher's right to publish; however, as an institution, the Laboratory does not endorse the viewpoint of a publication or guarantee its technical correctness. The submitted manuscript has been authored by an employee or employees of Triad National Security, LLC, operator of Los Alamos National Laboratory under Contract No. 89233218CNA000001 with the U.-S. Department of Energy. Accordingly, the U.S. Government retains an irrevocable, nonexclusive, royalty-free license to publish, translate, reproduce, use, or dispose of the published form of the work and to authorize others to do the same for U.S. Government purposes.

Contents

1	The Role of Nondestructive Assay in Safeguards, Security, and Safety . . . . . . W. H. Geist and J. Conner	1
2	The Origin of Gamma Rays . . . . J. Stinnett, P. J. Karpius, D. J. Mercer, T. D. Reilly, and D. T. Vo	7
3	Gamma-Ray Interactions with Matter . . . . P. J. Karpius and T. D. Reilly	27
4	Gamma-Ray Detectors . . . . M. Lombardi, M. H. Carpenter, D. J. Mercer, P. A. Russo, and D. T. Vo	43
5	Instrumentation for Gamma-Ray Spectroscopy . . . . D. T. Vo and J. L. Parker	59
6	Attenuation Correction Procedures . . . . D. J. Mercer, P. J. Karpius, J. L. Parker, P. A. Santi, and K. Schmidt	81
7	General Topics in Passive Gamma-Ray Assay . . . . P. J. Karpius and J. L. Parker	101
8	The Measurement of Uranium Enrichment . . . . D. T. Vo	149
9	Plutonium Isotopic Composition by Gamma-Ray Spectroscopy . . . . D. T. Vo, K. Koehler, and T. E. Sampson	169
10	Gamma-Ray NDA Applications & Techniques . . . . P. A. Santi, R. Hunneke, B. Jennings, P. J. Karpius, A. L. Lousteau, D. J. Mercer, T. D. Reilly, P. A. Russo, S. Smith, R. Venkataraman, D. T. Vo, and J. Younkin	185
11	Densitometry . . . . R. Venkataraman, R. McElroy, P. A. Russo, and P. A. Santi	237
12	X-Ray Fluorescence . . . . M. H. Carpenter, P. J. Karpius, M. C. Miller, and S. J. Pattinson	271
13	The Origin of Neutron Radiation . . . . M. T. Swinhoe and N. Ensslin	289
14	Neutron Interactions with Matter . . . . M. T. Swinhoe, J. D. Hutchinson, and P. M. Rinard	307
15	Neutron Detectors . . . . D. C. Henzlova, M. P. Baker, K. Bartlett, A. Favalli, M. Iliev, M. A. Root, S. Sarnoski, T. Shin, and M. T. Swinhoe	325

16	**Principles of Singles Neutron Counting**	359
	J. D. Hutchinson, K. Amundson, H. Kistle, J. Moussa, T. Shin, J. E. Stewart, and R. K. Weinmann-Smith	
17	**Principles of Neutron Coincidence Counting**	389
	M. T. Swinhoe, N. Ensslin, J. D. Hutchinson, M. Iliev, and K. Koehler	
18	**Principles of Neutron Multiplicity**	439
	D. C. Henzlova, N. Ensslin, A. Favalli, W. H. Geist, L. Holzleitner, K. Koehler, M. S. Krick, M. M. Pickrell, T. D. Reilly, J. E. Stewart, and K. D. Veal	
19	**Passive Neutron Instrumentation and Applications**	483
	R. K. Weinmann-Smith, T. J. Aucott, A. P. Belian, D. P. Broughton, M. Frankl, P. A. Hausladen, D. C. Henzlova, J. D. Hutchinson, R. D. McElroy, H. O. Menlove, T. P. Pochet, L. A. Refalo, M. A. Root, M. Nelson, J. K. Sprinkle, M. T. Swinhoe, and M. M. Watson	
20	**Active Neutron Instrumentation and Applications**	585
	M. T. Swinhoe, N. Ensslin, L. G. Evans, W. H. Geist, M. S. Krick, A. L. Lousteau, R. McElroy, M. M. Pickrell, and P. Rinard	
21	**Spent Fuel Measurements**	605
	A. C. Kaplan-Trahan, A. P. Belian, M. Croce, D. C. Henzlova, P. Jansson, G. Long, G. E. McMath, J. R. Phillips, E. Rapisarda, M. A. Root, and H. R. Trellue	
22	**Perimeter Radiation Monitors**	639
	M. G. Paff and J. Toevs	
23	**Principles of Calorimetric Assay**	657
	M. P. Croce, D. S. Bracken, R. N. Likes, C. R. Rudy, and P. A. Santi	
24	**Calorimetric Assay Instruments**	677
	M. P. Croce, D. S. Bracken, X. Brochard, R. N. Likes, C. R. Rudy, and P. A. Santi	
25	**Basics of Uncertainty**	693
	J. Stinnett, P. A. Santi, and M. T. Swinhoe	
26	**Nuclear Material Accounting and Control Measurements**	709
	G. E. McMath, A. L. Lousteau, and S. Smith	
27	**Nuclear Data for Nondestructive Assay**	719
	P. A. Santi and T. D. Reilly	

Correction to: Active Neutron Instrumentation and Applications ... C1
M. T. Swinhoe, N. Ensslin, L. G. Evans, W. H. Geist, M. S. Krick, A. L. Lousteau, R. McElroy, M. M. Pickrell, and P. Rinard

Index ... 731

Original Authors

The revised PANDA book was created using existing material from the original authors of PANDA and the 2007 Addendum.

M. P. Baker
D. S. Bracken
T. W. Crane
N. Ensslin
P. E. Fehlau
W. H. Geist
S. Hansen
M. Krick
R. N. Likes
M. C. Lucas
H. O. Menlove
M. C. Miller
G. W. Nelson
J. L. Parker
M. Pickrell
J. R. Phillips
T. D. Reilly
P. M. Rinard
C. Rudy
P. A. Russo
T. E. Sampson
H. A. Smith, Jr.
J. K. Sprinkle
J. E. Stewart
K. D. Veal
D. Vo

Contributing Authors

In addition to the authors of the original PANDA, the following people contributed to the revision, addition, and review of the current version of this book.

Los Alamos National Laboratory

K. Amundson
K. Bartlett
A. Belian
D. P. Broughton
T. L. Burr
M. H. Carpenter
J. Conner
M. P. Croce
D. Dale
J. S. Davydov
A. Favalli
W. H. Geist
O. Gillispie
D. C. Henzlova
J. D. Hutchinson
M. Illiev
B. Jennings
P. J. Karpius
H. Kistle
K. Koehler
S. P. LaMont
M. Lombardi
G. Long
A. Martin
G. E. McMath
P. Mendoza
D. J. Mercer
J. Moussa
G. Orlicz
M. G. Paff
S. J. Pattinson
M. A. Root
M. L. Ruch
P. A. Santi

R. E. Steiner
T. Shin
M. K. Smith
J. Stinnett
M. T. Swinhoe
J. Toevs
A. C. Trahan
H. R. Trellue
D. T. Vo
M. Watson
R. K. Weinman-Smith

Oak Ridge National Laboratory

L. G. Evans
P. Gibbs
P. Hausleden
R. Hunneke
A. L. Lousteau
R. McElroy
K. Schmitt
S. Smith
R. Venkataraman
J. Younkin

Sandia National Laboratories

M. W. Enghauser
A. A. Solodov

JRC-Karlsruhe

L. Holzleitner

Stockholm University

P. Jansson

Savannah River National Laboratory

T. Aucott
S. Branney
M. Howard

International Atomic Energy Agency

A. Belian
X. Brochard
M. Frankl
F. Mingrone
T. Pochet
E. Rapisarda
L. ReFalo

The Role of Nondestructive Assay in Safeguards, Security, and Safety

W. H. Geist and J. Conner

1.1 Introduction

The term *nondestructive assay* (NDA) refers to measurement techniques "applied to nuclear material and other items of safeguards interest to confirm their isotopic composition and quantity without destroying the items" [1]. NDA techniques are generally categorized as either "passive," measuring radiation that is spontaneously emitted from nuclear or other radioactive material, or "active," measuring induced emissions of radiation. Another method for characterizing material involves sampling material and analyzing the sample with destructive chemical procedures, known as destructive analysis (DA). Although NDA is usually less accurate than DA, NDA obviates the need for sampling, reduces operator exposure to radiation, is less expensive, and is much faster than DA.

Many factors are considered when selecting which measurement method to use. These factors include the desired measurement precision, the cost, the timeliness, and the form of the material to be characterized. NDA methods can be used to characterize materials in different ways, such as identifying if any radioactive material is present and, if so, determining the radionuclides present, quantifying the isotopic composition of elemental radionuclides, and determining the mass of the radioactive material contained in an item. For the nuclear fuel cycle, the nuclear materials of greatest interest are uranium and plutonium and, to a lesser degree, thorium and neptunium.

The original impetus for developing NDA techniques was primarily to support the application of International Atomic Energy Agency (IAEA) safeguards for nuclear material as a key component of international efforts to ensure the nonproliferation of nuclear weapons. As safeguards measures became widely applied to nuclear material at nuclear facilities around the world, the need emerged for the IAEA to deploy rapid measurement methods that would not alter the state of the nuclear material and that would cause minimal disruption to nuclear facility operations. The United States (U.S.) Nuclear Regulatory Commission and the U.S. Department of Energy initially supported the IAEA in addressing this need, and numerous other national and international organizations have also subsequently made significant contributions. NDA techniques are now an essential part of the toolkit used by IAEA inspectors to verify the inventories of nuclear material held worldwide.

Although most NDA techniques were developed first in support of safeguards applications, it became clear over time that NDA techniques could also be applied to nuclear material safety and security. In some cases, a single NDA measurement can be used for safeguards, security, and safety applications; typically, the measurement approach is tailored to meet a specific application that builds on the underlying fundamental measurement physics.

Los Alamos National Laboratory strongly supports academic freedom and a researcher's right to publish; as an institution, however, the Laboratory does not endorse the viewpoint of a publication or guarantee its technical correctness.

W. H. Geist (✉) · J. Conner
Los Alamos National Laboratory, Los Alamos, NM, USA
e-mail: wgeist@lanl.gov; jconner@lanl.gov

1.2 IAEA Safeguards

The Treaty on the Non-Proliferation of Nuclear Weapons (NPT; [2]), which requires non-nuclear-weapon States to accept IAEA safeguards measures on all nuclear material and facilities, is one of the most widely adopted, legally binding international treaties. At the core of its mission, the IAEA "verifies through its inspection system that States comply with their commitments, under the Non-Proliferation Treaty and other non-proliferation agreements, to use nuclear material and facilities only for peaceful purposes." Safeguards [3] are measures administered by the IAEA to verify that States do not divert nuclear material from peaceful uses to develop nuclear weapons or other nuclear explosive devices. The accounting of nuclear material is one of the primary methods that the IAEA uses to detect a diversion of material, and both NDA and DA methods are used to quantify nuclear material for accountancy. NDA techniques are used by facility operators and national regulatory authorities to demonstrate peaceful uses of nuclear material, and they are used by the IAEA and other international organizations to verify declarations of nuclear material inventories and derive safeguards conclusions.

1.2.1 Use of NDA for Safeguards at the Facility Level

To comply with NPT obligations and facilitate reporting to and inspections by the IAEA, the State's safeguards regulatory authority (SRA) will require each nuclear facility to have a system [4] in place to account for the nuclear material processed and stored at the facility. For facilities that process nuclear material, most of the nuclear material accountancy is based on DA measurements, with NDA measurements applied to those material types not suitable for DA methods.

At the end of a specified period of time, the facility will reconcile the nuclear material inventory to ensure that no material is missing (see Chap. 26). The inventory is reconciled by taking the beginning inventory, adding any material receipts, and subtracting the final inventory and any shipments out of the facility during the period. The resulting value is called the *inventory difference* (ID). The ID is also called *material unaccounted for* (MUF). Calculating the ID is straight-forward for nuclear facilities that do not process material but use only discrete items (such as fuel assemblies used and stored in nuclear power plants), and the ID should be zero. For bulk material processing facilities, such as reprocessing or fuel fabrication plants, the ID is typically non-zero because accountancy measurements in such facilities carry uncertainty. Therefore, to verify that no diversion of materials has occurred, the ID target value should be zero within uncertainty limits. The uncertainty of the ID (also called *sigmaMUF*) is determined by both the quantity of material of different types passing through the facility and the uncertainty of the measurement techniques used for each type. The SRA may set limits on the values for the ID and the associated uncertainty to satisfy international safeguards requirements; however, the SRA may also set different and, at times, more stringent requirements for national nuclear security purposes (see Sect. 1.3.2 and Chap. 26) but in a manner that also still facilitates meeting its safeguards obligations with the IAEA.

To ensure that the uncertainty of the ID is small, most accountancy values are based on DA, which yields very precise mass measurements. Material forms that are not suitable to DA—such as waste, holdup, and scrap—are measured with NDA methods.

1.2.2 Use of NDA for Safeguards at the State Level

Under safeguards agreements with the IAEA, States must establish and maintain "State or regional systems of accounting for, and control of, nuclear material and in making them more effective" (SSAC; [5]). An SSAC is a set of arrangements to account for and control nuclear material in a State; in particular, a "measurement system for the determination of the quantities of *nuclear material* received, produced, shipped, lost or otherwise removed from inventory, and the quantities on inventory" [6]. The SRA is responsible for collecting, verifying, and reporting to the IAEA the nuclear material inventory in the State. The SRA bases their reporting primarily on the accountancy conducted by the facility operator, which makes use of DA and NDA measurements. The SRA will review and verify the documentation provided by the facility operator to ensure completeness before reporting to the IAEA. Although the SRA is not required by the IAEA to perform measurements to verify the information received from the operator, if the State authority does perform such measurements, it would commonly use NDA.

1.2.3 Use of NDA for Safeguards by IAEA and Regional Authorities

At the international level, the IAEA is the primary organization tasked with ensuring that all States party to the NPT are fulling their safeguards obligations; however, regional authorities, such as the European Atomic Energy Community and the Brazilian-Argentine Agency for Accounting and Control of Nuclear Materials, work to verify peaceful uses of nuclear material at the regional level and may use NDA in support of that objective—either independently or in collaboration with the IAEA.

NDA or DA measurements of nuclear materials are used by the IAEA or by regional safeguards authorities to independently verify the State's declaration. International inspectors for these organizations perform these measurements in the facility at both regular and, in certain cases, random intervals to ensure that the IAEA goal of the timely detection of the diversion of a significant quantity of nuclear material is satisfied. To limit the impact to the facility, the measurements must be performed quickly, and in general, the measurement equipment must be easily portable.

Because it is impracticable to measure an entire inventory of nuclear material in a reasonable time, the inspector will measure a sample of the entire inventory. Items will be chosen to ensure that a diversion of nuclear material is detected with a defined detection probability. An inspector performs three types of measurements: gross defect, partial defect, and bias defect. *Gross defect* refers to an item that has been falsified to the maximum extent possible so that all or most of the declared material is missing. A gross defect measurement is performed using a nondestructive attribute measurement of the nuclear material, which is typically done by measuring characteristic gamma rays emitted for the item (see Sect. 10.2). *Partial defect* refers to an item that has been falsified to such an extent that some fraction of the declared amount of material is missing. To measure a partial defect, a quantitative measurement must be performed. NDA measurements based on gamma and/or neutron signatures are used because of the relatively short assay time and measurement precisions that can detect a partial defect. *Bias defect* refers to an item that has been slightly falsified so that only a small fraction of the declared amount of material is missing. Detection of a bias defect requires a measurement with a high precision; therefore, DA measurements are usually performed.

1.2.4 Arms Control Treaties

NDA measurements have been used, currently are used, and may be used under past, current, and hypothetical arms control treaties. The use of NDA to verify arms control agreements is driven by the same set of concerns as its use in safeguards. One advantage of using NDA techniques is that NDA does not require direct access to the nuclear material and limits the time and access needed by inspectors. The New Strategic Arms Reduction Treaty, currently extended to 2026, uses a lack of neutron detection from the approved radiation detection equipment to demonstrate that an item is not a nuclear weapon. A wide range of NDA techniques has been considered for various applications in arms control.

Proposed arms control measurements introduce complications that are rarely encountered in safeguards applications. Hypothetical future treaty scenarios assume various combinations of (1) direct measurements that confirm that a device is a nuclear weapon, (2) monitoring of weapons production facilities, and (3) verification of the dismantlement of nuclear weapons. All of these monitoring scenarios can be addressed by NDA techniques, but the NDA measurements may yield sensitive information that States may not be willing or, under the NPT, able to share. Information protection requirements need to be incorporated and often lead to difficulties in the authentication (ability of the inspector to trust the equipment) and certification (ability of the host to allow the equipment to be used). Much of the work needed to apply NDA techniques to arms control applications centers on solving these secondary problems.

1.3 Nuclear Security

1.3.1 Introduction

Nuclear security focuses on the prevention, detection, and response to criminal or intentional unauthorized acts involving or directed at nuclear material, other radioactive material, associated facilities, or associated activities [7]. Unlike safeguards, the IAEA is not mandated by treaty or other legally binding international instrument to carry out nuclear security responsibilities within a State. Instead, nuclear security is widely understood and accepted as the sovereign responsibility of each State. However, the IAEA is recognized by its Member States as having a central coordinating role in facilitating international cooperation on nuclear security, and it publishes recommendations [8] for and provides assistance, upon request, to States regarding establishing, implementing, and sustaining an effective nuclear security regime. According to IAEA guidance, an effective national nuclear security regime features, among other things, the following components: legal and regulatory framework, threat assessment, physical protection, security of material in transport, nuclear material accounting and control (NMAC), and detection of and response to nuclear and other radioactive material out of regulatory control (MORC). NDA measurements play a key role in two of these components—NMAC and the detection of and response to nuclear and other radioactive MORC.

1.3.2 Nuclear Material Accounting and Control

To support and enhance nuclear security—in particular, to prevent, detect, and respond to unauthorized removal of nuclear material—all nuclear facilities should implement an effective NMAC system. According to IAEA guidance, NMAC systems and measures focus, in particular, on mitigating the risk posed by insider threats, and in the event of loss or theft of nuclear material, NMAC systems should be able to identify the quantity and characteristics of the missing material [9]. NDA equipment is used to support the NMAC system by confirming nuclear material quantity and types within the accounting system. NMAC systems recommended by the IAEA for nuclear security purposes can fulfill many of the same purposes of the facility-level implementation of the SSAC system used for nuclear safeguards, and facilities may use the same accounting system in support of both the SSAC and NMAC. However, as noted previously, the national nuclear regulatory body for nuclear security—which oftentimes will serve also as the SRA—may set different and, at times, more stringent NMAC requirements for national nuclear security purposes. For example, whereas the IAEA will set measurement requirements within a facility to meet its goals for timely detection of the diversion of one significant quantity of nuclear material, the national regulatory body may require additional measurements to detect and respond to possible theft by an insider. NDA systems for NMAC are also designed to work in concert with administrative measures—such as implementation of a two-person rule and security background checks on personnel—to deter insider threats, whereas the IAEA does not establish requirements for such procedures as a part of safeguards. A more detailed description of NMAC systems and measures, including the role of NDA, is given in Chap. 26.

1.3.3 Detection and Response to Nuclear and Other Radioactive Material Out of Regulatory Control

As a part of an effective national nuclear security regime, States should also establish systems and measures to prevent, detect, and respond to a criminal or unauthorized act with nuclear security implications that involves nuclear and other radioactive MORC [10]. These systems and measures should combat illicit trafficking of nuclear and other radioactive material, including for cases of theft, illegal possession, transfer, or disposal of material—whether intentional or not—and for acts carried out either within or across national borders. NDA plays a major role in the detection of and response to illicit trafficking of nuclear material. Competent authorities for nuclear security (including front-line organizations, such as border police or customs) regularly use handheld and fixed NDA equipment, such as radiation portal monitors (RPMs), to scan people, vehicles, and cargo for MORC (Chap. 22). Typically, for example, if a vehicle RPM alarms at a border crossing point, a front-line officer will perform a secondary inspection of the vehicle using a handheld NDA instrument to identify any radionuclides present (see Sect. 10.3) and, taking into account the vehicle cargo manifest or declaration, determine whether an associated threat exists. For high-threat items, items for which a criminal nexus likely exists, or materials for which a desire exists to understand provenance, a nuclear forensic investigation should be initiated.

Nuclear forensics is an essential component of national response plans to events that involve MORC. It informs prevention, detection, and response and enhances the nuclear security of a state. *Nuclear forensic science*, or *nuclear forensics*, is the examination of nuclear or other radioactive material, or of evidence contaminated with radionuclides, in the context of legal proceedings under international or national law related to nuclear security [11, 12]. The goal of nuclear forensics is to discover linkages among people, places, materials, and events. Understanding the provenance of MORC allows a state to identify security weaknesses and strengthen them if the material originates from a facility within the State and to close smuggling pathways for material being transported through the State. A nuclear forensics examination in support of a nuclear security investigation may use both NDA and DA measurement techniques. NDA measurements performed usually include radionuclide identification, isotopic composition, and activity quantification (see Chaps. 8, 9, and 10). NDA measurements may provide enough information for prosecution or identification and understanding of provenance or ruling out potential origins.

For States that have an advanced forensics capability, DA measurements provide additional material characteristics and higher-precision results in cases where NDA results are inconclusive.

1.4 Safety and Compliance

1.4.1 Introduction

The same NDA measurements used to support safeguards and security can also be used to support safety and compliance. Safety concerns that can be addressed with NDA methods include radiation protection, criticality safety, and quantification of material at risk. Radiation protection often involves the measurement of dose (the energy of ionizing radiation absorbed per unit mass of material) to people and is outside the scope of this book, which focuses on the characterization of nuclear material. NDA techniques are often used to ensure the compliance of waste disposal limits. Because civilian nuclear facilities are commercial entities, nuclear material measurements can also serve as process controls to ensure that their finished product meets the consumer's specification. In addition, nuclear material has a monetary value and, to maximize profit, the facility operator wants to ensure that all nuclear material is accounted for.

1.4.2 Criticality Safety

NDA measurements are often used to support holdup measurements in nuclear facilities (Chap. 10, Sect. 10.4.3 and Chap. 19, Sect. 19.4.4). *Holdup* refers to the nuclear material that deposits or "holds up" in the processing equipment. If this holdup is ignored, over time, a large enough deposit of nuclear material could collect, resulting in a criticality hazard. To mitigate this hazard, facility operators perform measurements of the accumulated deposits in the processing equipment and associated piping, duct work, etc. These measurements are usually performed with gamma-ray-based NDA techniques [13], although some forms of nuclear material could be amenable to neutron-based techniques.

1.4.3 Waste Measurement

Waste repositories have strict requirements on the types and quantities of radionuclides in a waste container that the facility is licensed to accept. There are four main types of nuclear waste: high-level, transuranic, intermediate-level, and low-level waste. NDA measurements, along with acceptable knowledge from the waste generator, are often used to characterize the waste and to ensure that the waste acceptance criteria are satisfied. The Waste Isolation Pilot Plant in Carlsbad, New Mexico (WIPP), disposes of transuranic waste and performs NDA measurements [14] to ensure that the waste acceptance criteria [15] are satisfied.

References

1. *IAEA Safeguards Glossary*, International Nuclear Verification Series No. 3, Rev. 1 (International Atomic Energy Agency, Vienna, 2022)
2. *Treaty on the Non-Proliferation of Nuclear Weapons*, INFCIRC/140 (International Atomic Energy Agency, 1970)
3. J. Doyle (ed.), *Nuclear Safeguards, Security, and Nonproliferation: Achieving Security with Technology and Policy*, 2nd edn.., ISBN: 9780128032718 (Elsevier, 2019)
4. *Nuclear Material Accounting Handbook*, IAEA Service Series No. 15 (International Atomic Energy Agency, 2008)
5. *Guidance for States Implementing Comprehensive Safeguards Agreements and Additional Protocols*, IAEA Services Series No. 21 (International Atomic Energy Agency, Vienna, 2016)
6. *The Structure and Content of Agreements Between the Agency and States Required in Connection with the Treaty on the Non-Proliferation of Nuclear Weapons*, INFCIRC/153 (International Atomic Energy Agency, 1970)
7. *Objective and Essential Elements of a State's Nuclear Security Regime*, IAEA Nuclear Security Series No. 20 (International Atomic Energy Agency, Vienna, 2013)
8. *Nuclear Security Plan 2022–2025*, GC(65)/24 (International Atomic Energy Agency, 2021)
9. *Use of Nuclear Material Accounting and Control for Nuclear Security Purposes at Facilities*, IAEA Nuclear Security Series No. 25-G (International Atomic Energy Agency, Vienna, 2015)
10. *Nuclear Security Recommendations on Nuclear Material and Other Radioactive Material out of Regulatory Control* IAEANuclear Security Series No. 15 (International Atomic Energy Agency, Vienna, 2011)

11. K. Moody, P. Grant, I. Hutcheon, *Nuclear Forensic Analysis*, 2nd edn.., SBN 9780367778040 (CRC Press, 2021)
12. *Nuclear Forensics in Support of Investigations*, IAEA Nuclear Security Series No. 2-G, Rev.1 (International Atomic Energy Agency, Vienna, 2015)
13. P. Russo, *Gamma-Ray Measurements of Holdup Plant-Wide: Application Guide for Portable, Generalized Approach*, Los Alamos National Laboratory report LA-14206 (2005)
14. J. Veilleux, D. Cramer, *Comparison of Gamma and Passive Neutron Non-Destructive Assay Total Measurement Uncertainty Using the High Efficiency Neutron Counter*," Los Alamos National Laboratory report LA-UR-05-4296 (2005)
15. *Transuranic Waste Acceptance Criteria for the Waste Isolation Pilot Plant*, Rev. 10, DOE/WIPP-02-3122 (U.S. Department of Energy, Carlsbad Field Office, 2020)

Open Access This chapter is licensed under the terms of the Creative Commons Attribution 4.0 International License (http://creativecommons.org/licenses/by/4.0/), which permits use, sharing, adaptation, distribution and reproduction in any medium or format, as long as you give appropriate credit to the original author(s) and the source, provide a link to the Creative Commons license and indicate if changes were made.

The images or other third party material in this chapter are included in the chapter's Creative Commons license, unless indicated otherwise in a credit line to the material. If material is not included in the chapter's Creative Commons license and your intended use is not permitted by statutory regulation or exceeds the permitted use, you will need to obtain permission directly from the copyright holder.

The Origin of Gamma Rays

J. Stinnett, P. J. Karpius, D. J. Mercer, T. D. Reilly, and D. T. Vo

2.1 Gamma Rays and the Electromagnetic Spectrum

Gamma rays are high-energy electromagnetic radiation, or "photons," emitted from the de-excitation of the atomic nucleus. Electromagnetic radiation includes such diverse phenomena as radio, television, microwaves, infrared radiation, visible light, ultraviolet radiation, X-rays, and gamma rays. These radiations all propagate through vacuum at the speed of light. They can be described as wave phenomena that involves electric and magnetic field oscillations analogous to mechanical oscillations, such as water waves or sound. They differ from each other only in the frequency of oscillation. Although given different names, electromagnetic radiation actually forms a continuous spectrum, from low-frequency radio waves at a few cycles per second (hertz [Hz]) to gamma rays at 10^{18} Hz and above (see Fig. 2.1).

The parameters used to describe an electromagnetic wave frequency, wavelength, and energy are linked and may be used interchangeably; the Planck-Einstein relation is

$$E = h\nu = \frac{hc}{\lambda}, \tag{2.1}$$

where

E = Energy of the photon (joule [J])
h = Planck's constant, $6.62607015 \times 10^{-34}$ J/Hz
ν = Frequency of the photon (Hz)
c = Speed of light in a vacuum, 299,792,458 meters per second (m/s)
λ = Wavelength of the photon (meters [m])

A common practice is to use frequency or wavelength for radio waves, wavelength for visible or near-visible light, and energy for X-rays and gamma rays. Throughout this book, the electronvolt (eV) will be regularly used as the unit for energy.

Visible light can be emitted during changes in the chemical state of elements and compounds. These changes usually involve the least-tightly bound outer-shell atomic electrons. The colors of the emitted light are characteristic of the radiating elements and compounds and typically have energies of ~1 eV.[1] X-rays and gamma rays are very high-energy light with

---

Los Alamos National Laboratory strongly supports academic freedom and a researcher's right to publish; as an institution, however, the Laboratory does not endorse the viewpoint of a publication or guarantee its technical correctness.

[1] The electronvolt (eV) is a unit of energy equal to the kinetic energy gained by an electron accelerated through a potential difference of 1 V; 1 eV equals 1.602×10^{-19} J. This small unit and the multiple units keV (10^3 eV) and MeV (10^6 eV) are useful for describing atomic and molecular phenomena.

J. Stinnett (✉) · P. J. Karpius · D. J. Mercer · T. D. Reilly · D. T. Vo
Los Alamos National Laboratory, Los Alamos, NM, USA
e-mail: stinnett@lanl.gov; karpius@lanl.gov; mercer@lanl.gov; ducvo@lanl.gov

© The Author(s) 2024
W. H. Geist et al. (eds.), *Nondestructive Assay of Nuclear Materials for Safeguards and Security*,
https://doi.org/10.1007/978-3-031-58277-6_2

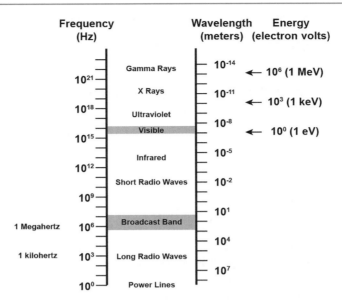

Fig. 2.1 The electromagnetic spectrum that shows the relative scale of different electromagnetic radiations, including gamma rays, X-rays, light waves, and radio waves

overlapping energy ranges of 10 keV and above. X-rays are emitted during changes in the state of more-tightly bound inner-shell electrons, whereas gamma rays are emitted during changes in the state of a nucleus. The energies of the emitted radiations are characteristic of the radiating elements and nuclides. Knowledge of these high-energy electromagnetic radiations began in Germany in 1895 with the discovery of X-rays by Wilhelm Röntgen. After observing that a zinc sulfide screen glowed when it was placed near a cathode-ray discharge tube, Röntgen found that the radiation that caused the glow was dependent on the electrode materials and the tube voltage, was not bent by electric or magnetic fields, and could readily penetrate dense matter. Natural radioactivity was discovered the following year in France by Henri Becquerel, who observed that uranium salts gave off a natural radiation that could expose (or blacken) a photographic plate. While studying these phenomena, Marie and Pierre Curie isolated and identified the radioactive elements polonium and radium. They determined that the phenomena were characteristic of the element, not its chemical form.

These "radioactive rays" were intensely studied in many laboratories. In 1899 in England, Ernest Rutherford discovered that 95% of the radiation of various natural uranium compounds was effectively stopped by 0.02 mm of aluminum, and 50% of the remaining radiation was stopped by 5 mm of aluminum or 1.5 mm of copper [1]. Rutherford named the first component "alpha" and the second, more penetrating radiation "beta." Both of these radiations were deflected by electric and magnetic fields, though in opposite directions; this fact indicated that the radiations carried an electrical charge. In 1900, Paul Villard and Henri Becquerel noted that a photographic plate was affected by radioactive materials even when the plate was shielded by 20 cm of iron or 2–3 cm of lead. They also noted that this penetrating radiation showed no magnetic deflection. In 1903, Rutherford named this component *gamma* and stated that "gamma rays are probably like Röntgen rays." Thus, the three major radiations were identified and named for the first three letters of the Greek alphabet: α, β, and γ.

As indicated by the brief description of their discovery, gamma rays often accompany the spontaneous alpha or beta decay of unstable nuclei. X-rays, like gamma rays, are energetic photons. They differ in that gamma rays are emitted through de-excitations of a nucleus, and X-rays are emitted during rearrangement of the atomic electron structure rather than the nuclear structure. X-rays are discussed in Sect. 2.3.2. X-ray energies are unique to each element but the same for different isotopes of one element. They frequently accompany nuclear decay processes, which can disrupt the atomic electron shell.

Gamma rays from spontaneous nuclear decay are emitted with a rate and energy spectrum that is unique to the decaying nuclear species. This uniqueness provides the basis for most gamma-ray assay techniques: by counting the number of gamma rays emitted with a specific energy, it is possible to determine the number of nuclei that emit that characteristic radiation, which can be used to compute a mass or activity of that radionuclide, determine the isotopic composition of a radiation source, and other applications to be explored throughout this book.

2.2 Characteristics of Nuclear Decay

2.2.1 Nuclear Decay Processes: General

The atomic nucleus is assumed to be a bound configuration of protons and neutrons. Protons and neutrons have nearly the same mass and differ principally in charge: protons have a positive charge of 1, and neutrons are electrically neutral. Different elements have nuclei with different numbers of neutrons and protons. The number of protons in the nucleus is called the *atomic number* and is given the symbol Z. In the neutral atom, the number of protons is equal to the number of electrons. The number of neutrons in the nucleus is given the symbol N. The total number of nucleons (protons and neutrons) in the nucleus is called the *atomic mass number* and is given the symbol A; (A = Z + N).

For all nuclear decay processes, the number of unstable nuclei of a given species is found to diminish exponentially with time:

$$n(t) = n_0 e^{-\lambda t}, \tag{2.2}$$

where

$n(t)$ = number of nuclei of a given species at time t
n_0 = number of nuclei at $t = 0$
λ = decay constant, the parameter characterizing the exponential.

Each nuclear species has a characteristic decay constant. Radioactive decay is most commonly discussed in terms of the nuclear half-life, $T_{1/2}$, which is related to the decay constant by

$$T_{1/2} = \frac{\ln(2)}{\lambda}. \tag{2.3}$$

The half-life is the time necessary for the number of unstable nuclei of one species to diminish by one-half. (Half-lives are commonly given in nuclear data tables.) The decay rate or specific activity can be represented in terms of the half-life as follows:

$$R = \frac{\ln(2)}{T_{1/2}} \times \frac{N_A}{A} = \frac{1.32 \times 10^{16}}{T_{1/2} \times A}, \tag{2.4}$$

where

R = rate in decays per second per gram
A = atomic mass in grams per mole
N_A = Avogadro's constant
$T_{1/2}$ = half-life in years.

Equation 2.4 is often used to estimate the activity per gram of a sample.

An alpha or beta decay of a given nuclear species is not always accompanied by gamma-ray emission. The fraction of decays that is accompanied by the emission of a specific energy gamma ray is called the *branching intensity*, also referred to as *branching ratio*, *yield*, *gamma yield*, or *photons per decay*. For example, the most intense gamma ray emitted by ^{235}U has an energy of 185.7 keV and a branching intensity of 57.2%. Uranium-235 decays by alpha-particle emission with a half-life of 7.038×10^8 years. Equation 2.4 thus implies an alpha emission rate of 7.98×10^4 α/s for 1 g of ^{235}U. Only 57.2% of the alpha particles are accompanied by a 185.7 keV gamma ray; therefore, the specific activity of this gamma ray is 4.6×10^4 γ/s.

Of the natural decay radiations, only the gamma ray is of interest for nondestructive assay (NDA) of bulk nuclear materials because the alpha- and beta-particle ranges are very short in condensed matter. Consider the following ranges in copper metal:

5 MeV α: 0:01 mm or 0.008 g/cm^2.
1 MeV β: 0.7 mm or 0.6 g/cm^2.
0.4 MeV γ: 12 mm or 10.9 g/cm^2 (mean free path).

Because of the limited range of alpha and beta particles, they are not generally transmitted through the container in which a material is located. Further, even for exposed material, only alphas and betas emitted from very near the surface can possibly escape the material.

2.2.2 Alpha Decay

The alpha particle is a doubly ionized (bare) ^{4}He nucleus. It is a very stable, tightly bound nuclear configuration. When a nucleus decays by alpha emission, the resulting daughter nucleus has a charge that is two units less than the parent nucleus and an atomic mass that is four units less. This generic reaction can be represented as follows:

$$^A_Z X \rightarrow ^{A-4}_{Z-2} Y + ^4_2 He. \tag{2.5}$$

The decay can occur only if the mass of the neutral parent atom is greater than the sum of the masses of the daughter and the ^{4}He atom. The mass difference between the parent and the decay products is called the *Q-value* and is equal to the kinetic energy of the decay products:

$$Q = (M_p - M_d - M_{He})c^2, \tag{2.6}$$

where

M_p = atomic mass of the parent atom
M_d = atomic mass of the daughter atom
M_{He} = atomic mass of a helium atom.

When the parent nucleus decays, most of the energy (Q) goes to the alpha particle because of its lower mass:

$$E_\alpha = \frac{Q(A-4)}{A}. \tag{2.7}$$

The remainder of the available energy goes into the recoil of the daughter nucleus.

Most of the approximately 750 known alpha emitters are heavy nuclei with atomic numbers greater than 82. The energy range of the emitted alpha particle is generally 2–12 MeV, and the half-lives largely vary from 10^{-8} s to 10^{10} years. The short-lived nuclei generally emit high-energy alpha particles when they decay, as shown in Fig. 2.2.

Immediately after the decay of the parent nucleus, the daughter nucleus can be either in the ground state or in an excited state. In the latter case, the nucleus can relax by either of two mechanisms: gamma-ray emission or internal conversion. The radiative relaxation leads to emission of one or more gamma rays (typically 10^{-14} s after the alpha emission) with discrete energies whose sum equals the original excitation energy. During internal conversion, the nucleus transfers the excitation energy directly to one of the most-tightly bound atomic electrons, usually a K-shell electron. The electron leaves the atom with an energy equal to the difference of the excitation energy and the electron binding energy. The filling of the electron shell vacancy leads to the emission of X-rays or electrons (called *Auger electrons*) with the characteristic energy spectrum of the daughter element. The probability of internal conversion increases strongly with atomic number (Z) and with decreasing excitation energy.

In some cases, the alpha decay leads to an excited state that lives much longer than 10^{-14} s. If the lifetime of this state is longer than approximately 10^{-6} s, it is called an *isomer* of the ground-state nucleus. An example of an isomer is the alpha decay of ^{239}Pu that leads to ^{235}U:

$$^{239}Pu \rightarrow ^{235m}U (99.96\%) \, [T_{1/2} = 26 \; min; decays \; to \; ^{235}U \; ground \; state]$$

$$^{239}Pu \rightarrow ^{235}U (0.04\%). \tag{2.8}$$

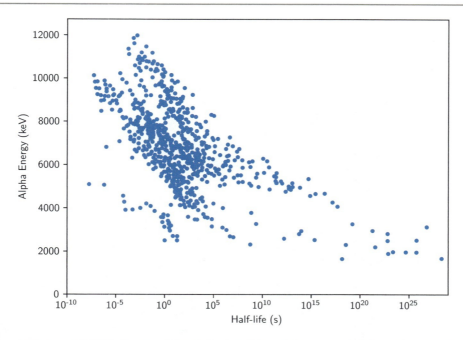

Fig. 2.2 Plot of alpha energies versus half-life. In general, longer-lived nuclides emit lower-energy alpha particles. Beryllium-8 (Q = 91.84 keV, Half-life = 8.19×10^{-17} s) is an interesting outlier in terms of the alpha decay energy, half-life, and Z; the next lowest known alpha emitter has Z = 52

The common decay mode of ^{239}Pu leads first to the isomer ^{235m}U, which has a half-life of 26 min. The direct decay to ^{235}U occurs only 0.04% of the time. Although isomers tend to be short-lived, there are exceptions: ^{91m}Nb has a half-life of 60.9 days. Interestingly, ^{180m}Ta is observationally stable, whereas ^{180}Ta has a half-life of just 8.15 h.

All of the alpha particles, gamma rays, internal conversion electrons, and X-rays emitted during the decay process have discrete, characteristic energies. The observation of these characteristic spectra showed that nuclei have discrete allowed states or energy levels analogous to the allowed states of atomic electrons. The various spectroscopic observations have provided information for developing the nuclear-level schemes presented in handbooks and online databases, such as the *National Nuclear Data Center NuDat database* [2]. An example appears in Fig. 2.3, showing the lower energy levels of ^{235}U populated during the alpha decay of ^{239}Pu. These levels give rise to the characteristic gamma-ray spectrum associated with the alpha decay of ^{239}Pu. Note that the characteristic gamma-ray spectrum is commonly associated with the parent or decaying nucleus even though the energies are determined by the levels of the daughter nucleus. Although this practice may seem confusing, it is universally followed for gamma rays. The confusion is further aggravated by the common use of X-ray nomenclature that associates the characteristic X-rays with the daughter element. Therefore, the alpha decay of ^{239}Pu leads to ^{235}U and is accompanied by the emission of ^{239}Pu gamma rays and uranium X-rays.

2.2.3 Beta Decay

In the beta decay process, the atomic number (Z) increases or decreases by one unit and the atomic mass number (A) stays constant. In effect, neutrons and protons change state. The three types of beta decay are β^-, β^+, and electron capture.

Beta-minus decay was the first detected process; the β^- particle was found to be a normal electron. During the decay process, the nucleus changes state according to the following formula:

$$^A_Z X \to ^A_{Z+1} Y + e^- + \bar{\nu}_e \tag{2.9}$$

The β^- decay process can be thought of as the decay of a neutron into a proton, an electron, and an electron antineutrino. This process is the common beta decay process for nuclei with high atomic number and for fission product nuclei, which usually have significantly more neutrons than protons. The decay is energetically possible for a free neutron (a neutron outside of a nucleus) and occurs with a half-life of 12.8 min.

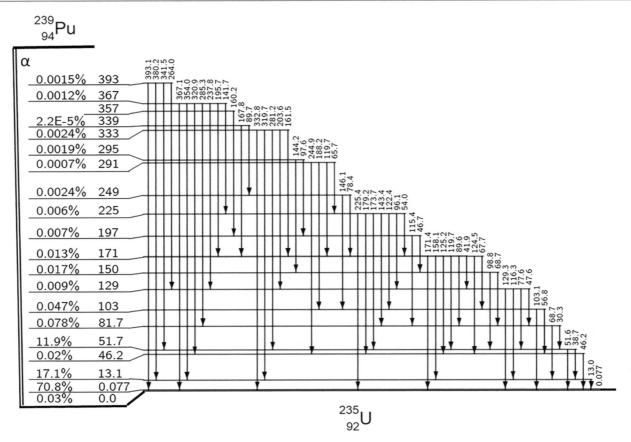

Fig. 2.3 Diagram of some of the nuclear energy levels of ^{235}U. These levels are populated during the alpha decay of ^{239}Pu and give rise to the characteristic gamma-ray spectrum of ^{239}Pu. (Figure adapted using data from [2])

During β^+ decay, a proton is converted to a neutron, a positron (also called an antielectron), and an electron neutrino; the nucleus changes state according to the following formula:

$$^{A}_{Z}X \rightarrow {}^{A}_{Z-1}Y + e^+ + \nu_e. \tag{2.10}$$

Electron capture competes with the β^+ decay process. The nucleus interacts with an inner atomic electron and, in effect, captures it, changing a proton into a neutron with the emission of a positron and an electron neutrino. The formula for this process is

$$^{A}_{Z}X + e^- \rightarrow {}^{A}_{Z-1}Y + \nu_e. \tag{2.11}$$

All unstable nuclei with an atomic number less than 82 decay by at least one of the three processes and sometimes by all three (see Fig. 2.4). Beta decay occurs whenever it is energetically possible—if the following conditions are met for the masses of the neutral parent (p) atoms and the potential daughter (d) atom:

$$\beta^- \ decay: M_p > M_d$$

$$\beta^+ \ decay: M_p > M_d + 2m_e$$

$$Electron\ capture: M_p > M_d. \tag{2.12}$$

Beta decay can be to the ground state or to an excited state in the daughter nucleus. In the latter case, the excited state decays by gamma-ray emission or internal conversion.

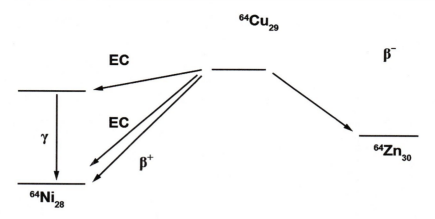

Fig. 2.4 Nuclear decay scheme of ^{64}Cu, showing three possible beta decay processes

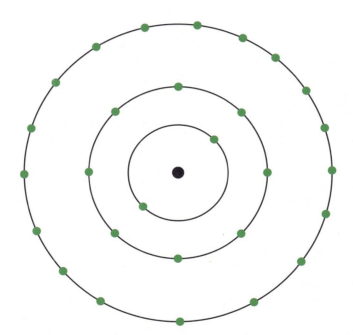

Fig. 2.5 The Bohr model of the atom. A dense nucleus of protons and neutrons is at the center, ringed by electron orbitals. The first orbital (K) contains 2 electrons, the second orbital (L) contains 8 electrons, the third (M) contains 18, and so on

2.3 X-Ray Production

2.3.1 The Bohr Model of the Atom

In the simple Bohr model of the atom (Fig. 2.5), the positive nucleus contains protons and neutrons and has an approximate radius of 1.4×10^{-15} ($A^{1/3}$) m and an approximate density of 2×10^{14} g/cm^3. The nucleus is surrounded by a cloud of negative electrons in discrete, well-defined energy levels or orbitals. The radii of these orbitals are in the range of 10^{-9} to 10^{-8} m. The original Bohr model had well-localized orbits and led to the familiar planetary diagram of the atom. Although this model is now understood to be an oversimplification that ignores much of modern physics, this concrete model is useful for explaining X-ray production.

The different energy levels of the atom are designated K, L_1, L_2, L_3, M_1, ... M_5, and so forth. (As an example, consider the K and L electron energy levels of uranium illustrated in Fig. 2.6.) The electric force between an electron and the positively

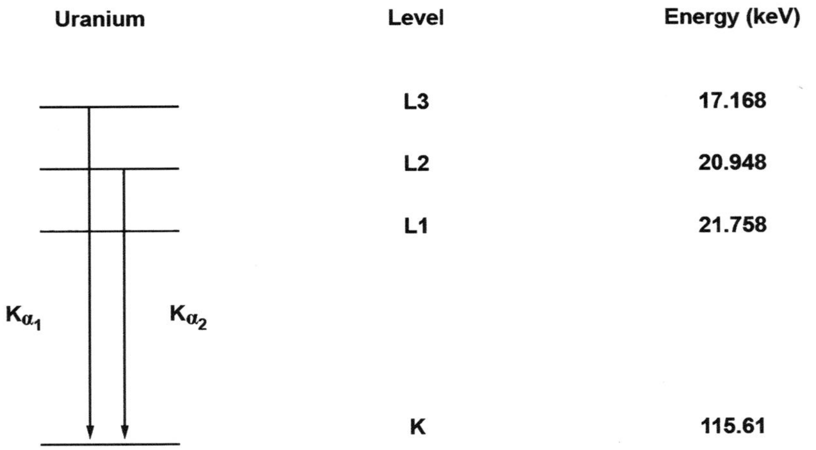

Fig. 2.6 Electron energy levels in uranium. Transitions between the levels shown give rise to the K-series X-rays

charged nucleus varies as the inverse square of the separation; therefore, the electrons closer to the nucleus have a higher binding energy (B). The binding energy is the energy required to remove the electron from the atom. The K-shell electrons are always the most-tightly bound. Quantum mechanics gives a good description of the energies of each level and how the levels fill up for different elements. The chemical properties of the elements are determined by the electron configuration.

In its normal resting configuration, the atom is stable and does not radiate. If an electron moves from a higher to a lower energy level, it radiates a characteristic X-ray to release the excess energy. Although different naming schemes are used for these X-rays, here the Siegbahn notation [3] is used. With this nomenclature, consider the $K_{\alpha1}$ X-ray: the electron fell into the K-orbital from one orbital out. Here, "α" means from one further orbital out, "β" would be two orbitals out, and so on. Finally, the "1" denotes the largest difference due to the difference in level energies.

For example, consider the electron energy levels in uranium, shown in Fig. 2.6. If an electron drops from the L_3 to the K state, the $K_{\alpha1}$ X-ray would be emitted, with an energy of $115.61 - 17.168 = 98.442$ keV.

2.3.2 X-Ray Production Mechanisms

Various interactions can remove an electron from being bound to an atom, a process called *ionization*. All energetic, charged particles interact with electrons as they pass through matter. X-ray and gamma-ray photons also interact with atomic electrons. Nuclear interactions, such as internal conversion or electron capture, can cause the ionization of atomic electrons.

When an electron leaves an atom, the atom is in an excited state with energy B_i by virtue of the vacancy in the ith electron level. This vacancy can be filled by a more-loosely bound electron from an outer orbital, the jth level. The change in energy level is accompanied by the emission of an X-ray with energy $B_i - B_j$ or by the emission of an Auger electron with energy $B_i - 2B_j$. In the latter case, the atom transfers its excess energy directly to an electron in an outer orbital. The fraction of vacancies in level i that result in X-ray emission is defined as the fluorescence yield W_i. Figure 2.7 shows the variation of the K-shell fluorescence yield with atomic number. X-ray emission is more probable for high-Z elements (for $Z > 70$, $W_K > 95\%$).

Bulk samples of radioactive material will emit characteristic X-rays of both the parent and the daughter. In high-Z materials, internal conversion is a probable decay mode that will lead to vacancies in the daughter's inner electron shell (usually K or L). The subsequent filling, from an outer shell electron, results in the emission of characteristic X-rays of the daughter. X-rays characteristic of the parent arise from energetic decay products (alpha, beta, gamma ray, or X-ray) that ionize nearby parent atoms.

Plutonium metal emits uranium X-rays by virtue of the internal conversion process that occurs after alpha decay. It also emits plutonium X-rays by virtue of X-ray fluorescence induced by alpha particles.

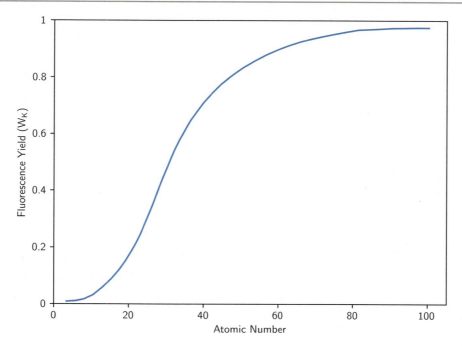

Fig. 2.7 Variation of the K-shell fluorescence yield, W_K, with atomic number

2.3.3 Characteristic X-Ray Spectra

Several possible causes exist for an electron to de-excite and thereby produce an X-ray:

- De-excitation could be the result of a decay process, such as internal conversion, during which a vacancy in an orbital is left after the decay. When the vacancy is filled, a characteristic X-ray is emitted.
- An external particle could cause ionization. When the resulting vacancy is filled, an X-ray is emitted. This process is called *induced fluorescence* and results in a characteristic X-ray.
- An energetic electron may be slowed down or deflected in the electric field around an atomic nucleus. This process is called *bremsstrahlung* and is discussed in Sect. 2.3.4. Unlike the previous options, the energies of the X-rays emitted from bremsstrahlung are a continuous spectrum and do not depend (in energy) on the material.

Each element has a characteristic X-ray spectrum. All elements have the same general X-ray pattern, but the X-ray energies are different. Because these characteristic X-rays are emitted when the electrons are reconfigured, and because the energy states of the electrons directly depend on how positively charged the nucleus is, these X-ray energies increase as the atomic number increases. Figure 2.8 shows the characteristic X-rays from lead and uranium.

Early investigators developed the system commonly used today for naming X-rays. A Roman letter indicates the final level to which the electron moves, and a Greek letter plus a number indicates the electron's initial energy level. (The Greek letter was originally related to the X-ray energy and the number to its intensity.) Table 2.1 gives the major K X-rays of uranium and plutonium. The L and M X-rays are of lower energy and can be found in the literature.

2.3.4 Bremsstrahlung (Braking Radiation)

Charged particles continuously decelerate as they move through condensed materials. As they decelerate, they emit photons with a continuous energy spectrum known as bremsstrahlung. These photons are of interest because their energies are often similar to those used for NDA.

Beta particles from nuclear decay often emit bremsstrahlung photons while stopping. Although beta particles have a very short range in condensed matter and rarely escape from the host material, the bremsstrahlung photons often escape and are

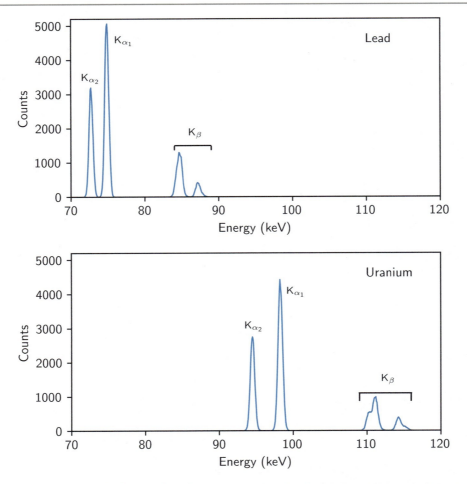

Fig. 2.8 Characteristic X-ray spectra from lead (top) and uranium (bottom). Note that the pattern is the same but shifted in energy

Table 2.1 Major K X-rays of uranium and plutonium[a] [4]

X-ray	Levels Final - Initial	Energy (keV; [2]) Uranium	Plutonium	Intensity[b] [2] Uranium	Plutonium
$K_{\alpha 2}$	K - L_2	94.65	99.52	62.5	63.2
$K_{\alpha 1}$	K - L_3	98.44	103.74	100	100
$K_{\beta 1}$	K - M_3	111.30	117.23	22.6	22.8
$K_{\beta 2}$	K - $N_{2,3}$	114.3, 114.56	120.44, 120.7	8.7	8.9
$K_{\beta 3}$	K-M_2	110.42	116.24	11.5	11.6

[a]Other X-rays in the K series are weaker than those listed here
[b]Relative intensity; 100 is maximum

detected along with the gamma rays of interest for NDA. Internal conversion electrons can also contribute to the production of bremsstrahlung radiation. The detected discrete gamma rays emitted by a decaying nucleus are superimposed on a continuous bremsstrahlung background. The electron linear accelerator uses the bremsstrahlung reaction to produce high-energy photons for nuclear research, nuclear medicine, and active NDA of nuclear materials [5]. An example of a gamma ray spectrum for beta emitters is shown in Fig. 2.9.

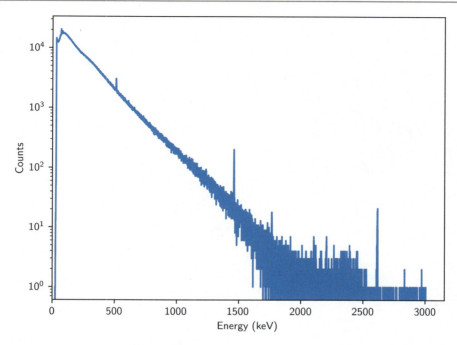

Fig. 2.9 High-resolution gamma-ray spectrum of ^{90}Y and ^{90}Sr, which are strong β^- emitters. As the β^- particles decelerate, bremsstrahlung is emitted. This continuum extends from 0 keV up to the maximum possible energy, which for ^{90}Y, is 2280 keV

2.4 Major Gamma Rays from Nuclear Material

2.4.1 Typical Spectra

Figures 2.10 through 2.15 show typical uranium, plutonium, and thorium gamma-ray spectra; these spectra, which are explained in detail in Sect. 4.3, were measured with high-resolution germanium detector systems (see Chap. 4). Figure 2.10 shows the spectrum of highly enriched uranium (HEU) in the 0–3 MeV range, with characteristic gamma rays from ^{235}U and the ^{238}U granddaughter ^{234m}Pa. The intense gamma rays in the 140–210 keV range are often used for the assay of ^{235}U. (Fig. 2.11 shows this region in more detail.) For comparison, Fig. 2.12 shows a spectrum of depleted uranium; the spectrum shows the ^{238}U daughter radiations often used for ^{238}U assay. Figures 2.13 and 2.14 show gamma-ray spectra of plutonium with approximate ^{240}Pu concentrations of 14% and 6%, respectively. Note the differences in relative peak heights between the two spectra; these differences are used to determine the plutonium isotopic composition (see discussion in Chap. 9, Sect. 9.4). Figure 2.15 shows the characteristic gamma-ray spectrum of ^{232}Th; all major gamma rays come from daughter nuclides.

2.4.2 Major Gamma-Ray Signatures for Nuclear Material Assay Fission-Product Gamma Rays

In principle, any of the gamma rays from nuclear material can be used to determine the mass of the isotope that produces them. In practice, certain gamma rays are used more frequently than others because of their intensity, penetrability, and freedom from interferences. The ideal signature would be an intense ($>10^4$ γ/g − s) gamma ray with an energy of several thousand keV. The mass attenuation coefficients of all materials show a broad minimum between 1 and 5 MeV, and very few natural gamma rays above 1 MeV can cause interference. Unfortunately, such gamma rays do not exist for uranium or plutonium. Table 2.2 lists the gamma rays most commonly used for the NDA of the major uranium and plutonium isotopes.

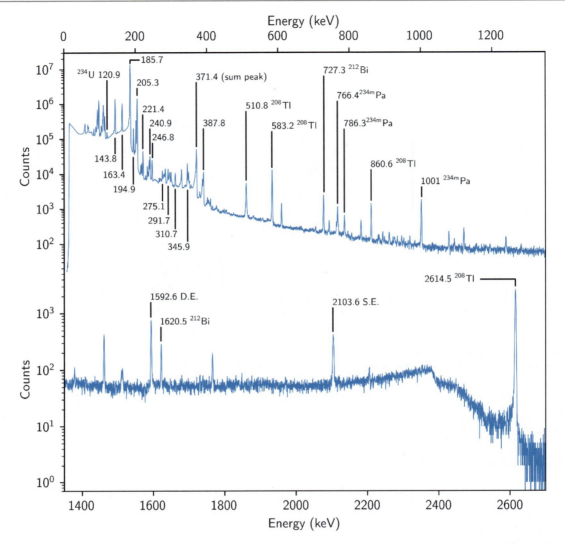

Fig. 2.10 High-resolution gamma-ray spectrum of HEU (93% ^{235}U). Energies given in keV. Labels S.E. and D.E. represent the single- and double-escape peaks of the 2614 keV gamma ray. Peaks not labeled with a specific isotope are from ^{235}U

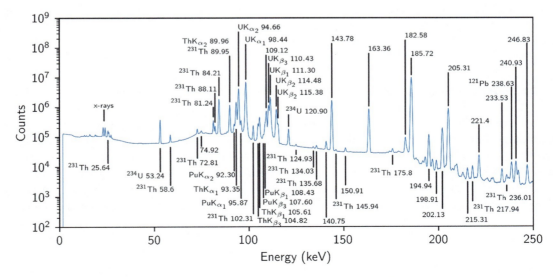

Fig. 2.11 Gamma-ray spectrum of HEU, showing the intense gamma rays often used for assay of ^{235}U. Peaks not labeled with a specific isotope are from ^{235}U

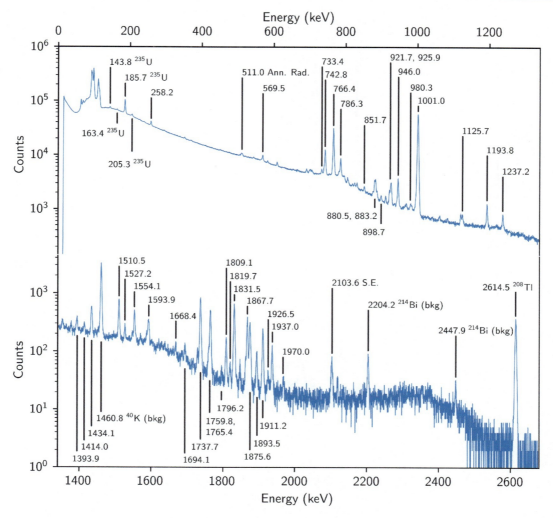

Fig. 2.12 Gamma-ray spectrum of depleted uranium (0.2% ^{235}U). The intense gamma rays at 766 and 1001 keV are from the ^{238}U daughter ^{234m}Pa and are often used for the assay of ^{238}U. Most of the weak gamma rays above 1001 keV also come from ^{234m}Pa. The label, Ann. Rad., is annihilation radiation; the small peak at 511 keV is due to positron annihilation. Peaks not labeled with a specific isotope are from the ^{238}U daughter ^{234m}Pa

2.4.3 Fission-Product Gamma Rays

Considerable interest has been shown in the measurement of irradiated fuel from nuclear reactors. The irradiated fuel has a high monetary value and a high safeguards value because of the plutonium produced during reactor operation. Gamma rays from the spontaneous decay of uranium and plutonium cannot be used for measurement of irradiated fuel because they are overwhelmed by the very intense gamma rays emitted by fission products that build up in the fuel during irradiation. The total gamma-ray intensity of the fission products from light-water-reactor fuel irradiated to 33,000 MWd/tU (megawatt days per ton of uranium) is approximately 2×10^{10} γ/g − s (g = gram of uranium) 1 year after removal of the fuel from the reactor, whereas the major uranium and plutonium gamma rays have intensities in the range of 10^3 to 10^4 γ/g − s. In some instances, the intensity of one or more fission products can be measured and related to the mass of the contained nuclear material.

Certain high-Z nuclei can fission or split into two or three medium-Z daughter nuclei. The fission process can occur spontaneously, or it can be induced when the parent nucleus absorbs a neutron. Spontaneous fission is more probable in nuclei with even atomic mass numbers (A). Induced fission can occur after absorption of either thermal or fast neutrons in nuclei with odd mass numbers; it occurs only after absorption of fast neutrons in even-numbered nuclei. The fission process was first discovered in 1939 by Otto Hahn and Freidrich Strassmann and correctly interpreted in the same year by Lise Meitner and Otto Frisch.

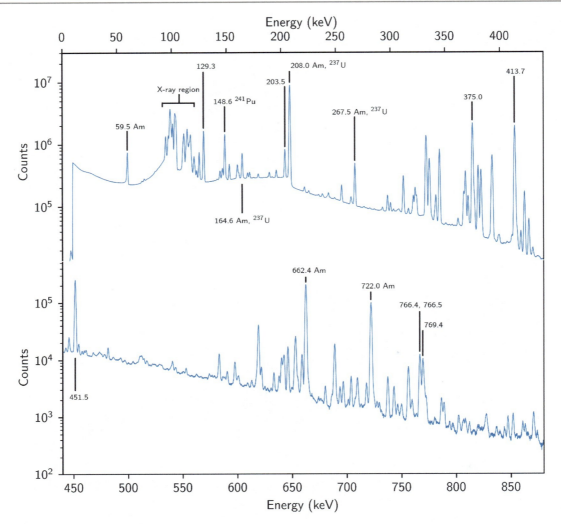

Fig. 2.13 Gamma-ray spectrum of plutonium with 14% ^{240}Pu and 1.2% ^{241}Am. Compare this spectrum with that of Fig. 2.14, noting the difference in relative intensities. Peaks not labeled with a specific isotope are from ^{239}Pu. Refer to the original PANDA book [6] for identification of additional peaks

The fission of a nucleus is a cataclysmic event when compared with the alpha decay and beta decay processes described in Sects. 2.2.2 and 2.2.3, respectively. The energy released in fission is approximately 200 MeV, whereas beta decays tend to release around 1 MeV and alpha decays generally release several MeV of energy. Most of the fission energy is carried as kinetic energy by the two (rarely three) daughter nuclei (called *fission products* or *fission fragments*). The fissioning nucleus also emits an average of two prompt neutrons and six prompt gamma rays at the instant it splits. A typical fission reaction is illustrated by the formula

$$n + {}^{235}_{92}U \rightarrow {}^{137}_{55}Cs + {}^{97}_{37}Rb + 2n. \tag{2.13}$$

This formula illustrates only one of the many possible fission reactions. The fission-product nuclei themselves are unstable—with an excess of neutrons—and decay by either neutron emission or β^- decay (frequently accompanied by gamma-ray emission); the neutron and gamma radiations from these reactions are called *delayed neutrons* and *delayed gamma rays*, respectively. The fission products have half-lives ranging from seconds to years. The gamma rays from fission products can be used to characterize irradiated fuel materials.

The most commonly measured fission-product gamma ray in irradiated fuel is from ^{137}Cs at 661.66 keV. This fission product has a high yield and a sufficiently long half-life (30.08 year) so that its concentration is proportional to the total number of fissions that have occurred in the fuel. (See Chap. 21 for a more complete discussion of the fission reaction and the measurement of irradiated fuel.)

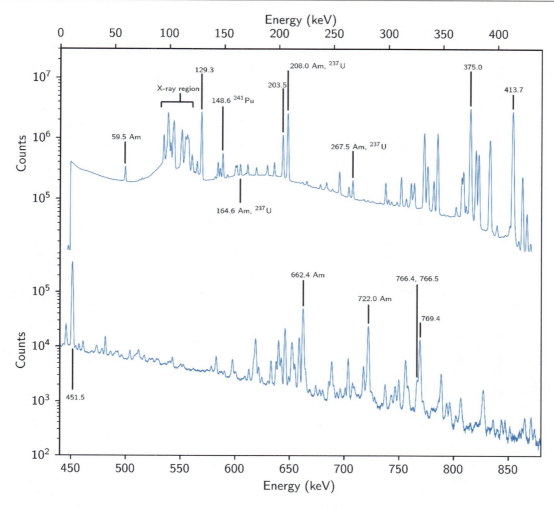

Fig. 2.14 Gamma-ray spectrum of low-burnup plutonium with approximately 6% ^{240}Pu. Peaks not labeled with a specific isotope are from ^{239}Pu. Refer to the original PANDA book [6] for identification of additional peaks

2.4.4 Background Radiation

All gamma-ray detectors will give some response even in the absence of a measurement sample. This response is due to the ambient background in the location of the detector. The ambient background consists of radiation from nuclear material in nearby storage areas, cosmic-ray interactions, and natural radioactivity in the local environment.

The radiation from the nuclear material stored nearby is often of the same nature as the radiation from the samples to be measured; for example, in a fuel production facility, the background generally consists of more uranium in addition to the naturally occurring radioactive background. This background spectrum often has a high Compton continuum (see Sect. 3.3.2), resulting from degradation and scattering in the materials that separate the detector from the storage area. Background radiation from nuclear material can be minimized with a judicious choice of detector location and shielding.

At the Earth's surface, cosmic rays consist primarily of high-energy gamma rays and charged particles. Although a neutron component exists, it has little effect on gamma-ray detectors. The charged particles are mostly muons but also include electrons and protons. The muon flux at sea level is approximately 0.038/cm^2 − s; at an altitude of 2000 m, the muon flux increases to approximately 0.055/cm^2 − s. The muon interacts with matter as though it were a heavy electron, and its rate of energy loss when passing through typical solid or liquid detector materials is approximately 8.6 MeV/cm. A typical penetrating muon deposits approximately 34 MeV in a 40 mm thick detector. Because this is much more energy than can be deposited by gamma rays from uranium or plutonium, muon interactions can overload or saturate the detector electronics. For a detector with a front surface area of 20 cm^2, the typical muon interaction rate at sea level is approximately 0.75/s.

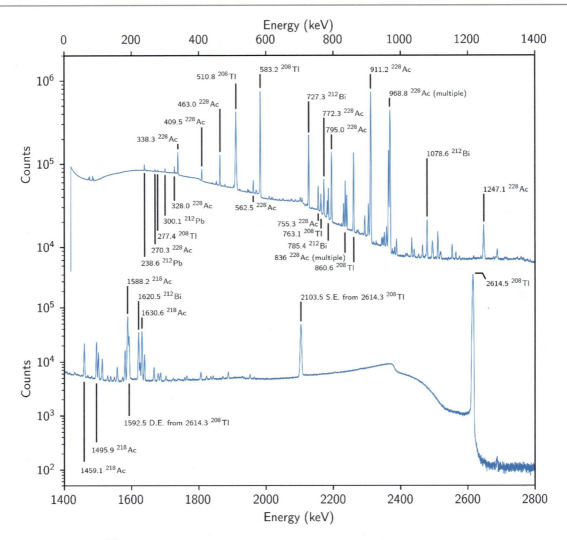

Fig. 2.15 Gamma-ray spectrum ^{232}Th and its daughter products. Thorium-232 emits no significant gamma rays of its own. The daughter nuclides grow into equilibrium with the ^{232}Th parent over a period of approximately 35 years

All materials have varying degrees of natural radioactivity. For example, the human body—and even some gamma-ray detectors—have some measurable natural radioactivity. Building materials, such as concrete, can be especially active. The major radioactive species in natural materials are ^{40}K, ^{232}Th and its daughters, and ^{235}U and ^{238}U and their daughters. Potassium-40 has a natural abundance of 0.0117% and decays by both electron capture (10.67%) and β^- decay (89.33%), with a half-life of 1.277×10^9 year. The electron capture is accompanied by the emission of a 1.461 MeV gamma ray that is evident in almost all background gamma-ray spectra taken on land. Potassium is present in most organic matter, with ^{40}K being the major source of radioactivity.

Thorium is a common trace element in many terrestrial rocks. Thorium-232 is the natural parent to the thorium decay series, which goes through 10 generations before reaching the stable nuclide ^{208}Pb. The most significant gamma rays associated with ^{232}Th are actually emitted by other radionuclides in its decay chain, particularly ^{228}Ac, ^{212}Bi, and ^{208}Tl.

Uranium is also found as a trace element in many rocks, although it is less common than thorium. The gamma-ray spectrum of unprocessed uranium ore is much different from that of uranium seen in the nuclear fuel cycle. Because of the long half-life of the daughter ^{230}Th (7.5×10^4 year), later generations take a long time to grow back into equilibrium after any chemical treatment that separates uranium daughters from the natural ore. Figure 2.16 shows a typical spectrum of uranium ore (compare with Fig. 2.12). Natural chemical processes in different rocks can often leach out some of the daughter nuclides and cause different ores to have different gamma-ray emissions. The natural sources discussed above are common and contribute

Table 2.2 Major nondestructive assay gamma-ray signatures

Isotope	Energy[a] (keV)	Activity[a] (γ/g − s)	Mean Free Path[b] (mm) (High-Z, ρ)	(Low-Z, ρ)
^{234}U	120.9	8.06E+04	1.0	69.9
^{235}U	143.8	8.76E+03	0.4	74.4
	185.7	4.56E+04	0.7	81.1
^{238}U	766.4[c]	3.94E+01	9.8	141.6
	1001.0[c]	1.05E+02	13.1	160.5
^{238}Pu	152.7	5.89E+06	0.5	76.0
	766.4	1.39E+05	9.8	141.6
^{239}Pu	129.3	1.45E+05	0.3	71.6
	413.7	3.37E+04	4.0	108.6
^{240}Pu	45.2	3.75E+06	0.1	33.1
	160.3	3.38E+04	0.5	77.3
	642.5	1.09E+03	7.9	130.8
^{241}Pu	148.6	7.12E+06	0.4	75.3
	208.0[d]	1.99E+07	0.9	84.3
^{241}Am	59.5	4.55E+10	0.2	47.0
	662.4	4.62E+05	8.2	132.6

[a]Data for energy and activity are extracted from [7]
[b]The mean free path is the absorber thickness that reduces the gamma-ray intensity to $1/e = 0.37$. The mean free path in uranium or plutonium oxide ($\rho = 10$ g/cm^3) is given for the high-density, high-atomic-number case (high-Z, ρ). The mean free path in aluminum oxide ($\rho = 1$ g/cm^3) is given for the low-density, low-atomic-number case (low-Z, ρ). Attenuation data are from [8]
[c]From the ^{238}U daughter ^{234m}Pa; equilibrium assumed
[d]From the ^{241}Pu daughter ^{237}U; equilibrium assumed

to the background gamma-ray spectrum in most locations. Other sources of background are occasionally encountered, such as materials contaminated by radioactive tracers. Slag from steel furnaces, which can have measurable levels of ^{60}Co, and uranium tailings are used as a concrete aggregate in some areas. The use of such materials in buildings can contribute to background radiation levels.

2.5 Additional Gamma-Ray Production Reactions

The discussion in Sect. 2.4 has been limited to gamma rays that come from the natural decay reactions of radioactive nuclides. These gamma rays provide the bulk of the signatures useful for NDA. This section discusses gamma rays produced in other nuclear reactions. Some of these radiations can interfere with NDA.

When nuclei interact with other particles, charged or neutral, the nuclei often emit gamma rays as products of the interaction. The neutron capture reaction (n,γ) is a classic example. Usually, the new nucleus is radioactive and is created in an excited state from which it can decay by gamma-ray emission. The following formulas illustrate the neutron-capture reaction that breeds plutonium in a fission reactor:

$$n + {}^{238}U \rightarrow \gamma + {}^{239}U \; [T_{1/2} = 23.45 \; min].$$

$$^{239}U \rightarrow e^- + \bar{\nu}_e + {}^{239}Np \, [T_{1/2} = 2.36 \; days].$$

$$^{239}Np \rightarrow e^- + \bar{\nu}_e + {}^{239}Pu \, [T_{1/2} = 24,110 \; year].$$

Photons from the capture reaction have discrete energies that are characteristic of the levels of the daughter nucleus. Their energies are typically several MeV for higher-atomic-number nuclei. A commonly measured example is the H(n,γ) at 2223.2 keV; the hydrogen in the ground (concrete, dirt, etc.) or in other objects (particularly humans) is often enough to make this line visible in gamma-ray measurements in an environment with a significant number of neutrons.

Inelastic scattering of neutrons (n,n'γ) is usually accompanied by gamma-ray emission. The gamma rays have discrete energies that are characteristic of the levels in the target nucleus. Gamma rays produced by this reaction are usually not of

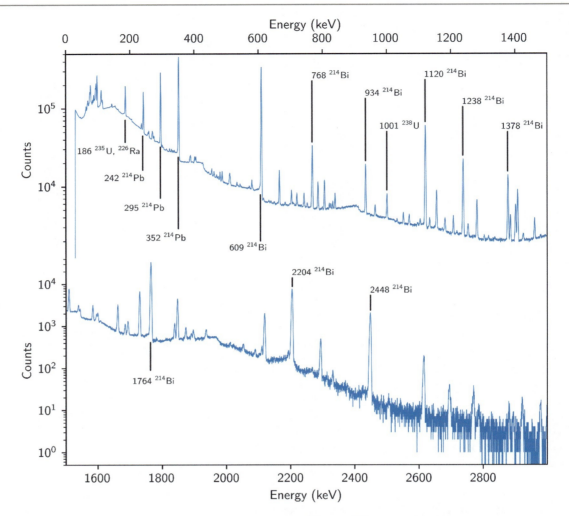

Fig. 2.16 Gamma-ray spectrum of uranium ore. Major radiations are from ^{214}Pb and ^{214}Bi. Compare to the spectrum of processed uranium in Fig. 2.12; most daughter products are removed during processing. Refer to the original PANDA book [6] for identification of additional peaks

interest for NDA, although they may complicate analysis and can provide information about materials present during a gamma-ray measurement in an environment with a significant neutron flux. For example, an 847 keV photon may indicate fast neutrons scattering on iron.

A major source of neutrons from plutonium compounds and UF$_6$ is the interaction of alpha particles from nuclear decay with low-atomic-number nuclei in the compound or surrounding matrix material (see Chap. 13 for more details). This effect can induce the emission of various particles; a common example in nuclear safeguards and security applications is the interaction of alpha particles with fluorine, which can undergo a few different reactions:

$$^{19}F + \alpha \rightarrow n + {}^{22}Na \; [T_{1/2} = 2.60 \; year]$$

$$^{19}F + \alpha \rightarrow p + {}^{22}Ne^*.$$

The fluorine reaction usually proceeds to the ground state of ^{22}Na, which does not result in prompt gamma-ray emission; however, the subsequent β^+ decay of ^{22}Na leads to gamma rays with energies of 511 and 1274.5 keV. These radiations are evident in samples of PuF$_4$ and ^{238}PuO$_2$ with even trace fluorine impurities and are not generally useful as assay signatures but are an indicator of chemical form or possibly impurities present in an item. Knowing that they are occurring is important for assay with neutron-measurement techniques.

References

1. E. Rutherford, VIII. Uranium radiation and the electrical conduction produced by it. The London, Edinburgh, and Dublin Philosophical Magazine and Journal of Science **47**(284), 109–163 (1899). https://doi.org/10.1080/14786449908621245
2. NuDat 3.0, https://www.nndc.bnl.gov/nudat3/. Accessed June 2023
3. M. Siegbahn, Relations between the K and L series of the high-frequency spectra. Nature **96**, 676 (1916)
4. G. Zschornack, *Handbook of X-Ray data* (Springer, Berlin/Heidelberg, 2007)
5. T. Gozani, *Active nondestructive assay of nuclear materials: Principles and applications*, NUREG/CR-0602 (U.S. Nuclear Regulatory Commission, Washington, DC, 1981)
6. D. Reilly, N. Ensslin, H. Smith Jr., S. Kreiner, *Passive nondestructive assay of nuclear materials (PANDA)* (U.S. Nuclear Regulatory Commission, 1991)
7. D.A. Brown, M.B. Chadwick, R. Capote, et al., ENDF/B-VIII.0: The 8th major release of the nuclear reaction data library with CIELO-project cross sections, new standards and thermal scattering data. Nuclear Data Sheets **148**, 1–142 (2018)
8. J.H. Hubbell, *Photon cross sections, attenuation coefficients, and energy absorption coefficients from 10 keV to 100 GeV* (U.S. Department of Commerce, National Bureau of Standards report NSRDS-NBS 29, 1969)

Open Access This chapter is licensed under the terms of the Creative Commons Attribution 4.0 International License (http://creativecommons.org/licenses/by/4.0/), which permits use, sharing, adaptation, distribution and reproduction in any medium or format, as long as you give appropriate credit to the original author(s) and the source, provide a link to the Creative Commons license and indicate if changes were made.

The images or other third party material in this chapter are included in the chapter's Creative Commons license, unless indicated otherwise in a credit line to the material. If material is not included in the chapter's Creative Commons license and your intended use is not permitted by statutory regulation or exceeds the permitted use, you will need to obtain permission directly from the copyright holder.

Gamma-Ray Interactions with Matter

P. J. Karpius and T. D. Reilly

3.1 Introduction

A knowledge of gamma-ray interactions is important to the nondestructive assay (NDA) practitioner to understand gamma-ray detection and attenuation. A gamma ray must interact with a detector to be recorded. Although the isotopes of uranium and plutonium emit gamma rays at fixed energies and emission probabilities, the gamma-ray intensity measured outside a distributed source is always attenuated because of gamma-ray interactions within the item. This attenuation must be carefully considered when conducting analyses of gamma-ray NDA measurements.

This chapter discusses the exponential attenuation of gamma rays in bulk materials and describes the major gamma-ray interactions and the application of gamma-ray shielding, filtering, and collimation. The discussion given here is necessarily brief. For more detailed discussions, see Refs. [1–3].

3.2 Exponential Attenuation

Gamma rays were first identified in 1900 by Villard [4] as a component of the radiation from radium that was unaffected by magnetic fields and had much higher penetrability than alpha and beta particles. In 1909, Soddy and Russell [5] found that gamma-ray attenuation followed an exponential law and that the ratio of the attenuation coefficient to the density of the attenuating material was nearly constant for all materials.

3.2.1 The Fundamental Law of Gamma-Ray Attenuation

Figure 3.1 illustrates a simple attenuation experiment.

When gamma radiation of intensity (I_0) is incident on an attenuator of thickness (x), the emerging intensity (I) of uncollided photons transmitted through the attenuator is given by the exponential expression

$$I = I_0 e^{-\mu_\ell x}, \tag{3.1}$$

where μ_ℓ is the linear attenuation coefficient (expressed in cm^{-1}). The ratio (I/I_0) is called the *transmission*.

Figure 3.2 illustrates exponential attenuation for three different gamma-ray energies and shows that the transmission increases with increasing gamma-ray energy and decreases with increasing attenuator thickness.

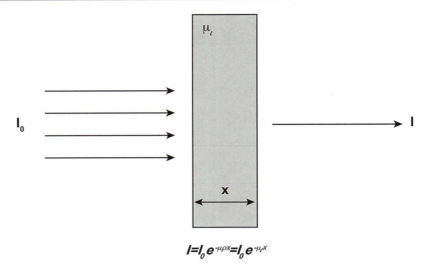

Fig. 3.1 The fundamental law of gamma-ray attenuation. The transmitted gamma-ray intensity (I) is a function of gamma-ray energy, attenuator composition, and attenuator thickness. The equation represents the intensity of gamma rays that transit the attenuator uncollided and are neither absorbed nor scattered

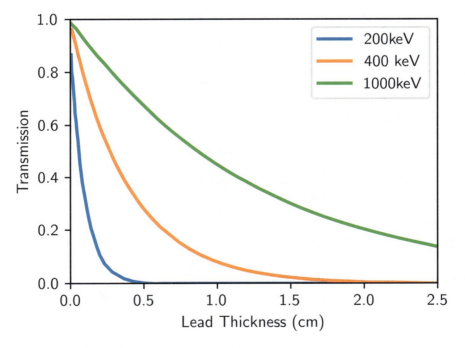

Fig. 3.2 Transmission of gamma rays through lead attenuators

Measurements with different sources and attenuators show that the linear attenuation coefficient (μ_ℓ) depends on the gamma-ray energy, the atomic number (Z), and the density (ρ) of the attenuator. For example, lead has a high density and a high atomic number and transmits a much lower fraction of incident gamma radiation without interaction than does a similar thickness of aluminum or steel. Figure 3.3 shows the linear attenuation of solid sodium iodide (NaI), a common material used in gamma-ray detectors. Note the dependence on the three different interactions (to be discussed later in this chapter): photoelectric absorption, Compton scattering, and pair production.

Alpha and beta particles have a well-defined range or stopping distance; however, as Fig. 3.2 shows, gamma rays do not have a unique range. The reciprocal of the attenuation coefficient ($1/\mu_\ell$) has units of length and is often called the *mean free*

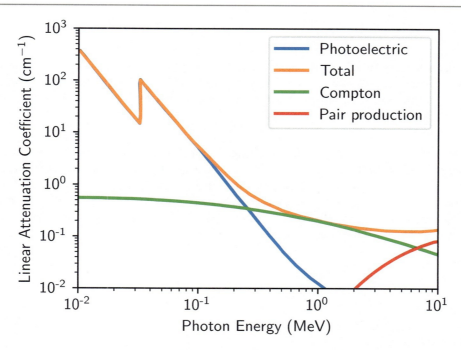

Fig. 3.3 Linear attenuation coefficient of NaI, showing contributions from photoelectric absorption, Compton scattering, and pair production

path. The mean free path can generally be thought of as the average distance a gamma-ray travels in a medium before interacting. Mathematically, it is the attenuator thickness that produces a transmission of 1/e, or ~ 0.37.

3.2.2 Mass Attenuation Coefficient

The linear attenuation coefficient is simple to measure experimentally, but care must be taken in its use because of its dependence on the density of the attenuating material. For example, at a given energy, the linear attenuation coefficients of water, ice, and steam are all different, even though the same elemental mixture of H_2O is involved in each case.

Gamma rays interact primarily with atomic electrons; therefore, the attenuation coefficient must be proportional to the electron density (P), which is proportional to the bulk density (ρ) of the absorbing material. However, for a given material, the ratio of the electron density to the bulk density is a constant (Z/A), independent of bulk density. The ratio (Z/A) is nearly constant for all but the heaviest elements and hydrogen.

$$P = Z\frac{\rho}{A}, \tag{3.2}$$

where

P = electron density
Z = atomic number
ρ = mass density
A = atomic mass

The ratio of the linear attenuation coefficient to the density ($\mu_{\ell/\rho}$) is called the *mass attenuation coefficient* (μ) and has the dimensions of area per unit mass (cm^2/g).[1] The units of this coefficient hint that one may think of it as the effective cross-sectional area of electrons per unit mass of the attenuator. The mass attenuation coefficient can be written in terms of a reaction cross section, $\sigma(cm^2)$,

---

[1] Note that some texts will call μ (sans subscript) the linear attenuation coefficient and μ/ρ the mass attenuation coefficient.

$$\mu = \frac{N_a \sigma}{A}, \tag{3.3}$$

where N_a is Avogadro's number (6.022×10^{23} atoms/mol), and A [g/mol] is the relative atomic mass of the attenuator material. The cross section is the probability that a gamma ray will undergo a single interaction. Chapter 14 gives a more complete definition of the cross-section concept. Using the mass attenuation coefficient, Eq. 3.1 can be rewritten as

$$I = I_0 e^{-\mu \rho x}. \tag{3.4}$$

The mass attenuation coefficient is independent of bulk density; for the example mentioned above, water, ice, and steam all have the same value of μ. This coefficient is more commonly tabulated than the linear attenuation coefficient because it quantifies the gamma-ray interaction probability for an individual element. References [6, 7] are widely used tabulations of the mass attenuation coefficients of the elements and were used, along with others, in compiling the National Institute of Standards and Technology Photon Cross Sections Database [8]. Equation 3.5 is used to calculate the effective mass attenuation coefficient for compound materials

$$\mu = \sum \mu_i m_i, \tag{3.5}$$

where

μ_i = mass attenuation coefficient of ith element
m_i = mass fraction of ith element.

The use of Eq. 3.5 is illustrated as follows for solid uranium hexafluoride (UF_6) at 200 keV:

μ_u = mass attenuation coefficient of U at 200 keV = 1.23 cm^2/g
μ_f = mass attenuation coefficient of F at 200 keV = 0.123 cm^2/g
m_u = mass fraction of U in UF_6 = 0.68
m_f = mass fraction of F in UF_6 = 0.32
ρ = density of solid UF_6 = 5.1 g/cm^3
$\mu = \mu_u m_u + \mu_f m_f = 1.23 \times 0.68 + 0.123 \times 0.32 = 0.88$ cm^2/g
$\mu_\ell = \mu \rho = 0.88 \times 5.1 = 4.5$ cm^{-1}.

3.3 Interaction Processes

The gamma rays of interest to NDA applications interact with detectors and attenuators via three major processes: photoelectric absorption, Compton scattering, and pair production. For NDA applications that involve nuclear safeguards, the typical energy range of interest extends from roughly 60 keV to 1001 keV. Because of the energy threshold on pair production, only the photoelectric effect and Compton scattering will play major roles in this range. For the application of nuclide identification—where it is necessary to extend to 3 MeV and beyond—pair production will be a commonly occurring interaction.

3.3.1 Photoelectric Effect

In 1887, Heinrich Hertz discovered the photoelectric absorption process [9] by observing that photons of ultraviolet light liberate electrons from a metal surface. In the equivalent process for gamma rays, the gamma ray loses all of its energy to an atomic electron in one interaction. The probability for this process depends very strongly on gamma-ray energy (E_γ) and atomic number (Z) of the material. A gamma ray may interact with a bound atomic electron in such a way that it loses all of its energy and ceases to exist as a photon (see Fig. 3.4). Some of the gamma-ray energy is used to overcome the electron binding energy, and most of the remainder is transferred to the freed electron as kinetic energy. To conserve momentum, a very small amount of recoil energy remains with the atom. Unscattered gamma rays from a sample that undergo photoelectric absorption

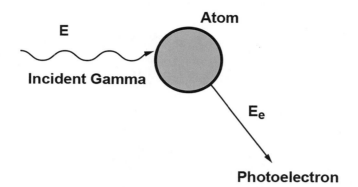

Fig. 3.4 A schematic representation of the photoelectric absorption process

in the detector contribute to the full-energy peak in a spectrum. These interactions are the most useful in NDA measurements. However, photoelectric absorption can also contribute to the region of a spectrum known as the *continuum* if the source gamma ray had a scattering event outside of the detector.

The probability of photoelectric absorption depends on the gamma-ray energy, the electron binding energy, and the atomic number of the atom. The probability is greater the more tightly bound the electron; therefore, K-shell electrons are most affected (more than 80% of photoelectric interactions involve K electrons) if the gamma-ray energy exceeds the K-electron binding energy. The probability is given approximately by Eq. 3.6, which shows that the interaction is more important for heavy nuclei—such as lead and uranium—and low-energy gamma rays.

$$\mu_{pe} \propto \frac{Z^n}{E^{3.5}}, \tag{3.6}$$

where μ_{pe} = photoelectric mass attenuation coefficient and the exponent (n) varies between 4 and 5 depending on the incident photon energy [10].

As the gamma-ray energy decreases, the probability of photoelectric absorption increases rapidly (see Fig. 3.3). Photoelectric absorption is the predominant interaction for low-energy gamma rays, X-rays, and bremsstrahlung.

The energy of the photoelectron (E_e) released by the interaction is the difference between the incident gamma-ray energy (E_γ) and the electron binding energy (E_b):

$$E_e = E_\gamma - E_b. \tag{3.7}$$

In most detectors, the photoelectron is stopped quickly in the active volume of the detector, which induces a response whose amplitude is proportional to the energy deposited by the photoelectron. The electron binding energy is not lost but appears as characteristic X-rays emitted in coincidence with the photoelectron. In most cases, these X-rays are eventually absorbed in the detector via photoelectric interaction in coincidence with the liberated photoelectron, and the resulting output pulse is proportional to the total energy of the incident gamma ray. For low-energy gamma rays that interact near the surface of detectors, a sufficient number of K-shell X-rays can escape from the detector to cause escape peaks in the observed spectrum; the peaks appear below the full-energy peak by an amount equal to the energy of the X-ray. This effect is illustrated in Fig. 3.5.

Figure 3.6 shows the photoelectric mass attenuation coefficient of lead. The interaction probability increases rapidly as energy decreases but then becomes much smaller at a gamma-ray energy just below the binding energy of the K-shell electron. This discontinuity is called the *K-edge*; below this energy, the gamma ray does not have sufficient energy to dislodge a K-shell electron. The discontinuity exists because, for energies below the K-edge, the K-shell electrons no longer play a part in photoelectric interactions, which causes a discreet change in the interaction probability across the edge. Below the K-edge, the interaction probability increases again until the energy drops below the binding energies of the L electrons; these discontinuities are called the L_I, L_{II}, and L_{III} *edges*. The presence of these absorption edges is important for densitometry and X-ray fluorescence measurements (see Chaps. 11 and 12).

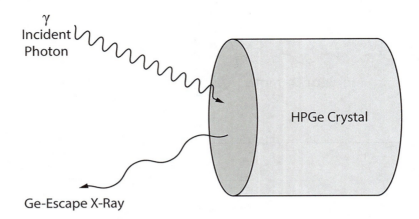

Fig. 3.5 An incident photon liberates an electron near the surface of a high-purity germanium (HPGe) crystal, and the resulting germanium X-ray escapes

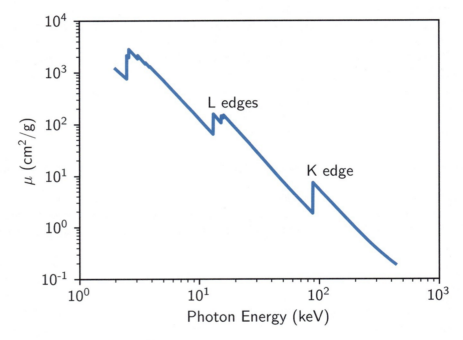

Fig. 3.6 Photoelectric mass attenuation coefficient of lead

3.3.2 Compton Scattering

Compton scattering is the process whereby a photon interacts with a bound or free electron and transfers part of its energy to the electron (see Fig. 3.7). When considering a bound electron, Compton scattering is more probable when the incident photon energy (E_γ) is much greater than the electron-binding energy (E_b).

Conservation of energy and momentum allows only a partial energy transfer when the electron is not bound tightly enough for the atom to absorb recoil energy. In this situation, the photoelectric effect is forbidden but Compton scattering is allowed. The probability for this process is weakly dependent on energy and is nearly independent of the atomic number (Z) because it depends on the electron density (N_a Z/A; electrons per unit mass of attenuating material), which varies little for all elements except hydrogen.[2]

---

[2] Hydrogen (>99% ^{1}H) is the only element without neutrons, so its electron density is about twice that of most nuclides.

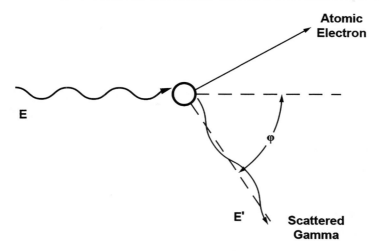

Fig. 3.7 A schematic representation of Compton scattering

The electron becomes a free electron with kinetic energy equal to the difference of the energy lost by the photon and the electron binding energy. Because the electron binding energy is very small compared with the incident photon energy, the kinetic energy of the electron is very nearly equal to the energy lost by the incident photon:

$$E_e = E_\gamma - E'_\gamma, \tag{3.8}$$

where

E_e = energy of scattered electron.
E_γ = energy of incident photon.
E_γ' = energy of scattered photon.

Two particles leave the interaction site: the freed electron and the scattered photon. The directions of the electron and the scattered photon depend on the amount of energy transferred to the electron during the interaction. Equation 3.9 gives the energy of the scattered photon,

$$E'_\gamma = \frac{m_0 c^2}{1 - \cos\phi + \frac{m_0 c^2}{E_\gamma}}, \tag{3.9}$$

where

$m_0 c^2$ = rest mass energy of electron = 511 keV
ϕ = angle between incident and scattered photons (see Fig. 3.7).

This energy is minimum for a collision where the photon is scattered 180° and the electron moves forward in the direction of the incident photon. For this case, the energy of the scattered photon is given by Eq. 3.10, and the energy of the scattered electron is given by Eq. 3.11:

$$E'_\gamma(min) = \frac{m_0 c^2}{2 + \frac{m_0 c^2}{E_\gamma}} \approx \frac{m_0 c^2}{2} = 256\, keV;\, if\, E_\gamma \gg \frac{m_0 c^2}{2} \tag{3.10}$$

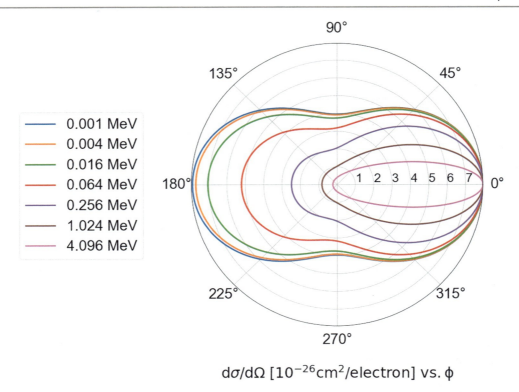

Fig. 3.8 A polar plot of the differential cross section for the number of photons (incident from the left) Compton scattered into a unit solid angle at a mean scattering angle ϕ. The curves represent incident photon energies [10]

$$E_e(max) = \frac{E}{1 + \frac{m_0 c^2}{2E}} \approx E_\gamma - 256\ keV;\ if\ E_\gamma \gg \frac{m_0 c^2}{2} \tag{3.11}$$

For very small angle scattering ($\phi \simeq 0°$), the energy of the scattered photon is only slightly less than the energy of the incident photon, and the scattered electron takes very little energy away from the interaction. The energy given to the scattered electron ranges from near zero to the maximum given by Eq. 3.11. The probability for Compton scattering of a photon from a free electron is given by the differential cross section first derived by Klein and Nishina [11]. This is shown in the polar plot of Fig. 3.8 for different incident photon energies.

The differential cross section (i.e., the interaction probability) with respect to solid angle for the scattered photon becomes strongly peaked in the forward direction (small photon-scattering angles) with increasing incident photon energy. Figure 3.9 shows the differential cross section (in electrons/MeV) for scattering a recoil electron into a certain kinetic energy interval as a function of that recoil energy. The multiple curves represent various incident photon energies. Note that especially as the incident photon energy increases, the distribution of recoil-electron energies is relatively flat but increases to a sharp maximum (the 'Compton Edge') at the highest possible recoil energy. Nelms [12] provides results in this regard for incident photon energies from 10 keV up to 500 MeV.

When Compton scattering occurs in a detector, the scattered electron is usually stopped in the detection medium, and the detector produces a response that is proportional to the energy lost by the incident photon. Compton scattering in a detector produces a spectrum of output pulses from zero up to the maximum energy given by Eq. 3.11. It is difficult to relate the Compton-scattering spectrum to the energy of the incident photon. Figure 3.10 shows the measured gamma-ray spectrum from a monoenergetic gamma-ray source (^{137}Cs). The full-energy peak at 662 keV is formed by interactions where the gamma-ray deposits all of its energy in the detector either by a single photoelectric absorption or by a series of Compton scatterings followed by photoelectric absorption.

The region of the spectrum below the full-energy peak is formed by Compton scattering, where the photon loses only part of its energy in the detector. The step near 470 keV corresponds to the maximum energy that can be transferred to an electron by a 662-keV gamma ray in a single Compton-scattering event. This step is called a *Compton edge*. Note that in the spectrum, the structure of the Compton edge is spread out and is most intense at a slightly lower energy than the maximum (free-

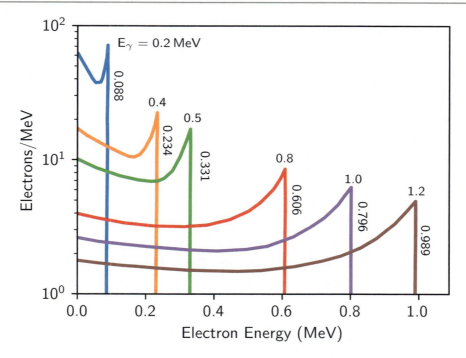

Fig. 3.9 Differential cross section (in electrons/MeV) for Compton scattering as a function of energy of scattered electrons and incident photon energy (E_γ). The sharp discontinuity corresponds to the maximum energy that can be transferred in a single scattering event

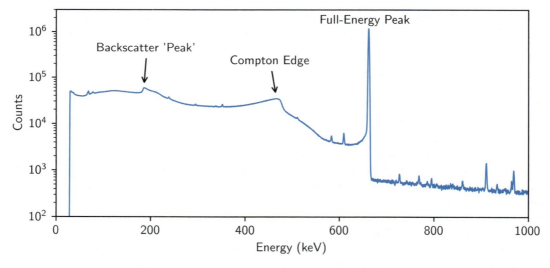

Fig. 3.10 High-resolution spectrum of ^{137}Cs showing full-energy peak, Compton edge, and backscatter peak from the 662 keV gamma ray. Events below the full-energy peak are caused by Compton scattering in the detector and surrounding materials

electron) calculated value. The limitations of detector resolution are insufficient to explain this, which is made clear in high-purity germanium spectra. The smearing of the edge is mainly due to the distribution of momentum of the bound electrons in the detection medium [13]. This effect is enhanced due to the tendency for large-angle scattering to occur from inner-shell electrons [14].

The small peak around 188 keV shown in Fig. 3.10 is called a *backscatter peak*. The backscatter peak is formed when the photon undergoes a large-angle scattering ($\simeq 180°$) in the material that surrounds the detector and then is absorbed in the detector by photoelectric interaction. The energy of the backscatter peak is given by Eq. 3.12, which shows that the maximum

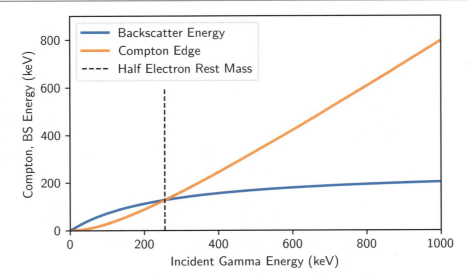

Fig. 3.11 Energy of the Compton edge and backscatter versus the energy of the incident gamma ray

energy is half the rest energy of the electron, i.e., 256 keV. The sum of the energy of the backscatter peak $E_e(180°)$ and the Compton edge $E'\gamma(180°)$ equals the energy of the incident photon,

$$E_e(180°) + E'_\gamma(180°) = E_\gamma. \tag{3.12}$$

Again, the backscatter gamma ray energy $\left(E'_\gamma(180°) = E_\gamma - E_e(180°)\right)$ is detected when the original photon first undergoes a 180° scatter outside the detector, followed by full-energy deposition of the scattered photon inside the detector. The Compton edge energy $\left(E_e(180°) = E_\gamma - E'_\gamma(180°)\right)$ is detected by 180° scattering of full-energy photons inside the detector, followed by escape of the scattered photon and full-energy deposition of the Compton electron in the detector.

If the incident full-energy photon undergoes Compton-edge scattering and is followed by additional scattering inside the detector before its escape, counts will accumulate between the Compton edge and the full-energy peak. Counts can also be contributed to this region by small-angle scattering of the incident photon before it reaches the detector. In this case, the scattered photon retains an energy greater than the Compton edge and deposits this energy in the detector via eventual photoelectric absorption.

Figure 3.10 illustrates the general features in a spectrum for incident photon energies well above half the rest energy of the electron. Figure 3.11 illustrates how the Compton-edge and backscatter energies depend on the incident photon energy. At half the rest energy of the electron (256 keV), the Compton-edge and backscatter energies are the same. Below that, the backscatter energy is higher than the Compton-edge energy.

When the incident photon energy is much larger than the electron binding energy, the electron can be considered essentially free in an approximate sense. The nucleus has only a minor influence, and the probability for Compton scattering is nearly independent of atomic number. The interaction probability depends on the electron density, which is proportional to Z/A, and is nearly constant for all materials. The Compton-scattering probability is a slowly varying function of gamma-ray energy (see Fig. 3.3).

3.3.3 Pair Production

A gamma ray with an energy of at least 1.022 MeV can create an electron-positron pair when it is under the influence of the strong Coulomb field of a nucleus (see Fig. 3.12). In this interaction, the nucleus receives a very small amount of recoil energy to conserve momentum, but the nucleus is otherwise unchanged, and the gamma ray disappears. This interaction has a threshold of 1.022 MeV, the minimum energy required to create the electron and positron, each of which have a rest energy of 0.511 MeV. If the gamma-ray energy exceeds 1.022 MeV, the excess energy is shared between the electron and positron as

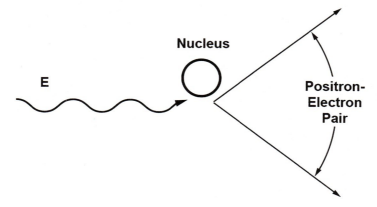

Fig. 3.12 A schematic representation of pair production

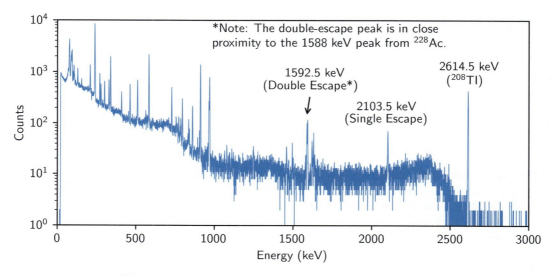

Fig. 3.13 Gamma-ray spectrum of a ^{232}Th spectrum showing single-escape (2103.5 keV) and double-escape (1592.5 keV) peaks that arise from pair-production interactions of 2614.5 keV gamma rays from ^{208}Tl in a germanium detector

kinetic energy. This interaction process is relatively unimportant for nuclear material assay when safeguards is the focus because most important gamma-ray signatures are below 1.022 MeV. However, when nuclide identification is the focus, we must consider the effects of pair production because this work often spans a multi-MeV energy range.

The electron and positron from pair production are rapidly slowed down in the attenuator. After losing its kinetic energy, the positron combines with an electron in an annihilation process, which releases two photons—each with an energy of 0.511 MeV. These photons can interact further with the attenuating material or can escape. In a gamma-ray detector, this interaction often gives three peaks for a high-energy incident gamma ray (see Fig. 3.13). When the pair production event occurs inside the detector, the kinetic energy of the electron and positron is absorbed in the detector. Analogous to the case of germanium X-ray escape discussed previously, none, one, or both of the annihilation photons can escape from the detector. If both annihilation photons are absorbed in the detector, the interaction contributes to the full-energy peak in the measured spectrum; if one of the annihilation photons escapes from the detector, the interaction contributes to the single-escape peak located 0.511 MeV below the full-energy peak; if both annihilation photons escape, the interaction contributes to the double-escape peak located 1.022 MeV below the full-energy peak. The relative heights of the three peaks depend on the energy of the incident gamma ray and the size of the detector. These escape peaks could arise when samples of irradiated fuel, thorium, and ^{232}U are measured because these materials possess important gamma rays above the pair-production threshold. A ^{232}Th

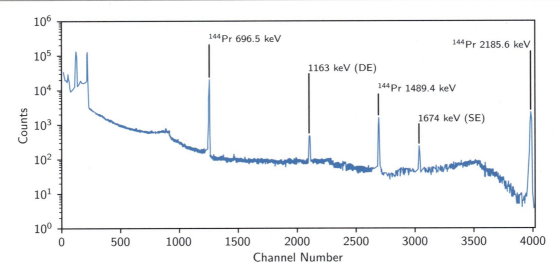

Fig. 3.14 Gamma-ray spectrum of the fission product ^{144}Pr showing single-escape (SE) (1674 keV) and double-escape (DE) (1163 keV) peaks that arise from pair-production interactions of 2186 keV gamma rays in a germanium detector

spectrum is shown in Fig. 3.13. The prominent 2614.5 keV gamma rays from the daughter ^{208}Tl are a common source of single- and double-escape peaks.

Irradiated fuel is sometimes measured using the 2186-keV gamma ray from from the short-lived daughter ^{144}Pr (17 minutes) of the fission product ^{144}Ce (285 days). The gamma-ray spectrum of ^{144}Pr shown in Fig. 3.14 shows the single-escape (SE) and double-escape (DE) peaks that arise from pair-production interactions of the 2186-keV gamma ray in a germanium detector.

Pair production is impossible for gamma rays with energy less than 1.022 MeV. Above this threshold, the probability of the interaction increases rapidly with energy (see Fig. 3.3). The probability of pair production varies approximately as the square of the atomic number (Z) and is significant in high-Z elements, such as lead or uranium. In lead, approximately 20% of the interactions of 1.5 MeV gamma rays are through the pair-production process, and the fraction increases to 50% at 2.0 MeV. For carbon, the corresponding interaction fractions are 2% and 4%.

3.3.4 Total Mass Attenuation Coefficient

All three interaction processes described above contribute to the total mass attenuation coefficient. The relative importance of the three interactions depends on incident photon energy and the effective atomic number of the attenuating material. Figure 3.15 shows a composite of mass attenuation curves that cover a wide range of energy and atomic number. It shows the dramatic variation of the three processes. All elements except hydrogen show a sharp, low-energy rise that indicates where photoelectric absorption is the dominant interaction. The position of the rise is very dependent on atomic number. Above the low-energy rise, the value of the mass attenuation coefficient decreases gradually, indicating the region where Compton scattering is the dominant interaction. The mass attenuation coefficients for all elements with atomic number less than 26 (iron) are nearly identical in the energy range of 200–2000 keV. The attenuation curves converge for all elements except hydrogen in the range of 1–2 MeV. The shape of the mass attenuation curve of hydrogen shows that it interacts with photons with energy greater than 10 keV almost exclusively by Compton scattering. The mass-attenuation coefficient curve for hydrogen does not converge with those of the other elements around 2 MeV because the lack of neutrons in hydrogen (^{1}H) gives it a (relative) electron density that is higher by a factor of 2. Above 2 MeV, the pair-production interaction becomes important for high-Z elements, and the mass attenuation coefficient begins to rise again. An understanding of the major features of Fig. 3.15 is very helpful to the understanding of NDA techniques.

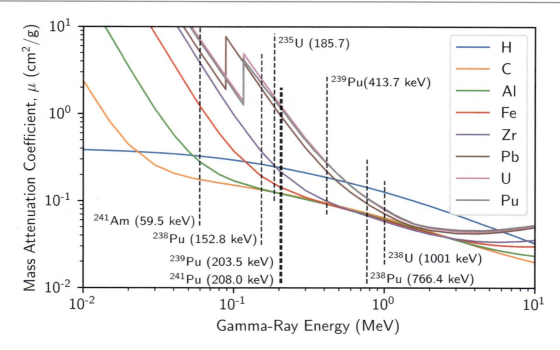

Fig. 3.15 Total narrow beam mass attenuation coefficients of selected elements. Also indicated are gamma-ray energies commonly encountered in NDA of uranium and plutonium

Table 3.1 Effect of a cadmium filter on a plutonium spectrum

	Plutonium signal (counts/s)[a]			
Absorber (cm)	60 keV	129 keV	208 keV	414 keV
0	3.57×10^6	1.29×10^4	8.50×10^4	2.02×10^4
0.1	2.40×10^4	0.67×10^4	6.76×10^4	1.85×10^4
0.2	1.86×10^2	0.34×10^4	5.37×10^4	1.69×10^4

[a] ^{241}Am = 0.135%; ^{239}Pu = 81.9%; ^{241}Pu = 1.3%. Signal is from a 2 g disk of plutonium metal, 1 cm diameter × 0.13 cm thick

3.4 Filters

In many assay applications, the gamma rays of interest can be measured more easily if lower-energy gamma rays can be absorbed by photoelectric interaction before they reach the detector. The lower-energy gamma rays can cause significant count-rate-related losses in the detector and spectral distortion if they are not removed. The removal process is often called *filtering*, although it is nothing more than selective attenuation. A perfect filter material would have a transmission of zero below the energy of interest and a transmission of unity above that energy, but as Fig. 3.15 shows, such a material does not exist. Useful filters can be obtained by selecting attenuators of appropriate atomic number such that the sharp rise in photoelectric cross section is near the energy of the gamma rays that must be attenuated but well below the energy of the assay gamma rays of interest.

Filtering is usually employed in the measurement of plutonium gamma-ray spectra. Except immediately after chemical separation of americium, all plutonium samples have significant levels of ^{241}Am, which emits a very intense gamma ray at ~60 keV. In most items, this gamma ray is the most intense gamma ray in the spectrum and must be attenuated so that the plutonium gamma rays can be accurately measured. A thin sheet of cadmium or tin is commonly used to attenuate these intense ^{241}Am gamma rays. Table 3.1 shows how a cadmium filter affects the spectrum of a 2 g disk of plutonium metal. In the absence of the filter, the 60 keV gamma ray dominates the spectrum and could even paralyze the detector. A 1–2 mm cadmium filter drastically attenuates the 60 keV activity but only slightly attenuates the higher-energy plutonium lines. The plutonium spectrum below 250 keV is usually measured with a cadmium filter. When only the ^{239}Pu 414 keV gamma ray is of interest, lead may be used as the filter material because it will—depending on its thickness—attenuate gamma rays in the 100–200 keV region and will stop most of the 60 keV gamma rays. It is interesting to note that at 60 keV, the mass attenuation coefficients of lead and cadmium are essentially equal in spite of the higher Z of lead (82) relative to cadmium (48). This effect occurs because the K-edge of lead appears at 88 keV, as discussed in Sect. 3.3.2.

A cadmium filter is often used when measuring ^{235}U because it attenuates gamma rays and X-rays in the 90–120 keV region and does not significantly affect the 186 keV gamma ray from ^{235}U. Filters may also be used for certain irradiated fuel measurements. The 2186 keV gamma ray from the fission products ^{144}Ce-^{144}Pr is measured in some applications as an indicator of the residual fuel material in leached hulls produced at a reprocessing plant (see Chap. 21). The major fission-product gamma-ray activity is in the 500–900 keV region and can be selectively reduced relative to the 2186 keV gamma ray using a 10–15 cm thick lead filter.

Graded filters with two or more materials are sometimes used to attenuate the characteristic X-rays from the primary filter material before they interact in the detector. When gamma rays are absorbed in the primary filter material, the interaction produces copious amounts of X-rays. For example, when the 60 keV gamma rays from ^{241}Am are absorbed in a thin cadmium filter, a significant flux of 23 and 27 keV cadmium X-rays can be produced. If these X-rays create a problem in the detector, they can be attenuated easily using a very thin sheet of copper. The K X-rays of copper are around 8 keV, which is usually a sufficiently low energy that does not interfere with the measurement. If the primary filter material is lead, cadmium is used to absorb the characteristic K X-rays of lead at 75 and 85 keV, and copper is used to absorb the characteristic K X-rays of cadmium at 23 and 27 keV. In graded filters, the lowest Z material is always placed closest to the detector.

3.5 Shielding

In NDA instruments, shields and collimators are required to limit the detector response to background gamma rays and to shield the operator and detector from transmission and activation sources. Gamma-ray shielding materials should be of high density and high atomic number so that they have a high total linear attenuation coefficient and a high photoelectric absorption probability. The most common shielding material is lead because it is readily available, has a density of 11.35 g/cm^3 and an atomic number of 82, and is relatively inexpensive. Lead can be molded into many shapes; however, because of its high ductility, it cannot be machined easily or hold a given shape unless supported by a rigid material.

In some applications, an alloy of tungsten (atomic number 74) is used in place of lead because it has significantly higher density (17 g/cm^3), can be more easily machined, and holds a shape well. Table 3.2 shows some attenuation properties of the two materials. The tungsten is alloyed with nickel and copper to improve its machinability.

Table 3.2 shows that at energies above 500 keV, the tungsten alloy has a significantly higher linear attenuation coefficient than lead because Compton scattering dominates in this energy range and is mainly dependent on density, which is higher for tungsten than lead. Thus, the same shielding effect can be achieved with a thinner shield. At energies below 500 keV, the difference between the attenuation properties of the two materials is less significant; the higher density of the tungsten alloy is offset by the lower atomic number. The tungsten alloy is used where space is severely limited or where machinability and mechanical strength are important. However, tungsten is 20–30 times more expensive than lead; therefore, it is used sparingly and is almost never used for massive shields. The alloy is often used to hold intense gamma-ray transmission sources or to collimate gamma-ray detectors. Note also that depleted uranium at atomic number 92 and density of 18.95 g/cm^3 is an even better gamma-ray shield than tungsten. The radioactivity of a depleted uranium shield (colloquially called a *pig*) is normally far less than the industrial radiation source it is designed to contain.

Table 3.2 Attenuation properties of lead and tungsten

Energy (keV)	Linear attenuation coefficient (cm^{-1})		Thickness (cm)[a]	
	Lead	Tungsten[b]	Lead	Tungsten[b]
1000	0.77	1.08	2.98	2.14
500	1.70	2.14	1.35	1.08
200	10.6	11.5	0.22	0.20
100	60.4	64.8	0.038	0.036

[a]Thickness of attenuator with 10% transmission
[b]Alloy: 90% tungsten, 6% nickel, 4% copper

References

1. R.D. Evans, *The Atomic Nucleus* (McGraw-Hill Book Co., New York, 1955)
2. G.F. Knoll, *Radiation Detection and Measurement*, 5th edn. (Wiley, New York, 2020)
3. G.R. Gilmore, *Practical Gamma-Ray Spectrometry*, 2nd edn. (Wiley, New York, 2008)
4. P. Villard, Sur la re'flexion et la re'fraction des rayons cathodiques et des rayons de'viables du radium. Comptes rendus **130**, 1010–1012 (1900)
5. F. Soddy, A. Russell, Lond. Edinb. Dubl. Phil. Mag. J. Sci. **18**(106), 620–649 (1909)
6. E. Storm and H. Israel, "Photon Cross Sections from 0.001 to 100 MeV for Elements 1 through 100," Los Alamos Scientific Laboratory report LA-3753 (1967)
7. J.H. Hubbell, *Photon Cross Sections, Attenuation Coefficients, and Energy Absorption Coefficients from 10 keV to 100 GeV*, U.S. Department of Commerce, National Bureau of Standards report NSRDS-NSB 29 (August 1969)
8. M.J. Berger, J.H. Hubbell, S.M. Seltzer, J. Chang, J.S. Coursey, R. Sukumar, D.S. Zucker, K. Olsen, *XCOM: Photon Cross Sections Database* (National Institute of Standards and Technology, 1998) https://www.nist.gov/pml/xcom-photon-cross-sections-database. Accessed Feb 2023
9. H. Hertz, Ueber einen Einfluss des ultravioletten Lichtes auf die electrische Entladung. Ann. Phys. **267**, 983–1000 (1887)
10. G.F. Knoll, *Radiation Detection and Measurement*, 3rd edn. (Wiley, New York, 2000)
11. O. Klein, Y. Nishina, Über die Streuung von Strahlung durch freie Elektronen nach der neuen relativistischen Quantendynamik von Dirac. Zs. f. Phys. **52**, 853–868 (1929)
12. A.T. Nelms, Graphs of the Compton energy-angle relationship and the Klein-Nishina formula from 10 Kev to 500 MeV. National Bureau of Standards Circular **542**, 18 (1953)
13. J. Felsteiner, S. Kahane, B. Rosner, Effect of the electron-momentum distribution on the shape of the Compton edge of Si(li) detectors. Nucl. Instrum. Methods **118**(1), 253–255 (1974)
14. L.V. East, E.R. Lewis, Compton scattering by bound electrons. Physica **44**(4), 595–613 (1969)

Open Access This chapter is licensed under the terms of the Creative Commons Attribution 4.0 International License (http://creativecommons.org/licenses/by/4.0/), which permits use, sharing, adaptation, distribution and reproduction in any medium or format, as long as you give appropriate credit to the original author(s) and the source, provide a link to the Creative Commons license and indicate if changes were made.

The images or other third party material in this chapter are included in the chapter's Creative Commons license, unless indicated otherwise in a credit line to the material. If material is not included in the chapter's Creative Commons license and your intended use is not permitted by statutory regulation or exceeds the permitted use, you will need to obtain permission directly from the copyright holder.

Gamma-Ray Detectors

M. Lombardi, M. H. Carpenter, D. J. Mercer, P. A. Russo, and D. T. Vo

4.1 Introduction

For a gamma ray to be detected, it must interact with matter, and that interaction must be recorded. Fortunately, the electromagnetic nature of gamma-ray photons allows them to interact strongly with the electrons in the atoms of all matter. The key process by which a gamma ray is detected is *ionization*, where it gives up part or all of its energy to an electron. The unbound energetic electrons collide with other atoms and liberate many more electrons. The liberated charge is collected, either directly (as with a proportional counter or a solid-state semiconductor detector) or indirectly (as with a scintillation detector), to register the presence of the gamma ray and measure its energy. The final result is an electrical pulse whose voltage is proportional to the energy deposited in the detecting medium. The resolution of a detector or its ability to distinguish incoming gamma rays of different energies depends on multiple factors, which are covered in some detail in Sect. 4.3.

In this chapter, we present some general information on types of gamma-ray detectors that are used in nondestructive assay (NDA) of nuclear materials. The electronic instrumentation associated with gamma-ray detection is discussed in Chap. 5. More in-depth treatments of the design and operation of gamma-ray detectors can be found in Refs. [1, 2].

4.2 Types of Detectors

Many different detectors have been used to register the gamma ray and its energy. In NDA, it is usually necessary to measure not only the amount of radiation emanating from a sample but also its energy spectrum. Therefore, the detectors of most use in NDA applications are those whose signal outputs are proportional to the energy deposited by the gamma ray in the sensitive volume of the detector. This chapter will give an overview of the most commonly used NDA gamma-ray detectors; detector selection is discussed in Chap. 5. Table 4.1 shows some of the properties of a few commonly used spectroscopic gamma-ray detectors, with microcal included for comparison.

4.2.1 Gas-Filled Detectors

Gas counters consist of a sensitive volume of gas between two electrodes (see Fig. 4.1). In most designs, the outer electrode is the cylindrical wall of the gas pressure vessel, and the inner (positive) electrode is a thin wire positioned at the center of the cylinder. In some designs (especially ionization chambers), both electrodes can be positioned in the gas, separate from the gas pressure vessel.

Los Alamos National Laboratory strongly supports academic freedom and a researcher's right to publish; as an institution, however, the Laboratory does not endorse the viewpoint of a publication or guarantee its technical correctness.

M. Lombardi (✉) · M. H. Carpenter · D. J. Mercer · P. A. Russo · D. T. Vo
Los Alamos National Laboratory, Los Alamos, NM, USA
e-mail: lombardi@lanl.gov; mcarpenter@lanl.gov; mercer@lanl.gov; ducvo@lanl.gov

Table 4.1 Typical gamma ray dector properties

Detector	NaI:Tl	LaBr$_3$:Ce[a]	CdZnTe	HPGe[b] (planar)	Microcal
Type	Scintillator[c]	Scintillator	Solid state	Solid state	Low temperature
Average Z	50	45	49	32	50
Density(g/cm^3)	3.67	5.06	5.76	5.32	7.3
Resolution:% FWHM[d] @ 662 keV	7%	<3%	3.2%[e,f]	0.2%[e]	N/A
Resolution:% FWHM @ 122 keV	13%	6%	6.3%	0.4%	<0.06%
Scintillation light peak λ (nm)[g]	415	380	N/A	N/A	N/A

[a]https://www.crystals.saint-gobain.com/products/standard-and-enhanced-lanthanum-bromide
[b]HPGe = high-purity germanium
[c]Intrinsic properties of scintillator materials are from Ref. [1] unless noted otherwise
[d]FWHM = full width at half maximum
[e]Resolution values taken from Ref. [3]
[f]Resolution at 662 keV improves to ~2.5% for thinner crystals
[g]Glass transmits down to 350 ηm. Quartz transmits down to 180 ηm. Commercial suppliers currently use glass-window photomultiplier tubes

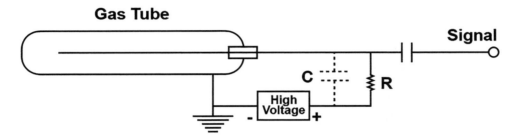

Fig. 4.1 The equivalent circuit for a gas-filled detector. The gas constitutes the sensitive (detecting) volume. The potential difference between the tube housing and the center wire produces a strong electric field in the gas volume. The electrons from ionizations in the gas travel to the center wire under the influence of the electric field, producing a charge surge on the wire for each detection event

An *ionization chamber* is a gas-filled counter for which the voltage between the electrodes is low enough that only the primary ionization charge is collected without any type of charge multiplication. The electrical output signal is not proportional to the energy deposited in the gas volume.

If the voltage between the electrodes is increased, the ionized electrons attain enough kinetic energy to cause further ionizations, which leads to charge multiplication within the detector. One then has a *proportional counter* that can be tailored for specific applications by varying the gas pressure and/or the operating voltage. The output signal is proportional to the energy deposited in the gas by the incident gamma-ray photon, and the energy resolution is intermediate between NaI scintillation counters and germanium (Ge) solid-state detectors. Proportional counters have been used for spectroscopy of gamma rays and X-rays whose energies are low enough (a few tens of keV) to interact with reasonable efficiency in the counter gas.

If the operating voltage is increased further, charge multiplication in the gas volume increases (avalanches) until the space charge produced by the residual ions inhibits further ionization. As a result, the amount of ionization saturates and becomes independent of the initial energy deposited in the gas. This type of detector is known as the *Geiger-Mueller (GM) detector*. A GM tube gas counter does not differentiate among the kinds of particles it detects or their energies; it counts only the number of particles that enter the detector. This type of detector is the basis of the conventional $\beta - \gamma$ dosimeter used in health physics.

Gas counters do not have much use in gamma-ray NDA of nuclear materials. The scintillation and solid-state detectors are much more desirable for obtaining the spectroscopic detail needed in the energy range typical of uranium and plutonium radiation (~100–1000 keV). Gas counters are described in more detail in Chap. 15 because they are more widely used for neutron detection.

4.2.2 Scintillation Detectors

The sensitive volume of a scintillation detector is a luminescent material (a solid, liquid, or gas) that is viewed by a device that detects the gamma-ray-induced light emissions, usually a photomultiplier tube (PMT) or a silicon photomultiplier (SiPM; [4]). The scintillation material may be organic or inorganic; the latter is more common. Examples of organic scintillators are

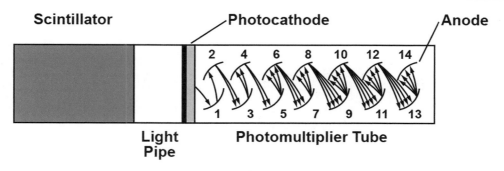

Fig. 4.2 Typical arrangement of components in a scintillation detector. The scintillator and PMT are often optically linked by a light pipe. The dynodes (1–13 in the figure) are arranged to allow successive electron cascades through the tube volume. The final charge burst is collected by the anode and is usually passed to a preamplifier for conversion to a voltage pulse

anthracene, plastic, and liquid. The latter two are less efficient than anthracene—the standard against which other scintillators are compared. Some common inorganic scintillation materials are sodium iodide (NaI), cesium iodide (CsI), zinc sulfide (ZnS), lanthinum tri-bromide (LaBr$_3$), and lithium iodide (LiI). The most common scintillation detectors are solid, and the most popular are the inorganic crystals NaI, LaBr$_3$, and CsI. Some new scintillators, such as CLYC, CLLBC, and SrI, are being explored in applications where better gamma resolution than NaI is desired. A comprehensive discussion of scintillation detectors can be found in Refs. [1, 2, 5, 6]. When gamma rays interact in scintillator material, ionized (excited) atoms in the scintillator material "relax" to a lower-energy state and emit photons of light. In a pure inorganic scintillator crystal, the return of the atom to lower-energy states via the emission of a photon is an inefficient process. Furthermore, the emitted photons are usually too high in energy to lie in the range of wavelengths to which the photomultiplier is sensitive. Small amounts of impurities, called *activators*, are added to all scintillators to enhance the emission of visible photons. Crystal de-excitations channeled through these impurities give rise to photons that can activate the photomultiplier. One important consequence of luminescence through activator impurities is that the bulk scintillator crystal is transparent to the scintillation light. A common example of scintillator activation encountered in gamma-ray measurements is thallium-doped sodium iodide [NaI(Tl)].

Scintillators such as CLYC can also be used for neutron detection. In this case, the ^{6}Li(n,α) reaction is used for thermal-neutron detection, and pulse-shape discrimination allows for separation of the gamma and neutron signals [7, 8]. This scintillator and others that are sensitive to both gamma rays and neutrons will be described in greater detail in Chap. 15.

The scintillation light is emitted isotropically; so, the scintillator is typically surrounded with reflective material (such as MgO) to minimize the loss of light and then is optically coupled to the photocathode of a PMT (see Fig. 4.2).

Scintillation photons incident on the photocathode liberate electrons through the photoelectric effect, and these photoelectrons are then accelerated by a strong electric field in the PMT. As these photoelectrons are accelerated, they collide with electrodes in the tube (known as dynodes) and release additional electrons. This increased electron flux is then further accelerated to collide with succeeding electrodes, causing a large multiplication (by a factor of 10^4 or more) of the electron flux from its initial value at the photocathode surface. Finally, the amplified charge burst arrives at the output electrode (the anode) of the tube. The magnitude of this charge surge is proportional to the initial amount of charge liberated at the photocathode of the PMT; the constant of proportionality is the gain of the PMT. Furthermore, by virtue of the physics of the photoelectric effect, the initial number of photoelectrons liberated at the photocathode is proportional to the amount of light incident on the phototube, which in turn, is proportional to the amount of energy deposited in the scintillator by the gamma ray (assuming no light loss from the scintillator volume). Thus, an output signal is produced that is proportional to the energy deposited by the gamma ray in the scintillation medium. However, as discussed above, the spectrum of deposited energies (even for a monoenergetic photon flux) is quite varied because of the competing processes of the photoelectric effect, the Compton effect, and various scattering phenomena in the scintillation medium, as well as the statistical fluctuations associated with all of these processes. This result is discussed in more detail in Sect. 4.3.

4.2.3 Solid-State Detectors

In solid-state detectors, the charge produced by the photon interactions is collected directly. The gamma-ray energy resolution of these detectors is dramatically better than that of scintillation detectors, so greater spectral detail can be measured and used for special nuclear material (SNM) evaluations. A generic representation of the solid-state detector is shown in Fig. 4.3. The

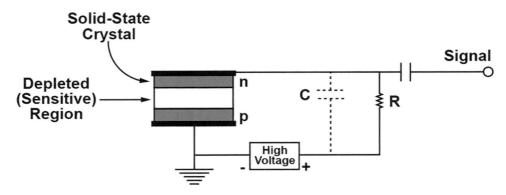

Fig. 4.3 Typical arrangement of components in a solid-state detector. The crystal is a reverse-biased p-n junction that conducts charge when ionization is produced in the sensitive region. The signal is usually fed to a charge-sensitive preamplifier for conversion to a voltage pulse (see Chap. 5)

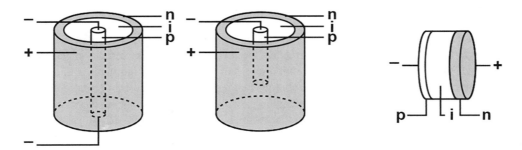

Fig. 4.4 Illustration of various solid-state detector crystal configurations: (**left**) open-ended cylindrical or true coaxial, (**middle**) closed-ended cylindrical, and (**right**) planar. The p-type and n-type semi-conductor materials are labeled accordingly. The regions labeled i are the depleted regions that serve as the detector sensitive volumes. In the context of semiconductor diode junctions, this region is referred to as the intrinsic region or a p-i-n junction

sensitive volume is an electronically conditioned region (known as the *depleted region*) in a semiconductor material in which liberated electrons and holes move freely. Germanium possesses the most ideal electronic characteristics in this regard and is the most widely used semiconductor material in solid-state detectors. As Fig. 4.3 suggests, the detector functions as a solid-state proportional counter, with the ionization charge swept directly to the electrodes by the high electric field in the semiconductor, produced by the bias voltage. The collected charge is converted to a voltage pulse by a preamplifier. The most popular early designs used lithium-drifted germanium [Ge(Li)] as the detection medium. The lithium served to inhibit trapping of charge at impurity sites in the crystal lattice during the charge-collection process. In recent years, manufacturers have produced high-purity (hyperpure) germanium (HPGe) crystals, essentially eliminating the need for the lithium doping and thus simplifying operation of the detector.

Solid-state detection media other than germanium and silicon have been applied to gamma-ray spectroscopy. In NDA measurements, as well as many other applications of gamma-ray spectroscopy, it would be advantageous to have high-resolution detectors that operate at room temperature, thereby eliminating the cumbersome apparatus necessary for cooling the detector crystal. Operation of room-temperature semi-conductor materials (such as CdTe, HgI2, and GaAs) has been extensively researched [9]. Their higher average atomic numbers provide greater photoelectric efficiency per unit-volume of material. The room-temperature semi-conductor most commonly used today is CdZnTe.

Solid-state detectors are produced mainly in two configurations: planar and coaxial. These terms refer to the detector crystal shape and the manner in which it is wired into the detector circuit. The most commonly encountered detector configurations are illustrated in Fig. 4.4. Coaxial detectors are produced either with open-ended (*true coaxial*) or closed-ended crystals (Fig. 4.4). In both cases, the electric field for charge collection is primarily radial, with some axial component present in the closed-ended configuration. Coaxial detectors can be produced with large sensitive volumes and therefore with large detection efficiencies at high gamma-ray energies. In addition, the radial electric field geometry makes the coaxial (especially the open-ended coaxial) solid-state detectors best for fast timing applications. The planar detector consists of a crystal of either rectangular or circular cross section and a sensitive thickness of 1–20 mm; for example, Fig. 4.4. The electric field is perpendicular to the cross-sectional area of the crystal. The crystal thickness is selected on the basis of the gamma-ray energy region relevant to the application of interest, with the small thicknesses optimum for low-energy measurements (e.g., in the L-X-ray region for SNM). Planar detectors usually achieve the best energy resolution because of their low capacitance; they are preferred for detailed spectroscopy, such as the analysis of the complex low-energy gamma-ray and X-ray spectra of uranium and plutonium.

In the quiescent state, the reverse-bias-diode configuration of a germanium solid-state detector results in very low currents in the detector (usually in the picoampere to nanoampere range). This leakage current can be further reduced from its room-temperature value by cryogenic cooling of solid-state medium, typically to liquid nitrogen temperature (77 K). This cooling reduces the natural, thermally generated electrical noise in the crystal but constitutes the main disadvantage of such detectors: the detector package must include capacity for cooling, which usually involves a dewar to contain the liquid nitrogen coolant. An alternative option is to use a mechanically cooled detector [10, 11]. A discussion on a less-common, popular solid-state detector material for photon spectroscopy, lithium-drifted silicon [Si(Li)] can be found in Sect. 3.3.2 of the original PANDA book [12].

4.2.3.1 Wide-Energy Detectors

Wide- or extended-energy germanium detectors—a type of planar germanium detector—have a point contact. They are more sensitive at lower energies than standard HPGe detectors, generally with an energy detection limit as low as 3 keV. These detectors often have thin carbon composite endcap windows. In addition, they lack the lithium dead layer that is standard on p-type HPGe detectors, enabling additional increased low-energy sensitivity. With thicknesses of up to 30 mm, these detectors not only have good resolution at low energy but also comparable resolution to coaxial HPGes at higher energies. Detection efficiency is maximized at low-to-mid energies. A discussion of the optimal HPGe detector selection, based on application, can be found in Ref. [13]. Further details on the design and use of solid-state detectors for gamma-ray spectroscopy can be found in Refs. [1, 2, 14].

4.2.3.2 Radiation Damage

Performance degradation due to radiation damage can occur in all gamma-radiation detectors but is particularly noticeable in high-resolution detectors. The amount of damage produced in the detector crystal per unit of incident flux is greatest for neutron radiation. Thus, in environments where neutron levels are high (such as accelerators, reactors, or instruments that have intense neutron sources), the most significant radiation damage effects will be observed. Furthermore, radiation damage effects can be of concern in NDA applications where large amounts of nuclear material are continuously measured with high-resolution gamma-ray spectroscopy equipment; for example, in measurements of plutonium isotopics in a high-throughput mode.

The primary effect of radiation damage is the creation of dislocation sites in the detector crystal. This effect increases the amount of charge trapping, reduces the amplitudes of some full-energy pulses, and produces low-energy tails in the spectrum photopeak; thus, the resolution is degraded, and spectral detail is lost. An example of this effect is shown in Fig. 4.5 (Ref. [15]). It has been generally observed that significant performance degradation begins with a neutron fluence of approximately 10^9 n/cm^2, and detectors become unusable at a fluence of approximately 10^{10} n/cm^2 [16]. However, n-type HPGe crystals are demonstrably less vulnerable to neutron damage. Procedures have been described in which the effects of the radiation damage can be reversed through warming (annealing) the detector crystal [17]. Additional information on radiation damage to high-resolution solid-state detectors can be found in Refs. [18–20].

4.2.4 Microcalorimeter Detectors

Microcalorimeter detectors (frequently referred to as microcal detectors) are sensitive devices that can produce extremely high-resolution spectra by exploiting material properties at very low temperatures (< 0.1 Kelvin). The technology has become quite mature and has been identified as an important technique when exceptionally precise gamma-ray measurements are required. A gamma-ray spectrum from a microcal detector can readily have a resolution an order of magnitude better than a HPGe detector, with a typical full width at half maximum (FWHM) of 75 eV for a peak near 100 keV [21]. However, because the detectors must operate at such cold temperatures, microcalorimeter instruments are typically more costly and complex than other gamma spectrometers. This section provides a brief overview of the technology; further information about microcal detectors for gamma-ray spectroscopy may be found in Refs. [21–23].

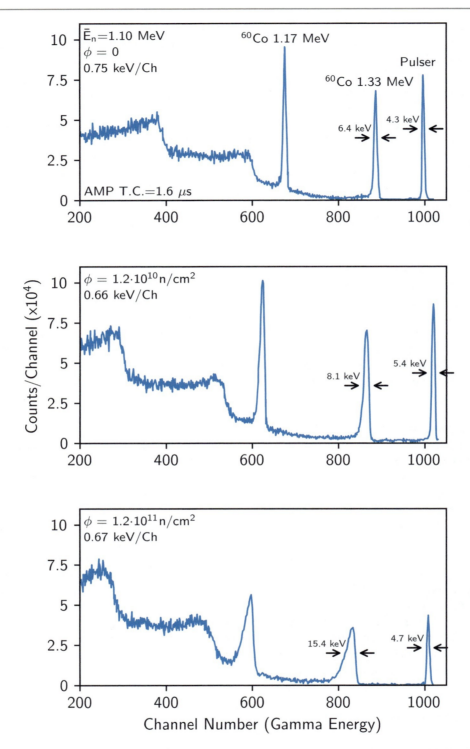

Fig. 4.5 The deterioration of a high-resolution, solid-state detector gamma spectrum with increasing neutron fluence(φ). The width of the 1.33 MeV ^{60}Co photopeak is indicated in each spectrum. Only the high-energy portion of the spectrum is shown. Also noted is the width of an electronic pulser peak. (Adapted from [15])

A microcalorimeter is a microfabricated device that consists of essentially two components: an absorber and a sensor. The absorber is engineered such that a single photon of the energy range of interest that strikes it will increase its temperature by a measurable amount. The sensor, thermally coupled to the absorber, is designed to be sensitive to the temperature increase of the absorber and convert the heat pulse into an electrical signal. The more energy the photon deposits in the absorber, the larger its temperature increase and the larger the electrical pulse produced in the sensor.

Microcalorimeters must operate at low temperatures for a few reasons. First, the absorber must have very low heat capacity so that the tiny amount of energy deposited by a photon leads to a measurable temperature change. Because the specific heat of most materials decreases with temperature, energy resolution is therefore inversely proportional to the operating temperature; however, the absorber must be large enough for a photon to have a high probability of being absorbed in the first place. Because the heat capacity of most metals decreases strongly with decreasing temperature, typically an absorber of practical size will have a useful heat capacity only when it is well below 1 Kelvin. The most common absorber is tin because of its favorable thermodynamic properties and reasonable absorption cross section. A typical tin gamma-ray absorber has an active area of 2.5 mm^2 and is 0.4–1 mm thick. Second, the low temperature is necessary to reduce thermal noise in the sensor and achieve the high-energy resolution. This effect is comparable to the need to cool solid-state detectors, such as HPGe detectors, below a certain temperature to eliminate thermally excited electron-hole pairs, where microcal detectors are sensitive to thermal excitations (phonons) instead of electron excitations.

The most common and mature sensor technology is the superconducting transition-edge sensor (TES)—a small superconducting structure with a superconducting transition temperature on the order of 100 millikelvin. Above this temperature, the superconductor behaves as a normal metal with finite electrical resistance; below this temperature, the superconductor carries no resistance. The transition between these phases is extremely sharp in temperature. The TES is voltage biased so that in this transition regime, a small temperature change will cause a large increase in resistance. This sensitivity allows a wide range of photon energies to be measured with high resolution. The TES is designed to then quickly cool again so that the pulse produced is on the order of 5 ms. Each TES is coupled to a device called a *SQUID* (*s*uperconducting *q*uantum *i*nterference *d*evice), which behaves as a highly sensitive cryogenic amplifier. Each signal is then measured using appropriate data acquisition electronics and software. After the signals are processed, a histogram of the pulse heights is made and calibrated into a spectrum.

Special refrigerators, or cryostats, are used to keep the microcalorimeters at the desired operating temperature. These cryostats are either adiabatic demagnetization refrigerators or dilution refrigerators. Both cryostat types typically use a pulse tube compressor to attain temperatures below 4 Kelvin—with no consumption of liquid cryogens—and have specialized stages to reach base temperatures of tens of millikelvin. Adiabatic demagnetization refrigerators exploit the cooling power of a paramagnetic salt placed within a high magnetic field, and dilution refrigerators use cooling properties of a mixture of liquid ^{3}He and ^{4}He. Both cryostats are typically large, expensive, and difficult to transport compared with HPGes.

Because they must be small, a single microcalorimeter device has poor efficiency compared with solid-state or scintillation detectors; therefore, a typical microcal detector system uses an array of hundreds of individual microcalorimeter sensor elements to increase detector efficiency, with typical active areas of around 50 mm^2. Even still, the detector efficiency is limited compared with scintillators and HPGe detectors. Current instruments are designed for the energy range of 50–250 keV needed for plutonium and uranium isotopic analysis and do not have useful efficiency at higher energies. Increases in the number of sensors and advancements in the ability to electronically read out many sensors promise to continue to improve detector efficiency. Large array modules, with more than 100 elements, have necessitated the technique of microwave multiplexing, which permits hundreds of signals modulated in frequency space to be transmitted along one coaxial cable. Current instruments operate at several thousand counts per second from a detector array of several hundred microcalorimeter sensor elements [21].

4.2.5 Comparison of Detector Types

There is often a trade-off between cost, ease of use, resolution, and efficiency when comparing various detector types. For example, NaI can be made very large; 2 × 4 × 16 in. or 4 × 4 × 16 in. NaI "logs" are commonly found in mobile search-type detectors. These detectors have much broader resolution than some of the common semiconductors but are unmatched in terms of efficiency. HPGe detectors have excellent resolution but must be operated at liquid-nitrogen temperatures. CZT falls in between, with good resolution and room-temperature operability; however, it cannot be grown very large as a single crystal and thus has lower efficiencies. Chapter 10, Sect. 10.4.1.2, discusses the trade-off between resolution and efficiency, with a focus on plutonium isotopics. A comparison of several commonly used semiconductor detector materials with NaI is presented in Table 4.2.

Table 4.2 Comparison of several semiconductor detector materials

Material	Atomic numbers	Energy pere-h Pair (δ)[a](eV)	Best γ-Ray energyresolution at 122 keV[b](keV)
Ge (77 K)	32	2.96	0.46
CdZnTe	48, 30, 52	5.0	Varies[c]
CdTe	48, 52	4.43	3.5
HgI$_2$	80, 53	4.3	3.2
GaAs	31, 33	4.2	2.6
NaI (300 K)[c]	11, 53	26.3[d]	14.2

[a]This quantity determines the number of charge carriers produced in an interaction (see Sect. 4.3.3)
[b]Representative resolution data, as tabulated in Ref. [24]. Energy resolution is discussed further in Sect. 4.3.3 and in Chap. 5
[c]Although not a semiconductor material, NaI is included in the table for convenient comparison
[d] Much work regarding reducing or compensating for hole trapping in CZT has greatly improved the energy resolution of this detector material [25]
[d] For NaI, energy required per scintillation photon (eV)

4.3 Characteristics of Detected Spectra

In gamma-ray spectroscopy applications, the detectors produce output pulses whose magnitudes are proportional to the energy deposited in the detecting medium by the incident photons. The measurement system includes some method of sorting all of the generated pulses and displaying their relative numbers. The basic tool for accomplishing this task is the multichannel analyzer, whose operation is discussed in Chap. 5. The end result of multichannel analysis is a histogram (spectrum) of the detected output pulses, sorted by magnitude. The pulse-height spectrum is a direct representation of the energy spectrum of the gamma-ray interactions in the detection medium and constitutes the spectroscopic information used in gamma-ray NDA.

4.3.1 Generic Detector Response

Regardless of the type of detector used, the measured spectra have many features in common. Consider the spectrum of a monoenergetic gamma-ray source of energy E_0. The gamma-ray spectrum produced by the decaying nuclei is illustrated in Fig. 4.6a. The gamma-ray photons originate from nuclear transitions that involve specific energy changes. There is a very small fluctuation in these energy values because of two effects: the quantum uncertainties in the energies of the transitions, called the *Heisenberg Uncertainty*; and recoil effects as the gamma-ray photons are emitted. These uncertainties are finite but negligible compared with the other energy-broadening effects (discussed as follows and so are not shown on the figure). Thus, the "ideal" monoenergetic gamma-ray spectrum from free decaying nuclei is essentially a sharp line at the energy E_0.

Because detected gamma-ray photons do not usually come from free nuclei but are emitted in material media, some photons undergo small-angle scattering before they emerge from the radioactive sample. This scattering leaves the affected photons with slightly less energy than E_0, and the energy spectrum of photons emitted from a material sample is slightly broadened into energies below E_0, as shown in Fig. 4.6b. The magnitude of this broadening is also quite small with respect to other effects discussed as follows and is exaggerated in Fig. 4.6b to call attention to its existence.

When the gamma ray enters the detection medium, it transfers part or all of its energy to an atomic electron, freeing the electron from its atomic bond. This freed electron then transfers its kinetic energy—in a series of collisions—to other atomic electrons in the detector medium.

The amount of energy required to produce electron-ion pairs in the detecting medium determines the amount of charge that will be produced from a detection event that involves a given amount of deposited energy (see Table 4.2). A photoelectric interaction transfers all of the incident photon's energy to a photoelectron; this electron subsequently causes multiple ionizations until its energy is depleted. The amount of charge produced from this type of event is therefore proportional to the original photon energy. A Compton-scattering interaction transfers only part of the incident photon's energy to an electron, and that electron subsequently causes ionization until its energy is depleted. The amount of charge produced from this type of event is proportional to the partial energy lost by the incident photon but conveys no useful information about the original photon energy. Multiple Compton-scattering events for a single photon can produce amounts of charge closer to the full energy of the original photon; however, the likelihood that a photon will undergo multiple Compton-scattering events within the detector depends on the initial photon energy and the physical size of the detector relative to the mean free paths of

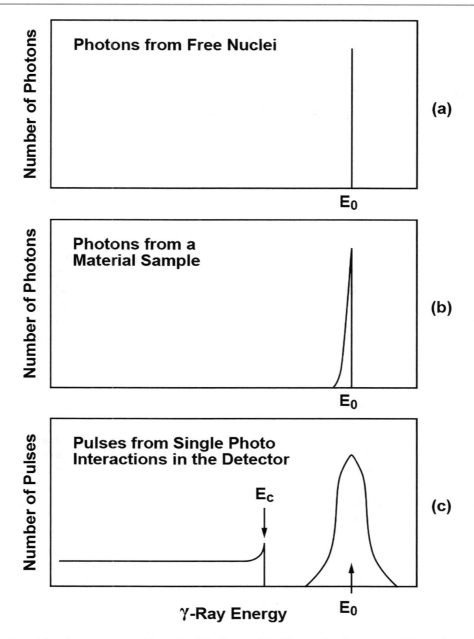

Fig. 4.6 An idealization of the photon spectrum (**a**) produced by free nuclei, (**b**) emerging from a material sample, and (**c**) displayed from interactions in a detecting medium

the scattered photons. The idealized detector response to the photoelectric and Compton interactions in the detection medium is shown in Fig. 4.6. The maximum energy that can be deposited in the detection medium from a Compton-scattering event comes from an event where the photon is scattered by 180°. The Compton-generated detector pulses are therefore distributed below this maximum energy, E_c in Fig. 4.6c, and constitute a source of "background" pulses that carry no useful energy information. This part of the spectrum is generally referred to as the Compton continuum. Note: Detected photons that undergo Compton scattering in materials between the source and the detector contribute to the continuum.

The full-energy peak in Fig. 4.6c is significantly broadened by the statistical fluctuation in the number of scintillation photons or electron-ion pairs produced by the photoelectron. This effect is the primary contributor to the width of the full-energy peak and is therefore the dominant factor in the detector energy resolution (see Sect. 4.3.3).

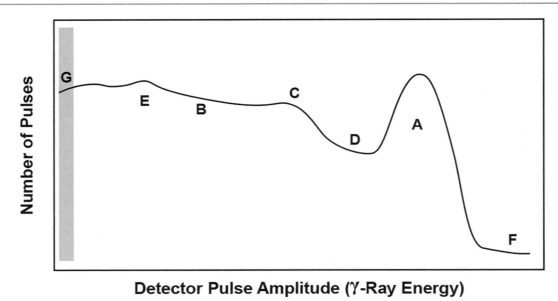

Fig. 4.7 A realistic representation of the gamma detector spectrum from a monoenergetic gamma source. The labeled spectral features are explained in the text

4.3.2 Spectral Features

A more realistic representation of a detector-generated gamma-ray spectrum from a monoenergetic gamma-ray flux is shown in Fig. 4.7. The spectral features labeled A through G are explained after the figure.

(A.) The Full-Energy Peak. This peak represents the pulses that arise from the full-energy, photoelectric interactions in the detection medium. Some counts also arise from single or multiple Compton-scattering events that are followed by a photoelectric absorption. Its width is determined primarily by the statistical fluctuations in the charge produced from the interactions plus a contribution from the pulse-processing electronics (see Sect. 4.3.3 and Chap. 5). Its centroid represents the photon energy E_0. Its net area, above background, represents the total number of full-energy interactions in the detector and is usually proportional to the activity/mass of the emitting nuclide.

(B.) Compton Continuum. These pulses, distributed smoothly up to a maximum energy E_c (see Fig. 4.6), come from interactions involving only partial photon energy loss in the detecting medium. Compton events from higher-energy photons are the primary source of background counts under the full-energy peaks in more complex spectra.

(C.) The Compton Edge. This region of the spectrum represents the maximum energy loss by the incident photon through Compton scattering. It is a broad asymmetric peak that corresponds to the maximum energy (E_c) that a gamma-ray photon of energy E_0 can transfer to a free electron in a single scattering event. This event corresponds to a "head-on" collision between the photon and the electron, during which the electron moves forward, and the gamma-ray scatters backward through 180° (see Chap. 3, Sect. 3.3.2). The energy of the Compton edge is given by Eq. 3.11.

(D.) The Compton Valley. For a monoenergetic source, pulses in this region arise from either multiple Compton-scattering events or from full-energy interactions by photons that have undergone small-angle scattering (in either the source materials or intervening materials) before entering the detector. Unscattered photons from a monoenergetic source cannot produce pulses in this region from a single interaction in the detector. In more complex spectra, this region can contain Compton-generated pulses from higher-energy photons.

(E.) Backscatter Peak. This peak is caused by gamma rays that have interacted by Compton scattering in one of the materials that surrounds the detector. Gamma rays scattered by more than ~115° will emerge with nearly identical energies in the 200–250 keV range. Therefore, a gamma ray source will give rise to many scattered gamma rays whose energies are near this minimum value (see Ref. [1] and Chap. 3, Sect. 3.3.2). The energy of the backscatter peak is given by Eq. 3.10.

(F.) Excess-Energy Region. With a monoenergetic source, events in this region are from high-energy gamma rays and cosmic interactions from the natural background and from pulse-pileup events if the count rate is high enough (see Chap. 5). In a more complex spectrum, counts above a given photopeak are primarily Compton events from the higher-energy gamma rays.

(G.) Low-Energy Rise. This feature of the spectrum, very near the "zero-pulse-height-amplitude" region, may contain contributions from low-amplitude electronic noise in the detection system that is processed in the same way as low-amplitude detector pulses. This noise tends to be at rather high frequency and so appears as a high-count-rate phenomenon. Electronic noise is usually filtered out of the analysis electronically (see Chap. 5), so this effect does not tend to be noticeable in the displayed spectrum.

In more complex gamma-ray spectra containing many different photon energies, the Compton-edge and backscatter-peak features tend to "wash out," leaving primarily full-energy peaks on a relatively smooth Compton background.

4.3.3 Detector Resolution

The resolution of a detector is a measure of its ability to resolve two peaks that are close together in energy. The parameter used to specify the detector resolution is the full width of the (full-energy) photopeak at half its maximum height (FWHM). If a standard Gaussian shape is assumed for the photopeak, the FWHM is given by

$$FWHM = 2\sigma\sqrt{2 \cdot \ln(2)}, \quad (4.1)$$

where σ is the width parameter for the Gaussian. High resolution (small FWHM) makes the distinction of close-lying peaks easier. In addition, the determination of the Compton continuum under the peak is more accurate without interference from overlapping peaks. The more complex a gamma-ray spectrum is, the more desirable it is to have the best energy resolution possible.

Both natural and technological limits apply to how precisely the energy of a gamma-ray detection event can be registered by the detection system. The natural limit on the energy precision arises primarily from the statistical fluctuations associated with the charge production processes in the detector medium. The voltage integrity of the full-energy pulses can also be disturbed by electronic effects (noise, pulse pileup, improper pole-zero settings, etc.). These electronic effects have become less important as technology has improved, but their potential effects on the resolution must be considered in the setup of a counting system. The electronic and environmental effects on detector resolution are discussed in more detail in Chap. 5.

The two types of detectors most widely used in gamma-ray NDA applications are the NaI(Tl) scintillation detector and the germanium solid-state detector. The NaI detector generates full-energy peaks that are much wider than their counterparts from the germanium detector. This result is illustrated in Fig. 4.8, where the wealth of detail evident in the germanium spectrum of plutonium gamma rays is all but lost in the corresponding NaI spectrum.

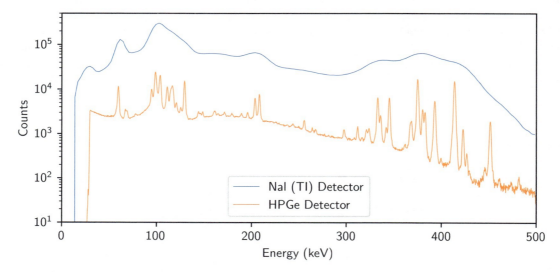

Fig. 4.8 Gamma-ray spectrum from a plutonium sample with 94.2% ^{239}Pu, taken with a high-resolution, HPGe solid-state detector and with a NaI scintillation counter

By considering the statistical limit on the energy precision, it is possible to understand the origin of the difference in the energy resolution achievable with various types of detectors. Ideally, the number of electron charges (n) produced by the primary detection event depends on the total energy deposited (E) and the average amount of energy required to produce an electron-ion pair (δ):

$$n = \frac{E}{\delta}. \tag{4.2}$$

The random statistical variance in n is the primary source of fluctuation in the full-energy pulse amplitude. However, for some detector types, this statistical variance is observed to be less than (that is, better than) the theoretical value by a factor known as the Fano factor [26, 27]:

$$\sigma^2(n) = Fn = \frac{FE}{\delta}. \tag{4.3}$$

This effect occurs because part of the energy lost by the incident photon goes into the formation of ion pairs, and part goes into heating the lattice crystal structure (thermal energy). The division of energy between heating and ionization is essentially statistical. Without the competing process of heating, all of the incident energy would result in ion-pair production, and there would be no statistical fluctuation in n (F = 0).

By contrast, if the probability of ion-pair production is small, then the statistical fluctuations would dominate (F $\cong$ 1). For scintillators, the factor (F) is approximately unity; for germanium, silicon, and gases, it is approximately 0.15 [1, 2]. Because the number of charge carriers (n) is proportional to the deposited photon energy (Eq. 4.2), the statistical portion of the relative energy resolution is given by

$$\frac{E_{stat}}{E} = \frac{2.35\sigma(n)}{n} = 2.35\left[\frac{F\delta}{E}\right]^{\frac{1}{2}}. \tag{4.4}$$

The electronic contribution to the energy fluctuations (ΔE_{elect}) is essentially independent of photon energy and determined largely by the detector capacitance and the preamplifier. Thus, the total energy resolution can be expressed as the combination of the electronic and statistical effects:

$$\Delta^2 E_{tot} = \Delta^2 E_{elect} + \Delta^2 E_{stat} = \alpha + \beta E. \tag{4.5}$$

In Fig. 4.9, the energy resolutions of scintillation, solid-state, and microcalorimeter detectors are compared for the measurement of gamma rays from a plutonium item. Techniques for measuring resolution are described in Chap. 5, Sect. 5.2.

The argument presented here assumes that the scintillation efficiency is the main factor influencing the number of electrons produced at the photocathode of a scintillation detector. Other factors, such as scintillator transparency, play important roles. To work effectively as a detector, a scintillating material must have a high transparency to its own scintillation light. In a similar vein, factors such as charge-carrier mobility play an important role in determining the resolution of a solid-state detector. This discussion is simplified by necessity, but it illustrates the primary reason why germanium detectors resolve so much better than scintillation detectors. See Ref. [1] for a more complete discussion of detector resolution.

4.3.4 Detector Efficiency

The basic definition of absolute photon detection efficiency is

$$\varepsilon_{tot} = \frac{total\ number\ of\ photons\ detected\ in\ the\ full\ energy\ peak}{total\ number\ of\ photons\ emitted\ by\ the\ source}. \tag{4.6}$$

For the following discussion, we will be concerned with only full-energy events and thus with the full-energy detection efficiency. This total efficiency can be expressed as the product of four factors, defined below. These factors will be covered in detail in Chaps. 5 and 6, although a short discussion on the final factor, intrinsic efficiency, is discussed here:

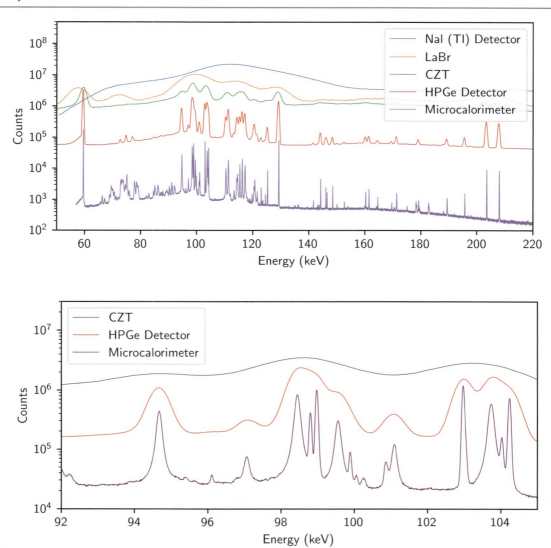

Fig. 4.9 Low-burnup plutonium spectrum obained with several different detectors. (top) Figure shows the energy spectrum in the low-energy region. (bottom) Figure shows an expanded view of the X-ray region. Counts have been arbitary normizalzed forcomparison. Good resolution in this region is critical in performing plutonium isotopic composition measurements

$$\varepsilon_{tot} = \varepsilon_{geom} \cdot \varepsilon_{absp} \cdot \varepsilon_{sample} \cdot \varepsilon_{int}. \tag{4.7}$$

The *geometric efficiency* ε_{geom} is the fraction of emitted photons intercepted by the detector.

The *absorption-efficiency* ε_{absp} is the fraction of emitted photons not absorbed in intervening materials.

The *sample efficiency* ε_{sample} is the fraction of emitted photons that escape the sample without being absorbed (the reciprocal of the sample self-absorption correction (CF_{atten}) discussed in Chap. 6).

The *intrinsic efficiency* ε_{int} is the probability that a gamma ray that enters the detector will interact and give a pulse in the full-energy peak. In simplest terms, this efficiency comes from the standard absorption formula

$$\varepsilon_{int} = 1 - exp(-\mu\rho x), \tag{4.8}$$

where μ is the photoelectric mass attenuation coefficient, and ρ and x are the density and thickness of the sensitive detector material. This simple expression underestimates the true intrinsic efficiency because the full-energy peak can also contain events from multiple Compton-scattering interactions. In general, ε_{int} is also a weak function of the radius of the detector because off-axis incident gamma rays are detected.

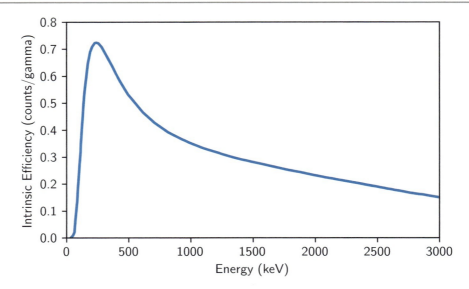

Fig. 4.10 Typical intrinsic efficiency curve for a generic HPGe detector

Figure 4.10 shows a generic intrinsic efficiency curve for a standard HPGe detector. HPGe detectors are frequently referred to by their relative efficiency, which by convention, is a comparison to the efficiency of a 3 × 3 in. NaI detector at 1332 keV (using a Co-60 gamma ray). This comparison will be discussed further in Chaps. 5 and 6.

References

1. G.F. Knoll, *Radiation Detection and Measurement*, 3rd edn. (Wiley, New York, 2000)
2. C.E. Crouthamel, F. Adams, R. Dams, *Applied Gamma-Ray Spectrometry* (Pergamon Press, New York, 1975)
3. D.T. Vo, P.A. Russo, *Testing the Plutonium Isotopic Analysis Code FRAM with Various CdTe Detectors* (Los Alamos National Laboratory Report LA-UR-02-0542, 2002)
4. M.C. Recker, E.J. Cazalas, J.W. McClory, J.E. Bevins, Comparison of SiPM and PMT performance using a $Cs_2LiYCl_6:Ce^{3+}$ (CLYC) scintillator with two optical windows. IEEE Trans. Nucl. Sci. **66**(8), 1959–1965 (2019). https://doi.org/10.1109/TNS.2019.2926246
5. J.B. Birks, *The Theory and Practice of Scintillation Counting*, 1st edn. (Pergamon Press, Oxford, 1964)
6. A. Giaz, V. Fossati, G. Hull, F. Camera, N. Blasi, S. Brambilla, S. Ceruti, N. Cherepy, B. Million, L. Pellegri, S. Riboldi, Characterization of new scintillators: SrI_2:Eu, $CeBr_3$, GYGAG:Ce and CLYC:Ce. J. Phys. Conf. Ser. **620**, 012003 (2005). https://doi.org/10.1088/1742-6596/620/1/012003
7. N. D'Olympia, P. Chowdhury, C.J. Lister, J. Glodo, R. Hawrami, K. Shah, U. Shirwadkar, Pulse-shape analysis of CLYC for thermal neutrons, fast neutrons, and gamma-rays. Nucl. Instrum. Methods Phys. Res., Sect. A **714**, 121–127 (2013)
8. C.M. Whitney, L. Soundara-Pandian, E.B. Johnson, S. Vogel, B. Vinci, M. Squillante, J. Glodo, J.F. Christian, Gamma-neutron imaging system utilizing pulse shape discrimination with CLYC. Nucl. Instrum. Methods Phys. Res., Sect. A **784**, 346–351 (2015)
9. E. Sakai, Present status of room temperature semiconductor detectors. Nucl. Instrum. Methods **196**(1), 121–130 (1982)
10. J.M. Marler, V.L. Gelezunas, Operational characteristics of a high purity germanium photon spectrometer cooled by a closed-cycle cryogenic refrigerator. IEEE Trans. Nucl. Sci. **20**, 522–527 (1973)
11. A. Lavietes, G. Mauger, E. Anderson, *Electromechanically-Cooled Germanium Radiation Detection System* (UCRL-JC-129695, 1998)
12. D. Reilly, N. Ensslin, H. Smith Jr., S. Kreiner, *Passive Nondestructive Assay of Nuclear Materials (PANDA)* (U.S. Nuclear Regulatory Commission, 1991)
13. I.D. Hau, K. Morris, W. Russ, Optimal HPGe detector selection for gamma-ray spectroscopy. Trans. Nucl. Sci. **60**(2), 1225–1230 (2013). https://doi.org/10.1109/TNS.2013.2251750
14. R.H. Pehl, Germanium gamma-ray detectors. Phys. Today **30**(11), 50 (1977)
15. H.W. Kraner, C. Chasman, K.W. Jones, Effects produced by fast neutron bombardment of Ge(Li) gamma-ray detectors. Nucl. Instrum. Methods **62**(2), 173–183 (1968)
16. P.H. Stelson, J.K. Dickens, S. Raman, R.C. Trammell, Deterioration of large Ge(li) diodes caused by fast neutrons. Nucl. Instrum. Methods **98**(3), 481–484 (1972)
17. R. Baader, W. Patzner, H. Wohlfarth, Regeneration of neutron-damaged Ge(Li) detectors inside the cryostat. Nucl. Instrum. Methods **117**(2), 609–610 (1974)
18. E.H. Seabury, C. Dew Van Siclen, J.B. McCabe, C.J. Wharton, A.J. Caffrey, Neutron damage in mechanically-cooled high-purity germanium detectors for field-portable Prompt Gamma Neutron Activation Analysis (PGNAA) systems, in *2013 IEEE Nuclear Science Symposium and Medical Imaging Conference (2013 NSS/MIC), Seoul, Korea (South)*, (2013), pp. 1–4. https://doi.org/10.1109/NSSMIC.2013.6829527

19. Y. Eisen, A. Shor, Fast neutron damage of a pixelated CdZnTe gamma ray spectrometer. IEEE Trans. Nucl. Sci. **56**(4), 1700–1705 (2009). https://doi.org/10.1109/TNS.2009.2020599
20. M. Koenen, J. Bruckner, M. Korfer, I. Taylor, H. Wanke, Radiation damage in large-volume n- and p-type high-purity germanium detectors irradiated by 1.5 GeV protons. IEEE Trans. Nucl. Sci. **42**(4), 653–658 (1995). https://doi.org/10.1109/23.467896
21. D.T. Becker, B.K. Alpert, D.A. Bennett, M.P. Croce, J.W. Fowler, J.D. Gard, A.S. Hoover, J. Young II, K.E. Koehler, J.A.B. Mates, G.C. O'Neil, M.W. Rabin, C.D. Reintsema, D.R. Schmidt, D.S. Swetz, P. Szypryt, L.R. Vale, A.L. Wessels, J.N. Ullom, Advances in analysis of microcalorimeter gamma-ray spectra. IEEE Trans. Nucl. Sci. **66**(12), 2355–2363 (2019)
22. J.N. Ullom, D.A. Bennett, Review of superconducting transition-edge sensors for X-ray and Gamma-ray spectroscopy. Supercond. Sci. Technol. **28**, 084003 (2015)
23. J.A.B. Mates, D.T. Becker, D.A. Bennett, B.J. Dober, J.D. Gard, J.P. Hays-Wehle, J.W. Fowler, G.C. Hilton, C.D. Reintsema, D.R. Schmidt, D.S. Swetz, L.R. Vale, J.N. Ullom, Simultaneous readout of 128 X-ray and Gamma-ray transition-edge microcalorimeters using microwave SQUID multiplexing. Appl. Phys. Lett. **111**, 062601 (2017)
24. P. Siffert, A. Cornet, R. Stuck, R. Triboulet, Y. Marfaing, Cadmium telluride nuclear radiation detectors. Trans. Nucl. Sci. **22**, 211 (1975)
25. M.D. Alam, S.S. Nasim, S. Hasan, Recent progress in CdZnTe based room temperature detectors for nuclear radiation monitoring. Progr. Nucl. Energy **140** (2021). https://doi.org/10.1016/j.pnucene.2021.103918. Accessed Feb 2023
26. U. Fano, On the theory of ionization yield of radiation in different substances. Phys. Rev. **70**, 44–52 (1946)
27. U. Fano, Ionization yield of radiations. II. The fluctuations of the number of ions. Phys. Rev. **72**, 26–29 (1947)

Open Access This chapter is licensed under the terms of the Creative Commons Attribution 4.0 International License (http://creativecommons.org/licenses/by/4.0/), which permits use, sharing, adaptation, distribution and reproduction in any medium or format, as long as you give appropriate credit to the original author(s) and the source, provide a link to the Creative Commons license and indicate if changes were made.

The images or other third party material in this chapter are included in the chapter's Creative Commons license, unless indicated otherwise in a credit line to the material. If material is not included in the chapter's Creative Commons license and your intended use is not permitted by statutory regulation or exceeds the permitted use, you will need to obtain permission directly from the copyright holder.

Instrumentation for Gamma-Ray Spectroscopy

D. T. Vo and J. L. Parker

5.1 Introduction

The subject of this chapter is the function and operation of the components of a gamma-ray spectroscopy system. In Chap. 4, it was shown that the output pulse amplitude is proportional to the gamma-ray energy deposited in most detectors. The pulse-height spectrum from such a detector contains full-energy peaks superimposed on a Compton continuum. The information that is most important for the quantitative nondestructive assay (NDA) of nuclear material is contained in the full-energy peaks of the gamma-ray spectrum. The electronic equipment that follows the detector serves to acquire an accurate representation of the pulse-height spectrum and to extract the desired energy and intensity information from that spectrum.

This chapter provides a relatively brief introduction to the wide variety of instruments used in the gamma-ray spectroscopy of nuclear materials. It emphasizes the function of each component and provides information about important aspects of instrument operation. For a detailed description of instrument operation, the reader should refer to the instruction manuals provided with each instrument. Because the state of gamma-ray spectroscopy equipment is rapidly advancing, the best sources of current information are often the manufacturers and users of the instruments. However, although the manufacturers are clearly the best source of information about the electronic capabilities of their equipment, those active in the application of gamma-ray spectroscopy to NDA are usually the best source of information on effective assay procedures and the selection of equipment for a given application. Books and reports on gamma-ray spectroscopy equipment are often out of date soon after they are published.

Gamma-ray spectroscopy systems can be divided into two classes according to whether they use analog or digital multichannel analyzers (MCAs). The main difference between the two types of MCAs is that for the analog MCA, the pulses from the preamplifier are processed by the analog amplifier and then analyzed for pulse height by the analog-to-digital converter (ADC) whereas the digital MCA digitizes the pulses from the preamplifier and then digitally processes the pulses using a quasi-trapezoid pulse shape. Figures 5.1 and 5.2 show block diagrams of the two classes. Both systems begin with a detector, where the gamma-ray interaction produces a weak electrical signal that is proportional to the deposited energy. They start to diverge downstream of the preamplifier.

Section 5.2 discusses the process of selecting an appropriate detector for different NDA applications. Sections 5.3 through 5.9 discuss the basic components of gamma-ray spectroscopy systems; the discussion of each component is presented in the order in which the electrical signal flows through the system. Section 5.10 presents auxiliary electronic equipment. Components other than those shown in Figs. 5.1 or 5.2 usually must be added to form a useful NDA system. Shields, collimators, sample holders, sample changers, scanning mechanisms, and source shutters are discussed in later chapters that describe specific assay techniques and instruments.

Los Alamos National Laboratory strongly supports academic freedom and a researcher's right to publish; as an institution, however, the Laboratory does not endorse the viewpoint of a publication or guarantee its technical correctness.

D. T. Vo (✉) · J. L. Parker
Los Alamos National Laboratory, Los Alamos, NM, USA
e-mail: ducvo@lanl.gov

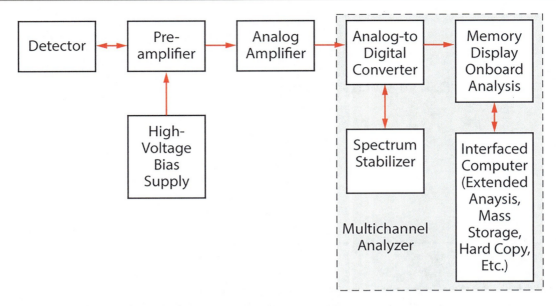

Fig. 5.1 Block diagram of an analog MCA-based gamma-ray spectroscopy system

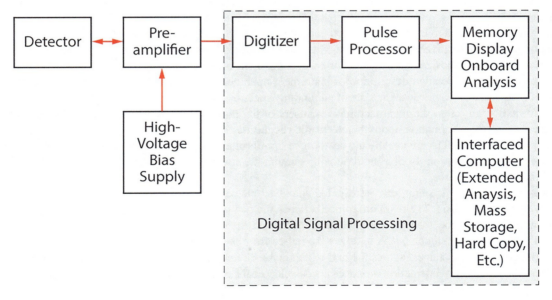

Fig. 5.2 Block diagram of a digital MCA-based gamma-ray spectroscopy system

5.2 Selection of Detector

This section contains some general guidance to the often-difficult matter of selecting an appropriate detector for a particular NDA application. Not only are there several generic types of detectors but also a myriad of variations of size, shape, packaging configuration, performance, and cost. The detector choice must be evaluated based on the technical requirements of a proposed application and the nontechnical but often overriding matter of budgetary constraints.

The first and most important detector parameter to consider is resolution. A detector with high resolution usually gives more accurate assays than one with low resolution. The resolution of a high-purity germanium (HPGe) detector is typically 0.5–2.0 keV for peak full width at half maximum (FWHM) in the energy range of interest for NDA applications (50–1000 keV), whereas the resolution of a NaI detector is 20–60 keV in the energy range of 180–1000 keV. It is common

5 Instrumentation for Gamma-Ray Spectroscopy

to see the resolution of the NaI detector expressed as percentage of the peak energy, such as 10% resolution at 186 keV or 7% resolution at 662 keV. (See Chap. 4 for more information about detector and detector resolution.) It is easier to accurately determine the area of full-energy peaks in a complex spectrum when the peaks do not overlap, and the probability of overlap is smaller with narrower peaks. The Compton continuum contributes less to the relative uncertainty on net full-energy-peak area for a high-resolution spectrum because it is a smaller fraction of the total area in the peak region. A detector with a large peak-to-Compton ratio will have a lower Compton continuum, which leads to more precise net peak areas. Full-energy-peak areas are easier to evaluate in high-resolution spectra because the interference from small-angle Compton scattering in the sample is reduced. Gamma rays that undergo small-angle scattering lose only a small amount of energy. If these scattered gamma rays still fall in the full-energy-peak region, the calculated full-energy-peak area is likely to be incorrect. (See Chap. 8 for discussions regarding the effect of scattered gamma rays on uranium enrichment measurements.) This problem is minimized by using a high-resolution detector, which provides narrow, full-energy peaks.

The complexity of the spectrum should influence the detector choice; the more complex the spectrum becomes, the more desirable high resolution becomes. Plutonium has a much more complex gamma-ray spectrum than uranium does; thus, HPGe detectors are used more often in plutonium assay applications than in uranium assay applications.

The second performance parameter to consider is efficiency, which determines the full-energy-peak count rates that can be expected as a function of energy, the time that is required to achieve a given precision, and the sensitivity that can be achieved. Higher efficiency always costs more for a given detector type, but a given efficiency is less expensive to obtain in a low-resolution NaI detector than in a high-resolution HPGe detector. There is considerable motivation to use a less-expensive, lower-resolution detector when it can give satisfactory assay results.

Other parameters, such as space and cooling requirements and portability, also must be considered. The selection of an appropriate detector is often difficult and could involve compromises among conflicting requirements. Once the selection is made, considerable care should be taken to specify all pertinent parameters to prospective vendors to ensure that the desired detector is obtained.

5.3 High-Voltage Bias Supply

All of the commonly used gamma-ray detectors require a high-voltage bias supply to provide the electric field that collects the charge generated by the gamma-ray interaction in the solid-state detector or the photomultiplier attached to a scintillator. The bias supply is not a part of the signal path but is required to operate the detector. It is usually the most reliable unit in a spectroscopy system and the easiest to operate.

HPGe detectors require very low currents, typically $<10^{-9}$ A. The voltage requirements range from about one thousand volts for a small planar HPGe detector to several thousand volts for a large coaxial HPGe detector. Bias supplies for HPGe detectors usually provide up to 5 kV and 100 µA. The voltage-resolution and low-frequency filtering requirements are modest because there is no charge amplification in the detector. The voltage is usually continuously variable in the range of 0–5 kV. In the past, it was necessary to vary the voltage very slowly (<100 V/s) when turning on or changing the detector bias because the field-effect transistor (FET) used in the first stage of the detector preamplifier is easily damaged by sudden voltage surges. However, the protection provided by the filter now included in all high-quality preamplifiers is so good that an FET is rarely destroyed for this reason.

The high-voltage bias supply requirements for photomultiplier tubes used with scintillation detectors are more stringent than for solid-state diode detectors. The required voltage is typically less than one thousand volts, but the required current handling is usually 1–10 mA. Because the gain of photomultiplier is a very strong function of the applied voltage, the stability and filtering must be excellent. The 100 µA supplies used with HPGe detectors will usually not operate a photomultiplier tube.

Bias supplies come in a variety of packages. The most common for laboratory setups is the nuclear instrumentation module (NIM), which is mated to a frame or bin (NIM bin) that supplies the necessary direct current (DC) voltages to power the module. NIM modules meet the internationally accepted standards for dimensions, voltages, wiring, and connectors and are widely used in NDA instrumentation. Other bias supplies fit in NIM bins but take power from the normal alternating-current main source. The high-current bias supplies used to power multiple photomultiplier-tube arrays are often mounted in standard 45.7 cm (18 in.) wide instrumentation racks.

NIM bias supplies frequently use an electronic switching device to generate the required voltage. The switching device generates a high-frequency noise that can find its way into the preamplifier and cause significant degradation of spectral quality. This problem can be minimized by careful grounding and cable positioning. The noise generated by the power supply can also introduce false signals into any pileup-rejection circuitry in use. Photomultiplier-tube bias supplies—even those that are not of the NIM type—can also be sources of high-frequency noise. As usual, an oscilloscope is a most useful aid in detecting the presence of interfering electrical noise from any source.

Field measurements often employ all-in-one MCAs (discussed later) instead of NIM bin systems. The high-voltage bias supplies for the all-in-one MCAs vary widely for different MCAs. Some connect to the MCA externally, and some connect to the MCA internally—the latter of which requires the user to remove the MCA cover to plug in the bias supply module. The high-voltage bias supplies are normally of either very high voltage (up to 5 kV) and low current (to be used with the HPGe detectors) or lower high voltage (up to 2 kV) and higher current (to be used with scintillation detectors). They also come with either positive bias or negative bias. Some all-in-one MCAs also have the built-in, high-voltage bias supply, and the user controls all of its aspects (high-voltage values, polarity, and HPGe or scintillator type) through software.

Note that detector bias supplies can be lethal. Caution is always required, particularly when working with the high-current supplies that power photomultiplier tubes. Persons who are accustomed to working with low-voltage, low-power, transistorized circuits must be made aware of the danger associated with the use of detector power supplies.

5.4 Preamplifier

Preamplifiers are required for HPGe detectors and improve the performance of NaI scintillation detectors. The detector output signal is usually a low-amplitude, short-duration current pulse; a typical pulse might be a few tenths of a microsecond long. The preamplifier converts this current pulse to a voltage pulse whose rise time is about the same as the duration of the current pulse, and the amplitude is proportional to the energy deposited in the detector during the gamma-ray interaction. To maximize the signal-to-noise ratio of the output pulse and preserve the gamma-ray energy information, the preamplifier must be placed as close to the detector as possible. The closeness of the preamplifier to the detector crystal minimizes the capacitance at the preamplifier input, thereby reducing the output noise level. The preamplifier also serves as an impedance-matching device between the high-impedance detector and the low-impedance coaxial cable that transmits the amplified detector signal to the analog amplifier in an analog system and to the flash ADC in a digital system. The amplifier and preamplifier may be separated by as much as several hundred meters.

Because the detector and preamplifier must be located closely to each other, the preamplifier is often positioned in an inconvenient location, surrounded by shielding, and inaccessible during use. Most preamplifiers have no external controls; the gain and pulse-shape adjustments are included in the analog amplifier in analog systems and in the filtering stage of digital systems, the latter of which is usually in a more convenient location close to other system electronics. Because it lacks external controls, the preamplifier occupies only a few hundred cubic centimeters. Its small volume is advantageous when the preamplifier must be located inside the detector shielding. For single NaI detectors, the preamplifier is often built into the cylindrical housing that holds the photomultiplier-tube socket. For HPGe detectors manufactured before the 1980s, the preamplifier was usually housed in a small rectangular box. Since then, preamplifiers for HPGe detectors have been packaged in an annular configuration behind the end cap of the detector cryostat. This configuration eliminates awkward boxes that protrude at right angles from cryostats and makes the detectors easier to shield. Figure 5.3 shows the basic preamplifier configurations just described.

Although preamplifiers have few controls, they have several connectors. Usually included are one or two output connectors and a test input through which pulses can be routed from an electronic pulser to simulate gamma-ray events for testing the performance of the preamplifier and the other signal-processing instruments in the system. (The simulated gamma-ray peak produced in the acquired spectrum can also give a good estimate of the electronic losses suffered by the system; see Chap. 7.) The detector high-voltage bias is often applied through a connector mounted on the preamplifier. A multi-pin connector is usually included to provide the power needed for operating the preamplifier; the power is often supplied by the main amplifier. Certain NaI preamplifiers generate the required low voltage from the detector bias voltage.

The preamplifier output pulse is a fast positive or negative step followed by a very slow, exponential decay. The rise time is a few tenths of a microsecond, and the decay time is ~100 μs or more (with ~50 μs decay constant). The amplitude of the fast step is proportional to the charge delivered to the preamplifier input and therefore proportional to the energy deposited in the detector. The long decay time means that a second pulse often occurs before the tail of the preceding pulse has decayed. This effect is seen in Fig. 5.4 (top), which shows an example of the preamplifier output from a large, coaxial HPGe detector. The amplitude of the fast-rising step can be distorted if the energy deposition rate becomes so high that the average DC level rises beyond the linear range of the preamplifier, as seen in Fig. 5.4 (bottom). In this situation, as count rate continues to increase, the voltage rises to such a level that a constant DC voltage equal to the maximum voltage of the preamplifier is output, and there are no pulses at all. The preamplifier is then paralyzed or saturated.

Fig. 5.3 Detectors that have cylindrical, annular, and rectangular preamplifiers

Most manufacturers offer several preamplifier models that are optimized for different detector types. Parameters such as noise level, sensitivity, rise time, and count-rate capability can be different for different models. The count-rate capability is usually specified as the maximum charge per unit time (C/s) delivered from the detector to the preamplifier input. For HPGe detectors, the equivalent energy per unit time (MeV/s) is often specified. When this number is divided by the average gamma-ray energy, the result is the maximum count rate that the preamplifier can handle.

Few choices usually can be made when selecting a preamplifier for a NaI detector; however, several significantly different options are available for HPGe detectors. The selection depends on the detector and the measurement application. Because HPGe detectors are always sold with an integral preamplifier, the selection must be made when the detector is purchased.

Because of its low noise, a FET is always the first amplifying stage in an HPGe detector preamplifier. When HPGe detectors were first produced, the FET was always operated at room temperature in the preamplifier enclosure; however, better resolution can be achieved when the FET is cooled along with the detector crystal. The improvement in resolution is especially significant at gamma-ray energies below 200 keV. The preamplifier feedback resistor and other associated circuit components could be located inside the cryostat with the FET and the detector crystal. The penalty for the improved performance is that if the FET fails and must be replaced, the cryostat must be opened, usually by the manufacturer at considerable expense to the user. However, preamplifiers that use cooled FETs are now so reliable and so well-protected from high-voltage surges that the transistors rarely fail. As a result, this type of preamplifier is now the most commonly used.

Most manufacturers also offer a high or low count-rate option. This option is needed because detector resolution cannot be optimized simultaneously for high and low count rates. Most detector-preamplifier units are optimized to operate at low count rates (<10,000 cps) because this option provides the best resolution possible. If the primary application will involve count rates greater than 50,000 cps, the manufacturer should be asked to optimize the detector for high-count-rate performance.

HPGe detector crystals are fabricated in planar, coaxial, semi-coaxial, or semi-planar geometries; the designation refers to the shape of the crystal and the location of the charge-collecting contacts. Because of their very low electrical capacitance, small planar detectors (<20 cm^3) have lower noise and better resolution than large detectors. To obtain the best possible resolution from small planar detectors, the feedback resistor is sometimes removed from the preamplifier; however, without the feedback resistor, the decay time of the output pulse is very long, and the output level increases with each successive pulse. Figure 5.5 shows the output of a preamplifier that does not have a feedback resistor. When the maximum allowable DC level is reached, the preamplifier must be reset using a pulsed-optical or transistorized method. The reset pulse can saturate the main amplifier for up to several hundred microseconds, and the data acquisition equipment must be disabled to avoid the analysis of invalid, distorted pulses. Pulsed-optical preamplifiers have a low reset threshold voltage, so they work effectively with only

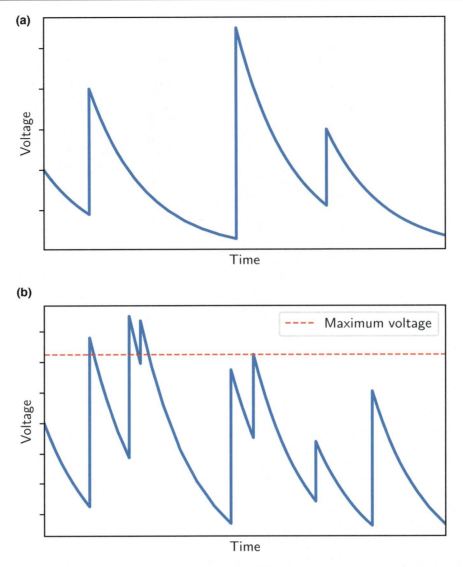

Fig. 5.4 (top) Normal output pulses from a typical HPGe detector preamplifier. The important energy information is contained in the amplitude of the fast-rising voltage step (~0.5 μs). The pileup of pulses on the long tails (~50 μs exponential decay constant) does not affect the validity of the energy information. (bottom) Output pulses from a high rate measurement. Some of the pulses are above the maximum range of the preamplifier and are cut off at the maximum voltage

X-rays and low-energy gamma rays. For higher-energy gamma rays and/or at high rates, a transistor reset preamplifier is needed.

In recent years, preamplifiers that use variations of the pulsed-optical method have been developed for high-count-rate applications. In one case, the optically coupled reset device is replaced by a transistor network. In another case, the reset is accomplished by optical means, but the preamplifier is reset after nearly every event, thereby reducing the amplifier saturation time.

5.5 Analog Amplifier

The sharp rise with long decay-time pulses of the preamplifier is not suitable for direct measurement of peak height. An ideal pulse gradually approaches a relatively flat top and then quickly falls away to the baseline voltage. The pulse should also be narrow to minimize the pileup.

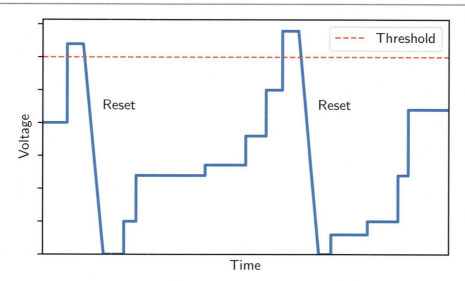

Fig. 5.5 Output pulses from a pulsed-optical preamplifier on an HPGe. Because there is no feedback resistor, the noise level is lower than in resistively coupled preamplifiers. The output level rises in a stairstep fashion and must be reset when the maximum allowable voltage is reached

After leaving the preamplifier, the gamma-ray pulses are amplified and shaped to meet the requirements of the pulse-height-analysis instrumentation that follows the main amplifier. Most spectroscopy-grade amplifiers are single- or double-width NIM modules. All-in-one MCAs have a built-in amplifier and are often adequate for the intended application.

The main amplifier accepts the low-voltage pulse from the preamplifier and amplifies it into a linear voltage range of 0–10 V for most high-quality amplifiers. Within the linear range, all input pulses are accurately amplified by the same factor. The amplification is nonlinear for output pulses that exceed the maximum range. The amplifier gain can be adjusted over a wide range, typically by a factor of 10–5000. Amplifiers usually have two gain controls (coarse and fine) to allow continuous gain adjustment.

The shaping function of the main amplifier is vital to the production of high-quality spectra. The amplified pulses are shaped to optimize the signal-to-noise ratio and to meet the pulse-shape requirements of the pulse-height-analysis electronics. Because MCAs measure the input pulse amplitude with respect to an internal reference (baseline) voltage, the amplifier output must return quickly to a stable voltage level—usually zero—between gamma-ray pulses. The stability of the baseline voltage level is extremely important because any baseline fluctuation perturbs the measurement of the gamma-ray pulse amplitude and contributes to the broadening of the full-energy peak.

A narrow pulse shape permits a quick return to baseline; however, the pulse must be wide enough to allow sufficient time to collect all of the charge liberated by the interaction of the gamma ray in the detector. The pulse shape should also provide a signal-to-noise ratio that minimizes the variation in output pulse amplitude for a given quantity of charge deposited at the preamplifier input. Unfortunately, the pulse width that provides the optimum signal-to-noise ratio is usually wider than that required for a quick return to baseline. At low count rates, the pulse can be wide because the probability is small that a second pulse will arrive before the amplifier output has returned to the baseline level. However, as the count rate increases, the probability that pulses occur on a perturbed baseline also increases, and the spectrum is distorted in spite of the optimum signal-to-noise ratio. A narrower pulse width than required for the optimum signal-to-noise ratio usually gives the best resolution at high count rates; however, the resolution is not as good as can be obtained at low count rates. Figure 5.6 shows the example of the preamplifier output pulses and the corresponding amplifier pulses.

The amplifiers used with high-resolution HPGe detectors employ a combination of electronic differentiation, integration, and active filtering to provide the desired pulse shape. Qualitatively, differentiation removes low frequencies from a signal, and integration removes high frequencies. Differentiation and integration are characterized by a time constant—usually expressed in units of microseconds—that defines the degree of attenuation as a function of frequency. The greater the time constant, the greater the attenuation of low frequencies by differentiation and the attenuation of high frequencies by integration. When both differentiation and integration are used, the low- and high-frequency components are strongly suppressed, and a relatively narrow band of middle frequencies is passed and amplified. Figure 5.7 shows schematics of the CR-RC (capacitor-resistor-resistor-capacitor) differentiation and integration filter.

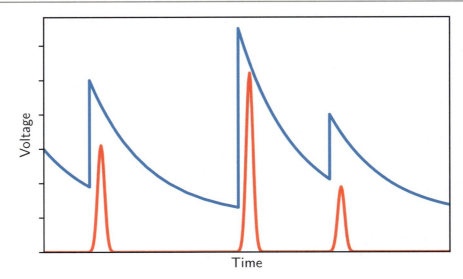

Fig. 5.6 Resistive feedback preamplifier pulses (blue) and the corresponding analog amplifier pulses (red). The heights of the amplifier pulses are proportional to the heights of the preamplifier pulses. The voltage scales of the preamplifier pulses and amplifier pulses are not the same; the amplitude of the latter is typically many times larger than the former

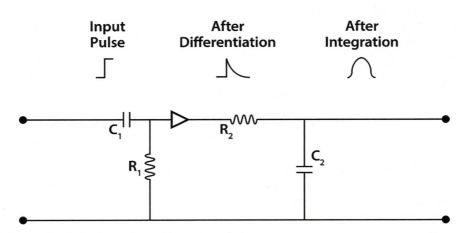

Fig. 5.7 Effects of the differentiation and integration filters

Because most spectroscopy amplifiers function best when the differentiation and integration time constants are equal, a single control usually selects time constants in the range of 0.25–12 μs. When the two time constants are equal, the amplifier output pulse is nearly symmetrical (see the unipolar pulse in Fig. 5.8). The total pulse width is approximately six times the time constant. At low count rates, large coaxial HPGe detectors usually have optimum resolution with time constants of 3–4 μs, and small planar HPGe detectors resolve best with time constants of 6–8 μs. The problem of pulse pileup is more severe when long time constants are used to exploit the intrinsically better resolution of the smaller detectors. The time constant used in a given situation depends on the detector, the expected count rate, and whether resolution or data throughput is of greater importance.

High-resolution HPGe detectors are relatively slow and require time constants longer than those needed for other types of detectors. NaI scintillation detectors, which have resolutions that are 10–20 times worse than those of HPGe detectors, operate well with time constants of 0.25–1.0 μs. Organic scintillation detectors, which have almost no energy resolution, can operate with time constants of only 0.01 μs; when energy resolution is not required but high-count-rate capability is, they are very useful. Unfortunately, no detector now available combines very high resolution with very high count-rate capability.

Spectroscopy amplifiers usually provide two different output pulse shapes: unipolar and bipolar. The bipolar pulse is usually obtained by differentiating the unipolar pulse. Figure 5.8 shows both unipolar and bipolar output signals from a typical

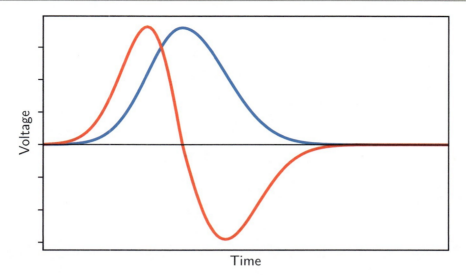

Fig. 5.8 Unipolar and bipolar output signals from a typical spectroscopy amplifier with differentiation/integration pulse shaping

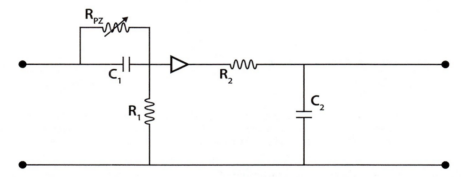

Fig. 5.9 Filter circuit with pole-zero compensation

spectroscopy amplifier. The unipolar output has a better signal-to-noise ratio and is usually used for energy analysis, whereas the bipolar output has superior timing information and overload recovery. The bipolar pulse shape is usually better for timing applications because the zero crossover point (the point where the bipolar pulse changes sign) is easily detected and is very stable. The crossover point corresponds to the peak of the unipolar output and is nearly independent of output pulse amplitude.

5.5.1 Pole-Zero Compensation Circuit

The CR-RC circuit shown in Fig. 5.7 assumes that the input pulse is a step pulse; however, a typical pulse from a normal resistive feedback preamplifier consists of a steep rise with a long falling tail. The output pulse from the CR (capacitor-resistor) differentiation filter will not be critically damped but will be underdamped. This underdamped "post-differentiation" pulse will lead to an underdamped "post-integration" output pulse.

Most amplifiers include a pole-zero compensation circuit to help maintain a stable baseline. The pole-zero circuit was introduced around 1967 and was the first major improvement in amplifier design after the introduction of transistors. It significantly improves amplifier performance at high count rates. The term *pole zero* arises from the terminology of the Laplace transform methods used to solve the simple differential equation that governs the circuit behavior. The circuit is very simple; it consists of an adjustable resistor in parallel with the analog amplifier input capacitor. Figure 5.9 shows the filter circuit with the pole-zero compensation.

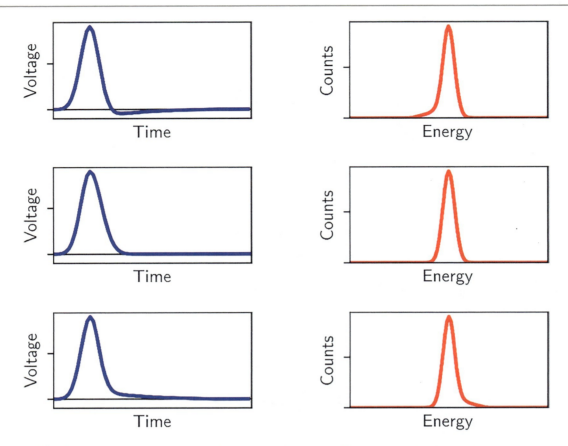

Fig. 5.10 The effect of pole-zero adjustment on the amplifier output (left) and the full-energy-peak shape (right). The upper frames show a pole-zero effect analogous to underdamping, which causes low-energy tailing on the MCA peak. The middle frames show correct pole-zero compensation (critical damping). The lower frames show pole-zero overdamping, which causes high-energy tailing on the MCA peak

In spite of the simplicity of the circuit, the proper adjustment of the pole-zero control is crucial for correct operation of most modern amplifiers. When the pole-zero control is properly adjusted, the amplifier output returns smoothly to the baseline level in the minimum possible time. When the control is incorrectly adjusted, the following conditions result: the output pulses exhibit an effect analogous to underdamping or overdamping that perturbs the output baseline and seriously degrades the amplifier performance at high count rates; and the full-energy peaks are broader and often have low- or high-energy tails depending on whether an underdamped or overdamped condition exists. Accurate determination of the full-energy peak areas is difficult. Figure 5.10 shows the amplifier pulse shapes and full-energy peak shapes that result from correct and incorrect pole-zero adjustment.

Adjustment of the pole-zero circuit is simple and is best accomplished using an oscilloscope to monitor the amplifier output pulse shape and following procedures found in the amplifier manual. The adjustment should be checked whenever the amplifier time constant is changed.

5.5.2 Baseline Restoration Circuit

Baseline restoration (BLR) circuits were added to spectroscopy amplifiers soon after the advent of pole-zero circuits. Like the pole-zero circuit, the BLR helps maintain a stable baseline. The pole-zero circuit is located at the amplifier input and is a very simple circuit; the BLR is located at the amplifier output and is often remarkably complex. The pole-zero circuit prevents underdamping or overdamping caused by the finite decay time of the preamplifier output pulse; the BLR suppresses the baseline shifts caused by the alternating current coupling between the analog amplifier and the circuitry that follows the unipolar output pulses. Although operation of the BLR is totally automatic in most modern amplifiers, some older amplifiers have several controls to optimize amplifier performance for different count rates and preamplifier types.

5.5.3 Pileup Rejection Circuit

Pileup rejection circuits have been added to many amplifiers to improve performance at high count rates. A pileup rejector (PUR; also pileup rejection) uses timing circuitry to detect and reject events where two or more gamma-ray pulses overlap. Such events have a combined pulse amplitude that is not characteristic of any single gamma ray and only increases the height of the background continuum in the acquired spectrum. Figure 5.11 shows how two gamma-ray pulses overlap to produce a pileup pulse. The PUR usually provides a logic pulse that can be used to prevent analysis of the pileup pulses. In high-count-rate situations, the PUR can provide better resolution and a lower background continuum; as a result, determination of the full-energy-peak areas is simplified. Figure 5.12 shows the improvement in spectral quality that can result from using a PUR. The figure also shows that the PUR can sharpen the appearance of sum peaks such that they may be mistaken for real full-energy peaks. This result is due to the limited ability of the PUR circuit to reject pulses that occur within the timing resolution as defined by the shaping time of the analog amplifier.

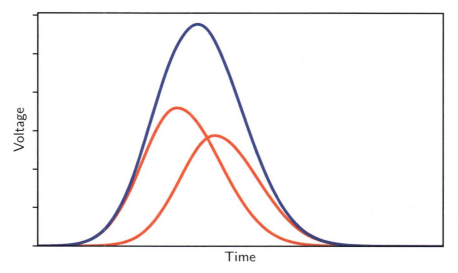

Fig. 5.11 The origin and effect of pulse pileup on the output of a spectroscopy amplifier. When two pulses (red) are separated by less than the amplifier rise time, the amplitude of the resulting sum pulse (blue) is not representative of either input pulse

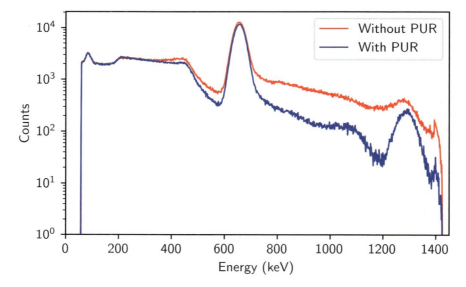

Fig. 5.12 A high dead-time (80%) NaI spectrum of ^{137}Cs that shows the improved spectral quality obtained with pileup rejection. Pileup rejection reduces the pileup continuum near the sum peak at 1323 keV by a factor of 10 but does not significantly reduce the sum-peak amplitude

The considerable benefits of pileup rejection are offset by increased complexity of operation and more stringent requirements for the preamplifier output pulse. The preamplifier output pulse must be free of high-frequency ringing that can cause false pileup signatures in the timing circuits. The output must also be free of high-frequency interference from power supplies, scalers, and computers. Such high-frequency pickup is usually filtered out in the main amplifier but can cause false pileup signatures in the pileup rejection circuit and lead to excessive rejection of good gamma-ray pulses that lead to spectral distortions. Considerable care must be used when adjusting pileup rejection circuits.

The proper use of pole-zero, BLR, and pileup rejection circuits can greatly improve the quality of the measured gamma-ray spectrum. Because an oscilloscope is virtually indispensable for adjusting these circuits for optimum performance, a good quality oscilloscope should be readily available to every user of a gamma-ray spectroscopy system. Some software has built-in oscilloscopes to allow the viewing of the input pulses from the preamplifier and/or the output pulses from the amplifier. Users should understand the operation of the oscilloscope as well as they understand the operation of the spectroscopy system. They can detect and/or prevent more difficulty through proper use of the oscilloscope than with any other piece of equipment.

5.5.4 Advanced Concepts in Amplifier Design

Advances in amplifier design have led to improvements in the ability of gamma-ray spectroscopy systems to operate at high count rates without excessive spectral degradation. The concept uses a narrow pulse shape to reduce pileup losses while preserving good peak shape, signal-to-noise ratio, and resolution.

In the design, a gated integrator is added to the output of a standard, high-quality amplifier. The amplitude of the integrator output pulse is proportional to the integral of the amplifier output pulse. The integrator output is digitized in the normal way by the ADC. For a given gamma-ray interaction, the charge collection time depends on the electric field strength in the detector and the location of the interaction. Charge carriers that are produced far from the collection electrodes or in regions where the electric field is weaker arrive later at the electrodes. Charge that is collected very late may not contribute to the information-carrying part of the preamplifier pulse; such charge is said to cause a *ballistic deficit*. If the amplifier time constants are comparable to the charge collection time, the integral of the amplifier output pulse is more nearly proportional to the collected charge than is the pulse amplitude. Qualitatively, the integration allows a longer period for charge collection and decreases the ballistic deficit. Shorter time constants can be used with the amplifier- integrator combination than can be used with the amplifier alone. The short time constants reduce pileup losses and increase data throughput. Figure 5.13 shows the amplifier and corresponding integrator output pulse.

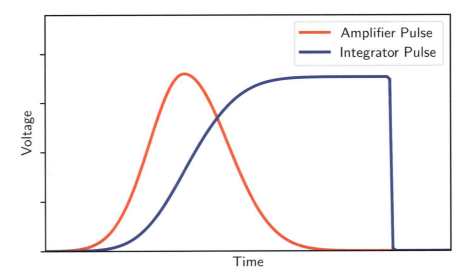

Fig. 5.13 The regular output and gated-integrator output from an amplifier that uses gated integration to reduce the ballistic deficit at short time constants

The gated integrator permits good resolution and lower pileup and dead-time losses at high input rates; however, its performance under normal conditions is likely to be slightly worse than that of semi-Gaussian shaping. Therefore, many gated-integrator amplifiers also include the standard analog amplifier output to be used in normal operating conditions.

5.6 Other Components

Some other components may or may not be part of a complete gamma-ray spectroscopy system. Those modules are single-channel analyzers (SCAs), counters, scalers, timers, and rate meters. They are described in Chap. 4 of the original PANDA book [1] and are omitted here.

5.7 Multichannel Analyzer

The functions listed inside the dashed line in Fig. 5.1 are usually performed by an MCA operating in the pulse-height-analysis mode. The MCA can operate in several modes, including pulse-height analysis, voltage sampling, and multichannel scaling. The MCA sorts and collects the gamma-ray pulses coming from the main amplifier to build a digital and visual representation of the pulse-height spectrum produced by the detector.

5.7.1 Analog-to-Digital Converter

The ADC performs the fundamental pulse-height analysis and is located at the MCA input. The ADC input is the voltage pulse from the analog amplifier; its output is a binary number that is proportional to the amplitude of the input pulse. The binary output number is often called an *address*. Other MCA circuits increment a storage register in the MCA memory that corresponds to the ADC address. The ADC performs a function that is analogous to that of the oscilloscope user's saying "five volts" when a 5 V pulse is applied to the oscilloscope input terminals. The ADC accepts pulses in a given voltage range, usually 0–8 V or 0–10 V, and sorts them into a large number of contiguous, equal-width voltage bins or channels. Because of the sorting function, the early MCAs were often called *kicksorters*. The original kicksorter developed by Otto Frisch worked by rolling ball bearings into different channels cut into a polystyrene block using an electrical "kick," which would form an electro-mechanical histogram of the pulse heights.

The number of channels into which the voltage range is divided is usually a power of 2 and is called the *ADC conversion gain*. In the mid-1950s, a high-quality ADC could divide 100 V into 256 channels. Now, ADCs routinely divide 10 V into as many as 32,768 channels. This capability is impressive; an individual channel is only 0.3 mV wide. The required conversion gain varies with detector type and with the energy range being examined. Figure 5.14 shows part of an 8192-channel

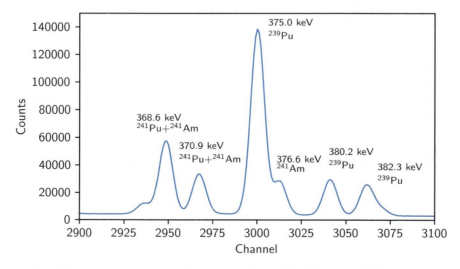

Fig. 5.14 A small portion of an 8192-channel plutonium spectrum taken with a high-quality coaxial HPGe detector at 0.125 keV/ch. The major peak is the 375.0 keV peak from ^{239}Pu decay

plutonium spectrum measured with a high-resolution HPGe detector. The full-energy peaks should contain enough channels to clearly define the structure of the spectrum. When peak fitting is required, 10 or more channels are needed to clearly define peak shape. Nominally ~3 FWHM would suffice.

The ADC sorts the amplifier output pulses according to voltage, which is proportional to the energy deposited in the detector during the gamma-ray interaction. Like the relationship between voltage and energy, the relationship between channel number and energy is nearly linear. An HPGe detector system is normally linear, with most of the nonlinearity coming from the amplifier and the ADC. A NaI detector system is not very linear, somewhat quadratic, with most of the nonlinearity coming from the scintillator itself. The relationship can be represented by Eq. 5.1:

$$E = mx + B, \tag{5.1}$$

where
E = energy in keV
X = channel number
m = slope in keV per channel
b = zero intercept in keV.

The slope (*m*) depends on the conversion gain and the amplifier gain; increasing the conversion gain or amplifier gain decreases the slope (*m*) and vice versa. Common values for slope (*m*) are 0.05 keV per channel (for planar HPGe detectors) to 1.0 keV per channel (for NaI detectors). Although it may seem logical to assume that zero energy corresponds to channel zero ($b = 0$), this assumption may not be the case. For HPGe detectors, the zero intercept is often very close to the origin—only a fraction of 1 keV. For NaI detectors, the zero intercept is most often between -15 and -10 keV. Due to the slightly quadratic relationship, the zero intercept of the NaI detector can change depending on the peaks used to determine the relationship. Most all-in-one MCAs (discussed later) do not allow adjustment of the zero intercept, but some single-module ADCs—such as those of the NIM system—do allow such adjustments. For such systems, the slope and zero intercept can be adjusted to fit the energy range of interest into the desired channel range. For example, plutonium measurements often use gamma rays in the 60–420 keV range. If the gains are adjusted to 0.1 keV per channel and the zero intercept to 10 keV, a 4096-channel spectrum covers the 10–419.6 keV energy range and includes the important gamma ray at 413.7 keV. In the example, the channel number can be converted easily to energy.

Because preamplifiers, amplifiers, and ADCs are not exactly linear, the relationship shown in Eq. 5.1 between energy and channel number is not exact; however, with good equipment, gamma-ray energies can be readily measured to a tenth of a keV by assuming a linear calibration. ADC linearity is usually specified with two numbers: integral and differential linearity. Integral nonlinearity is a slight curvature in the relationship between energy and channel number; differential nonlinearity is a variation in channel width. It is difficult to design an ADC that does not have differential nonlinearity. A common ADC problem that can influence assay results is a slow increase of the differential nonlinearity over time. The <1% differential nonlinearity of most ADCs is totally acceptable for most applications.

Two types of ADCs are in common use: the Wilkinson and the successive-approximation ADC. A Wilkinson ADC counts pulses from a fast oscillator for a time interval that is proportional to the amplitude of the amplifier pulse. The digitization time determines the channel number assigned to each pulse. A successive-approximation ADC examines the amplifier pulse using a series of analog comparators. The first comparator determines whether the pulse amplitude is in the upper or lower half of the ADC range. Each successive comparator determines whether the pulse amplitude is in the upper or lower half of the voltage interval determined by the previous comparator. Twelve comparators determine the pulse amplitude to one part in 2^{12} (or 4096) channels. The digitization time of a successive-approximation ADC is constant and independent of pulse height. Until the 1990s, Wilkinson ADCs dominated the gamma-ray spectroscopy field because they had superior differential linearity. Now, successive-approximation ADCs have comparable differential linearity and are becoming more popular because they are often faster than Wilkinson ADCs.

ADC speed is an important consideration for high-count-rate spectroscopy. While the ADC is processing one pulse, all other pulses are ignored. The pulse processing time, or *dead time*, can be a substantial fraction of the total acquisition time. A dead time of 25% means that 25% of the information in the amplifier pulse stream is lost. For both ADC types, the dead time per event is the sum of the digitization time and a fixed processing time (usually 2–3 μs). The 450–100 MHz oscillators used in Wilkinson ADCs require 12–43 μs to digitize and store a gamma-ray event in channel 4000. Successive-approximation ADCs (4096 channels) require 4–12 μs to analyze a gamma-ray event. A detailed comparison of ADC speed requires specification of the gamma-ray energy spectrum, the overall system gain, and the ADC range. In general, successive-approximation ADCs are faster than Wilkinson ADCs for spectra with 4096 channels or more. For spectra with few channels, the Wilkinson ADC may be faster. In a spectrum with an average channel number of 512, a 400-MHz Wilkinson ADC has an average dead time per event of 3 µs.

Several common features appear on most ADCs independent of type or manufacturer. Lower-level discriminators and upper-level discriminators determine the smallest and largest pulses accepted for digitization. The discriminators can be adjusted to reject uninteresting low- and high-energy events and reduce ADC dead time. The discriminator adjustment does not affect the overall gamma-ray count rate and cannot be used to reduce pulse pileup losses that occur in the detector, preamplifier, and amplifier. The discriminators form an SCA at the input to the ADC; most ADCs provide an SCA output connector. Most ADCs have coincidence and anticoincidence gates that allow external logic circuits to control the ADC. Pileup rejection circuits frequently provide an inhibit pulse that is fed to the anticoincidence gate to prohibit processing or storage of pileup pulses. The coincidence gate is also used to analyze gamma-ray events that are detected in two separate detectors. Most ADCs have an adjustable conversion gain and range; the range control determines the maximum channel number to be digitized. There is usually a dead-time indicator that displays fractional dead time. In computer-based MCAs, the ADC parameters often can be set under program control. Most small MCAs have a built-in ADC; large MCA systems use separate NIM- or rack-mounted ADCs.

5.7.2 Spectrum Stabilizers

For HPGe detectors, the relationship of energy and channel number changes with time even though the energy-to-charge collection factor is constant. The preamplifier, amplifier, and ADC are all subject to small but finite changes in gain and zero level caused by variation in temperature and count rate. Under laboratory conditions, the position of a full-energy peak at channel 4000 may shift only a few channels over a period of many weeks; however, even this small drift may be undesirable. Larger drifts may be encountered in the uncontrolled environment of production facilities. Spectrum stabilizers are electronic modules that fix the position of one or more full-energy peaks by adjusting a gain or DC level in the spectroscopy system to compensate for drift; they are especially recommended for gamma-ray spectroscopy systems that must be operated in uncontrolled environments by unskilled operators (as often required by routine production-plant assay systems). Stabilizers are also recommended whenever channel summation procedures are used to determine full-energy-peak areas.

The spectrum stabilizers used with HPGe detectors are usually digital circuits connected directly to the ADC. The stabilizer examines each gamma-ray event address generated by the ADC and keeps track of the number of counts in two narrow windows on either side of a selected full-energy-peak channel. The stabilizer generates a feedback signal for the ADC that is proportional to the difference in the number of counts in the two windows. The feedback signal adjusts the ADC gain or zero level so that the average number of counts in each window is the same; the adjustment fixes the position of the selected stabilization peak. Often two peaks are stabilized independently; a peak at the high-energy end of the spectrum is used to adjust ADC gain and another peak at the low-energy end is used to adjust the ADC zero level. With two-point stabilization, the spectroscopy system stability is often so good that no spectral peak shifts position by more than one-tenth of a channel over a period of many months. Digital stabilizers can be used to easily establish simple and convenient energy calibrations (e.g., $E = 0.1X$).

Stabilization peaks should be free from interference, adequately intense, and present at all times. Often one of the stabilization peaks comes from a gamma-ray source that is attached to the detector to provide a constant signal in the detected spectrum. Such a stabilization source is usually monoenergetic and provides the low-energy stabilization peak so that its Compton continuum does not interfere with other gamma-ray peaks of interest. In some cases, a very stable pulser may be connected to the test input of the preamplifier to provide an artificial stabilization peak. Peaks from pulsers or special stabilization sources may also be used to provide corrections for pulse-pileup and dead-time losses (see Chap. 7).

Digital stabilization is included in most all-in-one MCAs and is also available for some NIM modular ADCs. All stabilizers have controls to set the desired peak-centroid channel number and the width of the stabilization peak windows, and there is often a control to set the stabilizer sensitivity. The digital stabilizers of virtually every all-in-one MCA and some modular units are controlled by an external computer. This feature is useful when stabilization peaks must be changed during automatic assay procedures.

Digital stabilizers that have a small correction range are inadequate for use with NaI detectors. Because of the relatively greater instability of scintillator/photomultiplier detectors (as large as 1–2%/°C), spectrum stabilization is often more necessary for NaI detectors than it is for HPGe detectors.

Scintillation detector stabilizers are similar to digital stabilizers but operate with the amplifier rather than the ADC. The stabilizer compares the count rate on either side of the selected stabilization peak and generates a feedback signal that adjusts the amplifier gain to keep the two count rates equal. When a suitable stabilization peak is not available in the NaI spectrum, a pulser peak cannot be substituted because it can correct for only preamplifier and amplifier instability; the major drift in a NaI system occurs in the photomultiplier tube. Although an external gamma-ray source can provide a stabilization peak, the Compton background from the source can interfere excessively with the gamma rays of interest. An alternative solution is to use a detector with a built-in light pulser. NaI crystals can be grown with a small doping of an alpha-particle-emitting nuclide such as ^{241}Am. The alpha-particle interactions in the crystal provide a clean spectral peak with a fixed rate and gamma-ray-equivalent energy. Because the temperature dependences of alpha-particle-induced and gamma-ray-induced scintillation light are not identical, accurate stabilization over a large temperature range may require special temperature compensation circuitry.

5.7.3 Multichannel Analyzer Memory, Display, and Data Analysis

After the ADC converts the amplifier voltage pulse to a binary address, the address must be stored for later observation and analysis. All MCA systems have memory reserved for spectrum storage, and most have a spectral display and some built-in data analysis capability.

Although the most common memory size is 8192 channels, MCAs are available that have other memory sizes such as 1024, 4096, 16,384, and even 32,768 channels. The smaller memory size is adequate for NaI detector applications and for HPGe detector applications that involve a small energy region. To have sufficient channels in a full-energy peak, an overall system gain of 0.1 keV per channel is often required; however, with this gain, a 1024-channel MCA can collect data only in a 100 keV wide region.

The maximum number of counts that can be stored per channel is often an important consideration because this number sets a limit on the precision that can be obtained from a single measurement. Early transistorized MCAs often had a maximum capacity of 65,536 counts per channel. The present standard is typically 2×10^9 or 4×10^9 counts per channel. These numbers are probably more than will ever be required in any anticipated application.

A quick and useful way to obtain qualitative and semi-quantitative information from a spectrum stored in memory is to look at a plot of channel content versus channel number. Most MCAs have a spectral display, and many offer a wide range of display options. All displays offer several vertical and horizontal scale factors, and many offer both linear and logarithmic scales. Most displays have one or two cursors (visual markers) that can be moved through the spectrum; the channel number and contents of the cursor locations are displayed numerically on the screen. Most MCAs can intensify selected regions of interest or change the color of the regions of interest to emphasize particular spectral features. A good MCA can display two or more spectra simultaneously and can overlap spectra for careful visual comparison.

Until the 1990s, most MCA displays used cathode-ray tubes (CRTs) with electrostatic deflection, which are easily used for small screen displays—up to approximately 15 × 15 cm. Later, most MCAs have used magnetic deflection CRTs to allow a larger screen size. Modern systems are often integrated with the computer and its acquisition-controlling software and spectrum-analysis program. The computer will display the spectrum using its MCA emulator software. In all types, each channel is represented by a dot or bar whose vertical height is proportional to the channel contents.

Big-screen, magnetically deflected CRTs of the MCA and computer monitors in the 1990s made good pictures but had one annoying drawback: the horizontal oscillator in the magnetic-deflection circuit would generate bursts of electro-magnetic interference at a frequency of approximately 16 kHz. This interference was easily picked up on preamplifier signal lines and could cause significant degradation of spectral quality. Great care had to be taken in grounding, shielding, and routing signal cables to eliminate or minimize the problem. The monitors of modern computers—either desktop or laptop—are generally LED (light-emitting diode) or LCD (liquid crystal display) and do not generate electromagnetic interference that can degrade the spectral quality.

All modern, computer-integrated MCAs have some built-in data analysis functions. Common analysis functions determine the channel position and width of spectral peaks, the energy calibration, the number of counts in selected regions of interest, and the full-energy-peak areas. Other available functions might include smoothing, normalizing, and subtracting (stripping) a background spectrum. The computer can also control complete assay systems and execute complex analysis codes.

5.8 Digital Pulse Processing System

A digital pulse processing (DPP) system or digital MCA consists of the components in the dashed line in Fig. 5.2. Unlike the analog MCA system, where the pulses from the preamplifier are shaped and amplified by the amplifier before being sent to the ADC to be converted into numbers, the DPP system converts the pulses from the preamplifier into digitized models of the pulses and then analyzes the digitized pulses using a digital filter. The digital filters can be made to resemble the typical pulse shaping of the analog amplifier, such as the triangular or Gaussian pulse shape. They can also be made into different and more-effective pulse shapes that are not possible with traditional analog processing.

5.8.1 Flash ADC

The Wilkinson ADC and the successive-approximation ADC described in Sect. 5.7.1 convert the height of the output pulse from the amplifier into an n-bit number with maximum value of $2^n - 1$. The process is accurate but relatively slow, taking up to ~10 μs per pulse. A pulse from the preamplifier of an HPGe detector may have a rise time of several hundred nanoseconds and a fall time of several hundred μs. Digitizing the pulse would require an ADC with a measurement time less than the rise time of the pulse to accurately capture its rise time. The flash ADC (also known as a direct-conversion ADC) can satisfy this requirement. It is the fastest of all of the types of ADC, with a sampling rate capability of 10 MHz or larger (equivalent to conversion time of 100 ns or less). A flash ADC samples the signal using a bank of comparators in parallel, each one comparing the signal with a unique reference voltage. The output from the comparators is fed into a digital encoder, which then produces a binary output that represents the amplitude of the waveform at the respective address.

5.8.2 Pulse Processor

5.8.2.1 Filter

The output of the preamplifier can be described as fast rise followed by exponential decay. Figure 5.15 shows the section of the pulse near its beginning—the sharp rise. The height of the pulse corresponds to the absorbed gamma-ray energy. Determining the gamma-ray energy requires the measurement of the voltage step or the pulse height. The effect of the noise in the signal is minimized by filtering.

In Fig. 5.15, it is obvious that the pulse height can be determined by taking the average of some points after the step and subtracting the average of the points before the step; i.e., the averages are calculated from the two regions marked with L_{after}

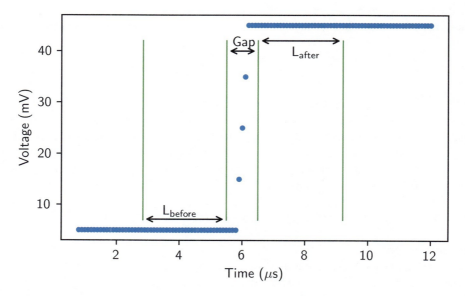

Fig. 5.15 The digitized pulse at the step

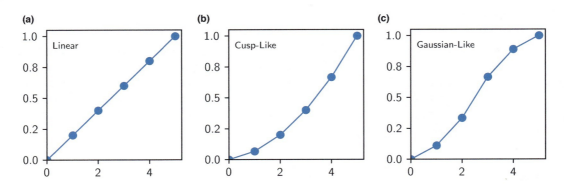

Fig. 5.16 Rise times (from left to right) produced by the sets of weights in the region L_{after}: (1,1,1,1,1,0), (5,4,3,2,1,0), and (1,2,3,2,1,0), respectively

and L_{before}. Their difference is the voltage step (V_s), which corresponds to the gamma-ray energy. The voltage step (V_s) can be expressed as

$$V_s = \frac{\sum_{i(after)} V_i W_i}{\sum_{i(after)} W_i} - \frac{\sum_{i(before)} V_i W_i}{\sum_{i(before)} W_i}, \tag{5.2}$$

where V_i and W_i are the voltage and weight of the data point (i), respectively. The values of the weights (W_i) and how the regions are selected determine the filter. For example, the lengths of the regions L_{after} and L_{before} are the rise time and fall time of the filter, respectively; the gap length is the flattop; constant weights (W_i) produce a linear rise time or fall time; larger weight values (W_i) for the region near the gap produce a cusp-like rise or fall time; larger weight (W_i) for the middle region of L_{after} or L_{before} produces Gaussian-like rise or fall time.

Figure 5.16 shows examples of three different rise times produced by three different sets of weights in the region L_{after}: (1,1,1,1,1,0), (5,4,3,2,1,0), and (1,2,3,2,1,0). The weight of zero of the last point in each set corresponds to the point on the right following the region L_{after}. This last point with no weight produces the first point (at the origin) in each graph. Similar shapes for the fall time also can be constructed using analogous weights for the region L_{before}.

To apply the filter, the difference of the two weighted, moving averages of Eq. 5.2 is calculated repeatedly over the entire data range. The first filter point requires (n) multiplication operations and ($2n - 4$) addition operations, where n is the total number of data points in L_{after} and L_{before}. Subsequent filter points require (n) multiplication operations and ($n + 2$) addition operations. This filtering operation is computationally intensive.

In reality, most digital pulse processors use weight (W_i) equal to unity with equal length L_{after} and L_{before}, which significantly simplifies the calculations and reduces the demand for computational power. Equation 5.2 then simplifies to

$$V_s = \frac{1}{L} \left(\sum_{i(after)} V_i - \sum_{i(before)} V_i \right), \tag{5.3}$$

where L is the number of data points (i) in L_{after} (also the same as number of data points (i) in L_{before}). The first filter point requires 1 multiplication operation and $2L - 2$ addition operations. Subsequent filter points require one multiplication operation and four addition operations. This filtering operation uses only a very small fraction of computing power like that of Eq. 5.2 and still produces an effective filter—the trapezoidal filter. When the gap is set to zero, then the filter becomes a triangular filter, which is equivalent to the analog triangular filter used for high rate measurements. A non-zero gap width is the flat top of the trapezoidal filter. This flat top width can be adjusted to significantly reduce or eliminate the ballistic deficit effects while not sacrificing throughput.

Figure 5.17 shows the original pulse from the preamplifier and the trapezoidal filter pulse.

5.8.2.2 Pole-Zero Correction

The pulse of the RC (resistor-capacitor) type (non-transistor/optical-reset) preamplifier decays exponentially to the baseline DC level of the preamplifier. Due to such decay, the filter output after the pulse is below the initial level, similar to the output pulse of the CR-RC filter (without pole-zero) of the analog amplifier. Figure 5.18 shows such a pulse and its filtered pulse.

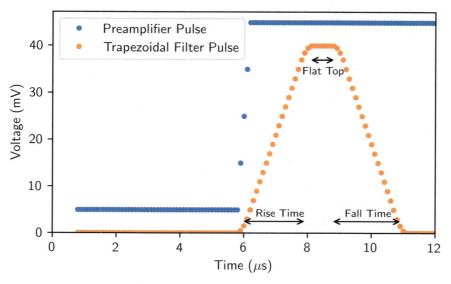

Fig. 5.17 Trapezoidal filtering of a pulse with rise time = fall time = 2 μs and flat top = 1 μs

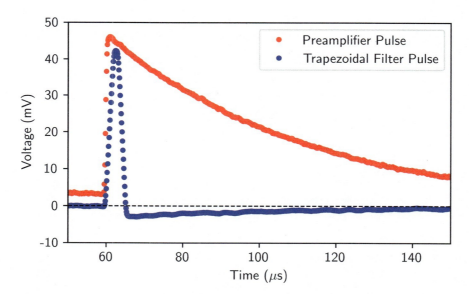

Fig. 5.18 A pulse displayed over a longer time period to show the effect of preamplifier decay time. The rise and fall times of the filter are 2.2 μs, and the flat top is 1.1 μs

The exponential decay of the preamplifier pulse can be corrected by applying the correction equation

$$V'_i = V'_{i-1} + \left[V_i - V_{i-1} e^{-\Delta t/\tau} \right], \tag{5.4}$$

where V_i and V'_i are the original voltage and corrected voltage above the preamplifier DC level at channel i, respectively, Δt is the time interval between two adjacent channels, and τ is the decay constant of the preamplifier pulse. The corrected pulse is then processed using the designed filter. Doing it correctly, the output pulse should not be underdamped nor overdamped. Figure 5.19 shows the comparison of the results without and with pole-zero correction.

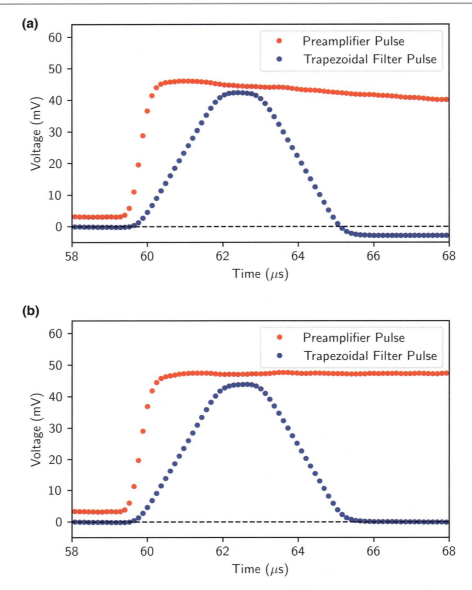

Fig. 5.19 Comparison of the results without and with pole-zero correction. (top) Figure shows the pulse from the preamplifier and the filtered output pulse; the output pulse is underdamped, going past the baseline. These are the same preamplifier and output pulses shown in Fig. 5.18, with 2.2 μs rise and fall times and 1.1 μs flat top. (bottom) Figure shows the corrected preamplifier pulse and the output pulse. The decay constant of 46 μs was used for the correction. The output pulse is critically damped, returning quickly to the baseline

5.9 All-in-One Multichannel Analyzer

All components of a modern gamma-ray spectroscopy system—including the low-voltage power supply for the preamplifier, the high-voltage power supply for the detector, the amplifier, and the traditional MCA but without the computer—are normally packed into a single box and are referred to as an all-in-one MCA, or simply an MCA. Some of these systems also include the data acquisition control and display, but most systems do not. All systems, including the ones with their own control and display, can be connected to and controlled by a computer.

Older, all-in-one MCAs of the 1990s or early 2000s communicate with the computer through the RS-232 serial connection or the parallel or printer port. Modern all-in-one MCAs, either analog or digital, connect to the computer by a USB (universal serial bus) or Ethernet cable. A portable MCA also has an internal rechargeable battery that can supply the power required to operate the system for several hours. Most of the gamma-ray spectroscopy systems used in nuclear safeguards today are all-in-one MCAs.

Going one step further, some systems combine both a detector and an all-in-one MCA system into one compact detector system. The detectors for these systems are of various kinds, including HPGe, CZT, $LaBr_3$, and NaI. The HPGe detector system is normally called an *electro-cooled HPGe* or *electrically cooled HPGe* because the detector crystal is cooled by an electro-mechanical cooler instead of liquid nitrogen. These systems can acquire and store the data, and the data can be downloaded to a computer for analysis. Some systems also have onboard computational capability to analyze the data and produce the results without an external computer.

5.10 Auxiliary Electronic Equipment

Figures 5.1 and 5.2 show only the basic components of gamma-ray spectroscopy systems. This section describes other instruments that may be used in addition to the basic components.

The oscilloscope is the most useful auxiliary instrument used with gamma-ray spectroscopy systems. It is virtually indispensable when setting up the spectroscopy system for optimum performance, monitoring system performance, detecting malfunctions or spurious signals, and correcting problems. An expensive oscilloscope is not required; a 50-MHz response, one or two vertical inputs, and an ordinary time base are usually quite adequate. Battery-powered, portable oscilloscopes can be carried easily to systems in awkward locations. Many all-in-one MCAs come with the software that can display the processed pulses.

Electronic pulsers are used to test system performance and correct for dead-time and pileup losses. Mercury-switch pulsers have excellent pulse amplitude stability but are quite slow and have limited pulse-shape variability. Other electronic pulsers often have high-repetition rate and very flexible pulse shaping but usually have neither great amplitude nor frequency stability. A few pulsers provide random intervals between pulses rather than the more common fixed intervals. Sliding pulsers are used to test ADC linearity; their pulse amplitude is modulated linearly with time.

Cameras were often used to take pictures of MCA and oscilloscope displays. Modern oscilloscopes today can digitize the waveforms into a digital form, store them, and transfer them to a computer to be displayed and/or processed. Pictures of waveforms help to document and diagnose problems; pictures of spectra provide a quick and useful way to record information in a notebook.

Many different instruments are available to provide information on gamma-ray pulse timing, usually to establish temporal relationships between two or more detectors. Timing-filter amplifiers sacrifice signal-to-noise performance and overall resolution to preserve timing information. Other instruments examine the preamplifier output, the bipolar output of the main amplifier, or the output of a timing-filter amplifier, and they generate a fast logic signal that has a fixed and precise temporal relationship to the gamma-ray events in the detector. The timing is determined using techniques such as fast leading-edge discrimination, constant-fraction discrimination, amplitude/rise-time compensation, and zero-crossover pickoff. The timing outputs are either counted or presented to coincidence circuits that determine whether specified time relationships are met by the events in two or more detectors. Depending on the type of detector, the coincidence gates can be as narrow as 100 picoseconds or less. The logic output of a coincidence circuit is either counted or used as a control signal. When more-detailed timing information is required, a time-to-amplitude converter can be used to generate an output pulse whose amplitude is proportional to the time interval between input pulses.

A linear gate can be used as a coincidence or control circuit at the input to an MCA. Linear gates pass analog signals with no change in amplitude or shape if they are gated by control signals that are derived from one of the timing circuits described previously. A linear stretcher generates a pulse with the same amplitude as the input pulse but with an adjustable length. A stretcher is occasionally used to condition the amplifier signal before subsequent processing in the ADC. Summing amplifiers, or mixers, produce outputs that are the linear sum of two or more input signals. A mixer can be used in connection with routing signals for collecting spectra from several detectors using a single ADC.

Compton suppression, a common procedure that improves the quality of gamma-ray spectra, uses some of the timing circuits previously described. A Compton suppression spectrometer usually includes a high-resolution detector that is surrounded by a low-resolution, annular detector. The scattered gamma ray from a less-than-full-energy interaction in the high-resolution detector is often detected in the annular detector. A coincidence event between the two detectors inhibits the storage of the high-resolution event in the MCA and reduces the Compton continuum between the full-energy peaks.

5.11 Concluding Remarks

The instrumentation described in this chapter can be assembled to form different gamma-ray spectroscopy systems for different NDA applications. Many instrument manufacturers can provide integrated spectroscopy systems that include all components—from the detector to the output printer. If the user has sufficient expertise, individual components can be procured from different manufacturers. In any case, careful consideration must be given to the requirements of the measurement application before selecting a spectroscopy system from the nearly endless array of options and configurations. References [2–6] provide detailed descriptions of the function and operation of gamma-ray spectroscopy instrumentation. For the user who is not active in gamma-ray spectroscopy, current information is best obtained from research reports, the commercial literature, and the developers and users of state-of-the-art instrumentation.

Gamma-ray assay systems that are dedicated to a particular operation can be very simple to operate. Conversely, vast versatility and flexibility are provided by combining the appropriate detector, amplifier, MCA, and analysis capability to make a large, modern gamma-ray spectroscopy system. Unfortunately, a complex, versatile instrument can never be truly simple to operate; a labor of several weeks is usually required to master the operation of the typical large system. However, the effort required is usually readily exerted to use instruments of truly amazing power. The power of modern gamma-ray spectroscopy systems is perhaps best appreciated by those who remember from personal experience when all spectral measurements were done with a NaI detector, an SCA, and a counter.

Gamma-ray spectroscopy equipment, as with virtually all electronics, has improved rapidly over the past 70 years as vacuum tubes were replaced by transistors and transistors were replaced by integrated circuits. The microprocessor chip has put greater capability into smaller and smaller volumes. The capability per dollar has increased in spite of inflation. The rate of improvement is still significant, particularly in the capability and flexibility of MCA memory, display, and data analysis. Spectral quality is not progressing as rapidly, although improvement is still occurring in pulse-processing electronics, especially when dealing with very high count rates (up to 10^6 cps) from high-resolution HPGe detectors. The technology of NaI and HPGe detectors is quite mature, and major improvements are not expected. Still, steady progress in all areas of gamma-ray spectroscopy technology will continue, and unexpected breakthroughs may indeed occur.

References

1. D. Reilly, N. Ensslin, H. Smith Jr., S. Kreiner, *Passive Nondestructive Assay of Nuclear Materials (PANDA)* (U.S. Nuclear Regulatory Commission, 1991)
2. G.F. Knoll, *Radiation Detection and Measurement*, 3rd edn. (Wiley, New York, 2000)
3. G.R. Gilmore, *Practical Gamma-Ray Spectrometry*, 2nd edn. (Wiley, New York, 2008)
4. P.W. Nicholson, *Nuclear Electronics* (Wiley, New York, 1974)
5. F. Adams, R. Dams, *Applied Gamma-Ray Spectroscopy* (Pergamon Press, Oxford, 1970)
6. W.J. Price, *Nuclear Radiation Detection*, 2nd edn. (McGraw-Hill Book Co., New York, 1964)

Open Access This chapter is licensed under the terms of the Creative Commons Attribution 4.0 International License (http://creativecommons.org/licenses/by/4.0/), which permits use, sharing, adaptation, distribution and reproduction in any medium or format, as long as you give appropriate credit to the original author(s) and the source, provide a link to the Creative Commons license and indicate if changes were made.

The images or other third party material in this chapter are included in the chapter's Creative Commons license, unless indicated otherwise in a credit line to the material. If material is not included in the chapter's Creative Commons license and your intended use is not permitted by statutory regulation or exceeds the permitted use, you will need to obtain permission directly from the copyright holder.

Attenuation Correction Procedures

D. J. Mercer, P. J. Karpius, J. L. Parker, P. A. Santi, and K. Schmidt

6.1 Introduction

Nuclear materials subject to nondestructive assay (NDA) are often macroscopic objects composed of highly attenuating material. In some situations, the objects under assay are also obscured by containers or shielding. Gamma rays emitted within the item matrix are subject to collisions in the material matrix or other intervening materials before reaching the detector. The loss of photons that occurs in the material of the sample is referred to as self-attenuation, and loss due to collisions in other intervening materials is external attenuation. These effects are key limiting factors for using gamma rays for the measurement of special nuclear material mass in extended sources.

Materials measured using NDA vary widely in size, but even a small sample with a high concentration of nuclear material will suffer significant self-attenuation. Although gamma-ray attenuation can frequently be ignored in filter papers or small vials encountered in radiochemical applications, it usually cannot be ignored in NDA measurements of nuclear material.

Attenuation can be thought of as a factor that degrades detector efficiency, which is particularly true in situations where the packaging or dead regions of the detector itself are significantly attenuating. In practice, it is sometimes convenient to calculate or measure detector efficiency, considering attenuation simultaneously.

Several methods are available for correcting for attenuation. Often the best method is to calibrate out the effects of attenuation and detector efficiency using standard sources. Alternatively, one can calculate or measure the detector efficiency for the counting geometry by considering the size, shape, and composition of the sample and any shielding, which can be done by using an analytical method that employs known attenuation coefficients, by measuring attenuation using gamma-ray sources at similar energies and interpolating, or by using software that performs radiation transport simulation of the geometry.

Sections 6.2 through 6.7 describe the nature and computation of the attenuation correction factor CF_{atten}. A more detailed discussion of this subject is given in Ref. [1]. In Chap. 10, Sect. 10.4 describes assay systems using transmission-corrected procedures.

Los Alamos National Laboratory strongly supports academic freedom and a researcher's right to publish; as an institution, however, the Laboratory does not endorse the viewpoint of a publication or guarantee its technical correctness.

D. J. Mercer (✉) · P. J. Karpius · J. L. Parker · P. A. Santi
Los Alamos National Laboratory, Los Alamos, NM, USA
e-mail: mercer@lanl.gov; karpius@lanl.gov; psanti@lanl.gov

K. Schmidt
Oak Ridge National Laboratory, Oak Ridge, TN, USA

6.2 Procedures

6.2.1 Preliminary Remarks

The procedures and methods described herein are best applied with high-resolution gamma-ray detectors. The merits of superior energy resolution and procedures for extracting counts from an individual peak are discussed in Chap. 7. The methods and correction factors may be used for assays with low-resolution detectors, but additional care must be exercised to achieve accurate peak integrations and background and continuum subtractions.

In addition to self-attenuation, rate-related losses also contribute to the loss of proportionality between detected count rate and sample mass. For the remainder of this chapter, it is assumed that rate-related losses have been accounted for as prescribed in Chap. 7.

6.2.2 General Description of the Attenuation Correction Factor

In determining CF_{atten}, the basic question is, what fraction of the gamma rays of interest that are emitted in the direction of the detector enter the detector at full energy? Fundamentally, the trajectories of gamma rays are interrupted by events, such as Compton scattering or photoelectric absorption, with probabilities that can be described using cross sections. Attenuation is calculated as the aggregate effect of those events for a large number of photons. Monte Carlo radiation transport models take the approach of simulating the paths of individual photons, taking into account the probabilities of scattering events stochastically. The aggregate effect can also be described analytically. A collimated beam of gamma rays traversing a material is attenuated according to the formula

$$T = exp(-\mu_\ell x), \qquad (6.1)$$

where

T = transmission
μ_ℓ = the linear attenuation coefficient (Chap. 3)
x = is the length of the path through the material.

The problem of calculating gamma-ray attenuation is then reduced to finding the attenuation coefficient for the material in question and accounting for the geometry of the problem appropriately.

The correction factor for attenuation can be represented as a product of the inverse transmissions associated with self-attenuation (T_{self}) and external attenuation (T_{ext}):

$$CF_{atten} = \frac{1}{T_{self}} \times \frac{1}{T_{ext}}. \qquad (6.2)$$

Because the attenuation coefficient depends on photon energy, CF_{atten} is also dependent on photon energy.

An issue comes in the notation. The linear attenuation coefficient is noted as the Greek letter μ_l and the mass attenuation coefficient is noted as only the Greek letter μ. Also, the literature sometimes states only the attenuation coefficient without indicating which type it is. A mass attenuation coefficient is simply the linear attenuation coefficient divided by the density of the material (ρ).

Plots of the attenuation coefficients of the elements expose the fundamental restrictions on the size, shape, composition, and density of samples that can be successfully assayed by gamma-ray methods. Figure 6.1 shows mass attenuation coefficients for selected elements ranging from hydrogen (Z = 1) to plutonium (Z = 94). Qualitatively, the information in the graph defines nearly all possibilities and limitations of passive gamma-ray NDA. Note that μ for uranium at 185.7 keV is nearly six times larger than that for plutonium at 413.7 keV, which means that the assay of ^{235}U by its 185.7 keV gamma ray is subject to considerably more stringent limitations on sample size, particle size, and uniformity than is the assay of ^{239}Pu by its 413.7 keV gamma ray. Below ~80 keV, the μ of most elements rises rapidly, making attenuation problems unmanageably severe for all but small samples of very small particle size.

6 Attenuation Correction Procedures

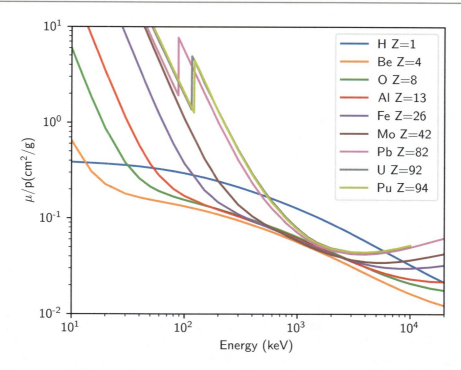

Fig. 6.1 Total mass attenuation coefficients (without coherent scattering contribution) versus energy for nine elements with an atomic number (Z) in the 1–94 range [2]

6.2.3 Necessary Assumptions for Determining an Accurate Self-Attenuation Correction Factor

Using the linear attenuation coefficient μ_ℓ, the fraction of gamma rays that escape without interaction in the item can, in general, be computed. Determining μ_ℓ is the key to determining CF_{atten}.

For samples where self-attenuation is a large effect, three conditions are necessary for accurate gamma-ray assays:

- The mixture of gamma-ray-emitting material and matrix material is reasonably uniform and homogeneous in composition and density.
- The gamma-ray-emitting particles are small enough that the self-attenuation within the individual particles is negligible.
- The sample must not be so thick that the contribution from material at the back of the sample is comparable to the uncertainty in the measured count rate.

The first two conditions guarantee that the linear attenuation coefficient of the material is single-valued on a sufficiently macroscopic scale that it can be used to accurately compute the gamma-ray-escape fraction. There are no restrictions on the chemical composition of the sample. All that is required is that μ_ℓ can be computed or measured. Unknown samples need not have the same or even similar chemical compositions as the calibration standards. There are also no basic assumptions about the size and shape of standards, although there are limitations.

The assumption of "reasonable" uniformity is admittedly vague and difficult to define. What constitutes reasonable uniformity depends on the gamma-ray energy, the chemical composition of the sample, and the accuracy required. Some sample types almost always satisfy the assumptions, and some almost never do so.

Figure 6.2 is given as an aid in estimating self-attenuation for individual particles based on Eq. 6.13 in Sect. 6.6.3. It gives the fraction of gamma rays that escape without interaction from spherical sources as a function of the product ($\mu\rho D$), where D is the diameter of the sphere. As an example, for a 200μm diameter, $\rho = 10$ g/cm^3 particle of UO_2, using Eq. 6.13 with $\mu_\ell = $ 13.1 1/cm at 186 keV, $\mu\rho D \simeq 0.26$, indicating that ~10% of the 185.7 keV gamma rays emitted are scattered with some energy loss or are completely absorbed within the particle.

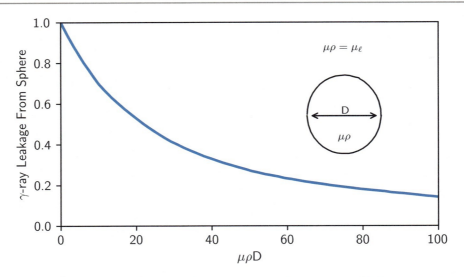

Fig. 6.2 Fraction of gamma rays that escape uncollided from spherical sources as a function of $\mu\rho$D. Coherent scattering has been neglected

Solutions meet the first two criteria for accurate gamma-ray assay, assuming that there are no contained particulates or precipitates. Pure powders (PuO_2, UO_2, U_3O_8, etc.) almost always are suitable, as are certain well-mixed scrap materials such as incinerator ash. High-temperature gas reactor (HTGR)-coated fuel particles and HTGR-type rods come close to meeting the requirements, but assay results are low by 5–10% unless correction is made for the self-attenuation in the particle kernels [3]. Small quantities of high-Z gamma emitters (<10 g) mixed with low-Z, low-density materials might meet the requirement if there are no agglomerations of the powder with significant self-attenuation. Large quantities of high-Z powders (greater than ~100 g) will almost surely create some significantly attenuating agglomerations when mixed with such materials. Among the worst cases are metal chips of high-Z, high-density metals or fuel pellets mixed with low-Z, low-density matrices; in these situations, assays may well be low by factors of 2 or 3 or even more. This fact causes one to be cautious about using gamma-ray methods to screen heterogeneous materials for possible criticality dangers or to perform mass measurements of nuclear material.

The third condition guards against the problem of "infinite" thickness. If activity from the back of the sample is shielded into the uncertainty of the count rate measurement, any amount of material can be placed there without measurable effect. This problem is most commonly encountered for samples of uranium. It is discussed in more detail in the context of enrichment measurements in Chap. 8, Sect. 8.3. Samples that are infinitely thick for low-energy peaks require the use of higher-energy peaks for mass assay.

It must be emphasized that the degree to which materials satisfy the three conditions is the most important factor in determining the potential accuracy of a gamma-ray assay. For example, experience indicates that small samples of solution (up to a few tens of cubic centimeters) may be assayed with accuracies of a few tenths of a percent. Samples of uniform, homogeneous powders of volumes up to a few liters have been assayed with accuracies approaching 1% in spite of significant density gradients. Larger containers of waste (for example, 30-gallon drums) rarely satisfy the assumptions well enough to allow errors of <10%, and the error will be much worse for the extremely heterogeneous cases.

Another important general fact about gamma-ray assay is that the results are almost always low when samples that do not satisfy the assumptions are assayed in conjunction with calibration standards that do satisfy the assumptions. The procedures that accurately determine the self-attenuation in acceptable samples underestimate the correction in samples that fail to satisfy the required conditions.

6.2.4 Methods for Determining the Correction Factor for Attenuation

To use gamma-ray methods to assay the amount of nuclear material within an item, the detection efficiency of the detector system must be well characterized. The recommended method for measuring the efficiency of the detector system is to use one or more standards and correct for the differences in attenuation between the standards and the items undergoing assay. Ideally, the standards would be similar to the sample so that the correction for differences in attenuation would be small. Several methods exist for determining CF_{atten} for the standard and the unknown, some assuming a slightly different procedure.

6.2.4.1 Calculating Correction Factor from Tabulated Values of μ_ℓ

The first and most obvious method, assuming sufficient knowledge of geometry and materials, is to compute the correction factor using tabulated values of $\mu\ell$ and the fundamental law of gamma-ray attenuation presented in Chap. 3, Sect. 3.2. Reference [2] provides the necessary mass attenuation coefficients. For simple cases, this method can be done analytically. Solutions for several example cases will be presented in Sect. 6.6. Correction factors for complex geometries can be computed as shown in Sect. 6.7 or by using ray-tracing methods (e.g., Refs. [3–5]). Prior knowledge about the geometry does not necessarily mean that the assay result is known in advance. In many cases, the correction factor is almost purely dependent on properties of external attenuators or matrix composition and mass, which are reasonably well known. When only verification measurements are required on well-characterized materials, the approach is useful even when self-attenuation dominates.

6.2.4.2 Determining Correction Factor from Transmission Measurements

An elegant and general method of obtaining the correction factor involves measuring the transmission through the sample of a beam of gamma rays from an external source. Solving for μ_ℓ in Eq. 6.1, we obtain

$$\mu_\ell = \frac{-\ln(T)}{x}. \tag{6.3}$$

Assuming that composition of the sample is uniform, we can use this relation to calculate the attenuation coefficient and then use an appropriate formula to calculate attenuation in the sample.

This method requires no knowledge of the chemical composition or density of the sample, just the basic assumptions on uniformity and particle size. In fact, it is often the preferred method even when some knowledge of the sample composition is available, particularly when the best obtainable accuracy is desired. The experimentally measured μ_ℓ includes all effects of chemical composition and density.

The transmission method can identify those samples for which accurate quantitative assays are impossible because of excessive self-attenuation. As the measured transmission decreases, its precision deteriorates along with the precision of the sample μ_ℓ, thus creating error in the computed value of CF_{atten}. The precision and accuracy of the measured transmission become unacceptable for transmissions between 0.01 and 0.001. Transmission values ≤ 0.001 (perhaps even calculated as negative) almost always indicate a sample that cannot be assayed. An illustrative example is included in Chap. 10.

6.2.4.3 Determining Correction Factor Using Representative Standards

A different (and older) way to approach the problem is to use several representative standards that span the configuration of the sample under assay. In this procedure, a set of calibration standards is prepared as nearly identical as possible in size, shape, and composition to the unknowns. The standards are counted in a fixed geometry to prepare a calibration curve, and the assay is accomplished by counting the unknowns in the same geometry and comparing the count directly with the calibration curve. This procedure produces good results only if the unknowns and standards are sufficiently similar that the same concentration of assay material in each gives rise to the same CF_{atten}. The representative standard procedure also assumes that the pileup and dead-time losses are equal for equal concentrations of assay isotopes. This method is applicable only when the nature and composition of the assay samples are well known and essentially unvarying.

6.2.4.4 Determining Correction Factor Based on Monte Carlo Simulations

The last method we consider is Monte Carlo simulation of radiation transport. The intricacies of this method are beyond the scope of this text, but its use is so ubiquitous and so adaptable that we cannot neglect it. Modern radiation transport calculations are typically performed using software packages or toolkits such as those listed or discussed in Refs. [6–9]. At least two ways of using Monte Carlo simulations exist to make corrections for the effects of gamma-ray attenuation.

First, it is possible to construct models for the purpose of calculating transmission for self-attenuation and/or external attenuation. The transport model propagates individual photons through the geometry and applies interaction cross sections as implemented in the model. Under this paradigm, effects such as coherent scattering and downscattering and complicated geometries can be implemented in a natural way, greatly reducing the need for simplifying assumptions.

Second, one could perform an end-to-end transport calculation, from gamma-ray emission in the sample to energy deposition in the sensitive volume of the detector (and even scintillation light transport to a photomultiplier in the case of scintillators). This calculation requires implementing a model of the sample, the detector, and all intervening materials. In this model, the effect of attenuation is integrated into a detector efficiency simulation, and a direct correlation is calculated between the number of photons emitted and the number detected.

One caveat for employing Monte Carlo simulations is that the results are only as good as the knowledge about the specific properties of the object being modeled. An inaccurate understanding of the object or measurement scenario being modeled can lead to an inaccurate determination of the effects of attenuation on the measurement. Therefore, it is important that physical measurements are performed to benchmark the simulations to establish their fidelity. Performing transmission measurements with sources is important to verify that the model represents physical reality accurately.

6.3 Differential Attenuation Methods

Some radioactive materials emit X-rays or gamma rays at multiple energies with well-known relative intensities. For these materials, if one uses a system with good energy resolution, multiple full-energy peaks will appear in a detected energy spectrum. In some cases, multiple peaks come from a single isotope, and in other cases, daughters of the major constituent have emissions that are useful if the history of the sample is known. Notable examples include the major actinides ^{235}U, ^{238}U (via daughters ^{234}Pa and ^{234m}Pa), and ^{239}Pu. The net areas of the peaks, together with the known relative intensities and some knowledge about the shape of the detector efficiency curve, can be used to characterize the attenuating materials. This method is often referred to as the Peak-Ratio method.

A detailed treatment of the many differential attenuation techniques is beyond the scope of this book, but a short discussion can give the reader insight for some of the possibilities and limitations. The basic idea is to determine μ_ℓ or CF_{atten} from the ratios of gamma-ray intensities at different energies. Consider a slab-shaped sample of thickness x that contains an isotope that emits gamma rays at energies E_1 and E_2; assume that the unattenuated emission rates are equal. Using Eq. 6.4, the peak area (A) ratio is.

$$\frac{A_2}{A_1} = \frac{CF_{atten}(E_1)}{CF_{atten}(E_2)} = \frac{\mu_\ell(E_1)}{\mu_\ell(E_2)} \text{ for } \mu_\ell x \gg 1. \tag{6.4}$$

If the composition of the attenuating material or an "effective Z" is known or assumed, it may be possible to use the measured ratio of attenuation coefficients to determine the individual coefficients and evaluate the correction factor for attenuation at energy $CF_{atten}(E)$. This idea is behind all ratio methods, namely that different energy gamma rays are attenuated differently and could carry information about the attenuation properties of the material they pass through.

Simple applications are illustrated by the example of germanium detector dead-layer measurements. These measurements are performed to indicate dead-layer thickness for germanium detectors using the 22 keV and 88 keV emissions from ^{109}Cd [10]. A collimated source illuminates a location near the center of a coaxial or planar detector. The intrinsic efficiency for photons that reach the sensitive volume is close to unity. Because the dead layer attenuates photons at 22 keV with a photoelectric cross section much higher than at 88 keV, the ratio of counts in the detector at 22 keV varies strongly with the dead-layer thickness, whereas the net counts at 88 keV are relatively unaffected. As a result, the ratio of counts for the 22 keV peak to counts in the 88 keV peak is highly sensitive to the thickness of the dead layer. In practice, the 22 keV peak is subject to significant attenuation from the source window, so this method is most applicable as a relative measurement. A single source is used to compare the dead-layer and housing thicknesses for detectors.

This simple treatment is possible at low energies, where the intrinsic efficiency of the detector is close to unity. The technique can be used for more highly attenuating plates with peaks at higher energies, but the nontrivial energy-dependent efficiency of the detector must be considered and requires an iterative procedure. Differential attenuation has been used for measuring isotopic mixtures from the multiple peaks from ^{239}Pu in the code FRAM [11], for example.

For self-attenuation, the gamma-ray ratio methods require the assumptions on uniformity and particle size discussed in Sect. 6.2.3 to give accurate results. If the assumptions are not met, the transmission-corrected methods give results that are usually low. Ratio methods give results that are generally greater than those from transmission methods but may overcompensate depending on the size of the emitting particles. In many cases, ratio methods can give a warning when the conditions on uniformity, particle size, and sample thickness are violated. Unfortunately, although the ratios can give a warning of potentially inaccurate assay situations, no way is currently known whereby the ratio methods can consistently correct for the problems detected. A combination of transmission and ratio methods gives the most information about self-attenuation in a given sample.

6.4 Formal Definition of the Correction Factor for Attenuation

6.4.1 The General Definition

Expressions for CF_{atten} can be formulated in various useful ways. The formulation adopted here is a multiplicative correction factor that gives a corrected count rate that is directly proportional to the quantity of isotope being measured.

$$R_{corr} = FEIR \times CF_{atten}, \tag{6.5}$$

where
R_{corr} = total corrected rate
$FEIR$ = full-energy interaction rate.

Procedures for determining the full-energy interaction rate (FEIR) based on the full-energy-peak area within the spectrum are discussed in detail in Chap. 7, Sect. 7.5.6.

It is useful to define CF_{atten} with respect to a specified geometrical shape, which is often simpler than the actual shape.

$$CF_{atten} = \frac{FEIR(\mu_\ell = 0; \text{Specified Shape})}{FEIR(\mu_\ell \neq 0; \text{Real Shape})}, \tag{6.6}$$

where

$FEIR(\mu_\ell = 0, \text{Specified Shape})$ = the FEIR that would have been measured if there were no attenuation and if the sample were changed to the specific shape
$FEIR(\mu_\ell \neq 0, \text{Real Shape})$ = the actual measured FEIR from the sample.

In practice, CF_{atten} is not computed from Eq. 6.6; it is determined using one of the methods described above. The remainder of this chapter is devoted to formulas for calculating CF_{atten} for self-attenuating sources from μ_ℓ, the geometrical configuration, and the position of the sample relative to the detector. Most often, the expressions for CF_{atten} are not integrable in terms of elementary functions, so numeric methods must be used.

For far-field scenarios, one can assume a point detector with equal efficiency for all angles of incidence, which considerably simplifies the computations. This assumption is usually good when the distance between sample and detector is at least several times the maximum dimension of either the detector or the sample. If the sample-to-detector distance must be kept small for reasons of efficiency and if the highest obtainable accuracy is required, the actual measured or calculated detector efficiency as a function of energy and position should be used. See Chap. 7, Sect. 7.8, for more detail on efficiency measurements.

6.4.2 Useful Specified Shapes

The most useful specified shapes are

- the actual sample shape,
- a point, and
- a line.

If one has many samples and standards of the same shape and size, then CF_{atten} may be computed with respect to a nonattenuating sample of the same shape. When the sample is sufficiently uniform and homogeneous and of reasonable size, let the detector view the whole sample, and use the CF_{atten} computed with respect to a nonattenuating point. This method allows the standards and the unknowns to be of different size, shape, and chemical composition; however, for such assays to be accurate, the entire contents must be reasonably well represented by a single μ_ℓ.

Samples often have vertical composition and density gradients, the natural consequence of filling relatively narrow containers from the top. The material tends to fall into the containers in layers. In such cases, a single μ_ℓ cannot adequately characterize the whole sample, but narrow layers or segments can be adequately characterized by a single μ_ℓ value. The assay accuracy can be improved by using a segmented scan, during which the detector views the sample through a collimator that defines relatively narrow horizontal segments in which μ_ℓ can be assumed constant. For such segmented scans, it is best to compute the CF_{atten} with respect to a nonattenuating line along the axis of the containers. In this way, cylindrical samples can be accurately assayed with respect to standards of quite different diameters.

6.5 Important Parameters of the Self-Attenuation Correction Factor

The correction factor for self-attenuation, CF_{atten}, is a function of many parameters. Those currently recognized as significant are listed in decreasing order of importance:

- the μ_ℓ of the sample material
- the volume and shape of the sample material
- the μ_ℓ of the sample container
- the size and shape of the sample container
- the position and orientation of the sample relative to the detector
- the size, shape, and efficiency of the detector

In many situations, the dependence of CF_{atten} on several of the parameters is mild. For example, when the sample-to-detector distance is at least several times the maximum dimensions of the detector, the dependence of CF_{atten} on the size, shape, and efficiency of the detector is often negligible. When the distance between a cylindrically shaped sample and the detector is at least several times the maximum dimension of either sample or detector, CF_{atten} is usually a strong function of the sample μ_ℓ and a mild function of the sample dimensions and distance from the detector and has negligible dependence on the detector size, shape, and efficiency.

The greatest simplifications occur in the far-field case, where the maximum dimensions of both sample and detector are negligible compared with their separation. All gamma rays reach the detector along essentially parallel paths. There is no dependence on detector size or shape, on small changes in the sample-to-detector distance, or on sample size except for the influence of size on the fraction of gamma rays that escape from the sample. Simple analytic expressions can be derived for several sample shapes. These expressions are often useful approximations for assay situations that are not truly far field. Indeed, the far-field situation is a useful reference case against which to compare near-field cases.

It is usually advantageous to plot CF_{atten} versus the parameter of strongest dependence (μ_ℓ) and to plot separate curves for specific values of other parameters. Because μ_ℓ is often found by measuring the gamma-ray transmission (T) and using the relationship $T = \exp(-\mu_\ell x)$, it is generally more convenient to plot CF_{atten} versus $\ln(T)$.

Consider the expression for CF_{atten} for the far-field assay of a slab-shaped sample viewed normal to a side,

$$CF_{atten} = \frac{\mu_\ell x}{[1 - exp(-\mu_\ell x)]}, \tag{6.7}$$

where x is the sample thickness along the normal to the detector. Using $T = \exp(-\mu_\ell x)$, we can write the simple expression

$$CF_{atten} = \frac{-ln(T)}{(1-T)}. \tag{6.8}$$

If $T \ll 1$, $CF_{atten} \simeq -\ln(T)$, so a plot of Cf_{atten} versus $\ln(T)$ is nearly linear. Figure 6.3 gives a plot of Eq. 6.8. It also shows CF_{atten} versus $\ln(T)$ for cylindrical and spherical samples, where T is measured across the sample diameter. All of the cases have the form $CF_{atten} \simeq -k \ln(T)$ for $T \ll 1$. This approximate $\ln(T)$ dependence exists for most assay geometries and is very useful to keep in mind.

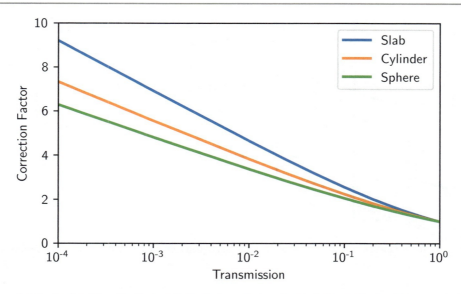

Fig. 6.3 Far-field correction factors for slab, cylindrical, and spherical samples as a function of transmission. The transmission is measured normal to the face of the slab sample and along a diameter of the cylindrical and spherical samples

Table 6.1 Far-field correction factors for slab, cylinder, and sphere as functions of transmission

Transmission	Slab[a]	Cylinder[b]	Sphere[b]
1.0000	1.000	1.000	1.000
0.8000	1.116	1.097	1.086
0.6000	1.277	1.231	1.202
0.4000	1.527	1.434	1.376
0.2000	2.012	1.816	1.701
0.1000	2.558	2.238	2.054
0.0500	3.153	2.692	2.431
0.0200	3.992	3.326	2.956
0.0100	4.652	3.826	3.370
0.0010	6.915	5.552	4.805
0.0001	9.211	7.325	6.288

[a]Transmission normal to surface
[b]Transmission along a diameter

6.6 Analytic Far-Field Forms for the Self-Attenuation Correction Factor

In general, the near-field integral expressions for CF_{atten} cannot be integrated in terms of elementary functions. However, far-field expressions have been derived for three simple sample geometries: box shaped (rectangular parallelepipeds), cylindrical, and spherical. Figure 6.3 gives the far-field CF_{atten} for all three sample shapes, and Table 6.1 gives numeric values for the three cases.

6.6.1 Slab-Shaped Samples

The slab-shaped sample is the only one for which a simple derivation exists. From Eq. 6.6, we can write CF_{atten} with respect to a nonattenuating sample (specified shape same as real shape) as

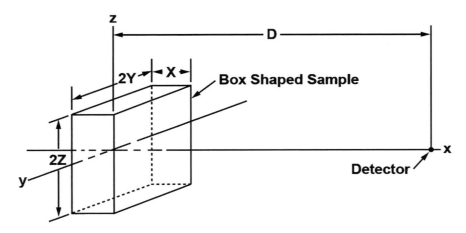

Fig. 6.4 Counting geometry for a slab-shaped sample with coordinates and dimensions for use in deriving the far-field correction factor

$$CF_{atten} = \frac{\int_v \rho I\varepsilon \, dV}{\int_v \rho I\varepsilon \exp(-\mu_\ell r) \, dV}, \qquad (6.9)$$

where
ρ = spatial density of the isotope being assayed (g/cm^3)
I = emission rate of the assay gamma ray (γ/g − s)
ε = absolute full-energy detection efficiency
μ_ℓ = linear attenuation coefficient of the sample
r = distance that gamma rays travel within the sample
dV = volume element.

The parameters ρ, I, and μ_ℓ are constant, whereas ε and r are functions of position. It is the exponential term in the denominator that, for most geometrical configurations, cannot be integrated in terms of elementary functions.

Consider the configuration shown in Fig. 6.4. The parameter (I) is a constant for a given isotope, and by virtue of the fundamental assumptions on uniformity, ρ and μ_ℓ are also constant. The far-field assumption is equivalent to assuming that ε is also a constant.

Because of the far-field assumption, only the integration in X is significant. After the obvious cancellations,

$$CF_{atten} = \int_0^X dx \Big/ \left(\int_0^X exp[-\mu_\ell (X-x)] \, dx \right). \qquad (6.10)$$

As in Eq. 6.7, this result evaluates to

$$CF_{atten} = \frac{\mu_\ell X}{1 - exp(-\mu_\ell X)}. \qquad (6.11)$$

6.6.2 Cylindrical Samples

For a cylindrical sample viewed along the diameter in the far field [12, 13],

$$CF_{atten} = \frac{1}{2} \frac{\mu_\ell D}{I_1(\mu_\ell D) - L_1(\mu_\ell D)}, \quad (6.12)$$

where

L_1 = modified Struve function of order 1
I_1 = modified Bessel function of order 1
D = sample diameter
μ_ℓ = linear attenuation coefficient of the sample.

The expression is very compact, but it is inconvenient to use because of the Struve and Bessel functions [14]. Equation 6.12 was used to generate the curve for a cylinder in Fig. 6.3. Note that the CF_{atten} for a cylinder is a little less than that for a slab-shaped sample. In the cylindrical sample, fewer gamma rays must penetrate the maximum thickness of material; therefore, the fraction escaping is greater, and the CF_{atten} is smaller.

6.6.3 Spherical Samples

For a spherical sample in the far field, the correction factor is [4]

$$CF_{atten} = \left(\frac{1.5}{\mu_\ell D} \left\{ 1 - \frac{2}{(\mu_\ell D)^2} + exp(-\mu_\ell D) \left[\frac{2}{\mu_\ell D} + \frac{2}{(\mu_\ell D)^2} \right] \right\} \right)^{-1}. \quad (6.13)$$

This expression is plotted in Fig. 6.3. The CF_{atten} for a sphere is smaller than that for either the parallelepiped or the cylinder. On the average, gamma rays travel shorter distances to escape from a sphere than from either a cylinder or a slab. Spherical samples are rarely met in practice, but the reciprocal of CF_{atten} gives the fraction of gamma rays that escape from spherical particles and is useful in deciding whether a sample meets the required assumption on particle size.

6.7 Numeric Computation in the Near Field

To calculate attenuation or detector efficiency in near-field scenarios, it is important to consider the depth of the detector and the sample. In these situations, it is typical to rely on an empirical efficiency measurement or Monte Carlo photon transport simulation. If neither option is practical, the efficiency of the system can still be calculated using simplified models and straightforward one-, two-, or three-dimensional numerical integration methods. The accuracy of gamma-ray NDA is usually determined more by the sample uniformity and homogeneity than by the accuracy of the CF_{atten} computation.

Approximate analytic forms exist that give adequately accurate values for CF_{atten} over reasonable ranges of transmission. A few such forms are described in the following sections. The adequacy of a particular expression can be determined by comparison with more accurate numeric computations. Approximate analytic forms often provide the capability to derive analytic expressions for the precision of CF_{atten}.

6.7.1 A Useful One-Dimensional Model

A common assay geometry is where a germanium detector views a bottle of solution from below. Both detector and sample can be approximated by right-circular cylinders. Assume that the axes of symmetry of the bottle and the detector coincide and that the detector radius is r_d, the sample radius is r_s, the sample depth is D, and the distance from the sample to the detector is d (Fig. 6.5). If d is a few times greater than both r_d and r_s, no gamma ray impinges on the detector at angles greater than ~10° to the common axis. In as much as $\cos\theta \geq 0.95$ for angles <19°, it is clear that no gamma ray travels more than a few percent greater distance on its way to the detector than those that travel parallel to the common centerline. Therefore, the assay situation can be described by a one-dimensional model that consists of a point detector and a line sample of "depth" (D) and

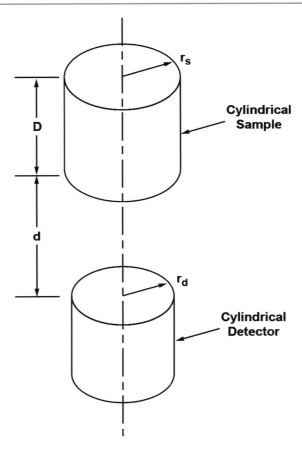

Fig. 6.5 Commonly used vertical assay geometry for which a one-dimensional model is appropriate for computing CF_{atten}

linear attenuation coefficient (μ_ℓ) separated from the detector by a distance (d) as indicated in Fig. 6.6. This model contains the effects of the inverse-square law and gamma-ray attenuation, which are the main influences on the CF_{atten}.

Using this model, CF_{atten} with respect to a non-attenuating sample is

$$CF_{atten} = \frac{\left[\int_0^D \frac{dx}{(d+x)^2}\right]}{\int_0^D \frac{[exp(-\mu_\ell x)]dx}{(d+x)^2}}, \tag{6.14}$$

where all constants that pertain to the detector efficiency and gamma-ray emission rates have cancelled. The numerator integrates to $D/[d(d+D)]$, but as simple as the denominator appears to be, it cannot be integrated in terms of elementary functions; however, it can be written as a sum in a simple way. The expression for CF_{atten} then becomes

$$CF_{atten} = \frac{\left[\frac{D}{d(d+D)}\right]}{\sum_{I=1}^{N} \frac{\{exp[-\mu_\ell(I-0.5)\Delta x]\}\Delta x}{[d+(I-0.5)\Delta x]^2}}, \tag{6.15}$$

where $\Delta x = D/N$, and N is the number of intervals for the numeric integration.

Generally, taking $N \simeq 100$ gives the result to <0.1%.

Equation 6.15 shows clearly the functional dependence of CF_{atten} on d, D, and μ_ℓ and the equivalence of the integral and the sum. The parameter D is well defined as the sample depth. However, the parameter d is less well defined because of its energy

6 Attenuation Correction Procedures

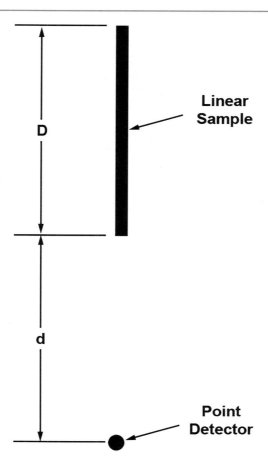

Fig. 6.6 One-dimensional model for computing CF_{atten}

dependence; high-energy gamma rays interact throughout the detector, whereas low-energy gamma rays tend to interact near the surface, so the average interaction depth is a function of energy. The parameter (d) implicitly includes the interaction depth. Experience shows that if the nominal value of d is at least a few times D, then with the help of a set of standards that cover a wide range of μ_ℓ, the value of d in Eq. 6.15 can be adjusted to give CF_{atten} such that the corrected rate per unit activity is nearly constant over a wide range of μ_ℓ. The adjustment of d compensates for the imprecisely known sample-to-detector distance and for deviation of the one-dimensional model from the actual three-dimensional assay geometry.

Figure 6.7 shows results of a measurement exercise using the procedure just described. The samples were 25 mL solutions of depleted uranium nitrate in flat-bottomed bottles of 10 cm² area (right circular cylinders 3.57 cm diameter and 2.5 cm deep). The uranium concentration varied from 5 to 500 g/L, and all samples were spiked with an equal amount of ^{75}Se. (The source material was ^{75}Se; uranium served as an absorber only.) The detector crystal was ~4.0 cm in diameter and ~4.0 cm long. For the 136.0 keV gamma ray of ^{75}Se, the corrections for electronic losses, $CF_{rateloss}$, varied by only ~10%, whereas the corrections for gamma-ray attenuation, CF_{atten}, varied by ~275%.

Because each sample had identical amounts of ^{75}Se, the corrected 136.0 keV rate should have been equal for all samples. The upper part of the figure gives the fractional deviation of the corrected rates from the average of all and indicates the typical precision of the measurements. All corrected rates are within about ±0.5% of the average. In this case, the actual distance of the sample bottom to the average interaction depth in the detector was ~8 cm, and the adjusted value was 9.0 cm. Qualitatively, the one-dimensional model gives values of CF_{atten} that are a little low compared with the correct three-dimensional model because the gamma rays pass through slightly greater thicknesses of solution than in the one-dimensional model. Increasing d increases CF_{atten} overall and also increases CF_{atten} more for lower values of T; therefore, the value of d used in computations is usually a little higher than the physical value.

If a set of solution samples has variable but determinable depths, one would prefer to compute CF_{atten} with respect to a nonattenuating point so that the corrected rates from all of the samples can be compared directly. The ratio between CF_{atten} with respect to the nonattenuating point and CF_{atten} with respect to the nonattenuating sample is $(1 + D/d)$, independent of μ_ℓ.

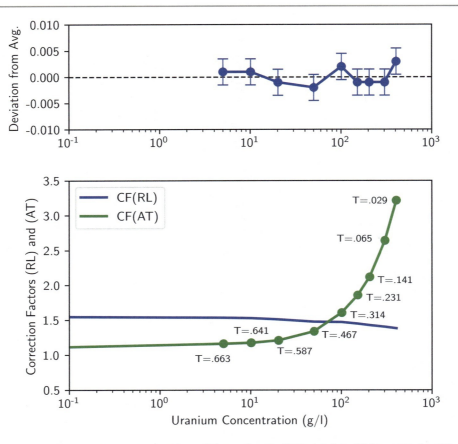

Fig. 6.7 Results of a measurement exercise designed to test the usefulness of a one-dimensional model for computing $CF_{atten} = CF_{atten}(T)$

All CF_{atten} values for both standards and unknowns should be computed with respect to the same nonattenuating shape so that the corrected rates are directly comparable.

6.7.2 A Useful Two-Dimensional Model

In another common assay geometry, a detector views a cylindrical sample from the side (Fig. 6.8). If the sample depth is less than the sample diameter and if the distance from the detector to the sample center is at least several times the sample diameter, then a simple two-dimensional model can often be used to compute CF_{atten}. The model is a point detector at a distance (D) from the center of a circular sample of radius (R), as shown in Fig. 6.9. The detector efficiency is essentially constant for gamma rays that originate at any point within the sample volume. The correction factor with respect to the non-attenuating sample can be written as

$$CF_{atten} \simeq \frac{\left(\frac{\pi}{2}\right) ln\left[1 - \frac{R^2}{D^2}\right]}{\sum_{m=1}^{M}\sum_{n=1}^{N}\left\{\frac{exp[-\mu_\ell t(m,n)]\Delta A(n)}{L^2(n,m)}\right\}}. \quad (6.16)$$

The derivation of this expression is given in Ref. [1]. The ratio of CF_{atten} calculated relative to the non-attenuating sample and CF_{atten} calculated relative to a non-attenuating point is $-(D^2/R^2) ln(1 - R^2/D^2)$. For a fixed value of T, CF_{atten} is a function only of the ratio D/R. Figure 6.10 gives CF_{atten} as a function of D/R for several values of T. The essential point is that CF_{atten} decreases slowly as D/R decreases; the larger changes occur for the smaller values of T. This behavior is a consequence of the inverse square law. For a given value of T, CF_{atten} asymptotically approaches a maximum as $D/R \rightarrow \infty$. The deviations from

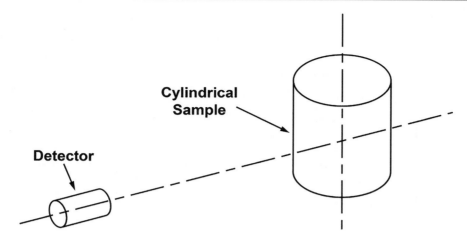

Fig. 6.8 Typical assay geometry for which a two-dimensional model for computing CF_{atten} is usually adequate

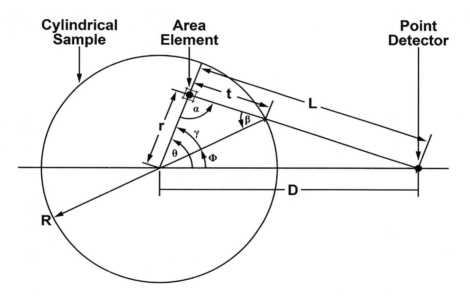

Fig. 6.9 Two-dimensional model for computing CF_{atten}, showing the distances that must be determined and the variables in terms of how they must be expressed. Note that $0 \leq \alpha \leq \pi/2$, and $0 \leq \gamma \leq \pi$

the far-field ($D/R = \infty$) case are plotted in Fig. 6.11. For $T > 0.001$ and $D/R \geq 50$, all deviations are $\leq 1\%$; therefore, $D/R \geq 50$ can be regarded as the far-field situation for most purposes. The variation of CF_{atten} with T is much stronger than the variation with the ratio D/R.

The results presented in Figs. 6.10 and 6.11 were obtained using values of $M = 200$ and $N = 200$, for which all results are within 0.1% of what the actual integrals would give. For a three-dimensional model, a modest extension in derivation and programming is required. If the third dimension is also given 200 increments, modern computers can still carry out the integration in a short amount of time.

6.7.3 A Three-Dimensional Model

As a final model for an assay geometry, consider the segmented assay of cylindrical samples. In the case of Fig. 6.12, a detector views the sample through a horizontal collimator, which defines sample segments that are assayed individually. The sample is usually as close to the collimator as possible and is usually rotated during the measurement to reduce the effect

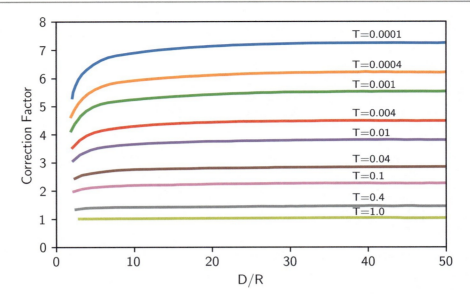

Fig. 6.10 Correction factors with respect to a nonattenuating sample as computed from the two-dimensional model. They are plotted versus the ratio D/R for various values of the transmission T, where D is the distance from the center of the cylindrical sample to the point detector, and R is the radius of the sample

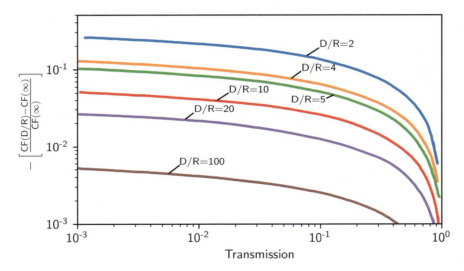

Fig. 6.11 Deviations of near-field values of CF_{atten} from the far-field values as a function of transmission for various values of D/R

of inhomogeneities. The detector is often a right circular cylinder of germanium ~5.0 cm in diameter and ~ 5.0 cm long. The inverse-square-law effects caused by the collimator must be considered explicitly; the two-dimensional model is not adequate.

The model consists of a perfect collimator (no leakage) and a vertical line detector centered at the rear of the collimator. The detector efficiency is assumed to be independent of either the position or angle at which the gamma rays strike the line detector. The distance from the emitting element to the detector is increased by a constant that is approximately equal to the average interaction depth in the detector. Inasmuch as materials are often packaged in metal containers that significantly attenuate the emitted gamma rays, the packaging is included in the model. The derivation of the three-dimensional model is outside the scope of this text; it is addressed fully in Ref. [1].

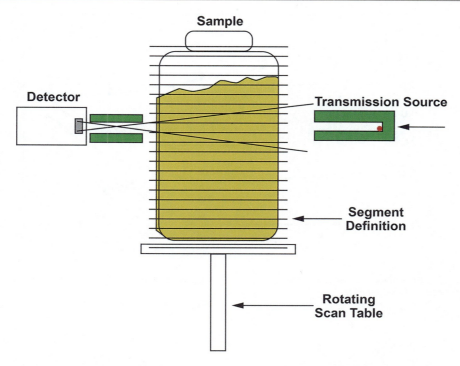

Fig. 6.12 Typical segmented assay situation for which a three-dimensional model for computing CF_{atten} is appropriate

6.7.4 Approximate Forms and Interpolation

The most accurate way to compute CF_{atten} for reasonable assay geometries is to use a simple mathematical model and numeric integration. However, because of lengthy execution times, it is often desirable to compute CF_{atten} for a few values of T (or μ_ℓ) and to use an interpolation scheme to find CF_{atten} for intermediate values. The interpolation problem can be approached in several ways.

Because CF_{atten} has an approximate $\log(T)$ dependence (see Fig. 6.3), it is reasonable to use a fitting function of the form

$$CF_{atten} = A + B\log(T) + C[\log(T)]^2. \tag{6.17}$$

The computer need only store the constants A, B, and C for each assay geometry. This scheme works very well over wide ranges of T. In a typical segmented scanning situation, A, B, and C can be determined to give values of CF_{atten} correct to $\leq 0.3\%$ for $0.008 \leq T \leq 0.30$.

A particularly simple scheme is based on the far-field form of CF_{atten} for a slab: $-\ln(T)/(1-T)$. Observing that a circle is not very different from a square, one is led to try

$$CF_{atten} \cong \frac{-\ln(T^k)}{(1-T^k)}, \tag{6.18}$$

with $k < 1$ as an approximate function for cylindrical samples even in the near-field situation. This form also has a $\ln(T)$ dependence for $T \ll 1$ and has only one constant to be determined. Figures 6.13 and 6.14 provide a feeling for how accurate the approximate form might be. Figure 6.13 gives the fractional deviation of Eq. 6.18 from the correct far-field values for a cylinder (Eq. 6.12) as a function of T and k. Figure 6.14 compares the approximate and correct values for a near-field assay of a cylindrical sample where $D/R = 5/1$. In Fig. 6.13, $k = 0.82$ gives CF_{atten} correct within $\pm 1\%$ for $0.01 \leq T \leq 1.0$, and in Fig. 6.14, $k = 0.75$ gives CF_{atten} correct within $\pm 1.5\%$ for $0.01 \leq T \leq 1.0$.

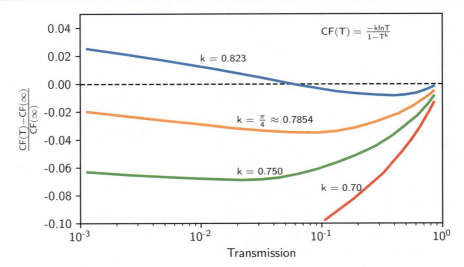

Fig. 6.13 Deviations of the values of CF_{atten} computed from the approximate expression $CF_{atten}(T) = -k \ln(T)/(1 - T^k)$ from the far-field values for a cylinder for several values of the parameter k

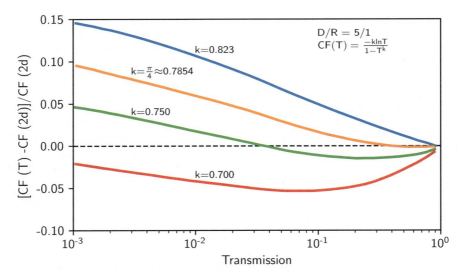

Fig. 6.14 Deviations of the values of CF_{atten} computed from the approximate expression $CF_{atten}(T) = -k \ln(T)/(1 - T^k)$ from the values from the two-dimensional model for cylindrical samples for D/R = 5. They are plotted as functions of the transmission T for several values of the parameter k

For the materials to be assayed, the choice of an interpolation procedure or approximate function for CF_{atten} depends on the accuracy desired or possible. For a field measurement of a heterogeneous drum that contains ^{235}U, the accuracy is determined far more by the heterogeneity of the sample material than by the function used for CF_{atten}. When ±25% accuracy is all that can be hoped for, it is wasteful to set up a model and do numeric integrations for CF_{atten}. On the other hand, if the samples are solutions—where careful modeling and computation can yield accuracies <1%—the effort is fully justified.

It is assumed that gamma-ray assay systems are calibrated with suitable physical standards. It is also assumed that CF_{atten} is determined for both the unknowns and the standards. Generally, CF_{atten} is mainly a function of the measured transmission T with some influence from the geometrical parameters.

It is easier to calibrate an assay system correctly for a narrow range of transmission (which usually implies a narrow range of concentration of the assay isotope) than for a broad range. Great care must be used in modeling the assay geometry and computing CF_{atten} if high accuracy is required over a wide range of concentrations.

Considering the challenges inherent in computing CF_{atten}, standards could be used to determine a variable calibration constant as a function of T. Although standards can be used only as a fine tuning of a system calibration, a variable calibration constant or nonlinear calibration curve attempts to account for only the physical aspects that are not understood about the assay arrangement.

6.7.5 Precision of Self-Attenuation Correction Factor and Total Corrected Rate

In a properly operating gamma-ray NDA system, the precision is almost totally a function of the random nature of the emission and detection of the gamma rays. The influence of electronic fluctuations and drifts should almost never influence the precision of the results at a level $> 0.1\%$. The dominant statistical component of the assay precision can usually be estimated from the full-energy peak areas and their precisions. The overall statistical precision, including any contribution from the equipment, is estimated from replicate assays. The electronic and mechanical stability of the assay system can be evaluated by comparing the overall precision with that estimated from peak areas and their precisions.

Consider the influence of the precision of CF_{atten} on the precision of the final assay. The assay is proportional to R_{corr}, which is given (see Eq. 6.5) as $R_{corr} = FEIR \times CF_{atten}$. The procedures used to derive expressions for $\sigma(R_{corr})$, $\sigma(FEIR)$, and $\sigma[CF_{atten}]$ are covered in detail in many sources. (Two relatively simple sources are Refs. [15, 16].) The intent here is only to emphasize a few points relative to obtaining a reasonable expression for $\sigma(R_{corr})$.

If R_{corr} can be written as an analytic function of the peak areas, then an expression for $\sigma(R_{corr})$ can be derived; however, when CF_{atten} is found by numeric procedures, $\sigma(R_{corr})$ cannot be computed directly. An approximate function for CF_{atten} can be used to derive an expression for $\sigma(R_{corr})$. The approximate forms for CF_{atten} are often not sufficiently accurate to compute R_{corr}, but they usually provide an adequate expression for $\sigma(R_{corr})$. In Sect. 6.6.2, a one-dimensional model was used to determine CF_{atten} for the assay of cylindrical samples. To derive an expression for $\sigma(R_{corr})$, one could use Eq. 6.8 or the modified form Eq. 6.18 for CF_{atten}. The proper value of k would be chosen by comparison with precisions computed from replicate assays. This procedure gives the accuracy provided by numeric integration of a more accurate model for CF_{atten} and still provides good estimates of $\sigma(R_{corr})$.

Although $\sigma(R_{corr})$ is the assay precision, $\sigma[CF_{atten}]$ alone is sometimes of interest. The expression for $\sigma[CF_{atten}]$ will always be simpler than that for $\sigma(R_{corr})$. If no peak areas are common to the expressions for FEIR and CF_{atten}, then

$$\sigma_r(R_{corr}) = \sqrt{\sigma_r^2(FEIR) + \sigma_r^2[CF_{atten}]}, \qquad (6.19)$$

where $\sigma_r(x) \equiv \sigma(x)/x$.

If there are peak areas common to the expressions for FEIR and CF_{atten}, Eq. 6.19 is not valid, and the expression for R_{corr} must be written as an explicit function of the peak areas concerned. Expressions for precision are frequently complex, but considerable simplification can usually be achieved by judicious approximations. The effort to make such simplifications reduces the computation time and provides a better understanding of the main source of imprecision.

References

1. J. L. Parker, *The Use of Calibration Standards and the Correction for Sample Self-Attenuation in Gamma-Ray Nondestructive Assay*. Los Alamos National Laboratory report LA-10045 (August 1984)
2. M.J. Berger, J.H. Hubbell, S.M. Seltzer, J. Chang, J.S. Coursey, R. Sukumar, D.S. Zucker, K. Olsen, *XCOM: Photon Cross Sections Database* (National Institute of Standards and Technology, 1998) https://www.nist.gov/pml/xcom-photon-cross-sections-database. Accessed Feb 2023
3. J. L. Parker, Correction for Gamma-Ray Self-Attenuation in Regular Heterogeneous Materials. Los Alamos National Laboratory report LA-8987-MS (September 1981)
4. J.P. François, On the calculation of the self-absorption in spherical radioactive sources. Nucl. Instrum. Methods **117**(1), 153–156 (1974)
5. J. L. Parker and T. D. Reilly, Bulk sample self-attenuation correction by transmission measurement, in *Proc. ERDA X- and Gamma-Ray Symposium*, Ann Arbor, Michigan, May 19–21, 1976 (Conf. 760,539)
6. MCNP Users Manual – Code Version 6.2, ed. C. J. Werner, Los Alamos National Laboratory, report LA-UR-17-29,981 (2017)
7. J. Allison, K. Amako, J. Apostolakis, et al., Recent developments in GEANT4. *Nucl. Instrum. Methods Phys. Res., Sect. A* **835**, 186–225 (2016)
8. S. M. Horne, G. G. Thoreson, L. A. Theisen, D. J. Mitchell, L. Harding, W. A. Amai, GADRAS-DRF 18.6 User's Manual, SAND2016–4345, Sandia National Laboratories technical report (May 2016)
9. R. Venkataraman, F. Bronson, V. Abashkevich, B.M. Young, M. Field, Validation of In Situ Object Counting System (ISOCS) mathematical efficiency calibration software. *Nucl. Instrum. Methods Phys. Res., Sect. A* **422**, 450–454 (1999)

10. ORTEC, Overview of Semiconductor Photon Detectors, https://www.ortec-online.com/-/media/ametekortec/other/overview-of-semiconductor-photon-detectors.pdf. Accessed Feb 2023
11. D. T. Vo, T. E. Sampson, *FRAM, Version 6.1 User Manual*, Los Alamos National Laboratory report LA-UR-20-21,287
12. Self-Shielding Correction for Photon Irradiation of Slab and Cylindrical Samples. Gulf General Atomic, Inc., Progress Report GA-9614 (July 1, 1968–June 30, 1969)
13. S. Croft, D. Curtis, M. R. Wormald, The calculation of self attenuation factors for simple bodies in the far field approximation. Mirion Technologies Inc. (2005). https://mirion.s3.amazonaws.com/cms4_mirion/files/pdf/technical-papers/calcselfatten-paper.pdf?1534969337. Accessed Feb 2023
14. *Handbook of Mathematical Functions with Formulas, Graphs, and Mathematical Tables*, ed. by M. Abramowitz, I. A. Stegun, National Bureau of Standards report, Applied Mathematics Series 55 (1970)
15. Y. Beers, *Introduction to the Theory of Error*, 2nd edn. (Addison-Wesley Publishing Co., Inc., Reading, MA, 1957)
16. P.R. Bevington, D. Keith Robinson, *Data Reduction and Error Analysis for the Physical Sciences*, 2nd edn. (McGraw-Hill Book Company, New York, 1992)

Open Access This chapter is licensed under the terms of the Creative Commons Attribution 4.0 International License (http://creativecommons.org/licenses/by/4.0/), which permits use, sharing, adaptation, distribution and reproduction in any medium or format, as long as you give appropriate credit to the original author(s) and the source, provide a link to the Creative Commons license and indicate if changes were made.

The images or other third party material in this chapter are included in the chapter's Creative Commons license, unless indicated otherwise in a credit line to the material. If material is not included in the chapter's Creative Commons license and your intended use is not permitted by statutory regulation or exceeds the permitted use, you will need to obtain permission directly from the copyright holder.

General Topics in Passive Gamma-Ray Assay

P. J. Karpius and J. L. Parker

7.1 Introduction

This chapter discusses general topics that apply to the gamma-ray assay techniques discussed in previous chapters. All of these topics must be understood if optimum results are to be obtained from any assay technique. Although many of the techniques are now automated by robust methods common to modern gamma-ray analysis software, they are discussed here for both completeness and for reference by developers. The topics include the following:

- Energy calibration and determination of peak position
- Energy resolution measurements
- Determination of full-energy-peak area
- Rate-related losses and corrections
- Effects of the inverse-square law
- Detector efficiency measurements
- Relative efficiency
- Nuclide-ratio measurements
- Nuclide-activity measurements

7.2 Energy Calibration and Determination of Peak Position

7.2.1 Introduction

The energy calibration of a gamma-ray spectroscopy system is the relationship between the energy deposited in the detector and the amplitude of the corresponding amplifier output pulse. For a system that employs a multichannel analyzer (MCA), we can take this further and state that the energy calibration is the relationship between the energy deposited in the detector and corresponding channel of the recorded spectrum. The energy calibration is used to determine the width and location of regions of interest (ROIs), to determine resolution as a function of energy, and to facilitate energy-based searches on unrecognized peaks when conducting nuclide identification. Also, it is often the case that gamma-ray analysis software assumes a certain set of energy calibration parameters.

Los Alamos National Laboratory strongly supports academic freedom and a researcher's right to publish; as an institution, however, the Laboratory does not endorse the viewpoint of a publication or guarantee its technical correctness.

P. J. Karpius (✉) · J. L. Parker
Los Alamos National Laboratory, Los Alamos, NM, USA
e-mail: karpius@lanl.gov

The energy calibration of a good spectroscopy system is nearly linear:

$$E = mx + b, \tag{7.1}$$

where
E = energy deposited in detector in keV
m = slope in keV per channel
x = channel number
b = zero intercept in keV.

The assumption of linearity is usually sufficient for high-resolution nondestructive assay (NDA) techniques. However, no system is exactly linear; each has small but measurable nonlinearities. When a more accurate relationship is necessary, a higher-order polynomial is used. For high-resolution detectors, gamma-ray energies can be determined to within 0.2 keV or better using a nonlinear calibration curve and several standard gamma-ray sources with energies known to better than 0.001 keV.

Low-resolution detectors (for example, NaI(Tl) scintillators) often use Eq. 7.1 with an intercept of about -10 to -15 keV, in part due to a direct current offset that exists regardless of energy deposition. This intercept is also a result of applying a linear energy calibration to the nonlinear relationship of energy and channel that exists for NaI detectors. A higher-order polynomial is usually applied to perform energy calibration on data from NaI detectors. For a very good germanium detector, a linear calibration will determine the peak energy to within one-tenth of a keV, which is adequate to identify the nuclides present in the measured item. For most of the nuclides of interest to NDA, the pattern of the gamma-ray spectrum is so distinctive that a visual examination of the MCA display by an experienced person is sufficient to identify the nuclides present, especially in the case of low-resolution spectra. Figure 7.1 shows the characteristic spectrum of low-burnup plutonium, and Fig. 7.2 shows the characteristic spectrum of low-enriched uranium (2% ^{235}U by mass).

The energy calibration procedure involves determining the channel location of peaks of known energy and fitting them to the desired calibration function. A common calibration source (or sources), such as ^{137}Cs, ^{57}Co, ^{60}Co, ^{152}Eu, or ^{232}Th, is often used to provide known gamma-ray peaks. In the absence of such a source, one might be able to use the gamma rays from the measured nuclear material item or the natural gamma-ray background, as shown in Fig. 7.3, to determine the energy calibration.

Plutonium spectra have interference-free peaks at 59.54, 129.29, 148.57, 164.57, 208.00, 267.54, 345.01, 375.04, and 413.71 keV. Similar internal calibrations are possible for many nuclides [1–3]. In a natural-background spectrum, it is helpful to be able to recognize the 1460.8 keV peak of ^{40}K and the 2614.5 keV peak of the ^{232}Th daughter ^{208}Tl.

Table 7.1 lists some of the most frequently used nuclides with the half-lives and energies of their principal emissions [3]. Most of the nuclides listed emit only a few gamma rays and are useful with both low- and high-resolution detectors. All

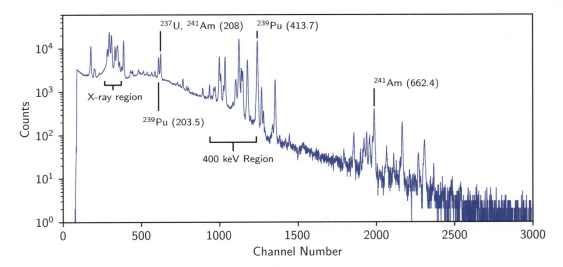

Fig. 7.1 A high-resolution spectrum of low-burnup plutonium (6% ^{240}Pu). Energies are given in keV

7 General Topics in Passive Gamma-Ray Assay

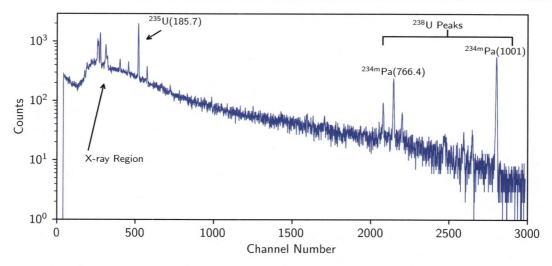

Fig. 7.2 A high-resolution spectrum of low-enriched uranium (2% ^{235}U). Energies are given in keV

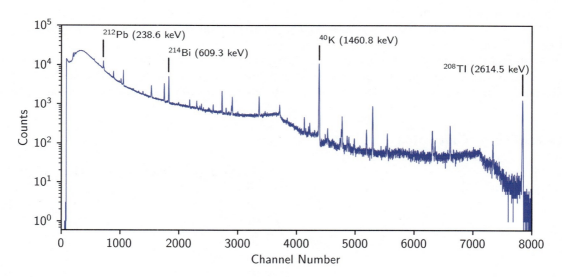

Fig. 7.3 A high-resolution spectrum of the natural gamma-ray background. The main components are ^{40}K and the progeny of ^{238}U and ^{232}Th

Table 7.1 Common nuclides used for energy calibration

Nuclide	Half-Life	Energy (keV)[a]	Remarks
^{241}Am	433 year	59.5	Many others but weaker by factors of 10^4 or greater
^{137}Cs	30.1 year	661.7	From daughter ^{137m}Ba, monoenergetic source + Ba X-rays
^{133}Ba	10.6 year	81.0, 276.4, 302.9, 356.0, 383.9	Several others, but much weaker
^{60}Co	5.27 year	1173.2, 1332.5	
^{22}Na	2.6 year	511.0, 1274.5	511.0 from annihilation radiation
^{109}Cd	1.27 year	88.0	Ag K X-rays at 22.16 keV and 24.91 keV
^{54}Mn	312 day	834.8	Monoenergetic source + Cr X-rays
^{65}Zn	244 day	511.0, 1115.5	511.0 from annihilation radiation
^{57}Co	272 day	122.1, 136.5	Two others of higher energy but much weaker
^{75}Se	120 day	121.1, 136.0, 264.7, 279.5, 400.7	Several others but much weaker
^{152}Eu	13.5 year	121.8, 344.3, 778.9, 964.1, 1085.8, 1112.1, 1408.0	Several others but weaker

[a]Energies are from the *NuDat 2.8* Database [3]

Table 7.2 Selected gamma-ray energies from the natural background

Nuclide	Parent Series	Energy (keV)[a]	Remarks
^{212}Pb	^{232}Th	238.6	Also present in other sources containing ^{228}Th (e.g., ^{232}U)
^{214}Pb	^{238}U	295.2	Also present in other sources containing ^{222}Rn (e.g., ^{226}Ra)
^{214}Pb	^{238}U	351.9	Also present in other sources containing ^{222}Rn (e.g., ^{226}Ra)
^{208}Tl	^{232}Th	583.2	Also present in other sources containing ^{228}Th (e.g., ^{232}U)
^{214}Bi	^{238}U	609.3	Also present in other sources containing ^{222}Rn (e.g., ^{226}Ra)
^{228}Ac	^{232}Th	911.2	Can distinguish ^{228}Th sources from ^{232}Th
^{214}Bi	^{238}U	1120.3	Also present in other sources containing ^{222}Rn (e.g., ^{226}Ra)
^{40}K	^{40}K	1460.8	Often the most intense peak in the background
^{214}Bi	^{238}U	1764.5	Also present in other sources containing ^{222}Rn (e.g., ^{226}Ra)
^{208}Tl	^{232}Th	2614.5	Also present in other sources containing ^{228}Th (e.g., ^{232}U)

[a]Energies are from the *NuDat 2.8* Database [3]

nuclides listed in the table are available from commercial vendors. Packaged sources usually contain a single nuclide and are produced in a wide variety of geometries. Source strengths between 0.1 and 100 μCi are usually adequate for energy calibration. Convenient sets of six to eight single-radionuclide sources are available from most vendors. Their use is required for setting up, testing, and checking many performance parameters of spectroscopy systems. The source sets are useful for determining energy calibration, testing detector resolution, measuring detector efficiency, setting the pole-zero adjustment, and correcting for rate-related counting losses.

Table 7.2 lists several gamma-ray peaks from the natural gamma-ray background that are useful for high-purity germanium (HPGe) energy calibration.

Gamma-ray standards are also available with several nuclides in one capsule. These multi-energy sources are used to define the energy calibration curve and efficiency curve of high-resolution detectors. Sources that are traceable by the National Institute of Standards and Technology Standard Reference Materials Program can be obtained for these and other purposes.

7.2.2 Linear Energy Calibration

Equation 7.1 describes the assumed functional form for a linear energy calibration. If the channels x_1 and x_2 of two full-energy peaks of energies E_1 and E_2 are known, m and b can be computed from

$$m = \frac{E_2 - E_1}{x_2 - x_1} \tag{7.2}$$

$$b = \frac{x_2 E_1 - x_1 E_2}{x_2 - x_1} \tag{7.3}$$

For a two-point calibration, the two calibration peaks should be near the low- and high-energy ends of the energy range of interest to avoid long extrapolations beyond the calibrated region. In practice, when calibrating over a wide energy range, a known high-energy peak is typically used to adjust the slope (gain), and a known low-energy peak is used to adjust the offset.

Often when an unacceptable degree of nonlinearity exists, several linear calibrations can be used over shorter energy intervals in a piecewise-linear fashion. The high-resolution spectrum of most plutonium items has nine well-resolved peaks between 59.5 keV and 413.7 keV so that eight linear calibrations can be constructed for the intervals between adjacent peaks; none of the intervals is greater than 78 keV. A series of short, linear calibrations often can be as accurate as a single quadratic or higher-order calibration function; however, modern routines do offer continuous polynomial-based functions for energy calibration.

When more than two peaks span the energy range of interest, least-squares fitting techniques can be used to fit a line to all of the peaks. This method can be used to obtain the following expressions for m and b for n peaks:

Table 7.3 Results of linear energy calibration using a high-quality plutonium spectrum (0.125 keV/ch)

Accepted energies (keV)	Peak positions (channels)	Energy difference (keV)[a]	
		Two-point calibration	Nine-point calibration
59.541	476.07	−0.001	−0.009
129.296	1034.00	−0.016	0.006
148.567	1188.18	−0.007	0.007
164.610	1316.39	−0.030	0.020
208.005	1663.75	−0.005	−0.005
267.540	2140.11	0.010	−0.010
345.013	2759.75	−0.013	0.003
375.054	3000.10	−0.004	0.004
413.713	3309.41	−0.003	0.003

[a]The tabulated numbers are the energies from the calibration minus the accepted energies. For the two-point calibration, m = 0.125 keV per channel and b = 0.032 keV. Energies are from the *NuDat 2.8* Database [3]

$$m = \frac{n \sum x_i E_i - \sum x_i \sum E_i}{\Delta} \quad (7.4)$$

$$b = \frac{\sum x_i^2 \sum E_i - \sum x_i \sum x_i E_i}{\Delta} \quad (7.5)$$

where $\Delta = n \sum x_i^2 - (\sum x_i)^2$.

A linear energy calibration is usually adequate for safeguards NDA applications, but it might not be adequate for tight-tolerance peak searches over the 3 MeV or greater range often required for nuclide identification. In such cases, more complex energy calibration methods are usually employed. Table 7.3 gives the results of two-point and nine-point linear calibrations of a high-quality plutonium spectrum. The nominal calibration, E (keV) = 0.1x + 20.0, was established by stabilizing the 59.536 keV gamma ray of ^{241}Am at channel 395.0 and the 413.712 keV gamma ray of ^{239}Pu at channel 3937.0. The second column of Table 7.3 gives the peak positions determined by fitting a Gaussian curve to the upper portion of the peaks. The third and fourth columns give the difference between the accepted energies and those obtained from the two-point and nine-point calibrations. Although there is a measurable curvature to the energy versus channel relation, the maximum error is only ~0.03 keV for the two-point calibration and ~ 0.017 keV for the nine-point calibration. The consistency of the results in Table 7.3 indicates that the peak positions have been located to within ~0.1 channel (~0.01 keV) and that the accepted energy values are consistent within ~0.01 keV.

7.2.3 Determination of Peak Position (Centroid)

Even with high-resolution detectors, full-energy peaks are usually at least several channels wide. The peaks are nearly symmetric, and the peak positions are chosen as the peak centers defined by the axis of symmetry. Full-energy peaks are usually well described by a Gaussian function of the form

$$y(x) = y_0 \, exp\left[-\frac{(x-x_0)^2}{2\sigma^2}\right], \quad (7.6)$$

where
$y(x)$ = number of counts in channel x
y_0 = peak amplitude
x_0 = peak centroid
σ^2 = variance.

References [4–6] provide detailed explanations of the properties of the Gaussian function. The function is symmetric about x_0, which is the peak centroid used in energy calibration. The parameter y_0 is the maximum value of the function and is nearly

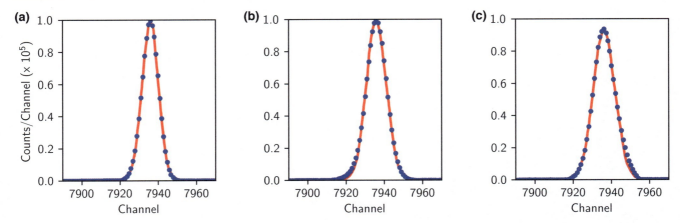

Fig. 7.4 The 1332.5 keV full-energy peak of ^{60}Co for three combinations of detector and count rate: (**a**) HPGe detector and low count rate; (**b**) germanium detector with poor peak shape; and (**c**) HPGe detector and high count rate

equal to the maximum counts per channel in the peak if the background under the peak is negligible. The parameter σ^2 (the variance) is related to the full width at half maximum (FWHM) of the function by

$$FWHM = 2\sqrt{2\ln(2)}\sigma = 2.35482\sigma. \tag{7.7}$$

The area under the Gaussian curve is given by

$$A = \sqrt{2\pi}\sigma y_0 = 2.50663\sigma y_0$$

$$= 1.06447(FWHM)y_0. \tag{7.8}$$

The constant in the second form of Eq. 7.8 is close to 1.0 because the area of a Gaussian is just a little greater than the area of an isosceles triangle with the same height and width at the half-maximum level.

Full-energy peaks are not exactly Gaussian shaped. For high-quality germanium detectors, the deviations are hardly visible, but for lower-quality detectors, the deviations are easily seen as an excess of counts on the low-energy side of the peak (called *tailing*). At very high rates or with poorly adjusted equipment, high-energy tailing is sometimes visible. The upper one-half to two-thirds of a peak is usually Gaussian, and the centroid determined by fitting a Gaussian to the upper portion of the peak is a well-defined measure of peak position. Figure 7.4a–c show the 1332.5 keV full-energy peak of ^{60}Co and the fitted Gaussian function. Figure 7.4a is from a high-quality germanium detector at low count rate with properly adjusted electronics. The deviations from the curve are hardly visible except for a very slight low-energy tailing. Figure 7.4b is from a detector with poor peak shape. The low-energy tailing is obvious. Figure 7.4c is from the same detector as Fig. 7.4a but at very high rates that cause distinct high-energy tailing and significant deviation from a true Gaussian shape. In all three situations, the Gaussian function fitted to the upper two-thirds of the peak gives a good peak location.

7.3 Detector Resolution Measurements

This section is devoted primarily to the measurement of detector resolution. The importance of good resolution and peak shape in obtaining unbiased NDA results cannot be overemphasized. A narrow, Gaussian peak shape simplifies area determination and minimizes the possibility of bias in assay results.

The FWHM (or FW.5M) is the basic measure of peak resolution. It is usually given in energy units (keV) for high-resolution detectors and expressed as a percentage of the measurement energy for low-resolution detectors. Resolution measured in energy units increases with energy: $FWHM^2 \approx a + bE + cE^2$ due to the nature of information (charge or photon) production by incident radiation. When expressed as a percentage, resolution decreases with energy.

Most detectors give full-energy peaks that are essentially Gaussian above the half-maximum level. The ratio of the full width at heights less than the half maximum to the FWHM has been used in the past to quantify the quality of the full-energy peak shape. Manufacturers measure the FWHM and its ratio to the full width at tenth maximum (FW.lM) to describe the peak shape; for many years, a value of FW.1M/FW.5M less than 1.9 was regarded as describing a good peak shape. It is also reasonable to specify FW.02M/FW.5M and even FW.01M/FW.5M when the best peak shape is required.

Table 7.4 gives the theoretical ratios for a Gaussian curve and the measured ratios for a high-quality coaxial germanium detector. The table shows that the actual peak shape closely approaches the Gaussian ideal. All ratios should be measured after background subtraction.

The pulse-height spectrum from a detector is continuous, whereas an MCA or single-channel analyzer groups the pulses in energy intervals. It is assumed that all events in an interval can be represented by the energy of the center of the interval. When a Gaussian is fitted to the center points of the intervals, the width parameter (σ) is slightly greater than that of the original continuous distribution. As discussed in Ref. [7], the grouped variance and the actual variance are related by

$$(\sigma^2)_G = (\sigma^2)_A + \frac{h^2}{12} \tag{7.9}$$

$$(FWHM^2)_G = (FWHM^2)_A + 0.462 h^2, \tag{7.10}$$

where
$(\sigma^2)G$ = grouped variance
$(\sigma^2)A$ = actual variance
h = group width (channel width).

For MCA spectra, h has units of keV per channel if FWHM is in keV and $h = 1.00$ if FWHM is in channels. Table 7.5 gives the ratio $(FWHM)_A/(FWHM)_G$. To measure the actual resolution to 0.1%, the system gain should be adjusted to provide more than 15 channels in $(FWHM)_G$. If $(FWHM)_G$ is three channels, the $(FWHM)_A$ is overestimated by ~3%. The correction has no practical bearing on full-energy-peak areas. The Gaussian function fitted to the binned points has the same area (to better than 0.01%) as the continuous distribution because the y_0 parameter is decreased by almost exactly the same factor as the width parameter is increased.

Table 7.4 Theoretical and measured resolution ratios

	$\dfrac{FW.1M}{FW.5M}$	$\dfrac{FW.02M}{FW.5M}$	$\dfrac{FW.01M}{FW.5M}$
Pure Gaussian	1.823	2.376	2.578
122.0 keV	1.829	2.388	2.599
	(1.003)[a]	(1.005)[a]	(1.008)[a]
1332.5 keV	1.856	2.428	2.640
	(1.018)[a]	(1.022)[a]	(1.024)[a]

[a] The ratio of the measured ratio to the theoretical ratio for a Gaussian

Table 7.5 The ratio of $(FWHM)_A$ to $(FWHM)_G$

FWHM$_G$ (channels)	Ratio of FWHM$_A$ to FWHM$_G$
3.0	0.9740
5.0	0.9907
10.0	0.9971
15.0	0.9990
20.0	0.9994
25.0	0.9996
30.0	0.9997
35.0	0.9998
40.0	0.9999

Various methods exist for determining the width of a peak. Modern gamma-ray analysis software provides options for width estimation. Peak-width estimation methods, such as graphical determination, analytical interpolation, and the second-moment method, are discussed in Chap. 5 of Ref. [8].

7.4 Determination of Full-Energy-Peak Area

7.4.1 Introduction

The gamma-ray pulse-height spectrum contains much useful information about gamma-ray energies and intensities. One of the most important concerns in applying gamma-ray spectroscopy is correct extraction of the full-energy peak areas and their associated uncertainties.

Full-energy peaks in gamma-ray pulse-height spectra rest on a continuum caused by the Compton scattering of higher-energy gamma rays. The most fundamental limitation in obtaining unbiased peak areas is the determination of the Compton continuum. When the continuum is small with respect to the peak, it can cause only a small fractional error in the peak area; however, when the ratio of the peak area to the continuum area becomes much less than 1.0, the statistical error on the net peak area increases, and the possibility of bias rises rapidly.

For many NDA applications, simple background-subtraction methods are adequate. Under certain circumstances, complex spectral fitting codes with long- and short-term tailing functions must be used. This tailing arises from small-angle Compton scattering and incomplete charge collection in germanium crystals. With low-resolution detectors, the problem of including small-angle-scattering events in the peak is severe, but computational corrections can sometimes be applied to resolve the problem [9]. Tailing can be an issue with enrichment-meter measurements using the 2-ROI method (see Chap. 8, Sect. 8.3).

7.4.2 Selection of Regions of Interest

The choice of ROI is as important as the choice of algorithms used to evaluate peak areas. Most procedures use two ROIs to define the continuum level on the low- and high-energy sides of a peak or multiplet. The average channel count of an ROI is taken as the continuum level at the center of the ROI. A third ROI defines the peak region. Advanced gamma-ray analysis software, such as FRAM [10], can use multiple ROIs to obtain net areas for one or more peaks with complex underlying continua.

For a Gaussian function, 99.96% of the area lies within a region centered at x_0 that is three times the FWHM of the function. The amplitude of the Gaussian function at $x_0 \pm 1.75$ FWHM is only 0.0082% of the maximum value at x_0, so continuum ROIs that begin at this point have minimal contributions from the peak. Thus, a peak ROI of 3 times the FWHM and continuum ROIs placed symmetrically 3.5–4.0 times the FWHM apart should obtain ~99.9% of the peak area.

In principle, the continuum is estimated more precisely if the continuum ROIs are quite wide; however, the possibility of systematic error increases as the energy interval increases. For most NDA applications, continuum ROIs of 0.5–1.0 times the FWHM are adequate. With an energy calibration of 0.1 keV per channel, typical background ROIs are three to five channels wide. When the continuum between neighboring peaks is very narrow, ROIs of one or two channels must be used. Peaks whose centers are separated by three times the FWHM can be considered resolved; usually a narrow ROI can be placed between them. It is better to sacrifice statistical precision than to introduce bias by using continuum ROIs that are too wide.

Spectra with significant low- or high-energy tailing may require a wider peak ROI than three times the FWHM. Because peak resolution deteriorates somewhat at high rates, the ROI should be set on a spectrum recorded at a high count rate, otherwise regions may be too narrow to properly encompass the broadening of peaks due to pileup at higher input rates. Better results usually are obtained if all ROIs are of equal width; therefore, the ROIs for low-energy peaks and reference pulser peaks are somewhat wider than three times the FWHM.

Software can be written to accurately and consistently choose ROIs. Digital stabilization can be used to keep the desired peaks within a single preselected set of ROIs for long time periods. Sometimes it is desirable to change the spectrum to fit a particular set of ROIs. Codes exist that can reshuffle the contents of a spectrum to give any desired energy calibration with little degradation of spectral quality.

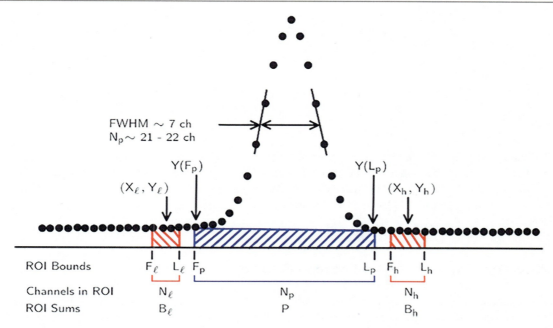

Fig. 7.5 Regions of interest and associated parameters used to compute the net area and the estimated standard deviation of a full-energy peak

7.4.3 Subtraction of Straight-Line Compton Continuum

It is often adequate to approximate the Compton continuum by a straight line between the high- and low-energy sides of well-resolved peaks or of overlapping peak groups. Figure 7.5 shows how the ROIs are selected and indicates the notation used in the background equations. Note that the continuum ROIs need not be symmetrically placed with respect to the peak ROI nor need they be of equal width. The background is the trapezoidal area beneath the continuum line given by

$$B = \left[Y(F_p) + Y(L_p)\right]\left(\frac{N_p}{2}\right), \tag{7.11}$$

where
$Y(F_p) = mF_p + b$
$Y(L_p) = mL_p + b$
and where
$m = (Y_h - Y_\ell)/(X_h - X_\ell)$
$b = (X_h Y_\ell - X_\ell Y_h)/(X_h - X_\ell)$.

The variance of the background B is

$$s^2(B) = \left(\frac{N_p}{2}\right)^2 \left[K^2 \frac{B_h}{N_h^2} + (2-K)^2 \frac{B_\ell}{N_\ell^2}\right], \tag{7.12}$$

where $K = \frac{(F_p + L_p - 2X_\ell)}{(X_h - X_\ell)}$.

Equation 7.12 assumes no uncertainty in the ROI bounds and is a function only of the statistical uncertainties in B_h and B_ℓ, which are estimated by $s^2(B_\ell) = B_\ell$ and $s^2(B_h) = B_h$. Equation 7.11 is correct when the background ROIs are not symmetrically placed relative to the peak ROI. If the continuum ROIs are placed symmetrically relative to the peak ROI, the expressions for both B and $s^2(B)$ are simplified. The symmetry requirement means that $(F_p - X_\ell) = (X_h - L_p)$ and $K = 1$, and so the expressions become

$$B = \left(\frac{Y_\ell + Y_h}{2}\right) N_p = \left(\frac{B_h}{N_h} + \frac{B_\ell}{N_\ell}\right) \frac{N_p}{2} \tag{7.13}$$

and

$$s^2(B) = \left(\frac{N_p}{2}\right)^2 \left(\frac{B_h}{N_h^2} + \frac{B_\ell}{N_\ell^2}\right). \tag{7.14}$$

Equations 7.13 and 7.14 frequently are used even when the symmetry requirement is not met, and if the net peak areas are much greater than the subtracted continuum, little error will result. However, when the peak areas are equal to or less than the subtracted continuum, the error may well be significant. In dealing with complex spectra (e.g., plutonium spectra), one is frequently forced to use asymmetrically placed ROIs. In the simplest situations, hand calculations may even be performed (see Chap. 5 of Ref. [8]).

7.4.4 Subtraction of Smoothed-Step Compton Continuum

The Compton continuum beneath a full-energy peak is not a straight line. Most of the continuum is caused by incomplete energy deposition in the detector. The part of the continuum under a peak caused by the gamma ray that generates the peak results from small-angle and multiple Compton scattering as well as full-energy interactions in the detector followed by incomplete charge collection. This contribution can be described by a smoothed step function, although in more subtle cases, these two processes contribute to low-energy tailing. Effects such as these are commonly modeled in modern gamma-ray analysis software.

Gunnink [11] devised the original procedure to generate a step-function continuum based on the overlying spectral shape. The procedure provides better results than the straight-line background approximation, especially for overlapping peak multiplets. For "clean" single peaks, the improvement is often negligible.

Figure 7.6 shows a logarithmic plot of a multiplet and a step-function background.

Using the notation of Fig. 7.5, the continuum at channel n is

$$B_n = Y_\ell - D \left[\sum_{i=X_\ell+1}^{i=n} \frac{(y_i - Y_h)}{\sum_{i=X_\ell+1}^{i=X_h}(y_i - Y_h)} \right], \tag{7.15}$$

where

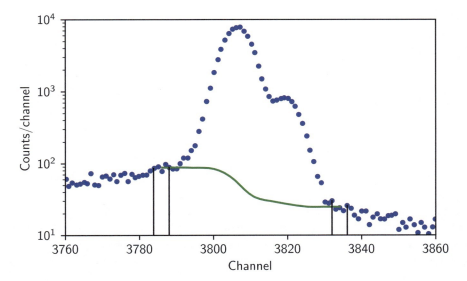

Fig. 7.6 A peak doublet with the estimated spectral continuum computed by the simple, smoothed-step algorithm

y_i = total counts in channel i
$D = Y_\ell - Y_h$
$B(X_\ell) = Y_\ell$
$B(X_h) = Y_h$.

The background Y_h is subtracted from every channel because the Compton events from higher-energy gamma rays cannot influence the shape of the smoothed step for lower-energy gamma rays. Equation 7.15 is usable when the continuum beneath a peak or multiplet has a slightly negative or zero slope.

A significant complication in using the smoothed-step procedure is that the expression for the precision of the net area becomes exceedingly complicated when derived from Eq. 7.15.

7.4.5 Subtraction of Compton Continuum Using a Single-Background Region of Interest

Estimating the background continuum from a single ROI is sometimes desirable or necessary. For example, a single ROI is often desirable when using a NaI detector to measure ^{235}U enrichment or ^{239}Pu holdup.

When the peak-to-continuum ratio is high, it may be adequate to assume a flat continuum. Here, the contribution of the continuum to the peak ROI is given by

$$B = \frac{N_p}{N_h} B_h \qquad (7.16)$$

$$s(B) \approx \frac{N_p}{N_h} s(B_h). \qquad (7.17)$$

Although this procedure is most often used with low-resolution scintillators, it is also used with germanium detectors when there is no convenient place for a background ROI on the low-energy side of a peak.

If the background continuum is not flat but can be assumed to have a constant slope over the energy range concerned, Eq. 7.16 may be modified to

$$B = K B_h, \qquad (7.18)$$

where K is a constant factor determined by experiment. Often K will change from item to item. Although these single-ROI procedures have limited accuracy, their use is preferred to ignoring the continuum problem entirely.

The two-standard method can be used to measure low-level radioactive contaminations in water or other fluids. See Chap. 8, Sect. 8.3, about uranium enrichment for more details on this procedure.

7.4.6 Using Region-of-Interest Sums to Measure Peak Areas

For well-resolved peaks, the simple summation of counts above the estimated background continuum is probably as good as any other method of finding the peak area. This method avoids any difficulty from imperfections in the peak-shape models of spectral fitting codes. The ROI-summation method is quite tolerant of small variations in peak shape and provides an accurate and straightforward estimate of the precision of the net peak area. This method may be applicable when the data are sufficiently sparse such that peak fitting is not feasible.

For all ROI-summation procedures, the peak area is given by

$$A = P - B, \qquad (7.19)$$

where P is the integral of the peak ROI, and B is the contribution from the background continuum. The expressions for $s^2(A)$, the estimated variance of the net area, vary according to the procedure used to estimate the background continuum.

When B is estimated by straight-line interpolation from continuum ROIs on either side of the peak ROI, the estimated variance of the peak area is

Table 7.6 Expressions for net full-energy-peak areas and estimated variances[a] (straight-line background assumed)[b]

Conditions on ROIs	Background B	Background Variance $s^2(B)$
Arbitrary position and width of background ROIs	$B = [Y(L_p) + Y(F_p)]\left(\frac{N_p}{2}\right)$ where $Y(L_p) = mL_p + b$ and $Y(F_p) = mF_p + b$ and where $m = (Y_h - Y_\ell)/(X_h - X_\ell)$ and $B = (X_h Y_\ell - X_\ell Y_h)/(X_h - X_\ell)$	$s^2(B) = \left(\frac{N_p}{2}\right)^2 \left[K^2 \frac{B_h}{N_h^2} + (2-K)^2 \frac{B_l}{N_l^2}\right]$ where $K = \frac{(F_p + L_p - 2X_l)}{(X_h - X_l)}$
Symmetric placement of background ROIs relative to peak ROI, $(F_p - X_\ell) = (X_h - L_p)$	$B = \frac{N_p}{2}(Y_h + Y_\ell) = \frac{N_p}{2}\left(\frac{B_h}{N_h} + \frac{B_\ell}{N_\ell}\right)$	$s^2(B) = \left(\frac{N_p}{2}\right)^2 \left(\frac{B_h}{N_h^2} + \frac{B_\ell}{N_\ell^2}\right)$

[a] $A = P - B$, $s^2(A) = P + s^2(B)$
[b] Notation summary as in Fig. 7.5; LE = low energy; HE = high energy: F_ℓ, L_ℓ = first and last channels of LE background ROI; F_p, L_p = first and last channels of peak ROI; F_h, L_h = first and last channels of HE background ROI; B_ℓ, P, B_h = integrals of LE background, peak, and HE background ROIs; N_ℓ, N_p, N_h = numbers of channels in LE background, peak, and HE background ROIs; $Y_h = B_h/N_h$ = average continuum level in HE background ROI; $Y_\ell = B_\ell/N_\ell$ = average continuum level in LE background ROI; X_h, X_ℓ = centers of background ROIs; $Y(F_p)$ and $Y(L_p)$ = ordinates of background line at F_p and L_p; m and b = slope and intercept of background line between (X_ℓ, Y_ℓ) and (X_h, Y_h)

$$s^2(A) = s^2(P) + s^2(B) = P + s^2(B). \tag{7.20}$$

Equations 7.11 through 7.18 give B and $s^2(B)$ for different conditions on the width and position of the background ROIs relative to the peak ROI. The expressions are summarized in Table 7.6. The simplest expressions are obtained when the background ROIs are symmetrically placed with respect to the peak ROI and have the appropriate widths. When adequate computational capacity is available, the most general form of the expressions should be used so that ROIs can be assigned without constraint

When the smoothed step function is used to estimate the background continuum, Equations 7.15 and 7.19 combine to give

$$A = P - \sum_{n=F_p}^{L_p} \left\{ Y_\ell - D\left[\sum_{i=X_\ell+1}^{n} \frac{y_i - Y_h}{\sum_{i=X_\ell+1}^{X_h}(y_i - Y_h)}\right] \right\}. \tag{7.21}$$

Because the continuum estimate is a function of the channel counts, the exact expressions for $s^2(A)$ become extremely complex. One of the estimates for $s^2(A)$ given in Table 7.6 should be used.

When a single background ROI is used, Eq. 7.19 holds for the net peak area, and the expression for $s^2(A)$ is based on Eq. 7.20. When the continuum is assumed to be flat, as in Eq. 7.16, the expressions for A and $s^2(A)$ become

$$A = P - \frac{N_p}{N_h} B_h \tag{7.22}$$

$$s^2(A) = P + \left(\frac{N_p}{N_h}\right)^2 B_h. \tag{7.23}$$

If a sloped continuum is assumed, as in Eq. 7.18, the expressions for A and $s^2(A)$ become

$$A = P - KB_h$$

$$s^2(A) = P + K^2 B_h \tag{7.24}$$

Although Equations 7.22 and 7.23 could correctly predict the repeatability of measurements, they do not predict any assay bias arising from the approximate nature of the single-ROI background estimate.

7.4.7 Using Simple Gaussian Fits to Measure Peak Areas

As shown in Sect. 7.2.3, the determination of σ and y_0 by a Gaussian fit also determines the peak area using Eq. 7.8. For cleanly resolved peaks, the areas obtained by fitting simple Gaussians are probably no better than those obtained from ROI sums and may be somewhat worse. This assertion is known to be true for germanium detectors. For NaI scintillators, a Gaussian fit may give more consistent peak areas than ROI methods. The simple Gaussian-fitting procedures do not provide straightforward ways to estimate peak-area precision.

In a few situations, Gaussian fitting is advantageous. When two peaks are not quite resolved such that the ROIs for the desired peaks overlap, a Gaussian can be fitted to one-FWHM-wide ROIs centered on each peak to determine the peak areas. When the centroid location and FWHM are the primary information desired from a Gaussian fit, the area estimate often comes with no extra effort. When a peak has significant low-energy tailing from Compton scattering in the item or shielding, a simple Gaussian fit to the middle FWHM of the peak can easily obtain the desired area.

When a Gaussian function is transformed to a line that is least-squares fit to obtain the parameters x_0 and σ, the parameter y_0 can also be determined using any of the original data points and Eq. 7.6 to solve for y_0. An average of the values of y_0 determined from several points near x_0 gives a satisfactory value for the area equation.

7.4.8 Using Known Shape Parameters to Measure Peak Areas in Multiplets

The previous discussion emphasizes well-resolved single peaks because most applications of gamma-ray spectroscopy to the NDA of nuclear material employ well-resolved peaks. However, to measure isotopic ratios from high-resolution plutonium spectra, it is necessary to analyze unresolved peak multiplets.

If the peak shape is described by an adequate mathematical model in which all of the parameters are known except the amplitude, unresolved multiplets can be analyzed quite simply by linear least-squares methods. For some purposes, the simple Gaussian function (Eq. 7.6) is adequate without any tailing terms. If the position and width parameters x_0 and σ are known, only the amplitude parameter y_0 is unknown. Frequently, the well-resolved peaks in spectrum can yield sufficient information to determine the x_0 and σ parameters for the unresolved peaks. If the gamma-ray energies are accurately known across the entire analysis region in a spectrum, then the energy calibration can be determined with sufficient accuracy to calculate the x_0 parameter for all unresolved peaks. The width parameter σ can be determined from the well-resolved peaks and interpolated to the unresolved peaks with the relation $\text{FWHM}^2 = a + bE$, which is quite accurate for germanium detectors above 100 keV. The well-resolved peaks can also yield information needed to determine the parameters of tailing terms in the peak-shape function.

The least-squares-fitting procedure for determining the peak amplitudes is most easily described by the following example, which assumes a three-peak multiplet where all of the peaks come from different nuclides. After the Compton continuum is subtracted from beneath the multiplet, the residual spectrum has only the three overlapping peaks, and the count in channel i may be written as

$$y_i = A1 \times F1_i + A2 \times F2_i + A3 \times F3_i \tag{7.25}$$

where A1, A2, and A3 are the amplitudes to be determined and F1, F2, and F3 are the functions that describe the peak shapes. Assuming that the peaks are well described by a pure Gaussian,

$$F1 = exp\left[K1(x_i - x1_0)^2\right]$$

$$F2 = exp\left[K2(x_i - x2_0)^2\right]$$

$$F3 = exp\left[K3(x_i - x3_0)^2\right] \tag{7.26}$$

where $x1_0, x2_0, x3_0 =$ known centroid positions

$$K1, K2, K3 = \frac{1}{2\sigma^2_{1,2,3}}$$

$$\sigma_i = \frac{(FWHM)_i}{2\sqrt{2\ln 2}}.$$

The least-squares fitting procedure determines A1, A2, and A3 to minimize the sum of the squared difference between the actual data points and the chosen function. With derivation, the expressions for A1, A2, and A3 are

$$A1 = \frac{1}{D}\begin{vmatrix} \sum yF1 & \sum F1F2 & \sum F1F3 \\ \sum yF2 & \sum F2^2 & \sum F2F3 \\ \sum yF3 & \sum F3F2 & \sum F3^2 \end{vmatrix}$$

$$A2 = \frac{1}{D}\begin{vmatrix} \sum F1^2 & \sum yF1 & \sum F1\,F3 \\ \sum F2\,F1 & \sum yF2 & \sum F2\,F3 \\ \sum F3\,F1 & \sum yF3 & \sum F3^2 \end{vmatrix}$$

$$A3 = \frac{1}{D}\begin{vmatrix} \sum F1^2 & \sum F1\,F2 & \sum yF1 \\ \sum F2\,F3 & \sum F2^2 & \sum yF2 \\ \sum F3\,F1 & \sum F3\,F2 & \sum yF3 \end{vmatrix}$$

$$D = \begin{vmatrix} \sum F1^2 & \sum F1F2 & \sum F1F3 \\ \sum F2F1 & \sum F2^2 & \sum F2F3 \\ \sum F3F1 & \sum F3F2 & \sum F3^2 \end{vmatrix} \tag{7.27}$$

The pattern of Eq. 7.26 can be followed for expanding to additional unknowns. The form of F1, F2, and F3 is not related to the solutions for A1, A2, and A3. The only requirement is that the functions are totally determined except for an amplitude factor. Tailing terms may be added to improve the accuracy of the peak-shape description. When two or more peaks in a multiplet are from the same nuclide, the known photon branching ratios[1] (yields), I_1, I_2 . . ., can be used to fit the peaks as a single component. If peaks one and two in the example are from the same nuclide, Eq. 7.24 becomes

$$y_i = A \times F_i + A3 \times F3_i, \tag{7.28}$$

where

$$F = exp\left[K1(x_i - x1_0)^2\right] + \left(\frac{I_2}{I_1}\right) exp\left[K2(x_i - x2_0)^2\right] \tag{7.29}$$

Equation 7.27 has only two unknowns: A and A3. Strictly speaking, the coefficients in F should be $1/E_1$ and I_2/I_1E_2, where E1 and E2 are the relative efficiencies at the two energies. If available, the efficiencies should be included, but often the related members of the multiplet are so close together in energy that E1 ≈ E2. When one of the related gamma rays is much more intense than the other, the errors in the intense components caused by assuming E1 = E2 are usually negligible.

---

[1] There are various names for photon emission probability. To stay consistent with the other chapters of this text, we will use the terminology "branching ratio" in lieu of "yield," undoubtedly to the chagrin of some authors.

7.4.9 Peak Fitting

A detailed description of peak-fitting techniques is given in Sect. 5.3 of Ref. [8]. Regarding a historical perspective, the techniques developed by Gunnink and coworkers at Livermore [12] have been widely used for both plutonium isotopic measurements and general gamma-ray spectroscopy. The GRPANL program [13, 14] was developed by Gunnink specifically to analyze the multiple peaks of the plutonium spectrum; it also forms the basis for the area determination routines of the GRPAUT program [15] that has been used at Mound Laboratory.

Both GRPANL and GRPAUT use a smoothed-step-function background and a Gaussian function with an exponential tail to describe the full-energy peak. The equation for the full-energy peak function is

$$Y_i = Y_0 \left\{ exp\left[\alpha(X_i - X_0)^2\right] + T(X_i) \right\} \quad (7.30)$$

where
Y_i = net counts in channel X_i for a single peak
Y_0 = peak amplitude
$\alpha = -4 \ln 2/(\text{FWHM})^2 = 1/2\sigma^2$, where σ is the standard deviation of the Gaussian function
X_0 = peak centroid
$T(X_i)$ = tailing function at channel X_i.

The tailing function is given by

$$T(X_i) = \{A \exp[B(X_i - X_0)] + C \exp[D(X_i - X_0)]\} \times \left\{1 - \exp\left[0.4\alpha(X_i - X_0)^2\right]\right\}\delta \quad (7.31)$$

where
A and C = short- and long-term tailing amplitude
B and D = short- and long-term tailing slope
$\delta = 1$ for $X_i < X_0$
$\delta = 0$ for $X_i \geq X_0$.

The second term brings the tailing function smoothly to zero at X_0, as shown in Fig. 7.7. For many applications, the long-term tail can be neglected (C = 0); for large multiplets with strong peaks, it should be included.

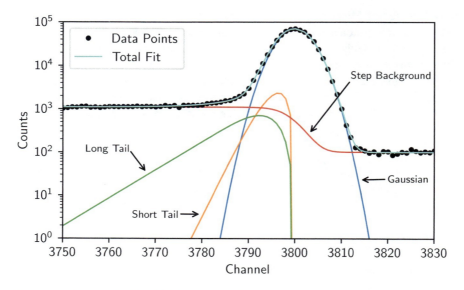

Fig. 7.7 Gamma-ray full-energy peak obtained with a HPGe detector showing Gaussian, short-term tailing, long-term tailing, and smoothed-step background contributions to spectral peak shape [16]

When all seven parameters in Equations 7.28 and 7.29 are treated as free parameters, the peak-fitting process is usually slow, although modern computers often permit a sufficiently fast analysis. Fortunately, many of the parameters can be predetermined. The peak positions (X_0) and width parameters (a) can be determined from two strong, isolated, reference peaks, such as the 148 and 208 keV peaks from ^{241}Pu and ^{237}U. Because the gamma-ray branching ratios are well known, the relative intensities of peaks from the same nuclide can be fixed.

Experience [12] shows that the short-term tailing slope (B) is constant for a given detector system and should be measured for a high-energy peak where tailing is large. The short-term tailing amplitude (A) is given by

$$\ln(A) = k_1 + k_2 E. \tag{7.32}$$

After B has been fixed, A can be determined from the two peaks that were used to determine the peak positions and width parameters.

If the long-term tailing is zero, the only free parameters are the peak amplitudes Y_0, and the fitting procedure is relatively fast. GRPANL allows other parameters to be free, but this method increases the analysis time. The step-function background is determined first and then subtracted from the acquired spectrum, GRPANL uses an iterative, nonlinear, least-squares technique [13, 14] to fit the residual full-energy-peak activity. Because the method is iterative, the analysis time depends on the number of peaks, the number of free parameters, and the type of computer system. Typically, analysis of a plutonium spectrum that contains more than 50 peaks in 15 groups in the 120–450 keV range takes ~10 min on a Digital Equipment Corporation PDP-11/23 computer or 3–4 min on a PDP-11/73. The analysis time is usually much shorter than the data accumulation time.

GRPANL can fit X-ray peaks that have a different intrinsic line shape (Lorentzian) than gamma rays [17]. This feature is necessary to fit peaks in the 100 keV region.

Much time has been invested in the development of computer codes to determine the peak areas from complex, overlapping peak multiplets. Multiple successful codes exist—FRAM [10], MGA [16], PeakEasy [18], and InterSpec [19]—along with many variations for special problems. Several of these codes are discussed in Chap. 10.

The codes describe full-energy peaks with a basic Gaussian shape plus one or two low-energy tailing terms (long- and short-term tailing) and sometimes a high-energy tailing term. The long-term tail is ascribed to small-angle Compton scattering within the item. The specific form of the tailing terms varies from code to code, although the results are often equivalent. The procedures to subtract the Compton continuum also vary; in general, the background subtraction procedures are most in need of improvement.

All fitting codes perform better on high-quality spectra with good resolution and minimal peak tailing. A fitting code cannot completely compensate for poor-quality detectors and electronics or for sloppy acquisition procedures. That said, an ounce of resolution is worth a pound of code.

7.4.9.1 Fitting Idiosyncrasies of Microcalorimeter Data

With the extremely high-energy resolution and pixelated nature of microcalorimeter spectrometers, summing spectra acquired simultaneously from the individual pixels could result in tails on the low- and high-energy side of the full-energy peak. This situation can arise when the pixels in a microcalorimeter array have a distribution of energy resolutions and is addressed by fitting the summed peak with either a Bortels function—a convolution of a Gaussian with an exponential (or two)—or the sum of two Gaussian functions fixed to the same centroid. In the latter case, one Gaussian describes the intrinsic detector response, and the second Gaussian describes the broadening and tailing that results from pixel summation. The single-low-energy-tailed Bortels is

$$Y_i = Y_0 \frac{\tau}{2} \exp\left(\tau\left(X_i - X_0 + \frac{(\sigma^2 \tau)}{2}\right)\right) \operatorname{erfc}\left(\frac{X_i - X_0 + \sigma^2 \tau}{\sqrt{2}\sigma}\right), \tag{7.33}$$

where

Y_i = net counts in channel X_i for a single peak
Y_0 = peak amplitude
σ = the standard deviation of the Gaussian function
X_0 = peak centroid
τ = tailing parameter on the low-energy side.

7 General Topics in Passive Gamma-Ray Assay

For a two-tailed Bortels function with tailing on both the low- and high-energy side, the function is

$$Y_i = \eta Y_0 \frac{\tau_L}{2} \exp\left(\tau_L\left(X_i - X_0 + \frac{\sigma^2 \tau_L}{2}\right)\right) \operatorname{erfc}\left(\frac{X_i - X_0 + \sigma^2 \tau_L}{\sqrt{2}\sigma}\right) \\ + (1-\eta) Y_0 \frac{\tau_R}{2} \exp\left(\tau_R\left(X_0 - X_i + \frac{\sigma^2 \tau_R}{2}\right)\right) \operatorname{erfc}\left(\frac{X_0 - X_i + \sigma^2 \tau_R}{\sqrt{2}\sigma}\right), \quad (7.34)$$

where
Y_i = net counts in channel X_i for a single peak
Y_0 = peak amplitude
σ = the standard deviation of the Gaussian function
X_0 = peak centroid
τ_L = tailing parameter on the low-energy side
τ_R = tailing parameter on the high-energy side
η = fraction between the two exponentials.

For X-rays, the peaks are fit with a Voigt function—a convolution of a Gaussian with a Lorentzian. The Lorentzian width (negligible for gamma rays) is defined by the intrinsic lifetime of the corresponding atomic level, which is much shorter in general than for nuclear lifetimes and is unique to each X-ray. The Gaussian component describes the detector response. The Voigt function is

$$Y_i = Y_0 \frac{F\left\{\frac{X_i - X_0 + i\Gamma}{\frac{\sigma}{\sqrt{2}}}\right\}}{\frac{\sigma}{\sqrt{2\pi}}}, \quad (7.35)$$

where
Y_i = net counts in channel X_i for a single peak
Y_0 = peak amplitude
σ = the standard deviation of the Gaussian function
X_0 = peak centroid
Γ = width for the Lorentzian component
F = real component of the Fadeeva function.

7.5 Rate-Related Losses and Corrections

7.5.1 Introduction

Rate-related losses can generally be divided into dead-time and pileup losses. As discussed previously, dead time is a measure of the fraction of the measurement time during which the system is busy processing incoming pulses from gamma-ray detection events. Pulse pileup occurs when two or more events occur within an interval less than the amplifier pulse width sum to give a pulse whose amplitude is not proportional to either of the original pulses.

Many texts discuss all counting losses in terms of two limiting cases (Ref. [20], for example). These cases are called *paralyzable* and *nonparalyzable*. In general, paralyzable behavior occurs when each pulse subsequent to the pulse being processed extends the dead time, whereas in the nonparalyzable case, the system ignores all subsequent pulses until the processing of the current pulse is complete, and the dead time is not extended.

Nonparalyzable behavior is typical of analog-to-digital converter (ADC) operation, wherein a busy signal is emitted during pulse digitization and memory storage while subsequent pulses have no effect on the system. In this case, the system can truly be considered "dead" with respect to subsequent pulses.

Paralyzable behavior is a result of pileup at the analog amplifier or digital pulse-processing stage. The terminology of *dead time*, as applied to this case, is unfortunate because no circuitry is dead during pileup; rather, events are lost from their proper channel because of the pulse distortion. At high count rates when multiple pulses overlap in time, several things can occur, resulting in counting losses (i.e., a reduction in spectrum throughput). For example, multiple pulses can be recorded as the sum of their amplitudes, counting only as a single event stored at a channel in the ADC that does not correspond to any individual pulse amplitude; or one or more overlapping pulses can be discarded if they arrive within the operational time resolution of a pileup-rejection circuit. In the absence of pileup rejection, a summation of pulses can have a resultant amplitude that exceeds the upper-level discriminator of the linear gate that precedes the ADC and then be discarded. All of these effects of pileup result in a reduction in throughput of the recorded spectrum even though no circuit is dead.

In an analog system, pileup can occur in the detector, the preamplifier, or the analog amplifier, but the overall effect is governed by the slowest component—usually the analog amplifier as previously described. Pileup always results in a loss of information; the degree of loss depends on the information sought and the gross count rates involved.

In high-resolution spectroscopy systems, the analog amplifier pulse width is often comparable to the ADC processing time, and the loss of information caused by pileup may be equal to or greater than the loss caused by dead time. Although an MCA can operate in a live-time mode and compensate for dead-time losses, it does not fully compensate for pileup losses. This effect can be understood if one considers that two photons that enter the detector simultaneously will result in a single pulse just as a single photon would. These two events in stand-alone fashion would relate to the same dead time, but the former is more likely in higher dead-time situations in the absence of true coincidence summing (see Sect. 7.6).

7.5.2 Counting Loss as a Function of Input Rate

In early systems, the dead-time losses were far higher than pileup losses, and the simple nonparalyzable model was quite adequate. In modern spectroscopy systems, counting losses are rarely completely described by the simple model of nonparalyzable dead time; however, the concept is described here for completeness.

We will examine the correction to the throughput rate (rate of photon events recorded in the spectrum) that yields the raw input rate (rate of photons that deposit energy in the detector). Because we are assuming that the raw input rate is for photons that have already reached the detector, the correction described below is independent of whether the source has been attenuated. See Chap. 6 for attenuation correction methods.

$$R_i = \frac{R_t}{f_L} = \frac{R_t}{1 - f_D}, \tag{7.36}$$

where

R_i = input rate of photons that deposit energy in the detector (time^{-1})
R_t = throughput rate of photon events recorded in the spectrum (time^{-1})
f_L = live-time fraction (unit-less)
f_D = dead-time fraction = $R_t \cdot \tau$ (unit-less)
τ = dead time per recorded event (time)

The dead time per event (τ) is typically on the order of tens of microseconds for modern HPGe detector systems (based on the pulse-processing time).

As an example, consider a dead time of 20% (dead-time fraction of 0.2). The corresponding live-time fraction is 0.8, which is 20% lower than a live-time fraction of 1.0 (i.e., the system is 100% live; no correction is needed because no data are being recorded). Using Eq. 7.36 and setting f_D to 0.2, a throughput rate of 1000 cps would correspond to an input rate of 1250 cps. Or conversely, losing 20% of events from an input rate of 1250 cps leaves a throughput of 1000 cps.

Note that the input rate is $R_i \to \infty$ as $R_t \to 1/\tau$. The term *nonparalyzable* arises because R_t rises monotonically toward the limit ($1/\tau$); however, for pileup losses, $R_t \to 0$ as $R_i \to \infty$, justifying the term *paralyzable*.

Although Wilkinson ADCs do not have a fixed dead time, Eq. 7.36 applies if τ is set equal to the average dead-time interval. Whether fixed or an average, the dead time (τ) is rarely determined directly because most users wish to correct for the combined rate-related losses.

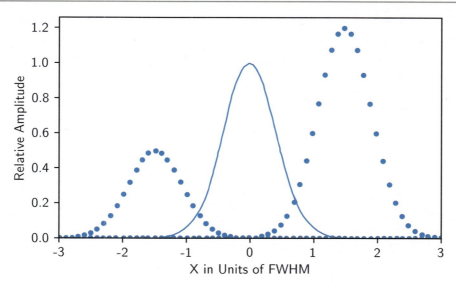

Fig. 7.8 Three Gaussian curves (amplifier pulses) separated by 1.5 × FWHM. As drawn, there is minimal pileup distortion to any pulse. If the pulse separation is reduced, the distortion will clearly increase

Pulse-pileup losses are important in high-resolution spectroscopy. The relatively long shaping times of the amplifier require optimum signal-to-noise-ratio pulses having widths up to 50μs, which increases the probability of pileup. A small amount of pileup can lead to a significant degradation of the resolution in the resulting gamma-ray spectrum. This result is especially evident for measurements taken with high-resolution HPGe systems.

Because NaI(Tl) systems operate with pulse-shaping time constants that are less than those for HPGe systems, they have lower pileup losses. Furthermore, because the NaI peaks are 10–20 times broader than germanium peaks, many events can suffer slight pileup and still remain within the full-energy peak; however, pileup losses in NaI(Tl) spectra are much harder to correct because of the broader peaks.

Figure 7.8 shows that if an amplifier pulse is preceded or followed by a pulse within approximately half the pulse width, its peak amplitude is distorted. The degree of distortion depends on the amplitude and timing of the interfering pulse relative to the analyzed pulse. Pileup-rejection circuitry frequently is used to detect and prevent analysis of the distorted events. Unfortunately, in rejecting pileup-compromised pulses, almost all systems reject some small fraction of non-distorted pulses.

7.5.3 The Poisson Nature of Counting Loss

If no other events occur within the dead-time interval (T) where pileup is possible, the pulse will be analyzed and stored in its proper location in the spectrum. The possibility for pileup depends on various factors, but the underlying nature is governed by Poisson statistics. Considering that nuclear decay is a Poisson process, the emission of gamma rays within a certain time interval (e.g., dead time per pulse [τ]) will also be Poisson. The fundamental expression from Poisson statistics [1, 3] that applies here is

$$P(N) = \frac{(R_i T)^N e^{-R_i T}}{N!}, \tag{7.37}$$

where P(N) is the probability of (N) events occurring within a time interval (T) if the average input rate is R_i. The probability that an event is not lost to pileup is obtained by setting N = 0 in Eq. 7.37:

$$P(0) = e^{-R_i T} \tag{7.38}$$

This probability is essentially the live-time fraction f_L. The fraction f_D of pulses lost during processing of a prior pulse (the dead-time fraction) is given by

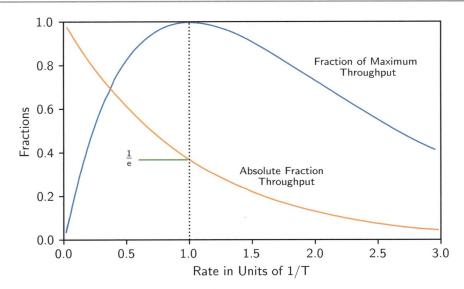

Fig. 7.9 The absolute fraction of events stored and the fraction of the maximum storage rate for a paralyzable system (with a time interval [T], where pileup is possible) as a function of input rate in units of 1/T

$$f_D = 1 - P(0) = 1 - e^{-R_iT}. \tag{7.39}$$

The measured throughput rate R_t is given by

$$R_t = R_i e^{-R_iT}, \tag{7.40}$$

where R_i is the input rate of photon events in the detector. Differentiation of Eq. 7.40 with respect to R_i gives

$$\frac{d}{dR_i}(R_t) = \frac{d}{dR_i}(R_i e^{-R_iT}) = 1 \cdot e^{-R_iT} + R_i \cdot \frac{d}{dR_i}(R_iT) \cdot e^{-R_iT} = e^{-R_iT}(1 + R_iT). \tag{7.41}$$

The maximum throughput (R_t) occurs when this derivative is zero (i.e., when $1 - R_iT = 0$). Therefore, R_t is maximized at $R_i = 1/T$, and the fraction of R_i stored at that rate is $1/e \approx 0.37$. The fraction of stored events is the live-time fraction of the measurement.

The fraction of the input rate that is stored (the live-time fraction) is e^{-R_iT}, and the stored rate (i.e., the throughput rate) as a fraction of the maximum stored rate $1/(eT)$ is given by $R_iT e^{1-R_iT}$. Both of these fractions are plotted in Fig. 7.9.

7.5.4 Throughput as a Function of Dead Time

Building on the discussion leading to Fig. 7.9, we consider throughput as a function of dead time. With appropriate sources and equipment, the throughput curve as a function of dead time can be determined experimentally. One could consider input count rate as the independent variable, but different systems will exhibit different throughput values for the same input count rate. For a discussion of throughput as a function of input rate, see Chap. 5 of Ref. [8].

For the same reported dead-time percentage, different systems will have different input count rates. The behavior of throughput versus input count rate will therefore be system dependent. But due to the Poisson nature of rate loss as described in Sect. 7.5.3, throughput as a function of dead time will be system independent.

From Eq. 7.41, we saw that the maximum throughput occurs when the input count rate R_i equals the inverse of the dead time per pulse, $1/T$. To calculate the dead-time percentage for maximum throughput, we set $R_i = 1/T$ in Eq. 7.39 for the dead-time fraction of pulses lost to pileup.

$$f_D\left(\frac{1}{T}\right) = 1 - e^{-\left(\frac{1}{T}\right)T} = 1 - e^{-1} = 1 - 0.37 = 0.63 \tag{7.42}$$

Equivalently and perhaps more intuitively, one can arrive at the same result by first writing the observed throughput rate as a function of dead time. Substituting $1 - f_D$ as the live-time fraction, we obtain the throughput rate as a function of dead time,

$$R_t = R_i e^{-R_i T} = R_i(1 - f_D). \tag{7.43}$$

We can rearrange the dead-time fraction of Eq. 7.39 and take the natural logarithm,

$$R_i = -\frac{1}{T} \ln(1 - f_D), \tag{7.44}$$

substitute this into Eq. 7.43 for throughput rate,

$$R_t = -\frac{1}{T} \ln(1 - f_D)(1 - f_D), \tag{7.45}$$

and take the derivative of the throughput rate with respect to dead-time fraction.

$$\frac{d}{df_D}(R_t) = \frac{d}{df_D}\left(-\frac{1}{T} \ln(1 - f_D)(1 - f_D)\right)$$

$$= \frac{1}{T}[1 + \ln(1 - f_D)]. \tag{7.46}$$

Setting this to zero yields the dead time for the maximum throughput,

$$f_D(max\ throughput) = 1 - e^{-1} = 0.63. \tag{7.47}$$

In other words, assuming the Poisson nature of pile up, the theoretical maximum throughput will occur at a dead time of 63%. An example of throughput versus dead-time percentage is shown in Fig. 7.10, where throughput is quantified as the net count rate in the 122-keV peak from a ^{57}Co source.

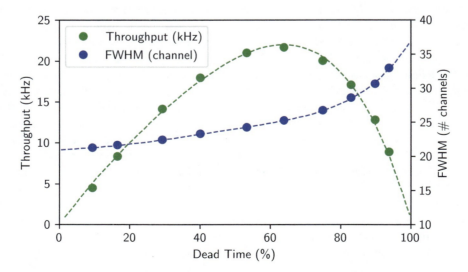

Fig. 7.10 Net count rate (throughput) and FWHM for the 122-keV peak of ^{57}Co as a function of dead time

In this experiment, 10 spectra were recorded at dead times in the range of ~10%–90%. The live time for each measurement was kept constant at 100 s. Note that the maximum throughput occurs near a dead time of just above 60%. The common Poisson nature of this relationship makes this outcome true for all systems. The degradation of resolution due to increasing pulse overlap can also be seen in Fig. 7.10 as the FWHM of the 122 keV peak plotted as a function of dead-time percentage.

7.5.5 General Comments on Data Throughput

The throughput rate is the rate at which events are stored in the spectrum. Figure 7.9 shows that the throughput rate peaks at surprisingly low values for ordinary high-resolution gamma-ray spectroscopy systems. For T = 50μs (a high though common value), $1/T = 20{,}000$ s^{-1}, and the maximum throughput is ~7350 s^{-1}. Where long time constants are necessary to produce the desired resolution, throughput must be sacrificed as the price for the highest resolution. In fact, the low-rate side of the throughput curve should be used when possible because it yields better resolution and peak shape. At a rate of 0.6(1/T), the throughput is 90% of maximum; at only 0.5(1/T), the throughput is still 82% of maximum.

Sometimes a spectroscopy system must be operated far beyond the throughput maximum. At a rate of 2/T (40,000 s^{-1} with T = 50μs), only ~14% of the information is stored, implying a correction for pileup losses of ~7. One important point is evident: to maximize system throughput and minimize the necessary corrections, T must be minimized, and some loss of resolution must always be accepted. Fortunately, much progress has been made to minimize T and still preserve resolution and peak shape (see Chap. 5 and Ref. [21]).

If Equations 7.38 through Eq. 7.40 are used to estimate throughput rates and loss fractions, R_i and T must be reasonably well known. The input rate (R_i) is usually easy to obtain. Many modern amplifiers include a provision for pileup rejection and have a fast-counting channel with a pulse-pair resolution of ~0.5–1.0μs and an output that can be counted with a scaler-timer. Equation 7.36 can be used to refine the value of R_i when there is significant loss in the fast-counting channel. Fast amplifiers and discriminators can be connected to the preamplifier output to measure the gross count rate.

The rejection or loss interval (T) is more difficult to estimate. If electronic pileup rejection is not used, T can be assumed to be approximately equal to the pulse width (see Fig. 7.8). An oscilloscope can be used to measure the width between the 1% or 2% amplitude points of the pulse. For many amplifiers, the pulse width is approximately six times the time constant (τ), but this assumption usually underestimates the pileup losses. After a pulse is analyzed, the amplifier output must fall below the ADC lower-level discriminator before another event is accepted. Because the discriminator level is usually low, a pulse preceded by another with less than a full pulse-width separation will not be analyzed. To compensate, T might be estimated at ~1.5 times the pulse width for systems without formal pileup rejection.

With electronic pileup rejection, different configurations have somewhat different values of T. One common procedure uses a fast-counting circuit to examine the intervals between preamplifier pulses and to generate an inhibit signal if an interval is less than a fixed value. The interval and inhibit signal length are approximately the width of an amplifier pulse. The inhibit signal is applied to the anticoincidence gate of the ADC to prevent analysis of pileup events. The value of T depends on the anticoincidence requirements of the ADC; usually a pulse is rejected if another pulse precedes it within the preset interval or if another pulse follows it before the ADC linear gate closes when digitization begins. Obviously, a good qualitative understanding of the operation of the ADC and pileup rejection circuitry is required to estimate T accurately. Additional losses caused by ADC dead time can often be ignored. For example, if the pulse width is 35μs (corresponding to the use of ~6μs time constants) and digitization takes 15μs or less beginning when the pulse drops to 90% of its maximum value, then the ADC completes digitization and storage before the pileup inhibit signal is released, and the ADC contributes no extra loss.

The fraction of good information stored is usually somewhat less than estimated. One reason is that rejection circuitry allows some pileup events to be analyzed, thus causing a loss of good events. Most pileup rejection circuitry has a pulse-pair resolution of 0.5–1.0μs. Pulses separated by less than the resolution time will pile up but are still analyzed, causing sum peaks in the spectrum (see Sect. 7.6). An example of this result is shown in Fig. 7.11. Note how the application of pileup rejection enhances the random sum peak at twice the energy of the incident 662 keV gamma rays as well as the associated summed continua, which occurs because many pulses that partially overlap in time and would otherwise broaden spectral features are, for the most part, thrown out.

Another cause of information loss is the generation of long-rise time preamplifier pulses. Usually preamplifier rise times are a few tenths of a microsecond; however, if the gamma-ray interaction occurs in a part of the detector where the electric field is weak or where an excess of trapping centers exists, it may take several microseconds to collect the liberated charge. The main amplifier produces a very long, low-amplitude pulse—often two or three times as long as normal. Good events that sum with these long, low-amplitude pulses are lost as useful information. The frequency with which such events are generated depends

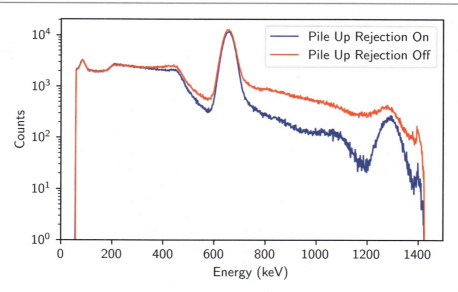

Fig. 7.11 Two NaI spectra of ^{137}Cs both recorded at a system-reported dead time of ~79%. The upper (red) spectrum was taken with pileup rejection off. The lower (blue) spectrum was taken with pileup rejection on

on detector properties and how the detector is illuminated with gamma rays. Gamma rays that fall on the detector edges where fields are often distorted and weak have a much greater chance of not being properly collected. In some applications, a detector performs better at high rates if the gamma rays can be collimated to fall only on the detector's center region. For a relatively poor detector under fully illuminated conditions, as many as 10% of the detected events can have long rise times, which result in a substantial loss of potential information. To achieve high throughput at high rates requires an excellent detector with minimum generation of the poorly collected, slow-rising pulses.

7.5.6 Correction Methods: General

The determination of the full-energy interaction rates (FEIR) of the gamma rays of interest is fundamental to many NDA procedures.

$$FEIR = \frac{A(\gamma)}{TT} CF(RL) \tag{7.48}$$

where
$A(\gamma)$ = full-energy-peak area
TT = real time of acquisition
$CF_{rateloss}$ = rate-related loss correction factor.

Three classes of correction procedures are discussed in this section. For detecting pileup events, the first procedure uses fast-counting electronics to measure the intervals between pulses. Corrections are made by extending the count time or by adding counts to the spectrum during acquisition. The second procedure adds an artificial peak to the spectrum by connecting a pulser to the preamplifier. The third procedure uses a gamma-ray source to generate the correction peak. The second and third procedures both use the variation in the correction-peak area to calculate a correction factor.

All three methods require the assumption that all peaks suffer the same fractional loss from the combined effects of pileup and dead time; in general, the assumption is good.

7.5.7 Pileup Correction Methods: Electronic

Methods that extend count time employ fast-counting circuits that operate directly from the preamplifier output; the time constants involved allow a pulse-pair resolution of 0.5–1.0μs. The time resolution is achieved at the sacrifice of energy resolution so that some small pulses analyzed by the ADC are lost to the timing circuitry. The circuitry can neither detect nor correct for pileup events where the interval is less than the circuit resolving time or where one of the events is below the detection threshold. When two or more pulses are closer together than the chosen pileup rejection interval but farther apart than the resolving time, the distorted event is not stored, and the count time is extended to compensate for the loss.

One method of extending the count time is to generate a dead-time interval that begins when a pileup event is detected and ends when the next good event has been processed and stored; this procedure is approximately correct. The procedure cannot compensate for undetected events; however, with a typical rejection-gate period of 20μs and a pulse-pair resolving time of ~1μs, the correction error could be only a few percent. For rates up to several tens of thousands of counts per second, the total error could be only 1% or less; the necessary circuitry is frequently built into spectroscopy amplifiers. The method requires live-time operation, so the assay period is not known a priori. The method also requires that the count rate and spectral shape are constant during the counting period; this limitation is of no consequence for the assay of long-lived nuclides, but it is important in activation analysis of very short-lived nuclides.

The activation analysis requirement to handle high count rates and rapidly changing spectral shapes has led to further advancement in dead-time pileup corrections. Such systems can handle input rates of hundreds of thousands of counts per second and accurately correct for losses in excess of 90%.

7.5.8 Pulser-Based Corrections for Dead Time and Pileup

This method uses a pulser to insert an artificial peak into the stored spectrum. The method has numerous variations depending on the type of pulser used. Most germanium and silicon detector preamplifiers have a TEST input through which appropriately shaped pulses can be injected. These pulses suffer essentially the same dead-time and pileup losses as gamma-ray pulses and form a peak similar to a gamma-ray peak. The pulser peak has better resolution and shape than the gamma-ray peaks because it is not broadened by the statistical processes involved in the gamma-ray detection process. The pulser peak area is determined in the same way as a gamma-ray peak area. The number of pulses injected into the preamplifier is easily determined by direct counting or by knowing the pulser rate and the acquisition time.

An advantage of the pulser method is that the artificial peak usually can be placed to avoid interference from gamma-ray peaks. In addition, because all pulser events are full energy, minimum extra dead time and pileup are generated. On the other hand, it is difficult to find pulsers with adequate amplitude stability, pulse-shaping capability, and rate flexibility.

Another common problem is the difficulty of injecting pulses through the preamplifier without introducing some underdamped behavior of the output pulse. Excessive underdamping is objectionable because gamma-ray pulses can pile up on the negative part of the pulse like they do on the positive part. The amplifier pole zero cannot compensate simultaneously for the different decay constants of the pulser and gamma-ray pulses, and compensation networks are rarely used at the TEST input because of probable deterioration in resolution. The underdamping problem can be minimized by using a long decay time on the pulser pulse (often as long as a millisecond), by using shorter amplifier time constants, and by using high baseline-restorer settings. Some sacrifice of overall resolution is usually required to adequately minimize the problem of underdamping.

The simplest approach is to use an ordinary fixed-period pulser in which the interval between pulses is constant and equal to the reciprocal of the pulse rate. The best amplitude stability comes from the mercury-switch pulser, which charges and discharges a capacitor through a resistor network by a mercury-wetted mechanical switch. The mechanical switch limits the useful rate of such pulsers to ≤100 Hz.

Assuming that the pulser peak and gamma-ray peaks lose the same fraction of events from dead time and pileup, the appropriate correction factor is

$$CF_{rateloss} = \frac{N_p}{A_p}, \tag{7.49}$$

where

N_p = number (rate) of pulses injected
A_p = area rate of pulser peak.

$CF_{rateloss}$ has a minimum value of 1.00 and is the reciprocal of the fraction of events stored in the peaks.

Equation 7.49 is not quite correct because pulser pulses are never lost as a result of their own dead time, nor do they pile up on one another. Thus the overall losses from gamma-ray peaks are greater than those from the pulser peak, although the difference is usually small. At moderate rates, the dead-time and pileup losses are nearly independent, and $CF_{rateloss}$ can be corrected with two multiplicative factors to obtain a more accurate result:

$$CF_{rateloss} = \frac{N}{A(P)}(1 + R_p T_D)(1 + R_p T), \tag{7.50}$$

where
R_p = pulser rate
T_D = dead time for pulser pulse
T = pileup interval.

The pulser dead time (T_D) can be adequately estimated from the speed of the ADC and, for Wilkinson ADCs, from the position of the pulser peak. The interval (T) is usually 1.5–2 times the pulse width. If R_p is ≤100 Hz, both factors are usually small. Assuming a typical value of 20μs for both T_D and T, the value of each additional factor is 1.002 so that the increase in $CF_{rateloss}$ is only ~0.4%. Larger corrections result if greater values of T_D or T are used. If R_p is increased to 1000 Hz to obtain high precision more quickly, the additional factors make a difference of several percent.

Concern about assay precision brings up a rather curious but useful property of periodic pulsers. The precision of the pulser peak is given by a different relation than that of gamma-ray peaks. In fact, the precision of a pulser peak is always better than the precision of a gamma-ray peak of the same area because gamma-ray emission is random and the generation of pulser pulses is not. The precision of gamma-ray peak areas is governed by Poisson statistics, whereas the precision of pulser peaks is governed by binomial statistics. In a given time interval, the pulser either fired a fixed number of times or it did not, unlike the theoretically infinite number of possible gamma-ray emissions to be detected during the same time interval.

The variance of a binomial distribution can be written in terms of the number of trials (n) and the probability for success (P) for each trial.

$$\sigma^2 = nP(1-P) \tag{7.51}$$

Assuming that the continuum under the pulser peak is negligible and that the peak area is A_p (equal to nP in Eq. 7.51), the variance and relative variance of A_p are given by

$$S^2(A_p) = A_p\left(1 - \frac{A_p}{N_p}\right)$$

$$S_r^2(A) = \frac{1}{A_p}\left(1 - \frac{A_p}{N_p}\right) \tag{7.52}$$

where N_p is the total number of pulses injected into the spectrum. Assuming again that the Compton continuum is negligible, the variance and relative variance of a gamma-ray peak of area (A) are given by

$$S^2(A) = A$$

$$S_r^2(A) = \frac{1}{A} \tag{7.53}$$

Figure 7.12 gives $S_r(A_p)$ versus A_p for several choices of A_p/N_p and demonstrates that by the time $A_p/N_P \approx 0.5$, the improvement in precision is quite negligible.

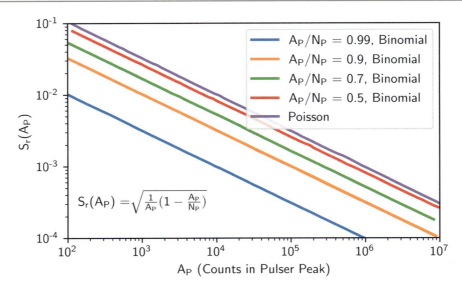

Fig. 7.12 The relative standard deviation $S_r(A_p)$ of a pulser peak area (A_p) as a function of A_p for several values of A_p/N_p, where N_p is the total number of pulses injected into the spectrum

When the pulser peak rests on a significant continuum, the expressions for $S(A_p)$ are more complex because of the random nature of the continuum. The pulser peak should be placed in a low-continuum (usually high-energy) portion of the spectrum so that the improved precision can be taken advantage of, and the simple Eq. 7.52 can be used.

The use of a high-energy pulser peak can complicate the minimization of underdamping of the amplifier pulse shape. An alternative approach is to use a rectangular pulse that is longer than the amplifier output; then the pole-zero problem disappears, and no underdamping occurs. Instead, a negative pulse is generated as the pulser output drops. Other events pile up on the negative pulse, but the pulse is cleanly defined and tends to throw pileup events out of their peak. The additional factor for the pileup losses can be written as $(1 + 2R_pT)$.

Pulser-based, dead-time pileup corrections are accurate only when both rate and spectral shape are constant throughout the counting period. When the count rate changes during a measurement, proper corrections cannot be made if the pulser operates at a fixed rate. In principle, a correction can be made using a pulser that operates at a rate that is a fixed function of the gross detector rate. Pulsers based on this concept have been built and used successfully [22]. They are used in activation analysis, half-life studies, accelerator experiments, and anywhere that variable rates with constant spectral shape might be encountered. The use of variable rate pulsers indicates the variety and ingenuity with which the fundamental idea of inserting a synthetic peak into a spectrum has been applied to the problem of dead-time pileup corrections.

7.5.9 Reference-Source Method for Dead-Time Pileup Corrections

The most accurate method for measuring the dead-time pileup correction uses a reference source fixed in position relative to the detector. The source provides a constant gamma-ray interaction rate in the detector. The reference peak performs the same function as the pulser peak.

Like the other methods, the reference-source method requires the assumption that all peaks suffer the same fractional loss from dead time and pileup. Given this assumption, the ratio of any peak area to the reference peak area is independent of such losses. Let $A(\gamma)$ and $FEIR(\gamma)$ represent, as usual, the full-energy-peak area and the FEIR of any gamma ray other than the reference gamma ray (R). If F is the fraction of stored events (common to all gamma rays in the spectrum) and TT is the true acquisition time, then the areas are

$$A(\gamma) = F \times FEIR(\gamma) \times TT$$

$$A(R) = F \times FEIR(R) \times TT. \qquad (7.54)$$

The ratio of the two expressions gives

$$\frac{A(\gamma)}{A(R)} = \frac{FEIR(\gamma)}{FEIR(R)}, \qquad (7.55)$$

which is independent of both F and TT. Gamma-ray assays often are based directly on the loss-independent ratios A(γ)/A(R) without ever explicitly determining CF$_{rateloss}$ or FEIR(γ). For the reference-source gamma ray, the correction factor becomes

$$CF_{rateloss} = FEIR(R) \times \frac{TT}{A(R)}. \qquad (7.56)$$

Combining this expression with Eq. 7.48 gives

$$FEIR(\gamma) = \frac{A(\gamma)}{A(R)} FEIR(R). \qquad (7.57)$$

If assay systems are calibrated with the help of standards, it is unnecessary to know FEIR(R) to obtain accurate assay values. In many assay procedures, the quantity sought, M (nuclide or element mass), is proportional to FEIR(γ). In Equations 7.58 through 7.60, K is the calibration constant, the subscript *s* denotes quantities pertaining to standards, and the subscript *u* denotes quantities pertaining to unknowns. Again, we are assuming that no attenuation corrections are required for the sources used.

$$M_u = \frac{\left[\frac{A(\gamma)_u}{A(R)_u}\right] \times FEIR(R)}{K} \qquad (7.58)$$

The calibration constant can be determined from a single standard:

$$K = \frac{\left[\frac{A(\gamma)_s}{A(R)_s}\right] \times FEIR(R)}{M_s} \qquad (7.59)$$

Combining Eq. 7.58 and Eq. 7.59 gives

$$M_u = \frac{\left[\frac{A(\gamma)_u}{A(R)_u}\right]}{\left[\frac{A(\gamma)_s}{A(R)_s}\right]} M_s, \qquad (7.60)$$

which is independent of FEIR(R).

Although an accurate value of FEIR(R) is not needed, it is useful in obtaining approximate values of FEIR(γ), FEIR(R), and CF$_{rateloss}$ so that actual rates of data acquisition are known, along with the fraction of information being lost to dead time and pileup. Having a calibration constant expressed as corrected counts per second per unit mass can be helpful when estimating required assay times.

A reasonably accurate value of FEIR(R) can be obtained by making a live-time count of the reference source alone and estimating a correction for the pileup losses. A more accurate value can be obtained by using a pulser to correct for dead-time pileup losses.

The reference-source method can be applied to any spectroscopy system without additional electronics. The method avoids problems caused by injecting pulser pulses into a preamplifier and by drift of the reference peak relative to the gamma-ray peaks. It also avoids the extra corrections required by a fixed-period pulser. Additionally, no error occurs because of the finite pulse-pair resolving time, and the reference peak is constantly present for digital stabilization and for checking system performance.

The most significant limitation to the procedure is that a reference source with appropriate half-life and energy is not always available. An additional limitation is that the reference source must have a significant count rate, which causes additional losses and results in poorer overall precision than that achievable using the same count time with other methods. The reference-source method, as well as the simpler pulser method, is applicable only when the count rate and spectral shape are constant.

The reference source should have a long half-life and an intense gamma ray in a clear portion of the spectrum. The energy of the reference gamma ray should be lower than the energy of the assay gamma rays so as not to add to the background beneath the assay peaks. Few sources meet all of the desired criteria, but several have proven adequate in many applications.

For ^{239}Pu assays based on the 413.7 keV gamma ray, ^{133}Ba is the most useful source. Its 356.0 keV gamma ray does not suffer interference from any plutonium or americium gamma ray, and it is within 60 keV of the assay energy. The 10.3 year half-life is very convenient. Although ^{133}Ba has several other gamma rays, they are all at energies below 414 keV.

For plutonium assays that make use of lower-energy gamma rays, ^{109}Cd is a useful reference source. The 88.0 keV gamma ray is its only significant emission except for the ~25 keV ^{109}Ag X-rays from electron capture, which are easily eliminated by a thin filter. Its half-life of ~453 days is adequate to give a year or two of use before replacement. Although no interfering gamma rays from plutonium or americium nuclides are present, a possible interference exists from lead X-rays fluoresced in the detector shielding. The lead $K_{\beta 2}$ X-ray falls almost directly under the 88.0 keV gamma ray from ^{109}Cd. Interference can be avoided by wrapping the detector in tin to absorb the lead X-rays and by using a sufficiently strong ^{109}Cd source so that any residual leakage of lead X-rays is overwhelmed. If different shielding material (e.g., iron or tungsten alloys) can be used, the problem disappears.

For assays of ^{235}U, the 122.0 keV gamma ray from ^{57}Co is used frequently. Its 271-day half-life is adequate although not as long as might be desired. The 122.0 keV gamma ray is approximately eight times more intense than the 136.5 keV gamma ray, which is the only other gamma ray of significant intensity. Note that in using ^{57}Co for assay of highly enriched uranium items, the 120.9 keV gamma ray from ^{234}U can be an annoying interference.

Often ^{241}Am can be used as a reference source for uranium or other assays. Although the 59.5 keV gamma ray from ^{241}Am is further removed than desirable from 186 keV, it can be used successfully, particularly if steps are taken to reduce the resolution difference. The half-life of 433.6 year is beyond fault. Americium-241 must be absolutely absent from any materials to be assayed. When using ^{169}Yb as a transmission source in densitometry or quantitative ^{235}U assay, ytterbium daughters emit X-rays that directly interfere with the 59.5 keV gamma ray; sufficient filtering combined with adequate source intensity can eliminate any possible difficulty. Note also that the $K\alpha 1$ X-ray emission from tungsten, a common attenuator material, is at 59.3 keV.

The current methods for dead-time pileup correction assume that all full-energy peaks suffer the same fractional loss. That assumption is not completely true primarily because the width and detailed peak shape are functions of both energy and count rate. In applying the reference-peak method, precautions can be taken to minimize the degree to which the assumption falls short. Four of those precautions, most of which apply to any of the correction methods, are listed as follows:

- Where possible, apply the procedure only over a narrow energy range.
- Keep the peak width and shape as constant as possible as functions of both energy and count rate, even if that effort slightly degrades the low-rate and low-energy resolution. Proper adjustment of the amplifier and the pileup rejection can help considerably.
- Avoid a convex or concave Compton continuum beneath important peaks, especially the reference peak. If possible, the ratio of the reference peak area to the continuum area should be ≥ 10.
- Exercise great care in determining peak areas. ROI methods may be less sensitive than some of the spectral fitting codes to small changes in peak shape.

Experimental results indicate that the reference-source method can correct for dead-time and pileup losses with accuracies approaching 0.1% over a wide count-rate range. Such accuracies can also be approached by pulser methods, particularly at lower rates and by some purely electronic methods.

7.6 Coincidence Summing in Gamma-Ray Spectra

Two or more gamma rays could enter the detector volume in a time interval that is very small compared with the charge collection times of germanium, silicon, or NaI(Tl) detectors. In this case, the multiple gamma rays are treated as a single interaction.

7.6.1 Random Coincidence Summing

Random coincidence summing occurs when two or more photons from different atoms deposit energy within the volume and resolving time of the detector. This process becomes more observable as dead time of the measurement increases. For example, consider a bare ^{137}Cs source of sufficient activity or at a source-to-detector distance such that the dead time is significant for a particular detector. If two 661.7 keV gamma rays deposit their full energy in the detector at roughly the same time, this process will contribute to counts in the spectrum at $2 \times 661.7 = 1323.4$ keV. If this effect occurs many times during a measurement, a "random sum peak" will occur at this energy. Of course, random summing can occur at all energies in all combinations, even creating summed versions of other spectral features such as Compton continua. In general, summed features will be less resolved because of the mixing of various energy combinations, which is illustrated in Fig. 7.13.

7.6.2 True Coincidence Summing

True coincidence summing occurs when two photons from the same decay event in an atom deposit energy within the volume and resolving time of the detector. True coincidence can result from gamma-gamma, gamma-X-ray (X-rays generated following internal conversion or electron capture), and gamma-annihilation photon coincidence. It is a function of the detection efficiency (counting geometry and detector) and the radionuclide's decay scheme (not a function of the overall count rate). Therefore, true coincidence can be reduced by moving the sample farther from the detector to reduce the subtended solid angle (see Sect. 7.7.2). Typically, true coincidence causes counts to be lost from the full-energy peaks, but it can also cause addition to full-energy peaks depending on the radionuclide's decay scheme. Common radionuclides susceptible to true coincidence are ^{60}Co, ^{88}Y, ^{152}Eu, ^{154}Eu, ^{125}Sb, ^{134}Cs, and ^{133}Ba [23]. Consider the example of ^{133}Ba, the level scheme of which is shown in Fig. 7.14.

Following the emission of a 356-keV gamma-ray from the 437-keV level, the nucleus reaches the 81-keV level, where the lifetime is only 6.3 ns. In terms of the microsecond-scale time resolution of typical gamma-ray detectors, the 356 and 81 keV gamma rays can be viewed as being emitted virtually at the same time. If both photons deposit energy in the detector, the result will be a count that represents the sum of the energies deposited. Due to the cascade-like nature of the decay, true coincidence

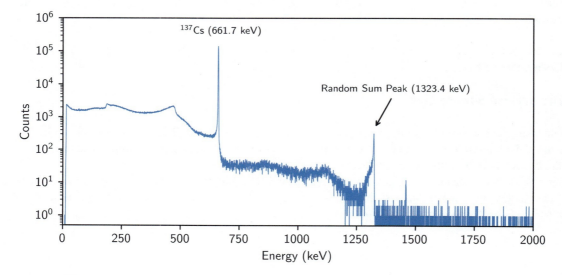

Fig. 7.13 Spectrum of ^{137}Cs at a dead time of 60%. A random sum peak occurs at 1323 keV. Note also the summed continuum

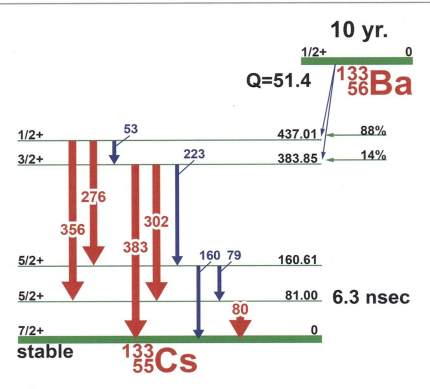

Fig. 7.14 Decay scheme for ^{133}Ba. Note the lifetime for the 81 keV level is only 6.3 ns

summing is often referred to as "cascade coincidence summing." Because both gamma rays emanate from the same atom, this effect is essentially independent of source activity and thus, dead time. In addition to true coincidence summing, random summing may occur between the 81-keV gamma ray of one atom and the 356-keV gamma ray of another atom.

Note that true coincidence events are not limited to decay cascades. Consider the decay of ^{22}Na by positron emission. The resulting 1275-keV gamma ray is emitted in coincidence with the positron. The annihilation of the positron will occur rapidly in matter, i.e., ~hundreds of picoseconds [24]. Therefore, the 511-keV annihilation radiation that results will be in coincidence with the 1275-keV gamma ray.

Cascade summing can result in subtractive or additive errors in the measured FEIRs. The problem becomes significant when the source is so close to the detector that the probability of detecting two or more cascade gamma rays simultaneously is large. If very short source-to-detector distances must be used to enhance sensitivity, cascade-summing problems must be carefully considered.

7.7 Geometric Effects

7.7.1 The Inverse Square Law

Radiation intensity decreases with the square of the distance from a point source. If the distance from a radiation source is at least three times the longest dimension of the source, the source can be treated as a point source, and the inverse square law will give the correct answer within a percent [15].

Consider a point source that emits (I) gamma rays per second. The gamma-ray flux (F) at a distance (R) is defined as the number of gamma rays per second that pass through a unit area on a sphere of radius (R) centered at the source. Because the area of the sphere is $4\pi R^2$, the expression for F is

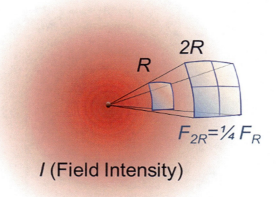

Fig. 7.15 Inverse-Square Law. The flux at a radius of 2R is one quarter of the flux at a radius of R. One would need a detector four times larger than the detector at R to record the same total count rate at 2R

$$F = \frac{I}{4\pi R^2}. \tag{7.61}$$

Essentially, as one doubles the source-to-detector distance, the recorded flux will drop by a factor of 4. In other words, to record the same count rate at a distance of 2R as one would record at R, a detector face four times larger would be required. This scenario is illustrated in Fig. 7.15.

7.7.2 Solid Angle

The inverse square law is directly related to the concept of solid angle. The maximum solid angle of 4π steradians can be covered only by a detector that completely surrounds the source, such as a well counter. A typical gamma-ray detector covers only a fraction of the Gaussian spherical surface through which the emitted photon flux will pass. If Ω is the solid angle subtended by the detector at the source, then $\Omega/4\pi$ is the solid-angle fraction covered by the detector, and it is the probability that a gamma ray from an isotropic source will be incident on the face of the detector. This outcome is illustrated in Fig. 7.16. Note that the maximum solid angle that could be covered in this situation is 2π steradians, which occurs when the source is in direct contact with the center of the detector face ($R = 0$). For a point source and a detector face with a circular aperture, the solid-angle fraction can be expressed in terms of the angle θ that is created by the source-to-detector distance (R) and the radius (r) of the detector face.

$$\frac{\Omega}{4\pi} = \frac{1}{2}(1 - cos\theta) \tag{7.62}$$

7.7.3 Extended-Source Geometries

The detector count rate is proportional to the incident flux and, as discussed, if the detector face can be approximated by a portion of a spherical surface centered at the source, the count rate has the same $1/R^2$ dependency as the flux. When low-intensity items are counted, a clear motivation exists to reduce the item-to-detector distance and increase the count rate. Unfortunately, when the item-to-detector distance is so small that different parts of the item have significantly different

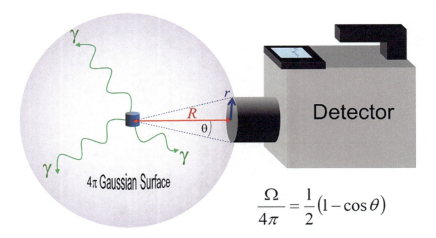

Fig. 7.16 Solid-angle fraction out of 4π steradians for a detector with radius (r) at a distance (R) from the source where $\theta = \tan^{-1}(r/R)$

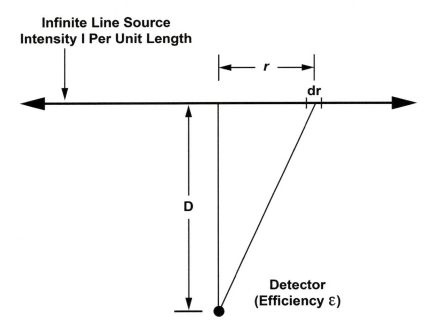

Fig. 7.17 Geometry for computing the response of a point detector to a line source

distances to the detector, the count rates from different parts of the item vary significantly. This variation can cause an assay error when the distribution of emitting material is nonuniform.

When the source-to-detector distance is on the order of or less than either the item or detector dimensions, the measurement can be considered to be in the near field. In this situation, the overall count rate from items of finite extent does not follow the simple inverse-square law; usually the variation is less strong than $1/R^2$. Knowledge of a few simple cases can help to estimate overall count rates and response uniformity.

The simplest extended source is a line, which is often an adequate model of a pipe that is carrying radioactive solution. Consider an ideal point detector with intrinsic efficiency (ε) at distance (D) from an infinitely long source of intensity (I) per unit length (Fig. 7.17).

7 General Topics in Passive Gamma-Ray Assay

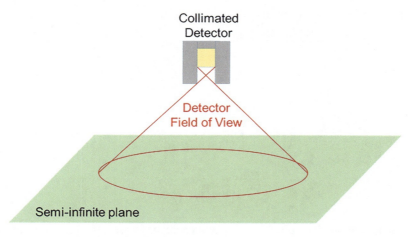

Fig. 7.18 Detector over a semi-infinite plane of uniform radioactive material

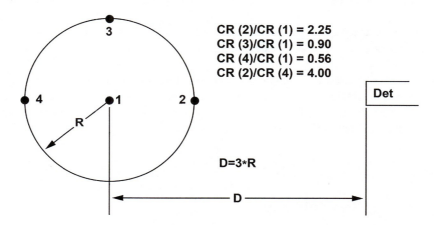

Fig. 7.19 Cross section through a cylindrical item and a point detector showing how count rate varies with position

The count rate from this source can be expressed as

$$CR = 2\int_0^\infty \frac{I\varepsilon\,dr}{r^2 + D^2} = \frac{\pi I \varepsilon}{D}. \tag{7.63}$$

In this ideal case, the count-rate dependence is 1/R rather than $1/R^2$; when pipes are counted at distances much smaller than their length, the count-rate variation will be approximately 1/R.

The count rate from a detector at a distance (R) from an infinite plane surface does not depend on R at all. Certain geometries can approximate an infinite plane. For a detector near a uniformly contaminated glove-box wall, count rates vary little with wall-to-detector distance changes. In this sense, the glove-box wall can be thought of as a semi-infinite plane. As the distance between the wall and the detector changes, the area of the wall within the detector field of view changes to compensate the observed count rate. This effect is illustrated in Fig. 7.18.

When possible, the variation of response with position inside an item should be minimized. The item-to-detector distance can be increased, but the penalty is a severe loss of count rate. A better strategy is to rotate the item. Consider the cross section of a cylindrical item of radius (R) whose center is at a distance (D) from a detector (Fig. 7.19). Unless D is much greater than R, the count rates for identical sources at positions 1, 2, 3, and 4 vary considerably. The figure shows that if D = 3R, the maximum count-rate ratio is CR(2)/CR(4) = 4. The ratio of the response of a rotating source at radius (R) to the response at the center (position 1 of Fig. 7.19) is

Table 7.7 The effect of item rotation on count-rate variation

R/D	CR(R)/CR(l)(Rotating)	CR(2)/CR(l)(Nonrotating)
1/2	1.33	4.0
1/3[a]	1.125	2.25
1/4	1.067	1.78
1/5	1.042	1.56
1/6	1.029	1.44
1/7	1.021	1.36

[a]See Fig. 7.19

$$\frac{CR(R)}{CR(1)} = \frac{1}{1 + \left(\frac{R}{D}\right)^2}. \tag{7.64}$$

The response is the same as that obtained for a uniform, nonattenuating, circular source of radius (R) whose center is at a distance (D) from a detector. Table 7.7 gives the value of this function for several values of R/D compared with CR(2)/CR(l) for the nonrotating source of Fig. 7.19. For relatively large values of R/D, rotation improves the uniformity of response. The response variation is even larger when attenuation is considered. Rotation only reduces $1/R^2$ effects; it does not eliminate them completely.

Rotation reduces response variations caused by radial positioning but does little to compensate for height variations. If the source height is less than one-third of the item-to-detector distance, the decrease in response is less than 10% relative to the normal position.

The choice of item-to-detector distance is a compromise between minimizing the response variations and maintaining an adequate count rate. A useful guideline is that the maximum count-rate variation is less than ±10% if the item-to-detector distance is equal to three times the larger of the item radius or the half-height. If an item cannot be rotated, it helps considerably to count it in four positions 180° apart.

7.8 Detector Efficiency Measurements

7.8.1 Absolute Full-Energy-Peak Efficiency

The absolute full-energy-peak efficiency is the fraction of gamma rays emitted by a source at a particular source-to-detector distance that produces a full-energy interaction in the detector. It is determined by measuring the efficiency at various energies using an unshielded point source (to minimize external and self-attenuation) and fitting the experimental data with an appropriate function.

$$\varepsilon_A(E_\gamma) = \frac{FEIR(E_\gamma)}{I(E_\gamma)}, \tag{7.65}$$

where
$\varepsilon_A(E\gamma)$ = absolute full-energy efficiency at the energy $E\gamma$
$FEIR(E\gamma)$ = full-energy interaction rate at the energy $E\gamma$
$I(E\gamma)$ = gamma-ray emission rate at the energy $E\gamma$.

Considering the more general case of an extended source shielded by a container or otherwise, we can also write the absolute detection efficiency in terms of the effects of self-attenuation and external attenuation, the intrinsic detector efficiency (see Sect. 7.8.2), and the solid-angle fraction that was discussed in Sect. 7.7.2.

$$\varepsilon_A(E_\gamma) = T_{Self}(E_\gamma) \cdot T_{Ext}(E_\gamma) \cdot \varepsilon_I(E_\gamma) \cdot \frac{\Omega}{4\pi}, \tag{7.66}$$

7 General Topics in Passive Gamma-Ray Assay

where

$T_{Self}(E\gamma)$ = self-transmission through material matrix at the energy $E\gamma$
$T_{Ext}(E\gamma)$ = transmission through external attenuators at the energy $E\gamma$
$\varepsilon_I(E\gamma)$ = intrinsic detector efficiency at the energy $E\gamma$ (see Sect. 7.8.2)
$\Omega/4\pi$ = solid-angle fraction covered by the detector

A discussion of the transmission terms $T_{Self}(E\gamma)$ and $T_{Ext}(E\gamma)$ can be found in Chap. 6.

7.8.2 Intrinsic Detector Efficiency

The intrinsic full-energy detector efficiency is the probability that a photon will deposit all of its energy in the detector volume, assuming that it has entered the detector volume. For far-field measurement geometries, photons are incident normal to the face of the detector. In this case, the intrinsic efficiency is independent of solid angle and thus, source-to-detector distance. For a far-field point source (no self-attenuation, $T_{Self} = 1$ for all energies) that is unshielded and in a negligible container (no external attenuation, $T_{Ext} = 1$ for all energies), the absolute full-energy efficiency ε_A and the intrinsic full-energy efficiency ε_I are related by the simplified version of Eq. 7.66:

$$\varepsilon_A(E_\gamma) = \frac{FEIR(E_\gamma)}{I(E_\gamma)} = \varepsilon_I(E_\gamma) \cdot \frac{\Omega}{4\pi} \qquad (7.67)$$

In general, as incident photon energy increases, the likelihood for full-energy deposition in the detector decreases. This result is especially true for thin detectors, illustrated in Fig. 7.20; however, the sensitive medium of a gamma-ray detector is usually enclosed in a housing that will attenuate photons but is not sensitive to the energy deposited. This attenuation will decrease the efficiency of detection for photons with low incident energies (e.g., $< \sim 60$ keV), as shown in Fig. 7.21, which is especially true for ruggedized detectors in which the housing is thicker.

The intrinsic full-energy efficiency is determined experimentally by measuring the absolute full-energy efficiency and solving Eq. 7.67 for ε_I. The source should be point-like in nature and set in a far-field configuration (to minimize self-attenuation and solid-angle geometry effects respectively) and in a thin-walled container of low atomic number (to minimize external attenuation). Such effects are addressed further in Sect. 7.9, which covers relative efficiency. An example intrinsic detector efficiency curve is shown in Fig. 7.22. This example is for an ORTEC Detective (HPGe), which has a ruggedized housing. The effect manifests itself as the drop in efficiency below approximately 130 keV. Below this energy, photons scatter or are absorbed in the nonsensitive housing before reaching the detector crystal. Above this energy, efficiency decreases as incident energy increases, and photons more likely scatter and escape the detector crystal or pass right through without depositing their full energy.

Based on Fig. 7.22, a photon with an energy of 130 keV that is incident on the detector face will have an approximate 45% chance of contributing a count to a full-energy peak at 130 keV. Again, this result is entirely independent of source-to-detector distance (i.e., solid-angle fraction).

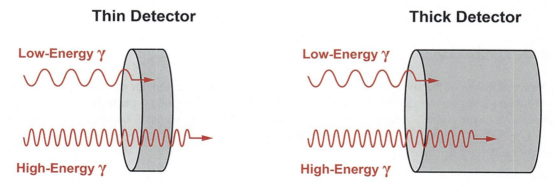

Fig. 7.20 In general, higher-energy photons will more likely pass through a detector without depositing their full energy. The intrinsic detector efficiency therefore decreases with increasing incident photon energy

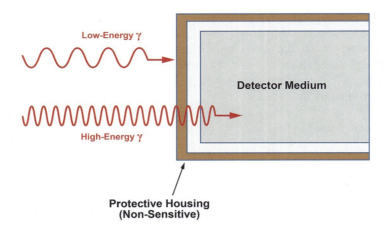

Fig. 7.21 The sensitive part of a gamma-ray detector is enclosed in a protective housing that attenuates but is not sensitive to photons, which decreases detection efficiency at low energies, especially for ruggedized detectors

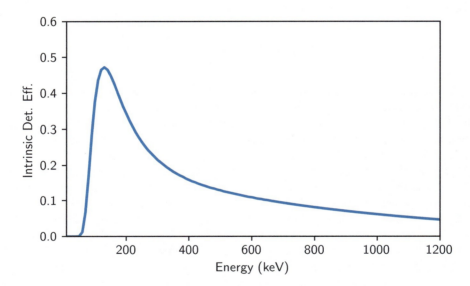

Fig. 7.22 Example of an intrinsic detector efficiency curve. This curve is for an ORTEC Detective (HPGe) detector

Note that the values of ε_I computed will depend slightly on the position of the source with respect the detector. Photons of different energies have different corresponding average interaction depths in the detector. If the source-to-detector distance is several times the detector thickness, then the differences in interaction depth as a function of energy can be ignored and the solid-angle fraction calculated according to Sect. 7.7.2.

7.9 Relative Efficiency

Frequently the actual values of the absolute or intrinsic full-energy-peak efficiency are not needed, and only the ratios of the efficiency at different energies are required. Relative efficiency differs from absolute efficiency only by a multiplicative constant that depends on the source activity and solid-angle fraction of the measurement. Exact gamma-ray emission rates are not required—only values proportional to the emission rates. Also, a relative efficiency curve is inherent in an absolute efficiency curve and defines its shape.

The emitted photon flux $I(E_\gamma)$ in Eq. 7.65 can be written as

$$I(E_\gamma) = \frac{N \cdot \ln(2) \cdot BR(E_\gamma)}{T_{1/2}} \qquad (7.68)$$

where
N = the number of nuclei available to emit the photon at energy Eγ
$T_{1/2}$ = the half life of the emitting nuclide
BR(Eγ) = the branching ratio for emitting the photon at energy Eγ

Substituting this in Eq. 7.65 and rearranging, we have.

$$\varepsilon_A(E_\gamma) = \left[\frac{T_{1/2}}{N \cdot \ln(2)}\right] \cdot \frac{FEIR(E_\gamma)}{BR(E_\gamma)} \qquad (7.69)$$

The term in brackets is independent of energy. We now remove that term and rewrite Eq. 7–69 as a proportionality relation. In doing so, we define relative efficiency.

$$\varepsilon_R \propto \frac{FEIR}{BR} \qquad (7.70)$$

We have removed the explicit energy dependence in Eq. 7.69 because it is common to all terms. The proportional nature of the relation illustrates that it is the shape of the relative efficiency curve as a function of energy that is important and not its absolute magnitude.

To reiterate, the shape of the relative efficiency curve will be governed by three things:

- intrinsic detector efficiency (Sect. 7.8.3),
- self-attenuation of the source matrix (Chap. 6), and
- external attenuation due to any intervening material between the source and detector (Chap. 6).

Relative efficiency must be considered when comparing measured count rates for gamma-ray peaks that are significantly separated in energy. "Significantly separated" is somewhat of a relative term itself because the slope of the relative efficiency curve can vary quite a lot for different energy regions of the spectrum.

7.9.1 Efficiency as a Function of Energy and Position

Usually NDA calibrations are done using standards that contain known amounts of the nuclides of interest in packages of appropriate shape and size. Because approximate detector efficiencies are used only to estimate expected count rates, little need exists to carefully characterize efficiency as a function of energy and position. A detector so characterized can assay without the use of standards although generally not with the same ease and accuracy as with them. When standards are not available or allowed in an area, verification measurements can be made of items of known geometry and content using the known detector efficiency to predict the FEIRs for a chosen detector-item configuration. If the measured FEIRs agree within the estimated error with the predicted rates, the item content is regarded as verified.

In constructing appropriate efficiency functions, the absolute efficiency is measured for many energies and positions and then fit to an adequate mathematical model. Cline's method [25] combines reasonable accuracy with a straightforward procedure for determining the efficiency parameters.

7.9.2 Toy Model for Relative Efficiency

To understand the general shape of a relative efficiency curve, it might help to first consider a toy model. Assume that a theoretical measurement is performed of a bare point source embedded in a perfect detector of infinite extent, as represented by Fig. 7.23.

The true point-like nature of the source would mean that no self-attenuation of emitted photons would exist. The absence of a container or other material around the source means that no external attenuation would exist. The idea that the detector is infinite in extent would mean that the full energy of every photon emitted by the source would be deposited in the detector. Therefore, the only reason for differences in the FEIR counts of the resulting spectrum, shown in Fig. 7.24, would be due to the branching ratio for each photon energy.

If we normalize the area of each peak by its respective branching ratio as in Eq. 7.70, the result would be a spectrum with all peaks of the same area, as shown in Fig. 7.25; that is, we have a flat relative efficiency curve for our toy model.

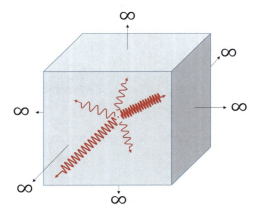

Fig. 7.23 Toy model of an infinitesimal point source embedded in a detector of infinite extent. Every gamma ray that is emitted by the source deposits its full energy in the detector

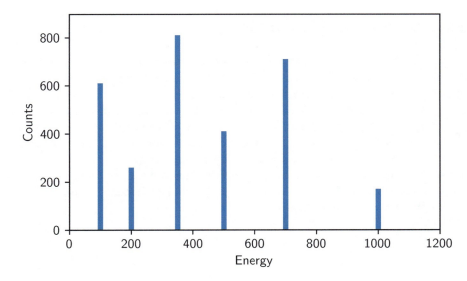

Fig. 7.24 Fictional spectrum (with arbitrary energy units) from the toy model of an infinitesimal point source embedded in an infinite detector. The only reason for the difference in recorded counts at each energy is the different branching ratios

7 General Topics in Passive Gamma-Ray Assay

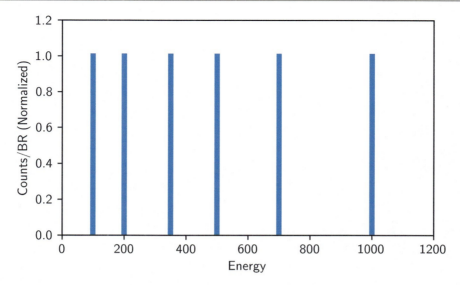

Fig. 7.25 Fictional relative efficiency (Counts/BR) from the toy model of an infinitesimal point source embedded in an infinite detector. Without the effects of realistic attenuation or intrinsic detector efficiency, the relative efficiency data are flat with respect to energy

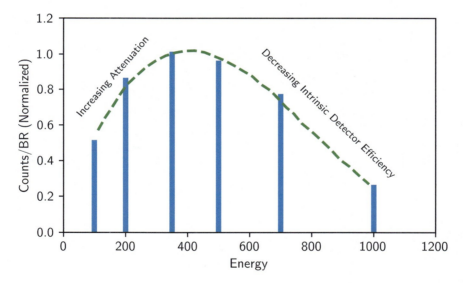

Fig. 7.26 Relative efficiency curve now including the effects of attenuation and intrinsic detector efficiency. Increasing attenuation reduces counts at lower energies, whereas decreasing intrinsic detector efficiency decreases counts at higher energies

We now apply the real-world effects of attenuation and intrinsic efficiency of a noninfinite detector to our toy model relative efficiency of Fig. 7.25. In general, attenuation will reduce counts at lower energies, whereas intrinsic detector efficiency will reduce counts at higher energies. This outcome is illustrated in Fig. 7.26.

7.9.3 Example Relative Efficiency Measurements

The simplest way to obtain a relative efficiency curve is to record net areas for several gamma-ray peaks from a single nuclide that span the energy range of interest and normalize the net areas by their respective branching ratios. An example of such a measurement is shown in Fig. 7.27 for a bare point source of ^{152}Eu. The lack of external and self-attenuation means that the intrinsic detector efficiency is the dominant effect in determining the shape of this curve.

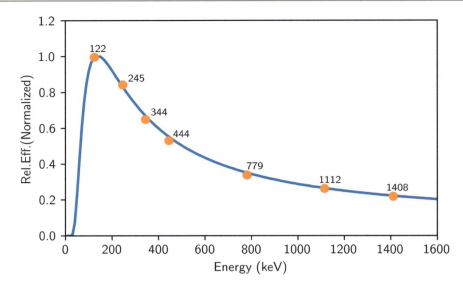

Fig. 7.27 Relative efficiency curve for a bare ^{152}Eu point source. The curve has been normalized to the maximum efficiency, which for this measurement, occurs for the 122-keV data point

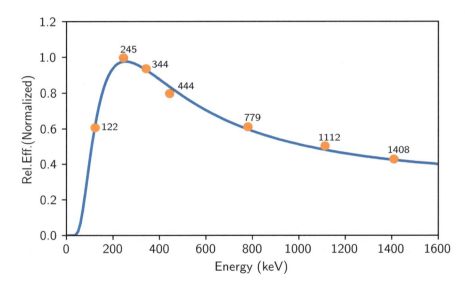

Fig. 7.28 Relative efficiency curve for a ^{152}Eu point source shielded by approximately 6 mm of steel. The curve has been normalized to the maximum efficiency, which for this measurement, occurs for the 245-keV data point

With the addition of external attenuation between the source and the sensitive medium of the detector, we begin to lose counts preferentially from peaks at lower energies. This effect is shown in Fig. 7.28, where a ^{152}Eu point source has been shielded by 6 mm of steel. The maximum relative efficiency shifts to higher energy (245 keV in this case) compared with the measurement of the bare source in Fig. 7.27. This effect is much more evident in Fig. 7.29, where a ^{152}Eu point source has been shielded by 50 mm of steel.

The data in Fig. 7.30 are governed by self-attenuation for a measurement of 10 g of weapons-grade plutonium, with some minor external attenuation provided by the thin steel container and a cadmium filter. The effect of decreasing intrinsic detector efficiency on the relative efficiency curve is evident for the 646-keV data point relative to the maximum near 414 keV.

It is important to note that degeneracies exist wherein various combinations of external and self-attenuation can produce curves that are similar in appearance.

7 General Topics in Passive Gamma-Ray Assay

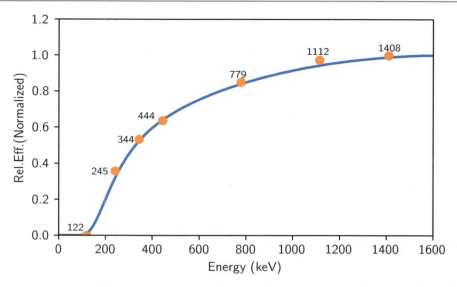

Fig. 7.29 Relative efficiency curve for a ^{152}Eu point source shielded by approximately 50 mm of steel. The curve has been normalized to the maximum efficiency, which for this measurement, occurs for the 1408-keV data point

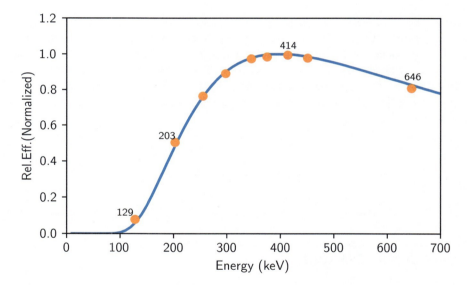

Fig. 7.30 Relative efficiency curve for a 10-gram ^{239}Pu (93%) source in a steel container approximately 2 mm thick, filtered by approximately 1 mm of cadmium. The curve has been normalized to the maximum efficiency, which for this measurement, occurs for the 414-keV data point

7.9.4 Relative Efficiency Curve Functions

When fitting relative efficiency curve data, two general approaches are often used. These approaches can be categorized as empirical forms and physical (or semi-empirical) forms. An example of an empirical form for fitting relative efficiency data from a single nuclide is

$$\ln(\varepsilon_R) = C_1 + C_2 E^{-2} + C_3 \ln(E) + C_4 (\ln(E))^2 + C_5 (\ln(E))^3 \tag{7.71}$$

The C_i factors are empirical fitting coefficients, and the formalism is in the natural logarithm of the relative efficiency to reduce the complexity of fitting what might otherwise be an exponential high-degree polynomial that often results in non-physical oscillatory behavior. Additional coefficients may be added to support the use of multiple nuclides of different concentrations in fitting the relative efficiency data.

Because Eq. 7.71 is a continuous function, it must not be used to fit data that cross the discontinuity created by any absorption edge because those are discontinuous by nature. Typically, the K-shell absorption edge of plutonium at ~121 keV is the highest energy edge that typically must be addressed. Therefore, a good practice is to use an empirical function, such as the above, only for data above this energy.

When the energy range crosses an absorption edge, a model for the relative efficiency function that incorporates attenuation coefficients for both external and self-attenuation must be used. Such an example is shown in Eq. 7.72 for a plutonium item shielded by iron and lead.

$$\varepsilon_R = \varepsilon_I \cdot [e^{-\mu\rho x}]_{Pb} \cdot [e^{-\mu\rho x}]_{Fe} \cdot \left[\frac{1 - e^{-\mu\rho x}}{\mu\rho x}\right]_{Pu} \tag{7.72}$$

The energy dependence is implicit in ε_I and the various mass attenuation coefficients (μ). It is sometimes the case that the intrinsic detector efficiency (ε_I) has been determined ahead of time. In that case, the only fitting parameters in a function such as this are thickness (x) of each of the external and self-attenuation factors. As with the empirical model, additional factors may be included for fitting data with multiple nuclides, etc.

7.9.5 Efficiency Relative to the Accepted Standard Size NaI(Tl) Detector

An alternative definition of the term *relative efficiency* pertains to the efficiency of a detector relative to a standard NaI (Tl) detector at a standard energy. The manufacturers of germanium detectors usually characterize the efficiency of coaxial detectors by comparing the efficiency with the absolute full-energy efficiency of a 3 × 3 in. NaI(Tl) detector. The comparison is usually made at 1332.5 keV with a source-to-detector distance of 25.0 cm and the efficiency expressed as a percentage of the NaI(Tl) efficiency. The efficiency of the germanium detector is measured, and the absolute full-energy efficiency of the NaI(Tl) detector is assumed to be 0.0012 for the stated energy and distance. The expression for computing the measured germanium-detector efficiency is

$$\varepsilon_{RNaI} = \frac{\frac{FEIR(1332.5)}{ER(1332.5)}}{0.0012} 100, \tag{7.73}$$

where FEIR(1332.5) includes corrections for dead-time pileup losses, and ER(1332.5) is the current emission rate.

7.10 Measurement of Nuclide Ratios

The full-energy peak count rate for any single gamma ray can be written as

$$\dot{C}\left(E_j^i\right) = \lambda^i N^i BR_j^i \varepsilon_A\left(E_j\right), \tag{7.74}$$

where

$\dot{C}\left(E_j^i\right)$ = full-energy-interaction rate (FEIR) of gamma ray j with energy E_j emitted from nuclide i
λ^i = decay constant of nuclide i ($\lambda^i = \ln(2)/T^i_{1/2}$, where $T^i_{1/2}$ is the half-life of nuclide i)
N^i = number of atoms of nuclide i
BR_j^i = branching ratio of gamma ray j from nuclide i
$\varepsilon_A(E_j)$ = absolute detection efficiency for full-energy peak detection of gamma ray with energy E_j. Includes detector intrinsic efficiency, geometry, item self-attenuation, and attenuation in materials between the item and detector.

The full-energy-interaction (or count) rate can be written also in terms of the mass of the nuclide as

$$\dot{C}\left(E_j^i\right) = \gamma_j^i M_i \varepsilon_A\left(E_j\right) \qquad (7.75)$$

where

γ_j^i = specific emission rate of gamma ray j from nuclide i in γ/s − g
M_i = mass of nuclide i (g).

These two equations can be used to give expressions for the atom and mass ratios of two nuclides. The atom ratio is given by

$$\frac{N^i}{N^k} = \frac{C\left(E_j^i\right)}{C\left(E_m^k\right)} \times \frac{T_{1/2}^i}{T_{1/2}^k} \times \frac{BR_m^k}{BR_j^i} \times \frac{\varepsilon_R(E_m)}{\varepsilon_R(E_j)} \qquad (7.76)$$

In Eq. 7.76, the absolute efficiency, $\varepsilon_A(E)$, has been replaced by the relative efficiency $\varepsilon_R(E)$ of Sect. 7.9 because the solid angle contribution in the former is common to both nuclides and drops out in the ratio. Also, the count rate for each gamma-ray peak has been replaced by counts because the live time for data acquisition will be the same for both and will also drop out in the ratio. The use of a relative efficiency curve does not correct for coincidence-summing effects.

The similar expression for the mass ratio is

$$\frac{M^i}{M^k} = \frac{C\left(E_j^i\right)}{C\left(E_m^k\right)} \times \frac{\gamma_m^k}{\gamma_j^i} \times \frac{\varepsilon_R(E_m)}{\varepsilon_R(E_j)} \qquad (7.77)$$

7.11 Estimation of Activity and Mass Using Gamma Rays

For situations where source self-attenuation is negligible, gamma-ray analysis can be used successfully to calculate source activity or mass, or both. When self-attenuation is non-negligible, especially if the item being assayed is infinitely thick with respect to the assay energies, gamma-ray analysis can be used only to place a lower bound on the activity or mass.

From Eq. 7.74, the activity $A = \lambda N$ of a nuclide can be written in terms of the count *rate*, branching ratio, and absolute efficiency for a single gamma-ray energy $E\gamma$.

$$Act = \frac{\dot{C}(E_\gamma)}{BR(E_\gamma) \cdot \varepsilon_A(E_\gamma)} \qquad (7.78)$$

A more robust analysis than the above would be to use an error-weighted average of activities, calculated using multiple gamma-ray peaks.

The nuclide mass M in an item can be obtained by scaling the activity by the atomic mass number A, the inverse of Avogadro's number, N_A, of 6.022149E23 atoms/mole, and the inverse of the decay constant $\lambda = \ln(2)/T_{1/2}$.

$$M = \frac{Act}{\lambda} \cdot \frac{A}{N_A} \qquad (7.79)$$

7.11.1 Absolute Efficiency Correction Using Relative Efficiency Analysis

The implicit difficulty in using gamma rays to estimate the activity and mass of an item is the determination of the absolute efficiency curve for cases where attenuation is non-negligible and unknown. We restate Eq. 7.66:

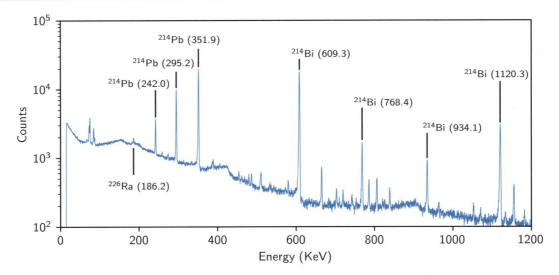

Fig. 7.31 Spectrum of ^{226}Ra shielded by 2–3 mm of lead. The spectrum was recorded with an ORTEC Detective (HPGe). The energy range for analysis has been limited to 186–1120 keV to mitigate uncertainties in detector efficiency over a multiple-MeV range

$$\varepsilon_A(E_\gamma) = \varepsilon_R(E_\gamma) \cdot \frac{\Omega}{4\pi} = T_{Self}(E_\gamma) \cdot T_{Ext}(E_\gamma) \cdot \varepsilon_I(E_\gamma) \cdot \frac{\Omega}{4\pi} \quad (7.80)$$

If the solid angle and intrinsic detector efficiency are known *a priori*, the remaining unknown in the absolute efficiency will depend on any attenuation that is present. The way to determine that attenuation is to conduct a *relative* efficiency (differential attenuation) analysis by first plotting the relative efficiency curve (counts normalized by branching ratio) and then normalizing that by the intrinsic detector efficiency curve. The proper correction for attenuation can be found by determining the attenuating parameters that will account for the residual nonlinearity and flatten this curve. This method will result in the relative efficiency data points being in a straight line with zero slope, which in effect, returns us to the toy model of Fig. 7.25.

This outcome is more easily illustrated with an example. Consider the case of a ^{226}Ra source shielded by a few millimeters of lead. The spectrum, recorded using an ORTEC Detective (HPGe) 25 cm from the source, is shown in Fig. 7.31. Note that the gamma-ray peaks chosen for analysis are from different nuclides in the ^{226}Ra decay chain. Any disequilibrium or independent attenuation of daughter nuclides would be manifested as data points that do not form a smooth relative efficiency curve.

Following Eq. 7.70, we extract the net count rate for several gamma-ray peaks and normalize each by the respective branching ratio to produce the relative efficiency curve shown in Fig. 7.32. Note that this curve has *not* been normalized by the maximum efficiency value because we will be conducting an activity quantification that, unlike an isotopic ratio calculation, is an absolute calculation.

Note that only statistical uncertainties are propagated in this example. The error bars are plotted but in most cases are too small to be visible.

The intrinsic detector efficiency curve we have for this detector was shown in Fig. 7.22. We now use this curve to obtain the intrinsic detector efficiency at the energies of interest in our example, as illustrated in Fig. 7.33.

We now normalize the relative efficiency data points in Fig. 7.32 by the intrinsic detector efficiency at each energy in Fig. 7.33, which results in the curve shown in Fig. 7.34.

Note that the slope of the higher-energy region of the curve in Fig. 7.34 approaches zero, which indicates an approach to a transmission value of unity as energy increases. The deviation from a zero slope at lower energies is mainly from the attenuation of the lead.

For simplicity, we will assume that any self-attenuation of the ^{226}Ra source is negligible. To account for the external attenuation of the lead, we must normalize the data points in Fig. 7.34 by the transmission ($e^{-\mu\rho x}$) through the lead. The only free parameter in our simple example is the thickness (x) of the lead attenuator.

A fitting routine was used to adjust the thickness of the lead until the normalization of the data points in Fig. 7.34 by the transmission through lead produced a flat slope. The resulting thickness of lead for this example was approximately 2.5 mm.

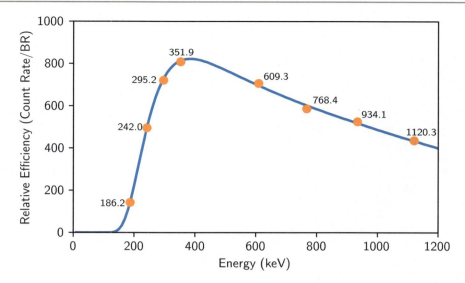

Fig. 7.32 Relative efficiency curve (un-normalized) for ^{226}Ra shielded by 2–3 mm of lead and recorded with an ORTEC Detective (HPGe). The energy range for analysis has been limited to 186–1120 keV to mitigate uncertainties in detector efficiency over a multiple-MeV range. The error bars are included

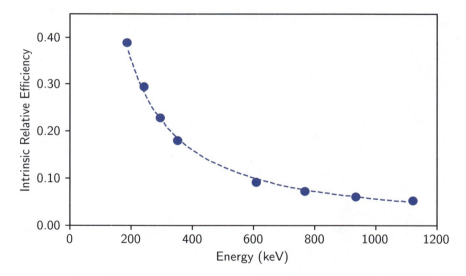

Fig. 7.33 Intrinsic detector efficiency for the energies of interest in our example of ^{226}Ra recorded with an ORTEC Detective (HPGe)

The curve for transmission through this thickness of lead is shown in Fig. 7.35. Note how it essentially follows the shape of Fig. 7.34. Hopefully it is clear that normalizing the relative efficiency curve of Fig. 7.34 by the transmission curve of Fig. 7.35 will produce a curve that is roughly linear with a near-zero slope, as shown in Fig. 7.36.

Finally, to obtain the activity, we normalize the ordinate values from Fig. 7.36 by the solid-angle fraction described in Fig. 7.16. For the measurement distance of 25 cm using an ORTEC Detective (HPGe) with a detector radius of 2.5 cm, the solid-angle fraction is 0.00248 steradians.

The activity as calculated for each data point is shown in Table 7.8.

As shown in Table 7.8 and as one might expect from Fig. 7.36, we obtain very similar activity results for all gamma-ray peaks. This result should be the case because all gamma rays used in this analysis emanate from the same source.

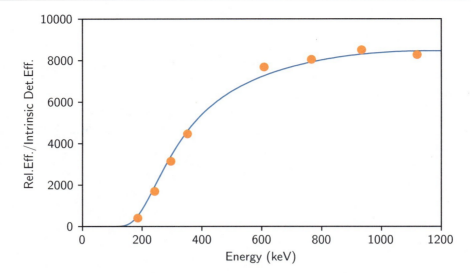

Fig. 7.34 Relative efficiency data for ^{226}Ra of Fig. 7.32 normalized by intrinsic detector efficiency of Fig. 7.33. This effectively represents the transmission

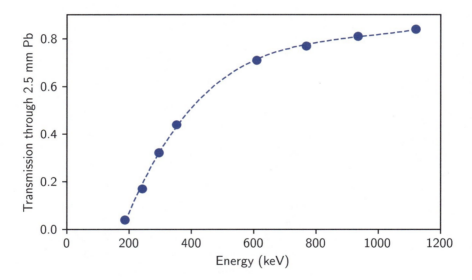

Fig. 7.35 Transmission through approximately 2.5 mm of lead for photon energies over the range of interest

We calculate the error-weighted average and uncertainty of the activity:

$$\overline{Act} = \frac{\sum (Act_i/\sigma_i^2)}{\sum (1/\sigma_i^2)} \text{ and } \sigma_{\overline{Act}} = \sqrt{\frac{1}{\sum (1/\sigma_i^2)}} \quad (7.81)$$

The error-weighted average of these data yield an activity of 4165 ± 9 kBq or approximately 112.6 ± 0.2μCi. The ground truth for the source activity in this case was ~104 ± 10μCi. Although the measurement result is different from the ground truth in a statistically significant manner, this result is still reasonable due to the unincorporated systematic uncertainties with using a generalized intrinsic detector efficiency function, for example. For some applications, an 8% difference from the true value might be unacceptable, but for rough nuclide quantification calculations, such as an initial assessment of interdicted radioactive material at a border crossing where being within a factor of 2 is often acceptable, an 8% difference such as this is quite acceptable.

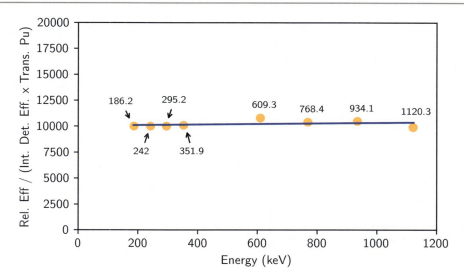

Fig. 7.36 Corrected relative efficiency curve for ^{226}Ra shielded by 2.5 mm of lead. The correction is a normalization by intrinsic detector efficiency and transmission through the lead attenuator

Table 7.8 Activity results for Radium-226 shielded by 2–3 mm of lead and recorded with an ORTEC Detective (HPGe)[a]

Energy	Activity [Bq]	Uncertainty [Bq]
186.2	4.03E+06	4.52E+05
242.0	4.04E+06	6.17E+04
295.2	4.03E+06	2.28E+04
351.9	4.07E+06	1.44E+04
609.3	4.35E+06	1.43E+04
768.4	4.20E+06	5.34E+04
934.1	4.24E+06	8.17E+04
1120.3	4.00E+06	3.02E+04

[a]Data have been corrected for relative efficiency and solid-angle fraction effects

References

1. R. Gunnink, J.E. Evans, A.L. Prindle, A Reevaluation of the Gamma-Ray Energies and Absolute Branching Intensities of ^{237}U, ^{283,239,240,241}Pu, and ^{241}Am. Lawrence Livermore Laboratory report UCRL-52139 (1976).
2. M.E. Anderson, J.F. Lemming, Selected Measurement Data for Plutonium and Uranium. Mound Laboratory report MLM-3009 (1982).
3. *NuDat 2.8*. https://www.nndc.bnl.gov/nudat2/chartNuc.jsp. Accessed Feb 2023.
4. R.D. Evans, *The Atomic Nucleus* (McGraw-Hill Book Co., New York, 1955) Chapters 23–25
5. P.R. Bevington, D. Keith Robinson, *Data Reduction and Error Analysis for the Physical Sciences*, 2nd edn. (McGraw-Hill Book Company, New York, 1992)
6. Y. Beers, *Introduction to the Theory of Error*, 2nd edn. (Addison-Wesley Publishing Co., Inc., Reading, 1957)
7. S.L. Meyer, *Data Analysis for Scientists and Engineers* (Wiley, New York, 1975), p. 37
8. D. Reilly, N. Ensslin, H. Smith Jr., S. Kreiner, *Passive Nondestructive Assay of Nuclear Materials* (PANDA) (U.S. Nuclear Regulatory Commission, January 1, 1991).
9. R.B. Walton, E.I. Whitted, R.A. Forster, Gamma-ray assay of low-enriched uranium waste. *Nucl. Technol.* **24**, 81–92 (1974)
10. T.E. Sampson, T.A. Kelley, D.T. Vo, Application guide to gamma-ray isotopic analysis using the FRAM software. Los Alamos National Laboratory report LA-14018 (2003).
11. R. Gunnink, Computer Techniques for Analysis of Gamma-Ray Spectra. Lawrence Livermore Laboratory report UCRL-80297 (1978).
12. R. Gunnink, J.B. Niday, Computerized Quantitative Analysis by Gamma-Ray Spectrometry, Volume 1, Description of the GAMANAL Program. Lawrence Livermore Laboratory report UCRL-51061, Vol. 1 (March 1972).
13. R. Gunnink, W.D. Ruhter, GRPANL: A Program for Fitting Complex Peak Groupings for Gamma and X-ray Energies and Intensities. Lawrence Livermore Laboratory report UCRL-52917 (January 1980).
14. J.G. Fleissner, R. Gunnink, GRPNL2: An Automated Program for Fitting Gamma and X-ray Peak Multiplets. Monsanto Research Corporation, Mound Facility report MLM-2807 (March 1981).

15. J.G. Fleissner, GRPAUT: A Program for Pu Isotopic Analysis (A User's Guide), Mound Facility report MLM-2799 (January 1981).
16. R. Gunnink, MGA: A Gamma-Ray Spectrum Analysis Code for Determining Plutonium Isotopic Abundances, Volume 1 Methods and Algorithms. Lawrence Livermore National Laboratory report UCRL-LR-103220, Vol. 1; ISPO-317 (1990).
17. R. Gunnink, An algorithm for fitting lorentzian-broadened K-series X-ray peaks of the heavy elements. Nucl. Instrum. Methods **143**, 145–149 (1977)
18. Los Alamos National Laboratory, computer program *PeakEasy V4.99.5*, Release LA-CC-13-040, https://PeakEasy.lanl.gov March 2023).
19. W.C. Johnson, E. Chan, E. Walsh, C. Morte, D. Lee, computer software *InterSpec v. 1.0.8.*, https://www.osti.gov//servlets/purl/1833849. Accessed Feb 2023.
20. G.F. Knoll, *Radiation Detection and Measurement*, 3rd edn. (Wiley, New York, 2000)
21. J.G. Fleissner, C.P. Oertel, A.G. Garrett, A high count rate gamma-ray spectrometer system for plutonium isotopic measurements. J. Nucl. Mater. Manage. **14**, 45–56 (1985)
22. H.H. Bolotin, M.G. Strauss, D.A. McClure, Simple technique for precise determination of counting losses in nuclear pulse processing systems. Nucl. Instrum. Methods **83**, 1–12 (1970)
23. M.W. Enghauser, FRMAC Gamma Spectroscopist Knowledge Guide. Federal Radiological Monitoring and Assessment Center, Sandia National Laboratories Unlimited Release SAND2019-9768 R (2019).
24. B. Barbiellini, P. Genoud, T. Jarlborg, Calculation of positron lifetimes in bulk materials. J. Phys.: Condens. Matter **3**(39), 7631–7640 (1991)
25. J.E. Cline, H.P. Yule (eds.), Technique of gamma-ray detector absolute efficiency calibration for extended sources, in *Proceeding of the American Nuclear Society Topical Conference on Computers in Activation Analysis and Gamma-Ray Spectroscopy*, Mayaguez, Puerto Rico, April 30–May 3, 1978 (Conf. 780421) (1978), pp. 185–196.

Open Access This chapter is licensed under the terms of the Creative Commons Attribution 4.0 International License (http://creativecommons.org/licenses/by/4.0/), which permits use, sharing, adaptation, distribution and reproduction in any medium or format, as long as you give appropriate credit to the original author(s) and the source, provide a link to the Creative Commons license and indicate if changes were made.

The images or other third party material in this chapter are included in the chapter's Creative Commons license, unless indicated otherwise in a credit line to the material. If material is not included in the chapter's Creative Commons license and your intended use is not permitted by statutory regulation or exceeds the permitted use, you will need to obtain permission directly from the copyright holder.

The Measurement of Uranium Enrichment

D. T. Vo

8.1 Introduction

Uranium and plutonium items are present in the nuclear fuel cycle in a wide variety of isotopic compositions, so the isotopic composition of an item is often the object of measurement (see Chap. 9). In this chapter, we consider a special case of isotopic analysis: the determination, by radiation measurement, of the fractional abundance of a specific isotope of an element. This measurement is most often applied to uranium items to establish the fraction of fissile ^{235}U, commonly referred to as the uranium enrichment. The term *enrichment* is used because the fraction of the item that is ^{235}U is usually higher than that in naturally occurring uranium.

Three isotopes of uranium are prevalent in nature (their isotopic atom abundances are shown in parentheses): ^{238}U (99.27%), ^{235}U (0.72%), and ^{234}U (0.006%). The ^{234}U comes from the alpha decay of ^{238}U:

$$^{238}U \xrightarrow{\alpha(4.5e^9 y)} {}^{234}Th \xrightarrow{\beta(24D)} {}^{234m}Pa \xrightarrow{\beta(1.2m)} {}^{234}U \tag{8.1}$$

Other uranium isotopes may be present if the item is reactor produced (recycled uranium); the isotopes include ^{232}U, ^{233}U, and ^{236}U. The ^{235}U *atom* fraction for uranium is defined as follows:

$$E_a(at\%) = \frac{\text{No. of atoms } ^{235}U}{\text{No. of atoms } U} \times 100. \tag{8.2}$$

The enrichment can also be expressed as a *mass* fraction:

$$E_w(wt\%) = \frac{\text{No. of grams } ^{235}U}{\text{No. of grams } U} \times 100. \tag{8.3}$$

The two ways of expressing uranium enrichment are related through the molar masses, $\frac{g}{mol}$, of the various isotopes present. Thus,

$$E_w = \frac{\alpha_{235} A_{235}}{\sum \alpha_x A_x} = E_a \frac{A_{235}}{A_U}, \tag{8.4}$$

where α_x and A_x are the fraction and atomic mass of the uranium isotope x, respectively, and $A_U = \sum \alpha_x A_x$ is the molar mass of the actual uranium composition present.

Los Alamos National Laboratory strongly supports academic freedom and a researcher's right to publish; as an institution, however, the Laboratory does not endorse the viewpoint of a publication or guarantee its technical correctness.

D. T. Vo (✉)
Los Alamos National Laboratory, Los Alamos, NM, USA
e-mail: ducvo@lanl.gov

Uranium enrichments in light-water-reactor fuel are typically in the range of a few percent. Canadian deuterium-uranium reactors use natural uranium, and materials test reactors use uranium with enrichments from almost 20% up to 93%. Determination of uranium enrichment in items is a key measurement for process or product control in enrichment and fuel fabrication plants, and it is very important in international safeguards inspections to verify that uranium stock is being used for peaceful purposes.

Enrichment measurement principles can be used to determine any isotopic fraction if a radiation signature is available and if a few specific measurement conditions are met. The following discussion describes various enrichment measurement techniques and their applications.

8.2 Radiations from Uranium Items

The isotopes of uranium emit alpha, beta, neutron, and gamma radiation. The primary radiation used in passive nondestructive assay of uranium items is gamma radiation, which is usually dominated by emissions from ^{235}U decay. However, in low-enriched-uranium items, X-ray radiation is the most intense component of the emission spectrum. The 185.7 keV gamma ray is the most frequently used signature to measure ^{235}U enrichment. It is the most prominent single gamma ray from any uranium item enriched above the natural ^{235}U level. Table 8.1 lists the most intense gamma rays from uranium isotopes of interest [1] and their progenies. Gamma-ray spectra from uranium items of varying degrees of enrichment are shown in Figs. 8.1 and 8.2 for high- and low-resolution gamma detectors, respectively.

Table 8.1 Intense gamma radiation from uranium isotopes

Isotope	Gamma-ray energy (keV)	Branching ratio [%]	Specific intensity (gamma/s-g of isotope)
^{232}U	57.8	2.00×10^{-1}	1.65×10^{9}
	129.1	6.82×10^{-2}	5.64×10^{8}
	270.2	3.16×10^{-3}	2.61×10^{7}
	327.9	2.83×10^{-3}	2.34×10^{7}
^{233}U	54.7	1.68×10^{-2}	5.99×10^{4}
	97.1	2.03×10^{-2}	7.24×10^{4}
	119.0	3.63×10^{-3}	1.29×10^{4}
	120.8	2.82×10^{-3}	1.01×10^{4}
	146.3	6.50×10^{-3}	2.32×10^{4}
	164.5	6.26×10^{-3}	2.23×10^{4}
	245.3	3.57×10^{-3}	1.27×10^{4}
	291.3	5.25×10^{-3}	1.87×10^{4}
	317.2	7.37×10^{-3}	2.63×10^{4}
^{234}U	53.2	1.23×10^{-1}	2.83×10^{5}
	120.9	3.50×10^{-2}	8.06×10^{4}
^{235}U	143.8	1.10×10^{1}	8.76×10^{3}
	163.4	5.08×10^{0}	4.06×10^{3}
	185.7	5.70×10^{1}	4.56×10^{4}
	202.1	1.08×10^{0}	8.64×10^{2}
	205.3	5.02×10^{0}	4.01×10^{3}
^{236}U	49.5	7.80×10^{-2}	1.87×10^{3}
	112.8	1.90×10^{-2}	4.55×10^{2}
^{238}U	63.3	3.70×10^{0}	4.60×10^{2}
In equilibrium with ^{234m}Pa	92.4	2.13×10^{0}	2.65×10^{2}
	92.8	2.10×10^{0}	2.61×10^{2}
	258.2	7.64×10^{-2}	9.50×10^{0}
	742.8	1.07×10^{-1}	1.33×10^{1}
	766.4	3.17×10^{-1}	3.94×10^{1}
	786.3	5.44×10^{-2}	6.77×10^{0}
	1001.0	8.42×10^{-1}	1.05×10^{2}
^{228}Th	238.6	4.35×10^{1}	5.41×10^{3}
In equilibrium with ^{208}Tl	583.2	3.06×10^{1}	3.81×10^{3}
	2614.5	3.59×10^{1}	4.46×10^{4}

8 The Measurement of Uranium Enrichment

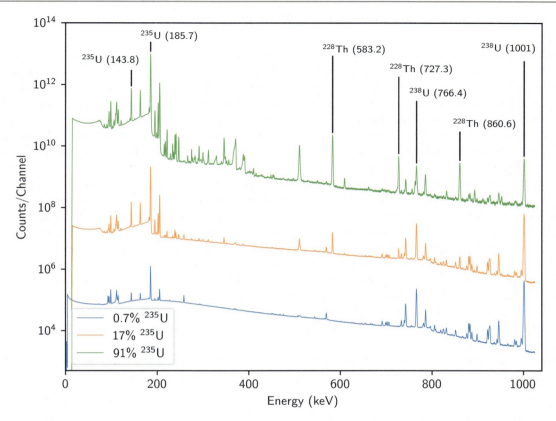

Fig. 8.1 Gamma-ray spectra from natural (bottom spectrum), 17% enriched (middle spectrum), and 91% enriched (top spectrum) uranium, measured with a 25% efficiency coaxial high-purity germanium detector. The spectra are shifted vertically for clarity. The peaks labeled ^{238}U are from the decay of ^{234m}Pa; the peaks labeled ^{228}Th are from the decay of its progenies. Note the dominance in the spectrum of the 186 keV peak from ^{235}U decay

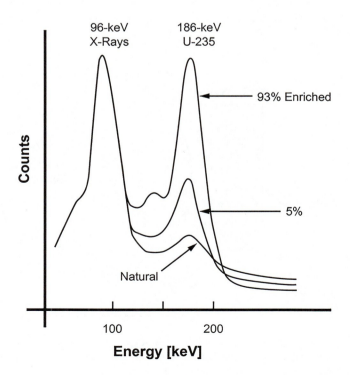

Fig. 8.2 Gamma-ray spectra from natural, 5% enriched, and 93% enriched uranium items, measured with a NaI(Tl) scintillation detector. As the ^{235}U enrichment increases, the 186 keV peak becomes more intense, and the background continuum (from the ^{238}U daughters) above the peak energy becomes weaker. (Figure from Ref. [2])

8.3 Infinite-Thickness Gamma Measurement Technique

The principle of gamma-ray uranium enrichment measurement states that under identical measurement conditions and geometry, the ^{235}U enrichment is proportional to the ^{235}U 186 keV peak rate. For measurements that are not identical, corrections can be applied to correct for the differences between measurement conditions and geometry. This enrichment measurement method is referred to as the *infinite thickness measurement* and is also known as the enrichment meter technique.

The principles of gamma-ray uranium enrichment measurement [3–5] were first applied to the measurement of UF$_6$ cylinders [6]. The basic measurement procedure involves viewing a uranium item through a collimated field of view with a gamma-ray detector (Fig. 8.3). The enrichment is deduced by comparing the 186 keV peak rate with a known reference rate. If the uranium item is large enough, the 186 keV gamma rays from only a fraction of the total item reach the detector because of the strong absorption of typical uranium-bearing materials at this energy. This "visible volume" of the item is determined by the collimator, the detector geometry, and the mean free path of the 186 keV radiation in the item material. Its size (illustrated in Fig. 8.3 by the dashed lines) is independent of the enrichment because the different uranium isotopes all have the same attenuation properties. If the depth of the item along the collimation axis is much larger than the mean free path of 186 keV photons in the item material, all items of the same physical composition present the same visible volume to the detector—called the *infinite-thickness* criterion. Gamma rays that originate at depths greater than the infinite thickness inside the item material have a very low probability of emerging out of the item material. Table 8.2 lists the mean-free-path and infinite-thickness values for the 186 keV gamma ray in commonly encountered uranium compounds. For many common uranium materials, the infinite-thickness criterion is satisfied with quite modest item sizes. However, because we see no deeper into the item than certain distances, as indicated in Table 8.2, gamma-ray-based enrichment measurements often interrogate only the surface of the uranium material. Then, for enrichment measurements to be meaningful for the entire item, the material must be isotopically uniform.

This technique measures the 186 keV peak area to determine the enrichment of the uranium item. It does not depend on peaks of ^{238}U or its progenies, so this technique can measure fresh uranium as well as aged uranium.

8.3.1 Uranium and Matrix Material

For a given detector/collimator geometry, all items of pure uranium metal have identical visible volumes because the mean free path of the 186 keV gamma ray is the same for each item. As a result, the detector views ^{235}U radiation from the same amount of total uranium regardless of the size of the metal item. Because the 186 keV intensity, although heavily absorbed, is still proportional to the number of ^{235}U atoms in the visible volume, it is proportional to the atom enrichment of the item.

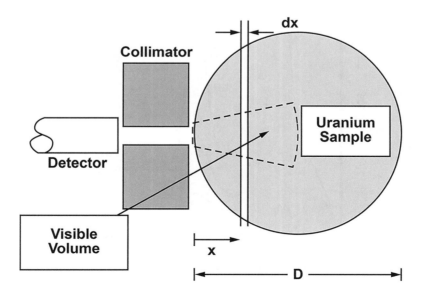

Fig. 8.3 The basic elements of a gamma-ray uranium-enrichment measurement setup. For purposes of illustration, the size of the visible volume compared with the detector and collimator is exaggerated. Normally the depth of the visible volume is much smaller than the source-to-detector distance

Table 8.2 Mean free paths and infinite thicknesses for 186 keV photons in uranium compounds

Uranium compound	Density (ρ) (g/cm^3)	Mean free path (cm)[a]	Infinite thickness (cm)[b]
Metal	18.7	0.04	0.26
UF$_6$ (solid)	4.7	0.20	1.43
UO$_2$ (sintered)	10.9	0.07	0.49
UO$_2$ (powder)	2.0	0.39	2.75
U$_3$O$_8$ (powder)	7.3[c]	0.11	0.74
Uranyl nitrate	2.8	0.43	3.04

[a] Equal to $1/\mu\rho$ at 186 keV for the material in question
[b] Defined as 7 mean free paths, the distance for which the error in assuming infinite-item size is less than 0.1% (see Eq. 8.8)
[c] Highly packed powder

The prototypical enrichment item consists of uranium and a (usually low-Z) matrix material instead of pure uranium metal. The measurement geometry is still the same as that shown in Fig. 8.3, but the absorption by the matrix material is an added factor in the measurement. Exhaustive summaries of the theory of this type of measurement have been published [7, 8]. Following is a summary of the key mathematical results necessary to analyze enrichment measurements.

Consider a gamma-ray measurement on a two-component item of thickness (D), where the item-to-detector distance is large compared with the depth of the visible volume. This feature permits the neglect of $1/r^2$ effects during integration over the item volume. The counting rate from an infinitesimal section of the item (see Fig. 8.3) is given by

$$dR = \varepsilon E_w S A \rho_U \, dx \, \exp(-\mu\rho x)\exp(-\mu_c\rho_c t_c), \tag{8.5}$$

where
ε = detection efficiency at the assay energy
E_w = uranium enrichment (*mass* percent; see Eq. 8.3)
S = specific activity of the gamma ray (185.7 keV; see Table 8.1)
A = collimator channel area
ρ_U = uranium density
$\mu_c\rho_c$ = linear photon absorption coefficient of the item container at the assay energy
t_c = single-wall thickness of the item container.

The quantity $\mu\rho$ represents the linear photon absorption coefficient of the combined uranium (U) and the matrix (m) at the assay energy,

$$\mu\rho = \mu_U\rho_U + \mu_m\rho_m. \tag{8.6}$$

Integration of Eq. 8.5 over the item thickness gives the total 186 keV count rate,

$$R = \varepsilon E_w S A \rho_U \exp(-\mu_c\rho_c t_c) \int_0^D \exp(-\mu\rho x)dx, \tag{8.7}$$

which reduces to

$$E_w = \left[\frac{\mu_U}{\varepsilon S A}\right] R \left[\frac{F\exp(\mu_c\rho_c t_c)}{1 - \exp(-\mu\rho D)}\right], \tag{8.8}$$

where

$$F = 1 + \left(\frac{\mu_m\rho_m}{\mu_U\rho_U}\right). \tag{8.9}$$

Table 8.3 Material composition correction factors (F/F_s)

Nuclear material of calibration standards (Factor F_s)	Nuclear material of items measured (Factor F)						
	U	UC	UC_2	UO_2	U_3O_8	UF_6	Uranyl nitrate
U (100% U)	1.000	1.004	1.009	1.011	1.015	1.039	1.056
UC (95% U)	0.996	1.000	1.004	1.007	1.011	1.034	1.051
UC_2 (91% U)	0.991	0.996	1.000	1.003	1.007	1.030	1.047
UO_2 (88% U)	0.989	0.993	0.997	1.000	1.004	1.027	1.044
U_3O_8 (85% U)	0.985	0.989	0.993	0.996	1.000	1.023	1.040
UF_6 (68% U)	0.963	0.967	0.971	0.974	0.977	1.000	1.016
Uranyl nitrate (60% U)	0.947	0.951	0.955	0.958	0.962	0.984	1.000

If the item thickness (D) is large enough, then the exponential in the denominator of Eq. 8.8 is negligible compared with 1, making variations in item dimensions unimportant. This is the origin of the infinite-thickness criterion. The first bracket in Eq. 8.8 contains factors that depend only on the instrument properties (ε and A) and the intrinsic properties of uranium (μ_U and S) and thus constitutes the basic calibration constant of the measurement. If the unknown items and calibration standards have identical containers, then the factor $\exp(\mu_c \rho_c t_c)$ can be absorbed into the calibration constant; otherwise, the factor must be used explicitly to correct for container attenuation (see Sect. 8.7).

The factor (F) in Eq. 8.8 reflects the matrix effects. For uranium metal, no other matrix material is present; therefore, its attenuation coefficient and density can be set to zero. If the calibration standards and the unknown items have the same matrix properties, then this factor can also be included in the calibration constant. If the item matrix factor (F) differs from the calibration matrix factor (F_s), then a small correction is also necessary for this difference. Table 8.3 gives values for this multiplicative correction (F/F_s) for various uranium compounds [2].

The main disadvantage of this technique—besides the infinite thickness and full field-of-view requirements—is that it requires knowledge of basically every parameter of the item, including the material composition, the container thickness, and the container density.

8.3.2 Instrumentation and Infinite-Item Technique: 2-ROI Method

The enrichment is proportional to the peak rate, which is the peak area divided by the acquisition live time. For a clean peak, such as the 186 keV peak of a high-purity germanium (HPGe) spectrum, a simple region of interest (ROI) on the peak would produce the needed net peak area. On the other hand, the 186 keV peak of a NaI spectrum is not clean and requires sophisticated software to fit and unravel the peak components to obtain the 186 keV peak area. Without the fitting software, an ROI method with two different calibration standards can be used to determine the enrichment of an item.

The basic measurement apparatus is a collimated gamma-ray detector and its associated electronics. Gamma-ray spectra are analyzed with two simple ROIs, with one ROI set on the 186 keV energy region (C1 in Fig. 8.4) and the second ROI set on a background continuum region above the assay peak energy (C2 in Fig. 8.4).

The uranium enrichment is proportional to the net 186 keV count rate (R in Eq. 8.8), which is given by

$$R = C1 - f\, C2. \tag{8.10}$$

This equation represents the subtraction of a background from the gross rate in the chosen 186 keV peak energy region. The major contribution to the background continuum comes from the higher-energy gamma rays of ^{238}U daughters that Compton scatter in the detector. Even though C2 is not actually in the assay energy region, it represents the background under the assay region to within a scale factor (f, to be determined by calibration). Because the enrichment (either atom or mass) is proportional to the net rate (R), we have

$$E = a\, R\, F\, \exp(\mu_c \rho_c t_c) = F\, \exp(\mu_c \rho_c t_c)(a\, C1 - a f\, C2). \tag{8.11}$$

The calibration constant (a) contains all of the geometric factors and the intrinsic uranium constants in Eq. 8.8. The matrix factor (F) and the container wall attenuation correction factor $\exp(\mu_c \rho_c t_c)$ have been displayed explicitly to emphasize their roles when standards and unknowns are made of different types of materials or packaged in different containers. If the

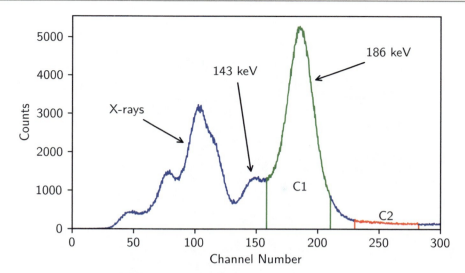

Fig. 8.4 A low-resolution uranium gamma-ray spectrum showing the two energy regions used in the enrichment measurement

measurement is performed on materials of the same type that are packaged in the same container, then $F \exp(\mu_c \rho_c t_c)$ can be included in the calibration constant. The enrichment is then written in terms of the measured data ($C1$ and $C2$):

$$E = a\,C1 + b\,C2. \tag{8.12}$$

The calibration constants a and b ($= -af$) now include the container attenuation and the matrix factor and are determined by measurement of two standards of known enrichments (E_1 and E_2).

Typical instrumentation [9] employs a portable, battery-powered, microprocessor-based multichannel analyzer with the NaI detector (Fig. 8.5). More recently, other detector types, such as the LaBr$_3$ and CZT, are increasingly being used. The instrument acquires a full uranium spectrum, integrates the counts in selected regions of interest that correspond to the count windows $C1$ and $C2$ (for example, as in Fig. 8.4), computes the enrichment and its statistical uncertainty, and presents the results on an alphanumeric display. The two-parameter, two-standard calibration procedure is also incorporated into the instrument software [10]. As with earlier measurements, many current applications still use NaI scintillation detectors [6]. However, when measuring the enrichment of recycled uranium samples (that have been through the nuclear fuel cycle), higher-resolution spectrometry with HPGe, CZT, or LaBr3 detectors is more effective in avoiding problems of interference from the 238.6 keV peak of ^{212}Pb, which is a ^{232}U progeny. Additionally, the higher-resolution detector is especially helpful where chemical processes have concentrated ^{238}U daughters in the deposit or in the uranium material itself. Some commercial processes have been observed to produce up to a ten-fold concentration of ^{238}U daughters. The radiation from the daughters produces a high Compton background in the detector, which can complicate the evaluation of the 185.7 keV peak area.

The 2-ROI technique with the NaI detector has one serious flaw: it assumes that the net area obtained in the procedure is from the full-energy interaction of ^{235}U gamma rays in the 186 keV region. This assumption is not quite correct. The net area also includes the counts from the low-angle, scattered gamma rays of the 186 keV gamma ray. The thicker container or filter will produce a lower-intensity, 186 keV peak with additional scattered gamma rays. The technique can correct for the transmission through the container and filter but cannot correct for the scattered gamma rays. When measuring an item with container thickness, bias will be introduced very differently from the container and filter thickness of the calibration standards. For this technique, it is suggested that the calibration should be done with the total container and filter thickness about the same as the thickness of the item to be measured. Then the scattered component from the calibration and measurement will cancel out, minimizing the bias from the scattered gamma rays.

Fig. 8.5 Gamma-ray uranium enrichment measurement equipment, including a portable, microprocessor-based multichannel analyzer and a NaI (Tl) gamma-ray detector incased in a collimator to define the field-of-view. This instrumentation is battery powered and is suited for mobile field applications

8.3.3 Instrumentation and Infinite-Item Technique: Peak-Fitting Method

The measurement system for this technique is the same as the one for the 2-ROI method: a collimated gamma-ray detector and its associated electronics. The 186 keV peak area can be determined by fitting the 186 keV region [11]. The 186 keV structure consists mainly of the three major peaks of ^{235}U (at 163.4, 185.7, and 205.3 keV), the background (continuum) from the Compton scattering of the higher-energy ^{238}U peaks, and the low-angle scatter gamma rays of the ^{235}U peaks in the structure. After the peak area is determined, the peak rate (R) is obtained by dividing the peak area by the measurement live time. This technique requires only one uranium standard to calibrate the system.

This fitting technique can analyze the spectra taken by a system designed for the 2-ROI method, such as that shown in Fig. 8.6. Modern instruments include a very portable, NaI or LaBr detector system shown in Fig. 8.7. This multipurpose system can perform enrichment measurements in addition to its capabilities to detect and locate, to determine the dose rate, and to identify radioactive sources. The enrichment is determined using the infinite thickness with peak-fitting technique. The software package "NaIGEM" (NaI Gamma Enrichment Measurement; Refs. [11, 12]), which is embedded into the instrument, is used to fit the 186 keV peak and the scattered components of the gamma-ray spectrum.

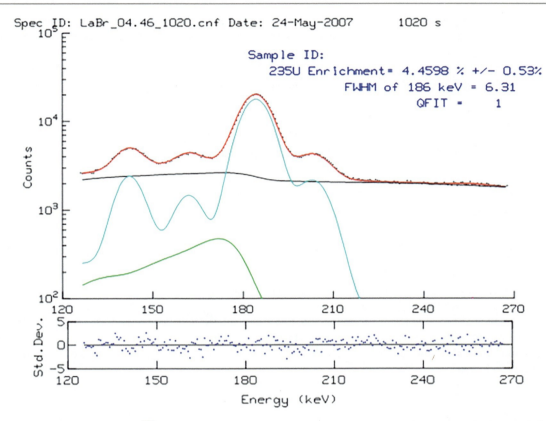

Fig. 8.6 The black data points show the ^{235}U gamma ray response recorded with a LaBr detector. The red curve is the overall fit to the whole 186 keV region, the cyan curve is a fit to the ^{235}U peaks, the black curve is a fit of the background, and the green curve is a fit of the scatter gamma rays. Photo is a screenshot from the NaIGEM software [11]

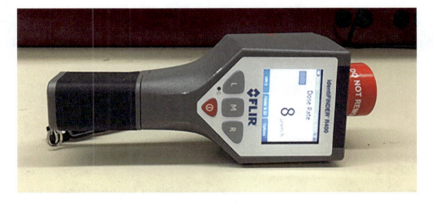

Fig. 8.7 HM-5 unit from FLIR. The unit can detect, locate, measure, and identify radioactive sources and determine the enrichment of uranium using the infinite thickness with peak-fitting method

8.4 Peak-Ratio Technique

For arbitrary, noninfinite uranium items (for example, thin foils, contamination deposits, or dilute solutions), it is difficult to correct the 186 keV gamma intensity for absorption-to-yield enrichment. This difficulty arises because the factor $[(1 - \exp(-\mu\rho D)]$ in Eq. 8.8 is difficult to estimate. The peak-ratio technique measures the areas of many peaks of different isotopes present in the sample and, from that, determines the ratios of uranium isotopes. It then uses the isotopic ratio information to determine the uranium composition. The technique needs to obtain the areas of many peaks, which requires detectors with very good energy resolution, such as the HPGe detectors. Some medium-resolution detectors, such as the LaBr$_3$ detector or the large CZT detector, can also be used with sophisticated software that can extract peak areas.

The technique is based on the ratios of same terms of different isotopes. (See Chap. 7, Sect. 7.10 for detailed discussion of the technique.) In principle, for a sample with a homogeneous isotopic composition, the atom ratio of two isotopes is given by the following equation:

$$\frac{N^i}{N^k} = \frac{C(E_j^i)}{C(E_m^k)} \times \frac{T_{1/2}^i}{T_{1/2}^k} \times \frac{BR_m^k}{BR_j^i} \times \frac{RE(E_m)}{RE(E_j)}, \tag{8.13}$$

where

$C\left(E_j^i\right)$ = photopeak area of gamma ray j with energy E_j emitted from isotope i

$T_{\frac{1}{2}}^i$ = half-life of isotope i

N^i = number of atoms of isotope i

BR_j^i = branching ratio (gamma rays/disintegration) of gamma ray j from isotope i

$RE(E_j)$ = relative efficiency for photopeak detection of gamma ray with energy E_j.

The half-lives $T_{1/2}$ and the branching ratios BR are known, published nuclear data. The photopeak counting intensity $C(E)$ is determined from the gamma-ray spectrum of the measured item.

After the isotopic ratios are measured, the isotopic fractions can be obtained. The sum of all isotopic fractions must equal unity. Neglecting ^{236}U, then

$$1 = f_{234} + f_{235} + f_{238}, \tag{8.14}$$

where f_i is the isotopic fraction of isotope i.

Dividing Eq. 8.14 by f_{235} and inversing, then

$$f_{235} = \left[\frac{f_{234}}{f_{235}} + 1 + \frac{f_{238}}{f_{235}}\right]^{-1}. \tag{8.15}$$

After the isotopic fraction f_{235} of ^{235}U is determined, the isotopic fraction of ^{234}U and ^{238}U are obtained from

$$f_i = f_{235} \times \left[\frac{f_i}{f_{235}}\right], i = 234, 238. \tag{8.16}$$

8.4.1 Determination of ^{236}U Concentration

Uranium-236 has only a couple of weak gamma rays, which are not visible in spectra of typical uranium items. Its content can be estimated using the empirical isotopic correlation with the other isotopic fractions. The correlation may have the form

$$^{236}U = A \times \left[\left(^{235}U\right)^B \times \left(^{238}U\right)^C \times \left(^{234}U\right)^D\right]. \tag{8.17}$$

The gamma rays of ^{234}U are generally weak, and the ^{234}U fraction normally has large uncertainty, which would significantly increase the ^{236}U uncertainty; therefore, the correlation may be simplified and made more robust by removing ^{234}U from the correlation. Its form will then become

8 The Measurement of Uranium Enrichment

Table 8.4 Decay characteristics for uranium isotopes and their progenies

Isotope	Half-Life (year)	Activity (dis/s − g)	Specific power (mW/g Isotope)
^{232}U	68.9 ± 0.4	8.274×10^{11}	717.6 ± 4.2
^{233}U	$(1.592 \pm 0.002) \times 10^5$	3.565×10^8	0.28039 ± 0.00036
^{234}U	$(2.455 \pm 0.006) \times 10^5$	2.302×10^8	0.17925 ± 0.00044
^{235}U	$(7.038 \pm 0.005) \times 10^8$	7.996×10^4	$(6.0959 \pm 0.0044) \times 10^{-5}$
^{236}U	$(2.342 \pm 0.004) \times 10^7$	2.393×10^6	$(1.7532 \pm 0.0030) \times 10^{-3}$
^{238}U	$(4.468 \pm 0.003) \times 10^9$	1.244×10^4	$(1.0220 \pm 0.0053) \times 10^{-5}$
^{228}Th	1.9125 ± 0.0009	3.033×10^{13}	$(1.6222 \pm 0.0008) \times 10^5$

$$^{236}U = A \times \left[\left(^{235}U\right)^B \times \left(^{238}U\right)^C \right]. \tag{8.18}$$

After the isotopic fraction of ^{236}U has been determined using the correlation, the other isotopic fractions can be corrected using

$$f_i^c = \frac{f_i}{1 + f_{236}}, i = 234, 235, 236, 238, \tag{8.19}$$

where f_i^c are the normalized isotopic fractions, including f_{236}^c. This correction renormalizes the fractions so that the sum over all uranium isotopes equals unity.

8.4.2 Decay Characteristics of Uranium Isotopes

All uranium items contain the isotopes ^{234}U, ^{235}U, and ^{238}U. Recycled uranium also contains ^{232}U, ^{233}U, ^{236}U, and ^{228}Th, which is the daughter of ^{232}U decay. Table 8.4 lists some of the decay characteristics of these important isotopes [1].

8.4.3 U-238 Secular Equilibrium

Uranium-238 decays to ^{234}Th, which then decays to the ^{234}Pa isomer ^{234m}Pa (1.17 m), which then quickly reaches secular equilibrium with ^{234}Th. Thorium-234 has a half-life of 24.1 days and will come into secular equilibrium with its ^{238}U parent in about 169 days (7 half-lives). After that, the gamma rays from the decay of ^{234}Th and ^{234}Pa can be considered to come from ^{238}U and can be used as a measure of ^{238}U. A uranium item that is chemically separated less than 169 days is considered fresh, and the gamma rays from ^{234}Th and ^{234}Pa cannot be used as a direct measure of ^{238}U. The amount of ^{238}U determined using those gamma rays will appear too small relative to those from ^{235}U, making the enrichment appear too large. The correct enrichment can still be obtained by applying correction using the time since chemical separation.

8.4.4 Spectra Interferences

Uranium enriched from uranium ore has only three uranium isotopes: ^{234}U, ^{235}U, and ^{238}U. Recycled uranium, which comes from reprocessing spent fuel, contains some additional uranium isotopes: ^{232}U, ^{233}U, and ^{236}U. Uranium-232 has a relatively short half-life (68.9 year) and quickly decays to ^{228}Th. Thorium-228 content will gradually increase until secular equilibrium with the parent ^{232}U is reached after slightly more than 13 year. Many enriched uranium items contain some fraction of recycled uranium, which comprises ^{228}Th. Therefore, ^{228}Th gamma rays can be used in the analysis to determine the isotopic composition of the material.

Soil, concrete—virtually everything—contains a small amount of ^{232}Th. Thorium-232 decays to ^{228}Ra, which then decays to ^{228}Ac. Actinium-228 then decays to ^{228}Th. Thus, the background radiation always has gamma rays emitted by ^{228}Th. For an unshielded or lightly shielded detector, gamma rays from ^{228}Th in the background can become significantly large and interfere with ^{228}Th gamma rays emitted from the uranium item, biasing the isotopic composition results or forcing the analysis to exclude the use of the ^{228}Th gamma rays.

8.4.5 Summing Effects

Many gamma rays from the decay of the ^{238}U daughter ^{234m}Pa are in coincidence with some other gamma rays [13]. When two gamma rays enter the detector at the same time, the resulting pulse from the detector will not correspond to the energy of either gamma ray but rather the sum of the two energies deposited by these gamma rays. Counts that contribute this sum are lost from the individual peaks.

Some of gamma rays and X-rays from ^{235}U decay also suffer similar summing [14]. Uranium-235 decays to ^{231}Th and releases gamma rays and intrinsic thorium X-rays. These intrinsic thorium X-rays, which follow internal conversion in the decay of ^{235}U, are in coincidence with many gamma rays from the decay. As a result, the thorium X-rays suffer coincidence summing and will record fewer counts in their respective peak areas.

To obtain accurate results, uranium analysis that employs the thorium X-rays or the ^{234m}Pa peaks that are affected by coincidence summing will need to be corrected for the loss of counts in the peaks.

8.4.6 Analytical Regions

An enrichment can be obtained by analyzing the region at 89–98 keV of the uranium spectrum [15]. This region contains the pair of gamma ray peaks at 92.4 and 92.8 keV from ^{238}U daughter ^{234}Th and the intrinsic thorium X-ray peak at 93.4 keV from ^{235}U decay. From these peaks, the ^{238}U/^{235}U ratio is obtained. The nearly identical energy of the peaks minimizes uncertainties in the energy-dependent relative efficiencies. The enrichment is calculated from the ^{238}U/^{235}U ratio by assuming that the uranium consists of only ^{235}U and ^{238}U.

$$E_a = f_{235} = \frac{N_{235}}{(N_{235} + N_{238})} = \left(1 + \frac{f_{238}}{f_{235}}\right)^{-1} \tag{8.20}$$

For full isotopic composition, a wider energy region is needed to measure many peak areas of many different isotopes [13, 16].

8.4.6.1 The 60–240 keV Region Analysis

This low-energy region is used mainly with spectra obtained by a planar HPGe detector, though a good semi-planar or coaxial HPGe detector with less than 0.7 keV full width at half maximum at 84 keV can be used. This region, shown in Fig. 8.8, includes peaks of ^{234}U, ^{235}U, and ^{238}U. Because of its low energy, the 63.3 keV peak of ^{238}U often is not visible when the material is moderately shielded, so a software code might not use this peak in its analysis.

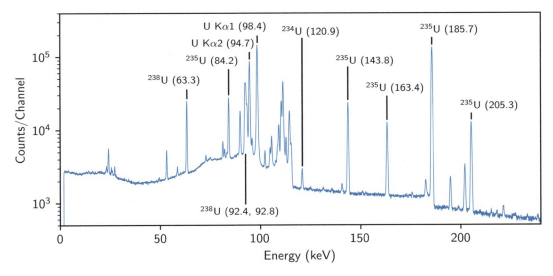

Fig. 8.8 Gamma-ray spectrum of a 4.46% ^{235}U item in the 60–240 keV region measured with a planar HPGe detector with resolution of 520 eV at 122 keV

8 The Measurement of Uranium Enrichment

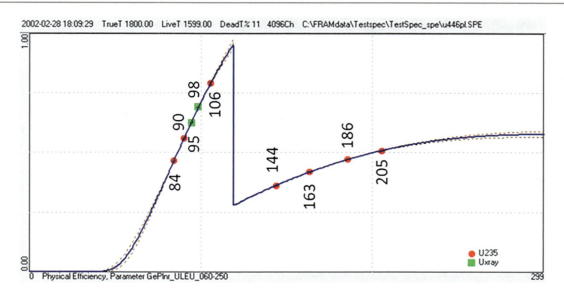

Fig. 8.9 An example of the relative efficiency curve of the analysis that employs the 60–240 keV energy region taken from the FRAM software

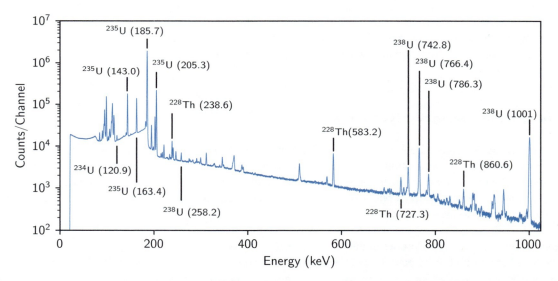

Fig. 8.10 Gamma-ray spectrum of a 20.1% ^{235}U item measured with a coaxial HPGe detector with resolution of 1.0 keV at 186 keV

Figure 8.9 shows the relative efficiency curve, determined by the FRAM software, from the main peaks in the 60–240 keV region.

8.4.6.2 The 120–1010 keV Region Analysis

This energy region (entirely above the complex X-ray region) is used mainly with spectra obtained by a coaxial HPGe detector. In situations where a coaxial HPGe detector is not available and the thick shielding prevents the low-energy peaks in the 100 keV region from getting out, a planar HPGe detector can be used to measure the peaks in this high-energy region, albeit with a longer measurement time due to small efficiency at high energy. This region, shown in Fig. 8.10, includes peaks of ^{234}U, ^{235}U, ^{238}U, and ^{228}Th. All of the intense ^{235}U peaks are below 206 keV, whereas most of the intense peaks of ^{238}U are above 740 keV. Uranium-238 has a relatively weak peak at 258 keV, connecting the low-energy region of the spectrum that consists of peaks from ^{235}U with the high-energy region that consists of peaks from ^{238}U. This 258 keV peak is sufficiently

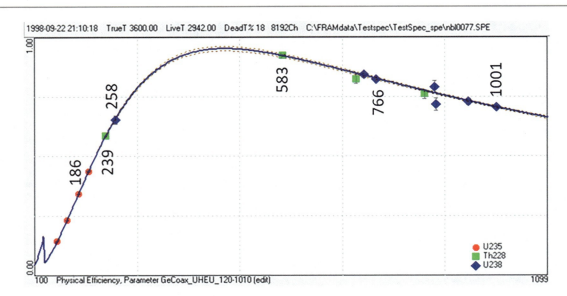

Fig. 8.11 An example of the relative efficiency curve of the analysis that employs the 120–1010 keV energy region taken from the FRAM software

intense for low-enriched uranium, but for highly enriched uranium, its intensity is weak; and for very highly enriched uranium, it may not even be present above the continuum. The peaks of ^{228}Th with energies at 239, 583, 727, and 861 keV then can help bridge the wide gap between the low-energy ^{235}U peaks and the high-energy ^{238}U peaks.

Figure 8.11 shows the relative efficiency curve, determined by the FRAM software, from the main peaks in the 120–1010 keV region.

8.4.7 Summary of Peak-Ratio Technique

The peak-ratio method described above has the advantage that the uranium enrichment can be determined without the use of enrichment standards or the determination of geometry-dependent calibration constants. In addition, the items do not need to satisfy the infinite-thickness criterion. Furthermore, the relative efficiency corrections are made for each item and include not only the absorption by the item material but also that by the item container and any external absorbers. The disadvantages of this technique are related to the following:

- The need for secular equilibrium between the ^{238}U and its daughters
- The need for a high-resolution detector (planar HPGe for X-ray region analysis and coaxial HPGe for higher-energy region analysis); advanced software code [16] allows the use of the technique with a medium-resolution detector, such as the large CZT detector or the LaBr$_3$ detector.
- The need for isotopic homogeneity in the item

The need for isotopic homogeneity manifests itself in cases where residual material from other sources could be in the container with the material currently being measured—for example, in the measurement of UF$_6$ cylinders in which uranium and/or its progenies from previous shipments could have deposited on the walls of the cylinder.

8.5 Visual Estimation of Enrichment

The peak ratio technique that employs the X-ray region uses the 92.4 and 92.8 keV peaks of ^{234}Th as a measure of the ^{238}U concentration of the item and could use the 93.4 keV thorium K$_{\alpha 1}$ peak alone or in addition to other peaks as a measure of the ^{235}U concentration. It just happens that the peak height relationship between the 93.4 keV thorium K$_{\alpha 1}$ peak and the ^{234}Th 92.4 keV peak can be used to estimate the ^{235}U enrichment. The relationship is simple and easy to remember. The enrichment in percent is expressed as

8 The Measurement of Uranium Enrichment

$$E \approx 10 \frac{H_{93.4}}{H_{92.4}} \quad (8.21)$$

where H is the height of the peak above the background. The background is a point in the middle of the valley on the left of the 92.4 keV peak. Figure 8.12 shows the 93 keV region of the spectra, as measured by a planar HPGe detector, where the horizontal lines are drawn through the counts at 91.1 keV (background), 92.4 keV (^{238}U), and 93.4 keV (^{235}U) to guide the eye.

By just eyeing the spectrum and mentally comparing the net height at 93.4 with the height at 92.4, a user can often estimate the enrichment of the item to within 20% of the actual value. If the user can access the spectrum, then the counts at 91.1, 92.4, and 93.4 can be obtained. The enrichment can then be simply calculated using the Eq. 8.21. The enrichment determined by this method can be accurate to within several percent for enrichment in the range of ~3%–50% ^{235}U. For enrichment outside the range, the technique still works, but the bias can be much larger than several percent. For the four spectra shown in

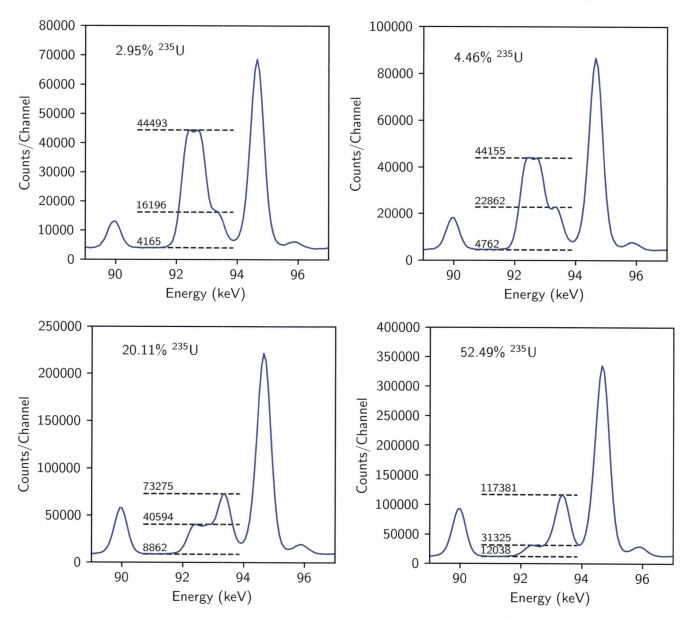

Fig. 8.12 The 93 keV region of the spectra as measured by an HPGe detector. The horizontal lines are drawn through the counts at 91.1 keV (background), 92.4 keV (^{238}U), and 93.4 keV (^{235}U) to guide the eye. The corresponding counts at 91.1, 92.4, and 93.4 keV are displayed above the lines

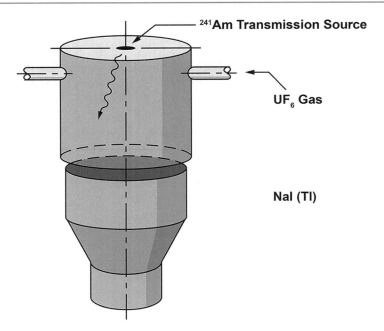

Fig. 8.13 The NaI(Tl)-based, gas-phase UF$_6$ enrichment monitor. The NaI detector views 60 keV gamma rays from the ^{241}Am source (for transmission measurement) and 186 keV gamma rays from the item chamber (for ^{235}U determination). (Figure from Ref. [18])

Fig. 8.12, the calculations using Eq. 8.21 give the enrichment results of 2.98%, 4.60%, 20.30%, and 50.81% ^{235}U, respectively. These values are reasonably close to the known values for these items at 2.95%, 4.46%, 20.11%, and 52.49% ^{235}U, respectively.

8.6 Gas-Phase Uranium Enrichment Measurement Techniques

An extreme case of performing enrichment measurements on a noninfinite item is the measurement of UF$_6$ in the gaseous phase. In one technique [17–20], the ^{235}U concentration was determined from a measurement of the 186 keV gamma-ray emission rate (R) from the decay of ^{235}U, and the total uranium concentration was determined by measuring the transmission (T$_{60}$) through UF$_6$ gas of 60 keV gamma rays from an external ^{241}Am source. Figure 8.13 shows the measurement system, with the orientation of the NaI(Tl) detector and the item chamber accompanied by the location of the ^{241}Am transmission source. The (atomic) enrichment (E$_a$) was related to the measured count rate (R) of the 186 keV rays by

$$E_a = {}^R\!/_{(C\,\ln(T_{60}))}, \tag{8.22}$$

where R was corrected for dead-time losses and attenuation in the gas, C was a calibration constant, and ln(T$_{60}$) was proportional to the total uranium in the item. Because the measurement accounted for variations in UF$_6$ density, the measured assay was independent of the UF$_6$ pressure. This method produced assay results with measured accuracies better than 1% relative over the range of UF$_6$ enrichments of 0.72%–5.4%, using a single-point calibration. For 1.0% enriched UF$_6$ at 700 torr, a 0.74% relative precision was obtained for a 1000 s counting time [19]. This technique was applied at relatively high UF$_6$ pressures, so the data signals were dominated by radiation from the UF$_6$ gas, making interferences from uranium deposits on the inner surface of the item chamber unimportant. In 1982, a NaI(Tl) gamma-ray detector was used during test and evaluation of this instrument at the Paducah product feed line of the Oak Ridge Gaseous Diffusion Plant. The instrument was modified for high-resolution gamma-ray detection and tested in 1983.

At lower UF$_6$ pressures (for example, tens of torr), the density of the UF$_6$ gas is not great enough for a transmission measurement to have sufficient sensitivity. For such a low pressure, the On Line Enrichment Monitor can be employed [21]. It determines the enrichment of UF$_6$ gas flowing through the header pipework by combining the 186 keV count rate with the gas pressure and temperature data. The 186 keV count rate was measured with a collimated NaI detector; the UF$_6$ gas pressure was measured by a pressure sensor, and the temperature was determined by temperature sensors attached to the outside of the pipe

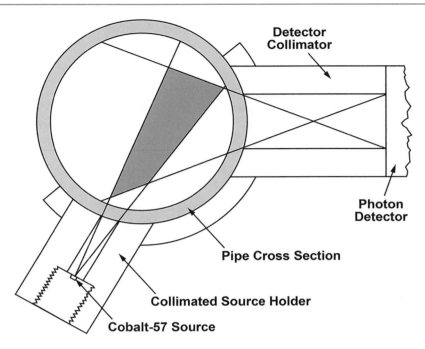

Fig. 8.14 The detector/collimator arrangement for the enrichment measurement of low-pressure UF$_6$ in pipework. The assembly consists of a collimated source holder and detector collimator rigidly connected to the pipework. The overlap of the two fields of view isolates a volume of gas in the middle of the pipe from the wall deposits. A tiny ^{57}Co source is used to fluoresce X-rays in the gas. (Figure from Ref. [24])

and thermally insulated from the room environment. The enrichment at time (t) was related to the 186 keV count rate, the pressure, and the temperature by

$$E(t) = {}^{CR(t)}/\rho(P,T,t), \tag{8.23}$$

where C is a calibration constant, R is the 186 keV count rate, and ρ is the UF$_6$ gas density as a function of the pressure P and temperature T.

Another technique that combines passive gamma-ray counting and (active) X-ray fluorescence (XRF) could verify the approximate enrichment of gaseous UF$_6$ at low pressures in cascade header pipework [22–26].

The enrichment measurement of UF$_6$ at low pressure is made more complicated by the radiation from material deposited on the container surfaces if it becomes a significant fraction of the total signal. Careful corrections for this interference are then required for accurate results. The correction for radiation from the uranium deposited on the inner surface of the pipework was established with gamma rays from ^{234}Th and ^{231}Th decays [25, 26]. The total mass of uranium in the gas was measured using XRF with the 122 keV gamma rays from a ^{57}Co excitation source. The ratio of the intensities of the 186 keV gamma rays from the UF$_6$ gas to the uranium K$_{\alpha 1}$ X-rays was calibrated to give a direct measurement of the gas enrichment. A variation of the correction for uranium deposits [24–26] determined the correction for the deposited uranium by passive gamma-ray measurements under two different collimation conditions (see Fig. 8.14). In both applications, the instruments could provide a "go/no-go" decision on whether the measured enrichment was less than or greater than 20%, thus providing the capability of detection of highly enriched uranium for enrichment plant safeguards.

8.7 Container Wall Attenuation Corrections

The standard relationship between the enrichment and the measured data (Eq. 8.8) includes the term $\exp(\mu_c \rho_c t_c)$ that corrects for the attenuation of the measured radiation by the walls of the item container, where t_c is the wall thickness. (See Chap. 6 for a detailed discussion on attenuation correction.) The attenuation may be included in the calibration if the calibration standards and the unknown items have the same type of container. In some cases, this simplification is not possible, and a container wall attenuation correction must be applied in each measurement. This section considers correction methods for an infinite-thickness enrichment measurement where the item matrix is constant. T_x, the transmission of one wall thickness of the unknown item container, is defined by

$$T_x = exp\left[-(\mu_c \rho_c t_c)_x\right] \tag{8.24}$$

The expression given in Eq. 8.24 is valid for normal incidence of gamma rays on the container wall. The condition of normal incidence is ensured by careful configuration of the measurement geometry using a collimated detector as shown in Fig. 8.3. If T_s is similarly defined as the container wall transmission in the calibration measurements, then the unknown enrichment is

$$E = KR(E_A)T_s/T_x, \tag{8.25}$$

where K is the calibration constant, and $R(E_A)$ is the net gamma-ray peak count rate from the unknown item at the assay energy (E_A = 186 keV), measured through the container wall. If an HPGe detector is used for the measurements, the transmission ratio T_s/T_x can be determined from the ratio of intensities of different-energy gamma rays from one isotope, assuming the item material is infinitely thick for the gamma rays measured (see Sect. 8.3). The technique is called *internal-line ratio technique*. (See Chap. 7 of Ref. [27] for a detailed discussion.)

The discussion that follows presents the direct measurement method for determining this container attenuation correction, T_s/T_x. In addition, the verification of UF$_6$ cylinders is discussed to provide an example of a class of measurements where this correction is especially critical.

8.7.1 Direct Measurement of Wall Thickness

If the container composition and wall thickness at the measurement point are known for both the calibration and item measurements, then T_s/T_x can be calculated directly from the exponential expression

$$\frac{T_s}{T_x} = exp\left[(\mu_c \rho_c t_c)_s - (\mu_c \rho_c t_c)_x\right], \tag{8.26}$$

where ρ_c represents the density of the container material, and μ_c is evaluated at the assay energy. The container wall thickness t_c can be measured directly using an ultrasonic thickness gauge (see Fig. 8.15). A burst of ultrasound is transmitted by the probe into the container material and travels until it reaches a material of substantially different physical character from the container material. The sound is then reflected back to the probe. The gauge electronics perform a precise measurement of the time needed for the ultrasound pulse to make the round trip in the container material and thereby determine the thickness of the material. Such thickness gauges are available commercially, and thickness results can be read usually to ±0.1 mm.

8.7.2 Measurement of UF$_6$ Cylinders

One of the most common container types in enrichment measurements is the large cylinder used to ship and store UF$_6$ in liquid or solid form. These cylinders vary in size and wall thickness. Table 8.5 gives some pertinent parameters for the most common cylinder types [28]. The thick, high-density walls of these cylinders have minor variations in thickness, which can result in significant variation in gamma-ray count rate. The relationship between the relative fluctuation of the enrichment result and the relative fluctuation of the wall thickness is obtained by differentiation of Eq. 8.8:

$$\frac{dE}{E} = \mu_c \rho_c dt_c = 1.12 dt_c, \tag{8.27}$$

where the factor 1.12 is applicable for steel (μ_c = 0.144 cm^2/g at 186 keV, ρ_c = 7.8 g/cm^3, and the unit for dt_c is cm). Thus, a 1 mm variation in cylinder wall thickness will cause an 11.2% bias in the corresponding enrichment measurement. Use of a thickness-gauge measurement of the wall thickness reduces the measurement error to a few tenths of one millimeter, essentially removing the wall thickness from consideration as a source of measurement bias. The UF$_6$ enrichment measurement apparatus for cylinders can be calibrated by using one or more cylinders as standards, which may then be used for

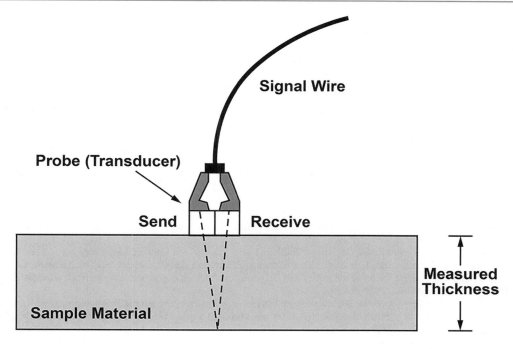

Fig. 8.15 Probe placement in an ultrasonic measurement of thickness. The probe must be acoustically coupled to the outer surface of the material for the ultrasound pulse to enter the material without being severely attenuated. The coupling is accomplished with a liquid compound (usually supplied with the thickness gauge) placed between the probe face and the material surface

Table 8.5 Physical characteristics of selected UF_6 storage and shipping cylinders [28]

	Cylinder type			
Characteristic	5A	8A	30B	48X
Nominal diameter (in.)	5	8	30	48
Nominal diameter (cm)	13	20	76	122
Nominal length (in.)	36	56	81	121
Nominal length (cm)	91	142	206	307
Material of construction	Monel	Monel	steel	steel
Wall thickness (in.)	1/4	3/16	1/2	5/8
Wall thickness (cm)	0.64	0.48	1.27	1.59

analysis. Alternatively, standards of U_3O_8 or UF_4 of known enrichment may be used, with the appropriate corrections for matrix differences (that is, the factor F/F_s in Table 8.3).

Early studies of enrichment measurements on type 30B and 5A cylinders with NaI scintillation detectors [6] achieved assay results with relative standard deviations of 5% for type 30B cylinders and <1% for type 5A cylinders. Count times were on the order of a few minutes, and the wall thickness measurement took only a few seconds. Due to the slower ultrasound speed in the paint, measuring the cylinder thickness through a painted surface would add to the steel thickness an extra thickness equal to 2.35 times the paint thickness. Good acoustic coupling between the thickness-gauge probe and the cylinder surface was obtained by sanding the paint off of a spot within the area viewed by the gamma-ray detector; the uncertainty in the thickness measurement was estimated at 0.4%.

References

1. *NuDat 2.8*, https://www.nndc.bnl.gov/nudat2/chartNuc.jsp. Accessed Feb 2023.
2. International Atomic Energy Agency, *An Introduction to Non-Destructive Assay Instrumentation: A Training Manual for the International Atomic Energy Agency Inspectorate* (International Atomic Energy Agency, Vienna, 1984)
3. J.T. Russell, *Method and Apparatus for Nondestructive Determination of ^{235}U in Uranium* (U.S. Patent No. 3,389,254, June 1968)

4. T.D. Reilly, R.B. Walton, J.L. Parker, The enrichment meter—A simple method for measuring isotopic enrichment, in *Nuclear Safeguards Research and Development Program Status Report, September–December 1970*, Los Alamos Scientific Laboratory report LA-4605-MS, ed. by G.R. Keepin, (1970), pp. 19–21
5. J.L. Parker, T.D. Reilly, The enrichment meter as a concentration meter, in *Nuclear analysis research and development program status report, September–December 1972*, Los Alamos Scientific Laboratory report LA-5197-PR, ed. by G.R. Keepin, (1972), pp. 11–12
6. R.B. Walton, T.D. Reilly, J.L. Parker, J.H. Menzel, E.D. Marshall, L.W. Fields, Measurement of UF_6 Cylinders with Portable Instruments. Nucl. Technol. **21**(2), 133–148 (1974)
7. L.A. Kull, R.O. Ginaven, *Guidelines for Gamma-Ray Spectroscopy Measurements of ^{235}U Enrichment* (Brookhaven National Laboratory report BNL-50414, 1974)
8. P. Matussek, *Accurate Determination of the ^{235}U Isotope Abundance by Gamma Spectrometry: A User's Manual for the Certified Reference Material EC-NRM-171/NBS-SRM-969* (Institut für Kernphysik report KfK 3752, Kernforschungszentrum, Karlsruhe, Federal Republic of Germany, 1985)
9. J.K. Halbig, S.F. Klosterbuer, R.A. Cameron, Applications of a portable multichannel analyzer in nuclear safeguards, in *Proceedings of IEEE 1985 Nuclear Science Symposium, San Francisco, CA, October 23–25, 1985*, (Los Alamos National Laboratory report LA-UR-85-3735, 1985)
10. *WinU235*, https://www.gbs-elektronik.de/. Accessed Feb 2023
11. R. Gunnink, A Guide for Using NaIGEM PC Version 2.1.3 for NaI and LaBr3 Detectors, (2010)
12. R. Gunnink, R. Arlt, Methods for evaluating and analyzing CdTe and CdZnTe spectra. Nucl. Instrum. Methods Phys. Res., Sect. A **458**(1–2), 196–205 (2001)
13. D. Vo, *Uranium Isotopic Analysis with the FRAM Isotopic Analysis Code* (Los Alamos National Laboratory report LA-13580, 1999)
14. A. Bosko and A. Berlizov, Evaluation of the True Coincidence Summing Effects on Uranium Enrichments Using a Monte Carlo Approach, ANS MC2105 – Joint International Conference on Mathematics and Computation (M&C), Supercomputing in Nuclear Application (SNA) and the Monte Carlo (MC) Method, Nashville, TN (April 19–23, 2015)
15. R. Gunnink, W.D. Ruhter, P. Miller, J. Goerten, M. Swinhoe, H. Wagner, J. Verplancke, M. Bickel, S. Abousahl, MGAU: A new analysis code for measuring U-235 enrichments in arbitrary samples, in *IAEA Symposium on International Safeguards, Vienna, Austria, March 8–14, 1994*, (Lawrence Livermore National Laboratory report UCRL-JC-114713, 1994)
16. D.T. Vo, *Improvements of FRAM Version 6.1* (Los Alamos National Laboratory report LA-UR-19-28449, 2019)
17. J.W. Tape, M.P. Baker, R. Strittmatter, M. Jain, M.L. Evans, Selected Nondestructive Assay Instruments for an International Safeguards System at Uranium Enrichment Plants. Nucl. Mater. Manage. **VIII, 719** (1979)
18. R.B. Strittmatter, J.N. Leavitt, R.W. Slice, *Conceptual Design for the Field Test and Evaluation of the Gas-Phase UF_6 Enrichment Meter* (Los Alamos Scientific Laboratory report LA-8657-MS, 1980)
19. R.B. Strittmatter, A gas-phase UF_6 enrichment monitor. Nucl. Technol. **59**, 355–362 (1982)
20. R.B. Strittmatter, L.A. Stovall, J.K. Sprinkle Jr., Development of an enrichment monitor for the Portsmouth GCEP, in *Proceedings of Conference on Safeguards Technology: The Process-Safeguards Interface, Hilton Head Island, SC, November 28–December 3, 1983 (Conf. 831106)*, (1984), p. 63
21. J. March-Leuba, J. Garner, J. Younkin, D.W. Simmons, *"On Line Enrichment Monitor (OLEM) UF_6 Tests for 1.5" Sch40 SS Pipe Revision 1* (Oak Ridge National Laboratory document, ORNL/LTR-2015/773, ORNL/PTS/60562, 2016)
22. D.A. Close, J.C. Pratt, H.F. Atwater, J.J. Malanify, K.V. Nixon, L.G. Speir, The measurement of uranium enrichment for gaseous uranium at low pressure, in *Proceedings of 7th Annual ESARDA Symposium on Safeguards and Nuclear Material Management, Liege, Belgium, May 21–23, 1985*, (1985), p. 127
23. D.A. Close, J.C. Pratt, J.J. Malanify, H.F. Atwater, X-ray fluorescent determination of uranium in the gaseous phase. Nucl. Instrum. Methods Phys. Res., Sect. A **234**, 556–561 (1985)
24. D.A. Close, J.C. Pratt, H.F. Atwater, Development of an enrichment measurement technique and its application to enrichment verification of gaseous UF_6. Nucl. Instrum. Methods Phys. Res., Sect. A **240**, 398–405 (1985)
25. T.W. Packer, E.W. Lees, Measurement of the enrichment of uranium in the pipework of a gas centrifuge plant, in *Proceedings of 6th Annual ESARDA Symposium on Safeguards and Nuclear Material Management, Venice, Italy, May 14–18, 1984*, (1984), p. 243
26. T.W. Packer, E.W. Lees, Measurement of the enrichment of UF_6 gas in the pipework of a gas centrifuge plant, in *Proceedings of 7th Annual ESARDA Symposium on Safeguards and Nuclear Material Management, Liege, Belgium, May 21–23, 1985*, (1985), p. 299
27. D. Reilly, N. Ensslin, H. Smith Jr., S. Kreiner, *Passive Nondestructive Assay of Nuclear Materials (PANDA)* (U.S. Nuclear Regulatory Commission, 1991)
28. Uranium Hexafluoride: Handling Procedures and Container Criteria. Oak Ridge Operations Office report ORO-651, Revision 4 (1977)

Open Access This chapter is licensed under the terms of the Creative Commons Attribution 4.0 International License (http://creativecommons.org/licenses/by/4.0/), which permits use, sharing, adaptation, distribution and reproduction in any medium or format, as long as you give appropriate credit to the original author(s) and the source, provide a link to the Creative Commons license and indicate if changes were made.

The images or other third party material in this chapter are included in the chapter's Creative Commons license, unless indicated otherwise in a credit line to the material. If material is not included in the chapter's Creative Commons license and your intended use is not permitted by statutory regulation or exceeds the permitted use, you will need to obtain permission directly from the copyright holder.

9 Plutonium Isotopic Composition by Gamma-Ray Spectroscopy

D. T. Vo, K. Koehler, and T. E. Sampson

9.1 Introduction

Interpreting the results of neutron coincidence counting or calorimetry measurements usually requires an accurate measurement of plutonium isotopic composition. Several methods have been developed for determining plutonium isotopic composition by gamma-ray spectroscopy; some of the early approaches are described in Refs. [1–5]. An American Society for Testing and Materials standard test method has been written for plutonium isotopic analysis using gamma-ray spectroscopy [6]. Different methods have been developed for different item types.

This chapter introduces the characteristics of plutonium spectra that influence isotopic measurements, describes useful spectral regions, and presents the principles of spectral analysis important to isotopic measurements. It includes descriptions of typical data collection hardware, details of data analysis methods, and descriptions of several implemented systems with examples of their accuracy and precision.

9.2 Background

9.2.1 Decay Characteristics of Plutonium Isotopes

Most plutonium items contain the isotopes ^{238}Pu, ^{239}Pu, ^{240}Pu, ^{241}Pu, and ^{242}Pu. Americium-241 and ^{237}U are always present as decay products of ^{241}Pu. Table 9.1 lists some of the decay characteristics of these important isotopes [7].

9.2.2 Decay Characteristics of ^{241}Pu

The decay of ^{241}Pu is shown in Fig. 9.1. Because of the long half-life of ^{241}Am, the concentration of the ^{241}Am daughter continues to increase for up to 75 years. Aged plutonium items often have very high ^{241}Am content, especially if the initial ^{241}Pu concentration was high.

Because of its short half-life, the ^{237}U daughter rapidly comes into secular equilibrium [8] with its ^{241}Pu parent. After approximately 47 days (seven half-lives), gamma rays from the decay of ^{237}U can be used as a measure of ^{241}Pu. Because ^{237}U has several strong gamma rays, it is especially useful for plutonium isotopic measurements. In this chapter, the terms ^{241}Pu-^{237}U *equilibrium* and *aged* refer to items where ^{237}U is in secular equilibrium with ^{241}Pu. Samples in which

Table 9.1 Decay characteristics for isotopes useful in plutonium isotopic measurements

Isotope	Half-life (year)	Specific activity (dis/s − g)	Specific power (mW/g isotope)
^{238}Pu	87.7 ± 0.1	6.336 × 10^{11}	567.79 ± 0.65
^{239}Pu	24,110 ± 30	2.295 × 10^{9}	1.9284 ± 0.0024
^{240}Pu	6561 ± 7	8.399 × 10^{9}	7.0721 ± 0.0075
^{241}Pu	14.290 ± 0.006	3.840 × 10^{12}	3.295 ± 0.026[a]
^{242}Pu	373,000 ± 3000	1.465 × 10^{8}	0.1170 ± 0.0009
^{241}Am	432.6 ± 0.6	1.268 × 10^{11}	114.58 ± 0.16
^{237}U	(6.75 ± 0.01 day)	9.485 × 10^{7} [b]	–

[a]Including the specific power of 237 U
[b]Uranium-237 activity was computed assuming ^{241}Pu-^{237}U equilibrium (see Fig. 9.1). The alpha branching ratio of ^{241}Pu is assumed to be 2.47 × 10^{-5}

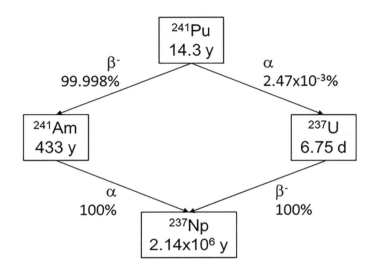

Fig. 9.1 Decay scheme of ^{241}Pu and its daughters

^{241}Pu-^{237}U equilibrium does not exist are called *freshly separated*. For those samples, ^{237}U gamma rays cannot be used as a measure of ^{241}Pu.

Figure 9.1 shows that both ^{241}Am and ^{237}U decay to the same isotope, ^{237}Np. Both isotopes can populate the same excited states in ^{237}Np and give rise to identical gamma rays; therefore, most of the useful ^{237}U gamma-ray peaks have a contribution from ^{241}Am that has grown in through the beta decay path of ^{241}Pu. The amount of this contribution depends on the particular gamma ray and the time since americium was last separated from the sample. Figure 9.2 shows the relative contributions for important ^{237}U gamma rays. To quantify the ^{241}Pu content, a correction should be made to ^{241}Pu-^{237}U peaks for their ^{241}Am content.

9.2.3 Determination of ^{242}Pu Concentration

Plutonium-242 has only a few gamma rays, similar in energy and branching ratio to those from ^{240}Pu; however, the long half-life of ^{242}Pu and its low abundance in most plutonium items make its detection by gamma-ray measurement impossible. Instead, empirical isotopic correlations [9, 10] are used to predict the ^{242}Pu content from the other isotopic fractions. Such predictions generally produce acceptable results (~20% uncertainty) for the concentration of ^{242}Pu (typically 0.03–5%) found in most plutonium if process batches have not been mixed and ^{241}Am has been neither added nor removed.

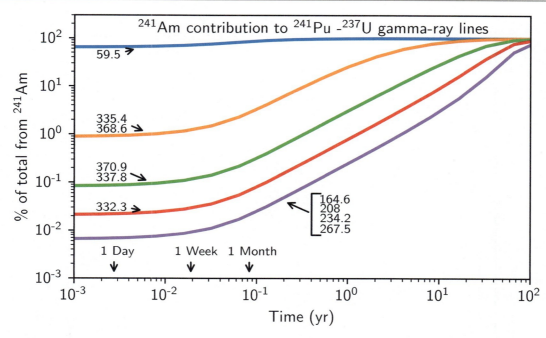

Fig. 9.2 Americium-241 contribution to ^{237}U gamma-ray peaks (energy in keV) as a function of time since separation. Uranium and americium concentrations are zero at t = 0

9.2.4 Spectral Interferences

Many regions of the gamma-ray spectrum can contain interfering gamma rays from other isotopes in the sample. For example, very high-burnup samples often contain ^{243}Am and its ^{239}Np daughter; aged samples can contain ^{237}Np and its daughter ^{233}Pa; and samples from reprocessed fuel can contain fission products. All possible interferences cannot be listed here; however, by knowing the history of a sample, the spectroscopist can anticipate possible spectral interferences.

9.2.5 Applications of Plutonium Isotopic Measurements

The principal application of plutonium isotopic measurements is to determine the quality of plutonium and to support other nondestructive assay methods in providing the total plutonium content of a sample. Two methods that use plutonium isotopic results are calorimetry and neutron coincidence counting.

Calorimetry uses the isotopic information to calculate the specific power P (W/g plutonium) of a sample from the measured isotopic fractions and the known specific power for each isotope (see Chaps. 23 and 24).

The response of neutron coincidence counters is a complicated function of all plutonium isotopes and ^{241}Am. An effective mass known as ^{240}Pu$_{eff}$ is determined using the response of a neutron coincidence counter. A measurement of isotopic composition is required to convert the coincidence counter response to individual plutonium isotopic masses (see Chap. 17).

9.3 Spectral Regions Useful for Isotopic Measurements

This section describes the spectral features that are important for the measurement of plutonium isotopic composition. The descriptions follow the practice used in works of Gunnink et al. [11] and Lemming and Rakel [12] in which the spectrum is divided into several different regions. The gamma-ray spectrum of plutonium varies greatly with isotopic composition and ^{241}Am concentration. Two sample spectra are shown in Figs. 9.3 and 9.4. Figure 9.3 represents low burnup and low ^{241}Am; Fig. 9.4 represents high burnup and relatively high ^{241}Am.

Table 9.2 lists most of the gamma rays that are useful for plutonium isotopic measurements [7]. The gamma rays are grouped into four separate classifications: 30–60 keV, 60–120 keV, 120–500 keV, and 500–800 keV. These four analytical

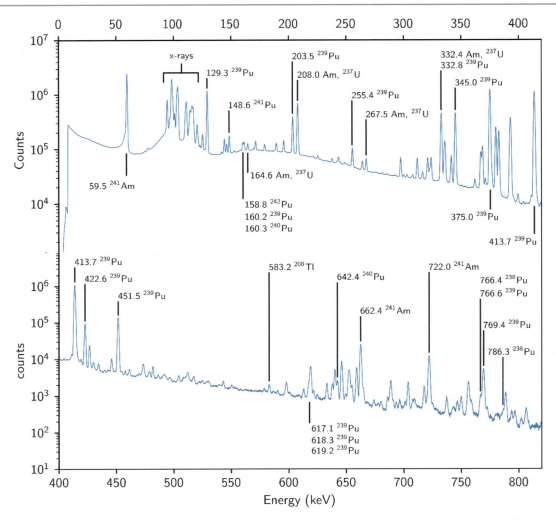

Fig. 9.3 Gamma-ray spectrum from low-burnup plutonium with approximate isotopic composition (wt%): ^{238}Pu, 0.012%; ^{239}Pu, 93.82%; ^{240}Pu, 5.90%; ^{241}Pu, 0.240%; ^{242}Pu, 0.02%; ^{241}Am, 0.063%. Peaks not labeled with a specific isotope are from ^{239}Pu

regions have one thing in common: each region has one—and only one—measurable gamma ray from both ^{238}Pu decay and ^{240}Pu decay. The list shows that the lower-energy gamma rays are more intense than those at higher energies. The lower-energy gamma rays should be used whenever possible.

Typically, a plutonium-bearing sample consists of two major isotopes: ^{239}Pu and ^{240}Pu. The sum of the fractions of these two isotopes is almost 100%. A bad measurement of one or both of these two isotopes will negatively affect the fractions of all other isotopes. Plutonium-239 has many intense peaks, so it is relatively easy to get good activity results; however, ^{240}Pu has only a few peaks that are not very intense and often have interferences with strong peaks from other isotopes. It is therefore very important to extract the ^{240}Pu peak areas accurately to have good isotopic results for all isotopes.

9.3.1 The 30–60 keV Region

The 30–60 keV region has been used mainly for analysis of freshly separated solutions from which ^{241}Am and ^{237}U have been removed. If too much ^{241}Am is present, its 60 keV gamma ray overwhelms all other peaks in the region. Usually this region is useful for 15–30 days after a separation of americium and uranium. A typical spectrum from a high-burnup reprocessing plant solution and a low-burnup spectrum are shown in Fig. 9.5. When the gamma rays can be measured, the 30–60 keV region is the most useful region for measuring ^{238}Pu, ^{239}Pu, and ^{240}Pu. The region does not have a measurable ^{241}Pu gamma ray. Small

9 Plutonium Isotopic Composition by Gamma-Ray Spectroscopy

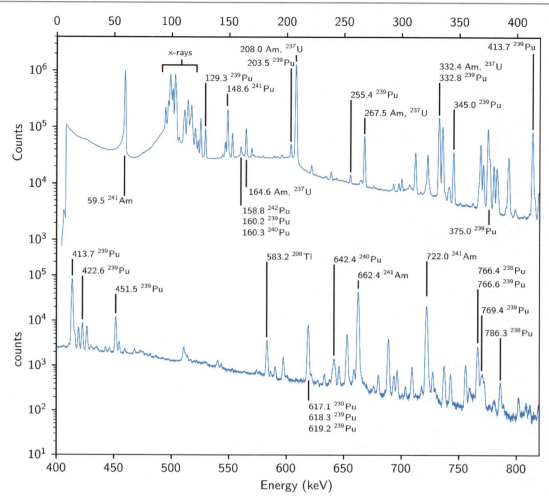

Fig. 9.4 Gamma-ray spectrum from high-burnup plutonium with approximate isotopic composition (wt%): ^{238}Pu, 0.202%; ^{239}Pu, 82.49%; ^{240}Pu, 13.75%; ^{241}Pu, 2.69%; ^{242}Pu, 0.76%; ^{241}Am, 1.18%. Peaks not labeled with a specific isotope are from ^{239}Pu

Table 9.2 Useful gamma rays in various energy regions

Region (keV)	^{238}Pu (keV)	(γ/s − g)	^{239}Pu (keV)	(γ/s − g)	^{240}Pu (keV)	(γ/s − g)	^{241}Pu (keV)	(γ/s − g)	^{241}Am (keV)	(γ/s − g)
30–60	43.50	2.55×10^8	38.66	2.46×10^5	45.24	3.87×10^6	–	–	59.54	4.71×10^{10}
			51.62	6.41×10^5						
60–120	99.85	4.74×10^7	–	–	104.23	6.18×10^5	101.06[a,b]	2.40×10^7	59.54	4.71×10^{10}
									98.97	2.66×10^7
									102.98	2.56×10^7
120–500	152.72	6.04×10^6	129.30	1.49×10^5	160.31	3.48×10^4	148.45	7.39×10^6	125.30	5.35×10^6
			203.55	1.34×10^4			164.61[a]	1.83×10^6	146.55	6.05×10^5
			345.01	1.31×10^4			208.00[a]	2.08×10^7	335.37	6.51×10^5
			375.05	3.66×10^4			267.54[a]	6.99×10^5		
			413.71	3.45×10^4			332.35[a]	1.18×10^6		
			451.48	4.46×10^3			370.94[a]	1.05×10^5		
500–800	766.39	1.43×10^5	645.94	3.57×10^2	642.35	1.12×10^3	–	–	662.40	4.78×10^5
			769.15	1.20×10^2					722.01	2.57×10^5
			769.37	1.60×10^2						

[a]Uranium-237 daughter of ^{241}Pu with ^{241}Pu-^{237}U equilibrium
[b]Neptunium X-ray from the decay of ^{237}U, daughter of ^{241}Pu

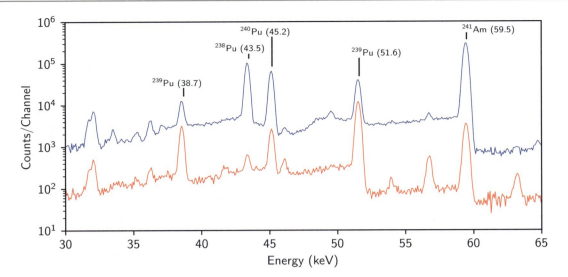

Fig. 9.5 Gamma-ray spectra in the 30–60 keV region from freshly separated solutions of plutonium. The top spectrum is shifted vertically for clarity. The top spectrum (blue line) isotopic composition (wt%): ^{238}Pu, 0.60%; ^{239}Pu, 70.42%; ^{240}Pu, 23.45%; ^{241}Pu, 3.20%; ^{242}Pu, 2.31%; ^{241}Am, 0.0011%. The bottom spectrum (red) line isotopic composition (wt%): ^{238}Pu, 0.01%; ^{239}Pu, 95.85%; ^{240}Pu, 3.80%; ^{241}Pu, 0.26%; ^{242}Pu, 0.07%; ^{241}Am, 0.0004%

contributions from ^{241}Pu and ^{237}U interfere with the ^{238}Pu peak at 43.5 keV, the ^{240}Pu peak at 45.2 keV, and the ^{239}Pu peak at 51.6 keV.

Several experimenters have used this region for solution measurements: Gunnink [13] and Russo [14] measured freshly separated solutions from a reprocessing plant; Umezawa [15] and Bubernak [16] applied these measurements to samples prepared in an analytical laboratory; and Li [17] measured submilligram-sized solid samples with modest ^{241}Am content. Gunnink used absolute counting techniques and calibrated with known solution standards; Umezawa used absolute counting with a calibrated detector; Bubernak calibrated with samples of known isotopic composition; and Russo and Li measured isotopic ratios independent of calibration standards, the methodology of which is discussed in Sect. 9.4.

9.3.2 The 60–120 keV Region

The 60–120 keV region includes the 100 keV region, which is the most complex region of the gamma-ray spectrum of plutonium. The uranium X-rays arise from plutonium decay. For mixed-oxide (MOX) samples, a second component of the uranium X-rays comes from the induced X-ray fluorescence (XRF) of uranium. The plutonium X-rays come from induced XRF of plutonium. By comparing the ratio of uranium and plutonium fluorescence X-rays, the uranium-to-plutonium ratio in a MOX sample can be determined.

The neptunium X-rays arise from the decay of ^{241}Am and ^{237}U. For aged plutonium, the neptunium X-rays from ^{237}U decay can be used to measure ^{241}Pu. The region does not have a measurable ^{239}Pu gamma ray; the most intense ^{239}Pu peak in this region at 98.78 keV sits on the shoulder of a much more intense uranium X-ray peak, and its area cannot be accurately extracted even for the low-burnup plutonium with greater than 90% ^{239}Pu.

The excessive overlapping and interfering nature of the spectrum in the 60–120 keV region is shown in Fig. 9.6. The 99.86 keV ^{238}Pu peak is intense for high-burnup plutonium, but for low-burnup plutonium, it is very weak and is made obscure by the intense plutonium fluorescence X-ray peak at 99.54 keV, rendering it unusable. The 103.68 keV ^{241}Pu peak and the intense 103.74 keV plutonium fluorescence X-ray peak interfere with the 104.24 keV ^{240}Pu peak. The intrinsic line shape of X-rays is different from that of gamma rays; the overall peak shape for an X-ray peak is the convolution of a Lorentzian distribution and a Gaussian distribution—called a *Voigt profile*—plus an exponential tail on the low-energy side. This different peak shape must be considered in the analysis of this region by peak-fitting or response-function methods [4, 18, 19].

9 Plutonium Isotopic Composition by Gamma-Ray Spectroscopy

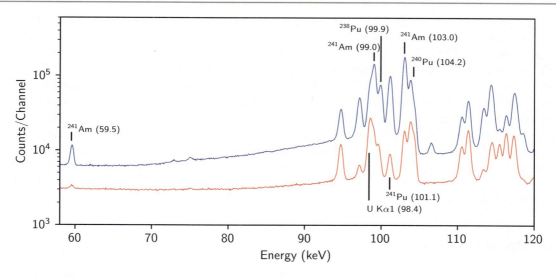

Fig. 9.6 Gamma-ray spectra of PuO$_2$ in the 60–120 keV region. The spectra are shifted vertically for clarity. The top spectrum (blue line) isotopic composition (wt%): ^{238}Pu, 1.10%; ^{239}Pu, 64.88%; ^{240}Pu, 26.33%; ^{241}Pu, 3.34%; ^{242}Pu, 4.35%; ^{241}Am, 5.01%. The bottom spectrum (red line) isotopic composition (wt%): ^{238}Pu, 0.01%; ^{239}Pu, 93.53%; ^{240}Pu, 6.31%; ^{241}Pu, 0.11%; ^{242}Pu, 0.04%; ^{241}Am, 0.22%

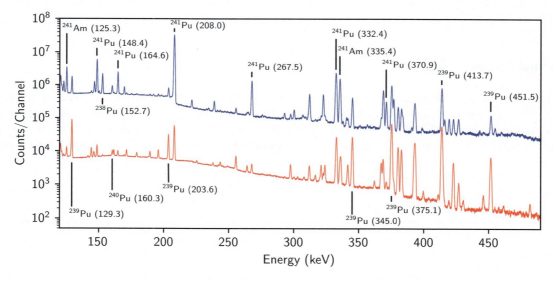

Fig. 9.7 Gamma-ray spectra of PuO$_2$ in the 120–500 keV region. The spectra are shifted vertically for clarity. The top spectrum (blue line) isotopic composition (wt%): ^{238}Pu, 1.10%; ^{239}Pu, 64.88%; ^{240}Pu, 26.33%; ^{241}Pu, 3.34%; ^{242}Pu, 4.35%; ^{241}Am, 5.01%. The bottom spectrum (red line) isotopic composition (wt%): ^{238}Pu, 0.01%; ^{239}Pu, 93.53%; ^{240}Pu, 6.31%; ^{241}Pu, 0.11%; ^{242}Pu, 0.04%; ^{241}Am, 0.22%

9.3.3 The 120–500 keV Region

The 120–500 keV region contains the gamma-rays of all plutonium isotopes (except ^{242}Pu) and ^{241}Am. Figure 9.7 shows this spectral region. High-burnup material (blue line) generally produces weaker ^{239}Pu peaks and stronger peaks from all other plutonium isotopes and ^{241}Am than low-burnup material (red line) does.

This region contains only one measurable gamma ray from ^{238}Pu decay at 152.72 keV and ^{240}Pu decay at 160.31 keV. The weak 152.72 keV ^{238}Pu peak is usually on a high background continuum and yields poor precision for low-burnup (0.01 wt% ^{238}Pu) material. The precision of the measurement can be as poor as 10% for a typical measurement.

The ^{240}Pu gamma ray at 160.31 keV has strong interferences from ^{241}Pu at 159.95 keV and ^{239}Pu at 160.19 keV. The quantification of the 160.31 keV peak is also made more challenging by the summing of the X-rays and gamma rays from the 100 keV region with the 59.54 keV ^{241}Am gamma rays that form summed peaks in the energy range from 158 keV to 163 keV. These summed peaks can interfere with the 160 keV complex and can be eliminated by using a filter to selectively absorb gamma rays and X-rays from the 60 keV and 100 keV regions before they interact with the detector. The plutonium spectrum of a planar high-purity germanium (HPGe) detector also includes one additional feature that makes the difficult-to-analyze region even harder to analyze: the backscatter structure of the 413.7 keV ^{239}Pu peak. Due to the thinness of the crystal, many high-energy gamma rays can pass through the crystal without interacting. Some of these gamma rays will bounce off of the copper rod cold finger of the detector or materials behind the crystal and get back into the crystal. These scattered gamma rays form the backscattered structure. The backscattered structure of the 413.7 keV gamma ray crests at 158 keV and introduces additional complexity to the already complex region, making it difficult to get a good 160.31 keV peak area result. The statistical precision of the ^{240}Pu component is seldom measured to better than ~1%.

9.3.4 The 500–800 keV Region

Figure 9.8 shows the characteristics of the 500–800 keV region, which is most useful for large or heavily shielded samples. The 645.94 keV, 769.15 keV, and 769.37 keV peaks are useful for measuring ^{239}Pu, and the 662.40 keV and 722.01 keV peaks are useful for measuring ^{241}Am. The 766.39 keV ^{238}Pu peak is intense for high-burnup plutonium but is very weak and difficult to measure for low-burnup plutonium. The fitting of the only ^{240}Pu peak in this region at 642.35 keV is made complicated by the nearby ^{239}Pu peak at 639.99 keV and the ^{241}Am peak at 641.47 keV. This region does not have any peaks from ^{241}Pu or its daughter ^{237}U.

9.3.5 Actual Analytical Regions

Of the four regions discussed in Sections 9.3.1 through 9.3.4, only the 120–500 keV region has peaks from all plutonium isotopes and ^{241}Am, so full isotopic analysis can be done with the peaks in this region alone. For the other three regions, each one lacks peaks from either ^{239}Pu or ^{241}Pu, making a full isotopic measurement using only the peaks in the region impossible. These regions need to be combined with peaks in other regions to fully determine the isotopic composition of plutonium.

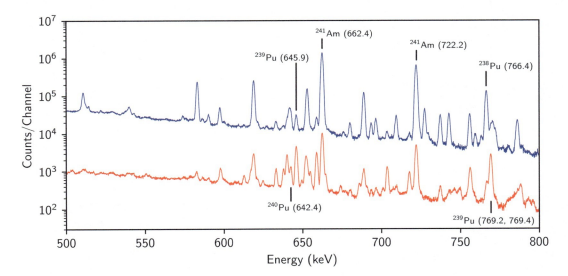

Fig. 9.8 Gamma-ray spectra of PuO$_2$ in the 500–800 keV region. The spectra are shifted vertically for clarity. The top spectrum (blue line) isotopic composition (wt%): ^{238}Pu, 1.15%; ^{239}Pu, 63.85%; ^{240}Pu, 25.93%; ^{241}Pu, 4.79%; ^{242}Pu, 4.28%; ^{241}Am, 3.48%. The bottom spectrum (red line) isotopic composition (wt%): ^{238}Pu, 0.01%; ^{239}Pu, 93.48%; ^{240}Pu, 6.31%; ^{241}Pu, 0.16%; ^{242}Pu, 0.04%; ^{241}Am, 0.17%

9 Plutonium Isotopic Composition by Gamma-Ray Spectroscopy

An analysis using the measurable peaks in all four regions from 30–800 keV would include many peaks and could be very accurate; however in practice, the results from the analysis of a very wide energy range might not be better than the analysis in a smaller energy range even though the wide energy range analysis uses more data. The reason for this outcome is that when the energy range is very large, the deviances in the efficiency curve could become large and could negatively affect the results. (See Chap. 7 and [10] for a description of efficiency curve.)

In the following sections are descriptions of four practical analytical regions that are normally used for plutonium isotopic analysis of HPGe spectra and one region that is used for analysis of microcalorimeter data.

9.3.5.1 The 30–210 keV Analytical Region

For freshly separated samples, peaks below 60 keV are visible and can be used for analysis. These peaks represent the most intense gamma rays for most of the plutonium isotopes except ^{241}Pu (and ^{242}Pu). Because this region lacks a ^{241}Pu peak, it is insufficient to use the 30–60 keV region alone for plutonium isotopic analysis. The region would need to be used in conjunction with another region that has a ^{241}Pu peak (see Table 9.2).

The 60–120 keV region has a neptunium X-ray at 101.06 keV (from the decay of ^{237}U) that can be used to measure ^{241}Pu in aged plutonium. For fresh plutonium, this X-ray peak cannot be used to quantify the ^{241}Pu, so a peak from the direct decay of ^{241}Pu is needed. The 120–500 keV region has the 148.45 keV peak from ^{241}Pu decay, so the 30–60 keV region can be combined with the 120–500 keV region for full isotopic analysis. In fact, the whole 120–500 keV region need not be used because that would make the combined 30–500 keV range too large for effective analysis. A combination of the 30–60 keV region and 120–210 keV region is sufficient [20]. This combination of two narrow regions is advantageous because the analysis avoids the 100 keV region, which is crowded and requires the capability of analyzing the X-ray peaks. For a better analysis, the 60–120 keV region can be included if capability exists to fit the X-ray peaks. The inclusion of this region does not extend the range of the efficiency curve but adds significant intense peaks to the analysis [19].

In practice, the analytical region is often extended to 230 keV to cover the 210 keV and 228 keV peaks from ^{243}Am and/or ^{243}Cm decay, which will give ^{243}Am to Pu and/or ^{243}Cm to Pu ratio results in addition to the plutonium isotopic composition. The 186 keV ^{235}U peak is often included to measure the ^{235}U to Pu ratio.

9.3.5.2 The 60–210 keV Analytical Region

The 60–210 keV analytical region can be used to analyze either fresh or aged plutonium. For fresh plutonium, only the 148.45 keV peak is used for ^{241}Pu activity. All of the peaks from the ^{241}Pu daughter ^{237}U (see Table 9.2) can be used to measure ^{237}U activity but not the ^{241}Pu activity. For aged plutonium, these ^{237}U peaks can be considered to come directly from ^{241}Pu decay and are used to measure ^{241}Pu. For a small plutonium item inside a thin container, this analytical region offers the best results because the peaks in the 100 keV region are intense and can give good results in a short measurement time. A drawback is the requirement of a detector with good resolution, such as that of a good planar detector, to de-convolute the many overlapping peaks. A detector with inferior resolution, such as a coaxial HPGe detector, can still give precise results due to intense peaks in the 100 keV region, but the bias can be significantly large due to the difficulty in resolving the overlapping peaks in the region.

Similar to the 30–210 keV region, the analytical region may be extended to 230 keV to cover the 210 keV and 228 keV ^{243}Am and/or ^{243}Cm peaks for ^{243}Am to Pu and/or ^{243}Cm to Pu ratio results. Also, the 186 keV ^{235}U peak is often included to measure the ^{235}U to Pu ratio.

9.3.5.3 The 120–420 keV Analytical Region

The 120–420 keV analytical region does not include the X-ray region, which makes analysis a bit easier because it does not require fitting the X-ray peaks. The drawback to using this region is its dependence on the weak 160.31 keV ^{240}Pu peak, which is a factor of 18 less intense than the ^{240}Pu peak at 104.23 keV (see Table 9.2). If the 451.48 keV peak is present in the spectrum, the analytical region can be extended to 460 keV to include this peak. The 186 keV peak is normally used for ^{235}U activity. If the ^{235}U to Pu ratio is large, then some other intense peaks from ^{235}U at 143.8, 163.4, and 205.3 keV can also be used.

9.3.5.4 The 200–800 keV Analytical Region

The 200–800 keV analytical region is most useful for large items and/or items shielded by thick and high-Z filter materials. The peaks from ^{238}Pu and ^{240}Pu in this analytical region are more than an order of magnitude weaker than those in the 120–420 keV analytical region. For large items or heavily shielded items, the low-energy peaks are significantly suppressed, which makes the weak peaks at high energies relatively more prominent. In practice, the lower bound of the analytical region is normally set at 180 keV to include the 186 keV ^{235}U peak, and the upper bound is extended to 1010 keV to include the 1001 keV ^{238}U peak so MOX items can be analyzed.

9.3.5.5 The 96–210 keV Analytical Region with Microcalorimeters

The efficiency of microcalorimeter gamma spectrometers is optimized for the 100 keV region, making the 96–210 keV analytical region optimal for determining plutonium isotopic composition with microcalorimeters. The high resolving power of microcalorimeters has been shown to result in reduced uncertainty and bias in quantitative analysis of plutonium isotope ratios as compared with HPGe measurements [21, 22].

For microcalorimeter regions, the high-energy resolution allows for more degrees of freedom in fitting for the regions of interest (ROIs), which is not possible when fitting the regions relevant to plutonium isotopic composition in the 96–210 keV region of HPGe data. For microcalorimeter data, because of the well-resolved lines, the gamma-ray energies can be allowed to vary up to 50 eV with few exceptions, which is 0.7–1.0 times the typical energy resolution. To achieve convergence with HPGe data in this analytical region, the energy position of the majority of peaks must be fixed relative to each other within an ROI with only a global shift considered. The uncertainty reported from peak minimization routines has significant dependence on these parameter constraints, and the fewer degrees of freedom artificially reduce uncertainties in peak areas while potentially introducing systematic error in the derived peak areas [21].

9.4 Measurement Principles

9.4.1 Measurement of Isotopic Ratios

The full-energy peak area for any single gamma ray can be written as

$$C(E_j^i) = \lambda^i N^i BR_j^i \varepsilon(E_j), \tag{9.1}$$

where

$C\left(E_j^i\right)$ = full-energy peak area of gamma ray j with energy E_j^i emitted from isotope i

λ^i = decay constant of isotope i; $\lambda^i = \frac{\ln(2)}{T_{1/2}^i}$, where $T_{1/2}^i$ is the half-life of isotope i

N^i = number of atoms of isotope i

BR_j^i = branching ratio (gamma rays/disintegration) of gamma ray j from isotope i

$\varepsilon(E_j)$ = total efficiency for full-energy peak detection of gamma ray with energy E_j; includes intrinsic detector efficiency, geometry (solid angle), item self-absorption, and attenuation in materials between the item and detector.

Equation 9.1 may be rearranged to give an expression for the atom ratio of two isotopes. The atom ratio is given by

$$\frac{N^i}{N^k} = \frac{C(E_j^i)}{C(E_m^k)} \times \frac{T_{1/2}^i}{T_{1/2}^k} \times \frac{BR_m^k}{BR_j^i} \times \frac{RE(E_m)}{RE(E_j)}. \tag{9.2}$$

In Eq. 9.2, the full-energy peaks areas $C(E)$ are measured, and the half-lives $T_{1/2}$ and branching ratios BR are either known or can be calculated from nuclear data. The total efficiency has been expressed in terms of the relative efficiency $RE(E)$. Geometric factors cancel, and the relative efficiency ratio includes only item self-absorption, attenuation in materials between the item and detector, and detector efficiency. The use of an efficiency ratio removes the need for reproducible geometry and makes the isotopic ratio method applicable to items of arbitrary size, shape, and composition.

A relative efficiency curve can be determined from the measured spectrum of every item. Equation 9.3 gives the following proportionality for gamma rays from a single isotope i:

$$\varepsilon(E_j) \propto RE(E_j) \propto \frac{C(E_j^i)}{BR_j^i}. \tag{9.3}$$

Because efficiency ratios are used in Eq. 9.2, only the shape of the relative efficiency curve, which is the ratios given in Eq. 9.3, is important.

After appropriate isotopic ratios are measured, it is usually desirable to calculate absolute isotopic fractions. The sum of all isotopic fractions must equal unity. Neglecting ^{242}Pu, this implies that

$$1 = f_{238} + f_{239} + f_{240} + f_{241}, \tag{9.4}$$

where f_i is the isotopic fraction of isotope i.

Dividing Eq. 9.4 by f_{239} gives

$$\frac{1}{f_{239}} = \frac{f_{238}}{f_{239}} + 1 + \frac{f_{240}}{f_{239}} + \frac{f_{241}}{f_{239}}. \tag{9.5}$$

Equation 9.5 expresses the isotopic fraction of ^{239}Pu (f_{239}) in terms of the three measured ratios f_{238}/f_{239}, f_{240}/f_{239}, and f_{241}/f_{239}. The remainder of the isotopic fractions are obtained from

$$f_i = f_{239} \times \left[\frac{f_i}{f_{239}}\right], i = 238, 240, 241. \tag{9.6}$$

Section 9.4.2 discusses the incorporation of ^{242}Pu into this analysis. If the ratio of ^{241}Am to any of the plutonium isotopes (usually ^{239}Pu) has been measured, then the absolute fraction of ^{241}Am can be calculated from

$$f_{Am} = f_i \times \left[\frac{f_{Am}}{f_i}\right]. \tag{9.7}$$

Note that this formula gives the weight or atom fraction of ^{241}Am in the sample with respect to total plutonium, not total sample.

The isotopic ratio method can be applied to items of arbitrary size, shape, and composition. The method works if the plutonium isotopic composition and the Am/Pu ratio are uniform throughout the item; the plutonium could be nonuniformly distributed within the item if the aforementioned uniformity is present. Two different methods were developed—one at Mound Laboratory [23] and the other at Los Alamos National Laboratory [24]—to measure electro-refining salt residues that have a nonuniform Am/Pu ratio.

9.4.2 The ^{242}Pu Isotopic Correlation

Plutonium-242 cannot be measured directly because of its low activity, low abundance, and weak gamma rays. Instead, isotopic correlation techniques [9] are used to predict the ^{242}Pu abundance from knowledge of the other isotopic fractions. It is well known that correlations exist among the plutonium isotopic abundances because of the nature of the neutron capture reactions that produce the plutonium isotopes in nuclear reactors. Because these correlations depend on the reactor type and details of the irradiation history, it is difficult—if not impossible—to find a single correlation that is optimal for all material. Gunnink [9] suggests that the correlation

$$f_{242} = \frac{K(f_{240})(f_{241})}{(f_{239})^2}, \tag{9.8}$$

where f is the isotopic fraction, is linear and relatively independent of reactor type. When the isotopic fractions are given in weight percent, the constant K equals 52. One disadvantage of using correlations that depend on ^{241}Pu is the relatively short half-life of ^{241}Pu, which decreases in absolute abundance by ~5% per year. The correlation works best if the ^{241}Pu abundance can be corrected to the time of fuel discharge from the reactor. When the discharge time is not known, a partial correction can be made by adding the quantities of ^{241}Am to the ^{241}Pu before computing the correlation. The total gives the ^{241}Pu content at the time of the last chemical separation. Reference [10] employs such correlation.

$$f_{242} = A\left[(f_{238})^B (f_{239})^C (f_{240})^D (f_{241} + f_{Am241})^E\right]. \tag{9.9}$$

The parameters A through E can be tuned specifically for materials of different reactor types. Setting parameter E to zero would make the correlation independent of ^{241}Pu. In more recent versions of the isotopic code Multi-Group Analysis (MGA) authored by Ray Gunnink, ^{242}Pu correlation is determined using the equation [25]

$$[^{242}Pu/^{239}Pu] = C1 \cdot [^{238}Pu/^{239}Pu]^{C2} \cdot [^{240}Pu/^{239}Pu]^{C3} \cdot [(^{241}Pu + Am)/^{239}Pu]^{C4} + C5 \cdot [^{238}Pu/^{239}Pu]^{C8} + C6$$
$$\cdot [^{240}Pu/^{239}Pu]^{C9} + C7 \cdot [(^{241}Pu + Am)/^{239}Pu]^{C10}. \tag{9.10}$$

The coefficients C1 to C10 are entered in the setup file for the MGA.

A correlation not involving ^{241}Pu has been suggested [9, 26]:

$$f_{242} = \frac{K(f_{240})^3}{(f_{239})^2}. \tag{9.11}$$

This correlation is linear for a given reactor type, but the slope K depends on reactor type.

Ruhter et al. [27] proposed a correlation formalism that does not depend on ^{241}Pu. It is given in the following equation:

$$\frac{^{242}Pu}{^{239}Pu} = C_0 \left[\frac{^{238}Pu}{^{239}Pu}\right]^{C_1} \cdot \left[\frac{^{240}Pu}{^{239}Pu}\right]^{C_2}. \tag{9.12}$$

This is equivalent to Eq. 9.9 with E = 0. The correlation given in Eq. 9.12 was tested and evaluated by Bignan et al. [28] using a theoretical and experimental approach. Based on extensive experimental data sets available from destructive analysis of pressurized-water reactor (PWR) and boiling-water reactor (BWR) fuels, Bignan et al. [28] determined that best values for the fitting coefficients C_1 and C_2 were 0.33 and 1.7, respectively. Rather than recommending a single value for the coefficient C_0, Bignan et al. [28] provided a table with recommended values of C_0 according to the plutonium characteristics, such as the range of initial ^{235}U enrichment and the range of burn-up. The average value of C_0 is 1.313 for PWR fuel and 1.117 for BWR fuel. According to Bignan et al. [28], using these average values for C_0 results in average error on ^{242}Pu evaluation of approximately 3% for PWR fuel and 7% for BWR fuel.

After the isotopic fraction of ^{242}Pu has been determined using a suitable correlation, known value, or stream average, the other isotopic fractions can be corrected using

$$f_i^c = f_i(1 - f_{242}), \tag{9.13}$$

where f_i^c are the normalized isotopic fractions (such that the sum of all plutonium isotope fractions is 100%) excluding f_{242}^c; the normalized ^{242}Pu fraction is the correlated fraction f_{242} determined from one of the Eqs. 9.4 through 9.8. This correction has a flaw in that if the ^{242}Pu fraction f_{242} happens to be greater than 1, then the normalized isotopic fractions f_i^c will become negative. A more robust equation to calculate the isotopic fractions is

$$f_i^c = \frac{f_i}{1 + f_{242}}, \tag{9.14}$$

where f_i^c are the normalized isotopic fractions, including f_{242}^c. This correction renormalizes the fractions so that the sum over all plutonium isotopes equals unity.

Table 9.3 Typical measurement precision

Region (keV)	Count time	^{238}Pu	^{239}Pu	^{240}Pu	^{241}Pu	^{241}Am	Specific power
30–210[a]	15–30 min	0.3–5%	0.05–0.5%	0.2–1.0%	1–5%	1–10%	0.1–1.0%
60–210	30–60 min	0.3–5%	0.05–0.5%	0.2–1.0%	0.2–0.8%	0.2–1.0%	0.1–1.0%
120–420	60–120 min	0.2–4%	0.05–0.5%	0.2–1.0%	0.2–0.8%	0.2–1.0%	0.1–1.0%
200–800	1–4 h[b]	1–10%	0.1–0.5%	1–5%	0.2–0.8%	0.2–10%	0.3–2%

[a]Fresh plutonium
[b]With high-count-rate systems, these precisions can be realized in less than 30 min

Table 9.4 Systematic and statistical uncertainties for plutonium isotopic composition from microcalorimeter and HPGe measurements[a]

Measurement	^{238}Pu/^{239}Pu	^{240}Pu/^{239}Pu	^{241}Pu/^{239}Pu	^{241}Am/^{239}Pu
Microcalorimeter (syst.) [%]	0.8–1	~1	~0.76	~1.2
HPGe (syst.) [%]	0.9–7	1–3	0.8–2	1–6
Scaling factor (stat.)	0.2–1	0.2–0.8	0.8–1	0.2–1

[a]Scaling factor is the microcalorimeter statistical uncertainty divided by the HPGe statistical uncertainty, assuming equal counts in the 96–210 keV region [21]

9.4.3 Summary of Measurement Precision (HPGe Detectors)

For all techniques discussed, the measurement precision is influenced most by the spectral region analyzed; a higher precision is obtained when measuring the lower-energy regions that have higher gamma-ray emission rates. Table 9.3 summarizes the measurement precision attainable for different energy regions. Measurement accuracy is usually commensurate with precision.

9.4.4 Summary of Measurement Precision (Microcalorimeters)

Microcalorimeters afford the opportunity to achieve lower total measurement uncertainty on determination of plutonium isotopic composition. They reduce (1) statistical errors on weak and/or overlapping peaks important for analysis and (2) systematic uncertainties that arise from uncertainties in the nuclear and atomic data used for isotopic analysis by both microcalorimeters and HPGe detectors. These systematic uncertainties currently limit the precision of plutonium isotopic determination to ~1%.

Existing codes such as FRAM do not include the uncertainties from nuclear and atomic data, such as branching ratios, half-lives, X-ray line widths, and energy calibration; uncertainties from empirical adjustments of these values are not calculated and propagated to isotopic ratios. The SAPPY code [29] based on SAP [21] does propagate these systematic sources of uncertainty in a consistent framework for both HPGe and microcalorimeter data. These systematic uncertainties limit the precision of gamma spectroscopy to approximately 0.8–7% in the 97–207 keV region, as can be seen in Table 9.4. For microcalorimetry, the dominant contribution to the systematic uncertainties arises from the branching ratio uncertainties (0.7–1.2%), and if these uncertainties were significantly improved, microcalorimetry results would be dominated by uncertainties in half-lives at the tenths-of-percent levels. Given improvements in branching ratios, HPGe analysis would still be dominated by the 0.2–7% uncertainties from the energies. Neither microcalorimeter nor HPGe analysis is sensitive to uncertainties in mass attenuation coefficients and material densities, variations of which produce relative uncertainties of less than 0.001% [21].

Assuming equal counts in the 96–210 keV region, the statistical uncertainties on microcalorimeter measurements of isotopic ratios can be up to a factor of 5 (0.2^{-1}) lower than HPGe results. Microcalorimeters achieve significantly better performance than HPGe detectors on materials with high ^{239}Pu content because the sensitivity of the detector allows for more precise measurements of the other less-abundant isotopes. For example, a 1-h HPGe measurement of PIDIE-1 yields uncertainties of ^{238}Pu/^{239}Pu, ^{240}Pu/^{239}Pu, ^{241}Pu/^{239}Pu, and ^{241}Am/^{239}Pu of 11%, 4%, 1%, and 1%, respectively. Using the scaling factors for statistical uncertainties (Table 9.4 and Table III in Ref. [2]), propagating these uncertainties for absolute fractions (e.g., ^{238}Pu/Pu) using the method described in Ref. [30], ignoring ^{242}Pu uncertainty, and ignoring covariance, the improvements in the uncertainties of the ^{238}Pu/Pu, ^{239}Pu/Pu, ^{240}Pu/Pu, ^{241}Pu/Pu, and ^{241}Am/Pu fractions are factors of 5.3,

5.0, 5.0, 1.0, and 3.2, respectively, with microcalorimeter data. This result would affect the specific power uncertainties for calorimetry measurement by a factor of 4.7, yielding a factor of 4.7 improvement in the calculation of total plutonium mass, assuming no uncertainty in the calorimetry measurement. A similar analysis for PIDIE-6 yields a factor of 1.1 improvement due to its higher burnup [21].

References

1. F.X. Haas, W.W. Strohm, Gamma-ray spectrometry for calorimetric assay of plutonium fuels. IEEE Trans. Nucl. Sci. **NS-22**(1), 734–738 (1975)
2. T. Dragnev, K. Schaerf, Non-destructive gamma spectrometry measurement of ^{239}Pu/^{240}Pu and Pu/^{240}Pu ratios. Int. J. Appl. Radiat. Isot. **26**(3), 125–129 (1975)
3. J.L. Parker, T.D. Reilly, Plutonium isotopic determination by gamma-ray spectroscopy, in Nuclear Analysis Research and Development Program Status Report, January–April 1974, ed. by G. Robert Keepin, Los Alamos Scientific Laboratory report LA-5675-PR (1974), pp. 11–13
4. R. Gunnink, J.B. Niday, P.D. Siemens, A system for plutonium analysis by gamma-ray spectrometry: part i: techniques for analysis of solutions. Lawrence Livermore Laboratory report UCRL-51577, Part I (April 1974)
5. R. Gunnink, Simulation study of plutonium gamma-ray groupings for isotopic ratio determinations. Lawrence Livermore Laboratory report UCRL-51605 (June 1974)
6. Standard test method for determination of plutonium isotopic composition by gamma-ray spectrometry. ASTM Standard Test Method C1030-10, American Society for Testing and Materials, Philadelphia (2018)
7. *NuDat 2.8.* https://www.nndc.bnl.gov/nudat2/chartNuc.jsp. Accessed Feb 2023
8. R.D. Evans, *The Atomic Nucleus* (McGraw Hill, New York, 1955), p. 484
9. R. Gunnink, Use of isotope correlation techniques to determine ^{242}Pu abundance. Nucl. Mater. Manage. **9**(2), 83–93 (1980)
10. T. Sampson, Plutonium isotopic analysis using PC/FRAM, Chapter 2, PANDA Addendum, Los Alamos National Laboratory report LA-UR-03-4403
11. R. Gunnink, J.E. Evans, A.L. Prindle, A reevaluation of the gamma-ray energies and absolute branching intensities of ^{237}U, $^{238, 239, 240, 241}$Pu, and ^{241}Am. Lawrence Livermore Laboratory report UCRL-52139 (October 1976)
12. J.F. Lemming, D.A. Rakel, Guide to plutonium isotopic measurements using gamma-ray spectrometry. Mound Facility report MLM-2981 (August 1982)
13. R. Gunnink, A.L. Prindle, Y. Asakura, J. Masui, N. Ishiguro, A. Kawasaki, S. Kataoka, Evaluation of TASTEX Task H: measurement of plutonium isotopic abundances by gamma-ray spectrometry. Lawrence Livermore National Laboratory report UCRL-52950 (October 1981)
14. P.A. Russo, S.-T. Hsue, J.K. Sprinkle, S.S. Johnson, Y. Asakura, I. Kando, J. Masui, K. Shoji (1982) In-plant measurements of gamma-ray transmissions for precise k-edge and passive assay of plutonium concentration and isotopic fractions in product solutions. Los Alamos National Laboratory report LA-9440-MS (PNCT 841–82-10) (1982)
15. H. Umezawa, T. Suzuki, S. Ichikawa, Gamma-ray spectrometric determination of isotopic ratios of plutonium. J. Nucl. Sci. Technol. **13**(6), 327–332 (1976)
16. J. Bubernak, Calibration and use of a high-resolution low-energy photon detector for measuring plutonium isotopic abundances. Anal. Chim. Acta **96**(2), 279–284 (1978)
17. T.K. Li, T.E. Sampson, S. Johnson, Plutonium isotopic measurement for small product samples, in *Proceeding of the 5th Annual ESARDA Symposium on Safeguards and Nuclear Material Management*, Versailles, France, April 19–21, paper 4.23 (1983)
18. R. Gunnink, Gamma spectrometric methods for measuring plutonium, in *Proceeding of the American Nuclear Society National Topical Meeting on Analytical Methods for Safeguards and Accountability Measurements of Special Nuclear Material*, Williamsburg, VA, May 15–17, 1978
19. D.T. Vo, T.K. Li, Plutonium isotopic analysis in the 30 keV to 210 keV range, 23rd Annual Meeting, Symposium on Safeguards and Nuclear Material Management, Bruges, Belgium, May 8–10, 2001, Los Alamos National Laboratory report LA-UR-01-2264
20. D.T. Vo, J.S. Hansen, T.K. Li., Plutonium isotopic analysis using low-energy gamma-ray spectroscopy, Sixth International Conference of Facility Operation-Safeguards Interface, Jackson Hole, Wyoming, September 20–24, 1999, Los Alamos National Laboratory report LA-UR-99.4890
21. A.S. Hoover, R. Winkler, M.W. Rabin, D.A. Bennett, W.B. Doriese, J.W. Fowler, J. Hayes-Wehle, R.D. Horansky, C.D. Reintsema, D.R. Schmidt, L.R. Vale, J.N. Ullom, K. Schaffer, Uncertainty of plutonium isotopic measurements with microcalorimeter and high-purity germanium detectors. IEEE Trans. Nucl. Sci. **61**(4), 2365–2372 (2014)
22. M. Croce, D. Becker, K.E. Koehler, J. Ullom, Improved nondestructive isotopic analysis with practical microcalorimeter gamma spectrometers, in *Proceeding of the Workshop on Isotopic Analysis of Uranium and Plutonium by Non-Destructive Assay Techniques*, February 15–19, 2021, J. Nucl. Mater. Manage. 49(3) (2021), pp. 108–113
23. J.G. Fleissner, Nondestructive assay of plutonium in isotopically heterogeneous salt residues, in *Proceeding of the Conference on Safeguards Technology*, Hilton Head Island, SC, Department of Energy publication CONF-831106 (1983), pp. 275–285
24. D.T. Vo, Improvements of FRAM Version 6.1, Los Alamos National Laboratory report LA-UR-19.28449 (2019)
25. Model S508 Multigroup Analysis (MGA) Stand-Alone Interface, V10.1 7064882B User's Manual Supplement, Mirion Technologies (Canberra), Inc. (2017)
26. H. Umezawa, H. Okashita, S. Matsurra, Studies on correlation among heavy isotopes in irradiated nuclear fuels, in Symposium on Isotopic Correlation and Its Application to the Nuclear Fuel Cycle held by ESARDA, Stresa, Italy (May 1978)
27. W.D. Ruhter, R. Gunnink, S. Baumann, S. Abeynaike, J. Verplancke, MGA and passive neutron measurements, in *Proceeding of the ESARDA Workshop on Passive Coincidence Counting*, Ispra (April 1993)
28. G. Bignan, H. Ottmar, A. Schubert, W. Ruhter, C. Zimmerman, Plutonium isotopic determination by gamma spectrometry: recommendations for the ^{242}Pu content evaluation using a new algorithm, ESARDA Bull 28 (January 1998)

29. D.T. Becker, B.K. Alpert, D.A. Bennett, M.P. Croce, J.W. Fowler, J.D. Gard, A.S. Hoover, J. Young II, K.E. Koehler, J.A.B. Mates, G.C. O'Neil, M.W. Rabin, C.D. Reintsema, D.R. Schmidt, D.S. Swetz, P. Szypryt, Advances in analysis of microcalorimeter gamma-ray spectra. IEEE Trans. Nucl. Sci. **66**(12), 2355–2363 (2019). https://doi.org/10.1109/TNS.2019.2953650
30. T. Sampson, R. Gunnink, The propagation of errors in the measurement of plutonium isotopic composition by gamma-ray spectroscopy. Los Alamos National Laboratory report LA-UR-83-1520 (1983)

Open Access This chapter is licensed under the terms of the Creative Commons Attribution 4.0 International License (http://creativecommons.org/licenses/by/4.0/), which permits use, sharing, adaptation, distribution and reproduction in any medium or format, as long as you give appropriate credit to the original author(s) and the source, provide a link to the Creative Commons license and indicate if changes were made.

The images or other third party material in this chapter are included in the chapter's Creative Commons license, unless indicated otherwise in a credit line to the material. If material is not included in the chapter's Creative Commons license and your intended use is not permitted by statutory regulation or exceeds the permitted use, you will need to obtain permission directly from the copyright holder.

Gamma-Ray NDA Applications & Techniques

10

P. A. Santi, R. Hunneke, B. Jennings, P. J. Karpius, A. L. Lousteau, D. J. Mercer, T. D. Reilly, P. A. Russo, S. Smith, R. Venkataraman, D. T. Vo, and J. Younkin

10.1 Introduction

Gamma-ray nondestructive assay (NDA) can provide a variety of useful information about a sample that contains nuclear or radiological material. The measurement of uranium enrichment (Chap. 8) and plutonium isotopic composition (Chap. 9) are two specific applications that have their own dedicated chapters. In this chapter, we will discuss several additional applications of gamma-ray NDA and their associated techniques, which can be divided into three categories: attribute measurements (Sect. 10.2), general nuclide identification (Sect. 10.3), and quantitative measurements (Sect. 10.4).

Attribute measurements are useful measurements that are intended to confirm the presence or absence of specific attributes of nuclear material. These rapid, low-cost, semiquantitative techniques can be used to supplement more rigorous (but less frequent) physical inventories, to allow an inspector to rapidly assess the material in a large facility, or to apply to process monitoring.

Nuclide identification refers to the process of analyzing a gamma-ray spectrum to determine which radionuclides are present in the sample. The instrumentation used to collect the spectrum determines which of several techniques are available for nuclide identification. These techniques are useful for identifying unknown or legacy items and in search and recovery operations.

Section 10.4, Quantitative Measurements, provides details on the applications and techniques for performing isotopic composition measurements, waste measurements, and holdup measurements.

10.2 Attribute Measurements

10.2.1 Introduction

Nuclear material measurements are usually quantitative assays in which the measurement goal is to fix a numerical value on the amount of nuclear material present. The assays are performed with the highest accuracy and precision possible, and prior knowledge about the samples could be extensive; however, several measurement challenges can be met with more qualitative information on samples, about which prior knowledge could vary widely. Some examples include

---

Los Alamos National Laboratory strongly supports academic freedom and a researcher's right to publish; as an institution, however, the Laboratory does not endorse the viewpoint of a publication or guarantee its technical correctness.

P. A. Santi (✉) · B. Jennings · P. J. Karpius · D. J. Mercer · T. D. Reilly · P. A. Russo · D. T. Vo
Los Alamos National Laboratory, Los Alamos, NM, USA
e-mail: psanti@lanl.gov; bjennings@lanl.gov; karpius@lanl.gov; mercer@lanl.gov; ducvo@lanl.gov

R. Hunneke · A. L. Lousteau · S. Smith · R. Venkataraman · J. Younkin
Oak Ridge National Laboratory, Oak Ridge, TN, USA
e-mail: lousteaula@ornl.gov; smithsk@ornl.gov; venkataramar@ornl.gov; younkinjr@ornl.gov

- characterization of unlabeled or mislabeled samples;
- go/no-go determination of nuclear material content for recovery, burial, transport, or criticality safety;
- rapid inventory verification to check consistency of declared values;
- confirmation of shipper values by the receiver;
- location of nuclear material holdup;
- process monitoring; and
- control of material movement.

Most of these tasks can be accomplished with qualitative or semiquantitative measurements that are rapid enough to save time, money, and personnel exposure.

NDA techniques are well suited to these types of measurements because they are usually fast, nonintrusive, and capable of measuring the item as a whole. If the nondestructive measurement is careful and accurate, it may be considered a material assay. If the measurement is completely qualitative and determines only some signature, fingerprint, quality, or characteristic of the material, it may be considered an attribute measurement. Between the extremes are semiquantitative measurements (such as waste characterization, monitoring of material movement, rapid inventory verification, and identification) and measurement of material holdup. These semiquantitative measurements are often very important to the day-to-day operation of nuclear fuel cycle facilities.

Section 10.2.2 summarizes nuclear material attributes and how they can be measured. Measurements that are more than attribute measurements but less than full quantitative assays include semiquantitative measurements of waste (Sect. 10.4.2) and holdup measurements (Sect. 10.4.3 and Chap. 19, Sect. 19.4.4). For discussions of two other measurement problems that fall into the category of semiquantitative measurements, see Chap. 21 on irradiated fuel assay and Chap. 22 on portal monitoring.

10.2.2 Measurement of Nuclear Material Attributes

The most fundamental task in measuring nuclear material attributes is simply to identify the presence or absence of particular radionuclides or nuclear materials. The primary radiation attributes (regardless of material type) are listed as follows:

- alpha radiation
- beta radiation
- gamma/X-ray radiation
- infrared radiation (heat)
- total neutron radiation
- coincident neutron radiation
- high-fission cross section for thermal neutrons (yielding prompt and delayed gamma rays and neutrons)

Information on the radiation emission rates of these attributes is summarized in Chaps. 2, 13, 23, and 27.

Nuclear material in elemental form is also very dense and strongly attenuates gamma radiation. A further attribute of uranium and plutonium is the discontinuities in their X-ray absorption cross section at the K- and L_{III}-absorption edges (Chap. 12).

Of the attributes listed above, only gamma ray spectroscopy provides unambiguous identification of the specific radionuclides that are present within the material in question, especially if the spectra are measured with high resolution. Although the other attributes mentioned are necessary features of nuclear material, they are not sufficient for unique identification.

In a full-fledged nuclear material assay, almost all attributes cited above are measured at one time or another. A simple way to view attributes measurements is to regard them as incomplete assays. The data are taken in the same way as for complete assays, but the measurements are made more quickly, with less precision, and often without any use of the absolute calibration of the instrument. Even semiquantitative confirmatory or verification measurements may involve only a determination of the relative magnitude of the attribute responses from sample to sample. Table 10.1 summarizes the measurement instruments that are commonly available in nuclear facilities and the attributes they can reveal. Some active assay instruments are included for completeness.

10 Gamma-Ray NDA Applications & Techniques

Table 10.1 Measurement instruments and the attributes they reveal

Instrument	Attribute
Visual inspection	Packaging, history, color
Scales	Weight, apparent density
Alpha counter	Presence of alpha emitting isotopes, contamination
Geiger counter	Gross beta/gamma activity; potential presence of U, Pu, or Am
Gamma spectrometer	Gamma-ray spectrum; U, Pu signature; enrichment; burnup
Radiograph	Density, distribution, shape
Densitometer	Density; X-ray absorption edges; U, Pu signature
Calorimeter	Heat output; presence of high alpha activity; warmth implies Pu, Am
Passive total neutron counter	Neutron emission; presence of spontaneous fissions or (α,n) reactions
Passive neutron coincidence counter	Spontaneous or induced fissions; presence of Pu or Cf likely
Active neutron coincidence counter	Induced fissions; presence of U or Pu likely
Californium shuffler	Delayed neutrons from induced fissions; presence of U or Pu likely
Fuel-rod scanner	Delayed neutrons or gamma rays; presence of U or Pu likely

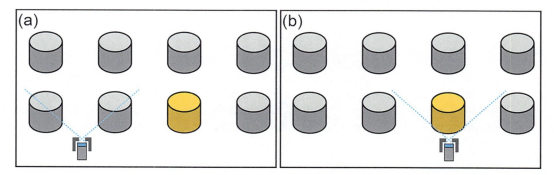

Fig. 10.1 Notional diagram of positioning a gamma-ray detector in measuring (**a**) the representative background and (**b**) the subsequent attribute measurement on the selected item (indicated in yellow). In this scenario, it is presumed that all of the items contain the same nuclear material and that the item in question cannot be readily moved from its current location

Attribute measurements can be a very effective tool for characterizing, verifying, or monitoring nuclear material. Measured one at a time, nuclear material attributes provide simple answers to inventory questions. Measured in combination, they can provide very reliable or even unique information with a minimum of effort.

10.2.3 International Atomic Energy Agency Verification of the Presence of Radionuclides

One of the standard methods for verifying the presence of nuclear material within a container is to verify that the container is emitting gamma rays that are characteristic of the declared nuclear materials. To perform this measurement, the count rate of the characteristic gamma ray must be statistically greater than the ambient background count rate. Because attribute measurements are performed within the nuclear facility and often with the item of interest located near other items containing that same material, the background count rate can be significant and can include the presence of the gamma ray of interest. To ensure that the observed count rate from the gamma ray of interest is coming from the item to be verified and not from neighboring items or the ambient environment, it is important to perform a background measurement that is as representative as possible of the background that is expected when measuring the item of interest. For instance, if the item to be verified (indicated in yellow in Fig. 10.1) were being stored with other items that contain similar types of material, positioning the gamma-ray detector so that the detector's field of view (FOV) was in between neighboring items, as indicated in Fig. 10.1a, would provide a representative background. By positioning the detector in this manner, the background measurement can provide a reasonable estimate for the contributions to the overall count rate from items that are next to and behind the item of interest.

Once an appropriate background count rate has been established, a measurement can be performed to determine whether the characteristic gamma ray is being emitted from the item of interest at a statistically significant higher rate than was observed from the measurement of the background. The determination of what is meant by "statistically significant" depends on the requirements associated with the confidence level of the measurement. In the case of an attribute measurement that is done for safeguards purposes, the minimum statistical confidence level that is needed to conclude that a characteristic gamma ray is being emitted from an item of interest is 99.7%, as shown in Eq. 10.1. To achieve a statistical confidence level of 99.7% or more in an attribute measurement, the characteristic gamma-ray count rate that was measured from the item must exceed the measured count rate in the ambient background by at least 3 times the total uncertainty of the net count rate between the item and the background measurements, as shown in Eq. 10.1.

$$R_M - R_B > 3\sqrt{\left(\frac{\sigma_M}{T_M}\right)^2 + \left(\frac{\sigma_B}{T_B}\right)^2} \qquad (10.1)$$

where
R_M = the count rate in the peak for the measurement
R_B = the count rate in the peak for the background
σ_M = the uncertainty in the measurement counts
σ_B = the uncertainty in the background counts
T_M = the measurement count time
T_B = the background count time

Given that the total uncertainty in the measurement is greatly dependent on the measurement time, one key aspect in performing a successful attribute measurement is to ensure that the count time for the item measurement is sufficiently long enough so that the uncertainty in the measurement is small enough to satisfy Eq. 10.1. In cases where the count rate in the peak from the ambient background is negligible, the condition for verifying an item using the attribute test can be simplified to determine if the net area (A) underneath the peak is greater than 3 times its uncertainty:

$$A > 3\sigma \qquad (10.2)$$

Often, the characteristic gamma-ray energy that is used to verify the presence of a specific radionuclide of interest is one that can be measured and identified using a low-resolution detector system, such as NaI. In the case of performing an attribute measurement on ^{235}U for instance, the 185.7 keV line is chosen as the characteristic gamma ray to verify its presence within the material of interest. Figure 10.2 shows an example NaI spectra with a region of interest (ROI) placed on the 186 keV

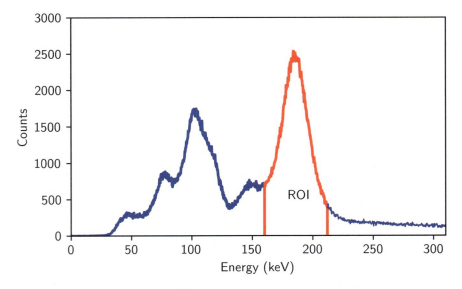

Fig. 10.2 NaI spectra associated with an attribute test of a ^{235}U item. The area under the 186 keV peak is located within the ROI

gamma-ray peak. In cases where the ambient background is negligible, once the measured area underneath the 186 keV peak becomes greater than 3σ, the ^{235}U attribute test will have been satisfied, verifying that the item of interest contains ^{235}U.

10.3 General Nuclide Identification

10.3.1 Introduction

The identification of unknown radionuclides (or "nuclide ID") is a task that encompasses measuring a gamma-ray spectrum from a sample that contains unknown radionuclides and determining all of the gamma-emitting isotopes that contributed to the measured spectrum. The gamma detector that is used to measure the spectrum will often determine the method that is most applicable. For example, an experienced gamma-ray spectroscopist can use a method known as *pattern recognition* to rapidly identify common gamma ray spectra from low-resolution detectors, such as NaI (Tl) or LaBr$_3$; but if presented with a spectrum from a high-resolution detector, the spectroscopist will likely rely on peak identification techniques to ensure that every full-energy peak in the spectrum is identified and accounted for. Alternatively, some radiation detector systems offer automated nuclide identification algorithms that apply computational analysis that is based on similar concepts to those used manually by a spectroscopist [1].

In an ideal situation, the gamma-ray detector system would measure every gamma ray emitted from the unknown item with perfect efficiency across all energies and sufficient resolution to have each gamma ray appear as a distinct peak. In practice, complications arise when using actual gamma-ray detectors to identify unknown radionuclides. Factors that complicate the use of gamma-ray spectroscopy to identify radionuclides include gamma rays present in the ambient background that interfere with the gamma rays being emitted by the item of interest, inability to distinguish gamma rays with similar energies due to the energy resolution of the detector being used, and uncertainty in the energy calibration of the gamma-ray detector, among other factors.

To perform nuclide ID, five practical steps must be taken to identify an unknown source using gamma-ray spectroscopy [2]. First, the detector being used must have an accurate energy calibration that spans the range of gamma-ray energies that are to be detected from the source. Appropriate calibration sources must be used in calibrating the detector, and they must emit known gamma rays over the appropriate energy range. Given that it is rare to have an energy calibration for a gamma-ray detector that is truly linear in nature, extrapolating an energy calibration beyond the range of known gamma-ray energies used to establish the calibration increases the uncertainty associated with the measured gamma-ray energy, which can make identifying a gamma-ray peak of interest within the spectrum more difficult.

Having calibrated the detector system, a background spectrum should be taken to determine how the ambient background is contributing to the detected unknown spectrum. Similar to the case of measuring an attribute described in Sect. 10.2.3, the background spectrum should be taken in the same environment as the unknown measurement will take place. As the background conditions can evolve over time, it is important that the background spectrum be taken as close in time as when the measurement of the unknown item will take place. In cases where a large or dense object is being measured, considerations must be made for how the presence of the item within the area would shield a portion of the background from the detector.

The background spectrum may be subtracted from the measured spectrum of the item of interest to reveal the features that remain within the spectrum that could be attributable to the unknown radionuclides. The process of subtracting background from the measured spectrum must consider any differences that occurred during these measurements, such as the live time of the two measurements.

However, directly subtracting the background can introduce complications into the process of nuclide identification. This process can cause artifacts in the net spectrum, particularly if there are any differences in background intensities (perhaps from a slightly different measurement location or due to shielding from the item of interest), energy calibration (such as from gain shift from temperature changes). Additionally, poor counting statistics may lead to missed features when analyzing background-subtracted spectra. When performing peak-based analysis, an alternative approach is to subtract net peak count rates extracted from both the spectrum of interest and the background.

The remaining features that are present after background subtraction must then be analyzed and compared against the signatures of known radionuclides. The specific features of a spectrum that are used in a radionuclide identification will depend on what analysis techniques are being used, with some techniques using only the peaks themselves while other techniques will use the entire spectrum, including all of the peaks, Compton edges, and Compton continuum [2].

Note that in many cases, to avoid the pitfalls of background subtraction, the live-time normalized representative background is often superimposed on the spectrum of the unknown item so that the two may be compared.

10.3.2 Manual Methods

Two manual methods for nuclide identification are presented here: peak search and lookup and pattern matching. Manual methods involve a gamma-ray spectroscopist who analyzes the raw gamma-ray spectrum collected by the instrument and manually analyzes it, either directly on the user interface of the instrument or, more commonly, by downloading the spectra from the instrument and analyzing it using an external spectral analysis tool.

Peak search and lookup is a manual nuclide identification technique that is performed primarily on high-resolution gamma-ray spectra, commonly collected with high-purity germanium (HPGe) detectors. To perform the best analysis, a minimum of three measurements is recommended:

- Background spectrum from the same location as the unknown sample
- A spectrum from a known gamma-ray source, ideally with multiple gamma-ray lines that can be used to confirm the energy calibration
- A spectrum from the unknown gamma-ray source

The background spectrum is useful to eliminate certain full-energy peaks from the analysis of the unknown spectrum, the known gamma ray spectrum is used to verify the energy calibration, and the unknown is what the analysis is performed on. Figure 10.3 shows a long dwell spectrum of a background collected with an HPGe detector, with the full-energy peaks labeled. Note that a good background spectrum can also be used to verify the energy calibration of the system.

To perform the peak search and lookup, the spectroscopist should first confirm the energy calibration by using either the known or background spectrum. After confirming that the energy calibration is correct, the spectroscopist should overlay and normalize (using the live time) the background spectrum on the unknown spectrum (see Fig. 10.4), which allows the spectroscopist to omit from the lookup process those peaks that are present with the same magnitude in both the background and unknown spectrum. This approach is often a less-error-prone operation than subtracting the background spectrum from the unknown.

Next, the spectroscopist must identify all other full-energy peaks that are present in the spectrum, noting the presence of any peaks that appear wider or of an unexpected shape that could indicate either multiple overlapped peaks or photons from other sources, such as electron-positron annihilation or other nuclear reactions. With all of the full-energy peaks identified in the spectrum, the spectroscopist must then look up the gamma emission lines from radionuclides to determine those nuclides present in the unknown sample. Because many more low-energy gamma rays are emitted than high-energy gamma rays, one

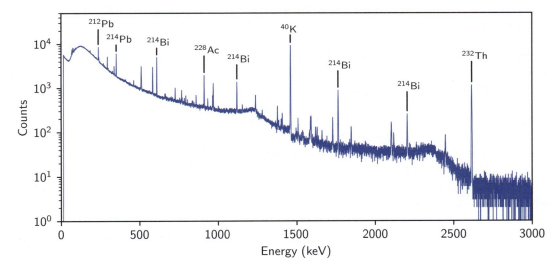

Fig. 10.3 Long dwell background spectrum with peak labels

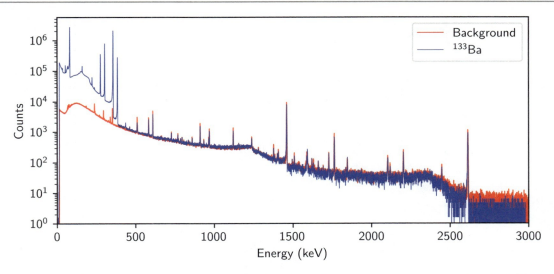

Fig. 10.4 Spectrum of Ba-133 source with background spectrum overlaid. All full-energy peaks at higher energy than 436 keV are present in the same magnitude in both the background and foreground measurement and therefore can be excluded from the analysis

Table 10.2 Isotopes that emit 384 keV gamma rays

Results	Energy (keV)	Difference	Phot/decay	Nuclide	Parent half-life
^{133}Ba	383.85	+0.15	8.940e−02	^{133}Ba	10.53 yr
^{133m}Xe	383.85	+0.15	2.515e−05	^{133}Xe	2.26 days
^{133}Xe	383.85	+0.15	2.400e−05	^{133}Xe	5.29 days
^{146}Gd	383.50	−0.20	4.600e−04	^{146}Gd	48.30 days
^{241}Am	383.81	+0.11	2.820e−07	^{241}Am	433.2 yr
^{227}Ac	383.52	−0.18	3.112e−04	^{227}Th	21.8 yr
^{227}Th	383.52	−0.18	4.674e−04	^{227}Th	18.50 days
^{108m}Ag	383.20	−0.50	8.349e−07	^{108}Ag	418.0 yr
^{239}Pu	383.81	+0.11	8.181e−08	^{241}Am	2.41e+004 yr
^{231}Pa	383.52	−0.18	1.352e−05	^{227}Th	3.28e+004 yr

strategy is to begin identifying unknown full-energy peaks from prominent high-energy peaks to low energy. Once a potential nuclide is identified, the spectroscopist should look to see if that nuclide emits any other gamma-ray lines and if those gamma lines are present in the spectrum. For the example shown in Fig. 10.4, the spectroscopist should start with the unknown full-energy peak at ~384 keV, which results in several possible nuclide candidates as seen in Table 10.2. From the list of possible radionuclides, selecting ^{133}Ba shows that several other gamma rays are emitted and present in the unknown spectrum. The other radionuclides that appeared in the search on the energy of 384 keV emit other gamma rays that do not appear in the unknown spectrum and can therefore be ruled out. Note that some peaks that appear in measured spectra may be single- or double-escape peaks, sum peaks, or X-rays; these should be considered when analyzing an unknown spectrum.

The second manual nuclide identification technique is pattern recognition, which relies on an experienced spectroscopist's recognizing the spectral shape, typically from a low-resolution gamma detector. To become proficient in using the pattern recognition technique, a spectroscopist must see and become familiar with a large number of spectra from various radionuclides that have different detector geometries and shielding. One way to gain this experience is by using spectral flash cards that contain an unlabeled spectrum on one side and the answer key on the back. One such example of spectral flash cards is the SpectraDeck [3]. Typically some level of pattern recognition is required to perform an energy calibration.

Figure 10.5 shows a common and easily recognizable low-resolution spectrum of Co-60. At a glance, an experienced spectroscopist will be able to quickly identify this spectrum as Co-60 without needing to identify the two full-energy peaks and look them up in a table. For radionuclides that emit gamma rays that are closer in energy, lower-resolution detectors might not allow the spectroscopist to identify the unique energies of all gamma rays that contribute to the spectrum and, therefore, pattern recognition becomes a more applicable manual identification method. Figure 10.6 shows a low-resolution spectrum of

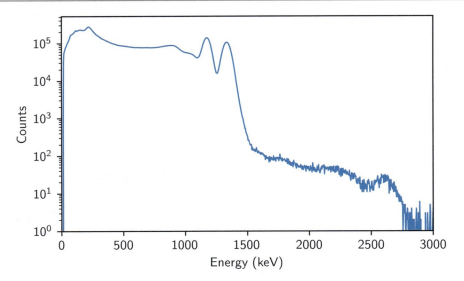

Fig. 10.5 NaI spectrum of unshielded Co-60 is easily identifiable by its two distinct peaks at 1173 and 1332 keV

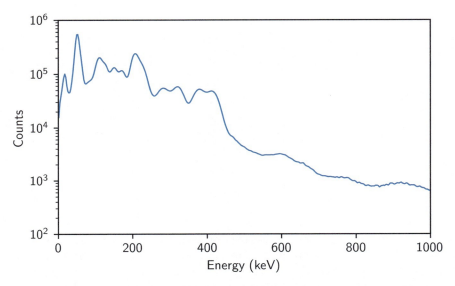

Fig. 10.6 NaI spectrum of unshielded Lu-177m has many gamma lines that cannot be resolved, making pattern recognition difficult. A manual nuclide identification technique could be required to identify the nuclide that creates this spectrum

Lu-177m in which individual full-energy peaks cannot be distinguished, but the overall spectrum has a distinct shape that can be recognized by a trained spectroscopist. Figure 10.7 shows a high-resolution spectrum of Lu-177m with the many gamma-ray lines.

10.3.3 Template Matching

Template-matching algorithms have been developed to automate the process of pattern recognition that is used when analyzing an unknown spectrum. Within a template-matching algorithm, the unknown spectrum is compared with a series of known template spectra with the quality of fit between the spectrum and the template quantified in some manner to ascertain whether the template is considered to be present within the unknown spectrum [2]. One common way to quantify the quality of fit is to use a reduced χ^2 method, which accounts for the differences between the template spectrum and the unknown spectrum on a channel-by-channel basis relative to the uncertainty associated at each channel in the unknown spectrum. This

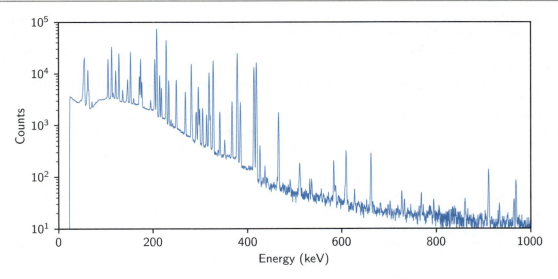

Fig. 10.7 HPGe spectrum of unshielded Lu-177m that shows many of the prominent gamma lines

method produces a value, which allows for the algorithm to determine whether the difference between the template and the unknown spectrum is statistically acceptable. The algorithm will determine which templates are the closest match to the unknown spectrum based on specific criteria associated with the quantification of the quality of fit [2].

To account for a wide range of detection conditions, the algorithm's database may contain several template spectra for a single radionuclide. Each template spectrum that is present within an algorithm's database will have either been calculated or measured with knowledge of the intervening materials between the source and the detector, the measurement geometry, and the detector medium [2]. Given that the template spectra are a product of both the radionuclide being detected and the conditions associated with the template, the matching of an unknown spectrum to a specific template spectrum can determine both the radionuclides present in the spectrum and the properties of the attenuating material between the source and the detector.

A complication that arises in template-matching algorithms is when analyzing an unknown spectrum that has more than one radionuclide present because the template spectra present within the algorithm's database are often produced from a single radionuclide source. To properly account for the possibility of more than one radionuclide's presence within the unknown spectrum, template-matching algorithms must either be able to dynamically and appropriately combine several template spectra together to compare with the unknown spectrum or have within their database special templates that were created based on the simultaneous measurement of different sources of radionuclides. Given the unconstrained number of measurement conditions that might need to be considered in analyzing an unknown spectrum (attenuation, distance to source, the presence of multiple sources, etc.), one of the drawbacks with using a template-matching algorithm is the lack of a clear template within the database that matches the unknown spectrum in a statistically significant manner [2]. In these cases, the spectroscopist would need to become more involved in the analysis of the spectrum either through working with the template-matching algorithm or through other means to identify the source.

In addition, implementations of template-matching (and other automated nuclide identification) algorithms generally use a relatively small number of radionuclides for their libraries. This practice reduces the computation needed—especially if mixtures are considered by the algorithm—and generally improves the probability of correct identification for nuclides that are in the library. However, the practice also renders these implementations unable to identify any radionuclide not included in the library. Radionuclide libraries are then generally selected for the application as well as the most likely radionuclides to be encountered [4].

10.3.4 Recent Software Applications

10.3.4.1 Spectral Analysis Software

Spectral analysis software encompasses any software that is used to view, analyze, and interpret gamma-ray spectra. This software could include the software and user interface that is incorporated directly into a radiation detection system, or it could be a standalone software that is used to perform gamma-ray spectroscopy on offline data. Although some differences exist in the features and functionality of each software package, the primary functionality of these applications is mainly the same. These applications include the ability to open spectra in many different file formats, zoom in or out, fit full-energy peaks, and have look-up libraries for gamma and X-rays. These applications commonly have additional features, such as summing multiple spectra, subtracting background spectra, adjusting the energy calibration, dose-to-activity conversion, and other useful tools for the analyst.

Spectral analysis software is specifically designed for analysis and is therefore often much easier for a spectroscopist to use when performing nuclide identification than any spectral viewing/analysis features included with the data acquisition software. Two commonly used spectral analysis applications are PeakEasy[1] and InterSpec.[2]

10.3.4.2 Detector Response and Inject Modeling Software

Detector-specific response and inject modeling software is used to synthetically generate gamma-ray spectra based on a source term that is modeled in a user-defined configuration. The source term can be modeled as a simple point source, or the software may allow for more complex problems, such as large, distributed sources, multilayered shielding, and sources dispersed in a matrix. These software applications are able to solve radiation transport problems to determine the gamma-ray flux incident on a detector. The software then applies a detector-specific model to account for the physical detector response to incident photons. This detector-specific modeling accounts for the intrinsic efficiency, energy resolution, and detector-specific pulse-processing electronics (dead time, pileup effects, etc.) to create an accurate, detector-specific spectrum for a specific source model.

Detector-specific response and inject modeling software is a powerful tool that can assist an analyst in multiple ways. For an unknown item in a complex geometry environment, a spectroscopist may use any of the previously discussed methods for performing the isotope identification. However, to gain a better understanding of the unknown source, the analysist may build the scenario with the modeling tool, tweaking parameters to determine the matrix, density, distribution, and other factors that would produce the measured spectrum from an unknown item. In other applications, these tools may be used to determine scenario-specific limits of detection, evaluate detector performance without the need for resource-intensive testing, or develop detector-specific training material for analysts and spectroscopists. One example software that contains all of the functionality described here is GAmma Detector Response and Analysis Software (GADRAS).[3]

10.4 Quantitative Measurements

10.4.1 Isotopic Composition Measurements

10.4.1.1 Introduction

An accurate measurement of uranium and plutonium isotopic composition is usually required to interpret the results of neutron coincidence or calorimetry measurements. Several methods have been developed for determining uranium and plutonium isotopic composition by using gamma-ray spectroscopy; some of the early approaches are described in Refs. [5–9]. An American Society for Testing and Materials standard test method has been written for plutonium isotopic analysis using gamma-ray spectroscopy [10]. Different methods have been developed for different sample types.

The principal application of plutonium isotopic measurements is to support other NDA methods in providing the total plutonium content of a sample. Two methods that use plutonium isotopic results are calorimetry and neutron coincidence counting.

Calorimetry uses the isotopic information to calculate the specific power P (W/g plutonium) of a sample from the measured isotopic fractions and the known specific power for each isotope (see Chapters 23 and 24).

The response of neutron coincidence counters is a complicated function of the plutonium isotopes and the ^{241}Am. A measurement of isotopic composition is required to convert the coincidence counter response to plutonium mass (see Chap. 17).

---

[1] Limited distribution, obtained from https://peakeasy.lanl.gov
[2] Open source; obtained from https://sandialabs.github.io/InterSpec/
[3] Limited distribution, obtained through https://rsicc.ornl.gov/

10.4.1.2 Data Acquisition Considerations for the Measurement of Uranium and Plutonium

10.4.1.2.1 Detectors

All data analysis methods benefit from the use of a detector that gives the best possible resolution and peak shape. These parameters are most important when selecting a detector for a plutonium isotopic system. A planar, HPGe detector is used in most applications. A detector with a front surface area of 200 mm^2 and a thickness of 10–13 mm gives a good trade-off between resolution and efficiency. Such detectors are available commercially with a resolution (full width at half maximum [FWHM]) better than 500 eV at 122 keV. A peak-shape specification of 2.55 or better for the ratio of full width at one-fiftieth maximum to FWHM at 122 keV helps ensure good peak shape. Good detectors give values of 2.5 or below for this parameter. The low efficiency of planar detectors restricts their use to regions below 400 keV. High-quality coaxial detectors can be used in the 100–400 keV region, but their lower resolution complicates the analysis of partially resolved peaks using channel-summation methods.

A coaxial detector with a relative efficiency of 10% or higher is recommended for measurements in the 600 keV region. Again, resolution is important. The very best resolution may negate the need to peak-fit the entire 600 keV region [11]. Resolutions of 1.7 keV or better at 1332 keV are available.

10.4.1.2.2 Filters

Filters must be used in nearly all situations that involve plutonium-bearing items to reduce the count rate from the 59.54 keV ^{241}Am gamma ray that dominates the unfiltered spectrum from any aged sample. If the detector is unfiltered, the americium peak will cause unnecessary dead time and will sum with X-rays and gamma rays in the 100 keV region to produce interferences in the 150–165 keV region. Typical filters use 0.15–0.30 cm of cadmium and 0.025 cm of copper to selectively absorb the 59.54 keV gamma ray. A reasonable rule of thumb is to design the filter to reduce the 60 keV peak height to just less than the peak heights in the 100 keV region. A thicker filter will unnecessarily reduce the intensity of the important plutonium peaks in the 120–200 keV area (see Chap. 9, Sect. 9.3.3). A further test for an adequate filter is to check that the region between 153 and 160 keV is flat and contains no sum peaks [12]. A more complete discussion of filter design is given in Chap. 3.

Little, if any, filtering is needed for freshly separated samples (no ^{241}Am or ^{237}U) when using the 100 keV region or the 40 keV region. If the detector is shielded with lead, the shield is often lined with approximately 0.25 cm of cadmium to suppress lead X-rays (72–87 keV) that would otherwise appear in the spectrum.

No filters are typically used or needed when measuring uranium items due to the lack of peaks that could interfere with the peaks from ^{235}U and ^{238}U.

10.4.1.2.3 Count Time

The count time required to produce the desired precision is a function of the spectral region studied. In the 40 and 100 keV regions, count times of 300–600 s are usually satisfactory. Count times of 1–2 h or longer are often necessary when using gamma rays above 120 keV to measure high-mass samples, although in some situations, samples as small as 10 g can be measured to better than 1% in less than 30 min. Small samples (1–2 g or less) could require overnight measurement times. For large samples, simple verification of the ^{239}Pu/^{241}Pu ratio may take only a few minutes.

10.4.1.2.4 Count Rate and Sample/Detector Geometry

Large plutonium samples have high neutron emission rates; 1 kg of plutonium emits 1 to 2 × 10^5 n/s. High neutron exposure is known to damage germanium detectors and degrade detector resolution. It is difficult to minimize this effect because, as the sample-to-detector distance is increased, the count time must be increased, and the neutron dose remains essentially constant.

10.4.1.2.5 Throughput and Resolution Trade-Off

For the same detector and electronic equipment (depending on the electronic setup and the measurement input count rate), the precision and accuracy of a measurement can be significantly different. The two mentioned parameters (electronic setup and input rate) are independent from each other, and a change in one or both parameters can affect the results of different analysis differently depending on the type of analysis and the analytical energy region.

Figure 10.8 shows examples of the resolution and throughput as a function of the input rate. The curves are generated from actual data of the 122 keV peak of a ^{57}Co source and the 1332 keV peak of a ^{60}Co source measured by a coaxial detector at three different rise times of an adjustable digital multichannel analyzer that uses a quasi-trapezoid pulse shape. The three rise times of 4, 6, and 12 μs pulse shaping in the figure correspond roughly to the shaping times of 2, 3, and 6 μs of the traditional analog amplifier.

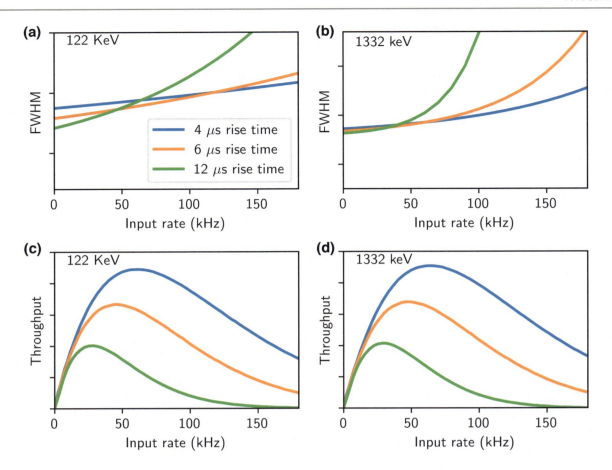

Fig. 10.8 Resolution and throughput as a function of the input rate. Both the FWHM and Throughput vertical axes have a minimum value of 0

For input rates less than 10 kHz, the throughput curves with different rise times are about the same, whereas the resolution of the 12 μs rise time is the best. As the input rate increases, smaller rise time would give better throughput, and for very large input rate, smaller rise time would also give better resolution. For a system that is used in low count-rate environments, a large rise-time setup is preferred. For a versatile system that measures sources of various intensities, a smaller rise-time setup is more designable.

The relationship between the resolution and throughput can also be displayed in a somewhat different viewpoint. Figure 10.9 shows the resolution and throughput as a function of the percent dead time.

The throughput curve shape is independent of the rise time, and its maximum occurs at a dead time of $1 - e^{-1} = 63\%$, as shown in Eq. 7.47. As Fig. 10.9 shows, there is trade-off between the resolution and throughput for dead time below 63%. Under no circumstances should the dead time of a measurement be allowed to exceed 63%. If the dead time exceeds 63%, then the user should increase the source-detector distance, add an extra filter, or reduce the rise time or shaping time of the data acquisition system to reduce the dead time to below 63%.

The shape and position of the FWHM curve in the figure can change somewhat depending on the detector and the electronics. It can be lower and/or level for a good system or higher and/or sloped for a poor system. With a good system, the dead time of a measurement can be pushed closer to 63%, but for a poor system, the trade-off from a small gain in throughput near the maximum is not worth the significant broadening of the peaks due to the large slope of the FWHM curve.

The peak broadening caused by increased dead time or input rate has different effects on different peaks. A clean peak on a low background will not be significantly affected; its uncertainty will increase only slightly. The increase in uncertainty of an isolated peak on a high background will be more noticeable. When a peak is joined with another peak and requires complicated fitting to extract its peak area, then a slight increase in its FWHM can translate into a large increase in the uncertainty.

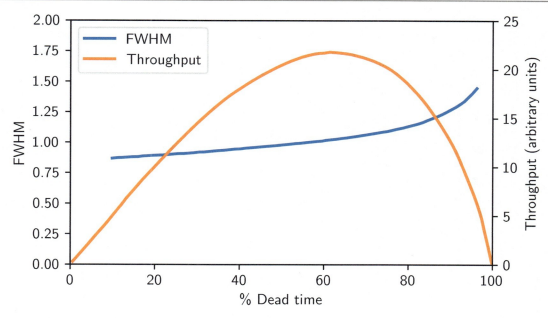

Fig. 10.9 Resolution and throughput as a function of the percent dead time

The uranium isotopic analysis employing the 120–1010 keV region uses peaks that are relatively clean. For this analysis, there is a balance between the increased uncertainty from the increased FWHM and the decreased uncertainty from the increase in throughput near the maximum. Therefore, the optimal statistical uncertainty can be achieved with a dead time slightly smaller than 63%.

For uranium analysis that employs the X-ray region, the ^{238}U activity is determined with a pair of ^{234}Th peaks at 92.4 and 92.8 keV. This pair is close to the thorium X-ray peak, at 93.4 keV from ^{235}U decay. A small increase of the FWHM can result in a large increase of the area uncertainties of this pair of peaks, which then translates to large uncertainty for ^{235}U and ^{238}U results.

For plutonium analysis, ^{240}Pu has only a few weak gamma ray peaks that can be measured and used for activity calculations. The analysis employs the low-, middle-, and high-energy regions using the ^{240}Pu peaks at 104.2 keV, 160.3 keV, and 642.5 keV, respectively. These peaks are weak and mixed in with other peaks; therefore, a small increase in the FWHM can significantly increase the uncertainties of these peaks. The dead time for optimal precision (smallest statistical uncertainty) can be far from the 63% dead time. Figure 10.10 shows the 208 keV peak area and FWHM and the Fixed-Energy Response-Function Analysis with Multiple Efficiency (FRAM)–reported uncertainties for ^{240}Pu effective and specific power as a function of the dead time of five plutonium spectra. The measurement time was 7200 s true time for each spectrum. The item was low burnup with 6.4% ^{240}Pu. The detector system used to collect these data consisted of a 25% relative efficiency coaxial HPGe detector together with the nuclear instrumentation module electronics. The high-energy (180–1010 keV) region analysis of FRAM was used to analyze the data.

The curves are the third order polynomial trendlines. The maximum of the 208 keV peak area is at 63.5% dead time, and the minima of the specific power and ^{240}Pu effective curves are at 45.5% and 49.4%, respectively. The detector system used to acquire these data was good, with only 7% broadening of the peaks when the dead time increased from 14% to 85%. For a system that is not as good, then the specific power and ^{240}Pu effective minima would occur at lower dead times.

The results in Fig. 10.10 show that when the dead time is reduced from 63.5% to 45.5%, the 208 keV area is reduced from 2.87×10^6 to 2.55×10^6 (11% reduction), the FWHM changed from 1.00 keV to 0.98 keV (2% improvement), and the specific power error is reduced from 0.42% to 0.34% (19% improvement). This result means that for this specific case, a small improvement of only 2% in resolution can significantly increase the precision despite the 11% loss in statistics. The trade-off will be different for analyses that employ different energy regions or different detector systems.

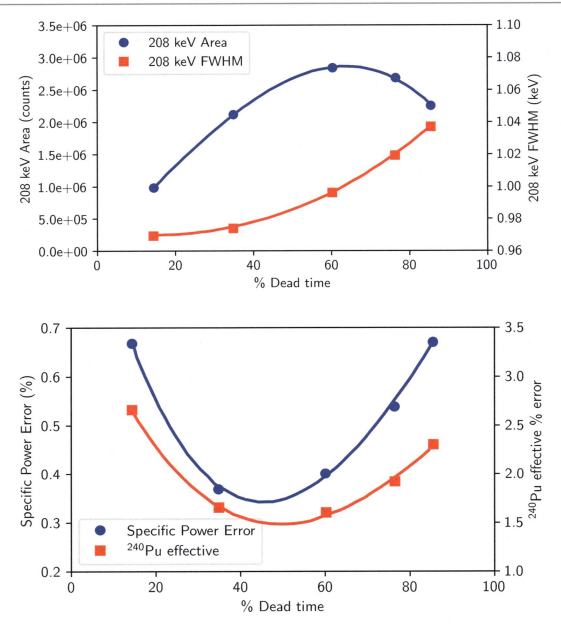

Fig. 10.10 Resolution, throughput, ^{240}Pu effective, and specific power as a function of the percent dead time. The 208 keV area curve peaks at 63.5%, and the specific power and ^{240}Pu effective curves bottom out at 45.5% and 49.4%, respectively

10.4.1.2.6 Detector Efficiency and Resolution Trade-Off

It is well known that for solid-state detectors such as HPGe and CZT, the larger detector would give better efficiency but worse resolution. The detector efficiency and resolution that are optimal for a measurement therefore depends on the measurement situation and what is being measured.

For example, if the source is a small 93% ^{235}U item, then the high-energy peaks of ^{238}U at 1001 keV are very weak due to the small fraction of ^{238}U in the item. A large coaxial detector is not a good choice because, for low-energy region analysis, its resolution in the 100 keV region could be insufficient, causing the areas of the pair of ^{234}Th peaks at 92.4 and 92.8 keV to have large uncertainties despite good statistics. The analysis in the high-energy region is not so good either because of the weak 1001 keV peak. In this case, the small planar detector is better because it can analyze the low-energy region well.

However, if the sample is large, then the scattered gamma rays from inside the large item will increase the background in the 100 keV region and reduce the precision of the analysis employing the low-energy region. In the meantime, the large item will prevent most of the low-energy peaks from getting out while still allowing the high-energy 1001 keV peak to get out of the item and to the detector. The overall effect is an increase of the 1001 keV peak intensity relative to the lower-energy peaks, which improves the analysis employing the high-energy region of the large coaxial. Combining with the decrease in the precision of the analysis employing the low-energy region of the planar detector, the high-energy region analysis with the coaxial detector is better.

Another example: If the source is very weak, then a large coaxial detector might be better than the small planar detector because it can get much better statistics. Conversely, if the source is intense, then the planar may be better because the optimal throughput for both systems will be the same, but the planar detector system will give better resolution.

Choosing a coaxial detector leaves the question of how large the crystal should be. For plutonium measurement, the highest useful peak is below 800 keV, whereas for uranium analysis, the highest useful peak is at 1001 keV. With these energies, a detector with highest efficiency is not necessary. A coaxial detector with a relative efficiency of approximately 25% is about right for all of the plutonium and uranium measurements with the coaxial detector.

10.4.1.2.7 Systematic Uncertainty

Discussion in the previous section was entirely about the statistics and precision of a measurement. When the input rate and/or the dead time of a measurement increases, the probability for the true and chance coincidence summing also increases. These coincidence events will decrease the counts of the involved individual peaks and create sum peaks where its energy is the sum of the energies of two individual gamma rays. The reduced counts of the individual peak areas, if not corrected, will affect the accuracy of the analysis. The creation of the sum peaks, which can be numerous, can either increase or decrease the area of a measured peak; if its energy is very close to the measured peak, then its counts will be added to that of the measured peak, and if its energy is right at the background region of the least-square fit, then it will increase the background and reduce the measured peak area.

Figure 10.11 shows the two spectra of a high-burnup plutonium item acquired at two different measurement conditions. The higher spectrum was acquired at 34% dead time and had the 60 keV peak height at 10% of the peaks in the 100 keV region. The lower spectrum was acquired at 40% dead time and had the 60 keV peak height at 120% of the peaks in the 100 keV region.

The top spectrum shows a relatively flat background below 159 keV and above 162 keV, whereas the bottom spectrum shows several peaks at 156.6, 158.5, and 162.5 keV. These are the sum peaks of the 59.5 keV peak of ^{241}Am summing with the 97.1 keV neptunium X-ray peak and the 98.9 keV and 103.0 keV gamma-ray peaks from ^{241}Am decay, respectively. Another intense sum peak occurs at 160.6 keV (59.5 keV + 101.1 keV neptunium X-ray) that is not resolved from the 160 keV peak structure. The inclusion of this 106.6 keV sum peak will increase the area of the 160.3-keV ^{240}Pu peak, whereas the 158.5 and 162.5 keV sum peaks would likely increase the background of the fit region and decrease the 160.3 keV peak area.

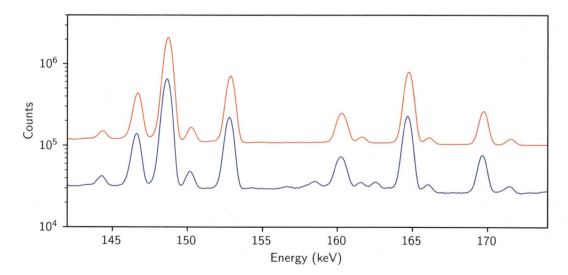

Fig. 10.11 Overlaid spectra of a high-burnup plutonium (26% ^{240}Pu). The higher spectrum (red) was acquired at 34% dead time and had the 60 keV peak height at 10% of the peaks in the 100 keV region. The lower spectrum (blue) was acquired at 40% dead time and had the 60 keV peak at 120% of the peaks in the 100 keV region

For a spectrum such as this, the analysis that employs the mid-energy region that uses the 160.3 keV peak would give incorrect ^{240}Pu activity. This situation can be remedied by either moving the source away from the detector to reduce the input rate and the dead time or by adding more filter to significantly reduce the 59.5 keV peak to only a fraction of the peaks in the 100 keV region.

10.4.1.3 Uranium Enrichment-Meter Techniques

Applications for determining the enrichment of uranium items via the net count rate of the 185.7 keV peak of ^{235}U can be broadly grouped under the enrichment-meter approach. In its simplest form, the enrichment of an item can be determined with the following formalism:

$$E = K \cdot \dot{C}_{186} \qquad (10.3)$$

Here, the enrichment E is related to the net count rate of the 186 keV peak of ^{235}U, $\dot{C}_{186}$, by a calibration constant K.

Such measurements are highly dependent on item and detector geometry and require representative calibration measurements to be made ahead of time. Changes in the geometry, detector FOV, or attenuation characteristics of the material matrix and intervening material with respect to that of the calibration can severely bias the enrichment results. Infinite thickness measurements are a subset of enrichment meter measurements, although many often equate the two. For the former, the item fills the FOV of the detector with a uranium matrix that is infinitely thick with respect to the 186 keV gamma rays from ^{235}U. This effect was shown in Fig. 8.3. The advantage of the infinite thickness approach is that it is easier to maintain the fixed-geometry requirement of the enrichment meter while shielding the detector from 186 keV gamma rays from uranium items that would otherwise be in the detector FOV. Although infinitely thick for 186 keV gamma rays, an item may not be infinitely thick for the 1001 keV and other associated gamma rays from ^{238}U progeny. Although such peaks are not considered directly in the enrichment meter approach, their scattering continuum can affect the relative uncertainty on the net-count rate 186 keV peak.

The first enrichment meter approach relied on two ROIs. This approach was designed for use with low-resolution detectors such as NaI, which did not allow direct determination of the background continuum underneath the 186 keV peak. This method required two calibration constants and two calibration measurements. Because this method has been mostly superseded by peak-fitting approaches, it is mentioned here only for historical purposes. For more information on the two-region enrichment meter, see Chap. 8, Sect. 8.3.2.

To apply the enrichment-meter technique using only a single calibration constant and without fitting the 186 keV peak requires the use of a high-resolution detector system (i.e., HPGe). Here, three ROIs are used with a region on both the low- and high-energy sides of the 186 keV peak that allow for direct determination of the underlying continuum, as shown in Fig. 10.12.

Modern applications of the enrichment meter to low-resolution data have generally been based on the approach of the NaIGEM code. NaIGEM (NaI Gamma Enrichment Measurements) is a code for determining uranium enrichment from low-resolution gamma-ray spectra [13]. NaIGEM, which is intended for use with NaI or LaBr$_3$ detectors, provides an approach to the enrichment-meter method that is an alternative to, and in many ways an improvement over, the two-region method discussed previously. As in the latter, the net-peak area of the 186 keV region of ^{235}U is related via calibration to the enrichment of items for a particular measurement configuration. However, instead of obtaining this net peak area using a simple relationship between the ROI gross counts as in the two-region method, NaIGEM makes use of least-squares fitting within a single ROI centered on the 186 keV peak. Although this approach is more complex, it offers advantages over the two-region method.

A result of NaIGEM's approach for obtaining the net area of the 186 keV peak of ^{235}U is that, in principle, only one calibration measurement is required for a fixed geometry. A caveat to this is that in the case of thick attenuators (e.g., 12–18 mm of steel) between the detector and the item, an additional calibration measurement will be needed to fine tune the fit of the resulting scattering component. However, this method is as opposed to the two-region method, which always requires two calibration measurements to solve for the two calibration constants relating the 186 keV and background-continuum ROIs.

The fitting components used by NaIGEM account for peaks from ^{235}U, the scattering continuum from higher-energy gamma rays of ^{238}U daughters, the scattering from the 186 keV peak itself, and interferences from ^{232}U daughters. A screenshot of a fitted 186 keV region is shown in Fig. 10.13.

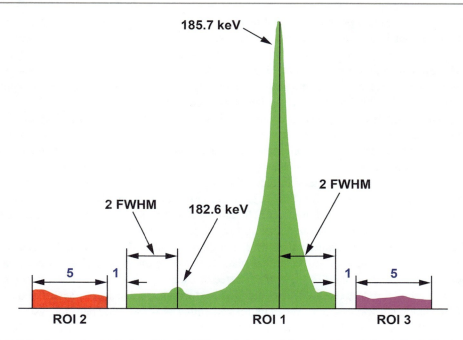

Fig. 10.12 Illustration of the three-region method used for the high-resolution, enrichment-meter technique. The ROIs above and below the ^{235}U peaks at 182.6 keV and 185.7 keV are used to directly determine the underlying Compton continuum

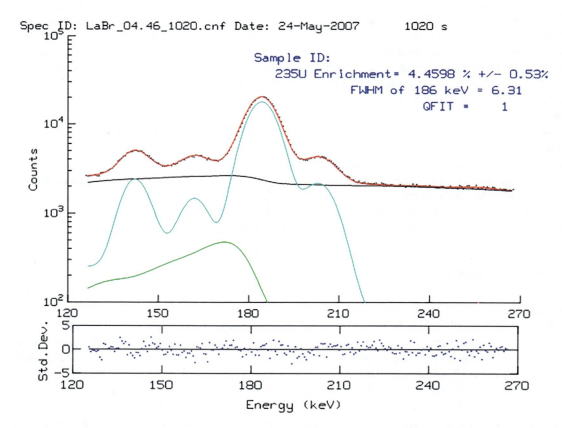

Fig. 10.13 Screenshot of NaIGEM fit of the 186 keV region of a uranium spectrum recorded with a LaBr detector [13]

By specifically fitting the continuum component that results from low-angle and multiple scattering of 186 keV gamma rays, NaIGEM circumvents an issue that can plague the two-region method. Due to the simple approach of applying a transmission factor $e^{-\mu\rho x}$ to correct for additional attenuation, the two-region method fails to account for the continuum that arises from scattering within the attenuator. This scattering component lies in part within the low-energy side of the 186 keV ROI and can result in an over-correction by the transmission factor, which in turn will bias the enrichment results. By fitting this scattering component directly, NaIGEM greatly mitigates this source of bias.

Uranium that has been through the fuel cycle at least once could become contaminated with ^{232}U, the decay chain of which contains ^{228}Th daughters that can complicate uranium gamma-ray spectra especially at higher enrichments. Therefore, the 238.6 keV peak of ^{212}Pb can become a source of interference in determining the continuum that underlies the 186 keV ROI. By including this contribution from the ^{232}U decay chain as a fitting component, NaIGEM is still able to accurately determine the background continuum and obtain the net area of the 186 keV peak in the presence of significant ^{232}U daughters.

Additionally, NaIGEM has some tolerance for shifts in gain and detector resolution. Because of this and the above reasons, NaIGEM has become a widely used tool in uranium-enrichment measurements as well as forming the basis for specialized enrichment-measurement codes to be discussed in a later section of this chapter.

In cases where the items are either not infinitely thick or do not fill the entire field-of-view of the detector (such as fresh fuel rods, fresh fuel pellets, or materials test reactor fresh fuel plates), a fixed-geometry method can be used where the item is placed at a fixed location relative to the field-of-view of the detector. Accurate enrichment measurements are possible using this method because the fresh fuel rods, pellets, or plates are produced with a specific set of physical properties (physical size, density, poison content, cladding thickness, etc.). When performed with a low-resolution detector system, a NaIGEM analysis is often used in extracting the count rate of the 186 keV peak from the spectrum. As was the case when NaIGEM was used for infinite thickness measurement, a single calibration measurement is needed to relate the 186 keV count rate to the uranium enrichment of the item.

In the case of fresh fuel rods and pellet measurements, differences between the physical dimensions of the items used to calibrate the system and the items being measured in the field can have a major impact on the accuracy and precision of the measurement results. This impact occurs because fresh fuel rods and pellets are made with various diameters, heights, cladding thicknesses, and neutron poison content, all of which will impact the count rate of the 186 keV peak independent of the enrichment of the item. Therefore, analysis of fixed-geometry enrichment measurements must include corrections for the physical size and properties of the item being measured relative to the items used to calibrate the system. These corrections are often generated using Monte Carlo methods, which calculate the response of the detector system to changes in the item characteristics.

Unlike infinite thickness measurements, where the item completely fills the detector's FOV, ^{235}U gamma rays that are present within the ambient background must be accounted for when performing fixed-geometry measurements because the items being measured are not infinitely thick and do not fill the detector's FOV. If possible, fixed-geometry measurements should be performed in as low an ambient background as possible. Before performing a measurement, a background run should be performed to ascertain the rate at which 186 keV gamma rays are being detected from the ambient background. A high ambient background can potentially reduce the accuracy and precision of the enrichment measurement if the rate of detecting the 186 keV gamma rays from the ambient background is comparable in size to the rate of gamma rays detected from the item.

10.4.1.4 Online Enrichment Monitor

The online enrichment monitor (OLEM) is the latest in the evolution of instruments used to determine the enrichment of UF_6 in process piping at gas centrifuge enrichment plants.

The OLEM was built specifically to perform unattended measurements on unit header pipes where the UF_6 gas pressures are higher than in the cascade. For example, for UF_6 product, the OLEM could be installed outside of the cascade hall after the unit header pump, where the UF_6 is denser, because it is under higher pressure (Fig. 10.14). This higher-density UF_6 gas provides a stronger gamma signature from the ^{235}U.

As shown in Fig. 10.14, the OLEM system often consists of two tamper-indicating enclosures because existing pressure transducers are often installed on secondary manifolds to facilitate calibration and maintenance. One enclosure mounted to the main unit header pipe contains the single-board computer, gamma spectrometer, and temperature sensor. This enclosure typically connects via tamper-indicating conduit to a secondary enclosure that surrounds the pressure sensor.

Initial OLEM prototypes were developed by Oak Ridge National Laboratory for use by the International Atomic Energy Agency. Some preliminary modeling is described in Ref. [14], and a complete description of the as-built units is available in Ref. [15].

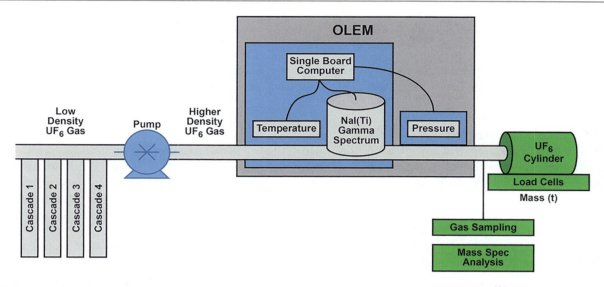

Fig. 10.14 In a unit header pipe, OLEM determines the density of the gas passively to determine the enrichment of ^{235}U in the gas as a function of time. OLEM uses the higher density found after the pump to improve the signal-to-noise ratio of the gamma measurements. OLEM may be used in combination with gas sampling or automated mass spectrometry analysis to provide infrequent, high-accuracy assay measurements. OLEM provides no mass flow information, but mass flow information could be acquired from load cells

OLEM determines the ^{235}U enrichment as a function of time by determining the density of ^{235}U compared with the total density of U, as shown in Eq. 10.4:

$$Enrichment\ (t) = \frac{Density_{U235}\ (t)}{Density_U\ (t)} \quad (10.4)$$

The density of ^{235}U is determined from gamma spectrometry, and the uranium density is determined from gas temperature and pressure measurements, unlike some of its enrichment-monitoring predecessors. The density of uranium in the UF$_6$ gas can be calculated empirically by Eq. 10.5 as provided by Refs. [16, 17], assuming the gas is pure UF$_6$. Equation 10.5 adjusts for deviation from the Ideal Gas Law due to stronger intermolecular forces of the UF$_6$ vapor. Equation 10.5 provides the density in terms of grams per cubic centimeter when pressure (P) is provided in atmospheres, and the temperature is provided in Kelvin. The pressure of the gas can be measured directly with access to the pipe, and the temperature of the gas is inferred by measuring the temperature of the pipe.

$$Density_U = \frac{4.291\ P}{T\left[1 + \left(-1.3769 \times 10^6\ \frac{P}{T^3}\right)\right]} \quad (10.5)$$

The density of ^{235}U (Density$_{U235}$) is proportional to gamma-ray emissions from ^{235}U, which are typically measured using the 186 keV peak. A calibration is needed to determine the proportionality constant (k) that accounts for the unit header pipe characteristics and measurement geometry. Monte Carlo N-Particle (MCNP) transport modeling can provide a first order approximation for k. The precision of k can be improved using UF$_6$ flow loop measurements or by comparing OLEM results with high-accuracy assay measurements such as those from a mass spectrometer.

The OLEM gamma measurement, like most measurements, is affected by background radiation sources. The background radiation may be from wall deposits of uranium inside the pipe or from other background sources in the area. The signal from the wall deposits can be determined by analyzing the 186 keV counting rate while plant operations cause pressure changes in the pipe; the 186 keV intensity measured from the gas is proportional to the gas pressure (at a fixed enrichment), whereas the wall deposit signal is independent of gas pressure. These pressure transients allow for the estimate wall deposit background. The OLEM measurement determines the enrichment (E) by using the counts from the gas (counts$_{186gas}$) by subtracting the counts from other background sources (counts$_{186bkg}$), as shown in in Eq. 10.6.

$$E = \frac{Density_{U235}}{Density_U} = k\frac{counts_{186gas}}{Density_U} = k\frac{(counts_{186gross} - counts_{186bkg})}{Density_U} \quad (10.6)$$

The OLEM measurement approach is capable of achieving approximately 1% relative uncertainty for an low-enriched uranium (LEU) product stream (i.e., 4.95 ± 0.05%), 3% for the feed stream (~0.71% ± 0.02%), and 10% for the tails stream (~0.20% ± 0.02%).

10.4.1.5 Isotopic Analysis Codes

Response-function analysis codes use the principles discussed in Chap. 7 to calculate the shape of the detector response to a particular isotope in a particular energy region. The peak-fitting procedure assigns a separate term with the form of Eq. 7.30 to each full-energy peak in the analysis region and allows some or all of the shape parameters to be free. The response-function analysis uses the same equation but fixes all shape parameters and the relative amplitudes Y_0 of all of the peaks from the same isotope; the only free parameters in the fitting procedure are the amplitudes of the isotopes that contribute peaks to the analysis region. The fitting procedure is reduced to a linear least-squares analysis that can be quickly solved.

The peak-shape characteristics of the detector must be known or determined from the spectrum of the sample under study. If the parameters are determined directly from each spectrum, variations that are due to different count rates or instrumental changes are automatically registered. (The procedure used to determine peak positions and shape parameters is discussed in Chap. 7, Sects. 7.2 and 7.4.) Given the shape parameters and positions for all gamma-ray peaks, it is easy to compute the response profile of each isotope in the analysis region. Response-function analysis has been used to fit the complex 100 keV region [8, 18, 19] and the many regions between 120 and 370 keV [20, 21]. The formalism of Ref. [22] should be used to compute X-ray line shapes when analyzing the 100 keV region.

10.4.1.5.1 Multi-group Analysis Program for Plutonium and Uranium Isotopic Abundance

The Multi-Group Analysis (MGA) software program for plutonium isotopic measurements was first developed in 1979 by Ray Gunnink at Lawrence Livermore National Laboratory [23]. The MGA code was initially developed to analyze the low-energy gamma-ray spectra (energies less than 300 keV) as measured using a high-resolution germanium detector to determine the plutonium isotopic composition of an item. One of the main advantages of analyzing peaks with energies less than 300 keV is that this region contains at least one intense gamma ray from all of the relevant plutonium isotopes except ^{242}Pu and ^{241}Am. Because the relevant gamma rays being analyzed are close in energy, the differences in both the attenuation and the detector efficiency for these gamma rays are small. To extract the area underneath individual gamma-ray peaks within the complex, MGA uses advanced analysis techniques to unfold overlapping peak groupings. The specific techniques used to extract these peak areas were previously discussed in Chap. 7, Sect. 7.4.9.

The MGA code represents one of the first implementations of using a relative or "intrinsic" efficiency curve (see Chap. 7, Sect. 7.9) to analyze a plutonium spectrum to extract plutonium isotopic fractions [23]. MGA determines the relative efficiency curve from the measured spectrum by using a functional form to describe the detector efficiency, the absorption by the steel container walls, and the self-attenuation of the gamma rays within the sample [24]:

$$I_j = \sum_{k=1}^{N}(p_{j,k} \cdot X_k)\left[\frac{e^{(-\mu_{con}(E_j)\cdot t_{con})}\cdot\left(1 - e^{(-\mu_{smp}(E_j)\cdot t_{smp})}\right)}{\mu_{smp}(E_j)\cdot t_{smp}}\right] \cdot \left[\varepsilon_j^0 \cdot (1 + bE_j + cE_j^2)\right] \quad (10.7)$$

where
I_j – intensity of selected peak in the spectrum from the isotope of interest
X_k – disintegration rate of isotope k (unknown)
$p_{j,k}$ – emission probability for peak j belonging to isotope k
μ_{con}, μ_{smp} – linear attenuation coefficients for the container and sample, respectively
t_{con}, t_{smp} – wall thickness for the container and sample, respectively (unknown)
ε_j^0 – estimated relative detector efficiency for peak j
b,c – unknown variables used to account for variations in efficiency as a function of Energy

The unknown variables in Eq. 10.7 (X_k, t_{com}, t_{smp}, ε_j^0, b, c) are solved by an iterative least-squares method [24]. Although the MGA code was initially developed to apply this methodology to address specific isotopic problems with limited input from the user, recent versions of the program allow for the user to specify the regions of a spectra to be analyzed, the specific X-ray and gamma-ray peaks that are present within the specific region, and relevant physical information associated with each peak, including its energy, emission probability, etc. [24]. In determining the ^{240}Pu$_{eff}$ fraction, MGA is typically able to achieve a precision of 2%–3% for a 300 s measurement.

The MGAHI program was developed to analyze plutonium spectra from items that are packaged with a sufficient container thickness to completely attenuate the 100 keV region. This program applies the MGA methodology to the measured gamma-ray peaks from 200 keV to 1 MeV [25]. For the MGAHI analysis to be successful, the 203 keV gamma-ray peak from ^{239}Pu must be present within the spectrum, and the energy resolution of the detector must be 1.1 keV or better at 208 keV. The plutonium spectra regions used in the MGAHI analysis are the 325–350 keV region, the 360–385 keV region, and the 635–670 keV region [26]. The detector efficiency, absorber, and sample thicknesses are calculated based on the known peaks from ^{239}Pu that are present within the spectrum above 200 keV [26].

In 1994, Gunnink and collaborators developed the Multi-Group Analysis Uranium (MGAU) code, which adapted the analysis methodology of the MGA code for plutonium to the analysis of uranium spectra in the 89–100 keV region [27]. As noted in Chap. 8, Sect. 8.4.6, the MGAU code determines a ratio of ^{238}U/^{235}U by analyzing peaks associated with the decay of ^{235}U and ^{238}U. Because the code contains only gamma-rays and X-rays associated with the decay of uranium isotopes, a requirement associated with using MGAU is that the item does not contain plutonium isotopes or significant amounts of other radionuclides [28]. As with all analysis codes that determine uranium enrichment using a ratio-based technique, the code requires that secular equilibrium is established between ^{238}U and its ^{234}Th daughter within the material. The typical accuracy of MGAU in determining the ^{235}U enrichment is to within 2%. The MGA analysis methodology is also used in the "U235" code to analyze uranium spectra [29]. One main difference between the MGAU and U235 codes is that MGAU limits its analysis to the 89–100 keV region whereas the U235 code has been set up to analyze peaks in the 0–300 keV range to determine the ^{234}U, ^{235}U, and ^{238}U ratios.

10.4.1.5.2 FRAM

The Fixed-Energy Response-Function Analysis with Multiple Efficiency (FRAM) software was developed and continues to be refined by Los Alamos National Laboratory. It was first developed in the mid-1980s, running on a MicroVAX computer using the Virtual Memory System (developed by Digital Equipment Corporation) [30]. The first version of the FRAM code that ran on PC/Windows 3.1—called *PC/FRAM*—was released in 1994 [31, 32]. Several upgraded versions of the FRAM code have been released since then, and the latest is FRAM version 6.1, released in March 2020 [33].

10.4.1.5.3 Plutonium Measurement and Analysis

FRAM's plutonium energy regions can be classified into four separate energy ranges: 30–60 keV, 60–120 keV, 120–500 keV, and 500–1010 keV. All of these regions have one thing in common: each region has one and only one measurable gamma ray from ^{240}Pu decay.

For the freshly separated samples such as spent-fuel dissolver solutions, all peaks above 30 keV can be used for analysis by FRAM. The peaks in the 30–60 keV range represent the most intense gamma rays for all plutonium isotopes except ^{241}Pu. Therefore, FRAM is not sufficient to use this region alone for complete plutonium isotopic analysis. This region will need to be used together with another region that has a ^{241}Pu peak.

For aged material, the 60 keV peak of ^{241}Am is very intense, and its corresponding Compton distribution would overwhelm all other peaks below it, making the region below 60 keV useless for isotopic measurements. Therefore, for aged plutonium, only the gamma rays in the region at and above 60 keV can be used.

The 60–120 keV region contains intense usable peaks from all of the plutonium isotopes except ^{239}Pu. Plutonium-239 has one fairly intense peak at 98.8 keV; however, it is sitting on the shoulder of a much more intense uranium X-ray peak and its area cannot be accurately extracted, so it is not used. As in the 30–60 keV region, the 60–120 keV region cannot be used by itself to obtain all isotopic information. The region must be combined with another region to obtain this information.

The 120–500 keV region contains peaks from all plutonium isotopes, so the isotopic ratios of all plutonium isotopes can be determined from just this region. However, the gamma rays in this region are about two and one orders of magnitude less intense than the peaks in the 30–60 keV and 60–120 keV regions, respectively.

The 500–1010 keV region does not have peaks from ^{241}Pu, so some gamma rays from the immediate lower-energy region will need to be included for a complete analysis. The gamma rays in this region are about one order of magnitude weaker than those in the 120–500 keV region. This energy region is best used with heavily shielded plutonium items where the low-energy gamma rays from other regions are only weakly visible, if at all.

Figure 10.15 shows an example of a plutonium spectrum with four regions separated by three vertical dashed lines. Each of these four regions has one, and only one, measurable gamma ray from ^{240}Pu decay. The four overlapping analytical regions

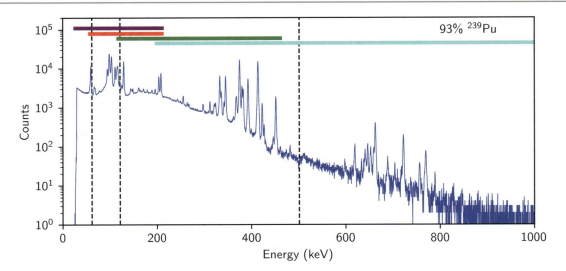

Fig. 10.15 A low-burnup plutonium spectrum. The three vertical, dashed lines separate the four regions. The four overlapping analytical regions that FRAM normally uses for the analysis are shown as four thick horizontal bars above the spectrum

that FRAM normally uses for the analysis are shown as four thick, horizontal bars above the spectrum. The top bar corresponds to the analytical region for freshly separated plutonium, called the *very low-energy analytical region*. The other three—low-, middle- or medium-, and high-energy analytical regions—are for aged plutonium.

The analyses of the first two analytical regions (which include the X-ray regions) with energy up to 210 keV require detectors with good resolution at low energy (<0.7 keV FWHM at 122 keV). A good planar detector is therefore normally used for these measurements. The resolution requirement for the analysis of the other two analytical regions with energy from 120 keV is less stringent and can be easily achieved using a standard planar or coaxial detector.

For standard plutonium analysis that employs the X-ray regions, the data are typically acquired with a planar detector in 4 K channels at 0.075 keV/ch. The data for very-low to medium analyses can also be obtained in 8 K channels at 0.075 keV/ch up to more than 600 keV so that the data could be analyzed in several different energy ranges: 30–210 keV (for fresh plutonium), 60–210 keV, and 120–500 keV.

For the coaxial detector, the spectra are normally acquired in 8 K channels at 0.125 keV/ch, covering the entire 0–1024 keV energy range. The coaxial detector's data can be analyzed employing either the 120–500 keV region (third analytical region in Fig. 10.15) or 200–1010 keV region (fourth analytical region in Fig. 10.15).

Some special coaxial detectors (such as the Lo-Ax from Ortec or wide-energy detectors from Canberra) have reasonable resolution at low energy and reasonable resolution and efficiency at high energy. With such a detector, the X-ray regions (with 0.125 keV/ch energy calibration) may also be employed in analyses in addition to employing the 120–500 keV and 200–1010 keV regions.

10.4.1.5.4 Uranium Measurement and Analysis

FRAM can also obtain a complete isotopic analysis for uranium. With the use of a planar detector, FRAM normally analyzes data in the 80–210 keV range. In low-enriched uranium (LEU), where the 63.3 keV peak from the ^{238}U decay-chain might be visible, FRAM may also use that peak. The data are typically acquired in 4 K channels at 0.075 keV/ch. If a planar detector is used in a measurement situation where uranium is shielded by a somewhat thick absorber (such as the walls of a UF_6 cylinder) and low-energy gamma rays and X-rays might not be visible, then FRAM may use the peaks in the range of 120–1010 keV for analysis. If a coaxial detector is used, the analytical energy range is normally in the 120–1010 keV range. The data for these higher-energy analyses are acquired in 8 K channels at 0.125 keV/ch up to 1024 keV.

Figure 10.16 shows an example of a uranium spectrum with two regions—one below the K-edge and one above the K-edge, separated by the dashed line. The two thick, horizontal bars above the spectrum represent the two overlapping (low and high) analytical regions that FRAM normally uses for the analysis.

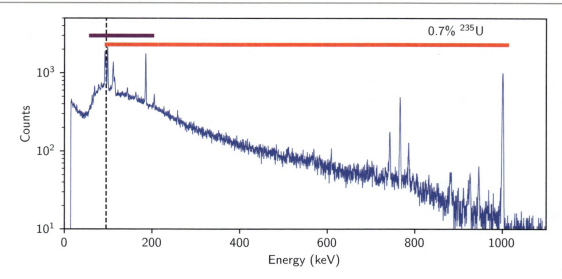

Fig. 10.16 A natural-uranium spectrum. The vertical dashed line separates the two regions—one below the K-edge and one above the K-edge. The two overlapping analytical regions that FRAM normally uses for the analysis are shown as two thick, horizontal bars above the spectrum

10.4.1.5.5 Analysis Technique

The FRAM code is structured to give the user as much control as desired over the analysis to increase the versatility and applicability of FRAM, which is accomplished via the use of the "parameter set." A parameter set contains all of the parameters required to carry out an analysis on a gamma-ray spectrum, including information on the isotopes to be analyzed; the specific gamma-ray peaks to use; the nuclear data for the isotopes and gamma-ray peaks; data acquisition conditions such as gain, zero, number of channels, and regions of the spectrum for analysis; and diagnostic test parameters. A parameter set in the database can be exported to a text file—called a *parameter file*—that can be sent to other users and imported into their FRAM database.

FRAM analyzes the data using the following steps:
1. Determine energy calibration.
2. Determine FWHM formula.
3. Determine peak shape formula.
4. Fit the regions to obtain peak areas.
5. Determine relative efficiency curve.
6. Calculate the relative activities of the isotopes.
7. Adjust the background of the analysis regions.
8. Repeat steps 4 through 7 several times (default: five iterations).
9. Estimate by correlation the ^{242}Pu or ^{236}U isotope if not measured or declared by the operator.
10. Calculate the final isotopic fractions of the isotopes.

Step 1: Determining Energy Calibration

The peaks are located using the default energy calibration in the parameter set. The algorithm locates the peak at the maximum count found in a region of $(6 + 0.0016 \times \text{peak centroid})$ channels on either side of the default peak position, which is located using the default energy calibration. After the peak is located, the peak centroid is found using the following method: (1) an 11 points smooth function is applied to the channels in the peak region, and the peak centroid is calculated from the channel with the largest smooth counts and its two adjacent channels using a quadratic function. If it fails, then (2) a least-squares fit of a quadratic function is applied to the logarithm of the counts. If both (1) and (2) fail, then the peak is not used in the calibration. A piecewise linear calibration between pairs of peaks is used for the energy calibration.

Step 2: Determining FWHM Calibration

The FWHM is calculated from a least-squares fit of a quadratic equation to the logarithm of the net counts. The fit is over a range of channels in which the counts exceed 75% of the peak maximum on the low-energy side and exceed 25% of the peak maximum on the high-energy side. The FWHM as a function of energy that is used in calculating the response function for an arbitrary fitted peak is found from a least-squares fit to the function,

$$FWHM(E) = \sqrt{\left(A_1 + A_2 E + \frac{A_3}{E}\right)}, \qquad (10.8)$$

where E is the peak energy in keV, and A_i are the fitted constants.

Step 3: Determining Peak Shape Calibration

The shape of a gamma-ray peak in the spectrum is described by a central Gaussian component with a single exponential tail on the low-energy side of the peak:

$$Y(E_i) = H \; exp\left[-\alpha(E_i - E)^2\right] + Tail(E_i), \qquad (10.9)$$

where

$Y(E_i)$ is the net count at energy E_i,
H is the peak height at the peak energy E, and.
$\alpha = 2.77259/FWHM^2$ is the peak width parameter.

The tailing parameter $Tail(E_i)$ is given by

$$Tail(E_i) = H \exp[(T_1 + T_2 E) + (T_3 + T_4 E)(E_i - E)] \cdot \left[1 - \exp(-0.4\alpha(E_i - E)^2)\right],$$

where T_i are the fitted constants.

Step 4: Determining Peak Areas Using Response Functions

The analysis starts by subtracting the background to get the net counts in a region. The background for the first iteration is available from the initial background calculation. The background for later iterations is from the adjusted background done in Step 7.

The process then fits the net counts in the channels in a region using the response functions. Each response function has the form $\sum_i f_i R_i(x)$, where each R_i is a unit area function that describes the shape of a photopeak and f_i is the associated area factor. The shape of a gamma-ray peak in the spectrum is described by a central Gaussian component with a single exponential tail on the low-energy side of the peak.

For HPGe spectra, the response function described can be fitted using linear least-squares fits. For a medium-resolution spectrum such as that of a LaBr$_3$ or CZT detector—due to the imprecision of the FWHM and tail calibration, the very broad peaks, and the overlapping of many peaks—the linear least-squares fit does not work well. Therefore, the nonlinear least-squares fit technique that combines the Powell's minimization method with the linear least-squares fit is used to fit medium-resolution spectra.

Step 5: Calculating relative efficiencies

There are two models for the relative efficiency curve: empirical relative efficiency and physical relative efficiency. For both models, let N be the number of isotopes that have peaks used for efficiency calculations and M be the number of efficiency functions. (M is 1 for homogeneous items and > 1 for inhomogeneous items.) Then the model for the empirical relative efficiency curve is

$$\ln(Area/BR) = c_1 + c_2/E^2 + c_3\ln(E) + c_4(\ln(E))^2 + c_5(\ln(E))^3 + c_i + c_j/E \quad (10.10)$$

where E is the energy in MeV. Each c_i is associated with isotopes beyond the first one; i ranges in value from 6 to $5 + (N - 1)$. Each c_j is associated with an efficiency function beyond the first one; j ranges in value from $6 + (N - 1)$ to $5 + (N - 1) + (M - 1)$.

The physical relative efficiency has two sub models that relate to how the heterogeneous materials are determined. For the true heterogeneous model, the design matrix is constructed as

$$RE = \frac{Area}{BR} = \left[\frac{(1 - \exp(-\mu_0 x_j))}{(\mu_0 x_j)}\right] \cdot [\exp(-\mu_1 x_1) \cdot \exp(-\mu_2 x_2) \cdot \exp(-\mu_3 x_3)]$$
$$\cdot [A_i] \cdot [Detector\ Efficiency] \cdot [Correction\ Factor]. \quad (10.11)$$

where E is the energy in MeV, μ_o refers to the absorption coefficient of plutonium or uranium, and x_j are the corresponding thicknesses of the nuclear materials that represent different efficiency curves. For $i = 1, 2,$ or 3, μ_i and x_i refer to the absorption coefficient and thickness of the ith absorber material. Each A_i is associated with isotopes beyond the first one; if any exist, i ranges in value from 7 to $6 + (N - 1)$. Each x_j beside x_o is associated with an efficiency function beyond the first one; if any exist, j ranges in value from $6 + (N - 1)$ to $5 + (N - 1) + (M - 1)$.

For the $\exp(C_j/E)$ model, the design matrix is constructed as

$$RE = \frac{Area}{BR} = \left[\frac{(1 - \exp(-\mu_0 x_0))}{(\mu_0 x_0)}\right] \cdot [\exp(-\mu_1 x_1) \cdot \exp(-\mu_2 x_2) \cdot \exp(-\mu_3 x_3)] \cdot [A_i] \cdot \left[\exp\left(\frac{c_j}{E}\right)\right]$$
$$\cdot [Detector\ Efficiency] \cdot [Correction\ Factor] \quad (10.12)$$

where x_o corresponds to the thickness of the nuclear material, and $\exp(c_j/E)$ is the term for multiple efficiency functions.

Step 6: Calculating relative activities.

The relative activity of a peak is $A_i = Area_i(BR_i)(RE_i)$, where $Area_i$ is the area of the peak i, BR_i is the branching ratio of the peak i, and RE_i is the relative efficiency at the peak i's energy. The relative activity of an isotope is found as the average in logarithm of the activities of the individual peaks of the isotope.

$$ln(A) = average(ln(A_i)) = \frac{\sum_i ln(A_i) w_i}{\sum_i w_i} \quad (10.13)$$

where $w_i = (A_i/\delta A_i)^2$ is the reciprocal of the square of the error associated with $\ln(A_i)$, and δA_i is the uncertainty of the relative activity of peak i.

Step 7: Adjusting the background

Each analysis region consists of one peak region and up to four background regions. In the initial background determination, the counts in each channel of a background region are set equal to the count in the channel of the spectrum. The whole background for the analysis region is calculated from these background regions.

After each iteration, the contribution of the fitted peaks to the background regions is subtracted from the initial background counts. The new background for the analysis region is then calculated from the adjusted background regions.

Step 8: Repeat analysis

After obtaining the new background for the HPGe spectra, the code repeats the analysis process (obtaining peak areas, fitting the relative efficiency curve, calculating the relative activities, and adjusting the background) several times by going back to Step 4. For the medium-resolution spectra, the code repeats both the calibration and analysis processes by going back to Step 1.

Step 9: Adding ^{242}Pu or ^{236}U

After the final iteration is complete, the final relative masses (relative to the first isotope in the isotope list) are combined to give the absolute isotopic fractions. If ^{242}Pu or ^{236}U is measured, then this combination step will be the final step, and Step 10 can be skipped. However, for virtually all of the analysis, these isotope fractions will be from either operator-declared values or from correlation with other isotopes.

The correlation used for ^{242}Pu is of the form

$$^{242}Pu = A \times \left[\left(^{238}Pu\right)^B \times \left(^{239}Pu\right)^C \times \left(^{240}Pu\right)^D \times \left(^{241}Pu + ^{241}Am\right)^E \right] \quad (10.14)$$

where the constants A, B, C, D, and E are Application Constants in the selected parameter set.

The correlation used for ^{236}U is of the form

$$^{236}U = A \times \left[\left(^{235}U\right)^B \times \left(^{238}U\right)^C \right] + D \times \left[\left(^{235}U\right)^E \times \left(^{228}Th\right)^F \right] \quad (10.15)$$

where the constants A, B, C, D, E, and F are application constants in the selected parameter set. If ^{228}Th is not present as an isotope in the parameter set, the correlation uses only the first term.

Step 10: Calculating the Final Isotopic Fractions

After the ^{242}Pu or ^{236}U fraction from either operator entry or correlation is added, then the total fraction will be greater than 1 (or 100%). The fractions are renormalized, accounting for the added ^{242}Pu or ^{236}U.

10.4.1.6 Special Measurement Cases

The following sections discuss the special cases that the user may encounter; many of the cases are from Ref. [34].

10.4.1.6.1 Uranium-Plutonium Mixed Oxide

For gamma-ray analysis, materials that contain a mixture of uranium and plutonium oxide (mixed oxide [MOX]) can be considered as a plutonium item with uranium as an impurity. This consideration is true even in those cases where the uranium fraction is many times larger than the plutonium fraction. Both the MGA and FRAM codes can analyze typical MOX items using the plutonium analysis (MGA) or the standard plutonium parameter sets (FRAM). FRAM also includes the standard parameter sets to analyze special MOX where the ^{235}U/Pu ratio is greater than 0.3.

Both the MGA code and the FRAM standard plutonium parameter sets include ^{235}U peaks at 143.8, 163.4, 185.7, 202.1, and 205.3 keV as interference peaks. The FRAM standard plutonium analysis in the high-energy region also includes the ^{238}U peaks at 766.4 and 1001.0 keV. Plutonium-238 and ^{238}U both contribute to peaks at 766.4 and 1001.0 keV, and all four peaks are used in the analysis. In the MOX samples, the 766.4 keV peak arises mainly from ^{238}Pu, and the 1001.0 keV peak arises mainly from ^{238}U. FRAM accounts for this by fixing the ^{238}U component at 766 keV to the ^{238}U component at 1001 keV and fixing the ^{238}Pu component at 1001 keV to the larger ^{238}Pu component at 766 keV. The iterative FRAM analysis then further refines this starting point.

Figure 10.17 shows a MOX spectrum (11.7% ^{240}Pu) with 1.1% uranium enrichment and U/Pu ratio of 6.0. The ^{235}U 185.7 keV peak is weak, whereas the 1001.0 keV peak of ^{238}U is intense and clearly seen in the spectrum.

10.4.1.6.2 Americium-243 and Neptunium-239 Impurities

Americium-243 may be present in plutonium samples. This isotope decays as shown in the following equation:

$$^{243}Am \xrightarrow{\alpha\ 7364y} {}^{239}Np \xrightarrow{\beta^-\ 2.356d} {}^{239}Pu \quad (10.16)$$

The gamma rays from the direct decay of ^{243}Am are too weak to be seen, but the gamma rays from the decay of ^{239}Np are very strong and easily identifiable in a plutonium spectrum. The intense peaks are at 106.1, 209.8, 228.2, 277.6, and 334.3 keV. These gamma rays, along with a few weaker neighboring gamma rays, are accounted for in analysis for ^{243}Am.

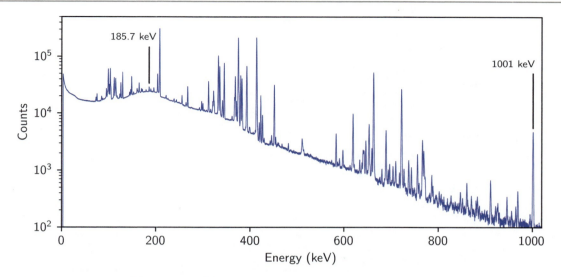

Fig. 10.17 A MOX spectrum (11.7% ^{240}Pu/Pu, 1.1% ^{235}U/U, and 6.0 U/Pu ratio)

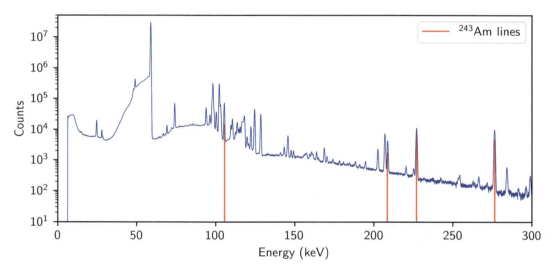

Fig. 10.18 A low-burnup plutonium spectrum with a significant amount of ^{243}Am. The 106.1, 209.8, 228.2, and 277.6 keV peaks of ^{243}Am are marked red

The standard FRAM parameter files easily analyze spectra from samples with "normal" concentrations of ^{243}Am. Very high concentrations are characterized by the peak height of the 209 keV ^{239}Np peak being an appreciable fraction of or greater than the peak height of the 208.00 keV ^{241}Pu-^{237}U peak. These cases should be examined more closely because it may be necessary to "tweak" the boundaries of some of the ROIs to obtain the best results. MGA can analyze spectra with "normal" concentrations of ^{243}Am but does not report it.

Figure 10.18 shows a low-burnup plutonium spectrum with a significant amount of ^{243}Am. The 106.1, 209.8, 228.2, and 277.6 keV peaks of ^{243}Am can be seen in the spectrum.

10.4.1.6.3 Heterogeneous Am/Pu

Pyrochemical purification processes, such as molten salt extraction and electrorefining, produce residues where the americium and plutonium are in different matrices. The americium is concentrated in a low-Z matrix in the form of $AmCl_3$, whereas small amounts of plutonium metal are imbedded in the low-Z matrix. The relative efficiency curve will be different for an americium gamma ray escaping from a chloride matrix compared with that of a plutonium gamma ray escaping from plutonium metal.

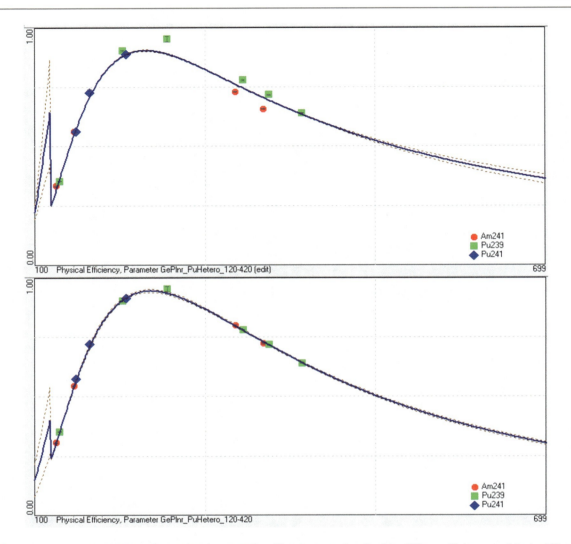

Fig. 10.19 Screenshot from the FRAM software showing examples of heterogenous data fitted by different efficiency models. (top) Data are fitted to a homogeneous efficiency curve. (bottom) Data are fitted to a heterogeneous efficiency curve

MGA cannot calculate heterogeneous models. The FRAM code calculates the different relative efficiency curve shapes for americium and plutonium based on actual physical heterogeneous models or a model proposed by Fleissner [35].

Figure 10.19 shows examples of heterogeneous data fitted by a homogeneous model efficiency (top), assuming americium and plutonium are in the same matrix and by a heterogeneous model efficiency (bottom), assuming americium is in a different matrix than plutonium.

10.4.1.6.4 Very High Pu-239/Very Low Pu-240

This material type can be defined roughly as having a ^{240}Pu content less than approximately 2%. Spectra from these materials usually have very little ^{241}Pu, so the ^{241}Pu-^{237}U peaks may not be strong enough to use for internal calibrations. Plutonium-239 lines are strong and suffer little interference because of the low ^{241}Pu and ^{238}Pu content in the sample. At these low levels, the 208 keV gamma ray from ^{241}Pu-^{237}U is less intense than the ^{239}Pu gamma ray at 203.5 keV, which is just the reverse of what is usually seen in plutonium spectra from higher-burnup samples.

Both MGA and FRAM—with their low-energy standard parameter sets—can analyze spectra with 99.9% ^{239}Pu and 0.05% ^{240}Pu. FRAM can analyze spectra with 1% ^{240}Pu with the medium- and high-energy parameter sets.

Figure 10.20 shows a very high ^{239}Pu/very low ^{240}Pu spectrum with 2.0% ^{240}Pu. The marked peaks are of the 203.5 and 208.0 keV peaks where the 203 keV peak is more than one order of magnitude larger than the 208.0 keV peak.

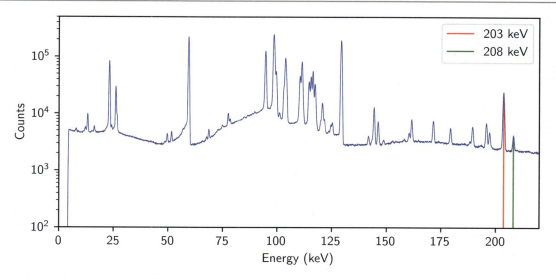

Fig. 10.20 A plutonium spectrum of 2.0% ^{240}Pu. The 203.5 (red) and 208.0 (green) keV peaks are marked

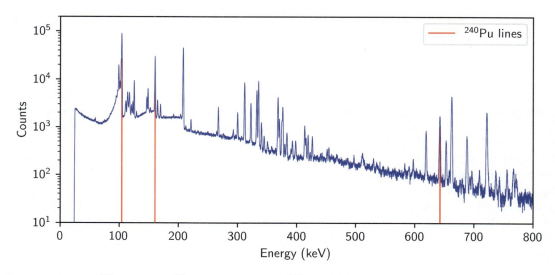

Fig. 10.21 Spectrum of a 1.0% ^{239}Pu and 94% ^{240}Pu. The three peaks of ^{240}Pu at 104.2, 160.3, and 642.5 keV are marked in red

10.4.1.6.5 Very Low Pu-239/Very High Pu-240

Samples with more than 90% ^{240}Pu are used in physics research applications and in safeguards applications for calibration of neutron coincidence counters. Such samples have three major peaks that span three analysis regions of FRAM at 104.2, 160.3, and 642.5 keV. Both MGA and FRAM—with their standard parameter sets in all energy regions—can analyze samples up to 95% ^{240}Pu (and 1% ^{239}Pu).

Figure 10.21 shows the spectrum of an item with 1.0% ^{239}Pu and 94% ^{240}Pu. The three most-intense peaks of ^{240}Pu at 104.2, 160.3, and 642.5 keV are marked in red.

10.4.1.6.6 Very High Neptunium-237

Neptunium-237 is present in every plutonium sample as a direct decay product of ^{241}Am and the ^{237}U daughter of ^{241}Pu. The levels vary as a function of the age of the sample, the burnup of the sample, and the chemical processing history.

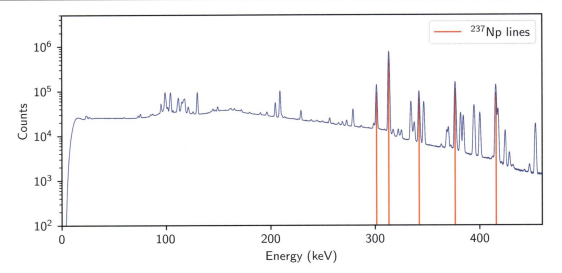

Fig. 10.22 A spectrum of a low-burnup plutonium with 2% ^{237}Np. The ^{237}Np peaks at 300.1, 311.9, 340.5, 375.4, 398.5, and 415.8 keV are marked in red

The concentration of ^{237}Np is determined by analyzing the gamma rays from its ^{233}Pa daughter. This isotope decays as

$$^{237}Np \xrightarrow{\alpha\ 2.144e6y} {}^{233}Pa \xrightarrow{\beta^-\ 26.975d} {}^{233}U. \tag{10.17}$$

The gamma-ray activity from ^{233}Pa will be in secular equilibrium with ^{237}Np after about 6 months. This condition does not present a problem except for materials fresh from chemical processing. The decay of ^{233}Pa produces intense gamma rays at 86.5, 300.1, 311.9, 340.5, 375.4, 398.5, and 415.8 keV that can be used for the quantification of the ^{237}Np/Pu ratio and/or must be considered as interferences for the plutonium isotopic composition measurement.

Normal ^{237}Np/Pu ratios range from 10^{-5} to 10^{-3}. FRAM can analyze the samples with ^{237}Np/Pu ratio up to 0.02 with the medium- and high-energy standard parameter sets. FRAM—with low-energy standard parameter sets—can analyze samples with a ^{237}Np/Pu ratio up to 0.1. Above a 0.1 ratio, FRAM can still analyze the sample using a parameter set tailored for very high ^{237}Np. With appropriate filtering to cut down the intense, low-energy gamma rays below 200 or 300 keV, spectra with ^{237}Np/Pu ratio up to 10 can be analyzed using the FRAM high-energy region analysis.

Figure 10.22 shows the spectrum of a low-burnup plutonium with 2% ^{237}Np. The ^{237}Np peaks at 300.1, 311.9, 340.5, 375.4, 398.5, and 415.8 keV are marked in red.

10.4.1.6.7 Very High Am-241

The concentration of ^{241}Am in normal plutonium ranges from 0 for fresh plutonium to 10% for very aged, high-burnup plutonium. Americium-241 concentrations above 10% relative to plutonium usually arise only in residues and wastes that contain ^{241}Am concentrated from purification processes. This is a very high concentration of ^{241}Am.

For very high ^{241}Am samples, the ^{241}Am gamma-ray activity will dominate the spectrum. The peaks from plutonium will ride on top of a large continuum from the americium gamma rays. The resulting signal/background ratio will be poor, and the plutonium gamma-ray peaks will have poor statistical precision. Also, the ^{241}Am peaks normally too weak to be visible above the background continuum will suddenly appear in the spectrum. These "new" peaks interfere with peak and background ROIs established for analyses of samples with "normal" americium concentrations.

Both MGA and FRAM—with their standard parameter sets in all energy regions—can analyze normal plutonium samples with up to 10% ^{241}Am. For very high ^{241}Am samples, special FRAM parameter sets can be created to analyze samples with a ^{241}Am/Pu ratio up to 3.

Figure 10.23 shows the low-burnup plutonium with 78% ^{241}Am. The intense peaks of ^{241}Am are marked in red. The intense, unmarked peaks at 300 and 312 keV are from ^{237}Np, which is the daughter of ^{241}Am.

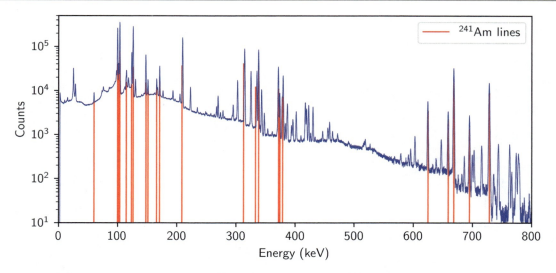

Fig. 10.23 Spectrum of low-burnup plutonium with 78% ^{241}Am. The intense peaks of ^{241}Am are marked in red

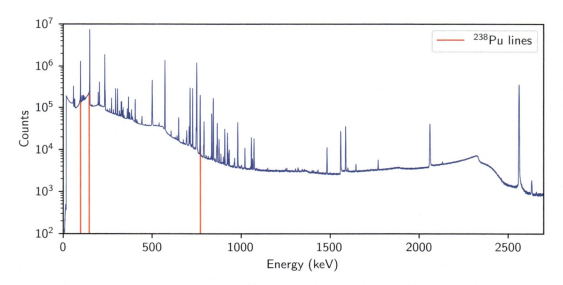

Fig. 10.24 Spectrum of a very high ^{238}Pu. The three intense peaks of ^{238}Pu are marked in red

10.4.1.6.8 Very High Pu-238 (Heat Source)

Heat-source-grade plutonium typically contains approximately 80% ^{238}Pu and 2%–4% ^{240}Pu, with the remainder being mostly ^{239}Pu. Plutonium-241 and ^{241}Am are also present and can be analyzed. The ^{236}Pu that is originally present in parts per million (ppm) amounts produces strong gamma-ray peaks from its thorium daughter decay products. The thorium daughter gamma-ray peaks are used to help define the relative efficiency curve.

Plutonium-238 has three intense peaks at 99.9, 152.7, and 766.4 keV. These peaks are very prominent in the spectrum. Old samples also have intense peaks from the thorium daughter. Neither MGA nor FRAM—with their standard parameter sets—can analyze the heat source spectra. Special FRAM parameter sets can be created to take the three major and some minor ^{238}Pu peaks at 201.0, 742.8, and 786.3 keV into account to analyze the data in all three energy regions: low, medium, and high. The analysis for ^{239}Pu, ^{241}Pu, and ^{241}Am is straightforward using the major higher-energy peaks. Plutonium-240 is the most difficult isotope to analyze because its peaks—at 104.2, 160.3, and 642.5 keV—are essentially undetectable. This issue results in very large percent relative standard deviation values for ^{240}Pu. The user must use prior knowledge or stream average data for ^{242}Pu because the standard FRAM isotopic correlation does not apply.

Figure 10.24 shows the spectrum of a very high ^{238}Pu. The three intense peaks of ^{238}Pu are marked in red. Most of the intense unmarked peaks in the spectrum are from ^{228}Th, the granddaughter of ^{236}Pu.

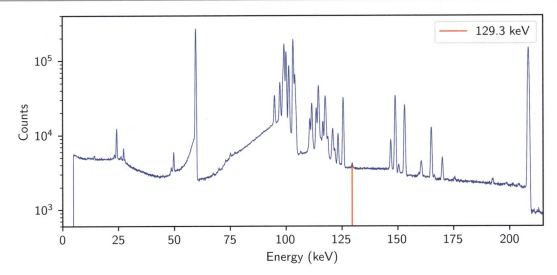

Fig. 10.25 Spectrum of a 95% ^{242}Pu item

10.4.1.6.9 Very High Pu-242

The concentration of ^{242}Pu in normal plutonium ranges from ~0.02% for very low-burnup plutonium to ~10% for very high-burnup plutonium. Plutonium-242 emits only few weak gamma rays, and those gamma rays are not measurable in normal plutonium spectra. However, for the material with very high ^{242}Pu concentration, the ^{242}Pu peaks could become sufficiently intense that they and the corresponding ^{242}Pu activity can be measured.

The three gamma rays emitted by ^{242}Pu are at 44.9, 103.5, and 159.0 keV. For fresh plutonium, the ^{241}Am has not had time to build up, and its 59.5 keV peak is not too intense, which allows the peaks below 60 keV to show and be measured. Plutonium-240 has an intense peak at 45.2 keV, which is less than one FWHM from the ^{242}Pu 44.9 keV peak, assuming that the detector is reasonably good with the FWHM ~0.4 keV at 45 keV. This closeness of the two peaks requires that the peak area of the ^{242}Pu 44.9 keV peak be at least 50% of the ^{240}Pu 45.2 keV peak area to get a decent uncertainty for the 44.9 keV peak area. This requirement translates into 95% or greater ^{242}Pu in the plutonium.

The 103.5 keV peak is next to the 103.67 keV peak of ^{241}Pu and the 103.74 keV plutonium fluorescence X-ray. Both peaks can be accounted for with one condition: that their areas are not much greater than the 103.5 keV peak area. The condition with the ^{241}Pu 103.67 keV peak requires greater than 95% ^{242}Pu. The condition with the 103.74 keV X-ray peaks requires that the plutonium concentration in the matrix is small, such as low-concentration solution, so that the plutonium fluorescence is small.

The 159.0 keV peak has smallest photoemission probability of the three, but it is far from some other peaks that can interfere with the extraction of its peak area. The peak closest to it is the ^{241}Pu 160.0 keV peak. A good planar HPGe detector may have a FWHM of 0.6 keV at 160 keV. This result is only 60% of the 1 keV separation between the ^{242}Pu 159.0 keV and the ^{241}Pu 160.0 keV peaks, and it allows accurate determination of the 159.0 keV peak area if it is measurable. To have reasonable 159.0 keV peak uncertainty, the ^{242}Pu/Pu ratio would need to be at least 85%.

Figure 10.25 shows the spectrum of a 95% ^{242}Pu item. The spectrum looks like that of a super high-burnup plutonium where even the most intense ^{239}Pu peak at 129.3 keV shows up only as a blip above the background.

Figure 10.26 shows a FRAM least-squares fit of the 160 keV region. The three intense and fitted peaks are the ^{242}Pu 159.0 keV peak, ^{241}Pu 160.0 keV peak, and ^{240}Pu 160.3 keV peak.

10.4.1.6.10 Very High Uranium-236

Similar to ^{242}Pu, ^{236}U has only a couple of weak gamma rays at 49.4 and 112.7 keV. The concentration of ^{236}U in normal uranium ranges from 0% for natural uranium and gradually increases to ~0.3% for very highly enriched uranium (HEU). This concentration is too small for its peaks to be visible to measure. However, for special uranium, where the ^{236}U concentration is ~20%, the intensity of the 49.4 keV peak is comparable to that of the 63.3 keV of the ^{238}U daughter ^{234}Th and of the 53.2 keV of ^{234}U with ~20% enrichment. If the enrichment is LEU, then the 53.2 keV peak intensity is less, but the 63.3 keV peak intensity increases, and vice versa for HEU. Regardless, a peak from either ^{234}U or ^{238}U at low energy will always extend the relative efficiency curve down to ~50 keV, allowing ^{236}U activity to be determined using the 49.4 keV peak intensity. Note

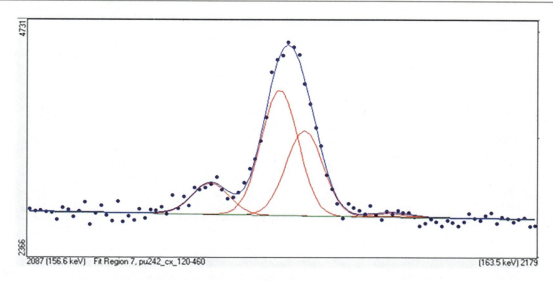

Fig. 10.26 Scrrenshot from the FRAM software showing a least-squares fit of the 160 keV region. The three intense peaks from left to right are of ^{242}Pu, ^{241}Pu, and ^{240}Pu

that ^{238}U has a peak at 49.55 keV, and the contribution of this peak needs to be accounted for when fitting the ^{236}U 49.4 keV peak.

Uranium-236 also has a peak at 112.7 keV. This peak is very close to the 112.8 keV peak of ^{234}Th (^{238}U's daughter). For LEU, this peak is overwhelmed by the 112.8 keV peak, and for HEU, it is not visible due to the high background in this region from ^{235}U. For uranium with greater than 20% ^{236}U, this 112.7 keV peak would become much larger and can be measured to determine the ^{236}U activity.

10.4.1.6.11 Freshly Separated Plutonium

Plutonium-241 has a short half-life (relative to other plutonium isotopes) and quickly decays to ^{241}Am. Americium-241 has a very intense peak at 59.5 keV. After several days, this peak may become the most intense peak in the spectrum. After several weeks, depending on the ^{241}Pu fraction, this peak may become so intense that its Compton background may obscure the peaks below it, preventing the analysis from using the peaks below 60 keV.

For very fresh plutonium, the region below 60 keV is the most intense and useful region for measuring ^{238}Pu, ^{239}Pu, ^{240}Pu, and ^{241}Am. The peaks are at 38.7 (^{239}Pu), 43.5 (^{238}Pu), 45.2 (^{240}Pu), 51.6 (^{239}Pu), and 59.5 (^{241}Am) keV. Figure 10.27 shows the spectrum of a fresh, high-burnup plutonium.

The region below 60 keV does not have a measurable ^{241}Pu peak; therefore, to get full isotopic composition, it will need to combine with other region(s). The 30–120 keV region would include peaks from all plutonium isotopes and ^{241}Am and can give complete isotopic composition without having to cross the plutonium K-edge. Analysis in this region would give good results.

There is a caveat when analyzing this region. The only intense ^{241}Pu peak in this region is at 103.7 keV. Also, the plutonium X-ray peak at 103.7 keV will need to be accounted for to accurately measure the ^{241}Pu peak. The plutonium X-ray peak is from fluorescence and from the decay of ^{239}Np, which is the daughter of ^{243}Am and can be present in fresh plutonium. The fluorescence plutonium X-ray intensity can be minimized by measuring a small amount of plutonium in solution. The plutonium X-ray from the decay of ^{239}Np can be reduced by measuring the sample 2 weeks or later after separation. Neptunium-239's half-life is 2.4 days, and only 1% will remain after six half-lives, which is 14 days.

The 100 keV region includes the X-ray peaks, which require Voigt (convolution of Lorentzian a Gaussian) peak-shape fitting. The fitting of this peak shape is complicated and not easily done. A combination of the 30–60 keV region with the 121–210 keV region will give complete isotopic composition without having to deal with the X-rays. The best analysis would be from the 30–210 keV region, including the 100 keV region.

Both the MGA and FRAM codes—with their low- and medium-region standard parameter sets—can analyze the freshly separated plutonium. Users can create FRAM parameter sets to include the peaks below 60 keV. The results will likely improve over that of the standard parameter sets.

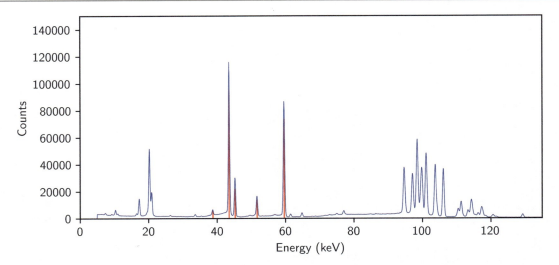

Fig. 10.27 Spectrum of a fresh, high-burnup (23.7% ^{240}Pu) plutonium sample. The five peaks marked in red are ^{239}Pu 38.7 keV, ^{238}Pu 43.5 keV, ^{240}Pu 45.2 keV, ^{239}Pu 51.6 keV, and ^{241}Am 59.5 keV

10.4.1.6.12 Freshly Separated Uranium

Uranium-238 decay emits a gamma ray at 49.55 keV. This peak is weak and would normally be overwhelmed by the high background from the Compton of the ^{235}U, ^{234}Th, and ^{234}Pa gamma rays. However, for fresh, low-enriched uranium, the gamma-ray signals and their Compton background from these three isotopes are weak and may allow this peak to be measured. To calculate the ^{238}U activity, the intensity of this peak needs to be compared with another peak nearby, or the relative efficiency curve can be extended to the location. Uranium-234 has a peak at 53.2 keV. If visible, this peak can extend the efficiency curve down to 50 keV and allow the determination of the ^{238}U activity. The ratio of the 49.55 keV peak to the 53.2 keV peak is ~1 for natural uranium and gradually reduces to 0.1 as the enrichment increases to 5%. For higher enrichment, the 49.55 keV peak would become very small and cannot be accurately measured. Figure 10.28 shows the spectrum of 1.0% enriched uranium 8 h after separation.

For uranium of any enrichment, the ^{238}U gamma rays that are used in the analysis of the MGA code and the standard parameter sets of FRAM are from the ^{234}Th daughter and/or ^{234}Pa granddaughter. Thorium-234 has a half-life of 24.1 days and will come into secular equilibrium with its ^{238}U parent in about 6 months. A uranium sample that has had its thorium and protactinium removed less than 6 months before is considered fresh. The amount of ^{238}U determined using ^{234}Th and ^{234}Pa gamma rays will appear too small relative to those from ^{235}U, which would make the enrichment appear too large. Correction using the time since chemical separation will need to be applied to bring the enrichment to its true value. FRAM has an option that allows the user to enter the separation date, and it will correct for the overestimated enrichment due to the lack of secular equilibrium between the ^{234}Th daughter and the ^{238}U parent. MGA does not have the option, and the correction must be done manually.

10.4.2 Waste Measurements

10.4.2.1 Introduction

The objective of radioactive waste assay is to measure waste items, quantify activities or masses of radionuclides present in the item and the total measurement uncertainties, compare the results with the regulatory limits, and adjudicate the appropriate disposal of the waste item. NDA techniques are very useful in radioactive waste assay. NDA systems that measure neutron, gamma, or heat signatures emitted by radioactive waste are commonly deployed. The NDA system that is best suited for a given waste stream will depend on the radionuclides expected to be present in the waste stream, data quality objective, acceptance criteria, fit for purpose, and cost, among other things.

Radioactive wastes are typically classified into several disposal categories. Although some variation in terminology exists from country to country, similar guidelines are generally followed throughout the world. Following are the four major categories of radioactive waste.

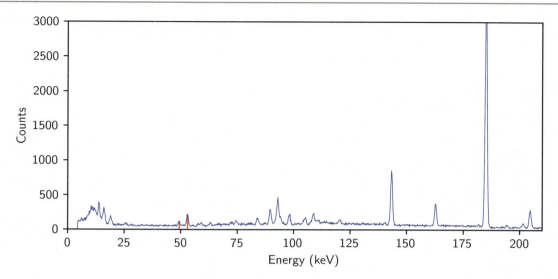

Fig. 10.28 Spectrum of a fresh 1.0% enriched uranium measured 8 h after removal of thorium and protactinium. The peak shown in red at 49.55 keV is from ^{238}U, and the one at 53.2 is from ^{234}U

10.4.2.1.1 Remote-Handled Waste

Remote-handled waste is typically defined as waste that has a dose rate of greater than 200 mR/h (2 mSv/h) on the surface of the container. This waste may or may not also be transuranic (TRU).

10.4.2.1.2 Transuranic Waste, Alpha Waste, and Intermediate-Level Waste

TRU waste is defined by U.S. Department of Energy waste acceptance criteria as waste that is contaminated with alpha-emitting TRU radionuclides with half-lives greater than 20 years and concentrations greater than 100 nCi/g (3700 Bq/g) at the time of assay. The term *intermediate level waste* is also used to describe similar types of waste internationally.

10.4.2.1.3 Low-Level Radioactive Waste

Low-level (radioactive) waste (LLW) is typically defined as waste that is not TRU and does not exceed certain concentration limits for specific nuclides. LLW includes items that have become contaminated with radioactive material or have become radioactive through exposure to neutron radiation. This waste typically consists of contaminated protective shoe covers and clothing, wiping rags, mops, filters, reactor water treatment residues, equipment, and tools. The radioactivity can range from just above background levels found in nature to very highly radioactive in certain cases, such as parts from inside the reactor vessel in a nuclear power plant.

10.4.2.1.4 Free-Release Waste

Free-release or very low-level waste criteria vary significantly from country to country and are subject to constant changes. Often country-specific limits and terminology apply; however, the measurement principles are the same. In Europe, the Basic Safety Standard 2013–59 legislation has established a release limit; some countries use 0.4–50 Bq/g for very low-level waste.

Depending on the radionuclides expected in the waste stream assayed, a neutron or gamma NDA system (or a calorimeter, which measures the heat from radioactive decay) is used.

10.4.2.1.5 Neutron Systems

Neutron waste assay systems are used to measure the uranium or plutonium content in waste items. They can be either passive (measuring spontaneous fission neutrons) or active (measuring induced fission neutrons). Active neutron systems, e.g., the ^{252}Cf Shuffler, are used to assay and quantify uranium. In these systems, a neutron-emitting interrogation source is used to induce fissions in ^{235}U contained in the waste, followed by the measurement of prompt fission neutrons or delayed neutrons. Neutron measurements are advantageous because lower-energy gamma rays (e.g., the 186 keV gamma ray emitted by ^{235}U) could be severely self-attenuated and/or attenuated in the waste matrix.

Passive neutron systems are used to assay and quantify plutonium by measuring the prompt neutrons. Measurement of the triples provides an additional piece of data that can be used to determine an additional unknown parameter; however, the efficiency of the neutron counter must be high enough so that good precision can be achieved in the triples measurement.

10.4.2.1.6 Gamma-ray Systems

Gamma-ray systems have the advantage of providing isotope-specific information. Waste items that consist of many fission products and activation products need to be assayed using a gamma NDA system that can identify the radionuclides and quantify their activities or masses. In NDA systems that measure only the gamma rays emitted by radionuclides (emission only), the geometry correction and the container and matrix attenuation are determined using a multi-density efficiency calibration. Some gamma NDA systems use transmission measurements to determine the container and matrix attenuations, which are then applied to the count rates in the full-energy peaks in the gamma-ray spectrum. Most gamma-based NDA systems can be used in both the emission only or the emission and transmission modes.

To completely characterize a waste item, both neutron and gamma measurements are often necessary. The ^{235}U mass or ^{240}Pu$_{equivalent}$ mass is determined using neutron counting. Gamma-ray measurements are necessary to determine the abundances of uranium and plutonium isotopes. Results from the neutron and gamma measurements are then combined to provide isotope-specific quantification for uranium or plutonium content of the waste.

10.4.2.1.7 Calorimetry

Calorimeters perform waste assay by measuring the thermal energy generated by radioactive decay. Calorimeters are typically used to measure waste materials that produce significant amounts of heat by alpha-particle decay (plutonium, americium), or beta-particle decay (tritium). Heat-flow calorimeters have been used routinely to assay items whose thermal power ranges from 0.001–135 W. Gamma-ray spectrometry is typically used to determine the plutonium isotopic composition and ^{241}Am-to-plutonium ratio.

10.4.2.2 ISOCS for Waste Measurements

When measuring waste, the In Situ Object Counting System (ISOCS) calibration software can be used to calculate the efficiency calibration when determining the activity and mass of nuclear material. This method of quantifying nuclear and radiological material is used with gamma spectroscopy to analyze measurements from an HPGe detector. The ISOCS efficiency calibration has been applied successfully to gamma assay of waste material for decades and has proven to be a well-established technique [36].

To determine an accurate efficiency calibration, the ISOCS software does not require the use of traditional calibration sources, which can lower the cost and time required for collection of waste measurements. Instead, this software uses the MCNP transport modeling code, mathematical geometry templates, and a few physical parameters input by the user that are specific to the detector, sample, and geometry used in the measurement [37]. The model created in ISOCS by the user can be adjusted to represent the waste sample and measurement geometry. To find the activity and mass of radioisotopes identified in the waste, the efficiency calibration from the ISOCS software is applied during spectral analysis of the measurement data.

Several tools can be used with ISOCS to provide the most accurate measurement values. The Line Activity Consistency Evaluator (LACE) tool can be used to optimize and validate these poorly known or unknown physical parameters. This tool is based on a fundamental rule of gamma spectroscopy in which all associated gamma energies for a given radionuclide have the same calculated activities. By plotting these calculated activities and providing a value for the slope, the user can determine if the material is over- or under-attenuated and can then adjust the model parameters as needed. This tool helps to improve accuracy and to remove measurement bias that is caused by differences between the model and the actual physical properties of the sample [38].

Another tool that can be used with the ISOCS software is the ISOCS Uncertainty Estimator (IUE), an automated method that iteratively creates and analyzes multiple models based on input by the user [38]. The user defines one or more physical parameters, the range of variation, and a probability distribution function [36]. The IUE allows the user to investigate the impact of varying one or more physical parameters and determines the average and standard deviation of efficiency values. This tool can also be used to simulate hotspots in a sample with random size and locations [36], which can be especially useful for waste measurements.

This method can be applied to a very wide range of nuclear material types and compositions and is sometimes the only technique available for certain applications. It is often a more cost-effective option because it eliminates the need for physical calibration sources. The ISOCS method for quantifying nuclear and radiological material has an extremely wide range of possible relative uncertainties. Another limitation is that creating the ISOCS models can be complex and will most likely require proper training and experience [39].

10.4.2.3 SNAP for Waste Measurements

The Spectral Nondestructive Assay Platform (SNAP) is a gamma-ray spectroscopy, point-kernel modeling routine owned by Pajarito Scientific Corporation.[4] The program imports gamma-ray spectrum photopeak data from any counting system to produce assay results for a wide variety of geometries and matrices. Nearly anything with gamma-ray-emitting contaminants that can be physically described can be analyzed by this method (waste, surfaces, holdup, etc.). The code is adaptable to any gamma-ray detector media (HPGe, $LaBr_3$, NaI, etc.) for which an intrinsic photopeak efficiency calibration can be obtained. SNAP is intended for an experienced spectroscopist because it provides the user with the opportunity to make specific choices during the analysis process to "self-direct" the routine to an optimal solution.

SNAP analyses follow a straightforward, four-step process: (1) nuclide identification of photopeaks in the spectrum report, (2) input of modeling data to describe the assay source/detector configuration, (3) photopeak selection for assay calculations of detected radionuclides as well as minimum detectable activities, and (4) fine-tuning of the final assay report. The nuclide identification process is library driven and is often completed in the automated mode. However, use of both semi-automated and manual modes allows the analyst to interact with the code so that correct choices are made regarding the radionuclides and interactions that are present. Coincidence peaks, escape peaks, neutron capture gamma rays, florescent X-rays, and other common phenomena can be properly identified. The user is also required to make photopeak choices during the assay calculation process to avoid inappropriate automated choices that frequently require post-analysis modification. Although user interactivity requires a minor amount of additional time during the analysis process, it helps avoid many of the common mistakes of purely automated routines.

SNAP assumes that the detector is pointed in the general direction of the item under investigation and calculates the detector response in a 180° FOV. This calculation is accomplished by a combination of empirically determining both the detector's intrinsic photopeak efficiency using sources normal to the detector face and applying the relative change in efficiency as a function of gamma-ray angle as it enters the detector (an angular response calibration). Both calibration subroutines are provided with the program. The intrinsic photopeak efficiency is established with measurements of National Institute of Standards and Technology–traceable sources over the desired energy range of interest (e.g., 50–3000 keV). The angular response calibration is derived from empirical measurements of a source at incremental angles through the forward-looking 180° space. The intrinsic photopeak efficiency measurements are made with the sources placed at common assay distances; the angular response measurements can be made with or without side-shielding/collimation of the detector. No limit exists to the number of efficiency and angular response calibrations that can be created and used.

The modeling routine requires the user to physically describe the item/sample under measurement and its geometric configuration to the detector (source-detector distance and vertical and horizontal offsets from the detector center). In addition to common simple geometries (e.g., rectangular box, right circular cylinder) the code can also produce results for area sources (dpm/100 cm^2) and linear sources (Ci/cm length). Furthermore, SNAP allows the user to selectively weight any number of point-kernels in the matrix to describe unique shapes (e.g., spheres, cones, curved pipes) and/or specific distributions of contaminants within the item/sample (e.g., bottom settling, accumulations in corners). Models of 15 special shapes and contaminant distributions are provided with the program.

More than 35 common materials can be used to describe the item's radioactive matrix, as well as the container walls and/or gamma-ray filters (e.g., cadmium, copper). Mass attenuation coefficients as a function of energy are provided for all materials to calculate attenuation losses of full-energy gamma rays headed toward the detector. Additional mass attenuation coefficients for other materials can be added by the user as needed. When the material in the radioactive matrix is unknown, gamma-ray attenuation can be determined using a transmission source measurement. To properly interpolate the transmission fraction, the transmission correction subroutine requires the use of a transmission source that has gamma-ray emissions on either side of the assay photopeak.

A graphical differential peak analysis allows the analyst to view the relative activities of different energy photopeaks from the same radionuclide. This tool enables a physics check on whether the item/sample has been properly modeled so that modifications can be made if necessary. Gamma-ray self-absorption corrections (aka *lump corrections*) of inherently dense nuclear materials, such as plutonium and uranium, can also be made when two or more photopeaks are present from the same isotope. These corrections are independent of the lump shape and will enhance the final accuracy of results for these materials.

SNAP includes a thorough radionuclide library that can be added to and/or modified by the user. It comes with 11 sublibraries of common radionuclide groups such as uranium isotopes, plutonium and other transuranics, and medical isotopes. New sublibraries can be created, and default sublibraries can be modified as necessary.

---

[4] Parajito Scientific Corporation, www.pajaritoscientific.com

A detailed error analysis is performed on all photopeaks used in the assay. Uncertainties are determined for random statistical errors in the net counts in photopeaks, polynomial curve fits to the intrinsic photopeak efficiency, and assumptions about the source distribution and attenuation losses. The default calculation assumes a uniform, homogeneous distribution of contaminants within the physical volume of the item/sample. However, the program will also calculate the difference in activity of a single point source within that volume at a near-worst-case location. This difference in activity is conservatively used as part of the total measurement uncertainty unless it is overridden by the analyst during the fine-tuning stage of the report process (e.g., it is known that the contaminants are uniformly distributed).

Not considering statistical uncertainties, the accuracy of results using SNAP (or any other modeling routine) are dependent on three primary factors: (1) how closely the modeling input parameters match the physical item/sample; (2) the physical size/volume and matrix density of the item/sample; and (3) the methods employed to collect the data (e.g., detector positioning, source rotation, count time, background interference). It is apparent that poor results will occur if the item/sample under investigation is not physically well known or properly modeled. Furthermore, large-volume items with higher-density matrix materials are much more susceptible to geometry and attenuation inaccuracies than smaller items with lower-density materials. However, it is also important to recognize that a well-positioned detector that measures all sides of an item equally will generally produce data that lends itself to more accurate results. The failure to measure all sides of an item equally (or to rotate it continuously during the count) means that potential hotspots will be either over- or underestimated during analysis. Positioning the detector close to a larger-volume item will also lead to excessive inaccuracies when the source distribution is heterogeneous as opposed to homogeneous (which is a common occurrence). Although modeling routines allow a great deal of flexibility in the ways that measurements with portable detectors can be conducted, poor choices in measurement methods can lead to undesirable results. Caution is advised.

10.4.2.4 Segmented Gamma-Ray Scanning/Tomographic Gamma-Ray Scanning

In the process of filling scrap and waste containers, vertical variations frequently occur in the volume densities of both source arid matrix materials. Radial inhomogeneities are often less pronounced, and their effects can be substantially reduced by sample rotation. The different layers could substantially meet the requirements on homogeneity even though large differences exist between layers. In such cases, the container may be assayed as a vertical stack of overlapping segments. The advantages of the segmented scanning procedure are gained at the loss of some degree of sensitivity; therefore, a system that employs segmented scanning would probably not be used on samples that contain <1 g of ^{239}Pu or ^{235}U.

Figure 10.29 shows the spatial relationships of detector, collimator, assay sample, transmission source, and reference source for a system tailored to the assay of ^{239}Pu in cylindrical containers <20 cm in diameter.

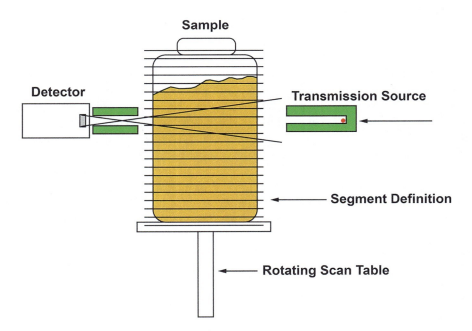

Fig. 10.29 General arrangement for segmented, transmission-corrected, gamma-ray assay

The sample container is positioned as close as possible to the collimator to maximize count rates and give the best segment resolution. The segment overlap is determined by the sample size, the collimator dimensions, and the relative positions of the segments. In Fig. 10.29, a collimator 1.25 cm high and 10 cm deep provides a reasonable trade-off in sensitivity and spatial resolution. For 30- and 55-gallon drums, a collimation 5 cm high and 20 cm deep is a reasonable choice. The spatial resolution cannot be as sharp in the latter case, but it is sufficient to provide useful information on the uniformity of material distribution. The choice of collimator material is usually lead. If space is a consideration, a tungsten alloy may be used.

To maximize count rates, the detector is as close as possible to the collimator. For the plutonium measurement, a filter of lead (~1.5 mm) and cadmium (~0.8 mm) reduces the rate of low-energy events from ^{241}Am and the X-rays of both plutonium and lead. For ^{235}U assay, the cadmium alone should suffice because no 60 keV ^{241}Am flux is found in plutonium materials.

For ^{239}Pu assay, ^{75}Se is the transmission source, and ^{133}Ba is the reference source. About 10 mCi of ^{75}Se provides usable intensity for at least 1 year in spite of the relatively short, 120-day half-life. Sources of this strength must be encased in a collimator shield to avoid undue personnel exposure. The 356.0 keV gamma ray of ^{133}Ba provides the reference; it can also be used for spectrum stabilization because it is always present in the acquired spectrum. The 10.4-year half-life is convenient; a single source usually lasts the useful life of a germanium detector. A source of ~10 µCi may be positioned anywhere on the front side of the detector end cap; however, a slightly better peak shape results when the source is mounted on the front of the end cap along the crystal axis.

Segmented scans may be accomplished in several ways, described as continuous or discrete scans. In a continuous scan, the rotating sample moves past the collimator at a constant speed. The count time is often chosen as the time required for the container to move the height of the collimator. In a discrete scan, the sample is positioned vertically, counted, repositioned, counted again, etc. This mode of operation avoids detector microphonics caused by vibration in the vertical drive system. In practice, a segment spacing equal to one-half of the collimator height works well and might be recommended as a rule of thumb. The continuous mode probably gives a somewhat better value for the average transmission within a segment. The discrete scan is usually easier to achieve. It also lends itself to two-pass assays during which the container is counted once with the transmission source exposed and once with it shuttered off. The two-pass scheme is useful when the utmost sensitivity and accuracy is desired and is particularly useful when ^{169}Yb is used as a transmission source in ^{235}U assays. Other variations in the application of the segmented scanning procedure are possible and are described in Refs. [40, 41].

An increasing demand to assay a class of materials that was difficult or impossible to assay with standard NDA techniques, including segmented gamma-ray scanning (SGS), led to the development of tomographic gamma-ray scanning (TGS) technology [42]. Unlike SGS, which measured the content within vertical segments of an item, TGS technology uses a collimation system that divides the samples into small-volume elements called *voxels*. Measuring the transmission in each voxel provides a three-dimensional map of the sample that indicates the density of material that exists within each voxel. This map provides a means for measuring and correcting for the attenuation that is occurring within the item as a function of the position within the item. By measuring the gamma-ray emission rate from each voxel and applying the attenuation corrections from the measured three-dimensional density map, a three-dimensional image of the nuclear material content within the item can be produced, which allows for an accurate assay of nuclear material within the item. Examples of TGS three-dimensional image reconstructions of 208 L drums that contain various matrices are shown in Fig. 10.30 [42].

Figure 10.31 shows a conceptual configuration for the TGS. During a scan, the item undergoes a continuous motion consisting of both rotation and translation to maximize throughput of items. The translation consists of moving the item both vertically and horizontally. The gamma rays from ^{75}Se are used to determine attenuation coefficient images within each voxel as a function of gamma-ray energy. The distribution of gamma-ray-emitted material is then determined based on the emission analysis, with the emitted gamma rays corrected for attenuation. From this information, the mass of nuclear material within the item is determined based on mass calibrations that were performed on the TGS for the primary isotope that is to be measured.

To illustrate the capabilities of a TGS system, Fig. 10.32 shows the results of assays performed by a portable TGS system at the Rocky Flats Environmental Technology Site on electro-refining salts that were destined for disposal at the Waste Isolation Pilot Plant in comparison with assay results on those same items using a combination of calorimetry and gamma-ray isotopic (Cal/ISO) measurements [42, 43]. The fact that the data presented in Fig. 10.32 does not significantly deviate from the solid line that indicates a slope of unity shows no significant bias in the TGS data. An analysis of the TGS data indicated that an uncertainty per measurement was 9.4%, with a bias between the TGS and Cal/ISO data of $-0.04 \pm 0.99\%$.

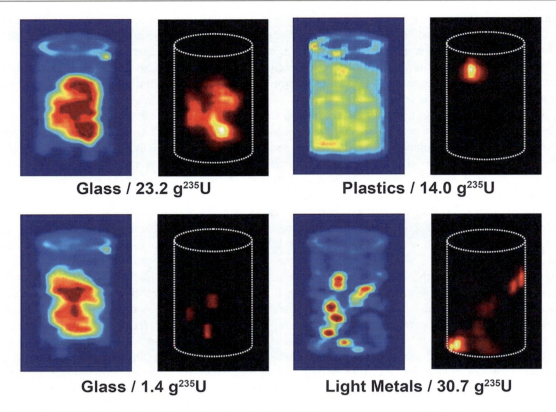

Fig. 10.30 Transmission (blue background) and emission (black background) image reconstructions of 208 L drums in varying matrices [42]

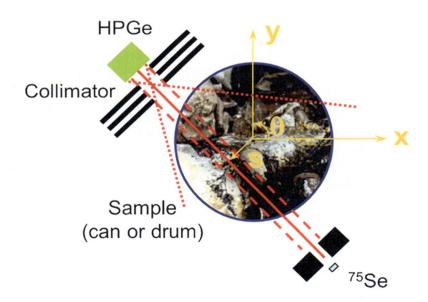

Fig. 10.31 Conceptual configuration for a TGS system

10.4.3 Holdup Measurements

10.4.3.1 Introduction

Process holdup is defined as the residual nuclear material retained in process equipment after cleanout [44]. The effective characterization and quantification of process holdup is critical for plant operations and important to many plant programs,

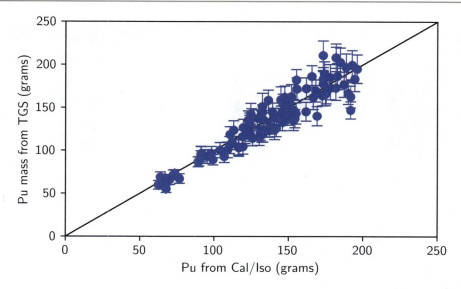

Fig. 10.32 Assay results from TGS measurements on electro-refining salts performed at Rocky Flats Environmental Technology Site as a function of assay results using Cal/Iso, showing no significant bias in the results. For reference, the diagonal line in the graph indicates a 1-to-1 correspondence between the TGS and Cal/Iso results [42, 43]

including material control and accountability, nuclear criticality safety, and environmental management. These in-situ measurements are most often performed using NDA methods. Holdup measurements present significant challenges due to the unique and nonideal nature of the deposits. The distribution (i.e., geometry) and thickness of a given deposit is largely unknown. The concentration and isotopic information are assumed based on process knowledge or samples measured at a specific time in the process. Additional challenges include the lack of representative calibration standards, the large number of holdup deposit locations and relatively short count times available, high or variable background environments, and poor accessibility to measurement locations. As a result, large measurement uncertainties are expected.

Several holdup measurement methods have been developed and used through the years for both uranium and plutonium applications. One of the earliest approaches—the generalized geometry holdup (GGH) method—is still used today for most uranium applications. More modern techniques, such as ISOCS or ISOTOPIC-32, are currently being applied for plutonium applications. The most common methods and the accompanying instrumentation are described in the following sections.

10.4.3.2 GGH Method, Introduction, Calibrations, and Corrections

10.4.3.2.1 Introduction

Traditionally, holdup measurements are performed using a robust methodology—the GGH method [44–47]. This method is commonly applied to uranium holdup applications and can account for geometric differences in deposits, correct for both wall attenuation and self-attenuation, and be used in various radiological environments. The basis of the GGH method is to force each holdup deposit to fit the GGH model definition of a point, line, or area geometry with respect to the collimated detector FOV. The assumption is that the detector can be positioned such that the holdup deposit will assume one of the geometries shown in Fig. 10.33.

- Point: Point in the center of the detector FOV and small enough in diameter to occupy less than 3% of the FOV
- Line: Line extending across the FOV in one direction but occupying less than 3% FOV in the opposite (1D) direction (narrow)
- Area: Uniform distribution that extends beyond the FOV in all directions

Through the development of representative models and application of corrections for wall attenuation, finite source geometries, and self-attenuation, a measured mass for point deposits, a linear density for line deposits, or an areal density for area deposits can be determined. It should be noted that the GGH method is applied to most solid deposits in process equipment where self-attenuation dominates the attenuation of the signal. This method is not applicable for materials where attenuation is driven by other factors [46].

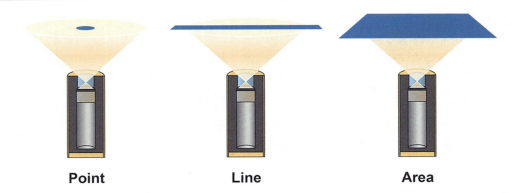

Fig. 10.33 Diagram depicting the geometries considered in the GGH method

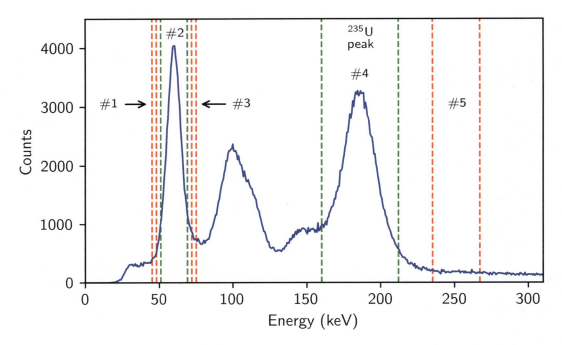

Fig. 10.34 Example uranium spectrum and assigned ROIs for uranium holdup using GGH

10.4.3.2.2 Determining Net Counts for Analysis

The GGH method assumes that all deposits can be modeled as points, lines, or areas with respect to the detector FOV and measured as such to determine the isotopic mass of the deposit. The measurement itself is a simple ROI technique that focuses on the 186 keV gamma ray from ^{235}U for uranium applications or the 414 keV gamma ray from ^{239}Pu for plutonium applications [47]. The GGH method begins with analysis of the net count rate in the ROI indicative of the isotope of interest.

An example spectrum showing the traditional ROIs used for uranium analysis is given in Fig. 10.34. ROIs #1 through #3 are related to the reference peak (^{241}Am in the following scenario and usually ^{109}Cd for plutonium applications) and are used for quality control checks. The reference peak and associated backgrounds are monitored to ensure that the detector is stable and operating properly. ROI #4 is the analysis ROI that encompasses the 186 keV peak. ROI #5 is referred to as the Compton ROI and serves as a measure of the Compton continuum that is subtracted from the integral of ROI #4 to determine the 186 keV net peak area.

It is important to note that room background can have a significant impact on the GGH analysis and results and therefore must be carefully considered. Room background is generally measured alongside the process equipment under inspection, with the detector located at the same position and either pointed away from the deposit or with a high-Z plug inserted into the collimator. The net room background for the isotope of interest is determined as previously described.

The net room background rate is then subtracted to obtain the net count rate attributed to the source. This net count rate is referred to as the *uncorrected, room-background-subtracted, net count rate* because it has yet to be corrected for wall or self-attenuation.

$$\dot{C}_{unc} = \left\{ \frac{1}{t_S} \left[\sum_{i=Beg_A}^{End_A} S_A(i) - \sum_{i=Beg_C}^{End_C} S_C(i) \right] \right\} - \left\{ \frac{1}{t_B} \left[\sum_{i=Beg_A}^{End_A} B_A(i) - \sum_{i=Beg_C}^{End_C} B_C(i) \right] \right\} \quad (10.18)$$

where
Beg_A = first channel marking the analysis ROI
End_A = last channel marking the analysis ROI
Beg_C = first channel marking the Compton ROI
End_C = last channel marking the Compton ROI
$S_A(i)$ = counts in the ith channel of the analysis ROI from the source (cts)
$S_C(i)$ = counts in the ith channel of the Compton ROI from the source (cts)
$B_A(i)$ = counts in the ith channel of the analysis ROI from the background (cts)
$B_C(i)$ = counts in the ith channel of the Compton ROI from the background (cts)
t_S = count time for the source measurement (s)
t_B = count time for the background measurement (s)

This uncorrected, room-background-subtracted, net count rate is the basis for further analyses including determination of the calibration constants and measurements of holdup deposits. Once the room-background-subtracted, net count rate is determined and appropriate corrections are applied, the isotopic mass of the deposit can be determined.

10.4.3.2.3 Determining Calibration Constants

Detector calibrations are carefully performed using a single, well-characterized calibration standard to understand the detector response to the generalized geometries (i.e., point, line, or area) and distributions of the deposit. Acceptable calibration sources must be representative of the material to be measured (thin, deposit-like standards are preferred to thicker or spherical standards), in a stable form, have a known geometry and packaging, and have a known self-attenuation value.

The calibration source is measured at the center of the FOV along the detector axis at a fixed distance (r_0) to determine the response to a point source. (Refer to the point-source geometry with respect the detector axis shown in Fig. 10.33.) The source is then measured at small radial increments (s) across the FOV to determine the one-dimensional radial response or the detector response to a distributed line-geometry deposit, as shown in Fig. 10.35. For a cylindrical collimator, rotational symmetry is a good assumption; therefore, the response at each of the radial locations is assumed to be constant around the concentric rings of the detector FOV (also shown in in Fig. 10.35) and determines the response to a distributed area-geometry deposit.

This series of calibration measurements is first used to determine the geometric correction factors for a line (L) and an area (A).

$$L = s \left[2 \frac{\sum_i C_{Radial}(i)}{C_{Radial}(0)} - 1 \right] \quad (10.19)$$

$$A = \pi \frac{s^2}{4} + \pi \cdot 2s^2 \frac{\sum_i [|i| C_{Radial}(i)]}{C_{Radial}(0)} \quad (10.20)$$

where
L = geometric correction factor for a line (cm)
A = geometric correction factor for an area (cm^2)
s = incremental radial measurement spacing (cm)
i = off-axis measurement positions
$C_{Radial}(i)$ = net count rate at position i from the radial response measurements (cps)

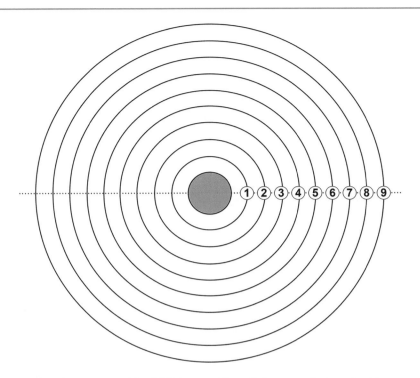

Fig. 10.35 Diagram depicting the spatial response of the GGH detector. The radial response is determined by measuring the calibration (point) source at each of the off-axis positions. The distance between each position is s

The calibration constants are then calculated using the following equations:

$$K_P = \frac{m_{U235}}{\dot{C}_{unc} \cdot CF_{wall} \cdot CF_{self} \cdot r_0^2} \qquad (10.21)$$

$$K_L = \frac{m_{U235}}{\dot{C}_{unc} \cdot CF_{wall} \cdot CF_{self} \cdot L \cdot r_0} \qquad (10.22)$$

$$K_A = \frac{m_{U235}}{\dot{C}_{unc} \cdot CF_{wall} \cdot CF_{self} \cdot A} \qquad (10.23)$$

where
K_P = calibration factor for point-type deposits (g–s/cm²)
K_L = calibration factor for line-type deposits (g–s/cm²)
K_A = calibration factor for area-type deposits (g–s/cm²)
m_{U235} = isotopic mass of the calibration standard (g)
$\dot{C}_{unc}$ = uncorrected, room-background-subtracted, net count rate from the isotope of interest (cps)
CF_{Wall} = correction factor for wall attenuation
CF_{Self} = correction factor for self-attenuation in the deposit itself
L = geometric correction factor for a line (cm)
A = geometric correction factor for an area (cm²)
r_0 = measurement distance from the detector face (including the collimator offset) to the calibration standard (cm)

The correction factors for the calibration standard should be well known. Alternatively, they can be calculated using the following equations:

$$CF_{wall} = \exp(\mu_{wall} \cdot \rho_{wall} \cdot x_{wall}) \qquad (10.24)$$

$$CF_{self} = \frac{\mu_{act} \cdot \rho_{act} \cdot x_{act}}{1 - \exp(\mu_{act} \cdot \rho_{act} \cdot x_{act})} \qquad (10.25)$$

where

μ_{wall} = mass attenuation coefficient for the container wall material for the gamma-ray energy of interest (cm^2/g)
ρ_{wall} = density of wall material (g/cm^3)
x_{wall} = thickness of the container wall (cm)
μ_{act} = mass attenuation coefficient for the actinide compound mixture at the energy of interest (cm^2/g)
ρ_{act} = density of actinide compound mixture (g/cm^3)
x_{act} = thickness of the actinide compound material (cm)

Once the calibration constants are determined for each geometry, measurements can be performed to obtain a total gram quantity of a given holdup deposit.

10.4.3.2.4 Analysis of Holdup Deposits

Like the calibration calculations, the holdup deposit analysis begins by determining the uncorrected, room-background-subtracted, net count rate in the ROI. However, an area background correction factor (CF$_B$) is now applied for area geometries to compensate for attenuation of the background signal by the process equipment: $CF_B = CF_{Wall}$ for a single-walled deposit and $CF_B = CF_{Wall}^2$ for double-walled deposits. For points and lines: CF$_B$ = 1.

$$\dot{C}_{unc} = \left\{ \frac{1}{t_S} \left[\sum_{i=Beg_A}^{End_A} S_A(i) - \sum_{i=Beg_C}^{End_C} S_C(i) \right] \right\} - \frac{1}{CF_B} \left\{ \frac{1}{t_B} \left[\sum_{i=Beg_A}^{End_A} B_A(i) - \sum_{i=Beg_C}^{End_C} B_C(i) \right] \right\} \qquad (10.26)$$

where

Beg$_A$ = first channel marking the analysis ROI
End$_A$ = last channel marking the analysis ROI
Beg$_C$ = first channel marking the Compton ROI
End$_C$ = last channel marking the Compton ROI
S$_A$ = counts in the analysis ROI from the source measurement at channel I (cts)
S$_C$ = counts in the Compton ROI from the source measurement at channel I (cts)
B$_A$ = counts in the analysis ROI from the background measurement at channel I (cts)
B$_C$ = counts in the Compton ROI from the background measurement at channel I (cts)
t$_S$ = count time for the source measurement (s)
t$_B$ = count time for the background measurement (s)
CF$_B$ = area background correction factor

The importance of background measurements should also be noted. Background measurements should be performed frequently. When background is unknown, unstable, or rapidly changing over the measurement campaign, it might be necessary to perform background measurements before every deposit measurement. In more-stable environments, a single background measurement may be used for multiple subsequent deposit measurements. Walk-downs of the facility and reliable process knowledge should be considered when deciding how to proceed with background measurements.

10.4.3.2.5 Corrections

For holdup deposits, three corrections are typically applied: wall attenuation correction, finite source correction, and self-attenuation correction. Without application of these correction factors, a systematic negative bias will result. The wall-attenuation correction is straightforward and easily applied to gamma-ray measurements. The wall-attenuation correction is again determined using Eq. 10.24.

Next, the finite source correction must be applied to account for the fact that point and line deposits have a non-negligible width that fills some fraction of the detector FOV at the given measurement distance. The portions of the deposit that are off-axis from the center of the detector FOV have a lower detection efficiency that must be corrected for. The finite source correction factor (CF_{FS}) is dependent on the radial response of the detector and an empirical width parameter (w). This width parameter is effectively the estimated diameter of a point or the estimated width of a line deposit. For example, w for a point can be estimated as the diameter of an elbow in a pipe where material has accumulated. For a line, w can be estimated by the width of a pipe. The finite source correction is not applied to area geometries because they are assumed to fill the detector FOV completely.

$$w_0 = w \cdot \frac{r_0}{r} \tag{10.27}$$

$$G = \exp\left[-\frac{1}{2} \cdot \left(\frac{2\sqrt{2 \cdot \ln(2)} \cdot \frac{w_0}{2}}{FWHM}\right)^2\right] \tag{10.28}$$

$$CF_{FS} = \left[\frac{1+G}{2}\right]^{-n} \tag{10.29}$$

where

w_0 = relative width (for lines)/diameter (for points) scaled by the ratio of the calibration distance at i = 0 to the measurement distance to the deposit (cm)
w = empirical parameter for point diameter or line width of the deposit (cm)
r_0 = calibration distance at position i = 0 (including the collimator offset) (cm)
r = measurement distance from the detector face (including the collimator offset) to the deposit (cm)
G = calculated radial response at the edge of the deposit relative to the center
FWHM = full width half maximum of the Gaussian fit of the radial response (cm)
CF_{FS} = finite source correction factor
n = geometry order (n = 2 for a point, n = 1 for a line, n = 0 for an area)

Measurement distances can be carefully chosen to reduce the effects of the finite source; however, environmental factors, such as room background and physical access to measurement locations, generally drive the selection of the measurement distance.

Both the wall-attenuation correction and the finite source correction are applied linearly such that an intermediate specific mass of the isotope can be calculated using Eq. 10.30.

$$m_{int} = (K \cdot r^n)(\dot{C}_{unc} \cdot CF_{FS} \cdot CF_{Wall}) \tag{10.30}$$

where

m_{int} = intermediate specific mass (wall attenuation and finite source corrections applied) (g, g/cm, g/cm^2)
K = calibration constant associated with the deposit geometry ($K = K_P$, K_L, or K_A depending on the deposit geometry) (g–s/cm^2)
r = measurement distance from the detector face (including the collimator offset) to the deposit (cm)
n = geometry order (n = 2 for a point, n = 1 for a line, n = 0 for an area)
$\dot{C}_{unc}$ = uncorrected, room-background-subtracted, net count rate in the analysis peak for the measured deposit (c/s)
CF_{FS} = finite source correction factor
CF_{wall} = wall attenuation correction factor

Standard uncertainty propagation techniques can be applied for calculations thus far because all are considered linear [46].

Once the finite source and wall-attenuation corrections have been applied, the nonlinear self-attenuation correction can be applied. The self-attenuation correction must be applied to account for the gamma rays that are attenuated by the (high-Z) deposit itself. The most common application of the self-attenuation correction is provided herein and depends on the measured areal density of the deposit [48]. Alternative methods have been developed to address self-attenuation [49].

Because self-attenuation depends on the actinide density, conversion of the specific mass of the isotope to the areal density of the actinide is required. This conversion calls for process knowledge of both the isotopic distribution and the actinide matrix.

$$m_E = \frac{m_{int}}{f_I \cdot f_E} \qquad (10.31)$$

where

m_E = elemental or actinide mass of the deposit (g, g/cm, g/cm^2)
m_{int} = intermediate specific mass (wall-attenuation and finite source correction applied) (g, g/cm, g/cm^2)
f_I = isotopic fraction by weight (enrichment)
f_E = elemental fraction by weight (based on deposit compound)

The measured areal density of the actinide deposit is then determined by dividing the specific actinide mass by the area of the point deposit or the width of a line deposit. The areal density for an area deposit is directly measured.

$$(\rho x)_{Meas} = \frac{m_E}{\pi \left(\frac{w}{2}\right)^2}, \text{for a point} \qquad (10.32)$$

$$(\rho x)_{Meas} = \frac{m_E}{w}, \text{for a line} \qquad (10.33)$$

$$(\rho x)_{Meas} = m_E, \text{for an area} \qquad (10.34)$$

where
$(\rho x)_{Meas}$ = measured areal density (g/cm^2)
w = empirical parameter for point diameter or line width

The true area density as a function of the measured areal density can then be determined using Eq. 10.35. Note that the self-attenuation correction is now applied.

$$(\rho x) = -\frac{1}{\mu} \cdot \ln[1 - \mu \cdot (\rho x)_{Meas}] \qquad (10.35)$$

where
(ρx) = true areal density of the actinide (g/cm^2)
μ = mass attenuation coefficient for actinide deposit material at energy of interest (cm^2/g).

The self-attenuation corrected areal densities must then be converted back to specific mass of the isotope:

$$m_P = (\rho x) \cdot \pi \left(\frac{w}{2}\right)^2 \cdot f_I \cdot f_E \qquad (10.36)$$

$$m_L = (\rho x) \cdot w \cdot f_I \cdot f_E \qquad (10.37)$$

$$m_A = \begin{cases} (\rho x) \cdot f_I \cdot f_E, for\ points, lines, and\ single-walled\ deposits \\ \frac{1}{2}(\rho x) \cdot f_I \cdot f_E, for\ double-walled\ area\ deposits \end{cases} \qquad (10.38)$$

where
m_P = final specific mass for a point (g)
m_L = final specific mass for a line (g/cm)
m_A = final specific mass for an area (g/cm^2)

Note that for a double-walled deposit, the specific mass is divided by 2 to account for the fact that the deposits on both walls are contributing to the signal.

Finally, to determine the total grams for a given deposit,

$$M = m \cdot D \tag{10.39}$$

where
M = total grams of the isotope of interest in the deposit (g)
m = specific mass for the deposit (g, g/cm, g/cm2)
D = total deposit area or length depending on geometry (1 for a point) (cm^2, cm)

For completeness, the resulting self-attenuation correction factor is

$$CF_{SELF} = \frac{(\rho x)}{(\rho x)_{Meas}} \tag{10.40}$$

where CF_{Self} = self-attenuation correction factor.

Note that the uncertainty for the self-attenuation correction factor cannot be calculated using standard propagation techniques but instead must be propagated nonlinearly [44].

The holdup deposit width parameter (w) impacts the finite source correction (CF_{FS}) and the self-attenuation correction (CF_{SELF}) in opposite ways. If the estimated width parameter is larger than the actual deposit, then CF_{FS} is overestimated, and CF_{SELF} is underestimated. The converse is also true; if the estimate of the width parameter is smaller than the actual deposit, then CF_{FS} is underestimated, and CF_{SELF} is overestimated. Therefore, any bias caused by an inaccurate estimate of w is largely cancelled and its magnitude much smaller than the overall uncertainty. The practical takeaway from this case is that the NDA practitioner who is making holdup measurements with large enough r (measurement distance from the detector face to the deposit) such that the deposits resemble those depicted in Fig. 10.33 can simply take the process equipment dimension as the deposit width, saving valuable time by eliminating the need to measure deposit widths using handheld dose-rate meters.

The application of these correction factors is critical for the measurement of process holdup. Although the bias introduced for a single measurement could be small, the amount of nuclear material in process holdup throughout an entire facility could be significant. Without application of these corrections, holdup measurement results will always be biased low.

10.4.3.2.6 Hardware Considerations

Because the GGH method relies on simple ROIs, low-resolution gamma detectors are generally acceptable. Advantages of low-resolution systems, such as NaI(Tl), include room-temperature operation, low cost, and the lightweight (easily maneuverable) nature. The use of NaI(Tl) and the ROI method can be complicated or biased due to interferences, such as ^{232}U/^{232}Th decay (resulting from reprocessing of reactor returns) or the relatively low sensitivity to thin deposits in the presence of high-energy gamma rays from room background [47, 50].

Because the GGH method assumes a radially symmetric response across the detector FOV, the detectors used for this method should be cylindrical and must be side-shielded to reduce the effects of room background and collimated to ensure rotational symmetry. The NaI(Tl) detectors commonly used include a 1-inch collimator. For uranium applications, a ^{241}Am seed is placed between the detector face and the tin filters. This setup provides a stable and constant reference source for the holdup measurements. For plutonium applications, a ^{109}Cd source is often employed. Tin filters are often placed in front of the detector to eliminate low-energy gamma rays and X-rays from interfering with the reference peak. A diagram that illustrates the primary detector components is provided in Fig. 10.36.

The data acquisition system for holdup measurements includes a modern multichannel analyzer (MCA) to store spectra and a handheld computer or similar device to control the measurements. The current MCA used for NaI(Tl) systems is the GBS Elektronik MCA-527.

10.4.3.2.7 Benefits/Applications

The GGH method has many advantages. A single calibration source can be used to establish the response of the measurement system. The robust geometry models are flexible enough to accommodate most holdup deposits. If the crude geometry assumptions are reasonable, this method can provide a very good estimate of the mass of the deposit in a reasonably short

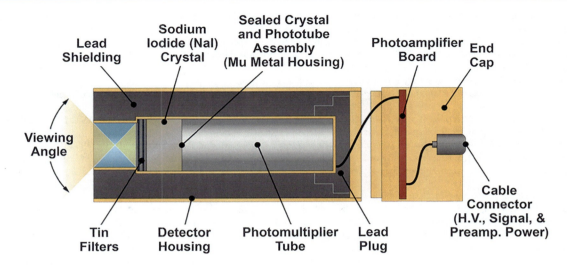

Fig. 10.36 Diagram of the traditional NaI(Tl) detector used for GGH measurements

amount of time, which allows a high volume of measurements to be performed in the relatively short inventory periods. Finally, because high resolution is not required for most applications of the GGH method, NaI detectors with high efficiency are often employed. NaI is lightweight, can be operated at room temperature, and is relatively inexpensive; therefore, it is cost efficient to have multiple measurement systems operating around a facility.

10.4.3.2.8 Limitations

GGH is not without its limitations. The method relies on many assumptions and user inputs (including process knowledge related to actinide mixture, enrichment, and geometric distributions of the source). The GGH method is only as good as the assumptions that go into the models; thus, measurement experience is invaluable. The better the assumptions, the better the results. Another significant limitation is the inability to handle deposits that approach infinite thickness. Because GGH relies on the ROI method, certain interferences (especially those that have gamma rays in the Compton ROI) may affect the measurement results.

10.4.3.3 ISOCS for Holdup Measurements

The ISOCS calibration software can be used to take more-accurate holdup measurements. As discussed in Sect. 10.4.2.2, ISOCS has the capability to model the measurement geometry and physical parameters associated with the holdup material.

Information about the location, composition, and density that corresponds to holdup material is often limited; therefore, the LACE and IUE tools are especially useful for these types of measurements. The LACE tool indicates whether the associated gamma energies for a given radionuclide have the same calculated activities, and the IUE tool allows the user to investigate the impact of varying one or more physical parameters in the model.

ISOCS is becoming more commonly used for holdup measurements, including by the International Atomic Energy Agency for safeguards verification activities that involve uranium holdup measurements [39]. If the information regarding the holdup characteristics is not well known, the uncertainties may not be the most precise or accurate [39]; however, it can provide a low measurement uncertainty if the measurement parameters are well known.

References

1. T. Burr, M. Hamada, Radio-isotope identification algorithms for NaI γ spectra. Algorithms **2**(1), 339–360 (2009)
2. D. Mayo et al., *Nuclear Reachback Reference Manual*, Los Alamos National Laboratory report LA-UR-18-28694 (2018)
3. S.E. Garner, *Spectra Deck*, Los Alamos National Laboratory report LA-UR-08-00502 (2008)
4. "American National Standard Performance Criteria for Handheld Instruments for the Detection and Identification of Radionuclides, ANSI N42.34-2021 (American National Standards Institute, Inc. New York, 2021)
5. F.X. Haas, W.W. Strohm, Gamma-ray spectrometry for calorimetric assay of plutonium fuels. IEEE Trans. Nucl. Sci. **NS-22**(1), 734–738 (1975)

6. T. Dragnev, K. Schaerf, Non-destructive gamma spectrometry measurement of ^{239}Pu/^{240}Pu and Pu/^{240}Pu ratios. Int. J. Appl. Radiat. Isot. **26**(3), 125–129 (1975)
7. J.L. Parker, T.D. Reilly, Plutonium isotopic determination by gamma-ray spectroscopy, in *Nuclear Analysis Research and Development Program Status Report, January–April 1974*, Los Alamos Scientific Laboratory report LA-5675-PR, ed. by G.R. Keepin, (1974), pp. 11–13
8. R. Gunnink, J.B. Niday, P.D. Siemens, *System for Plutonium Analysis by Gamma-Ray Spectrometry: Part I: Techniques for Analysis of Solutions*, Lawrence Livermore Laboratory report UCRL-51577, Part I (1974)
9. R. Gunnink, *Simulation Study of Plutonium Gamma-Ray Groupings for Isotopic Ratio Determinations*, Lawrence Livermore Laboratory report UCRL-51605 (1974)
10. *Standard Test Method for Determination of Plutonium Isotopic Composition by Gamma-Ray Spectrometry*, ASTM Standard Test Method C1030-10, American Society for Testing and Materials, Philadelphia (2018)
11. J.G. Fleissner, *GRPAUT: A Program for Pu Isotopic Analysis (A User's Guide)*, Mound Facility report MLM-2799 (1981)
12. J.G. Fleissner, J.F. Lemming, J.Y. Jarvis, Study of a two-detector method for measuring plutonium isotopics, in *Measurement Technology for Safeguards and Materials Control Proceedings of American Nuclear Society Topical Meeting, Kiawah Island, SC, November 26–30, 1979*, (1979), pp. 555–561
13. R. Gunnink, *A Guide for Using NaIGEM*, Version 2.1.4 (Oct 2010)
14. L.E. Smith, A.R. Lebrun, Design, modeling and viability analysis of an online uranium enrichment monitor, in *2011 IEEE Nuclear Science Symposium Conference Record, Valencia, Spain*, (2011), pp. 1030–1037. https://doi.org/10.1109/NSSMIC.2011.6154314
15. J. Younkin, L.E. Smith, J. March-Leuba, J. Garner, *On-Line Enrichment Monitor: Field Prototype Design Description*, IAEA document SG-RM-13445 (2013)
16. R. DeWitt, *Uranium Hexafluoride: A Survey of the Physicochemical Properties*, Report no. GAT-280 (Goodyear Atomic Corp, Portsmouth, 1960)
17. B. Weinstock, E.E. Weaver, J.G. Malm, Vapour-pressures of NpF$_6$ and PuF$_6$; thermodynamic calculations with UF$_6$, NpF$_6$ and PuF$_6$. J. Inorg. Nucl. Chem. **11**(2), 104–114 (1959), ISSN 0022-1902). https://doi.org/10.1016/0022-1902(59)80054-3
18. R. Gunnink, A.L. Prindle, Y. Asakura, J. Masui, N. Ishiguro, A. Kawasaki, S. Kataoka, *Evaluation of TASTEX Task H: Measurement of Plutonium Isotopic Abundances by Gamma-Ray Spectrometry*, Lawrence Livermore National Laboratory report UCRL-52950 (1981)
19. R. Gunnink, Gamma spectrometric methods for measuring plutonium, in *Proceedings of American Nuclear Society National Topical Meeting on Analytical Methods for Safeguards and Accountability Measurements of Special Nuclear Material*, (Williamsburg, VA, 1978)
20. W.D. Ruhter, D.C. Camp, *A Portable Computer to Reduce Gamma-Ray Spectra for Plutonium Isotopic Ratios*, Lawrence Livermore Laboratory report UCRL-53145 (1981)
21. W.D. Ruhter, *A Portable Microcomputer for the Analysis of Plutonium Gamma-Ray Spectra*, Lawrence Livermore National Laboratory report UCRL-53506, Vols. I and II (1984)
22. R. Gunnink, An algorithm for fitting Lorentzian-broadened K-series X-ray peaks of the heavy elements. Nucl. Instrum. Methods **143**, 145–149 (1977)
23. R. Gunnink, *Plutonium Isotopic Analysis of Nondescript Samples by Gamma-Ray Spectrometry*, Conference on Analytical Chemistry in Energy Technology, Gatlinburg, Tennessee, October 6–8, 1981
24. W.D. Ruhter, Experience with a General gamma-ray isotopic analysis approach, in *Proceedings of 44th Annual Meeting of the INMM*, Phoenix, AZ, July 13–17, 2003, Lawrence Livermore National Laboratory report UCRL-JC-151542 (2003)
25. D. Clark, T.-F. Wang, W. Romine, W. Buckley, K. Raschke, W. Parker, W. Ruhter, A. Friensehner, S. Kreek, Uranium and plutonium isotopic analysis using MGA++, in *Proceedings of 39th Annual Meeting of the INMM, Naples, FL, July 26–30, 1998*, Lawrence Livermore National Laboratory report UCRL-JC-131168 (1998)
26. W.E. Parker, T.-F. Wang, D. Clark, W.M. Buckley, W. Romine, W.D. Ruhter, Plutonium and Uranium Isotopic Analysis: Recent Developments of the MGA++ Code Suite, in *Proeedings of. 6th International Meeting on Facilities Operations-Safeguards Interface*, Jackson Hole, WY, September 20–24, 1999, Lawrence Livermore National Laboratory report UCRL-JC-133130 (1999)
27. R. Gunnink, W.D. Ruhter, P. Miller, J. Goerten, M. Swinhoe, H. Wagner, J. Verplancke, M. Bickel, S. Abousahl, *MGAU: A New Analysis Code for Measuring U-235 Enrichments in Arbitrary Samples*, IAEA Symposium on International Safeguards, Vienna, Austria, March 8–14, 1994, Lawrence Livermore National Laboratory report UCRL-JC-114713
28. W.D. Ruhter, R. Gunnink, *Measurement of Plutonium and Uranium Isotopic Abundances by Gamma-Ray Spectroscopy*, 11th International Workshop on Accurate Measurements in Nuclear Spectroscopy, Sarov, Russia, September 2–6 1996. Lawrence Livermore National Laboratory report UCRL-JC-123412
29. D. Clark, *U235: A Gamma Ray Analysis Code for Uranium Isotopic Determination*, Lawrence Livermore National Laboratory report UCRL-ID-125727
30. T.E. Sampson, G.W. Nelson, T.A. Kelley, *FRAM: A Versatile Code for Analyzing the Isotopic Composition of Plutonium from Gamma-Ray Pulse Height Spectra*, Los Alamos National Laboratory report LA-11720-MS (1989)
31. T.E. Sampson, T.A. Kelley, T.L. Cremers, *PC/FRAM: New Capabilities for the Gamma Ray Spectrometry Measurement of Plutonium Isotopic Composition*, Los Alamos National Laboratory document LA-UR-95-3287 (1995)
32. T.A. Kelley, T.E. Sampson, D. DeLapp, *PC/FRAM: Algorithms for the Gamma-Ray Spectrometry Measurement of Plutonium Isotopic Composition*, Los Alamos National Laboratory document LA-UR-95-3326 (1995)
33. D. Vo, Improvements of FRAM version 6.1, in *Proceedings of Workshop on Isotopic Analysis of Uranium and Plutonium by Non-Destructive Assay Techniques* (virtual event), February 15–19, 2021, J. Nucl. Mater. Manage. 49 (3), 19–30 (2021)
34. T.E. Sampson, T.A. Kelley, D.T. Vo, *Application Guide to Gamma-Ray Isotopic Analysis Using the FRAM Software*, Los Alamos National Laboratory report LA-14018 (2003)
35. J.G. Fleissner, Nondestructive assay of plutonium in isotopically heterogeneous salt residues, in *Proceedings of Conference on Safeguards Technology*, Hilton Head Island, SC, Department of Energy publication CONF-831106 (1983), pp. 275–285
36. A. Bosko, R. Venkataraman, F.L. Bronson, G. Ilie, W. R. Russ, Waste Characterization Using Gamma Ray Spectrometry with Automated Efficiency Optimization – 13404, *WM2013 Symposium*, February 24–28, 2013, Phoenix, AZ (2013)
37. *Model S573 ISOCS™ Calibration Software*, Mirion Technologies (Canberra), Inc., C40166 (2013)

38. *Technical Advantages of ISOCS™/LabSOCS™*, Mirion Technologies Application Note C39530 (2012)
39. E. Braverman, A. Lebrun, V. Nizhnik, F. Rorif, In Situ Object Counting System (ISOCS™) Technique: Cost-Effective Tool for NDA Verification in IAEA Safeguards, in *Symposium on International Safeguards: Preparing for Future Verification Challenges*, November 1–5, 2010, Vienna, Austria (2010) IAEA-CN-184/047 https://inis.iaea.org/collection/NCLCollectionStore/_Public/42/081/42081489.pdf?r=1
40. E.R. Martin, D.F. Jones, J.L. Parker, *Gamma-Ray Measurements with the Segmented Gamma Scan*, Los Alamos Scientific Laboratory report LA-7059-M (December 1977)
41. *Standard Test Method for Nondestructive Assay of Special Nuclear Material in Low-Density Scrap and Waste by Segmented Passive Gamma-Ray Scanning*, ASTM Standard ASTM-C-1133-96/XAB (American Society for Testing and Materials, West Conshohocken, PA, 1996)
42. J. Steven Hansen, *Application Guide to Tomographic Gamma Scanning of Uranium and Plutonium*, Los Alamos National Laboratory report LA-UR-04-7014 (2004)
43. D.J. Mercer, S. Hansen, J.P. Lestone, T.H. Prettyman, TGS measurements of pyrochemical salts at rocky flats, in *Proceedings of 42nd Annual Meeting of the INMM*, Indian Wells, CA, July 15–19, 2001, Los Alamos National Laboratory report LA-UR-01-3509 (2001)
44. U.S. Nuclear Regulatory Commission, *In-Situ Assay of Enriched Uranium Residual Holdup*, NRC Regulatory Guide 5.37, Rev. 1, Washington, D.C. (October 1983)
45. *Nondestructive Assay of Special Nuclear Materials Holdup*, Los Alamos National Laboratory training manual LA-UR-99-2597 (August 1999 or earlier versions, beginning in February 1991)
46. P.A. Russo, *Gamma-Ray Measurements of Holdup Plant-Wide: Application Guide for Portable, Generalized Approach*, Los Alamos National Laboratory report LA-14206 (2005)
47. J.K. Sprinkle, R.A. Cole, M.L. Collins, S.-T. Hsue, P.A. Russo, R. Siebelist, H.A. Smith, R.N. Ceo, S.E. Smith, *Low-Resolution Gamma-Ray Measurements of Process Holdup*, Los Alamos National Laboratory report LA-UR-96-3482 (1996)
48. P.A. Russo, T.R. Wenz, S.E. Smith, J.F. Harris, *Achieving Higher Accuracy in the Gamma-Ray Spectroscopic Assay of Holdup*, Los Alamos National Laboratory report LA-13699-MS (2000)
49. R.B. Oberer, C.A. Gunn, L.G. Chiang, *Improved Background Corrections to Uranium Holdup Measurements*, Y-12 National Security Complex report Y/DK-2107 (2004)
50. P.A. Russo, T.H. Prettyman, *Inter-Comparison of Detectors for Portable Gamma-Ray Spectrometry at Y-12*, Los Alamos National Laboratory report LA-UR-99-0199 (1999)

Open Access This chapter is licensed under the terms of the Creative Commons Attribution 4.0 International License (http://creativecommons.org/licenses/by/4.0/), which permits use, sharing, adaptation, distribution and reproduction in any medium or format, as long as you give appropriate credit to the original author(s) and the source, provide a link to the Creative Commons license and indicate if changes were made.

The images or other third party material in this chapter are included in the chapter's Creative Commons license, unless indicated otherwise in a credit line to the material. If material is not included in the chapter's Creative Commons license and your intended use is not permitted by statutory regulation or exceeds the permitted use, you will need to obtain permission directly from the copyright holder.

Densitometry

R. Venkataraman, R. McElroy, P. A. Russo, and P. A. Santi

11.1 Introduction

The term *densitometry* refers to measurement of the density of a material by determining the degree to which that material attenuates electromagnetic radiation of a given energy. Chapter 3 details the interaction of electromagnetic radiation (specifically X-rays and gamma rays) with matter. Because electromagnetic radiation interacts mainly with atomic electrons, densitometry measurements are element specific, not isotope specific. Two phenomena occur during a densitometry measurement: first, part of the incident radiation energy is absorbed; and second, the ionized atoms emit characteristic X-rays as they return to their ground states. This latter process, known as *X-ray fluorescence* (XRF), is a powerful method for element-specific assays. (See Chap. 12 for details of the XRF technique.) In some cases, gamma-ray transmission measurements can provide information not only on the bulk density of a sample but also on its composition. Because the attenuation of low-energy photons (primarily by the photoelectric effect) is a strong function of the atomic numbers (Z) of the elements in the sample, a measurable signature is provided on which an assay can be based.

This chapter describes various densitometry techniques that involve measurement of photon absorption at a single energy and at multiple energies and measurement of differential photon attenuation across absorption edges. Applications using these techniques are discussed, and measurement procedures with typical performance results are described.

All densitometry measurements discussed in this chapter are based on determination of the transmission of electromagnetic radiation of a given energy by the sample material. The mathematical basis for the measurement is the exponential attenuation relationship between the intensity (I_0) of photon radiation of energy (E) incident on a material and the intensity (I) that is transmitted by a thickness (x) of the material:

$$I = I_0 e^{-\mu \rho x} \tag{11.1}$$

where ρ is the mass density of the material, and μ is the mass attenuation coefficient, which is evaluated at the photon energy (E). The incident and transmitted intensities are the measured quantities. Their ratio (I/I_0) is called the *transmission* (T) of the material at the photon energy of interest. If any two of the three quantities in the exponent expression are known from other data, the third quantity can be determined by the transmission measurement. A strong advantage of a procedure that measures photon transmissions is that the data are handled as a ratio of two similarly measured quantities, thereby removing many bothersome systematic effects that often complicate the measurement of absolute photon intensity.

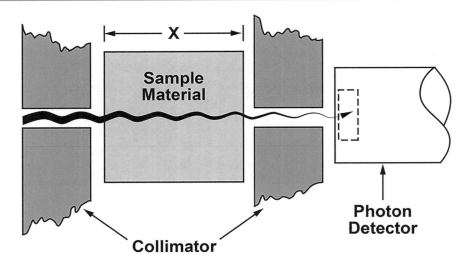

Fig. 11.1 Key components of a densitometry measurement

The measured electromagnetic radiation can originate from an artificial X-ray source (that emits photons with a continuous-energy spectrum) or from a natural gamma-ray source (that emits gamma rays with discrete energies). The sample material is placed between a photon source and a photon detector (see Fig. 11.1). The transmission of the sample is determined by measuring the photon intensity of the source both with (I) and without (I_0) the sample material present.

11.2 Single-Line Densitometry

11.2.1 Concentration and Thickness Gauges

If a sample is composed of one type of material or of a mixture of materials whose composition is tightly controlled except for one component, then the sample transmission at one gamma-ray energy can be used as a measure of the concentration (density ρ) of the varying component. Normally, discrete-energy gamma-ray sources are used. For example, consider a two-component system such as a solution of uranium and nitric acid, whose components have respective densities ρ and ρ_0 and mass attenuation coefficients μ and μ_0 at a given gamma-ray energy. The natural logarithm of the photon transmission at that energy is given by

$$\ln(T) = -(\mu\rho + \mu_0\rho_0)x. \qquad (11.2)$$

With ρ as the unknown concentration,

$$\rho = -\left(\frac{1}{\mu x}\right)\ln(T) - \frac{\mu_0\rho_0}{\mu}. \qquad (11.3)$$

Equation 11.3 may be applied as a gauge for the concentration of an unknown amount of substance (ρ) in a known, carefully controlled solvent concentration (ρ_0). In applying the concentration gauge to special nuclear material (SNM; uranium and plutonium) solutions, it is critical that both of the mass attenuation coefficients be well characterized for the solvent (μ_0) and for the SNM (μ) and that the solvent composition (ρ_0) be well known and constant from sample to sample. The sample solutions should not be vulnerable to contamination because contamination would cause variations in the effective values of ρ_0 and μ_0.

Single-line measurement can also be applied as a thickness gauge for materials of known and tightly controlled composition. Online measurement of the transmitted photon intensity at one energy through metals and other solids in a constant measurement geometry is a direct measure of the thickness (x) of those materials. Such information is useful for timely control of some commercial production processes.

11.2.2 Measurement Precision

Consider the case of a single-line concentration measurement in which no significant fluctuations are present in the solvent composition. The measurement precision of the unknown quantity (ρ) is determined by the statistical variance of the transmission (T). The relative precision of the density measurement is obtained by propagating the statistical uncertainty in the transmission ratio (T).

$$\frac{\sigma(\rho)}{\rho} = \left(\frac{1}{\ln(T)}\right)\left[\frac{\sigma(T)}{T}\right] \qquad (11.4)$$

In writing Eq. 11.4, it is assumed that transmission (T; equal to I/I_0) is due to the SNM component only. The transmission through the matrix component is accounted for as part of I_0; i.e., I_0 is the transmission through a blank solvent solution with no SNM.

$$I_0 = e^{-\mu_0 \rho_0 x}, \qquad (11.5)$$

$$I = e^{-(\mu\rho + \mu_0 \rho_0)x}, \qquad (11.6)$$

$$T = \frac{I}{I_0} = e^{-\mu\rho x}. \qquad (11.7)$$

Therefore, $\ln(T) = -\mu\rho x$.

This expression shows a range of transmission values over which the relative precision of the density measurement is smaller than that of the transmission, as illustrated in Fig. 11.2, which shows the favorable precision regime, $|\ln(T)| > 1$ or $T < 0.37$. For larger transmission values, the relative precision of the density is larger than that of the transmission, and the measurement suffers accordingly. Note that when T approaches 1, the expression for the relative precision diverges because of the factor $1/(\ln(T))$. Because the sample material is absorbing none of the incident radiation, there is no assay signal.

The range of useful transmission values can also be related to a characteristic concentration, $\rho_c = 1/\mu x$. When $\ln(T) > 1$, then $\rho > \rho_c$, and the measurement is in the favorable precision regime; but when $\rho > \rho_c$, the assay signal is too small, and the precision is unfavorable. By determining the favorable operating range from the point of view of this characteristic

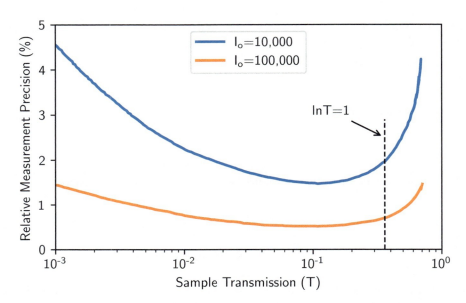

Fig. 11.2 Precision of single-energy densitometry as a function of sample transmission for two values of incident photon total counts (from Eq. 11.8). The optimum transmission is that which gives the smallest relative measurement precision, which corresponds to a concentration that is greater than the characteristic concentration, ρc, where $\ln(T) = 1$. (Note the logarithmic scale on the horizontal axis)

concentration, one can choose a reasonable sample thickness (x) given the intrinsic properties of the sample material to be measured (μ) and the expected range of sample concentrations.

Because of the symmetry in ρ and x in Eqs. 11.1, 11.2, and 11.3, Eq. 11.4 also expresses the relative precision of a thickness measurement. For a thickness measurement, the precision can be enhanced by a judicious choice of photon energy.

There are limits on both how high and how low the sample transmission should be for optimum measurement precision. Because $T = I/I_0$ and the intensities are statistically varying quantities, Eq. 11.4 can be rewritten as

$$\frac{\sigma(\rho)}{\rho} = \frac{1}{\ln(T)} \left(\frac{T+1}{I_0 T} \right)^{\frac{1}{2}}. \tag{11.8}$$

In Fig. 11.2, a plot of this relationship shows the deterioration of the measurement precision at the high- and low-concentration extremes. The optimum range of T is below the point where $\ln(T) = 1$, in keeping with the definition of ρ_c. The range of T over which the quantity $\sigma(\rho)/\rho$ is near a minimum determines the instrument design features (sample thickness, measurable concentration range, and photon energy). These features are also important in the more complex densitometry measurements described in Sects. 11.3 and 11.4.

Note that the above discussion deals with the measurement precision determined by counting statistics alone. Generally, other factors can cause added fluctuations in the measurement results; they include variations in the matrix material (solvent) and possible instrumental fluctuations. As a result, the precision of an assay instrument should be determined by making replicate measurements of known standards that represent the full range of sample and solvent properties.

11.3 Multiple-Energy Densitometry

Measurement of photon transmission at one energy allows for the assay of only one substance or component of a mixture; the concentration of the other components must be kept constant. Measurement of photon transmission at two energies allows for the assay of two components of a mixture. Such a compound measurement stands the greatest chance for success the more the attenuation coefficients of the two components differ from one another. Analysis of the concentration of a high-Z element in a low-Z solvent is an excellent example of a two-energy densitometry measurement.

11.3.1 Analysis of Two-Energy Case

Consider a mixture of two components with (unknown) concentrations ρ_1 and ρ_2. Let the mass attenuation coefficient of component i measured at energy j be given by

$$\mu_i^j = \mu_i(E_j) \tag{11.9}$$

and define the transmission at energy j as

$$T_j = e^{-\left(\mu_1^j \rho_1 + \mu_2^j \rho_2\right)x}. \tag{11.10}$$

The measurement of two transmissions gives two equations for the two unknown concentrations:

$$\begin{aligned}\frac{-\ln(T_1)}{x} &= M_1 = \mu_1^1 \rho_1 + \mu_2^1 \rho_2 \\ \frac{-\ln(T_2)}{x} &= M_2 = \mu_1^2 \rho_1 + \mu_2^2 \rho_2\end{aligned} \tag{11.11}$$

By attributing the measured absorption to the two sample components, we are actually defining the incident radiation to be the intensity transmitted by an empty sample container. The solution to the above equations is

$$\rho_1 = M_1\mu_2^2 - \frac{M_2\mu_2^1}{D}$$
$$\rho_2 = M_2\mu_1^1 - \frac{M_1\mu_1^2}{D} \tag{11.12}$$
$$D = \mu_1^1\mu_2^2 - \mu_2^1\mu_1^2.$$

For Eq. 11.11 to have a solution, the determinant of the coefficients (D) must be nonzero. This condition is virtually assured if the mass attenuation coefficients for the two components have significantly different energy dependences. Physically, this condition means that the assay is feasible if the components can be distinguished from one another by their absorption properties. This criterion further suggests two possible choices of photon energies. First, if two widely differing energies are used, the different slopes of μ versus E for the high-Z and the low-Z components suffice to differentiate between them. Second, by choosing photon energies near and on either side of an absorption edge for the heavier (higher-Z) component, the energy dependence for the mass attenuation coefficient of the higher-Z material will appear to have the opposite slope to that of the low-Z component, making the two components easily distinguishable. This approach is especially promising in assays of SNM in low-density matrices or in assays of two SNM components.

11.3.2 Measurement Precision

The primary source of random measurement uncertainty is the statistical variance of the transmission measurements. The expression for the relative precision of each component's concentration is given by

$$\frac{\sigma(\rho_1)}{\rho_1} = \frac{1}{\mu_2^1 \ln(T_2) - \mu_2^2 \ln(T_1)} \left\{ \left[\mu_2^2 \frac{\sigma(T_1)}{T_1}\right]^2 + \left[\mu_2^1 \frac{\sigma(T_2)}{T_2}\right]^2 \right\}^{\frac{1}{2}}$$
$$\frac{\sigma(\rho_2)}{\rho_2} = \frac{1}{\mu_1^2 \ln(T_1) - \mu_1^1 \ln(T_2)} \left\{ \left[\mu_1^2 \frac{\sigma(T_1)}{T_1}\right]^2 + \left[\mu_1^1 \frac{\sigma(T_2)}{T_2}\right]^2 \right\}^{\frac{1}{2}}. \tag{11.13}$$

Because the assay result varies inversely with the sample thickness (see Eq. 11.11), the sample thickness (x) must be very well known or held constant within close tolerance.

11.3.3 Extension to More Energies

In principle, the multiple-energy densitometry technique can be extended to three or more energies to measure three or more sample components. In practice, such a broadening of the technique undermines the sensitivity of the measurement for some sample components because it is extremely difficult to select gamma-ray energies that can sample different energy dependences of the absorption of each of the components. Accordingly, multiple-energy densitometry is rarely extended beyond the two-energy case.

11.4 Absorption-Edge Densitometry

Absorption-edge densitometry is a special application of two-energy densitometry. The photon energies at which the transmissions are measured are selected to be as near as possible to and on opposite sides of the absorption-edge discontinuity in the energy dependence of the mass attenuation coefficient for the unknown material [1]. Both the K and the L_{III} absorption edges have been used in nondestructive assay (NDA) of SNM (see Sect. 11.7 for specific applications). Figure 11.3 shows the attenuation coefficients for plutonium, uranium, and selected low-Z materials and includes the K- and L-edges for the heavy elements.

Absorption-edge densitometry involves the measurement of the transmission of a tightly collimated photon beam through the sample material. The collimation defines the measurement geometry and reduces interference from radiation emitted by

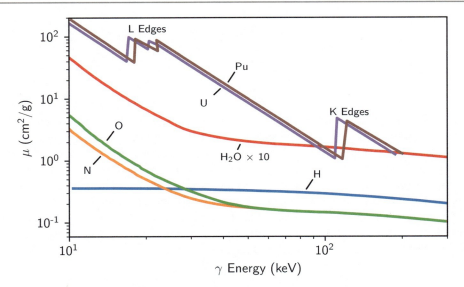

Fig. 11.3 Energy dependence of the photon mass attenuation coefficients for uranium, plutonium, and selected low-Z materials. Note the absorption edge discontinuities for uranium and plutonium in the 17–20 keV (L-edge) and 115–122 keV (K-edge) energy regions

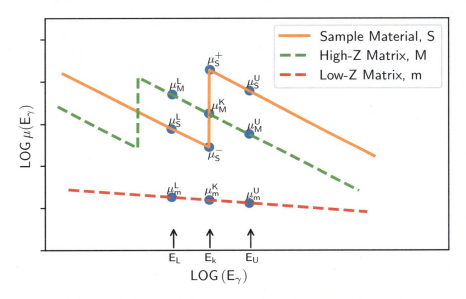

Fig. 11.4 Expanded schematic of the mass attenuation coefficient as a function of photon energy. Curves are shown for a sample material (s) assumed to be a heavy element, a heavy-element matrix component (M), and a light-element matrix component (m)

the sample material. Because the collimation selects only a small fraction of the sample volume, the sample must be highly uniform for the assay to be representative of all of the material. As a result, the absorption-edge technique is best suited for solution assays, although it has been used for assays of solids [2–4].

11.4.1 Description of Measurement Technique

Consider the typical case of a high-Z (SNM) component in a low-Z (solvent) matrix. Figure 11.4 depicts the attenuation coefficients and measurement energies above (U) and below (L) an absorption edge. (The discussion emphasizes K-edge measurements, but the analysis is similar in the L-edge region as well.) The subscript S refers to the measured element, and the

11 Densitometry

Table 11.1 Absorption-edge energies and discontinuities for selected SNM components

Property	Uranium	Plutonium
E(K)	115.6 keV	121.8 keV
E(L$_{III}$)	17.2 keV	18.0 keV
$\Delta\mu$(K)	3.7 cm^2/g	3.4 cm^2/g
$\Delta\mu$(L$_{III}$)	55.0 cm^2/g	52.0 cm^2/g

subscripts M and m refer to the high- and low-Z matrix elements, respectively. The magnitudes of the attenuation coefficient discontinuities and the edge energies of interest are given in Table 11.1.

Equation 11.14 gives the transmission of photons through the solution at the two measurement energies E_U and E_L.

$$\ln(T_L) = -(\mu_s^L \rho_s + \mu_m^L \rho_m)x$$
$$\ln(T_U) = -(\mu_s^U \rho_s + \mu_m^U \rho_m)x \quad (11.14)$$

To solve for the measured element concentration,

$$\rho_s = \frac{1}{\Delta\mu x} \ln\left(\frac{T_L}{T_U}\right) + \rho_m \left(\frac{\Delta\mu_m}{\Delta\mu}\right) \quad (11.15)$$

where

$$\Delta\mu = \mu_s^U - \mu_s^L > 0$$
$$\Delta\mu_m = \mu_m^L - \mu_m^U > 0 \quad (11.16)$$

The second term in Eq. 11.15 expresses the contribution from the solvent matrix. Because the transmissions are measured relative to an empty sample container, the transmission of the sample container does not influence Eq. 11.15. Note the similarity of Eq. 11.15 to the single-line case (Eq. 11.3), with μ's replaced by $\Delta\mu$'s.

Because the matrix term in Eq. 11.15 is independent of SNM concentration and sample cell geometry, it can be applied to any absorption-edge densitometry measurement for which the solution transmissions are measured relative to an empty sample container. Ideally, if $E_L = E_U = E_K$, then $\Delta\mu_m = 0$, and the measurement is completely insensitive to any effects from the matrix. However, in practice, the two measurement energies differ by a finite amount, so some residual matrix correction may be necessary. In cases where the matrix contribution can be significant, it can be determined empirically by assaying a solution that contains only the matrix material, or its effect can be deduced analytically. For further discussion of matrix corrections for absorption-edge densitometry, see Sect. 11.4.4.

The ratio of the two transmissions at the two measurement energies ($R = T_L/T_U$) is the measured quantity, and $\Delta\mu$ and x are constants that can be evaluated from transmission measurements with calibrated standards of well-defined concentrations. With judiciously chosen photon energies, this technique will provide very reliable, nearly matrix-independent assays of specific elements whose absorption edges lie between the transmission source energies.

11.4.2 Measurement Precision

Differentiation of Eq. 11.15 gives the relative precision of a densitometry measurement:

$$\frac{\sigma(\rho_s)}{\rho_s} = \left(\frac{1}{\Delta\mu\rho_s x}\right)\left[\frac{\sigma(R)}{R}\right] = \left(\frac{1}{\ln(R)}\right)\left[\frac{\sigma(R)}{R}\right]. \quad (11.17)$$

The fractional error in R is determined by the counting statistics of the transmission measurements. In analogy with the discussion of Eq. 11.4, the choice of measurement parameters can be guided by reference to a characteristic concentration, $\rho_c = 1/\Delta\mu x$. When $\rho > \rho_c$, the measurement is in the favorable precision regime, where $\sigma(\rho)/\rho < \sigma(R)/R$. But if the SNM

Table 11.2 Characteristic concentrations for uranium and plutonium

Characteristic concentration	Uranium (g/L)	Plutonium (g/L)
$\rho_c(K)$	270	294
$\rho_c(L_{III})$	18	19

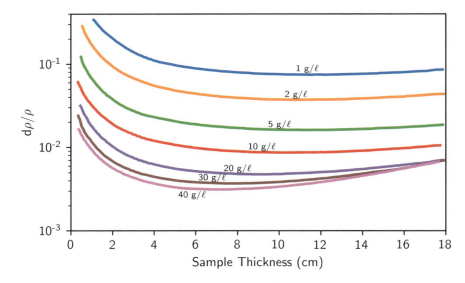

Fig. 11.5 Calculated relative statistical uncertainty in plutonium concentration (g/L) by K-edge densitometry as a function of sample cell thickness (transmission path length). The empty cell transmission counts below the K-edge (I_{0L}) were taken to be 2×10^6 in the 121.1 keV photopeak

concentration is too far above ρ_c, the excessive absorption deteriorates the measurement precision, primarily because of the enhanced absorption of the transmission gamma rays above the absorption edge. The statistical fluctuations of the very small, transmitted intensity at E_U is then overpowered by the statistical fluctuations of the background in that energy region.

Table 11.2 shows the values of these characteristic concentrations for a 1 cm transmission path length (x = 1 cm). The table implies, for example, that for a 1 cm sample cell thickness, K-edge assays of plutonium concentrations greater than 300 g/L would be in the favorable precision regime. For assays of 30 g/L solutions, the sample-cell thickness should be greater than 0.5 cm for L_{III}-edge assays and greater than 9 cm for K-edge assays.

A more analytical approach can be used to optimize measurement parameters. Figure 11.5 shows the calculated statistical measurement precision (Eq. 11.17) as a function of transmission path length (x) for a variety of SNM concentrations. The figure shows, for example, that a densitometer designed for 30 g/L SNM solution assays should have a sample cell thickness of 7–10 cm.

The final test in evaluating the design of a densitometer is empirical determination of the assay precision. Figure 11.6 shows the precision of a series of measurements on a K-edge densitometer (KED) designed for low- to medium-concentration plutonium solution assays with a 7 cm thick sample cell [5]. Figure 11.6 agrees well with the theoretical curve shown in Fig. 11.5.

Calculations of measurement precision are helpful in determining design parameters for optimum instrument performance. Figure 11.7 shows the results of such calculations for both K- and L_{III}-edge densitometers [6]. The ranges of plutonium concentrations over which the relative measurement precision is better than 1% are shown for different sample thicknesses (x).

11.4.3 Measurement Sensitivity

A useful parameter in the specification of an NDA instrument is its "minimum detectable limit," which is that quantity of nuclear material that produces an assay signal significantly above background in a reasonable count time [7, 8]. For nuclear waste measurements where the minimum detectable limit is an important instrument specification, an assay signal that is three

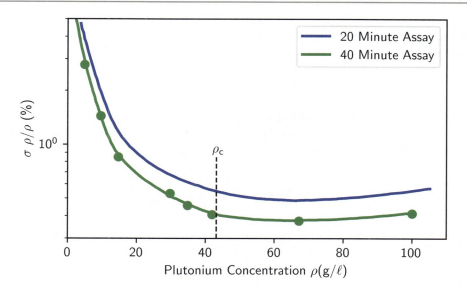

Fig. 11.6 Relative statistical precision achieved in a plutonium concentration measurement by K-edge densitometry using ^{50}mCi of ^{75}Se and 25 mCi of ^{57}Co as a function of sample concentration for a sample cell thickness of 7 cm. At 200 g/L, the precision increased to 5% for a 40 min count time. Curves are shown for two count times [5]

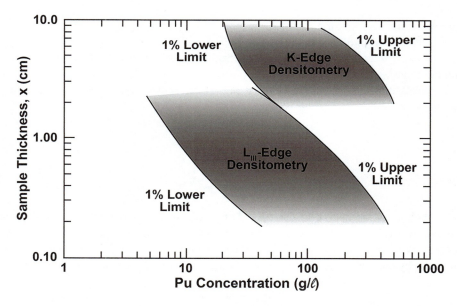

Fig. 11.7 Indications of sample transmission thicknesses (x) over which plutonium concentration assays can be performed by absorption-edge densitometry to better than 1% statistical precision. Shaded regions for the K- and L_{III}-edge techniques show plutonium concentration ranges over which this precision is achievable

standard deviations (99% confidence level; [9]) above background is considered to be significant. This limit can also be regarded as a measurement sensitivity because it characterizes a lower limit of SNM than can be detected with some level of confidence.

Because absorption-edge densitometers are usually built for specific assay applications in well-defined SNM solution concentration ranges, the minimum detectable limit is not particularly important. However, the measurement sensitivity can serve as a convenient quantity for comparing design approaches and other factors that influence instrument performance.

To obtain an expression for the measurement sensitivity of an absorption-edge densitometer, the assay background must be defined so that the minimum detectable assay signal can be determined. The statistical uncertainty in the measured density is given in Eq. 11.17. The ratio (R) of the two transmissions above and below the absorption edge is composed of raw gamma-ray (or X-ray) photon intensities that vary according to the usual statistical prescriptions. When the SNM concentration is zero, the solution is entirely matrix material (typically acid) and

$$T_U \approx T_L = T = exp(-\mu_m \rho_m x). \qquad (11.18)$$

When the SNM concentration is zero, R = 1. Starting from Eq. 11.15, assuming that the solvent concentration ρ_m does not fluctuate and recognizing that $T_L = I_L/I_{0L}$, $T_U = I_U/I_{0U}$, and $T_L \approx T_U = T$, it can be shown that the uncertainty in the concentration of SNM is calculated as follows:

$$\sigma(\rho_s) = \frac{1}{\Delta \mu x} \left(\frac{1}{I_{0U}} + \frac{1}{I_{0L}} + \frac{1}{TI_{0U}} + \frac{1}{TI_{0L}} \right)^{\frac{1}{2}}. \qquad (11.19)$$

Equation 11.19 expresses the uncertainty in the background. The 3σ criterion provides an expression for the minimum detectable limit (or sensitivity, s) for an absorption-edge densitometer:

$$s = \frac{3}{\Delta \mu x} \left[\left(\frac{1}{I_{0U}} + \frac{1}{I_{0L}} \right) \left(\frac{T+1}{T} \right) \right]^{\frac{1}{2}} (g/L), \qquad (11.20)$$

where the units of $\Delta \mu x$ are cm^3/g.

Equation 11.20 shows that the measurement sensitivity is affected by several measurement parameters:

- The sensitivity suffers in low-transmission samples.
- Long counts of the unattenuated photon intensities (I_0) improve measurement sensitivity.
- L-edge measurements, with their larger $\Delta \mu$, are more sensitive than K-edge measurements (if all other measurement parameters remain the same).
- An increase in sample cell thickness may improve the measurement sensitivity, but the accompanying decrease in T will compete with that improvement.

11.4.4 Matrix Effects

The absorption-edge technique is insensitive to the effects of matrix materials if both transmissions are measured at the absorption edge. However, with a finite energy separation of the transmission gamma rays, the matrix contribution is nonzero and is represented by the second term in Eq. 11.15. This term can become significant for low SNM concentrations, ρ_s, or when the spacing between the assay energies, E_L and E_U, becomes large; either condition threatens the validity of the inequality,

$$\Delta \mu_m \rho_m \ll \Delta \mu \rho_s. \qquad (11.21)$$

The natural width of the absorption edge (less than 130 eV) and the energy resolution of the detection system (typically 500 eV or more) are intrinsic limitations to the design of an instrument that attempts to minimize the effects of matrix attenuation by using closely spaced assay energies. The limited availability of useful, naturally occurring radioisotopes also leads to compromises in the choice of transmission sources. One very useful technique for reducing the matrix effect is an extrapolation procedure applied to the measured transmission data [2, 5]. The procedure attempts to extrapolate the measured transmissions to the energy of the absorption edge. This extrapolation is possible because the energy dependence of the mass attenuation coefficients over narrow energy ranges is known to be a power law:

$$\log \mu(E) = k \log E + B. \qquad (11.22)$$

11 Densitometry

Table 11.3 Slopes (k) and intercepts (B) for the linear dependence of log $\mu(E)$ versus log (E) for various materials of interest in the 100–150 keV energy region (log to base 10) [10]

Solution Component	k	B
Plutonium (above K-edge)	−2.48	5.83
(below K-edge)	−2.56	5.42
Uranium (above K-edge)	−2.49	5.82
(below K-edge)	−2.71	5.65
Tungsten	−2.50	5.65
Tin	−2.45	5.12
Iron	−1.57	2.70
Aluminum	−0.500	0.227
Water	−0.306	−0.153
Nitric acid	−0.314	−0.171

The slope parameter (k) is essentially the same for elements with Z > 50, with an average value of approximately −2.55 near the uranium and plutonium K-edges [2]. Table 11.3 gives the extracted values for the slopes and intercepts of several substances of interest to SNM assay [10].

As an example of a general assay case, consider a solution of SNM in a low-Z solvent with possible additional heavy-element (Z > 50) matrix contaminants. Equation 11.15 generalizes to

$$\rho_s = \left(\frac{1}{\Delta\mu x}\right) ln\left(\frac{T_L}{T_U}\right) + \rho_M \left(\frac{\Delta\mu_m}{\Delta\mu}\right) + \rho_m \left(\frac{\Delta\mu_m}{\Delta\mu}\right). \tag{11.23}$$

The subscript M refers to the high-Z matrix contaminant, and the subscript m represents the low-Z matrix (solvent); and in analogy with Eq. 11.16,

$$\Delta\mu_M = \mu_M^L - \mu_M^U. \tag{11.24}$$

(See Fig. 11.4.) The measured transmissions are then extrapolated to the SNM K-edge using the energy dependence of $\mu(E)$ for the heavy elements. Because the slope parameters (k) for Z > 50 are all essentially the same, the SNM and high-Z matrix absorption coefficients can be transformed with the same k (for example, the average value, −2.55). As a result, the transformed $\Delta\mu_M$ vanishes, and the assay result becomes

$$\rho_s = \left(\frac{1}{\Delta\mu_\pm x}\right) ln\left(\frac{T_L^a}{T_U^b}\right) + \left(\frac{c\mu_m^K \rho_m}{\Delta\mu_\pm}\right), \tag{11.25}$$

where $\Delta\mu_\pm$ (which equals $\mu_s^+ - \mu_s^-$; see Fig. 11.4) is now defined across the absorption edge (in this case, the K-edge) rather than between the energies E_L and E_U.

The constants a, b, and c are defined as

$$\begin{aligned} a &= (E_K/E_L)^k \\ b &= (E_K/E_U)^k \\ c &= \left(\frac{E_K}{E_L}\right)^{k-k'} - \left(\frac{E_K}{E_U}\right)^{k-k'}, \end{aligned} \tag{11.26}$$

where k = −2.55 and k′ = −0.33 (the average value of k for elements with atomic numbers less than 10). This procedure renders the assay essentially independent of the heavy-element matrix but still leaves a residual correction for the light-element matrix. It is not possible to remove the effects of both the light- and heavy-element matrix materials because k ≠ k′. The transmissions must be corrected for the acid matrix contributions because the transmissions are measured relative to an empty sample cell. If the reference spectra (the I_0 intensities) were taken with the cell full of a representative acid solution, no acid matrix correction would be necessary. However, any fluctuation in acid molarity would bias the measurement of an actual sample. The density of nitric acid (ρ_m) and the acid molarity (M) are related [11] by

$$\rho_m = 1 + 0.03\,M. \tag{11.27}$$

For plutonium K-edge assays in which the K-edge is closely bracketed by ^{57}Co and ^{75}Se gamma rays (see Sect. 11.4.6), this low-Z matrix correction is small but may be important at low plutonium concentrations. For example, the correction term in Eq. 11.23 for 3 M nitric acid is equivalent to approximately 0.87 g plutonium/L [5]. Equation 11.25 shows that fluctuations in acid molarity cause fluctuations in the acid matrix correction that are only 3% as large, so careful control of the acid molarity is important only at very low SNM concentrations.

For uranium K-edge assays with a ^{169}Yb transmission source ($E_L = 109.8$ keV, $E_U = 130.5$ keV), the extrapolation procedure greatly improves the quality of the assay results. This effect is demonstrated graphically in Ref. [2], where assays of uranium solutions with varying tin concentrations were shown to be matrix-independent with the extrapolation correction. Several other matrix effects studies are described in Ref. [12].

11.4.5 Choice of Measurement Technique

Because of differences in the $\Delta\mu$ values at the K-edge versus the L_{III}-edge, the measurement sensitivity (defined in Eq. 11.20) is more than an order of magnitude larger at the L_{III}-edge than at the K-edge, other parameters being equal (see also Table 11.1). However, because of the higher penetrability of photons at the K-edge energies, thicker samples can be used for the K-edge measurements, thereby enhancing K-edge sensitivity.

If significant quantities of lower-Z elements (such as yttrium and zirconium) are present in a sample, the K-edges of these elements cause discrete interferences that bias the L_{III} assays of uranium and plutonium [12]. Furthermore, detector resolution at L_{III} energies limits the ability to perform L_{III}-edge assays in the presence of significant amounts of neighboring elements (uranium with protactinium or neptunium; plutonium with neptunium or americium). The K-edge measurements are not subject to such interferences. In addition, the higher photon energies required for the K-edge transmission measurements permit the use of thicker or higher-Z materials for sample cell windows, an important practical consideration for in-plant operation. Finally, more flexibility exists in the availability of discrete gamma-ray transmission sources for K-edge measurements.

11.4.6 Transmission Sources

The most versatile transmission source is the bremsstrahlung continuum produced by an X-ray generator. The intensity of this source can be varied to optimize the count rate for a variety of sample geometries, concentrations, and thicknesses.

The X-ray generator voltage, which determines the assay energy range, can be adjusted and the spectrum tailored appropriately for the assay of specific elements. Furthermore, matrix effects can be minimized by extrapolation of the measured transmissions to the absorption edge. Commercial units are available with power supplies that are highly stable and X-ray tubes that are long-lived for long-term reliable operation in either the K- or the L_{III}-edge energy regions.

The use of discrete gamma-ray lines that bracket the absorption edge—the alternative to the continuum transmission sources—has been demonstrated successfully in several instruments. This technique is appropriate for K-edge assays. Discrete gamma rays are not available as primary emissions in the L_{III}-edge energy region. This approach depends on the availability of relatively slowly decaying radioisotopes that emit gamma rays of appropriate energies and sufficient intensities. For example, a convenient combination for the K-edge assay of plutonium ($E_K = 121.8$ keV) is the 121.1 keV gamma ray from ^{75}Se (half-life = 120 days) and the 122.1 keV gamma ray from ^{57}Co (half-life = 270 days). The proximity of both energies to the plutonium absorption edge minimizes the effects of the matrix and enhances the sensitivity of the assay [1]. Because of the different half-lives, accurate decay corrections or frequent measurements of the unattenuated intensities (I_0) are required. The use of ^{169}Yb (half-life = 32 days) for uranium assay at the K-edge [2, 3] has the advantage of requiring no decay correction because both gamma rays come from the same source. However, the larger energy separation ($E_L = 109.8$ keV, $E_K = 115.6$ keV, $E_U = 130.5$ keV) introduces a larger matrix sensitivity (larger $\Delta\mu_m$) and a smaller assay sensitivity (smaller $\Delta\mu$). Furthermore, whether using ^{75}Se or ^{169}Yb, to maintain acceptable counting statistics, the source must be replaced frequently because both sources have short half-lives. The extrapolation technique discussed in Sect. 11.4.4

is especially effective in reducing the matrix sensitivity. A detailed discussion of convenient radioisotopic sources for absorption-edge densitometry appears in Ref. [1]. Several variations on these two basic transmission source configurations are discussed in Ref. [12].

11.5 Single-Line Densitometers

The measurement of photon transmission at a single energy has been applied using low-resolution detectors for assay of SNM in solution and in reactor fuel elements. These instruments use low-energy gamma-ray transmission sources to minimize the ratio μ_0/μ (see Eq. 11.3) and thus reduce the sensitivity to variations in the low-Z matrix.

One instrument uses an ^{241}Am transmission source mounted in the center of an annular cell that contains SNM solution [13, 14]. The cell is surrounded by a 4π plastic scintillator. The instrument separates the transmitted 60 keV gamma ray from the background sample radiation by modulating the source with a rotating, slotted, tungsten shield. Designed to assay high concentrations (>200 g/L) of SNM, the instrument is sensitive to 1% changes in SNM concentration at the 95% confidence level.

A single-line densitometer has been used to determine the density of SNM in pelleted and compacted ceramic fuel elements [15]. The 67 and 76 keV gamma rays of ^{171}Tm and the 84 keV gamma ray of ^{170}Tm are detected by a 1 in. diameter NaI (Tl) detector. The gross detector signals are counted in the multichannel scaling mode as the fuel elements are scanned to give the SNM density profile. The sensitivity of the instrument to SNM is 0.2 g/cm^3 at the 95% confidence level.

11.6 Dual-Line Densitometers

Dual-line densitometry has application to solids (fuel elements) and to solutions. Low- and high-resolution gamma spectrometers have been used to assay a low- and a high-Z component as well as to assay two high-Z components.

A dual-line densitometer has been used to determine the densities of the low-Z (silicon and carbon) and high-Z (thorium and uranium) components in high-temperature, gas-cooled reactor fuel pellets [16, 17]. The transmission source provides two widely differing gamma-ray energies (122 keV from ^{57}Co and 1173 and 1332 keV from ^{60}Co) so that the sensitivity to the two components is based on the different slopes of μ versus E at low and high Zs. Equation 11.12 applies in this case. Fuel pellet cakes that contain 92–95% thorium and 5–8% ^{238}U with a low-Z-to-heavy-Z weight ratio of 1.6–2.4 were assayed in 2 min measurement periods. The sensitivity to changes in the weight of either component was 3% or better at the 95% confidence level.

Dual-line solution densitometry has also been applied to the assay of two SNM components by measuring transmissions at two low gamma-ray energies [18, 19]. The transmission energies were chosen to bracket the L-absorption edges of the higher-Z component (element 2) in such a way that in Eq. 11.11, $\mu_2^1 = \mu_2^2$ and $\mu_1^1 > \mu_1^2$. Thus, Eq. 11.11 can be solved to give the concentration of element 1, independent of element 2:

$$\rho_1 = \left(\frac{1}{\Delta\mu_1 x}\right) ln\left(\frac{T_1}{T_2}\right), \tag{11.28}$$

where $\Delta\mu_1 = \mu_1^2 - \mu_1^1$. The measured T_2 and ρ_1 are then used to obtain the concentration of element 2:

$$\rho_2 = \left(\frac{\ln(T_2)}{\mu_2^2}\right) - \frac{\rho_1 \mu_1^2}{\mu_2^2}. \tag{11.29}$$

Dual-line densitometry has been applied to thorium and uranium assay using secondary sources of niobium and iodine K$_\alpha$ X-rays (at 16.6 and 28.5 keV, respectively) fluoresced by a 100 mCi ^{241}Am source. These X-rays bracket the L-edges of uranium; however, 16.6 keV is just above the L$_{III}$ absorption edge of thorium (at 16.3 keV). Measurements were performed using low-resolution [18] and high-resolution [19] gamma-ray spectroscopy. The high-resolution experiments used reference solutions containing mixtures of thorium and uranium with total SNM concentrations between 35 and 70 g/L. In the range of $0.25 \leq \rho_{Th}/\rho_U \leq 4.0$, the precision of the thorium and uranium concentration assay was 1% or better for 4000 s count periods.

11.7 Absorption-Edge Densitometers

Assay of uranium and plutonium solutions by the absorption-edge densitometry technique has been demonstrated in field tests of several instruments that perform K-edge or L_{III}-edge measurements. The instruments were designed for solution scrap recovery or reprocessing applications. Each instrument uses a high-resolution gamma-ray spectrometer (typically high-purity germanium [HPGe] for K-edge assays and Si(Li) for L_{III}-edge assays) and a computer-based multichannel analyzer. The measurement precision achieved in each case approaches the statistical prediction, which is typically 0.5% or better for short (30 min) count periods.

The transmission sources used by the K-edge instruments are discrete gamma-ray sources or bremsstrahlung continuum (X-ray) sources. The 109.8 and 130.5 keV gamma rays of ^{169}Yb are used for discrete K-edge assays of uranium, and the 121.1 and 122.1 keV gamma rays of ^{75}Se and ^{57}Co are used for discrete K-edge assays of plutonium. Only X-ray generators have been used in the L_{III}-edge instruments.

The absorption-edge assay relies on Eq. 11.15. The assay precision (Eq. 11.17) depends on several variables, including $\Delta\mu$, x, solution concentration, count time, and incident beam intensity. It is therefore convenient in comparing various instruments to use the characteristic concentration parameter [$\rho_c = (1/\Delta\mu x)$] for each instrument. The instrument relative precision is defined as the precision measured at the optimum concentration for a fixed count period. This optimum concentration is that for which the relative precision [$\sigma(\rho)/\rho$] is a minimum (see Figs. 11.5 and 11.6).

Tables 11.4 and 11.5 list the K-edge and L_{III}-edge densitometers that have undergone field testing. The characteristic concentration (ρ_c) and the empirically determined optimum concentration (shown in parentheses beneath ρ_c) are given for each instrument. The tables specify the solutions used to obtain the data and to quote the measured precisions at the optimum concentrations in specified count periods. Detailed discussions of the instruments listed in Tables 11.4 and 11.5 are given in Sects. 11.10 and 11.11.

11.7.1 K-Absorption-Edge Densitometers

Several KEDs have been tested and evaluated under actual or simulated in-plant environments. Table 11.4 summarizes the performance data for the instruments.

Table 11.4 K-absorption-edge densitometers

Instrument test location	ρ_c ρ optimum (g/L)	SNM	Solution type	Precision 1σ (%)	Live time (s)	References
Los Alamos National Laboratory Los Alamos, New Mexico (USA)	135 (300)	U	HEU[a] SR[b] miscellaneous	0.5	1000	[12, 20, 21]
Oak Ridge National Laboratory Y-12, Oak Ridge, Tennessee (USA)	55 (100)	U	HEU[a] SR[b] miscellaneous	0.5	600	[22, 23]
Allied General Nuclear Services Barnwell, South Carolina (USA)	80 (200)	Pu	Prepared (fresh, aged)	0.2	1200	[12, 24, 25]
Power Reactor and Nuclear Fuel Development Corporation, Tokai (Japan)	150 (300)	Pu	RP[c] product (fresh, aged)	0.2	2000	[12, 26–28]
Savannah River Plant[d] Aiken, South Carolina (USA)	40 (60)	Pu	RP[c] product (fresh)	0.2	2000	[5, 12, 29]
International Atomic Energy Agency Seibersdorf (Austria) (portable)	150 (300)	Pu	Prepared	0.3	500	[30]
Kernforschungszentrum Karlsruhe (KfK)[e] Karlsruhe (Germany) (continuum source)	150 (300)	U Pu U (+Pu) Pu (+U)	Prepared RP[c] product ["RP feed" "U : Pu :: 3 : 1"]	0.2 0.2 0.2 1.0	1000 1000 1000 1000	[12, 31, 32] [33]

[a] *HEU* highly enriched uranium
[b] *SR* scrap recovery
[c] *RP* reprocessing plant
[d] Present-day Savannah River Site
[e] Present-day Karlsruhe Institute of Technology

Table 11.5 L_{III}-absorption-edge densitometers

Instrument test location	ρ_c ρ optimum (g/L)	SNM	Solution type	Precision 1σ (%)	Live time (s)	References
Savannah River Laboratory[a] Aiken, South Carolina (USA)	16 (50)	U or Pu U (+Pu) Pu (+U)	RP[b] product RP product U:Pu::2:1	0.3 0.2 1.0	1000	[12, 21, 34]
Argonne National Laboratory New Brunswick Laboratory[c] Chicago, Illinois (USA)	15 (50)	U or Pu U (+Pu) Pu (+U)	Prepared Prepared U:Pu::2:1	0.3 0.2 0.9	1000 2000 2000	[35]
Allied General Nuclear Services Barnwell, South Carolina (USA)	19 (35)	U	U, natural enrichment (flowing)	0.7	250	[36]
Los Alamos National Laboratory Los Alamos, New Mexico (USA) (Compact)	16 (60)	U	U, natural enrichment	0.2	1000	[37]

[a]Present-day Savannah River Site
[b]RP reprocessing plant
[c]Present-day NBL Program Office

11.8 The Hybrid K-Edge Densitometry/X-Ray Fluorescence System

Based on the principles of continuous source K-edge absorption densitometry and XRF, a combined system was developed and tested by Herbert Ottmar and Heinrich Eberle at KfK in Karlsruhe, Germany, in the late 1970s and 1980s. Compared with destructive analysis, the hybrid K-edge densitometry/X-ray fluorescence (HKED) method is rapid. The HKED method requires minimal sample preparation and handling and generates a minimal amount of radioactive waste. Since its development, the HKED method has become an important tool for fuel fabrication, process control, quality control, material control and accountancy, and safeguards in nuclear fuel reprocessing plants.

Figure 11.8 is a schematic diagram of an HKED system that uses separate containers for K-edge densitometry and XRF.

The bremsstrahlung continuum source served both as a transmission source for K-edge assay of uranium and as a fluorescing source for XRF assay of the plutonium/uranium concentration ratio. The intensity of the continuum source allows the highly restrictive sample collimation required for K-edge and XRF assays while greatly reducing the passive count rate from the samples, which contain high levels of fission products. The concentration of the major element is determined by measuring the transmission of the incident X-ray beam across the characteristic K-edge energy of the element. Simultaneously, the intensities of prominent K_α X-rays fluoresced by the incident X-ray beam are used to determine the element ratios of actinides present in the sample. The minor element concentration is calculated from the product of the densitometry value for the major element concentration and the elemental ratio of the actinides from the XRF measurement.

HKED instruments can be operated in three different modes: KED only, where uranium or plutonium (or both) are measured in the 50–400 g/L concentration range; hybrid mode, where the concentration of uranium is determined using KED and the U:Pu ratio (100:1 typically) ratio is determined using XRF measurements; and stand-alone XRF for low uranium and plutonium concentrations (0.2–50 g/L). The KED and XRF measurements are performed using HPGe detectors—capable of detecting low energies—coupled to digital signal processors. Three replicate measurements of 1000 s live time each are performed. The instrument is capable of assaying actinide concentrations in solutions to yield a precision of $\pm 0.3\%$ or better for the major element (the actinide element present in high concentrations, typically 50–400 g/L) and a precision of $\pm 1\%$ for the minor element (concentration of 1 g/L). The major and minor actinide elements are typically uranium and plutonium, respectively. The random and systematic uncertainties from KED and HKED can be found in Ref. [38] (Table 4a for uranium element concentration measurements and Table 4b for plutonium element concentration measurements).

The KED measurement depends on the effective path length of the X-ray beam through the solution. For meeting the international target values, this geometrical parameter must be carefully controlled because its fractional uncertainty propagates directly into the fractional uncertainty of the uranium or plutonium concentration measurement. The uncertainty on the path length in this case must be small compared with other sources of uncertainty (0.01% typically). The preferred type

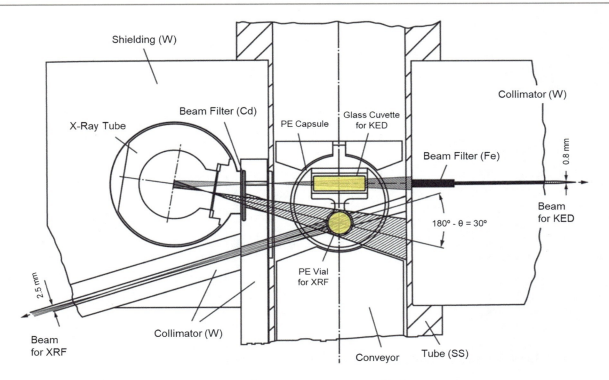

Fig. 11.8 Hybrid KED/XRF system that uses separate sample containers for K-edge and XRF

of sample vials is spectroscopy cells, whose thickness is known to a precision less than 0.01%. The HKED instrument may use a single cylindrical vial for both the KED and XRF measurements or separate sample containers for KED and XRF. For systems that use separate sample containers, a rectangular cuvette of well-known dimensions is used for KED, and a cylindrical vial is used for XRF measurement.

11.8.1 Determination of Elemental Concentration Using KED Branch of HKED Instrument

A typical KED spectrum from a uranium sample is shown in Fig. 11.9. X-ray and gamma-ray peaks from the decay of a ^{109}Cd source are used for energy calibration and to gain stabilization of the electronics. The 22.1 and 25.0 keV Ag X-ray peaks and the 88.03 keV gamma-ray peak are used for energy calibration. The 22.1 keV Ag X-ray and the 88.03 keV gamma ray are used for gain stabilization. Tungsten X-rays evident in the spectrum originate from the X-ray tube and from the tungsten used for collimation and shielding.

The shape and intensity of the low-energy region of the bremsstrahlung spectrum is tailored using beam filters (typically 1 mm thick cadmium and 20 mm thick iron). However, the count rate on the low-energy side does not drop to zero because a significant fraction of the photons that reach the HPGe undergo inelastic scattering, thus leaving a partial energy deposition in the detector. This deposition then accumulates on the low-energy side of the spectrum. As the photon energy increases, the mass attenuation coefficient decreases, and the count rate registered in the pulse-height spectrum increases with increasing channel number. At the characteristic K-edge energy of the high-Z element, the mass attenuation coefficient increases abruptly, and the count rate drops. The drop or jump in the count rate is proportional to the concentration of the high-Z element in the sample.

The abbreviated mathematical approach is described in this section. See Ref. [39] for a detailed description.

The current implementation of KED analysis uses an approach based on region of interest (ROI) to determine the ratio of transmission across the K-edge and therefore determine the elemental concentration. A physics-based fitting approach has been developed by McElroy et al. [40] that has improved the precision by a factor of 2. The physics-based fitting approach is discussed in Sect. 11.9.

The transmitted counts in the channels in the ROI below and above the K-edge are fit as a function of energy. Conventionally, a linear fit of a natural log-log function is used.

11 Densitometry

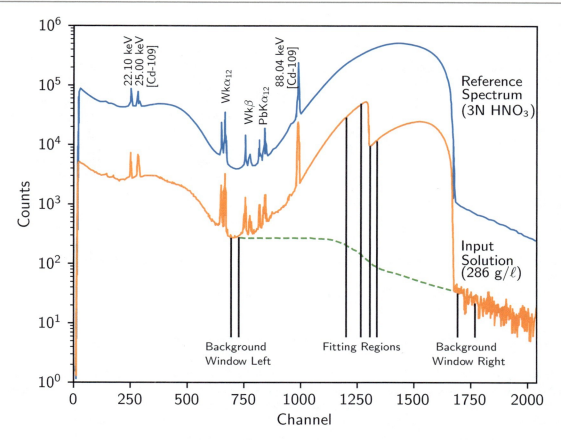

Fig. 11.9 Typical K-edge spectrum from uranium sample with regions of interest indicated. [39]

$$\ln\left[\ln\left(\frac{1}{T}\right)\right] = a - b\ln(E) \quad (11.30)$$

In Eq. 11.28, T is the transmission at energy E, and a and b are fitting coefficients specific to each ROI. Following the discussion given in Ref. [39], the uranium or plutonium concentration, ρ, in the sample at the limits of the ROI windows below and above the K-edge can be written as follows:

$$\rho = \frac{\ln(T_{E-}) - \ln(T_{E+})}{\Delta\mu \cdot d} \quad (11.31)$$

where

E− = upper-energy boundary of the ROI below the K-edge
E+ = lower-energy boundary of the ROI above the K-edge
Δμ = difference in mass attenuation coefficients at energies E− and E+
d = length of the sample vial.

With the accurately known value of the sample path length and by determining the transmission values at energies E− and E+ using calibration standards of known concentrations (ρ), one can experimentally determine the quantity (Δμ). The Δμ value at the K-edge energy can also be determined by extrapolating the fitting functions used below and above the K-edge to the K-edge energy. Therefore, the parameter (Δμ) is the calibration parameter for a given HKED system. If the system has been configured and calibrated accurately, the experimentally determined value of Δμ must agree with the theoretical value (to within an uncertainty of 1σ) derived using mass attenuation coefficients from a reliable database, such as the National Institute of Standards and Technology (NIST) XCOM database [41].

Table 11.6 Summary of uncertainties for K-edge densitometry measurement for a two-sample container system (rectangular cuvette for KED and cylindrical vial for XRF)

Uncertainty component	Magnitude of uncertainty (%)	Comment
Counting precision (3 times 1000 s live time)	0.15%	Concentration range: 150–300 g/L
Cell length	0.01%	For individual cuvette
	<0.1%	Variation for a production batch of cuvettes
Cell positioning	<0.1%	Determined by dimensional tolerances for sample holder
Sample matrix	<0.2%	Can be considered in calibration
Uranium isotopic composition	0.013%	Per percent change of ^{235}U enrichment
Sample temperature	0.05%	Per degree centigrade of sample temperature
Calibration	0.2%	Uncertainty in reference concentration from chemical analysis
Nonlinearity	<0.2%	Concentration range: 150–300 g/L
Instrument variability	<0.3%	Monitored from control charts
Total	0.5%	Summed in quadrature

When an unknown sample is measured, the concentration (ρ) can be derived from the calibration parameter ($\Delta\mu$) and the path length (d) in Eq. 11.29.

The uncertainty components for a KED measurement are given in Table 11.6.

11.8.2 Determination of Ratio of Major to Minor Element Concentrations Using XRF Branch of HKED

The fluoresced X-rays from the heavy elements in the sample are measured using a second HPGe detector optimized for low energy that is also coupled to a multichannel analyzer. The net peak areas of the $K_{\alpha 1}$ and $K_{\alpha 2}$ peaks from the major and minor actinide elements are quantified, and the ratio of major/minor element concentrations is determined. In HKED measurements, the major element is typically uranium, and the minor element is plutonium. In the current implementation of the hybrid analysis, the X-ray peak areas are determined using an ROI approach. As mentioned in Sect. 11.8.2, McElroy et al. [42] have developed a physics-based method that also fits the XRF spectra.

Figure 11.10 shows an XRF spectrum from the bremsstrahlung irradiation of a dissolver solution sample with uranium and plutonium concentrations of 276 and 1.85 g/L, respectively [39]). The K_α and K_β X-ray peaks are evident, as are the tungsten X-rays. The passive spectrum taken with the bremsstrahlung source turned off is also shown in the figure. The fission-product gamma-ray peaks are evident in the passive spectrum.

In the schematic drawing of the HKED geometry shown in Fig. 11.8, the HPGe detector measuring the XRF is configured at an approximate angle of 150° because, for a maximum incident photon energy of 150 keV, the energy of the inelastically scattered photons is less than 100 keV for scattering angles ≥150°. Therefore, configuring the HKED system geometry such that the XRF detector is at an angle close to 150° relegates the inelastically scattered photons from energy regions where the characteristic K X-rays from uranium and plutonium occur. The signal-to-continuum ratio of the spectrum at these X-ray energies is drastically improved as a result, thus improving the precision of the results and the sensitivity of the system. Figure 11.11 illustrates the energy of inelastically scattered photons as a function of the scattering angle. The incident photon energy is 150 keV.

The XRF measurement and analysis in the hybrid mode entails the determination of the uranium-to-plutonium concentration ratios. The U/Pu ratio is determined by calculating the ratios of the $K_{\alpha 1}$ X-ray peaks from uranium and plutonium, determining the relative efficiency curve in the energy range of interest, and factoring in the excitation probabilities for the X-ray emission and the atomic weights of uranium and plutonium. This estimate includes the intrinsic efficiency of the detector as well as attenuation through the sample, the container wall, and any absorber between the sample vial and the detector. Usually, the geometrical factors are held constant between calibration and assay so that the efficiency ratio is based solely on the solution concentration. The governing equation is

$$\frac{U}{Pu} = \frac{A(U)}{A(Pu)} \cdot \frac{UK_{\alpha 1}}{PuK_{\alpha 1}} \cdot \frac{\varepsilon(PuK_{\alpha 1})}{\varepsilon(UK_{\alpha 1})} \cdot \frac{1}{R_{\frac{U}{Pu}}}. \tag{11.32}$$

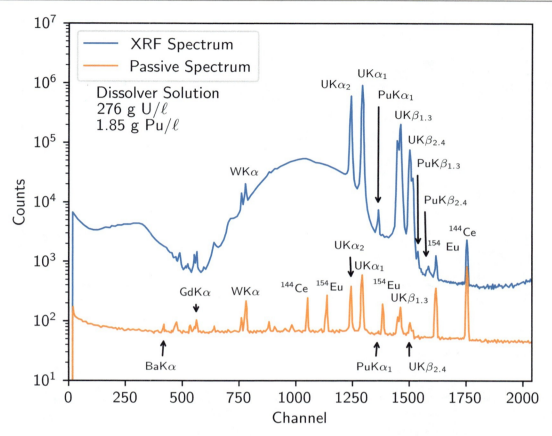

Fig. 11.10 Example of an X-ray fluorescent spectrum from a uranium-plutonium dissolver solution sample. The spectrum with the bremsstrahlung source turned off is shown

In Eq. 11.29,

A = atomic weight of element
$UK_{\alpha 1}$ = net peak area ratio of the $UK_{\alpha 1}$ X-ray peak
$PuK_{\alpha 1}$ = net peak area ratio of the $PuK_{\alpha 1}$ X-ray peak
$\varepsilon(Pu\ K_{\alpha 1})/\varepsilon(U\ K_{\alpha 1})$ = ratio of overall relative efficiency for $K_{\alpha 1}$ X-rays.

The factor $R_{U/Pu}$ is the ratio of the excitation probabilities for the emission of uranium $K_{\alpha 1}$ and plutonium $K_{\alpha 1}$ X-rays. $R_{U/Pu}$ depends on the spectral distribution of the incident photon beam used for fluorescence. Because the spectral distribution is modified by the attenuation properties of the sample, the factor $R_{U/Pu}$ is a weak function of sample composition. It is the calibration factor for the XRF portion of the system. The calibration factor is determined over the applicable range of uranium and plutonium concentrations and U/Pu ratios (100:1 typically), where the HKED technique is in common use.

The uncertainty components for the U/Pu ratio determined using HKED XRF measurements are shown in Table 11.7.

11.8.3 Stand-Alone XRF for Measuring Low Concentrations of U and Pu

In the stand-alone XRF-only mode, the normalized full-energy peak rates of the $K_{\alpha 1}$ line of uranium and plutonium are measured and related to a concentration (ρ) through a fitting function. A separate function is used for uranium and plutonium. The typical functional form is given by Eq. 11.33:

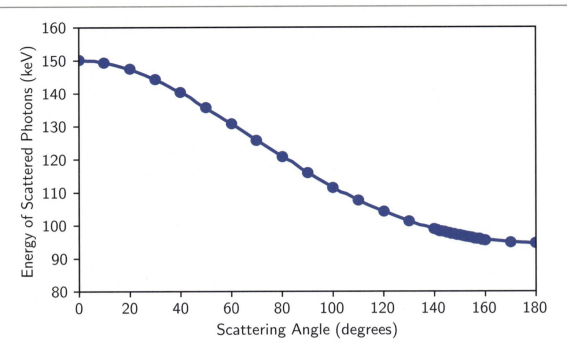

Fig. 11.11 Energy of an inelastically scattered photon as a function of scattering angle

Table 11.7 Summary of uncertainties for U/Pu ratio measurement for a two-sample container system (rectangular cuvette for KED and cylindrical vial for XRF)

Uncertainty component	Uncertainty magnitude
Counting statistics (1 h assay)	0.5%
Self-irradiation	0.3%
Atomic weights	<0.1%
Calibration	0.15%
Instrument variability	<0.3%
Total	0.7%

$$\rho = a_0 \cdot C \cdot e^{(a_1 \cdot C)}. \qquad (11.33)$$

In Eq. 11.33, a_0 and a_1 are fitting coefficients, and C is the net peak count rate of the $K_{\alpha 1}$ X-ray peak.

The stand-alone XRF mode of HKED is used for uranium or plutonium samples with concentrations up to 50 g/L typically. At higher concentrations, the self-attenuation of the fluoresced X-rays is significant enough such that the net peak count rate of the $K_{\alpha 1}$ X-ray is not proportional to the uranium or plutonium sample concentration.

11.9 Physics-Based Fitting Approach to KED Analysis

In the KED analysis approach described in Sect. 11.8.1, the ultimate uncertainty in the KED results for uranium or plutonium concentration is dependent on the accuracy of the calibration standards that were prepared and used. To lower the uncertainties in KED results and therefore enhance safeguards verification, a physics-based fitting approach was developed by McElroy et al. [40]. The approach is described in detail in the following sections.

The spectral fitting methodology takes full advantage of the information content of the transmission spectra, allowing determination of the major actinide and matrix concentration for the sample solution. The K-edge transmission measurement is effectively a first-principles measurement and should therefore require no calibration other than the energy calibration of the HPGe detector. However, the KED spectrum is remarkably complex due to scattering and electronic effects. To reduce the complexity of the analysis, we limit the fitting function to encompass the energy range that spans the ^{109}Cd peak at 88 keV to

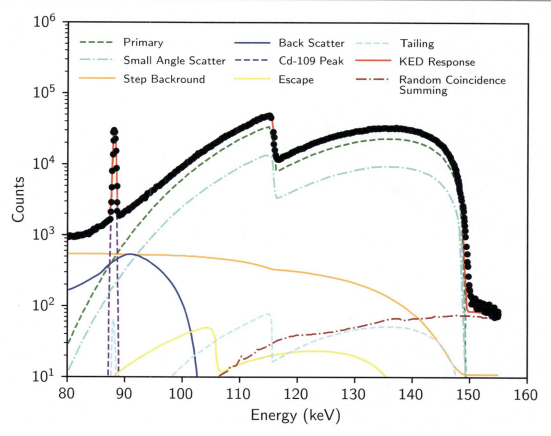

Fig. 11.12 The fitted K-edge response function and various contributors to the response superimposed on a measured spectrum from a 322 g U/L solution standard. The deviation near the high-voltage cutoff is tentatively attributed to jitter in the high-voltage supply in the X-ray generator

Table 11.8 Fit parameters for the KED transmission measurement

Sample parameters	Detector parameters	X-ray generator parameters
U concentration	Energy offset	Endpoint energy
Np concentration	Energy slope	X-ray intensity
Pu concentration	Gaussian width	Shape parameter [4, 5]
Am concentration	Fast tail decay	Small-angle scatter fraction
Cm concentration	Fast tail intensity	Detector backscatter intensity
Matrix concentration	Slow tail decay	
	Slow tail intensity	

just below the X-ray generator endpoint energy (i.e., 85–147.5 keV). Even with this limitation, the list of potential free parameters is rather lengthy, as illustrated in Fig. 11.12 and summarized in Table 11.8.

The KED response function uses Shaltout's representation [43] of the X-ray source term modified by the assembly of attenuating layers (e.g., beam filters, sample vial, detector housing), the sample contents (i.e., the actinides and sample matrix), and detector response function. Examination of experimental KED spectra supplemented by Monte Carlo N-Particle [44] modeling showed that the transmitted spectrum includes a substantial contribution from small-angle scattering within the long, narrow tungsten collimator (120 × 0.8 mm) as well as a small backscatter peak from the detector's copper cold finger assembly.

In the representation of the KED response function, the mass attenuation data from the NIST XCOM Photon Cross Sections Database [41] was incorporated and assumed for the present that these values have no uncertainty. These cross-section tables provide a delta function step at the K-edge transition, whereas others assume that the K-edge transition is Lorentzian broadened [45]. It was discovered that a mixture of the two assumptions provides the best empirical representation

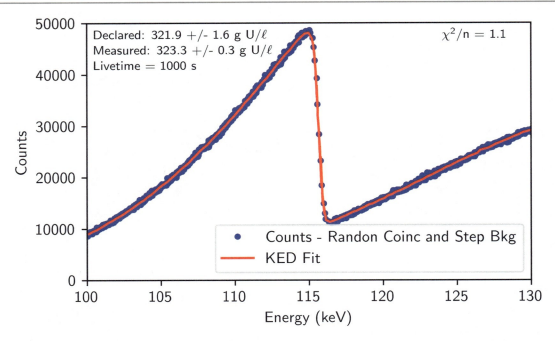

Fig. 11.13 K-edge spectrum obtained from a solution containing approximately 322 g uranium per liter. The plot shows the fit to the measured spectrum after correction for background and random coincidence summing. The relative residuals were all within 3 standard deviations, which illustrates the fidelity of the K-edge fitting process

for the KED response function; that is, for energies below the K-edge transition, a Voigt broadening of the mass attenuation function is required, but above the K-edge, a simple Gaussian broadening provides the best representation. The shape of high-energy response is potentially explained as an artifact of the fine structure in the mass attenuation function above the edge that is smoothed out by the instrumental response; however, more detailed cross-section data will be needed to verify this conjecture. The resulting response function provides an excellent representation of the observed data, as illustrated in Fig. 11.13. Many components are used to calculate the response function to fit the measured data. These components of the response function, described in detail in Ref. [39], include the X-ray source term, attenuation due to system hardware, sample attenuation, small-angle scattering, fission-product background, the detector response, and calculation of mass attenuation coefficients. The detector response includes Gaussian broadening, exponential tailing, stepped background, detection efficiency, germanium escape peaks, backscatter, and random coincidence summing.

11.10 Examples of K-Edge Densitometers XRF and Hybrid K-Edge Systems

This section contains information about KEDs and hybrid K-edge systems previously and currently in use throughout the world.

1. Los Alamos National Laboratory [12, 20, 21].
   The Los Alamos uranium solution assay system (USAS) was a hybrid assay instrument used offline at the Los Alamos HEU scrap recovery facility. The USAS measurement head is shown in Fig. 11.14.
   The USAS applied three distinct gamma-ray methods to assay uranium concentration in 20 or 50 mL uranium solution samples (in disposable plastic sample vials) in three concentration ranges. Waste solutions with uranium concentrations in the range of 0.001–0.5 g/L were counted for 2000 s with no transmission correction. Process solutions with concentrations in the range of 1–50 g/L were measured using a ^{169}Yb transmission source. The highest range, 50–400 g/L, which corresponded to product solutions, was assayed by the K-edge method using a ^{169}Yb transmission source. Accuracies of 0.7–1.5% were achieved in measurement times of 400–2000 s.
   The assay results were used for process control and nuclear material accounting. The instrument was in routine use in the scrap recovery facility from January 1976 until August 1984, when the facility was closed.

11 Densitometry

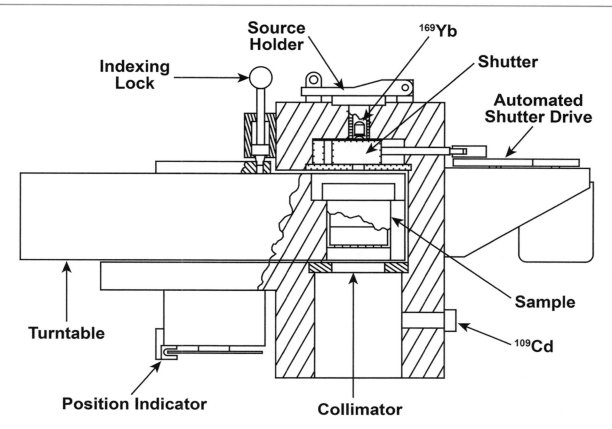

Fig. 11.14 Scale line drawing of the uranium solution assay system measurement head. The solution thickness in the transmission path is 2 cm

2. Oak Ridge Y-12 Plant [22, 23]

   The HEU scrap recovery facility at the Y-12 Plant uses a solution assay system (SAS) that is analogous to the USAS. The K-edge method is used to assay 50 mL uranium solution samples in the concentration range of 50–200 g/L. The samples include the product, and the SAS uses a ^{169}Yb transmission source and disposable plastic sample vials. The system was put into routine use at Y-12 in October 1981 for process control and materials accounting.

3. Allied General Nuclear Services [12, 24, 25]

   A discrete-source KED was evaluated for plutonium assay at the Allied General Nuclear Services facility in Barnwell, South Carolina, during 1977–78. The hybrid instrument performed passive and K-edge measurements on prepared 10 mL solution samples of typical light-water-reactor plutonium in a fixed quartz sample cell. The transmission source was a combination of ^{75}Se and ^{57}Co. The results were reported for cells of different transmission path lengths.

4. Power Reactor and Nuclear Fuel Development Corporation [12, 26–28]

   A discrete-source KED operated in the Tokai-Mura Reprocessing Plant analytical laboratory of the Power Reactor and Nuclear Fuel Development Corporation in Japan. Freshly separated and aged plutonium solution samples of the products of boiling-water-reactor and pressurized-water-reactor fuel reprocessing were assayed by the K-edge method in a two-cycle assay (first with a ^{75}Se transmission source, then with ^{57}Co). Figure 11.15 shows the location of the measurement station under the glove box at the Tokai-Mura plant laboratory. Figure 11.16 is a scale line drawing of the measurement head, which included a well that extends down from the glove-box floor. The instrument performed an isotopics assay on the fresh solutions in a third cycle. The solution samples were assayed in a well that was an extension of a glove box. The gamma-ray detector and the transmission sources were external to the glove box. The sample cells were disposable plastic vials that required approximately 10 mL of solution.

   The Tokai instrument was installed in November 1979 and was operated through 1980 in an evaluation mode. The instrument was in routine use in the facility starting in early 1981. Figure 11.17 is a plot of the percent difference between K-edge assay results and the reference values (from destructive analysis) obtained throughout 1981 during routine facility use of the instrument [28]. The densitometer was available for facility use by International Atomic Energy Agency inspectors from 1982.

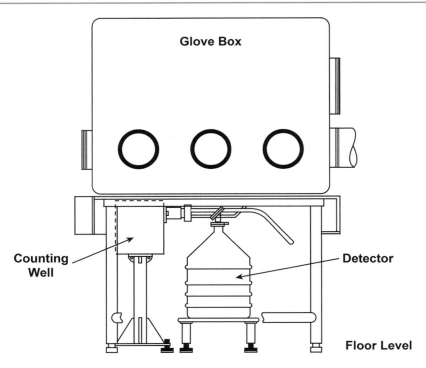

Fig. 11.15 The Tokai KED measurement station beneath the laboratory glove box

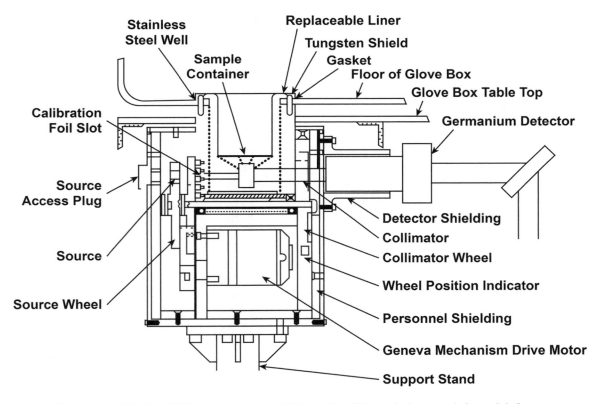

Fig. 11.16 Scale line drawing of the Tokai KED measurement head. The solution thickness in the transmission path is 2 cm

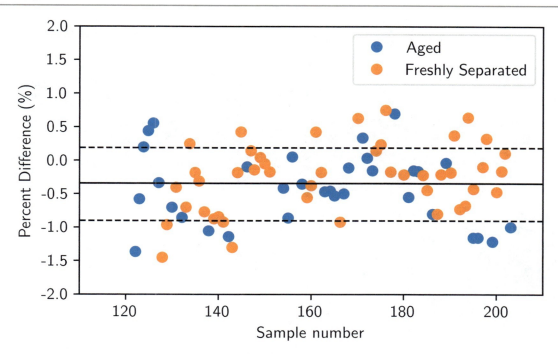

Fig. 11.17 Percent difference between 600 s K-edge and destructive assays/or plutonium concentration, plotted as a function of sample identification number. The solid line is the average relative result of −0.36%. This apparent bias is the result of calibration (in 1979) using only a small number of reference samples [27]

5. Savannah River Plant [5, 12, 29]
   A discrete-source K-edge plutonium solution densitometer was designed for inline testing at the Savannah River Plant (present-day Savannah River Site). A flow-through stainless steel sample cell was plumbed into a bypass loop on process solution storage tanks and resided in a protrusion of the process containment cabinet. The detector and transmission sources were located outside the containment cabinet on either side of the protrusion for measurement of the gamma-ray transmissions through the solution-filled cell. The measurements were performed on approximately 100 mL of static solution, after circulation of the tank solution through the bypass loop. (The freshly separated plutonium in the solutions is produced during reprocessing of low-burnup fuel.) The K-edge transmission measurements were performed in two cycles, as with the Tokai instrument, and a third cycle determined the plutonium isotopic composition. The instrument was also used to investigate the measurement of plutonium concentration in the presence of uranium admixtures. The extrapolation procedure described in Sect. 11.4.4 was used on solutions with uranium-to-plutonium ratios greater than 2:1 [5]. The offline testing of the instrument took place at the plant from April 1980 until December 1981. Figure 11.6 (see Sect. 11.4.2) is a plot of the measurement precision versus concentration (over the range of 5–200 g/L) obtained in this testing phase [5]. The inline testing began in December 1982 and ended in June 1983.
6. International Atomic Energy Agency Safeguards Analytical Laboratory [30]
   A portable KED has been designed for testing as an inspection tool to authenticate the concentrations of plutonium samples inside glove boxes. The densitometer consists of hardware to hold and shield the detector and transmission sources and a portable multichannel analyzer equipped with electronics for the analog signal processing. The hardware slides inside the glove of the glove box so that a plutonium solution sample in a disposable plastic vial can be mounted and clamped in a holder between the detector and transmission source for the two-cycle K-edge assay. Figure 11.18 shows the measurement head of the portable K-edge instrument inserted in a glove-box glove.
7. Kernforschungszentrum Karlsruhe [12, 31–33]
   A continuum-source KED has been tested at KfK in Karlsruhe, Federal Republic of Germany, since 1978. The detector and X-ray head reside outside of a glove box, and the samples and collimators are located inside the glove box. The instrument has been used to assay reprocessing product solutions and fast-breeder-reactor reprocessing feed solutions for concentrations of both uranium and plutonium. A hybrid version of this instrument was used for assaying light-water-reactor feed solutions in which the plutonium content was approximately 1% of the uranium content. The continuum source served both as a transmission source for K-edge assay of uranium and as a fluorescing source for XRF assay of the

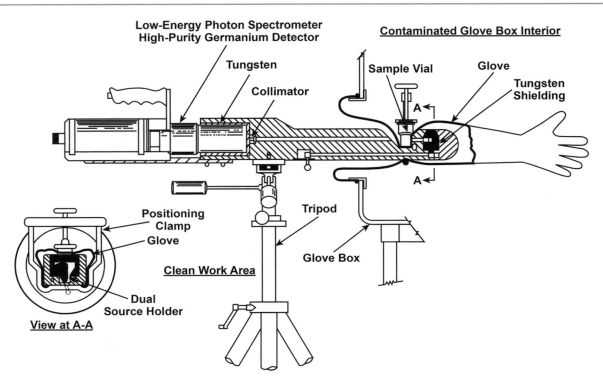

Fig. 11.18 Scale line drawing of the measurement head of the portable KED inserted into a glove-box glove. The sample cell thickness in the transmission path is 2 cm

plutonium/uranium concentration ratio. The intensity of the continuum source allows the highly restrictive sample collimation required for K-edge and XRF assays while greatly reducing the passive count rate from the samples, which contain high levels of fission products. Figure 11.19 is a line drawing of the measurement head for the hybrid instrument.

8. Oak Ridge National Laboratory HKED System

   Recent changes in safeguards objectives and the evolution of the nuclear fuel cycle have resulted in an expansion of the list of actinides of interest and a need for algorithm enhancements for the HKED analysis. New material-processing regimes and evolving fuel cycles have made it necessary to accommodate much higher levels of plutonium in the solutions in which the relative concentrations can be on the order of one to one (i.e., U:Pu = 1:1). These new processes will result in higher concentrations of the minor actinides, such as americium and neptunium. Recent increased emphasis on the tracking of americium and neptunium and the increase in the ratio of plutonium relative to uranium in proposed processing solutions has necessitated validation of the HKED software for these applications. To support these development activities, a commercial HKED system has been installed at the Oak Ridge National Laboratory (ORNL) Radiological Engineering Development Center, a multipurpose, Category II, radiochemical processing facility that provides the capability to fabricate a broad range of mixed actinide solution standards for testing and future developments of the HKED technique. The ORNL HKED system, shown in Fig. 11.20, is a single cylindrical sample-vial measurement system. In other words, the KED and XRF measurements are both performed on the solution sample contained in a single cylindrical vial.

11.11 Examples of L_{III}-Absorption-Edge Densitometers

This section contains descriptions of several L_{III}-edge densitometers that have been tested and evaluated under actual or simulated in-plant environments. The first three L_{III}-edge densitometers described were designed to be equivalent, mechanically and electronically. Figure 11.21 is line drawing of the measurement head for the Allied General Nuclear Services instrument and represents all three instruments. Table 11.5 summarizes the performance data for the instruments.

11 Densitometry

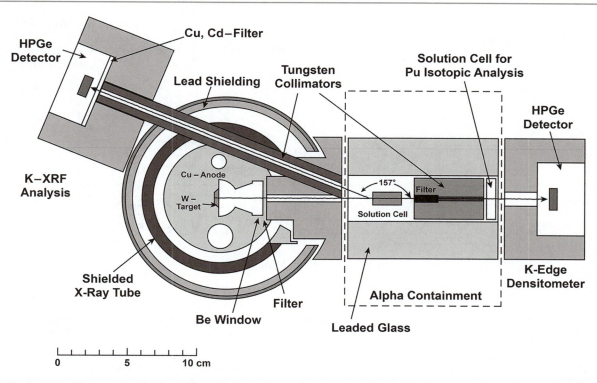

Fig. 11.19 Cross-sectional view of the KfK combined K-edge/XRF system. The size of the alpha containment, a standard glove box, is not shown to scale. The sample cell thickness in the transmission path is 2 cm

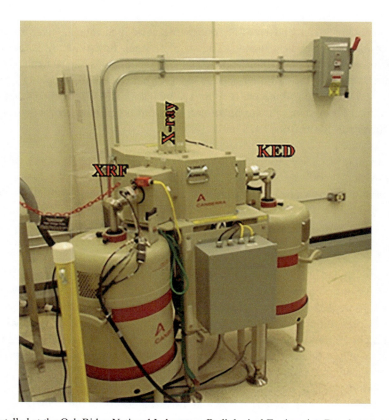

Fig. 11.20 HKED system installed at the Oak Ridge National Laboratory Radiological Engineering Development Center

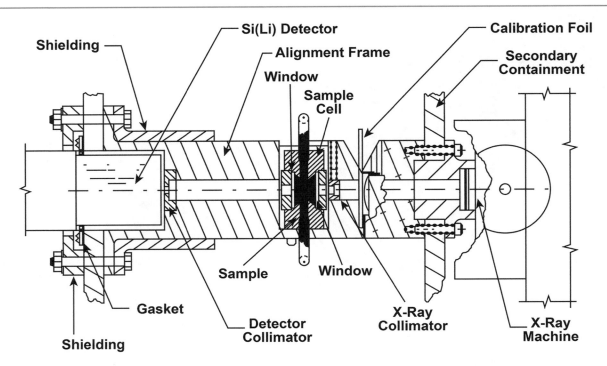

Fig. 11.21 Scale line drawing of the L_{III}-edge densitometer measurement head for the Savannah River Laboratory, New Brunswick Laboratory, and Allied General Nuclear Services instruments. The flow-through sample cell is shown cut off at the inlet and outlet tubes. The darkened area indicates the solution in the cell (1 cm transmission path length). The materials for secondary containment, shielding, frame, sample cell, and collimator are polycarbonate (Lexan), stainless steel, aluminum, stainless steel (with Kel-F windows), and brass, respectively

1. Savannah River Laboratory [12, 21, 34].

    The L_{III}-edge densitometer at the Savannah River Laboratory (present-day Savannah River Site) was tested in conjunction with a solution coprocessing demonstration facility. The stainless-steel, flow-through, solution sample cell (fitted with plastic windows) was plumbed into the glove box that housed the coprocessing setup so that solution from various points in the process could be introduced into the cell for L_{III}-edge assay of either uranium or plutonium or both. The instrument measured 15 mL static solutions in the cell; before each assay, the instrument was flushed several times with the solution. The assay precisions obtained for pure uranium or plutonium solutions are plotted versus concentration in Fig. 11.22. The instrument operated at Savannah River from 1978 until 1980.

2. New Brunswick Laboratory [35].

    The L_{III}-edge densitometer at the U.S. Department of Energy New Brunswick Laboratory (NBL) at Argonne was designed to reproduce the measurement geometry and assay method of the Savannah River Laboratory instrument. Prepared reference solutions of uranium, plutonium, and mixed solutions were used in a carefully controlled evaluation of the precision and accuracy of this instrument. The NBL assay results are compared with reference values for pure uranium solutions in Fig. 11.23. The sensitivity to matrix contaminants with low-, intermediate-, and high-Z elements was examined for contamination levels up to 10% (of SNM) by weight. This evaluation took place during 1980–1981.

3. Allied General Nuclear Services [36].

    An L_{III}-edge densitometer designed to perform continuous assays of uranium concentration in flowing process streams was tested in 1981 at the Allied General Nuclear Services Barnwell facility. The stainless-steel flow-through cell was plumbed into a line that continuously sampled the product stream of a solvent extraction column. The instrument operated for 7 days without interruption, providing uranium concentration results every 5 min, analyzing flowing solutions from startup (essentially zero uranium concentration) to steady-state levels of approximately 40 g/L. The instrument provided a hard copy of the near-real-time results automatically to a materials control and accounting computer programmed to draw near-real-time material balances using readouts from process-monitoring equipment.

4. Los Alamos National Laboratory [37].

    A compact L_{III}-edge densitometer was tested in 1984 at Los Alamos National Laboratory. This instrument used a commercial X-ray generator designed for portable applications. Figure 11.24 shows the measurement head. Although a

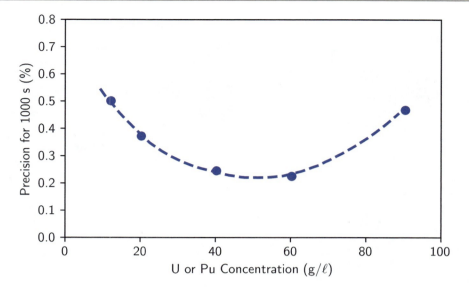

Fig. 11.22 Precision measured at Savannah River Laboratory for 1000 s L_{III}-edge assays of uranium or plutonium plotted versus concentration. The dashed line is the calculated standard deviation, based on counting statistics, using Eq. 11.17

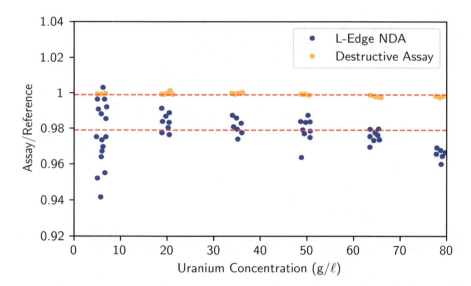

Fig. 11.23 Comparison of single 1000 s L_{III}-edge assays and destructive analysis of solution samples introduced into the sample cell of the New Brunswick Laboratory instrument. Groups of three destructive assay data points plotted horzontally represent repeated assays of the same sample. These data were used to establish the calibration of the New Brunswick Laboratory instrument

standard rack of electronics was employed for analog-signal processing and for data acquisition and analysis, the portable multichannel analyzer used in the compact KED at Seibersdorf (see Sect. 11.10) could have been employed in the L-edge densitometer at Los Alamos National Laboratory, allowing portable applications to be considered for L_{III}-edge measurements. The performance of the compact densitometer with prepared reference solutions of uranium was equal to that of the L_{III}-edge instruments tested previously.

5. Combined Procedure for Uranium Concentration and Enrichment Assay (COMPUCEA) [46–48].

   COMPUCEA is a transportable system for accurate onsite analytical measurements of uranium elemental assay and enrichment during the physical inventory verification in European low-enriched uranium fuel fabrication plants. COMPUCEA makes use of X-ray absorption edge spectrometry (K- and L-edge) for the uranium element analysis and high-resolution gamma spectrometry for the ^{235}U abundance determination, in addition to involving careful analytical procedures, such as quantitative sample dissolution, solution density measurements, and quantitative aliquoting.

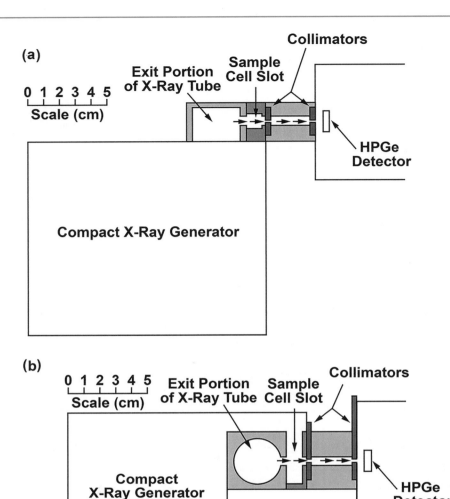

Fig. 11.24 Scale line drawings of the compact L_{III}-edge densitometer measurement head at Los Alamos National Laboratory: (**a**) view from above in the horizontal plane of the X-ray transmission path; (**b**) side view. The solution thickness in the transmission path is 1 cm

COMPUCEA uses a combination of chemistry and spectrometry measurements to determine uranium content; equipment for the uranium concentration measurements is based on the following hardware components:

- A compact X-ray generator with maximum ratings of 30 kV and 100 μA (model Eclipse-II/III from AMPTEK Inc.) with a Ag target transmission tube and associated controller as a radiation source.
- A Peltier-cooled, 10 mm^2 × 0.5 mm Si Drift Detector system with associated preamplifier/amplifier and power supply (model AXAS-SSD10 from KETEK GmbH) for high pulse rates, offering an energy resolution of 142 eV at 5.9 keV at a pulse shaping time constant of 0.5 μs Gaussian equivalent.
- A digital signal processor (Canberra, model DSA 1000).
- A shielding/collimation assembly with exchangeable sample adapters for L-edge densitometry and optional XRF measurements; for the L-edge measurements, a fixed flow-through cell with a path length of 2 mm is used, which is rinsed and dried between measurements. In this configuration, the path length of the photon beam through the sample cell remains constant, minimizing error sources.

Figure 11.25 shows the schematic setup of the L-edge densitometer, together with an L-edge absorption spectrum obtained from a uranium solution. For energy calibration and instrument control, part of the X-ray beam is directed onto a Ti/Ge target;

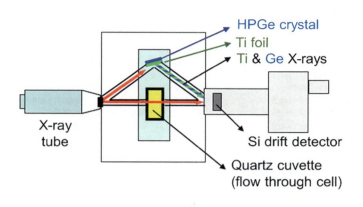
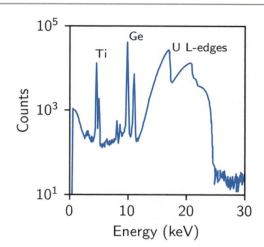

Fig. 11.25 Schematic of the COMPUCEA L-edge densitometer (left) and L-edge spectrum of a uranium solution (right). The K X-ray peaks of Ti and Ge are used for energy calibration

the characteristic K X-rays of Ti and Ge can be seen in the spectrum. These reference lines were chosen because they are in an energy range where they do not interfere with the uranium L-edge spectrum.

The total weight of the COMPUCEA system, including the shielding for the X-ray tube, is approximately 6 kg (without laptop computer), which meets the design goal of true portability.

References

1. T.R. Canada, J.L. Parker, T.D. Reilly, Total plutonium and uranium determination by gamma-ray densitometry. Trans. Am. Nucl. Soc. **22**, 140 (1975)
2. T.R. Canada, D.G. Langner, J.W. Tape, Nuclear safeguards applications of energy-dispersive absorption-edge densitometry, in *Nuclear Safeguards Analysis*, Series No. 79, ed. by E.A. Hakkila, (American Chemical Society, Washington, DC, 1978), p. 96
3. T.R. Canada, S.-T. Hsue, D.G. Langner, D.R. Martin, J.L. Parker, T.D. Reilly, J.W. Tape, Applications of the absorption-edge densitometry technique to solutions and solids. Nucl. Mater. Manage. **6**(3), 702 (1977)
4. R.L. Bramblett, Passive and active measurements of SNM in 55-gallon drums. Nucl. Mater. Manage. **4**, 137 (1975)
5. H.A. Smith Jr. et al., Test and Evaluation of the In-line Plutonium Solution K-Absorption-Edge Densitometer at the Savannah River Plant: Phase I, Off-line Testing. Los Alamos National Laboratory report LA-9124-MS (1982)
6. S.-T. Hsue, Densitometry Design. Personal communication to R. B. Walton, Los Alamos document Q-1-80-243 (May 1980)
7. L.A. Currie, Limits for quantitative detection and quantitative determination. Anal. Chem. **40**, 586 (1968)
8. C.E. Crouthamel et al., A compilation of gamma-ray spectra (NaI detector), in *Applied Gamma-Ray Spectrometry*, ed. by F. Adams, R. Dams, (Pergamon Press, Braunschweig, 1970)
9. T.W. Crane, Detectability Limits and Precision for Shufflers. Los Alamos National Laboratory report LA-10158-MS (1984)
10. E. Storm, H.I. Israel, Photon Cross Sections from 0.001 to 100 MeV for Elements 1 Through 100. Los Alamos Scientific Laboratory report LA-3753 (1967)
11. R.C. Weast (ed.), *Handbook of Chemistry and Physics*, 55th edn. (Chemical Rubber Company Press, Cleveland, 1975)
12. P.A. Russo, S.-T. Hsue, D.G. Langner, J.K. Sprinkle Jr., Nuclear safeguards applications of energy-dispersive absorption-edge densitometry. Nucl. Mater. Manage. **9**, 730 (1981)
13. F. Brown, D.R. Terry, J.B. Hornsby, R.G. Monk, F. Morgan, J. Herrington, P.T. Good, K.C. Steed, V.M. Sinclair, Application of instrumental methods to the determination of nuclear fuel materials for safeguards, in *Safeguards Techniques, Proceedings of the IAEA Karlsruhe Symposium*, vol. II, (IAEA, Vienna, July 1970), p. 125
14. D.R. Terry, A.P. Dixon, A Portable Gamma Absorptiometer for Safeguards Use in Nuclear Fuel Reprocessing Plants. Atomic Weapons Research Establishment report AWRE/44/88/28 Cos 28, Aldermaston, England (1975)
15. J.E. Ayer, D.R. Schmitt, A gamma-ray absorptiometer for nuclear fuel evaluation. Nucl. Technol. **21**, 442 (1975)
16. T. Gozani, H. Weber, Y. Segal, A gamma-ray transmission gauge for determination of heavy and light metal loading of fuel elements. Nucl. Mater. Manage. **2**(3), 139 (1973)
17. T. Gozani, *Active Nondestructive Assay of Nuclear Materials: Principles and Applications*, NUREG/CR-0602 (U.S. Nuclear Regulatory Commission, Washington DC, 1981), p. 118
18. G. Bardone, M. Aparo, F.V. Frazzoli, Dual-energy X-ray absorptiometry for the assay of mixed special nuclear material in solution, in *Proceedings of the Second Annual Symposium on Safeguards and Nuclear Materials Management*, Edinburgh, Scotland, March 26–28, 1980 (European Safeguards Research and Development Association, Brussels, 1980), p. 270

19. M. Aparo, B. Mattia, F.V. Frazzoli, P. Zeppa, Dual-energy X-ray absorptiometer for nondestructive assay of mixed special nuclear material in solution, in *Proceedings of the Fifth Annual Symposium on Safeguards and Nuclear Materials Management*, Versailles, France, April 19–21, 1983 (European Safeguards Research and Development Association, Brussels, 1983), p. 271
20. J.K. Sprinkle, Jr., H.R. Baxman, D.G. Langner, T.R. Canada, T.E. Sampson, The in-plant evaluation of a uranium NDA system, in *Proceedings of the American Nuclear Society Topical Conference on Measurement Technology for Safeguards and Materials Control*, Kiawah Island, SC, November 26–28, 1979 (National Bureau of Standards, Washington, DC, 1980), p. 324
21. T.R. Canada, J.L. Parker, P.A. Russo, Computer-based in-plant nondestructive assay instrumentation for the measurement of special nuclear materials, in *Proceedings of the American Nuclear Society Topical Conference on Computers in Activation Analysis and Gamma-Ray Spectroscopy*, Mayaguez, Puerto Rico. April 30–May 3, 1978 (US DOE Technical Information Center, Washington, DC), p. 746
22. H.H. Hogue, S.E. Smith, Off-line nondestructive analysis at a uranium recovery facility, in *Safeguards Technology: The Process-Safeguards Interface, Proceedings of the American Nuclear Society-INMM Topical Conference*, Hilton Head Island, SC, November 28–December 2, 1983 (Conf. 831106, 1984)
23. H.H. Hogue, S.E. Smith, Nondestructive Analysis at the Oak Ridge Y-12 Plant. Oak Ridge Y-12 Plant report Y-2297 (1984)
24. K.J. Hofstetter, G.A. Huff, R. Gunnink, J.E. Evans, A.L. Prindle, On-line measurement of total and isotopic plutonium concentrations by gamma-ray spectrometry, in *Analytical Chemistry in Nuclear Fuel Reprocessing*, ed by W. S. Lyon (Science Press, Princeton, 1978), p. 266, and University of California report UCRL-52220 (1977)
25. K.J. Hofstetter, G.A. Huff, On-line Isotopic Concentration Monitor. Allied General Nuclear Services report AGNS-1040-2.3-52 (1978)
26. L.R. Cowder, S.-T. Hsue, S.S. Johnson, J.L. Parker, P.A. Russo, J.K. Sprinkle, Jr., Y. Asakura, T. Fukuda, I. Kondo, An instrument for the measurement of Pu concentration and isotopics of product solutions at Tokai-Mura, in *Proceedings of the Second Annual Symposium on Safeguards and Nuclear Materials Management*, Edinburgh, Scotland, March 26–28, 1980 (European Safeguards Research and Development Association, Brussels, 1980), p. 119
27. P.A. Russo, S.-T. Hsue, J.K. Sprinkle, Jr., S.S. Johnson, Y. Asakura, I. Kondo, J. Masui, K. Shoji, In-Plant Measurements of Gamma-Ray Transmissions for Precise K-Edge and Passive Assay of Plutonium Concentration and Isotopic Fractions in Product Solutions. Los Alamos National Laboratory report LA-9440-MS (1982)
28. Y. Asakura, I. Kondo, J. Masui, K. Shoji, P.A. Russo, S.-T. Hsue, J.K. Sprinkle Jr., S.S. Johnson, In-plant measurements of gamma-ray transmissions for precise K-edge and passive assay of plutonium concentration and isotopic abundances in product solutions at the Tokai reprocessing plant. Nucl. Mater. Manage. **11**, 221 (1982)
29. H.A. Smith Jr., T. Marks, L.R. Crowder, C.O. Shonrock, S.S. Johnson, R.W. Slice, J.K. Sprinkle, Jr., K.W. MacMurdo, R.L. Pollard, L.B. Baker, Development of in-line plutonium solution NDA instrumentation at the savannah river plant, in *Proceedings of the Second Annual Symposium on Safeguards and Nuclear Materials Management*, Edinburgh, Scotland, March 26–28, 1980 (European Safeguards Research and Development Association, Brussels, 1980), p. 123
30. L.R. Cowder, S.F. Klosterbuer, R.H. Augustson, A. Esmailpour, R. Hawkins, E. Kuhn, A compact K-edge densitometer, in *Proceedings of the Sixth Annual Symposium on Safeguards and Nuclear Materials Management*, Venice, Italy, May 14–18, 1984 (European Safeguards Research and Development Association, Brussels, 1984), p. 261
31. H. Eberle, P. Matussek, H. Ottmar, I. Michel-Piper, M.R. Iyer, P.P. Chakraborty, Nondestructive elemental and isotopic assay of plutonium and uranium in nuclear materials, in *Nuclear Safeguards Technology 1978*, vol. II, (International Atomic Energy Agency, Vienna, 1979), p. 27
32. H. Eberle, P. Matussek, I. Michel-Piper, H. Ottmar, Assay of uranium and plutonium in solution by K-edge photon absorptiometry using a continuous X-ray beam, in *Proceedings of the Second Annual Symposium on Safeguards and Nuclear Materials Management*, Edinburgh, Scotland, March 26–28, 1980 (European Safeguards Research and Development Association, Brussels, 1980), p. 372
33. H. Eberle, P. Matussek, I. Michel-Piper, H. Ottmar, Operational experience with K-edge photon absorptiometry for reprocessing feed and product solution analysis, in *Proceedings of the Third Annual Symposium on Safeguards and Nuclear Materials Management*, Karlsruhe, Federal Republic of Germany, May 6–8, 1981 (European Safeguards Research and Development Association, Brussels, 1981), p. 109
34. P.A. Russo, T.R. Canada, D.G. Langner, J.W. Tape, S.-T. Hsue, L.R. Crowder, W.C. Moseley, L.W. Reynolds, M.C. Thompson, An X-ray L_{III}-edge densitometer for assay of mixed SNM solutions, in *Proceedings of the First Annual Symposium on Safeguards and Nuclear Materials Management*, Brussels, Belgium, April 25–27, 1979 (European Safeguards Research and Development Association, Brussels, 1979), p. 235
35. W.J. McGonnagle, M.K. Holland, C.S. Reynolds, N.M. Trahey, A.C. Zook, Evaluation and Calibration of a Los Alamos National Laboratory L_{III}-Edge Densitometer. U.S. Department of Energy New Brunswick Laboratory report NBL-307 (1983)
36. P.A. Russo, T. Marks, M.M. Stephens, A.L. Baker, D.D. Cobb, Automated On-Line L-Edge Measurement of SNM Concentration for Near-Real-Time Accounting. Los Alamos National Laboratory report LA-9480-MS (1982)
37. M.L. Brooks, P.A. Russo, J.K. Sprinkle, Jr., A Compact L-Edge Densitometer for Uranium Concentration Assay. Los Alamos National Laboratory Report LA-10306-MS (1985)
38. International Target Values for Measurement Uncertainties in Safeguarding Nuclear Materials. IAEA STR-368, Rev 1.1 (International Atomic Energy Agency, Vienna, September 2022). https://nucleus.iaea.org/sites/connect/ITVpublic/Resources/International%20Target%20Values%20for%20Measurement%20Uncertainties%20in%20Safeguarding%20Nuclear%20Materials.pdf
39. H. Ottmar, H. Eberle, The Hybrid K-Edge/K-XRF Densitometer: Principles – Design – Performance; Report KfK 4590. Karlsruhe (1991)
40. R.D. McElroy Jr., S. Croft, S.L. Cleveland, G.S. Mickum, Spectral fitting approach to the hybrid k-edge densitometer: preliminary performance results, in *Proceedings of the 56th Annual Meeting of the INMM*, Indian Wells, CA, July 12–16, 2015
41. M.J. Berger, J.H. Hubbel, S.M. Seltzer, J.S. Coursey, R. Sukumar, D.S. Zuker, K. Olsen, XCOM: Photon Cross Sections Database, National Institute of Standards and Technology (1998), https://www.nist.gov/pml/xcom-photon-cross-sections-database. Accessed Feb 7, 2022
42. R.D. McElroy, Performance Evaluation of the ORNL Multi-elemental XRF (MEXRF) Analysis Algorithms, Oak Ridge National Laboratory Technical Report ORNL/TM-2016/594, Oak Ridge, TN (May 2016)
43. A.A. Shaltout, On the X-ray tube spectra, the dependence on the angular and electron energy of X-rays from the targets. Eur. Phys. J. Appl. Phys. **37**, 291–297 (2007)
44. C.J. Werner (ed.), MCNP Users Manual – Code Version 6.2. Los Alamos National Laboratory, Report LA-UR-17-29981 (2017)
45. T. Materna, J. Jolie, W. Mondelaers, B. Masschaele, Near K-edge measurement of the X-ray attenuation coefficient of heavy elements using a tuneable X-ray source based on an electron LINAC. Radiat. Phys. Chem. **59**, 449–457 (2000)

46. H. Ottmar et al., COMPUCEA: a high-performance analysis procedure for timely on-site uranium accountancy verification in LEU Fuel Fabrication Plants, in *Proceedings of the 48th Annual Meeting of the INMM*, Tucson, AZ, July 8–12, 2007
47. N. Erdmann, H. Ottmar, P. Amador, H. Eberle, H. Schorlé, P. van Belle, Validation of COMPUCEA 2nd Generation. JRC-ITU-TN-2008/37 (2009)
48. N. Erdmann et al., COMPUCEA: a high-performance analysis procedure for timely onsite uranium accountancy verification in LEU fuel fabrication plants. ESARDA Bull. **43**, 30–39 (2009)

Open Access This chapter is licensed under the terms of the Creative Commons Attribution 4.0 International License (http://creativecommons.org/licenses/by/4.0/), which permits use, sharing, adaptation, distribution and reproduction in any medium or format, as long as you give appropriate credit to the original author(s) and the source, provide a link to the Creative Commons license and indicate if changes were made.

The images or other third party material in this chapter are included in the chapter's Creative Commons license, unless indicated otherwise in a credit line to the material. If material is not included in the chapter's Creative Commons license and your intended use is not permitted by statutory regulation or exceeds the permitted use, you will need to obtain permission directly from the copyright holder.

12 X-Ray Fluorescence

M. H. Carpenter, P. J. Karpius, M. C. Miller, and S. J. Pattinson

12.1 Introduction

The potential use of X-rays for qualitative and quantitative elemental assay was appreciated soon after X-rays were discovered. The early applications used Geiger-Mueller tubes and elaborate absorber arrays or crystal diffraction gratings to measure X-rays. Later, advances in semiconductor detectors and associated electronics opened up the field of energy-dispersive X-ray fluorescence (XRF) analysis for general elemental assay.

XRF analysis is based on the fact that the X-rays emitted from an ionized atom have energies that are characteristic of the element involved. The X-ray intensity is proportional to both the elemental concentration and the strength of the ionizing source. Photon ionization, which is achieved using either an X-ray tube or radionuclide, is most directly applicable to the nondestructive assay (NDA) of nuclear material. Other methods of ionization (e.g., via electron or heavy ion beam) are generally prohibitive because of the physical size and complexity of the ionization source and the physical form of the material.

XRF analysis is a complementary technique to densitometry (Chap. 11). Densitometry measures photons that are transmitted through the sample without interaction, whereas XRF measures the radiation produced by photons that interact within the sample. As indicated by Fig. 12.1, densitometry is usually better suited for measuring samples with high concentrations of the element of interest, whereas XRF is the more useful technique for measuring samples with lower concentrations.

The literature on XRF analysis includes several general Refs. [1–5] that provide a thorough discussion of the method, with extensive bibliographies and information on attenuation correction procedures and both energy- and wavelength-dispersive XRF.

12.2 Theory

12.2.1 X-Ray Production

Section 2.3 in Chap. 2 contains a brief discussion of X-ray production. X-rays originate from atomic electron transitions and are element-specific. In the stable atom, electrons occupy discrete energy levels that are designated (in order of decreasing binding energy) K, L_1, L_2, L_3, M_1, ..., M_5, N_1, ..., N_7, and so forth. The binding energy is the energy that must be expended to remove an electron from a given orbit. The vacancy thus created is filled by an electron from an outer orbit. The resultant loss in potential energy may appear as an X-ray whose energy is equal to the difference in the binding energies of the two

---

Los Alamos National Laboratory strongly supports academic freedom and a researcher's right to publish; as an institution, however, the Laboratory does not endorse the viewpoint of a publication or guarantee its technical correctness.

M. H. Carpenter (✉) · P. J. Karpius · M. C. Miller · S. J. Pattinson
Los Alamos National Laboratory, Los Alamos, NM, USA
e-mail: mcarpenter@lanl.gov; karpius@lanl.gov; mike.miller@inl.gov; spattinson@lanl.gov

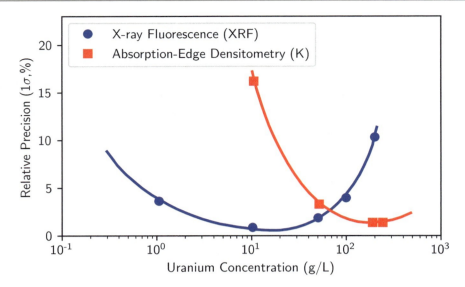

Fig. 12.1 Solution assay precision versus uranium concentration for typical XRF measurements (blue circles) and absorption-edge densitometry measurements (red squares)

electron states. For example, if a uranium K electron is removed from the atom and an electron from the L_3 level falls into its place, the energy of the emitted X-ray is 98.428 keV (115.591 keV minus 17.163 keV). The X-ray produced by this transition is designated $K_{\alpha 1}$. The K-series X-rays are produced by outer electrons that fill a K-shell vacancy.

Each X-ray transition has a specific probability or intensity for a given shell. Comparison of intensities between shells (e.g., between lines originating from L_3 vs. L_2 vacancies) must consider the excitation conditions that generate the distribution of vacancies and is not generalizable. The spectral distribution of the X-ray generator or source, the geometry and density of the item being measured, and the different excitation cross section versus energy for each shell all factor into the ratio of vacancies and the observed outer shell line intensities. As a result, the observed inter-shell L-line intensities are different when generated as a result of filling a K-shell vacancy versus excited directly by a low-energy X-ray generator. In the K-series, the K-to-L_3 transition is the most probable, and other intensities are usually expressed relative to $K_{\alpha 1}$.

Figure 12.2 depicts the transitions involved in the production of the most prominent K and L X-rays. Table 12.1 presents the major K and L lines of uranium and plutonium, along with their relative intensities normalized to each subshell. Figures 12.3 and 12.4 show the K and L X-ray spectra of uranium. References [6] and [7] provide detailed emission line and level energies referenced to multiple sources for all elements. Reference [6] also provides a more detailed discussion on the origin and physics of X-ray production.

12.2.2 Fluorescence Yield

All ionizations do not result in X-ray emission. The Auger effect is a competing mechanism of atomic relaxation. In this process, the atom regains energy stability by transferring energy directly to an outer-shell electron, which is then emitted with this energy less its binding energy. The ratio of the number of emitted X-rays to the total number of ionizations is called the *fluorescence yield* (ω_i), where i designates the shell involved. Fluorescence yield increases with atomic number and is greater than 95% for K X-rays of elements with Z > 78 (see Fig. 12.5). For a given element, the fluorescence yield decreases from the K series to the L and M series. The fluorescence yield can be approximated by Ref. [1],

$$\omega_i = \frac{Z^4}{A_i + Z^4}, \qquad (12.1)$$

where A_i is approximately 10^6 for the K shell and 10^8 for the L shell.

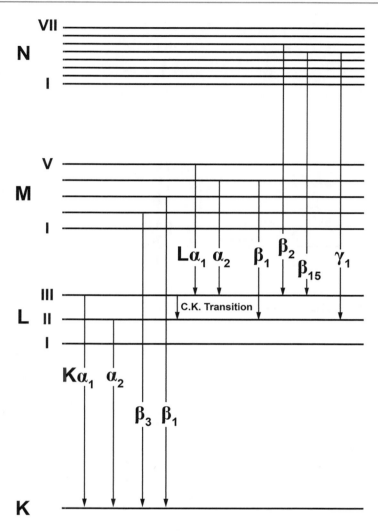

Fig. 12.2 Diagram of energy levels showing the atomic transitions that produce the major K and X-rays (*C.K.* Coster Krönig)

Table 12.1 Energies and relative intensities of the major K and L X-rays of uranium and plutonium

Line	Transition (Final–Initial)	Energies in keV[a] Uranium (%)[b]	Plutonium (%)[b]
$K_{\alpha 1}$	K–L_3	98.44 (100)	103.74 (100)
$K_{\alpha 2}$	K–L_2	94.65 (62.5)	99.52 (63.2)
$K_{\beta 1}$	K–M_3	111.30 (22.6)	117.23 (22.8)
$K_{\beta 3}$	K–M_2	110.42 (11.5)	116.24 (11.6)
$K_{\beta 2}$	K–$N_{2,3}$	114.33, 114.56 (8.7)	120.44, 120.70 (8.9)
$L_{\alpha 1}$	L_3–M_5	13.62 (100)[c]	14.28 (100)[c]
$L_{\alpha 2}$	L_3–M_4	13.44 (11.4)	14.08 (11.4)
$L_{\beta 2}$	L_3–N_5	16.43 (21)	17.26 (21.3)
L_{ℓ}	L_3–M_1	11.62 (16.7)	12.12 (7.2)
$L_{\beta 1}$	L_2–M_4	17.22 (100)[c]	18.29 (100)[c]
$L_{\gamma 1}$	L_2–N_4	20.17 (22.6)	21.42 (23)
$L_{\beta 4}$	L_1–M_2	16.58 (100)[c]	17.56 (100)[c]
$L_{\beta 3}$	L_1–M_3	17.45 (87)	18.54 (82)

[a]Energies from Ref. [6]
[b]Intensities relative to either $K_{\alpha 1}$, $L_{\alpha 1}$, and $L_{\beta 4}$ in percent, calculated from Ref. [6]
[c]Note for L lines, vacancies in L1, L2, L3 are not necessarily proportional. Different intensity ratios between shells will be observed depending on excitation conditions

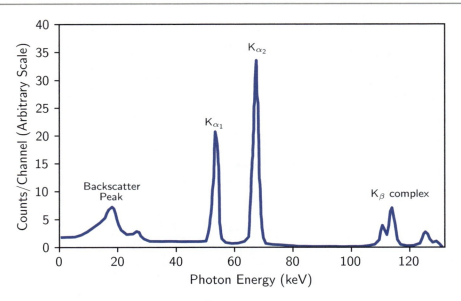

Fig. 12.3 K X-ray spectrum of uranium. The excitation source is ^{57}Co

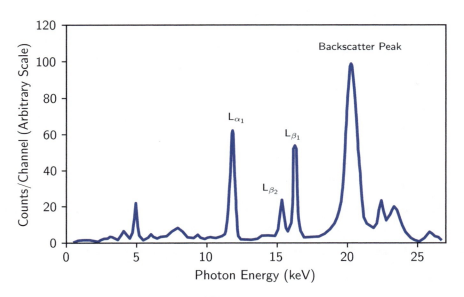

Fig. 12.4 L X-ray spectrum of uranium. The excitation source is ^{109}Cd

12.2.3 Photon Transmission

For a photon to eject an electron, the photon energy must be greater than or equal to the electron binding energy. For example, to ionize K electrons of plutonium, the energy of the excitation photon must be at least 121.82 keV.

The fraction of photons (F) that interact with the atomic electrons of a particular material is given by

$$F = 1 - e^{-\mu \rho x}, \tag{12.2}$$

where

μ = mass attenuation coefficient
ρ = density of sample
x = thickness of sample.

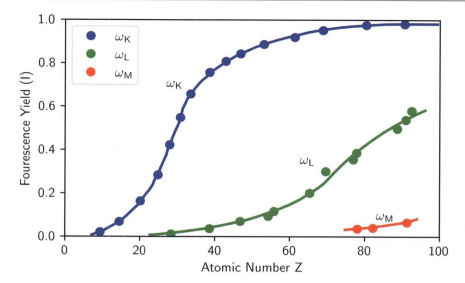

Fig. 12.5 Fluorescence yield for K, L, and M X-rays as a function of atomic number

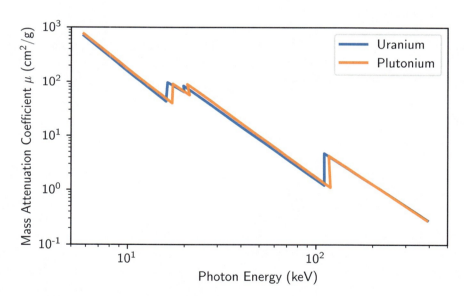

Fig. 12.6 Mass attenuation coefficient versus energy for uranium and plutonium

If one plots the mass attenuation coefficient versus photon energy for a given element, sharp discontinuities (known as *absorption edges*) are observed. Figure 12.6 shows the mass attenuation coefficient for uranium and plutonium. The edges indicate the sudden decrease in the photoelectric cross section for incident photon energies just below the binding energy of that electron state. The photoelectric interaction is the dominant process involved in photon-excited X-ray excitation.

Attenuation limits the sample size that can be analyzed by X-ray transmission techniques. Figure 12.7 shows the mean free path of 20, 100, and 400 keV photons in water and in a 50 g/L uranium solution. In general, transmission techniques are applicable for samples whose transmission path lengths are less than four or five mean free paths.

Equation 12.2 is useful when comparing K XRF and L XRF. For L XRF, μ is larger, and more of the excitation flux interacts with the sample. For K XRF, μ is smaller, and both the excitation photons and X-rays are attenuated less (relative to L XRF). This attenuation difference implies that L XRF is more sensitive (more X-rays produced per unit excitation flux and cross-sectional area) than K XRF. Conversely, K XRF allows greater flexibility regarding the choice of sample container and intervening absorbers.

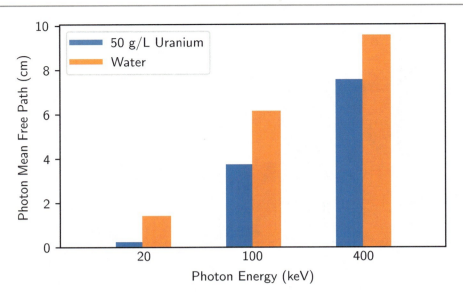

Fig. 12.7 Mean free path of 20, 100, and 400 keV photons in water ($\rho = 1$ g/cm^3) and in a 50 g/L uranium solution

12.2.4 Measurement Geometry

The choice of geometry is very important in an XRF system. Although photoelectric interactions of the excitation photons with analyte atoms are of primary interest, other intentions—particularly Compton backscatter interactions—must be considered. The energy of a Compton-scattered gamma ray is

$$E' = \frac{511}{\left(1 - \cos\phi + \frac{511}{E}\right)}, \tag{12.3}$$

where

E', E = scattered, incident photon energy in keV
ϕ = angle between incident and scattered photons.

(See Chap. 3, Sect. 3.3.2, and Ref. [8].)

The energy E' is a minimum when $\phi = 180°$, and photons that have scattered at or near this angle can produce a backscatter peak in the measured spectrum. For 122 keV photons from ^{57}Co (a suitable source for K XRF of uranium or plutonium), the backscatter peak is at 82.6 keV. If the scattering angle ϕ is 90°, E' is 98.5 keV, which is in the middle of the K X-ray spectrum from uranium and plutonium. If ^{57}Co is used as an excitation source, the measurement geometry should be arranged such that ϕ is close to 180° for most of the scattered gamma rays that reach the detector. This arrangement puts the backscatter peak and the Compton continuum of scattered photons below the characteristic X-rays and minimizes the background under the X-ray photopeaks (see Fig. 12.3). The annular source described later in the chapter provides this favorable geometry. For L X-rays, the geometry is not as critical because $E'(180°)$ is 20.3 keV for 22 keV silver X-rays from ^{109}Cd (a good L XRF source for uranium), and the backscatter peak is above the X-ray region of interest. Scattering materials near the detector must be carefully controlled to minimize the magnitude of the backscatter peak. Some investigators [9] use excitation sources that have energies much higher than the binding energy of interest, thereby minimizing the scattering effects in the spectral region of the induced X-rays. This approach requires higher-intensity excitation sources (by an order of magnitude or more) to produce sufficient X-ray activity due to diminishing ionization probability at energies far away from the absorption edge.

The detector must be shielded from the excitation source and other background radiation to reduce dead time and pileup losses. Detector collimation is usually necessary to limit the interference from unwanted sources. To stabilize the X-ray response, the relative positions of the source, sample, and detector must be fixed; often these components are physically connected. Figure 12.8 shows a possible geometry for a transmission-corrected XRF analysis.

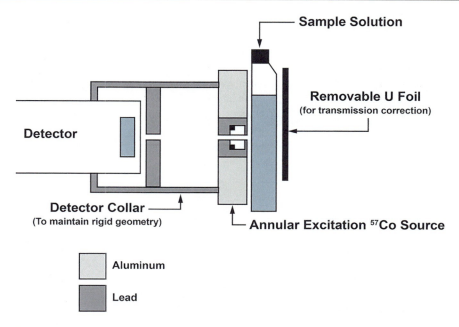

Fig. 12.8 Cross-sectional view of geometry for a transmission-corrected assay using an annular excitation source

12.3 Types of Sources

Two types of sources are commonly used: discrete gamma-ray or X-ray sources and continuous sources, such as X-ray generators. Each type has advantages and disadvantages. The selection of a suitable source involves consideration of type, energy, and strength. It is most efficient to choose a source whose energy is above but as close as possible to the absorption edge of interest. As shown by the graph of μ versus photon energy in Fig. 12.6, the value of the mass attenuation coefficient is greatest just above an absorption edge.

Cobalt-57 emits a gamma ray at 122 keV, an efficient energy for K-shell ionization of either uranium or plutonium. X-ray generators are available for K XRF of uranium and plutonium, but they are too bulky for portable applications. A good discrete source for L XRF of uranium and plutonium is ^{109}Cd, which emits silver K X-rays ($K_{\alpha 1}$ energy = 22 keV). X-ray generators are available that are small enough for portable applications that require photons in the 25 keV energy range.

Discrete line sources are small, extremely stable, and operationally simple, making them attractive for many XRF applications. Their major disadvantage is that they decay with time and require periodic replacement. (Two commonly used sources, ^{57}Co and ^{109}Cd, have half-lives of 272 days and 453 days, respectively.) Another disadvantage is that discrete sources cannot be turned off, causing transportation and handling difficulties. Because the source strength is often 1 mCi or greater, both personnel and detector must be carefully shielded. Table 12.2 lists some radionuclides that can be used for XRF analysis of uranium and plutonium. The geometry of the annular source shown in Fig. 12.9 is commonly used because it shields the detector from the excitation source and minimizes backscatter interference.

X-ray generators produce bremsstrahlung by thermionic emission of electrons from a filament. The electrons are accelerated through a high-voltage potential into a target of high atomic number. The maximum X-ray energy emitted by the generator is a function of the accelerating voltage, whereas the intensity is a function of the beam power (accelerating voltage × beam current). Because they require a high-voltage supply and a means of dissipating the heat produced in the target, X-ray generators can be bulky, especially for higher operating potentials. Small generators that operate below 70 keV and portable generators—with power ratings up to 50 W—are available that do not require elaborate cooling systems. For a given power rating, higher maximum operating voltage is achieved at the expense of lower available current.

Table 12.2 Excitation sources suitable for uranium and plutonium assay

Radionuclide	Half-life	Decay mode	Useful emissions Type	Energy (keV)
^{57}Co	270 d	Electron capture	Gamma rays	122
^{109}Cd	453 d	Electron capture	Ag K X-rays	22
^{75}Se	120 d	Electron capture	Gamma rays	121
^{144}Ce	285 d	Beta decay	Pr K X-rays	36
^{125}I	60 d	Electron capture	Te K X-rays	27
^{147}Pm-Al	2.6 yr	Beta decay	Continuum	12–45[a]

[a]End point of the bremsstrahlung spectrum

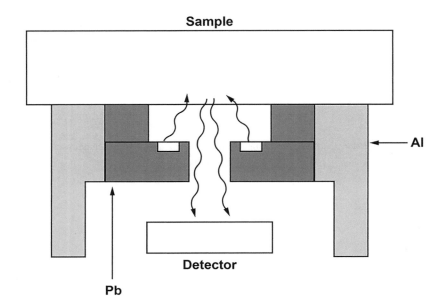

Fig. 12.9 Annular excitation source

The spectrum from an X-ray generator spans the energy range from the accelerating potential of the generator to the transmission cutoff of the X-ray window. The shape I(E) and total intensity (I) of this distribution is given by [5]),

$$I(E) \propto iZ(V-E)E$$
$$I \propto iZV^2, \tag{12.4}$$

where

i = tube current
V = operating voltage
Z = atomic number of target.

Figure 12.10 shows the output spectrum from an X-ray generator. In addition to the continuous spectrum, the characteristic X-rays of the target material are produced. These X-rays may cause an interference, which can be removed with filters. The filter chosen should have an absorption edge just below the energy to be attenuated. A more comprehensive discussion of X-ray generator spectrum shape and features may be found in Ref. [10].

X-ray generators can be switched on and off, and their energy distribution and intensity can be varied as desired. They typically provide a more intense source of photons than radionuclide sources (~10^{12} photons/s or greater); however, their flexibility is possible only at the expense of simplicity and compactness. Because an X-ray generator is an electrical device, system failures and maintenance problems are possible concerns. The assay precision is dictated by the stability of the X-ray

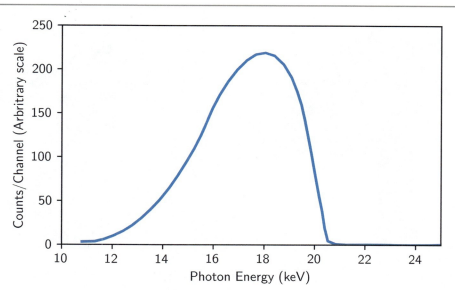

Fig. 12.10 Typical X-ray generator spectrum. The generator target is tungsten, and the operating potential is 20.4 kV

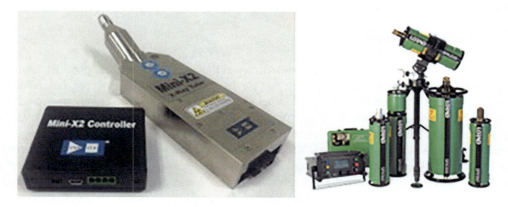

Fig. 12.11 Portable X-ray generators. (left) Mini-X2 from AMPTEK, up to 70 kV accelerating voltage with silver anode suitable for U/Pu L excitation [11]. (right) SPX range of portable X-ray generators from Spellman HV, up to 300 kV accelerating voltage for U/Pu K excitation [12]. Note the size and complexity scaling with energy range

tube. Modern generators exhibit less than 0.1% fluctuation for short-term stability and 0.2–0.3% for long-term stability. Figure 12.11 shows two different classes of portable X-ray generators suitable for actinide L-shell and K-shell excitation.

These standalone X-ray generators may be incorporated into commercial all-in-one XRF systems that integrate an X-ray detector and analysis software. Portable XRF systems are widely available from major manufacturers (Fig. 12.12) and are routinely used in the field, as well as the laboratory, for a wide range of material composition analysis. These units lack sufficient power and accuracy for high-precision assay of nuclear material concentration, but they are useful in situations that require triage or pre-screening of samples for more careful analysis with more precise instrumentation.

Other sources may be used for XRF. A secondary fluorescent source uses a primary photon source to excite the characteristic X-rays of a target, and the target X-rays are used to excite the sample to be analyzed. The primary excitation source can be discrete or continuous. Depending on the target material, this scheme can produce a great variety of monoenergetic excitation photons. The major drawback is the need for a high-intensity primary source. If the primary source is a radionuclide, radiation safety may be an important concern. It is possible to make a bremsstrahlung source using a radionuclide rather than an X-ray generator. Such a source consists of a beta-decaying isotope mixed with a target material (for example, ^{147}Pm-Al, with aluminum being the target material).

Fig. 12.12 Portable integrated XRF measurement systems with integrated X-ray generator. (left) Bruker TRACER 5G [13]. (right) Thermo Scientific Niton XL5 [14]

12.4 Correction for Sample Attenuation

12.4.1 Effects of Sample Attenuation

As in passive gamma-ray assays, sample attenuation is a fundamental limitation to the accuracy of XRF analysis. Attenuation corrections are required for the X-rays leaving the sample and for the gamma rays or X-rays from the excitation source. XRF analysis is unsuitable for large, solid samples because the attenuation is too large to be accurately treated with any correction procedure. For example, the mean free path of 122 keV gamma rays in uranium metal is approximately 0.013 cm. The low penetrability of this radiation means that XRF can be accurately used to assay solids only if the sample is smooth and homogenous. This limitation is even more true for L XRF using 22 keV photons. XRF can be used to accurately assay dilute uranium solutions because the mean free path of photons in water is approximately 6.4 cm at 122 keV and 1.7 cm at 22 keV. Because the excitation source energy is above the absorption edge and the energies of the characteristic X-rays are just below the absorption edge, the attenuation of the excitation radiation is higher and determines the range of sample thickness that can be accurately assayed. Figure 12.13 plots the mean free path of 122 keV gamma rays as a function of uranium concentration (uranyl nitrate in 4-M nitric acid).

Attenuation considerations also affect the choice of sample containers. Because the K X-rays of uranium and plutonium are in the 100 keV range, low- to medium-atomic-number metal containers such as aluminum and steel can be tolerated, and K XRF can be applied to inline measurements. However, L X-rays are severely attenuated by even thin metal containers and can be measured only in low-Z containers, such as plastic or glass.

12.5 General Assay Equation

For quantitative analysis, the X-ray emission rate must be related to the element concentration and corrected for rate losses. The desired relation, as presented in Chap. 7, Sect. 7.5, is

$$\rho = \frac{RR \times CF_{rateloss} \times CF_{atten}}{K}, \quad (12.5)$$

where

ρ = element concentration
RR = raw rate of X-ray detection

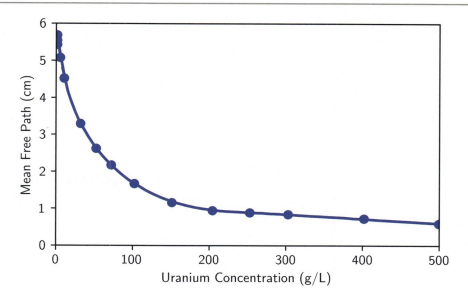

Fig. 12.13 The 122 keV photon mean free path versus uranium concentration (uranyl nitrate in 4-M nitric acid)

$CF_{rateloss}$ = correction factor for rate-related losses
CF_{atten} = correction factor for attenuation
K = calibration constant.

$CF_{rateloss}$ can be determined using either pulser or radionuclide normalization (Chap. 7, Sect. 7.5). The attenuation correction has two parts: one for excitation radiation and one for fluoresced X-rays.

Consider a far-field measurement geometry where the sample is approximated by a slab and the excitation source is monoenergetic (Fig. 12.14). The flux (F_γ) of excitation photons at a depth (x) in the sample is given by

$$F_\gamma = I_\gamma \ exp(-\mu^\gamma X / cos \phi). \tag{12.6}$$

The variables in Eqs. 12.6, 12.7, 12.8, 12.9 and 12.10 are defined in Table 12.3; note that μ here denotes linear attenuation coefficient rather than mass attenuation coefficient, as shown in Fig. 12.6. The number of excitation photons that interact in the volume dx and create a $K_{\alpha 1}$ X-ray is

$$F_x \ dx = F_\gamma \ \tau \rho \omega B \ \frac{dx}{cos \phi}. \tag{12.7}$$

The fluoresced X-rays are attenuated in the sample according to

$$F_x(out) = F_x \ exp\left(\frac{-\mu^x X}{cos \theta}\right). \tag{12.8}$$

Note that as the fluoresced X-ray energies are strictly less than the energy of the corresponding absorption edge (e.g., K X-rays lie below the K absorption edge), they are in a local minimum of the material attenuation coefficient, and they have a relatively long path length compared with X-rays farther away from the edge (either higher or lower in energy). As a result, the path length of the excitation photons that enter the sample may differ significantly from that of the fluoresced X-rays. Combining and integrating Eqs. 12.6, 12.7, 12.8, 12.9 and 12.8 yield the following expression or the X-ray rate at the detector surface:

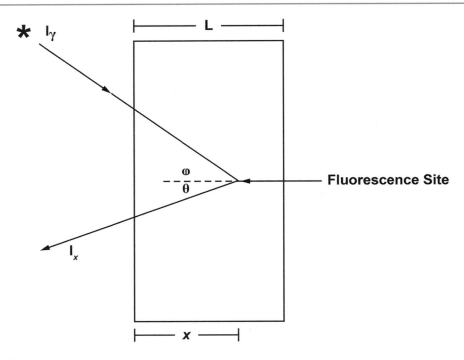

Fig. 12.14 General XRF slab geometry

Table 12.3 Variables in Eqs. 12.6, 12.7, 12.8, 12.9 and 12.10

I_0	Excitation flux at sample surface
τ	Photoelectric cross section, K shell, energy
ρ	Concentration of given element
ω	K fluorescence yield
B	Branching ratio for $K_{\alpha 1}$
Ω	Detector solid angle
$\mu^\gamma = \sum \mu_i^\gamma \rho_i$	Linear attenuation coefficient, γ energy, element i
$\mu^x = \sum \mu_i^x \rho_i$	Linear attenuation coefficient, x energy, element i
ϕ	Incident angle of excitation
θ	Exiting angle of X-ray
L	Slab thickness

$$I_x = \frac{I_\gamma \, \tau \rho \omega B \Omega}{4\pi[(\cos\theta/\cos\phi)\mu^\gamma + \mu^x]} \left\{ 1 - exp\left[-\left(\frac{\mu^\gamma}{\cos\phi} + \frac{\mu^x}{\cos\theta}\right)L \right] \right\}. \tag{12.9}$$

The factor $(\Omega/4\pi)\cos\phi/\cos\theta$ has been added for normalization. If an X-ray generator is used as the excitation source, Eq. 12.9 must be integrated from the absorption edge to the maximum energy of the generator.

When the sample is infinitely thick for the radiation of interest, Eq. 12.9 becomes

$$I_x = \frac{I_\gamma \, \tau \rho \omega B \Omega}{4\pi[(\cos\theta/\cos\phi)\mu^\gamma + \mu^x]}. \tag{12.10}$$

This equation is similar to that of the enrichment meter (see Chap. 8, Sect. 8.3). The result is very important for XRF analysis because it implies that the X-ray rate is directly proportional to the concentration of the fluoresced element.

In plutonium and highly enriched uranium materials, the self-excitation of X-rays by the passive gamma rays can complicate the assay. For mixed U/Pu materials, the dominant signals are passive X-rays from the alpha decay of plutonium. When the excitation source can fluoresce both plutonium and uranium (as can ^{57}Co and ^{109}Cd), additional uranium fluorescence is caused by the plutonium X-rays. A separate passive count is usually required to correct for this interference.

12.5.1 Attenuation Correction Methods

The most effective XRF methods account for sample attenuation. The simplest approach uses calibration curves derived from chemically similar standards. The method is effective only if the standards are well characterized, match the samples chemically, and span the concentration range to be assayed in sufficient numbers to define the calibration curve. Changes in matrix composition may require recalibration with new standards.

A procedure that is less sensitive to matrix variation is the transmission-corrected assay [15–17] in which a transmission measurement is made for each sample to correct for attenuation. Consider the attenuation correction factor for the situation shown in Fig. 12.15 (assume that $\theta = 0$). The expression for CF_{atten} has the functional form for a slab that was discussed in Chap. 6:

$$CF_{atten} = \frac{-\ln(\alpha)}{1-\alpha}, \qquad (12.11)$$

where

$$\alpha = exp\left[-\left(\frac{\mu^\gamma}{\cos\phi} + \mu^x\right)L\right].$$

A measurement of the transmissions of the excitation and the fluoresced X-rays can be used to determine α. For this method, a foil of the element being measured is placed behind the sample and the induced X-ray signal is measured with and without the sample. An additional measurement (see Fig. 12.15) is made with the sample only (no foil), and α is computed from

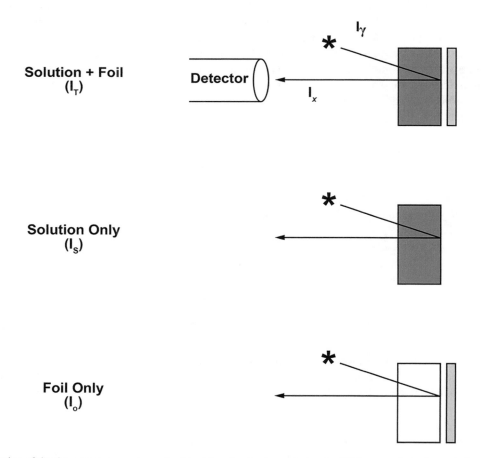

Fig. 12.15 Explanation of the three measurements required to determine the transmission for XRF assay of uranium solutions

$$\alpha = \left(\frac{I_T - I_S}{I_0}\right), \quad (12.12)$$

where

I_T = fluoresced X-ray intensity with foil plus sample
I_S = fluoresced X-ray intensity with sample only
I_0 = fluoresced X-ray intensity with foil only.

This measurement includes the attenuation of the excitation source and of the induced X-ray signal. Although there are advantages to using the same element in the transmission foil as that being assayed, other elements can be used if their characteristic X-rays are sufficiently close to those of the assay element. For example, thorium metal has been used successfully for the measurement of uranium solutions.

A suitable number of standards is needed to evaluate the calibration constant (K) in Eq. 10.5. Eqs. 12.11 and 12.12 are exact only for a far-field geometry, and most XRF measurements are made in a near-field geometry; therefore, even with rate and sample attenuation corrections, it is important to use several standards to evaluate the calibration constant K.

12.6 Applications and Instrumentation

Instrumentation used in XRF analysis is similar to that of other gamma-ray assay systems: detector, associated electronics, multichannel analyzer, and excitation source. This instrumentation is discussed in detail in Chap. 5.

XRF analysis has been used in analytical chemistry laboratories for many years. In most cases, an X-ray generator is used as an excitation source rather than a radionuclide. Low- to intermediate-Z elements are measured, and sample preparation is a key factor in the analysis. Many techniques require that the sample be homogenized and pressed before analysis. When the sample can be modified to optimize the assay, XRF analysis is very sensitive. Short count times (<1000 s) can yield accurate and precise data, with sensitivities in the nanogram range [1]. Complete XRF systems are available commercially.

Several XRF measurement techniques are used for materials that contain uranium or plutonium. John et al. [18] used a ^{57}Co source to excite uranium X-rays in solutions and simultaneously observed the 185.7 keV gamma ray from ^{235}U as a measure of enrichment. The ratio of the fluoresced X-ray emission to the 185.7 keV gamma-ray intensity was found to be independent of uranium concentration in the 8–20 g/L range for enrichments of 0.4%–4.5% ^{235}U. Accuracies of better than 1% were reported.

Rowson and Hontzeas [19] proposed a Compton-scattering-based correction for sample attenuation to measure uranium ores. An annular 50 mCi ^{241}Am source was used to excite characteristic X-rays from a molybdenum foil (~17.4 keV), which can excite only the L_{III} subshell in uranium, thereby considerably simplifying the L X-ray spectrum. Matrix corrections were determined from the ratio of the molybdenum K_α backscatter to the uranium L_α X-rays. Canada and Hsue [20] give a good theoretical description of this method and suggest an improvement that involves additional ratios using the K_β or L_β lines to further minimize matrix effects.

Baba and Muto [21] used an X-ray generator to excite the L X-ray spectrum of uranium- and plutonium-bearing solutions. An internal standard was used to determine the matrix attenuation correction. A known and constant amount of lead nitrate was added to all solutions, and the uranium or plutonium L_α X-ray activity was normalized to the rate of the lead L_α X-ray. A linear calibration (to within 1% for uranium and 2% for plutonium) was obtained for 200 s measurements. The solution concentration range was 0.1–200 g/L of uranium and up to 50 g/L of plutonium. Mixed U/Pu solutions were also measured.

Karamanova [22] investigated the use of beta-particle-induced XRF in addition to gamma-ray excitation for K XRF analysis of uranium and mixed U/Th oxides. A ^{57}Co source was used for gamma-ray excitation, and a ^{90}Sr ^{90}Y source was used for beta-particle-induced fluorescence. Because the attenuation cross sections of the matrix and the heavy element are similar for beta-particle excitation, the volume of sample is essentially constant, and the net X-ray signal is proportional to the concentration of the element being assayed. Samples with uranium concentrations of 0.5%–88% were measured with precisions of 0.1%.

Several investigators have measured reprocessing plant solutions. These solutions can contain fission products and have high U/Pu ratios, making them very difficult to assay by passive techniques. They require either extensive chemical separations or a sensitive technique, such as XRF; Pickles and Cate [23] employed XRF using X-ray generator. Samples

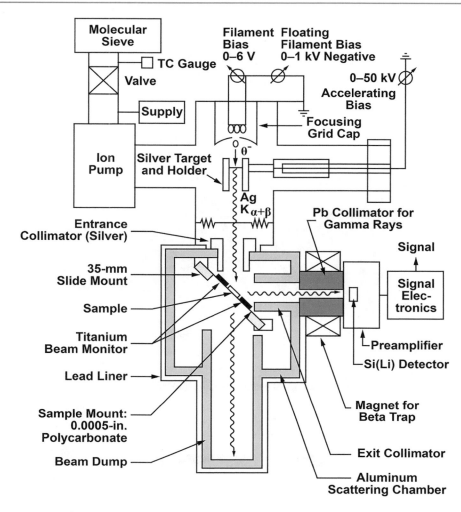

Fig. 12.16 Schematic drawing of system used by Pickles and Cate [23]

with U/Pu ratios of up to 400 and fission product activities of 2 Ci/g were analyzed with a precision of 1% and an accuracy of 2% in the mass range of 1–58 µg using a 10 min count. The samples were evaporated onto a thin polycarbonate film to minimize sample attenuation. Figure 12.16 shows the instrumental configuration. The sample chamber contained titanium sheets on either side of the sample mount. X-rays from the titanium provided a rate-loss correction. A magnetic beta-particle trap and a lead collimator were employed to reduce the passive signal at the detector.

Camp et al. [24, 25] use a ^{57}Co excitation source to fluoresce K X-rays from uranium and plutonium in product streams at reprocessing plants. Total heavy element concentrations of 1–200 g/L are assayed using a nonlinear polynomial calibration. The self-attenuation correction uses the incoherently scattered 122 keV gamma rays from ^{57}Co [24] or an actual transmission measurement using the 122 keV gamma ray [25]. In samples that contain both uranium and plutonium, a passive count is made to correct for passive X-ray emission.

Andrew et al. [26] investigated the feasibility of measuring uranium and U/Th solutions with uranium concentrations of 10–540 g/L. For solutions that contain both uranium and thorium, errors in the U/Th ratio were 0.4%. Uncertainties in concentration measurements were 0.5% for single-element solutions and 1% for mixed solutions. This work was extended to U/Pu solutions from reprocessing plants [27, 28]. Uranium and plutonium solutions in the range of 1–10 g/L with a fission product activity of 100 µCi/mL were measured by tube-excited K XRF. The data analysis was similar to that used for U/Th solutions, and the authors suggested combining K XRF and K-edge densitometry to obtain absolute element concentrations.

Ottmar et al. [29, 30] investigated the combination of K XRF and K-edge densitometry using an X-ray generator. The two techniques are complementary and produce a measurement system with a wide dynamic range. Light-water-reactor dissolver solutions with activities of ~100 Ci/L and U/Pu ratios of ~100 have been accurately measured with this system. The major

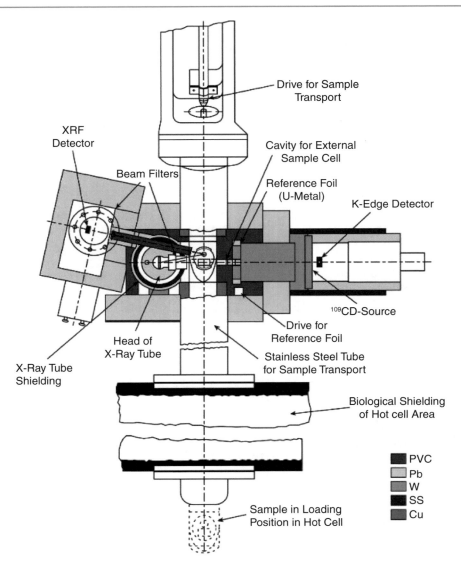

Fig. 12.17 Schematic drawing of hybrid K-edge/K XRF system (Ref. [29]; courtesy of H. Ottmar)

component, uranium (~200 g/L), is determined using absorption-edge densitometry, whereas the U/Pu ratio is determined by XRF. Precisions of 0.25% for the uranium concentration and 1% for the U/Pu ratio are obtained in 1000 s count times. Two sample cells are employed: a 2 cm glass cuvette (whose dimensions are known to ±2 μm) for the densitometry measurement and a 1 cm diameter polyethylene vial for the XRF measurement. The XRF measurement is made at a back angle of ~157° to maximize the signal-to-background ratio. Figure 12.17 shows this hybrid system.

Lambert et al. [31] employed secondary-excitation L XRF to measure the U/Pu ratio in mixed-oxide fuel pellets. A pellet with 25% PuO_2 and 75% UO_2 gave a precision of ~0.5% in a 3 min count time. The desired X-rays were excited using selectable secondary target foils (rhodium). The method requires good sample homogeneity because the portion of sample analyzed (~30 μm depth of analysis) is relatively small. Misra et al. [32] have studied XRF methods for U/Th mixed-oxide solids using both tube and radionuclide X-ray sources, demonstrating that—although radionuclide sources may give around 2X higher precision than tube sources due to reduced background (0.8% versus 2% precision)—tube sources are preferred for routine material analysis because of much shorter measurement times owing to higher source brightness.

With the development of the silicon drift detector (SDD), investigators have developed L XRF that uses SDDs to take advantage of their high-throughput capabilities at low energies compared with Si(Li) detectors. Park et al. [33] have commissioned a hybrid L-edge densitometer/L XRF instrument with SDDs for nuclear material assay with the capability to measure materials from 0.05 to >0.2 g/cm^3 within international target value limits in 10 min count times.

XRF methods are also suitable for assay of trace actinides in waste streams, often in the presence of conflicting contaminants and deployed in the field. Matsuyama et al. [34] developed methods using a portable total reflection X-ray fluorescence (TXRF) spectrometer from OURSTEX Corporation to characterize trace uranium and thorium contaminants in contaminated effluent from the Fukushima Daiichi nuclear power plant disaster. This portable unit uses a tungsten target with a maximum tube voltage and current of 40 kV and 0.2 mA, respectively, yielding a power of 8 W. These energies and powers are well matched to analysis of actinide L lines between 10 and 20 keV. An SDD with 7 mm^2 active area enabled reasonable measurements in 180 s acquisition time. This portable system was benchmarked against a more complex benchtop TXRF instrument (NANOHUNTER-II from Rigaku Co.), which uses a monochromator to select a narrow excitation energy to decrease the reflected background. This instrument has a molybdenum target and uses the molybdenum K_α line at 17.17 keV for efficient excitation of the uranium L lines. Matsuyama et al. achieved a minimum detection limit (MDL) for uranium of 0.22–0.25 ppm with the portable TXRF and 1.4 ppb with the benchtop TXRF, demonstrating efficacy for rapid uranium concentration measurements in effluent without the need to evaporate solutions and assay using alpha spectroscopy.

The International Atomic Energy Agency is also invested in the development of XRF for safeguards NDA. Examples include recent work by Docenko et al. [35] to implement online uranium ore assay using a commercial XRF analyzer and Balaji Rao et al. [36], who developed an online process stream XRF system for concentration measurements. Using the Bruker P2 Con-X XRF with a Si SDD, Docenko et al. achieved a MDL of 60–200 ppm in a 5 min measurement time in solid ore samples—with a dynamic range up to 80% concentration uranium. This result is sensitive to typical uranium and thorium concentrations of order 100 ppm found in rutile sands, as well as 0.1–4% typical uranium concentrations in ores. The methods developed by Balaji Rao et al. targeted concentrations of 0.1–10 g/L uranium using uranium L lines—with a demonstrated residual of 1–5%. This work used a portable XRF (ALPHA-4000 from Innov-X systems) with a Ta anode and a maximum operating range of 40 kV and 80 μA voltage and current, detected with a silicon PiN diode detector. Both of these studies illustrate the range of applications of XRF to analysis of nuclear fuel cycle process products, yielding accurate and timely concentration measurements for process verification.

References

1. E.P. Bertin, *Principles and Practice of X-Ray Spectrometric Analysis* (Plenum Press, New York, 1975)
2. R. Jenkins, R.W. Gould, D. Gedcke, *Quantitative X-Ray Spectrometry* (Marcel Dekker, Inc, New York, 1981)
3. R.E. Van Grieken, A. Markowicz, *Handbook of X-Ray Spectrometry*, 2nd edn. (Marcel Dekker Inc., New York, 1993)
4. R. Tertian, F. Claisse, *Principles of Quantitative X-Ray Fluorescence Analysis* (Heyden & Son, Inc, Philadelphia, 1982)
5. G.R. Lachance, F. Claisse, *Quantitative X-Ray Fluorescence Analysis, Theory and Application* (Wiley, New York, 1995)
6. G. Zschornack, *Handbook of X-Ray Data* (Springer, Berlin, Heidelberg, 2007)
7. A.C. Thompson, *X-Ray Data Booklet*, 3rd edn. (Lawrence Berkeley National Laboratory, University of California, 2009) https://xdb.lbl.gov/. Accessed Oct 2021
8. G.F. Knoll, *Radiation Detection and Measurement*, 2nd edn. (Wiley, New York, 1979), p. 310
9. P. Martinelli, Possibilities of plutonium analysis by means of X-ray fluorescence with an iridium-192 radioactive source. Analysis **8**(10), 499–504 (1980)
10. N.A. Dyson, *X-Rays in Atomic and Nuclear Physics*, 2nd edn. (Cambridge University Press, Cambridge, 1990)
11. AMPTEK, AMPTEKMini-X2 X-Ray Tube: Specifications. https://www.amptek.com/products/mini-x-x-ray-tube. Accessed Feb 2022
12. Spellman High Voltage Electronics Corporation, SPX Portable X-Ray NDT Systems, 128121-001 REV. F. https://www.spellmanhv.com/-/media/en/Products/SPX.pdf. Accessed Feb 2022
13. Bruker, Tracer 5 Portable XRF Analyzer, Brochure. https://www.bruker.com/en/products-and-solutions/elemental-analyzers/handheld-xrf-spectrometers/TRACER-5/_jcr_content/root/sections/highlights/sectionpar/twocolumns/contentpar-1/calltoaction_copy.download-asset.pdf/secondaryButton/TRACER-5_Brochure.pdf. Accessed Feb 2022
14. ThermoFisher Scientific, Niton™ XL5 Plus Handheld XRF Analyzer. Spec Sheet, https://www.thermofisher.com/document-connect/document-connect.html?url=https://assets.thermofisher.com/TFS-Assets%2FCAD%2FDatasheets%2Fniton-xl5-plus-spec-sheet.pdf. Accessed March 2022
15. R. Strittmatter, M. Baker, P. Russo, Uranium solution assay by transmission-corrected X-ray fluorescence, in *Nuclear Safeguards Research and Development Program Status Report, May–August 1978*, Los Alamos Scientific Laboratory report LA-7616-PR, ed. by S.D. Gardner, (1979), pp. 23–24
16. T.R. Canada, D.C. Camp, W.D. Ruhter, Single-Energy Transmission-Corrected Energy-Dispersive XRF for SNM-Bearing Solutions. Los Alamos National Laboratory report LA-UR-82-557 (1982)
17. P. Russo, M.P. Baker, T.R. Canada, Uranium-Plutonium Solution Assay by Transmission-Corrected X-Ray Fluorescence, in *Nuclear Safeguards Research and Development Program Status Report, September–December 1977*, ed. by J.L. Sapir, Compiler, Los Alamos Scientific Laboratory report LA-7211-PR (1978), pp. 22–28
18. J. John, F. Sebesta, J. Sedlacek, Determination of uranium isotopic composition in aqueous solutions by combined gamma spectrometry and X-ray fluorescence. J. Radioanal. Chem. **78**(2), 367–374 (1983)

19. J.W. Rowson, S.A. Hontzeas, Radioisotopic X-ray analysis of uranium ores using Compton scattering for matrix correction. Can. J. Spectrosc. **22**(1), 24–30 (1977)
20. T.R. Canada, S.-T. Hsue, A note on the assay of special nuclear materials in solution by X-ray fluorescence. Nucl. Mater. Manage. **XI**(2), 91 (1982)
21. Y. Baba, H. Muto, Determination of uranium and plutonium in solution by energy-dispersive X-ray fluorescence analysis. Bunseki Kagaku **32**, T99–T104 (1983)
22. J. Karamanova, Development of an Express NDA Technique (Using Radioisotopic Sources) for the Concentration Measurements of Nuclear Materials, 1 November 1974–30 June 1977. International Atomic Energy Agency report IAEA-R-1557R (1977)
23. W.L. Pickles, J.L. Cate, Jr., Quantitative Nondispersive X-Ray Fluorescence Analysis of Highly Radioactive Samples for Uranium and Plutonium Concentrations. Lawrence Livermore Laboratory report UCRL-7417 (1973)
24. D.C. Camp, W.D. Ruhter, Nondestructive, energy-dispersive X-ray fluorescence analysis of product stream concentrations from reprocessed nuclear fuels, in *Proceedings of the American Nuclear Society Topical Conference on Measurement Technology for Safeguards and Materials Control*, Kiawah Island, SC, November 26–28, 1979 (National Bureau of Standards, Washington, DC, 1980), p. 584
25. D.C. Camp, W.D. Ruhter, K.W. MacMurdo, Determination of actinide process- and product-stream concentrations off-line or at-line by energy-dispersive X-ray fluorescence analysis, in *Proceedings of the Third Annual Symposium on Safeguards and Nuclear Materials Management*, Karlsruhe, Federal Republic of Germany, May 6–8, 1981 (European Safeguards Research and Development Association, Brussels, 1981), p. 155
26. G. Andrew, B.L. Taylor, B. Metcalfe, Estimation of Special Nuclear Materials in Solution by K X-Ray Fluorescence and Absorption Edge Densitometry. United Kingdom Atomic Energy Authority report AERE-R9707 (1980)
27. G. Andrew, B.L. Taylor, The Measurement of Pu and U in Reprocessing Plant Solutions by Tube-Excited K X-Ray Fluorescence. United Kingdom Atomic Energy Authority report AERE-R9864 (1980)
28. G. Andrew, B.L. Taylor, The Feasibility of Using K-XRF for the On-Line Measurement of Pu/U Ratios of Highly Active Dissolver Solutions. United Kingdom Atomic Energy Authority report AERE-M3134 (1980)
29. H. Ottmar, H. Eberle, P. Matussek, I. Michel-Piper, Qualification of K-Absorption Edge Densitometry for Applications in International Safeguards. International Atomic Energy report IAEA-SM-260/34 (1982)
30. H. Ottmar, H. Eberle, P. Matussek, I. Michel-Piper, How to Simplify the Analytics for Input-Output Accountability Measurements in a Reprocessing Plant. Kernforschungszentrum Karlsruhe report KfK 4012 (1986)
31. M.C. Lambert, M.W. Goheen, M.W. Urie, N. Wynhoff, An Automated X-Ray Spectrometer for Mixed-Oxide Pellets. Hanford Engineering Development Laboratory report HEDL-SA1492 (1978)
32. N.L. Misra, S.S. Kumar, S. Dhara, A.K. Singh, G.S. Lodha, S.K. Aggarwal, A comparative study on determination of uranium and thorium in their mixed oxides by EDXRF using tube and radioisotope X-ray sources. X-Ray Spectrom. **40**, 379–384 (2011)
33. S. Park, S. Joung, J. Park, Nuclear fuel assay through analysis of uranium L-shell by hybrid L-edge/XRF densitometer using a surrogate material. EPJ Web Conf **170**, 08007 (2018)
34. T. Matsuyama, Y. Izumoto, K. Ishii, Y. Sakai, H. Yoshii, Development of methods to evaluate several levels of uranium concentrations in drainage water using total reflection X-ray fluorescence technique. Front. Chem. **7**, 152 (2019)
35. D. Docenko, V. Gostilo, A. Sokolov, A. Rozite, *On-Line Measurement of Uranium in Ores Using XRF Analyzer P2 Con-X," IAEA-TECDOC-CD--1739* (International Atomic Energy Agency, Vienna, 2014)
36. Y. Balaji Rao, B.V.V. Ramana, P. Gayathri Raghavan, R.B. Yadav, Determination of uranium in process stream solutions from uranium extraction plant employing energy dispersive X-ray fluorescence spectrometry. J. Radioanal. Nucl. Chem. **294**, 371–376 (2012)

Open Access This chapter is licensed under the terms of the Creative Commons Attribution 4.0 International License (http://creativecommons.org/licenses/by/4.0/), which permits use, sharing, adaptation, distribution and reproduction in any medium or format, as long as you give appropriate credit to the original author(s) and the source, provide a link to the Creative Commons license and indicate if changes were made.

The images or other third party material in this chapter are included in the chapter's Creative Commons license, unless indicated otherwise in a credit line to the material. If material is not included in the chapter's Creative Commons license and your intended use is not permitted by statutory regulation or exceeds the permitted use, you will need to obtain permission directly from the copyright holder.

The Origin of Neutron Radiation

M. T. Swinhoe and N. Ensslin

13.1 Introduction

The nuclear materials that are accounted for in the nuclear fuel cycle emit neutrons as well as gamma rays. For most isotopes, the neutron emission rate is very low compared with the gamma-ray emission rate. For other isotopes, the neutron emission rate is high enough to provide an easily measurable signal. If the sample of interest is too dense to permit the escape of gamma rays from its interior, then assay by passive neutron detection may be the preferred technique.

Neutrons are emitted from nuclear materials with a wide spectrum of energies. As neutrons travel through matter, they interact and change their energy in a complex manner (see Chap. 14). However, neutron detectors (see Chap. 15) do not usually preserve information about the energy of the detected neutrons. Consequently, neutron assay consists of counting the number of emitted neutrons without knowing their specific energy. (This result is in sharp contrast with gamma-ray assay, where gamma rays of discrete energy are emitted by specific radioactive isotopes.) How then can the assayist obtain a neutron signal that is proportional to the quantity of the isotope to be measured?

This chapter describes the production of neutrons by spontaneous fission, by neutron-induced fission, and by reactions with alpha particles or photons. In many cases, these processes yield neutrons with unusually low or high emission rates, distinctive time distributions, or markedly different energy spectra. This information can be used to obtain quantitative assays of a particular isotope if the sample's isotopic composition is known and only a few isotopes are present.

The discussion of neutron radiation in this chapter emphasizes features that can be exploited by the assayist. Chapter 16 focus on Singles-neutron-counting techniques that exploit high emission rates or unusual energy spectra, Chap. 17 on coincidence counting techniques, and Chap. 18 on multiplicity counting. Chapters 19 and 20 highlight passive and active neutron instrumentation and analysis techniques.

13.2 Spontaneous and Induced Nuclear Fission

The spontaneous fission of uranium, plutonium, or other heavy elements is an important source of neutrons. An understanding of this complex process can be aided by visualizing the nucleus as a liquid drop (Fig. 13.1). The strong, short-range nuclear forces act like a surface tension to hold the drop together against the electrostatic repulsion of the protons. In the heaviest elements, the repulsive forces are so strong that the liquid drop is barely held together. A small but finite probability exists that the drop will deform into two droplets connected by a narrow neck (saddle point). The two droplets may spontaneously separate (scission) into two fragments. Within 10^{-13} s of scission, each of the two fragments emits multiple prompt neutrons and gamma rays. The fragments are usually unequal in size, with mass distributions centered near atomic numbers 100 and

---

Los Alamos National Laboratory strongly supports academic freedom and a researcher's right to publish; as an institution, however, the Laboratory does not endorse the viewpoint of a publication or guarantee its technical correctness.

M. T. Swinhoe (✉) · N. Ensslin
Los Alamos National Laboratory, Los Alamos, NM, USA
e-mail: swinhoe@lanl.gov

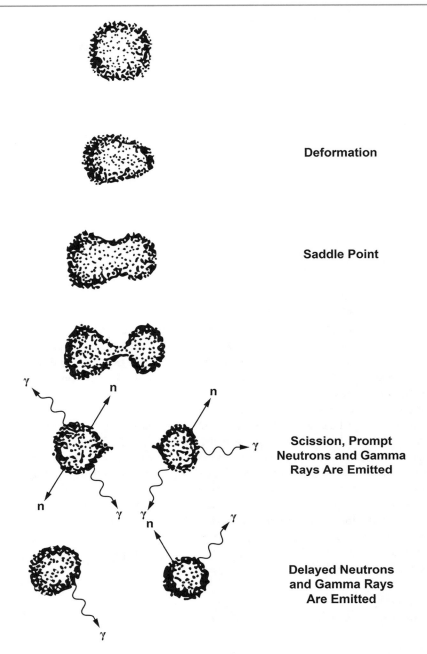

Fig. 13.1 Spontaneous fission of a nucleus represented as the breakup of a liquid drop

140 (see Fig. 21.1 in Chap. 21). These fission fragments carry away the majority of the energy released in fission (typically 170 MeV) in the form of kinetic energy. Also, within milliseconds or seconds, many of the fragments decay by beta-particle emission into other isotopes that may emit delayed neutrons or gamma rays.

Spontaneous fission is a quantum mechanical process that involves penetration of a potential barrier. The height of the barrier, and therefore the fission rate, is a very sensitive function of atomic number Z and atomic mass A. The fission yields of some heavy isotopes are summarized in Table 13.1 [1–8]. For thorium, uranium, and plutonium, the fission rate is low compared with the rate of decay by alpha-particle emission, which dominates the total half-life. For californium and even heavier elements, the fission rate can approach the alpha decay rate. The fission yield of ^{240}Pu—1020 n/s − g [4, 5]—is the most important single yield for passive neutron assay of plutonium because ^{240}Pu is usually the major neutron-emitting plutonium isotope present.

13 The Origin of Neutron Radiation

Table 13.1 Spontaneous fission neutron yields

Isotope	Number of protons Z	Number of neutrons N	Total half-life [1]	Spontaneous fission half-life [6] (year)	Spontaneous fission yield [2] (n/s − g)	Spontaneous fission multiplicity [2, 7, 8] ν	Induced thermal fission multiplicity [7] ν
^{232}Th	90	142	1.40×10^{10} year	1.2×10^{21}	1.01×10^{-7}	2.13	1.9
^{232}U	92	140	68.9 year	$>6.8 \times 10^{15}$	1.43×10^{-2}	1.71	3.13
^{233}U	92	141	1.59×10^{5} year	$>2.7 \times 10^{17}$	3.70×10^{-4}	1.76	2.4
^{234}U	92	142	2.45×10^{5} year	1.5×10^{16}	6.82×10^{-3}	1.81	2.4
^{235}U	92	143	7.04×10^{8} year	1×10^{19}	1.05×10^{-5}	1.86	2.41
^{236}U	92	144	2.34×10^{7} year	2.5×10^{16}	4.24×10^{-3}	1.89	2.2
^{238}U	92	146	4.47×10^{9} year	8.20×10^{15}	1.34×10^{-2}	1.98	2.3
^{237}Np	93	144	2.14×10^{6} year	$>1.0 \times 10^{18}$	1.14×10^{-4}	2.05	2.70
^{238}Pu	94	144	87.74 year	4.75×10^{10}	2.56×10^{3}	2.19	2.9
^{239}Pu	94	145	2.41×10^{4} year	8×10^{15}	1.49×10^{-2}	2.16	2.88
^{240}Pu	94	146	6.56×10^{3} year	1.14×10^{11}	1.04×10^{3a}	2.154^{a}	2.8
^{241}Pu	94	147	14.4 year	$<6 \times 10^{16}$	2.06×10^{-3}	2.25	2.8
^{242}Pu	94	148	3.75×10^{5} year	6.77×10^{10}	1.74×10^{3}	2.149	2.81
^{241}Am	95	146	432.7 year	1.2×10^{14}	1.14	2.5	3.09
^{242}Cm	96	146	163 days	7×10^{6}	1.98×10^{7}	2.54	3.44
^{244}Cm	96	148	18.1 year	1.32×10^{7}	1.11×10^{7}	2.71	3.46
^{249}Bk	97	152	320 days	1.8×10^{9}	1.0×10^{5}	3.40	3.7
^{252}Cf	98	154	2.645 year	85.5	2.29×10^{12}	3.757	4.06

^{a}References [4] and [5]

The strong dependence of spontaneous fission rates on the number of protons and neutrons is important for assay considerations. The fission rate for odd-even isotopes is typically 10^3 lower than the rate for even-even isotopes, and the fission rate for odd-odd isotopes is typically 10^5 lower. These large differences are due to nuclear spin effects [9]. As the fissioning nucleus begins to deform, the total ground-state nuclear spin must be conserved; however, the quantized angular momentum orbits of the individual neutrons or protons have different energies with increasing deformation. The lowest energy orbit of the undeformed nucleus may not be the lowest energy orbit in the deformed nucleus. In the case of heavy even-even nuclei whose total ground-state spin is zero, the outermost pairs of neutrons and protons can simultaneously couple their spins to zero while shifting to the lowest energy orbits. In the case of odd nuclei, a single neutron or proton must occupy the orbit that conserves total nuclear spin even though extra energy is required [10, 11]. This effect raises the fission barrier and makes odd-even and odd-odd isotopes more rigid against spontaneous fission than even-even isotopes.

Among the even-even isotopes with high spontaneous fission yields are ^{238}U, ^{238}Pu, ^{240}Pu, ^{242}Pu, ^{242}Cm, ^{244}Cm, and ^{252}Cf. Isotopes with odd neutron numbers or odd proton numbers do not have high spontaneous fission yields as described above. However, isotopes with odd neutron numbers can easily be induced to fission if bombarded with low-energy neutrons; absorption of an extra neutron yields an unbounded neutron pair whose pairing energy is now available to excite the compound nucleus to an energy near the fission barrier. Among the even-odd isotopes that can be fissioned by neutrons of zero energy but have low spontaneous fission yields are ^{233}U, ^{235}U, and ^{239}Pu. These isotopes are called *fissile*. Even-even isotopes, such as ^{238}Pu and ^{240}Pu, that are not easily fissioned by low-energy neutrons are called *fertile*. This term comes from reactor theory and refers to the fact that through neutron capture, these isotopes are fertile sources of fissile isotopes. Examples of induced fission cross sections for fertile and fissile isotopes are given in Chap. 14.

13.3 Neutrons and Gamma Rays from Fission

Prompt neutrons and gamma rays emitted at the time of scission are the most useful for passive assay because of their intensity and penetrability. Many passive assay instruments, such as coincidence counters, are designed to detect prompt fission neutrons and are often also sensitive to gamma rays. For this reason, this section describes both neutron and gamma-ray emissions.

Figure 13.2 shows an energy spectrum of the neutrons emitted during the spontaneous fission of ^{252}Cf [12, 13]. The mean energy is 2.14 MeV. The spectrum depends on many variables, such as fission-fragment excitation energy and average total

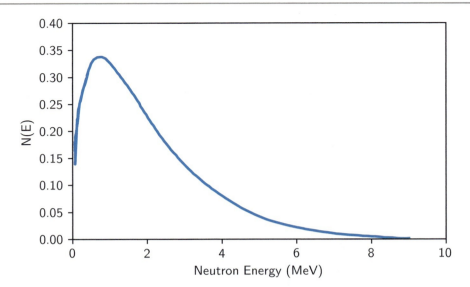

Fig. 13.2 Prompt neutron energy spectrum from the spontaneous fission of ^{252}Cf, as calculated from a Maxwellian distribution with "temperature" T = 1.43 MeV

Table 13.2 Measured prompt fission multiplicity distributions

Probability distribution	^{235}U induced fission [17]	^{238}Pu spontaneous fission [8]	^{239}Pu induced fission [17]	^{240}Pu spontaneous fission [8]	^{242}Pu spontaneous fission [8]	^{252}Cf spontaneous fission [8]	^{244}Cm spontaneous fission [8]
P(0)	0.033	0.056	0.011	0.063	0.068	0.002	0.015
P(1)	0.174	0.210	0.101	0.231	0.229	0.026	0.115
P(2)	0.335	0.380	0.275	0.333	0.334	0.125	0.300
P(3)	0.303	0.222	0.324	0.253	0.248	0.274	0.334
P(4)	0.123	0.105	0.199	0.099	0.0997	0.305	0.182
P(5)	0.028	0.026	0.083	0.018	0.0182	0.185	0.044
P(6)	0.003		0.008	0.002	0.0031	0.066	0.0085
P(7)						0.014	0.0004
P(8)						0.0018	0.00002
P(9)						0.0001	
P(10)						5 E-08	
$\bar{\nu}$	2.406	2.19	2.875	2.154	2.149	3.757	2.71
$\overline{\nu(\nu-1)}$	4.626	3.874	6.75	3.79	3.809	11.95	5.941
$\overline{\nu(\nu-1)(\nu-2)}$	6.862	5.417	12.81	5.21	5.35	31.67	10.112

fission energy release but can be approximated by a Maxwellian distribution N(E), where N(E) varies as $\sqrt{E}\exp\left(-\frac{E}{1.43}MeV\right)$. This spectrum is proportional to $\sqrt{E}$ at low energies; it then falls exponentially at high energies. The neutron spectra for spontaneous fission of ^{240}Pu and thermal-neutron induced fission of ^{233}U, ^{235}U, and ^{239}Pu can also be approximated by Maxwellian distributions, with spectrum parameters 1.32, 1.31, 1.29, and 1.33 MeV, respectively [14, 15].

The number of neutrons emitted in spontaneous or induced fission is called the *neutron multiplicity*. Average neutron multiplicities $\bar{\nu}$ are included in the last two columns of Table 13.1. For neutron-induced fission, the multiplicity increases slowly and linearly with the energy of the incoming neutron [16]. The multiplicities given in the last column of Table 13.1 are approximately correct for thermal- or low-energy incident neutrons.

From one fission to another, the neutron multiplicity may vary from 0 to 6 or more depending on the distribution of excitation energy among the fission fragments. Table 13.2 [17–19] lists the measured prompt neutron multiplicity distributions P(ν) of some important isotopes for spontaneous or thermal neutron-induced fission. Uncertainties on the individual probabilities vary 1–5% near the maxima to 30–50% near the end points.

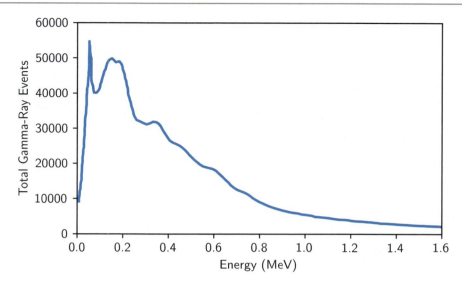

Fig. 13.3 Prompt gamma-ray spectrum that accompanies spontaneous fission of ^{252}Cf, as recorded by a NaI detector [22]

Terrell [20] has shown that the multiplicity distributions for both spontaneous and thermal-neutron-induced fission can be approximated by a Gaussian distribution centered at $\bar{\nu}$, the mean multiplicity:

$$P(\nu) = \frac{1}{\sqrt{2\pi\sigma^2}}\, e^{-(\nu-\bar{\nu})^2/2\sigma^2}. \tag{13.1}$$

A distribution width σ of 1.08 can be used as an approximation for all isotopes except ^{252}Cf, where 1.21 should be used.

Information about the neutron multiplicity distribution in fission is used in the analysis of coincidence counting and neutron multiplicity counting (see Chaps. 17 and 18). One question that has arisen in this regard is whether the neutron multiplicity and the mean neutron energy are correlated. In other words, if the number of neutrons emitted in a fission is above average, will the mean neutron energy be below average? The available experimental evidence indicates that the mean neutron emission energy is approximately constant and that the number of neutrons emitted increases with the amount of available energy [21]. Thus, the mean energy may be approximately independent of the multiplicity.

After a nucleus undergoes fission, prompt neutrons are evaporated from the fission fragments until the remaining excitation energy is less than the neutron binding energy. At this point, prompt gamma rays carry away the remaining energy and angular momentum. On the average, 7–10 prompt gamma rays are emitted, with a total energy of 7–9 MeV [9]. Figure 13.3 shows the prompt gamma-ray spectrum that accompanies spontaneous fission of ^{252}Cf, as recorded by a NaI detector [22]. A spectrum obtained with a high-resolution detector might reveal many discrete transitions, although the transitions would be Doppler-broadened by the recoil of the fission fragments. Prompt gamma rays from fission are of much lower intensity than the gamma rays that follow alpha decay (see Chap. 2). Therefore, they are not useful for passive assay despite their relatively high energy; however, prompt gamma rays from fission are useful for coincidence counting, where their high multiplicity can lead to a strong signal.

This section includes brief descriptions of the delayed neutrons and gamma rays emitted after fission. In passive assay systems, the delayed neutrons and gamma rays are usually masked by the stronger prompt emissions; however, the time delay is often used by active assay systems to discriminate between the interrogation source and the induced fission signal. For additional details on these delayed signals, see Ref. [23].

Delayed neutrons originate from some of the isotopes produced during beta decay of fission fragments. They are emitted by highly excited isotopes as soon as the isotopes are created by the beta decay of their precursors; therefore, delayed neutrons appear with half-lives characteristic of their precursors. Although there are many such isotopes, delayed neutrons can be categorized into six groups, with decay half-lives ranging from 200 ms to 55 s [24]. The neutron yield of each group is different for each uranium or plutonium isotope. In principle, active assay systems can use this variation as an indication of the isotopic composition of the irradiated sample [23], but in practice, this use is difficult to implement. Delayed neutron energy spectra are highly structured in comparison to the smooth Maxwellian distributions of prompt neutrons. Also, the average energy of delayed neutrons is only 300–600 keV as opposed to the 2 MeV average of prompt neutrons. Most importantly, the number of delayed neutrons is typically only 1% of the number of prompt neutrons, so delayed neutrons contribute to passive neutron measurements, but their effect is not large.

Delayed gamma rays from fission have a higher intensity and slower emission rate than delayed neutrons. Their average multiplicity and energy is comparable to that of prompt gamma rays: 6–8 gamma rays, each with an average energy close to 1 MeV. There is no clear-cut distinction between the emission time of prompt and delayed gamma rays like there is for prompt and delayed neutrons. Gozani [23] has used a time of 10^{-9} s after fission as a convenient demarcation. The delayed gamma rays so defined are then emitted over times of several seconds or minutes. The intensity of these gamma rays is two orders of magnitude above the intensity of delayed neutrons.

13.4 Neutrons from (α,n) Reactions

Nuclei can decay spontaneously by alpha- or beta-ray emission as well as by fission. Alpha particles are helium nuclei with two protons and two neutrons, and beta particles are energetic free electrons. In principle, all nuclei of atomic mass greater than 150 are unstable toward alpha decay; however, alpha decay is a quantum mechanical barrier penetration process like spontaneous fission, and the Coulomb barrier is high enough to make alpha decay unlikely for all but the heaviest elements. Table 13.3 [1, 2, 25–27] lists the alpha decay rates of some heavy elements. The total half-lives of the isotopes listed in the table are almost the same as the alpha decay half-lives except for ^{241}Pu and ^{249}Bk, where beta decay dominates, and ^{252}Cf, where the spontaneous fission rate is ~3% of the alpha decay rate.

The alpha decay process leads to the emission of gamma rays from unstable daughters (see Chap. 2). Also, the alpha particles can produce neutrons through (α,n) reactions with certain elements. This source of neutrons can be comparable in intensity to spontaneous fission if isotopes with high alpha decay rates, such as ^{233}U, ^{234}U, ^{238}Pu, or ^{241}Am, are present. This section describes the production of neutrons by (α,n) reactions and provides some guidelines for calculating the expected neutron yield.

Following are two examples of (α,n) reactions that occur in many nuclear fuel cycle materials:

$$\alpha + {}^{18}O \rightarrow {}^{21}Ne + n$$
$$\alpha + {}^{19}F \rightarrow {}^{22}Na + n$$

The alpha particle is emitted from uranium or plutonium with energies in the range of 4–6 MeV. Because ^{234}U is the dominant alpha emitter in enriched uranium, the average energy for alpha particles emitted from uranium is 4.7 MeV (see Table 13.3).

For plutonium, an average energy of 5.2 MeV is typical. In air, the range of alpha particles from uranium is 3.2 cm, and the range of alpha particles from plutonium is 3.7 cm. The range in other materials can be estimated from the Bragg-Kleeman rule [28].

$$range = 0.00032 \frac{\sqrt{A}}{density(g/cm^3)} \times range\ in\ air, \qquad (13.2)$$

where A is the atomic weight of the material. The range in uranium and plutonium oxide is roughly 0.006 and 0.007 cm, respectively; therefore, the alpha particles lose energy very rapidly when traveling through matter. In many cases, this short range means that the alpha particle can never reach nearby materials in which (α,n) reactions could take place. However, if elements such as oxygen or fluorine are intimately mixed with the alpha-emitting nuclear material, an (α,n) reaction may take place because the alpha particle can reach these elements before it loses all of its energy.

When the alpha particle arrives at another nucleus, the probability of a reaction depends on the Q-value, the threshold energy, and the height of the Coulomb barrier. The Q-value is the difference in binding energies between the two initial nuclei and the two final reaction products. A positive Q-value means that the reaction will release energy. A negative Q-value means that the alpha particle must have at least that much energy in the center-of-mass reference frame before the reaction can proceed. If this minimum energy requirement is transformed to the laboratory reference frame, it is called the *threshold energy*:

13 The Origin of Neutron Radiation

Table 13.3 (α,n) reaction neutron yields

Isotope	Total half-life [1]	Alpha decay half-life [1]	Alpha yield [1] (α/s − g)	Average alpha energy [1] (MeV)	(α,n) yield in oxide [2] (n/s − g)	(α,n) yield in UF$_6$ [25, 26]/PuF4 [27] (n/s − g)
^{232}Th	1.40×10^{10} year	1.40×10^{10} year	4×10^3	4.00	2.2×10^{-5}	
^{232}U	68.9 year	68.9 year	8.0×10^{11}	5.30	1.49×10^4	2.6×10^6
^{233}U	1.59×10^5 year	1.59×10^5 year	3.57×10^8	4.82	4.8	7.0×10^2
^{234}U	2.45×10^5 year	2.45×10^5 year	2.3×10^8	4.76	3.0	5.8×10^{2a}
^{235}U	7.04×10^8 year	7.04×10^8 year	8.0×10^4	4.40	7.1×10^{-4}	0.122
^{236}U	2.34×10^7 year	2.34×10^7 year	2.4×10^6	4.48	2.4×10^{-2}	3.96
^{238}U	4.47×10^9 year	4.47×10^9 year	1.2×10^4	4.19	8.3×10^{-5}	0.014
^{237}Np	2.14×10^6 year	2.14×10^6 year	2.6×10^7	4.77	3.4×10^{-1}	
^{238}Pu	87.74 year	87.74 year	6.3×10^{11}	5.49	1.34×10^4	2.2×10^6
^{239}Pu	2.41×10^4 year	2.41×10^4 year	2.3×10^9	5.15	3.81×10^1	5.6×10^3
^{240}Pu	6.56×10^3 year	6.56×10^3 year	8.4×10^9	5.15	1.41×10^2	2.1×10^4
^{241}Pu	14.329 year	5.90×10^5 year	9.3×10^7	4.89	1.3	1.7×10^2
^{242}Pu	3.75×10^5 year	3.76×10^5 year	1.5×10^8	4.90	2.0	2.7×10^2
^{241}Am	432.7 year	432.7 year	1.3×10^{11}	5.48	2.69×10^3	4.4×10^5
^{242}Cm	163 days	163 days	1.2×10^{14}	6.10	3.76×10^6	
^{244}Cm	18.1 year	18.1 year	3.0×10^{12}	5.80	7.73×10^4	
^{249}Bk	330 days	6.1×10^4 year	8.8×10^8	5.40	1.8×10^1	
^{252}Cf	2.645 year	2.731 year	1.9×10^{13}	6.11	6.0×10^5	

^{a}Considerable uncertainty

$$\text{Threshold energy} = -Q\left(1 + \frac{4}{A}\right) \text{ if } Q \text{ is negative} \quad (13.3)$$
$$\text{Threshold energy} = 0 \text{ if } Q \text{ is positive.}$$

The Coulomb barrier is the strength of the electrostatic repulsion that the alpha particle must overcome to enter the target nucleus and react.

$$\text{Coulomb barrier (MeV)} = \frac{Z_1 Z_2 e^2}{r_0\left(A_1^{\frac{1}{3}} + A_2^{\frac{1}{3}}\right)}, \quad (13.4)$$

where $Z_1 = 2$, $A_1 = 4$, $e^2 = 1.44$ MeV − fm, $r_0 = 1.2$ fm, and Z_2 and A_2 refer to the target nucleus [29]. Thus, an (α,n) reaction is energetically allowed only if the alpha particle has enough energy to (1) overcome or penetrate the Coulomb barrier and (2) exceed the threshold energy. (Note that the two energy requirements are not additive.) Table 13.4 [28, 30] summarizes these properties for a series of low-mass isotopes.

Table 13.4 shows that (α,n) reactions with 5.2 MeV alpha particles are possible in 11 low-Z elements. In all elements with atomic number greater than that of chlorine, the reaction is energetically not allowed. The observed yield of neutrons from (α, n) reactions is given in Table 13.5 [31–35] for thick targets. A thick target is a material that is much thicker than the range of the alpha particle and one in which the alpha particles lose energy only in the target element. From Eq. 13.2, the range of alpha particles in solids is on the order of 0.01 cm.

(Alpha,n) reactions can occur in compounds of uranium or plutonium (such as oxides or fluorides) and in elements (such as magnesium or beryllium) that could be present as impurities. The neutron yield per gram of source nuclide in pure oxides and fluorides is given in the last two columns of Table 13.3. In other materials, the yield will depend very sensitively on the alpha activity of the nuclear isotopes, the alpha-particle energy, the reaction Q-values, the impurity concentrations, and the degree of mixing (because of the short range of the alpha particle).

The thick target yield depends on the incident energy, which is why the (α,n) yields shown in Table 13.3 are different for each nuclide shown. The thick target yields for commonly encountered low-Z elements are shown in Table 13.5. This yield is the number of neutrons emitted per 10^6 alpha particles and is given both at 4.7 MeV (average alpha energy for uranium) and 5.2 MeV (average alpha energy for plutonium).

Table 13.4 (Alpha,n) Q-values, threshold energies, and Coulomb barriers

Nucleus	Natural abundance (%)	Q-value [30] (MeV)	Threshold energy [30] (MeV)	Coulomb barrier (MeV)	Maximum neutron energy for 5.2 MeV alpha [28]
^{4}He	100	−18.99	38.0	1.5	
^{6}Li	7.5	−3.70	6.32	2.1	
^{7}Li	92.5	−2.79	4.38	2.1	1.2
^{9}Be	100	+5.70	0	2.6	10.8
^{10}B	19.8	+1.06	0	3.2	5.9
^{11}B	80.2	+0.16	0	3.2	5.0
^{12}C	98.9	−8.51	11.34	3.7	
^{13}C	1.11	+2.22	0	3.7	7.2
^{14}N	99.6	−4.73	6.09	4.2	
^{15}N	0.4	−6.42	8.13	4.1	
^{16}O	99.8	−12.14	15.2	4.7	
^{17}O	0.04	+0.59	0	4.6	5.5
^{18}O	0.2	−0.70	0.85	4.6	4.2
^{19}F	100	−1.95	2.36	5.1	2.9
^{20}Ne	90.9	−7.22	8.66	5.6	
^{21}Ne	0.3	+2.55	0	5.5	7.6
^{22}Ne	8.8	−0.48	0.57	5.5	4.5
^{23}Na	100	−2.96	3.49	6.0	1.8
^{24}Mg	79.0	−7.19	8.39	6.4	
^{25}Mg	10.0	+2.65	0	6.4	7.7
^{26}Mg	11.0	+0.03	0	6.3	5.0
^{27}Al	100	−2.64	3.03	6.8	2.2
^{29}Si	4.7	−1.53	1.74	7.2	3.4
^{30}Si	3.1	−3.49	3.96	7.2	1.4
^{37}Cl	24.2	−3.87	4.29	8.3	1.0

Table 13.5 Thick-target yields from (α,n) reactions (error bars estimated from scatter between references)

Element (natural isotopic composition)	Neutron yield per 10^6 alphas of energy 4.7 MeV (^{234}U)	Neutron yield per 10^6 alphas of energy 5.2 MeV (avg Pu)	References	Avg neutron energy (MeV) for 5.2 MeV alphas [31]
Li	0.16 ± 0.04	1.13 ± 0.25	[30]	0.3
Be	44 ± 4	65 ± 5	[31]	4.2
B	12.4 ± 0.6	17.5 ± 0.4	[29, 30, 33]	2.9
C	0.051 ± 0.002	0.078 ± 0.004	[29–31]	4.4
O	0.040 ± 0.001	0.059 ± 0.002	[29–31]	1.9
F	3.1 ± 0.3	5.9 ± 0.6	[29, 30, 33]	1.2
Na	0.5 ± 0.5	1.1 ± 0.5	[32]	
Mg	0.42 ± 0.03	0.89 ± 0.02	[29–31]	2.7
Al	0.13 ± 0.01	0.41 ± 0.01	[29–31]	1.0
Si	0.028 ± 0.002	0.076 ± 0.003	[29–31]	1.2
Cl	0.01 ± 0.01	0.07 ± 0.04	[32]	

The (α,n) yields in Table 13.5 are accurate to 5–10% for the best-measured elements. The oxide (α,n) yields in Table 13.3 are known to 10% or better. Thus, the neutron yield calculations are accurate to 10% at best, even with perfect mixing.

The thick target yield is different from the yield observed in a mixture or compound because the energy deposited by the alpha particles is split between the constituents. The (α,n) yield from compounds and mixtures (assuming perfect mixing) can be calculated from the general equation given in Ref. [36].

13 The Origin of Neutron Radiation

Table 13.6 Mass stopping forces for commonly encountered elements [37][a]

Element	Mass stopping force (C_j) @4.7 MeV (U) MeV/(g.cm^{-2})	Stopping force relative to oxygen for U (C_{j1})	Mass stopping force (C_j) @5.2 MeV (Pu) MeV/(g.cm^{-2})	Stopping force relative to oxygen for Pu (C_{j1})
Li	824	1.062	768	1.059
Be	774	0.997	725	1.000
B	808	1.041	750	1.034
C	846	1.090	788	1.087
N	815	1.050	760	1.048
O	776	1.000	725	1.000
F	723	0.932	675	0.931
Na	674	0.869	627	0.865
Mg	672	0.866	628	0.866
Al	627	0.808	589	0.812
U	227	0.293	216	0.298
Pu[b]	227	0.293	210	0.290

[a]The nuclear component is negligible and is ignored here
[b]Extrapolated using the ratio of U/Th values

$$Y_{mixture} = \frac{\sum_j a_j A_j C_{j1} Y_j}{\sum_j a_j A_j C_{j1}}, \quad (13.5)$$

where a_j is the number of atoms of type j of atomic weight A_j in the mixture. C_{j1} is the mass stopping power (force) of element j relative to element 1. Y_j is the thick target yield of element j. This equation weights the yield of each element by the amount of alpha energy deposited in that element. Table 13.6 gives the stopping forces for 4.7 MeV alphas (average alpha energy for uranium) and for 5.2 MeV alphas (average energy for plutonium) and the values relative to oxygen (C_{j1}). Note that the product $a_j A_j$ is proportional to the mass, m_j, of element j in the material, and so Eq. 13.5 can be written as

$$Y_{mixture} = \frac{\sum_j m_j C_j Y_j}{\sum_j m_j C_j}, \quad (13.6)$$

which makes Eq. 13.6 easier to apply in the case of mixtures. Note that we have also changed to the actual mass stopping force (power) of each element rather than the value relative to that of oxygen.

In the special case of a two-component mixture where only one element has an (α,n) yield (such as PuO$_2$), Eq. 13.6 simplifies to

$$Y_{mixture} = \frac{a_1 A_1 Y_1}{a_1 A_1 + a_2 A_2 C_{21}} \text{ or } Y_{mixture} = K\, Y_1, \quad (13.7)$$

where K is a "reduction factor." In the case of PuO$_2$, $a_1 = 2$, $A_1 = 16$, $a_2 = 1$, $A_2 = 239$, and $C_{21} = 0.29$, giving a value of K of 0.316. Thus the (α,n) emission of PuO$_2$ is reduced by this factor compared with the thick target yield ($0.059 \times 0.316 = 0.0186$).

The actual (α,n) emission from an item can be calculated in multiple ways. For PuO$_2$ or PuF$_4$ (and UO$_2$ and UF$_6$), for example, the specific (α,n) production rates can be used from Table 13.3 and summed for all isotopes.

$$Y_{oxide} = \sum_i m_i Y_i, \quad (13.8)$$

where m_i is the mass in grams of the ith isotope and Y_i is the neutron yield per gram of each alpha-emitting nuclide as given in Table 13.3. The summation over i should include ^{241}Am, which is a strong alpha emitter.

Secondly, for PuO$_2$, the "alpha" value of the material can be calculated as described in Chap. 17 of this book by using Eq. 17.62. This equation uses the values from Tables 13.1 and 13.3 to calculate the ratio of (α,n) neutron emission to spontaneous fission neutron production for PuO$_2$. The (α,n) emission of a PuO$_2$ item can then be calculated by multiplying the alpha value by the spontaneous fission rate. The alpha value can be calculated for other materials by multiplying by an "alpha weight" value, which is equal to the (α,n) yield of the material relative to that of PuO_2. For plutonium metal, the alpha weight is 0.0. For other materials, the alpha weight can be calculated using the general method as follows.

For any material, Eq. 13.7 or Eq. 13.8 can be used directly and then multiplied by the alpha-particle production rate of the item (using the α emission rates given in Table 13.3). This equation is simple to apply for typical uranium and plutonium isotopes but needs additional corrections for alpha emitters with different alpha energies. One approximate way to deal with this issue is to approximate the thick target yield in a particular light element (e.g., F) from a new isotope[1] (e.g., ^{242}Cm) by taking the thick target yield in fluorine from plutonium (from Table 13.5) and multiplying by the ratio of the yield in oxide from ^{242}Cm to the yield in oxide from plutonium (from Table 13.3). This formula works because, over a limited range of energy, the (α,n) thick target yields vary in roughly the same way with energy. With this method, approximate results can be obtained from all of the nuclides listed in Table 13.3 and all of the light elements listed in Table 13.5. Example: The thick target yield of fluorine from ^{242}Cm is approximately given by

$$Y_{242Cm} = Y_{Pu} \times \frac{\left(\frac{n}{\alpha} oxide\right)_{242Cm}}{\left(\frac{n}{\alpha} oxide\right)_{Pu}}$$

$$Y_{F,242Cm} = 0.014 \times \frac{\frac{3.76E6}{1.2E14}}{\frac{38.1}{2.3E9}} = 0.026 \text{ n}/10^6 \text{ } \alpha$$

(13.9)

where the values are from Tables 13.3 and 13.5. We have used the values for ^{239}Pu as a convenient measure for the average for Pu.

When N materials are in a mixture with I α-emitters and J light elements (N = I + J), the total (α,n) contribution is

$$N_\alpha = \sum_{i=1}^{I} m_i S_{\alpha i} \frac{\sum_{j=1}^{J} m_j C_j Y_{ji}}{\sum_{j=1}^{N} m_j C_j},$$

(13.10)

where $S_{\alpha i}$ is the specific α emission of material i (α/s/g), and Y_{ji} is thick target yield of light element j from α-emitter i (obtained using the above approximation, where appropriate).

The stopping-force results obtained with SRIM (Stopping Range of Ions in Matter computer code; Ref. [37]) have been extrapolated to larger Z values by fitting a power function to the results from Z = 50 to 92 (see Fig. 13.4). The results are given in Table 13.7.

A more accurate method of calculating (α,n) reaction rates (and spectra) is to use the code SOURCES4c [38], which uses libraries of alpha particle spectra, stopping powers, and (α,n) cross sections to calculate the (α,n) emission for multiple different geometries.

The energy of the neutron emitted in an (α,n) reaction depends on the energy that the alpha particle has at the time of the reaction and on the Q-value of the reaction in the isotope. Average thick-target neutron energies are given in Table 13.5. Maximum neutron energies are given in the last column of Table 13.4. AmBe and AmLi spectra are given in Figs. 13.5 and 13.6.

Another important characteristic of neutrons from (α,n) reactions is that only one neutron is emitted in each reaction. These events constitute a neutron source that is random in time with a multiplicity of 1. This characteristic is exploited by neutron coincidence counters (Chaps. 17 and 18), which can distinguish between spontaneous fission neutrons and neutrons from (α, n) reactions.

Note that both (α,n) and (α,p) reactions may leave the nucleus in an excited state from which the nucleus decays to the ground state by emitting one or more gamma rays. For example:

---

[1] This method also gives improved results for unusual plutonium isotopic compositions, such as mainly ^{238}Pu.

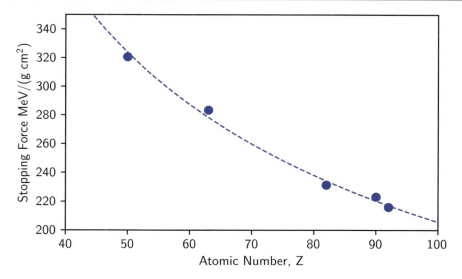

Fig. 13.4 Stopping force as a function of atomic number

Table 13.7 Stopping force for selected elements with extrapolation by power fit aZ^b ($a = 4188.7$, $b = -0.654$)

Element	Z	SRIM 5.2 MeV/(g/cm^{-2})	Power fit
Li	3	768	
Be	4	725	
B	5	750	
C	6	788	
N	7	760	
O	8	725	
F	9	675	
Na	11	627	
Mg	12	628	
Al	13	589	
Si	14	595.3	
Ge	32	394.3	
Sn	50	320.6	323.7
Pb	82	231.1	234.2
Eu	63	283.6	278.3
Th	90	222.4	220.3
U	92	216	217.2
Np	93		215.7
Pu (extrap)	94		214.2
Am (extrap)	95		212.7
Cm (extrap)	96		211.2
Bk (extrap)	97		209.8
Cf (extrap)	98		208.4

$$^{19}F(\alpha, n)^{22}Na^* \xrightarrow{\gamma} {}^{22}Na \xrightarrow[2.6\text{yr}]{\beta^+} {}^{22}Ne^* \xrightarrow{\gamma(1275\text{keV})} {}^{22}Ne$$

$$^{19}F(\alpha, p)^{22}Ne^* \xrightarrow{\gamma} {}^{22}Ne^* \xrightarrow{\gamma(1275\text{keV})} {}^{22}Ne.$$

Here, the asterisk refers to a nucleus in an excited state. Because many of the gamma rays from these reactions are of high energy and are often emitted nearly simultaneously with the neutron, they can affect the response of total neutron counters or neutron coincidence counters that contain detectors that are sensitive to gamma rays.

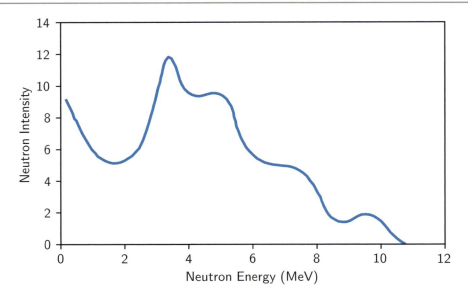

Fig. 13.5 Typical neutron spectrum of an AmBe source

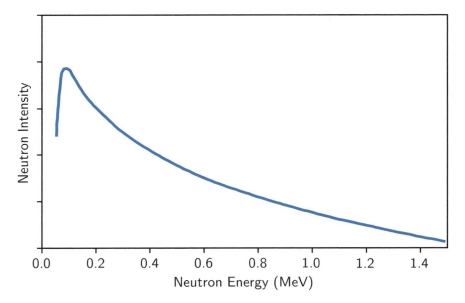

Fig. 13.6 Typical neutron energy spectrum of an AmLi source

13.5 Neutrons from Other Nuclear Reactions

Spontaneous fission, induced fission, and (α,n) reactions are the primary sources of neutrons observed in passive measurements; however, other reactions, such as (γ,n), (n,n′), and (n,2n), may take place in the sample or detector assembly and contribute slightly to the observed count rate. This section describes these reactions briefly; examples are given in Table 13.8. These reactions are more important in active nondestructive assay measurements; details can be found in Ref. [23].

The (γ,n) reaction can produce neutrons in any element if the gamma-ray energy is high enough. The typical minimum threshold energy (~8 MeV) is much higher than the energies of gamma rays emitted from radioactive nuclides; however, the (γ,n) threshold energies for beryllium (1.66 MeV) and deuterium (2.22 MeV) are anomalously low. Therefore, it is possible to

Table 13.8 Other nuclear reactions that may affect passive neutron counting

Radiation source	Threshold energy (MeV) [30]	Target material	Reaction	Outgoing radiation	Outgoing energy [30]
Gamma	1.665	Beryllium	(γ,n)	Neutron	8(Eγ − 1.665)/9 MeV
Gamma	2.224	Deuterium	(γ,n)	Neutron	(Eγ − 2.224)/2 MeV
Neutron	0	Hydrogen	(n,γ)	Gamma	2.224 MeV
Proton	1.880	Lithium-7	(p,n)	Neutron	≥30 keV
Proton	1.019	Tritium	(p,n)	Neutron	≥0 keV
Neutron	0.1–1.0	Lead	(n,n′)	Neutron	
Neutron	0.1–1.0	Tungsten	(n,n′)	Neutron	
Neutron	0.1–1.0	Uranium	(n,n′)	Neutron	
Neutron	1.851	Beryllium	(n,2n)	Neutrons	
Neutron	3.338	Deuterium	(n,2n)	Neutrons	
Neutron	5.340	Tungsten	(n,2n)	Neutrons	

create a photoneutron source by surrounding relatively intense, long-lived, high-energy gamma-ray sources (such as ^{124}Sb or ^{226}Ra) with a mantle of beryllium or D_2O. A detailed list of photoneutron sources is given in Table 4.3 of Ref. [23]. For passive assay applications, it is necessary to keep in mind that prompt fission gamma rays or gamma rays from some (α,n) reactions can produce extra neutrons only if the detector assembly contains beryllium or deuterium. Or conversely, neutrons can be captured in hydrogen to produce deuterium and 2.22 MeV gamma rays. These possibilities are included in Table 13.8.

Inelastic neutron scattering (n,n′) can occur in heavy nuclei with neutron energies of roughly 0.1–1.0 MeV or higher. This reaction is possible if the target nucleus has energy levels low enough to be excited by the neutron. The probability of this reaction is not high, and the number of neutrons present is not altered; however, the average energy of neutrons in the material will decline somewhat faster than would be expected from elastic scattering alone.

The (n,2n) reaction can increase the number of neutrons present, but the threshold energy in most elements is in the range of 10 MeV. The thresholds are lower for deuterium, beryllium, and tungsten, but the number of extra neutrons produced is likely to be small. The possibility of (n,2n) reactions should be considered only when the neutrons are known to have high energy; when deuterium, beryllium, or tungsten are present; and when the coincidence count rates to be measured are very low. In such cases, the observed response may be enhanced.

13.6 Neutrons from Cosmic Rays

High-energy cosmic rays that enter the atmosphere create showers of secondary particles that can give rise to counts in neutron detectors. These secondary particles affect neutron-counting instruments by interactions with the material of the neutron detector or the item being counted. These interactions can create events with extremely high multiplicities, resulting in larger contributions to the higher moments (Doubles, Triples, etc.; see Chap. 17) than to the Singles. The flux of secondary particles, and thus the rate of occurrence of these events, decreases strongly with altitude. The doubles rate from cosmic rays at 2200 m is roughly 10 times that at sea level. The number of interactions for a given secondary flux also depends strongly on the atomic number (Z) of the material concerned. For example, the number of interactions in an item composed primarily of uranium will be much higher than in an item of the same mass that is composed mainly of iron. This background from cosmic ray interactions is normally a concern only for the measurement of small amounts of material because the average counting rate from cosmic rays is small; however, in these cases it is difficult to determine the magnitude of the cosmic background contribution for a couple of reasons. Firstly, the background depends on the item being measured. For example, when determining the amount of uranium contamination on large pieces (tons) of metal process equipment, the background counting rate will be different before and after the equipment is placed in the instrument. Secondly, because these are relatively rare events, it is difficult to make measurements with reliable uncertainties. One approach to reducing the contribution of cosmic background events has been to divide the measurement time into many small intervals and reject data from any interval that has anomalously high counting rates. A discussion on cosmic ray effects can be found in Ref. [39].

13.7 Isotopic Neutron Sources

Compact, portable neutron sources are useful for laboratory work, for verifying the proper operation of assay instruments, and for irradiating samples to obtain other induced signals. For accountability or safety purposes, it is often important to have sources that contain little or no plutonium or uranium. Such sources can be manufactured by using other isotopes that emit neutrons by spontaneous fission or by taking advantage of (α,n) reactions between strong alpha-particle-emitting isotopes and low-Z materials.

Californium-252 is the most commonly used spontaneous fission neutron source; it can be fabricated in very small sizes and still provide a strong source for a practical period of time. Table 13.9 summarizes some of the properties of ^{252}Cf; Figs. 13.2 and 13.3 give the prompt neutron and gamma-ray spectra. For some applications, it is important to remember that ^{252}Cf neutrons are emitted with an average multiplicity of 3.757; therefore, they are strongly correlated in time and will generate coincidence events.

Sources that emit random, uncorrelated neutrons can be manufactured by mixing alpha emitters, such as ^{238}Pu or ^{241}Am, with beryllium, lithium, fluorine, or other elements in which (α,n) reactions are possible. Table 13.10 [1, 28, 40] summarizes the characteristics of some common (α,n) sources. One important feature for practical applications is the half-life of the heavy element that emits the alpha particles. All of the sources listed in Table 13.10 have long half-lives with the exception of ^{210}PoBe. Another important feature is the neutron energy spectrum obtained from the source. In some cases, it is important to have a high-energy, highly penetrating neutron source. In other cases, it might be important to avoid neutron energies high enough to fission plutonium or uranium isotopes (that is, the source must provide subthreshold interrogation) or high enough to excite (n,2n) reactions.

Two common (α,n) sources in use today are ^{241}AmBe and ^{241}AmLi. Typical neutron energy spectra for these two sources are given in Figs. 13.5 and 13.6. The energy spectra can vary somewhat because of impurity elements or imperfect mixing. (Also, (α,n) spectra can change their shape somewhat in time depending on the source construction and the particular isotopes involved.) Note that AmLi sources are usually fabricated by mixing ^{241}AmO$_2$ with lithium oxide and that (α,n) reactions in the oxide contribute a high-energy tail to the spectrum.

The ^{241}AmBe sources are compact and relatively inexpensive and do not require much gamma-ray shielding; however, the high-energy spectrum allows (n,2n) reactions that will produce coincidence counts. The ^{241}AmLi sources are less compact and more expensive and require tungsten shields. Because of their low-energy neutron spectra, they are the most widely used sources for subthreshold interrogation in active assay and for random-neutron check sources in passive coincidence counting. For the latter application, one should be aware of the possibility of plutonium contamination in the americium, which can yield spurious coincidence counts from spontaneous fission.

(Alpha,n) sources also emit gamma and beta radiation, and in many cases, the dose observed outside the container is dominated by gamma radiation. (For comparison, the dose from 10^6 n/s is about 1 mrem/h at 1 m.) The neutron yield of an (α, n) source relative to its total radiation output in curies may thus be an important selection criterion. This ratio is given in the

Table 13.9 Characteristics of ^{252}Cf

Total half-life	2.646 year
Alpha half-life	2.731 year
Spontaneous fission half-life	85.5 year
Neutron yield	2.34×10^{12} n/s $-$ g
Gamma-ray yield	1.3×10^{13} γ/s $-$ g
Alpha-particle yield	1.9×10^{13} α/s $-$ g
Average neutron energy	2.14 MeV
Average gamma-ray energy	1 MeV
Average alpha-particle energy	6.11 MeV
Neutron activity	4.4×10^9 n/s $-$ Ci
Neutron dose rate	2300 rem/h $-$ g at 1 m
Gamma dose rate	140 rem/h $-$ g at 1 m
Conversion	558 Ci/g
Decay heat	38.5 W/g
Avg spontaneous fission neutron multiplicity	3.757
Avg spontaneous fission gamma multiplicity	8

13 The Origin of Neutron Radiation

Table 13.10 Characteristics of some isotopic (α,n) sources

Source	Half-life (year) [1]	Average alpha energy (MeV) [1]	Average neutron energy (MeV) [1]	Maximum neutron energy (MeV) [28]	Gamma dose in mrem/h at 1 m/(10^6 n/s) [40]	Curies per gram[a]	Yield in 10^6 n/s − Ci[c] [40]
^{210}PoBe	0.38	5.3	4.2	10.9	0.01	4490	2–3
^{226}RaBe	1600	4.8	4.3	10.4	60	1	0–17
^{238}PuBe	87.74	5.49	4.5	11.0	0.006	17	2–4
^{238}PuLi	87.74	5.49	0.7	1.5	~1	17	0.07
^{238}PuF$_4$	87.74	5.49	1.3	3.2	~1	17	0.4
^{238}PuO$_2$	87.74	5.49	2.0	5.8	~1	17	0.003
^{239}PuBe	24,120	5.15	4.5	10.7	6	0.06	1–2
^{239}PuF$_4$	24,120	5.15	1.4	2.8	~1	0.06	0.2
^{241}AmBe	433.6	5.48	5.0	11.0	6	3.5	2–3
^{241}AmLi	433.6	5.48	0.3	1.5	2.5	3.5	0.06
^{241}AmB	433.6	5.48	2.8	5.0		3.5	
^{241}AmF	433.6	5.48	1.3	2.5		3.5	

[a](Alpha yield/s − g)/(3.7×10^{10} dps/Ci)

Table 13.11 Reactions for producing neutrons (from Ref. [41]). The most commonly used reactions are shown in **bold**

Reaction	Threshold incident particle energy (MeV)	Q value (MeV)	Threshold neutron energy (MeV)
^{2}H(d,n)^{3}He	**0**	3.266	**2.448**
^{3}H(p,n)^{3}He	1.019	−0.764	0.0639
^{3}H(d,n)^{4}He	**0**	17.586	**14.064**
^{9}Be(α,n)^{12}C	0	5.708	5.266
^{12}C(d,n)^{13}N	0.328	−0.281	0.0034
^{13}C(α,n)^{16}O	0	2.201	2.07
^{7}Li(p,n)^{7}Be	1.882	−1.646	0.0299

last column of Table 13.10. Because of their high gamma-ray output, some (α,n) sources should be encapsulated in shielding material. For example, ^{241}AmLi sources are enclosed in 1/4–3/8 in. thick tungsten to shield against the intense 60 keV gamma rays from americium decay.

13.8 Neutron Generators

Neutrons can also be produced by neutron generators. These generators consist of an ion source, an acceleration mechanism, and a target. An International Atomic Energy Agency report gives a detailed description of the history and operating principles of such generators and includes commonly used neutron-producing reactions, which are given in Table 13.11 [41].

The most common neutron generators either accelerate deuterium ions into a deuterium target, giving 2.5 MeV neutrons from the ^{2}H(d,n)^{3}He (DD) reaction, or accelerate deuterium and tritium ions into a target that contains deuterium and tritium, giving predominantly 14 MeV neutrons from the ^{3}H(d,n)^{4}He (DT) reaction. The target material is usually contained in a metal target. In both cases, the beam continuously reloads the target. These two reactions have significant cross sections in the low-energy region and require ion beams with energies of only 100–500 keV to produce neutrons. The cross section for the DT reaction is about 100 times that of the DD reaction and so produces a correspondingly higher yield. Table 13.12 lists available neutron generators from a recent review [42] that use the DD or DT reaction.

Some neutron generators allow the detection of the charged particle that is created at the same time as the neutron. The angle of the charged particle determines the angle of emission of the neutron, providing a neutron cone. The time between the charge particle detection and the neutron (or gamma) detection gives an estimate of the neutron interaction distance. This information can be used for imaging in active interrogation.

Table 13.12 A list of available neutron generators from Ref. [42]

Model	Maximum neutron yield (n/s)	Typical tube lifetime	Operating mode
Thermofisher Scientific Inc.			
API 120 NG	2×10^7	1200 h @ 1×10^7 n/s	Continuous
D 711 NG	2×10^{10}	1000 h @ 1×10^{10} n/s	Continuous
MP 320 NG	1×10^8	1200 h @ 1×10^8 n/s	Continuous and pulsed
P 211 NG	1×10^8	Up to 500 h or more	Continuous and pulsed
P 385 NG	5×10^8	4500 h @ 1×10^8 n/s	Continuous and pulsed
Sodern			
GENIE 16	2×10^8	8000 h @ 5×10^8 n/s	Continuous and pulsed
GENIE 35	1×10^{10}	2000 h @ 1×10^{10} n/s	Continuous and pulsed
Adelphi Technology Inc.			
DD108	1×10^8	2000 h	Continuous and pulsed
DD109.1	1×10^9	2000 h	Continuous and pulsed
DD109.4	4×10^9	2000 h	Continuous and pulsed
DD-109 M	4×10^9	2000 h	Continuous
DD10MB	2×10^{10}		Continuous
DT108API	1×10^8		Continuous
DT110-14 MeV	1×10^{10}		Continuous
All Russia Research Institute of Automatics VNIIA			
ING-013	5×10^9	1600 h @ 1×10^8 n/s	Pulsed
ING-03	3×10^9	1600 h @ 1×10^8 n/s	Pulsed
ING-031	2×10^{10}	1600 h @ 1×10^8 n/s	Pulsed
ING-07	1×10^9	1600 h @ 1×10^8 n/s	Continuous and pulsed
ING-17	3×10^8		Continuous and pulsed
ING-27	1×10^8		Continuous
NG-14	2×10^{10}		Continuous
ING-10	5×10^8		Pulsed
ING12	2×10^9		Pulsed

13.9 Conclusions

What properties of neutron radiation can be used by the assayist to measure the quantity of specific isotopes? Several important features are summarized as follows:

1. The odd-even effect in spontaneous fission means that only fertile isotopes, such as ^{238}U, ^{238}Pu, ^{240}Pu, and ^{242}Pu, are strong emitters of high-energy (2 MeV average) neutrons. For metallic samples of plutonium, the total neutron emission rate is usually directly related to the masses of the even isotopes that are present, which is also true for metallic uranium, although kilogram quantities are required for practical assays because of the lower neutron emission rate.
2. The prompt neutron multiplicity ($\nu = 2$–3) means that coincidence-counting techniques can provide a nearly unique signature for the presence of the even isotopes; however, the multiplicity does not vary enough from one isotope to another to permit discrimination between them.
3. The detection of prompt fission gamma rays along with the neutrons can greatly enhance instrument sensitivity; however, the different behavior of neutrons and gamma rays in the sample matrix and in the detector increases the difficulty of relating the measured response to the sample mass. Therefore, this approach is not recommended for most applications. The use of prompt gamma rays alone is an almost untouched field, but relating measured response to sample mass is again likely to be a difficult problem.
4. Delayed neutron yields are too low for passive assay. Delayed gamma rays usually are not detected by neutron detectors, but they contribute to the response of scintillators. Both delayed neutrons and delayed gamma rays are very important for active assay but not for passive assay.

5. Fissile isotopes, such as ^{235}U and ^{239}Pu, are assayed either by active techniques or indirectly by passive assay of adjacent fertile isotopes if the isotopic composition of the sample is known.
6. (Alpha,n) reactions allow good passive assays of compounds such as ^{238}PuO$_2$ and ^{234}UF$_6$. Again, the quantity of other isotopes can be inferred from the known isotopic composition. (Alpha,n) reactions also can yield unwanted passive emissions that complicate the assay. Neutron coincidence counting is often used to discriminate against (α,n) reactions.

The principles and applications of these techniques are described in Chaps. 16, 17, 18, 19 and 20.

References

1. *NuDat 3.0*, https://www.nndc.bnl.gov/nudat3/. Accessed 15 June 2023.
2. R.T. Perry, W.B. Wilson, Neutron Production from (α,n) Reactions and Spontaneous Fission in ThO$_2$, UO$_2$, and (U,Pu)O$_2$ Fuels. Los Alamos National Laboratory report LA-8869-MS (June 1981)
3. D.L. Johnson, Evaluation of neutron yields from spontaneous fission of transuranic isotopes. Trans. Am. Nucl. Soc. **22**, 673–675 (1975)
4. P. Fieldhouse, D.S. Mather, E.R. Culliford, The spontaneous fission half-life of ^{240}Pu. J. Nucl. Energy **21**(10), 749–754 (1967)
5. C. Budtz-Jørgensen, H.-H. Knitter, Neutron-induced fission cross section of plutonium-240 in the energy range from 10 keV to 10 MeV. Nucl. Sci. Eng. **79**(4), 380–392 (1981)
6. N.E. Holden, D.C. Hoffman, Spontaneous fission half-lives for ground-state nuclide. Pure Appl. Chem. **72**(8), 1525–1562 (2000)
7. *Evaluated Nuclear Data File ENDF/B-V* (available from and maintained by the National Nuclear Data Center at Brookhaven National Laboratory)
8. P. Santi, M. Miller, Reevaluation of prompt neutron emission multiplicity distributions for spontaneous fission. Nucl. Sci. Eng. **160**(2), 190–199 (2008)
9. E.K. Hyde, *The Nuclear Properties of the Heavy Elements III: Fission Phenomena* (Dover Publications, New York, 1971)
10. J.O. Newton, Nuclear properties of the very heavy elements. Prog. Nucl. Phys. **4**, 234 (1955)
11. J.A. Wheeler, *Niels Bohr and the Development of Physics* (Pergamon Press, London, 1955), p. 166
12. J.W. Boldeman, D. Culley, R. Cawley, The fission neutron spectrum from the spontaneous fission of ^{252}Cf. Trans. Am. Nucl. Soc. **32**, 733 (1979)
13. W.P. Poenitz, T. Tamura, Investigation of the prompt neutron spectrum for spontaneously-fissioning ^{252}Cf, in *Proceedings of the International Conference on Nuclear Data Science and Technology*, Antwerp, Belgium, September 1982, p. 465
14. D.G. Madland, J.R. Nix, New calculation of prompt fission neutron spectra and average prompt neutron multiplicities. Nucl. Sci. Eng. **81**(2), 213–271 (1982)
15. J. Terrell, Neutron yields from individual fission fragments. Phys. Rev. **127**(3), 880–904 (1962)
16. W.G. Davey, An evaluation of the number of neutrons per fission for the principal plutonium, uranium, and thorium isotopes. Nucl. Sci. Eng. **44**(3), 345–371 (1971)
17. J.W. Boldeman, M.G. Hines, Prompt neutron emission probabilities following spontaneous and thermal neutron fission. Nucl. Sci. Eng. **91**(1), 114–116 (1985)
18. D.A. Hicks, J. Ise Jr., R.V. Pyle, Probabilities of prompt-neutron emission from spontaneous fission. Phys. Rev. **101**, 1016 (1956)
19. M.S. Zucker, N. Holden, Parameters for several plutonium nuclides and ^{252}Cf of safeguards interest, in *Proceedings of the 6th Annual ESARDA Symposium on Safeguards and Nuclear Material Management*, Venice, Italy (1984), p. 341
20. J. Terrell, Distributions of fission neutron numbers. Phys. Rev. **108**, 783 (1957)
21. A. Gavron, Z. Fraenkel, Neutron correlations in spontaneous fission of ^{252}Cf. Phys. Rev. C **9**, 632–645 (1974)
22. H.R. Bowman, S.G. Thompson, J.O. Rasmussen, Gamma-ray spectra from spontaneous fission of ^{252}Cf. Phys. Rev. Lett. **12**(8), 195 (1964)
23. T. Gozani, *Active Nondestructive Assay of Nuclear Materials: Principles and Applications*, NUREG/CR-0602 (U.S. Nuclear Regulatory Commission, Washington, DC, 1981)
24. G.R. Keepin, *Physics of Nuclear Kinetics* (Addison-Wesley Publishing Co., Inc, Reading, 1965)
25. T.E. Sampson, Neutron yields from uranium isotopes in uranium hexafluoride. Nucl. Sci. Eng. **54**(4), 470–474 (1974)
26. W.B. Wilson, J.E. Stewart, R.T. Perry, Neutron production in UF$_6$ from the decay of uranium nuclides. Trans. Am. Nucl. Soc. **38**, 176 (1981)
27. W.B. Wilson, Los Alamos National Laboratory memorandum T-2-M-1432 to N. Ensslin (1983)
28. R.D. Evans, *The Atomic Nucleus* (McGraw-Hill Book Co., New York, 1955)
29. J.B. Marion, F.C. Young, *Nuclear Reaction Analysis* (Wiley, New York, 1968), p. 108
30. R.J. Howerton, Thresholds and Q-Values of Nuclear Reactions Induced by Neutrons, Protons, Deuterons, Tritons, ^{3}He Ions, Alpha Particles, and Photons. Lawrence Livermore National Laboratory report UCRL-50400, vol. 24 (1981)
31. G.J.H. Jacobs, H. Liskien, Energy spectra of neutrons produced by α-particles in thick targets of light elements. Ann. Nucl. Energy **10**(10), 541–552 (1983)
32. J.K. Bair, J. Gomez del Campo, Neutron yields from alpha-particle bombardment. Nucl. Sci. Eng. **71**(1), 18–28 (1979)
33. D. West, A.C. Sherwood, Measurements of thick-target (α,n) yields from light elements. Ann. Nucl. Energy **9**(11/12), 551–577 (1982)
34. J. Roberts, Neutron Yields of Several Light Elements Bombarded with Polonium Alpha Particles. U.S. AEC report MDDC-731 (1947)
35. W.B. Wilson, R.T. Perry, Thick-Target Neutron Yields in Boron and Fluorine. Los Alamos National Laboratory memorandum T-2-M-1835 to N. Ensslin (1987)
36. D. West, The calculation of neutron yields in mixtures and compounds from the thick target (α,n) yields in the separate constituents. Ann. Nucl. Energy **6**, 549–552 (1979)
37. J.F. Zeigler, SRIM-2013. http://www.srim.org/SRIM/SRIMLEGL.htm. Accessed 1 Feb 2023

38. W.B. Wilson, R.T. Perry, E.F. Shores, W.S. Charlton, T.A. Parish, G.P. Estes, T.H. Brown, M. Bozoian, T.R. England, D.G. Madland, J.E. Stewart, SOURCES 4C: A Code for Calculating (α,n), Spontaneous Fission, and Delayed Neutron Sources and Spectra. American Nuclear Society/Radiation Protection and Shielding Division, 12th Biennial Topical Meeting, Santa Fe, NM, April 14–18, 2002, Los Alamos National Laboratory report LA-UR-02-1839
39. N. Ensslin, W. Geist, J. Lestone, D.R. Mayo, H.O. Menlove, Cosmic ray background analysis for a cargo container counter, in *Proceedings of the 42nd Annual Meeting of the INMM,* Indian Wells, CA, July 15–19, 2001, Los Alamos National Laboratory report LA-UR-01-339 (2001)
40. Joint Publication Research Center, *Neutron Sources*, JPRS-48421 (JPRS, Washington, DC, July 1969)
41. International Atomic Energy Agency, Neutron Generators for Analytical Purposes. IAEA Radiation Technology Reports No. 1, Vienna, Austria (2012)
42. N. Marchese, A. Cannuli, M.T. Caccamo, C. Pace, New generation non-stationary portable neutron generators for biophysical applications of neutron activation analysis. Biochim. Biophys. Acta B **1861**(1), 3661–3670 (2017)

Open Access This chapter is licensed under the terms of the Creative Commons Attribution 4.0 International License (http://creativecommons.org/licenses/by/4.0/), which permits use, sharing, adaptation, distribution and reproduction in any medium or format, as long as you give appropriate credit to the original author(s) and the source, provide a link to the Creative Commons license and indicate if changes were made.

The images or other third party material in this chapter are included in the chapter's Creative Commons license, unless indicated otherwise in a credit line to the material. If material is not included in the chapter's Creative Commons license and your intended use is not permitted by statutory regulation or exceeds the permitted use, you will need to obtain permission directly from the copyright holder.

Neutron Interactions with Matter

M. T. Swinhoe, J. D. Hutchinson, and P. M. Rinard

14.1 Introduction

How neutrons interact with matter affects the ways that assays can be performed with neutrons. Neutron interactions with the assay material affect the interpretation of neutron measurements and limit the amount of fissile material that the assay instrument can contain safely. A neutron detector is based on some neutron interaction with the material in the detector. Also, the use of shielding materials during neutron measurements may be necessary to protect radiation workers.

This chapter provides fundamental information about neutron interactions that are important to nuclear material measurements. The first section describes the interactions on the microscopic level, where individual neutrons interact with other particles and nuclei. The concepts are then extended to macroscopic interactions with bulk compound materials.

14.2 Microscopic Interactions

14.2.1 The Cross-Section Concept

The probability that a particular event will occur between a neutron and a nucleus is expressed through the concept of the cross section. If a large number of neutrons of the same energy are directed into a thin layer of material, some can pass through with no interaction, others can have interactions that change their directions and energies, and still others can fail to emerge from the sample. A probability exists for each of these events. For example, the probability that a neutron will not emerge from a sample (that is, being absorbed or captured) is the ratio of the number of neutrons that do not emerge to the number originally incident on the layer. The cross section for being absorbed is the probability of neutrons being absorbed divided by the areal atom density (the number of target atoms per unit area of the layer). Therefore, the cross section has the dimensions of area; it must be a small fraction of a square centimeter because of the large number of atoms involved. Because this type of cross section describes the probability of neutron interaction with a single nucleus, it is called the *microscopic cross section* and is given the symbol σ. (A macroscopic cross section for use with bulk matter is defined in Sect. 14.3.)

Another approach to understanding the concept of the microscopic cross section is to consider the probability of a single neutron that attempts to pass through a thin layer of material that has an area A and contains N target nuclei, each of cross-sectional area s. The sum of the areas of all of the nuclei is Ns. The probability of a single neutron hitting one of these nuclei is roughly the ratio of the total target area Ns to the area of the layer A. In other words, the probability of a single neutron having a collision with a nucleus is Ns/A or (N/A)s, the areal target density times s; however, on the atomic level, cross sections for neutron interactions are not simply the geometrical cross-sectional area of the target. By replacing this s by the σ of the preceding paragraph, σ might be thought of as an effective cross-sectional area for the interaction. The cross section for the interaction retains the dimensions of area that s had.

The physical cross-sectional area s of a heavy nucleus is ~2 × 10^{-24} cm². Interaction cross sections for most nuclei are typically between 10^{-27} and 10^{-21} cm². To avoid the inconvenience of working with such small numbers, a different unit of area is used: the *barn*, denoted by the symbol b. It is defined to be 10^{-24} cm², so that the physical cross-sectional area of a heavy nucleus is ~2 b. Many neutron interaction cross sections range between 0.001 and 1000 b.

Each type of event has its own probability and cross section. The probability of each type of event is independent of the probabilities of others, so the total probability of any event occurring is the sum of the individual probabilities. Similarly, the sum of all of the individual cross sections is the total cross section.

14.2.2 The Energy-Velocity Relationship for Neutrons

Cross-section magnitudes are strong functions of neutron energy, as discussed in Sect. 14.2.4. As an introduction to that discussion, this section describes the relationship between neutron energy and velocity. This connection is important not only for understanding cross sections but also for estimating the time that neutrons are present in regions, such as those found in assay instruments.

The classical expression for kinetic energy, $E = mv^2/2$, is sufficiently accurate because even a kinetic energy of 100 MeV is still only about one-tenth of the rest mass energy of a neutron (939.55 MeV). For velocity (v) in meters per second and kinetic energy (E) in MeV,

$$E = 5.227 \times 10^{-15} v^2 \tag{14.1}$$

and

$$v = 1.383 \times 10^7 E^{\frac{1}{2}} \tag{14.2}$$

Figure 14.1 shows a graph of these equations for ready use. The graph shows, for example, that a 1 MeV neutron has a speed of 1.383×10^7 m/s and therefore will cross a 15 cm sample region in a typical assay instrument in ~11 ns. A thermal neutron with an energy of 0.025 eV (see Sect. 14.2.3) has a speed of 2187 m/s and will cross the same 15 cm region in ~70 μs.

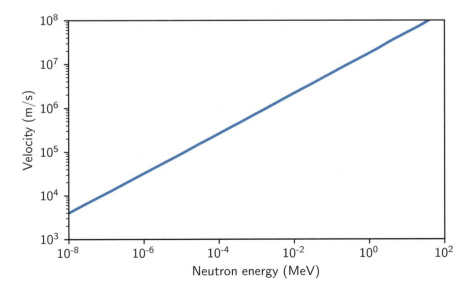

Fig. 14.1 Graph showing the relationship between a neutron's speed and its kinetic energy

14.2.3 Types of Interactions

A neutron can have many types of interactions with a nucleus. Figure 14.2 shows the types of interactions and their cross sections. Each category of interaction in the figure consists of those linked below it. The total cross section σ_t expresses the probability that any interaction will occur.

A simple notation can be used to give a concise indication of an interaction of interest. If a neutron n impinges on a target nucleus T—forming a resultant nucleus R and the release of an outgoing particle g—this interaction is shown as T(n,g)R. The heavy nuclei are shown outside the parentheses. To denote a type of interaction without regard for the nuclei involved, only the portion in parentheses is shown. An example of an (n,p) reaction is $^5B(n,p)^5Be$.

An interaction can be one of two major types: scattering or absorption. When a neutron is scattered by a nucleus, its speed and direction change, but the nucleus is left with the same number of protons and neutrons that it had before the interaction. The nucleus will have some recoil velocity, and it can be left in an excited state that will lead to the eventual release of radiation. When a neutron is absorbed by a nucleus, a wide range of radiations can be emitted, or fission can be induced.

Scattering events can be subdivided into elastic and inelastic scattering. In elastic scattering, the total kinetic energy of the neutron and nucleus is unchanged by the interaction. During the interaction, a fraction of the neutron's kinetic energy is transferred to the nucleus. For a neutron of kinetic energy E that encounters a nucleus of atomic weight A, the average energy loss is $2EA/(A + 1)^2$. This expression shows that to reduce the speed of neutrons (that is, to moderate them) with the fewest number of elastic collisions, target nuclei with small A should be used. By using hydrogen, with A = 1, the average energy loss has its largest value of E/2. A neutron with 2 MeV of kinetic energy will have (on the average) 1 MeV left after one elastic collision with a hydrogen nucleus, 0.5 MeV after a second such collision, and so on. To achieve a kinetic energy of only 0.025 eV would take a total of about 27 such collisions. (A neutron of energy 0.025 eV is roughly in thermal equilibrium with its surrounding medium and is considered a "thermal neutron." From the relation E = kT, where k is Boltzmann's constant, an energy E of 0.025 eV corresponds to a temperature T of 20 °C.) In general, after n elastic collisions, the neutron's energy is expected to change from E_0 to $E_n = E_0[(A^2 + 1)/(A + 1)^2]^n$. To reach E_n from E_0 thus requires n = $\log(E_n/E_0)/\log[(A^2 + 1)/(A +1)^2]$ collisions, on average. Table 14.1 gives examples of the number of collisions required to "thermalize" a 2 MeV neutron in some materials.

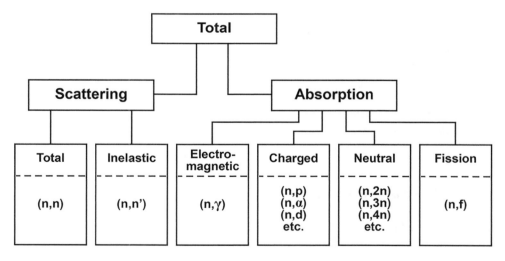

Fig. 14.2 Various categories of neutron interactions. The letters separated by commas in the parentheses show the incoming and outgoing particles

Table 14.1 Average number of collisions required to reduce a neutron's energy from 2 MeV to 0.025 eV by elastic scattering

Element	Atomic weight	Number of collisions
Hydrogen	1	27
Deuterium	2	31
Helium	4	48
Beryllium	9	92
Carbon	12	119
Uranium	238	2175

Inelastic scattering is like elastic scattering except that in the former, the nucleus undergoes an internal rearrangement into an excited state from which it eventually releases radiation. The total kinetic energy of the outgoing neutron and nucleus is less than the kinetic energy of the incoming neutron; part of the original kinetic energy is used to place the nucleus into the excited state. It is no longer easy to write an expression for the average energy loss because it depends on the energy levels within the nucleus. But the net effect on the neutron is again to reduce its speed and change its direction. If all excited states of the nucleus are too high in energy to be reached with the energy available from the incoming neutron, inelastic scattering is impossible. In particular, the hydrogen nucleus does not have excited states, so only elastic scattering can occur in that case. In general, scattering moderates or reduces the energy of neutrons and provides the basis for some neutron detectors (for example, proton recoil detectors).

Instead of being scattered by a nucleus, the neutron can be absorbed or captured. A variety of emissions can follow, as shown in Fig. 14.2. The nucleus can rearrange its internal structure and release one or more gamma rays. Charged particles can also be emitted; the more common ones are protons, deuterons, and alpha particles. The nucleus can also rid itself of excess neutrons. The emission of only one neutron is indistinguishable from a scattering event. If more than one neutron is emitted, the number of neutrons now moving through the material is larger than the number present before the interaction; the number is said to have been *multiplied*. Finally, a fission event could lead to two or more fission fragments (nuclei of intermediate atomic weight) and more neutrons (see Chap. 13).

Many safeguards instruments have neutron detectors that use an absorption reaction as the basis of the detection technique. The lack of an electric charge on the neutron makes direct detection difficult, so the neutron is first absorbed by a nucleus, which then emits a charged particle (such as a proton or deuteron). Helium-3, uranium-235, and boron-10 are commonly used in detectors because they have large absorption cross sections for the production of charged particles with low-speed neutrons.

When moderation alone is desired, absorption should be avoided. For example, hydrogen is a better moderator than deuterium (i.e., it requires fewer collisions to achieve a particular low speed), but it also has a larger absorption cross section for neutrons. The net effect is that deuterium will yield more thermal neutrons than hydrogen and may be the preferred moderating material.

The cross sections associated with the various interactions described above can be designated by the following notation:

σ_t = total cross section ($\sigma_s + \sigma_a$)
σ_s = total scattering cross section ($\sigma_{el} + \sigma_i$)
σ_{el} or $\sigma_{n,n}$ = elastic-scattering cross section
σ_i or $\sigma_{n,n'}$ = inelastic-scattering cross section
σ_a or σ_c = absorption or capture cross section
σ_{ne} = nonelastic cross section $\sigma_t - \sigma_{el}$
$\sigma_{n,\gamma}$ = radiative capture cross section
σ_f or $\sigma_{n,f}$ = fission cross section
$\sigma_{n,p}$ = (n,p) reaction cross section
$\sigma_{n,\alpha}$ = (n,α) reaction cross section.

14.2.4 Energy Dependence of Cross Sections

All of the cross sections described previously vary with neutron energy and with the target nucleus, sometimes in a dramatic way. This section gives generalizations about the energy dependence of cross sections and shows data [1] for a few important nuclei.

Figure 14.3 illustrates the total cross section for ^{239}Pu for incident neutrons of 0.001 eV to 10 MeV energy. As a general rule, the cross section decreases with increasing energy. At low energies (below 1 MeV), the elastic cross section is nearly constant, whereas the inelastic-scattering cross section and absorption cross sections are proportional to the reciprocal of the neutron's speed (i.e., 1/v). So at low energies, the total cross section can be nearly constant or decreasing with energy depending on which type of event dominates. For example, in ^{239}Pu, the inelastic cross section dominates, and the total cross section varies as 1/v. Similar behavior is observed for most light- and intermediate-weight nuclei as well. Figures 14.4 and 14.5 illustrate the low-energy total cross-section behavior of boron and cadmium. The unusually high absorption cross sections of these two materials make them useful as thermal-neutron poisons.

At higher energies, the cross section can have large peaks superimposed on the 1/v trend. These peaks are called *resonances* and occur at neutron energies where reactions with nuclei are enhanced. For example, a resonance will occur if

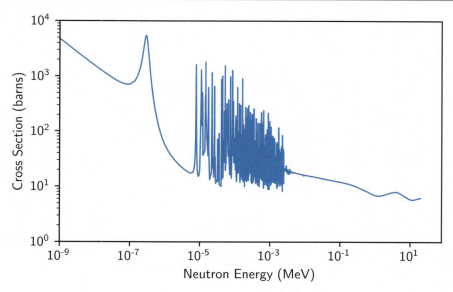

Fig. 14.3 Total neutron cross section of ^{239}Pu

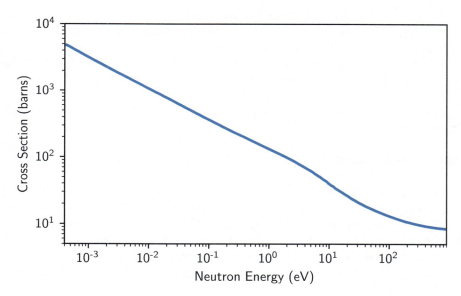

Fig. 14.4 Low-energy total neutron cross section of boron [1]

the target nucleus and the captured neutron form a compound nucleus and the energy contributed by the neutron is close to that of an excited state of the compound nucleus.

In heavy nuclei, large and narrow resonances appear for neutron energies in the eV range. For energies in the keV region, the resonances can be too close together to resolve. In the MeV region, the resonances are more sparse and very broad, and the cross sections become smooth and rolling. For light nuclei, resonances appear only in the MeV region and are broad and relatively small. For nuclei with intermediate weights (such as cadmium, nickel, and iron), resonances can be found below 1 keV. These resonances have heights and widths between those of light and heavy nuclei.

Some exceptions to the general trends exist in ^{1}H and ^{2}H where there are no resonances at all and in nuclei with "magic" numbers of protons or neutrons where the behavior can be similar to that of light nuclei despite the actual atomic weight. In practice, it is necessary to rely on tables of cross sections for the nuclei of interest because there is no convenient way to calculate cross sections.

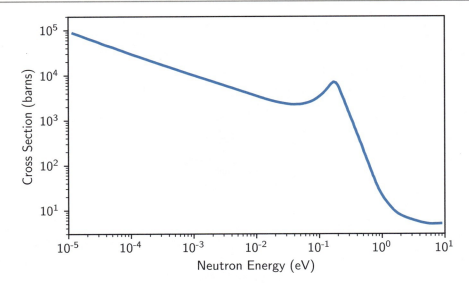

Fig. 14.5 Low-energy total neutron cross section of cadmium [1]

Some neutron-induced fission cross sections important for nondestructive assay are shown in Fig. 14.6. The fissile isotopes ^{235}U and ^{239}Pu have large cross sections (~1000 b) for fission by thermal or near-thermal neutrons. For fission by fast neutrons (10 keV to 10 MeV), these cross sections are reduced to 1–2 b. The fertile isotopes ^{238}U and ^{240}Pu have negligible fission cross sections for low-energy neutrons but exhibit a threshold near 1 MeV neutron energy. Above 1 MeV, the fission cross sections of the fertile isotopes are comparable to those of the fissile isotopes.

14.3 Macroscopic Interactions

14.3.1 Macroscopic Cross Sections

Although study of the interactions of a neutron with a single nucleus on the microscopic scale provides a basis for understanding the interaction process, measurements are actually performed with thick samples that often contain a mixture of elements. These additional features are described by using the macroscopic cross sections appropriate for bulk materials.

The definition of the macroscopic cross section arises from the transmission of a parallel beam of neutrons through a thick sample. The thick sample can be considered a series of atomic layers; for each layer, we can apply the results found with the microscopic cross-section concept. By integrating through enough atomic layers to reach a depth x in the sample, the intensity I(x) of the uncollided neutron beam is

$$I(x) = I_0 e^{-N\sigma_t x}, \tag{14.3}$$

where I_0 is the intensity of the beam before it enters the sample, N is the atom density, and σ_t is the total cross section. Figure 14.7 shows the uncollided intensity remaining in a parallel beam as it passes through a thick layer of matter. Note that the fraction transmitted without collisions, $I(x)/I_0$, depends on the energy of the neutrons through the energy dependence of the microscopic total cross section σ_t.

An expression similar to Eq. 14.3 is used for gamma-ray attenuation. In that case, low-energy gamma rays are very likely to be absorbed and thus removed not only from the parallel beam but from the material entirely. With neutrons at low energies, elastic scattering is the most likely event. Although Eq. 14.3 gives the intensity of the neutrons that have had no interaction up to a depth x, the actual number of neutrons present that can be detected can be much larger because of multiple scattering, multiplication, or finite detector acceptance angle.

The total macroscopic cross section is $\Sigma_t = N\sigma_t$. Σ_t has dimensions of cm^{-1} (see Eq. 14.3) and is analogous to the linear attenuation coefficient for gamma rays. If only a particular type of interaction is of interest, a macroscopic cross section for it

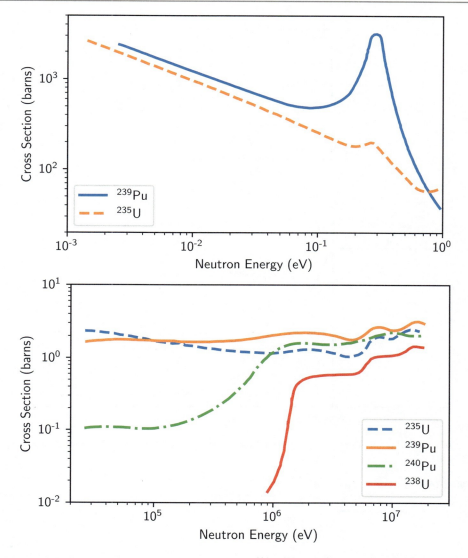

Fig. 14.6 Neutron induced fission cross sections for some important fissile (^{235}U, ^{239}Pu) and fertile (^{238}U, ^{240}Pu) isotopes [1]

alone can be defined using its microscopic cross section in place of the total cross section. For quantitative calculations, the concept of macroscopic cross section is less used than the analogous gamma-ray linear attenuation coefficient because of the complications of multiple scattering and other effects mentioned in the previous paragraph.

If the sample is a compound instead of a simple element, the total macroscopic cross section is the sum of the macroscopic cross sections of the individual elements:

$$\Sigma = \Sigma_1 + \Sigma_2 + \Sigma_3 \ldots \tag{14.4}$$

The atom density N_i of each element i is given by

$$N_i = \rho N_a \frac{n_i}{M}, \tag{14.5}$$

where ρ is the density of the compound; M is the molecular weight of the compound; N_a is Avogadro's number, 6.022×10^{23} atoms/mole; and n_i is the number of atoms of element i in one molecule. From Eqs. 14.4 and 14.5, the general form of the macroscopic cross section can be written as

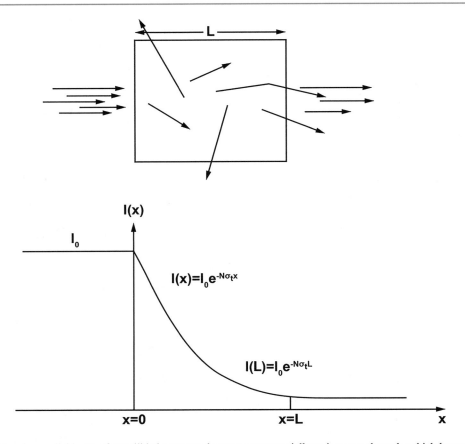

Fig. 14.7 The intensity of a parallel beam of uncollided neutrons decreases exponentially as it passes through a thick layer of matter

Table 14.2 Nuclear data for $^{nat}UO_2$

Isotope	n_i	σ_t at 1 MeV(b)
^{235}U	0.007	6.84
^{238}U	0.993	7.10
^{16}O	2.000	8.22

$$\Sigma = \frac{\rho N_a}{M}(n_1\sigma_1 + n_2\sigma_2 + n_3\sigma_3 + \ldots) \tag{14.6}$$

As an illustration of these equations, the total macroscopic cross section for 1 MeV neutrons in $^{nat}UO_2$ (density 10 g/cm^3, molecular weight 270) is calculated from the data shown in Table 14.2.

$$\Sigma_t = \frac{(10)(0.6022)}{270}[(0.007)(6.84) + (0.993)(7.10) + 2(8.22)] = 0.525 cm^{-1} \tag{14.7}$$

Powers of 10^{24} and 10^{-24} have been cancelled in Avogadro's number and in the cross-section values. These cross-section values were taken from Table 14.3 [2], which is a compilation of microscopic and macroscopic cross-sections at two neutron energies, 0.025 eV (thermal) and 1 MeV.

14 Neutron Interactions with Matter

Table 14.3 Neutron cross sections of common materials [2]

Material	Atomic or molecular weight	Density (g/cm³)	Cross sections[a] E = 0.0253 eV				E = 1 MeV			
			$\sigma_t(b)$	$\sigma_a(b)$	$\Sigma_t(cm^{-1})$	$\Sigma_a(cm^{-1})$	$\sigma_t(b)$	$\sigma_a(b)$	$\Sigma_t(cm^{-1})$	$\Sigma_a(cm^{-1})$
Al	27	2.7	1.61	0.232	0.097	0.014	2.37	0.000	0.143	0.000
B	10	2.3	3845	3843	533	532	2.68	0.189	0.371	0.0262
B	11	2.3	5.28	0.005	0.665	0.0006	2.13	0.000	0.268	0.000
Be	9	9.0	6.35	0.010	3.82	0.0060	3.25	0.003	1.96	0.0018
C	12	1.9	4.95	0.003	0.472	0.0003	2.58	0.000	0.246	0.000
Nat ca	40.08	1.55	3.46	0.433	0.081	0.101	1.14	0.004	0.027	0.0001
Cd	112	8.7	2470	2462	115.5	115.2	6.50	0.058	0.304	0.0027
Nat cl	35.45	Gas	50.2	33.4	Gas	Gas	2.30	0.0005	Gas	Gas
Nat cu	63.55	8.94	12.5	3.80	1.06	0.322	3.40	0.011	0.288	0.0009
F	19	Gas	3.72	0.010	Gas	Gas	3.15	0.000	Gas	Gas
Fe	56	7.9	14.07	2.56	1.19	0.217	5.19	0.003	0.441	0.0003
Nat Gd	157.25	7.95	49,153	48,981	1496	1491	7.33	0.223	0.223	0.0068
H	1	Gas	30.62	0.33	Gas	Gas	4.26	0.000	Gas	Gas
H	2	Gas	4.25	0.000	Gas	Gas	2.87	0.000	Gas	Gas
He	3	Gas	5337	5336	Gas	Gas	2.87	0.879	Gas	Gas
He	4	Gas	0.86	0.000	Gas	Gas	7.08	0.000	Gas	Gas
Li	6	0.534	938	937	50.3	50.2	1.28	0.230	0.069	0.0123
Li	7	0.534	1.16	0.036	0.053	0.0017	1.57	0.000	0.072	0.000
Nat Mg	24.31	1.74	3.47	0.063	0.150	0.0027	2.66	0.001	0.115	0.000
Mn	55	7.2	14.5	13.2	1.14	1.04	3.17	0.003	0.250	0.0002
N	14	Gas	12.22	1.9	Gas	Gas	2.39	0.021	Gas	Gas
Na	23	0.971	3.92	0.529	0.100	0.0134	3.17	0.000	0.081	0.000
Ni	59	8.9	23.08	4.58	2.10	0.416	3.66	0.0008	0.322	0.0001
O	16	Gas	3.87	0.000	Gas	Gas	8.22	0.000	Gas	Gas
Pb	204	11.34	11.40	0.18	0.381	0.0060	4.39	0.0033	0.147	0.0001
Pu	238.05	19.6	599.3	562.0	29.72	27.87	6.66	0.190	0.330	0.0094
Pu	239.05	19.6	1021	270	50.4	13.3	7.01	0.026	0.346	0.0013
Pu	240.05	19.6	294	293	14.5	14.4	7.15	0.108	0.352	0.0053
Pu	241.06	19.6	1390	362	68.1	17.7	7.98	0.117	0.391	0.0057
Pu	242.06	19.6	26.7	18.9	1.30	0.922	7.31	0.098	0.357	0.0048
Nat Si	28.09	2.42	2.24	0.161	0.116	0.0084	4.43	0.001	0.230	0.0001
Th	232	11.3	20.4	7.50	0.598	0.220	7.00	0.135	0.205	0.0040
U	233.04	19.1	587	45.8	29.0	2.26	6.78	0.069	0.335	0.0034
U	234.04	19.1	116	103	5.70	5.07	8.02	0.363	0.394	0.0178
U	235.04	19.1	703	96.9	34.3	4.74	6.84	0.117	0.335	0.0057
U	236.05	19.1	13.3	5.16	0.648	0.251	7.73	0.363	0.377	0.0177
U	237.05	19.1	487.5	476.4	23.6	23.1	6.72	0.135	0.326	0.0066
U	238.05	19.1	11.63	2.71	0.562	0.131	7.10	0.123	0.343	0.0059
Nat U	238.03	19.1	16.49	3.39	0.797	0.1637	7.01	0.120	0.343	0.0058
Nat W	183.85	19.3	23.08	18.05	1.459	1.141	6.95	0.057	0.439	0.0036
CH_2	14	0.94			2.68	0.027			0.449	0.0000
H_2O	18	1.0			2.18	0.022			0.560	0.0000
D_2O	20	1.1			0.410	0.000			0.420	0.0000
	Average fission products of:									
^{235}U	117		4496	4486			7.43	0.00036		
^{239}Pu	119		2087	2086			7.48	0.00093		

[a] A zero value means zero to the number of figures shown

14.3.2 Mean Free Path and Reaction Rate

A very descriptive feature of the transmission of neutrons through bulk matter is the mean free path length, which is the mean distance that a neutron travels between interactions. It can be calculated from Eq. 14.3, with $N\sigma_t$ replaced by Σ_t. The mean free path length λ is

$$\lambda = \frac{1}{\Sigma_t}, \qquad (14.8)$$

the reciprocal of the macroscopic cross section. For the case of 1 MeV neutrons in UO_2 calculated in Eq. 14.8, a macroscopic cross section of 0.525 cm^{-1} implies a mean free path length of 1.9 cm.

The mean free path length has many qualitative applications in assay instruments and shielding.

- If the mean free path length of neutrons emitted by a sample in a passive assay instrument is long compared with the dimensions of the sample, it is likely that most of the neutrons will escape from the sample and enter the detection region.
- If the number of collisions required to thermalize a neutron is known, the necessary moderator thickness of a shield can be estimated.
- If the thickness of a shield is many times the mean free path length of a neutron that tries to penetrate the shield, then the shield fulfills its purpose. (Because the mean free path length is a function of the neutron's energy, the actual calculation is not so simple.)

A closely related concept is the reaction rate. When traveling with a speed v, a neutron has an average time between interactions of λ/v. The reaction rate is the frequency with which interactions occur, v/λ, or $v\Sigma_t$. In uranium oxide, for example, a 1 MeV neutron will have a reaction rate of 7.26×10^8 per second (from Eqs. 14.2 and 14.7). However, this calculation does not mean that in 1 s, that many reactions will occur; it means that with each collision, the neutron's energy decreases, and the cross section changes, thereby altering the instantaneous reaction rate.

Neutron flux, ϕ, is defined as the number of neutrons that cross unit cross-sectional area (in all directions) in unit time. We can use this concept to write a general expression for the reaction rate for any particular reaction—for example, fission—as

$$\textit{Fission rate per sec} = \phi \Sigma_f V, \qquad (14.9)$$

where ϕ is the neutron flux (n/cm^2/sec), Σ_f is the macroscopic fission cross section (cm^{-1}), and V is the volume (cm^3) over which we want to calculate the reaction rate. Note that $\phi \Sigma_f V$ can also be written $\phi \sigma_f N_{tot}$, where N_{tot} is the total number of atoms of the nuclide. This expression works for beams (with monodirectional neutrons as in Eq. 14.3) as well as reactors (with neutrons that move in all directions). Of course, the limitation is that the flux and the cross section must be uniform over the relevant volume. If we were to put a very small amount of fissile material into a region of fixed flux, we could calculate the reaction rate using Eq. 14.9. However, if we were to put larger and larger amounts of fissile material into the region, the reaction rate would not increase linearly with mass (as Eq. 14.9 implies) because the average neutron flux over the sample volume is reduced by the reactions that take place in the fissile material. This effect is referred to as *self-shielding* and is a very important concept in understanding the operation of neutron instrumentation (as well as nuclear reactors).

The paths of neutrons in matter can be simulated with Monte Carlo calculations. Figure 14.8 shows a few paths for neutrons that have 1 MeV of energy entering cylinders of different materials. The mean free path length depends on both the type of material and the energy of the neutron. After each collision, the energy is decreased, and the mean free path length is affected accordingly. Figure 14.8 shows that a cylinder of polyethylene is more effective in preventing the transmission of neutrons than a cylinder of heavy metal. A neutron loses most of its energy by colliding with the light elements in polyethylene, and then the mean free path length becomes small as the cross sections increase. An important effect of polyethylene is that it seemingly retains a large fraction of the neutrons near a certain depth; these neutrons have had enough collisions to lose nearly all of their kinetic energy. If a thermal-neutron detector is placed in this region, the chance of detecting neutrons is optimized.

14 Neutron Interactions with Matter

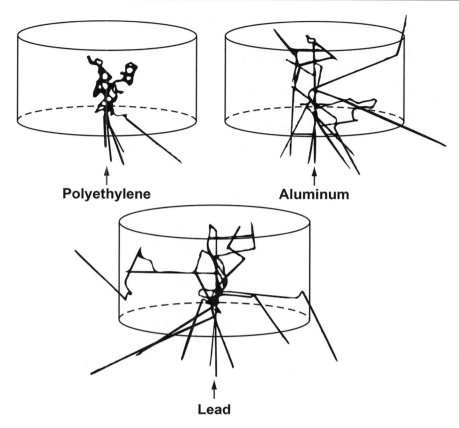

Fig. 14.8 Neutrons with 1 MeV of kinetic energy are shown entering cylinders of material from the bottom and then being scattered or absorbed. The paths were calculated using a Monte Carlo technique

14.4 Effects of Moderation in Bulk Matter

It is often a design goal to reduce or moderate the speed of neutrons in the sample region or in the detector region or in both. Recalling the general 1/v trend of interaction cross sections (Figs. 14.3 through 14.6), the purpose of the reduction in speeds is to increase the probability of an interaction. In other regions, it may be desirable to hinder interactions by choosing materials that are poor moderators or by adding low-energy neutron absorbers to remove neutrons once they become moderated.

For example, in the assay of plutonium, moderation is not a desirable effect in the sample region. High-speed neutrons are more able to penetrate the sample, and they have lower fission cross sections so that multiplication is less than with low-speed neutrons. Conversely, in the detector region, moderation increases the detection efficiency for detectors such as ^{3}He proportional counters. By placing hydrogenous material (such as polyethylene) around the detectors, the neutrons can be counted with more efficiency. Also needed is a filter that will let high-speed neutrons enter the detector region where they can become moderated but prevent the moderated neutrons' return to the sample region where they could produce additional fissions. A layer of material with a large absorption cross section for slow neutrons (such as cadmium, Fig. 14.5) placed between the sample region and the detector region is effective in this regard.

A standard basis for comparing moderating abilities of different materials is the moderating power. If one material has a larger moderating power than another, less of that material is needed to achieve the same degree of moderation. Two factors are important: the probability of a scattering interaction and the average change in kinetic energy of the neutron after such an interaction. To be an effective moderator, both the probability of an interaction and the average energy loss in one scatter should be high. The moderating power is defined as $\xi\Sigma_s$, where Σ_s is the macroscopic scattering cross section and ξ is the average logarithmic energy decrement in a scatter. This decrement is $\ln(E_{before}) - \ln(E_{after})$. When elastic collisions in an element with atomic weight A dominate the scattering process, the decrement becomes

Table 14.4 Moderating powers and ratios of selected materials [4]

Moderator	Moderating power (1 eV to 100 keV)	Moderating ratio (Approximate)
Water	1.28	58
Heavy water	0.18	21,000
Helium at STP	0.00001	45
Beryllium	0.16	130
Graphite	0.064	200
Polyethylene (CH$_2$)	3.26	122

STP standard temperature and pressure

$$\xi = 1 - \frac{(A-1)^2}{2A} \ln \frac{(A+1)}{(A-1)}. \tag{14.10}$$

For $A > 2$, ξ can be approximated by $2/(A + 0.67)$ [3]. The moderating power of a compound is given by

$$\xi \Sigma_s = \frac{\rho N_a}{M}(n_1 \sigma_1 \xi_1 + n_2 \sigma_2 \xi_2 + \ldots), \tag{14.11}$$

where ρ is the density of the compound, M is its molecular weight, N_a is Avogadro's number, n_i is the number of atoms of element i in one molecule, σ_i is the microscopic scattering cross section for element i, and ξ_i is the logarithmic energy decrement for element i.

A material with a large moderating power might nevertheless be useless as a practical moderator if it has a large absorption cross section. Such a moderator would effectively reduce the speeds of those neutrons that are not absorbed, but the fraction of neutrons that survive may be too small to be used in a practical manner. A more comprehensive measure of moderating materials is the moderating ratio, $\xi \Sigma_s/\Sigma_a$. A large moderating ratio is desirable; it implies not only a good moderator but also a poor absorber. For a compound, the moderating ratio is given by Eq. 14.10 with each σ_i replaced by σ_s/σ_a for element i.

Table 14.4 gives the moderating powers and ratios for some common moderator materials for neutrons in the 1 eV to 100 keV energy range [4]. Ordinary water has a higher moderating power than heavy water because the atomic weight of hydrogen is half that of deuterium. But the hydrogen nucleus (a proton) can absorb a neutron and create deuterium much more readily than a deuterium nucleus can absorb a neutron and create tritium. This difference in absorption cross sections gives heavy water a much more favorable moderating ratio; however, because of its availability and low cost, ordinary water is often preferred. The solid materials given in the table have a higher moderating ratio than ordinary water and can have fabrication advantages. Polyethylene is commonly selected as a moderator because of its high moderating power and moderating ratio.

14.5 Effects of Multiplication in Bulk Matter

When a neutron interaction yields more than one neutron as a product, a multiplication event has occurred. More neutrons will be present in the material after the interaction than before. The most widely known multiplication event is fission, but other absorption interactions, such as (n,2n), can be important contributors to multiplication.

Of the neutrons in a given material at a given moment, some will eventually escape, and the others will be absorbed. Additional neutrons can originate in the material as products of the absorptions. The definition of the multiplication M is the total number of neutrons that exist in the sample divided by the number of neutrons that were started. If 100 neutrons are started in the sample and an additional 59 are found to be created from multiplication events, the multiplication is 1.59. Only a fraction of the first generation of 100 neutrons produces additional neutrons through multiplication events; the others escape or are absorbed by other types of interactions. The same fraction of the second generation produce a third generation and so on. The number of neutrons remaining in the sample steadily decreases until it is zero, and the total number of neutrons produced by all of the multiplication events is 59.

A related concept more commonly used is the multiplication factor, which relates the numbers of neutrons in successive generations. Two categories of multiplication factors apply to different physical sizes of the material involved. If the material

Table 14.5 Example of neutron population decline

Generation	Average no. of neutrons for $k_{eff} = 0.37$
1	100
2	37
3	14
4	5
5	2
6	1
7	0
	159

is infinite in extent, the multiplication factor is written k_∞ and is defined as the ratio of the number of neutrons in one generation to the number in the previous generation. Because of the infinite size of the material, all neutrons of a generation become absorbed. Thus k_∞ is also the ratio of the number of neutrons produced in one generation to the number absorbed in the preceding generation.

If the material is not infinite in size, some neutrons in a generation may escape through the surface and not be absorbed; these are "leakage" neutrons. The multiplication factor for this more practical situation is called k_{eff}. It is defined as the ratio of the number of neutrons produced in one generation to the number either absorbed or leaked in the preceding generation. The multiplication factor k_{eff} is the more practical ratio for safeguards work because instruments are often made small to comply with size and weight constraints.

As an example of k_{eff} and its connection with the multiplication M, consider the case described earlier. The original 100 neutrons would constitute the first generation. If the original neutrons create 37 neutrons through reactions, the 37 neutrons would be the next generation. The multiplication factor in this case is thus 0.37. With k_{eff} less than one, the number of neutrons in succeeding generations decreases, eventually reaching zero. As Table 14.5 indicates, it takes seven generations to reduce the number of neutrons from 100 to about zero.

The number of neutrons in one generation is found by multiplying the number in the previous generation by the multiplication factor k_{eff}, which in this example is 0.37. This is a statistical process, of course, and the exact number in any generation cannot be exactly known, but for large numbers of neutrons, the ratio of populations in successive generations is very nearly constant.

The multiplication M is readily connected to the multiplication factor k_{eff} when k_{eff} is less than one. By adding together all of the numbers of neutrons in all of the generations, the geometric sum can be found.

$$M = \frac{1}{1 - k_{eff}}, k_{eff} < 1. \qquad (14.12)$$

With the multiplication of 1.59 in Table 14.5, the multiplication factor is 0.37, showing that the number of neutrons is decreasing in successive generations.

As k_{eff} approaches one, the value of M becomes larger and larger, as shown in Fig. 14.9. When $k_{eff} = 1$, the formula shows that there is no limit to the number of neutrons that will be produced; in practice, there is a finite number of nuclei that can produce neutrons, so the number of neutrons created is finite but extremely large. Criticality is said to be reached when $k_{eff} = 1$. If k_{eff} is larger than 1, the sample is supercritical; with k_{eff} less than 1, the sample is subcritical.

The term *fission chain* is often used to refer to neutrons that have a common ancestor. As an example, assume that a neutron induced a fission that produced two neutrons. And then assume that those two neutrons each induced fission that also each produced two neutrons (as shown in Fig. 14.10). In this example, all seven neutrons (the starting neutron, the two neutrons created by the first fission, and the four neutrons created from the subsequent fissions) are part of the same fission chain. The probability of having a fission chain of some length will increase as the multiplication of the system increases. Similarly, the average length of fission chains in the system will also increase as the multiplication increases. See [5] for more information on fission chains.

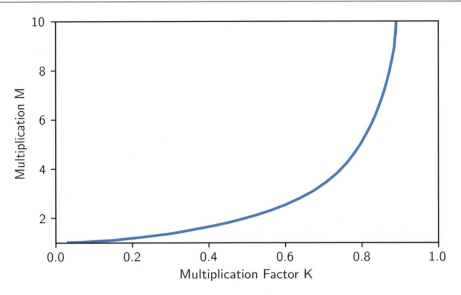

Fig. 14.9 The multiplication M is shown as a function of the multiplication factor k_{eff}. Only subcritical values (k_{eff} less than 1) are included here

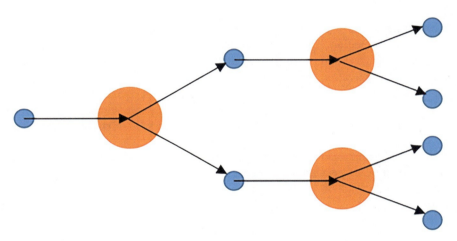

Fig. 14.10 Example of a fission chain. The blue circles are neutrons, and the orange circles are nuclei of a fissionable nuclide (such as ^{235}U). All seven neutrons in this example are part of the same fission chain

14.6 Neutron Shielding

To protect personnel from the biological effects of neutrons and to reduce background counts, neutron shielding is often necessary. The selection and arrangement of shielding materials vary with the circumstances. Some general principles can be derived from the neutron interactions with matter described earlier in this chapter.

Thermal neutrons that have energies of 0.025 eV or less are absorbed with great effectiveness by thin layers of boron or cadmium, as suggested by the large cross sections shown in Figs. 14.4 and 14.5. Boron is often used in the form of boron carbide (B_4C) or boron-loaded solutions. One commonly used material is boral, a mixture of boron carbide and aluminum, which is available in sheets of varying thickness. Cadmium has the disadvantage of emitting high-energy gamma rays after neutron capture, which may necessitate additional gamma-ray shielding. A comparison of Figs. 14.4 and 14.5 shows that cadmium is more effective than boron for absorbing thermal neutrons, whereas boron is more effective for absorbing epithermal neutrons (energy range of 0.1–10 eV).

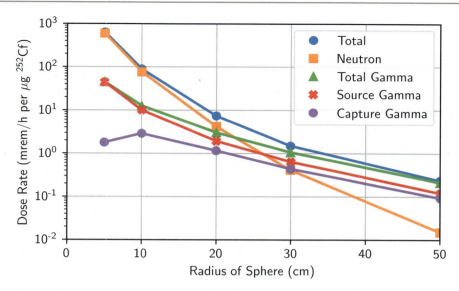

Fig. 14.11 Various dose rates on a spherical surface from a point ^{252}Cf source in polyethylene. (Data supplied by G. E. Bosler, Los Alamos, New Mexico)

High-speed neutrons are more difficult to shield against because absorption cross sections are much lower at higher energies; therefore, it is first necessary to moderate high-speed neutrons through elastic or inelastic-scattering interactions. Inelastic scattering or absorption could again produce potentially hazardous gamma rays; for example, neutron capture in hydrogen releases a 2.224 MeV gamma ray. An effective radiation shield consists of a combination of materials: hydrogenous or other low-A materials to moderate neutrons; thermal-neutron absorbers; and high-Z materials to absorb gamma rays. Examples of hybrid shielding materials are polyethylene and lead, concrete that contains scrap iron, and more exotic materials such as lithium hydride.

In safeguards work with small samples of fissionable materials or weak neutron sources, shielding may be restricted to several centimeters of polyethylene. The shielding properties of polyethylene are illustrated in Fig. 14.11, which gives dose rates on the surface of a sphere of polyethylene with a ^{252}Cf source in its center. The source emits neutrons with a high-energy fission spectrum comparable to most uranium and plutonium isotopes. Also produced are fission gamma rays and additional 2.22 MeV gamma rays from neutron capture in polyethylene. Neutrons provide most of the dose for spheres less than 22 cm in radius; beyond that, source gamma rays are the major contributors, followed by gamma rays from capture reactions. By increasing the radius from 5 to 12 cm or from 20 to 37 cm, the total dose rate can be reduced by a factor of 10. A rule of thumb for neutron dose reduction is that 10 cm of polyethylene will reduce the neutron dose rate by roughly a factor of 10.

More effective shields can sometimes be obtained by adding boron, lithium, or lead to polyethylene. The addition of boron or lithium results in a lower capture gamma-ray dose than that provided by pure polyethylene; lead effectively attenuates the source and the capture gamma-ray flux because it is a heavy element. However, because neutrons provide most of the dose up to a radius of 22 cm, the addition of these materials has little effect until the shield becomes substantially thicker than 22 cm. Figure 14.12 shows the effects on the total dose of adding other materials to polyethylene; these effects are important only for shields thicker than 30 cm. Boron- lithium-, or lead-loaded polyethylene is substantially more expensive than pure polyethylene, so the additional cost is an important consideration.

14.7 Transport Calculational Techniques

Neutron histories are difficult to determine because of the large number of different interactions possible in materials. This difficulty is further increased when the composition of matter changes frequently along the path of a neutron, as it often does throughout the volume of an assay instrument. Techniques for calculating the behavior or transport of neutrons and gamma rays in such circumstances are important for the design of assay instruments, the interpretation of measurements, and the development of shielding configurations. Two techniques for calculating the transport of neutrons in matter are described briefly in Sects. 14.7.1 and 14.7.2.

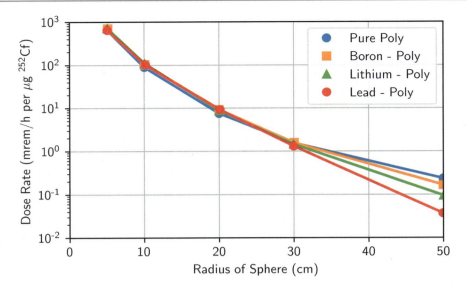

Fig. 14.12 Total dose rate on a spherical surface from a point ^{252}Cf source in various materials. (Data supplied by G. E. Bosler, Los Alamos, New Mexico)

14.7.1 Monte Carlo Techniques

The probability that a neutron interaction could occur is an important feature in the description of neutrons traveling through matter. Instead of trying to predict what an individual neutron might do, one can use procedures to predict what fraction of a large number of neutrons will behave in some manner of interest. Calculational techniques that, in simplistic terms, predict neutron events with "rolls of dice" (actually the generation of random numbers in a computer) are called *Monte Carlo methods*. The response of an assay system can often be calculated from the transport of many individual neutrons despite the inclusion of a few improbable neutron histories that deviate drastically from the average behavior.

The Monte Carlo method can allow a detailed, three-dimensional geometrical model to be constructed mathematically to simulate a physical situation. A neutron can be started at a selected location with a certain energy and direction. It travels distances that are consistent with the mean free path lengths in the materials, with random variations from the expected mean. At the end of each step in the neutron's path, a decision may be made to simulate a certain interaction, with the decision based on the cross section for the interaction with that material at that neutron energy. If an interaction is selected, the results of the interaction are simulated and its consequences followed. Eventually, a point is reached where no further interest in the neutron exists, and its history is terminated. This result might occur with the escape of the neutron or its moderation to very low energy. The neutron might be absorbed, followed by the emission of a gamma ray of no interest, or it might undergo a multiplication event. If a multiplication event occurs, the histories of the new neutrons are followed. In principle, the history of a simulated neutron is one that might actually occur with a real neutron.

By repeating this procedure for many thousands of neutrons and by keeping tallies of how many enter a detector region, how many cause fissions, how many escape through a shield, or whatever else is of interest, an average behavior and its uncertainty are gradually deduced. Many specialized techniques can be used to get good average values with the fewest number of neutrons, but cases exist where even a fast computer cannot provide enough histories within the constraints of time and budget. Nonetheless, Monte Carlo techniques provide essential assistance in design work by closely modeling the actual geometry of a problem and by having imaginary neutrons that simulate the motions and interactions of real ones. Examples of the results of Monte Carlo calculations are the shielding calculations in Figs. 14.11 and 14.12 and the coincidence counter design calculations described in Chaps. 17, 19, and 20.

The main codes used for neutron transport calculation for safeguards are the Monte Carlo N-Particle (MCNP) code [6] and GEANT [7]. The MCNP code has many included capabilities that make it easy to use for the general user, but to make specific modifications requires specialized skill and knowledge [8]. A book that details the use of the MCNP code for safeguards-relevant detector simulations was published in 2022 [9]. A special version of the MCNP code (MCNP-PoLiMi, [10]) provides some additional capabilities relevant to safeguards (for example, scintillator modeling). GEANT has fewer built-in capabilities, and the user needs to provide some of the functionality, but it is easier than the MCNP code to make modifications to include specialty capabilities.

14.7.2 Discrete Ordinates Techniques

Analytical transport equations exist that describe the exact behavior of neutrons in matter; however, only approximate numerical solutions to these equations can be obtained for complicated systems. Procedures for obtaining these numeral solutions are classified as discrete ordinates techniques.

Some important differences distinguish discrete ordinates techniques from Monte Carlo techniques. Only one- or two-dimensional geometries are generally practical with a discrete ordinates process, and the neutrons are considered to be at discrete locations instead of moving freely through a three-dimensional geometry. In a two-dimensional discrete ordinates case, for example, it is as if the surface material were covered by a wire mesh, and the neutrons existed only at the intersections of the wires. Furthermore, the energy of a neutron at any time must be selected from a finite set—in contrast to the continuously varying energy of a neutron in the Monte Carlo method.

Despite these disadvantages, discrete ordinates techniques can produce useful results in many cases. For problems that involve large volumes and amounts of materials (such as reactor cores), the Monte Carlo technique can be too cumbersome and slow; a discrete ordinates solution might be feasible. The previous sentence was written around 1990 and, despite the astounding progress in computing capability since then, is still true to some extent. The SCALE suite of codes [11] for nuclear safety analysis and design, for example, offers both discrete ordinates and Monte Carlo capabilities to cover cases where Monte Carlo methods are still too slow.

References

1. D.I. Garber, R. R. Kinsey, Neutron cross sections, vol. II, curves, Brookhaven National Laboratory report BNL 325 (1976)
2. Evaluated Nuclear Data File ENDF/B-V, available from and maintained by the National Nuclear Data Center of Brookhaven National Laboratory, https://www.nndc.bnl.gov/endf/. Accessed Mar 2023
3. J.R. Lamarsh, *Introduction to Nuclear Reactor Theory* (Addison-Wesley, Reading, 1966)
4. S. Glasstone, A. Sesonske, *Nuclear Reactor Engineering* (D. Van Nostrand, Princeton, 1967)
5. S.D. Nolen, The chain-length distribution in subcritical systems, PhD thesis, Los Alamos National Laboratory report LA-13721-T (2000)
6. The MCNP® Code., https://mcnp.lanl.gov for latest information. Accessed Feb 2023
7. J. Allison et al., Recent developments in GEANT4. Nucl. Instrum. Methods Phys. Res., Sect. A **835**, 186–225 (2016)
8. See webpage mcnp.lanl.gov for latest information
9. J.S. Hendricks, M.T. Swinhoe, A. Favalli *Monte Carlo N-Particle Simulations for Nuclear Detection and Safeguards—An Examples-Based Guide for Students and Practitioners* (Springer open access, 2022) ISBN 978-3-031-04128-0, https://doi.org/10.1007/978-3-031-04129-7. Accessed Feb 2023
10. S.A. Pozzi, E. Padovani, M. Marseguerra, MCNP-PoLiMi: A Monte Carlo code for correlation measurements. Nucl. Instrum. Meth. A **513**, 550–558 (2003)
11. SCALE 6.2.4, available from and maintained by Oak Ridge National Laboratory, https://www.ornl.gov/scale. Accessed Feb 2023

Open Access This chapter is licensed under the terms of the Creative Commons Attribution 4.0 International License (http://creativecommons.org/licenses/by/4.0/), which permits use, sharing, adaptation, distribution and reproduction in any medium or format, as long as you give appropriate credit to the original author(s) and the source, provide a link to the Creative Commons license and indicate if changes were made.

The images or other third party material in this chapter are included in the chapter's Creative Commons license, unless indicated otherwise in a credit line to the material. If material is not included in the chapter's Creative Commons license and your intended use is not permitted by statutory regulation or exceeds the permitted use, you will need to obtain permission directly from the copyright holder.

15. Neutron Detectors

D. C. Henzlova, M. P. Baker, K. Bartlett, A. Favalli, M. Iliev, M. A. Root, S. Sarnoski, T. Shin, and M. T. Swinhoe

15.1 Mechanisms for Neutron Detection

Mechanisms for detecting neutrons in matter are based on indirect methods. Neutrons, as their name suggests, are neutral. Also, they do not interact directly with the electrons in matter as gamma rays do. The process of neutron detection begins when neutrons, interacting with various nuclei, initiate the release of one or more charged particles. The electrical signals produced by the charged particles can then be processed by the detection system.

Two basic types of neutron interactions with matter occur. First, the neutron can be scattered by a nucleus, transferring some of its kinetic energy to the nucleus. If enough energy is transferred, the recoiling nucleus ionizes the material surrounding the point of interaction. This mechanism is efficient only for neutrons that interact with light nuclei. In fact, only hydrogen and helium nuclei are light enough for practical detectors. Second, the neutron can cause a nuclear reaction. The products from these reactions, such as protons, alpha particles, gamma rays, and fission fragments, can initiate the detection process. Some reactions require a minimum neutron energy (threshold), but most take place at thermal energies. Detectors that exploit thermal reactions are usually surrounded by moderating material to take maximum advantage of this feature.

Detectors that employ either the recoil or reaction mechanism can use solid, liquid, or gas-filled detection media. Although the choice of reactions is limited, the detecting media can be quite varied, leading to many options. This chapter describes gas-filled proportional counters, fission chambers, ^{10}B-lined chambers, scintillators, and other types of neutron detectors. In Sects. 15.4.1 through 15.4.4, gas detectors are discussed in the order of their frequency of use. Organic scintillators are discussed in more detail in Sect. 15.5, which also includes principles of scintillation signal generation and processing. Other types of scintillation detectors and neutron detectors are discussed in Sects. 15.5.7 and 15.6, respectively.

The energy information obtained in neutron-detection systems is usually poor because of the limitations of the available neutron-induced reactions. Recoil-type counters measure only the first interaction event. The full neutron energy is usually not deposited in the detector, and the only energy information obtained is whether a high- or low-energy neutron initiated the interaction. Reaction-type counters take advantage of the increased reaction probability at low neutron energies by moderating the incoming neutrons, but knowledge of the initial neutron energy before moderation is lost. The energy recorded by the detector is the reaction energy (plus, perhaps, some of the remaining initial neutron energy). Thus, in general, neutron detectors provide information only on the number of neutrons detected and not on their energy. Information on the range of detected neutron energies can usually be inferred from the detector type and the surrounding materials. If information on the neutron energy spectrum is needed, it can sometimes be obtained indirectly, as discussed in Sect. 15.7.

Los Alamos National Laboratory strongly supports academic freedom and a researcher's right to publish; as an institution, however, the Laboratory does not endorse the viewpoint of a publication or guarantee its technical correctness.

D. C. Henzlova (✉) · M. P. Baker · K. Bartlett · A. Favalli · M. Iliev · M. A. Root · S. Sarnoski · T. Shin · M. T. Swinhoe
Los Alamos National Laboratory, Los Alamos, NM, USA
e-mail: henzlova@lanl.gov; kbartlett@lanl.gov; afavalli@lanl.gov; metodi@lanl.gov; margaret@lanl.gov; sarnoski@lanl.gov; thshin@lanl.gov; swinhoe@lanl.gov

15.2 General Properties of Gas-Filled Detectors

Gas-filled detectors were among the first devices used for radiation detection. They can be used to detect either thermal neutrons via nuclear reactions or fast neutrons via recoil interactions. After the initial interaction with the neutron has taken place, the remaining detection equipment is similar, although changes in high-voltage or amplifier gain settings can be made to compensate for changes in the magnitude of the detected signal. Figure 15.1 shows a typical setup for neutron counting with a gas-filled detector. Figure 15.2 (top) shows some commonly used detectors with a range of diameters between 1.9 cm and 5 cm (0.75 in. and 2 in.) and aluminum or stainless steel shells. Figure 15.2 (bottom) shows a photograph of a cut-out of a typical ^{3}He detector tube, with a view of the top and bottom segments of anode wire connection inside the tube.

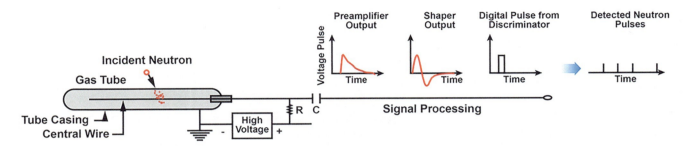

Fig. 15.1 Typical setup for gas-filled neutron detectors

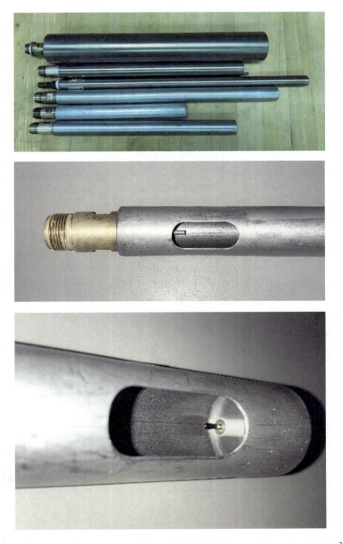

Fig. 15.2 (top) Various sizes of gas-filled neutron detectors. (middle and bottom) Cut-out views of a typical ^{3}He detector tube

The exterior appearance of a gas detector is that of a metal cylinder with an electrical connector at one end (and occasionally at both ends for position-sensitive measurements). The choice of connector depends on the intended application. For safeguards, instrumentation most typically involves an HN connector, as seen in Fig. 15.2; a super high voltage connector, shown on the longest tube in Fig. 15.2 (top); or simple wire leads. Detector walls are manufactured from either stainless steel or aluminum and are ~0.25–0.5 and 0.8 mm thick for stainless steel and aluminum, respectively. Generally, stainless steel walls are slightly thinner due to the higher structural strength of iron compared with aluminum. The performance of either material is quite satisfactory, with only slight differences in neutron transmission or structural strength. Steel walls absorb ~3% of the neutrons; aluminum walls, ~0.5%. Thus, aluminum tubes are usually preferred because of their higher detection efficiency. However, steel tubes have some small advantages over aluminum tubes for certain applications: they require less-careful handling during assembly, the connecting threads are less susceptible to galling, and impurities can be kept lower. The melting point of aluminum is much lower than that of stainless steel, limiting the temperature at which the aluminum tubes can be heat treated, which can lead to problems with outgassing. For this reason, some counter manufacturers use a lining of graphite (referred to as an MG [activated charcoal] coating) to remove gas impurities [1]. Graphite coatings have been shown to inhibit radiation damage and improve resolution [2]. In very low count-rate applications, a background of about 2–4 counts per day has been observed [3]. The MG coating is also sometimes used in tubes operated in high neutron or gamma-ray fields. The activated charcoal serves to absorb electronegative gases that build up during irradiation. For example, in a BF_3-filled detector, three fluorine atoms are released with each neutron capture. The fluorine atoms will combine with electrons released in subsequent neutron captures. Similar build-up of electronegative gases occurs in ^{3}He-filled tubes due to dissociation of polyatomic quench gas additives (such as CO_2, CH_4 or CF_4) under prolonged irradiation. Initially, this process reduces the electric pulse amplitude, and eventually, output pulses are eliminated altogether [2]. Additional details on the design of gas-filled detectors are given in Ref. [1, 4].

The central anode wire shown in Fig. 15.2 (bottom) is typically 0.03 mm thick, gold-plated tungsten. Tungsten provides tensile strength for the thin wire, and the gold plating offers improved electrical conductivity. The wire is held in place by ceramic insulators at the top and the bottom of the ^{3}He tube, as seen in Fig. 15.2 (bottom). Although the area around the ceramic insulators is filled with counting gas, it does not contribute to the measured signal and is often referred to as the *dead region*. An active length is therefore usually specified, which corresponds to the length of the anode wire not including the dead regions. This distinction is of particular importance for Monte Carlo N-Particle modeling, as discussed in Chap. 19, Sect. 19.5.3.

The front-end, signal-processing electronics take the detector signal as input and produce a logic (transistor-transistor logic [TTL]) pulse as output for each detected neutron. Several processes inside the detector described later in this chapter contribute to the production of a charge pulse with certain magnitude and time distribution that depend on initial deposited charge, gas multiplication, charge transport properties, distribution of charge within the detector volume, and electric-field strength through the detector volume. The purpose of the signal-processing electronics is to extract enough signal from the detector pulse to be able to determine when a neutron had been detected and, following that determination, to be ready for the next detector pulse as quickly as possible. These functions are performed by three functional elements—called the *preamplifier*, the *shaper*, and the *discriminator*—schematically depicted in Fig. 15.1.

The detector output signal, which is quite weak, undergoes initial amplification in the preamplifier. The preamplifier usually outputs an amplified voltage pulse proportional to the integrated charge from the detector with a decay constant that returns the output voltage to baseline. The amplified integrated signal goes through a filter—the shaper—which shaper is optimized to produce a pulse that is as short as possible while having an amplitude proportional to the energy deposited by the initial reaction, rejecting noise and unused parts of the signal (such as long components), and minimizing random pulse variation effects (such as pulse-shape variability and discrete charge effects). Oftentimes the trade-off between these requirements is dictated by the application. Typically, ^{3}He detectors need robust neutron-gamma discrimination (instead of high-fidelity energy information) while minimizing dead time; therefore, a fast time constant of the shaper is more important because it ensures reliable detection of a neutron pulse while enabling fast recovery to detect subsequent pulses.

The two classes of commonly used shapers are unipolar shapers and bipolar shapers. The example pulse shown in Fig. 15.1 is produced by a bipolar shaper (i.e., the step response of the shaper has two lobes, positive and negative, with equal areas). Even though unipolar shapers have somewhat better dead-time and noise performance, most ^{3}He electronics use bipolar shapers because a unipolar shaper requires a precisely tuned pole-zero compensation and baseline restorer circuits to avoid signal baseline shift at high count rates, whereas bipolar shapers are inherently more immune to the long time constant and imperfect tuning [5].

The third component of the signal-processing chain is the discriminator, which has a calibrated amplitude threshold and produces a logic (TTL) pulse for each shaper pulse above this threshold. The duration of this logic pulse commonly

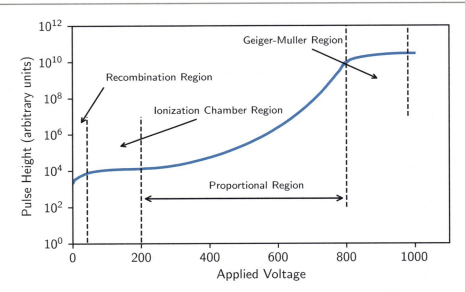

Fig. 15.3 Pulse-height versus applied-voltage curves to illustrate ionization, proportional, and Geiger-Mueller regions of operation

corresponds to time above threshold; however, in typical ^{3}He-based instruments where the outputs of multiple discriminators are daisy-chained together, the logic pulse is shortened to few tens of nanoseconds (typically 40–50 ns) to minimize the pileup with other logic pulses in the same chain and limit additional dead time.

These three elements (preamplifier, shaper, and discriminator) are combined in a compact set of electronics that connects to the detector. Oftentimes the whole front-end, signal-processing chain is colloquially abbreviated simply as *the amplifier* or even *the preamplifier* even though it includes all three elements. More details and photographs of amplifier modules commonly used in nondestructive assay (NDA) applications are provided in Chap. 19.

As described in Sect. 15.1, the detection of neutrons requires the transfer of some or all of the neutrons' energy and reaction Q-value to charged particles. The charged particle will then ionize and excite the atoms along its path until its energy is exhausted. In a gas-filled detector, approximately 30 eV is required to create an ion pair. The maximum number of ion pairs produced is then $E/30$ eV, where E is the kinetic energy of the charged particle(s) in eV. For example, an energy transfer of 765 keV—typical in ^{3}He-gas detectors—will release a total positive and negative charge of $\sim 8 \times 10^{-15}$ coulombs.

If little or no voltage is applied to the tube, most of the ions will recombine, and no electrical output signal is produced. If a positive voltage is applied to the central wire (anode), the electrons will move toward it, and the positively charged ions will move toward the tube wall (cathode). An electrical output signal will be produced, whose magnitude depends on the applied voltage, the geometry of the counter, and the fill-gas type and pressure. These parameters determine whether the detector operates in the ionization region, the proportional region, or the Geiger-Mueller region. These different operating regions are shown in Fig. 15.3.

In the ionization region, enough voltage has been applied to collect nearly all of the electrons before they can recombine. At this point, a plateau is reached, and further small increases in voltage yield no more electrons. Detectors operated in this region are called *ion chambers*. The charge collected is proportional to the energy deposited in the gas and independent of the applied voltage.

The region beyond the ionization region is called the *proportional region*. Here, the electric field strength is large enough so that the primary electrons can gain sufficient energy to ionize the gas molecules and create secondary ionization. If the field strength is increased further, the secondary electrons can also ionize gas molecules. This process continues rapidly as the field strength increases, thus producing a large multiplication of the number of ions formed during the primary event. This cumulative amplification process is known as *avalanche ionization*. When a total of "A" ion pairs result from a single primary pair, the process has a gas amplification factor of A, which will be unity in an ionization chamber where no secondary ions are formed and as high as 10^3 to 10^5 in a well-designed proportional counter. Note that in the proportional region, the charge collected is also linearly proportional to the energy deposited in the gas.

For the amplification process to proceed, an electron must acquire sufficient energy in one or more mean free paths to ionize a neutral molecule. The mean free path is the average distance that the electron travels between collisions in proportional-counter gas, which equals approximately 1–2 µm. For amplifications of 10^6 fewer than 20 mean free paths are necessary, which indicates that only a small region around the wire needs to be involved in the amplification process. In the rest of the volume, the electrons drift toward the anode without inducing charge multiplication. Because the amplification process requires a very high electric field, an advantage of the cylindrical detector design is the high electric field near the inner wire. The total amplification will depend on the electric field traversed, not the distance traversed.

At the same time that the electrons are drifting toward the anode, the positive ions are drifting toward the cathode. In a proportional counter, the drift velocity of the electrons is several orders of magnitude larger than the drift velocity of the positive ions. The electrons that have a larger drift velocity are collected in an extremely short time interval; the slower-drifting positive ions are collected on the cathode over a much longer time interval. The pulses observed have an initial fast rise time because of the motion of the electrons and fast positive ions near the anode wire in the intense electric field and a subsequent slower rise time because of the motion of the positive ions in the remaining volume of the counter [4]. A typical rise time for a ^{3}He proportional counter is on the order of 1–2 microseconds, followed by positive ion collection on the order of hundreds of microseconds [6]. This behavior can be compared with ^{10}B-lined proportional counters, which generally exhibit much faster electron rise times—on the order of 50 ns [7, 8]—due to the short-ranged and dense ionization processes. In addition, because the positive ions are initially formed close to the anode and must drift across the entire anode-to-cathode gap, the pulse amplitude is largely due to the drifting of the positive ions. The pulse reaches full amplitude only when the positive ions are fully collected. For a typical proportional counter, this collection process may take 200μs. Through differentiation, the pulse can be made much shorter without a substantial loss of pulse height so that rapid counting is possible. It is possible to approach the time dispersion caused by the variation in the drift time of the primary electrons from the interaction site to the anode wire. This time dispersion depends on tube voltage and diameter and has been reported as 1.1μs (CH_4), 2.5μs (^{3}He), and 17μs (^{4}He) for some typical gas-filled tubes [9].

As the applied voltage is increased further, the proportionality between the primary charge deposited and the output signal is gradually lost, primarily due to saturation effects at the anode wire. As the primary ions reach the high field regions near the anode wire, the avalanche process begins and quickly grows to a maximum value as secondary electrons create additional avalanches axially along the wire. Unlike operation in the proportional region where the avalanche is localized, the avalanche now extends the full length of the anode wire, and the multiplication process terminates only when the electrostatic field is sufficiently distorted to prevent further acceleration of secondary electrons. For weakly ionizing primary events, amplification factors of up to 10^{10} are possible. Detectors operated in this region are called *Geiger-Mueller counters*, which comprise very simple electronics and form the basis for rugged field inspection instruments. Because they are saturated by each event, Geiger-Mueller counters cannot be used in high-count-rate applications, but this limitation does not interfere with their use as low-level survey meters.

Depending on the associated electronics, neutron counters operated in either the ionization or proportional mode can provide an average output current or individual pulses. Measuring only the average output current is useful for radiation dosimetry and reactor power monitors. For assay of nuclear material, it is customary to operate neutron counters in the pulse mode so that individual neutron events can be registered.

Gas-filled detectors used in neutron counting typically employ ^{3}He, ^{4}He, BF_3, or CH_4 as the primary constituent at pressures of less than 1 to ~20 atm, depending on the application. Conventional inert counting gases (such as argon with a small concentration of N_2 or argon-methane) are used in the coated detectors where the neutron interactions occur in a thin layer of neutron-absorbent material coated on the detector walls (^{10}B-lined counters or fission chambers) as opposed to directly in the counting gas. In the case of ^{3}He and ^{4}He, other gases are often added to the primary constituent gas to improve detector performance. For example, a heavy gas such as argon can be used to reduce the range of the reaction products so that more of their kinetic energy is deposited within the gas, thereby improving the output pulse-height resolution. Adding a heavy gas also speeds up the charge collection time but has the adverse effect of increasing the gamma-ray sensitivity (Ref. [6, 10] and Sect. 15.4.1).

A polyatomic gas may also be added to proportional counters to serve as quench gas. The rotational degrees of freedom available to polyatomic gas molecules serve to limit the energy gained by electrons from the electric potential, thus helping to dampen and shorten the avalanche process and improve the pulse-height resolution. Gases such as BF_3 and CH_4 are already polyatomic gases and require no additional quench gas. Tubes filled with ^{3}He and ^{4}He often have a small quantity of CH_4, CO_2, or CF_4 added. Detectors filled with these gases typically require higher operating voltages. Also, the relatively large quantity of polyatomic gas restricts the intercollisional energy gain so that these detectors are usually not operated at fill pressures as high as those used for detectors filled with monoatomic gases. The presence of a small component of a polyatomic gas also improves electron-drift velocity because it increases inelastic collisions, which reduce the agitation energy and consequently increase the collision time during which the field drifts the electrons [6].

Generally, it is not possible to operate a ^{3}He counter without a small amount of quench gas. Helium as is a noble gas and, following irradiation, it produces metastable states that can cause breakdowns by interactions of deexcitation photons with the cathode wall (photoionization and photoelectric effect). One role of quench gas is to deexcite these states to prevent breakdowns [1]; however, use of some polyatomic quench gas mixtures can lead to dissociation under prolonged irradiation and, as a consequence, formation of electronegative impurities (such as oxygen in the case of CO_2 or fluorine in the case of CF_4). If the gas contains high amounts of electronegative impurities such as oxygen, then the electrons formed in the initial neutron interaction can attach themselves to these gas atoms and form slow-moving negative ions, which will essentially remove them from the fast-electron-drift process. Choice of quench gas is therefore dependent on the target application of the proportional counter. Typical quench gas choices for the majority of nuclear safeguards applications include $Ar + CH_4$, CF_4, and CO_2 or N_2. The latter two are especially useful for high-gamma-dose environments (see Sect. 15.4.1.). The former are typically favored for high neutron-count-rate applications.

Finally, the impact of neutron interaction location within the proportional counter volume needs to be considered. The location of the interactions and orientation of the ionization track of the charged particles in the detector volume can induce a spread of the electron collection time and, consequently, the rise time of detected pulses. The charged particle products of neutron interaction in the proportional counter gas can traverse the proportional counter in various directions, causing spread in the rise times of pulses from the detector. The shortest rise times correspond to ionization tracks parallel to the anode wire. The charge pulse spread results in variation of shapes and amplitudes of current pulses, which can reveal complex structures with multiple humps, as illustrated in Fig. 15.4 for a ^{3}He counter. These structures are particularly apparent in the case of tracks oriented perpendicularly to the anode wire. The minimization of this effect can be accomplished by choosing a quench

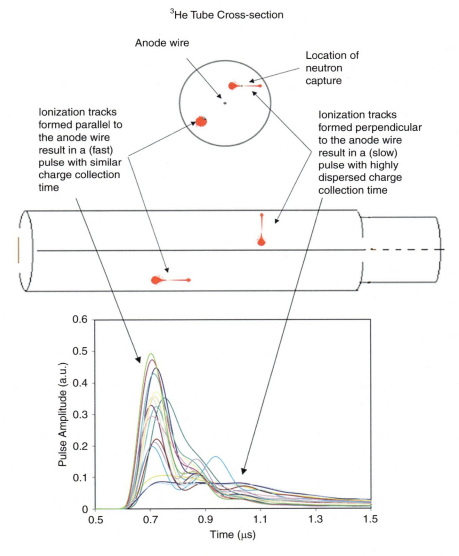

Fig. 15.4 (top) Simplified visualization of charged-particle ionization tracks inside a proportional counter; (bottom) pulse shapes of slow and fast pulses that correspond to these different orientations of ionization tracks within a ^{3}He detector

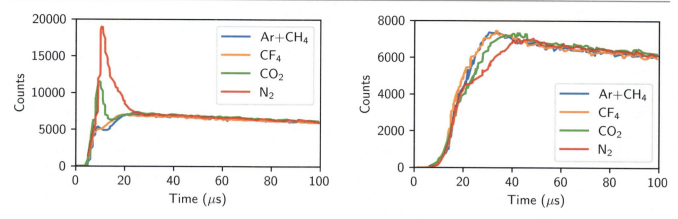

Fig. 15.5 Rossi-α distribution for Ar + CH$_4$, CF$_4$, CO$_2$ and N$_2$ quench-gas admixtures in a 1 in. ^{3}He tube with AMPTEK-A111 (left) and PDT-10A (right) amplifier modules

gas with high stopping power (thus reducing the track length) and high electron-drift velocities, which will reduce the counter dead time [6]. Ar + CH$_4$ is often considered for this reason.

The overall performance of a proportional counter is a combination of the counter-gas properties and the signal-processing electronics. PDT [11], AMPTEK [12], and KM200 [13] are the most common amplifier/discriminator modules used in present-day applications (see Chap. 19). Due to the short time constant of the AMPTEK shaping amplifier (~190 ns), its output pulse is dominated by the structures present in the detector pulse. Therefore, a re-triggering of discriminator on the complex pulse structures can be expected, which causes false events (double-pulsing), as illustrated in Rossi-α (also written as *Rossi-alpha*) distribution in Fig. 15.5 (left). Such discriminator retriggering can be eliminated by use of a longer shaping time, as shown in Fig. 15.5 (right), in the case of a PDT-10A amplifier module with a shaping time of ~500 ns. In addition, the effect is significantly reduced for Ar + CH$_4$ and CF$_4$ quench gas admixture, as shown in Fig. 15.5 (left), due to fast electron-drift velocities for these quench-gas types. Using faster gas mixtures together with faster electronics results in a detector system that has lower dead time and no double pulsing. Having control over the shaping time of the electronics and properly tuning the electronics to the detector reponse therefore ensures an optimum detection performance for a given application.

15.3 Gamma-Ray Sensitivity of Neutron Detectors

The neutron detectors described in this chapter are sensitive in some degree to gamma rays as well as neutrons. Because most nuclear material emits 10 or more times as many gamma rays as neutrons, the gamma-ray sensitivity of the neutron detector is an important criterion in its selection. For measurements of spent fuel or some reprocessed materials where gamma-ray fluxes of 1000 R/h or more are encountered, the gamma-ray sensitivity of the detector may dominate all other considerations.

In any detector, gamma-rays can transfer energy to electrons by Compton scattering (see Chap. 3) just as neutrons can transfer energy to other nuclei by scattering or nuclear reactions. Compton scattering can take place in the detector walls or in the fill gas, yielding a high-energy electron that in turn produces a column of ionization as it traverses the detector. In some detectors, electronic pulses induced by gamma-rays are comparable in size to neutron-induced pulses; in other detectors, they are much smaller but can pile up within the resolving time of the electronics to yield pulses comparable to neutron pulses. Four factors should be considered when evaluating the relative magnitudes of the neutron and gamma-ray signals.

- The presence of gamma-ray shielding has a substantial effect on the relative magnitude of the signals. For example, for a detector exposed to 1 MeV fission neutrons in the presence of 1 MeV fission gamma-rays, 5 cm of lead shielding absorbs roughly 0.1% of the neutrons and 90% of the gamma rays.
- Some detector materials and designs favor the absorption of neutrons. Table 15.1 gives examples for thermal- and fast-neutron detectors. From the table, it is clear that thermal neutrons can be absorbed with much higher probability than gamma rays. For fast-neutron detection, the neutron and gamma-ray interaction probabilities are comparable.
- In some detectors, neutrons deposit more energy than gamma-rays do. Neutrons can induce a nuclear reaction that releases more energy than the Compton scattering of the gamma ray imparts to the electron. (The average energy imparted by a

Table 15.1 Neutron and gamma-ray interaction probabilities in typical thermal- and fast-neutron detectors

	Interaction probability	
Thermal detectors	Thermal neutron	1 Mev gamma ray
^{3}He (2.5 cm diameter, 4 atm)	0.77	0.0001
Ar (2.5 cm diameter, 2 atm)	0.0	0.0005
BF$_3$ (5.0 cm diameter, 0.66 atm)	0.29	0.0006
Al tube wall (0.8 mm)	0.0	0.014
	Interaction probability	
Fast detectors	1 MeV neutron	1 MeV gamma ray
^{4}He (5.0 cm diameter, 18 atm)	0.01	0.001
Al tube wall (0.8 mm)	0.0	0.014
Scintillator[a] (5.0 cm thick)	0.78	0.26

[a]Performance is similar for any typical organic scintillator material

Table 15.2 Neutron and gamma-ray energy deposition in typical thermal- and fast-neutron detectors

Thermal detectors	Alpha or proton range (cm)	dE/dx for 400 keV electron (keV/cm)	Average neutron reaction energy deposited (keV)	Electron energy deposited (keV)[a]	Ratio of neutron to electron energy deposition
^{3}He (2.5 cm diameter, 4 atm)	2.1	1.1	~500	4.0	125
^{3}He (2.5 cm diameter, 4 atm) + Ar (2 atm)	0.5	6.7	~750	24.0	30
BF$_3$ (5.0 cm diameter, 0.66 atm)	0.7	3.6	~2300	25.7	90
Fast detectors					
^{4}He (5.0 cm diameter, 18 atm)	0.1	6.7	1000	48	20
Scintillator[b] (5.0 cm thick)	0.001	2000	1000	400	2.5

[a]This calculation assumes a path length of $\sqrt{2}$ × tube diameter
[b]Performance is similar for any typical organic scintillator material

1 MeV gamma ray is roughly 400 keV). Also, in gas detectors, the range of the electron is typically much longer than the range of the heavy charged particles produced by neutron interactions. When the gas pressure is chosen to just stop the heavy charged particles, the electrons will escape from the tube after depositing only a small fraction of their energy in the gas. Table 15.2 gives some numerical examples of these effects. The table also shows that for fast-neutron detection by plastic scintillators, the relative neutron and gamma-ray energy deposition are comparable.

- The charge collection speeds for neutron and gamma-ray detection can be different. This effect is very dependent on the choice of fill gas or scintillator material. In gas detectors, the long range of the electron produced by a gamma-ray interaction means that energy will be deposited over a greater distance, and more time will be required to collect it. An amplifier with fast differentiation will then collect relatively less of the charge released by a gamma-ray interaction than a neutron interaction. In scintillators, less distinction exists between the two kinds of events; however, in some circumstances, pulse-shape discrimination (PSD) between neutrons and gamma rays can be achieved (see Sect. 15.5).

To achieve good gamma-ray discrimination, it is often necessary to use materials or material densities that are not optimum for neutron detection. The result can be a reduced neutron detection efficiency. Table 15.3 lists the neutron-detection efficiency and approximate gamma-ray radiation limit for various types of neutron detectors. The detection efficiency is for a single pass through the detector at the specified energy. The actual efficiency for a complete detector system depends on the geometry; the obtainable efficiency can be lower than the estimate given in Table 15.3. Additional details on these detectors and their gamma-ray sensitivity are given in the following sections.

Table 15.3 Typical values of efficiency and gamma-ray sensitivity for some common neutron detectors

Detector type	Size	Neutron active material	Incident neutron energy	Neutron-detection efficiency[a] (%)	Gamma-ray sensitivity (R/h)[b]
Plastic scintillator	5 cm thick	^{1}H	1 MeV	78	0.01
Liquid scintillator	5 cm thick	^{1}H	1 MeV	78	0.1
Loaded scintillator	1 mm thick	^{6}Li	Thermal	50	1
Hornyak button	1 mm thick	^{1}H	1 MeV	1	1
Methane (7 atm)	5 cm diam	^{1}H	1 MeV	1	1
^{4}He (18 atm)	5 cm diam	^{4}He	1 MeV	1	1
^{4}He (100–200 atm)	5.2 cm diam	^{4}He	Thermal	5	0.02 [14]
^{3}He (4 atm), Ar (2 atm)	2.5 cm diam	^{3}He	Thermal	77	1 [10]
^{3}He (4 atm), CO_2 (5%)	2.5 cm diam	^{3}He	Thermal	77	10–100 [10, 15]
^{3}He (7.5 atm), CO_2 (5%)	2.5 cm diam	^{3}He	Thermal	N/A	20 [10]
^{3}He (9 atm), Ar + CO_2, MG coating	2.5 cm diam	^{3}He	Thermal	N/A	1 [16]
BF_3 (0.66 atm)	5 cm diam	^{10}B	Thermal	29	10
BF_3 (1.18 atm)	5 cm diam	^{10}B	Thermal	46	10
^{10}B-lined chamber	0.2 mg/cm^2	^{10}B	Thermal	10	10^3 [17, 18]
Fission chamber	2.0 mg/cm^2	^{235}U	Thermal	0.5	10^6 to 10^7

[a]Interaction probability for neutrons of the specified energy that strike the detector face at right angles
[b]Approximate upper limit of gamma-ray dose that can be present with the detector still providing usable neutron output signals

15.4 Gas-Filled Detectors

15.4.1 Helium-3 and BF$_3$ Thermal-Neutron Detectors

Gas-filled thermal-neutron detectors use most typically ^{3}He, but detectors with BF$_3$ have also been used. In the case of BF$_3$, the gas is enriched in ^{10}B. Helium-3 is only ~1 ppm of natural helium, so it is usually obtained by separation from tritium produced in reactors. The nuclear reactions that take place in these gases are

$$^3\text{He} + n \rightarrow {}^3\text{H} + {}^1\text{H} + 765 \text{ keV}$$
$$^{10}\text{B} + n \rightarrow {}^7\text{Li}^* + {}^4\text{He} + 2310 \text{ keV} \tag{15.1}$$

and

$$^7\text{Li}^* \rightarrow {}^7\text{Li} + 480 \text{ keV} \tag{15.2}$$

These reactions are exothermic and release energetic charged particles into the gas. The counters are operated in the proportional mode, and the ionization produced by these particles initiates the multiplication process that leads to detection. The amount of energy deposited in the detector is the energy available from the nuclear reaction.

In the case of ^{3}He, the neutron causes the breakup of the nucleus into a tritium nucleus (triton), ^{3}H, and a proton (^{1}H). The triton and the proton share the 765 keV reaction energy. In the case of ^{10}B, the boron nucleus breaks up into a helium nucleus (alpha particle) and a lithium nucleus, with 2310 keV shared between them. During 94% of occurrences, the lithium nucleus is left in an excited state, from which it subsequently decays by emitting a 480 keV gamma ray. This gamma ray is usually lost from the detector; therefore, only 2310 keV is deposited. About 6% of the time, the lithium nucleus is left in the ground state, so 2790 keV is deposited in the detector. This double-reaction mode yields an additional small full-energy peak in the pulse-height spectrum of BF$_3$ tubes.

The cross section for the ^{3}He and ^{10}B reactions shown in Eq. (15.1) is 5330 b and 3840 b respectively for thermal neutrons. Both reaction cross sections are strongly dependent on the incident neutron energy E and have roughly a $1/\sqrt{E}$ dependence. Figure 15.6 illustrates these cross sections [4]. As an example, a 2.54 cm diameter tube with 4 atm of ^{3}He has a 77% efficiency for thermal neutrons (0.025 eV). This configuration is nearly optimum for thermal neutrons; more ^{3}He would give relatively little additional efficiency and would usually not be cost effective. But higher-pressure (up to 10 atm) ^{3}He tubes are currently

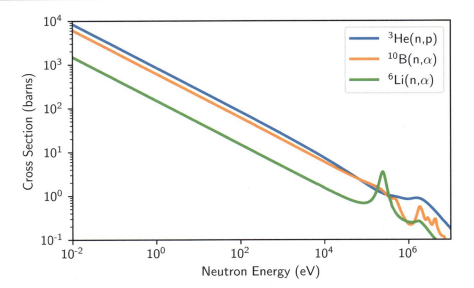

Fig. 15.6 Helium-3 (n,p), $^{10}B(n,\alpha)$, and $^{6}Li(n,\alpha)$ cross sections as a function of incident neutron energy

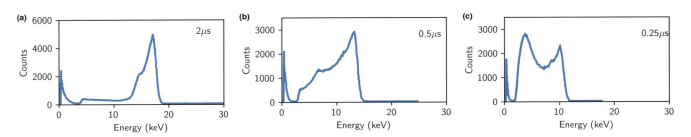

Fig. 15.7 Differential pulse-height spectrum for thermal neutrons detected by a ^{3}He-filled counter for three different amplifier time constants

in use in applications that require high overall neutron detection efficiency, such as multiplicity counting (see Chaps. 18 and 19). At 100 eV, the efficiency is roughly 2%; at 10 keV, roughly 0.2%; and at 1 MeV, roughly 0.002%. Because of this strong energy dependence, it is customary to embed ^{3}He or BF$_3$ detectors in approximately 10 cm of polyethylene or other moderating materials to maximize their counting efficiency (see Chap. 19 for additional details on detector design).

Figure 15.7 (top) is a typical pulse-height spectrum from a ^{3}He proportional counter. The shape of this spectrum is due primarily to the kinematics of the reaction process and the choice of amplifier time constants. The full-energy peak in the spectrum represents the collection of the kinetic energy of both the proton and the triton. (It should be emphasized that this peak represents the 765 keV released in the reaction and is not a measure of the incident neutron energy.) If one or the other particle enters the tube wall, less energy is collected in the gas, which results in a low-energy tail. Because the two charged particles are emitted back-to-back, one of the two is almost certain to be detected; thus, a minimum collection energy—with a wide valley below and then a low-energy increase—results from noise and piled-up gamma-ray events. If the discriminator is set in the valley, small changes in tube voltage or amplifier gain will not affect the count rate. The result is a very stable detection system; stability of 0.002% was reported under constant environmental conditions and approximately 0.06%/°C with temperature variation [19].

The choice of amplifier time constant determines the degree of charge collection from the tube. Time constants of 2µs or greater result in nearly complete charge collection and yield spectra, such as the spectrum shown in Fig. 15.7 (top). Time constants of 0.5–0.1µs cause gradual complete loss of the peak shape as illustrated in Fig. 15.7 (middle and bottom) but allow counting at higher rates with less noise pickup and gamma-ray interference. A 0.5µs time constant is a commonly used compromise between good resolution and high-count-rate capability. Amplifier modules used in the majority of current safeguards applications (PDT, AMPTEK, KM200) typically employ short time constants on the order of 0.2–0.5µs.

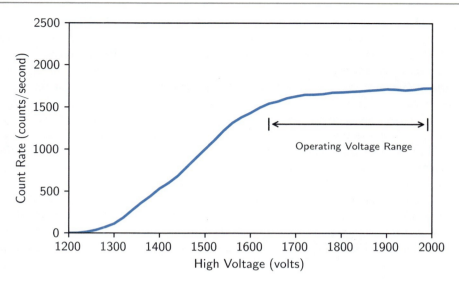

Fig. 15.8 Typical plateau curve shape measured for a 4 atm, gas-filled, ^{3}He slab counter

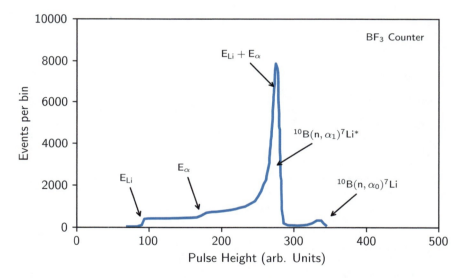

Fig. 15.9 Differential pulse-height spectrum for thermal neutrons detected by a ^{10}BF$_3$-filled counter

Helium-3 tubes are usually operated in the range of +1600–1800 V. Over this range, the increase in counting efficiency with voltage caused by improved primary charge collection is very slight, ~1%/100 V [19]. (A typical plateau curve is shown in Fig. 15.8.) However, the total charge collected (due to multiplication in the gas) changes rapidly with voltage, by ~100%/100 V. When ^{3}He tubes are used in multiple detector arrays, it is important to specify good resolution (on the order of 5% full width at half maximum [FWHM] and uniform gas mixture so that the position and width of the full-energy peak will be the same for all tubes.

A pulse-height spectrum for a BF$_3$ proportional counter is shown in Fig. 15.9 [20]. For BF$_3$ tubes, the resolution is in the range of 5–30% (FWHM) but is usually not as good as for ^{3}He. Gas pressures are in the range of 0.2–2 atm. To help compensate for the lower pressure, tube diameters are usually 5 cm. Operating voltages are in the range + 1400–2800 V, higher than for ^{3}He. Plateau curves are similar to those of ^{3}He. BF$_3$ gas is less expensive than ^{3}He, so manufacturing costs are less; however, helium is an inert gas and BF$_3$ is toxic, which presents concerns for deployment in nuclear facilities and has been discouraged by the International Atomic Energy Agency [15]. Nevertheless, mitigation approaches have since been developed and evaluated to minimize BF$_3$ risks, such as the concept that implements a reservoir for a chemical reactant such as sodium carbonate (Na_2CO_3) to neutralize the BF_3 gas in case of leakage [14]. In [14], a BF_3-based, high-level neutron coincidence counter -type prototype counter was also proposed. (For information on additional ^{3}He-alternative coincidence counters, see Chap. 19). U.S. Department of Transportation regulations place detectors with more than 2 atm fill pressure in the high-pressure compressed gas category, so ^{3}He-filled detectors are often more difficult to ship. Boron trifluoride is also a hazardous gas, and transportation of BF_3 is subject to strict U.S. Department of Transportation regulations and may be prohibited in some locations.

Helium-3 detectors find many applications in passive and active neutron assay because they are very stable, efficient, and relatively gamma insensitive. The detection efficiency for thermal neutrons is high, and the interaction probability for gamma rays is low, as indicated in Table 15.1. Also, much more energy is deposited in the gas by neutron interactions than by gamma-ray interactions, as indicated in Table 15.2. However, if the gamma dose is more than that emitted by typical plutonium and uranium samples, the response of ^{3}He and BF_3 detectors will be affected.

As an example, Fig. 15.10 shows the effect of increasing gamma-radiation fields on a ^{3}He detector (4 atm and 9 atm) that contains CO_2 quench gas [16]. The practical operating limit is on the order of 1–100 R/h. The figure illustrates the increased gamma-ray sensitivity with increasing gas pressure. For moderate dose rates (up to ~20 R/h), the benefit of increased neutron-detection efficiency through increasing the fill-gas pressure outweighs the detriment of the increased number of target atoms

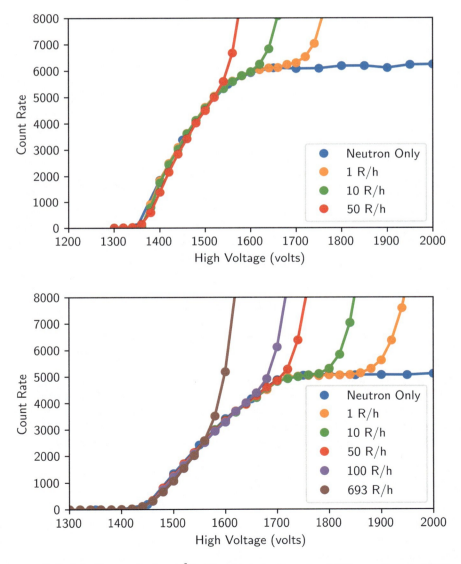

Fig. 15.10 (top) Gamma sensitivity for a 20 cm active length ^{3}He tube with 4 atm pressure and CO_2 quench gas and (bottom) a 9 atm pressure and Ar + CO_2 quench gas

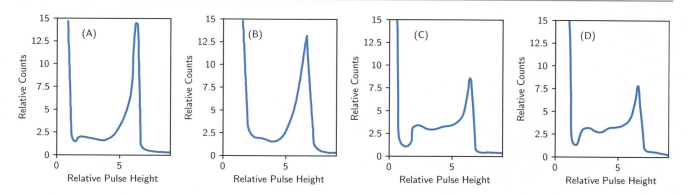

Fig. 15.11 Differential pulse-height spectra for various ^{3}He neutron detectors. The amplifier time constant was set at 0.5 μs. (**a**) ^{3}He + Ar + CH$_4$ mixture with neutron source; (**b**) ^{3}He + Ar + CH$_4$ mixture with neutron plus 1-R/h gamma-ray source; (**c**) ^{3}He + 5% CO$_2$ mixture with neutron source; (**d**) ^{3}He + 5% CO$_2$ mixture with neutron plus 1-R/h gamma-ray source

for gamma-ray interactions. However, as the dose rate increases beyond this value, the higher fill pressure aggravates the gamma-ray pileup problem and results in the reduction of neutron detection efficiency [10].

The use of argon in the quench gas, which is common in modern ^{3}He proportional counters, also results in a slight increase in gamma-ray sensitivity due to argon's high atomic number. A comparison between CO$_2$ and Ar + CH$_4$ is illustrated in Fig. 15.11 [21]. However, use of argon improves the relative size of the full-energy peak because the reaction products now have shorter ranges and deposit more of their energy in the gas. Also, the shorter ranges lead to faster charge collection and roughly 35% shorter electronic dead times, which make Ar + CH$_4$ quench gas desirable for high count-rate applications. Use of CF$_4$ quench gas is also increasingly considered in nuclear safeguards due to its high count-rate performance. The use of CF$_4$ increases the electron-drift velocity and localizes the volume where the ionization is produced. The ionization from the reaction products is thus produced in a shorter time; a shorter amplifier shaping time can be used, which improves the overall counting and dead-time characteristics. As stated earlier, argon gas is expected to be more sensitive to gamma pileup due to its higher Z. However, similar gamma-ray sensitivity was observed for Ar + CH$_4$ and CF$_4$ quench gas admixtures [22], which suggests that the effect is dominated by the interactions in the counter walls.

For applications that require tolerance to high gamma dose rates, such as measurements of spent fuel materials, Ar + CH$_4$ quench gas is typically not the ideal choice [10]; use of CO$_2$ or N$_2$ is preferred due to their lower Z. Additionally, polyatomic quench gas admixtures (CH$_4$ or CF$_4$) suffer from radiolytic decomposition under prolonged gamma exposures, which results in gain shift and, as a consequence, count-rate loss in the ^{3}He tube. Therefore, CO$_2$ and (especially) N$_2$ quench gases are preferred in long-term gamma-ray exposure applications [1, 10]. An alternative implementation using ^{10}B coating on the walls of ^{3}He-gas-filled tubes was considered [10] for neutron counting in very high gamma dose environments, which combines the benefits of high ^{3}He neutron detection efficiency and extended gamma-ray tolerance of ^{10}B-coated proportional counters (see Sect. 15.4.2). Such a hybrid tube provides capability to function as a ^{3}He detector for low-to-moderate gamma dose rates (up to ~20 R/h) and as ^{10}B-coated detectors for very high gamma dose rates (>200 R/h). For a 4 atm ^{3}He gas fill pressure, the performance improvement from the ^{10}B coating was shown to provide better neutron detection performance than an equivalent 6 atm ^{3}He tube without the ^{10}B coating, which means that the hybrid ^{3}He-filled/^{10}B-coated design can provide slight performance improvement in high gamma dose rate applications with reduced ^{3}He gas need. Note that for any high gamma dose rate application, the choice of detector design must be based on a range of factors specific to the application, such as the type of exposure (i.e., periodic or chronic), the actual dose on the detector, the gamma-ray energy, and the amplifier module time constant, economics.

The gamma-ray sensitivity of BF$_3$ detectors is comparable to but perhaps slightly better than ^{3}He. The ^{10}B reaction deposits more energy in the gas than the ^{3}He reaction, but gamma-ray interactions also deposit more energy (see Table 15.2). The ^{3}He reaction has a higher cross section than the BF$_3$ reaction. The cross section for a gamma-ray interaction will depend on the relative amounts of ^{3}He (including type of quench gas) and BF$_3$ and on the relative tube-wall thicknesses (see Table 15.1). BF$_3$ detectors can operate in gamma-radiation fields up to 10 R/h, which is better than the performance of ^{3}He + argon counters; however, the performance of ^{3}He + CO$_2$ counters is comparable to that of BF$_3$.

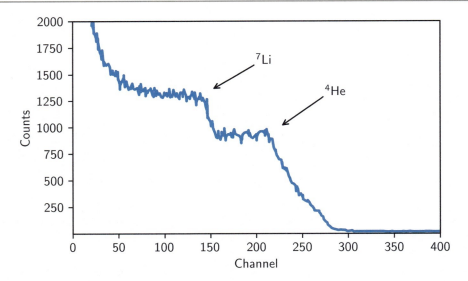

Fig. 15.12 Differential pulse-height spectrum of a ^{10}B-lined proportional counter

15.4.2 ^{10}B-Lined Detectors

Detectors lined with ^{10}B lie between ^{3}He and BF$_3$ proportional counters and fission chambers in terms of neutron-detection efficiency and gamma-ray insensitivity. Structurally, ^{10}B-lined detectors are similar to fission chambers—with the neutron-sensitive material, boron (in form of enriched ^{10}B or ^{10}B$_4$C), plated in a very thin layer (~0.2–0.5 mg/cm^2) on the walls of the detector [23].

The ^{10}B-lined detectors rely on the nuclear reaction given in Eq. (15.1) to detect neutrons. Either the alpha particle or the lithium nucleus enters the detection gas (not both; they are emitted back to back), and the detection process is initiated. Because the range of the alpha particle is ~1 mg/cm^2 in boron, the coating thickness must be optimized to ensure maximum neutron-detection efficiency. Typically, the optimum thickness depends on the system design and configuration and may be different for a single detector than for a system of several boron-lined proportional counters [23]. The detection principle of boron-lined detectors is similar to BF$_3$; however, because the nuclear reaction does not take place in the fill gas, the gas can be optimized for fast timing [24]. Argon at subatmospheric pressure, with a small admixture of CO$_2$, is one common choice. The boron-lined counter is operated in the proportional mode at a voltage of +600–1000 V [14].

Figure 15.12 shows the pulse-height spectrum of a ^{10}B-lined chamber [25]. The stepped structure of the spectrum results because either the alpha particle or the lithium nucleus can enter the gas. Because the lighter alpha particle carries more of the energy, the step that results from the alpha particle is shown farther to the right. The large number of low-energy pulses is due to the energy loss of the particles in the boron coating of the walls. The detector threshold is usually set above these low-energy pulses. Because there is no well-defined "valley" to set the threshold in, the count-rate plateau curve exhibits a slight slope of roughly 10%/100 V, contrary to 1–2%/100 V for ^{3}He [19].

Despite the lower thermal-neutron reaction cross section compared with ^{3}He (Fig. 15.6), careful optimization of boron coating thickness, boron coating surface area, and moderator design led to development of boron-lined detectors with performance comparable to ^{3}He proportional counters to provide a viable alternative to ^{3}He neutron-detection technologies [14]. Examples of such technologies and full-scale neutron coincidence counters based on such technologies are discussed in more detail in Chap. 19.

The ^{10}B-lined counter also provides a very useful technology in applications where it is necessary to detect neutrons in the presence of high gamma-ray fields [10]. With proper electronics, the detector can be operated in gamma-ray dose rates of nearly 1000 R/h with only ~30% reduction in neutron-detection efficiency (Fig. 15.13), resulting from the higher discriminator setting required to reject piled-up gamma events [18]. The reduced gamma-ray sensitivity of the ^{10}B-lined counter relative to the ^{3}He counters provides an essential extension of neutron-detection capability for challenging applications, such as pyroprocessing or advanced nuclear reactors (see also Chap. 19, Sect. 19.6.2).

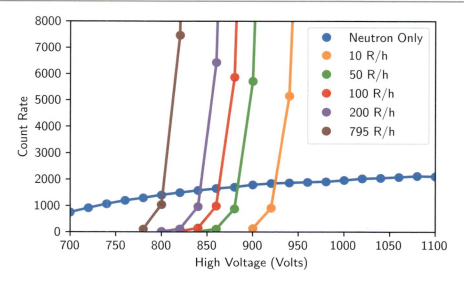

Fig. 15.13 High-voltage response of boron-lined, high-dose neutron detector for range of gamma dose rates

15.4.3 Helium-4 and CH₄ Fast-Neutron Detectors

Helium-4 and CH₄ fast-neutron detectors rely on the recoil of light nuclei to ionize the gas in the tube. The interaction is the elastic scattering of the neutron by a light nucleus. If the recoiling nucleus is only a hydrogen nucleus (proton), the maximum possible energy transfer is the total neutron kinetic energy E. For heavier elements, the maximum energy transfer is always less. For a nucleus of atomic weight (A), the maximum energy transfer is [4]

$$E(\max) = \frac{4AE}{(A+1)^2} \tag{15.3}$$

For a single scattering event, the actual energy transferred to the recoiling nucleus lies between 0 and $E(\max)$, depending on the scattering angle, and has equal probability for any value in this range.

Equation 15.3 shows that the target nucleus must have low atomic weight to receive a significant amount of energy from the neutron. Hydrogen is the most obvious choice; it can be used in a gaseous form or, more commonly, in liquid or plastic scintillators (see Sect. 15.5). Popular gas detectors usually employ methane (CH₄), which has a high hydrogen content, or ⁴He, which has a maximum energy transfer of 0.64 $E(n)$. (Helium-3 gas is also a suitable candidate by these criteria, but it is usually not used because of the stronger thermal reaction described in Sect. 15.4.1.) Figure 15.14 illustrates the elastic-scattering cross sections for ¹H and ⁴He, showing that they match the shape of the fission-neutron energy spectrum fairly well [26]. Note that the cross sections are substantially lower than those given in Fig. 15.6 for ³He and ¹⁰B. The efficiency for detecting a fast neutron by an elastic-scattering interaction is about 2 orders of magnitude lower than the efficiency for capture of a thermal neutron (see Table 15.1). Thus, a single ⁴He or CH₄ tube has an intrinsic efficiency of ~1.5% [26].

These gas counters are operated as proportional counters, with voltages in the range of +1200–2400 V. Gas fill pressures are typically 10–20 atm. Relative to helium, the polyatomic gases CH₄ or H₂ again require higher operating voltages, have slightly lower efficiencies, are limited to lower pressures, and exhibit faster signal rise times. The gamma-ray sensitivity of the two types of counters is comparable. Neutron counting can be done in gamma-radiation fields of roughly 1 R/h if a moderately high threshold is set [27].

Figure 15.15 shows a pulse-height spectrum from a ⁴He proportional counter collected with a ²⁵²Cf neutron source. The observed spectrum shape is the convolution of the following effects:

- the ²⁵²Cf spontaneous fission neutron spectrum, as illustrated in Fig. 13.2;
- the probability of transferring an energy between 0 and $E(\max)$ to the recoiling nucleus;
- the probability of multiple neutron scatters and the probability of losing recoiling nuclei in the tube walls; and
- the detection of low-energy noise pulses and gamma-ray pileup events.

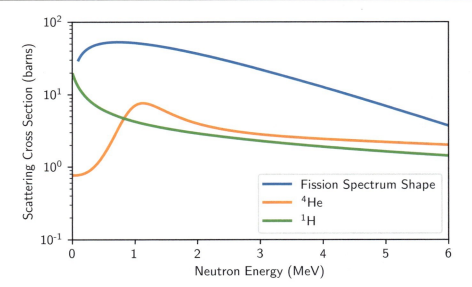

Fig. 15.14 Hydrogen and ^{4}He elastic-scattering cross sections, with a fission spectrum shape (not drawn to scale) superimposed

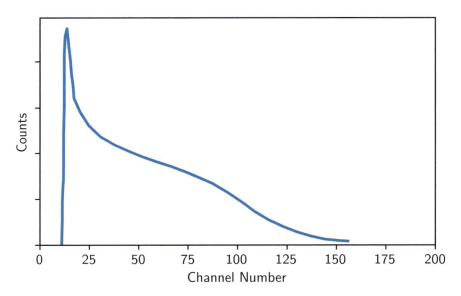

Fig. 15.15 Differential pulse-height spectrum of ^{4}He proportional counter for a ^{252}Cf source

Because of these effects, the pronounced peak in the initial neutron energy spectrum can be lost—or nearly lost—as indicated in Fig. 15.15. Nevertheless, some energy information remains, and more can be obtained by attempting to unfold the previously listed effects. It is customary to set a threshold high enough to reject low-energy noise and gamma-ray events but low enough to collect many of the medium- and high-energy neutron events. Because the threshold must be set on a sharply falling curve, a recoil detector is not as stable as a thermal detector.

Despite the apparent disadvantages of recoil-type detectors in terms of lower efficiency and stability, the detection process takes place without prior thermalization of the incident neutron. Thus, the neutron is detected very rapidly, and some information on its initial energy is preserved. Fast-neutron counters can detect neutrons in the energy range of 20 keV to 20 MeV, and some are useful for fast-coincidence counting, with 10–100 ns resolving times. It is also possible to set a threshold that will reject gamma rays and low-energy neutrons, a feature that is particularly suitable for active assay systems.

15.4.4 Fission Chambers

Fission chambers are a variation of the gas-filled counters previously described. They detect neutrons that induce fissions in fissionable material coated on the inner walls of the chamber. Often the exterior appearance of fission chambers is quite similar to that of other gas counters, although they are also available in smaller diameters and in other shapes. For neutron-counting applications, the fissionable material is usually uranium highly enriched in ^{235}U or ^{239}Pu (less common) because of its high cross section for thermal neutrons. However, other nuclides can be used for specific applications, such as ^{238}U, when a measurement of fast neutrons in the presence of thermal neutrons is desired. The coating-material form typically includes metallic uranium, UO_2, or U_3O_8 [28]. A very thin layer (0.02–2 mg/cm^2 surface thickness) is electroplated (sometimes evaporated or painted) on the inner walls. The thin layer is directly exposed to the detector gas. After a fission event, the two fission fragments travel in nearly opposite directions. The ionization caused by the fission fragment that entered the gas is sensed by the detector; the fragment traveling in the opposite direction is absorbed in the detector wall.

The two fragments share ~160 MeV of energy [29], but their range is quite short. For a typical plating material such as uranium, the average fission-fragment range is only ~7μm, equivalent to ~13 mg/cm^2 of coating thickness. Consequently, fission fragments that are produced at a depth of more than 7μm in the detector wall cannot reach the gas to cause ionization. Furthermore, most fragments exit at a grazing angle, so their path length is longer than the minimum needed to escape [30]. Because the coating must be kept thin to allow the fission fragments to enter the gas, the fission chamber uses only a small quantity of fissionable material and has a low detection efficiency. The maximum cross section for ^{235}U fission is ~580 b for thermal neutrons, and moderators are used to maximize the efficiency of the detector system (similar to ^{3}He systems). The intrinsic efficiency is typically 0.5–1%. Fast neutrons can also be detected but with even lower efficiency because the fast-fission cross section for fissioning nuclides is on the order of 100 times less than the thermal cross section of ^{235}U.

Fission chambers are operated in the ion chamber mode because the ionization caused by the fission fragments is sufficient, and no further charge multiplication within the detector is necessary. The electronics configuration shown in Fig. 15.1 is frequently used with an applied voltage in the +200–800 V range. A mixture of 90% argon and 10% methane is a common fill gas. At this pressure, the range of fission fragments is ~2 cm.

Figure 15.16 shows a pulse-height spectrum from a fission chamber [28]. If energy losses in the coating or in the walls are not too great, the double-hump shape caused by light and heavy fission fragments (near 70 and 100 MeV) is visible. Also, an alpha--particle background is present at low energies because nearly all fissionable material contains alpha-emitting isotopes. The alpha-particle energy is typically 5 MeV, whereas the fission-fragment energy is an order of magnitude larger. Thus, the threshold setting of the counting electronics can be set above the alpha-induced signal. At this threshold setting, some of the low-energy fission-fragment pulses will be lost. Plutonium has a much higher alpha activity than uranium; as a consequence,

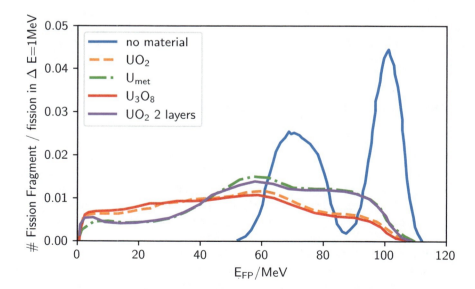

Fig. 15.16 Number of fission fragments per fission in a 1 MeV energy bin as a function of the fission-fragment energy. The plot shows the fission fragments as they emerge from the layer surface and enter the detector volume for no material, UO_2 (1.6μm), metallic U-U_{met} (0.8μm), U_3O_8 (2.2μm), and two layers of UO_2 (0.84μm)

more alpha pulses pile up, and the threshold for plutonium-lined fission chambers must be set higher than for uranium-lined chambers.

Because of the high-energy and high-ionization state of fission fragments, they deposit energy at a high rate. Thus, the size of the gas region where the energy is deposited can be small—less than the full fission-fragment range. Some designs consist of concentric cylinders that are only a few millimeters apart, with uranium coated on both electrode walls [28]. This process limits the energy that can be deposited in the region by alpha particles and gamma rays; therefore, fission chambers have the highest insensitivity to gamma rays (roughly 10^6 R/h) of any of the neutron detectors. They are the only detectors capable of direct unshielded neutron measurement of spent reactor fuel and vitrified waste. The inherently low efficiency of fission chambers is typically compensated for in these applications by the large number of neutrons available for counting.

15.5 Scintillation Detectors

The initial sections of this chapter (15.5.1 through 15.5.6) are dedicated to organic scintillation detectors, followed by descriptions of several other scintillator types of relevance for neutron-counting applications in Sect. 15.5.7. Sections 15.5.2 through 15.5.5 describe principles of scintillation signal generation, signal-processing electronics, and PSD.

15.5.1 Background

Organic scintillators are often used for fast-neutron detection because of their fast response and modest cost. Fast response is particularly beneficial for coincidence counting applications where the ratio of real-to-accidental coincidence events can have a significant impact on the statistical precision of measurement (see Chap. 17, Sect. 17.8.1). Although organic scintillators have response times of a few nanoseconds, the coincidence-resolving time for assay applications is usually dictated by the dynamic range of neutron flight times (tens of nanoseconds) from the sample to the detectors. (A 500 keV neutron will traverse a flight path of 1 m in ~100 ns). However, the resolving times of coincidence counting systems that moderate fast neutrons before detection are dominated by the dynamic range of times (tens of microseconds) required for thermalization.

Although several advantages exist for using organic scintillators for coincidence and multiplicity counting, one major disadvantage specifically in NDA applications is their high gamma-ray sensitivity. Detection probabilities for neutrons and gamma rays are comparable, and the energy spectra that results from monoenergetic radiation of both types are broad and overlapping. Therefore, pulse height alone yields little information about particle type; however, in certain organic scintillators, PSD techniques can be used to effectively distinguish between neutron and gamma-ray interactions. Additionally, long-term environmental stability of the detection electronics needs to be appropriately addressed for reliable facility deployment. Figure 15.17 illustrates the steps involved in detecting radiation using a typical organic scintillator detector [31].

15.5.2 Particle Interaction and Scintillation Mechanisms

Organic scintillators are able to detect both neutrons and gamma rays. Neutrons directly interact through elastic and inelastic scattering on nuclei present within the scintillator (typically C and H nuclei); the most useful scintillation light comes from elastic scatters on recoiling hydrogen nuclei (i.e., protons). This interaction occurs because of two primary reasons: a neutron can transfer 100% of its energy to a recoiling proton, but only 28% can be transferred to a recoiling ^{12}C nucleus in a single-scatter interaction; and ^{12}C recoil nuclei produce significantly less light compared with a recoiling proton [4, 32–34]. As an example, the ^{12}C recoil nuclei produce less than 2% relative to a fast electron in EJ 309 [35]. Gamma rays interact through Compton scattering (Chap. 3), producing a fast electron. The kinetic energy of the recoiling particles (protons and fast electrons for neutron and gamma-ray interactions, respectively) is absorbed by the scintillator and is then converted to heat and scintillation light [31]. Due to the different interaction types for neutrons and gamma rays, the resulting scintillation light is characteristically different from one another, allowing for particle identification for each detection.

The different amount of scintillation light produced for the recoiling protons and electrons is due to the ionization densities along the slowing-down paths; recoil protons will produce significantly less light along its track relative to a fast electron given the same deposited energy. The response to electrons is considered to be linear for particle energies above ~125 keV [4],

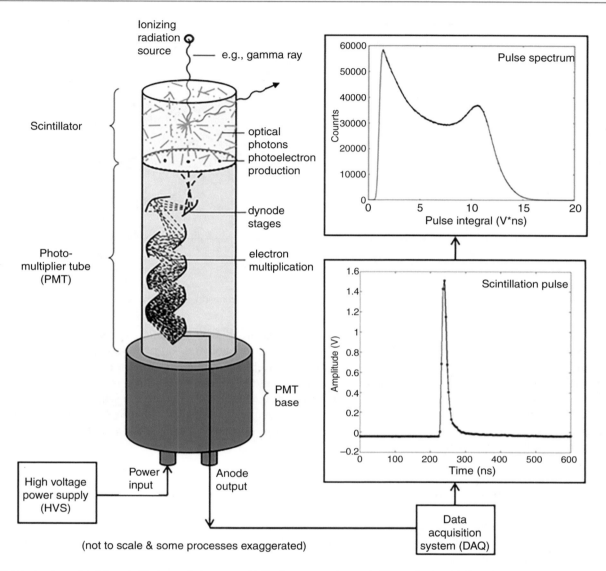

Fig. 15.17 A schematic of the scintillation and electron multiplication process for a scintillator-photomultiplier tube assembly

whereas recoil proton response is nonlinear and is always less than that of the electron response. To establish a unit that describes the light yield in an absolute sense, the unit of MeV electron equivalent (MeVee) is typically used where the amount of energy to produce 1 MeVee of light is 1 MeV for fast electrons but significantly higher for recoil protons depending on the scintillator material properties. Furthermore, the time-dependent emission of the scintillation is also distinctly different; recoil protons exhibit a longer-lived emission profile due to delayed fluorescence where a higher fraction of the total light appears in the latter region of the pulse. This effect can be leveraged to determine the incident particle type (neutron or gamma ray) that deposits equal energy in the detector.

Although the mechanism by which fast neutrons transfer their kinetic energy to protons in an organic scintillator is identical to that in a hydrogen or methane recoil proportional counter, various features of the overall detection process are markedly different. This distinction is largely due to the differences in physical properties of organic scintillators and gases. For example, the density of gas in a recoil proportional counter is on the order of 10^{-3} g/cm^3, whereas that of an organic scintillator is on the order of unity.

This difference in density means that for a given detection path length in the two materials, the probability of interaction for both neutrons and gamma rays will be substantially higher in the scintillator than in the proportional counter. Figure 15.18 shows the energy dependence of the interaction probability (expressed as attenuation coefficients) for neutron and gamma-ray

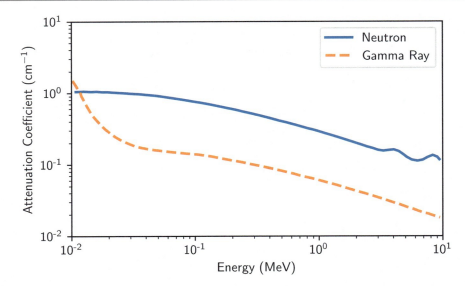

Fig. 15.18 Attenuation coefficients as a function of incident energy for neutron and gamma-ray interactions in EJ 301

interactions in EJ 301. For example, this figure shows that a 1 MeV neutron has an interaction probability of ~78% in a 5 cm thick EJ 301 liquid scintillator, whereas a 1 MeV gamma ray has an interaction probability of ~26%.

In addition, the ranges of the recoiling protons and electrons will be substantially shorter in the scintillator than in the proportional counter. Except for events that occur near the boundaries of the detectors, this fact is of little importance when considering the recoiling protons. However, the shortened range of the electrons in the organic scintillators has a profound effect because high-energy electrons can stop inside the detection volume. For example, a 500 keV electron can deposit all of its energy in a scintillator but deposit only a small fraction in a gas proportional counter, resulting in the higher gamma-ray sensitivity of scintillation detectors.

15.5.3 Detector Response: Energy, Timing, and Pulse-Shape Information

The detector responses to neutrons and gamma rays have characteristically different features due to the mechanisms described in Sect. 15.5.2. For most safeguards applications, three important quantities are processed for each detected signal: energy (i.e., light yield), timing, and pulse shape.

15.5.3.1 Energy

The energy deposited by an incident neutron or gamma ray can be directly related to the absolute light yield produced within the detector. Gamma-ray response retains a linear relationship between the energy deposited and the light yield, whereas neutron response exhibits a nonlinear relationship and is always less than that of a gamma-ray detection with equal energy deposition [33]. Energy information is particularly important for applications that aim to use the obtained signals to unfold the incident neutron spectrum.

Figure 15.19 shows the detector response for neutrons and gamma rays from a trans-stilbene scintillator, demonstrating the significantly higher light produced by fast electrons relative to that of recoil protons [35]. Note that trans-stilbene exhibits anisotropic light yield relative to the crystalline axis [36, 37]; Fig. 15.19 shows the light yield from a single axis merely to demonstrate the significantly less light produced by recoil protons relative to fast electrons.

The gamma-ray energy distribution produced by an organic scintillator will consist only of a Compton continuum and a Compton edge; the low-Z composition of the detector makes photoelectric absorptions improbable. Energy calibration, which provides the conversion from measured signal (e.g., analog to digital units) to light yield (e.g., MeVee), is acquired by characterizing the response to monoenergetic gamma-ray sources. Rather than using the photopeak, the Compton edge must be used, and the determination of the Compton edge location can be established either experimentally [38] or through iterative optimization of Monte Carlo simulations [39]. Figure 15.20 shows an example of an experimentally characterized energy distribution for 662 keV photons from a ^{137}Cs source using a backscatter measurement technique.

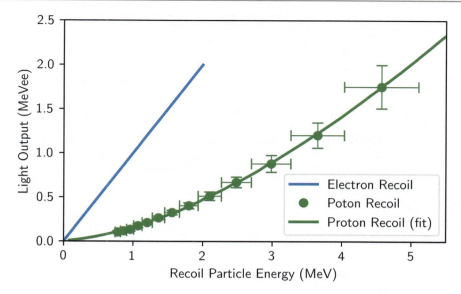

Fig. 15.19 Scintillation light yield as a function of particle energy deposition for electrons and protons in trans-stilbene

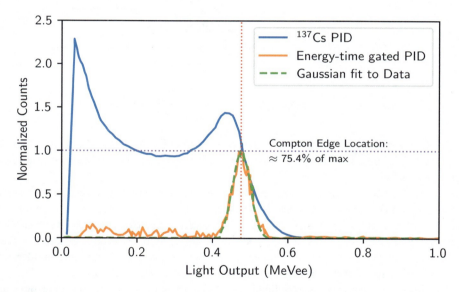

Fig. 15.20 Backscatter experiment in a 2 in. × 2 in. stilbene crystal for 662 keV photons from ^{137}Cs, showing the typical gamma-ray response (blue), along with the energy-time-gated distribution that isolates the approximate Compton edge location

To characterize the nonlinear neutron response as shown previously in Fig. 15.19, multiple different experiments can be performed using direct or indirect beam measurements [40], indirect measurements employing a coincidence scatter method [41], a double time-of-flight technique [42], and a time-of-flight method with edge characterization [35, 43], among others.

15.5.3.2 Time

The time response of the detected signals is primarily governed by the fast leading edge of the pulse as well as the digitization process of the pulse itself. Conventional time pick-off methods include leading edge triggering, crossover timing, and constant fraction timing, among others. A commonly used approach with modern digital electronics is the digital constant fraction discrimination (DCFD) method, derived from the concept of constant fraction timing where the detection time is determined to be the time the leading edge has passed a constant fraction of the peak amplitude [4]. DCFD is a digital implementation where the input signal is first attenuated and then inverted as well as delayed; the attenuated and the inverted, delayed signal

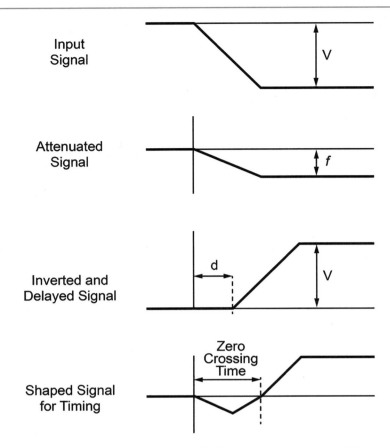

Fig. 15.21 A schematic that demonstrates the signal processing for establishing detection time using DCFD

are then summed to produce a bipolar signal, and the zero crossing time is taken as the detection time. Figure 15.21 is an example schematic of the DCFD process that is implemented in CAEN waveform digitizers commonly used in scintillator signal-processing applications [44].

The time response following a neutron detection in organic scintillators is very fast compared with that of thermal-neutron detectors (less than a nanosecond). Furthermore, because organic scintillators do not require intervening moderating material between the source and detector head to thermalize neutrons emitted from nuclear material, a very fast system response can also be achieved. The system coincidence resolving time for assay applications is usually dictated by the dynamic range of neutron flight times (typically on the order of tens of nanoseconds), whereas systems that require moderation before detection are dominated by the dynamic range of thermalization and diffusion times (on the order of tens of microseconds). The drastic decrease in the resolving times using an organic-scintillator-based system, in turn, allows for higher-precision, time-correlated measurements to be made for fixed acquisition times compared with thermal-neutron systems [45] for applications where good neutron-gamma discrimination can be assured. This aspect has been further evaluated in support of correlated neutron-counting applications [14]. Figure 15.22 shows a measured time-interval distribution of a trans-stilbene-based system using a ^{252}Cf spontaneous fission source, where the resolving time was calculated to be approximately 6.29 ns.

15.5.3.3 Pulse Shape

Determining information pertaining to the pulse shape of detected signal is essential for any application that uses organic scintillators to perform PSD to identify neutron and gamma-ray detections. Figure 15.23 demonstrates the time dependence of scintillation pulses in stilbene for gamma rays, neutrons, and alpha particles [4].

To exploit the pulse-shape differences that arise from incident neutron and gamma rays that deposit the same energy, a discriminating metric that contains information of the pulse shape is used. The most common approach is the tail-to-total ratio, where a ratio of the integral of the tail region and the total region of the pulse is calculated. Comparing the tail-to-total ratio for all detected pulses as a function of the measured light yield can then be used to separate neutron- and gamma-ray-induced

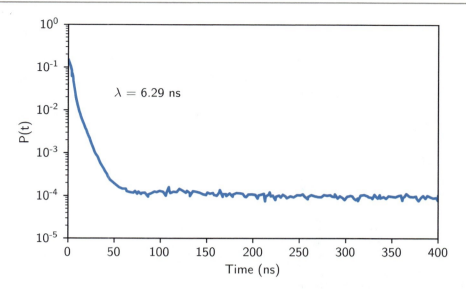

Fig. 15.22 Measured resolving time of a trans-stilbene-based, fast-neutron multiplicity counter that demonstrates a resolving time of approximately 6.29 ns

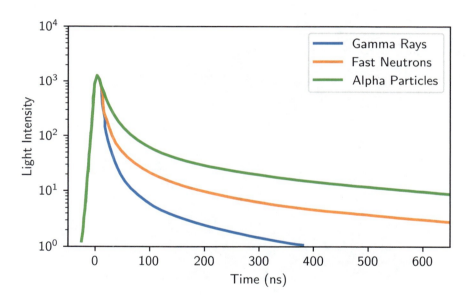

Fig. 15.23 The time dependence (in nanoseconds) of scintillation pulses in stilbene when excited by different incident particle types

signals. Figure 15.24 shows a histogram of the tail-to-total ratio, where two distinct distributions can be seen corresponding to neutron and gamma-ray signals [38].

The canonical figure of merit (FOM) to describe PSD performance for any organic scintillator is calculated as the ratio of the distance between the mean of the neutron and gamma-ray distributions to the sum of the FWHM for each distribution [46]. A two-dimensional histogram showing the tail-to-total ratio as a function of the light yield reveals two distinct clusters of data that correspond to neutron and gamma-ray signals, as shown in Fig. 15.25. To establish a discrimination line between the two data clusters, one can implement optimization algorithms, such as a slice-based optimization [47] and a template-based optimization [48].

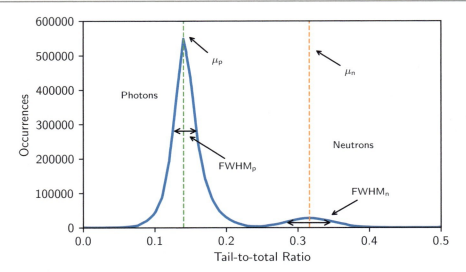

Fig. 15.24 Illustration of the figure of merit for measuring performance of PSD systems

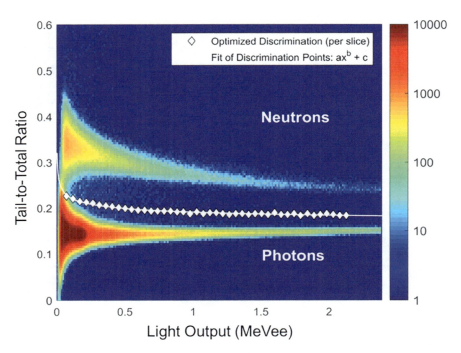

Fig. 15.25 Two-dimensional histogram of the tail-to-total ratio that shows distinct clusters of data that represent neutrons (top cluster) and photons (bottom cluster) for stilbene. Algorithms such as the one detailed in Ref. [44] can be used to optimize the discrimination line that separates the two particle types

15.5.4 Light Collection

The process of collecting the scintillation light produced in the detector involves creating an electrical signal that can subsequently be digitized without adding significant noise to the signal. This process can be achieved using light-collection electronics, such as photomultiplier tubes (PMTs) and silicon photomultipliers (SiPMs), among others. The primary function of these light-collection electronics is to greatly amplify the signal while ideally conserving the relationship between the energy deposited and the scintillation light produced in a linear manner [4].

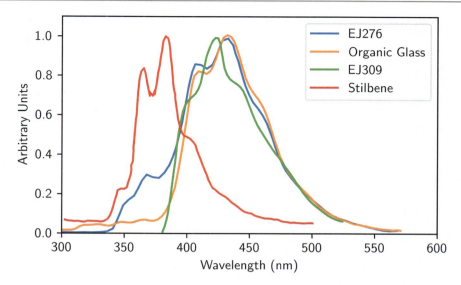

Fig. 15.26 Comparison of emission spectra for commonly used organic scintillators scaled to their maximum scintillation wavelength

The main function of PMTs is to convert the optical photons into an amplified signal, which is done through two major components that include the photosensitive layer—called the *photocathode*—and an electron-multiplying mechanism. The photocathode converts the incident optical photons into photoelectrons, which are then passed through the multiplication process where the photoelectrons will traverse through a series of dynodes. The overall multiplication factor is known as the *gain* and is typically on the order of 10^6 for commonly used PMTs. The signal produced by the multiplied electrons is collected at the anode placed after the series of dynodes and provides a usable electrical current pulse related to the number of original photoelectrons produced within the photocathode.

In comparison, SiPMs are designed to function as an array of small-dimension Geiger mode avalanche photodiodes. Typical Geiger mode avalanche photodiodes, as the name implies, lose all proportionality to the detected optical photons; however, SiPMs are specifically designed such that the proportionality between the initial electron-hole pairs created by the incident radiation is retained. This result is achieved by designing photodiode cells that are small enough to be affected only by a single optical photon, where the number of cells that produce an avalanching effect is proportional to the incident scintillation photons. The individual cells are summed to create an output that is proportional to the total number of optical photons detected throughout the entire array.

Spectral matching between the scintillation material and the photosensitive material in a PMT or SiPM is important to achieve maximum light-collection efficiency in a given detector assembly. Figure 15.26 shows the wavelength emission spectra for commonly used organic scintillators [49]. The wavelength of maximum scintillation light emission is typically ~400 nm. At that wavelength, light attenuation lengths are in the range of 1–5 m. Because light can travel relatively long distances in the scintillator material without significant attenuation, organic scintillators with dimensions on the order of 1 m are not uncommon because they are able to be manufactured to such sizes.

Figure 15.27 shows the quantum efficiency for commonly used SiPM (SensL J-series) and PMT (Electron Tube 9214B), demonstrating that the sensitivity range for these light-collection electronics corresponds to the range of the emission wavelengths given by typical scintillators.

Several environmental effects should be considered for practical use of detector assemblies that use either PMTs or SiPMs, including operational stability, temperature dependence, and magnetic interference. The latter is more significantly pronounced in PMTs but can be mitigated by placing a magnetic shield (e.g., mu-metal) around the entire dynode chain. Operational stability is typically dictated by the electronics associated with the detector assembly, including the high-voltage supply, analog-to-digital converter (ADC), and cable interferences. Thoughtful system design and engineering can also mitigate stability issues to provide a system that is robust and reliable for long-term use. Ambient temperature effects can be a large source of uncertainty in many organic-scintillator-based systems, where both the scintillating detector material and the light-readout electronics can have appreciable performance variability. Of course, these effects can be offset by using well-characterized systems in temperature-controlled facilities but should be considered in cases with unknown or dynamic temperature environments and should be subsequently corrected for with frequent gain calibration measurements.

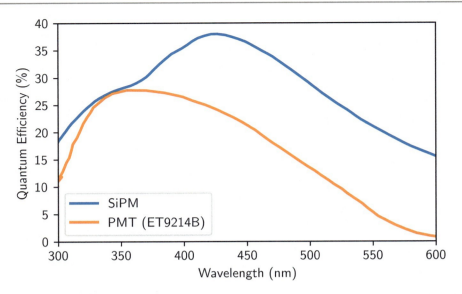

Fig. 15.27 Quantum efficiency of commercially available SiPM (SenSL J-Series) and PMT (ET 9214B)

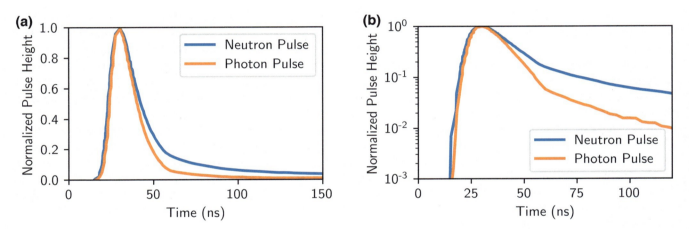

Fig. 15.28 Digitized scintillation pulses in stilbene using a CAEN V1730 digitizer with 14-bit resolution and 500 MHz sampling rate (2 ns bins) (**a**), logarithmic scale (**b**)

15.5.5 Signal Acquisition with Modern Digital Electronics

The state-of-the-art method for retrieving the current pulse produced by the detector assembly involves using fast ADC electronics to digitize the analog signal and has allowed for efficient storage of large data for offline processing while also providing the means for modular system designs. The fast ADC electronics typically has several field-programmable gate arrays (FPGAs), which sample the analog waveform with a frequency that is on the order of the inverse time scale of the scintillation pulse. For example, the sampling rate typically ranges from 250 MHz to greater than 1 GHz, which corresponds to sampled time bins of 4 ns to 1 ns, respectively. The collection of several FPGAs placed in a single digitizing unit is often referred to as a *waveform digitizer*, which can have varying specifications—such as sampling rate, vertical bit resolution, and number of channels—all of which should be chosen appropriately for a given detector assembly and system. Figure 15.28 shows the digitized pulse using a CAEN V1730 waveform digitizer with 14-bit resolution and 500 MHz sampling rate, which corresponds to 2 ns time bins [38, 44].

Rather than saving each waveform for post-processing, waveform digitizers also have the ability to save relevant information pertaining to each pulse—including the total light yield, timing, and pulse-shape parameters—significantly reducing the data throughput. Total light yield is calculated using a *long-gate* integration, where the entirety of the waveform is integrated. A *short-gate* integration is also performed to provide the quantities to calculate the tail-to-total ratio for PSD [44]. Lastly, timing information is extracted using a DCFD or other time-pickoff technique [44].

15.5.6 Commonly Used Organic Scintillators

A good scintillator for neutron detection has relatively high efficiency for converting recoil particle energy to fluorescent radiation, good transparency to its own radiation, and good matching of the fluorescent light spectrum to the light-collection electronics while also being physically robust and impervious to environmental effects. Although it is impossible to find organic scintillators that exhibit all of the aforementioned qualities, several commercially available scintillators are widely used. Four such scintillators worth mentioning here are liquid organic scintillators, EJ-309, [50], plastic organic scintillators, EJ-276, [51], amorphous organic glass scintillators [35, 52–54], and crystalline organic scintillators, trans-stilbene, [35–37, 55]. Table 15.4 provides a brief overview on the performance qualities pertaining to the four scintillator types.

Figure 15.29 shows a comparison of neutron light output responses for EJ-309, EJ-276, organic glass scintillator [51] and trans-stilbene [35] materials. The light output is given relative to the 477 keV electron response (i.e., location of Compton edge from 662 keV photons emitted by ^{137}Cs; Ref. [56]).

Another important performance characteristic of a candidate organic scintillator is the PSD performance; discriminating between neutron- and photon-induced signals is essential for ensuring that the data consist of only neutron signatures. Figure 15.30 compares the canonical PSD FOM.

Table 15.4 Summary of the properties for commonly used organic scintillators

Scintillator	State	Scintillation efficiency	Decay constant	Scintillation peak
Trans-stilbene	Solid	15,000	4.5 ns	390 nm
EJ-276	Solid	8,600	13 ns	425 nm
Organic glass	Solid	17,400	1.5 ns	425 nm
EJ-309	Liquid	12,300	3.5 ns	424 nm

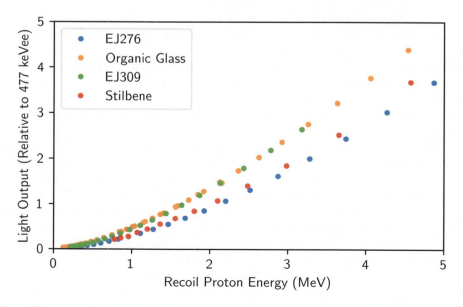

Fig. 15.29 A comparison of the measured neutron light yield response as a function of the neutron-equivalent energy (i.e., recoil proton energy) for the four commonly used organic scintillators [35, 51]. Note that the light output is shown relative to the light given by a 477 keV fast electron, which offsets any potential influence due to nonlinearities in the electron light yield [56]. Also note that, as described in Fig. 15.19, the light yield for trans-stilbene shown here is for a single crystalline axis and can vary appreciably depending on the axis orientation [36, 37]

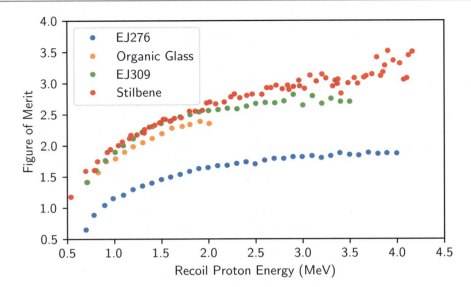

Fig. 15.30 Comparison of pulse-shape discrimination performance for commonly used organic scintillators as a function of the neutron-equivalent energy (recoil proton energy)

15.5.7 Other Scintillator Types

15.5.7.1 ^{6}Li Scintillators

Some scintillators are manufactured with neutron-active material added to achieve enhanced thermal neutron-detection capability. The purpose is to achieve more localized and more rapid detection of neutrons than is possible with thermal-neutron gas counters. Gadolinium, ^{10}B, and ^{6}Li are typical materials "loaded" into the scintillator. The neutron-active material initiates the light production by releasing energetic charged particles or gamma rays when the neutron is captured. After the initial interaction with the neutron occurs, the detection process is the same as if the light were produced by a gamma ray. Because the scintillator is also a gamma-ray detector, its gamma-ray sensitivity is generally very high; however, several possible configurations exist with good neutron-detection efficiency and low gamma-ray sensitivity.

One useful configuration for thermal-neutron counting consists of lithium-loaded scintillators. ZnS(Ag) crystals in a glass medium or cerium-activated silicate glasses are available. Thermal neutrons interact with ^{6}Li via the (n,α) reaction, and the heavy alpha particles excite the scintillator. Detectors of this type are available in sheets with thicknesses of ~1 mm. For thermal neutrons, efficiencies of 25–99% are possible in gamma-ray fields on the order of 1 R/h. The low gamma-ray sensitivity is due to the high thermal-neutron capture cross section, the large 4.78 MeV energy release in the reaction, and the thinness of the detector [4]. Lithium-coated scintillating fibers were also designed with alternating layers of lithium fluoride/zinc sulfide compounds, ^{6}LiF/ZnS(Ag), with wavelength-shifting fibers or flat light-guide layers to extract the scintillation light. A recent application of this technology includes ^{6}LiF/ZnS blades made of bulk polyvinyltoluene (PVT) wavelength-shifting guide coated with LiF/ZnS(Ag) neutron screens on both sides and surrounded by light-tight packaging [57]. Solid-state readout SiPMs and compact pulse-processing electronics are mounted at the top of the detector active area, providing a slim neutron detector. A proprietary PSD algorithm is used to maximize neutron efficiency while optimizing gamma-ray rejection. A prototype coincidence counter was developed based on this technology [14].

If ZnS(Ag) crystals are dispersed in Lucite, detection of fast neutrons is possible. The interaction mechanism is the elastic scattering of neutrons by hydrogen. The recoiling proton deposits its energy in the scintillator, and by transfer reactions, the ZnS(Ag) crystals are excited. ZnS(Ag) offers good gamma-ray insensitivity because relatively high energies are required to excite the light-emitting property of the zinc sulfide crystals. Detectors that consist of ZnS(Ag) crystals dispersed in Lucite are called *Hornyak buttons* [58]; their efficiency is low (on the order of 1%) because the poor light-transmission properties of this material limit its use to small sizes. Thin sheets have also been used for measurements of waste crates at the Rocky Flats Plant [59]. Hornyak buttons can operate in gamma fields up to ~1R/h because of the properties of ZnS(Ag) and because the thinness of the detector limits the gamma-ray-induced energy deposition.

Another recent development includes heterogeneous neutron detectors that comprise ^{6}Li scintillator (GS20) glass particles with a few mm^3 volume embedded in an organic, nonscintillating light-guide matrix [60, 61]. A prototype module is shown in

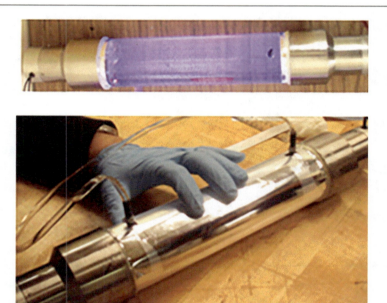

Fig. 15.31 (top) Lithium-6 glass scintillator rods assembled in a quartz tube, immersed in mineral oil, coupled with two PMTs; (bottom) detector prototype with light reflector installed

Fig. 15.31. The matrix serves to moderate the incident fission neutrons to thermal energies, thus enhancing the capture efficiency by the scintillator; absorb the energy from secondary electrons produced by gamma-ray interactions without producing light, thus limiting gamma pulse amplitudes; and transmit the light emitted by the scintillator particles. Preliminary experimental results and modelling have shown that the ^{6}Li-glass technology achieves about twice the neutron-detection efficiency per unit volume and about one order of magnitude higher count-rate capability (resulting in dead-time reduction of the same order) than traditional, equivalent, 4 atm ^{3}He neutron detectors while still maintaining the separation of the neutron and gamma signals.

15.5.7.2 Helium-4 Gas Scintillators

High-pressure, ^{4}He gas-filled scintillation detectors with pressures on the order of 100–200 atm use scintillation light for detection of neutrons [14, 62]. The appearance of these detectors is very similar to ^{3}He tubes. The typical detection mechanism involves elastic scattering of fast neutrons in the pressurized fill gas; addition of a ^{6}Li coating on the detector walls can also provide thermal neutron-detection capability. The recoil and/or charged particles produced in these interactions result in production of scintillating light, which is collected by PMT or SiPM. The elastic-scattering cross section of ^{4}He is several orders of magnitude smaller in the low-energy region than the neutron-capture cross section on ^{3}He; however, it exhibits a peak that is located at roughly 1 MeV, matching the emission energy of fission neutrons. For the high-pressure ^{4}He scintillation detectors, gamma rejection of ~0.01 R/h was reported while preserving 5% intrinsic neutron detection efficiency [63]. The applicability of these detectors for safeguards is discussed in further detail in Ref. [14].

15.5.7.3 Elpasolites

In recent years, a particular class of dual neutron and gamma-ray-sensitive inorganic scintillators, known as *elpasolites*, have become commercially available. These types of scintillators are of particular interest to the safeguards community for applications that would benefit from either neutron detection or dual-neutron and gamma-ray detection in a single volume, given their ability to differentiate detected incident radiation using PSD techniques.

In general, scintillators belonging to the elpasolite class of chemical compounds carry a similar structure that follows the pattern of $A_2B\Gamma\Delta_6$. Here A, B, Γ, and Δ are elements from the heavier alkali metals, lighter alkali metals, lanthanides or transition metals, and halogen groups of the periodic table, respectively. Additionally, cerium is added as a dopant to tune the scintillation characteristics. Examples of commercially available elpasolites include CLYC (Cs_2LiYC_6:Ce), CLLB ($Cs_2LiLaBr_6$:Ce), and CLLBC ($Cs_2LiLaBr_{4.8}C_{1.2}$:Ce)—the most mature candidate elpasolites for potential safeguards applications. Other candidate elpasolites are available for purchase from various manufacturers; however, these candidates are typically regarded as being more suitable for research applications given their lower level of technical readiness.

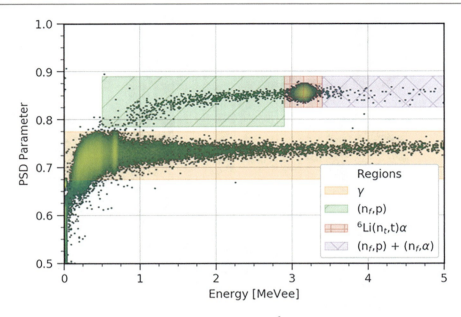

Fig. 15.32 A sample pulse-shape discrimination versus energy spectra from a 1 cm^3 sample of CLYC with the various reaction regions highlighted. Recorded response was captured using a combination of a ^{137}Cs and a ^{241}AmBe source

Neutron sensitivity in elpasolites is accomplished using two mechanisms. First, with the introduction of enriched ^{6}Li during the scintillator growth process, the presence of ^{6}Li in the final chemical structure of the scintillator allows for efficient thermal-to-epithermal neutron detection via the ^{6}Li(n,t)α reaction, where the alpha-triton pair propagate in the crystal structure, exciting a scintillation response. Second, via low-efficiency, fast-neutron reactions (n,p) and (n,α) with the chlorine and bromine elements [64], which are typically a few orders of magnitude lower in cross section compared with the ^{6}Li thermal reaction. The use of elpasolites for fast-neutron detection should not be the primary motivation for a potential application, given their low efficiency and the availability of other more-sensitive technologies such as organic scintillators, as discussed in Sect. 15.5.6. Gamma-ray sensitivity comes from the presence of the cerium dopant. Elpasolites have a moderate scintillation light yield in the 20–45 kph/MeV range, depending on crystal type. Scintillation pulse decay lengths range from 1 to 3μs. At 662 keV, typical gamma-ray energy resolutions are in the 3%–6% range and depend on the particular type of elpasolite used [65].

An attractive property of elpasolite scintillators is the ability to employ PSD techniques to separate out the respective neutron and gamma-ray response. Similar to PSD spectra from organic scintillators, elpasolites produce PSD spectra with distinct bands for neutrons and gamma rays (see Fig. 15.32). The alpha and triton products from the ^{6}Li capture of thermal neutrons show up as a cluster of events centered on 3.1–3.5 MeVee at room temperature, which is due to the quenched response of the Q = 4.78 MeV from the ^{6}Li reaction.

Other properties include being hydroscopic—similar to NaI scintillators—which requires a hermetic enclosure around the scintillator crystal. Several manufacturers offer hermetically sealed, preassembled systems that use either PMTs or SiPMs as light-collection devices and range in size as large as 3 in. diameter × 4 in. long. Examples of vendors established in the field include Radiation Monitoring Devices, Saint Gobain Crystals, and Berkeley Nucleonics Corp. Elpasolites are known to have a temperature dependence [66], which could be a consideration for specific safeguard applications that require operation outside of traditional laboratory conditions. Additionally, they are known to have only moderate radiation hardness [67], which also might be of concern for particular applications, for example in close proximity to high-flux reactors.

15.6 Other Types of Neutron Detectors

This section describes several neutron-detector technologies that are based on mechanisms different than proportional counting or scintillation detection that were considered or developed for nuclear material assays but have not yet found widespread use. Like the other detectors described in this chapter, they rely ultimately on either recoil interactions or direct nuclear reactions to detect neutrons.

15.6.1 Semiconductor Detectors

Desire to develop viable ^{3}He alternative technologies to mitigate concerns with limited ^{3}He supply (as discussed in more detail in Chap. 19) led to development of a range of alternative neutron detection technologies, including semiconductor-based detectors [14]. Such technology typically involves semiconducting wafers doped or lined with materials that contain ^{6}Li or ^{10}B. Similar to scintillator-based technologies, one of the key challenges for this technology involves optimization of neutron-detection efficiency performance in the presence of gamma-ray backgrounds typical for safeguards applications. In addition, the small size of the individual detectors (typically less than ~100 cm^2) will require multiple detector configurations to achieve system dimensions typical for a nuclear safeguards environment. This limitation brings additional challenges that involve the setting of signal-processing electronics to match the response of individual components and ensure the overall long-term and environmental stability of the complete system. The semiconductor technologies currently under development for nuclear safeguards typically involve small subsystems of several individual detectors in the form of planar or cylindrical arrangements to represent a full-scale single detector module [14, 68].

15.6.2 Activation or Optically Stimulated Response

Most neutron detectors combine neutron-sensitive material and detection electronics into one inseparable unit; however, it is possible to employ a detection system that is more compact and portable by using only the neutron-sensitive material. This material is first placed at the point of interest, then removed to measure the actual neutron flux by observing isotopic or crystalline structure changes. An example is the use of thermoluminescent dosimeters (TLDs), which consist of crystals that when heated, emit an amount of light proportional to the dose received. TLDs are primarily used for gamma-ray measurements, but one common crystal, LiF, can be made to be neutron sensitive by increasing the enrichment of ^{6}Li [4].

Activation foils—used for criticality safety and low-level detection—provide other examples of the use of neutron-sensitive material. One application has been the use of thin copper sheets to monitor plutonium migration in soil near nuclear waste storage sites [69]. This technique relies on neutron capture in ^{63}Cu to yield ^{64}Cu, which decays to ^{64}Ni + e$^+$ with a 12.7-hour half-life. The foils are buried long enough to achieve an equilibrium level of ^{64}Cu and then retrieved to permit measurement of the positron emission rate. Plutonium concentrations as low as 10 nCi/g have been monitored.

Optically stimulated luminescence (OSL) fiber dosimetry represents a new, passive, cost-effective radiation-measurement technique with great potential in radiation measurement applications. OSL fibers are inexpensive, passive, as thin as fishing line, and can be additively manufactured into other structures. OSL fibers absorb radiation, which excites electrons into a metastable state. The electrons remain trapped in this metastable state in crystal defects or holes until they are released by an external stimulus. When interrogated by an external stimulus, the electrons de-excite to their ground state and release a luminescence signal. A laser will emit a red light into the OSL fiber and release the electrons back to their ground state, subsequently releasing blue light that can be measured. The fiber is optically reset each time it is read, allowing a life cycle as follows: radiation exposure → measurement → readout (at which time the cycle restarts with no effects from the previous measurement). Investigations are ongoing into whether these fibers can be used to detect neutrons, in addition to their known gamma and X-ray detection capabilities [70, 71]. Neutron detection capabilities could open up the possibility of OSL fiber use for safeguards applications in the future.

15.7 Measurement of Neutron Energy Spectra

15.7.1 Background

As noted in Sect. 15.1, passive neutron assays are usually based on counting neutrons without regard to their energy because radioactive materials emit neutrons with broad energy spectra that are very similar from one isotope to another and because neutron detection is an indirect process that preserves little information about the incident neutron energy. This chapter has shown that neutron detection usually produces a broad spectrum of events that is only indirectly related to the neutron energy. A partial exception is found in the case of recoil detectors, such as ^{4}He-gas-filled counters and plastic scintillators; however, none of the detectors described in this chapter can distinguish nuclear isotopes on the basis of their neutron energy.

As a consequence, passive neutron assay is usually based on the counting of thermal or fast neutrons, with perhaps some tailoring of the detector or its surroundings to favor a particular broad-energy interval. Detectors are also chosen on the basis of their ability to produce fast (10–100 ns) or slow (10–100µs) output signals for coincidence counting. Some detectors are also designed to have a detection efficiency that is nearly independent of neutron energy.

15.7.2 Techniques

Although measurement of neutron energy spectra is not necessary for passive neutron assay, it is sometimes important for research or instrument development activities. Such a measurement is difficult but possible through a variety of techniques, including proton recoil spectrometers, neutron time-of-flight measurements, and ^{3}He spectrometers. An example of the use of ^{3}He spectrometers in measuring neutron energy spectra follows.

The ^{3}He spectrometer developed by Shalev and Cuttler [72, 73] has been used to measure delayed neutron energy spectra. (The AmLi neutron spectrum given in Fig. 13.6 was measured with an instrument of this type.) The spectrometer is a gas-filled proportional counter that contains ^{3}He, argon, and some methane. Neutrons are detected via the ^{3}He (n,p) reaction in the energy range of 20 keV to 2 MeV, where reaction cross section is smooth and nearly flat, declining from roughly 10 to 1 b. To detect these fast neutrons, the tube is not enclosed in moderating material; rather, it is wrapped in cadmium and boron sheets to reduce the contribution of the much stronger thermal ^{3}He (n,p) reaction (5330 b). Also, a lead shield is often added to reduce the effects of gamma-ray pileup on the neutron energy resolution. The intrinsic efficiency is low—on the order of 0.1%.

The energy spectrum of a ^{3}He spectrometer includes a full-energy peak at the neutron energy E_n + 765 keV, a thermal-neutron capture peak at 765 keV, and a ^{3}He (n,n′) elastic-scattering recoil spectrum with a maximum at 0.75 E_n (from Eq. 15.3). To emphasize the full-energy peak at E_n + 765 eV, long-charge collection time constants of 5–8µs are used, which favor the slower proton signals from the (n,p) reaction over the faster signals from recoiling ^{3}He nuclei. It is also helpful to collect data in a two-dimensional array of charge collected versus signal rise time to obtain more pulse-shape discrimination. In this way, a neutron energy spectrum can be obtained, although it must be carefully unfolded from the measured data.

The Shalev and Cuttler spectrometer [72] was used in Ref. [74] to measure the fast-neutron spectra of a deuterium-deuterium neutron generator and ^{252}Cf spontaneous fission neutron sources—two typical neutron sources used in safeguards applications.

In addition to the Shalev and Cuttler [72], the work was based on the ring ratio as described in Chap. 18, Sect. 18.2.6.2, and Chap. 19, Sect. 19.3.4.2. Because of the larger penetration of higher-energy neutrons, the ratio of counts in an outer ring of detectors to an inner ring depends on the initial neutron energy. A five-ring, ^{3}He-based multiplicity counter [75] was used in Ref. [76] to constrain the neutron energy spectrum of an AmLi neutron source typically used in active interrogation systems such as active well coincidence counter and Neutron Collar detectors (Chap. 20). Knowledge of the AmLi neutron energy spectrum is, in fact, necessary for accurate simulations of active systems such as neutron collar detectors [77].

References

1. D. Mazed, S. Mameri, R. Ciolini, Design parameters and technology optimization of ^{3}He-filled proportional counters for thermal neutron detection and spectrometry applications. Radiat. Meas. **47**(8), 577–587 (2012)
2. A.E. Evans et al., Radiation damage to ^{3}He proportional counter tubes. Nucl. Instrum. Methods **133**, 577–578 (1976)
3. M. Reginatto, Determination of the background of ^{3}He-filled proportional counters used for low-level neutron measurements. Radiat. Prot. Dosim. **180**(1–4), 403–406 (2018)
4. G.F. Knoll, *Radiation Detection and Measurement*, 4th edn. (Wiley., ISBN: 978-0-470-13148-0, 2010)
5. E. Fairstein, "Bipolar pulse shaping revisited," in IEEE Trans. Nucl. Sci. 44(3), 424–428 (June 1997)
6. A. Ravazzani et al., Characterisation of ^{3}He proportional counters. Radiat. Meas. **41**(5), 582–593 (2006)
7. J.L. Lacy et al., *Boron Coated Straw Detectors as a Replacement for ^{3}He*, in IEEE Nuclear Science Symposium Conference Record (NSS/MIC), Orlando, FL, (2009), pp. 119–125
8. D. Henzlova, H.O. Menlove, *High-Dose Neutron Detector Development for Measuring Alternative Fuel Cycle Materials*, in Proceedings of the global 2017, September 24–29, 2017, Seoul (Korea), Paper A-018, Los Alamos National Laboratory report LA-UR-17-23325 (2017)
9. T. L. Atwell, H. O. Menlove, *Measurement of the Time Resolution of Several ^{4}He and CH_4 Proportional Counters*, in Nuclear Analysis Research and Development Program Status Report, September–December 1973, ed. by G. Robert Keepin. Los Alamos Scientific Laboratory report LA-5557-PR (February 1974), pp. 12–14
10. D.H. Beddingfield, N.H. Johnson, H.O. Menlove, ^{3}He neutron proportional counter performance in high gamma-ray dose environments. Nucl. Instrum. Methods Phys. Res., Sect. A **455**(3), 670–682 (2000)
11. https://pdt-inc.com/

12. J.E. Swansen, Deadtime reduction in thermal neutron coincidence counter. Nucl. Instrum. Methods Phys. Res., Sect. B **9**(1), 80–88 (1985)
13. M. Iliev, K. Ianakiev, M. Swinhoe, *KM-200 Front-End Electronics for Thermal Neutron Detectors*, in Proceedings of the 57th Annual Meeting of the INMM, Atlanta, GA, 24–28 Jul 2016
14. D. Henzlova et al., Current status of ^{3}He alternative technologies for nuclear safeguards, Los Alamos National Laboratory report LA-UR-15-21201 (2015)
15. M. Pickrell, A. Lavietes, V.A. Gavron, D. Henzlova, M.J. Joyce, R.T. Kouzes, H.O. Menlove, The IAEA workshop on requirements and potential technologies for replacement of ^{3}He detectors in IAEA safeguards applications. J. Nucl. Mater. Manag. **41**(2), 14–29 (2013)
16. A. Sagadevan, D. Henzlova, H.O. Menlove, M. Croce, M. Dion, R. Morris, B. Bevard, Experimental validation of nondestructive assay capabilities for molten salt reactor safeguards FY21 report, Los Alamos National Laboratory report LA-UR-21-29908 (2021)
17. L. Oscar Lindquist, E.J. Dowdy, Neutron detector counting capabilities for ^{10}B-lined and ^{235}U fission chambers in high gamma-ray fluxes, Los Alamos National Laboratory report LA-10376-MS (1985)
18. D. Henzlova, H. Menlove, Report on operational test of high dose neutron detector, Los Alamos National Laboratory report LA-UR-17-29204 (2017)
19. L.G. Evans, D. Henzlova, H.O. Menlove, M.T. Swinhoe, S. Croft, J.B. Marlow, R.D. McElroy, Nuclear safeguards ^{3}He replacement requirements. J. Nucl. Mater. Manag. **40**, 88–96 (2012)
20. R. Nolte, Detection of neutrons. Lecture notes, https://ejc2014.sciencesconf.org/conference/ejc2014/pages/Lecture_notes_R_Nolte.pdf. Accessed Feb 2023
21. T.W. Crane, *Gas Mixture Evaluation for ^{3}He Neutron Detectors*, in Nuclear Safeguards Research and Development Program Status Report, May–August 1977, Joseph L. Sapir, Compiler, Los Alamos Scientific Laboratory report LA-7030-PR (March 1978), p. 39
22. D. Henzlova, H. Menlove, Quench gas and preamplifier selection influence on ^{3}He tube performance for spent fuel applications. ESARDA Bull. **47**, 10 (2012)
23. K. S. McKinny, T. R. Anderson, and N. H. Johnson, "Optimization of coating in Boron-10 lined proportional counters," IEEE Trans. Nucl. Sci. 60(2), 860–863 (April 2013)
24. D. Henzlova, H.O. Menlove, High-dose neutron detector development, Los Alamos National Laboratory report LA-UR-15-27500 (2015)
25. E.R. Siciliano, R.T. Kouzes, Boron-10 lined proportional counter wall effects. Pacific Northwest National Laboratory report PNNL-21368 (May 2012)
26. T. Gozani, *Active Nondestructive Assay of Nuclear Materials: Principles and Applications*, NUREG/CR-0602 (U.S. Nuclear Regulatory Commission, Washington DC, 1981)
27. M.L. Evans, NDA technology for uranium resource evaluation, January 1–June 30, 1978, Los Alamos Scientific Laboratory report LA-7617-PR (1979), pp. 36–41
28. A. Borella, R. Rossa, K. van der Meer, Modeling of a highly enriched ^{235}U fission chamber for spent fuel assay. Ann. Nucl. Energy **62**, 224–230 (2013)
29. D.G. Madland, Total prompt energy release in the neutron-induced fission of ^{235}U, ^{238}U, and ^{239}Pu. Nucl. Phys. A **772**(3–4), 113–137 (2006)
30. S.V. Chuklyaev et al., A method for determining the average charge inside a fission chamber. Instrum. Exp. Tech. **44**(2), 153–159 (2001)
31. C. S. Sosa, *The Importance of Light Collection Efficiency in Radiation Detection Systems that Use Organic Scintillators*, Doctoral dissertation (University of Michigan, Ann Arbor, 2018)
32. F.D. Brooks, Development of organic scintillators. Nucl. Instrum. Methods **162**(1–3), 477–505 (1979)
33. J.B. Birks, *The Theory and Practice of Scintillation Counting* (Pergamon Press, New York, 1964)
34. M.A. Norsworthy et al., Light output response of EJ-309 liquid organic scintillator to 2.86–3.95 MeV carbon recoil ions due to neutron elastic and inelastic scatter. Nucl. Instrum. Methods Phys. Res., Sect. A **884**, 82–91 (2018)
35. T.H. Shin et al., Measured neutron light-output response for *trans*-stilbene and small-molecule organic glass scintillators. Nucl. Instrum. Methods Phys. Res., Sect. A **939**, 36–45 (2019)
36. P. Schuster, E. Brubaker, Characterization of the scintillation anisotropy in crystalline stilbene scintillator detectors. Nucl. Instrum. Methods Phys. Res., Sect. A **859**, 95–101 (2017)
37. R.A. Weldon et al., Characterization of stilbene's scintillation anisotropy for recoil protons between 0.56 and 10 MeV. Nucl. Instrum. Methods Phys. Res., Sect. A **977**, 164178 (2020)
38. T.H. Shin, *Fast-Neutron Multiplicity Counting Techniques for Nuclear Safeguards Applications*, Doctoral dissertation (University of Michigan, Ann Arbor, 2019)
39. G. Dietze et al., Gamma-calibration of NE 213 scintillation counters. Nucl. Instrum. Methods Phys. Res. **193**(3), 549–556 (1982)
40. T.A. Laplace et al., Simultaneous measurement of organic scintillator response to carbon and proton recoils. Phys. Rev. C **104**, 014609 (2021)
41. R.A. Weldon et al., Measurement of EJ-228 plastic scintillator proton light output using a coincident neutron scatter technique. Nucl. Instrum. Methods Phys. Res., Sect. A **953**, 163192 (2020)
42. J.A. Brown et al., Proton light yield in organic scintillators using a double time-of-flight technique. J. Appl. Phys. **124**, 045101 (2018)
43. N. V. Kornilov et al., "Total characterization of neutron detectors with a ^{252}Cf source and a new light output determination," Nucl. Instrum. Methods Phys. Res., Sect. A 599(2–3), 226–233 (2009)
44. CAEN DPP-PSD Control Software Manual, https://www.caen.it/products/dpp-psd-control-software/. Accessed Feb 2023
45. A. Di Fulvio et al., Neutron Rodeo Phase II Final Report, Argonne National Laboratory technical report ANL-18/46 148938 (2019)
46. F. D. Brooks, R. W. Pringle, and B. L. Funt, "Pulse shape discrimination in a plastic scintillator," IRE Trans. Nucl. Sci. 7(2/3), 35–38 (June 1960)
47. J.K. Polack et al., An algorithm for charge-integration, pulse-shape discrimination and estimation of neutron/photon misclassification in organic scintillators. Nucl. Instrum. Methods Phys. Res., Sect. A **795**, 253–267 (2015)
48. M.M. Bourne et al., Neutron detection in a high-gamma field using solution-grown stilbene. Nucl. Instrum. Methods Phys. Res., Sect. A **806**, 348–355 (2016)
49. T.A. Laplace et al., Comparative scintillation performance of EJ-309, EJ-276, and a novel organic glass. J. Instrum. **15**, P11020 (2020)
50. J.W. Downs et al., Organic-glass scintillators. Nucleonics **16**, 94–96 (1958)
51. A. Tomanin et al., Characterization of a cubic EJ-309 liquid scintillator detector. Nucl. Instrum. Methods Phys. Res., Sect. A **756**, 45–54 (2014)

52. J.S. Carlson et al., Taking advantage of disorder: Small-molecule organic glasses for radiation detection and particle discrimination. J. Am. Chem. Soc. **139**(28), 9621–9626 (2017)
53. J.S. Carlson, P.L. Feng, Melt-cast organic glasses as high-efficiency fast neutron scintillators. Nucl. Instrum. Methods Phys. Res., Sect. A **832**, 152–157 (2016)
54. N. Zaitseva et al., Scintillation properties of solution-grown *trans*-stilbene single crystals. Nucl. Instrum. Methods Phys. Res., Sect. A **789**, 8–15 (2015)
55. J. J. Manfredi et al., "Proton light yield of fast plastic scintillators for neutron imaging," IEEE Trans. Nucl. Sci. 67(2), 434–442 (February 2020)
56. L.M. Bollinger, G.E. Thomas, Measurement of the time dependence of scintillation intensity by a delayed-coincidence method. Rev. Sci. Instrum. **32**, 1044 (1961)
57. M. Schear et al., *Monte Carlo Modelling and Experimental Evaluation of a 6LiF:ZnS(Ag) Test Module for Use in Nuclear Safeguards Neutron Coincidence Counting Applications*, in Proceedings of the 2014 IAEA Symposium on Nuclear Safeguards, Vienna, Austria, IAEA report no. IAEA-CN-220, 20–24 Oct 2014
58. W.F. Hornyak, A fast neutron detector. Rev. Sci. Instrum. **23**(6), 264 (1952)
59. R.A. Harlan, *Uranium and Plutonium Assay of Crated Waste by Gamma-Ray, Single Neutron, and Slow Coincidence Counting*, in Proceedings of the American Nuclear Society Topical Conference on Measurement Technology for Safeguards and Materials Control, *Kiawah Island, SC*, November 26–28, 1979 (National Bureau of Standards Publication 582, 1980), p. 622
60. K.D. Ianakiev et al., Neutron detector based on particles of 6Li glass scintillator dispersed in organic lightguide matrix. Nucl. Instrum. Phys. Res., Sect. A **784**, 189–193 (2015)
61. B.W. Wiggins et al., Computational investigation of arranged scintillating particle composites for fast neutron detection. Nucl. Instrum. Phys. Res., Sect. A **915**, 17–23 (2019)
62. R. Chandra et al., Fast neutron detection with pressurized ^{4}He scintillation detectors. J. Instrum. **7**, C03035 (2012)
63. R. Chandra et al., Results from Noble gas scintillation detectors with solid state light readout. Int. J. Mod. Phys. Conf. Ser. **27**, 1460137 (2014)
64. J. Glodo et al., Fast neutron detection with Cs_2LiYCl_6. IEEE Trans. Nucl. Sci. **60**(2), 864–870 (2013)
65. J. Glodo et al., Selected properties of Cs_2LiYCl_6, $Cs_2LiLaCl_6$, and $Cs_2LiLaBr_6$ scintillators. IEEE Trans. Nucl. Sci. **58**(1), 333–338 (2011)
66. B.S. Budden, L.C. Stonehill, J.R. Terry, A.V. Klimenko, J.O. Perry, Characterization and investigation of the thermal dependence of Cs_2LiYCl_6:Ce_3+ (CLYC) waveforms. IEEE Trans. Nucl. Sci. **60**(2), 946–951 (2013)
67. K.E. Mesick et al., Effects of proton-induced radiation damage on CLYC and CLLBC performance. Nucl. Instrum. Methods Phys. Res., Sect. A **948**, 162774 (2019)
68. R.G. Fronk et al., High-efficiency microstructured semiconductor neutron detectors for direct ^{3}He replacement. Nucl. Instrum. Methods Phys. Res., Sect. A **779**, 25–32 (2015)
69. L.E. Bruns, Capability of field instrumentation to measure radionuclide limits, Rockwell Hanford report RHO-LD-160 (1981), p. 45
70. B. Marcheschi, B.L. Justus, A.L. Huston, Optically Stimulated Luminescence Dosimetry Using Doped Lithium Fluoride Crystals, U.S. Patent No. 91219481, 1 Sept 2015
71. A.L. Huston et al., Optically stimulated luminescent glass optical fibre dosemeter. Radiat. Prot. Dosim. **101**(1–4), 23–26 (2002). https://doi.org/10.1093/oxfordjournals.rpd.a005974
72. S. Shalev, J.M. Cuttler, The energy distribution of delayed fission neutrons. Nucl. Sci. Eng. **51**(1), 52–66 (1973). https://doi.org/10.13182/NSE73-A23257
73. H. Franz et al., Delayed-neutron spectroscopy with ^{3}He spectrometers. Nucl. Instrum. Methods **144**(2), 253–261 (1977)
74. D.L. Chichester, J.T. Johnson, E.H. Seabury, Fast-neutron spectrometry using a ^{3}He ionization chamber and digital pulse shape analysis. Appl. Radiat. Isot. **70**(8), 1457–1463, https://doi.org/10.1016/j.apradiso.2011.12.045. Epub 2011 Dec 30. PMID: 22728128 (August 2012)
75. D.G. Langner et al., Neutron multiplicity counter development, ESARDA Meeting, Avignon, France, May 14–16, 1991, Los Alamos National Laboratory report LA-UR-91-1569 (1991)
76. R. Weinmann-Smith et al., Variations in AmLi source spectra and their estimation utilizing the 5 ring multiplicity counter. Nucl. Instrum. Methods Phys. Res., Sect. A **856**, 17–25 (2017)
77. D.P. Broughton et al., Sensitivity of the active neutron coincidence collar response during simulated and experimental fresh fuel assay. *Nucl. Instrum. Methods Phys. Res., Sect. A* **1001**, 165243 (2021). https://doi.org/10.1016/j.nima.2021.165243

Open Access This chapter is licensed under the terms of the Creative Commons Attribution 4.0 International License (http://creativecommons.org/licenses/by/4.0/), which permits use, sharing, adaptation, distribution and reproduction in any medium or format, as long as you give appropriate credit to the original author(s) and the source, provide a link to the Creative Commons license and indicate if changes were made.

The images or other third party material in this chapter are included in the chapter's Creative Commons license, unless indicated otherwise in a credit line to the material. If material is not included in the chapter's Creative Commons license and your intended use is not permitted by statutory regulation or exceeds the permitted use, you will need to obtain permission directly from the copyright holder.

Principles of Singles Neutron Counting

J. D. Hutchinson, K. Amundson, H. Kistle, J. Moussa, T. Shin, J. E. Stewart, and R. K. Weinmann-Smith

16.1 Introduction

Singles neutron counting indirectly measures the neutron emission rate. This neutron-counting technique uses the number of detected neutron events to infer information about an item. As discussed in Chap. 13, neutrons are primarily produced by three mechanisms: spontaneous fission, (α,n) reactions, and induced fission. If the neutron production rate, the detector efficiency, the multiplication, and the measured singles count rate are all known, the mass of nuclear material can be determined. Neutron counting is often paired with gamma spectroscopy measurements to determine the item's isotopic composition and can be effective for providing the spontaneous fission portion of the neutron production rate. Different terminology is often used for the same concepts within different disciplines (safeguards compared with reactor physics for example); Ref. [1] contains tables that link the historical terminology. The term *Singles* is simply used to distinguish these measurements from coincidence and multiplicity measurements, discussed in Chaps. 17 and 18.

Historically, ^{3}He and BF_3 neutron proportional counters were the most common types of detectors used for singles neutron counting, and examples of these types of systems are given throughout the chapter; Chap. 15 discusses other types of detector systems used for neutron detection.

Singles neutron-counting systems are very simple and often consist only of the components shown schematically in Fig. 16.1 for proportional counters. All detector events that produce an amplifier output pulse with an amplitude greater than a threshold set by the integral discriminator are counted for a set time in a scaler. The usual choice of discriminator setting is high enough to reject low-amplitude pulses produced by gamma rays and electronic noise and low enough to count all neutron-induced pulses. It is important to note that the pulse-height spectra of ^{3}He and BF_3 neutron proportional counters contain very little information about the energy of the detected neutrons; however, some information can be obtained using physical characteristics of the counters that moderate and absorb materials. The scalar and timer count the number of detections for the specified measurement time. This concept can be implemented many ways; for example, in list-mode electronics (discussed in Chap. 17), the times of each detection are recorded and stored so the data may be reanalyzed, and the singles rate in any time window can be calculated.

This chapter covers basic principles that are important in using singles neutron counting for passive assay of materials that contain uranium and plutonium. The example of polyethylene-moderated ^{3}He proportional counters is used to explain the basic principles because such detectors are routinely used for a wide variety of neutron-counting applications in nuclear facilities throughout the world. The organization of this chapter follows the primary production of neutrons in an item, their interactions in the item, and then detection by the instrument. Factors that affect each step are discussed. The chapter concludes with discussion of the effects of background and room return neutrons (neutrons scattered off room surfaces), which complicate and possibly increase the uncertainty or the absolute bias in the measurement.

Los Alamos National Laboratory strongly supports academic freedom and a researcher's right to publish; as an institution, however, the Laboratory does not endorse the viewpoint of a publication or guarantee its technical correctness.

J. D. Hutchinson (✉) · K. Amundson · H. Kistle · J. Moussa · T. Shin · J. E. Stewart · R. K. Weinmann-Smith
Los Alamos National Laboratory, Los Alamos, NM, USA
e-mail: jesson@lanl.gov; kamundson@lanl.gov; hadyn@lanl.gov; jmoussa@lanl.gov; thshin@lanl.gov; rweinmann@lanl.gov

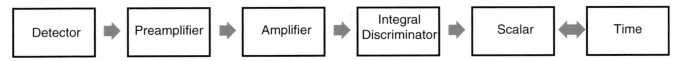

Fig. 16.1 The basic components of a simple singles neutron-counting system

16.1.1 Theory of Singles Neutron Counting

The measured singles neutron-counting rate is given by the simple formula

$$S = \varepsilon M_L I, \qquad (16.1)$$

where S = measured singles neutron count rate (counts/s)
ε = absolute detection efficiency (counts/n emitted)
M_L = item leakage multiplication due to induced fission (dimensionless)
I = item primary neutron production (n/s); I is often specified as $F + a$, where F is the spontaneous fission neutron emission rate and a is the (α,n) neutron emission rate.

The absolute detection efficiency ε is the number of counts produced by the detector per neutron emitted from the item. The item leakage multiplication M_L is the number of neutrons emitted from the outer surface of the item per neutron born inside the item. The item neutron production is the number of neutrons per second created from spontaneous fission (F) and (α,n) reactions. The uncertainty in the singles count rate can be determined through the application of Eq. A-19 from Chap. 25.

16.2 Primary Neutron Production Sources

The first of the three important factors in Eq. (16.1) that affects singles neutron counting is primary neutron production in the item (I). Primary neutron production is from (α,n) reactions (sometimes referred to as A) and spontaneous fission (sometimes referred to as F); secondary neutron production is from induced fission. Sometimes the neutrons from primary production sources are referred to as *starting neutrons* or *starter neutrons*. Chap. 13, Sects. 13.2 through 13.4, describe physical processes of primary neutron production and give spontaneous fission and (α,n) reaction neutron yields from actinide isotopes of interest for passive neutron assays. Yields from (α,n) reactions are also given for oxides and fluorides.

This section describes those features of neutron production in compounds of uranium and plutonium that affect assays based on singles neutron counting. General calibration principles to convert from a singles count rate to nuclear material mass are discussed. Initially no multiplication is assumed, ignoring neutron production by induced fission and neutron loss by absorption. These topics are introduced in Sect. 16.4.

This section includes many equations aimed at understanding the primary neutron production term I in Eq. (16.1), including the concept of determining an "effective mass" that is representative of the neutron source intensity for a specific measurement (depending on the nuclides involved and the form of the material). A difference exists in coefficients used for effective mass for singles counting compared with coincidence counting. For this reason, some of the equations might be similar to those in Chap. 17, but the coefficients are different. If coincidence counting is being used, the coefficients and equations in Chap. 17 should be used instead of those presented here.

16.2.1 Plutonium Compounds

To apply singles neutron counting as a signature for one or more isotopes of uranium or plutonium requires knowledge of the chemical form and isotopic composition of the item. Neutron production from spontaneous fission depends on the isotopic composition of the item but not on its chemical form. Neutron production from (α,n) reactions depends on both. This point is well illustrated with examples. Consider 100 g items of plutonium in the form of metal, PuO_2, and PuF_4 with four plutonium isotopic compositions, including weapons-grade plutonium and compositions representative of low-, medium-, and high-burnup fuel from light-water reactors. Table 16.1 gives neutron production rates for each isotope, from spontaneous fission

16 Principles of Singles Neutron Counting

Table 16.1 Primary neutron production rate from spontaneous fission and from (α,n) in PuO$_2$, and PuF$_4$ for four plutonium isotopic compositions

Isotope	Primary neutron production rate for 100 g of Pu (n/s)			
	Amount (wt%)	Spontaneous fission	(α,n) PuO$_2$	(α,n) PuF$_4$
^{238}Pu	0.02	52	268	44,000
^{239}Pu	93.735	2	3571	524,916
^{240}Pu	5.95	6069	839	124,950
^{241}Pu	0.2685	0	0	46
^{242}Pu	0.028	48	0	8
^{241}Am	0.0557	0	150	24,599
Totals		**6171**	**4828**	**718,519**
^{238}Pu	0.024	62	322	52,800
^{239}Pu	89.667	2	3416	502,135
^{240}Pu	9.645	9838	1360	202,545
^{241}Pu	0.556	0	1	95
^{242}Pu	0.109	187	0	29
^{241}Am	0.327a	0	880	144,417
Totals		**10,089**	**5979**	**902,021**
^{238}Pu	0.059	153	791	129,800
^{239}Pu	82.077	2	3127	459,631
^{240}Pu	16.297	16,623	2298	342,237
^{241}Pu	1.231	0	2	209
^{242}Pu	0.336	578	1	91
^{241}Am	0.162a	0	436	71,546
Totals		**17,356**	**6655**	**1,003,514**
^{238}Pu	1.574	4077	21,092	3,462,800
^{239}Pu	57.342	1	2185	321,115
^{240}Pu	24.980	25,480	3522	524,580
^{241}Pu	10.560	0	14	1795
^{242}Pu	5.545	9537	11	1497
^{241}Am	1.159a	1	3118	511,863
Totals		**39,096**	**29,942**	**4,823,650**

a^{241}Am wt% relative to plutonium

and from (α,n) reactions in PuO$_2$ and PuF$_4$. The total singles neutron output for a 100 g item (assuming no multiplication) is the sum of spontaneous fission output and the (α,n) output for the chemical form of the material, assuming no multiplication from induced fission. Plutonium metal is often assumed to be pure plutonium without light elements, so the (α,n) production will be 0, and therefore only the spontaneous fission column in this table is used. The rates were computed from yields in Chap. 13, Tables 13.1 and 13.3.

Equation 16.1 could be rewritten to address the neutron production for each isotope individually. Note that most measurement methods used in this chapter cannot distinguish which isotope emitted a neutron. In this full form, every isotope is included, and the coefficients describe the specific neutron production for spontaneous fission and (α,n) reactions.

$$I = F + A = (a^F_{238} + a^\alpha_{238})(^{238}Pu) + (a^F_{239} + a^\alpha_{239})(^{239}Pu) + (a^F_{240} + a^\alpha_{240})(^{240}Pu) + (a^F_{241} + a^\alpha_{241})(^{241}Pu)$$
$$+ (a^F_{242} + a^\alpha_{242})(^{242}Pu) + (a^F_{241Am} + a^\alpha_{241Am})(^{241}Am), \tag{16.2}$$

where a^F_{xyz} is the specific neutron production rate from spontaneous fission (n s^{-1} g^{-1}) for ^{xyz}Pu, and (^{xyz}Pu) is the mass of ^{xyz}Pu in grams. Note that although a^F_{xyz} is a nuclear data constant, the (α,n) specific neutron production coefficient a^α_{xyz} depends on the chemical form of the Pu and surrounding materials.

For many items, some production mechanisms are nonexistent or insignificantly small, and Eq. (16.2) may be simplified, which is illustrated for an item with only ^{238}Pu and ^{239}Pu:

$$I = F + A = (a^F_{238} + a^\alpha_{238})(^{238}Pu) + (a^F_{239} + a^\alpha_{239})(^{239}Pu). \tag{16.3}$$

Another simplification can be performed by combining the specific production from spontaneous fission and from (α,n) into a single coefficient. A single equation with simplified coefficients is sufficient to convert from measured singles count rate to the item's nuclear material mass.

$$I = F + A = (a_1)(^{238}Pu) + (a_2)(^{239}Pu). \tag{16.4}$$

If the isotopic composition is constant, then the equation can be simplified even further to remove dependency on the isotopics.

$$I = F + A = a_3(Pu). \tag{16.5}$$

Finally, if the detector efficiency is known, the measured singles rate (from Eq. 16.1) can be determined using a single calibration coefficient.

$$S = \varepsilon I = \varepsilon(F + A) = a_4(Pu). \tag{16.6}$$

Note that the a_4 term might need to include different efficiency values for differences in spontaneous fission and (α,n) neutron energies (discussed in Sect. 16.2.4). Detector efficiency is discussed in greater detail in Sect. 16.4.

A commonly used simplification of Eq. (16.2) is for plutonium metal, in which 99.9% of the neutrons is produced by ^{238}Pu, ^{240}Pu, and ^{242}Pu spontaneous fission,

$$I = (a_5)(^{238}Pu) + (a_6)(^{240}Pu) + (a_7)(^{242}Pu), \tag{16.7}$$

where the multipliers a_5 through a_7 are the specific spontaneous fission neutron yields given in Table 13.1 (n/s − g). Values for these constants are given in Table 16.4.

Plutonium-240 often dominates primary neutron production (98%, 98%, 96%, and 65% for the four plutonium isotopic compositions in Table 16.1). Thus, the concept of ^{240}Pu effective mass (^{240}Pu$_{eff}$) is used. Plutonium-240 effective mass is the mass of ^{240}Pu that would result in the same neutron emission as the actual neutron emission from all nuclides in the sample. Although items do contain other isotopes, assuming that the item is effectively only ^{240}Pu allows the neutron measurement results to focus on a single mass value. This concept is useful because the item isotopics are sometimes unknown. Plutonium-240 effective mass can be combined with a measurement of the item's isotopics to determine the total plutonium mass and the mass of each isotope.

The ^{240}Pu$_{eff}$ for plutonium metal is given by

$$^{240}Pu_{eff} = a_8(^{238}Pu) + (^{240}Pu) + a_9(^{242}Pu). \tag{16.8}$$

The constants a_8 and a_9 are 2.43 and 1.69, respectively, and account for the greater specific (per gram) spontaneous fission neutron production in ^{238}Pu and ^{242}Pu relative to ^{240}Pu. In other words, 1 g of ^{238}Pu produces 2.43 times the neutrons of 1 g of ^{240}Pu. If an item had exactly 2 g of ^{238}Pu, the ^{240}Pu$_{eff}$ mass would be 4.86 g, and the item would produce the same number of neutrons as 4.86 g of ^{240}Pu. These constants are derived from Table 13.1. As previously mentioned, these constants are different than the values used for ^{240}Pu$_{eff}$ mass in coincidence counting; see Chap. 17 for information on ^{240}Pu$_{eff}$ mass for coincidence counting.

For PuO$_2$, the ratio of (α,n) to spontaneous fission neutron production in Table 16.1 is 0.78, 0.59, 0.38, and 0.77 for the four plutonium isotopic compositions. Note that the ratio of the last (high-burnup, 24.980 wt% ^{240}Pu) composition did not decrease because of the larger amount of ^{238}Pu compared with the other examples, which have a large (α,n) yield, as shown in Table 13.3. In most typical isotopic compositions encountered, ^{238}Pu, ^{239}Pu, ^{240}Pu, and ^{241}Am are significant contributors to (α,n) neutron production in PuO$_2$. The total neutron production in PuO$_2$ (n/s) is described by a generic equation of the form

$$I = a_{10}(^{238}Pu) + a_{11}(^{239}Pu) + a_{12}(^{240}Pu) + a_{13}(^{242}Pu) + a_{14}(^{241}Am), \tag{16.9}$$

where the multipliers a_{10} through a_{14} are specific neutron yields (n/s − g) for both spontaneous fission and (α,n) reactions in PuO$_2$ and are given in Table 16.4.

The multipliers are determined from plutonium isotopics and specific (α,n) (Table 13.3) and spontaneous fission yields (Table 13.1) for each isotope.

For PuF$_4$, (α,n) reactions produce more than 98% of the neutrons for the four isotopic compositions. Neutron production in PuF$_4$ is described as in Eq. (16.10) with new neutron emission coefficients

$$I = a_{15}\left(^{238}Pu\right) + a_{16}\left(^{239}Pu\right) + a_{17}\left(^{240}Pu\right) + a_{18}\left(^{242}Pu\right) + a_{19}\left(^{241}Am\right). \tag{16.10}$$

The multipliers a_{15} through a_{19} are specific neutron yields (n/s) for (α,n) reactions in PuF$_4$ (Table 13.3), with small components from spontaneous fission (Table 13.1) and are given in Table 16.4.

For the examples shown in Table 16.2, ^{241}Am values were taken at the time of analysis. Americium-241 is a daughter of ^{241}Pu. Americium-241 mass increases over the first 70 years because its half-life exceeds that of its parent. After 70 years, the ^{241}Pu mass has reduced enough that the ^{241}Am decay outpaces additional growth due to ^{241}Pu. At that point, the ^{241}Am mass will start to decrease. The ^{241}Am content is often expressed as a mass percent of the total plutonium. Accurate estimation of ^{241}Am at the time of the analysis is especially important for PuF$_4$ and all high (α,n) production systems, evidenced by the very large value of a_{19} in Table 16.4 for ^{241}Am.

16.2.2 Uranium Compounds

Just as for plutonium compounds, singles neutron counting of uranium compounds requires prior knowledge of chemical form and isotopic composition. Examples of primary neutron rates in uranium forms and compositions frequently found in the nuclear fuel cycle are given in Tables 16.2 and 16.3. Considered are 10 kg items of uranium in the form of metal, UO$_2$, UO$_2$F$_2$, and UF$_6$. Note that 10 kg was used for uranium in these tables (as opposed to 100 g for plutonium in Table 16.1) due to the much lower neutron emission in uranium versus plutonium. Similar to Table 16.1, this table also assumes no additional multiplication present in the item ($M_L = 1$). Table 16.2 references enrichments of 0.2%, 0.7%, 3.0%, and 18.2% (low-enriched uranium [LEU]); Table 16.3 references enrichments of 31.7%, 57.4%, 69.6%, and 97.6% (highly enriched uranium [HEU]). Spontaneous fission and (α,n) neutron rates are given by isotope and form for each enrichment. The rates

Table 16.2 Primary neutron production rates in uranium metal, UO$_2$, UO$_2$F$_2$, and UF$_6$ for four uranium isotopic compositions (depleted uranium, natural uranium, and low-enriched uranium)

Isotope	Amount (wt %)	Neutron production rate for 10 kg of U (n/s)			
		Metal (spontaneous fission)	UO$_2$ (α, n)	UO$_2$F$_2$ (α, n)	UF$_6$ (α,n)
^{234}U	0.0005	0	0	9	29
^{235}U	0.1977	0	0	1	2
^{236}U	0.0036	0	0	0	1
^{238}U	99.8	136	1	175	279
Totals		**136**	**1**	**185**	**311**
^{234}U	0.0049	0	1	90	284
^{235}U	0.7108	0	0	3	6
^{236}U	–	–	–	–	–
^{238}U	99.28	135	1	174	278
Totals		**135**	**2**	**267**	**568**
^{234}U	0.0244	0	7	449	1415
^{235}U	3.001	0	0	11	24
^{236}U	0.0184	0	0	2	5
^{238}U	96.96	132	1	170	271
Totals		**132**	**8**	**632**	**1715**
^{234}U	0.0865	0	26	1592	5017
^{235}U	18.15	1	1	65	145
^{236}U	0.2313	0	1	28	67
^{238}U	81.53	111	1	143	228
Totals		**112**	**29**	**1828**	**5457**

Table 16.3 Primary neutron rates in uranium metal, UO_2, UO_2F_2, and UF_6 for four uranium isotopic compositions (highly enriched uranium)

Isotope	Amount (wt %)	Neutron production rate for 10 kg of U (n/s)			
		Metal (spontaneous fission)	$UO_2(\alpha,n)$	$UO_2F_2(\alpha,n)$	$UF_6(\alpha,n)$
^{234}U	0.1404	0	42	2583	8143
^{235}U	31.71	1	2	114	254
^{236}U	0.3506	0	1	42	102
^{238}U	67.80	92	1	119	190
Totals		93	46	2858	8689
^{234}U	0.2632	0	79	4843	15,265
^{235}U	57.38	2	4	207	459
^{236}U	0.5010	0	1	60	145
^{238}U	41.86	57	0	73	117
Totals		59	84	5184	15,986
^{234}U	0.3338	0	100	6142	19,360
^{235}U	69.58	2	5	250	557
^{236}U	0.5358	0	1	64	155
^{238}U	29.55	40	0	52	83
Totals		42	106	6508	20,155
^{234}U	1.032	1	310	18,989	59,856
^{235}U	97.65	3	7	352	781
^{236}U	0.2523	0	1	30	73
^{238}U	1.07	1	0	2	3
Totals		5	318	19,373	60,713

were computed from yields in Tables 13.1 and 13.3 (Chap. 13) and from yields shown in Ref. [2] (UO_2F_2). A note of interest is that the (α,n) emission is very dependent on the amount of ^{234}U present.

Uranium-238 spontaneous fission dominates neutron production in uranium metal for ^{235}U enrichments below ~70%, which allows ^{238}U assay based on singles neutron counting of large uranium metal items for all but the highest ^{235}U enrichments (when $M_L = 1$). Because of the low neutron rates relative to plutonium metal, longer count times and larger mass items/samples are required for acceptable precision. Linear calibrations of singles neutron count rate versus ^{238}U mass are used for total uranium-metal determinations with low-enrichment material.

For UO_2, total neutron production [spontaneous fission plus (α,n)] is very low for LEU. With increasing enrichment, the spontaneous fission component decreases as the (α,n) component increases. At enrichments more than 60%, the (α,n) component grows rapidly. Uranium-234 alpha decay is the dominant source of (α,n) reactions in UO_2 for enrichments of 3% or greater. Enrichment processes affect the concentration of ^{234}U even more than ^{235}U, so ^{234}U increases with uranium enrichment. Total neutron production in UO_2 is generally described by

$$I = a_{20}\left(^{234}U\right) + a_{21}\left(^{235}U\right) + a_{22}\left(^{236}U\right) + a_{23}\left(^{238}U\right), \tag{16.11}$$

where a_{20} through a_{23} are specific neutron yields (n/s − g) for both spontaneous fission and (α,n) reactions in UO_2 and are given in Table 16.4.

Uranium oxide as usually found in the nuclear fuel cycle (cans, rods, finished assemblies) suggests assay by active neutron techniques rather than passive because the relatively low passive singles neutron rates for standard item sizes must compete with room background. Active methods typically do not use singles counting but rather the correlated signals from ^{235}U-induced fission as discussed in Chap. 20; however, large UO_2 items may lend themselves to assay by passive singles neutron counting depending on item characteristics and measurement goals.

For UO_2F_2 (a chemical reaction product of UF_6 and water), neutron production is dominated by the (α,n) component. Neutron production rates increase with ^{235}U enrichment. Equations analogous to Eq. (16.12) describe the passive singles-neutron-counting calibration of UO_2F_2 with known isotopic composition. Passive singles neutron counting has been used to quantify UO_2F_2 deposits inside process equipment in gaseous diffusion enrichment plants [3].

Table 16.4 Coefficients for calculation of neutron production in Eqs. 16.4 through 16.12

Equation number	Coefficient	Unit	Value	Process	Nuclide(s)	Note
16.4	a_1	n/s − g	*	F + a	^{238}Pu	^{238}Pu and ^{239}Pu
16.4	a_2	n/s − g	*	F + a	^{239}Pu	
16.5	a_3	n/s − g	*	F + a	Pu	Pu
16.6	a_4	Cts/s − g	*	F + a	Pu	Pu w/ ε
16.7	a_5	n/s − g	2.59E+03	F	^{238}Pu	Pu metal
16.7	a_6	n/s − g	1.02E+03	F	^{240}Pu	
16.7	a_7	n/s − g	1.72E+03	F	^{242}Pu	
16.8	a_8	−	2.43	F	^{238}Pu/^{240}Pu	^{240}Pu$_{eff}$
16.8	a_9	−	1.69	F	^{242}Pu/^{240}Pu	
16.9	a_{10}	n/s − g	1.60E+04	F + a	^{238}Pu	PuO$_2$
16.9	a_{11}	n/s − g	3.81E+01	F + a	^{239}Pu	
16.9	a_{12}	n/s − g	1.16E+03	F + a	^{240}Pu	
16.9	a_{13}	n/s − g	1.72E+03	F + a	^{242}Pu	
16.9	a_{14}	n/s − g	2.69E+03	F + a	^{241}Am	
16.10	a_{15}	n/s − g	2.20E+06	F + a	^{238}Pu	PuF$_4$
16.10	a_{16}	n/s − g	5.60E+03	F + a	^{239}Pu	
16.10	a_{17}	n/s − g	2.20E+04	F + a	^{240}Pu	
16.10	a_{18}	n/s − g	1.99E+03	F + a	^{242}Pu	
16.10	a_{19}	n/s − g	4.42E+05	F + a	^{241}Am	
16.11	a_{20}	n/s − g	1.84E+02	F + a	^{234}U	UO$_2$
16.11	a_{21}	n/s − g	3.61E-02	F + a	^{235}U	
16.11	a_{22}	n/s − g	1.22E+00	F + a	^{236}U	
16.11	a_{23}	n/s − g	3.11E-02	F + a	^{238}U	
16.12	a_{24}	n/s − g	5.80E+02	F + a	^{234}U	UF$_6$
16.12	a_{25}	n/s − g	8.03E-02	F + a	^{235}U	
16.12	a_{26}	n/s − g	2.91E+00	F + a	^{236}U	
16.12	a_{27}	n/s − g	4.16E-02	F + a	^{238}U	

*Dependent on material form

For UF$_6$ (the standard process material for uranium enrichment), ^{234}U alpha decay and the subsequent ^{19}F(α,n)^{22}Na reaction dominate neutron production. The material is similar to UO$_2$F$_2$, with the (α,n) component being more dominant because of additional fluorine. For arbitrary enrichment, total UF$_6$ neutron production can be described by

$$I = a_{24}\left(^{234}U\right) + a_{25}\left(^{235}U\right) + a_{26}\left(^{236}U\right) + a_{27}\left(^{238}U\right), \qquad (16.12)$$

where a_{24} through a_{27} are specific neutron yields (n/s − g) for both spontaneous fission and (α,n) reactions in UO$_2$F$_2$ and are given in Table 16.4.

Passive singles neutron counting can be used for verification of UF$_6$ cylinders of all sizes and is routinely used on cylinders that contain low-enriched UF$_6$. This assay method is fast, simple, and inexpensive. Small cylinders are counted in nearly 4π geometry ("well" counters). For large, low-enriched cylinders, simpler calibration expressions are obtained by assuming a constant ^{238}U weight fraction and a constant ratio of ^{235}U to ^{234}U.

16.2.3 Primary Neutron Production Summary

The constants for the equations in the last two sections are shown in Table 16.4.

The data in Tables 16.1, 16.2, and 16.3 are plotted in Fig. 16.2. Compared are total spontaneous fission plus (α,n) neutron production rates for plutonium metal, PuO$_2$, PuF4, uranium metal, UO$_2$, UO$_2$F$_2$, and UF$_6$. The specific (per gram uranium or plutonium) rates span eight orders of magnitude—from uranium metal to PuF$_4$. Rates are plotted as a function of wt% ^{235}U and ^{239}Pu. The plot is useful for visual comparison and for estimating counting statistics with known detectors and geometries.

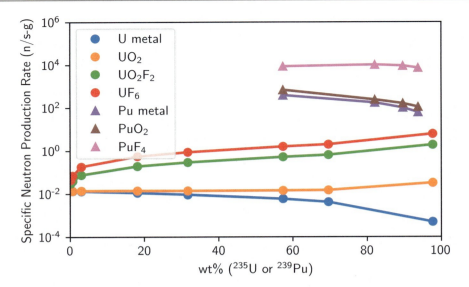

Fig. 16.2 Specific neutron production rates for uranium and plutonium metals, oxides, and fluorides. Data are from Tables 16.2 through 16.4

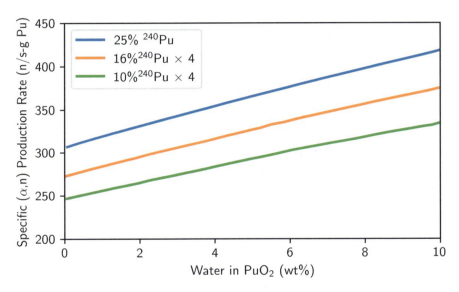

Fig. 16.3 Specific neutron production rates from (α.n) reactions in PuO_2 versus added moisture for ~10% ^{240}Pu, ~16% ^{240}Pu, and ~ 25% ^{240}Pu plutonium isotopic compositions

16.2.3.1 Impurities

Rarely are the plutonium and uranium compounds found in the nuclear fuel cycle completely free of impurities. These impurities can significantly alter primary neutron production. As an example, consider PuO_2 with H_2O added. Water is commonly found in PuO_2, with a nominal value of 1 wt%. Figure 16.3 shows the dependence of (α,n) neutron production (a) on weight percentage of water for three plutonium isotopic compositions. The quantity a for the 25% ^{240}Pu material is roughly a factor of 5 higher than for the 16% and 10% materials (because of the high ^{238}Pu fraction) regardless of moisture content. Figure 16.4 displays the relative increase in a for wet relative to dry PuO_2 as a function of wt% H_2O. Note that the moisture effect is independent of plutonium isotopic composition; i.e., adding 1 wt% moisture yields the same relative change in a for 10%, 16%, and 25% ^{240}Pu, namely +4.4%, which indicates that small changes in initial alpha-particle energies that result from changes in plutonium isotopics do not significantly affect neutron production in wet PuO_2.

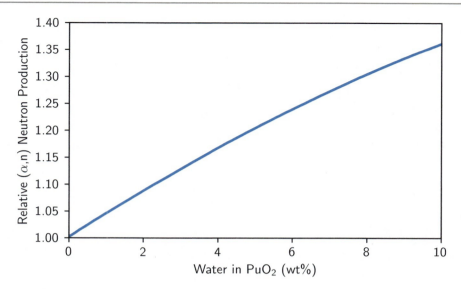

Fig. 16.4 Relative increase in neutron production from (α,n) reactions in wet relative to dry PuO_2 as a function of added moisture. The 10%, 16%, and 25% ^{240}Pu isotopic compositions fall on a single curve

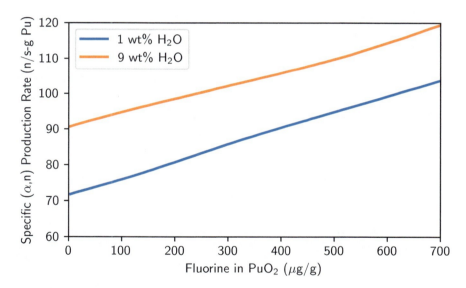

Fig. 16.5 Specific neutron production rates from (α,n) reactions in PuO_2 versus fluorine contamination for 16% ^{240}Pu plutonium isotopic composition. Both 1 and 9 wt% H_2O cases are shown

Although many trace contaminants are found in PuO_2, fluorine is usually the most significant for altering a. Figure 16.5 displays a versus F contamination in PuO_2 (16% ^{240}Pu), with 1 and 9 wt% H_2O. Figure 16.6 shows the a values for PuO_2 (16% ^{240}Pu), with fluorine impurities relative to pure PuO_2 versus fluorine concentration. Cases are shown for both 1 and 9 wt % moisture. The relative change in a from F contamination is greater for dry than for wet PuO_2.

If H_2O and F concentrations are known, data such as those shown in Figs. 16.3 through 16.6 can be used to adjust calibration parameters for singles-neutron-counting assays. These data were calculated using the Los Alamos SOURCES code [4, 5]. Chapter 13, Sect. 13.4, gives approximate formulas for calculating contributions to a from impurities in uranium and plutonium oxides.

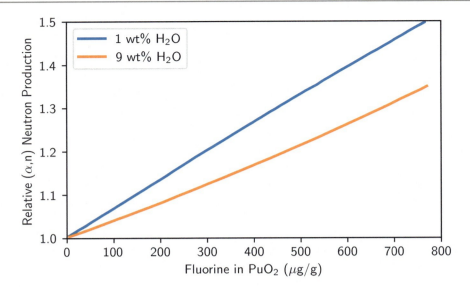

Fig. 16.6 Neutron production from (α,n) reactions in PuO$_2$ with fluorine impurities relative to pure PuO$_2$ versus fluorine concentration for 16% ^{240}Pu. Both 1 wt% and 9 wt% H$_2$O cases are shown

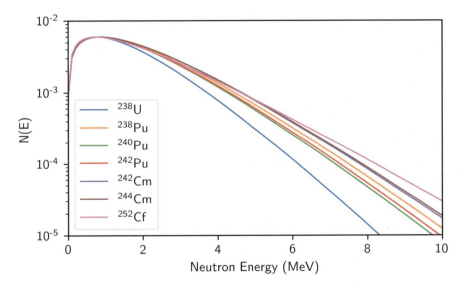

Fig. 16.7 Specific spontaneous fission neutron production spectrum for nuclides listed in Table 16.5. (Developed using Eq. 16.13)

16.2.3.2 Neutron Energy Spectrum Effects

Neutrons are produced in spontaneous fission and (α,n) reactions with characteristic energy distributions or spectra, which are important in the design of neutron counters that use polyethylene-moderated ^{3}He detectors.

For plutonium compounds, the spontaneous fission spectrum depends slightly on plutonium isotopic composition and not on item chemistry. The spectrum is dictated by the nuclear kinematics of the spontaneous fission disintegration process, which differ slightly for ^{238}Pu, ^{240}Pu, and ^{242}Pu. Figure 16.7 is a plot of the neutron energy distribution, N(E), for the spontaneous fission nuclides listed in Table 16.5. It can be seen that the different nuclides disagree more at higher energies. The spontaneous fission spectrum is described well by the Watt fission spectrum, which uses the form [6].

$$N(E) \propto e^{-\frac{E}{A}} sinh\left(\sqrt{BE}\right), \tag{16.13}$$

Table 16.5 Watt fission spectrum constants for spontaneous fission nuclides

Nuclide	A (MeV)	B (MeV^{-1})
^{238}U	0.648318	6.81057
^{238}Pu	0.847833	4.16933
^{240}Pu	0.794930	4.68927
^{242}Pu	0.819150	4.36668
^{242}Cm	0.887353	3.89176
^{244}Cm	0.902523	3.72033
^{252}Cf	1.180000	1.03419

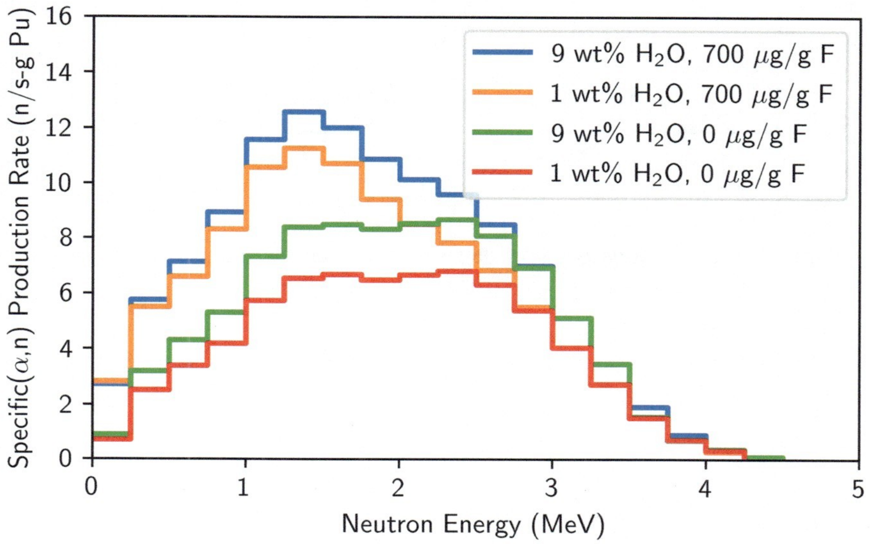

Fig. 16.8 Specific (α,n) neutron production rate spectra of PuO$_2$ (16% ^{240}Pu) with variable moisture and fluorine content calculated from SOURCES [7]

where E = neutron energy (MeV) and A (MeV) and B (MeV^{-1}) are constants. Table 16.5 shows the A and B constants for several spontaneous fission nuclides [6].

The laboratory spectrum of neutron energies from (α,n) reactions in plutonium compounds depends on item chemistry, impurity levels, and slightly on plutonium isotopic composition, all of which determine the slowing-down spectrum of alpha-particle energies and the (α,n) reaction cross-section dependence on alpha-particle energy. Figure 16.8 displays four specific (α,n) neutron production spectra for PuO$_2$ (16% ^{240}Pu) with variable moisture and fluorine contamination. An increase in moisture increases the spectrum slightly, and an increase in fluorine decreases the spectrum. Figure 16.9 shows the four total neutron spectra that correspond to the moisture and fluorine concentrations of Fig. 16.8. The average energies of these four spectra are not substantially different. This similarity indicates that spectral shape differences are generally not major factors that affect singles neutron counting of PuO$_2$ with moisture and fluorine contamination levels in the ranges used. These ranges are typical of a wide variety of PuO$_2$ items.

Figure 16.10 displays normalized neutron production spectra from (α,n) reactions in ^{234}UF$_6$. Spectra for two ^{22}Na level branching schemes are shown. The ^{234}UF$_6$ (α,n) spectrum is softer than that for PuO$_2$, with an average neutron energy of approximately 1.2 MeV. For PuO$_2$, the average is approximately 2.0 MeV.

16.2.3.3 Thin-Target Effects

In the previous sections, primary neutron production by (α,n) reactions was assumed to take place in items that qualify as "thick targets"—materials in which alpha particles lose all of their energy in the item. If the item density is low enough, neutron production is reduced because alpha particles escape before they undergo (α,n) reactions with target isotopes. Neutron production by ^{234}U alpha particles that drive the ^{19}F(α,n)^{22}Na reaction in gaseous UF$_6$ is a "thin target" situation. In this case, alpha particles may escape the gas volume at energies above the ^{19}F(α,n) cross-section threshold. Reference [8] presents

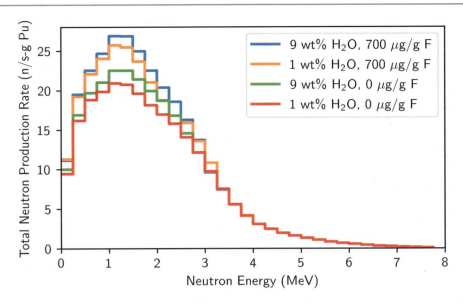

Fig. 16.9 Total neutron production rate spectra of PuO$_2$ (16% ^{240}Pu) with variable moisture and fluorine content calculated from SOURCES [7]

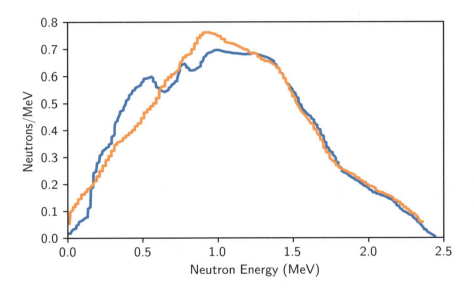

Fig. 16.10 Normalized (α,n) neutron production spectra of ^{234}UF$_6$ with two ^{22}Na level branching models content calculated from SOURCES [7]

methods for calculating thin-target neutron production in UF$_6$ gas. Figure 16.11 (from Ref. [8]) displays UF$_6$ neutron production versus projected range. The projected range is the product of atom density and target thickness. Neutron production saturates above approximately 6×10^{19} atoms/cm^2 at the thick-target neutron production value for ^{234}UF$_6$. At approximately 1.3×10^{19} atoms/cm^2, the number of neutrons produced per alpha particle is half of the thick-target value.

16.2.3.4 Neutron Transport in the Item

After specifying the primary neutron production in the item, the next logical step in using passive singles neutron counting for assay is to consider the number of neutrons that escape the item per neutron produced. The description of the transport of neutrons in the item volume (including all of the processes of neutron creation and loss) is complex and depends on many factors. Chapter 14 describes many of the basic principles of neutron transport. Here, we present a simple formula for item leakage multiplication, discuss numerical results, and describe the use of the formula.

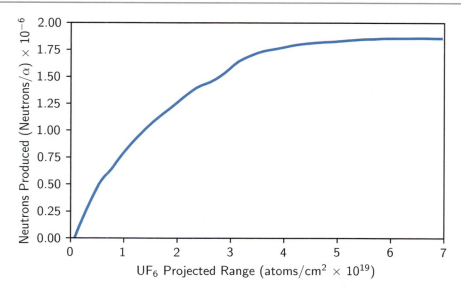

Fig. 16.11 Neutron production by a 4.77 MeV alpha particle from ^{234}U decay versus projected range in UF$_6$. The range in cm can be obtained by dividing the abscissa by the UF$_6$ atom density (atoms/cm^3)

16.2.3.5 Leakage Multiplication

The number of neutrons that escape the item (and therefore are available for counting) per primary neutron produced is defined as the leakage multiplication of the item, M_L, and is related to the induced fission probability in the item. Note that leakage multiplication is often assumed to be 1, which is a good assumption for items with small mass and/or large leakage (such as in fuel pins). Some of the examples used throughout this chapter do assume that the multiplication is 1 (and it is stated when done). The quantity M_L differs from the total multiplication M defined in Chap. 14. The quantity M is the average number of neutrons created in the item per starting neutron. The quantity M_L includes only those neutrons that escape the item. Therefore, M_L is more pertinent to singles neutron counting than M because only neutrons that escape the item can be measured. The two quantities are closely related as described below. The following definitions apply.

$\bar{\nu}$ = the average number of neutrons created by induced fission
p = probability that a neutron will induce a fission
p_c = probability that a neutron will be captured (and not fission)
p_L = probability that a neutron will escape the item (leakage probability)

A neutron of a given generation can induce a fission with probability p and is lost with probability $1 - p$. Loss mechanisms include capture and leakage,

$$p + p_c + p_L = 1. \tag{16.14}$$

The number of fissions produced in a given generation per fission in the previous generation is $p\bar{\nu}$. The quantity $p\bar{\nu}$ is the multiplication factor k (often referred to as k_{eff}) from reactor physics discussed in Chap. 14. Induced fission requires a neutron to be absorbed, and therefore the net neutron profit per fission is $\bar{\nu} - 1$. Values of $\bar{\nu}$ for various induced fission nuclides are given in Tables 13.1 and 13.2.

Given an item governed by Eq. 16.14, a single source neutron will cause p fissions and create $p\bar{\nu}$ new neutrons with a net neutron profit of $p(\bar{\nu} - 1)$. These created neutrons, $p\bar{\nu}$, will go on in a second generation to cause $p(p\bar{\nu})$ fissions, with $(p\bar{\nu})^2$ new neutrons and a net profit of $2p\bar{\nu}(\bar{\nu} - 1)$. This multiplication process is shown in Table 16.6 for the first few neutron generations.

One exercise that can be performed to better understand the multiplication process is to insert a constant number of neutrons into an item of known k_{eff} and calculate the number of neutrons present in the item. An example of this calculation is shown in Table 16.7. Here, 1000 neutrons are inserted into each generation. (These neutrons are from primary production mechanisms, such as spontaneous fission or (α,n) discussed in the previous section.) A k_{eff} of 0.5 is used for this example.

Table 16.6 The neutron multiplication process through the fourth fission generation

Generation	Number of fissions	Neutrons created	Net neutron profit
0	–	1 (source)	1
1	p	$p\bar{\nu}$	$p(\bar{\nu} - 1)$
2	$p(p\bar{\nu})$	$(p\bar{\nu})^2$	$p(p\bar{\nu})(\bar{\nu} - 1)$
3	$p(p\bar{\nu})^2$	$(p\bar{\nu})^3$	$p(p\bar{\nu})^2(\bar{\nu} - 1)$
4	$p(p\bar{\nu})^3$	$(p\bar{\nu})^4$	$p(p\bar{\nu})^3(\bar{\nu} - 1)$

Table 16.7 Number of neutrons present in the multiplication process through 10 generations for an item with a $k_{eff} = 0.5$ that adds 1000 neutrons per generation

	G0	G1	G2	G3	G4	G5	G6	G7	G8	G9	G10	Sum (to infinity)
G0	1000	500	250	125	63	31	16	8	4	2	1	2000
G1		1000	500	250	125	63	31	16	8	4	2	2000
G2			1000	500	250	125	63	31	16	8	4	2000
G3				1000	500	250	125	63	31	16	8	2000
G4					1000	500	250	125	63	31	16	2000
G5						1000	500	250	125	63	31	2000
G6							1000	500	250	125	63	2000
G7								1000	500	250	125	2000
G8									1000	500	250	2000
G9										1000	500	2000
G10											1000	2000
Sum (to current generation)	1000	1500	1750	1875	1938	1969	1984	1992	1996	1998	1999	2000

Given the definition of k_{eff} (the number of neutrons in a generation divided by the number of neutrons in the previous generation), the neutron population will be cut in half during each generation (decreasing from 1000 to 500 to 250 . . .). But during each generation, a new 1000 neutrons are also inserted. As seen in Table 16.7, the sum over generations is 2000. Because 1000 neutrons are inserted, the multiplication is 2 (2000/1000). Generally, the multiplication can be described using the equation

$$M = \frac{1}{1 - p\bar{\nu}} = \frac{1}{1 - k_{eff}}; k_{eff} < 1. \tag{16.15}$$

Once enough generations have occurred and no new neutrons exist from the initial 1000 that were introduced, the columns in Table 16.7 add up to 2000, corresponding to a multiplication of two. In addition, the sum over the rows (within a single generation) also equals 2000 once the population has reached an equilibrium.

Figure 16.12 shows the data for $k_{eff} = 0.5$ and extends it to other values of k_{eff}.

Note that Fig. 16.12 confirms Eq. (16.15). For instance, at $k_{eff} = 0.75$, 4000/1000 = 4, the same result given by Eq. (16.15). Also seen in Fig. 16.12 is that as the k_{eff} of the item increases, the number of generations required to a steady-state neutron population also increases; however, this outcome does not have implications in most passive assays for two reasons. First, the generation time (from ns scale for fast items to msec for thermal items) is often very short compared with the assay time (generally at least 100 s). Second, this buildup in population has already happened before the start of the assay (for passive measurements); however, for some pulsed neutron experiments, this outcome may matter.

For all generations, the sum of the net neutron profit is the total net neutron profit per source neutron $(1 - p)/(1 - p\bar{\nu})$. Not all of the net neutron profit will escape the item—some will be captured. The leakage multiplication M_L is the total net neutron profit per source neutron × the probability of escape divided by the probability of disappearance; that is,

$$M_L = \left(\frac{1-p}{1-p\bar{\nu}}\right)\left(\frac{p_L}{p_L + p_c}\right) = \frac{p_L}{1 - p\bar{\nu}} = \frac{1 - p - p_c}{1 - p\bar{\nu}}. \tag{16.16}$$

Finally, from Eqs. (16.15 and 16.16), the relationship between M_L and M is

16 Principles of Singles Neutron Counting

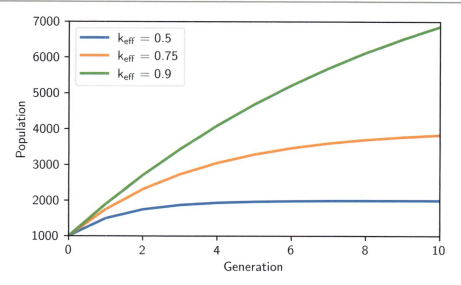

Fig. 16.12 Number of neutrons present in the multiplication process through 10 generations for systems with an assumed source of 1000 neutrons per generation

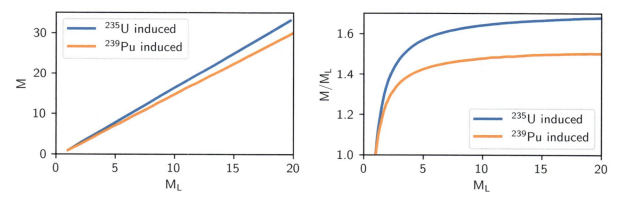

Fig. 16.13 (left) Total multiplication (M) versus leakage multiplication (M_L); (right) ratio of M/M_L versus M_L.

$$M_L = p_L M. \tag{16.17}$$

If the probability of capture p_c is small,

$$M_L \approx (1-p)M = \frac{1-p}{1-p\bar{\nu}}. \tag{16.18}$$

Equations 16.15 and 16.18 can be rearranged to solve for total multiplication values given leakage multiplication (still assuming p_c is small).

$$M = \frac{\bar{\nu} M_L - 1}{\bar{\nu} - 1} \tag{16.19}$$

If both p_c and p are small, M_L and M are approximately the same. The quantity M is always greater than or equal to M_L. Figure 16.13 shows the relationship between M and M_L; the $\bar{\nu}$ data from Table 13.2 are used in Eq. (16.19) to generate these plots. Figure 16.14 shows multiplication as a function of k_{eff}, generated using Eqs. (16.15 and 16.17) (again with $\bar{\nu}$ data from Table 13.2).

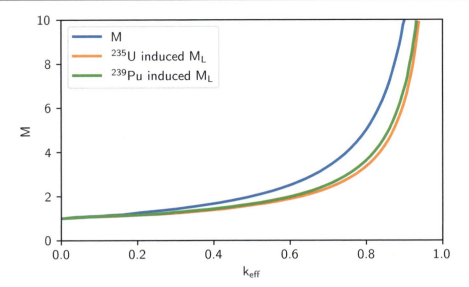

Fig. 16.14 Total multiplication (M) and leakage multiplication (M_L) versus k_{eff}

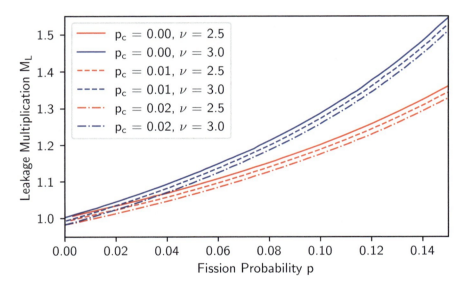

Fig. 16.15 Leakage multiplication (M_L) versus fission probability for two values of the average fission neutron multiplicity and three values of the capture probability

Equation 16.16, the expression for M_L, depends on three parameters: ν, p, and p_c. These quantities depend on the energy of the neutron that is inducing a fission or is being captured and the nuclide that undergoes fission; therefore, the quantities are understood to represent averages over the spectrum of neutron energies in the item. The term *MAGIC MERV* is often used in the criticality safety community as a pneumonic device to remember all of these parameters that affect criticality (and therefore k_{eff}, M, M_L, p, and p_c): mass, absorption, geometry, interaction, concentration, moderation, enrichment, reflection, volume (and sometimes temperature is included as well).

Figure 16.15 shows plots of the leakage multiplication M_L versus the fission probability p; Eq. (16.16) was used to generate the plots. Two sets of curves are shown. The lower set of three curves is representative of uranium-bearing items with $\nu = 2.5$. The upper set of three curves with $\nu = 3.0$ is representative of plutonium-bearing items. In each set of curves, the capture probability (p_c) varies from 0 to 0.02, which covers a wide range of items.

To explore what masses of nuclear material correspond to various multiplication values, two examples are presented. In the first example, shown in Fig. 16.16, the total and leakage multiplication of various masses of HEU and plutonium metal are

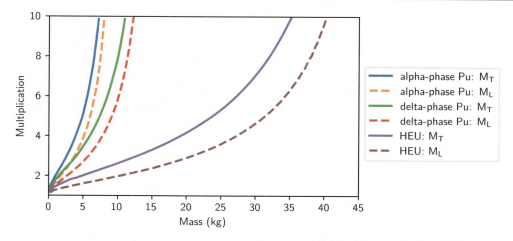

Fig. 16.16 Leakage multiplication (ML) and total multiplication (M) versus mass; assumes bare (unreflected) material in a spherical geometry

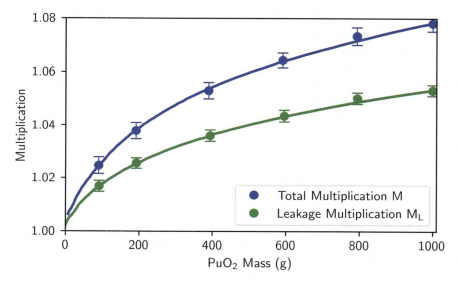

Fig. 16.17 (Lower curve) leakage multiplication (M_L) and (upper curve) total multiplication (M) versus PuO_2 mass in a container with an 8.35 cm inside diameter. The plutonium is 10 wt% $^{240}Pu_{eff}$. The item PuO_2 has a density of 1.3 g/cm^3 and is 1 wt% H_2O. The fill height increases as mass is added

shown. These results are based on previous critical experiments [9]. Two sets of results are shown for plutonium; one set of curves is for α-phase plutonium (which has a density of ~19.6 g/cm^3), whereas the other curves are for δ-phase plutonium (which has a lower density of ~16 g/cm^3). These results assume HEU with 93% ^{235}U and plutonium with 6% ^{240}Pu. It can be seen that the multiplication is much higher for a given mass in plutonium than the same mass in HEU, which is not surprising because the number of neutrons emitted in ^{239}Pu fission is greater than in ^{235}U. Note that this example assumes a spherical geometry with no reflection. Use of reflection or moderation could greatly increase the multiplication for the same amount of mass.

For the second example, Fig. 16.17 shows plots of both total and leakage multiplication versus PuO_2 mass in a cylindrical volume with a diameter of 8.35 cm. The ^{240}Pu (effective) is 10 wt% of plutonium. The PuO_2 has a density of 1.3 g/cm^3 and contains 1 wt% H_2O. The item fill height increases as PuO_2 mass increases. The leakage multiplication values were calculated using the Monte Carlo transport code (Monte Carlo N-Particle [MCNP]; Ref. [6]). The total multiplication was calculated using Eq. (16.19), with $\bar{\nu} = 3.13$. For the items described by Fig. 16.17, the capture probability p_c is negligible. The plots clearly show the difference in total multiplication M and leakage multiplication M_L. Reference [10] contains additional information on the MCNP code calculations of M_L and information on coincidence multiplication corrections.

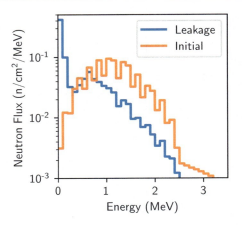 

Fig. 16.18 (left) Initial and leakage neutron spectra from a 2230 kg UF$_6$ cylinder (3.0% ^{235}U, density 5.1 g/cm^3), Monte Carlo uncertainty smaller than line width. (right) Cross section of UF$_6$ cylinder where 25% of UF$_6$ is bound to cylinder walls, and 75% of UF$_6$ is pooled at bottom of cylinder (typically referred to as X = 25) [11]

Although the bulk of neutrons produced from induced fissions are prompt neutrons, a fraction of neutrons are caused by inverse beta decay of fission products and are called *delayed neutrons*. These neutrons, discussed in Sect. 16.2.3.5, can be important in multiplicity analysis, but they are essentially not important in singles counting. For the purposes of singles counting, they are included in the M_L term in Eq. 16.1 and are therefore detected or not based on the same efficiency term. The energy of delayed neutrons is lower than that of prompt neutrons on average, so technically, a slightly different efficiency will be associated with delayed neutrons. Given the small percentage of neutrons that are delayed neutrons (~0.65% for ^{235}U-induced fission and ~0.2% for ^{239}Pu-induced fission), delayed neutrons essentially have no impact on singles counting.

16.2.3.6 Leakage Spectra–Nuclear Material

Neutrons that escape from the nuclear material have a lower average energy than those produced in the initial fission or (α,n) reactions. Neutrons lose energy in the nuclear material through scattering events, during which elastic collisions are more prevalent with light nuclei and inelastic collisions with heavy nuclei. Effects of self-moderation become more important to consider, especially for large samples.

An example of neutron energy losses in the item, as calculated with the Monte Carlo Code MCNP 6.2 [6], is shown in Fig. 16.18 as a histogram energy-distribution spectrum (width of energy bins is 0.05 MeV). This figure shows a histogram distribution of the simulated starting and leakage spectra of neutrons from a type 30B UF$_6$ storage cylinder containing 2230 kg at a ^{235}U enrichment of 3% (isotopics and emission rates as specified in Ref. [11]). For this calculation, independent simulations of the spontaneous fission and (α,n) sources were simulated and added in post processing. The ^{19}F(α,n)^{22}Na thick-target energy spectrum (Fig. 16.10) was based on a SOURCES4C calculation [7]. The change between the initial and leakage spectra in Fig. 16.18 shows that the neutron moderation within this large UF$_6$ cylinder is significant. The average energy of the leakage neutrons is 0.505 MeV compared with the average initial source neutron energy of 1.20 MeV (1.51 MeV, including induced fissions). For this case, 21.2% of the source neutrons does not escape from the cylinder because of capture.

Another leakage spectrum calculation, performed with the MCNP Code [6], is shown in Fig. 16.19. The item modeled is 800 g of PuO$_2$, with a density of 1.3 g oxide/cm^3. The item contains 706 g of plutonium (10% ^{240}Pu) and 1 wt% water. The cylindrical item is 8.35 cm diameter and 11.24 cm high. The smooth curve in Fig. 16.19 is the ^{240}Pu spontaneous fission neutron emission spectrum (Fig. 16.7 and Eq. 16.11). The calculated item neutron leakage spectrum is the histogram distribution, with 1 σ error bars shown. The average energy of the leakage spectrum is 1.91 MeV compared with 1.93 MeV for the emission spectrum, which implies very little moderation of the source spectrum by this PuO$_2$ item. The small buildup in the leakage spectrum between 0.6 and 1.0 MeV is from inelastic scattering by plutonium nuclides and elastic scattering by oxygen. The slight buildup between 10 keV and 100 keV is from elastic scattering by hydrogen and oxygen. This buildup would increase with added moisture. For this item, the leakage multiplication is ~1.04.

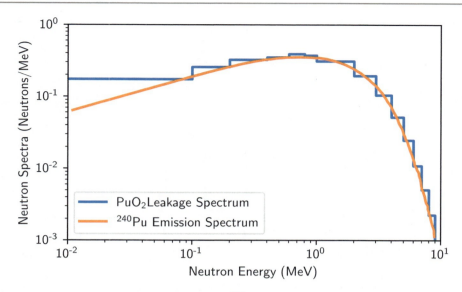

Fig. 16.19 Neutron leakage spectrum from an 800 g PuO2 item (10% ^{240}Pu), assuming an energy spectrum characteristic of ^{240}Pu spontaneous fission. Source neutrons were uniformly distributed in the item volume

16.2.3.7 Leakage Spectra–External Materials

Materials can be present in between the nuclear material and a detection system, which could significantly alter the neutron spectra. Examples include nuclear material normally encapsulated in a metal container that can have significant thickness or stored in containers with varying thicknesses of metals or polyethylene. This encapsulation can moderate the spectrum and affect its energy in the same ways that the nuclear material does. For example, the standard encapsulation around a ^{252}Cf neutron source has a radial thickness of ~0.5 cm stainless steel and reduces the average neutron energy by 3%. Material containers in nuclear facilities can be much thicker. Hydrogenous materials, such as plastic and water, can have much stronger neutron moderation effects because the average energy lost in a collision is much greater than for metals. This effect on the ^{252}Cf neutron spectrum, as calculated using the MCNP code, is shown in Fig. 16.20 [12]. This figure shows the relative neutron intensity, which is the intensity per source particle adjusted for the bin width. Modeling with a varying bin width results in a loss of quantitative information but a more accurate qualitative visualization of the neutron intensity of different regions. The polyethylene spectrum is a bimodal distribution with a group of thermalized neutrons and a group of fast neutrons. Seven percent of the total neutron population exists between 0 and 100 eV.

The shape of this spectrum can be important in determining detection efficiency, as described in Sect. 16.4.

16.2.3.8 Detector Efficiency

This chapter describes the chain of events and related physics in the production of neutrons by an item that contains nuclear material. The final steps in singles neutron counting are detection by a neutron-detection instrument to relate the neutron-detection rate to the amount of nuclear material.

16.2.4 Leakage Spectrum Effects on Detector Efficiency

Assuming a fixed neutron detector, different neutron leakage spectra will have different detection efficiencies, which occurs because higher-energy neutrons will penetrate further through polyethylene before thermalizing for absorption in ^{3}He. For example, the neutron-detection efficiency as a function of neutron energy in the active well coincidence counter (AWCC; Chap. 20, Sect. 20.5.1) is plotted in Fig. 16.21. This neutron counter is a two-ring design; the inner ring is centered at a polyethylene depth of 4 cm and the outer ring at 7.5 cm. The peak efficiencies of the inner ring, the outer ring, and the total occur at ~0.6 MeV, 2.1 MeV, and 1.2 MeV, respectively. In the example from the previous section, the 0.5 cm stainless steel encapsulation around most ^{252}Cf sources increases the relative detection efficiency in the AWCC by 2% compared with ^{252}Cf without encapsulation [12].

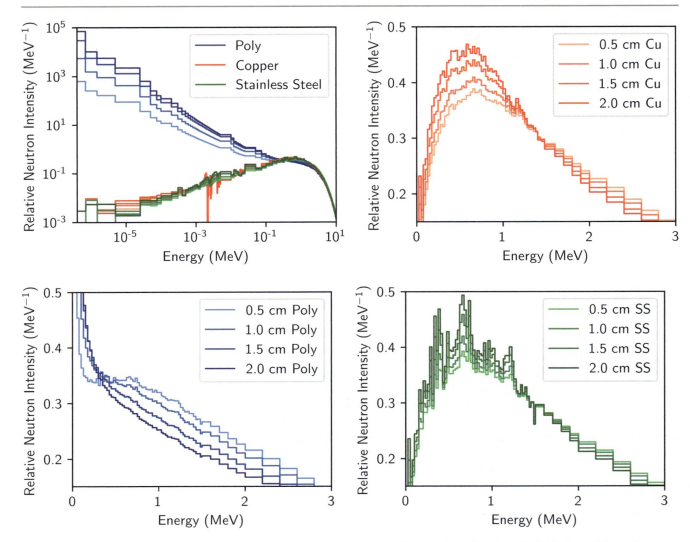

Fig. 16.20 Californium-252 leakage spectra in various encapsulations. Note the log-log scale in the first figure. Individual materials are shown in a range of 0–3 MeV on a linear scale [12]. The relative neutron intensity was calculated by tallying the current in each energy bin that summed to 1. Then the current was divided by the bin width to account for smaller bins at the lower energy levels, which results in a loss of quantitative information but a more accurate qualitative visualization of the neutron intensity of different regions. The polyethylene spectrum is a bimodal distribution with a group of thermalized neutrons and a group of fast neutrons. The combination of the manner in which the data were normalized and the usage of the log scaling for both the X and Y axes in the upper left graph results in the thermal neutrons appearing to dominate the spectrum in the case of the poly encapsulation. In reality, the thermal neutrons in this case are essentially a spike of neutrons at a narrow energy, whereas the rest of the distribution is continuous over a wide energy range. Seven percent of the total neutron population exists between 0 and 100 eV

To further illustrate this concept, a parametric study was performed to compare the detector efficiency of various starter neutron nuclides (those given in Table 16.5 and Fig. 16.7) using MCNP 6.2. In this study, a spherical shell of polyethylene was present around the source with variable thickness. Then a shell of ^{3}He with variable pressure surrounded the polyethylene (with a fixed thickness of 100 cm). The detector efficiency was simulated and normalized to the efficiency for a ^{252}Cf source (for the same polyethylene thickness and ^{3}He pressure). Each nuclide is compared with ^{252}Cf because sources of ^{252}Cf are often used to calibrate neutron detectors. The results for this study are shown in Fig. 16.22.

As shown in Fig. 16.22, ^{238}U has the greatest deviation in detection efficiency from a ^{252}Cf source. These deviations ranged from −10% to 14% depending on the configuration. The greatest deviation occurs at configurations with low ^{3}He pressure (less than 15 atm) and a thin polyethylene reflector (less than 4 cm). The deviations in detection efficiency are more sensitive to the variations in polyethylene thickness as opposed to ^{3}He pressure. For a given polyethylene thickness, the deviation in detection efficiency remains approximately constant over various pressures. For other isotopes, the deviations from a ^{252}Cf source were not as drastic; for example, ^{240}Pu deviations range only from −3.2% to 3.2%.

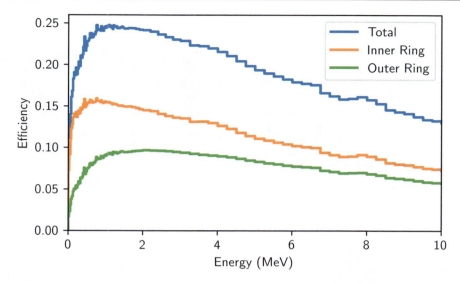

Fig. 16.21 MCNP-modeled neutron-detection efficiency as a function of emitted neutron energy in the AWCC [12]

16.3 Neutron Detector Design and Efficiency

Neutron detectors can be designed for a specific application, or an existing commercial off-the-shelf (COTS) detector can be used. This section describes some of the considerations in detector design for singles neutron counting. First, the nuclear physics mechanisms that relate a measured neutron count rate to the amount of nuclear material should be considered. For example, when measuring UF_6 cylinders, this mechanism is $^{19}F(\alpha,n)^{22}Na$ reactions. As the previous sections have shown, many physics considerations can depend on the material's element, isotopics, chemical form, and container. Detector systems are often designed to maximize detection efficiency for neutrons produced by the main mechanisms encountered for a particular application. Other considerations include background minimization, repeatability, cost, and required performance. To minimize design costs, a COTS system may be chosen instead of a custom optimized instrument, although some analysis is still required to choose the system and calibrate it. COTS systems have already been optimized for common materials. For some other nuclear materials, a COTS system would function nearly as well as an optimized detector, but for others, no COTS systems are capable of the measurement at all.

Equation 16.1 defines *detector efficiency* ε as the number of detected neutrons per neutron emitted from the item, which is referred to as *absolute efficiency*, not to be confused with *intrinsic efficiency*; intrinsic efficiency is the number of neutrons detected per neutron that exits the cavity for a well counter or that enters the detector system (for an external system). The absolute efficiency is the product of the geometric and intrinsic efficiency. This section describes important factors that affect ε for neutron detectors, including item placement within the cavity (for well counters, which is equivalent to the detector distance/placement for external systems), the use of moderating and/or shielding materials and their design, the item leakage spectrum, and background/room return neutrons.

16.3.1 Geometric Efficiency

The geometric efficiency is the probability that a neutron leaving the item will reach the detection system. Similar to the detection of gamma rays, the geometric efficiency of a neutron detector system will be largely dependent on the relationship between the location of the detector and the source (given in Eq. 7.62). For systems that are external to the source, getting closer to the source will increase the geometric efficiency. Similar to Eq. (7.66) (for gamma detection systems), the absolute efficiency is the product of the geometric and intrinsic efficiency. So, if the detector is closer to the source, both the geometric and absolute efficiency will increase. Note that for external neutron detectors, the efficiency may or may not directly scale by $1/r^2$ like gamma-ray detectors, which is largely due to room return effects discussed in Sect. 7.7.1.

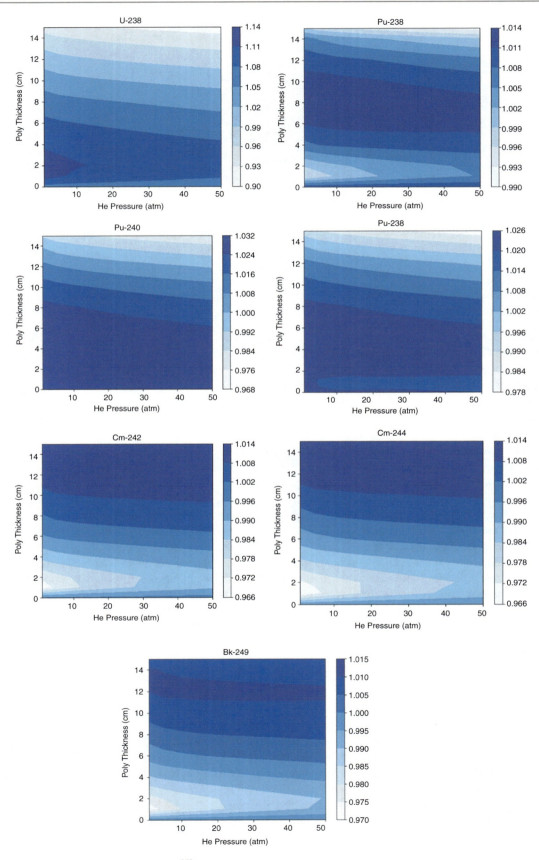

Fig. 16.22 Neutron-detection efficiency relative to ^{252}Cf [13]

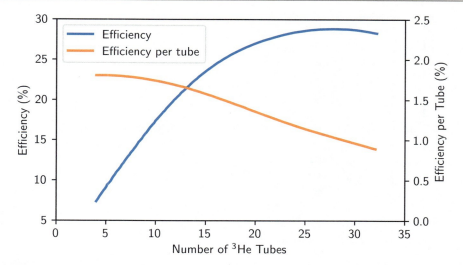

Fig. 16.23 Results of Monte Carlo calculations of singles-neutron-counting efficiency versus number of ^{3}He counting tubes for the configuration described in Sect. 16.3.2

16.3.2 Counter Arrangement

For detector systems that are focused on fast-neutron detection (such as organic scintillators), the counter arrangement is generally simple and is focused on maximizing the geometric efficiency. Detector systems that use thermal-neutron detection (such as ^{3}He detectors) require the use of moderating materials. These moderating materials need to be composed of low-Z materials, as described in Chap. 14, Sect. 14.2.3, to maximize the probability of neutron thermalization. High-density polyethylene (often referred to as simply "polyethylene") is a natural choice for the moderating material in detector systems due to its high hydrogen density, low cost, and easy machinability.

For a fixed moderator geometry, the location and number of ^{3}He proportional counters strongly affects counting efficiency. As an example, a series of efficiency calculations were performed using the MCNP Code [6] for a variable number of ^{3}He counters placed within a 1 m tall annulus of polyethylene. The counting tubes had a diameter of 1 in. and 4 atm of ^{3}He (77% thermal-neutron intrinsic counting efficiency). The internal and external diameters of the annulus were 17.8 cm (7 in.) and 38.1 cm (15 in.), respectively. The ^{3}He counters were evenly spaced within the annulus on a circle 27.9 cm (11 in.) in diameter. A 1 MeV monoenergetic source of neutrons was assumed for the calculations.

The two curves in Fig. 16.23 show the results of the calculations. The curve associated with the left ordinate is the absolute counting efficiency versus the number of ^{3}He counters. This curve shows a peak absolute counting efficiency of ~29% for 28 ^{3}He counters. The peak decreases after 28 ^{3}He counters because not enough polyethylene is present between the ^{3}He counters and, therefore, fewer thermal neutrons are being absorbed in the ^{3}He. This trend would continue if additional counters were added—going to a very low value because of less moderation between tubes—and eventually only fast neutrons, which have a very low capture cross section in ^{3}He, would be detected. Because ^{3}He proportional counters are expensive, absolute counting efficiency is sometimes compromised, and systems are designed that do not reach the peak maximum efficiency. The curve associated with the right ordinate is the counting efficiency per ^{3}He counter (an index of cost effectiveness) versus the number of counters.

A reasonable compromise for this example would be around 15 ^{3}He counters. The choice of the number of ^{3}He counters would also depend on item neutron emission intensity and desired counting precision.

16.3.3 Moderator Design

With a fixed number and arrangement of ^{3}He counters, the amount and location of polyethylene moderator can also strongly influence counting efficiency. This section looks at a specific example; a similar approach should be used when designing any thermal-neutron-detection system, using constraints associated with the specific application. For this example, calculations were done for a polyethylene slab 25.4 cm (10 in.) long and 15.2 cm (6 in.) tall with variable thickness (see Fig. 16.24). Two

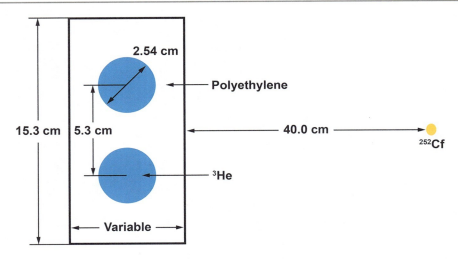

Fig. 16.24 The geometry used for a series of Monte Carlo calculations of detector efficiency. In the calculations, the thickness of polyethylene in front of and behind the two ³He counters was varied

Table 16.8 Relative efficiency of a simple slab detector for variable polyethylene moderator in front of and behind the ³He counters

Front polyethylene thickness (cm)													
Back polyethylene thickness (cm)		2	2.5	3	3.5	4	4.5	5	5.5	6	6.5	7	7.5	8
	2	0.181	0.271	0.334	0.430	0.489	0.521	0.560	0.608	0.622	0.639	0.659	0.615	0.575
	2.5	0.242	0.351	0.419	0.508	0.577	0.625	0.661	0.700	0.706	0.717	0.687	0.690	0.644
	3	0.301	0.426	0.494	0.596	0.645	0.665	0.713	0.757	0.782	0.765	0.747	0.733	0.692
	3.5	0.374	0.497	0.565	0.668	0.717	0.762	0.779	0.821	0.825	0.839	0.807	0.771	0.742
	4	0.425	0.547	0.611	0.719	0.769	0.812	0.830	0.872	0.867	0.866	0.826	0.799	0.773
	4.5	0.469	0.597	0.650	0.766	0.797	0.847	0.871	0.886	0.908	0.901	0.862	0.839	0.792
	5	0.511	0.628	0.688	0.798	0.844	0.887	0.896	0.919	0.946	0.919	0.889	0.849	0.808
	5.5	0.547	0.658	0.723	0.833	0.860	0.911	0.917	0.943	0.963	0.944	0.894	0.848	0.810
	6	0.564	0.682	0.745	0.843	0.874	0.921	0.941	0.952	0.973	0.965	0.898	0.858	0.815
	6.5	0.580	0.696	0.759	0.858	0.895	0.936	0.940	0.962	0.986	0.955	0.910	0.875	0.820
	7	0.604	0.703	0.755	0.874	0.909	0.935	0.961	0.975	0.977	0.963	0.921	0.873	0.832
	7.5	0.600	0.703	0.773	0.873	0.908	0.945	0.957	0.978	0.989	0.968	0.920	0.873	0.826
	8	0.613	0.722	0.775	0.872	0.913	0.954	0.968	0.983	1.000	0.969	0.929	0.883	0.835

³He counters are embedded within the slab with their axes parallel to the slab's long axis and separated by 5.3 cm (2.1 in.). The counters have a 2.54 cm (1 in.) diameter, 25.4 cm (10 in.) active length, and 4 atm fill pressure. A ²⁵²Cf neutron source is located 40.0 cm (15.75 in.) from the slab on a line perpendicular to the plane of the ³He counters. The Monte Carlo code MCNP [6] was used to calculate detection efficiency for various polyethylene thicknesses in front of and behind the ³He counters (relative to the source). The Watt fission spectrum was used for the calculations (Eq. 16.13) with $A = 1.025$ and $B = 2.926$ for ²⁵²Cf.[1] These parameters and this slab geometry represent typical values encountered in actual neutron detectors.

Results of the calculations are shown in Table 16.8 and Fig. 16.25. The uncertainty of the calculated relative efficiencies is approximately ±1%. The total slab thickness can be obtained by adding front and back polyethylene thicknesses because thickness is measured from the ³He tube centers. The moderator configuration for highest detection efficiency is 6 cm of polyethylene in front of the ³He counters and 8 cm behind. Note that for fixed rear polyethylene thickness, efficiency peaks and then decreases with increasing front polyethylene thickness. The decrease is due to neutron capture in hydrogen. For fixed

---

[1] These values are very different than those given in Table 16.5 for ²⁵²Cf. The values given in this paragraph were from the MCNP5 manual, whereas the values in the table are from the MCNP6.2 manual. From discussions with nuclear data experts, it is noted that the values in the table (not those in this paragraph) should be used. That being said, it is unlikely that using the values here would have a large impact on the results presented because the polyethylene study has a much larger impact than that related to the neutron emission energy.

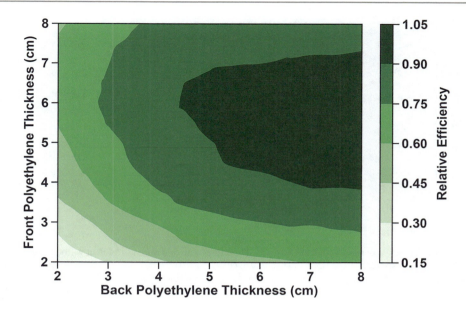

Fig. 16.25 Relative 252Cf neutron-detection efficiency as a function of front and back polyethylene thickness for the small slab detector illustrated in Fig. 16.24

front polyethylene thickness, efficiency approaches an asymptotic value with increasing rear polyethylene thickness because of neutron reflection from the rear polyethylene. If additional polyethylene were added, the efficiency would continue to increase slightly until the true asymptotic values were reached at the infinite reflector thickness (~10 cm); after that, additional polyethylene would no longer increase the efficiency because the neutrons would not make it back to the ^{3}He counters. Often, size and weight constraints limit the total polyethylene moderator thickness to be used. In the U.S., polyethylene is typically purchased in slabs 10.2 cm (4 in.) thick. For ease of fabrication, if the total slab thickness is constrained to this value, the ^{3}He counters should be placed with 4.6 cm of polyethylene in front and 5.6 cm behind for optimum efficiency.

16.4 Background and Room Return Neutrons

This section discusses neutrons that affect singles neutron counting other than those coming directly from the item. Items measured with a slab-type counter are often more sensitive to background neutrons than items measured with a well-type counter. With a well counter, one can measure the background with an empty well and simply subtract the background. This method works favorably in areas where the background is contact. In addition, well counters usually have higher detection efficiency than slab counters such that the background will be a small fraction of the measured signal. For a slab counter, the neutron source to be measured often affects the background through room scattering.

Typically background neutrons are largely caused by cosmic-ray events. The neutrons are more prevalent at lower energies, as shown in Fig. 16.26 (the peak energy is at ~0.1 eV). In multiplicity counting, it may be possible to remove some contribution from cosmic neutrons (those that have higher multiplicity than clearly exhibited by the sample), but removing some contribution is not possible for singles counting.

Room return neutrons occur due to neutron scattering. The most common occurrence would be neutrons that leave the item, scatter in the ground (such as a concrete floor), and then enter the detector volume. This effect has been previously described [16], which showed

$$\frac{D_S}{D_0} = 1 + C \frac{\frac{S^2}{D}}{\frac{S^2}{D} + \left(2\frac{R}{D_1}\right)^2}, \tag{16.20}$$

where D_S is the detected count rate (the same as S in Eq. 16.1), D_0 is the count rate that would be detected with no contribution from room return neutrons, S/D is the distance from the source to the detector, R/D_1 is the distance from a reflector (such as the

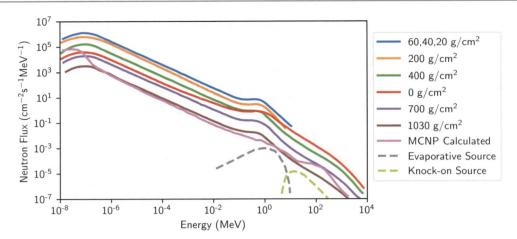

Fig. 16.26 Cosmic-ray neutron energy spectrum from Refs. [14, 15]. The values were measured in 1959 at various altitudes. The MCNP value was calculated using MCNP's cosmic ray source, as described in [15], where cosmic rays were transported through the atmosphere to the elevation and location of Los Alamos National Laboratory on April 1, 2013. The air density at this elevation is 810 g/cm^2

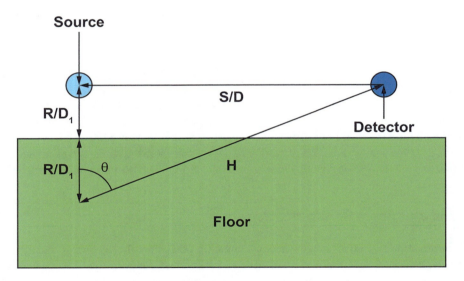

Fig. 16.27 Room return example with one scattering surface. (From [16])

ground) to the detector, and C is a constant. Figure 16.27 shows the setup for these definitions. It is assumed in this section that the center of the detector and the center of the item/source are at the same height above the ground. The hypotenuse (H in Fig. 16.27) is the square root of the denominator in Eq. (16.20) ($H = \sqrt{\frac{S^2}{D} + \left(2\frac{R}{D_1}\right)^2}$).

The example given above is for a single reflective surface, but this concept can be applied to any number of reflective surfaces. This example has previously been extended to five reflective surfaces in Ref. [16], as shown in Fig. 16.28.

$$\frac{D_S}{D_0} = 1 + C \sum_{i=1}^{5} \frac{\frac{S^2}{D}}{\frac{S^2}{D} + \left(2\frac{R}{D_1}\right)^2} \tag{16.21}$$

Note that one must use prior data to determine the appropriate value of C, which will depend on the detector system used as well as the composition of the reflective surfaces. Using a value of C = 0.6904 (the value given for the detector system in [16]), the effect of room return is plotted for a single reflective surface (such as the ground) in Fig. 16.29. It can be seen in this figure that the contribution from room return neutrons (D_S/D_0) increases as the source-to-detector distance (S/D) increases. As

16 Principles of Singles Neutron Counting

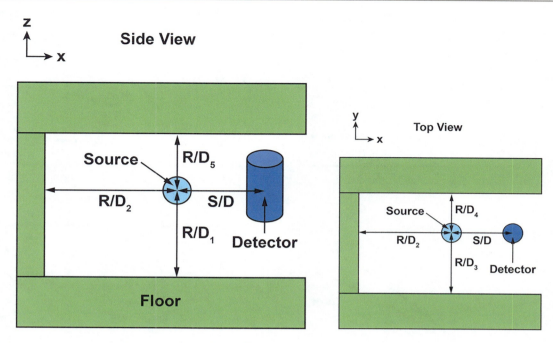

Fig. 16.28 Room return example with five scattering surfaces. (From [16])

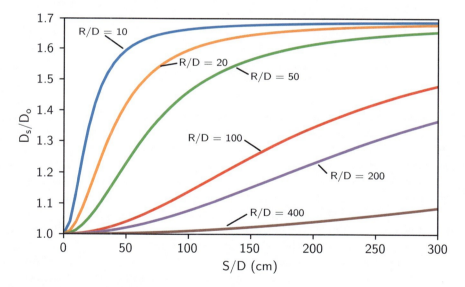

Fig. 16.29 Room return results with one reflective surface

S/D approaches 0, there are no room return neutrons ($D_S/D_0 = 1$). As the distance from the reflector (R/D) decreases, the contribution from room return neutrons increases as expected. Figure 16.29 shows that as R/D increases, the S/D distance at which the asymptote of D_S/D_0 is reached increases. For the specific constant used, it can be seen that at large S/D and small R/D, as much as 70% of the neutrons is caused by room return.

Figure 16.30 shows results when multiple reflective surfaces are present. For this study, an assumption was made that the distance to each of the reflector surfaces (R/D) was the same, but in practice, this scenario is generally not the case. For many measurements, one reflective surface (generally the floor) will be much closer than the other reflective surfaces. But it can be seen in this figure that if many close reflective surfaces are present, the total count rate can be dominated by room return neutrons (with five close reflective surfaces, it is seen that D_S/D_0 is greater than 4).

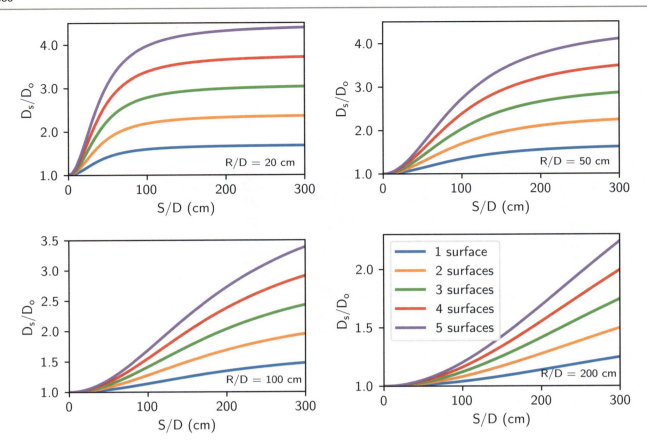

Fig. 16.30 Room return results with multiple reflective surfaces. All surfaces are assumed to have the same reflector-to-detector (R/D) distance in this figure

Both background and room return neutrons are unwanted sources of neutrons during a measurement. They complicate Eq. (16.1) and will result in incorrect estimates of the neutron source intensity (I) and/or leakage multiplication (M_L). Therefore, a best practice is to try to correct for and/or eliminate these neutrons.

In the case of background neutrons, it might be possible to measure the background at the same (or a similar) location before or after performing measurements with the item. Doing so would allow one to subtract out these neutrons (simply take the difference of the item plus background measurement and the background-only measurement). In some cases, the use of shielding may be possible to prevent background neutrons from reaching the active volume of the detector system. In other situations, the background neutrons will be negligible; for example, with measurements on items that have high neutron emission, removing or accounting for the background is generally not needed.

For room return neutrons, one can correct for these neutrons using Eq. (16.20) or (16.21). Correction simply requires performing measurements to determine the appropriate constant. In addition, it may also be possible to reduce room return neutrons depending on the control one has for the measurement setup. If possible, having the detector system closer to the item and having the item as far as possible from reflective surfaces can drastically reduce the amount of room return neutrons. In addition, it is also possible to use shielding to ensure that fewer room return neutrons are detected.

Neutrons emitted by other nuclear material could contribute to singles counting even though they are not emitted by the item. The neutron source could include neutron emission via spontaneous fission, induced fission, or (α,n). One example would include a nearby vault of nuclear material or staging locations in processing facilities. Although these neutrons can provide useful information about the items from which they are emitted, they could be a source of unwanted background associated with measurements of a separate item nearby. Such neutrons could negatively impact the measurement and yield inaccurate results. If possible, moving the item being measured from the additional sources of neutrons and/or using shielding to reduce the number of neutrons from the other nuclear material (that reach the detection system) could help improve the results.

References

1. T. Shin et al., A note on the nomenclature in neutron multiplicity mathematics. Nucl. Sci. Eng. **193**, 663–679 (2019)
2. W.B. Wilson, R.T. Perry, J.E. Stewart, *Neutron Production in UO_2F_2 from the Spontaneous-Fission and Alpha Decay of U Nuclides and Subsequent $^{17,18}O$ (α,n) and ^{19}F (α,n) Reactions*, in Applied Nuclear Data Research and Development Progress Report, April 1–June 30, 1981, P. G. Young, Compiler, Los Alamos National Laboratory report LA-09060-PR (December 1981), p. 50
3. R.H. Augustson, R.B. Walton, W. Harbarger, J. Hicks, G. Timmons, D. Shissler, R. Tayloe, S. Jones, R. Harris, L. Fields, *Measurements of Uranium Holdup in an Operating Gaseous Diffusion Enrichment Plant*, in Proceedings of the ANS/INMM Conference on Safeguards Technology: The Process-Safeguards Interface, Hilton Head Island, SC, November 28–December 2, 1983, U.S. DOE New Brunswick Laboratory, Conference No. 831106 (August 1984), pp. 77–88
4. W.B. Wilson, R.T. Perry, J.E. Stewart, T.R. England, D.G. Madland, E.D. Arthur, *Development of the SOURCES Code and Data Library for the Calculation of Neutron Sources and Spectra from (α,n) Reactions, Spontaneous Fission, and β-Delayed Neutrons*, in Applied Nuclear Data Research and Development Semiannual Progress Report, October 1, 1982–March 31, 1983, E. D. Arthur, Compiler, Los Alamos National Laboratory report LA-9841-PR (August 1983), p. 65
5. W.B. Wilson, Calculations of (α,n) neutron production in PuO_2 with variable moisture and fluorine contamination, Los Alamos National Laboratory memorandum T-2-M-1581, to J. E. Stewart (March 1985)
6. C.J. Werner (editor), MCNP users manual—code version 6.2, Los Alamos National Laboratory, report LA-UR-17-29981 (2017)
7. W. B. Wilson, R.T. Perry, E.F. Shores, W.S. Charlton, T.A. Parish, G.P. Estes, T.H. Brown, M. Bozoian, T. R. England, D. G. Madland, J. E. Stewart, *SOURCES 4C: a code for calculating (α,n), spontaneous fission, and delayed neutron sources and spectra*, American Nuclear Society/Radiation Protection and Shielding Division, 12th Biennial Topical Meeting, Santa Fe, NM, April 14–18, 2002, Los Alamos National Laboratory report LA-UR-02-1839 (2002)
8. J.E. Stewart, Neutron production by alpha particles in thin uranium hexafluoride, Los Alamos National Laboratory report LA-9838-MS (July 1983)
9. H.C. Paxton, N.L. Pruvost, Critical dimensions of systems containing ^{235}U, ^{239}Pu, and ^{233}U, 1986 revision, Los Alamos National Laboratory report LA-10860-MS (1987)
10. N. Ensslin, J. Stewart, J. Sapir, Self-multiplication correction factors for neutron coincidence counting. Nucl. Mater. Manag. **8**(2), 60–73 (1979)
11. D.P. Broughton, M.S. Grund, G. Renha, S. Croft, A. Favalli, Sensitivity of the simulation of passive neutron emission from UF_6 cylinders to the uncertainties in both $^{19}F(α,n)$ energy spectrum and thick target yield of ^{234}U in UF_6. Nucl. Instrum. Methods Phys. Res. A **1009**, 165485 (2021). https://doi.org/10.1016/j.nima.2021.165485
12. R. Weinmann-Smith et al., Changes to the ^{252}Cf neutron spectrum caused by source encapsulation. ESARDA Bull. **54**, 45–53 (2017)
13. J. Moussa, J. Hutchinson, G. McKenzie, Impact of spontaneous fission neutron emission energy on neutron detector response. Trans. Am. Nucl. Soc. **125**(1), 614–617 (December 2021)
14. W. Hess et al., Cosmic-ray neutron energy spectrum. Phys. Rev. **116**, 445 (1959)
15. R. Weinmann-Smith, M.T. Swinhoe, J.S. Hendricks, Measurement and simulation of cosmic rays effects on neutron multiplicity counting. Nucl. Instrum. Methods Phys. Res., Sect. A **814**, 50–55 (April 2016)
16. J. Hutchinson, D. Loaiza, B. Rooney, Correcting for room return neutrons using an empirical relationship. Trans. Am. Nucl. Soc. **96**(1), 453–454 (2007)

Open Access This chapter is licensed under the terms of the Creative Commons Attribution 4.0 International License (http://creativecommons.org/licenses/by/4.0/), which permits use, sharing, adaptation, distribution and reproduction in any medium or format, as long as you give appropriate credit to the original author(s) and the source, provide a link to the Creative Commons license and indicate if changes were made.

The images or other third party material in this chapter are included in the chapter's Creative Commons license, unless indicated otherwise in a credit line to the material. If material is not included in the chapter's Creative Commons license and your intended use is not permitted by statutory regulation or exceeds the permitted use, you will need to obtain permission directly from the copyright holder.

17 Principles of Neutron Coincidence Counting

M. T. Swinhoe, N. Ensslin, J. D. Hutchinson, M. Iliev, and K. Koehler

17.1 Introduction

Neutrons emitted from an item result (mainly) from a combination of spontaneous fission, induced fission, and (α,n) reactions, which makes it hard to quantify the mass of a particular element by simply measuring the number of neutrons emitted. (For example, roughly half of the neutrons emitted from a typical plutonium oxide item come from (α,n) reactions.) It is difficult to distinguish between these types of neutrons using energy spectroscopy because their energy spectra are similar. Because fission neutrons are emitted in bursts and (α,n) neutrons are emitted singly, it is possible to distinguish between these two types by their time distribution. Using the time distribution of pulses from a neutron detector is commonly referred to as *neutron coincidence counting*. In many neutron applications, we do not measure simple coincidences (such as between two gamma detectors for Compton suppression or cosmic veto); rather, we look at the statistical properties of the pulse train in a way that could be more properly called an *autocorrelation measurement*. A pulse train detected from a pure (α,n) source will have a purely random Poisson distribution of pulses, whereas a fission source will give a non-Poisson distribution. This neutron coincidence technique allows us to measure the quantity of uranium or plutonium present in bulk items of metal, oxide, mixed oxide, fuel rods, etc., nondestructively.

Table 13.1 (Chap. 13) summarizes the spontaneous fission neutron yields and multiplicities of many isotopes important in the nuclear fuel cycle. For plutonium, the table shows that ^{238}Pu, ^{240}Pu, and ^{242}Pu have large spontaneous fission yields. No large yields exist for uranium; however, ^{238}U in kilogram quantities will have a measurable yield. Spontaneous fission is usually accompanied by the simultaneous emission of more than one neutron. Thus, a measurement technique that is sensitive only to coincident neutrons will be sensitive only to these isotopes. The quantity of these particular isotopes can be determined even if the chemical form of the material yields additional single neutrons from (α,n) reactions. Then, if the isotopic composition of the material is known, the total quantity of plutonium or uranium can be calculated.

For a plutonium item that contains ^{238}Pu, ^{240}Pu, and ^{242}Pu, the observed coincidence response will be due to all three isotopes; however, ^{240}Pu is usually the major even isotope present in both low-burnup plutonium (~6% ^{240}Pu) and high-burnup, reactor-grade plutonium (~15–25% ^{240}Pu). For this reason, it is convenient to define an effective ^{240}Pu mass for coincidence counting by

$$^{240}Pu_{eff} = m_{240} = 2.52\,^{238}Pu + {}^{240}Pu + 1.68\,^{242}Pu, \tag{17.1}$$

where ^{xxx}Pu represents the mass of that isotope in the item. The ^{240}Pu$_{eff}$ is the mass of ^{240}Pu that would give the same coincidence response as that obtained from all of the even isotopes in the actual item. In Eq. 17.1, the ^{240}Pu$_{eff}$ quantity in an item is expressed in grams and is also denoted m_{240}. (The ^{240}Pu$_{eff}$ of an item can also be expressed as a percentage of the total

---

Los Alamos National Laboratory strongly supports academic freedom and a researcher's right to publish; as an institution, however, the Laboratory does not endorse the viewpoint of a publication or guarantee its technical correctness.

M. T. Swinhoe (✉) · N. Ensslin · J. D. Hutchinson · M. Iliev · K. Koehler
Los Alamos National Laboratory, Los Alamos, NM, USA
e-mail: swinhoe@lanl.gov; jesson@lanl.gov; metodi@lanl.gov; kkoehler@lanl.gov

plutonium mass.) Typically, ^{240}Pu$_{eff}$ is 2–20% larger than the actual ^{240}Pu content of the item. The coefficients 2.52 and 1.68 are determined by the relative spontaneous fission half-lives of each isotope (Chap. 13, Table 13.1), the relative neutron multiplicity distributions of each isotope (Chap. 13, Table 13.2), and the manner in which these multiplicities are processed by the coincidence circuitry (see Ref. [1], for example). The relative spontaneous fission yields are the dominant effect. The coefficients given above are appropriate for the shift register circuitry described later in this chapter but would change only slightly for other circuits. The values for the coefficients given in Eq. 17.1 are those that have been traditionally used with equipment fielded by Los Alamos National Laboratory for several decades. As better nuclear data become available, these coefficients could be updated. (For example, see discussions in Refs. [2, 3].) However, it is important to note that for instruments with existing calibrations that were calculated using the original values for the coefficients, it is likely that improving the values by using new nuclear data would have the effect of worsening the measurement results. As we shall see, this result is because the calibration constants include—to some extent—the values of the ^{240}Pu$_{eff}$ coefficients, and so decisions to change these coefficients should be made with care. (For a fixed plutonium isotopic composition, the coefficients have no effect on a calibration curve and are important only when the isotopic composition varies widely.)

Passive counting of spontaneous fission neutrons is the most common application of neutron coincidence counting. However, because fission can be induced—particularly in fissile isotopes such as ^{239}Pu and ^{235}U—an item that contains large quantities of fissile isotopes can be assayed by coincidence counting of induced fissions. The induced coincidence response will be a measure of the quantity of fissile isotopes present. If the fissions are induced by an (α,n) neutron source, the coincidence circuit can statistically discriminate the induced correlated signal from the uncorrelated source.

Passive and active neutron coincidence counters have found many applications in domestic and international safeguards, as described in Chaps. 19 and 20. Coincidence counters are usually more accurate than detectors that rely on single-neutron counting because they are not sensitive to neutrons from (α,n) reactions or room background; however, the single-neutron count rate can provide information that complements the coincidence information. For a wide range of material categories, it is generally useful to measure both the coincidence response and the total neutron response.

17.2 Characteristics of Neutron Pulse Trains

As an aid to understanding coincidence counting, it is helpful to consider the train of electronic pulses produced by the neutron detector. These electronic pulses—each representing one detected neutron—constitute the input to the coincidence (or autocorrelation) circuit. This input can be thought of either as a distribution of events in time or as a distribution of time intervals between events, whichever is more convenient. In any case, the observed distribution is produced by some combination of spontaneous fissions, induced fissions, and (α,n) reactions; and external background events. As mentioned in Sect. 17.1, fission events usually yield multiple neutrons that are correlated (close together in time) with each other, whereas (α,n) reactions and background events yield single neutrons that are uncorrelated or random in time.

17.2.1 Ideal and Actual Pulse Trains

An ideal neutron pulse train that contains both correlated and uncorrelated events might look like train (a) in Fig. 17.1. An actual pulse train detected by a typical neutron coincidence counter will look more complex, as shown by train (b) in Fig. 17.1 because the neutron coincidence counter design affects the pulse train in several ways.

Fig. 17.1 Neutron pulse trains as they might appear on a time axis. (**a**) An idealized pulse train that contains correlated and uncorrelated events. (**b**) An actual pulse train observed at high counting rates using a detector with typical efficiency and die-away-time characteristics

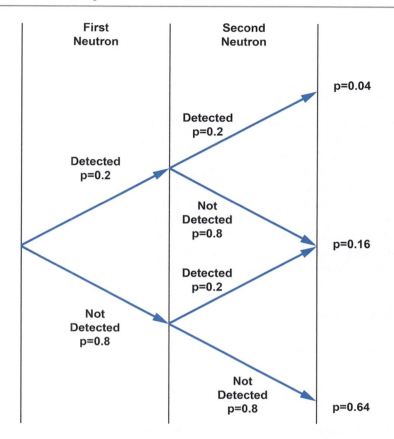

Fig. 17.2 Probability of detecting 0, 1, or 2 neutrons when 2 neutrons are emitted in a detector with neutron-detection efficiency of 20%

One kilogram of plutonium that contains 20% ^{240}Pu will emit ~200,000 n/s from spontaneous fission. If a coincidence counter has a typical detection efficiency of 20%, the total neutron count rate (singles) will be 40,000 n/s, and therefore, the mean time interval between detected events will be 25 μs. The typical efficiency $\varepsilon = 20\%$ of a coincidence counter is substantially less than 100% so that the majority of emitted neutrons is not detected. Most spontaneous fissions are also not detected. If n coincident neutrons are emitted, the probability of detecting k is given by the binomial distribution:

$$P(n,k) = \frac{n!}{(n-k)!k!}\varepsilon^k(1-\varepsilon)^{n-k} \qquad (17.2)$$

In this example, if two neutrons were emitted from a fission event (close to the mean spontaneous fission multiplicity of 2.15 for ^{240}Pu), the probability P(2,0) of detecting no neutrons is 0.64. See Fig. 17.2.

The probability P(2, 1) of detecting one neutron is 0.32, and the probability P(2,2) of detecting two neutrons is 0.04. (We call a detected pair of neutrons a *double*.) Therefore, more than half of all fission events are never detected, and most of those that are detected register only one neutron. Detecting bursts of two of more neutrons is relatively rare, occurring only 4% of the time in the above example. Nevertheless, in our example, there are about 100,000 spontaneous fissions/s. If each fission gave exactly two neutrons, the result would be about 3800 detections of two neutrons every second. (Taking the actual multiplicity distribution into account, we have about 6000 detections of two or more neutrons every second). Many of the apparent coincidences in the observed pulse train will be due to accidental overlaps of background events, background and fission events, or different fission events.

Another important effect in thermal-neutron-detector systems is the finite thermalization and detection time of the neutrons in the polyethylene body of the well counter. The process of neutron moderation and scattering within the counter can require many microseconds of time. At any moment, the process can be cut short by absorption in the polyethylene, the detector tube, or other materials or by leakage out of the counter. The effective neutron lifetime in the counter can also be prolonged by neutron-induced fission, leading to additional fast neutrons that undergo moderation and scattering before they, in turn, are absorbed. As a consequence of all of these processes, the neutron population in the counter dies away with time in a complex, gradual fashion after a spontaneous fission occurs. To a good approximation, this die-away can be represented by a single exponential,

$$N(t) = N(0)e^{-\frac{t}{\tau}}, \quad (17.3)$$

where N(t) is the neutron population at time t, and τ is the mean neutron lifetime in the counter, the die-away time. Die-away times are determined primarily by the size, shape, composition, and efficiency of the neutron coincidence counter but are also slightly affected by scattering, moderation, or neutron-induced fission within the sample being assayed. Typical values for most counter geometries are in the range of 20–100 μs; thus, the finite die-away time of the neutron coincidence counter causes the detection of prompt fission neutrons to be spread out over many microseconds. For large samples and typical counters, the mean lifetime may be comparable to, or longer than, the mean time interval between detected events.

As a result of the effects described, an actual observed pulse train may contain relatively few "real" doubles events among many "accidental" doubles events. Also, the real events will not stand out in any obvious way from the background of accidental events in the pulse train, as illustrated in Fig. 17.1, train (b). To visualize and quantify real and accidental events, it is helpful to use the interval distribution or the Rossi-α distribution.

17.2.2 The Interval Distribution

The interval distribution is the distribution of time intervals between detected events. This distribution is given by [4]:

$$I(t) = exp\left[-\int_0^t Q(t)d(t)\right]. \quad (17.4)$$

I(t) is the probability of detecting an interval of length *t*, and *Q(t)* is the probability of a second event as a function of time following a first event at t = 0. For a random neutron source, the probability of a second event is constant in time. If the singles count rate is S counts/s, the normalized interval distribution is $I(t) = Se^{-St}$. In this case, the interval distribution is exponential, and the most likely time for a following event to occur is immediately after the first event. On a semilogarithmic scale, the interval distribution will be a straight line. If real coincidence events are present in addition to random events, the interval distribution is given by a more complex equation [5]. Figure 17.3 illustrates an interval distribution that contains both coincidence and random events. As the system multiplication increases, the data deviate more from the straight line.

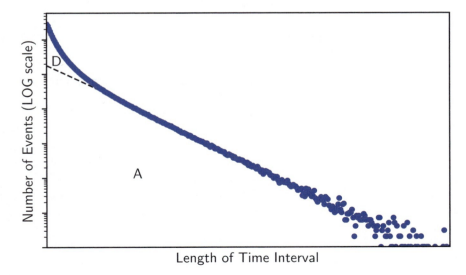

Fig. 17.3 An interval distribution formed by real coincidence events D and accidental events A. The slope of the accidental distribution on this semilogarithmic scale is the singles count rate S

17.2.3 The Rossi-α Distribution

The Rossi-α approach [6, 7], developed for reactor noise analysis by Bruno Rossi during the Manhattan project [8], uses another useful distribution. Similar to the interval distribution, the Rossi-α is also based on time differences between events recorded in the pulse train but differs in how the time differences are calculated. Figure 17.4 summarizes three ways that these time differences are obtained in analysis. Type I binning is used most often in data analysis and consists of storing the difference between the first recorded event in the pulse train and all subsequent events until the time difference exceeds a reset time—selected by the user to cover the region of interest—as shown in Fig. 17.5. Then the process is repeated, starting from

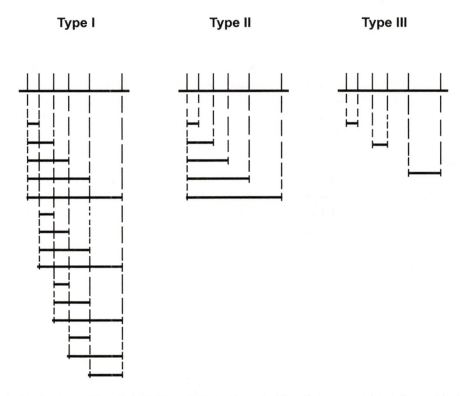

Fig. 17.4 Three Rossi-α binning types. (From Ref. [9]. The pulse train (six vertical lines that represent detected events) is shown at the top, and the time differences used for analysis are shown at the bottom

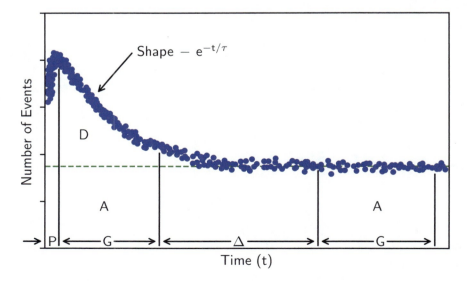

Fig. 17.5 A Rossi-α distribution that shows detected neutron events as a function of time following an arbitrary starting event. D represents real coincidence events (Doubles), and A represents accidental coincidence events. P = predelay, G = prompt and delayed gates, Δ = long delay, and τ = die-away time

the second recorded event in the pulse train. Type II binning follows the same procedure as Type I, the differences begin from the first pulse, but subsequent differences begin with the first recorded event outside of the selected reset time. Type III binning determines the time difference between the first two recorded events, ignores the interval between events 2 and 3, and then moves on to determine the time difference between the third and fourth recorded event, and so on, until all of the recorded events are examined. Note that Type III binning is very similar to the interval distribution that includes the difference between each detected event.

The Rossi-α distribution is the distribution in time of events that follow after an arbitrarily chosen starting event. If only random events are being detected, the distribution is constant as a function of time. If real coincidence events are also present, the Rossi-α distribution is given by

$$RA(t) = A + De^{-\frac{t}{\tau}}, \tag{17.5}$$

where RA(t) is the height of the distribution at time t; A is the accidental, or random, count rate; D is the real coincidence count rate; and τ is the detector die-away time (see additional discussion related to this term below). Figure 17.5 illustrates a Rossi-α distribution with D, A, and other labeled variables (defined later). The exponential die-away is clearly seen in this distribution.

When using moderated ^{3}He systems, the detector die-away time will be much longer than the prompt neutron lifetime in the item itself, which is why the detector die-away (τ) is given in Eq. 17.5. For unmoderated ^{3}He systems, measurements of intermediate or thermal systems near critical may allow for measurement of the true prompt neutron decay constant [10–15]. Similarly, if a faster detection system, such as an organic scintillator, is used for Rossi-α analysis [16–18], then the true prompt neutron decay constant (α) of the system can be measured for any energy system.

For systems where significant reflection is present, some of the assumptions of the point kinetics model used to develop Eq. 17.5 have been violated [19], and it has been shown that the fit obtained using Eq. 17.5 could be inadequate [20]. To compensate for these issues, Eq. 17.5 can be expanded to include a second exponential term [18, 21],

$$RA(t) = A + D\rho_1 e^{r_1 t} + D\rho_2 e^{r_2 t}, \tag{17.6}$$

where the coefficients are given by

$$\rho_1 = \frac{(1-R)^2}{r_1} + \frac{2(1-R)R}{r_1 + r_2} \tag{17.7}$$

$$\rho_2 = \frac{R^2}{r_2} + \frac{2(1-R)R}{r_1 + r_2} \tag{17.8}$$

and the exponent terms are given by

$$r_j = (-1)^j \sqrt{\frac{1}{\tau_r}(f' + \alpha) + \frac{1}{4}\left(\frac{1}{\tau_r} - \alpha\right)^2} + \frac{1}{2}\left(\alpha - \frac{1}{\tau_r}\right), \tag{17.9}$$

where τ_r is the mean time in the reflector (mean time outside of the core), R is a number between 0 and 1 that weights the exponents to determine α

$$\alpha = r_1(1-R) + r_2 R \tag{17.0}$$

and

$$f' = -\frac{(\alpha - r_1)(\alpha - r_2)}{\alpha - r_1 - r_2}. \tag{17.11}$$

Use of this two-exponential fit has been shown to give a better representation of the true prompt neutron decay constant in the item when reflection is present [17].

17 Principles of Neutron Coincidence Counting

When analyzing list-mode data (as described in Chap. 18), the user must select the reset time and the bin width used to plot/assemble the data before fitting. These times will depend on the average neutron energy of the system being measured. Work has been performed to estimate the uncertainty associated with Rossi-α measurements [16, 21].

One use of both Rossi-α and interval distribution analysis measurements is estimating the dead time of detector systems. As shown in Fig. 17.5, the peak in the data does not occur for the smallest possible time difference, which is what one would expect from theory. This outcome occurs because of the detector dead time (described in Sect. 17.3.2). Rossi-α and interval distribution analysis can be used to estimate the detector system dead time (see Sect. 17.6.5). One way to estimate dead time is to perform simulations with varying artificial dead-time values and compare these with measurements to bound the possible dead time of the detector system [22].

17.3 Basic Features of Coincidence Circuits

This section contains a brief description of some effects that will be useful for later discussions.

17.3.1 Electronic Gates

Coincidence circuits often contain electronic components—called *one-shots* or *gate generators*—that produce an output pulse of fixed duration whenever an input pulse is received. Gate generators used to convert the input pulses from the neutron detector into very short output pulses are called *triggers*. Gate generators used to convert the input pulses into long output pulses are called *gates*. Such gate generators—as well as amplifiers, detectors, and other circuits—exhibit an electronic dead time before they can function again. This dead time is at least the length d of the gate. Depending on the design, this dead time can be nonupdating or updating.

17.3.2 Updating and Nonupdating Dead Times

A nonupdating, or nonparalyzable, dead time is illustrated in pulse train (a) in Fig. 17.6. Of the four events, Events 1, 2, and 4 initiate pulse processing, but Event 3 does not and is lost. Using Eq. 17.4, it can be shown that for a true random input rate S, the measured output rate S_m is

$$S_m = \frac{S}{1 + d \cdot S}. \tag{17.12}$$

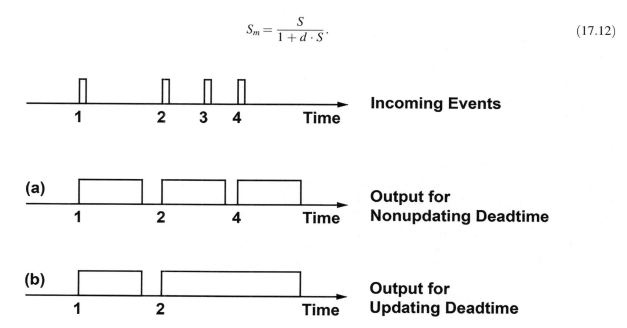

Fig. 17.6 Two gate generators with different electronic dead-time characteristics: (**a**) non-updating dead time; (**b**) updating dead time

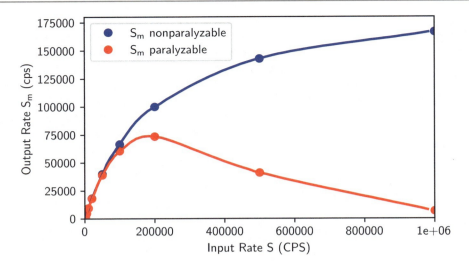

Fig. 17.7 Comparison between updating dead time and nonupdating dead time for a dead-time constant of 5 μs

As the input rate becomes very large, the output rate will approach the limiting value $1/d$, where d is the dead time.

An updating, or paralyzable, dead time is illustrated in pulse train (b) in Fig. 17.6. The appearance of Event 3 causes the gate produced by Event 2 to be extended or updated. Consequently, Event 4 does not generate a new pulse. Only Events 1 and 2 initiate pulses, and Events 3 and 4 are lost. Using Eq. 17.4, it can be shown that for random events

$$S_m = S e^{-d \cdot S}. \tag{17.13}$$

As the input rate increases, the output rate increases up to a maximum value (which occurs when the input rate is $1/d$) and then declines toward 0 (approaches paralysis) as the input rate continues to increase.

The behavior of the two types of dead time is shown in Fig. 17.7 for a dead time of 5μs.

For input rates that are small, identical dead-time corrections are obtained from Eqs. 17.12 and 17.13. In practice, real ^{3}He counting systems are neither fully updating nor fully nonupdating [23].

17.3.3 Cross-Correlation and Autocorrelation Circuits

Electronic one-shots or gate generators can be combined with accumulators in many possible ways to create coincidence circuits. Each combination will be subject to different electronic dead times and will require different equations for analysis. For neutron counting, cross-correlation or autocorrelation circuits are the most useful [24]. A simple cross-correlation measurement is shown in circuit (a) in Fig. 17.8. Trigger pulses from Detector 1 are compared with gates generated from Detector 2. This type of circuit is most useful for very fast detector pulses and short gates because discrimination against detector noise is good and because very few accidental coincidences are produced.

Circuit (b) in Fig. 17.8 illustrates an idealized autocorrelation measurement. Both detector inputs are first combined into one pulse train. Then every pulse in the train generates both a short trigger and a long gate so that every pulse can be compared with every following pulse. Autocorrelation circuits are best suited for thermal-neutron counters because many detector banks can be summed together for high efficiency and because the substantial die-away time of the neutrons causes many overlaps between detector banks. Gate lengths are chosen to be comparable to the die-away time, and a separate parallel circuit with a delayed trigger or gate is usually used for the subtraction of accidental coincidences (see Sects. 17.4 and 17.5). The autocorrelation circuits described in Sects. 17.4 and 17.5 are the most important circuits for neutron coincidence counting.

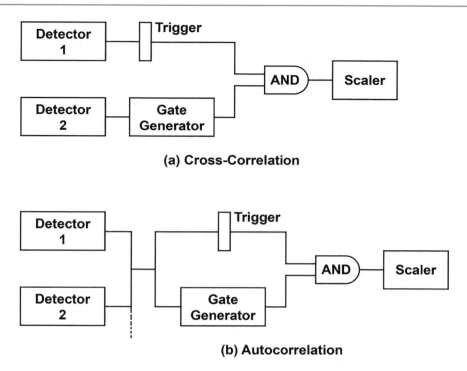

Fig. 17.8 Two types of coincidence circuits: (**a**) cross-correlation and (**b**) autocorrelation

17.4 Feynman Variance-to-Mean Method

Like other neutron noise methods, the Feynman variance-to-mean method uses the property of fission in which multiple prompt neutrons may be born simultaneously from the same fission event (described in Chap. 13, Sects. 13.2 and 13.3). If no neutron multiplication occurs, the emission (and also detection) of the number of neutrons within a time gate will obey a Poisson distribution. As multiplication increases, so does the deviation of the emission from a purely Poisson distribution.

This section will assume that the Feynman variance-to-mean analysis will start from list-mode data as described in Chap. 18. Similar to the other coincidence and multiplicity methods described, the analysis could also be performed using hardware to perform the initial binning instead of starting from list-mode data. Many of the examples presented in this section use data from the MC-15 detector system, which is described in detail in Chap. 19, Sect. 19.5.4.

17.4.1 Feynman Variance-to-Mean History

In 1944, Feynman, de Hoffman, and Serber introduced the method now known as the *Feynman variance-to-mean method* [25]. The first measurements that used this method were performed on the Los Alamos Water Boiler (also known as LOPO) during the Manhattan Project [8, 26]. As stated in Ref. [27]:

> In the 1950s and 1960s a great deal of theoretical work was devoted to multiplication or reactivity measurements. This was aided by: (1) advancements in nuclear instrumentation and (2) nuclear material availability. The former allowed for scientists to measure the number of neutrons in small time gates, allowing for one to apply correlated neutron techniques. The latter provided the means to test the theory at facilities such as the Los Alamos Pajarito Site, the Oak Ridge Critical Experiments Facility, and Argonne National Laboratory.

Theoretical work in the 1980s expanded the Feynman variance-to-mean method; of particular note are the Hansen-Dowdy method from Los Alamos [28] and the Hage-Cifarelli method from Italy [29]. Both of these are moments-based methods used to solve for multiplication, spontaneous fission rate, and/or the (α,n) emission rate. A great deal of research has been performed on this method through the years (a sampling is given in Refs. [7, 30–59]). Much of the description in this section focuses on the Hage-Cifarelli formalism and comes from Refs. [49, 57]. This section does not include uncertainties in the listed terms; uncertainties are given in Refs. [49, 57]. Application of this method requires many assumptions (namely the point-source assumption); a list of many additional assumptions is presented in Ref. [57] and other publications.

The Feynman variance-to-mean method is also referred to as the Feynman-α method, and these two terms can be used interchangeably. It should be noted that different terminology and nomenclature are often used for the same concepts within different disciplines (safeguards compared with reactor physics for example); Ref. [60] contains tables that link the terminology that has historically been used. The Feynman variance-to-mean and Rossi-α (described in Sect. 17.5) methods are linked to one another. One can obtain the Feynman variance-to-mean terms by performing a double integration of the Rossi-α distribution (Eq. 17.5) as described in Refs. [50, 59, 60].

17.4.2 Binning

Like other time-domain neutron noise methods, Feynman variance-to-mean starts by binning the pulse train into time gates. Multiple binning methods may be used as described in Ref. [45]; this section will focus on the sequential binning because it is the easiest to describe. Using sequential binning, the entire pulse train is divided into gates of equal width G, as shown on the left side of Fig. 17.9. The size of the gate width will depend on both the detection system and the timing characteristics of the system being measured. For ^{3}He systems, these gate widths are often on the order of 1 μs to tens of msec; however, for organic scintillators, these gate widths are often on the order of 1 to hundreds of ns. The number of detected neutrons in each gate width is recorded and plotted to create a histogram, as shown on the right side of Fig. 17.9. In this histogram, the x-axis is the number of detected neutrons (n), and the y-axis is the total number of gates (C_n) that recorded exactly n neutrons. The simple example of Fig. 17.9 contains only three gates. In the first gate, two neutrons were recorded; therefore, in the histogram, the multiplet of 2 is incremented from 0 to 1 (represented by the blue box). Three neutrons were then recorded in the second and third gates, which increments the multiplet of 3 from 0 to 2 (represented by the green and purple boxes). This entire process is repeated many times (generally hundreds of thousands or millions of times). Note that the total number of gates depends only on the counting time and gate width; it does not depend on the emission rate of the item.

Figures 17.10 and 17.11 show examples of Feynman histograms for low- and high-emission rate systems measured using two MC-15 detectors. The two detectors were connected (such that counts from all 30 ^{3}He tubes are recorded in the data file), and the gate width for the following figures is at 2048 μs. Starting with the low-emission-rate systems shown in Fig. 17.10, these systems all had less than 50 cps in the MC-15 detectors. It can be seen that the most probable value of n is always 0, which is to be expected because the emission rate is very low. (As the emission rate increases, the value of n—where the peak occurs—will increase.) Starting with the background in green, every gate observed either 0 or 1 event, and it can be seen that the data (green histogram) agrees with the Poisson distribution (green line). The other two sets of data were measured using the Rocky Flats hemishells, which are a set of nesting highly enriched uranium (HEU) hemishells. In particular, RF3–30 hemishells were used, which have a mass of 21.8 kg. The orange data are on the bare Rocky Flats hemishells, which have $M_L = 2.1$. It can clearly be seen when comparing the orange data versus the orange line that the data definitely deviate from a Poisson distribution. This result can be seen most clearly at n of 3 or 4. (Typically it is always easiest to see the deviation at the highest values of n that are present for that data.) Using this plot alone, one can conclude that multiplication is present in this system. Lastly, the blue data use the same HEU core, but here it is reflected by 2.5 in. thick, high-density polyethylene (HDPE). This HDPE reflector increases M_L to 9.8. It can be seen when comparing the blue data that it is very different than the

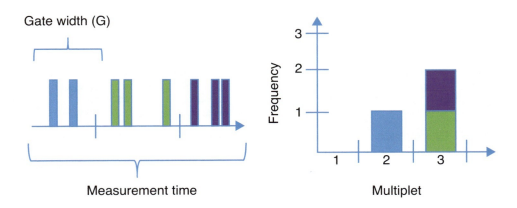

Fig. 17.9 Sequential binning to create Feynman histograms. (From Ref. [52])

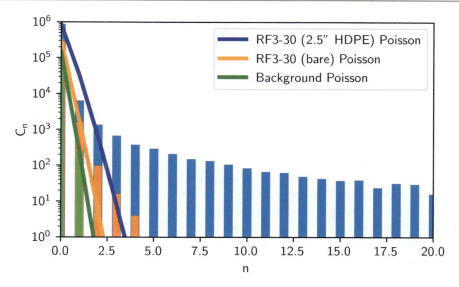

Fig. 17.10 Feynman histograms for three low-count-rate systems measured with the MC-15. The three systems shown are background, bare Rocky Flats 3–30 hemishells (21.8 kg highly enriched uranium, $M_L = 2.1$), and the same highly enriched uranium reflected with 2.5 in. thick polyethylene ($M_L = 9.8$)

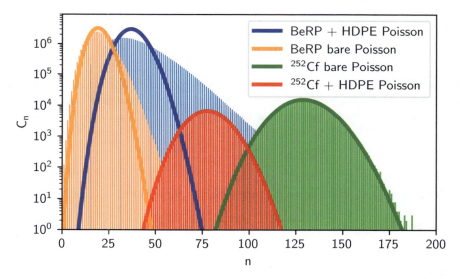

Fig. 17.11 Feynman histograms for three medium–/high-count-rate systems measured with the MC-15, including a ^{252}Cf source and the BeRP ball (a ~ 4.5 kg sphere of weapons-grade plutonium). Both are shown bare and reflected by polyethylene. (The solid lines show the equivalent Poisson distribution)

Poisson distribution. A really large deviation such as this is an indication that significant multiplication is present. Additional information on the Rocky Flats hemishells measurements is given in Ref. [61].

Similar comparisons can be performed for medium/high-count-rate systems, as shown in Fig. 17.11. Here, the count rate of the two MC-15 detectors was always greater than 10,000 cps. First, results from a bare ^{252}Cf source are shown in green. This source had an emission rate of 2,850,000 n/s when these measurements took place, and it can be seen that the count rate is very high (because every gate recorded many neutrons, with the peak occurring around $n = 130$), but the data are very similar to a Poisson distribution. A very slight deviation is seen at high n, but compared with the plutonium (in blue and orange), there is clearly very little deviation. From this curve alone, one can conclude that essentially no multiplication occurred (and therefore both M and M_L are very close to 1). The same conclusions are drawn for the ^{252}Cf source reflected by 4 in. of HDPE (in red). The count rate decreased, but the data clearly still follow a Poisson distribution. The plutonium cases in blue and orange are more interesting. Here, the beryllium-reflected plutonium (BeRP) ball (a ~ 4.5 kg sphere of weapons-grade α-phase plutonium) was used. Although the name includes beryllium, no beryllium was present for these measurements. (The BeRP ball was named after an experiment in the 1980s in which the ball was reflected by beryllium). Many publications describe the BeRP ball in more detail; see Refs. [48, 51, 56] for additional information. The orange data show the bare BeRP ball (which has $M_L = 3.31$, as given in Ref. [56]). Although the bare BeRP ball has a lower count rate than the ^{252}Cf source, it shows clear signs of multiplication, and it is very easy to see that it differs greatly from the Poisson distribution. When HDPE is added around the BeRP ball (in blue) the deviation from the Poisson distribution increases just like in the HEU example.

For all Feynman histograms, the data and Poisson distribution must integrate to the same value. As discussed previously, it is easiest to see deviation from Poisson by looking at the data above the curve for high n. In addition, the data are also above the curve at very low n (as seen in the orange and blue data in Fig. 17.10), which means that the data must be lower than the Poisson distribution near the peak—indeed the case in the orange and blue curves shown in Fig. 17.11. Another way to visualize this scenario is to plot the detection time versus event number, as shown in Fig. 17.12. When random neutrons are emitted, it is expected that the time should be proportional to the event number. As seen in the top plot of Fig. 17.12, for a ^{252}Cf source, the data fall close to this linear fit and have a very high R^2 value. As the multiplication increases, the data deviate further from this linear fit, which can be seen both visually (on the two lower graphs) and in their respective R^2 values.

Although Figs. 17.10 and 17.11 were not normalized, it is common practice to normalize the Feynman histograms, which is important if comparing measured data with different measurement times. This normalization is traditionally represented as

$$p_n = \frac{C_n}{\sum_{n=0}^{\infty} C_n} \qquad (17.14)$$

where C_n is the number of gates (y-axis in Figs. 17.10 and 17.11) that contained n events (x-axis in Figs. 17.10 and 17.11). From this outcome, the moments of the counting distribution can be determined using

$$\overline{C_r} = \sum_{n=0}^{\infty} n^r p_n \qquad (17.15)$$

where r is the index (that indicates the order of the moment). The reduced factorial moments are also often used and are very similar to the moments in Eq. 17.15; the main difference is that here, n is decremented due to the loss of a neutron when inducing fission. The reduced factorial moments are described by

$$m_r = \frac{\sum_{n=0}^{\infty} n(n-1)\ldots(n-r+1)p_n}{r!} \qquad (17.16)$$

Table 17.1 shows the first four moments and reduced factorial moments. From this table, it can be seen how these moments are related to one another.

Uncertainties of these terms are discussed in detail in Refs. [49, 57].

17.4.3 Excess Variance

From these moments, the excess variance (Y) can be calculated using

$$Y = \frac{\overline{C_2} - \overline{C_1}^2}{\overline{C_1}} - 1 = \frac{2m_2 + m_1 - m_1^2}{m_1} - 1. \qquad (17.17)$$

Note that Y and Y_m are often used interchangeably in publications. Equation 17.17 demonstrates why the method is referred to as "variance-to-mean." Here, $\overline{C_1}$ is the mean, and "$\overline{C_2} - \overline{C_1}^2$" is the variance. If no multiplication occurs in the item, then the data of the Feynman histogram will exhibit behavior of a Poisson distribution, in which the variance is equal to the mean and therefore, Y is equal to 0 (there is no "excess" in the variance). As the multiplication of the item increases, the excess variance increases. This concept is revisited in additional detail in Sect. 17.4.5.

It is common within the Hage-Cifarelli formalism to use the reduced factorial moments for a different expression of excess variance. First, a recursive formula is used

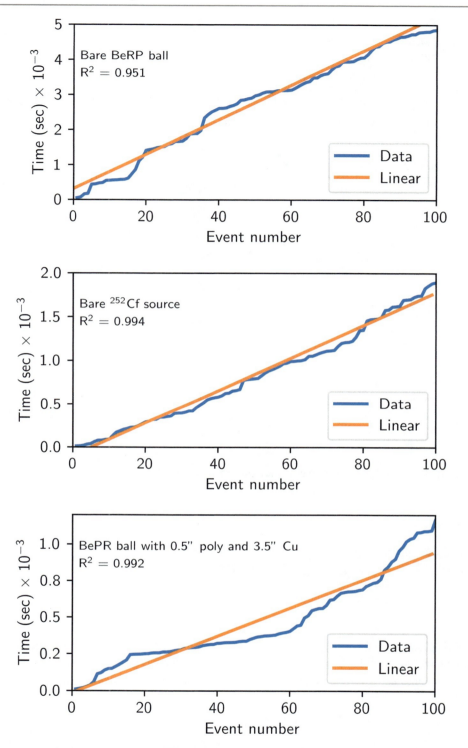

Fig. 17.12 Detection time for the first 100 events. Bare ^{252}Cf source shown at top, bare BeRP ball ($M_L = 3.31$) shown in the middle, and the BeRP ball with 0.5 in. thick polyethylene and 3.5 in. thick copper shown at bottom (SCRαP configuration 15, $M_L = 12.97$). (All data are from Ref. [56])

Table 17.1 First four moments

Order	Moment	Reduced factorial moment
1	$\overline{C_1} = \sum_{n=0}^{\infty} n p_n = m_1$	$m_1 = \sum_{n=0}^{\infty} n p_n$
2	$\overline{C_2} = \sum_{n=0}^{\infty} n^2 p_n = 2m_2 + m_1$	$m_2 = \frac{\sum_{n=0}^{\infty} n(n-1) p_n}{2!}$
3	$\overline{C_3} = \sum_{n=0}^{\infty} n^3 p_n = 6m_3 + 6m_2 + m_1$	$m_3 = \frac{\sum_{n=0}^{\infty} n(n-1)(n-2) p_n}{3!}$
4	$\overline{C_4} = \sum_{n=0}^{\infty} n^4 p_n = 24m_4 + 36m_3 + 14m_2 + m_1$	$m_4 = \frac{\sum_{n=0}^{\infty} n(n-1)(n-2)(n-3) p_n}{4!}$

Table 17.2 First four excess variance terms

Order	Excess variance term
1	$Y_1 = \frac{m_1}{G}$
2	$Y_2 = \frac{1}{G}\left(m_2 - \frac{1}{2}m_1^2\right)$
3	$Y_3 = \frac{1}{G}\left(m_3 - m_2 m_1 + \frac{1}{3}m_1^3\right)$
4	$Y_4 = \frac{1}{G}\left(m_4 - m_3 m_1 + m_2 m_1^2 - \frac{1}{2}m_2^2 - \frac{1}{4}m_1^4\right)$

Table 17.3 First four shaping terms

Order	Excess variance shaping term
1	$\omega_1 = 1$
2	$\omega_2 = 1 - \frac{1-e^{-\lambda G}}{\lambda G}$
3	$\omega_3 = 1 - \frac{1}{2\lambda G}\left(3 - 4e^{-\lambda G} + e^{-2\lambda G}\right)$
4	$\omega_4 = 1 - \frac{1}{6\lambda G}\left(11 - 18e^{-\lambda G} + 9e^{-2\lambda G} - 2e^{-3\lambda G}\right)$

$$m_\mu = \sum_{r=0}^{\mu-1} \frac{\mu - r}{\mu} m_r m^*_{(\mu-r)}, \qquad (17.18)$$

with $m_0 = 1$. Given Eq. 17.18, the following excess variance equation can be used

$$Y_\mu = \frac{m^*_\mu}{G}. \qquad (17.19)$$

The first four terms of this equation are shown in Table 17.2 (from Ref. [49]).

It should be noted that Y and Y_2 are proportional to one another by

$$Y_2 = \frac{Y \overline{C_1}}{2G}. \qquad (17.20)$$

Often, either Y or Y_2 is plotted as a function of gate width, as shown in Fig. 17.13. Note several interesting features from this figure. First, it can be seen that as multiplication increases, the Y_2 value increases due to both the increase in excess variance and the increase in the count rate ($\frac{\overline{C_1}}{G}$). Also of note is the shape of this curve. This shape follows

$$\omega_\mu = \sum_{r=0}^{\mu-1} \binom{\mu-1}{r} (-1)^r \frac{1 - e^{-\lambda G r}}{\lambda G r} \qquad (17.21)$$

where λ is the prompt neutron decay constant (and is sometimes used interchangeably with the α from the Rossi-α method). The first four terms of this equation are shown in Table 17.3 (from Ref. [49]).

Two interesting features can be noted when comparing the red data (C16, the BeRP ball with 4 in. thick polyethylene) and the green data (C11, the BeRP ball with 4 in. thick copper) in Fig. 17.13. First, note that the shapes of the curves look very different from one another. In particular, the red data do not reach an asymptote until much later in time (and the rise is shallower). This result is due to having a much longer prompt neutron lifetime and therefore different λ (from Ref. [56]: C11 has $1/\lambda = 41.64$ μs; C16 has $1/\lambda = 280.02$ μs). When ^{3}He systems that include polyethylene moderation are used, the

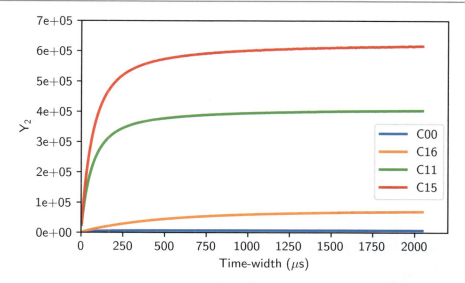

Fig. 17.13 Y_2 versus gate width for SCRαP configurations [56]. In order of increasing multiplication: C00 is the bare BeRP ball ($M_L = 3.31$), C16 is the BeRP ball with 4 in. thick polyethylene ($M_L = 9.69$), C11 is the BeRP ball with 4 in. thick copper ($M_L = 10.87$), and C15 is the BeRP ball with 0.5 in. thick polyethylene and 3.5 in. thick copper ($M_L = 12.97$)

minimum prompt neutron lifetime ($1/\lambda$) is related to the slowing-down time inside the polyethylene. For the MC-15 system, this time is 35–40 μs—one benefit of scintillation systems if this timing information is useful. The second interesting feature of Fig. 17.13 is the difference in magnitude of the green and red data. One might expect these to be similar with one another because they have similar multiplication ($M_L = 9.69$ for the red curve; 10.87 for the green curve). But this difference in magnitude is due to the difference in the count rates, which is driven by the fact that many neutrons are absorbed in the polyethylene reflector and do not make it to the detection system.

Given the excess variance (Y_μ) and shaping (ω_μ) terms, the counting rates are simply

$$R_\mu = \frac{Y_\mu}{\omega_\mu}, \tag{17.22}$$

so the detector count rate is simply $R_1 = Y_1$, which is also the rate at which a single neutron from a single fission chain is detected. R_2 is known as the *doubles counting rate*, which is the rate at which two neutrons from a single fission chain are detected (can be extended to higher order).

17.4.4 Determination of Leakage Multiplication and Mass

Once the counting rates have been determined, they can be used to infer information on the measured item using the following equation:

$$R_\mu = \varepsilon^\mu \left(b_{\mu 1} N_{sf} + b_{\mu 2} N_\alpha \right), \tag{17.23}$$

where ε is the absolute detection efficiency (counts/n), N_{sf} is the spontaneous fission rate of the system, and N_α is the (α,n) neutron emission rate. The first three terms of this equation are given in Table 17.4.

The coefficients are given in Tables 17.5 and 17.6.

Here, $\bar{\nu}$ is the moment of the prompt neutron multiplicity distribution. The subscripts *S* and *I* refer to *spontaneous fission* and *induced fission*, respectively. The numbered subscripts refer to the order of the reduced factorial moment of the distribution (1 for the first moment, 2 for the second, . . .) given by

Table 17.4 Counting rate terms

Order	Counting rate term
1	$R_1 = \varepsilon(b_{11}N_{sf} + b_{12}N_\alpha)$
2	$R_2 = \varepsilon^2(b_{21}N_{sf} + b_{22}N_\alpha)$
3	$R_3 = \varepsilon^3(b_{31}N_{sf} + b_{32}N_\alpha)$

Table 17.5 Counting rate coefficients

Order	$b_{\mu 1}$ Coefficient
1	$b_{11} = M_L \overline{\nu_{S1}}$
2	$b_{21} = M_L^2 \left(\overline{\nu_{S2}} + \frac{M_L - 1}{\overline{\nu_{I1}} - 1} \overline{\nu_{S1}}\, \overline{\nu_{I2}} \right)$
3	$b_{31} = M_L^3 \left[\overline{\nu_{S3}} + \frac{M_L - 1}{\overline{\nu_{I1}} - 1} \left(\overline{\nu_{S1}}\, \overline{\nu_{I3}} + 2\overline{\nu_{S2}}\, \overline{\nu_{I2}} \right) + 2\left(\frac{M_L - 1}{\overline{\nu_{I1}} - 1} \right)^2 \overline{\nu_{S1}}\, \overline{\nu_{I2}}^2 \right]$

Table 17.6 Counting rate coefficients

Order	$b_{\mu 2}$ Coefficient
1	$b_{12} = M_L$
2	$b_{22} = M_L^2 \left(\frac{M_L - 1}{\overline{\nu_{I1}} - 1} \overline{\nu_{I2}} \right)$
3	$b_{32} = M_L^3 \frac{M_L - 1}{\overline{\nu_{I1}} - 1} \left[\overline{\nu_{I3}} + 2\left(\frac{M_L - 1}{\overline{\nu_{I1}} - 1} \right) \overline{\nu_{I2}}^2 \right]$

Table 17.7 Prompt neutron multiply reduced factorial moments

Order	Reduced factorial moment
1	$\overline{\nu_1} = \sum_{\nu=0}^{\infty} \nu P_\nu$
2	$\overline{\nu_2} = \frac{1}{2} \sum_{\nu=0}^{\infty} \nu(\nu - 1) P_\nu$
3	$\overline{\nu_3} = \frac{1}{6} \sum_{\nu=0}^{\infty} \nu(\nu - 1)(\nu - 2) P_\nu$

$$\overline{\nu_\mu} = \sum_{\nu=\mu}^{\infty} \binom{\nu}{\mu} P_\nu, \quad \delta_{n,\mu} = \begin{cases} 1 & \text{for } n = \mu \\ 0 & \text{for } n \neq \mu \end{cases}. \quad (17.24)$$

This equation results in the moments given in Table 17.7.

It should be noted that this first moment is often referred to as simply $\overline{\nu}$ and is the average number of neutrons released per fission, which are the same values given for various nuclides in Chap. 13, Tables 13.1 and 13.2. Note that for the second order moment, the values in Table 13.2 should be divided by 2 (and the third order moment should be divided by 6); for example, $\overline{\nu_2}$ for ^{240}Pu is 1.9125 (given by 3.825/2). Sensitivity to these terms is demonstrated in Sect. 17.4.5.

Note that the singles rate (R_1) in Tables 17.3 through 17.5 is essentially the same as the singles rate (S) in Eq. 16.1, in which the item neutron source intensity term is given by

$$I = \overline{\nu_{S1}} N_{sf} + N_\alpha. \quad (17.25)$$

Assuming that it is known what nuclides are undergoing spontaneous and induced fission (and therefore the $\overline{\nu_\mu}$ terms are known), one is left with four unknowns in Tables 17.4 through 17.6. Three of these terms are specific to the item being measured: the spontaneous fission rate (N_{sf}), the (α,n) neutron emission rate (N_α), and the leakage multiplication (M_L). The fourth unknown is the detector efficiency (ε), which is specific to the detection system (and the setup of the detection system). To solve for all four of these unknowns, one would need to have four equations (for the singles, doubles, triples, and quad counting rates), which may be possible for very high-efficiency systems (such as a well counter) using items with a high emission rate, but this scenario is not possible for portable systems such as the MC-15. It can be seen in Table 17.4 that the order of the rate is proportional to the power of the efficiency (so doubles is proportional to efficiency squared, triples is proportional to efficiency cubed, etc.). For these reasons, it will be impractical to have reliable data for triples and quads for many detector systems. The resulting data are obviously extremely noisy and therefore should not be used.

Table 17.8 Known and unknown terms

Equations used	Known/assumed terms	Unknown terms
R_1, R_2	ε, N_α	N_{sf}, M_L
R_1, R_2	ε, M_L	N_α, F_S
R_1, R_2	ε, N_{sf}	N_α, M_L
R_1, R_2	N_α, N_{sf}	ε, M_L
R_1, R_2	N_α, M_L	N_α, N_{sf}
R_1, R_2	N_{sf}, M_L	ε, N_α
R_1, R_2, R_3	ε	N_α, N_{sf}, M_L
R_1, R_2, R_3	N_α	ε, N_{sf}, M_L
R_1, R_2, R_3	M_L	$\varepsilon, N_\alpha, N_{sf}$
R_1, R_2, R_3	N_{sf}	$\varepsilon, N_\alpha, M_L$
R_1, R_2, R_3, R_4	None	$\varepsilon, N_\alpha, N_{sf}, M_L$

Because it may not be possible to use these higher-order counting rates, some of the other parameters are often determined via other means. Table 17.8 shows the various ways in which one can apply these equations. Note that one extreme example would be to use Singles only and assume that the other three parameters are known. This assumption is not discussed here because it does not use correlated neutrons and is already discussed in Chap. 16. Many of the possibilities listed in Table 17.8 are used in practice, whereas others would rarely or never be used. In this section, we will focus on the top row (known/assumed ε and N_α) because it is both used often and can result in relatively easy equations.

Here we will assume that the detector efficiency is known, which could be true in practice for a few different reasons, including an independent replacement measurement with a neutron source of known strength or from simulations. In addition, we will also assume that the (α,n) neutron emission rate is zero, which is a fairly good assumption for items that include only fissionable material in metal form. When $N_\alpha = 0$, the terms in Table 17.4 simplify to

$$R_1 = \varepsilon M_L \overline{\nu_{S1}} N_{sf} \tag{17.26}$$

$$R_2 = \varepsilon^2 M_L{}^2 N_{sf} \left(\overline{\nu_{S2}} + \frac{M_L - 1}{\overline{\nu_{I1}} - 1} \overline{\nu_{S1}}\, \overline{\nu_{I2}} \right) \tag{17.27}$$

Here, Eq. 17.26 is rearranged to solve for N_{sf}

$$N_{sf} = \frac{R_1}{\varepsilon M_L \overline{\nu_{S1}}} \tag{17.28}$$

Equation 17.28 is then substituted into Eq. 17.27, resulting in the quadratic eq.

$$0 = a_1 M_L{}^2 + a_2 M_L + a_3 \tag{17.29}$$

with coefficients

$$a_1 = \frac{\overline{\nu_{S1}}\, \overline{\nu_{I2}}}{\overline{\nu_{I1}} - 1} \tag{17.30}$$

$$a_2 = \overline{\nu_{S2}} - \frac{\overline{\nu_{S1}}\, \overline{\nu_{I2}}}{\overline{\nu_{I1}} - 1} \tag{17.31}$$

$$a_3 = -\frac{R_2 \overline{\nu_{S1}}}{R_1 \varepsilon} \tag{17.32}$$

Solving this quadratic equation gives leakage multiplication

$$M_L = \frac{-a_2 + \sqrt{a_2^2 - 4a_1 a_3}}{2a_1} = \frac{-a_2 + a_4}{2a_1} \qquad (17.33)$$

with

$$a_4 = \sqrt{a_2^2 - 4a_1 a_3} \qquad (17.34)$$

After obtaining leakage multiplication, one can simply plug this result into Eq. 17.28 to solve for the spontaneous fission rate. Once the spontaneous fission rate is known, one can simply use the spontaneous fission yields in Table 13.1 to solve for the mass of the spontaneous fission nuclide. As an example, say that Eq. 17.28 resulted in $N_{sf} = 1.28e5$ fissions/s (the rate given for the BeRP ball in Ref. [56]). Using a $\overline{\nu_{S1}}$ value of 2.156 for ^{240}Pu (from Table 13.2), the result is 2.77e5 neutrons/s (1.28e5 × 2.156). Then, if we use a spontaneous fission yield of 1020 n/s − g (from Table 13.1), the result is a ^{240}Pu mass of 272 g (2.77e5/1020). One would then have to know the isotopic composition of the item to get the total plutonium mass. If we assume 6% ^{240}Pu, then the result is a total mass of 4.5 kg (272/0.06), which is quite close to the reported mass [56]. Of course, this result can be used in conjunction with the ^{240}Pu$_{eff}$ equation (Eq. 17.1) to further refine the mass estimate.

17.4.5 Additional Considerations and Notes

Now that it has been demonstrated how to use measured counting rates to solve for leakage multiplication and the spontaneous fission rate, a few additional considerations can be explored.

In Sect. 17.4.3, it was discussed that the excess variance (Y) increases as the leakage multiplication increases. This concept is shown in Fig. 17.14, for which it was assumed that $N_\alpha = 0$, and values from Table 13.2 were used, assuming that spontaneous fission is from ^{240}Pu and induced fission is from ^{239}Pu. Note that these results are independent of the spontaneous fission rate. It can be seen from this figure that the excess variance does indeed increase as leakage multiplication increases.

In Sect. 17.4.3, it was also discussed that for a random source ($M_L = 1$), the excess variance should be equal to 0; however, it is seen in Fig. 17.14 that when $M_L = 1$, the value of Y is not 0. (Note that the points on the far left of Fig. 17.14 are $M_L = 1$.) This concept is further explored in Fig. 17.15. Here, M_L is always equal to 1, and Y is shown as a function of efficiency for plutonium and ^{252}Cf. The excess variance is only truly 0 when the efficiency is also 0. The plutonium and ^{252}Cf data come from Table 13.2.

As discussed in Sect. 17.4.4 (and in particular, Table 17.8), the following parameters were treated as unknowns: ε, N_α, N_{sf}, M_L. It was assumed that the $\overline{\nu_\mu}$ terms were known, which may or may not be a good assumption depending on what information about the item is known and what types of detection systems, such as gamma detection, are available. Depending on the number of unknowns and equations available, one can of course solve for the $\overline{\nu_\mu}$ terms. However, here we

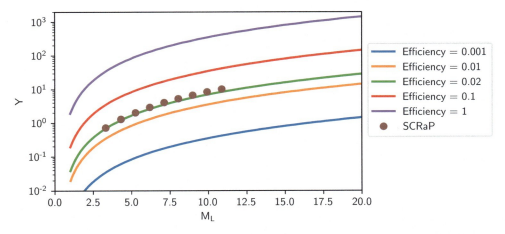

Fig. 17.14 Y versus leakage multiplication (M_L) for various detector efficiencies. Several measured data points from the SCRαP experiment [56] are shown. Those experiments had an efficiency very close to 0.02, as described in Ref. [56]

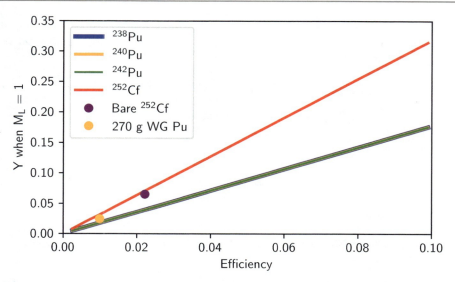

Fig. 17.15 Y versus efficiency for Pu and ^{252}Cf. Note that all plutonium nuclides give the exact same curve (so only the line for ^{242}Pu can be seen). Also shown are measured data points of ^{252}Cf [56] and 270 g of Pu [62]

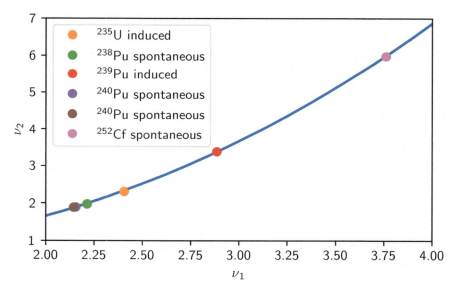

Fig. 17.16 $\overline{\nu_2}$ versus $\overline{\nu_1}$ for several nuclides. (Data are from Table 13.2)

will look at the sensitivity of multiplication and spontaneous fission rate results to the $\overline{\nu_\mu}$ terms for situations in which they are unknown.

As mentioned in Sect. 17.4.4, the $\overline{\nu_\mu}$ terms are given in Table 13.2. As previously noted, to get $\overline{\nu_2}$, one must divide the $\overline{\nu(\nu-1)}$ column in Table 13.2 by 2. Figure 17.16 shows the $\overline{\nu_1}$ and $\overline{\nu_2}$ terms from Table 13.2, in which it can be seen that one can fit these data (to provide an estimate of $\overline{\nu_2}$ given $\overline{\nu_1}$). Note that this exercise was also done in Ref. [29].

Using the fit from Fig. 17.16, we can study how M_L and N_{sf} are affected by the $\overline{\nu_\mu}$ terms. Both views in Fig. 17.17 show two curves; one of the curves uses constant $\overline{\nu_S}$ with varying $\overline{\nu_I}$, and the other curve uses constant $\overline{\nu_I}$ with varying $\overline{\nu_S}$. When constant values are used, they are from Table 13.2. When variable values are used, they span a range of 2–4 (which can be seen to encompass uranium, plutonium, and californium nuclides in Fig. 17.16). The $\overline{\nu_2}$ terms for both spontaneous and induced fission were determined from the equation given from the fit in Fig. 17.16. For both views in Fig. 17.17, the rest of the parameters are those given in Ref. [56] for the bare BeRP ball measured with two MC-15 detectors, and M_L and N_{sf} are calculated using Eqs. 17.33 and 17.28, respectively. It can be seen in the top view that for leakage multiplication, the results

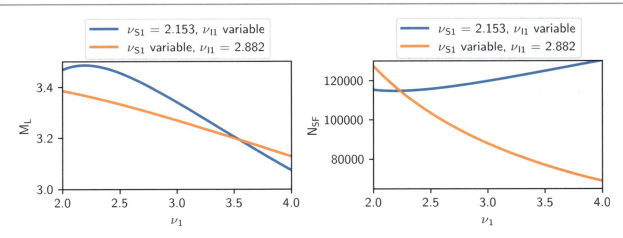

Fig. 17.17 (left) M_L versus $\overline{\nu_I}$ terms; (right) N_{sf} versus $\overline{\nu_I}$ terms. Both are for the bare BeRP ball configuration [56]

are not particularly sensitive to the $\overline{\nu_\mu}$ terms. When induced fission $\overline{\nu_\mu}$ is varied, the leakage multiplication values varied from ~3.05–3.5 (~14%), which may or may not matter depending on the knowledge of $\overline{\nu_{I1}}$ and the needed accuracy of the measurement. When spontaneous fission $\overline{\nu_\mu}$ was varied, the results mattered even less—from ~3.1–3.4 (~10%). On the right in Fig. 17.17, the sensitivity to the spontaneous fission rate (which as above, directly affects the mass) is shown. When induced fission $\overline{\nu_\mu}$ is varied, the change in F_S is not very dramatic (again ~12%); however, when spontaneous fission $\overline{\nu_\mu}$ is varied, the results change dramatically (almost by a factor of 2). This outcome is not surprising because N_{sf} and $\overline{\nu_{S1}}$ are inversely proportional, as shown in Eq. 17.28. From this equation, it can be seen that if it is desired to estimate N_{sf} (or mass, which is derived from this term), then $\overline{\nu_{S1}}$ must be fairly well known.

17.5 The Shift Register Coincidence Circuit

17.5.1 Principles of Shift Register Operation

Classical coincidence circuits require large corrections for electronic dead time because coincidence analysis begins with one event at t = 0 and continues until t = G, the gate length. If n events arrive within a time G, the first event will start the gate and the other n − 1 will be detected. A second gate cannot be started until a time of length G has passed, thus creating a dead time of that length.

An alternative approach is to store the incoming pulse train for a time G so that every event can be compared with every other event for a time G. In effect, every pulse generates its own gate; it is not necessary for one gate to finish before the next can start. This storage of events eliminates the dead-time effect described above and allows operation at count rates of several hundred kilohertz or more.

It is possible to store incoming pulses for a time G by means of an integrated circuit called a *shift register*. The circuit consists of a series of clock-driven flip flops linked together in stages. For example, a 64-stage shift register driven by a 2 MHz clock (0.5 μs/stage) defines a gate G of length 32 μs. Incoming pulses "shift" through the register one stage at a time (on each clock cycle), and the whole process takes 32 μs.

This dead-time-free shift register concept was introduced by Boehnel [24]. Versions of the circuitry have been developed by Stephens, Swansen, and East [63] and improved by Swansen [64, 65] and Lambert [66]. Currently, the shift register circuit is the most commonly used circuit for domestic and international coincidence-counting applications. Examples are given in Chap. 19. (Modern circuits do not use actual shift registers but instead use a pointer to locations in memory; however, to understand the operation, it is clearer to continue to describe the system as a shift register. Actual shift registers are the reason why gate lengths are often given in binary steps, such as 16 μs, 32 μs, and 64 μs, whereas they can now easily be set to any numerical value.)

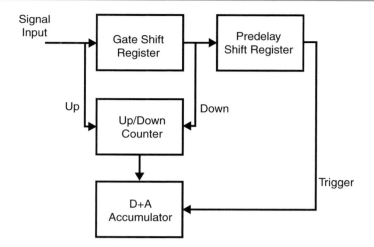

Fig. 17.18 A simplified block diagram of a shift register coincidence circuit that measures real + accidental (D + A) double events

Table 17.9 Number of recorded coincidences for different numbers of events in a group

For the following number of closely spaced events...	... the number of recorded coincidences will be
0	0
1	0
2	1
3	3
4	6
n	$n(n-1)/2$

17.5.2 The D + A Gate

Operation of the shift register coincidence circuit is best visualized by referring to the Rossi-α distribution of Fig. 17.5. This figure shows a prompt gate of length G that collects real and accidental doubles (D + A) and a delayed gate of length G that collects only accidental doubles (A). The two gates are separated by a long delay Δ. Note that coincidence counting does not begin until a short time interval P (the predelay) has passed. During this time—typically 1.5–6 μs for thermal-detector-based systems—the Rossi-α distribution is perturbed by pulse pileup and electronic dead times in the amplifiers, and the true coincidence count rate cannot be measured. Even more importantly, the accidental gate is not affected by these effects, and so the subtraction of (D + A) − (A) is biased negatively if a predelay is not used (see Sect. 17.6.2). After the predelay, the prompt D + A gate is defined by a shift register that is typically 32–64 μs long.

A simplified diagram of a shift register circuit that measures D + A events is illustrated in Fig. 17.18 [67]. The input is the logical OR of all of the amplifier-discriminator outputs, thus creating an autocorrelation circuit. Every input event passes through the gate shift register (length G) and then through the predelay shift register (length P). Every event that enters the gate shift register increments an up-down counter, and every event that leaves the gate shift register decrements the up-down counter. Thus, the up-down counter keeps an instantaneous tally of the current number of events in the gate shift register. Every pulse that leaves the predelay shift register causes the addition of the up-down counter tally to the D + A accumulator (trigger action).

The above sequence of events ensures that isolated, widely spaced events will never be registered in the D + A accumulator. However, if two events appear with a time separation greater than P but less than P + G, then one event will be in the shift register (and the up-down counter will have a count of 1) when the other event triggers the addition of the up-down counter tally into the D + A accumulator. Thus, a coincidence will be recorded, as required by Fig. 17.5. Note that if three or more events are present within the prescribed time interval, the counting algorithm will record all possible pairs of coincidences between events. See Table 17.9 for an example.

The possible permutations in counting twofold coincidences (Doubles) can exceed the number of events. A burst of four closely spaced pulses leads to six recorded Doubles.

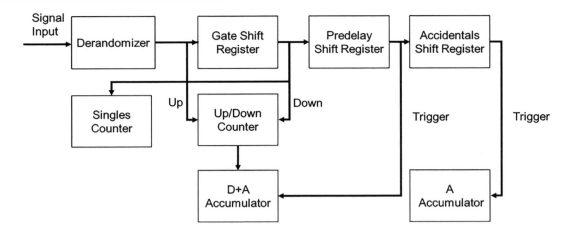

Fig. 17.19 Block diagram of a complete shift register coincidence circuit including singles, D + A, and A accumulators

The coincident events discussed above can represent two or more neutrons from one spontaneous fission (real fission event) or just the random overlap of background neutrons or neutrons from different fissions (accidental events). Thus, the counts accumulated by the circuit described above are called *D + A counts*.

17.5.3 The A Gate

Real doubles events (D) can be determined indirectly by adding a second complete shift register circuit that separately measures only accidental events (A). This circuit is identical to the D + A circuit except that a long delay (length Δ) is introduced between the shift register that defines the A gate and the input event that triggers the contents of the up-down counter into the A accumulator. The long delay is usually long compared with the detector die-away time so that no neutrons that are correlated to any event near t = 0 are still present, as illustrated in Fig. 17.5. A common choice for Δ is approximately 1000–4000 μs, which is very long compared with typical die-away times of 30–100 μs. When the long delay is many times the die-away time, the A accumulator will record only accidental coincidences, including random background events, uncorrelated overlaps between fission and background events, and uncorrelated overlaps between different fission events. The number of accidental events registered in the A accumulator will be—within random counting fluctuations—the same as the number of accidental events registered in the D + A accumulator if both the A and the D + A shift registers are exactly the same length in time. Then the net difference in counts received by the two accumulators is the net real doubles count (D), which is proportional to the fission rate in the sample.

In practice, the circuit that measures accidentals can be formed by introducing a second delayed shift register circuit or by introducing a second delayed trigger. The latter approach is used in recent circuit designs [64–66] for simplicity and because it ensures A and D + A gates of the same length. Figure 17.19 shows a block diagram of a recent shift register coincidence circuit design that includes singles (S), D + A, and A accumulators. The contents of the original up/down counter are added to the A accumulator whenever a pulse leaves the accidentals shift register. The Singles, defined in Chap. 16, are the total number of individual counts collected during the measurement or the equivalent rate. If a shift register coincidence circuit is operating with a 4 MHz clock, then each stage of the shift register corresponds to 250 ns. If two input pulses arrive within 250 ns, then one of them would be lost because one stage can hold only one pulse. To prevent this loss, a derandomizer [68] is used to "smooth out" bursts of pulses by spreading them out in time. (The pulses are spread out over a very small time compared with the detector die-away time, so the results are not significantly perturbed.) The derandomizer is placed at the input to store pulses (for example, up to 16) until space is available in the shift register. In the case of the advanced multiplicity shift register (AMSR), the derandomizer records pulses at up to 16 MHz and puts them into the shift register at 4 MHz [68]. With a derandomizer, a conventional shift register can operate at count rates approaching 2 MHz with virtually no synchronizer counting losses.

The A accumulator records accidental coincidences between the total neutron events recorded. The following relationship is true within random counting fluctuations:

17 Principles of Neutron Coincidence Counting

$$A = GS^2, \quad (17.35)$$

where A and S are expressed as count rates, and G is the coincidence gate length [69]. This relationship can be understood by considering that the average number of pulses in the gate is SG and that they are accumulated at the trigger rate S. This nonlinear relationship shows that A will exceed S when the total count rate is greater than 1/G. By means of Eq. 17.36, it is possible to calculate A rather than measure it; however, it is better to measure A with the circuit described previously because this method corrects continuously and automatically for any change in the singles neutron count rate during the assay. Equation 17.36 can then be used later as a diagnostic check for count-rate variations or instrument performance.

The AMSR introduced a possibility to improve the precision of the measurement of the accidental rate, which was done by decoupling the sampling of the D + A and A gates. As described above, the D + A and A gates are traditionally sampled at the input counting rate, which is necessary for the D + A gate but not for the A gate. In the AMSR, the A gate can be sampled at the clock speed, much faster than the sampling of the D + A gate, which improves the precision of the A rate, thereby reducing count times for all coincidence and multiplicity measurements. The count times are reduced by roughly a factor of 2 for a given precision. See Sect. 17.8.1 for details of the effect on measurement uncertainty.

17.5.4 Net Coincidence Response D

The net coincidence response D, or Doubles, contains a contribution from any group of pulses that number more than 1 in the gate period. They are the second moment of the frequency distribution directly accumulated in the shift register. This second moment can also be calculated from the measured multiplicity distribution in a multiplicity shift register (see Chap. 18). The two methods of calculation should lead to identical values.

From Fig. 17.4 and the above discussion, the ideal shift register coincidence response is related to the measured accumulator outputs by the equation

$$D = \frac{(D+A)\text{scaler} - (A)\text{scaler}}{e^{-\frac{P}{\tau}}\left(1 - e^{-\frac{G}{\tau}}\right)\left[1 - e^{-\frac{\Delta+G}{\tau}}\right]}. \quad (17.36)$$

The doubles gate fraction, f_D, is given by $f_D = e^{-P/\tau}(1 - e^{-G/\tau})$ and represents the fraction of the area under the whole Rossi-α curve (Sect. 17.2.3) that is recorded in the shift register. (Only the interval from P to P + G is recorded.)

(Corrections for amplifier dead times are given in Sect. 17.6. The term $[1 - e^{-(\Delta + G)/\tau}]$ should be very close to unity if the long delay Δ is much longer than the detector die-away time τ. Consequently, this term will be dropped in the following discussions.)

In Eq. 17.36, D represents the total number of coincidence counts that could be obtained if finite predelays, gate lengths, or delays were not required. In practice, it is customary to keep P, G, and Δ fixed and allow for their effects in the process of calibration with known standards. Then D = (D + A) accumulator—(A) accumulator is considered to be the true, observed coincidence response. An important equation that relates D to the physical properties of the sample (without multiplication), the detector, and the coincidence circuit can be derived from Eqs. 17.2, 17.12, and 17.14 [24, 70]:

$$D = m_{240}I_{240}\varepsilon^2 f_D \sum_\nu P(\nu)\frac{\nu(\nu-1)}{2!} = m_{240}I_{240}\varepsilon^2 e^{-P/\tau}\left(1 - e^{-G/\tau}\right)\sum_\nu P(\nu)\frac{\nu(\nu-1)}{2!}, \quad (17.37)$$

where

D = true coincidence count rate
m_{240} = ^{240}Pu-effective mass of the sample
I_{240} = the specific fission rate of ^{240}Pu (around 473 fissions/s/g),
ε = absolute detector efficiency
f_D = Doubles gate fraction
ν = spontaneous fission neutron multiplicity
$P(\nu)$ = multiplicity distribution
P = predelay

G = coincidence gate length
τ = detector die-away time.

Equation 17.37 illustrates again that the response of the shift register circuit to ν closely spaced events is proportional to $\nu(\nu-1)/2$. In Sect. 17.5.2, it was shown that the expression $\nu(\nu-1)/2$ represents the sum of all two-fold coincidences for ν closely spaced events; thus, the shift register collects all possible valid coincidences. However, the sample self-multiplication effects described in Sect. 17.10 affect shift register circuits more than conventional circuits, so the shift register circuits require larger correction factors.

Equation 17.37 provides a means of determining the detector die-away time τ. If the same sample is assayed in the same way at two different gate settings G_1 and G_2 (where G_2 is twice G_1) with the coincidence results D_1 and D_2, respectively, then

$$\tau = -\frac{G_1}{\ln\left(\frac{D_2}{D_1}-1\right)}. \tag{17.38}$$

17.6 Dead-Time Corrections for the Shift Register

In the preceding section, it was shown that the coincidence gate length G does not introduce dead times into the shift register circuit, which permits operation at count rates well above 100 kHz; however, at such high rates, multiple smaller dead times associated with the analog and digital parts of the circuitry become apparent. These times include

- detector charge collection time
- amplifier pulse-shaping time
- amplifier baseline restoration time
- losses in the discriminator OR gate, and
- shift register input synchronization losses.

These dead-time effects can be studied with time-correlated californium neutron sources, with uncorrelated AmLi neutron sources, and with digital random pulsers [71]. Even though the dead times can often be studied singly or together, the total effect is difficult to understand exactly because each dead time perturbs the pulse train and alters the effect of the dead times that follow. This section summarizes what is presently known about these dead times. Overall empirical correction factors are given, and several electronic improvements that reduce dead time are described.

17.6.1 Detector and Amplifier Dead Times

For most shift-register-based coincidence counters in use today, the analog electronic components consist of gas-filled proportional counters, charge-sensitive preamplifiers, amplifiers, and discriminators. As described in Chap. 15, Sect. 15.2, a charge signal can be obtained from the gas counter within an average time of 1–2 μs after the neutron interaction. This time dispersion is limited by variations in the spatial position of the interaction site and is not actually a dead time; however, the ability of the detector to resolve two separate pulses will be comparable to the time dispersion. The preamplifier output pulse has a rise time of ~0.1 μs, and the amplifier time constant is usually 0.15 or 0.5 μs. If all of the electrical components listed above are linked so that one preamplifier and one amplifier with 0.5 μs time constant serve seven gas counters, a total dead time of ~5 μs is observed [72]. In practice, this dead time is reduced by using multiple preamplifier-amplifier chains, as described in Sect. 17.6.4.

The amplifier output enters a discriminator that consists of a level detector and a short one-shot. The one-shot output is 50–150 ns long.

17.6.2 Bias Resulting from Electronic Effects

In addition to actual dead times, the electrical components can produce a bias in the shift register output. Bias is defined as the difference between the D + A and A counting rates when a random source, such as AmLi, is used. For a random source, the difference (D + A)—A should be zero; if it is not, the percent bias is 100 D/A. Possible sources of bias include electronic noise, uncompensated amplifier pole zero, shift register input capacitance, a dead time longer than the predelay P, or amplifier baseline displacement following a pulse, which is the most important source of bias if the electronic components are properly adjusted to minimize the other sources. Any closely following pulses that fall on the displaced baseline before it is fully restored to zero have a different probability of triggering the discriminator. Bias that results from pulse pileup is proportional to the square of the count rate and can become noticeable at high count rates. If the baseline is not fully restored in a time less than the predelay time, the effect will extend into the D + A gate, and a bias will result.

Figure 17.20 [73] illustrates a bias measurement as a function of predelay. The measurement used a coincidence counter with six amplifier channels. The observed bias was reduced to an acceptable value of 0.01% or less for predelay settings of 4.5 μs or more. These results are typical for well-adjusted electronics. For some high-efficiency and long die-away-time counters that operate at rates above 100 kHz, a conservative predelay setting of 6–8 μs may be warranted, but in general, 4.5 μs is sufficient. Modern (twenty-first century) electronics can operate with a predelay of 1.5 μs. At high count rates, D is typically on the order of 1% of A; a pulse pileup bias of 0.01% in D/A implies a relative bias of 1% in D—a bias that is only barely acceptable.

17.6.3 OR Gate and Synchronizing Dead Times

Because of the dead time in the amplifier-discriminator chain, it is customary to divide the detector outputs of a coincidence counter among four to six amplifiers. Each amplifier channel may serve three to seven detectors. The discriminator outputs of each channel are then "OR-ed" together before they enter the shift register. Now the dead time after the OR gate is much less than before if the two events are from different channels. The dead-time contribution of the OR gate itself can be calculated under the assumptions that no losses occur within a channel because of the longer preceding amplifier dead time and losses between channels are due to pulse overlap.

$$OR\ gate\ overlap\ rate = \frac{n(n-1)}{2!} 2(disc.output\ width)\left(\frac{S}{n}\right)^2, \qquad (17.39)$$

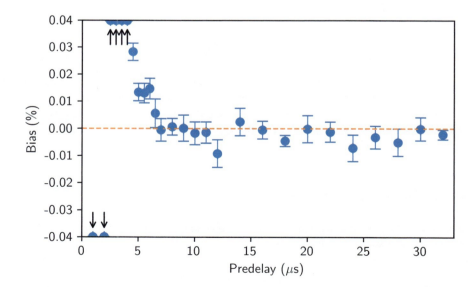

Fig. 17.20 Shift register coincidence bias D/A as a function of predelay P for electronics with 0.5μs time constant, as measured with strong random AmLi neutron source. For this measurement, bias was minimized by using optimum values of 100 kΩ for the amplifier pole-zero resistance and 68 pF for the shift register input synchronizer capacitor. Sensitivity to any remaining bias was maximized by using an 8μs coincidence gate G for the measurements [73]

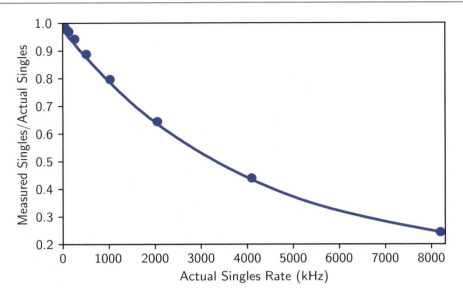

Fig. 17.21 Shift register synchronizer dead time as measured with a digital random pulser attached directly to the synchronizer input. The shift register clock period is 500 ns, and the digital random pulser has a pulse-pair resolution of 60 ns

where n is the number of channels and S is the total count rate. The ideal dead time for an OR gate that is accepting 50 ns wide pulses is then

$$OR\ gate\ deadtime = \frac{n-1}{n}(50\ ns). \tag{17.40}$$

This dead time is for total events; the coincidence dead time has not been calculated but would be larger. This OR-gate dead time can be eliminated by use of a multi-input derandomizer [68].

The output of the OR gate is a digital pulse stream that enters the shift register. At this point, the 50 ns wide pulses must be synchronized with the 500 ns wide shift register stages. The limit of one pulse per stage means that some closely following pulses will be lost unless a derandomizing buffer (Sect. 17.6.6) is used. These losses have been measured with a digital random pulser, as illustrated in Fig. 17.21. The ratio of measured Singles (S_m) to input Singles (S) is given by

$$\frac{S_m}{S} = \frac{1-e^{-pS}}{p}, \tag{17.41}$$

where p is the shift register clock period (500 ns in this case), and S is the total input rate [74]. At low rates, Eq. 17.41 yields a non-updating dead time of p/2; at high rates, the dead time approaches p. The coincidence dead time is on the order of 2p, as described in Ref. [74]. In general, the synchronizer dead time is small compared with the amplifier dead time, but it can be appreciable at high count rates. For example, at 256 kHz, the totals losses will be 6%, and the corresponding coincidence losses will be larger.

17.6.4 Empirical Dead-Time-Correction Formulas

Under the assumption that the electronic components have been adjusted so that bias is negligible, as discussed in Sect. 17.6.2, commonly used empirical dead-time-correction equations are

$$S(corrected) = S_m e^{\delta_S S_m} \tag{17.42}$$

$$D(corrected) = D_m e^{\delta_D S_m} \qquad (17.43)$$

where S_m is the measured totals rate, and D_m is the measured coincidence rate, (D + A) accumulator—(A) accumulator. Note that in Eqs. 17.42 and 17.43, the argument of the exponential contains S_m instead of the corrected rate S that appears in Eq. 17.13. The use of S_m is a convenient approximation at rates up to ~100 kHz. At higher rates, this approximation forces δ_s and δ_D to become functions of the count rate rather than constants. In typical implementations, the following expressions have been used,

$$\delta_S = \frac{A + BS_m}{4} \text{ and } \delta_D = A + BS_m \qquad (17.44)$$

where A and B are constants. The factor of 4 between the expressions has often been shown in simulations, but no sound theoretical basis exists. Constants A and B can be found experimentally. To have the same dead-time correction factor as a function of count rate, we need the exponential term d·S (from Eq. 17.13) to be equal to (A + B·S_m)·S_m/4 (from Eq. 17.42) (where d = A/4). With first-order series expansions, we obtain an approximate relationship between A and B:

$$B = \frac{A^2}{4} \qquad (17.45)$$

17.6.5 Dead-Time Determination

17.6.5.1 Addition of Random Sources

The total effect of the analog and digital dead times described above has not been calculated but can be determined empirically with californium and AmLi neutron sources. The coincidence dead time δ_D can be determined by placing a californium source in a fixed location inside a well counter and measuring the coincidence response while stronger and stronger AmLi sources are introduced. During these measurements, it is important to center the sources (so that all detector channels observe equal count rates) and to keep the sources well separated (so that scattering effects are minimized). The result of such a measurement is shown in Fig. 17.22. Within measurement uncertainties, the overall coincidence dead time is well represented by the updating dead-time equation (Eq. 17.13). The singles dead time δ_S can be measured by the source addition technique, where two

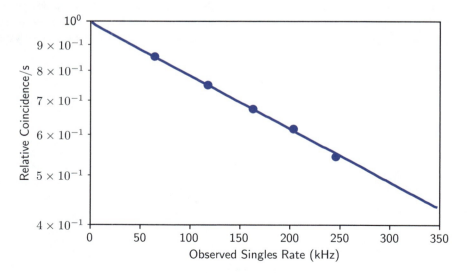

Fig. 17.22 Semilogarithmic plot of relative coincidence response from a californium source as a function of increasing totals count rate that results from additional AmLi sources. The points are measured values; the line is a least-squares fit to an exponential with dead-time coefficient $\delta_D = 2.4 \mu s$ [69]

californium or AmLi sources are measured in the counter, separately and then together. An updating dead-time equation also works well for the singles count-rate correction. Bias can be measured by placing only random AmLi sources in the counter.

17.6.5.2 Paired Sources

In this method, two sources (usually ^{252}Cf) are measured separately and together. The sum of the counting rates (both Singles and Doubles) of the individual sources will be larger than the counting rates from the combined sources due to dead time. After dead-time correction, the rates should be equal. The dead-time constants can be found by fitting the results to Eqs. 17.42, 17.43, and 17.44. This measurement needs to be carried out very carefully to get reliable results. For the measurement of each source separately, a dummy source capsule should be used in the position of the other source to ensure that any scattering effects are reproduced. One useful trick is to first measure source A and a dummy capsule together, then replace the dummy capsule with source B for the combined measurement, and finally, replace source A with the dummy capsule. In this way, the true sources never need to be repositioned. (The result is less sensitive to the precise positioning of the dummy capsule). The sources should be placed as symmetrically as possible to reproduce the relative counting rates seen in each preamplifier channel in an actual measurement.

17.6.5.3 Ratio of Two Known Sources

If the intensity ratio of two sources is well known, then the dead-time-corrected counting rates can be adjusted to obtain the known ratio.

17.6.5.4 Rossi-α Distribution

The recovery of the counting rate after a pulse is recorded in the very early times of the Rossi-α distribution as described in Sect. 17.2.3. For a system with ideal dead time, the recovery would be a step function following the initial pulse after a time d. A more realistic picture would be as shown in Fig. 17.23, where d is the length of the equivalent step function. The dead time, d, shown on these figures is the singles dead time (equivalent to the first term in Eq. 17.45, A/4).

This method of dead-time determination has the advantage over the other methods because it can be carried out at very low counting rates—perhaps even without the presence of a source—although in that case, the measurement time may be extremely long.

Figure 17.24 shows the Rossi-α distribution for a two-channel system. The second channel is not affected by the presence of a pulse on the first channel, so its counting rate is unaffected. Similar pictures can be obtained for n-channel systems. (At very early times, a small additional dead time (not shown here) could occur depending on the electronics [OR-gate or derandomizer].)

17.6.5.5 Dead-Time Measurement by Switching

A convenient method of dead-time measurement [23] can be built into the detector electronic design. By incorporating a switch into the preamplifier inputs of a system of 2 N preamplifier channels, the full input rate can be switched to only N preamplifier channels, and the measured count rates in the two cases give two equations that give the system dead time. One

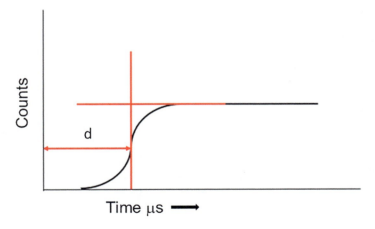

Fig. 17.23 The counting rate of a single-channel counting system following a pulse, which indicates the Singles dead time

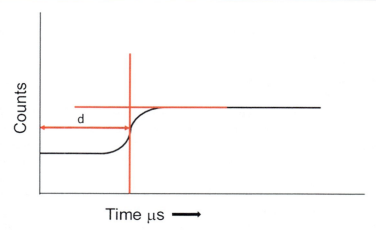

Fig. 17.24 The counting rate of a dual channel counting system following a pulse, which indicates the singles dead time

important feature of this method for high-count-rate items is that the dead time can be determined using the actual count rate of the item (rather than extrapolating from the [smaller] counting rates of ^{252}Cf sources).

17.6.5.6 Zero Dead-Time Method

As mentioned in Sect. 17.6.1, the charge collection time in the detector is not really a dead time. Two simultaneous events will deposit roughly twice as much charge as a single event, but the standard processing electronics treat this aggregate pulse as a single event over threshold. An alternative approach [75] is to use a gated integrator to measure the amount of deposited charge and generate the corresponding number of pulses to enter into the pulse train (in a similar way to a derandomizer; see Sect. 17.6.6). This *zero-dead-time* method eliminates a large fraction of the dead time of conventional processing electronics. It would work exactly only if the pulse-height spectrum of the detector was a delta function. Experimental tests are promising.

17.6.6 Typical Electronics and Derandomizing Buffer

Many typical neutron coincidence instruments use the AMPTEK A-111 preamplifier manufactured by AMPTEK, Inc., of Bedford Massachusetts. AMPTEK A-111 hybrid circuits have a charge-sensitive preamplifier, an amplifier with a bipolar output, a discriminator set to provide 50 ns output pulses, and a pulse-shaping circuit. They provide sufficient gain and signal-to-noise ratio if the ^{3}He tubes are operated at around 1680 V. The AMPTEK time constant is set to ~150 ns, which is the minimum recovery time before they can provide another output pulse. Each AMPTEK is mounted on a small circuit board that also provides an output pulse to drive a light-emitting diode that flashes when a neutron is detected and connections for "OR-ing" multiple channels together. Six channels of A-111 units can be operated with a reduced predelay of 3 μs with less than 0.01% bias.

Another more recent electronic package is the KM200 [76], which is a configurable preamplifier/discriminator unit suitable for use with a wide range of neutron detectors. The shaping time and sensitivity of the device can be optimized for demanding applications, such as spent fuel, and thus could improve the measurement performance of existing systems where conventional electronics do not perform well. The KM200 includes an analog front end (high-voltage decoupling, charge sensitive preamplifier, and bipolar shaper) and a sophisticated discriminator, which includes a comparator with threshold adjustment, double-pulsing suppression circuit, output logic pulse-width adjustment, OR logic input for daisy-chaining of devices, and an output logic pulse buffer. The KM200 dual-channel architecture design option expands the measurement capabilities of existing front-end electronics and detectors by increasing the counting-rate range and providing real-time diagnostics. The design is based on modern bipolar (no complementary metal oxide semiconductor technology) off-the-shelf components in standard footprints used in the communications industry, which mitigates the risk of components becoming obsolete or scarce.

To get a rough picture of dead time in a multiple-channel system, we can take the following example. If we have six identical channels—each with an input rate of 10,000 cps—and we assume that because the charge collection time in the preamplifier pulses closer than 1 μs cannot be separated, then we will have an output rate (in each channel) of 9900.5 cps ($10,000 \exp(-1 \times 10^{-6} \times 10,000)$), giving a total output counting rate over six channels of 59,403 cps. Treating these channels as a single system with an input of 60,000 cps and an output of 59,403, this system has a singles dead time of 0.167 µs. Each original channel has a singles dead-time effect of ~1%, so the summed system also has a dead-time effect of ~1%; however, because the counting rate in the overall system is six times higher, the effective dead-time constant is six times lower. From the discussion following Eq. 17.45, the coincidence dead time is four times higher, and so we have a value for the coincidence dead-time constant δ_D of $4 \times 0.167 = 0.67$ µs. This value also corresponds to A in Eq. 17.45. The coincidence dead-time effect on such a system at 100 kHz is ~7%.

The dead-time coefficient depends weakly on the detector gas mixture and strongly on the number of amplifier channels available. The number of detector tubes per amplifier channel has no measurable effect on the coefficients, although this situation may change if the detector tubes are subject to count rates in excess of ~20 kHz per tube.

The derandomizing buffer holds pulses that are waiting to enter the shift register, thus eliminating the input synchronization losses described in Sect. 17.6.3. Input pulses separated by less than 0.5 µs—the shift register clock period—are stored in a 16-count buffer until the shift register can accept them. This circuit eliminates the coincidence dead time of roughly 1.0 µs associated with the shift register input and permits counting at rates approaching 2 MHz with virtually no synchronizer counting losses. However, as the derandomizing buffer stretches pulse strings out in time, it may create strings longer than the predelay and thereby produce a bias. Because the AMPTEK A-111 amplifier requires a predelay of 1.5–3 µs, the maximum recommended singles rate for less than 0.01% bias is 500 kHz, which permits passive assays of almost any plutonium item—with criticality safety of the item in the well being the only limit.

17.7 Effect of Double Pulsing

Double pulsing is a phenomenon that occurs when an extra pulse that does not correspond to a neutron arrival is recorded. These spurious pulses are usually temporally correlated with a real neutron detection. The most common origin of double pulsing is due to a mismatch between the timing of the electron-ion collection within the ^{3}He tube and the pulse-shaping electronics. Depending on the path of the proton-triton pair after neutron capture by the ^{3}He, the resultant charge collection from their ionization can produce a two-peak structure in the pulse, resulting in a double trigger if the time constant of the shaping amplifier is shorter than the temporal separation of the two peaks [77]. Although increasing the time constant of the shaping amplifier to be the same length as or longer than the pulse duration would remove this source of double pulsing, it would also negatively impact the dead time of the system. A second source of double pulsing can be due to electronic feedback in the preamplifier or cross-talk between different channels if the transistor-transistor logic pulse from a preamplifier is routed near the sensitive input side. This second source of double pulsing can be entirely eliminated with appropriate detector and electronics design.

Double pulsing has been identified in commercial systems as early as 1990 [78] and continues to be observed in coincidence counters [79–83]. Because most origins of double pulsing are due to the charge-collection mechanism, double pulsing varies from system to system depending on preamplifier settings, ^{3}He tube dimensions, gas mix and pressure, applied high voltage, and gain. A common way to explore double-pulsing effects in a system is to increase the high voltage, which effectively increases the gain, making the discriminator more likely to retrigger on structures in the pulse shape.

Observed double pulsing usually occurs approximately 1 µs after the first trigger on the pulse. In a Rossi-α distribution, this effect manifests as a peaked distribution of time intervals around 1 µs, as seen in Fig. 17.25 (left). The area of this peak (A_p) can be used to determine the double-pulsing fraction, r, defined as the fraction of pulses that do not correspond to an actual neutron detection but are triggered by one

$$r = \frac{A_p}{S_m - A_p} = \frac{A_p}{S_0}, \qquad (17.46)$$

where S_m and S_0 are the singles rates with and without double pulsing, respectively [80].

Despite attempts to identify and quantify double pulsing with standard shift register approaches [82], it is difficult without list-mode data or at least a Rossi-α distribution. Assumptions that using a predelay longer than the characteristic double-pulsing time will eliminate any measurement effects are unsubstantiated, which is evident when considering a neutron pulse train with double-pulsing events inserted, as in Fig. 17.25 (right). The spurious double-pulse events increase both the measured singles rate and the measured doubles rate, even when a predelay is used.

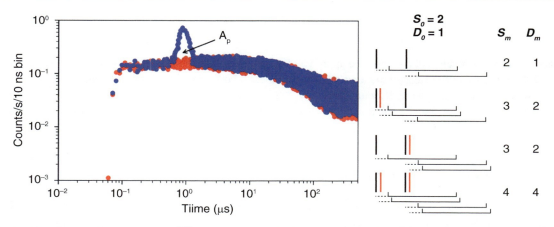

Fig. 17.25 (left) Two Rossi-α distributions of a ^{252}Cf source measured with an active well coincidence counter at normal operating voltage (1680 V) and a higher voltage to induce double pulsing (1760 V). The Rossi-α distribution of the measurement at 1760 V shows the characteristic double-pulsing peak with area Ap. (right) A pulse train with inserted double-pulsing (red) affects measured singles and doubles rates despite a predelay [80]

Equations that describe the effect double pulsing has on coincidence rates have been derived from first principles [82]. These principles can be used to correct measured singles and doubles rates (S_m and D_m) to the inherent rates without double-pulsing effects (S_0 and D_0) by using the double pulse fraction, r.

$$S_m = S_0(1+r) \tag{17.47}$$

$$D_m = D_0(1+r)^2 \approx D_0(1+2r). \tag{17.48}$$

These equations imply that a double-pulsing rate of 1% increases the observed singles and doubles rates by 1% and 2%, respectively. Double-pulsing rates up to 6% have been observed in systems when operating at high voltages [80, 83], which increases the measured singles and doubles rates by 6% and 12%, respectively. The effect on higher multiplicities is even greater [80], but no relationships have been derived for these multiplicities.

These equations do not consider the need to correct for both dead time and double pulsing, although reasonable results have been achieved by correcting for dead time using the standard methods and then proceeding with a double-pulsing correction [80, 82] by using the equations above.

17.8 Uncertainties Resulting from Counting Statistics

In principle, the effect of counting statistics on the coincidence response is very complex because the input pulse train contains both random and correlated events and because correlated events can overlap in many ways. Some of the complicating factors are described briefly in this section. For practical coincidence counters, these factors are not large, and it is usually possible to calculate measurement uncertainties for coincidence counting with the simple Eq. 17.50 given in Sect. 17.8.1.

The major factor that complicates measurement uncertainties is the nonrandom distribution of neutrons from spontaneous fission. Random neutrons from background or (α,n) events follow a Poisson distribution: for n counts, the variance is n and the relative error is $\sigma_n/n = \frac{\sqrt{\text{var}(n)}}{n} = \frac{1}{\sqrt{n}}$. However, if a spontaneous fission source emits a total of N neutrons in I fissions, with $N = \bar{\nu}I$, where $\bar{\nu}$ is the mean fission multiplicity, the relative error is $1/\sqrt{I}$ rather than $1/\sqrt{N}$. The number of spontaneous fissions follows a Poisson distribution, but the total number of single neutrons does not because the emission of more than one neutron per fission does not provide any more information to reduce the measurement uncertainty.

Boehnel [24] has shown that counting n spontaneous fission neutrons with an efficiency ε has a variance

$$\frac{\mathrm{var}(n)}{n} = 1 + \varepsilon \frac{\overline{\nu^2} - \overline{\nu}}{\overline{\nu}}. \tag{17.49}$$

If $\overline{\nu}$ approaches 1 or ε approaches 0, the variance approaches the Poisson distribution value of var.(n) = n but always remains larger. Eq. 17.49 implies that the measurement uncertainty will depend on the multiplicity of the fission source, the fraction of random events ($\overline{\nu} = 1$) present, and the detector efficiency. Other complicating factors are the detector die-away time and the total count rate, which affect the degree to which events overlap. Coincidence counting will then introduce additional complications.

17.8.1 Statistical Uncertainty for the Shift Register

As we saw in Sect. 17.5.4, the Doubles is calculated as the difference of the D + A accumulator and D accumulator. If the (D + A) and (A) registers are assumed to be uncorrelated and to follow the Poisson distribution, the relative error would be

$$\frac{\sigma_D}{D} = \frac{\sqrt{(D+A)+A}}{D} = \frac{\sqrt{D+2A}}{D}, \tag{17.50}$$

where D, (D + A) and A represent the accumulated counts in the accumulators. However, the correlated nature of the pulse train can modify this expression significantly [84], which leads to the inclusion of a scaling factor, ξ_D, that increases the doubles uncertainty.

$$\frac{\sigma_D}{D} = \frac{\xi_D \sqrt{D+2A}}{D}, \tag{17.51}$$

where the scaling factor for the doubles uncertainty, ξ_D, is given by

$$\xi_D = 1 + \frac{8D}{S} \frac{\gamma}{f_D},$$

with γ given by

$$\gamma = 1 - \frac{\left(1 - e^{-G/\tau}\right)}{G/\tau}. \tag{17.52}$$

Over the range of detection efficiency seen in safeguards instruments, the factor ξ_D varies from 1 to 5.

A shift register method was developed to improve the precision of the accidentals by a technique called *fast accidentals*, in which the D + A gate is sampled conventionally at the input rate, but instead of sampling the accidentals rate at the input pulse rate as shown in Fig. 17.19, it is sampled at the internal clock speed of the electronics. At the end of the measurement, the measured accidental sum is multiplied by the input rate and divided by the clock speed. The precision of the accidentals is then much smaller than the precision of the D + A in Eq. 17.50 and can be neglected, resulting in an improvement of a factor of $\sqrt{2}$ in the doubles uncertainty in most cases.

As an alternative to these theoretical expressions, the statistical uncertainty on D + A, A and D can be determined by multiple repeat measurements—often the normal procedure in typical measurements. The observed standard deviation (sample standard deviation) can be used as an estimate of the uncertainty on each of these quantities as an alternative to the theoretical expressions outlined above. Comparison of the sample standard deviations with these theoretical expressions confirms the validity of the equations. The sample standard deviations include variations (such as background changes) that are not included in the theoretical expressions and thus represent a more realistic value. It is good practice to compare both types of uncertainty to ensure that the observed uncertainty is not too much ($\lesssim$ 1.5 times) larger than expected. These sample standard deviations automatically account for techniques such as fast accidentals in the electronics.

(As an aside, the uncertainty on the accidentals is not usually needed to determine the measurement uncertainty, but Ref. [84] also contains a scaling factor for the uncertainty on the accidentals. This factor can increase the uncertainty by more than an order of magnitude over the square root of the number of accidentals as counting rates increase.)

Using Eq. 17.35, the approximate uncertainty Eq. 17.50 can be rewritten as

$$\frac{\sigma_D}{D} = \frac{\sqrt{D + 2GS^2}}{D\sqrt{t}} \qquad (17.53)$$

where D and S are dead-time-corrected count rates (Eqs. 17.42 and 17.43), and t is the count time. Because D is proportional to $(1 - e^{-G/\tau})$, the optimum value of gate length G that minimizes the relative error for a given die-away time τ can be derived by differentiating Eq. 17.53. The result is

$$G = \frac{\tau}{2}\left(e^{\frac{G}{\tau}} - 1\right) \approx 1.257\tau. \qquad (17.54)$$

The uncertainty as a function of gate width drops steeply from zero gate width, and then there is a wide, shallow minimum. The uncertainty rises only slowly as the gate width increases; thus, the uncertainty is not very sensitive to the actual value of the gate, especially if the gate is longer than the optimum value.

17.8.2 Uncertainties for Passive Counting

In passive neutron coincidence counting, the measured total response is proportional to $\varepsilon m_{240}t$, and the measured coincidence response is proportional to $\varepsilon^2 m_{240}t$, where m_{240} is the ^{240}Pu-effective mass, and t is the count time. The statistical measurement uncertainty (Eq. 17.50 or 17.53) is then proportional to

$$\frac{\sigma_D}{D} \propto \frac{\sqrt{k_1 m_{240} + 2k_2 G m_{240}^2}}{\varepsilon m_{240}\sqrt{t}}, \qquad (17.55)$$

where k_1 and k_2 are two constants of proportionality. For very small items, the relative error is proportional to $1/\sqrt{m_{240}}$; for large items, the relative error is independent of item mass. In either case, the relative error is proportional to $1/\varepsilon$, which implies that the efficiency of the passive well counter should be as high as possible.

The overall uncertainty on the measured mass includes the uncertainty on the plutonium isotopic composition because it is used to calculate the ^{240}Pu$_{eff}$ and the α ratio in some techniques (Sect. 17.11). The most reliable source of plutonium isotopic data is mass spectrometry. Larger uncertainties occur with high-resolution-gamma-spectroscopy-measured isotopic compositions (Chap. 9, Sect. 9.4). In all cases, the isotopic composition must be updated to the measurement date [85].

17.9 Measurement of ^{240}Pu$_{eff}$ Mass: Passive Calibration Assay

The most straightforward way to measure the ^{240}Pu$_{eff}$ mass of an item is to make a calibration curve of doubles rate versus known ^{240}Pu$_{eff}$ mass. The doubles rate is corrected for dead time and background. It is important for calibration purposes to use similar items that cover the mass range of interest (see Fig. 17.26). The Doubles plotted on the figure show the effect of neutron multiplication increasing the doubles rates beyond a linear response at higher masses; however, if the items used for calibration are truly representative of the item to be assayed, the results can be very reliable. The plutonium mass is obtained by dividing the ^{240}Pu$_{eff}$ mass by the ^{240}Pu$_{eff}$ fraction of the material, calculated from the isotopic composition. It is important to note that the doubles counting rate depends on the type of material (plutonium metal or oxide, for example); the geometry of the item (short, fat items have more neutron multiplication than tall, thin ones); the item density; and for items with high neutron multiplication, the presence of impurities that can give (α,n) neutrons. The method is less affected by the presence of small amounts of impurities than the known alpha technique (see Sect. 17.11).

The passive calibration curve method is one of several methods that have been incorporated in commonly used nondestructive assay software such as International Neutron Coincidence Counting (INCC; Ref. 86). The nonlinear functions that are used to fit the calibration data include polynomials, power curves, Padé functions, and exponentials. Depending on the calibration function used, the uncertainty on the measured ^{240}Pu$_{eff}$ mass can be determined by propagation from the statistical uncertainty on the Doubles [87].

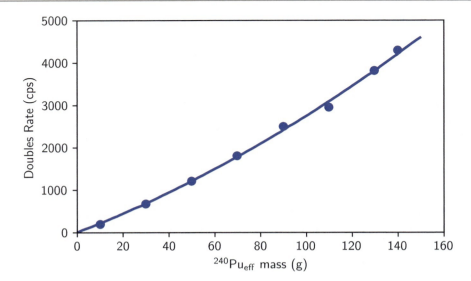

Fig. 17.26 Passive calibration curve from passive neutron doubles rate [70]

For cans of plutonium oxide powder, the results of this method were seen to change if the containers were shaken because the density of the powder—and therefore the neutron multiplication—had changed. This change led to the development of the known alpha technique (see Sect. 17.11), which makes a correction for the effect of neutron multiplication based on the singles and doubles counting rates from the item.

All of the calibration curves used in coincidence counting assume that the detector is operating correctly (as of the time of calibration). To check for correct operation, a *normalization measurement* is carried out. Usually, the doubles counting rate from a reference ^{252}Cf source is measured and compared with the doubles rate that was measured at the time of calibration, corrected for source decay. This ratio $K = D_{ref}/D_{now}$ is known as the normalization constant. It is usual to check that this measured ratio has sufficient statistical precision (e.g., <0.5%) and is within established limits (e.g., <3σ) to confirm that the detector is operating correctly. In some circumstances, it is appropriate to correct the singles count rate by $\sqrt{K}$ and the doubles count by K.

17.10 Effects of Sample Self-Multiplication

Among the effects that may perturb passive coincidence counting, self-multiplication of the coincidence response that results from induced fissions within the sample is usually dominant. This self-multiplication takes place in all plutonium samples and—to a lesser extent—in all uranium samples. Passive coincidence counters respond to induced fissions as well as to spontaneous fissions. Thus, the response from a given amount of spontaneously fissioning material is multiplied and appears to indicate more nuclear material than is actually present. This section describes the magnitude of this effect for plutonium and provides a self-multiplication correction factor that is useful for some assay situations.

17.10.1 Origin of Self-Multiplication Effects

Two common internal sources of neutrons induce fissions. One source is the spontaneously fissioning isotopes themselves. For example, neutrons emitted by ^{240}Pu may be captured by ^{239}Pu nuclei and induce these nuclei to fission. The spontaneous fission multiplicity ν_{sf} is 2.16, and the thermal-neutron-induced fission multiplicity ν_i is 2.88 (from Chap. 13, Table 13.1). The coincidence circuitry cannot, in practice, distinguish between these two multiplicities, so the total of both types of fissions is detected.

The other common source of neutrons is from (α,n) reactions with low-Z elements in the matrix. For example, in plutonium oxide, alpha particles from ^{238}Pu may react with ^{17}O or ^{18}O to create additional neutrons that may induce fissions in ^{239}Pu. The (α,n) neutrons, with multiplicity 1, do not in themselves produce a coincidence response; however, the induced fission

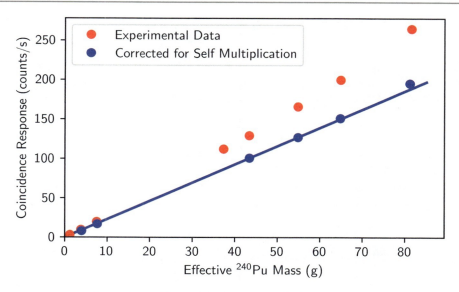

Fig. 17.27 Coincidence response of PuO$_2$ standards. The upward curvature in the data is due to self-multiplication in PuO$_2$. Monte Carlo calculations described in the text were carried out for all but the first and fourth samples to correct for self-multiplication, yielding a linear fit to the data

neutrons, with multiplicity $v_i = 2.88$, do. The magnitude of this coincidence response depends on the alpha emitter source strength, the low-Z element density, the degree of mixing between alpha emitters and low-Z elements, the fissile isotope density, and the geometry of the sample, and in general, is not proportional to the quantity of the spontaneously fissioning isotopes that are to be assayed.

The multiplication of internal neutron sources by induced fission is the same process that eventually leads to criticality. What is surprising is the appearance of multiplication effects in the assay of relatively small samples whose mass is far from critical. Even 10 g samples of plutonium metal show 5% enhancements in the coincidence response. At 4000 g of plutonium metal—considerably closer to criticality—the multiplication of the total neutron output is roughly a factor of 3, and the multiplication of the coincidence response is roughly a factor of 10.

The magnitude of self-multiplication effects on the passive coincidence assay of PuO$_2$ cans is illustrated in Fig. 17.27 [87]. The data show a definite upward curvature, and the deviation from a straight line determined by the smallest samples amounts to ~38% at the largest sample, 779 g of PuO$_2$. The following sections discuss other features of Table 17.10 that describe self-multiplication corrections applied to the data.

17.10.2 Calculational Results

Self-multiplication within a sample can be calculated by Monte Carlo techniques. The results of calculations done for the samples listed in Table 17.10 are given in Columns 5 through 9. These calculations were carried out with the Monte Carlo code described in Ref. [87]; however, the detector itself was not modeled in detail because it was necessary to obtain only the net leakage multiplication across a surface surrounding the sample. The Monte Carlo code selected initial (α,n) or spontaneous fission neutrons according to the ratio

$$\alpha = \frac{N_a}{\nu_s N_{sf}}, \tag{17.56}$$

where N_α is the number of (α,n) reactions, and N_{sf} is the number of spontaneous fissions. The values for α, obtained from Eq. 17.56 or 17.59, are given in Column 4 of Table 17.10. Each neutron-induced fission chain was followed to its end. The Monte Carlo code calculated the leakage multiplication M_L (defined in Chap. 16), which is related to the probability of fission, p, by Eq. 16.14 in that chapter.

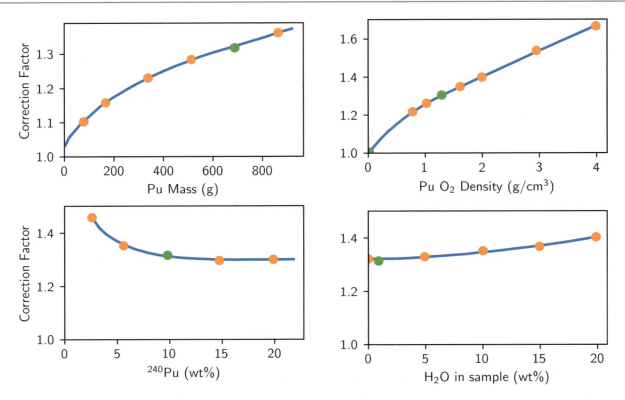

Fig. 17.28 Monte Carlo calculations of self-multiplication effects of various parameters for coincidence counting of PuO_2. The green data points denote the nominal calculation

The calculated values of M_L are given in Column 5 of Table 17.10. These values are the ratios by which the total (single) neutron count is enhanced by multiplication, with consideration for leakage, absorption, fission, and reflection. For simplicity, the leakage multiplication is denoted by M in the remaining discussion.

The Monte Carlo code also calculated a coincidence correction factor

$$CF = 1 + f_{sf} + f_{\alpha n}, \qquad (17.57)$$

where $1 + f_{sf}$ is the coincidence correction for net multiplication of spontaneous fission neutrons, and $f_{\alpha n}$ is the additional correction for net multiplication of (α,n) neutrons. In Table 17.10, Columns 6 and 7 show the relative size of these two induced-fission multiplication effects for the PuO_2 items measured. Column 8 shows the overall correction factor (CF), and Column 9 demonstrates that the corrected coincidence response per gram is now nearly constant.

Using the code described above, a series of reference calculations were made to determine the effect of sample mass, density, isotopic composition, and water content on the coincidence correction factor. The results are plotted in Fig. 17.28 [87]. All calculations represent variations about an arbitrary nominal sample of 800 g PuO_2, with a density of 1.3 g oxide/cm^3. This sample contains 706 g of plutonium at 10% $^{240}Pu_{eff}$ and 1 wt% water in a container that has an 8.35 cm inside diameter. For each calculation, only one parameter was varied from the nominal values. For the mass and density variations, the fill height was adjusted to conserve mass. For the H_2O content variation, the sample density was adjusted to conserve volume. Figure 17.28 shows that coincidence correction factors are appreciable even at low mass and low density.

The curves in Fig. 17.28 can be used to estimate coincidence correction factors for other similar plutonium oxide samples. The exact range of applicability is not known. For the samples in Table 17.10, the correction factors were calculated directly by the Monte Carlo code except for the first and fourth samples, which were extrapolated from Fig. 17.28 with consistent results.

Table 17.10 Self-multiplication correction factors for the plutonium oxide samples in Fig. 17.27

1	2	3	4	5[a]	6[a]	7[a]	8[a]	9[a]	10[b]	11[b]
Sample mass (g)	^{240}Pu effective (%)	Coincidence response (g − s)	α	Leakage multiplication (M_L)	f_{sf} see equation 17.57	$f_{αn}$ see equation 17.57	Correction factor (CF)	Corrected response (g − s)	From D/S ratio M_L	CF
20	6.0	2.35(2)	0.66				1.02(1)	2.31(3)		
60	6.4	2.42(2)	1.43	1.005	0.024	0.020	1.04(1)	2.32(3)	1.003	1.03
120	6.4	2.53(2)	1.36	1.010	0.049	0.035	1.08(1)	2.33(3)	1.012	1.08
480	7.8	2.99(3)	0.74				1.28(1)	2.34(4)	1.044	1.26
459	9.5	2.98(3)	0.64	1.046	0.192	0.068	1.26(1)	2.36(4)	1.048	1.28
556	9.9	3.03(3)	0.62	1.049	0.215	0.075	1.29(1)	2.35(4)	1.043	1.25
615	10.6	3.08(3)	0.60	1.056	0.260	0.084	1.34(1)	2.30(4)	1.052	1.30
779	10.4	3.26(3)	0.61	1.061	0.285	0.095	1.38(1)	2.36(4)	1.070	1.41

[a]Columns 5 through 9 are based on Monte Carlo calculations
[b]Columns 10 and 11 are based on the D/S ratio
[c]CF correction factor

17.10.2.1 Effects on Shift Register Response

It is possible to write expressions for the effects of sample self-multiplication on the shift register response. The total neutron count rate S, after subtraction of the background count rate, is given by

$$S = m_{240} I_{240} \varepsilon M \overline{\nu_{sf}} (1 + \alpha) \tag{17.58}$$

where m_{240} is the effective ^{240}Pu mass, I_{240} is the specific fission rate of ^{240}Pu (~473 fissions/s/g), ε is the detector efficiency, M is the leakage multiplication, $\overline{\nu_{sf}}$ is the spontaneous fission multiplicity, and α is defined by Eq. 17.56. If all other quantities are known, α can be determined by inverting Eq. 17.58:

$$1 + \alpha = \frac{S}{m_{240} I_{240} \varepsilon M \overline{\nu_{sf}}}. \tag{17.59}$$

The coincidence count rate D is given by the following equations [88]:

$$D = \frac{m_{240} I_{240} \varepsilon^2 (1 + \alpha \overline{\nu_{sf}}) \overline{\nu(\nu-1)}}{2} f_D, \tag{17.60}$$

where $\overline{\nu(\nu-1)}$ is the average moment per starting event [fission or (α,n)], given by

$$\overline{\nu(\nu-1)} = M^2 \left[\frac{\overline{\nu_{sf}(\nu_{sf}-1)}}{1 + \alpha \overline{\nu_{sf}}} + \frac{(M-1)}{\overline{\nu_i}-1} \frac{(1+\alpha)}{1 + \alpha \overline{\nu_{sf}}} \overline{\nu_{sf}} \, \overline{\nu_i(\nu_i-1)} \right], \tag{17.61}$$

and where $\overline{\nu_{sf}(\nu_{sf}-1)}$ and $\overline{\nu_i(\nu_i-1)}$ are the second moments of the spontaneous and induced fission multiplicity distributions.

Note that f_D represents the fraction of coincidences measured, $e^{-P/\tau}(1 - e^{-G/\tau})$. These equations assume that all fission chains produced from the original fission appear to be simultaneous within the resolving time of the coincidence counter. This assumption, called the *superfission concept* [24], is valid for thermal-neutron counters because of their long die-away time.

From Eqs. 17.58 and 17.60 for S and D and from Columns 5 and 8 of Table 17.2, it is apparent that sample self-multiplication affects coincidence counting (Doubles) more than singles counting. As a simple example of this effect, suppose that a spontaneous fission releases two neutrons, one of which is captured by a fissile nucleus that, in turn, releases three neutrons upon fissioning. The total number of neutrons has increased from two to four (M = 2); however, the coincidence response has increased from one to six (CF = 6). Thus, the ratio D/S has increased with multiplication. Laboratory measurements have shown that D/S can be used as a measure of multiplication. This ratio is the basis of the simple self-multiplication correction described in the following section.

17.11 Measurement of ^{240}Pu$_{eff}$ Mass: Known Alpha Assay

Because it is usually not possible to perform a Monte Carlo calculation to determine the self-multiplication of each sample to be assayed, a strong need exists for a self-multiplication correction that can be determined for each sample from the measured parameters D and S. As mentioned earlier, the ratio D/S is sensitive to sample multiplication so that it is possible to use D for the assay and D/S for a multiplication correction. The procedure for calculating this correction for plutonium samples follows:

Step 1: Assay a small 10 g or 20 g reference sample that, as an approximation, can be considered as non-multiplying. Use the same physical configuration and electronic settings as those to be used in Step 2 for assay of larger samples. This measurement yields the values D_0, S_0, and α_0. If the non-multiplying sample is pure metal, $\alpha_0 = 0$. Otherwise, α_0 can be determined from Eq. 17.59 with M = 1. (A multiplying reference sample can also be used if it is sufficient to obtain relative correction factors [89].

Step 2: Now assay an unknown multiplying sample that requires a self-multiplication correction. This measurement yields D and S. If the sample is pure metal, $\alpha = 0$. If the sample is of the same composition as the small reference sample used in Step 1, then $\alpha = \alpha_0$. If the sample is pure plutonium oxide, then by using Tables 13.1 and 13.3, it is possible to calculate

$$\alpha = \frac{13{,}400 f_{238} + 38.1 f_{239} + 141 f_{240} + 1.3 f_{241} + 2.0 f_{242} + 2690 f_{Am-241}}{1020(2.54 f_{238} + f_{240} + 1.69 f_{242})} \tag{17.62}$$

if the isotopic fraction f of each plutonium isotope and of ^{241}Am are known. Eq. 13.10 (Chap. 13) can be used to correct the calculated value of α for the presence of major impurities that have high (α,n) cross sections if the concentrations of these impurities are known. For inhomogeneous or poorly characterized plutonium oxide, scrap, or waste where α cannot be determined by one of the above methods, this self-multiplication correction cannot be used; thus, the name *known alpha* method:

Step 3: Calculate the ratio

$$r = \frac{\frac{D}{S}}{\frac{D_0}{S_0}} \frac{(1+\alpha)}{(1+\alpha_0)}. \tag{17.63}$$

The denominator of this equation is often referred to independently as ρ_0:

$$\rho_0 = \frac{D_0(1+\alpha_0)}{S_0} \tag{17.64}$$

The ratio r will be larger than 1 for multiplying samples with M > 1 because sample self-multiplication increases D more than S. The ratio r is independent of detector efficiency, die-away time, and coincidence gate length. Note that all count rates in Eq. 17.63 should be corrected for background and electronic dead times.

Step 4: The leakage multiplication M is obtained by solving the quadratic equation

$$K(1+\alpha)M^2 - [K(1+\alpha) - 1]M - r = 0, \tag{17.65}$$

where K is given by

$$K = \frac{\overline{\nu_{sf}\nu_i(\nu_i - 1)}}{\overline{\nu_s(\nu_s - 1)}\overline{\nu_i} - 1} \approx 2.062 \, (for\, plutonium)$$

Equation 17.65 can be derived from Eqs. 17.60 and 17.61 [90, 91].

Step 5: The coincidence counting correction factor for self-multiplication is M r. To summarize,

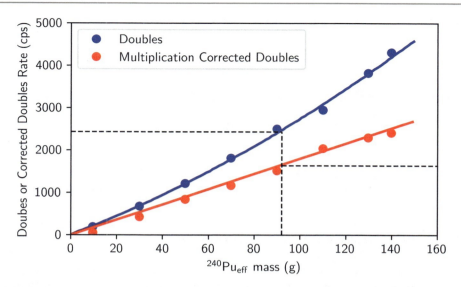

Fig. 17.29 Data from the high-level neutron coincidence measurement in Fig. 17.26 (blue) corrected for multiplication. A numerical example is given in the text

Table 17.11 Isotopic composition of a 538-gram plutonium item

Nuclide	^{238}Pu	^{239}Pu	^{240}Pu	^{241}Pu	^{242}Pu	^{241}Am
Percentage	0.055%	82.93%	16.4%	0.27%	0.344%	1.09%

$$S(corrected\ for\ mult.) = \frac{S}{M} \tag{17.66}$$

$$D(corrected\ for\ mult.) = \frac{D}{Mr}. \tag{17.67}$$

This self-multiplication correction has no adjustable parameters and is geometry independent. For example, when two plutonium samples are brought closer and closer together, M will increase, the induced-fission chain lengths will increase, the mean effective multiplicity will increase, and D/S will increase. Equation 17.65 will yield larger values of M, and Eqs. 17.66 and 17.67 will automatically yield larger correction factors. Examples are given in Ref. [87] and in Fig. 18.28. Figure 17.29 shows the result of applying this procedure to the data from Fig. 17.26, leading to the corrected points (*multiplication-corrected doubles*) lying on a straight line.

The following gives a numerical worked example. Take an item of pure PuO$_2$ with a plutonium mass of 538 g. The isotopic composition is given in Table 17.11.

Using Eqs. 17.1 and 17.62, we can calculate the m$_{240}$ = 92 g and α = 0.525. The measurement was made in a high-level neutron coincidence counter (see Chap. 19, Sect. 19.4.2), which has ρ$_0$ = 0.103. The dead-time-corrected and background-subtracted measured counting rates were S = 26,176 and D = 2434 cps. A value of 2434 cps on the doubles passive calibration curve corresponds to a m$_{240}$ mass of 92 g (Fig. 17.29). Solving Eq. 17.65 for M, we obtain a multiplication of 1.08 and a multiplication-corrected doubles value of 1634 cps. This value corresponds to a mass, m$_{240}$, of 92 g on the multiplication-corrected ("known-α") calibration curve (Fig. 17.29).

An important benefit arises from the possibility of a mass measurement from both the passive calibration curve (Sect. 17.9) and a result from the known alpha method based on the same measurement data. Agreement between the results from these two methods gives additional confidence in the result. If the results disagree, the cause should be understood before relying on either answer. Some potential reasons for perturbations to both methods are given in Sect. 17.13.

When Eqs. 17.66 and 17.67 are used to linearize the calibration curve so that

$$m_{240} = \frac{D}{kMr},\qquad(17.68)$$

then Eqs. 17.58, 17.63, and 17.68 require that the calibration constant k and the detector efficiency ε be related by

$$k = \varepsilon \bar{\nu}_s I_{240} \frac{D_0}{S_0}(1+\alpha_0).\qquad(17.69)$$

This relation is not important in practice because k is usually obtained by calibration, but it might provide a diagnostic to indicate whether the detector efficiency or the small reference sample have been properly measured.

17.12 Applications and Limitations of Known Alpha Assay

Although the self-multiplication correction factors given by Eqs. 17.66 and 17.67 provide a complete correction with no adjustable parameters, the following assumptions were made in the derivation to obtain simple equations:

- It was assumed that detector efficiency was uniform over the sample volume. This assumption is not always the case but is becoming easier to realize with instruments such as the upgraded High-Level Neutron Coincidence Counter (HLNCC-II) described in Chap. 19, Sect. 19.4.2.
- It was assumed that (α,n) neutrons and spontaneous fission neutrons had the same energy spectra so that the detection efficiency ε, fission probability p, and induced-fission multiplicity ν_i would be the same for both neutron sources. In general, this assumption is not the case, although for plutonium oxide, the (α,n) and spontaneous fission neutrons have similar mean energies (2.03 MeV and 1.96 MeV, respectively) but different spectrum shapes.
- It was assumed that all fission chains are simultaneous within the die-away time of the detector. This assumption is not true for neutrons that re-enter the sample from the detector (reflected neutrons) [24].

These approximations introduce errors into the correction. Values for M given by Eq. 17.65 may differ from values obtained from Monte Carlo codes. Values for the coincidence correction factor CF = Mr are usually better, presumably because some errors cancel in the use of ratios. The correction usually gives best fits of 2–3% to the data, which is good but is larger than the measurement relative standard deviation (on the order of 0.5%).

Applications of the simple self-multiplication correction are given in Table 17.10, Columns 10 and 11, and in Fig. 19.27 for PuO_2. Good results have been reported for PuO_2 in Ref. [66], for plutonium metal in Refs. [88, 91], and for breeder fuel-rod subassemblies in Ref. [92].

The above applications show that good results, typically 2–3%, can be obtained with the self-multiplication correction for well-characterized material despite the assumptions made in the derivation. However, the need to know α—the ratio of (α,n) to spontaneous fission neutrons—for each sample to be assayed poses a severe limitation on the applicability of the technique. For scrap, waste, impure oxide, or metal with an oxidized surface, α cannot be determined. Any error in the choice of α leads to an error of comparable size in the corrected assay value. In such cases, the multiplication correction should be used only as a diagnostic for outliers. For many classes of oxide, where α may be somewhat uncertain but sample density and geometry are fixed, Krick [93] has found that two-parameter calibration curves without self-multiplication corrections provide the best assay accuracy.

The fundamental limitation of the simple multiplication correction is that only two parameters are measured by each assay: D and S. The number of unknown variables is at least three: the sample mass, the sample self-multiplication, and the (α,n) reaction rate. Further improvements in multiplication corrections can be made if coincidence counters are built that provide a third measured parameter, such as triple coincidences (Ref. [67, 87] and Chap. 18 of this book).

17.13 Other Matrix Effects

The dominant matrix effect in passive neutron coincidence counting is usually the self-multiplication process described in Sect. 17.10. If corrections for electronic count-rate losses and self-multiplication can be properly applied, the coincidence response is usually linear with sample mass. However, other matrix effects can affect the assay and might be overlooked at times. These effects are summarized in this section, which is based in part on Ref. [94].

- (α,n) contaminants
  For plutonium samples, the most important (α,n) emitters are oxygen and fluorine. Fluorine concentrations of 10–400 ppm are typical, and oxygen (in water) may be as high as several percent. The calculated effect of fluorine and water on the total neutron count rate is given in Chap. 16, Sect. 16.2. Such (α,n) contaminants may bias the coincidence assay by a few percent. If their concentrations are known, the effects can be accounted for in the self-multiplication correction.
- Hydrogen content. The hydrogen in water affects the neutron coincidence response by shifting the neutron energy spectrum (see Chap. 16, Sect. 16.2). This effect increases the detector efficiency and the sample self-multiplication. The former effect can be minimized by careful detector design, and the latter is taken into account by the multiplication correction.
- Container wall effects. Neutron scattering and reflection by the container wall can increase detection efficiency and sample self-multiplication. An increase in the coincidence count rate of up to 7% has been observed. Container effects can be estimated by measuring a californium source with and without an empty sample can.
- Influence of uranium on plutonium assay. The addition of uranium to plutonium (as in mixed oxide) has the following effects: additional multiplication in ^{235}U; decrease in plutonium multiplication due to a "dilution" of the plutonium; and additional fast multiplication in ^{238}U. Despite different ^{239}Pu, ^{235}U, and ^{238}U fission multiplicities, the multiplication correction works well for mixed plutonium and uranium if no additional unknown (α,n) sources exist.
- Neutron moderation and absorption (self-shielding). In plutonium nitrate solutions, moderation that leads to increased neutron absorption has been observed. In active coincidence counting of uranium, neutron absorption and self-multiplication are both strong and opposing effects. The presence of both effects often yields nearly straight calibration curves (Chap. 20, Fig. 20.2).
- Neutron poisons. Boron, cadmium, and some other elements have high thermal-neutron capture cross sections and can absorb significant numbers of neutrons. Problems with neutron poisons have been observed in the active assay of fresh light-water-reactor fuel assemblies.
- Sample geometry. If the detection efficiency is not uniform over the sample volume, then the coincidence response can vary with sample geometry. Passive counters are now usually designed so that the whole sample will be in the region of uniform efficiency. For active coincidence counters, the source-to-sample distance is very important, and consistent positioning of samples is essential.
- Sample density. Variations in plutonium oxide density due to settling or shaking during shipping and handling can affect the passive coincidence response by as much as 10%.[1] The multiplication correction can take these variations into account for samples of similar composition if the samples are within the detector's uniform efficiency region. For active coincidence counting, density variations affect both self-multiplication and self-shielding. No correction is available.
- Scrap and waste matrices. Here, it is helpful to know what the matrix is and to know which of the above-mentioned effects might be present. For plutonium-bearing materials, the coincidence response is usually more reliable than the singles response but may provide only an upper limit on the quantity of ^{240}Pu. In general, it is useful to measure both the singles and the coincidence response (Doubles) and to use the singles response or the doubles/singles ratio as a diagnostic to help interpret the coincidence response.

---

[1] Private communication from F. J. G. Rogers, Harwell, United Kingdom, 1984.

17.14 Known M Assay

The known multiplication (M) method [95] is appropriate for the measurement of items with impurities that increase α. First, a ^{239}Pu-effective mass is defined in a similar way to the ^{240}Pu$_{eff}$ (Eq. 17.1) as follows:

$$^{239}Pu_{eff} = m_{239} = 0.786\,^{238}Pu + {}^{239}Pu + 0.515\,^{240}Pu + 1.414\,^{241}Pu$$
$$+ 0.422\,^{242}Pu + 0.545\,^{241}Am + 0.671\,^{235}U + 0.082\,^{238}U. \quad (17.70)$$

This quantity is a measure of the effective mass for induced rather than spontaneous fission. The constants depend on the fission cross section and multiplicity distribution of each isotope and therefore, the neutron spectrum.

This method assumes that α is unknown but that the multiplication is a known function of the effective ^{239}Pu mass. This function can be obtained by calibration with known standards or by Monte Carlo simulation.

Eliminating α from the point-model equations for the Singles and Doubles (Eqs. 17.58 and 17.60), we obtain

$$D = \frac{\varepsilon f_D M}{2}\left[k_M m_{239} I_{240} \varepsilon M \overline{\nu_{sf}(\nu_{sf}-1)} + \frac{M-1}{\nu_i - 1}\overline{\nu_i(\nu_i-1)}S\right] \quad (17.71)$$

where k_M is the ratio of the ^{240}Pu$_{eff}$ mass to the ^{239}Pu$_{eff}$ mass (m_{240}/m_{239}), calculated from the isotopic composition of the item, and the other symbols have their usual meanings. One possible formulation for the known M function is

$$M = 1 + b \cdot m_{239} + c \cdot m_{239}^2, \quad (17.72)$$

where b and c are calibration constants. Using this relationship, we can solve Eq. 17.71 iteratively for m_{239}, and then we can calculate m_{240} (from k_M), M (from Eq. 17.72), and α from

$$\alpha = \frac{S}{m_{240} I_{240} \varepsilon M \overline{\nu_{sf}}} - 1. \quad (17.73)$$

In addition to its use for impure items, the known M method can be used together with the known α technique to verify the isotopic composition of the item. For example, if the ^{241}Am amount was deliberately overstated, then the known α method would give a value for m_{240} that was higher than the true value (potentially allowing diversion of plutonium), but the known M method would not be affected, and the results (for both m_{240} and α) would be inconsistent, indicating an anomaly. The principal disadvantage of the method is the need to establish the relationship between M and m_{239}; however, it could be worthwhile in the case of items of fixed, well-defined geometry, such as mixed-oxide fuel assemblies.

17.15 Active Coincidence Counting

17.15.1 Introduction

For the assay of items where the passive neutron output is small (such as uranium items), an external neutron source can be used to induce fission. This process is called *active neutron interrogation*. A classical exposition is given in Ref. [96], and a more modern treatment is given in Ref. [97]. A very wide variety of applications exist using (pulsed) neutron generators or radioactive sources, together with many different kinds of detectors. In this section, we will limit ourselves to those methods that detect fission neutrons with neutron coincidence counting as described above. (Active neutron techniques for spent fuel are described in Chap. 21.)

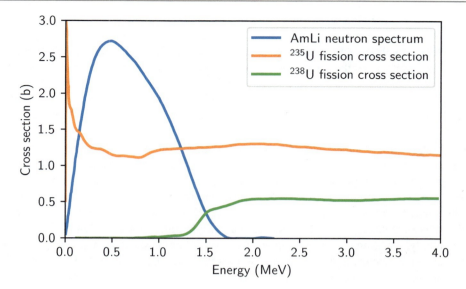

Fig. 17.30 AmLi neutron spectrum overlaid on uranium fission cross sections

17.15.2 Choice of Interrogation Source

The classical application of active neutron interrogation to safeguards has been the use of AmLi sources to induce fission because these sources emit neutrons singly, and the large majority of neutrons emitted are below the fission threshold energy of ^{238}U (see Fig. 17.30).

In addition, the americium has a very long half-life, although the source yield does depend on the intimate mixing of the AmO_2 with Li_2O (for example), and the yield may be anisotropic [98] and may change if the source is subject to rough treatment. The alpha activity of typical AmLi sources (that emit 10^4 to 10^5 n/s) is very large (~1 Ci /~10^{11} Bq), and the sources are usually housed in a tungsten capsule to reduce the potential dose from the intense 60 keV gammas. More recently [99], and partly because AmLi sources are difficult to procure, ^{252}Cf has been used. The disadvantages of ^{252}Cf are that it has a much higher fraction of source neutrons above the fission threshold of ^{238}U, a much shorter half-life, and the potential of ^{250}Cf buildup, which makes calculation of the source strength more uncertain for older sources. Another disadvantage is that the spontaneous fission neutrons from a ^{252}Cf interrogation source create a doubles background in the detector, which needs to be accounted for. However, there is a potential benefit: any interrogation source can emit a neutron that causes an induced fission in the item, and the neutrons that result have a certain probability of producing coincidence counts. However, when the interrogation source emits multiple neutrons simultaneously, the probability of a coincidence count can be increased in multiple ways. For example, one of the neutrons can induce a fission, and another neutron from the same spontaneous fission can be detected directly. In this case, the number of neutrons detected is then one greater than that of the induced fission event alone, which increases the probability of a coincidence event being recorded. The multiplicity of the event might change from 1 to 2 or from 2 to 3. Another case could be that two of the source neutrons could each cause a fission, which would lead to a high-multiplicity event. This enhancement of the coincidence probability compared with a single-neutron-emitting source is called *time-correlated induced fission* [100].

17.15.3 Fast and Thermal Modes

An important feature of active interrogation is the coupling between the external source and the fissionable material. Because the fission cross section for ^{235}U is much higher for thermal neutrons than for higher-energy neutrons, the induced fission rate can be greatly increased by moderating the neutron source and/or allowing thermal neutrons to enter the item, for example, from the polyethylene body of the detector. This operating mode—called *thermal mode*—is suitable for items that have low masses or low density of uranium. As the amount and/or density of the uranium increases, the induced fission rate becomes saturated because of self-shielding (see Chap. 14, Sect. 14.3.2), and the interior regions of the item are not fully probed. In such cases, the detector is configured to prevent thermal neutrons from entering the item (for example, by placing a cadmium layer around the item). This absorption reduces the induced fission rate but ensures a more uniform interrogation of larger mass items. Using the detector when it is configured to prevent thermal neutrons from entering the item is called *fast mode*.

A key feature of the fast mode is that by reducing the thermal-neutron fission rate in the item, the doubles rate is less sensitive to the presence of thermal-neutron absorbers in the item. This outcome is particularly important for the measurement of fresh, low-enriched fuel assemblies, where burnable poisons are intentionally added for fuel-performance reasons. A fast mode measurement is less dependent on the amount of burnable poison declared by the operator of the fuel fabrication facility.

17.15.4 Active Calibration Curve

In the case of active methods, the doubles rate measured in a detector follows the same form as Eq. 17.37 but with the spontaneous fission rate replaced by the primary induced fission rate. The resulting fission neutrons multiply in the same way as in the passive case, and so the Doubles increase nonlinearly as the item mass increases. In active coincidence counting, there is no net contribution to the doubles rate from (α,n) neutrons produced in the item, and so the α term is zero. The passive counting rate of the item may be subtracted from the doubles count. (The singles counting rate is not used in the analysis.) The resulting relationship between Doubles and ^{235}U mass is used as a calibration curve in the same way as the passive calibration curve of Sect. 17.9. Similarly, the accuracy depends on whether the calibration standards are representative of the assay items.

When a ^{252}Cf (or other spontaneous fission) source is used as the interrogation source, the doubles background from the source should be subtracted.

One widely used example of active coincidence counting is the active well coincidence counter [101], which will be described in Chap. 20, Sect. 20.5.1. This detector uses two AmLi sources to induce fission in uranium metal or oxide in cans. The instrument has a removable cadmium liner to allow operation in fast or thermal mode. The calibration curve is typically a third-order polynomial to consider self-shielding (see Chap. 14, Sect. 14.3.2) and multiplication. A special insert allows the active well coincidence counter (AWCC) to be used with one or two AmLi source(s) for the assay of materials test reactor fuel elements.

The uranium neutron collar (UNCL) is a second example of a widely used active interrogation instrument [102]. The instrument—described in Chap. 20, Sect. 20.5.3—has three polyethylene sides that contain ^{3}He detectors and a fourth side that contains an AmLi source. For most of the instruments described in this book, the calibration curve is generated by measuring calibration standards in the detector. These standards can be verified by destructive analysis and weighing of a sample. In the case of the collar detector, it is not possible to verify a calibration standard fuel assembly by taking a sample for destructive analysis. Therefore, a reference calibration was made with a reference collar and the AmLi source, and a family of detectors was established by cross referencing the collar efficiencies and AmLi source strengths [102]. In a fuel assembly measurement, the passive doubles counting rate from the item is measured. Then the interrogation source is added, and the active plus passive doubles rate is measured. (The counting rates in a typical collar measurement are quite small, and dead-time corrections are not applied.) The net doubles rate (from subtraction), D_m, is then adjusted for various factors to put it on the same scale as the reference calibration:

$$D_{corrected} = k_0 k_1 k_2 k_3 k_4 k_5 D_m, \tag{17.74}$$

where

k_0 = a correction for the source strength of the AmLi source used

k_1 = a relative measurement of a normalization source (see Sect. 17.9); in this case, the normalization measurement uses the AmLi source in an empty collar rather than a ^{252}Cf source, as in the case of passive counters

k_2 = the relative efficiency of the collar being used compared with the reference collar

k_3 = a correction for the declared burnable poison content of the measured fuel assembly

k_4 = a correction for the heavy metal linear density of the measured fuel compared with the reference fuel assembly used in the calibration

k_5 = a correction factor to account for other differences between the measurement and the calibration conditions (such as cardboard or plastic liners on the measured assembly).

The resulting $D_{corrected}$ is then used with the reference (active) calibration curve in the normal way. (In INCC, $D_{corrected}$ is referred to as *collar doubles*.)

An essential feature of both the AWCC and UNCL is that they are designed to minimize the singles counting rate that comes directly from the source, which is important because the accidental rate (being proportional to the square of the singles rate) makes a large contribution to the uncertainty. In both cases, the AmLi sources are embedded in polyethylene to reduce the direct contribution to the counting rate. Using the same principle in more recent collar designs [103, 104], some of the ^{3}He tubes near the AmLi source were omitted because the overall uncertainty was reduced despite the loss of efficiency.

A recent development is a neutron collar using liquid scintillator detectors instead of ^{3}He detectors [105]. The direct detection of fast neutrons (>0.1 MeV) allows the coincidence gate to be reduced by a factor of 1000 (from microseconds to nanoseconds) and virtually eliminates the uncertainties due to the accidental coincidence rate. The statistical precision is somewhat better than a modern fast-mode ^{3}He system [106], and the measured sensitivity to burnable poisons is less, resulting in better performance for the same measurement time.

17.15.5 Uncertainties for Active Counting

For the AWCC, the statistical measurement uncertainty on the Doubles is again given by Eq. 17.53. The coincidence response is proportional to $\varepsilon^2 m_{235} t I_{AmLi}$, where m_{235} is the ^{235}U mass, t is the count time, and I_{AmLi} is the AmLi source strength. Although the singles response is increased by these induced fissions, the effect is small in practice and, for error calculations, it is reasonable to assume that the singles response is directly proportional to $\varepsilon I_{AmLi} t$. Then

$$\frac{\sigma_D}{D} \propto \frac{\sqrt{k_1 \varepsilon m_{235} I_{AmLi} + k_2 I_{AmLi}^2}}{k_1 \varepsilon m_{235} I_{AmLi} \sqrt{t}}, \qquad (17.75)$$

where k_1 and k_2 are two constants of proportionality. For large uranium masses and weak sources, the relative error is proportional to $1/\sqrt{m_{235} I_{AmLi}}$, as expected. For strong sources, the relative error is proportional to $1/\varepsilon m_{235}$.

This last relationship has several interesting consequences. First, the relative error is independent of source strength for sources large enough to ensure that D is much less than A. This feature has the advantage that the sources need only be large enough to meet this criterion, which in practice, has been measured as 2×10^4 n/s for two sources, assuming negligible background and no passive signal from the sample [100]. However, this feature has the disadvantage that assay precision cannot be improved by introducing larger sources. Once G, ε, k_1, and k_2 are determined by the design of the well counter, the assay precision can be varied only by varying the counting time. Second, the absolute assay precision is almost independent of sample mass and is determined primarily by the accidental coincidence rate. For the AWCC described in Ref. [101], the absolute assay precision in the "fast configuration" for 1000 s count times is equivalent to 18 g of ^{235}U.

17.15.6 Advanced Analysis of Collar Data to Solve for Burnable Poison and Uranium Linear Density Simultaneously

Burnable poisons were introduced in the late 1980s to increase the lifetime of commercial light-water-reactor (LWR) fuel. The high thermal-neutron absorption by burnable poisons complicates the verification of ^{235}U content in reactor fuel when using the UNCL. Two primary methods are typically used to make more accurate measurements when burnable poisons are present [102]. Fast-mode measurements with thermal-neutron-absorbing cadmium liners are used to remove the thermal-neutron signal; however, these measurements are time consuming—requiring hour-long measurements versus 10 min measurements in thermal mode without cadmium liners. Burnable poison correction factors based on the declared number of burnable poison rods in an assembly can be used to obtain more accurate measurements with the detector in thermal mode; however, the correction factors rely on operator declarations for the burnable poison content. Experimental and simulated data were used to study the UNCL response to variations in burnable poison content and enrichment, eventually yielding an advanced analysis method that can be used to accurately measure the burnable poison content and the linear density of ^{235}U (g ^{235}U/cm) of fresh LWR fuel simultaneously in an inexpensive and timely manner.

The technique requires Monte Carlo simulations to generate the multiple calibration curves that result from changes in ^{235}U content and burnable poison content in fuel assemblies as seen in Fig. 17.31 [107].

Multiple curve fits are used to eventually generate equations for Singles and Doubles as functions of the burnable poison content and the linear density of ^{235}U in the fuel. The "knowns," Singles and Doubles, can be used to solve for the "unknowns," burnable poison content and linear density of ^{235}U. The Triples (T) can also be used in some cases, depending on the counting statistics. This technique has been demonstrated using simulated data for an AmLi interrogation source, a ^{252}Cf interrogation source, and for passive measurements with the passive version of the UNCL (the fast-neutron passive

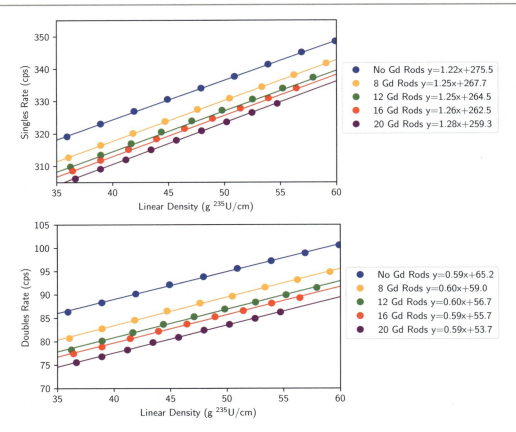

Fig. 17.31 Simulated singles and doubles rates as functions of linear density and number of Gd_2O_3 poison rods using the fast-neutron passive collar [107]

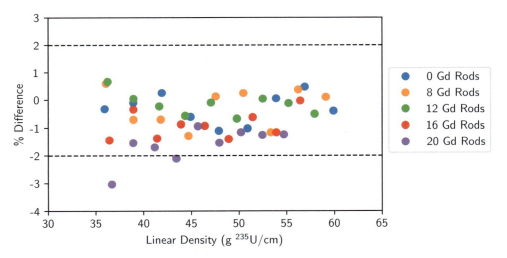

Fig. 17.32 Percent difference between calculated and actual linear density values for various combinations of low-enriched uranium and Gd_2O_3 rods. The calculated values are determined using Singles and Doubles. In the legend, Gd_2O_3 is simplified to Gd [109]

collar [FNPC]; Refs. [107–109]). See Ref. [107] for a full description of the technique using Singles and Doubles and the FNPC; Ref. [108] for a full description of the technique using Singles, Doubles, and Triples with a ^{252}Cf interrogation source; and Ref. [109] for an abbreviated description of the technique using Singles, Doubles, and Triples with an AmLi source.

The technique showed low uncertainties in the calculated linear density of the fuel in all cases. An example of the results of using this technique with the FNPC is shown in Fig. 17.32. In the range of typical commercial reactor fuel (~3–5% enriched in

^{235}U or ~ 40–60 g ^{235}U/cm), the errors in the calculated linear density were nearly all below 2%, which is comparable to uncertainties seen in routine measurements by inspectorates [102].

References

1. M.S. Krick, *^{240}Pu-Effective Mass Formula for Coincidence Counting of Plutonium with Shift Register Electronics*, in Nuclear Safeguards Research and Development Program Status Report, May–August 1977, J. L. Sapir, Compiler, Los Alamos Scientific Laboratory report LA-7030-PR (1978), p. 16
2. S. Croft, S. Cleveland, A. Nicholson, *Calculation of the ^{240}Pu-Effective Coefficients for Neutron Correlation Counting*," in Proceedings of the 56th Annual Meeting of the INMM, Indian Wells, CA, 12–16 Jul 2015
3. H. Ottmar, P. van Belle, S. Croft, P.M.J. Chard, L.C-A. Bourva, U. Blohm-Hieber *An Empirical Measurement of the Specific ^{240}Pu-Effective Mass of ^{238}Pu and ^{242}Pu*, in Proceedings of the ESARDA Symposium, Seville, Spain, 1999
4. C.H. Westcott, A study of expected loss rates in the counting of particles from pulsed sources. Proc. R. Soc. Lond. A **194**(1039) (1948)
5. C.H. Vincent, The pulse separation spectrum for the detection of neutrons from a mixture of fissions and single-neutron events. Nucl. Instrum. Methods **138**(2), 261–266 (1976)
6. N. Pacilio, *Reactor Noise Analysis in the Time Domain*, AEC critical review series, report no. TID 24512 (1969). https://doi.org/10.2172/4786612
7. R.E. Uhrig, *Random Noise Techniques in Nuclear Reactor Systems*, U.S. Atomic Energy Commission, ISBN 1114265659 (Ronald Press Co., 1970)
8. R.P. Feynman, F. de Hoffmann, R. Serber, "Dispersion of the neutron emission in U-235 fission," J. Nucl. Energy *(1954)* 3(1–2), 64–69 (1956)
9. G. Hansen, Rossi alpha method, Los Alamos National Laboratory report LA-UR-85-4176 (1985)
10. G.E. McKenzie IV, J.D. Hutchinson, W.L. Myers, Prompt neutron decay constant measurements on a polyethylene-reflected sphere of HEU. Trans. Am. Nucl. Soc. **117**, 97–100 (2017)
11. G. McKenzie et al., *Prompt Neutron Decay Constant Measurements on a Lead Moderated, Copper Reflected Weapons Grade Plutonium System*, in Proceedings of the ANTPC 2018, November 11–15, 2018, Orlando, FL (2018)
12. G. McKenzie et al., Prompt neutron decay constant measurements on the KRUSTY cold critical configuration. Trans. Am. Nucl. Soc. **119**, 822–825 (2018)
13. G. McKenzie et al., Prompt neutron decay constants in a highly enriched uranium-lead copper reflected system. Trans. Am. Nucl. Soc. **115**, 931–934 (2016)
14. G. McKenzie et al., Prompt neutron decay constants in a highly enriched uranium copper reflected system. Trans. Am. Nucl. Soc. **110**, 303–305 (2014)
15. G. McKenzie, Modern Rossi Alpha Measurements, Master's thesis, University of Illinois (2014)
16. M.Y. Hua et al., Measurement uncertainty of Rossi-alpha neutron experiments. Ann. Nucl. Energy **147**, 107672 (2020)
17. M.Y. Hua et al., Rossi-alpha measurements of fast plutonium metal assemblies using organic scintillators. Nucl. Instrum. Methods Phys. Res., Sect. A **959**, 163507 (2020)
18. M.Y. Hua et al., Validation of the two-region Rossi-alpha model for reflected assemblies. Nucl. Instrum. Methods Phys. Res., Sect. A **981**, 164535 (2020)
19. G. McKenzie, *Limits on Subcritical Reactivity Determination using Rossi-α and Related Methods*, Doctoral dissertation (University of Illinois, 2018)
20. J. Hutchinson et al., Prompt neutron decay constant fitting using the Rossi-alpha and Feynman variance-to-mean methods. Trans. Am. Nucl. Soc. **117**(1), 986–989 (2017)
21. M. Hua et al., Derivation of the two-exponential probability density function for rossi-alpha measurements of reflected assemblies and validation for the special case of shielded measurements. Nucl. Sci. Eng. **194**(1), 56–68 (2020)
22. T. Cutler et al., Copper and polyethylene-reflected plutonium-metal-sphere subcritical measurements, in *International Handbook of Evaluated Criticality Safety Benchmark Experiments*, NEA/NSC/DOC/(95)03/I, FUND-NCERC-PU-HE3-MULT-003 (2019)
23. K.I. Ianakiev, M. Iliev, M.T. Swinhoe, A.M. Lafleur, C. Lee, *Self-Calibration Method for Dead time Losses in Neutron Counting*, in Proceedings of the 39th ESARDA Conference, Dusseldorf, Germany, May 16–18 2017, Los Alamos National Laboratory report LA-UR-17-23751 (2017)
24. K. Boehnel, Determination of Plutonium in Nuclear Fuels Using the Neutron Coincidence Method, KFK2203, Karlsruhe, 1975, and AWRE Translation 70 (54/4252), Aldermaston (1978)
25. F. de Hoffman, R. Feynman, R. Serber, Intensity fluctuations of a neutron chain reactor, Los Alamos Scientific Laboratory report LA-DC-256 (1944)
26. F. de Hoffman, R. Feynman, R. Serber, Statistical fluctuations in the water boiler and the dispersion of neutrons emitted per fission, Los Alamos Scientific Laboratory report LA-101 (1944)
27. J. Hutchinson et al., *Subcritical Multiplication Experiments & Simulations: Overview and Recent Advances*, in ANS Advances in Nonproliferation Technology and Policy Conference, Santa Fe, NM (September 2016)
28. A. Robba, E. Dowdy, H. Atwater, Neutron multiplication measurements using moments of the neutron counting distribution. Nucl. Instrum. Methods **215**, 473–479 (1983)
29. D. Cifarelli, W. Hage, Models for a three-parameter analysis of neutron signal correlation measurements for fissile material assay. Nucl. Instrum. Methods Phys. Res., Sect. A **251**, 550–563 (1986)
30. G. Keepin, *Physics of Nuclear Kinetics* (Addison-Wesley, 1965)
31. I. Pázsit, L. Pál, *Neutron Fluctuations: A Treatise on the Physics of Branching Processes* (Elsevier Science, 2008). https://doi.org/10.1016/B978-0-08-045064-3.X5001-7

32. K. Hashimoto et al., Experimental investigation of dead-time effect on Feynman-α method. Ann. Nucl. Energy **23**(14), 1099–1104 (1996)
33. Y. Yamane et al., Formulation of data synthesis technique for Feynman-α method. Ann. Nucl. Energy **25**(1–3), 141–148 (1998)
34. Y. Kitamura et al., General formulae for the Feynman-α method with the bunching technique. Ann. Nucl. Energy **27**(13), 1199–1216 (2000)
35. I. Pázsit, M. Ceder, Z. Kuang, Theory and analysis of the Feynman-α method for deterministically and randomly pulsed neutron sources. Nucl. Sci. Eng. **148**(1), 67–78 (2004)
36. R. Soule et al., Neutronic studies in support of accelerator-driven systems: The MUSE experiments in the MASURCA facility. Nucl. Sci. Eng. **148**(1), 124–152 (2004)
37. J. Kloosterman, Y. Rugama, Feynman-alpha measurements on the fast critical zero-power reactor MASURCA. Prog. Nucl. Energy **46**(2), 111–125 (2005)
38. I. Pázsit et al., Calculation of the pulsed Feynman-alpha formulae and their experimental verification. Ann. Nucl. Energy **32**, 986–1007 (2005)
39. J. Mattingly, Polyethylene-Reflected Plutonium Metal Sphere: Subcritical Neutron and Gamma Measurements, SAND2009–5804 (2009)
40. Y. Rana, S. Degweker, Feynman-alpha and rossi-alpha formulas with delayed neutrons for subcritical reactors driven by pulsed non-poisson sources with correlation between different pulses. Nucl. Sci. Eng. **169**(1), 98–109 (2011)
41. J. Muñoz-Cobo et al., Feynman-α and Rossi-α formulas with spatial and modal effects. Ann. Nucl. Energy **38**(2–3), 590–600 (2011)
42. S. Croft, A. Favalli, D.K. Hauck, D. Henzlova, P.A. Santi, Feynman variance-to-mean in the context of passive neutron coincidence counting. Nucl. Instrum. Methods Phys. Res., Sect. A **686**, 136–144 (2012)
43. M.K. Prasad, N.J. Snyderman, Statistical theory of fission chains and generalized poisson neutron counting distributions. Nucl. Sci. Eng. **172**(3), 300–326 (2012)
44. A. Chapelle et al., Joint neutron noise measurements on metallic reactor caliban. Nucl. Data Sheets **118**, 558–560 (2013)
45. T. Cutler et al., Deciphering the binning method uncertainty in neutron multiplicity measurements. Trans. Am. Nucl. Soc. **111**, 846–849 (2014)
46. E.C. Miller et al., Computational evaluation of neutron multiplicity measurements of polyethylene-reflected plutonium metal. Nucl. Sci. Eng. **176**(2), 167–185 (2014)
47. W. Noonan, Neutrons: It is all in the timing—The physics of nuclear fission chains and their detection. J. Hopkins APL Tech. Dig. **32**(5), 762–773 (2014)
48. B. Richard et al., Nickel-reflected plutonium-metal-sphere subcritical measurements, in *International Handbook of Evaluated Criticality Safety Benchmark Experiments*, FUND-NCERC-PU-HE3-MULT-001 (2014)
49. M.A. Smith-Nelson, T.L. Burr, J.D. Hutchinson, T.E. Cutler, Uncertainties of the Yn parameters of the Hage-Cifarelli formalism, Los Alamos National Laboratory report LA-UR-15-21365 (2015), https://www.osti.gov/biblio/1170708
50. S. Degweker, R. Rudra, On the relation between Rossi alpha and Feynman alpha methods. Ann. Nucl. Energy **94**, 433–439 (2016)
51. B. Richard et al., Tungsten-reflected plutonium metal sphere subcritical measurements, in *International Handbook of Evaluated Criticality Safety Benchmark Experiments*, FUND-NCERC-PU-HE3-MULT-002 (2016)
52. J. Arthur et al., Development of a research reactor protocol for neutron multiplication measurements. Prog. Nucl. Energy **106**, 120–139 (2018)
53. J. Arthur et al., Validating the performance of correlated fission multiplicity implementation in radiation transport codes with subcritical neutron multiplication benchmark experiments. Ann. Nucl. Energy **120**, 348–366 (2018)
54. P. Humbert, *Rossi and Feynman-Alpha Formulas Including Prompt Neutron Decay and Detector Die Away Time Constants*, in Proceedings of the Topical Meeting of the Radiation Protection and Shielding Division of ANS, Santa Fe, NM, August 26–31 (2018)
55. J. Arthur et al., Genetic algorithm for nuclear data evaluation applied to subcritical neutron multiplication inference benchmark experiments. Ann. Nucl. Energy **133**, 853–862 (2019)
56. T. Cutler et al., Copper and polyethylene-reflected plutonium-metal-sphere subcritical measurements, in International Handbook of Evaluated Criticality Safety Benchmark Experiments, FUND-NCERC-PU-HE3-MULT-002 (2019)
57. J. Hutchinson et al., Validation of statistical uncertainties in subcritical benchmark measurements: Part I—Theory and simulations. Ann. Nucl. Energy **125**, 50–62 (2019)
58. J. Hutchinson et al., Validation of statistical uncertainties in subcritical benchmark measurements: Part II—Measured data. Ann. Nucl. Energy **125**, 342–359 (2019)
59. M. Nelson et al., *Comparison of the Feynman-Y and the Rossi-α Methods for Subcritical Systems* (PHYSOR, Cancun, 2019)
60. T. Shin et al., A note on the nomenclature in neutron multiplicity mathematics. Nucl. Sci. Eng. **193**(6), 663–679 (2019)
61. J. Gomez et al., Results of three neutron diagnosed subcritical experiments. Nucl. Sci. Eng. **193**(5), 537–548 (2019)
62. J. Hutchinson et al., Subcritical sensitivity measurements using the Thor core. Trans. Am. Nucl. Soc. **109**, 819–822 (2013)
63. M. Stephens, J. Swansen, L. East, Shift register neutron coincidence module, Los Alamos Scientific Laboratory report LA-6121-MS (1975)
64. J. Swansen, N. Ensslin, M. Krick, H. Menlove, *New Shift Register for High Count Rate Coincidence Applications*, in Nuclear Safeguards Research and Development Program Status Report, September–December 1976, Joseph. L. Sapir, Compiler, Los Alamos Scientific Laboratory report LA-6788-PR (1977), p. 4
65. J. Swansen, P. Collinsworth, M. Krick, Shift-register coincidence electronics system for thermal neutron counters. Nucl. Instrum. Methods **116**, 555 (1980)
66. K. Lambert, J. Leake, A. Webb, F. Rogers, A passive neutron well counter using shift register coincidence electronics, Atomic Energy Research Establishment report 9936 (1982)
67. M. Krick, J. Swansen, Neutron multiplicity and multiplication measurements. Nucl. Instrum. Methods Phys. Res. **219**(2), 384–393 (1984)
68. J.E. Stewart, S.C. Bourret, M.S. Krick, M.R. Sweet, T.K. Li, A. Gorobets, New shift-register electronics for improved precision of neutron coincidence and multiplicity assays of plutonium and uranium mass, Los Alamos National Laboratory report LA-UR-99-4927 (1999)
69. M. Krick, H. Menlove, The high-level neutron coincidence counter (HLNCC): User's manual, Los Alamos Scientific Laboratory report LA-7779-M (1979)
70. M.S. Krick, *Calculations of Coincidence Counting Efficiency for Shift-Register and OSDOS Coincidence Circuits*, in Nuclear Safeguards Research and Development Program Status Report, May–August 1977, Joseph L. Sapir, Compiler, Los Alamos Scientific Laboratory report LA-7030-PR (1978), p. 14
71. J. Swansen, N. Ensslin, A digital random pulser for testing nuclear instrumentation. Nucl. Instrum. Methods Phys. Res. **188**(1), 83–91 (1981)
72. E. Adams, Dead time measurements for the AWCC, Los Alamos National Laboratory memorandum Q-1-82-335 to H. Menlove (29 Apr 1982)

73. J. Swansen, N. Ensslin, *HLNCC Shift Register Studies*, in Nuclear Safeguards Research and Development Program Status Report, April–June 1980, ed. by G. Robert Keepin, Los Alamos Scientific Laboratory report LA-8514-PR (February 1981), p. 11
74. C.H. Vincent, Optimisation of the neutron coincidence process for the assay of fissile materials. Nucl. Instrum. Methods **171**(2), 311–317 (1980)
75. K.D. Ianakiev, M.L. Iliev, A. Favalli, M.T. Swinhoe, *Reduction of Dead Time Losses in Neutron Multiplicity Counting Systems*, in *Proceedings of the IEEE Nuclear Science Symposium*, Manchester UK, 2019, Los Alamos National Laboratory report LA-UR-19-24210 (2019)
76. K.D. Ianakiev, M. Iliev, M.T. Swinhoe, Fact sheet for KM200 front-end electronics, Los Alamos National Laboratory report LA-UR-15-24536 (2015)
77. A. Fazzi, V. Varoli, Signal shaping optimization with ^{3}He tubes in high rate neutron counting. IEEE Trans. Nucl. Sci. **46**(3), 342–347 (1999)
78. N. Dytlewski, N. Ensslin, J.W. Boldeman, A neutron multiplicity counter for plutonium assay. Nucl. Sci. Eng. **104**(4), 301–313 (1990)
79. T.C. Nguyen, J. Huszti, Q.V. Nguyen, Effect of double false pulses in calibrated neutron coincidence collar during measuring time-correlated neutrons from PuBe neutron sources. Nucl. Instrum. Methods Phys. Res., Sect. B **358**, 168–173 (2015)
80. K.E. Koehler, V. Henzl, S.S. Croft, D. Henzlova, P.A. Santi, Characterizations of double pulsing in neutron multiplicity and coincidence counting systems. Nucl. Instrum. Methods Phys. Res., Sect. A **832**, 279–291 (Oct 2016)
81. A.T. Simone, S. Croft, J.P. Hayward, L.G. Worrall, Using the JCC-71 neutron coincidence collar as a benchmark for detector characterization with PTR-32 list mode data acquisition. Nucl. Instrum. Methods Phys. Res., Sect. A **908**, 24–34 (Nov 2018)
82. A.T. Simone, S. Croft, C.L. Britton, R.D. McElroy, M.N. Ericson, J.P. Hayward, Investigating the non-ideal behavior of the AMPTEK A111 charge sensitive preamplifier & discriminator board—^{3}He proportional counter combination used in common safeguards neutron coincidence counters. Nucl. Instrum. Methods Phys. Res., Sect. A **954**, 161355 (Feb 2020)
83. A.S. Moore, S. Croft, R.D. McElroy, J.P. Hayward, Methods for diagnosing and quantifying double pulsing in a uranium neutron collar system using shift register logic. Nucl. Instrum. Methods Phys. Res., Sect. A **941**, 162333 (Oct 2019)
84. N. Dytlewski, M.S. Krick, N. Ensslin, Measurement variances in thermal neutron coincidence counting. Nucl. Instrum. Methods Phys. Res., Sect. A **327**(2–3), 469–479 (April 1993)
85. T.E. Sampson, J.L. Parker, Equations for plutonium and americium-241 decay corrections, Los Alamos National Laboratory report LA-10733-MS (1986)
86. J.F. Longo, W.H. Geist, W.C. Harker, M.S. Krick, INCC software users manual, Los Alamos National Laboratory report LA-UR-10-06227 (2010)
87. N. Ensslin, J. Stewart, J. Sapir, Self-multiplication correction factors for neutron coincidence counting. Nucl. Mater. Manag. **8**(2), 60–73 (1979)
88. K. Böhnel, The effect of multiplication on the quantitative determination of spontaneously fissioning isotopes by neutron correlation analysis. Nucl. Sci. Eng. **90**(1), 75–82 (1985)
89. M.T. Swinhoe, Multiplication effects in neutron coincidence counting: uncertainties and multiplying reference samples, U.K. Atomic Energy Commission report AERE-R 11678, Harwell (March 1985)
90. N. Ensslin, *A Simple Self-Multiplication Correction for In-Plant Use*, in Proceedings of the 7th ESARDA Symposium on Safeguards and Nuclear Material Management, Liege, Belgium, 21–23 May 1985
91. W. Hage, K. Caruso, An Analysis Method for the Neutron Autocorrelator with Multiplying Samples, Joint Research Centre, Ispra, Italy, report EUR 9792 EN (1985)
92. G. Eccleston, J. Foley, M. Krick, H. Menlove, P. Goris, A. Ramalho, "Coincidence measurements of FFTF breeder fuel subassemblies," Los Alamos National Laboratory report LA-9902-MS (1984)
93. M.S. Krick, Neutron multiplication corrections for passive thermal neutron well counters, Los Alamos Scientific Laboratory report LA-8460-MS (1980)
94. M.S. Krick, R. Schenkel, K. Boehnel, Progress in Neutron Coincidence Counting Techniques, report of the IAEA Advisory Group Meeting, Vienna, Austria, IAEA Department of Safeguards General Report STR-206 (7–11 Oct 1985)
95. H.O. Menlove, R. Abedin-Zadeh, R. Zhu, The analyses of neutron coincidence data to verify both spontaneous-fission and fissionable isotopes, Los Alamos National Laboratory report LA-11639-MS (1989)
96. T. Gozani, *Active Nondestructive Assay of Nuclear Materials: Principles and Applications*, NUREG/CR-0602 (U.S. Nuclear Regulatory Commission, Washington DC, 1981)
97. I. Jovanovic, A. Erickson (eds.), *Active Interrogation in Nuclear Security* (Springer, 2018)
98. M. Looman, R. Jaime, P. Peerani, P. Schillebeeckx, S. Jung, P. Schwalbach, M.T. Swinhoe, P.J. Chare, W.F. Kloeckner, *The Effect of Gadolinium Poison Rods on the Active Neutron Measurement of LEU Fuel Assemblies*, in 23rd Annual Meeting, Symposium on Safeguards and Nuclear Material Management, Bruges, Belgium, May 2001, pp. 568–573
99. M.A. Root, H.O. Menlove, R.C. Lanza, C.D. Rael, K.A. Miller, J.B. Marlow, Technical basis for the use of a correlated neutron source in the uranium neutron coincidence collar. Nucl. Technol. **197**(2), 180–190 (2017)
100. H.O. Menlove, S.H. Menlove, C.D. Rael, The development of a new, neutron, time correlated, interrogation method for measurement of ^{235}U content in LWR fuel assemblies. Nucl. Instrum. Methods Phys. Res., Sect. A **701**, 72–79 (2013)
101. H.O. Menlove, Description and operation manual for the active well coincidence counter, Los Alamos Scientific Laboratory report LA-7823-M (1979), p. 25
102. H.O. Menlove, J.E. Stewart, Z. Qiao, T.R. Wenz, G.P.D. Verrecchia, Neutron collar calibration and evaluation for assay of LWR fuel assemblies containing burnable neutron absorbers, Los Alamos National Laboratory report LA-11965-MS (1990)
103. L.G. Evans, M.T. Swinhoe, H.O. Menlove, P. Schwalbach, P. DeBaere, M.C. Browne, A new fast collar for safeguards inspection measurements of fresh low enriched uranium fuel assemblies containing burnable poison rods. Nucl. Instrum. Methods Phys. Res., Sect. A **729**, 740–746 (2013)
104. A. Favalli, H.O. Menlove, P. Polk Jr., M.T. Swinhoe, A new fast neutron collar for fresh boiling water reactor fuel assemblies, Los Alamos National Laboratory report LA-UR-16-24169 (2016)
105. J. Beaumont, T. Lee, M. Mayorov, J.K. Jeon, A. Bonino, M. Grund, M. Dutra, G. Renha, S.H. Ahn, K.H. Kim, J.S. Park, G.H. Yang, Field testing of a fast-neutron coincidence collar for fresh uranium fuel verification. Nucl. Instrum. Methods Phys. Res., Sect. A **962**, 163682 (2020)

106. M.T. Swinhoe, H.O. Menlove, P. De Baere, D. Lodi, P. Schwalbach, C.D. Rael, M. Root, A. Tomanin, A. Favalli, A new generation of uranium coincidence fast collars for assay of LWR fresh fuel assemblies. Nucl. Instrum. Methods Phys. Res., Sect. A **1009**, 165453 (2021)
107. M. Root, S. Sarnoski, H. Menlove, W. Geist, Advanced Analysis to Determine Both the U-235 and Gadolinium Content in Fresh PWR Fuel Assemblies," Attachment to "Experimental Evaluation of the Fast-Neutron Passive Collar (FNPC) Performance and Comparison with Scintillation Based Collars," Final Report for FNPC for NA-24, Los Alamos National Laboratory report LA-UR-19-31922 (2019)
108. M.A. Root, H.O. Menlove, R.C. Lanza, C.D. Rael, K.A. Miller, J.B. Marlow, J.G. Wendelberger, Using the time-correlated induced fission method to simultaneously measure the ^{235}U content and the burnable poison content in LWR FUEL ASSEMBLIEs. Nucl. Technol. **203**(1), 34–47 (2018). https://doi.org/10.1080/00295450.2018.1429112
109. M.A. Root, H.O. Menlove, J.G. Wendelberger, *An Advanced Analytical Technique for the Determination of Burnable Poison Content in Fresh LWR Fuel Assemblies*, in Proceedings of the 59th Annual Meeting of the INMM, Baltimore, MD, 22–26 Jul 2018

Open Access This chapter is licensed under the terms of the Creative Commons Attribution 4.0 International License (http://creativecommons.org/licenses/by/4.0/), which permits use, sharing, adaptation, distribution and reproduction in any medium or format, as long as you give appropriate credit to the original author(s) and the source, provide a link to the Creative Commons license and indicate if changes were made.

The images or other third party material in this chapter are included in the chapter's Creative Commons license, unless indicated otherwise in a credit line to the material. If material is not included in the chapter's Creative Commons license and your intended use is not permitted by statutory regulation or exceeds the permitted use, you will need to obtain permission directly from the copyright holder.

18. Principles of Neutron Multiplicity

D. C. Henzlova, N. Ensslin, A. Favalli, W. H. Geist,
L. Holzleitner, K. Koehler, M. S. Krick, M. M. Pickrell,
T. D. Reilly, J. E. Stewart, and K. D. Veal

18.1 Introduction

A nondestructive assay (NDA) technique for plutonium, called *passive neutron multiplicity counting*, was developed during the 1990s as an extension of neutron coincidence counting [1]. This technique has led to the design and fabrication of neutron multiplicity counters, accompanied by advances in data-processing electronics, analysis algorithms, and data-analysis software, discussed in Chap. 19. Altogether, this technology has led to significantly better measurement accuracy for plutonium metal, oxide, scrap, and residues without the need for empirical calibration curves as required for coincidence counting (see Chap. 17). An NDA technique for bulk, highly enriched uranium (HEU)—called *active neutron multiplicity counting*—has been developed as an extension of *passive* neutron multiplicity counting [2–4]. *Passive* multiplicity analysis usually uses neutron singles, doubles, and triples count rates to solve for plutonium mass, multiplication, and alpha. *Active* multiplicity analysis was developed to provide a similar capability for uranium, and the doubles and triples count rates are used to solve for the sample multiplication and the induced fission rate. This chapter describes the principles of *passive* (Sect. 18.2) and *active* (Sect. 18.3) multiplicity counting.

Passive multiplicity counting has applications in various different areas: improved materials accountability measurements, verification measurements, confirmatory measurements, and weapons materials inspections for verification of disarmament treaties. Although the historical motivation for developing the technique was improved accountability measurements of impure plutonium, new applications have arisen in the areas of verification and confirmation because the technique does not require prior knowledge of the item or prior calibration with standards. For similar reasons, multiplicity counting is in use by the International Atomic Energy Agency (IAEA) for verification of materials that have unknown matrices or limited production records or when coincidence counting calibrations are not available. Multiplicity counting can be used for all plutonium items, but the additional information is beneficial primarily for impure items. For some material categories, especially very impure items, multiplicity might not be helpful because of the limited precision of the triple coincidences. These materials include some plutonium-bearing wastes or process residues that contain sufficiently significant low-Z material impurities in which the high (α,n) reaction rate ruins the precision of the triples. Limitations of multiplicity counting for materials with high (α,n) and the most recent advancements to improve performance for materials with high (α,n) are discussed in Sect. 18.2.8. For pure plutonium metal or oxide, conventional coincidence counting provides better precision and sufficient accuracy. However, if any doubts exist about the plutonium purity, the multiplicity and conventional coincidence

---

Los Alamos National Laboratory strongly supports academic freedom and a researcher's right to publish; as an institution, however, the Laboratory does not endorse the viewpoint of a publication or guarantee its technical correctness.

D. C. Henzlova (✉) · N. Ensslin · A. Favalli · W. H. Geist · K. Koehler · M. S. Krick · M. M. Pickrell · T. D. Reilly · J. E. Stewart · K. D. Veal
Los Alamos National Laboratory, Los Alamos, NM, USA
e-mail: henzlova@lanl.gov; afavalli@lanl.gov; wgeist@lanl.gov; kkoehler@lanl.gov; mpickrell@lanl.gov; kevin.veal@nnsa.doe.gov

L. Holzleitner
European Commission, Joint Research Center, Karlsruhe, Germany
e-mail: Ludwig.HOLZLEITNER@ec.europa.eu

results can be compared, and the more accurate result can be used. Additional information on multiplicity applications and expected performance is provided in Chap. 19, Sect. 19.5.

Active neutron multiplicity counting has potential applications in shipper/receiver confirmation, inventory verification, or accountability measurements of uranium in difficult material forms. U.S. Department of Energy (DOE) facilities contain many metric tons of uranium in varied materials forms, including mixed oxides and metals, weapons components in shielded storage drums, high-density scrap/waste, remote-handled waste, and non-self-protecting irradiated fuels. The capability for direct accountability, inventory verification, or shipper/receiver verification of the ^{235}U is often not available because of the difficult measurement geometries involved. The use of active multiplicity counting can reduce inventory differences, expedite materials consolidation, and help reduce security and storage costs. Despite the active multiplicity-counting technique development that dates back to the 1990s, to date, only limited literature and field experience are available. The purpose of the active chapter is to document what is known to date and to give the reader a sense for the expected measurement performance of this technique.

18.2 Passive Neutron Multiplicity Counting

Historically, the benefit of passive neutron counting has been the great penetrability of neutrons. Neutrons are often the only way to rapidly assay large, dense samples. Neutrons can usually measure the entire volume of the item, and neutrons are not easily shielded except by hydrogenous materials with or without neutron poisons, such as boron. The first neutron instruments used only the total neutron count rate; however, very few plutonium materials could be accurately assayed, as implied by the long list of potential unknowns discussed in Sect. 18.2.2.

The next development was neutron coincidence counting, described in Chap. 17, which uses the spontaneous fission signature and is not affected directly by (α,n) neutrons. Coincidence counting has had wide application for international safeguards inspections. The coincidence-counting technique has had a more limited application domestically because large errors can occur when measuring impure materials. The fundamental limitation of coincidence counting is that it measures only two parameters. For a typical item, at least the first three unknowns are listed in Sect. 18.2.2 (i.e., spontaneous fission rate, which is linked to the plutonium mass and is the goal of the assay; induced fission or self-multiplication in an item; and the (α, n) reaction rate). Therefore, it is usually not possible to obtain accurate assays of impure samples—where the (α,n) reaction rate can be high—using conventional coincidence counting. One must assume that either the (α,n) rate is known and solve for mass and self-multiplication (Chap. 17, Sect. 17.10) or that self-multiplication is known and solve for mass and (α,n) rate (Sect. 17.13). If the assumed information is incorrect, large errors can occur. In fact, for many impure or heterogeneous samples, neither the multiplication nor the (α,n) yield can be known beforehand.

Based on the need for better accuracy, the goal of neutron multiplicity analysis is to correctly assay in-plant materials without prior knowledge of the item matrix, which is possible for many materials—including moist or impure plutonium oxide, oxidized metal, and some categories of scrap and waste—by the availability of a third measured parameter.

18.2.1 Definition of Neutron Multiplicity Counting

As discussed in Chap. 13, the important passive NDA signature for plutonium is spontaneous fission, which leads to the nearly simultaneous emission of multiple, indistinguishable neutrons. The number of neutrons emitted in spontaneous fission can vary from zero to eight. The distribution of the number of neutrons is called the *multiplicity distribution*. The multiplicity distributions of plutonium isotopes were introduced in Chap. 13, Table 13.2. Figure 18.1 provides a graphical illustration of multiplicity distributions for spontaneous fission in ^{240}Pu and the 2 MeV-neutron-induced fission for ^{239}Pu.

Multiplicity counting sums up separately the number of measured neutrons (0, 1, 2, 3, 4, 5, 6, 7, etc.) within the coincidence gate width of the data acquisition electronics (see Chap. 17, Sect. 17.5, for the definition of *coincidence gate*). This technique measures the multiplicity distribution of neutrons that are emitted, detected, and counted within the gate. For this reason, the word *multiplicity* is specifically associated with the extension of conventional coincidence counting to the collection of higher-order multiples of neutrons. However, we also associate the word *multiplicity* with a special neutron counter design (multiplicity counter; see Chap. 19, Sect. 19.5) and with the mathematics of the data analysis process as described in Sect. 18.2.4.

In practice, multiplicity data analysis is usually not based directly on the observed multiplicity distribution but on its factorial moments. The first moment is the Singles, the second factorial moment is the Doubles, and the third factorial moment

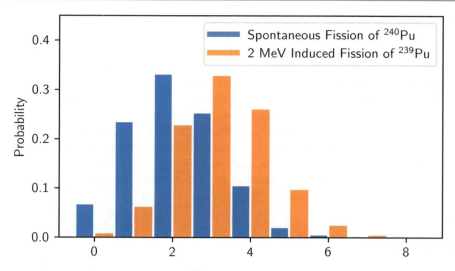

Fig. 18.1 The spontaneous fission multiplicity distribution for ^{240}Pu and the 2 MeV neutron-induced fission multiplicity distribution for ^{239}Pu

is the Triples, all of which are explained in more detail in Sect. 18.2.4.2. Neutron multiplicity analysis works with all three of these moments, whereas conventional coincidence counting uses only the Singles and Doubles, as discussed in Chap. 17. Thus, when we use the word *multiplicity*, we really mean that we add a third measured parameter: triple coincidences. Extensions to include higher-order multiplicity moments (Quads, Pents, and beyond) have also been explored [5, 6] and will be briefly discussed in Sect. 18.2.9, although currently they are not yet implemented in practical applications.

18.2.2 Basic Principles of Neutron Multiplicity Counting

Coincidence counting is an NDA technique that extracts quantitative information from the neutrons emitted by plutonium. Ideally, this information should determine the actual grams of ^{240}Pu$_{eff}$ in the item, defined as the mass of ^{240}Pu that would give the same double coincidence response as that obtained from all of the even isotopes in the item (also see Chap. 17, Sect. 17.1):

$$^{240}Pu_{eff} = m_{240} = 2.52\,^{238}Pu + \,^{240}Pu + 1.68\,^{242}Pu. \tag{18.1}$$

Gamma-ray spectroscopy or mass spectroscopy is then used to obtain the isotopic composition of the plutonium, which makes it possible to obtain the total plutonium mass (m_{Pu}):

$$m_{Pu} = \frac{^{240}Pu_{eff}}{(2.52 f_{238} + f_{240} + 1.68 f_{242})}, \tag{18.2}$$

where f_{238}, f_{240}, and f_{242} are the fractions of the plutonium isotopes present in the item.

In practice, the neutron flux emitted by the item can be affected by various—usually unknown or incompletely known—item or detector properties. The list of potentially unknown properties includes the following:

- spontaneous fission rate—the goal of the assay;
- induced fission, or self-multiplication in an item, and its variation across the item;
- the (α,n) reaction rate in the item;
- the energy spectrum of the (α,n) neutrons;
- spatial variation of the neutron multiplication;
- spatial variation in neutron-detection efficiency;
- energy spectrum effects on detection efficiency;
- neutron capture in the item; and
- the neutron die-away time in the detector.

Clearly, more unknowns potentially exist than conventional coincidence counting can determine. We need N measured parameters to solve for N unknown parameters. Conventional coincidence counting provides only two measured parameters: Singles and Doubles. The basic principle of neutron multiplicity counting is the use of a third measured parameter, the Triples, so that one can solve for three unknown item properties, typically the fission rate (mass of $^{240}\mathrm{Pu}_{\mathrm{eff}}$), item leakage multiplication, and the (α,n) reaction rate. The fourth and fifth unknowns are discussed in Sect. 18.2.8. The sixth and seventh unknown parameters—related to detection efficiency—are usually eliminated as unknowns by careful counter design and characterization (Chap. 19, Sect. 19.5). The other potential unknowns are usually less important and are assumed to be small or constant.

Because we are solving three equations for three unknowns, the solution is exact, complete, and self-contained. This effort has some interesting consequences. For items that meet the point-model assumptions presented in Sect. 18.2.4.1, the assay is bias free and accurate within counting statistics; however, if an item does not meet the assumptions, the assay could be biased. In principle, calibration with physical plutonium standards is not needed, but it is nevertheless important to use standards for validation, bias reduction, and measurement control.

18.2.3 Multiplicity Electronics

The electronics for thermal-neutron multiplicity counters are similar to those used for conventional coincidence counters. They are based on the detection of neutrons with ^{3}He proportional counters embedded in a polyethylene moderator [1] as discussed in Chap. 19, Sect. 19.5. Although non-helium-3-based counters have also been considered [7, 8], the systems currently used in practice still use ^{3}He. Modern thermal coincidence and multiplicity counters use preamplifier modules (AMPTEK, Ref. [9]; PDT, Ref. [10]; or KM200, Ref. [11]) to amplify the ^{3}He output pulses from the ^{3}He tubes and convert the pulses above a discriminator threshold to digital pulses (see Chap. 17, Sect. 17.6.6; and Chap. 19, Sect. 19.2). To provide very short electronic dead times, multiplicity counters usually have 80–130 ^{3}He tubes and many more preamplifier modules than coincidence counters.

18.2.3.1 Multiplicity Shift Register

To extract the fission rate and multiplicity information from the detected neutron pulses (i.e., neutron pulse train, Chap. 17, Sect. 17.2), we need to extract the correlated neutrons from the background of uncorrelated ones. The multiplicity shift register circuit achieves this task in an elegant fashion by separating the incoming pulse train into correlated and uncorrelated events (D + A and A gates), just like the standard coincidence shift register introduced in Chap. 17, Sect. 17.5. However, there is more information in a neutron pulse train than single and double neutron events, as recorded in the standard coincidence shift register. Multiplicity counting looks at the distribution of 0's, 1's, 2's, 3's, etc., to deduce the multiplicity distribution. Special electronics are required to measure the neutron multiplicity distributions in the D + A and A coincidence gates. These electronics record the number of times each multiplicity (i.e., number of pulses) occurs in the gates. For example, if seven neutron pulses are in the D + A gate when another neutron arrives, then "1" is added to the D + A multiplicity counter that tallies sevens because seven pulses are in the gate correlated with the arriving neutron pulse. Examples of current multiplicity shift registers can be found in Chap. 19.

Figure 18.2 shows a simplified circuit diagram of the multiplicity shift register (MSR) electronics with fast accidentals. This circuit includes a derandomizer (Chap. 17, Sect. 17.5), which spaces very close pulses far enough apart to correspond to the MSR clock rate (typically 4 or 50 MHz) and eliminates shift register electronics dead time. Pulses then enter a gate shift register; the up-down counter is incremented when a pulse enters the gate shift register and decremented when the pulse exits the gate shift register. Thus, the up-down counter keeps an instantaneous tally of the current number of pulses in the gate shift register. The predelay shift register delays the accumulation of the value from the up-down counter by the length of the predelay, typically 1–10 ms. The writing of data from up-down counter to the D + A accumulator is triggered (D + A trigger in Fig. 18.2) only when a predelay time elapses after the pulse left the gate shift register. This predelay blocks storing of any pulses that arrived within a fixed time of the triggering pulse. The predelay is selected to correspond to the timeframe affected by the dead time of the detector and the electronics and provides a fixed delay that masks pulses that cannot dependably be used in the multiplicity count rate calculations. The D + A trigger causes a data write action (red arrow in Fig. 18.2) to store the tally from the up-down counter in the D + A accumulator. The A accumulator is populated from the same up-down counter as the D + A accumulator but with a different trigger. Fast accidentals sampling is included in modern MSRs and is accomplished by triggering the write action into the A accumulator at the clock rate of approximately 4 MHz [12]. Conventional implementation—where the A accumulator is triggered at the same rate as the D + A accumulator but after a sufficiently

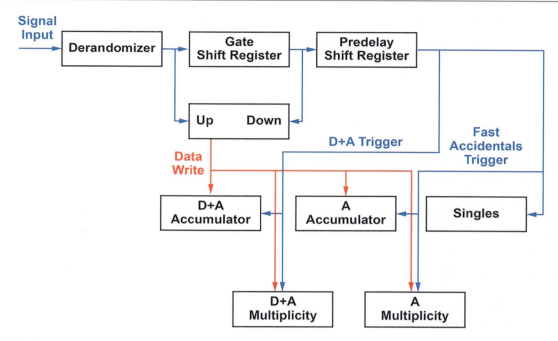

Fig. 18.2 Multiplicity shift register block diagram with fast accidentals

long delay of typically 1000–4000μs (Chap. 17, Sect. 17.5)—is also included in modern MSRs, but use of fast accidentals is typically preferred because it results in better statistics for the accidentals as discussed in this section.

The information contained in the D + A and A accumulator is equivalent to the information obtained from the standard coincidence shift register (Chap. 17, Sect. 17.5). The key distinction of MSR functionality is that it also populates D + A and A multiplicity scalers (counters), as shown at the bottom of Fig. 18.2. This function is accomplished using the same triggers that populate the D + A and A accumulators, which simultaneously trigger a write action into the D + A and A multiplicity bins. During this write action (shown by red arrows in Fig. 18.2), a bin in the D + A or A multiplicity (i.e., address of the multiplicity memory) that corresponds to the number of counts stored in the up-down counter gets incremented by 1. Thus, if 3 counts are in the up-down counter at the time of the D + A trigger, the D + A multiplicity bin 3 gets incremented by 1, so the multiplicity distributions for D + A and A gates get populated. The values accumulated in D + A and A accumulators are used for consistency checks in International Neutron Coincidence Counting (INCC) software, called *checksum tests*, to confirm the validity of data stored in the D + A and A multiplicity distributions [13]. These checksum tests compare the total number of events counted in D + A multiplicity distribution with the content of the D + A accumulator. The same comparison is performed for the total number of events counted in A multiplicity and the content of the A accumulator. For a correctly functioning MSR, these values should be identical, and the checksum tests therefore provide assurance that the MSR hardware is operating as expected.

As shown above, separate multiplicity distributions are recorded for the D + A and A gates. Table 18.1 shows typical D + A and A distributions obtained from a 60 g plutonium oxide sample measured in a multiplicity counter with roughly 56% efficiency. Each distribution contains the number of times each multiplicity occurred in the corresponding gate. As an example from this table, seven neutron pulses were found 183 times in the D + A coincidence gate and 42 times in the A coincidence gate. For the conventional A multiplicity distribution sampling, the A multiplicity is triggered by each detected pulse, and therefore, the sum of all multiplicities in the A distribution (37,153,097) is the total number of detected pulses. The sum of all multiplicities in the D + A distribution (37,153,123) is not always exactly equal to the total number of triggers because the D + A gate interval is shifted in a typical case by a long delay of 4096μs from that of the A gate. (For a purely random pulse stream, the two distributions are the same within statistical errors.) For a correlated pulse stream, the D + A distribution has more high-multiplicity events, and the A distribution has more low-multiplicity events.

The (D + A) and A distributions in Table 18.1 can be analyzed to obtain the number of single, double, and triple neutron pulses. Note that the number of 1's, 2's, and 3's in Table 18.1 does not correspond to the singles, doubles, and triples counts. Instead, the singles count corresponds to the sum of all of the multiplicities (in the case of fast accidentals normalized by

Table 18.1 Multiplicity distribution for a 60-g plutonium oxide sample

Multiplicity	Counts in (D + A) Gate, i.e., f(i) distribution	Counts in A Gate, i.e., b(i) distribution
0	26,804,360	29,731,130
1	8,187,530	6,222,207
2	1,772,831	1,016,603
3	325,270	157,224
4	53,449	22,387
5	8231	3093
6	1237	402
7	183	42
8	30	8
9	2	1
10	0	0
SUM	37,153,123	37,153,097

trigger frequency), although in practice, Singles is obtained directly by counting all pulses as shown in Fig. 18.2. The Doubles is the sum of all of the multiplicities times the mean of the D + A distribution minus the mean of the A distribution, which is the same as a conventional shift register output. The triples is a more complex unfolding of the D + A and A distributions. The individual expressions for obtaining the count rates from the recorded multiplicity distribution are discussed in more detail in Sect. 18.2.4.2.

The complete multiplicity distributions are used to compute the first, second, and third factorial moments, which can be related to Singles, Doubles, and Triples. The reason that very high multiplicities—8's, 9's, 10's, etc.—must be measured is that the average number of events inside the gate of the shift register is the singles count rate times the gate width. For example, if the singles rate is 100,000 cps and the gate width is 64μs, the average number of events in either the D + A or the A gate sampled at random times is 6.4. So even for a purely random neutron source, the D + A and A multiplicity distributions will range from 0's to ~20's, with their peak around 6 or 7; very high count-rate situations can be encountered with multiplicities in excess of 128. Note that the upper limit of multiplicity distribution is dictated by hardware and software implementation, which currently enables multiplicity distributions in excess of 500 (Chap. 19).

In the case of fast accidentals sampling, the A multiplicity is triggered at the clock speed of the shift register electronics, which results in much faster sampling than the sampling of the D + A gate. When using fast accidentals, the sum of all multiplicities in the A distribution is no longer equal to the total number of detected pulses because the A multiplicity is now triggered with much higher frequency than the incoming neutron detections. To correct for this outcome, the A multiplicity must be normalized by the sampling clock frequency. This correction is typically performed in INCC software, and the measured A multiplicity distributions obtained from INCC therefore look very similar for fast accidentals as well as for conventional sampling using a long delay. The advantage of fast accidentals is in improved precision of the accidentals rate (A) and consequently of the doubles count rate (D), thereby reducing count times for all coincidence and multiplicity measurements. As discussed in Chap. 17, Sect. 17.8.1, the doubles relative error is defined as

$$\frac{\sigma_D}{D} = \frac{\sqrt{(D+A)+A}}{D} = \frac{\sqrt{D+2A}}{D}. \tag{18.3}$$

For typical safeguards measurement situations, the doubles count rate $D \ll A$ and Eq. 18.3 collapse to

$$\frac{\sigma_D}{D} = \frac{\sqrt{2A}}{D}. \tag{18.4}$$

In the case of fast accidentals, the variance of A becomes small compared with the variance of (D + A) and the doubles relative error, still assuming the limit of $D \ll A$, reduces to [14]:

$$\frac{\sigma_D}{D} \sim \frac{\sqrt{A}}{D}. \tag{18.5}$$

From Eqs. 18.4 and 18.5, it can be seen that fast accidentals lead to reduction of count times by roughly a factor of 2 for a given precision of doubles count rates. Similar impact on triples count rates can be expected, based on the triples uncertainty expression in Ref. [14]. Nevertheless, caution should be applied when using the fast accidentals option in situations where variations of count rates can be expected throughout the duration of the assay (such as variations due to nuclear material moves in the vicinity of the assay system or during long-term unattended measurements). Under such conditions, fast accidentals do not correctly reproduce the actual variation of accidentals during the measurement interval and can lead to incorrect doubles and triples count rates. As discussed in more detail in Chap. 19, for conventional accidentals, such changes are correctly detected in data consistency checks performed by software (INCC), but these tests will not be valid when fast accidentals are used.

18.2.3.2 List Mode

Another approach to collect and extract multiplicity distributions and singles, doubles, and triples count rates is to use list-mode hardware. List-mode capability has been proposed since the early 2000s due to its more complete information about the assayed item, state-of-health monitoring capability, and natural falsification-resistant features. Today, list mode still represents a novel capability in safeguards; however, with a dedicated list-mode hardware for safeguards applications becoming commercially available, the use of list mode is gaining increasing interest. The Hungarian Centre for Energy Research developed the list-mode module, PTR-32, which has been authorized by the IAEA for safeguards applications [15], and other units have been developed to support safeguards needs. A complete overview of current list-mode hardware is provided in Chap. 19.

The key advantage of list mode is in recording and storage of all of the neutron-detection times (i.e., the entire pulse train). Recall that in an MSR, the neutron pulse train is processed in the MSR hardware to extract the D + A and A multiplicity distributions, but the individual neutron-detection times are not saved; therefore, list-mode data represent the most complete information that can be obtained from thermal-neutron counters and as such, provide the most comprehensive information about the counter as well as the assayed material. The list-mode data can be collected from a total output of a neutron counter or from the individual preamplifier modules within the counter (called *channel-by-channel recording*). In the former case, the list-mode hardware essentially serves as a direct replacement of the MSR without the need for any further modification of the neutron counter. It can be used for coincidence- as well as multiplicity-counting applications to provide the additional detail contained in the complete pulse train. The latter implementation (channel-by-channel recording) requires a modified neutron counter to enable recording of signals from individual preamplifier modules. Traditionally, only a single (total count rate) output is used from the neutron counter; however, several counters have been developed (Ref. [16] and Chap. 19, Sect. 19.6.1) or modified, such as in the case of Los Alamos National Laboratory epithermal neutron multiplicity counter (ENMC), [17] to support the channel-by-channel recording capability. Although most of these counters are research instruments, they provide a valuable insight into the full potential of the list-mode capability. The count rates recorded from individual channels within a neutron counter can be used to deduce additional information, such as spatial distribution of nuclear material in the measured area. Such techniques have been proposed for improved assay of waste in large drums or crates as well as for next-generation holdup measurements inside glove boxes in plutonium reprocessing facilities (Chap. 19, Sect.19.4.4). Additionally, information from individual channels provides a direct state-of-health monitoring capability, and the multiple signals are much more difficult for a potential diverter to falsify than a single output. An example of state-of-health monitoring is shown in Fig. 18.3. Background count rates from a set of 22 channels are shown over a period of 15 months, where a malfunctioning channel can be clearly identified because its rate is much larger than all of the other channels (Fig. 18.3, top). Figure 18.3, bottom, illustrates how the list mode can be used to simultaneously collect high-voltage plateaus from individual preamplifier modules in the counter. Such information can be used to identify potential detector or preamplifier module issues, as shown in the reduced gain observed on one of the channels. (The high-voltage plateau shown in the figure was obtained using a boron-lined proportional counter, which has a lower operating high voltage than a standard ^{3}He tube.)

The pulse train from a full counter as well as individual channels can be analyzed to extract Rossi-α or time-interval distributions. Rossi-α distributions are produced by creating a histogram of times that follow after an arbitrarily chosen starting neutron detection, and time-interval distributions are produced by creating a histogram of times between subsequent neutron detections. For more details on origins of these distributions, see Chap. 17, Sects. 17.2.2 and 17.2.3. These distributions provide a direct access to time-dependent components of the measured signal and expand assay capabilities for a variety of scenarios, such as spent fuel measurements [18–20]. Finally, pulse-train data can be analyzed using other algorithms different from shift register implementation, such as Feynman variance-to-mean (Chap. 17, Sect. 17.4; and Chap. 18, Sect. 18.2.9.2) or the randomly triggered inspection (RTI) technique (Chap. 18, Sect. 18.2.9.3). These algorithms provide additional flexibility in data analysis and can be used to expand current analysis techniques (Chap. 18, Sect. 18.2.9).

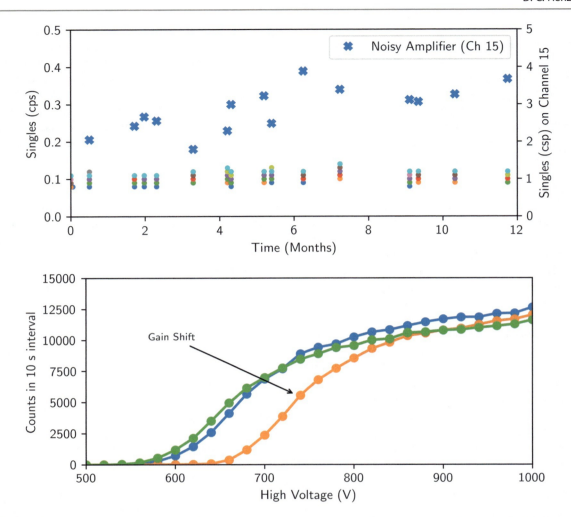

Fig. 18.3 Examples of list-mode use for state-of-health monitoring. (top) Noisy preamplifier module can be clearly identified in a system of 22 identical detectors; (bottom) reduced gain can be seen in high-voltage plateau on one of the channels

The recorded pulse train can be analyzed either online, in a similar fashion as a shift register (but with the shift register algorithms implemented in software), or stored for subsequent offline analysis. The online implementation is the preferred option for field use and has been incorporated into the INCC6 software (Chap. 19, Sect. 19.2.6). The offline analysis option provides added flexibility for future re-analysis, for example, to investigate observed verification inconsistencies; however, it can result in large amounts of data and high demands on data storage, especially for high-count-rate (high-mass) applications. For example, a 1000 s measurement of 600 g of PuO_2 in a high-efficiency neutron multiplicity counter, such as ENMC, would result in ~0.5–0.9 gigabytes of data, depending on the list-mode hardware. Options to store the list-mode data in a more compact form while preserving some of its comprehensive information—such as the potential to store Rossi-α distributions instead of a full pulse train—have been proposed. The advantage of such implementation would be a significant reduction of data storage requirements while preserving the time-dependent information.

To extract the singles, doubles, and triples count rates from list mode, the raw data must first be reduced to multiplicity distributions in the same format as obtained from the multiplicity shift register. This step is performed in software as described Sect. 18.2.4.2.

18.2.4 Passive Multiplicity Mathematics

Once the Singles, Doubles, and Triples have been measured, mathematical equations are used to directly relate those rates to ^{240}Pu mass and other item parameters. The starting point for these equations is the spontaneous fission process in plutonium, which provides the assay signature for multiplicity counting as it does for conventional coincidence counting. Most plutonium items also emit neutrons from (α,n) reactions with matrix materials. Neutrons from either spontaneous fission or (α,n) reactions can induce fissions in the item and result in self-multiplication. These processes are described in detail in Chap. 14.

The passive multiplicity-counting equations are based on fundamental assumptions about the fission process and the item, which set limits on their accuracy and range of applicability. The extraction of item parameters from the measured Singles, Doubles, and Triples requires two sets of equations—one set of equations to calculate Singles, Doubles, and Triples from the measured multiplicity distributions and the second set of equations to relate the measured singles, doubles, and triples count rates to the properties of the item (i.e., ^{240}Pu mass, multiplication, and α). The two sets of equations are described in Sects. 18.2.4.2 and 18.2.4.3 and contain everything that is needed for plutonium mass assay: how Singles, Doubles, and Triples can be extracted from the multiplicity counter measurement; and how Singles, Doubles, and Triples are related to the item properties.

18.2.4.1 Point-Model Assumptions

The equations given in the subsequent sections are based on some important assumptions and mathematical models about the process of neutron emission and detection. To the extent that the theoretical model matches the plutonium items, the measured singles, doubles, and triples rates provide an exact solution for the ^{240}Pu$_{eff}$ mass, its multiplication, and its (α,n) rate. To the extent that the model is not a perfect match, we can expect to encounter some biases or limitations in the multiplicity technique; therefore, the following assumptions are important to remember [21]:

1. *It is assumed that all induced fission neutrons are emitted simultaneously with the original spontaneous fission or (α,n) reaction (the superfission concept).* This concept is generally well satisfied for thermal-neutron counters due to the long lifetime of neutrons (die-away time) in the detector system.
2. *It is assumed that the detector efficiency and the probability of fission are uniform over the item volume.* This assumption is called the *point model* because it is equivalent to assuming that all neutrons are emitted at one point. The weighted point-model and spatial multiplication models described in Sect. 18.2.8 provide some improvements to this assumption for items with high multiplication.
3. *It is assumed that (α,n) neutrons and spontaneous fission neutrons have the same energy spectrum so that the detection efficiency, the fission probability, and the induced-fission multiplicity are the same for both neutron sources.* This assumption is not valid for many (α,n) sources. The two-energy point model described in Sect. 18.2.8.4 provides some improvements to this assumption.
4. *It is assumed that the neutron die-away time in the sample/detector combination is well approximated by a single exponential time constant.* This assumption is generally well satisfied in practical point-model applications for the majority of well coincidence and multiplicity counter designs. Advanced analysis based on Rossi-α distribution (Sect. 18.2.9.4) was developed for situations such as spent fuel measurements, where multiple die-away components contribute to the measured result.

18.2.4.2 Measured Singles, Doubles, and Triples Count Rates

This section will describe how the singles, doubles, and triples count rates are obtained from the multiplicity counter measurement using information from a multiplicity shift register or from list-mode data.

The multiplicity shift register measures the foreground multiplicity distribution $f(i)$ in the D + A gate and the background distribution $b(i)$ in the A gate, as given in Table 18.1. The software (typically INCC, Ref. [13]) computes the factorial moments $f_1, f_2, b_1,$ and b_2 of these distributions,[1] as described in Ref. [1]. The factorial moments are normalized to D + A and A gate trigger rates, respectively. The singles rate (S) is the total number of detected pulses per unit time and is provided directly by multiplicity shift register, as shown in Fig. 18.2. In terms of the factorial moments, the singles rate (S), the doubles rate (D), and the triples rate (T) are given by

$$S = S \cdot f_0 \qquad (18.6)$$

---

[1] $f_1 = \sum_{i=1}^{Nmax} i \cdot f(i)$ $f_2 = \sum_{i=2}^{Nmax} \frac{i(i-1)}{2} f(i)$ $b_1 = \sum_{i=1}^{Nmax} i \cdot b(i)$ $b_2 = \sum_{i=2}^{Nmax} \frac{i(i-1)}{2} b(i)$, where N_{max} is the maximum recorded multiplicity.

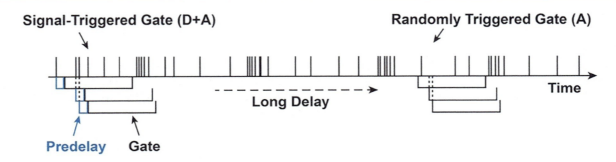

Fig. 18.4 Gates used to calculate the multiplicity distribution of a pulse train

$$D = S(f_1 - b_1) \tag{18.7}$$

$$T = \frac{S(f_2 - b_2 - 2b_1(f_1 - b_1))}{2}. \tag{18.8}$$

Note that $f_0 = 1$ because the distributions are normalized to 1. The Doubles are the difference between the first moment of the foreground multiplicity distribution and the first moment of the background multiplicity distribution (times the singles rate), which represents a subtraction of purely accidental coincidences from all possible coincidences detected in the D + A gate. This formulation is equivalent to the subtraction of the D + A and A accumulator in the conventional shift register (Chap. 17, Sect. 17.5). The Triples equation is not so intuitively obvious because of the cross terms; however, it can be viewed as subtraction of purely accidental triple coincidences occurring in A gate (term b_2) and those that originate from counting a correlated pair and an uncorrelated neutron as an accidental triple coincidence, term $b_1(f_1 - b_1)$, from all possible triple coincidence in the (D + A) gate (f_2 term).

The extraction of singles, doubles, and triples count rates from list mode is performed in software and can be implemented using the standard shift register algorithm. As discussed in Sect. 18.2.9, other implementations (using Feynman or RTI algorithms) are also possible. Here, we focus on the shift register algorithm, which is the most relevant for current practical implementation.

The first step in extraction of Singles, Doubles, and Triples from list mode is calculation of multiplicity distributions, which is accomplished by reproducing the multiplicity shift register operation in software by imposing a fixed-length predelay P, followed by gate G, and counting all pulses that fall within this gate, as shown in Fig. 18.4. The foreground multiplicity distribution within the D + A gate is obtained by binning all pulses within this gate based on their multiplicity. The D + A gate in the software, called the *signal-triggered gate*, is opened by every recorded neutron pulse. The background multiplicity distribution is determined from A gates that follow each detected pulse as before but with a sufficient time delay (typically 4095μs) so that any correlation in the pulse train has died away, as in a conventional multiplicity shift register. Fast accidentals can also be implemented in software by opening the A gate with predefined frequency independent of the frequency of detected neutron pulses. The resulting fast accidentals background multiplicity distributions have to be normalized to the sampling frequency similarly to the case of a multiplicity shift register, as described in Sect. 18.2.3.1. The reduced factorial moments of the background and foreground multiplicity distributions ($b(i)$ and $f(i)$, respectively) can then be calculated using the method previously described. These reduced factorial moments are used to determine the singles, doubles, and triples rates using Eqs. 18.6 through 18.8. Because Eqs. 18.7 and 18.8 combine the reduced factorial moments of both foreground and background multiplicity distributions, this approach is sometimes called *mixed* in literature [22].

When extracting multiplicity distributions from list mode, the measurement time should be calculated carefully because it impacts the resulting count rate (S, D, T) normalization and potentially leads to discrepancies with results obtained from a multiplicity shift register. Reference [23] has introduced some of the concepts to correctly extract the measurement time in list-mode analysis to minimize such inconsistencies. Additionally, different definitions of predelay and gate interval edges in software can have a slight impact on the extracted count rates, as shown in Ref. [17]. When multiplicity distributions are extracted from raw list-mode data in software, each individual pulse gets compared with predelay and gate interval edges to determine whether it falls within its limits. The resulting multiplicity distributions will be slightly different depending on whether the first (or the last) pulse within a gate interval is counted. Although the impact was found to be small, it illustrates the potential reasons for subtle differences between the results obtained from multiplicity shift register and list mode.

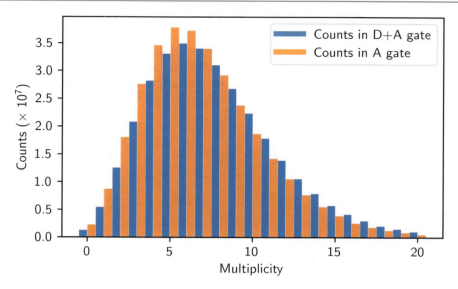

Fig. 18.5 Multiplicity distribution for a 3.8 kg plutonium metal sample measured in a 56% efficient multiplicity counter

However, as evaluated in Refs. [17, 24], for correctly implemented list-mode reduction in software, the agreement with multiplicity shift register results was found to be within 3 σ or better than 1% for the singles, doubles, and triples count rates.

Figure 18.5 shows a histogram of the multiplicity distributions in the D + A and A gates from a 3.8 kg plutonium metal item [1]. The high singles rate from this item yields multiplicities as high as 20. At these high rates, the D + A and A distributions do not look very different to the eye, so the use of factorial moments and Eqs. 18.7 and 18.8 are essential to unfold the correct doubles and triples count rates. These rates must also be carefully corrected for background count rates and electronic dead times.

18.2.4.3 Analytical Definition of Singles, Doubles, and Triples Count Rates

After having extracted the singles, doubles, and triples count rates from the measured multiplicity distributions, we also need analytical expressions that relate these count rates to the item properties. These equations were derived by Böhnel [21] and Cifarelli and Hage [25] based on the point-model and other assumptions described earlier and are described in more detail in Ref. [1].

$$S = F\varepsilon M \nu_{s1}(1 + \alpha) \tag{18.9}$$

$$D = \frac{F\varepsilon^2 f_d M^2}{2}\left[\nu_{s2} + \left(\frac{M-1}{\nu_{i1}-1}\right)\nu_{s1}(1+\alpha)\nu_{i2}\right] \tag{18.10}$$

$$T = \frac{F\varepsilon^3 f_t M^3}{6}\left[\nu_{s3} + \left(\frac{M-1}{\nu_{i1}-1}\right)[3\nu_{s2}\nu_{i2} + \nu_{s1}(1+\alpha)\nu_{i3}] + 3\left(\frac{M-1}{\nu_{i1}-1}\right)^2\nu_{s1}(1+\alpha)\nu_{i2}^2\right] \tag{18.11}$$

where F = spontaneous fission rate,
ε = neutron-detection efficiency,
M = neutron leakage multiplication,
α = (α,n) to spontaneous fission neutron ratio,
f_d = doubles gate fraction,
f_t = triples gate fraction,
$\nu_{s1}, \nu_{s2}, \nu_{s3}$ = first, second, and third factorial moments of the spontaneous fission neutron distribution, and
$\nu_{i1}, \nu_{i2}, \nu_{i3}$ = first, second, and third factorial moments of the induced fission neutron distribution.

Because the gate width used to extract correlated doubles and triples is finite, only a fraction of all fission neutron correlations is counted by the multiplicity electronics. To relate the measured doubles and triples count rates to the item's physical properties (mass, multiplication, α), these count rates must be corrected to include the coincidences that fall outside the finite gate to establish the complete correlated signature. The doubles and triples gate fractions (f_d and f_t, respectively) represent the fractions of the detected Doubles and Triples that are actually counted inside the gate width and are used to correct for the finite gate width. For a neutron detector with a die-away time characterized by a single exponential with time constant τ, f_d is given by

$$f_d = e^{-\frac{P}{\tau}}\left(1 - e^{-\frac{G}{\tau}}\right), \tag{18.12}$$

where τ = detector die-away time,
G = shift register gate width, and
P = shift register predelay.

The triples gate fraction is given by $f_t = f_d^2$ in theory but is usually determined experimentally, as discussed in Sect. 18.2.5.2, because the die-away curve for a real detector is not exactly a single exponential.

The F, M, and α in Eqs. 18.9 through 18.11 are the three unknowns that we need to solve for. The other parameters in the equations are either known physics parameters (ν_{i1-3}, ν_{s1-3}; Ref. [1]) or detector parameters (ε, f_d, f_t) that are established in detector characterization and calibration measurements (Sects. 18.2.5 and 18.2.6).

18.2.4.4 Final Solution for Sample Mass, Multiplication, and α

Equations 18.9 through 18.11 relate singles, doubles, and triples rates to the unknown item parameters, and Eqs. 18.6 and 18.8 calculate the singles, doubles, and triples rates from the multiplicity distributions; these are the relationships needed for multiplicity analysis. For measurements of large mass items in small containers, the detection efficiency ε is usually assumed to be a known parameter obtained from careful measurement of a californium reference source (see Sect. 18.2.5.1). Then Eqs. 18.9 through 18.11 can be solved for m_{240}, α, and M. The solution for M is obtained first by solving the following equation:

$$a + bM + cM^2 + M^3 = 0, \tag{18.13}$$

where the coefficients are functions of S, D, and T:

$$a = \frac{-6T\nu_{s2}(\nu_{i1} - 1)}{\varepsilon^2 f_t S(\nu_{s2}\nu_{i3} - \nu_{s3}\nu_{i2})} \tag{18.14}$$

$$b = \frac{2D[\nu_{s3}(\nu_{i1} - 1) - 3\nu_{s2}\nu_{i2}]}{\varepsilon f_d S(\nu_{s2}\nu_{i3} - \nu_{s3}\nu_{i2})} \tag{18.15}$$

$$c = \frac{6D\nu_{s2}\nu_{i2}}{\varepsilon f_d S(\nu_{s2}\nu_{i3} - \nu_{s3}\nu_{i2})} - 1. \tag{18.16}$$

Once M is determined, the item fission rate, F, is given by

$$F = \frac{\dfrac{2D}{\varepsilon f_d} - \dfrac{M(M-1)\nu_{i2}S}{\nu_{i1} - 1}}{\varepsilon M^2 \nu_{s2}}. \tag{18.17}$$

The second term in the numerator of Eq. 18.17 represents the effect of item self-interrogation due to induced fission, which must be subtracted from the emitted doubles to obtain the spontaneous fission rate. Once F is obtained, the item's ^{240}Pu effective mass, m_{240}, is given by

$$m_{240} = \frac{F}{\left(473.5\frac{\text{fissions}}{s \cdot g}\right)}. \tag{18.18}$$

Note that the value of ^{240}Pu specific spontaneous fission rate (473.5 fissions/s/g) has been recently re-evaluated and updated to 474.7 fissions/s/g [26]. The original value of 473.5 fissions/s/g is used here for consistency with the value traditionally used in the INCC software.

The item's α value is given by

$$\alpha = \frac{S}{F\varepsilon\nu_{s1}M} - 1 \tag{18.19}$$

If the item's isotopic composition is known, the total Pu mass, m_{Pu}, can be calculated from

$$m_{Pu} = \frac{m_{240}}{2.52f_{238} + f_{240} + 1.68f_{242}}, \tag{18.20}$$

where f_{238}, f_{240}, and f_{242} are usually obtained by mass or gamma-ray spectroscopy.

For measurements of low-Pu-mass items in large containers such as waste drums, the detection efficiency ε may vary from item to item. In this situation, it might be a good approximation to assume that sample self-multiplication, M, equals 1. Then M can be considered a known parameter, and one can solve Eqs. 18.9 through 18.11 for m_{240}, α, and ε. The equations for this case are given in Ref. [1].

18.2.5 Multiplicity Detector Characterization Measurements

A series of detector characterization measurements are required to establish the detector parameters used in Eqs. 18.9 through 18.11 and are necessary for routine operation. These measurements are typically performed in a well-defined laboratory environment using well-characterized ^{252}Cf and, if available, Pu standards.

18.2.5.1 Efficiency

The purpose of this procedure is to determine the neutron-detection efficiency of the multiplicity counter from the measured singles rate of a ^{252}Cf source that has a known neutron yield (neutrons/s). The singles rates from the californium reference source is a convenient choice if the californium source strength is known. A ^{252}Cf source has a half-life ($T_{1/2}$) of 2.645 yr., so the expected yield (Y) is

$$Y = Y_0 e^{-\left(\frac{\ln(2)\Delta T}{T_{\frac{1}{2}}}\right)}, \tag{18.21}$$

where ΔT is the time from the date that the original source strength Y_0 was determined and $\ln(2) = 0.6931$. The source should be placed at a reproducible position in the center of the assay chamber. The detector efficiency ε is then determined from the singles count rate, S, with the singles background subtracted.

$$\varepsilon = \frac{S}{Y} \tag{18.22}$$

Note that the efficiency determined with a californium source may be slightly different than the value needed for actual material measurements and can be further tuned in subsequent multiplicity counter calibration described in Sect. 18.2.6.

18.2.5.2 Gate Fractions

Because multiplicity counters can have more than one significant component to their die-away curves, the calibration process outlined here does not rely directly on the gate fractions f_d, f_t that are determined from the measured die-away time using

Eq. 18.12. Instead, we empirically determine the actual fractions of the Doubles and Triples that are counted within the gate width G. These doubles and triples gate fractions are calculated from the singles, doubles, and triples rates measured with the ^{252}Cf reference source. For this procedure, the strength of the californium source does not need to be known, but it requires a source with no multiplication (M = 1) and no (α,n) reactions (α = 0). Then from Eqs. 18.9 through 18.11, the following equations for the doubles gate fraction f_d and the triples gate fraction f_t are obtained.

$$f_d = \frac{2\nu_{s1}D}{\varepsilon \nu_{s2}S} \qquad (18.23)$$

$$f_t = \frac{3f_d \nu_{s2}T}{\varepsilon \nu_{s3}D} \qquad (18.24)$$

In these equations, ν_{s1}, ν_{s2}, and ν_{s3} are the factorial moments of the ^{252}Cf source distribution from Ref. [1] or elsewhere. Typical values are ν_{s1} = 3.757, ν_{s2} = 11.952, ν_{s3} = 31.668 (Table 13.2 and Ref. [27]).

18.2.5.3 Die-Away Time, Predelay, and Gate Width

The preceding sections describe the determination of detector parameters used in Eqs. 18.9 through 18.11. Several additional parameters need to be determined for routine measurements, including die-away time τ, predelay P, and gate width G.

The gate width is typically determined as a multiple of detector die-away time. The die-away time is extracted in series of measurements of ^{252}Cf performed using a pair of gate widths that increase by a factor of 2 expected for the multiplicity counter, typically 16μs and 32μs or 32μs and 64μs. These measurements verify that the detector die-away time is as expected and that the coincidence gate width (G) is correctly set for this die-away time. From Eq. 18.12, the ratio of the doubles at a gate length of 64μs to the doubles at 32μs (for the same predelay, die-away time τ, and measurement time) is

$$\frac{D_{64}}{D_{32}} = \frac{1 - e^{-\frac{64}{\tau}}}{1 - e^{-\frac{32}{\tau}}}. \qquad (18.25)$$

Solving this equation for τ gives

$$\tau = \frac{-32}{ln\left(\frac{D_{64}}{D_{32}} - 1\right)}. \qquad (18.26)$$

Some drum-sized multiplicity counters may have more than one significant component to their die-away curves because the distribution of polyethylene and ^{3}He is not uniform throughout the detector. Multiple significant components to die-away curves will also be observed in cases where the measured item has a longer die-away time than the detector itself, such as in the case of spent fuel (see Chap. 21, Sect. 21.4.5). Therefore, if the calculation of the die-away time is repeated with other gate lengths, a somewhat different die-away time can be obtained.

The optimum gate length for a given coincidence counter corresponds to the gate length that gives the lowest relative error for coincidence counting, which occurs at approximately 1.27τ. This optimum is a very broad and shallow minimum so that setting the gate width to the nearest multiple of 2 is usually sufficient. At high count rates, it may also be beneficial to set the gate width to a smaller value to reduce dead-time effects on the triples count rate. Optimum gate width can be obtained experimentally by acquiring a series of measurements using a ^{252}Cf source or plutonium standard for a range of gate widths and plotting doubles and triples relative standard deviation (RSD) as a function of gate width [28]. Using a shift register, this task can be accomplished by multiple measurements each with different gate width value (typically increased in factor-of-2 increments). This effort is significantly simplified using list-mode acquisition because a single measurement can be post-processed using multiple gate widths. An example of such analysis [29] is illustrated in Fig. 18.6 for a plutonium standard measured in the epithermal neutron multiplicity counter (ENMC). Reference [29] also provides additional evaluation of impacts of gate-width setting for a typical multiplicity counter.

For typical neutron well-counters, the value of predelay has traditionally been selected in the range of 1.5μs and 4.5μs, with the lower value used in multiplicity counters with a high number of preamplifier modules (i.e., low dead time). The predelay can also be determined experimentally through a series of measurements with a fixed gate-width value and range of predelay values (typically 1–5, in 0.5μs increments). Both shift register and list mode can be used to acquire such data, with list-mode

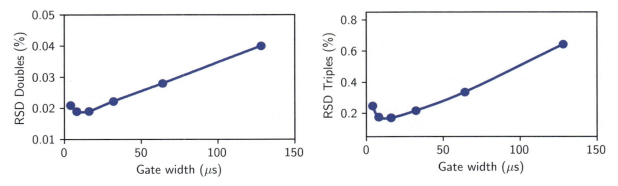

Fig. 18.6 Doubles and Triples RSD as a function of gate width measured for a ^{252}Cf source in ENMC [29]

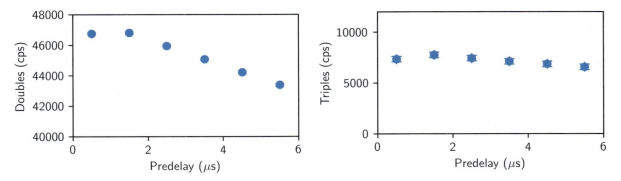

Fig. 18.7 Doubles and Triples as a function of predelay measured for a ^{252}Cf source in an AWCC

acquisition providing the advantage of requiring a single measurement that can be subsequently re-analyzed using multiple predelay values. The doubles and triples count rates that correspond to different predelay values are then plotted as a function of predelay, as illustrated in Fig. 18.7 for a ^{252}Cf source measured in active well coincidence counter (AWCC). Figure 18.7 illustrates that as predelay is shortened, the doubles and triples count rates increase linearly up to the point where electronics and dead-time effects become significant, and count rates begin to deviate from the linear trend. The optimum predelay setting would be selected at predelay values before this nonlinearity occurs to ensure that the results are not affected by these dead-time and electronics effects. In the example shown in Fig. 18.7, the optimum predelay would correspond to ~3μs.

18.2.5.4 Dead-Time Coefficients

Lastly, multiplicity analysis requires that careful dead-time corrections be applied to the singles, doubles, and especially the triples count rates. At high neutron count rates, some of the neutron pulses are lost as a result of detector and electronic dead time. If two neutrons are detected too close together, their pulses overlap and appear as one pulse in the counting circuits. These effects can occur in the ^{3}He tubes, in the amplifier/discriminator circuits, or in the OR-ing circuits that combine the pulses for input into the multiplicity shift register or list-mode acquisition (Chap. 17, Sect. 17.6). The derandomizer was implemented in some multiplicity counters to remove the impact of the OR-ing circuit, but the other contributions still remain. It should be emphasized that although multiplicity counting uses a fixed predelay to compensate for some of the dead-time effects, the dead time is still present in the data because it is present throughout the pulse train and affects the pulses that are inside the D + A or A gates. The main role of predelay is to minimize discrepancies between the D + A and A gates that would otherwise result in biased doubles and triples rates (Chap. 17, Sect. 17.6.2).

The singles and doubles rates can be corrected for dead time according to the same equations as in coincidence counting:

$$S_0 = S_m e^{\frac{\delta S_m}{4}} \tag{18.27}$$

$$D_0 = D_m e^{\delta S_m}, \tag{18.28}$$

where $\delta = A + BS_m$, A and B are the dead-time coefficients.

The subscript "m" means *measured*, and "0" refers to the quantity corrected for dead time. The coefficients A and B depend on the multiplicity counter, particularly on the number of preamplifier modules. One way to determine A and B is to reproducibly measure californium sources of different strengths and adjust A and B to obtain the same doubles/singles ratio for all sources.

The complex equations used for correcting the triples count rate are built into the INCC code [1, 13] and use a constant parameter δ_{mult} called the *multiplicity dead time*. The parameter δ_{mult} can be determined by measuring a weak and a strong ^{252}Cf source. The ratio of Triples to Doubles should be independent of the ^{252}Cf source strength after dead-time corrections, so the dead time can be determined by adjusting δ_{mult} to give the same Triples/Doubles ratio. The triples dead-time correction can be expressed as

$$T = e^{\delta_{mult} \, S_m} S_m T_\delta (1 + c S_m), \quad (18.29)$$

where T_δ represents the triples count rate extracted from dead-time-corrected multiplicity distributions based on the approach developed by Dytlewski [30]. The additional exponential factor is needed to fully correct for the dead-time effects. The parameter c is typically set to 0 and was developed to enable additional empirically determined dead-time coefficients for very high-count-rate situations [31]. Typical values of the A, B, and δ_{mult} for a range of existing multiplicity counters can be found in Chap. 19.

Other software implementations are used in the facility environment, such as NDA2000 [32], which includes similar dead-time implementations.

18.2.6 Multiplicity Calibration Procedure

The calibration procedure for neutron multiplicity counting does not require a series of representative physical standards to determine a curve of instrument response versus ^{240}Pu effective mass as is the case for coincidence counting (Chap. 17, Sect. 17.9). Multiplicity counters are used to assay a wide range of impure plutonium items, and representative physical standards are usually not available. It is possible to calibrate the counter directly by solving the singles, doubles, and triples equations for M, α, and m_{240} for a given class of items by using three detector parameters: ε, f_d, and f_t and Eqs.18.13 through 18.20. To the extent that the plutonium items satisfy the assumptions of the point model, this method provides an accurate assay; however, whenever possible, traceable physical standards should be used to validate this procedure or to remove remaining biases caused by idealizations of the point-model assumptions.

Initial determination of the detector parameters needed for multiplicity assay can be done with a ^{252}Cf source alone, as described in the previous section. However, multiplicity assays of plutonium based on the parameters determined from ^{252}Cf alone can be biased because of uncertainties in the nuclear data parameters for ^{252}Cf and plutonium, differences in detection efficiency between ^{252}Cf and plutonium fission neutrons, and differences between the actual items to be assayed and the assumptions of the point model. As a result, additional adjustments of the detector parameters are typically performed to calibrate the counter to the type of items to be assayed.

For instance, the initial determination of the detection efficiency using a ^{252}Cf reference source can be biased because of uncertainties in the neutron yield of the source and slight differences in energy between ^{252}Cf and plutonium fission neutrons. The efficiency should be corrected with Monte Carlo calculations using plutonium fission spectrum representative of items to be assayed. If Monte Carlo calculations show a significant difference in neutron-detection efficiency between a ^{252}Cf point source and the actual plutonium items, the efficiency can be adjusted accordingly. The magnitude of the adjustment will depend on the detector, but it will typically be in the range of 1–2%.

The doubles and triples gate fractions are calculated from the singles, doubles, and triples rates measured with the ^{252}Cf reference source using Eqs. 18.23 and 18.24. If one or more physical standards are available, the calibration can usually be improved by adjusting f_t to obtain the best assays for the standards. This method corrects for uncertainties in the nuclear data parameters of ^{252}Cf and plutonium and for differences between the actual items assayed and the point model assumptions. The adjustment to f_t may be on the order of 10%.

If M or α for the physical standard is known, it may also be helpful to vary ε and obtain the best agreement with the known M, α, and mass values. This approach can be helpful only if M or α is well known; otherwise, it will introduce a bias into the assays that will increase as M or α increases. In general, if no independent information is available on M or α for the standards, changes to ε are not advisable unless they are based on Monte Carlo calculations. An exception is where various well-known samples are available with characteristics very similar to the unknowns. In this case, all three detector parameters can be varied to provide the best assays.

After calibration, it is also helpful to verify the applicability of the multiplicity-counting technique by measuring some items to which the technique will be applied. The measurements should be verified relative to calorimetry or some other traceable process. If new material categories that might not be appropriate for multiplicity counting need to be measured, some fraction of the measurements should undergo a verification process.

18.2.6.1 Multiplication Bias Correction

If large plutonium metal or dense oxide items are to be measured, a variable bias correction is needed to correct multiplicity assays for the nonuniform probability of fission inside large multiplying items [33]. This correction is applied to the measured ^{240}Pu effective mass and is implemented in the multiplicity software analysis code INCC [13] as follows:

$$CF = 1 + a(M-1) + b(M-1)^2, \qquad (18.30)$$

where M is the measured item multiplication, and CF is a multiplicative factor that increases the calculated value of m_{240} in Eq. 18.18. An empirical set of coefficients appropriate for metal items in different multiplicity counters is $a = 0.0794$ and $b = 0.1386$ [34]. These empirical coefficients were derived from known plutonium items and are not necessarily applicable to all plutonium metal items and all counters. The correction factor approaches 1 as M approaches 1, so it can be left on even if the counter is used to assay only non-metallic items.

The weighted point-model and spatial multiplication model equations discussed in Sects. 18.2.8.1 and 18.2.8.2 provide a more robust bias correction within the framework of the multiplicity analysis equations. Although the weighted point model is currently implemented in standard safeguards software (INCC), the spatial multiplication model represents a newer development that is yet to be fully experimentally tested. The performance of the weighted point model is illustrated in Fig. 18.8, which compares the performance of the weighted point model on 232 metal samples measured in the Savannah River FB-Line multiplicity counter with the performance on 66 MCNP simulated metal cylinders—ranging in shape from pancakes to rods—that were used to determine the weighted point-model doubles and triples weighting factors [35]. The Savannah River bias correction coefficients were determined to be $a = 0.1460$ and $b = 0.1485$ and were used as additional parameters in the INCC analysis using the weighted point-model option. Because the Savannah River and Monte Carlo N-Particle eXtended (MCNPX) curves in Fig. 18.8 are similar, these coefficients may be less dependent on plutonium geometry, impurities, or counter cavity size. Measurements of impure plutonium metal show that the coefficients still vary with high levels of impurities and that the weighted point model can have an alpha-dependent bias for items with α values of 0.5 or more.

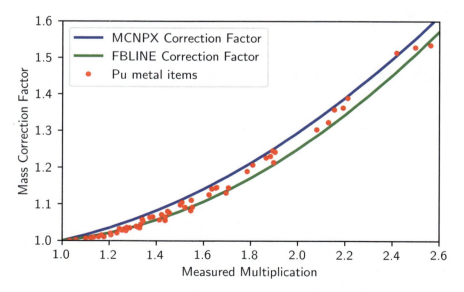

Fig. 18.8 Plutonium-240 effective mass correction factor versus measured multiplication for two sets of plutonium metal samples. The measured set consists of 232 plutonium metal items (upper curve). The computational set consists of 66 plutonium metal cylinders used to determine the doubles and triples weighting factors set. (Lower curve; Ref. [35])

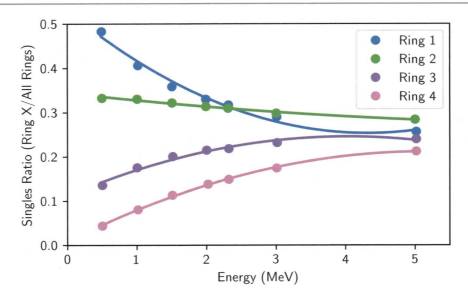

Fig. 18.9 Relative count rates for the four rings in the pyrochemical multiplicity counter as a function of neutron energy. Ring 1 is the inner ring

18.2.6.2 Energy Bias Correction

As discussed in Chap. 19, Sect. 19.3.4, Monte Carlo calculations and laboratory measurement campaigns have shown that the ratio of the Singles in the inner and outer rings is a good indicator for neutron energy spectrum shifts that may bias the assay. For example, Fig. 18.9 plots the relative responses of the four ^{3}He tube rings in the pyrochemical multiplicity counter as a function of energy. Low-energy neutrons are preferentially detected in the inner ring and high-energy neutrons in the outer rings. The neutron energy spectrum affects the overall efficiency, the probability of induced fission, and the induced fission multiplicity distribution. These effects are usually not large, but they can cause energy-dependent biases in plutonium measurements of items with large amounts of (α,n)-emitting impurities or neutron moderators.

Most multiplicity counters provide separate outputs for the inner and outer rings; these outputs can be fed into multiplicity shift registers using the auxiliary connectors. The reported singles count rates for these rings can thereby provide an estimate of the average energy of the (α,n) neutrons. A correction procedure based on ring ratios is discussed in Sect. 18.2.8.4. Even if the correction is not used, the ring ratio provides a useful diagnostic that can warn of high levels of impurities or moderators that may bias the assay.

18.2.7 Passive Multiplicity Counting Performance and Areas of Applications

To use passive multiplicity counting in practical measurements, it is important to understand the area of its applicability as well as its expected performance. This section summarizes expected performance and provides an overview on areas of use of passive multiplicity counting. Passive multiplicity counting is compared with coincidence counting for range of material types.

18.2.7.1 Expected Assay Precision

Multiplicity counter precision is determined primarily by the statistical uncertainty in the singles, doubles, and triples count rates and the reproducibility of item placement. The dominant uncertainty is usually in the triples and is determined primarily by detector efficiency, die-away time, count time, count rate, neutron multiplication, and the (α,n) rate. The propagated uncertainty in the plutonium mass is estimated by the analysis software (INCC) in one of two ways: sample standard deviation or theoretical standard deviation. The sample standard deviation is determined from the statistical scatter between the repeat runs that make up a single assay. A typical assay is performed in cycles (10–30 s long) of multiple repeat runs that sum up to the desired total measurement time. The accuracy of the sample standard deviation is determined by the number or repeats. The theoretical standard deviation is calculated from theoretical methods that have been benchmarked against measurements of the observed scatter [36]. The theoretical calculation is the default setting in INCC, but whenever possible, use of sample

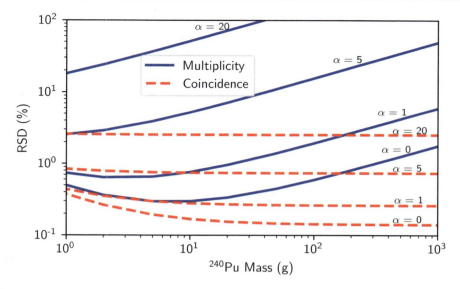

Fig. 18.10 Estimated precision for multiplicity and conventional coincidence assay using a multiplicity counter with a detector efficiency of 50%, a gate width of 64μs, a die-away time of 50μs, and a predelay of 3μs. The background is 100 cps, and the counting time is 1000 s

standard deviation is recommended to accurately reflect the actual statistical scatter in the measurement. For the code to give a reliable value for the sample standard deviation, a reasonable number of individual cycles should be recorded for each measurement. Ten cycles will give an uncertainty of ~5% on the standard deviation. In addition to these standard methods, a Bayesian framework for uncertainty quantification was developed in Ref. [37] to provide a comprehensive approach to uncertainty quantification in multiplicity counting. This method enables a complete uncertainty propagation that considers uncertainty of neutron detector calibration parameters as well as nuclear data to estimate final plutonium mass and provides enhanced capability to accurately reflect individual uncertainty contributions for plutonium mass assay results.

Figure 18.10 provides rough estimates of the predicted assay repeatability due to counting statistics for plutonium metal ($\alpha = 0$), oxide ($\alpha = 1$), scrap ($\alpha = 5$), and residues ($\alpha = 20$) for a high-efficiency multiplicity counter [36]. The α values of such materials vary; the values selected here are representative. Note that the repeatability due to counting statistics is always better for conventional coincidence counting than for multiplicity analysis.

18.2.7.2 Typical Assay Bias

Assay bias for multiplicity counting is very low for samples that meet the mathematical assumptions of the point model; however, in practice, container and matrix factors may yield noticeable biases. Table 18.2 provides a summary of typical performance for multiplicity assay of nuclear materials commonly found in DOE facilities and can be used to estimate performance for other similar applications. The observed repeatability and bias estimates include the uncertainties from neutron counting, gamma-ray isotopic analysis of the ^{240}Pu effective fraction, and reference values based on calorimetry/isotopics or destructive analysis. Results for metal measurements were corrected for multiplication bias.

To provide a standard reference on expected performance of typical neutron multiplicity counters in use by IAEA, IAEA evaluates and compiles historical verification data to establish target performance based on instrument and material category. These international target values (ITVs) are provided in Ref. [38] and serve as industry standard. ITVs are evaluated and updated regularly. An ITV release was made in 2022.

One important question for safeguards personnel is when to use multiplicity counting versus conventional coincidence counting. Factors to be considered in selecting either conventional or multiplicity counting vary with material type. Factors include plutonium mass, (α,n) reactions, available detector efficiency, self-multiplication, neutron energy effects, spatial distribution of fissile material, other matrix effects, available counting time/required precision, and container size and shape. For impure items with unknown multiplication and α, the accuracy for multiplicity counting is usually much better; however, if the conventional coincidence and multiplicity results are the same within counting statistics, then it might be better to use the more precise conventional results. Other considerations for several major material types are given in the following sections.

Table 18.2 Typical multiplicity counter performance on various nuclear materials measured in multiplicity counters with roughly 50%–55% efficiency [39, 40]

Nuclear material category	No. items measured	Pu Mass (g)	(α,n)/sf rate α	Count time (s)	RSD (%)	Bias (%)
Pu metal	13	200–4000	0–1.3	1800	4.6	1.3
	14	1500–5000	0	1800	2.7	−0.1
	283	5–2300	0.2–1.7	1000	2.8	0.4
	32	2000	0		0.5	−0.8
	6	1100–2100	0–0.6		2.2	2.0
Calex Std. [41]	8	398	1	1800	1.3	0.3
	150	398	1	1000	1.4	0.8
Pu oxide	45	500–5000	1	1800	2.2	0.0
Impure Pu	12	20–875	0.7–4.3	1000	2–3	0.8
Oxide	261	4–1800	0.2–3.2	1800	2.3	0.3
	7	20–800	0.6–0.8	5000	0.7	0.5
Pu scrap	16	80–1175	1–6	3600	5.7	−1.6
	24	2000	1–6	1800	5.8	−1.0
	17	300–1400	2–30	1000	11.4	−2.1
Pu residues	10	40–300	13–29	3000	4.8	0.9
	8	160–340	7–34	3600	18.8	−9.2
	17[b]	0–10	−1–82	1000	24	−2.3
	21[b]	10–40	−4–29	1000	22	1.8
	16[b]	40–400	0.4–13	1000	17	−1.4
Pu waste	est.	1	5	1000	10	2–5
U/Pu oxide	8	200–800	1–2	1000	1–2	1–3
Pu inventory	106	1000–4000	1–6	1800	4.2	0.6
Verification	67	300–1000	1–10	1200	8	0.0
	90	1–4000	1–6	6–12 (h)	10	−0.5
	150		0.8–4.5	1000	5	0.2

18.2.7.3 Plutonium Metal

Pure plutonium metal has $\alpha = 0$, so conventional coincidence counting (known-α; Chap. 17, Sect. 17.10) will give assays with better precision. In reality, most metal items contain some impurities, and their surface is usually oxidized. Actual α values range from 0.1 to ~1.0, which would produce unacceptable biases in conventional coincidence counting.

Plutonium metal buttons are dense, compact items for which the theoretical point model does not correctly describe multiplication. The variable-multiplication correction in the weighted point model described in Sect. 18.2.8.1 is usually needed to obtain good assay results, as illustrated in Fig. 18.8. The weighted point-model equations have provided good results at Savannah River, as shown in Ref. [35]. Additional information on multiplicity-counting performance for plutonium metal can be found in Ref. [1].

18.2.7.4 Plutonium Oxide

Pure plutonium oxide yields neutrons from spontaneous fission and from (α,n) reactions on oxygen. Depending on burnup, α is in the range of 0.4–0.8. If some impurities are present, it is conservative to estimate that $\alpha = 1$. Multiplicity information is not needed if the oxide is so pure that α can be calculated and the known-α approach can determine the mass and the multiplication from the singles and doubles rates. Some oxides in existing facilities are impure, with α values between 1 and 4; for these oxides, multiplicity counting is significantly more accurate than coincidence counting because of induced fissions caused by (α,n) neutrons and changes in multiplication caused by density variations.

Figure 18.11 compares coincidence and multiplicity assays of eight pure oxide samples and seven impure oxide samples measured in a five-ring multiplicity counter [42]. The coincidence results include the known-α correction for (α,n) reactions in oxide. The pure oxide samples have an average assay/reference ratio of 1.004 ± 1.4% by coincidence counting and 1.006 ± 0.66% by multiplicity counting. The impure oxide samples are 1.039 ± 8.2% by coincidence counting and 1.005 ± 0.68% by multiplicity counting. Additional examples of multiplicity-counting performance for PuO_2 can be found in Ref. [1].

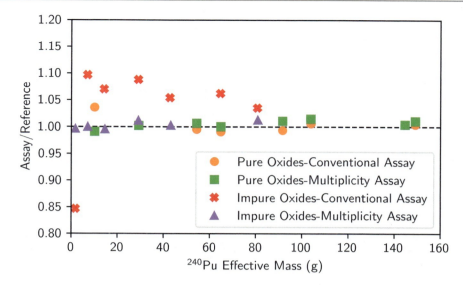

Fig. 18.11 Comparison of conventional coincidence and multiplicity assays of pure and impure plutonium oxide samples in the five-ring neutron multiplicity counter [42]

18.2.7.5 Plutonium Scrap

Scrap is recyclable plutonium-bearing material left over from processing activities. It can include relatively pure metal or oxide or materials with large quantities of matrix elements such as fluorine and beryllium. For multiplicity counting, we define scrap as *items with α in the range of 1–6*. The best assay technique for scrap (highest precision and lowest bias) depends on the nature of the item. An impure metal item is best assayed with multiplicity counting, but an item with very low multiplication and a very high (α,n) rate, such as waste, is best assayed with coincidence counting. The high α values will result in loss of multiplicity-counting precision due to the contribution of a high number of accidental coincidences into the doubles and especially triples count rates (Sect. 18.4). The loss of counting precision with increasing α was also illustrated in Ref. [43]. The selection of multiplicity or coincidence counting depends on whether the reduced bias with multiplicity assay, which can correct for induced fissions, outweighs the loss of precision. Additionally, the coincidence counting will require standards representative of the material to be assayed to establish a calibration dependence. Conventional coincidence counting usually provides an upper mass limit because it undercorrects for multiplication.

If the scrap contains moderating materials or if it emits enough (α,n) neutrons with energies much different from fission neutrons, then the detection efficiency can vary from item to item and result in biased multiplicity assay. The ratio of count rates in the inner and outer rings is sensitive to neutron energy, and very impure scrap items display dramatic changes in their ring responses. Figure 18.12 illustrates the measured ratio of Ring 1 to Ring 4 as a function of mean energy in the in-plant multiplicity counter [44]. The ring ratio can be a valuable tool in identifying items that contain gross impurities and in distinguishing metal from oxide or other impure plutonium. Often the multiplicity assay results can be corrected for bias from (α,n) neutrons using an empirical technique described in Ref. [45].

18.2.7.6 Plutonium Residues

Residues are plutonium-bearing materials retained during production operations. They can include ash, combustibles, inorganics, salts from pyrochemical processes, and wet items. Residues are often packaged in large cans, have significant quantities of nuclear material, and are very heterogeneous and difficult to measure. Many residues contain large quantities of elements, such as fluorine and beryllium, and therefore they may exhibit α values of 10–30 or more.

Multiplicity counting is the only feasible passive neutron option, but long count times are typically needed to assay plutonium residues. The multiplicity technique is generally not well suited for items that produce many more (α,n) neutrons than spontaneous fission neutrons because the errors from counting statistics are large. To get good precision on the triples, such materials should be assayed only with extremely long count times. This method is suggested by the curve for $\alpha = 20$ in Fig. 18.10. However, for plutonium items with unknown (α,n) rates, multiplicity analysis will be far less biased than conventional coincidence counting, particularly if a ring-ratio correction is used to compensate for changes in the neutron energy spectrum.

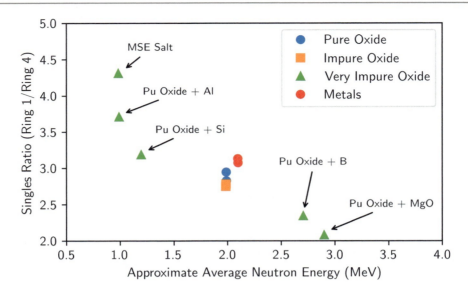

Fig. 18.12 The measured ratio of Ring 1 to Ring 4 as a function of mean neutron energy for various samples in the in-plant multiplicity counter [44]

An example is the measurement of plutonium-bearing salts using a 30-gallon multiplicity counter [40] at the Livermore Nuclear Materials Facility. The results are given in Table 18.2 for three mass ranges and are dependent on the success of the chemical separation process that yielded the salts. All multiplicity measurements were compared with much slower but very accurate calorimetry/isotopics measurements. Even at high α, the multiplicity results appear unbiased but with a larger RSD. For measurements of high-α plutonium scrap and residue items, the epithermal neutron multiplicity counter (ENMC; Chap. 19) is a good approach for obtaining much shorter counting times. In Ref. [46], it was shown that the ENMC reduced assay times by factors of ~20 relative to a plutonium scrap multiplicity counter (PSMC; Chap. 19).

18.2.7.7 Plutonium Waste

Multiplicity counting can improve the assay of plutonium waste in 208 L drums or standard waste boxes (~1900 L capacity) even though the waste contains only a few grams of plutonium. The additional information can flag the presence of shielding materials, detect highly multiplying items, or correct for (α,n)-induced fissions or detector efficiency variations. The expected assay precision for multiplicity analysis of waste drums has been estimated using a figure-of-merit (FOM) code [36, 47], included in Table 18.2. Multiplicity assay will have poor precision relative to conventional coincidence counting but may be more accurate because the bias from (α,n) induced fissions is corrected. Figure 18.13 plots the FOM code estimated precision for neutron coincidence and multiplicity assay of waste as a function of α if the multiplicity information is used to solve for mass, M, and α [47]. The assumed detector efficiency is 35%. The figure shows that the multiplicity assay of relatively clean plutonium waste that has a value between 0.5 and 5 will have an RSD in the range of 2–10%. This result is not the same high precision and accuracy that multiplicity counters can achieve for the assay of small cans of plutonium, but it does show that multiplicity of waste is feasible with high-efficiency drum counters. Waste with high fluoride content may have α values of 20–150, and the RSD will rise to 20–100% or more. Even with this poor precision, multiplicity assay may be much more accurate than conventional coincidence counting because the bias caused by (α,n) induced fissions is corrected.

Note that when using multiplicity analysis to solve for detector efficiency rather than item multiplication, as described in Sect. 18.2.4.4, the RSD increases by a factor of 3–4 over the entire mass range and is 5–15% at best. Because the use of multiplicity analysis to solve for detector efficiency significantly increases the RSD, alternative techniques for determining efficiency—such as segmented add-a-source [48] or ring-ratio analysis [44]—should be employed. For multiplicity assay of waste, the drum counter should be operated with a room background in the range of 10–100 counts/s. Higher values will seriously degrade both coincidence and multiplicity assay RSD over the low mass range of 1–50 g of plutonium. For screening at the transuranic-waste detectability limit, multiplicity counting usually does not have sufficient precision.

18.2.7.8 Mixed Uranium/Plutonium Oxide

Mixed oxides do not meet the assumptions used in the multiplicity mathematics and must be assayed with caution. The induced fission multiplicity distributions, fission cross sections, and capture cross sections in uranium are different from those

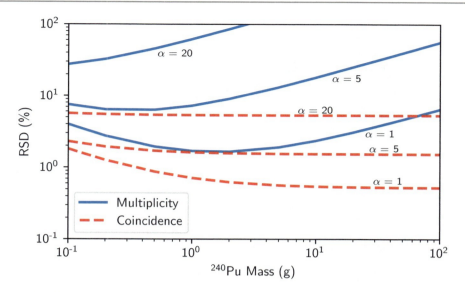

Fig. 18.13 Estimated precision for neutron coincidence and multiplicity assay of waste as a function of ^{240}Pu mass for various α values if the multiplicity equations are solved for mass, M, and α. Efficiency = 35%, gate width = 64μs, die-away time = 100μs, counting time = 1000 s, and background = 10 counts/s

in plutonium. If the calibration parameters appropriate for plutonium are used to assay plutonium oxides that have a large uranium concentration relative to their plutonium content, the assay results tend to bias low [33]. If the coefficients are adjusted to fit a particular mixed-oxide (MOX) material with a fixed U/Pu ratio, then the multiplicity performance can be good. Assay of high-burnup MOX items that contain a few hundred grams of plutonium using the PSMC in Japan gives 1–2% precision in 1000 s measurements [49].

18.2.7.9 Plutonium Inventory Verification

Multiplicity counting can be used successfully for inventory verification. The technique provides a better verification than is possible with coincidence counting because it requires less initial inventory information. This section provides some general guidelines on inventory verification measurements and includes past results as examples.

For inventory verification, it is helpful to segregate items into categories, such as calibration and measurement control standards, plutonium metal, low-α plutonium (impure oxides and scrap), and high-α plutonium (residues with α > 6). These categories can be defined by the observed sample multiplication, mass, α, or measurement precision. For low-α plutonium, count times of 1000–1800 s are usually sufficient to eliminate counting statistics as a significant contribution to the overall precision. For high-α plutonium, multiplicity counting may not be the preferred option because of the long count times required. For metal items, the data analysis should include a multiplication bias correction.

For example, Fig. 18.14 compares known-α coincidence assay and multiplicity assay for IAEA physical inventory verification (PIV) measurements at Rocky Flats using a 30-gallon multiplicity counter [50]. Multiplicity improved the average agreement between declared and assay by nearly a factor of 2 over the best conventional approach. Multiplicity verified 61% of the samples to within ±3% and 100% to within ±18%. Overall, 1σ agreement between multiplicity and site declarations was 4.2% for all items. Additional examples of application of multiplicity-counting performance to PIV can be found in Ref. [1].

Experience suggests that a small fraction of the inventory has multiplicity assays well outside the reasonable expected limit of error because of unknown matrix effects that do not meet the point-model assumptions. These outliers require calorimetry and/or gamma-ray isotopics to resolve. However, multiplicity counting can substantially reduce the number of items that require these techniques and thereby allow an increase in measurement throughput.

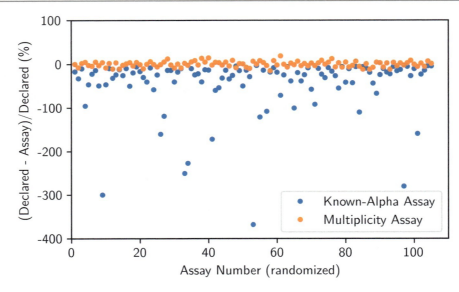

Fig. 18.14 Comparison of known-α conventional coincidence assay and multiplicity assay for physical inventory verification measurements using a 30-gallon neutron multiplicity counter [50]

18.2.8 Improved Point-Model Developments

This section summarizes some extensions to the conventional point model that were developed to mitigate impacts of some of point-model assumptions on the assay results for certain assay scenarios (such as large, dense plutonium materials or items with high α) and to expand observables of the original point-model equations (e.g.. point model with quads and pents rates) to improve the assay accuracy. The weighted point model was developed during the 1980s; the spatial multiplication model and two-energy point model represent more recent developments (~2010s). The latter two models are currently not implemented in standard safeguards software or practiced but are presented here for completeness on the available capability. In addition, an extension of the point model that incorporates delayed neutron contribution is discussed briefly for its value in understanding the common practice of neglecting delayed neutrons in the analysis.

18.2.8.1 Point-Model Equation to Higher Order

The traditional point model uses equations for three measured, correlated count rates (Singles, Doubles, and Triples) that relate the properties of the assayed items, the detector parameters, and the nuclear data to the measured singles, doubles, and triples count rates. Extensions of the point-model equations to higher-order count rates (i.e., beyond triples) were developed, and explicit expressions up to fifth order were reported in Ref. [6]. The corresponding count rates are called *quads* (for the fourth order) and *pents* (for the fifth order). The extension of the point model to higher-order correlations is further discussed in Sect. 18.2.9.1 in the context of dead-time correction, which is a prerequisite to fully use the higher-order correlated count rates (quads and pents) in practical assay applications.

18.2.8.2 Weighted Point-Model Equations

The point-model assumptions of constant neutron energy and uniform multiplication can cause significant biases in measurements of impure plutonium metal and large, dense plutonium oxides. For example, multiplication inside such items is not uniform but is actually much larger in the center and falls off toward the surface. To correct for these biases, weighted point-model equations [35] have been developed to add variable multiplication weighting factors to the expressions in Eqs. 18.10 and 18.11 and implemented in INCC [13]. The weighting factors are obtained from Monte Carlo simulations using the Monte Carlo N-Particle (MCNP) code [51], which supports the simulation of spontaneous fission sources and can tally the source and detected neutron multiplicity distributions.

The weighting factors are introduced for both the spontaneous fission and (α,n) contributions to the doubles and triples rates to account for multiplication variations. The MCNP code is used to model items with widely varying geometries, densities, and impurities to obtain simulated singles, doubles, and triples count rates. The weighting factors are then selected to correct for the biases observed in the simulated assays. For unknown samples, the weighted point-model equations are used to compute M, α, and m_{240} but require solving a fifth-order equation for multiplication. Performance of the weighted point model to correct for multiplication bias is described in Sect. 18.2.6.1.

18.2.8.3 Spatial Multiplication Model

Similar to the weighted point model, the spatial multiplication model was developed to correct for point-model assumption on constant neutron multiplication over the physical extent of an item. In the conventional point model, multiplicity rates are described as functions of the neutron leakage multiplication raised to an integer power M_L^n; i.e., the singles rate is a function of M_L, the doubles rate is a function of M_L^2, the triples rate is a function of M_L^3, etc. The spatial multiplication model makes the assumption that the distribution of the leakage multiplication is spherical: $M_L(r)$ [52] and modifies the point model by substituting $\langle M_L(r) \rangle^n$ with $\langle M_L(r)^n \rangle$.

Figure 18.15 demonstrates the effect of such a substitution visually and illustrates that, assuming the leakage multiplication is a cosine distribution (Fig. 18.15 left), the point model assumes that the doubles rate is a function of $\langle \cos(r) \rangle^2$ (grey line in Fig. 18.15 right), whereas the spatial multiplication model assumes that the doubles rate is a function of $\langle \cos^2(r) \rangle$ (black line in Fig. 18.15 right). The point-model assumption leads to an underestimate in the leakage multiplication in highly multiplying items. The traditional point-model approach evaluates Singles correctly but underestimates the leakage multiplication dependence for higher multiplicity rates, making the singles rates seem inflated. This result leads to an overestimate of (α, n) rates and an underestimate of the assayed mass and multiplication within the point model (Fig. 18.16). The weighted point

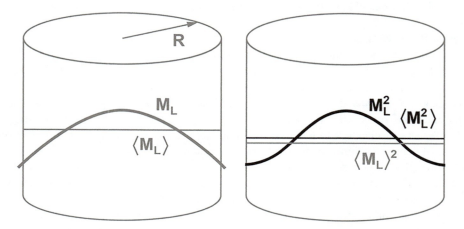

Fig. 18.15 Comparison of effects of leakage multiplication varying across an item. (left) Leakage multiplication assuming a cosine distribution (grey curve); (right) point-model assumption (grey line) and spatial multiplication model assumption (black line). The point-model assumption leads to an underestimate in the leakage multiplication in highly multiplying items [52]

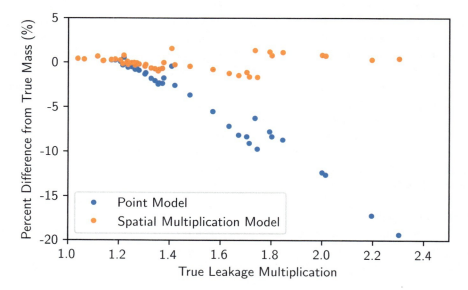

Fig. 18.16 Comparison of point model with the spatial multiplication model results assuming a Bessel function spatial distribution and known M_R [52]

model [53] was created to account for this effect by weighting the individual $\langle M_L(r) \rangle^n$ terms, whereas the spatial multiplication model aims at an analytical solution.

Several even functions have been explored for the spatial distribution of the leakage multiplication, including cosine and Bessel functions [35, 52, 54]. If one assumes the leakage multiplication has a cosine distribution,

$$M_L(r) = M_A \cos(ar), \qquad (18.31)$$

where M_A is the multiplication amplitude, and $M_R = aR$ (where R is the radius of the item) is the spatial multiplication extent. Then, the spatial multiplication model has four unknown parameters: mass, (α,n) rates, M_A, and M_R. These parameters can be determined by using multiplicity rates up to quadruples. Approaches to extend multiplicity counting up to Quads and Pents were developed and are discussed in Sect. 18.2.9. Alternatively, one can use the known M_R approach that uses Singles, Doubles, and Triples to solve for three item properties: mass, (α,n) rate, and average neutron multiplication M_L, using Eqs. 18.9 through 18.11. In this approach, a relationship between M_R and M_L is known to reduce the number of parameters [52]. Results with a known-M_R spatial multiplication model (three-parameter) approach have shown significant improvements on the point model up to a leakage multiplication of 2.5 (see Fig. 18.16 and Ref. [52]). Masses determined with the four-parameter spatial multiplication model have been shown to be accurate up to $M_L = 1.3$ [52, 54].

18.2.8.4 Two-Energy Point Model

The implications of point-model assumption that (α,n) neutrons and spontaneous fission neutrons have the same energy spectrum were evaluated in Ref. [55], which considers an energy-dependent bias in the assay of impure plutonium dioxide by neutron multiplicity counting and suggests that the ring ratio might be used as a spectral index from which the effective energy of the (α,n) neutrons might be estimated. This work introduced (α,n) neutron energy dependence via modification of the original point-model equations by making the following quantities functions of (α,n) neutron energy: neutron-detection efficiency, probability for an (α,n) neutron to induce a fission relative to that for fission spectrum neutron, and first through third factorial moments of neutron multiplicity distribution from fissions induced by (α,n) neutrons. Reference [55] provided revised expressions for Singles, Doubles, and Triples and evaluated bias resulting from point-model assumption of single-neutron energy for a range of impure plutonium dioxide items and a range of typical neutron multiplicity counters.

The equations and methodology from Ref. [55] are implemented in INCC code [13] as a dual-energy model option; however, to date, the implementation has not been extensively experimentally verified. The INCC implementation assumes that measurements are performed with a multiplicity counter that provides ring-ratio capability, i.e., a counter that is equipped with multiple rings (or rows) of detectors and provides individual transistor-transistor logic signals from two of its rings (rows). The INCC implementation requires the user to specify the neutron-detection efficiency of the counter, its ring-ratios, and relative induced fission probability (for (α,n) neutron versus fission neutron) for the expected range of (α,n) neutron energies. The energy dependence of neutron detection efficiency and ring ratios can be obtained experimentally using plutonium oxide materials that contain various impurities or other semi-monoenergetic sources (such as AmLi) or by MCNP modeling. The relative fission probability is typically obtained from MCNP. During assay using the dual energy model option, the INCC uses measured count rates from two of the counter rings to obtain a measured ring ratio, which is then used to extract the corresponding neutron energy, the neutron-detection efficiency, and the relative induced fission probability and to calculate the energy-dependent, induced-fission factorial moments. These quantities are then used in the two-energy point-model equations adopted from Ref. [55] to calculate the mass of the assayed impure item.

Additionally, an energy-dependent weighted point model was evaluated in Ref. [56], where extension to quads was explored to correct for (α,n) neutron energy dependence. This model uses quads as an additional observable to iteratively solve the weighted point-model equations with (α,n) neutron energy as an adjustable parameter. The method demonstrated a significant reduction in plutonium mass bias for items with α ≥0.5 contribution using MCNP calculations; however, this method requires additional evaluation for more realistic detector configurations.

18.2.8.5 Point Model with Delayed Neutrons

A further extension of the traditional point-model equations was developed to include the contribution of the delayed neutrons [57]. The traditional point-model equations are derived under the *superfission assumption*, where all of the neutrons are assumed to be generated simultaneously (Sect. 18.2.4.1). However, delayed neutrons cannot be accommodated, by definition, into the concept of superfission, and an extension of the point model was developed to account for this effect. This extension has proven that the influence of delayed neutrons is small (within the typical measurement uncertainty) for the types of deeply subcritical assay problems that concern the nuclear safeguards community. Such results justify the common practice of neglecting the delayed neutron contributions in point-model equations.

18.2.9 Passive Multiplicity-Counting Developments

This section describes additional developments of passive neutron multiplicity-counting algorithms that were pursued to further improve multiplicity-counting capability. They include improved dead-time treatments; extension to higher-order correlations beyond Triples; approaches to extract Singles, Doubles, and Triples and higher-order correlated count rates that do not rely on traditional shift register implementation; and use of list mode to extract instrument or item properties from Rossi-α distribution. Several advanced dead-time-correction algorithms were developed to improve dead-time-correction performance in very high-count-rate situations, where traditional empirical expressions (Eqs. 18.27 through 18.29) were shown to begin to fail [58]. These dead-time-correction algorithms also opened a possibility to extend multiplicity counting toward higher-order correlations, including correlations of four, five, and more neutrons (quads, pents, and higher), as described in Sect. 18.2.9.1. Such extension was not previously readily available because appropriate dead-time corrections for these higher-order correlated count rates did not exist. Although quads count rates equations were developed previously in the context of weighted point model [35] and are also implemented in INCC, they were not corrected for dead time. Alternative methods to extract correlated count rates from the measured multiplicity distributions provide additional flexibility and can lead to improved assay uncertainty (Sect. 18.2.9.3).

18.2.9.1 Advanced Dead-Time Algorithm and Multiplicity Counting Beyond Triples

Conventionally, dead time is associated with counting losses at high event rates; however, in contrast to random neutron sources, dead-time losses in time-correlated neutron pulse streams do not vanish as the count rate approaches zero because neutrons are emitted in time-correlated "bursts" from fission events and are detected with a high instantaneous count rate. Each burst is subject to intrinsic dead-time losses even when individual neutron bursts are widely distributed in time. Further, dead-time losses are increased for overlapping chains. For example, multiplicative items in which the presence of fissile material leads to fission chains have increased neutron multiplicity (number of neutrons per chain) and, therefore, higher dead-time losses. These examples illustrate the need for accurate methods for correcting correlated neutron data for dead time.

The traditional dead-time-correction technique (Eqs. 18.27 through 18.29) relies on semi-empirical dead-time-correction factors, which assume that the neutrons are emitted randomly in time [59]. Depending on the degree of neutron correlation in an item (i.e., the mass and amount of neutron multiplication from induced fission), the random source assumption that is built into the current dead-time-correction technique can strongly affect the dead-time-correction accuracy. The random source assumption also strongly impacts the ability to use higher-order correlated rates that could act as additional observables from the correlated neutron data. From a practical standpoint, the application of fourth (quads) and higher-order correlations have so far been hampered by the lack of a dead-time-correction scheme.

To address this shortcoming, three advanced dead-time-correction algorithms were developed in the 2010s [60–64]. The first approach, the Dytlewski-Croft-Favalli (DCF) approach [60], introduces multiplicity distribution-based dead-time correction, where each entry of multiplicity distribution is corrected before the factorial moments and correlated count rates are calculated. It represents an iterative improvement on the work by Dytlewski [30], which is what current dead-time corrections for triples (Eq. 18.29) is based on. The second approach involves the creation of a complete dead-time model for a single-channel, paralyzable system (correlated neutron dead-time model [CNDTM]; Refs. [61, 62]). The paralyzable (also called *updating*) dead-time model is a typical assumption in correlated (i.e., coincidence and multiplicity) neutron counting and represents a scenario where each neutron pulse that arrives within the dead-time interval extends the dead time regardless of whether the pulse is counted or suppressed. The third approach, the cross-channel comparison (CCC) method [63, 64], is a self-calibrating dead-time correction (using multichannel list-mode hardware) that compares pulses recorded on different preamplifier modules (i.e., channels) against each other. It uses features of Rossi-α distribution measured from each channel in the neutron counter to estimate and recover pulse losses due to dead time. The first two methods require a single dead-time-correction parameter that needs to be established in a separate calibration. The third method is completely self-calibrating and therefore does not require a prior calibration of dead-time coefficient. All three methods enable extension of dead-time correction to higher-correlated count rates beyond triples.

The DCF approach develops dead-time corrections for Singles, which were not included in the original Dytlewski framework, and extends the original work of Dytlewski [30] to any correlated order beyond Triples [60]. Figure 18.17 (left) illustrates the performance of the DCF method on a series of ^{252}Cf sources measured in an epithermal neutron

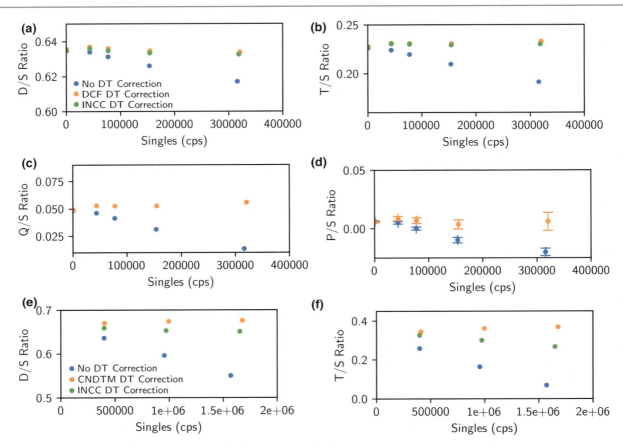

Fig. 18.17 (**a**)-(**d**) DCF-corrected D/S through P/S ratios for a set of ^{252}Cf sources measured in ENMC as a function of dead-time-corrected singles rate compared with INCC dead-time correction (Eqs. 18.27 through 18.29) and uncorrected count rate ratios; (**e**)-(**f**) the same comparison for CNDTM up to Triples [65]

multiplicity counter (ENMC). The displayed count-rate ratios for ^{252}Cf should be independent of the singles count rate if properly dead-time corrected. Figure 18.17 (left) illustrates a good performance of the DCF algorithm in comparison with the empirical correction (Eqs. 18.27 through 18.29) implemented in INCC software. It also demonstrates its extension to and capability to correct for dead time for Quads and Pents. The key advantage of this method is in its computational simplicity, allowing direct extension to higher-order correlated rates and straightforward software implementation.

The CNDTM approach considers the correlated nature of the neutrons that are being emitted by a fissioning item by using joint probability functions to provide a mathematical relationship between the *emitted* neutron pulse train and the *measured* pulse train. It uses three assumptions regarding the measurements that are common to the currently deployed dead-time correction (a single electronics chain for processing the detected signals, the die-away of the neutrons of the system that are described by a single exponential, and paralyzable dead time). The derivation of the CNDTM dead-time model is extensive and is presented in its entirety in Ref. [62]. The model uses a *modified distribution function* (function for the expectation of detection of *n* neutron counts after dead time) to predict the multiplicity moments that are measured after dead-time effects, based on the underlying detection rates (the rates that would be measured if there were no dead-time effects). The modified distribution functions are integrated to find the measured multiplicity moments. The formulas for the measured multiplicity moments can be inverted to directly extract dead-time-corrected detection rates. The corresponding mathematical expressions thus require inversion (through numerical integration) to obtain the dead-time-corrected rates. The model was developed for any multiplicity order but fully formulated up to and including quads. Figure 18.17 (right) illustrates the performance of the CNDTM method on a series of ^{252}Cf sources measured in ENMC compared with the standard empirical dead-time correction implemented in INCC and dead-time uncorrected count rates. The key feature of the CNDTM method is that it represents the closest attempt that has been made to a fully fundamental correction or description of dead time in a temporally correlated counting system. The disadvantage is in its mathematical complexity that must be considered when developing software implementation.

The CCC method [63, 64] makes use of neutron-detection information from each preamplifier module (channel) within the counter. Using multichannel list-mode hardware, neutron-detection time and corresponding channel information can be recorded simultaneously from each preamplifier module. This process enables correction of measured D + A and A multiplicity distributions for dead-time losses using statistical comparison between the pulse arrival times on individual channels within the counter. Because the calculation of correlated count rates is derived directly from the D + A and A distributions, this dead-time-correction method works in principle for any order (Singles, Doubles, Triples and higher). This approach compares detection rates of a preamplifier module that had recently detected a neutron and is therefore under dead time with the rest of the preamplifier modules unaffected by the dead time associated with this detection. The method is self-calibrating, which is done by calculating the probabilities of dead-time loss at specific channels using the actual measured data and basic properties of Rossi-α distribution. The key advantage of the CCC dead-time-correction method is that it does not require any prior calibration or dead-time coefficients and can be implemented live during each measurement. The limitation is in its requirement to record data from each preamplifier module separately, which needs to be implemented in the detector design stage to prevent later hardware modifications because most of the current instruments provide only a single signal output from all of the channels.

All of the dead-time-correction techniques discussed here enable an extension of multiplicity counting to higher-order correlated count rates, which represents the most direct way of extracting additional information from the measured data. Although none of these techniques are currently implemented in safeguards practice, they represent advancements with a potential to improve the accuracy and capabilities of neutron multiplicity counting into the future. As discussed in Sect. 18.2.8, the use of the additional measured quantities (Singles through Quads or Pents)—if correctly corrected for dead time—can lead to improved determination of the physical properties of the assayed item for items that violate some assumptions of the point model.

18.2.9.2 Feynman Multiplicity Histograms

In safeguards applications, it is common to use the shift-register-based analysis and Singles, Doubles, and Triples to extract item properties as discussed throughout this Chapter. In the area of criticality studies, the item properties are extracted using Feynman analysis as described in Chap. 17, Sect. 17.4.1. Both Feynman analysis and traditional multiplicity-counting analysis described in this chapter use measurement gates in time-correlation analysis. The moments of the gate-extracted multiplicity distributions are then related to material properties, such as spontaneous fission rate and neutron multiplication. In Feynman analysis, it is conventional to use series of consecutive gates—equivalent to A gates opened in tight, non-overlapping sequence—to extract moments and calculate the Y-value, which is then related to physical properties of the item. The conventional multiplicity analysis discussed in this chapter uses a combination of detected pulse-triggered D + A gates and random A gates (called *mixed* gates).

However, the Feynman and conventional multiplicity analysis approaches are not limited to their conventional associated gate structure, and both approaches are equally applicable using either purely random gates (A gates) or *mixed* gates (combination of D + A gates and A gates). This method is illustrated in Sect. 18.2.9.3 for the case of multiplicity-counting analysis. In Ref. [66], the Feynman and multiplicity analysis approaches were shown to be mathematically identical. The relationship between Feynman analysis and conventional multiplicity analysis highlights that there are, in fact, multiple ways to extract correlated count rates from multiplicity distributions. Therefore, the primary potential improvements to current analysis methods within point-model assumptions lie in optimization of gate structure and different methods of finding multiplicity rates from the gate moments, as discussed in the following section.

18.2.9.3 Alternative Methods to Extract Correlated Count Rates

The motivation to expand conventional multiplicity analysis beyond the shift register approach is to explore potential improvements of the assay performance. Two primary considerations exist for improving performance: precision of the correlated count rates and the degree to which the correlated count rates are subject to dead-time effects, which is a major component of accuracy. This section focuses on alternative methods of extracting the correlated count rates (Singles, Doubles, Triples, and beyond) using non-traditional gate structures and the degree to which they are affected by dead time.

The two basic types of gate structures result in three different analysis schemes: (1) randomly generated gates (i.e., A gates) opened either with a delay following the detected neutron pulse or at high-speed clock frequency, (2) pulse-triggered D + A gates, and (3) traditional multiplicity-counting analysis.

The first analysis scheme uses the randomly triggered gates only and is called *randomly triggered inspection* (RTI). In RTI counting, a gate of fixed time length (G) is opened at random intervals, and the number of neutron counts that occur within the gate is counted. The multiplicity shift register or software (in the case of list mode) analyzes the number of times each count

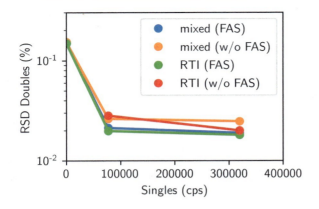 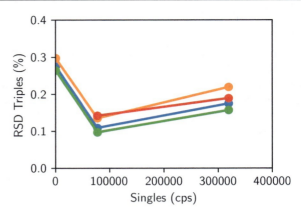

Fig. 18.18 Doubles and triples count rate precision as a function of singles count rate for the different analysis techniques; mixed techniques with (blue) and without (orange) fast accidentals (FAS) are compared with RTI techniques with (green) and without (red) fast accidentals [67]

total occurred within the gate to produce a *background* multiplicity distribution (A multiplicity). The *background* multiplicity distribution alone is then used to calculate the Singles, Doubles, Triples, and beyond [22].

The second analysis scheme uses the pulse-triggered gates only. In pulse-triggered counting (also called *signal-triggered inspection*; STI) each neutron detection opens a gate preceded by a predelay (P), and the *foreground* multiplicity distribution (D + A multiplicity) is generated. A system of equations was derived in Ref. [22] to calculate Singles, Doubles, Triples, and higher-order correlated count rates using *foreground* multiplicity distribution alone. However, it was found that because the *foreground* multiplicity distributions capture a combination of genuine and accidental coincidences within the STI-only formalism, it is not possible to get the accidental coincidence fraction in isolation experimentally and extract gate fractions as described in Sect. 18.2.5.2. To account for this contribution, theoretical relationships must be used, which make the STI-only formalism impractical for routine implementation.

The third analysis scheme is the traditional multiplicity-counting analysis, which uses a combination of RTI and STI gates, called the *mixed gate* approach. This technique forms the core of the standard neutron multiplicity counting as described in this chapter, based on Eqs. 18.6 through 18.8. Note that both RTI and mixed analysis schemes can use high-speed clock frequency to generate A gates, i.e., fast accidentals, as described in Sect. 18.2.3.1.

It was shown in Refs. [61, 67] that extracting correlated count rates using the RTI-only analysis and conventional mixed gates analysis results in different dead-time effects. Additionally, as shown in Ref. [67], these different analysis schemes affect the precision of the measured correlated count rates, illustrated in Fig. 18.18 on a range of ^{252}Cf sources with increasing yields measured in ENMC.

Figure 18.18 shows doubles and triples RSD comparison for the RTI and mixed analysis schemes discussed above, with and without fast accidentals. It demonstrates that the measurement precision is increasingly more affected by the choice of gate structure with increasing item emission rate. In the case of an item with low yield, the doubles precision varies by less than a few percent depending on the gate structure; in the case of high yield, the variation corresponds to tenths of a percent. The use of the RTI technique with a high frequency of A gates (fast accidentals) results in improvement in precision by a factor of ~1.4 for doubles and triples rates with respect to the mixed technique without fast accidentals, which is implemented in the older shift registers. Improvement similar to that of the RTI technique is achieved using the mixed technique with fast accidentals (also discussed in Sect. 18.2.3.1). This improvement corresponds to reducing the required measurement time by approximately a factor of 2. The mixed technique with fast accidentals represents the current standard in safeguards measurements and, as shown in Fig. 18.18, provides one of the highest measurement precisions. However, as described in Sect. 18.2.3.1, to prevent potentially incorrect results, care should be taken to ensure constant measurement conditions when using fast accidentals.

18.2.9.4 Rossi-α Analysis

Rossi-α distribution (RAD) is another advanced analysis technique that became directly accessible with use of list-mode data acquisition. RAD can be used to provide insight into some general neutron counter characteristics and also troubleshooting. Chapter 17, Sect. 17.2.3, describes the overall features of RAD and how it can be used to extract information on detector dead time. Section 17.7 also discusses the use of RAD to evaluate neutron counter signal for potential double-pulsing artifacts.

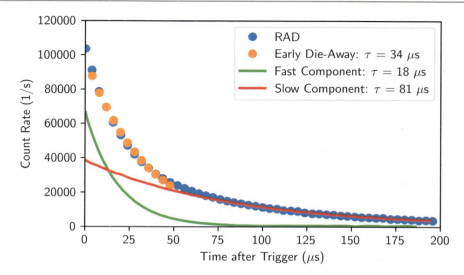

Fig. 18.19 Example of a Rossi-α distribution broken into two additive components (fast and slow) and fit with the single exponential early die-away curve

RAD provides several practical advantages in analysis and precision of neutron multiplicity measurements. Although shift registers require gate selection in advance of a measurement, RAD contains information from all possible combinations of predelay and gate width, enabling optimization of gate selection during post processing. In the unique case where the die-away time of the measured item is longer than that of the detector system, it may be possible to extract item multiplication from the distribution via the early die-away time technique [68]. The early die-away time is a single exponential fit to the early time domain of the RAD distribution that encompasses both the die-away times of individual components and their relative magnitudes. An example of the early die-away fit is shown in Fig. 18.19. The early die-away time and its connection to multiplication in spent fuel is discussed further and in the context of spent fuel measurements in Chap. 21, Sect. 21.4.5.

18.3 Active Neutron Multiplicity Counting

Active multiplicity analysis was developed to provide a multiplicity-counting capability for uranium-bearing materials [2–4]. Active multiplicity counting uses doubles and triples count rates to solve for the sample multiplication and the induced fission rate. An active well coincidence counter (AWCC; see Chap. 20) or similar instrument is used to induce fissions in the ^{235}U. The singles rate is not used because it is dominated by the large background from the interrogation sources (typically AmLi, but ^{252}Cf may also be used). However, active multiplicity counting is more complex than passive multiplicity counting because of an additional new parameter: the interaction of the interrogation source neutrons with the assay item, which is called the *coupling*. The ^{235}U mass depends on both the induced fission rate and the coupling, and this parameter must also be determined.

Active neutron coincidence counting using an AWCC equipped with AmLi interrogation sources, discussed in Chap. 13, is a standard technique for assay of ^{235}U, as described in Chap. 20. Calibration curves are normally obtained by counting uranium standards and plotting the double coincidence rate versus the ^{235}U mass. Because of neutron multiplication and absorption in the uranium, the calibration curves are nonlinear and are sensitive to the geometry and ^{235}U density of the item. The calibration curve approach works extremely well for many situations and often yields nearly linear calibrations because of cancellation between multiplication and absorption effects.

To obtain accurate results with the calibration curve approach, the calibration standards must be representative of the assay items. Some of the important sample characteristics are the enrichment, density, geometry, and material composition. However, appropriate calibration curves are not always available, either because suitable standards are not available or because the important characteristics of the items to be assayed are not well known. For example, uranium metal and uranium oxide items cannot be assayed with the same calibration curve, so representative physical standards of both material types are needed. For large-mass HEU metal items, variations in the geometry of the items or how they are stacked in the can will change the self-multiplication dramatically and cause large biases in the assay result. In general, the measurement bias will increase as the characteristics of the calibration standards and assay items diverge.

The goal of active neutron multiplicity counting, described in this chapter, is to use both the double and triple coincidence rates to provide a new measured parameter that solves for sample self-multiplication and thereby removes most of the bias caused by irregular sample geometry and density. This technique is much less sensitive to differences between the standards and the unknowns and has multiple potential areas of application.

The following sections describe the equations used for active multiplicity, how detectors are characterized, and how measurements are made.

18.3.1 Active Multiplicity Counter Configuration

Active multiplicity-counting electronics is the same as for the passive multiplicity-counting applications described in Sect. 18.3.1. The AmLi sources do not produce a very high count rate, so a variety of existing multiplicity shift registers (including those without derandomizer circuits) or list-mode modules may be used for this application.

18.3.2 Active Multiplicity Equations

The analysis equations for active multiplicity counting are derived from those for passive multiplicity counting, as given in Sect. 18.2, and are similar in form. They are again based on the measured neutron singles, doubles, and triples count rates. The most important difference is the use of an active well counter equipped with interrogating sources to induce fissions in the ^{235}U. Thus, the fission rate to be solved for is not a passive rate but an induced one. Also, the measured Singles are dominated by the interrogation source and are usually not useful for assay. For the same reason, the parameter "alpha," a measure of (α, n)-induced neutrons, is not relevant. The equations are no longer self-contained but now require additional information to determine the coupling between the interrogation source neutrons and the assay item.

The active multiplicity analysis equations are also based on the assumptions described in Sect. 18.2 for passive multiplicity analysis. If actual measurement geometries and items do not meet all of these assumptions, we can expect to encounter some biases or limitations. Specifically, the point-model assumption that the neutron detector efficiency and the probability of fission are uniform over the item volume is not always valid. The neutron counters used for active multiplicity analysis, like the AWCC (see Chap. 20), were not designed to provide a completely flat efficiency profile across the item volume. More importantly, the assumption of constant fission probability across the item volume is not valid for bulk HEU, and a multiplication bias correction has not yet been developed for active multiplicity analysis. Relationships for coupling as a function of multiplication, described in Sect. 18.3.5, tend to compensate for these effects but not completely.

18.3.3 Analytical Definition of Singles, Doubles, and Triples Count Rates

The neutron multiplicity electronics and software yield the first three reduced factorial moments of the measured neutron multiplicity distribution, which are the singles, doubles, and triples count rates. Theoretically, the singles, doubles, and triples count rates (S, D, T) are given by

$$S = S_0 + B + S_s + FM\varepsilon_f\nu_{s1} \tag{18.32}$$

$$D = \frac{F\varepsilon_f^2 f_D \nu_{s2}}{2} \cdot C_d \tag{18.33}$$

$$T = \frac{F\varepsilon_f^3 f_t \nu_{s3}}{6} \cdot C_t, \tag{18.34}$$

where.
S_o = Singles count rate from the AmLi sources without an item present,
B = background singles rate,
S_s = change to S_0 due to scattering and absorption of AmLi neutrons by the item,
F = induced fission rate in item,
M = neutron multiplication,

ε_f = efficiency for detecting induced fission neutrons, and

$\nu_{s1}, \nu_{s2}, \nu_{s3}$ = first, second, and third reduced factorial moments for AmLi-induced fissions in ^{235}U.

(Note that these symbols have a different definition for active versus passive multiplicity counting.)

C_d = a correction factor for self-multiplication of Doubles,

C_t = a correction factor for self-multiplication of Triples,

f_d = doubles gate fraction, and

f_t = triples gate fraction.

These gate fractions are as given in Sect. 18.2.5.2.

For a neutron detector with a die-away time characterized by a single exponential with a time constant τ, the doubles gate fraction f_d is given by

$$f_d = e^{-P/\tau}\left(1 - e^{-G/\tau}\right) \tag{18.35}$$

$$f_t = f_d^2 \tag{18.36}$$

but is usually determined experimentally. Note that ε_f is the efficiency for detecting induced fission neutrons. A different (lower) efficiency ε_a also exists for detecting AmLi source neutrons that pass through the shielded endcaps and still get detected.

In principle, the induced singles count rate could be used to determine the ^{235}U mass. In practice, the Singles are dependent on AmLi neutron source scattering, neutron absorption, and background fluctuations. For large ^{235}U masses of well-defined geometry, useful assays have been obtained by singles counting if varying background is not a problem. But in general, the active multiplicity approach described in Sect. 18.3.5—based on doubles and triples count rates—is much more accurate.

18.3.4 Calculation of Sample/Item Self-Multiplication

Expressions for the correction factors C_d for doubles self-multiplication and C_t for triples self-multiplication can be derived from the passive multiplicity equations. In terms of the self-multiplication M, the expressions are as follows [2, 4]:

$$C_d = M^2\left[1 + \frac{(M-1)\nu_{s1}\nu_{i2}}{\nu_{s2}(\nu_{i1}-1)}\right] \tag{18.37}$$

$$C_t = M^3\left[1 + \frac{(M-1)(3\nu_{s2}\nu_{i2} + \nu_{s1}\nu_{i3})}{\nu_{s3}(\nu_{i1}-1)} + \frac{(M-1)^2 3\nu_{s1}\nu_{i2}^2}{\nu_{s3}(\nu_{i1}-1)^2}\right], \tag{18.38}$$

where $\nu_{i1}, \nu_{i2}, \nu_{i3}$ = first, second, and third reduced factorial moments for subsequent generations of fission neutron-induced fissions in ^{235}U.

Using Eqs. 18.37 and 18.38 for the self-multiplication correction factors and Eqs. 18.33 and 18.34 for D and T, it is possible to solve for item self-multiplication without knowing the fission rate F, which cancels out in the ratio T/D. The self-multiplication, M, can be obtained by solving the following cubic equation [2]:

$$\begin{aligned}M^3 &+ M^2\left[\frac{3\nu_{s2}(\nu_{i1}-1)\nu_{i2} + \nu_{s1}(\nu_{i1}-1)\nu_{i3} - 6\nu_{s1}\nu_{i2}^2}{3\nu_{s1}\nu_{i2}^2}\right] \\ &+ M\left[\frac{\nu_{s3}(\nu_{i1}-1)^2 + 3\nu_{s1}\nu_{i2}^2 - 3\nu_{s2}(\nu_{i1}-1)\nu_{i2} - \nu_{s1}(\nu_{i1}-1)\nu_{i3} - (3T/D\varepsilon_f)(f_d/f_t)\nu_{s1}(\nu_{i1}-1)\nu_{i2}}{3\nu_{s1}\nu_{i2}^2}\right] \\ &+ \frac{3T}{D\varepsilon_f}\frac{f_d}{f_t}\left[\frac{\nu_{s1}(\nu_{i1}-1)\nu_{i2} - \nu_{s2}(\nu_{i1}-1)^2}{3\nu_{s1}\nu_{i2}^2}\right] = 0.\end{aligned} \tag{18.39}$$

The triples/doubles ratio provides a good measure of the neutron multiplication, as documented in Ref. [69]. A value for the multiplication is needed to perform active multiplicity assays and is useful by itself to help authenticate uranium items.

18.3.5 Definition of Item Coupling

Equations 18.32 through 18.34 are similar to those used in passive multiplicity counting; however, for active multiplicity counting, F is the rate at which neutrons from the AmLi interrogation sources induce first-generation fissions in ^{235}U rather than the ^{240}Pu-effective spontaneous fission rate. Therefore, active multiplicity analysis requires a new parameter called the *coupling*, which describes the induced fission rate, F, in terms of the AmLi source strength, Y, and the mass, m, of ^{235}U in the assay item:

$$F = CmY, \qquad (18.40)$$

where.
C = coupling,
F = induced fission rate in item from the AmLi neutrons,
Y = total output of both AmLi sources (neutrons/s), and
m = mass of ^{235}U in grams.

The coupling depends on the item's geometry, ^{235}U density, chemical and isotopic composition, and location in the assay chamber. In broader terms, it depends on the solid angle between the item and the AmLi sources, the neutron multiplication in the item, and other neutron moderation, scattering, and absorption effects. For these reasons, the coupling will not be linear with item mass.

If we substitute Eq. 18.40 into Eqs. 18.33 and 18.34 for the doubles and triples count rates, the product Cm appears together in both equations. Thus, we cannot solve for the sample mass m without using some additional information to obtain an equation for the coupling C. During the development of active multiplicity counting, several different expressions for the coupling have been derived.

The original uranium metal measurements at Savannah River Site and Y-12 [3] used a coupling equation of the form

$$CY = a + \frac{b}{m^{1/3}}, \qquad (18.41)$$

where a and b are calibration constants. This equation was selected so that the fission rate F in Eq. 18.40 would have the form $am + bm^{2/3}$. One term is proportional to item mass, as expected for high-energy interrogation neutrons, and one term is proportional to the item's surface area, as expected for low-energy interrogation neutrons.

Later work revealed an empirical relationship between the coupling and the multiplication [4, 69], which was an important discovery because the multiplication can be obtained from the triples/doubles ratio independent of sample/item mass or coupling, as described in the previous section. Monte Carlo calculations for a series of metal and oxide standards [70] showed that the coupling was inversely related to the multiplication because of the increasing penetration of the AmLi source neutrons at lower ^{235}U densities. The coupling was also found to be nearly independent of ^{235}U mass and density. A good fit to the data was obtained with the following equation:

$$C = a - \frac{b(M-1)}{1 + c(M-1)}, \qquad (18.42)$$

where a, b, and c are calibration coefficients. This relationship between the coupling and the multiplication is illustrated in Fig. 18.20 [69].

With the coefficients shown in Fig. 18.20, Eq. 18.42 was applied to a series of very impure "skull oxide" items measured at Y-12 [69]. Because these items had known masses, fill heights, enrichments, and diameters and were relatively uniform, it was possible to calculate their multiplication using Monte Carlo. The calculated values compared very well with those obtained using the triples/double ratios. The coupling and the multiplication correction factor tend to compensate each other so that the product is insensitive to ^{235}U mass over a relatively large range. This result provides a quantitative explanation for why an almost linear calibration curve of double coincidences versus ^{235}U mass works well for AWCC measurements of many metal and oxide items (see Chap. 20). For active multiplicity analysis, the use of calibration curves of coupling versus multiplication looks promising for the assay of uranium items whose detailed characteristics are not known, such as irregular pieces of metal and impure oxide. Note that during the coupling calibration measurements, it is important to carefully control the standard

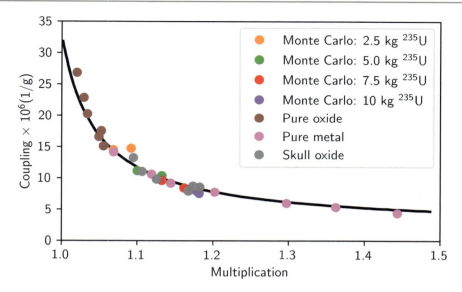

Fig. 18.20 Coupling as a function of multiplication for three categories of uranium samples [69]

positioning because the shape and magnitude of the coupling curve will depend on whether the items are always placed on a fixed stand—independent of their fill height—or whether they are positioned at different heights to keep the uranium itself in the center of the assay chamber.

A mathematical relationship for the coupling as a function of the solid angle between the sample/item and the AmLi sources and the item multiplication has also been developed [70, 71]. This relationship can be written as

$$C = k\phi(\Omega)\phi(M), \tag{18.43}$$

where $\varphi(\Omega)$ represents the dependence of the coupling on the source-item geometry, $\varphi(M)$ represents the dependence of the coupling on the flux depression within the item, and k is a scaling factor determined from calibration standards. To obtain an expression for $\varphi(M)$, this approach derives the *loss-to-fission ratio* of first-fission-generation neutrons that are leaking out of the item as

$$\frac{P(loss)}{P(fission)} = \frac{\nu_{i1}M}{M-1} - 1. \tag{18.44}$$

If leakage is the main loss mechanism, then this ratio is the probability that a fission neutron will leak from the item divided by the probability that it will induce further fission. And if fission is the main absorption mechanism, then the flux depression of the AmLi interrogation neutrons is simply the ratio of the probability that they will pass through the item to the probability that they will be absorbed in fission. If a similar functional form is assumed to apply to AmLi neutrons that enter the item, the coupling can be written as

$$C = k\phi(\Omega)\left(\frac{\nu_{i1}M}{M-1} - 1\right). \tag{18.45}$$

Once the coupling is determined from calibration curves of coupling versus multiplication using physical standards and/or Monte Carlo calculations, as illustrated in Figs. 18.20 and 18.21, the product of the scaling constant k and $\varphi(\Omega)$ for the set of standards is calculated by dividing the coupling by the $\varphi(M)$ term. This approach provides a calibration of the $k\,\varphi(\Omega)$ term for use in the unknown item measurements, where Eq. 18.45 can then be used to calculate C from the measured M (from T/D). The approach works well for disks placed on top of each other in different configurations, as described in Ref. [71].

Monte Carlo calculations have also been used to study the coupling for uranium geometries and matrices for which standards are not available [72]. The calculations used the MCNPX code, which is capable of modeling the induced fission coincidence rate in uranium from an AmLi interrogation source [73]. The MCNPX results were benchmarked to

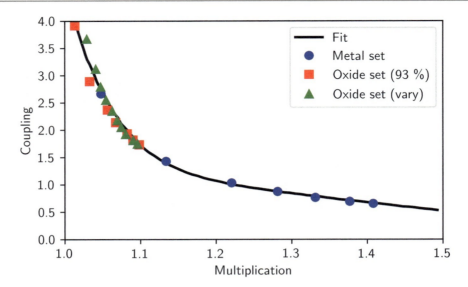

Fig. 18.21 The Monte Carlo–modeled relationship between the multiplication and the coupling for three material types: uranium metal, uranium oxide of constant enrichment, and uranium oxide of variable enrichment. The black line is an empirical fit to all of the data points [72]

measurements made with an AWCC, and then the code was used to model one set of metal items and two sets of oxide items in a cylindrical geometry. For cases where item diameter was constrained so that solid-angle effects were minimized, the relationship between coupling and multiplication is nearly collapsed to a single curve, as shown in Fig. 18.21.

The relationship shown in Fig. 18.21 is almost independent of item characteristics, with significant divergence only at multiplications below 1.05. The fit shown through the data was determined with a complex combination of linear and exponential functional forms and is a specific empirical fit to the detector and item geometries selected for the study. It is recommended that this approach be used by each facility that is undertaking active multiplicity measurements until such time as a more rigorous approach becomes available. Each facility should recalculate the curve of coupling versus multiplication with their active well counter and their available physical standards and supplement with MCNP calculations, if necessary, to obtain the curve needed to analyze their data.

18.3.6 Determination of Uranium Mass

Once the item multiplication M is obtained from the ratio of T/D using Eq. 18.39, the rate F of AmLi-induced fissions in the sample/item is obtained from the doubles count rate (Eq. 18.33), corrected for self-multiplication using Eq. 18.37. A relationship for coupling as a function of M and/or m must be selected using one of the approaches described in Sect. 18.3.5. The ^{235}U mass, m, is then given by

$$m = \frac{F}{CY}. \tag{18.46}$$

If the uranium enrichment is known, the total uranium mass is the result m divided by the enrichment.

18.4 Physics of High Accidentals Multiplicity Counting

This chapter will address the limitations of the traditional thermal-neutron multiplicity counting observed for items that are difficult to measure because the coincidence and multiplicity accidentals rate is too high. The high accidentals rate causes a large uncertainty in the counting statistics, which degrades the measurement precision. Traditionally, these measurement problems have been so severe that, in some cases, neutron assay by multiplicity or coincidence counting became impractical. The problem categories include

- Passive multiplicity measurements of impure plutonium items with moderately high alpha values[2]
- Active/passive measurements of HEU/Pu
- Active multiplicity assay of many forms of HEU

A common physics feature exists with all of these "difficult" problems: all of these measurements have a high neutron accidentals rate. The accidentals rate applies to either doubles (coincidence) counting or triples (multiplicity) counting. In both cases, it refers to those cases where multiple, uncorrelated neutrons are detected and counted within the time window of the coincidence gate. These counts appear as correlated but are indeed merely chance coincidence. The multiplicity formalism separates out the chance coincidences (accidentals) from the true coincidences using a statistical model. In the mathematics of neutron multiplicity counting, described in Chap. 18, Sect. 18.2, the accidentals counts form the background distribution, $b(i)$ and are counted in the accidentals or A gate, as described in Sect. 18.2.3.

This section will describe two approaches to illustrate the effects of the accidentals on the multiplicity precision. The first method will be to use a numerical calculation of the neutron multiplicity rates (for Singles, Doubles, and Triples) and their associated variances, as well as the resultant multiplicity assay precision, which considers the combined variances of all of the measured quantities and the inversion of the point model equations. This numerical calculation follows the theoretical formalism developed by Ensslin et al. [36]. The second approach will be analytical, with an approximate calculation of the expected variance of the doubles counts. The mathematics for the triples variance is more complicated and does little to elucidate the underlying physics. For that, we will rely on the Ensslin figure-of-merit (FOM) numerical calculations from Ref. [36]; however, for the doubles, the analytical result does illustrate the basic dependence of variance on the accidentals.

The effect of accidentals substantially changes the variance behavior for neutron multiplicity counting compared with, for example, counting individual neutrons or counting any individual, uncorrelated particle or event. From classical probability, if a measurement is made of any system that satisfies the conditions for a Poisson process, then relative error of the count (percent error) scales as $1/\sqrt{n}$, where n is the number of counts. Therefore, if singles, doubles, and triples counts in multiplicity all obeyed Poisson statistics, the relative standard deviation would be expected to scale the same way for all of them; however, that is not the case. The Singles do indeed follow this relationship, nearly, but the behavior of the Doubles and Triples is quite different.

Figure 18.22 illustrates some of these points. Using the Ensslin FOM code, we calculate the variance for the singles, doubles, triples, and multiplicity assay. To vary the accidentals rate strongly, the assay mass was varied over a large range – 0.01 – 100,000 g. The parameter α (α is the ratio of (α,n) reaction neutrons to fission neutrons) is fixed at zero. The item leakage multiplication is fixed at unity. The assumed detector efficiency is 35%, the die-away time is 50μs, the predelay is 3μs, and the gate width is 50μs, and all of these measurements are fixed.

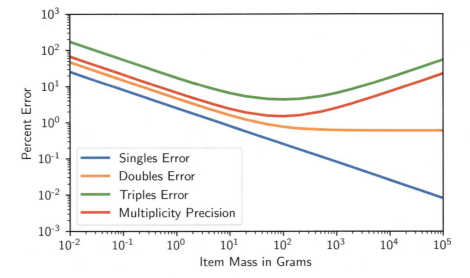

Fig. 18.22 Measurement precision versus item mass for different multiplicity orders

---

[2]For multiplicity measurements of impure Pu, the increase in α with increasing impurities hurts the counting precision on triple coincidences, eventually making the required counting time too long.

The results of Fig. 18.22 demonstrate the effects on variance from the accidentals. The singles relative standard deviation does indeed scale as $1/\sqrt{n}$, as is expected (note the graph is a log-log plot); however, the behavior of the doubles and triples RSD is quite different. At low masses, the doubles RSD scales as $1/\sqrt{n}$, but at modest mass values, the scaling stops, and the doubles RSD remains nearly constant. By contrast, the triples RSD scales as $1/\sqrt{n}$ for low masses, but above ~100 g and for this particular case, the triples RSD actually increases at a rate that is nearly proportional to the count rate. The total multiplicity assay RSD depends on the variances of Singles, Doubles, and Triples, and it also increases after about the 100 g level. This general behavior can also be addressed analytically using the point-model Eqs. 18.9 through 18.11.

Consider first the case of the Singles, which are simply the total number of neutrons detected. There are no accidentals to subtract from the total number of neutron detections because there are no coincidences, accidentals or otherwise. Using the point-model equation for the Singles, we note that the singles variance scales approximately as $1/\sqrt{n}$ because there are no accidentals to change this scaling.

Next consider the variance for the Doubles, which are genuine coincidences. These equations are best understood in two limiting cases: the case where the plutonium mass (m_{240}) is low, and the case where it is high. "Low" and "high" will be defined in the equations. For the case of coincidence counting (Doubles only), an estimate of the variance is, as discussed in Chap. 17, Sect. 17.8.1:

$$\sigma_D = \sqrt{(D+A) + A}, \tag{18.47}$$

where $(D + A)$ is the sum of all events in the D + A gate, and A is the sum of all events in the A gate. In the INCC software, these values are calculated as the factorial moment summations over the foreground, $f(i)$, distribution and the background, $b(i)$, distribution. The accidentals rate can be calculated from the singles rate. The total accidentals value is the accidentals rate times the count time:

$$A = G \cdot \dot{S}^2 \cdot t, \tag{18.48}$$

where G is the coincidence gate width and $\dot{S}$ is the singles rate. Note that the singles rate and the doubles rate, $\dot{D}$, are proportional to the plutonium mass, but the accidentals rate, $\dot{A}$, is proportional to the mass squared:

$$\dot{S} \sim m_{240} \tag{18.49}$$

$$\dot{A} \sim m_{240}^2 \tag{18.50}$$

$$\dot{D} \sim m_{240}. \tag{18.51}$$

Combining these equations, and solving for the RSD results in

$$\frac{\sigma_D}{D} = \frac{\sqrt{D + 2GS^2}}{D\sqrt{t}}. \tag{18.52}$$

Now, consider the two limiting cases. In the case of low mass, the second term in the numerator, $2GS^2$, which depends on the square of the mass, becomes arbitrarily smaller than the first term, which is linear with the mass. Therefore, at sufficiently low mass, the second term is negligible, and the equation for the RSD simplifies to

$$\frac{\sigma_D}{D} \simeq \frac{1}{\sqrt{t\varepsilon m_{240}}}, \tag{18.53}$$

which scales as $1/\sqrt{n}$, just as the FOM calculations show in Fig. 18.22, for masses less than 100 g of plutonium. The variance is limited by the statistical precision of the doubles counting. The triples behave in the same fashion.

By contrast, consider the case of large mass, namely m_{240}, which is sufficiently large that the second term in the numerator in Eq. 18.52 is much larger than the first term: $2GS^2 >> D$. Then, the equation for the RSD scales as

$$\frac{\sigma_D}{D} \sim \frac{\sqrt{\tau}}{\varepsilon \sqrt{t}}. \tag{18.54}$$

In this case, the total variance is dominated by the accidentals, and the scaling of the variance with efficiency, die-away time, and plutonium mass is quite different than for the low-mass case. The derivation of Eq. 18.54 assumed that the system is optimized so that the gate is proportional to the die-away time, τ. Indeed, there is no change in variance with mass; the usual benefit of improved statistics is lost. Note also that the RSD improves with short die-away time and large detection efficiency. A similar but more complicated calculation can be done for the triples count rate, which shows that the triples RSD actually increases with mass, just as the FOM calculation illustrated in Fig. 18.22.

The clear result is that in the case of a high accidentals rate, the variance induced by the accidentals can be the dominant contribution to the measurement variance. In the limit of large masses, and therefore large accidentals, the variance contribution from the Poisson statistics of the doubles rate is negligible compared with the variance contribution of the accidentals. Other measurement problems that also have a high accidentals rate will be affected in the same way.

Another example of a high-accidentals-rate measurement problem is active neutron multiplicity counting. The significance of active multiplicity counting is that the interrogating source provides a high level of single neutrons but no real doubles or triples counts (unless ^{252}Cf interrogation is used). Because the accidentals rate scales as $A = GS^2$, where G is the gate width and S is the singles rate, the accidentals rate is very high as well. Figure 18.23 illustrates this point for a nominal set of active measurement conditions: ^{235}U mass ranges from 100 to 10,000 g; a doubles rate of 10 counts per gram; an interrogation neutron rate of 10^6; and the same detector parameters as before. Figure 18.23 illustrates that the high accidentals rate caused by the interrogating source has substantially increased the doubles and triples uncertainty. By contrast, the singles variance is not affected as much. These RSD values are much higher than for the passive multiplicity case, where no interrogating source is used.

Another example is the case of materials with a high α. The effect of α is similar to that of an interrogating source. The (α,n) reaction neutrons appear as Singles, and they increase the measured singles rate, which increases the accidentals rate. Let's consider a nominal case that is similar to the calculation plotted in Fig. 18.22, (the detector parameters are identical), hold the mass fixed at 1000 g, and vary the α from 0.1 (essentially zero) to 5. The results are plotted in Fig. 18.24. Note that the RSD for the Doubles and Triples increases because the singles rate increases, and therefore, the accidentals rate increases. However, the RSD for the Singles decreases because the count rate increases, and this result improves the statistical precision.

In all of these cases, the neutron measurement problem is made challenging by a high accidentals rate for the doubles and triples counts. In these instances, the variance contribution of the accidentals dominates the variance of the measurement. Knowing that the accidentals have such a profound impact, it is possible to consider design changes in the neutron detectors that can reduce the variance contribution from accidentals. The evaluation proceeds from Eq. 18.54, which approximates the

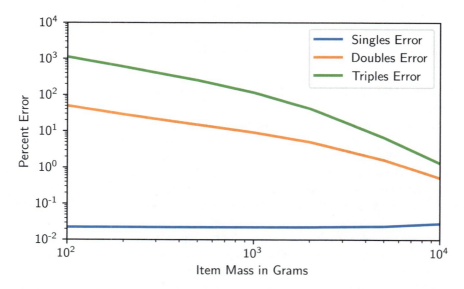

Fig. 18.23 Counting errors for active multiplicity counting versus item mass

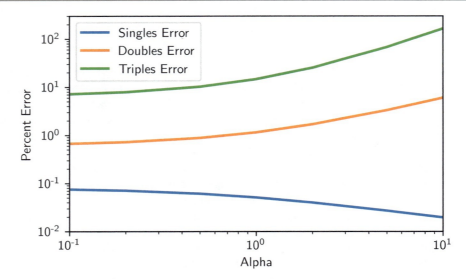

Fig. 18.24 Counting errors versus alpha (α)

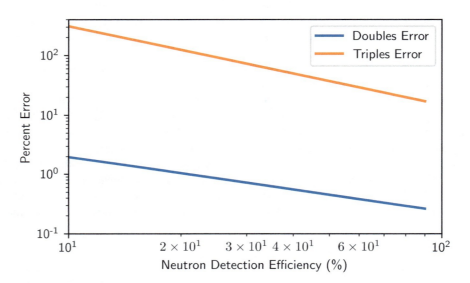

Fig. 18.25 FOM calculation: Varying efficiency for item with mass of 100,000 g of plutonium

doubles variance in the limit of high accidentals (i.e., large mass); it is not necessary to consider the case of low accidentals for this part of the discussion.

An increase in efficiency improves the precision of the multiplicity and coincidence measurements, which can be demonstrated with a FOM calculation. Figure 18.25 shows the case where efficiency is increased. The other detector parameters are the same as for Fig. 18.22; however, the mass value is chosen to be 100,000 g, which from Fig. 18.22, is well into the regime of accidentals-dominated variance for the Doubles and Triples. In this calculation, the efficiency varies from 20% to 80%. Figure 18.25 shows, consistent with Eq. 18.54, that the efficiency improves the variance for multiplicity counting in both the low mass (limited by real coincidence-counting statistics) and the high mass (limited by the accidentals rate) cases.

For completeness, it should also be noted that the variance for the opposite case—when the mass is low—is also improved as efficiency is increased. Die-away time has little effect, though. The significance is that a counter with high efficiency and low die-away time will improve the variance over the entire spectrum of measurement cases, from the limiting case where accidentals are the dominant contribution to variance at the high end to the limiting case where counting statistics of the real

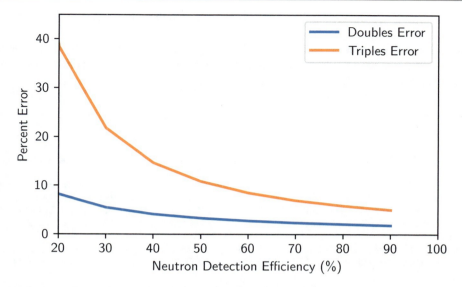

Fig. 18.26 FOM calculation: Varying efficiency for item with mass of 1 g of plutonium

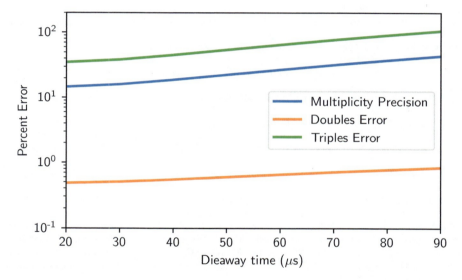

Fig. 18.27 FOM calculation: Variation of die-away time

coincidences dominates at the low end. These effects can also be seen from the FOM calculations. Figure 18.26 is identical to Fig. 18.25 except that the mass value is 1 g instead of 100,000 g of plutonium. Again, the efficiency varies from 20% to 80%.

Finally, the FOM code is used to calculate the improvement in detector performance (i.e., reduction in variance) when the die-away time is decreased. Figure 18.27 plots the same case as Figs. 18.25 and 18.26 except that the efficiency is held constant at 35%, the mass is 100,000 g, the gate width is reduced to 35μs, and the die-away time is varied from 20μs to 60μs. For the case of Doubles and Triples, the variance is improved when the die-away time is reduced and the efficiency is increased. Although we did not develop the analytical expressions for triples, the effect is more pronounced.

These effects can be understood from the basic principles of correlation counting and can be illustrated for doubles using a shift register. The D + A gate in a shift register measures both the genuine coincidences and the accidentals. To obtain the correlated neutron count rates (doubles), the accidentals contribution must be subtracted off. The accidentals are measured by opening up a second gate, the A gate. The A gate measures only uncorrelated (i.e., accidental) neutrons. Because these two numbers are subtracted, their variances must add; therefore, the total variance includes contributions from both the genuine coincidences and the accidentals. When the die-away time is reduced, the gate width is reduced proportionately. As the gate width is reduced, the number of accidentals that are counted is reduced, and the contribution to the variance is correspondingly reduced.

The conclusions from this discussion are that high accidentals—caused by high mass, high α, or active interrogation—significantly degrade the precision of multiplicity measurements. However, a neutron detector that has a high efficiency and a low die-away time can significantly mitigate this problem.

References

1. N. Ensslin, W. Harker, M. Krick, D. Langner, M. Pickrell, J. Stewart, Application guide to neutron multiplicity counting, Los Alamos National Laboratory report LA-13422-M (Nov 1998)
2. N. Ensslin, M.S. Krick, D.G. Langner, M. C. Miller, *Active Neutron Multiplicity Counting of Bulk Uranium*, in Proceedings of the 32nd Annual Meeting of the INMM, New Orleans, LA, July 28–31, 1991, Los Alamos National Laboratory report LA-UR-91-2470 (1991)
3. N. Ensslin, M.S. Krick, W.C. Harker, M.C. Miller, R.D. McElroy, P.A. McClay, W.L. Belew, R.N. Ceo, L.L. Collins, P.K. May, *Analysis of Initial In-Plant Active Neutron Multiplicity Measurements*, in Proceedings of the 34th Annual Meeting of the INMM, Scottsdale, AZ, July 18–21, 1993, Los Alamos National Laboratory report LA-UR-93-2631 (1993)
4. M.S. Krick, N. Ensslin, D.G. Langner, M.C. Miller, R.N. Ceo, P.K. May, L.L. Collins, Jr., *Active Neutron Multiplicity Analysis and Monte Carlo Calculations*, in Proceedings of the 35th Annual INMM Meeting, Naples, FL, July 17–20, 1994, Los Alamos National Laboratory report LA-UR-94-2440 (1994)
5. P. Santi, A. Favalli, D.K. Hauck, V. Henzl, D. Henzlova, Derivation of higher order moments for neutron multiplicity analysis methodologies, Los Alamos National Laboratory report LA-UR-15-20496 (2015)
6. A. Favalli, S. Croft, P. Santi, Point-model equations for neutron correlation counting: Extension of Böhnel's equations to any order. Nucl. Instrum. Methods Phys. Res., Sect. A **795**, 370–375 (2015)
7. J. Dolan, A. Dougan, D. Peranteau, S. Croft, An international view on ^{3}He alternatives for nuclear safeguards. J. Nucl. Mater. Manag. **43**(3), 17–29 (Spring 2015)
8. D. Henzlova, R. Kouzes, R. McElroy, P. Peerani et al., Current status of Helium-3 alternative technologies for nuclear safeguards, Los Alamos National Laboratory report LA-UR-15-21201, Rev. 3 (2015)
9. AMPTEK., https://www.amptek.com/products/charge-sensitive-preamplifiers/preamplifier-discriminators. Accessed Feb 2023
10. PDT., http://pdt-inc.com/. Accessed Feb 2023
11. K.D. Ianakiev, M. Iliev, M.T. Swinhoe, Fact sheet for KM200 front-end electronics, Los Alamos National Laboratory report LA-UR-15-24536 (2015)
12. J.E. Stewart, S.C. Bourret, M.S. Krick, M.R. Sweet, T.K. Li, A. Gorobets, New shift-register electronics for improved precision of neutron coincidence and multiplicity assays of plutonium and uranium mass, Los Alamos National Laboratory report LA-UR-99-4927 (2009)
13. J.F. Longo, W.H. Geist, W.C. Harker, M.S. Krick, INCC software users manual, Los Alamos National Laboratory report LA-UR-10-06227 (2010)
14. S. Croft, P. Blanc and N. Menaa, *Precision of the Accidentals Rate in Neutron Coincidence Counting*, WM '10, March 07 – March 11, 2010, in Phoenix, AZ (2010)
15. J. Huszti, Development of a Pulse-Train Recorder for Safeguards, *Proceedings of the 2014 IAEA Symposium on Nuclear Safeguards: Linking Strategy, Implementation and People*, IAEA-CN-220, Vienna, Austria, 20–24 Oct 2014
16. A. Trahan, G.E. McMath, P. Mendoza, H.R. Trellue, U. Backstrom, et al., Results of the Swedish spent fuel measurement field trials with the differential die-away self-interrogation instrument. Nucl. Instrum. Methods Phys. Res., Sect. A **955**, 163329 (Mar 2020)
17. D. Henzlova, H.O. Menlove, M.T. Swinhoe, J.B. Marlow, I.P. Martinez, C.D. Rael, *Neutron Data Collection and Analysis Techniques Comparison for Safeguards*, in Proceedings of the IAEA Symposium on International Safeguards: Preparing for Future Verification Challenges, IAEA-CN-184/178, Vienna, Austria, 1–5 Nov 2010
18. A.C. Kaplan, V. Henzl, H.O. Menlove, M.T. Swinhoe, A.P. Belian, M. Flaska, S.A. Pozzi, *Utilizing Simulated Rossi-Alpha Distributions to Develop New Methods of Characterizing Spent Nuclear Fuel*, in PHYSOR Conference Proceedings, Kyoto, Japan, 29 Sept–3 Oct 2014
19. A.C. Trahan, *Utilization of the Differential Die-Away Self-Interrogation Technique for Characterization of Spent Nuclear Fuel* (University of Michigan, Ann Arbor, 2016)
20. C. Thompson, G. McMath, V. Henzl, P. Mendoza, C. Rael, M. Root, A. Trahan, W.S. Charlton, U. Backstrom, A. Sjoland, H.R. Trellue Improved evaluation of safeguards parameters from spent fuel measurements with the Differential Die-Away (DDA) Instrument. *Nucl. Instrum. Methods Phys. Res., Sect. A* **1029**, 166462, Los Alamos National Laboratory report LA-UR-21-26890 (2022)
21. K. Böhnel, The effect of multiplication on the quantitative determination of spontaneously Fissioning isotopes by neutron correlation analysis. Nucl. Sci. Eng. **90**(1), 75–82 (1985)
22. S. Croft, D. Henzlova, D.K. Hauck, Extraction of correlated count rates using various gate generation techniques: Part I theory. Nucl. Instrum. Methods Phys. Res., Sect. A **691**, 152–158 (2012)
23. D.H. Beddingfield, M.T. Swinhoe, J. Huszti, M.R. Newell, A prescription for list-mode data processing conventions, Los Alamos National Laboratory report LA-UR-15-27846 (2015)
24. M. Newell, R. Rothrock, D. Henzlova, *Demonstration of the Advanced List Mode Module*, in Proceedings of the 58th Annual Meeting of the INMM, Indian Wells, CA, July 16–20, 2017, Los Alamos National Laboratory report LA-UR-17-25577 (2017)
25. D.M. Cifarelli, W. Hage, Models for a three-parameter analysis of neutron signal correlation measurements for fissile material assay. Nucl. Instrum. Methods Phys. Res., Sect. A **251**(3), 550–563 (1986)

26. S. Croft, A. Favalli, Review and evaluation of the spontaneous fission half-lives of ^{238}Pu, ^{240}Pu, and ^{242}Pu and the corresponding specific fission rates. Nucl. Data Sheets **175**, 269–287 (Jul–Aug 2021)
27. P. Santi, M. Miller, Reevaluation of prompt emission multiplicity distributions for spontaneous fission, Los Alamos National Laboratory report LA-UR-07-6229 (2007)
28. S. Croft, D. Henzlova, A. Favalli, D.K. Hauck, P.A. Santi, The optimum choice of gate width for neutron coincidence counting. Nucl. Instrum. Methods Phys. Res., Sect. A **764**, 322–329 (2014)
29. D. Henzlova, H.O. Menlove, S. Croft, A. Favalli, P. Santi, The impact of gate width setting and gate utilization factors on plutonium assay in passive correlated neutron counting. Nucl. Instrum. Methods Phys. Res., Sect. A **797**, 144–152 (2015)
30. N. Dytlewski, Dead-time corrections for multiplicity counters. Nucl. Instrum. Methods Phys. Res., Sect. A **305**(2), 492–494 (1991)
31. H.O. Menlove, Multiplicity dead time corrections, Los Alamos National Laboratory (unpublished) memo N-1-92-742 (1992)
32. NDA 2000, https://www.mirion.com/products/nda-2000-non-destructive-assay-software. Accessed Feb 2023
33. D.G. Langner, M.S. Krick, J.E. Stewart, N. Ensslin, *The State-of-the-Art of Thermal Neutron Multiplicity Counting*, in Proceedings of the 38th Annual Meeting of the INMM, Phoenix, AZ, July 20–24, 1997, Los Alamos National Laboratory report LA-UR-97-2734 (1997)
34. W.H. Geist, *Multiplication Dependent Correction Factors for Multiplicity Assay of Plutonium Items*, in Proceedings of the 47th Annual Meeting of the INMM, Nashville, TN, July 16–20, 2006, Los Alamos National Laboratory report LA-UR-06-4211 (2006)
35. M.S. Krick, W.H. Geist, D.R. Mayo, A weighted point model for the thermal neutron multiplicity assay of high mass plutonium samples, Los Alamos National Laboratory report LA-14157 (2005)
36. N. Ensslin, M.S. Krick, N. Dytlewski, Assay variance as a figure of merit for neutron multiplicity counting. Nucl. Instrum. Methods Phys. Res., Sect. A **290**(1), 197–207 (1990)
37. A. Favalli, T.L. Burr, S. Croft, D. Henzlova, B. P. Weaver, Comprehensive bayesian uncertainty quantification for neutron correlation counting of special nuclear material, Los Alamos National Laboratory report LA-UR-20-27531 (2020)
38. K. Zhao, M. Penkin, C. Norman, S. Balsley, K. Mayer, P. Peerani, C. Pietri, S. Tapodi, Y. Tsutaki, M. Boella, G. Renha Jr., E. Kuhn, International target values 2010 for measurement uncertainties in safeguarding nuclear materials, IAEA STR-368 (International Atomic Energy Agency, Vienna, Austria, Nov 2010)
39. D. Dearborn, M. Mount, *Lawrence Livermore National Laboratory Experience with the 30-Gallon Drum Neutron Multiplicity Counter*, in Proceedings of the 44th Annual Meeting of the INMM, Phoenix, AZ, 13–17 Jul 2003
40. D.M. Dearborn, S.C. Keeton, *Lawrence Livermore National Laboratory Experience Using 30-Gallon Drum Neutron Multiplicity Counter for Measuring Plutonium-Bearing Salts*, in Proceedings of the 45th Annual Meeting of the INMM, Orlando, FL, 18–22 Jul 2004
41. S.M. Long, S.S. Hildner, D. Guiterrez, C.J. Mills, W. Garcia, C. Gurule, Fabrication of 12% ^{240}Pu calorimetry standards, Los Alamos National Laboratory report LA-UR-95-2334 (1995)
42. D.G. Langner, M.S. Krick, N. Ensslin, G.E. Bosler, N. Dytlewski, *Neutron Multiplicity Counter Development*, in Proceedings of the ESARDA Symposium Safeguards & Nuclear Material Management, Avignon, France, May 14–16, 1991, Los Alamos National Laboratory report LA-UR-91-1569 (1991)
43. J.E. Stewart, M.S. Krick, J. Xiao, R.J. Lemaire, V. Fotin, L. McRae, D. Scott, G. Westsik, *Assay of Scrap Plutonium Oxide by Thermal Neutron Multiplicity Counting for IAEA Verification of Excess Materials from Nuclear Weapons Production*, in Proceedings of the 37th Annual Meeting of the INMM, Naples, FL, July 28–31, 1996, Los Alamos National Laboratory report LA-UR-96-2515 (1996)
44. D.G. Langner, M.S. Krick, D.W. Miller, The use of ring ratios to detect sample differences in passive neutron counting. Nucl. Matls. Manag. (Proceedings Issue) **21**, 790–797 (1992)
45. W.H. Geist, D.G. Langner, N. Ensslin, Analysis of FB-line neutron multiplicity data, Los Alamos National Laboratory report LA-UR-00-5792 (2000)
46. J.E. Stewart, H.O. Menlove, D.R. Mayo, W.H. Geist, L.A. Carillo, G.D. Herrera, Epithermal neutron multiplicity counter design and performance manual: More rapid plutonium and uranium inventory verifications by factors of 5–20, Los Alamos National Laboratory report LA-13743-M (2000)
47. N. Ensslin, M.S. Krick, H. O. Menlove, *Expected Precision of Neutron Multiplicity Measurements of Waste Drums*, LA-UR-95-2275, in Proceedings of the 36th Annual Meeting of the INMM, Palm Desert, CA, 9–12 Jul 1995
48. H.O. Menlove, D.H. Beddingfield, M.M. Pickrell, D.R. Davidson, R.D. McElroy, D.B. Brochu, *Design of a High Efficiency Neutron Counter for Waste Drums to Provide Optimized Sensitivity for Plutonium Assay*, Los Alamos National Laboratory report LA-UR-96-4585, in Proceedings of the 5th Nondestructive Assay and Nondestructive Examination Waste Characterization Conference, Salt Lake City, UT, 14–16 Jan 1997
49. H.O. Menlove, J. Baca, M.S. Krick, K.E. Kroncke, D.G. Langner, Plutonium scrap multiplicity counter operation manual, Los Alamos National Laboratory report LA-12479-M (ISPO-349) (Jan 1993)
50. D.G. Langner, J.B. Franco, J.G. Fleissner, V. Fotin, J. Xiao, R. Lemaire, Performance of the 30-gallon drum neutron multiplicity counter at rocky flats environmental technology site, Los Alamos National Laboratory report LA-UR-96-2569 (1996)
51. J.S. Hendricks et al., MCNPX extensions Version 2.5.0, Los Alamos National Laboratory report LA-UR-04-0570 (2004)
52. D.K. Hauck, V. Henzl, "Spatial multiplication model as an alternative to the point model in neutron multiplicity counting, Los Alamos National Laboratory report LA-UR-14-21991 (2014)
53. S. Croft, E. Alvarez, P.M.J. Chard, R.D. McElroy, S. Philips, *An Alternative Perspective on the Weighted Point Model for Passive Neutron Multiplicity Counting*, in 48th Annual Meeting of the INMM, Tucson, AZ, 8–12 Jul 2007
54. V. Henzl, D.K. Hauck, K.E. Koehler, P.A. Santi, *Simulation Study to Develop Spatial Multiplication Model in Neutron Multiplicity Counting*, in Proceedings of the 56th Annual Meeting of the INMM, Indian Wells, CA, 12–16 Jul 2015
55. M.S. Krick, D.G. Langner, J.E. Stewart, Energy-dependent bias in plutonium verification measurements using thermal neutron multiplicity counters, Los Alamos National Laboratory report LA-UR-97-3427 (1997)
56. P. Santi, W. Geist, *Energy Dependent Bias in the Weighted Point Model*, in Proceedings of the 46th Annual Meeting of the INMM, Phoenix, AZ, July 10–14, 2005, Los Alamos National Laboratory report LA-UR-05-4287 (2005)
57. S. Croft, A. Favalli, Incorporating delayed neutrons into the point-model equations routinely used for neutron coincidence counting in nuclear safeguards. Ann. Nucl. Energy **99**, 36–39 (2017)

58. K.E. Koehler, V. Henzl, D. Henzlova, W. Geist, *The Badlands of Neutron Multiplicity Counting*, in Proceedings of the 57th Annual Meeting of the INMM, Atlanta, GA, July 24–28, 2016, Los Alamos National Laboratory report LA-UR-16-25460 (2016)
59. M.S. Krick, Monte Carlo deadtime simulations for thermal neutron multiplicity counters using shift-register electronics, Los Alamos National Laboratory report LA-UR-07-1764 (2007)
60. S. Croft, A. Favalli, Extension of the Dytlewski-style dead time correction formalism for neutron multiplicity counting to any order. Nucl. Instrum. Methods Phys. Res., Sect. A **869**, 141–152 (2017)
61. D.K. Hauck, S. Croft, L.G. Evans, A. Favalli, P.A. Santi, J. Dowell, Study of a theoretical model for the measured gate moments resulting from correlated detection events and an extending dead time. Nucl. Instrum. Methods Phys. Res., Sect. A **719**, 57–69 (2013)
62. D.K. Hauck, An exact dead time model for correlated neutron counting based on the joint probability function, Los Alamos National Laboratory report LA-UR-12-24194 (2012)
63. L. Holzleitner, D. Henzlova, M. Swinhoe, V. Henzl, *Estimation of Dead time Loss for high Neutron Count-Rates and associated Multiplicity Correction using Multi-Channel List-Mode Data*, Powerpoint presentation, Symposium on International Safeguards: Building Future Safeguards Capabilities, Vienna, Austria, 5–8 Nov 2018
64. L. Holzleitner, M. Swinhoe, Dead time correction for any multiplicity using list mode neutron multiplicity counters: A new approach – Low and medium count-rates. Radiat. Meas. **46**(3), 340–356 (2011)
65. D. Henzlova, T. Cutler, A. Favalli, W. Geist, V. Henzl, K. Koehler, M. Lockhart, C. McGahee, R. Parker, P. Santi, Improving neutron measurement capabilities – Final report, Los Alamos National Laboratory report LA-UR-17-29756 (2017)
66. S. Croft, A. Favalli, D.K. Hauck, D. Henzlova, P.A. Santi, Feynman variance-to-mean in the context of passive neutron coincidence counting. Nucl. Instrum. Methods Phys. Res., Sect. A **686**, 136–144 (2012)
67. D. Henzlova, S. Croft, H.O. Menlove, M.T. Swinhoe, Extraction of correlated count rates using various gate generation techniques: Part II Experiment. Nucl. Instrum. Methods Phys. Res., Sect. A **691**, 159–167 (2012)
68. A.C. Kaplan, V. Henzl, H.O. Menlove, M.T. Swinhoe, A.P. Belian, M. Flaska, S.A. Pozzi, Determination of spent nuclear fuel assembly multiplication with the differential die-away self-interrogation instrument. Nucl. Instrum. Methods Phys. Res., Sect. A **757**, 20–27 (2014)
69. M.S. Krick, N. Ensslin, R.N. Ceo, P.K. May, *Analysis of Active Neutron Multiplicity Data for Y-12 Skull Oxide Samples*, in Proceedings of the 37th Annual Meeting of the INMM, Naples, FL, July 28–31, 1996, Los Alamos National Laboratory report LA-UR-96-0343 (1996)
70. C.A. Beard, N. Ensslin, W.H. Geist, Relationship between source-sample coupling and multiplication for active multiplicity assay, Nuclear Engineering Teaching Laboratory, University of Texas at Austin, unpublished technical report PRC 159, R9000 (1998–2000)
71. W.H. Geist, N. Ensslin, L.A. Carrillo, C.A. Beard, *Advanced Analysis Techniques for Uranium Assay*, in Proceedings of the 42nd Annual Meeting of the INMM, Indian Wells, CA, July 15–19, 2001, Los Alamos National Laboratory report LA-UR-01-3331 (2001)
72. W.H. Geist, K. Frame, *Modeling Active Neutron Coincidence Counters with MCNPX*, in Proceedings of the 45th Annual Meeting of the INMM, Orlando, FL, July 18–22, 2004, Los Alamos National Laboratory report LA-UR-04-4355 (2004)
73. J. S. Hendricks et al., Monte Carlo neutron-photon extended code, Los Alamos National Laboratory report LA-UR-04-0570 (2004)

Open Access This chapter is licensed under the terms of the Creative Commons Attribution 4.0 International License (http://creativecommons.org/licenses/by/4.0/), which permits use, sharing, adaptation, distribution and reproduction in any medium or format, as long as you give appropriate credit to the original author(s) and the source, provide a link to the Creative Commons license and indicate if changes were made.

The images or other third party material in this chapter are included in the chapter's Creative Commons license, unless indicated otherwise in a credit line to the material. If material is not included in the chapter's Creative Commons license and your intended use is not permitted by statutory regulation or exceeds the permitted use, you will need to obtain permission directly from the copyright holder.

Passive Neutron Instrumentation and Applications

R. K. Weinmann-Smith, T. J. Aucott, A. P. Belian, D. P. Broughton, M. Frankl, P. A. Hausladen, D. C. Henzlova, J. D. Hutchinson, R. D. McElroy, H. O. Menlove, T. P. Pochet, L. A. Refalo, M. A. Root, M. Nelson, J. K. Sprinkle, M. T. Swinhoe, and M. M. Watson

19.1 Introduction

Passive neutron measurements are a nondestructive assay (NDA) technique to determine the amount of nuclear material in an item of interest. Neutron NDA is used in many applications related to nuclear industries, including process control, nuclear material control and accounting, decontamination and decommissioning, and international nuclear safeguards. A complete neutron-detection system consists of the neutron detector instrument, electronics to process the detector pulses into count rates (most typically a shift register), and optionally a computer with software to operate the shift register and store the data.

A complete NDA system typically combines neutron measurement results with information from gamma measurements or known from outside information. Neutron measurements determine how much nuclear material is present but cannot determine what the nuclear material is because they can't determine what isotope or element the neutron is from. They determine only that more neutrons correspond to more of the unknown isotope or element. Gamma measurements or other information determine the isotopic composition of the item, and these results are combined to fully characterize the item.

Passive neutron measurements are a powerful tool because the neutrons are emitted from the nuclear material itself, so they need only be detected passively by some instrument. The neutrons easily penetrate most materials, which allows measurement of nuclear material sealed in a container without opening it. Neutrons and their characteristics are a unique signature that is difficult to spoof either accidentally or deliberately. A typical measurement takes on the order of 10 min, with a total measurement uncertainty on the order of 5%. A typical instrument can operate for decades with little maintenance and no consumables, reducing costs beyond the large, upfront cost for the instrument. Because these instruments have been widely used since the 1980s, they are commercially available for common materials.

However, initial setup for neutron measurements can be significant. A new ^{3}He-based instrument costs between tens and hundreds of thousands of dollars depending on the size and capabilities. In contrast with the procedure used in chemical analysis where the sample is modified to "fit" the instrument, in NDA, the instrument is modified to fit the sample. Commercial systems are available for standard applications. Designing and calibrating a custom instrument for unique materials can also cost hundreds of thousands of dollars. The associated electronics can cost thousands of dollars. The need for gamma measurements can add additional complexity and cost for the capability to fully characterize an item.

Passive neutron instruments are ideal when the nuclear material of interest emits neutrons, which may be measured to determine the amount of nuclear material. Typical measurement uncertainties for different applications can be found in the most recent International Target Values, published about every 10 years by the International Atomic Energy Agency (IAEA). The neutron emission rate must be sufficiently above background and high enough to allow statistical precision in an acceptable measurement time. The neutron emission rate must sufficiently correlate to quantity of interest; for example, the rate must increase and decrease with nuclear material mass instead of some other quantity. Neutron NDA engineers design instruments to achieve these two principles as efficiently as possible. The principles are discussed in greater detail in Chaps. 15–18. In practice, plutonium is most often measured passively because of its large spontaneous fission emission rate. Uranium has a much lower radioactive decay rate and a corresponding lower spontaneous fission and alpha decay rate and is measured passively in only a few cases, which are discussed in this chapter.

Passive neutron measurements are categorized as singles, coincidence, and multiplicity counting. The neutron count rate is the most basic measurement and is known as *totals* or *singles counting*. In most applications, the material of interest undergoes fission. Fission emits multiple neutrons at the same time, so groups of correlated neutrons are a more direct signature of the amount of nuclear material. Thus, pairs of two or groups of three neutrons are measured and are called Doubles (or coincidence) and Triples. Measurements that analyze only the Doubles rates are referred to as *coincidence counting* and measurements that use the Singles, Doubles, and Triples count rates are referred to as *multiplicity counting*.

Singles measurements are most often used when the detection efficiency is low or the nuclear material produces neutrons from (α,n) reactions instead of fission. Singles measurements work in the broadest range of applications but are most susceptible to background neutrons. Coincidence counting is the most common form of passive neutron measurements. It uses the doubles rate, although some advanced analysis also uses the singles rate to determine additional characteristics of the nuclear material. The doubles rates are much less susceptible to background and are much more difficult to spoof. Multiplicity counting uses the singles, doubles, and triples rates in a complex analysis that enables the measurement of more complex material, including impure material.

Passive neutron counting measures materials that typically produce neutrons through three mechanisms: spontaneous fission, induced fission, and (α,n). These mechanisms correspond to three characteristics of the material: nuclear material mass, multiplication, and alpha. Alpha is the ratio of (α,n) to spontaneous fission neutron production and is defined that way for use in multiplicity-counting analysis. In this brief discussion, it can be considered essentially the amount of (α,n) neutron production. In most applications, the objective is to determine the nuclear material mass. The item's multiplication and alpha are features that obscure the measurement by also affecting neutron production without a corresponding change in mass and must be accounted for by assuming their value, including knowing it to be negligible or measuring it as part of the analysis. Table 19.1 is a matrix that describes this idea for the various analysis methods where the mechanisms are either assumed or determined by measuring the Singles, Doubles, or Triples (S, D, T).

Singles, doubles, and triples relate to the grouping of neutrons. The probability of detecting (n) neutrons scales with the detector efficiency (ϵ) raised to the nth power, ϵ^n. For a fixed detector efficiency, the Singles will reach a given statistical uncertainty much faster than the Doubles and the Doubles much faster than Triples. An instrument designed for a measurement analysis technique that uses only Singles can achieve sufficient performance with a much lower efficiency and corresponding low cost. Instruments designed for coincidence counting using Doubles have higher efficiency and cost, and multiplicity counters are designed with very high efficiencies for sufficient triples uncertainty and the largest cost. The detection performance also scales inversely to die-away time for Doubles and Triples. The die-away time does not directly

Table 19.1 Method of determination of the characteristics of an item for various analysis methods

Analysis method	Item characteristics		
	Mass / spontaneous fission	Multiplication / induced fission	Alpha / (α,n)
Single	S	Assumed[a]	Assumed[a]
Coincidence	D	Assumed[a]	Assumed[a]
Known alpha (advanced coincidence)	D	S	Assumed[a]/calculated
Known multiplication (advanced coincidence)	D	Assumed[a]	S
Multiplicity	S, D, T	S, D, T	S, D, T

[a]*Assumed* to be consistent with the calibration

affect the singles performance (although it does affect singles dead time losses). The die-away time can be reduced at the cost of more ^{3}He, which increases the detector cost. Detailed technical analysis of this issue is discussed in Chap. 17 and 18. Typical detection efficiencies are <10%, 10–30%, and 60% for singles, coincidence, and multiplicity counters, respectively. Typical die-away times are 40–60 μs for typical coincidence counters and ~20 μs for typical multiplicity counters. Technically, any typical neutron instrument can measure Singles, Doubles, and Triples (and even Quads, Pents, and so on). Instruments are typically categorized according to the application for which they were intended.

A typical workflow in deploying a passive neutron counter is described as follows: The application, performance requirements, and budget must be identified. The objective of the measurement and how it integrates into the overall facility system or need should be understood. Any other facility constraints, such as available space, personnel, and equipment should be known. The nuclear material to be measured should be evaluated for its neutron-production characteristics. Some physics mechanism that allows a relation of neutron rate to quantity of interest must be identified. Many other practical considerations must be made. A neutron-detection instrument optimized for this application can be chosen and purchased from commercial vendors, or if none exists, it can be designed and built. Finally, the instrument must be calibrated to the material to be measured. If the material is simple and consistent, calibration may be simple and can be performed by measuring some known materials of similar composition. In more complex cases, bounding conditions must be determined and simulations used to supplement measurements to calibrate the system for all materials expected to be measured. However, in multiplicity counting, no calibration is needed, in principle, because the analysis solves for the three unknowns from first principles—one of the major advantages of multiplicity counting.

Helium-3-based neutron detectors have been the gold standard instrument used since the 1980s because of their many advantages. This chapter, unless otherwise noted, assumes that neutron-detection instruments are ^{3}He-based thermal-neutron counters. In 2020, the large majority of deployed systems are ^{3}He based. Other technologies are being developed to replace ^{3}He detectors, and if they achieve widespread adoption, then some of the specific assumptions made in this book will not apply because the new detectors may use different physics to detect the neutrons. However currently, these alternatives have downsides that make ^{3}He still preferable. The most promising of these new technologies are described at the end of this chapter and include alternative thermal-neutron counters—such as boron—and fast-neutron detectors—such as scintillators.

This chapter structure begins with this introduction, capturing big-picture concepts and context for choosing and deploying passive neutron-counting instruments. Electronics are discussed, which are common to all neutron counters. Then, singles, coincidence, and multiplicity instruments are discussed and are sorted by their application. This chapter seeks to describe most instruments in use today so that a reader may identify a solution to their NDA challenge or learn a general understanding of how an existing instrument functions, with references to further technical details. At least one example of each current analysis technique is captured for completeness. Then, instruments based on technology other than ^{3}He are described. Finally for completeness, neutron-detection methodologies used for applications other than determining nuclear material mass are described, including neutron direction and energy determination.

Some terminology is reviewed because the dominant language has changed over the years. *Singles*, *Doubles*, and *Triples* are the count rates from neutron measurements. *Singles* and *totals* are interchangeable and signify the total neutron count rate. *Doubles* and *real coincidence rate* are the same. *Coincidence* is an individual event of two neutrons detected at the same time, and there are *real* and *accidental* coincidences, but *Doubles* is the rate of real coincidences only (see Chap. 18). Coincidence counters are designed to measure Doubles, and multiplicity counters are designed to measure Triples. Singles, coincidence, and multiplicity counters are often called *thermal-neutron counters*, *^{3}He detectors*, or *well counters* and describe the physics, technology, or geometry, respectively, instead of the analysis.

Finally, summary tables of the various instruments reviewed in this chapter are presented in Tables 19.2, 19.3, and 19.4.

19.2 Neutron Acquisition Electronics

The data acquisition electronics take the raw signal output from the ^{3}He tubes and converts it through many steps to usable singles, doubles, and triples rates. The key steps are as follows:

1. Neutron absorption in ^{3}He produces energetic charged particles, which create ionization in the gas in a detector tube. A tube-wall cathode and inner anode wire with applied high voltage multiplies this ionization and collects the corresponding charge.

Table 19.2 Singles-counting instrument overview

Name	Application	Cavity size	Tube rings/rows	No. of ³He tubes	Tube pressure (atm)	Amplifier modules	High voltage (V)	Efficiency (%)
Shielded neutron assay probe	Portable measurements	Any	1	1	10	1		Application dependent
Portable handheld neutron counter	Facility inspections	Any	1	4	4	1	1680	7.4 in collar geometry
High-dose neutron detector	High gamma dose	Any	6	6 boron detection cells	N/A		Variable	Application dependent
Mini high-dose neutron detector	High gamma dose	Any	3	3 boron detection cells	N/A		Variable	Application dependent
Plutonium inventory monitoring system	Glove box plutonium inventory	Glove-box dependent	N/A	Glove-box dependent, 142 in Japan				Glove-box dependent
Vitrified canister assay system	Vitrified waste Pu content	Vitrified canisters						
Rokkasho hulls measurement system	Pu content of reprocessing hulls	Hull drums						
Hybrid induced fission based Pu-accounting instrument	U/TRU ingots from reprocessing		2	6	4	6	1680	7.8
Reactor gate discharge monitor	Fuel discharge from reactor to cooling pond	N/A	1	1		1		N/A, moving source

U/TRU uranium/transuranic

Table 19.3 Doubles-counting instrument overview

Name	Application	Cavity size	No. of tube rings	No. of ³He tubes	Tube pressure (atm)	Amplifier modules	High voltage (V)	Pd (μs)	Gate width (μs)	Eff. (%)	Die-away (μs)
High-level neutron coincidence counter	Bulk Pu cans	17.5 cm D × 41 cm H	1	18	4	6	1680	4.5	64	17.5	43
Plutonium canister assay system	MOX canisters	Version dependent									
Waste drum assay system	55-gallon waste drums	55 gal.	1	60	4	10	1680	4.5	128	18.6	80
Chernobyl drum assay system	55-gallon waste drums	55 gal.	1	15	4		1680				
High-efficiency neutron counter	55-gallon waste drums	55 gal.	2	113	4	16	1680	3	128	32	50
Super high-efficiency neutron counter	Standard waste boxes	SWB	2	260	10	32	1720			40	
Glove box unattended assay and monitoring system	Glove-box holdup	Glove-box dependent	1	16–20 per glove box	4		1680			<1%	
Fast neutron passive coincidence collar	Fresh fuel with high gadolinium content	23.5 cm × 23.5 cm	1	28	6	4	1680	2	200	21	70
Passive neutron coincidence collar	Fresh fuel	23.5 cm × 23.5 cm	1	28	6	4	1680	2	300	24	70
Uranium fluoride coincidence collar	²³⁵U mass and enrichment in UF₆ cylinders	14.7 cm D	1	12	4	6	1680	2	128	18	75

MOX mixed oxide, *SWB* standard waste box

Table 19.4 Multiplicity-counting instrument overview

Multiplicity counter name	Facility	Application	No. of tube rings	No. of ³He tubes	Amplifier modules	Derandomizer	Multiplicity dead time (ns)	Eff. (%)	Die-away time (μs)	Cavity size
Five-ring multiplicity counter	Los Alamos	Technology development	5	130	34	Yes	36	53	49	16.5 cm D × 25.4 cm H
Three-ring multiplicity counter	Los Alamos	Technology development	3	60	12	Yes	83	45	63	20 cm D × 30 cm H
Pyrochemical multiplicity counter	Los Alamos	In-plant metals, oxides	4	126	36	No	90	57	47	24 cm D × 38 cm H
Plutonium scrap multiplicity counter	Hanford and Japan	Pu inventory verification	3½	80	19	No	121	55	47	20 cm D × 41 cm H
ARIES neutron counter	Los Alamos	Pu metals and residues	3½	80	20	Yes	60	55	47	20 cm D × 41 cm H
FB-line multiplicity counter	Savannah River	Metal, oxide inventory	4	113	24	Yes	50	58	50	20 cm D × 41 cm H
30-gallon multiplicity counters	Rocky flats and Livermore	Inventory verification	3	126	54	Yes	25	42	55	30-gallon drum
High-efficiency neutron counter	Los Alamos	Waste assay	2	113	16	No	171	32	50	55-gallon drum
Epithermal neutron multiplicity counter	Los Alamos	Pu inventory verification	4	121 10 atm	27	Yes	37	65	22	20 cm D × 43 cm H
Epithermal neutron multiplicity counter with inventory sample counter (INVS)	Los Alamos	Low mass, nuclear data measurements	6	142 10 atm	30	ENMC–yes, INVS–no	100	80	19	5 cm D × 15 cm H
Mini-epithermal neutron multiplicity counter	Los Alamos	Pu inventory verification	4	104 10 atm	24	Yes	38	62	19	15.9 cm D × 38 cm H or larger
KAMS neutron multiplicity counter	Savannah River	Receipts verification	3	198 10 atm	54	Yes	19	52	37.3	55-gallon drum
SuperHENC	Rocky flats	Standard waste box assay	2	260 10 atm	32	Yes		40.3		Standard waste box
AVIS	Japan MOX fuel fabrication facility	MOX powders and pellets	4	74 10 atm	14	Yes	73	67	22	6 cm D × 12.5 cm H
MC-15 (slab instead of well geometry)	Los Alamos	Portable measurements	3	15 10 atm	15 (KM-100, not AMPTEK)	Yes				

ARIES advanced recovery and integrated extraction system, *ENMC* epithermal neutron coincidence counter, *INVS* inventory sample counter, *KAMS* K-area material storage, *HENC* high-efficiency neutron counter, *AVIS* advanced inventory verification sample (system)

2. A preamplifier[1] integrates the raw detection pulse generated in the ³He tube.
3. A shaping amplifier[1] further shapes the analog pulse.
4. A discriminator[1] rejects small pulses from gamma-energy deposition and produces a digital transistor-transistor logic (TTL) pulse for each neutron pulse.
5. The TTL outputs from each module may be combined in a derandomizer circuit, which ensures that TTL pulses that arrive at the same time from different modules are not lost to dead time.
6. The derandomizer output is fed to the shift register (or list-mode data acquisition system), which processes the pulse train to produce the reals+accidentals and accidentals multiplicity distributions to calculate the singles, doubles, and triples rates.

---

[1] *These three components are integrated into a single module, which is referred to as an "amplifier module" or casually as a "preamplifier." Between one and a few ³He tubes are connected to a single amplifier module; therefore, most detectors have multiple amplifier modules.*

Software is used to control data acquisition and to perform calculations for analysis techniques.

These steps can be accomplished by a range of equipment that is custom made or procured from commercial vendors. This section describes the commonly used equipment, which is primarily commercial.

The choice of electronics affects several aspects of detector performance. The number and layout of amplifier modules affects the resolution of list-mode data. The layout also affects dead time and gamma pileup, essentially the maximum neutron and gamma radiation in which the detector can function. More modern amplifiers have reduced dead time, and newer shift registers have improved capabilities, including higher maximum count rate capability and the ability to measure the multiplicity histogram, which allows calculation of the triples.

19.2.1 Amplifier Modules

The AMPTEK A111 preamplifier/discriminator chip, shown in Fig. 19.1, is the basis for multiple electronics boards made by various manufacturers for neutron assay systems. AMPTEK-based boards are widely used for their excellent performance. The various mounting and electrical connection options in the following photographs are chosen based on the application and are not specific to the amplifier module.

Precision Data Technology, Inc. (PDT) [1] makes neutron-monitoring modules that include a preamplifier, an amplifier, and a discriminator, shown in Fig. 19.2. The modules are conveniently contained in a housing that protects the fragile electronics within and allows them to be easily screwed onto the top of ^{3}He tubes instead of more complex mounting methods. Their low-voltage, high-voltage, and signal connectors can be *flying leads* that extend from the module a few centimeters for easy connection. This configuration allows the modules to be quickly attached or removed from ^{3}He tubes and quickly connected to data acquisition systems. The modules can have both TTL-out and TTL-in connections to string multiple units together for one overall detector output.

The KM-200 amplifier module, shown in Fig. 19.3, was designed at Los Alamos National Laboratory to incorporate new advances in signal processing [2]. The KM-200s are more compact, allowing one amplifier module per ^{3}He tube in densely packed detector designs. They also have unique features that reduce their dead time and increase their capabilities. Unique features include customizable signal-processing options such as the shaping time and sensitivity for optimization required in demanding applications such as spent fuel.

19.2.2 Scalar Instruments

The most basic data acquisition system, which is sufficient only for singles counting, is a scalar counter. These modules simply count the number of pulses detected. They can be coupled with timers to measure for a specified measurement time. They are older technology and less common but still used in some locations. They are often part of the nuclear instrumentation module (NIM) framework—a standard used in many nuclear applications beginning in the 1970s. Scalar counters may be preferred in applications where analog processing is required due to processing speed or security.

Fig. 19.1 A-111 board. The A-111 chip is the circular silver component

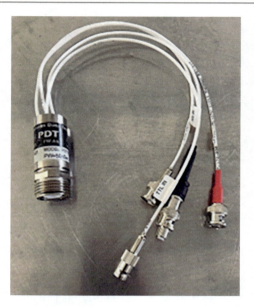

Fig. 19.2 PDT 10A preamplifier with flying leads

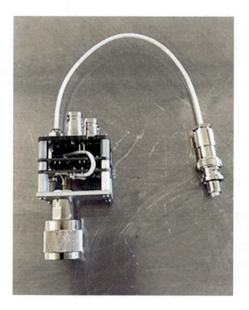

Fig. 19.3 KM-200 basic board stack

19.2.3 Coincidence Shift Registers

19.2.3.1 JSR-12

The Jomar Shift Register-12, shown in Fig. 19.4, is a shift register designed to measure only coincidence counts, so it is not capable of recording triples. Like most shift registers, it can be controlled from a computer or by using buttons and a screen on the front of the module. It has a 4 MHz internal clock.

Fig. 19.4 JSR-12 Shift Register

Fig. 19.5 AMSR

19.2.4 Multiplicity Shift Registers

19.2.4.1 JSR-14

The JSR-14 neutron analysis shift register [3] improved on the JSR-12 by including multiplicity capabilities. It is also based on a Los Alamos National Laboratory design and is manufactured by Canberra (now part of Mirion Technologies). It has the same form factor as the Canberra InSpector Multichannel Analyzer and a sampling rate of 4 MHz.

19.2.4.2 Advanced Multiplicity Shift Register

The advanced multiplicity shift register (AMSR), shown in Fig. 19.5, was designed by Los Alamos National Laboratory (LANL) and commercially manufactured by ANTEC/Ortec as the AMSR-150 [4]. The device was optimized for use with the IAEA neutron coincidence counting (INCC) software. It was the choice shift register until manufacture of the device was discontinued in 2015. The JSR-15 has since replaced the AMSR, although the AMSR is still used today if available.

The AMSR incorporated fast accidental sampling to improve measurement precision. Neutron coincidence and multiplicity measurement precision was limited by the level of accidental coincidence pulses (A). Achieving acceptable measurement precision required long count times, which limited the quality and quantity of neutron measurements. The AMSR was developed to improve neutron measurement precision and reduce count times using a digital signal-processing technique called *fast accidental* (FA) sampling.

FA differs from traditional shift register electronics by decoupling the A gate from the R + A gate. In FA, the R + A gate is triggered in the same manner as traditional shift register electronics, but the A gate is sampled according to the clock rate and

Fig. 19.6 JSR-15 multiplicity shift register (or handheld multiplicity register)

independently of the R + A gate trigger. The clock rate is typically much greater than the pulse rate, so the A gate is sampled more often than in traditional shift register electronics during a measurement. Increased sampling of the A gate decreased the measurement uncertainty on the Doubles by a factor of approximately $\sqrt{2}$ and count time by a factor of 2.

19.2.4.3 JSR-15

The JSR-15, shown in Fig. 19.6, was designed by LANL as the handheld multiplicity register and was commercially manufactured by Canberra Industries as the JSR-15 [5]. As of 2021, it is the state-of-the-art commercial shift register. The JSR-15 has a front panel display that allows the user to fully control the measurement conditions without the use of an additional computer. It displays results, measurement parameters, and measured multiplicity distributions. It has an internal battery for handheld operation and a USB (universal serial bus) interface for software control. The JSR-15 is greatly improved over previous shift registers, having a 50 MHz sampling rate that allows much higher detector count rates in applications such as spent fuel measurements and large-mass items. It allows the user to switch between fast and standard accidentals sampling [6].

19.2.4.4 Unattended Multiplicity Shift Register

The unattended multiplicity shift register (UMSR), shown in Fig. 19.7, was designed by Los Alamos National Laboratory and was state of the art in 2021. It was designed for integration into unattended continuous measurement systems. It is compact, produces 5 and 12 V low voltage, and can be controlled via web browser (internet connection not needed) or by INCC software.

19.2.5 List-Mode Data Acquisition Hardware

List-mode data collection represents the most advanced and complete information that can be obtained from a neutron-detection instrument. *List mode* refers to collection and recording the time of occurrence of all of the neutron detections within an NDA instrument. The data can be recorded for individual modules or for the sum of all modules. The data are typically stored in binary files and contain the full set of neutron-detection times (called *pulse train*). List-mode data acquisition represents an advanced neutron data acquisition option that gained interest primarily due to expanded analysis options (access to time-based information such as time-interval distribution or Rossi-α distribution), increased complexity to resist signal tampering, advanced troubleshooting possibilities for NDA instrument design, and straightforward access to a high density of input channels. Several list-mode data acquisition modules are commercially available; the key features of these modules are summarized in the following sections. List-mode data acquisition is described in Chap. 18, Sect. 18.2.3.2.

Fig. 19.7 UMSR

Fig. 19.8 (left) Photograph of 32-channel, BNC-input converter box; (right) photograph of PTR-32 module

19.2.5.1 PTR-32

The PTR-32 (Pulse Train Recorder 32; Fig. 19.8, left) was developed by the Hungarian Centre for Energy Research and, as the name suggests, can accommodate 32 input channels with an additional Bayonet Neill–Concelman (BNC)-input converter box (Fig. 19.8, right). The PTR-32 accepts standard TTL pulses from neutron-detection instrument and records time intervals (follow-up times) between the consecutive neutron-detection times in 10 ns increments as well as the corresponding input channel number [7]. The device also provides high voltage (up to 2000 V) and includes a 5 V DC (direct current) supply. A performance comparison focused on multiplicity counting (Singles, Doubles, and Triples) was performed between the PTR-32 and standard shift register hardware (JSR-15) with a high-efficiency neutron counter (epithermal neutron coincidence counter [ENMC]) and a range of ^{252}Cf sources with emission rates up to 5×10^5 n/s. This comparison indicated good agreement with JSR-15 for detected correlated count rates for measured singles count rates up to 2.7×10^5 cps [8].

19.2.5.2 Advanced List-Mode Module

The advanced list-mode module (ALMM), shown in Fig. 19.9, represents a next-generation list-mode hardware [9] developed by LANL. The ALMM is based on an earlier LANL list-mode module—the list-mode multiplicity module (LMMM)—but provides a more compact and robust design. The ALMM is a 32-channel, high-speed, TTL pulse event networked recorder. The ALMM records the pulse arrival times in 100 ns increments as well as channel number for each of the 32 input channels. Measurement results are streamed out of the Ethernet port to the user's computer or stored in the internal memory. The ALMM also includes high-voltage supply (0–2000 V) and both 12 V and 5 V DC supplies. ALMM performance for multiplicity-counting applications was evaluated in comparison with standard shift register hardware (JSR-15) using the high-efficiency neutron counter ENMC and a range of ^{252}Cf sources and plutonium standards. This evaluation confirmed reliable ALMM performance and excellent agreement with JSR-15 for measured singles, doubles, and triples count rates for singles

Fig. 19.9 Photograph of the ALMM front (left) and rear connections (right)

count rates up to 4×10^5 cps, corresponding to the maximum source strength available in the evaluation [9]. The ALMM can be operated in both attended and unattended modes.

19.2.6 Software

A single measurement can be taken by simply using the shift register without any software to analyze the results. The shift register displays the results, and the user could manually store and process the results. However, software is normally used to control the shift register more easily; to record results; and to keep records of measurement metadata such as date, time, facility, etc., and for subsequent analysis and analysis of multiple measurements such as creating calibration curves and then applying the calibration to assay of unknown items. These types of analysis could be performed by hand or in programs such as Microsoft Excel, but a few commercial software packages have been produced for this application. Three common software packages are INCC, NDA2000, and list-mode software.

19.2.6.1 IAEA Neutron Coincidence Counting Software

INCC software is a Microsoft Windows–based program often used to take measurements for international safeguards and other measurements as well. At the time of publication, the most widely used version of INCC5 is available by obtaining a license from LANL.[2] INCC5 was developed and has been maintained by LANL since about 1980. The software, used by IAEA inspectors, is tailored to international safeguards and is robust. Many features needed for applied NDA measurements have been incorporated. It can control the shift register and receive the measurement data from the shift register. It can store operating parameters for multiple detectors and organizes measurements based on detector and material type. It is excellent at record keeping, with the facility, material balance area, inspector name, and all measurement and detector details being stored and reported in each measurement result. INCC can save and export individual measurement files or the entire measurement database, making it easy to transfer measurement results between computers and to organize historical records of results.

INCC performs complex statistical calculations automatically for the user. For a given measurement, the user can specify that INCC break the measurement into multiple smaller cycles—for example, 30 cycles of 30 sec, for a total measurement time of 15 min—which allows INCC to automatically calculate the statistical uncertainty of the Singles, Doubles, and Triples from the distribution about the average count rates. INCC propagates this uncertainty in the fitting of calibration curves and their application to assay measurements. INCC also robustly handles passive and active backgrounds and can subtract them automatically for the appropriate measurements and propagate the resulting uncertainties. INCC automatically handles radioactive decay corrections.

INCC applies various quality control (QC) tests to the measured data. Once sufficient measurement cycles are collected, each cycle is compared with the average and standard deviation of the population of cycles, and outliers are rejected. This rejection is recalculated after every cycle in real time. Because the average and standard deviation are recalculated, cycles near the outlier limit can bounce back and forth between pass and fail as the measurement proceeds. This outlier test feature protects against scenarios such as large changes in background from cosmic ray bursts or the movement of nearby nuclear material.

---

[2]To obtain a license for INCC5, contact licensing@lanl.gov and specify LANL Computer Code LA-CC-10-092.

Another test is the accidentals/singles test. This test checks that the singles count rate remains (statistically) constant throughout the measurement, which is especially important for analysis methods that use the singles rates alone or in combination with the Doubles or Triples. The accidental register accumulates the average number of counts in the gate (S*G) every time the gate is opened (S times per second), where S in both cases is the instantaneous singles rate. Thus, the expected measured accidentals count rate is $\overline{S^2}G$ counts/s. The "calculated accidentals" is calculated after the run from the average singles rate for the run and is equal to $\overline{S}^2 G$. These two quantities are equal if the singles rate is constant over the cycle (because then, the mean of the square of the Singles is equal to the square of the mean of the Singles). As an example, if the singles rate is 10^5 counts/s and the gate length is 64 μs, then the average number of pulses in the gate is $(10^5)(64 \times 10^{-6}) = 6.4$. The calculated accidentals rate is then 6.4×10^5 counts/s. The INCC quality control test checks that this calculation matches the measured accidentals rate and, if not, rejects the cycle. This test always passes for shift registers that use "fast accidental" sampling (see Chap. 17, Sect. 17.8) because the measured accidentals rate is C*S*G normalized by S/C, where C is the clock frequency, and S is the average singles rate over the run. This measured fast accidentals is equal to the mean Singles squared, $\overline{S^2}G$, which is equal to the calculated accidentals.

INCC is capable of many common passive and active NDA analysis techniques. Passive techniques include calibration curve, known alpha, known multiplication, multiplicity counting, add-a-source, curium ratio, and truncated multiplicity. Within some of these methods, different sub-methods are available, such as moisture corrections in the known alpha method. INCC has built-in data needed for the methods, such as the fission neutron emission moments used in multiplicity counting. INCC performs other advanced functions for real-world scenarios.

19.2.6.2 NDA2000

NDA2000 is a complete acquisition, analysis, and archival package from Canberra/Mirion [10]. NDA2000 can be used for gamma and neutron measurements. The software provides full control of data acquisition electronics and automated assay system operation and is capable of various counter arrangements, detector arrangements, analysis sequences, hardware control, and report generation.

19.2.6.3 INCC6 and List-Mode Software

Typically, the various list-mode hardware devices described in Sect. 19.2.5 are operated by device-specific data acquisition software, which provides main communication controls (such as the capability to define measurement duration, high-voltage setting, etc.) and records complete pulse trains in device-specific binary file formats. To provide streamlined communication and software control in support of *attended* nuclear safeguards measurement needs, the INCC software was upgraded to include the capability to accommodate list-mode hardware. This functionality is incorporated in INCC6 [11], which provides control, data collection, and analysis capability for list-mode devices. INCC6 currently supports PTR-32, ALMM, and MCA527 (1 channel) direct control. To support list-mode data collection, INCC6 can be operated in two modes: as a virtual shift register or to provide advanced list-mode capability.

In the first case, the complete pulse train from a list-mode device is collected by INCC6. The data are directly analyzed to extract multiplicity distributions and singles, doubles, and triples count rates in the same way as from shift registers. The goal of this mode is to provide a seamless transition for the INCC users, where the list-mode hardware can be used as a direct drop-in replacement for existing shift registers. In this mode, the INCC6 standard functionality—familiar from traditional shift registers (such as calibration, calculation of item mass, declaration-assay difference)—is available to the user with unchanged interface.

In the second case, the advanced list-mode functionality is provided to the user. This mode was developed to provide access and full benefit of the additional information contained in complete pulse trains to the user. In this advanced list mode, INCC6 can be used to generate time-interval or Rossi-α distributions; to calculate correlated count rates (S, D, T) for unlimited sets of predelay and/or gate width values; to provide a matrix of count rates recorded on different channels; and to extract correlated count rates using alternative analysis techniques (see Chap. 18, Sect. 18.2.9). This added functionality provides expanded flexibility to the user and direct access to advanced analysis approaches and additional information contained in the list-mode data otherwise not available from shift register hardware. The advanced list-mode implementation of INCC6 represents the latest development and attempt to make full list-mode capability accessible to the user.

In addition to INCC6 and *attended* safeguards measurements, list-mode data collection is also fully compatible with *unattended* applications. Dedicated unattended list-mode software components are maintained at the IAEA and at LANL. The unattended software continuously acquires list-mode data, simultaneously analyzing the raw data at a preselected acquisition interval using a configurable virtual shift register. Original list-mode data and accumulated analyses results are preserved and updated continuously.

19.3 Singles Counting Instruments

Singles counting is a straightforward measurement of the neutron count rate. Singles counting can be successful with low-efficiency measurement configurations or detectors (in contrast with doubles counting because the doubles rate is proportional to ε^2). A singles measurement may be preferred to minimize detector cost if other restrictions limit the efficiency of the detector or measurement configuration, if measurement time or statistical precision are limiting, if the neutron emission rate is low, or if the neutron emission rate is high and a low efficiency is required to reduce detector dead time.

Singles counting measurement analysis is often simple and straightforward but in some applications is complex, such as in the hybrid induced-fission-based plutonium-accounting instrument (HIPAI; see Sect. 19.3.3). In singles counting, all sources of neutrons are detected. Singles measurements are more susceptible to undesired changes, with common examples being changes in background and neutrons from (α,n) reactions that change with the chemical form of the material, not necessarily the amount of nuclear material. Still, when the material is well understood and consistent, singles counting is a valuable assay technique.

19.3.1 Portable and Limited Access Applications

19.3.1.1 Shielded Neutron Assay Probe

The shielded neutron assay probe (SNAP) is one of the simplest and most portable of the neutron detectors described in this chapter. Several versions of the SNAP detector have been made at LANL. (At the time of this writing, the SNAP-IV is just finishing development, so this section will focus mostly on the SNAP-III.) All references for this section are from Refs [12–14]. unless noted otherwise. Information on the SNAP-II can be found in the previous version of the PANDA book [15]. Photographs of the SNAP-III are shown in Fig. 19.10, and cross-sectional views are shown in Fig. 19.11.

The SNAP-III consists of one ^{3}He counter (2.54 cm diameter, 10.16 cm active length, 170.5 psi fill pressure) embedded in a 7.62 cm diameter, high-density polyethylene (HDPE) cylinder. Details on the ^{3}He tube are given in Table 19.5. The HDPE cylinder is wrapped in a thin cadmium sheet, outside of which is another cylinder of HDPE with an outer diameter of 20.4216 cm. This outer cylinder is not present on all sides, as shown in Fig. 19.11. The detector was designed such that it can be set up with the outer HDPE present on the sides and back (but not present between the item being measured and the source). The combination of this outer HDPE (roughly 3 in. thick) and cadmium are used to maximize the number of detected neutrons that come from the item being measured and minimize the number of neutrons detected from other sources (either other nuclear material or room return neutrons). As seen in Fig. 19.11, HDPE above and below the ^{3}He tube further reduces neutrons from other sources, which results in the SNAP being a very directional detector. The inner cylinder HDPE thickness was chosen to maximize the detection efficiency of fission neutrons. The overall weight of the SNAP-III is only 11 kg. The SNAP detector is used to measure the neutron count rate of an item and infer the neutron emission. For items in which the neutron multiplication is known, the mass of the item can be estimated.

Fig. 19.10 Photographs of the SNAP-III detector

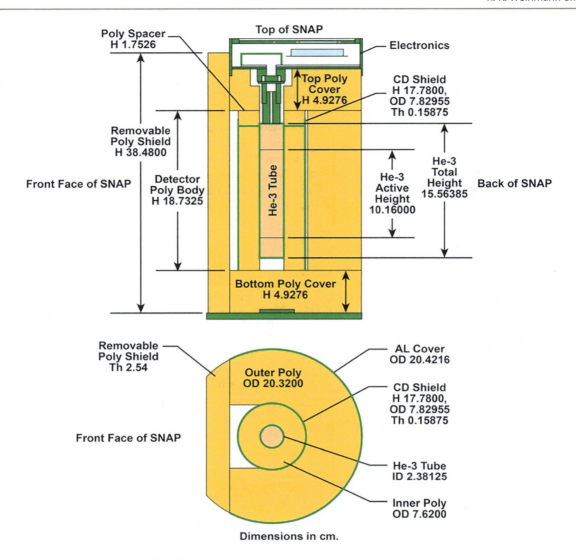

Fig. 19.11 Cross-sectional views of the SNAP-III detector [13]

Table 19.5 SNAP ^{3}He tube specifications [13]

Manufacturer	Reuter stokes
Model number	SA-P4-0804-104
Body material	Aluminum 1100
External diameter	1.00 in.
Thickness	1/32 in.
Height (including cladding)	15.7 cm
^{3}He pressure	170.5 psia
Active length	4.00 in.

A removable slab of HDPE can be placed on the front of the detector, as shown in Fig. 19.12. This slab has a thickness of 2.54 cm. The addition of this slab will greatly change the efficiency to detect thermal and intermediate energy neutrons. Figure 19.13 shows the simulated response of the SNAP-III detector using the Monte Carlo N-Particle (MCNP) code [16]. Measurements with the SNAP-III detector are often performed with and without the removable HDPE. Performing such measurements can give insight into the energy of the majority of neutrons that reach the detector.

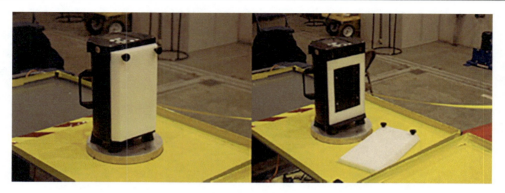

Fig. 19.12 Removable front HDPE on the SNAP-III detector [17]

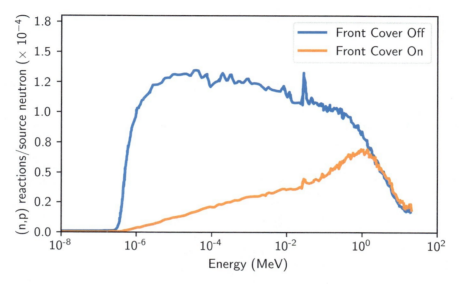

Fig. 19.13 Simulated SNAP-III detector response function as a function of energy [18]

Given the simplicity of the SNAP-III design (a single ^{3}He tube with thick HDPE shielding to reduce contribution from room return and other sources), it is somewhat easy to determine the detector efficiency for this system. Characterization measurements have been performed to determine the detector absolute (including both geometry and intrinsic) efficiency. Measurements were performed at a variety of source-to-detector (S/D) and reflector-to-detector (R/D) distances. The term *R/D* is often used in conjunction with the height above the floor, and other potential reflectors are often ignored. Figure 19.14 and Table 19.6 show measured results for the SNAP-III efficiency. Note that these values are shown in percent; using Table 19.2 as an example, the minimum S/D of 25 cm and R/D of 14 cm (which is the value with the SNAP on the floor) is 0.1010%, which shows how small the efficiency of the SNAP is, given the small ^{3}He tube that it contains; however, studies have been performed with reflection by up to five nearby surfaces (essentially a tunnel; see [19]). See Chap. 16, Sect. 16.4, for an example of how this technique is applied.

Many characterization measurements have also been performed in which various thicknesses of HDPE are present between a source (or a plutonium item) and the SNAP-III detector [12–14, 17, 19, 20]. The term *transmission* is used to describe the ratio of the count rate for a reflected configuration divided by the count rate for a bare source. Figure 19.15 shows the transmission both with and without the removable HDPE shield when the source is surrounded by up to 6 in. of HDPE. Results from the same measurements are shown in Fig. 19.16, but here, the count rate without the removable HDPE shield is divided by the count rate with the shield. From this figure, it can be seen why the SNAP-III can provide only limited neutron energy information. When the ratio value is in the 1.2–1.7 range, it can provide a unique HDPE thickness; however, above 1.7, the solutions are degenerate (and the results would be similar for any HDPE thickness).

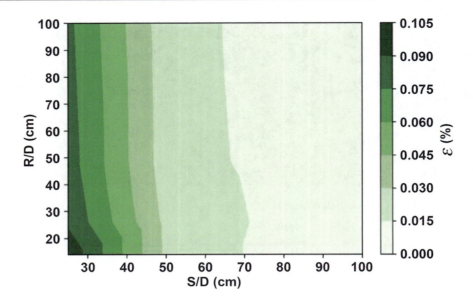

Fig. 19.14 SNAP-III measured efficiency (ε) for different source-to-detector (S/D) and reflector-to-detector (R/D) distances

Table 19.6 SNAP-III measured efficiency (ε) for different source-to-detector and reflector-to-detector distances

		R/D (cm)				
		14 (%)	26 (%)	48.5 (%)	100 (%)	400 (%)
S/D (cm)	25	0.1010	0.0877	0.0820	0.0787	0.0758
	50	0.0270	0.0262	0.0223	0.0216	0.0210
	75	0.0117	0.0130	0.0112	0.0101	0.0099
	100	0.0075	0.0077	0.0067	0.0060	0.0059
	125	0.0045	0.0045	0.0047	0.0040	0.0039
	175	0.0024	0.0025	0.0024	0.0022	0.0021
	225	0.0015	0.0015	0.0016	0.0015	0.0013
	275	0.0011	0.0009	0.0010	0.0010	0.0008
	325	0.0009	0.0007	0.0007	0.0007	0.0006

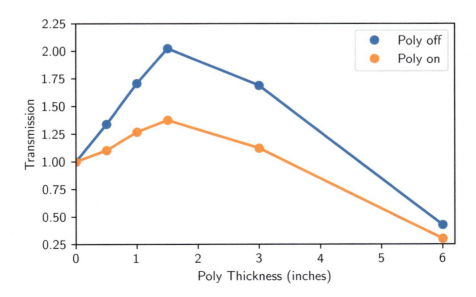

Fig. 19.15 SNAP-III measured transmission (count rate of a configuration divided by the count rate for the bare source) versus HDPE thickness

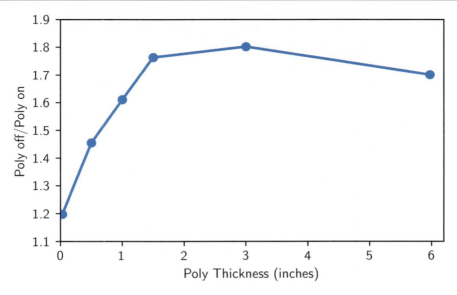

Fig. 19.16 The SNAP-III measured count rate without the removable HDPE shield versus HDPE thickness

The SNAP-III has often been used in measurements with the beryllium-reflected plutonium (BeRP) ball. The BeRP ball is a 4.5 kg, weapons-grade, α-phase plutonium sphere at the National Criticality Experiments Research Center (NCERC). These measurements include transmission on other materials as well as simulated results for the SNAP-III detector. See Refs. [13, 14, 17, 19, 20], and for more information on these measurements.

As previously mentioned, an effort is ongoing to develop a SNAP-IV detector. The SNAP-IV has the same physical design (same ^{3}He tube, HDPE, and cadmium design) as the SNAP-III, so all of the information above is relevant for the SNAP-IV; however, the major differences are in the detector electronics and the data acquisition. The SNAP-IV records list-mode data and outputs data in the .lmx format (identical to the ALMM and MC-15), which technically allows for the performance of correlated or multiplicity analysis of the SNAP-IV data. However, given the low detector efficiency, the SNAP-IV likely will not be used for that purpose very often. The benefit of this approach is to make data acquisition easier and improve the quality of the data analysis.

19.3.1.2 Slab Detectors

Slab detectors contain an array of thermal-neutron detectors inserted into a slab of moderating material. They are larger and heavier than SNAP detectors but provide higher detection efficiencies and better directionality if heavier shielding and collimators are added. Slab detectors often consist of ^{3}He tubes placed parallel to each other in a moderator block [21]. The size and number of tubes vary with the application. Figure 19.17 shows a slab detector that contains nine ^{3}He tubes (2.5 cm diameter, 4 atm fill pressure) embedded in a 10 cm thick polyethylene slab. The junction box holds the tubes rigidly and encloses the high-voltage buss wire and amplifier modules in an air-tight, electrically shielded space. For some applications, the polyethylene slab is covered with a thin cadmium sheet to absorb thermal neutrons.

Slab detectors have been used for plutonium holdup measurements after a cleanout operation [22]. The model used to interpret the results assumed that the geometry of the source material could be approximated by a uniform plane near the floor. This assumption could be investigated by moving the detector and repeating the measurement. The detector was 0.23 m^2 in frontal area and unshielded. The cadmium absorbers were removed to obtain maximum sensitivity to low-energy neutrons. The advantages of using neutron techniques over gamma-ray-based techniques are that shielding of the source material is less of a problem and room return tends to enhance the counter response to material at the edge of a room. These two effects reduced the number of measurement positions required to adequately survey the facility. The room-to-room shielding was good; consequently, cross talk between rooms was not a problem. The chemical and isotopic form of the plutonium was known so that the response per gram could be determined. The detector response was averaged over several measurement positions to reduce the detector's sensitivity to hot spots (or to determine if they existed). Although this procedure yields 50% results at best, such results are adequate for applications that involve a few hundred grams of plutonium spread thinly over large areas.

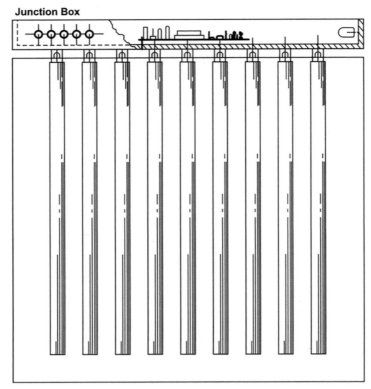

Fig. 19.17 Side view of slab detector with nine 2.5 cm diameter ^{3}He proportional counters. The junction box contains an amplifier module board

Other slab detectors [23] have been used for measurements on containers of spent fuel. The device uses a pair of slab detectors separated by a polyethylene shield so that the two counting rates give directional information on signal and background fluxes.

19.3.1.3 Portable Handheld Neutron Counter

Implementation of the Additional Protocol and complimentary access increased and broadened the scope of inspection activities in nuclear facilities. Novel instruments needed to be designed and manufactured to meet the growing demands on the IAEA. Among the instruments needed was a fieldable instrument capable of operating under non-ideal conditions, performing large-area ad-hoc field inspections, and quantifying nuclear material. The portable handheld neutron counter (PHNC) was designed to meet these needs [24].

The PHNC is a lightweight and portable neutron-detection system that consists of two HDPE slabs and two detector modules. Each detector module is a slab of HDPE with four ^{3}He tubes embedded inside. The detector modules are manufactured by PDT, who developed a novel method of embedding the ^{3}He tubes into the high-voltage and amplifier (PDT-20A) package to reduce the size and weight of the unit—only 3.8 kg. The ^{3}He tubes have an active length of 17.8 cm and a pressure of 4 atm. The small number and size of the ^{3}He counters mean that the efficiency for the system is low, at ~7.4% for two slabs. The shift register electronics are battery operated for portability purposes. The system is small enough to fit into a briefcase and weighs ~10 kg. The detector module can also be carried using a shoulder strap, with the electronics package carried by a belt strap.

The PHNC operates in two different modes for field measurements and for mass quantification measurements. For field measurements, the PHNC is operated in survey mode. The purpose of the survey mode for the PHNC is to locate hidden plutonium sources. Only singles counts are measured in short intervals between 1 and 600 s. The detector has limited directionality, so the user needs to be cognizant of directions. No quoted uncertainty exists for this mode because it is used for identification of neutron sources and not quantification. The signal is collected in real time, and an alarm is triggered when a threshold is crossed.

The mass quantification mode turns the PHNC into a collar detector. The two slabs are removed from the briefcase and are coupled with HDPE slabs to surround the source in question. The PHNC can measure plutonium mass from less than 1 g to hundreds of grams in containers made of stainless steel, aluminum, or glass; however, each container type requires a unique calibration. A coincidence measurement is collected using a shift register for 10–15 min. A measurement of pure oxide sample can obtain precision to 2–3%, although other sample compositions will have a much higher uncertainty.

19.3.1.4 Plutonium Inventory Monitoring System

A plutonium inventory monitoring system (PIMS; Refs. [25, 26]) consists of a permanently installed set of ^{3}He tubes deployed around one or more glove boxes. The response of the system is a set of counting rates that correspond to the locations of plutonium in the glove boxes. Calibration consists of determining a matrix, whose elements consist of the response of each detector to unit plutonium mass at many different locations. The plutonium distribution is obtained by multiplying the count rate vector by the inverse of the response matrix. PIMS installations exist in the United Kingdom and in the Rokkasho Reprocessing Plant in Japan. To give an idea of scale, the latter comprises 142 ^{3}He tubes. The system can be used to detect changes in inventory on a near-real-time basis to verify cleanout and measure any residual material.

19.3.2 Assay of Low-Level Waste

Passive neutron counting is often used for measuring nuclear waste material because neutrons can penetrate large waste containers much better than gamma rays can, particularly if the waste contains dense, high-Z materials. Waste containers are typically 55 gallons or larger in volume, and passive gamma-ray detectors tend to underestimate the amount of nuclear material because of gamma-ray absorption in the matrix. In contrast, neutron measurements tend to overestimate the amount of nuclear material; (α,n) reactions in the matrix or moderation followed by induced fission create "extra" neutrons. Neutron coincidence counting can substantially reduce this matrix sensitivity by discriminating between source fission neutrons and matrix (α,n) neutrons; however, total neutron counting may be more sensitive to small quantities of nuclear material if (α,n) reactions increase the neutron emission rate. For example, it is several orders of magnitude more sensitive to the fluorides PuF$_4$ or UF$_6$. Total neutron counting is often used for discard/save decisions rather than for quantitative assays.

19.3.2.1 Detection Sensitivity

The neutron emission rates of some common nuclear materials are listed in Tables 13.1, 13.3, and 13.5 (see Chap. 13). From these tables, it is clear that low-Z materials that allow (α,n) reactions can significantly increase the neutron output. Because the neutron output of the plutonium compounds is much higher than the neutron output of the uranium compounds, the sensitivity to plutonium is much better than the sensitivity to uranium. Here, the sensitivity of the assay is defined as ΔC/C, where ΔC is the 1-standard-deviation error in the counts C.

An important concept for the measurement of low-level waste is the detectability limit, which is that quantity of material that produces a signal larger than background by the ratio d = C/ΔC. For a background rate b much less than the signal rate, the detectability limit m (in grams) is given by

$$m = d^2/At_1 \tag{19.1}$$

For a background rate b much larger than the signal rate,

$$m = \frac{d}{A}\sqrt{\frac{b}{t_1} + \frac{b}{t_2}} \tag{19.2}$$

where A is the response rate of the instrument in counts per second per gram, and t1 and t2 are the signal and background count times, respectively. Detectability limits at 3σ above background (d = 3) are 23 mg for low-burnup plutonium, 0.5 mg for PuF$_4$, 170 g for natural uranium (NU), and 30 g for natural UF$_6$. These calculations assume 1000 s counting times in a large 4π counter with 15% absolute efficiency. The 4π counter is recommended for assaying low-level waste because of the weak emission rate and heterogeneous nature of the waste.

The calculated detectability limits show that passive neutron counting of low-level waste is usually practical only for plutonium. For 55-gallon drums that contain 100 kg of non-absorbing matrix materials, the plutonium limit of 23 mg

Table 19.7 Physical characteristics of Rocky Flats 55-gallon drum standards

Description	Matrix	Matrix composition (wt%)	Matrix avg net weight (kg)	Matrix avg density (g/cm^3)	Plutonium (as PuO$_2$) loadings (g Pu)
Graphite molds	60-mesh graphite	100	110	0.53	60, 145, 195
Dry combustibles	Carbon	90	24	0.12	10, 165, 175
	Plastics	5	N/A	N/A	N/A
	Cellulose	5	N/A	N/A	N/A
Wet combustibles	Cellulose	80	51	0.25	28.5, 166
	Water	15	N/A	N/A	N/A
	Plastics	5	N/A	N/A	N/A
Washables	Polyvinyl	42	32	0.15	10, 90, 160
	Lead gloves	28	N/A	N/A	N/A
	Polyethylene	20	N/A	N/A	N/A
	Cellulose	7	N/A	N/A	N/A
	Surgical gloves	3	N/A	N/A	N/A
Raschig rings	Pyrex glass with 12% boron as B$_2$O$_3$	100	82	0.39	40, 95, 185
Resin	Dowex 1 × 4	100	–	–	25, 110
Benelex-Plexiglas	–	–	–	–	75

corresponds to ~23 nCi/g. This detectability limit can easily increase by an order of magnitude for actual drums that contain significant quantities of moderators or neutron poisons; however in most cases, passive neutron assay overestimates the quantity of nuclear material present because of (α,n) reactions in the matrix. Unless the chemical and isotopic form of the waste is known, no quantitative conclusion can be drawn about the nuclear content of a barrel other than an upper limit.

19.3.2.2 Assay of 55-Gallon Drums

Total neutron counting of 55-gallon drums that contain PuO$_2$-contaminated waste has been investigated at LANL [27]. Measured were a set of 17 standards constructed at the Rocky Flats Plutonium Processing Facility. The standards were designed to simulate the contaminated process materials and residues routinely assayed in the Rocky Flats drum counter; Table 19.7 summarizes the characteristics of these drums. A standard deviation of ±16% was obtained for the 17 drums. Because the plutonium isotopics and chemical form were both fixed and well known, the total neutron signal was a reasonable measure of the plutonium content.

19.3.2.3 Assay of Large Crates

A neutron counter large enough to assay 1.2 × 1.2 × 2.1 m waste crates was developed and used extensively at the Rocky Flats facility [28]. This 4π counter uses twelve 30 cm diameter ZnS scintillators spaced around the sample chamber. Because these fast-neutron detectors also exhibit some gamma-ray sensitivity, the discriminator thresholds are set above the 1.3 MeV ^{60}Co gamma-ray energy. Most of the neutron signal is also discriminated out, and the measured efficiency of the counter is 0.1%. Although the desired signal is from spontaneous fission neutrons, total counting is preferred to coincidence counting because of the low detection efficiency.

The major sources of inaccuracy for the crate counter are variable matrix effects and the unknown chemical form of the plutonium. Some comparisons with a 55-gallon-drum counter show that the crate counter tends to overestimate the plutonium content. Typical crate loadings are less than 100 g; the counting times are 200 s. The results typically agree with tag values within a factor of 4.

The crate counter is used to flag crates that need to be opened and checked more carefully. The counter periodically identifies crates that have been labeled incorrectly and is also used in conjunction with passive gamma-ray counting if more quantitative results are desired.

19.3.3 Assay of Curium-Bearing Materials

Curium is produced by neutron absorption in nuclear fuel. It is found in spent nuclear fuel and is a transuranic that is usually removed with processing to remove fission products. Table 19.8 list properties of the two most common curium isotopes,

Table 19.8 Common curium isotope properties

Isotope	Half-life	Spontaneous fission neutron production yield (n/s − g)	Spontaneous fission multiplicity	Induced thermal fission multiplicity
^{242}Cm	163 d	2.10×10^7	2.54	3.44
^{244}Cm	18.1 year	1.08×10^{12}	2.72	3.46

^{242}Cm and ^{244}Cm. Both isotopes decay by spontaneous fission and have large specific neutron production rates. The half-life of ^{242}Cm is 163 days. For recently irradiated spent fuel, the neutron production can be dominated by ^{242}Cm. For older fuel, ^{242}Cm neutron production may be negligible, although it is also produced by decay of the long-lived ^{242m}Am [29]. The half-life of ^{244}Cm is 18.1 years, so it can contribute a significant fraction of the total neutron emission of spent nuclear fuel for decades after the fuel was last irradiated. The amount of both curium isotopes is also highly dependent on the burnup of the fuel.

Neutron assay of curium-bearing nuclear material is challenging because the objective of the measurements is often to measure the plutonium content by its neutron emission. The curium adds significant neutron background, and for typical fuel, curium can produce 99% of the spontaneous fission neutron emission.

Instruments used to measure nuclear material that contains curium often must be able to handle high neutron rates from the curium and high gamma rates from fission products. Shielding, low efficiencies, advanced electronics, and high-dose detectors (such as fission chambers for neutrons and ion chambers for gamma) are used to address these issues. Certain analysis techniques may be preferred—for example, techniques using only the singles rates—because high count rates will cause large doubles and triples uncertainties, or the efficiency may be prohibitively low.

Several analysis approaches have been investigated. One option is to determine the fractional curium content through destructive assay and include it as additional isotopes in the ^{240}Pu$_{eff}$ calculations, although that adds significant complication, expense, and time delays over NDA. Determining the fractional curium content through gamma measurements would be a preferred approach, but this method is often difficult due to few useful curium gamma rays and high gamma emission rates from fission products. Another approach that focuses on whole-facility nuclear material tracking instead of individual measurements and assumes that the plutonium is always intimately mixed with the curium is to measure and track the curium instead as an indirect measurement of the plutonium [30]. Several NDA techniques have been investigated, and specific instruments are described in the following sections. The techniques include self-interrogation (curium neutrons induce fission in the plutonium, where the multiplication is measured) and the passive neutron albedo reactivity technique [31, 32]. Multiplicity counting also has some applications; for example, if the neutron emission rates from curium and plutonium are similar (which is rare), then the ratio of the curium and plutonium can be determined from the detected neutron multiplicity distribution; some fraction will be produced from curium and (1− the fraction) from plutonium.

19.3.3.1 Vitrified Canister Assay System

The vitrified canister assay system (VCAS; Refs. [33, 34]) is used to measure the plutonium content of canisters that contain the material that resulted from the vitrification of fission product solutions. The large inventory of fission products causes these canisters to emit a very intense gamma radiation, leading to a high gamma dose rate at the detector location. For this reason, ^{235}U fission chambers (rather than ^{3}He tubes) are used to measure the neutron emission. Polyethylene is used as a moderator to increase the efficiency of the fission chambers. Tungsten gamma shielding is used to reduce the gamma dose to the detectors, electronics, and cables.

The main signal recorded by fission chambers is the neutron emission rate of the canister. Most (greater than 95%) of this signal comes from the ^{244}Cm in the canister, and so the system is calibrated in terms of the singles rate as a function of curium mass. The plutonium mass (the real quantity of interest) is obtained by multiplying the curium mass by the plutonium/curium ratio obtained from destructive analysis of samples taken from the fission product stream before vitrification.

Additional fission chambers (^{235}U and ^{238}U) are used to check the fill height and to ensure that the neutron spectrum corresponds to the expected combination of fission and (α,n) neutrons and that the amount of thermal-neutron emission corresponds to vitrified waste matrix rather than to an aqueous solution.

19.3.3.2 Rokkasho Hulls Measurement System

The Rokkasho hulls measurement system (RHMS; Refs. [34, 35].) is used to verify the plutonium content of drums of hulls and end pieces that result from the shearing of fuel assemblies before the dissolver stage at the Rokkasho Reprocessing Plant. In each of the two identical systems, three ^{3}He tubes are arranged around the drum. The system is collocated with an active interrogation instrument of the operator. The operator's system uses a 14 MeV neutron generator and many more ^{3}He tubes.

The RHMS itself measures the passive neutron emission from the drum and uses the singles rate to determine the ^{244}Cm mass (as in the case of VCAS). The plutonium mass is obtained by multiplying the curium mass by the plutonium/curium ratio taken from destructive analysis of samples from the accountability tank.

19.3.3.3 Hybrid Induced-Fission-Based Plutonium-Accounting Instrument

The Hybrid Induced-Fission-Based Plutonium-Accounting Instrument (HIPAI) is a prototype designed to measure the uranium/transuranic products (U/TRU) from a pyroprocessing fuel cycle. In this fuel cycle, spent reactor fuel is processed to separate the fission products from the uranium (still mixed with transuranics) for reintroduction into a reactor as recycled fuel. The transuranics include the plutonium present in the fuel and also the curium isotopes. Because the U/TRU product contains uranium and plutonium, it is an important object for nuclear material accountancy and international nuclear safeguards. HIPAI was designed to measure the amount of plutonium in the U/TRU ingot to facilitate this accountancy. The primary reason that a custom solution was needed is because of the curium present in the ingot. Curium-244 is a strong neutron emitter from spontaneous fission and, in a typical fuel composition, it emits about 1000 times more neutrons than plutonium. Most neutron NDA systems aim to measure spontaneous fission neutrons from plutonium, but for this material, the doubles rate actually indicates the amount of curium that is not the quantity of interest for nuclear material accountancy and safeguards. Instead, HIPAI measures the plutonium directly by measuring the multiplication of the ingot. If more plutonium is in the ingot, the multiplication will be higher; if less plutonium, the multiplication will be lower. This relationship assumes that the ingot has the same dimensions and density so that plutonium content is the only variable that affects multiplication. The plutonium isotopics must also be known from the isotopics of the spent fuel input into the process. The uranium in the ingot has a negligible effect on the multiplication. For nuclear material accountancy, the amount of uranium present in the ingot must be determined from some other means—likely gamma measurements—to determine the ratio of uranium to plutonium.

HIPAI measures the item's multiplication through a technique called *passive neutron albedo reactivity* (PNAR; see Ref. [36]). The concept of PNAR is that the item is placed in a typical thermal-neutron well counter that surrounds the item with polyethylene. Some fraction of the neutrons passively emitted by the item become thermalized in the polyethylene and are reflected back into the sample cavity, where they cause induced fission in the item. The item's multiplication affects the number of induced fissions, the neutron production rate, and the subsequent detector count rate. A cadmium liner can be placed around the cavity to absorb these thermal neutrons before they can return to the cavity, thus preventing this multiplication. PNAR works by taking measurements of the item with and without the cadmium liner. Dividing the no-cadmium measurement (with multiplication) by the cadmium measurement (without multiplication) gives a ratio that indicates the multiplication of the item. The ratio accounts for changes in the passive neutron emission rate. If 10% more neutrons are emitted, for example, by an increase in curium content, the measurements with and without the cadmium will both increase by 10%, which will be canceled out in the ratio. Thus, the ratio is a direct measurement of the item's multiplication, which directly and exclusively corresponds to the plutonium mass.

The use of the ratio is especially important for neutralizing curium's large neutron production rate. Many effects change the number of neutrons produced by the curium—including the simple passage of time—so using the ratio is necessary for the measurement to be independent of these changes. The curium content depends on the fuel's burnup and cooling time. Curium-242 has a half-life of 160 days, so in the first few years, the neutron emission of the item is dominated by the ^{242}Cm. After sufficient decay of ^{242}Cm, the dominant neutron emitter becomes ^{244}Cm, with a half-life of 18.1 years. The neutron emission from curium is about 1000 times greater than that of plutonium for typical spent fuel, so the measurement must be completely independent from the curium neutron production rate.

The HIPAI instrument consists of two rows of three ^{3}He tubes each, embedded in polyethylene. The location of the tubes and other features were optimized [37]. The tubes use PDT110 amplifiers, which are wired so that the count rate of each ring can be recorded independently. The overall detection efficiency is 7.8%, with the inner and outer ring efficiencies of 7 and 0.8%, respectively [38]. The detection efficiency is deliberately low to reduce count rate, and thus dead time from the high neutron-emission rate is due to curium. A photo of the HIPAI is shown in Fig. 19.18. The tall, central cylinder is the sample chamber, which is designed to be easily removed for addition and removal of the cadmium liner. Simulations showed that HIPAI was able to measure the plutonium mass within 2.6% ± 2%, on average [37].

Fig. 19.18 Hybrid induced-fission-based plutonium-accounting instrument

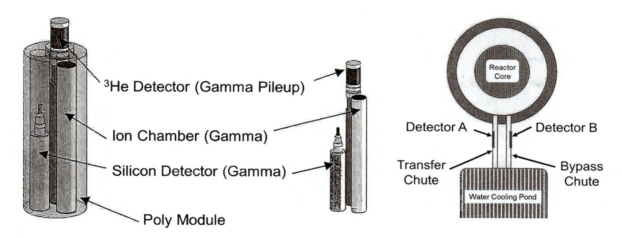

Fig. 19.19 Detector module detail of the RGDM (left) and example reactor configuration (right; Ref. [39])

19.3.4 Other Singles-Based Analysis Techniques

19.3.4.1 Reactor Gate Discharge Monitors and Direction of Motion

Reactor gate discharge monitors (RGDMs), gate monitors, and radiation monitors are used to monitor fueling activities at nuclear reactors, including confirming the time of fuel discharge from the reactor to the cooling pond and the fuel's direction of motion. They measure total neutron count rate, gamma dose, and potentially coarse gamma energy as a function of time. A gate monitor and its configuration in a reactor facility is shown in Fig. 19.19.

In the RGDM, the ^{3}He detector is used for neutron measurements and can also be operated in gamma pileup mode [39, 40]. The ion chamber and silicon detector perform gamma measurements. Tungsten shielding is used to reduce the intensity of the gamma dose and to protect the detector electronics. Directionality is achieved by placing the shielding on the bottom and not the top of the instrument, as shown in Fig. 19.20. As fuel approaches from the top, the count rate increases relatively slowly as a function of geometric attenuation of $1/r^2$. As the fuel passes the detector and moves away from the bottom, the lower thick shielding causes a fast drop-off in measured radiation greater than the effect of $1/r^2$ alone, as shown in Fig. 19.21. Tungsten shielding can also be used in the front half of the gamma detector as a measurement of the gamma energy. The ratio of the gamma dose measured with and without the radiation passing through the fixed thickness of tungsten

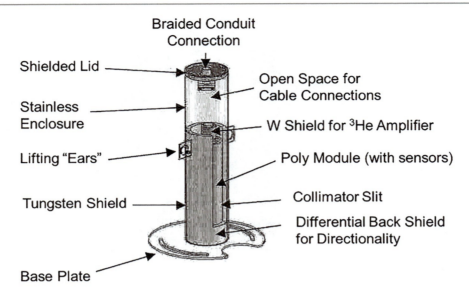

Fig. 19.20 RGDM shielding [39]

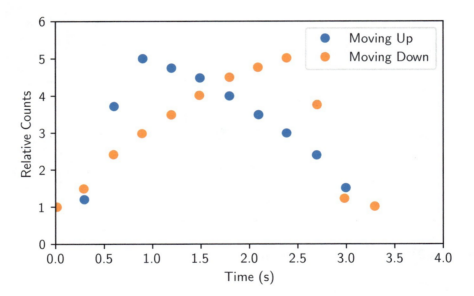

Fig. 19.21 Simulated RGDM signal showing directionality [40]

is a function of the gamma energy. Overall, a similar instrument design can be used in any application that requires direction of motion or a coarse measurement of gamma energy.

19.3.4.2 Ring-Ratio Analysis

The ring-ratio technique uses ^{3}He tubes embedded at different polyethylene depths to measure the average neutron energy of a sample for a variety of analyses. This technique can be used by neutron well counters with a central sample cavity and rings of ^{3}He tubes, by slab counters with rows of ^{3}He tubes, or by any other geometry with a similar configuration. In this description, we will focus on well counter geometry, although the physics is largely the same in other configurations. The rings are located at different polyethylene depths, and the neutron count rate can often be read individually for each ring. The count rate in each ring depends on the average neutron energy because higher-energy neutrons will penetrate through more polyethylene before thermalization and absorption, whereas lower-energy neutrons will be absorbed sooner, in the more-inner rings. This principle is a similar principle to Bonner spheres, which have long been used to measure neutron energies. The ratio of the ring count

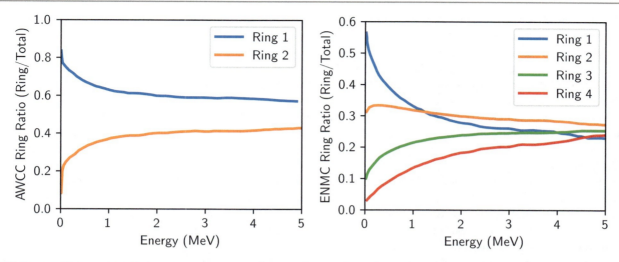

Fig. 19.22 AWCC (left) and ENMC (right) ring ratios as a function of energy. Ring 1 is the inner ring

rates is taken to normalize for changes in source strength. The ring ratio can be used in many applications—from scientific measurements of a source's average neutron energy to application-based uses such as determining the moisture content in a sample for detector efficiency corrections or for determining the concentration of a specific (a,n) emitter [41, 42]. The ring ratios and the detector Singles, Doubles, and Triples can be read out simultaneously, so the ring ratios can be combined with more-traditional analysis.

Ring ratios are a low-resolution measurement of neutron energy. Neutrons from a single isotope or (a,n) reaction are emitted in a spectra of energy. The average energy may be well known, but an individual neutron could be several MeV above or below the average. Then, the neutrons undergo a random number of scatters, losing a random amount of energy each time. For example, neutrons of the exact same single energy might be detected in the middle ring most often, but a significant portion will be detected in more inner or outer rings as well. The combination of those two mechanisms results in ring ratios that represent average characteristics of the neutron spectra. This effect is in contrast to gamma measurements, which precisely measure the characteristic discrete energy emitted by an isotope. A ring ratio can be precise because the statistical uncertainty of a count rate and the corresponding calculated ring-ratio uncertainty can be very low. For example, the ring ratio measurably changes when placing a thin plastic bag around a source [43]. Ring ratios reflect relatively small changes for large changes in neutron energy. For example, simulations of the active well coincidence counter (AWCC) outer ring/inner ring ratio show that doubling the neutron energy (from 1 to 2 MeV) changes the ratio from 0.58 to 0.67 [44]. The change in ring ratio can be larger in detectors that have more rings. One aspect of this small change is that neutrons lose an average of one-half of their energy in a collision with the hydrogen in polyethylene. It takes an average of 27 collisions to go from an average fission energy of ~2 MeV to an average thermal energy of ~0.00000002 MeV. It takes only one collision on average to go from 2 to 1 MeV, which is only 4% of the collisions needed to reach thermal energy and be detected. Simulated ring ratios for the AWCC and the ENMC as a function of neutron energy are shown in Fig. 19.22.

Ring ratios can be used in several applications. The ratio approximately measures the average neutron energy, which can be used as an indication of any characteristic that affects neutron energy. Knowledge of this characteristic can then be used in the analysis. These characteristics can include the source position in the cavity, the source geometry, and even the energy spectrum—even if the average energy remains constant. In fact, because so many source characteristics affect the ring ratio, the characteristics must be constant or controlled for the ring ratio to indicate only one characteristic. As a specific example, the ratio can be benchmarked to the moisture content in an item. The moisture changes the neutron energy, the amount of self-shielding due to increased thermal-neutron fission absorption in the nuclear material, the overall multiplication, and the absorption in the hydrogen in the moisture and in the cadmium liner. These effects can be corrected by calibrating the ring ratio to an empirical efficiency correction coefficient. Overall, the ring ratio can be a powerful tool to constrain the results of conventional singles, coincidence, or multiplicity counting, reducing measurement uncertainties, and enabling measurements of unique and challenging nuclear material.

19.3.4.3 Head-End Monitoring Systems, Input Spent-Fuel Verification System, and Integrated Head-End Verification System

To provide safeguards assurance on nuclear material in spent fuel assemblies that enter a reprocessing facility, it is essential to monitor spent fuel assemblies at the receipt into the facility and when they are moved into the shearing process. The nuclear material contained in intact spent fuel assemblies is in the item form from a safeguards perspective, and therefore, the verification activities rely on item-monitoring techniques, which include a combination of containment and surveillance measures and NDA techniques. Optical cameras combined with neutron and gamma counting are used to detect the presence of nuclear material and direction of motion of spent fuel assemblies. Examples of such implementation are the input spent-fuel verification system (ISVS) and the integrated head-end verification system (IHVS), both installed at the Rokkasho Reprocessing Facility.

ISVS is used to verify the number of spent fuel assemblies unloaded from transport casks into the spent fuel storage pools and for verification of empty returning shipping casks. It uses passive neutron and gamma detection as well as a combination of cameras. The system consists of two pair of neutron/gamma detection systems mounted on each side of the spent fuel transfer canal. The detection systems provide information on the number of spent fuel assemblies unloaded as well as confirm the direction of motion into the storage pool. IHVS is a similar system installed in the transfer path of spent fuel assemblies moving into the shearing process and also employs a system of cameras, ^{3}He-based neutron detectors, and ion chambers for neutron and gamma monitoring. To provide continuous monitoring capability, both systems are operated in an unattended mode, where the data are automatically collected in Multi-Instrument Collect (MIC) software (see Sect. 19.7.4) for unattended monitoring) and are available for review by inspectors. These measurement systems represent a good example of "safeguards by design" because the detector systems were incorporated into the wall of the shielded cells before the concrete fabrication.

19.3.4.4 Nuclear Material Transfer Monitor

Nuclear material under International Atomic Energy Agency (IAEA) safeguards is required to be verified before being transferred to retained waste. In certain circumstances, where the measurable radiation sources from this material are well understood, continuous unattended systems can be developed for quantitative assay of nuclear material. One such need was identified for uranium-bearing material to be transferred to a waste repository by means of a 40-foot-long steel sea/land shipping container carried on a 53-foot-long flatbed trailer. The several metric tons of material in each batch of transferred waste required the development of a large unattended system to assay each truckload prior to unloading into the repository.

The waste material to be disposed is held in 2.7 m^3 lift bags and consists of depleted uranium/titanium (DU/Ti) oxide which contains between 20% and 80% uranium by weight. Because of the size and mass of material total neutron counting was selected as the assay method.

The unattended system designed and implemented to quantify the mass of uranium in each truckload is the Nuclear Material Transfer Monitor (NMTM) [45]. The NMTM configuration for this particular use case is a portal with six neutron slab detectors in an overhead arrangement as seen in Figs. 19.23 and 19.24.

Each neutron slab detector consists of six helium-3 (^{3}He) tubes—each 1 meter in active length and filled with 4 atmospheres of ^{3}He—embedded in polyethylene (Fig. 19.25). The preamplifiers are PDT100 series and the outputs of all six tubes from a slab are daisy-chained to a single output. Data acquisition is based on the ADM2 pulse counting module from BOT Engineering Ltd.

The major contributors to the neutron count rate in the detectors from the DU/Ti oxide waste stream are spontaneous fission of ^{238}U and induced fission in ^{235}U. Calculations with SOURCES 4C code [46] determined that (α,n) reactions on oxygen are negligible. A review of historic facility records provided information on the make-up of the DU/Ti oxide waste stream: uranium weight percentage ranged from 20% to 80% $\pm$ 6% (relative), ^{235}U/^{238}U ratio average value 0.002 $\pm$ 5% (relative), and the bulk density ranged from 1.7 to 4.0 g/cm^3 $\pm$ 20%. The bulk density range is assumed to be freely flowing powder at the low end and likely non-flowing hardened or compressed powder at the high end.

Waste is transported to the site in three to five lift bags per truckload. The calibration coefficients were determined via Monte Carlo simulation using the MCNP code [16]. The benchmarking of the model has been described previously 0. The calibration curves for DU/Ti oxide and the associated calibration constants are seen in Fig. 19.26. The two curves correspond to the minimum and maximum density range. The non-linearity of the calibrations comes from multiplication in ^{235}U. For example, induced fission in ^{235}U accounts for nearly 50% of the estimated count rate from the high-density calibration simulations regardless of the uranium wt%. In the low-density simulations, the ^{235}U induced fission contribution is 3% for simulations at 20 wt% uranium and 14% for simulations at 80 wt% uranium.

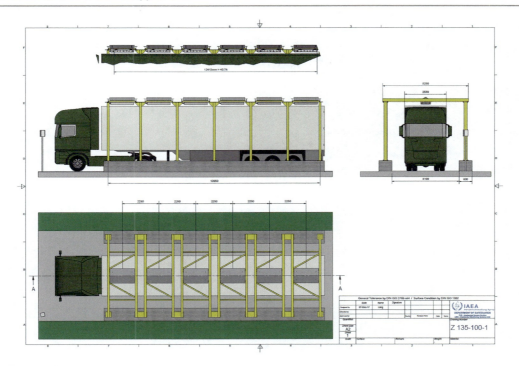

Fig. 19.23 NMTM configuration as an overhead portal

Fig. 19.24 Finished neutron portal

Several tens of loads of DU/Ti oxide, in a configuration that resembles the assumptions used to develop the calibration, have passed through the unattended portal since its installation. The average declared uranium mass of these was 11.4 metric tons (t) with a minimum of 5.3 t and a maximum of 13.8 t.

Knowledge of the lift bag capacity and the average load mass indicates that the material density is in the low range. Using the low-density calibration, the average assayed uranium mass was 12.2 t with a minimum of 6.2 t and a maximum of 15.0 t. A comparison of measured vs. declared uranium mass of all assayed loads is shown in Fig. 19.27. The dashed blue line fit to the data shows that the calibration generally overpredicts, assuming the operator's declaration is correct. Figure is a graph of the operator—inspector difference (OID) in metric tons. The average OID is -0.7 t and the average of the absolute value

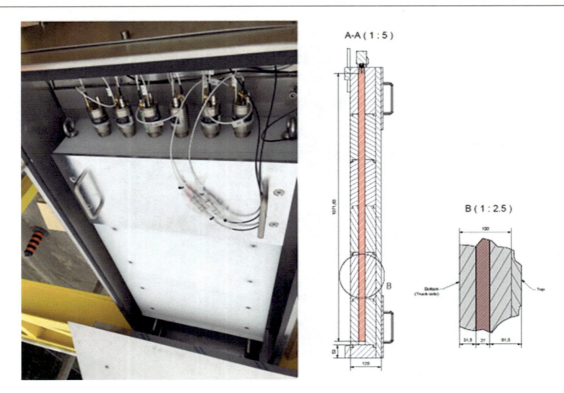

Fig. 19.25 Helium-3 based neutron slab detector photo and cutaway drawing (dimensions in mm)

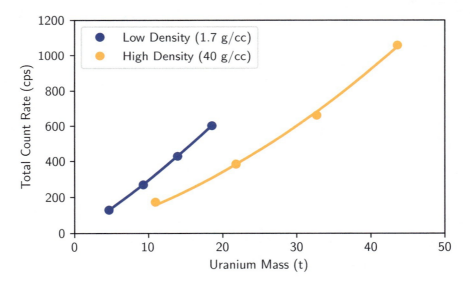

Fig. 19.26 Calibrations determined for DU + Ti oxide via Monte Carlo simulation

of all OID is 0.8 t or 7%. The single data point at the low mass of 5.3 t (declared) is not primarily responsible for the slope of the trendlines in Figs. 19.27 and 19.28. They remain largely unchanged if this data point is removed.

The NMTM has the capability to accurately assay the uranium mass in large loads of depleted uranium and titanium oxide. Monte Carlo methods work well to determine proper calibration constants, as long as the assumptions in the model are a fair representation of reality and the neutron source term is relatively simple. In this application, the predominant source of neutrons is spontaneous fission in ^{238}U. The assay of material streams with varying concentrations of impurities and/or more

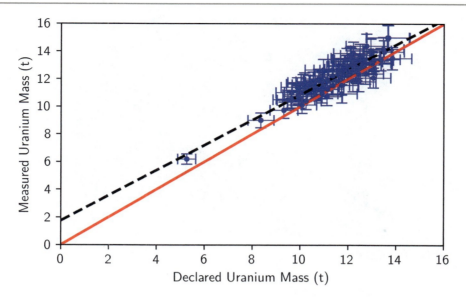

Fig. 19.27 Assayed mass vs. declared mass for the DU/Ti oxide transported in lift bags

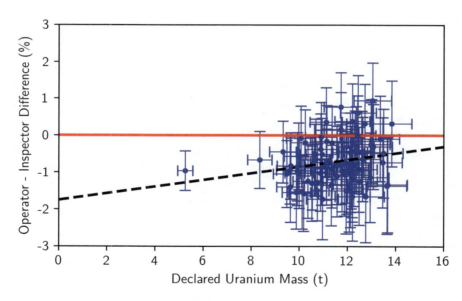

Fig. 19.28 Operator—inspector difference in metric tons for DU/Ti oxide transported in lift bags

complex neutron source terms, including higher contributions from induced fission and large sources of (α,n) production, will be less accurate.

The unattended NMTM system has been demonstrated to be durable and robust in the outdoor environment in which it is installed. The portal was installed in the summer of 2018 and continues to operate unattended despite a seasonal average temperature variation from +26 °C in summer to −13 °C in winter, seasonal extreme temperatures of ±35 °C, and accumulations of snow in winter.

It is likely that additional NMTM installations will be needed for other applications involving the transfer of large bulky materials between sites. Similar methods for benchmarking and calibration via Monte Carlo simulations will be utilized given the likelihood that certified sources will not be attainable. One such use of the NMTM will be at the New Safe Confinement at Chernobyl Unit 4, where it will be used to verify the absence of nuclear material in loads of construction waste being removed from the Unit 4 sarcophagus.

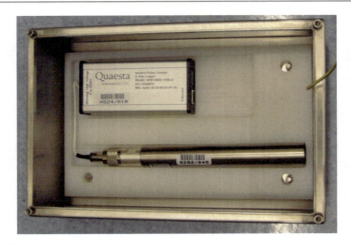

Fig. 19.29 RDL1 connected to ^{3}He detector

19.3.4.5 Mobile Unit for Neutron Detection

Developed by the IAEA and first used in the early 2000s, Mobile Unit Neutron Detection (MUND) systems are deployed along the nuclear fuel cycle in facilities throughout the world for mobile monitoring of nuclear material transfers. Whereas initially MUNDs were used in CANDU reactors to provide neutron detection during the spent fuel transfer process from the reactor facility to the dry storage silo area, nowadays the MUND's principal purpose has broadened to provide continuity of knowledge on neutron-emitting material during a transfer process.

The nuclear material transfer processes monitored by MUNDs are without external power or other connections; thus, the device was designed to operate stand-alone. Typically, two MUNDs are installed to provide redundancy. The MUND enclosure is sealed as well as being sealed to a specific location, both with tamper-indicating IAEA seals.

The main parts of a MUND are the ^{3}He detector for neutron detection, pulse counting hardware and a battery to support counting during the transfer process.

Due to the increasing need for MUND devices for dry storage programs around the world and seeking improvements in several areas, the IAEA began development of an improved MUND that integrates the majority of the system's separate components into a single small component and enhances data logging and counting capabilities. In 2011, the IAEA collaborated with Quaesta Instruments to develop the Radiation Data Logger (RDL1), which is primarily a neutron pulse monitor with integrated data logger. By now a field-tested, central component in MUND systems, the RDL1 has proven to simplify MUND operation and improve its performance (Fig. 19.29).

The microcontroller-based RDL1 has the following primary elements: high-voltage supply, charge-sensitive preamplifier, adjustable gain amplifier, discriminator, counter, waveform capture and peak detector, multichannel analyzer, real-time clock, redundant SD cards (internal and external), and an Ethernet interface, shown in Fig. 19.30. Additional features include temperature and humidity sensors, digital input monitor line, supply voltage monitor, and LED indicator.

The RDL1, shown in Fig. 19.31, has a range of features that make it suitable for mobile applications. The device has a small form factor of 8.7 cm × 6.2 cm × 3.4 cm. At 1500 V and room temperature, its average current consumption is less than 30 mA, which allows it to operate for extended periods on modest battery power. Overall power consumption during operation is reduced because the Ethernet interface may be powered down most of the time and turned on by an external control signal only when data transfer is required. Reliability is enhanced with redundant SD cards, one internal and one externally accessible. The maximum signal count rate is 6.25 kHz, a substantial increase above the previous maximum rate of 100 Hz.

Included within the RDL1 design is a multichannel analyzer (MCA) for the neutron pulse. The MCA is the primary information used for setting up the high voltage, gain, and upper and lower discriminators. As an IAEA component, the RDL1 underwent extensive testing. Since the component is utilized outdoors in a variety of weather conditions, one of the primary tests needed for use was temperature testing, where the RDL1 showed proper operation from −20 to +60 °C with no errors.

The RDL1-based MUND design [47] includes the detector, RDL1, battery, and VPN hardware. The power design is typically done such that the MUND can operate on batteries for several weeks between charging cycles. The neutron detector commonly used is an RS-P4-0805-223 ^{3}He tube; however, other detectors have been used depending on the detector sensitivity required for the material to be monitored.

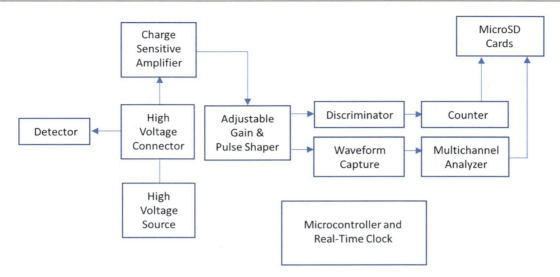

Fig. 19.30 RDL1 functional block diagram

Fig. 19.31 RDL1 MUND

Figure 19.32 shows the data from two MUND devices. The blue line shows the data from a transport device which holds two spent fuel assembly containers. The red line shows the data from a transfer device which loads and unloads the two spent fuel assembly containers in the transport device. The combination of data from the two MUNDs provides complete coverage of the transfer process.

19.4 Coincidence Counting Instruments

Coincidence or doubles counting is widely used for nuclear material assay. The usefulness of the technique is due primarily to the good penetrability of fast neutrons and the uniqueness of time-correlated neutrons to the fission process and thus the nuclear material content. It uses the doubles rate to assay the nuclear material mass of interest. The doubles rate is the real coincidence rate and comes from neutrons that are emitted at the same time, which in most applications, means fission. Because fission is a unique property of special nuclear material, the doubles rate is a more exclusive signature than the singles rate. The doubles rate is significantly less sensitive to (α,n) neutrons, which produce singles counts and whose production rate depends on the chemical form of the sample, potentially leading to a change in the assay results due to chemical form instead

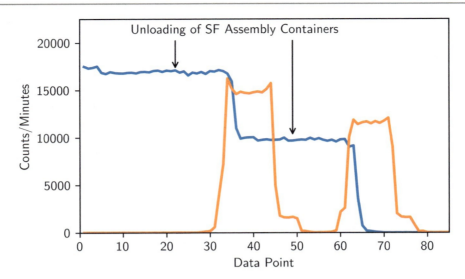

Fig. 19.32 MUND data from transfer

of nuclear material mass. The doubles rate is also significantly less susceptible to background sources. The natural background is primarily Singles and contributes very few Doubles. Nearby sources also contribute primarily Singles and not Doubles because the Singles is proportional to efficiency (ε), Doubles is proportional to efficiency squared (ε^2), and external sources have low detection efficiency.

The simplest doubles analysis technique is the calibration curve where the measured doubles rate is proportional to the mass quantity of interest; for example, if the sample has more $^{240}Pu_{eff}$ mass, the detector will measure a higher doubles rate. The quantities that contribute to the doubles rate must be well known and, if not an objective of the assay, controlled. For example, the even isotopes of plutonium decay by spontaneous fission and contribute to the passive doubles rate. The $^{240}Pu_{eff}$ fraction (isotopic composition) of the plutonium must be known to convert the measured doubles rate to the samples plutonium mass. The shape of the plutonium affects the Doubles by multiplication or induced fission, so the shape must be controlled. This effect is described in more detail in Chap. 17.

Doubles counting can be used in more complex analysis techniques that solve for other unknowns in addition to the nuclear material mass. Because doubles counting instruments can also measure Singles, two values are measured, which allows two unknowns to be solved for. In "known alpha" analysis, the sample's (α,n) production must be known, and the sample's multiplication is solved for, so the shape and other aspects that affect the multiplication can change freely without affecting the assay of mass. In "known multiplication" analysis, the multiplication must be known or held constant, and the (α,n) production can be solved and accounted for in addition to solving for the nuclear material mass. For more detail, see Chap. 17, Sect. 17.14.

19.4.1 Neutron Coincidence System Design Principles

In general, neutron coincidence counters need higher detection efficiency than total-neutron-counting systems because of the requirement to count at least two neutrons in a short time window. This requirement makes the coincidence counting rate proportional to the square of the detector efficiency. The high efficiency is usually accomplished by good geometric coupling between the sample and the detector (for example, a 4π or well counter) and by the use of efficient thermal-neutron detectors based on ^{3}He, although other technologies are discussed at the end of this chapter.

Coincidence counter designs generally seek to maximize the detection efficiency and minimize the die-away time. Performance targets are typically 10%–30% efficiency and 40–60 μs die-away time. The instrument is designed to minimize cost once other goals have been met. A standard figure of merit is used to generally compare the performance of different detectors: ε^2/τ, where ε is efficiency, and τ is die-away time. A higher ε^2 increases the probability of detecting coincidences, and a lower die-away time allows a shorter gate width to minimize the counting of accidental coincidences to minimize the statistical uncertainty. See Chap. 17 for more detail.

The polyethylene thickness is optimized for the expected neutron energy. As more polyethylene is added between the ^{3}He tubes and the source, the detection efficiency will increase as the fraction of thermalized neutrons increases. After some thickness, the efficiency will peak and then decrease because most of the neutrons are already thermal, and additional polyethylene is only adding more absorber and not increasing the fraction of thermal neutrons. This effect is called *overmoderated*. An undermoderated configuration has less polyethylene than needed for maximum efficiency. Polyethylene behind the ^{3}He tubes also increases the efficiency by reflecting neutrons that would otherwise escape back into the ^{3}He tubes for detection. Adding more polyethylene behind the ^{3}He will always reflect more neutrons, although the effectiveness decreases, and the detector size and weight increases, so normally only a couple of inches is used.

Die-away and efficiency are controlled by the spacing and gas density in the ^{3}He tubes. The trade-off to improved performance is increased cost, with an empty tube and the ^{3}He gas itself each having significant cost. As a rule of thumb, in a typical instrument, neutrons have a one-third probability of being absorbed and detected in the ^{3}He, being absorbed in the polyethylene, or escaping the detector. Adding additional rows of ^{3}He tubes minimizes leakage. Decreasing space between tubes minimizes absorption in polyethylene. Neutrons thermalize in polyethylene in ~5 μs and then diffuse through the polyethylene, with a die-away time from polyethylene or ^{3}He absorption of ~50 μs. If thermal neutrons reach a ^{3}He tube, they are very likely to be absorbed, so increasing tube density also reduces die-away time.

Although not often done for coincidence counting, increasing the ^{3}He pressure also significantly reduces die-away time and increases detection efficiency. At higher pressures, the macroscopic cross section (the overall probability of absorption: density of ^{3}He × microscopic cross section) increases, and the probability of detecting neutrons at epithermal (nearly thermal, with lower microscopic cross section) energies increases, so they are absorbed before reaching the 50 μs of diffusion through the detector. This effect is illustrated by comparing two multiplicity counters of the same design with different tube pressures. Changing the plutonium scrap multiplicity counter (PSMC; see Sect. 19.5.4) from 4-atm tubes to 10-atm tubes changes the efficiency from 55% to 62% and the die-away time from 49 to 36 μs.

Other aspects may also be considered. The background is normally measured and subtracted. It is often reduced by an outer layer of cadmium to absorb thermal neutrons and can be reduced further by the addition of external polyethylene to increase the effectiveness of the external cadmium at the expense of size and weight. Coincidence counters can achieve efficiency goals with a single row of ^{3}He tubes, but additional rows reduce changes in detection efficiency due to variable neutron emission energy because inner rows can be undermoderated while outer rows are overmoderated. Single-row detectors are slightly undermoderated to account for moderation in the item itself. In well configurations, end plugs are used to reflect neutrons into the detectors to minimize changes in efficiency as a function of sample vertical position in the cavity.

Computer simulations, most often the MCNP code [16], are used to optimize the design of ^{3}He neutron coincidence detector systems. A range of detector designs can be modeled along with the expected nuclear material to be assayed. The final design can be optimized relatively quickly and inexpensively, without the need for experiments. The accuracy of the simulation efficiency and die-away time is usually within ~5% of real-world systems. The simulations are also useful to evaluate the underlying nuclear physics to better understand the measurement and optimize the detector design. They can even be used to model variations in samples to better understand measurement uncertainties. Additional detail on MCNP modeling is given at the end of this chapter.

19.4.2 Bulk Pu Measurements

19.4.2.1 High-Level Neutron Coincidence Counter II

The high-level neutron coincidence counter II (HLNCC-II) contains 18 ^{3}He tubes but in a cylindrical polyethylene body [48]. AMPTEK (Sect. 19.2.1) amplifiers have been incorporated into the electronics, and the detector body has a design to flatten the vertical response. The vertical extent of the uniform efficiency counting zone is three times longer than that of the original unit without an increase in size or weight. Figure 19.33 shows a cross-sectional view of the HLNCC-II and a photograph of the complete system.

A primary design goal for the HLNCC-II was to obtain a uniform or flat counting response profile over the height of the sample cavity while still maintaining a portable system. This configuration was achieved by placing rings of polyethylene as "shims" at the top and bottom of the detector to compensate for leakage of neutrons from the ends. In addition to these outside rings, the interior end plugs were designed to increase the counting efficiency at each end. The end plugs were constructed of polyethylene with aluminum cores to give a better response than plugs made of either material alone. Also, the sample cavity has a cadmium liner to prevent thermal neutrons from reflecting back into the sample and inducing additional fissions.

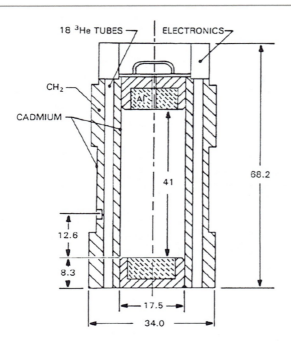

Fig. 19.33 Cross-sectional view and photo of the upgraded HLNCC-II

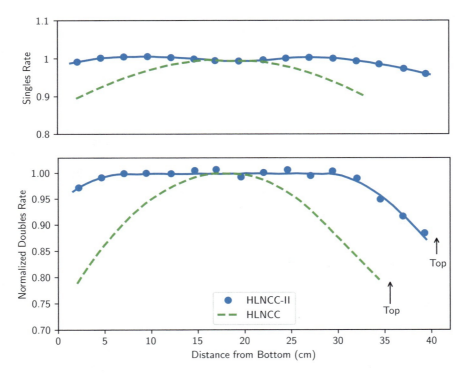

Fig. 19.34 Normalized response profiles/or total and coincident neutron counting for the HLNCC (dashed lines) and the upgraded version HLNCC-II (solid lines), showing a three-times-longer flat response profile for the HLNCC-II

Because the cadmium liner does not extend into the region of the end plugs, the polyethylene in the walls of the end plugs becomes an integral part of the moderator material for the ^{3}He tubes.

The totals and coincidence response profiles of the new counter were measured by moving a ^{252}Cf source along the axis of the sample cavity. The normalized response profiles are shown in Fig. 19.34, where the dashed curves refer to the original HLNCC. The improvement in response is apparent. Table 19.9 compares some of the key features of the HLNCC and the upgraded HLNCC-II.

Table 19.9 Detector parameter comparison for the HLNCC and the HLNCC-II

Parameter	HLNCC	HLNCC-II
Cavity diameter	17.5 cm	17.5 cm
Cavity height	35.0 cm	41.0 cm
Outside diameter	32–36 cm	34.0 cm
System weight	48 kg	43 kg
^{3}He tubes		
Number	18	18
Active length	50.8 cm	50.8 cm
Diameter	2.5 cm	2.5 cm
Gas fill	4 atm	4 atm
Gas quench	Ar + CH$_4$	Ar + CH$_4$
Efficiency	12%	17.50%
Die-away time	33 μs	43 μs
Cadmium liner	Fixed	Removable
Flat counting zone		
Coincidence, 2% from max	11.0 cm	30.5 cm
Totals, 1% from max	10.5 cm	33.5 cm

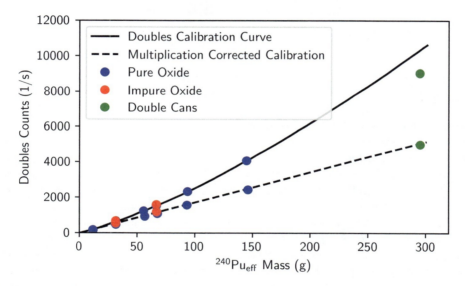

Fig. 19.35 Coincidence response of the HLNCC-II with the new, faster electronics for a variety of large PuO$_2$ samples, both with and without multiplication correction

The counting electronics package developed in parallel with the HLNCC-II was based on the AMPTEK A-111 hybrid charge-sensitive preamplifier/discriminator [49]. Pulses that result from neutron events are discriminated on the basis of pulse height from noise and gamma-ray events at the output of the preamplifier. This approach eliminates the need for additional pulse-shaping circuitry and allows a maximum counting rate of ~1300 kHz—about four times higher than previously attainable. The electronic dead time is also a factor of 4 lower than that of the HLNCC.

This electronics package is capable of measuring samples of significantly larger mass, usually limited only by criticality considerations. The small preamplifier/discriminator circuit is placed directly next to the base of the ^{3}He tubes inside a sealed box to enhance the signal-to-noise ratio. Under laboratory conditions, the singles counting stability was measured to be 0.002% over a 2-week counting period—the best stability ever observed with NDA systems.

The HLNCC-II has been used to assay PuO$_2$, PuF$_4$, mixed oxide, and other plutonium compounds. An example of the response of the system for PuO$_2$, both with and without multiplication corrections, is shown in Fig. 19.35. The highest mass point—at ~300 g ^{240}Pu-effective—corresponds to two cans of PuO$_2$ stacked on top of each other. The air gap between the two plutonium masses reduces the geometric coupling compared with that of a single can with the same total mass. This reduction in coupling results in less neutron multiplication and causes the double-can data point to lie below the calibration curve. After

Fig. 19.36 Photograph of iPCA2, with the load cell assembly on top and three HPGe detectors in its side

the multiplication is corrected for (as described in Chap. 17, Sect. 17.10), the double-can data point lies on the straight line defined by the single-can data.

19.4.2.2 Plutonium Canister Assay System Instruments

Plutonium canister assay system (PCAS) instruments represent a series of instruments that have been developed for verification of plutonium content in MOX storage canisters in reprocessing and MOX fuel fabrication facilities. Several generations of these instruments currently exist and support MOX canister verification activities in reprocessing and fuel fabrication facilities in Japan. The initial instrument design (PCAS), which is currently installed in the Plutonium Fuel Production Facility (PFPF) in Japan, was further improved to support measurements of higher-mass canisters at large-scale reprocessing plants. This next-generation instrument, called *iPCAS* (improved plutonium canister assay system), is used for measurements of MOX canisters at Rokkasho Reprocessing Facility in Japan. The newest instrument, iPCAS2 (improved plutonium canister assay system 2) was developed in 2009 to support MOX canister verification in the J-MOX fuel fabrication facility in Japan. All of these systems provide a similar basic functionality in capability to perform plutonium verification using neutron coincidence counting and confirmation of plutonium isotopic composition by built-in, high-purity germanium (HPGe) detector systems. The instruments are operated in unattended mode and use the NDA measurements in combination with camera surveillance to assure continuity of knowledge on the measured canisters. The latest instrument design (iPCAS2) also includes a load cell and a built-in camera to record canister inside diameter. Typical measurement times correspond to 15–30 min.

The iPCAS2 instruments consist of a ^{3}He-based passive neutron well counter equipped with one ring of ^{3}He tubes to measure the doubles rate and an inner ring of bare tubes to measure the moisture content in the MOX. The system includes an integrated HPGe system that comprises three individual HPGe detectors. The plutonium content is determined using a known-alpha analysis of the passive neutron coincidence count rate from plutonium spontaneous fission neutrons. The measurements are performed on MOX canisters that contain three MOX cans stacked vertically so that each can is simultaneously measured by one HPGe detector. The PCAS instruments are designed to assay kilogram quantities of plutonium in high gamma-ray backgrounds emitted by the measured material. A photograph of the newest improved plutonium canister assay system 2 (iPCA2) instrument is shown in Fig. 19.36.

19.4.2.3 Passive Neutron Coincidence Collar

The verification of the plutonium mass of light-water reactor (LWR) MOX assemblies can be done using the passive analysis methods of the HLNC with a detector that looks like the uranium neutron collar (UNCL). The original UNCL-I was

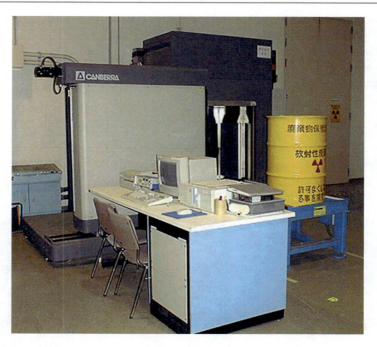

Fig. 19.37 Photograph of the WDAS [52]

configurable for active measurements of both boiling-water reactor and pressurized-water reactor (PWR) fuel and also had another detector side that could be configured into a four-sided detector for passive measurement of MOX assemblies [50]. Other detectors have been specifically built for this application.

The Passive Neutron Coincidence Collar (PNCL) measures the plutonium mass/cm of the fuel assembly. An active length verification (for example, with a handheld gamma detector) is necessary to verify the active length of the fuel assembly.

19.4.3 Waste and Scrap Measurements

19.4.3.1 Waste Drum Assay System

The waste drum assay system (WDAS; Fig. 19.37) is a waste assay system developed for assay of 200-L waste drums [51]. The system has six banks of ^{3}He tubes—one in each of the four sides and one each on the top and bottom. It includes a motor-driven door, where waste drums are inserted. The basic WDAS design includes an additional ~10 cm thick slab of high-density polyethylene (HDPE) for shielding from external neutrons to improve the sensitivity of the system for low-background applications. The system uses passive neutron coincidence counting, which in combination with plutonium isotopic composition, provides the plutonium content of the waste drum. The plutonium isotopic composition can be operator declared or measured by an optional high-resolution gamma spectroscopy (HRGS) system that can be combined with the WDAS instrument. The system also uses an add-a-source technique (see Sect. 19.4.8) to determine the perturbation and correction due to the waste matrix. The waste assay using the WDAS therefore consists of an add-a-source measurement with an external ^{252}Cf source (~3 min) and a passive measurement with the external ^{252}Cf source retracted (~10 min). The ^{252}Cf is automatically transferred from a shielded location to the measurement location on the bottom or side of the drum. The detection limit depends on the instrument location (i.e., background in the measurement area) and typically corresponds to ~1–2 mg of ^{240}Pu effective. The WDAS is used in attended mode with standard shift register electronics.

Various models of these counters have been built by commercial vendors (Pajarito Scientific Corporation and Canberra) and are used at facilities around the world. Typical performance characteristics taken from Ref. [51] are summarized in Table 19.10.

19.4.3.2 Chernobyl Drum Assay System

The Chernobyl drum assay system (CDAS) was designed to measure radioactive debris in 55-gallon drums from excavation work related to the construction of the New Safe Confinement (NSC) at the Chernobyl Nuclear Power Plant (ChNPP; see Ref. [53]). As the NSC construction occurs, radioactive debris from the accident is expected to be uncovered, stored, and

Table 19.10 Summary neutron performance parameters for the Waste Drum Assay System [51]

Performance parameter	Value
Sensitivity (^{240}Pu effective)	0.19
Operating high voltage	1680 V
Dead time coefficients	
A	0.71 µs
B	0.23 µs^2
C	0
Gate	128 µs
Predelay	4.5 µs

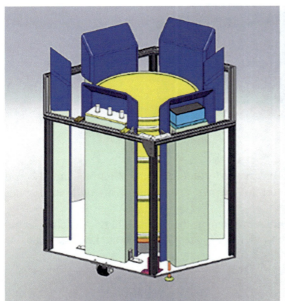

Fig. 19.38 CDAS simulation model (left) and CDAS on top of a drum container (right)

safeguarded. The radioactive debris is fuel from the reactor, which contains special nuclear material. CDAS performs a passive neutron measurement and an active add-a-source (see the discussion in this chapter) measurement and incorporates a gamma measurement in the analysis, which is taken earlier in the drum-processing procedure. CDAS is essentially a large well counter that surrounds the drum. A simulation model and the as-built system are shown in Fig. 19.38.

Because the 55-gallon drums are highly radioactive, they are stored and transported in thick casks to reduce the external radiation dose. CDAS sits on top of these containers, and a crane is lowered through CDAS to pull the drum out of the container and up into CDAS. The crane and container are seen in Fig. 19.38 (right). CDAS consists of six slabs of polyethylene and ^{3}He. Half of the slabs are close against the drum surface, and half are moved back by 15 cm to leave space for the crane. To optimize the use of ^{3}He, the close slabs contain three ^{3}He tubes each, and the offset slabs contain two. A ^{252}Cf source is used as the external source for the add-a-source technique. It is inserted in a tray underneath CDAS. The tray is made of polyethylene, with 1 in. nickel reflector underneath the source.

The CDAS analysis is complex. Overall, CDAS uses the point-model equations (used in multiplicity counting) for Singles and Doubles with an assumption that multiplication is negligible because the nuclear material mass is small. Thus, Singles and Doubles can solve for mass and multiplication. The isotopic composition is needed to solve some variables of the equations and is calculated from the results of a gamma measurement. The ratio of cesium isotopes was benchmarked to various burnups, and the isotopic compositions at the burnups are known from simulations. The gamma measurement is taken by ChNPP earlier in the drum-processing process, and the information is transferred into the CDAS analysis. Then, an add-a-source measurement is taken by introducing a strong ^{252}Cf source and comparing the expected value with no drum with the measured value with the drum, which accounts for the drum contents effects on efficiency and die-away time compared with an empty drum. Finally, CDAS measures the spontaneous fission neutrons emitted by the ^{238}U, ^{240}Pu, and ^{244}Cm. Combining all of this information, the total mass of nuclear material is determined from solving the point model and is used to calculate the mass of the relevant isotopes from the known isotopic composition.

The measurement uncertainty is typical of waste measurement systems, and an overall uncertainty in special nuclear material (SNM) mass is estimated to be less than 30%.

19.4.3.3 High-Efficiency Neutron Counter

The high-efficiency neutron counter (HENC; see Ref. [54]) was developed to address the needs and challenges associated with impure waste material stored in 200 L drum-type containers. The prior measurement systems, such as the WDAS (see beginning of this section and Ref. [51]), had inadequate efficiency to make multiplicity neutron measurements that were needed for samples that contained nuclear material with enough fissile mass that multiplication made the results difficult to interpret. However, for criticality safety reasons, 200-L waste drums normally contain plutonium at such a low level that neutron multiplication is negligible. The matrix materials in the drums are heterogeneous and cause significant changes in the detection efficiency, so the primary measurement challenge is the correction for the change in efficiency. Multiplicity measurements imply the use of the point-model equations [55] to use the singles, doubles, and triples rates to solve for the plutonium, the alpha, and the multiplication; for waste drums with negligible multiplications, the equations can be used to estimate the efficiency; however, the size and heterogeneity of the waste drums make the application of the point-model equations problematic, and the approach was not used for the HENC deployment at LANL Technical Area 54.

The actual measurement approach used by the HENC and several other drum measurement systems (WDAS, WCAS [waste crate assay system], SuperHENC) is the passive doubles measurement where the multiplication is negligible, and the efficiency change from the matrix material is determined by ^{252}Cf source add-a-source (AAS) measurement [56]. The triples rate is used to help eliminate cosmic-ray source neutron coincidence counts that have a higher multiplicity distribution than neutrons from spontaneous fission.

The HENC shown in Fig. 19.39 was fabricated from HDPE with 10 cm thick detector banks backed by 30 cm thick HDPE shielding. For fabrication reasons, the cavity size was set to be the same as the Canberra WM3100 system [57]. The sample cavity size is 82 cm wide × 109 cm high × 77 cm deep. The system contains 113 ^{3}He tubes, with uniform coverage on the four sides, the top, and the bottom. Doors on both the front and back of the system allow for the drums to pass through the system. The electronics cabinet contains the computer, printer, and the CI2150 multiplicity electronics. The system uses 16 AMPTEK

Fig. 19.39 Photograph of the Canberra High Efficiency Neutron Counter system showing the drum-loading system and the add-a-source shielding cube on the top

Table 19.11 HENC operating parameters

Operating parameter	Value
Efficiency (^{240}Pu energy)	0.31
Die-away time	50 µs
Dead time coefficients	
A	0.50 µs
B	0.16 µs^2
C	0
Multiplicity dead time	171 ns
Gate	128 µs
Predelay	3.0 µs
Doubles calibration coefficient	53.8 counts/s·g ^{240}Pu
Multiplication constant (p_0)	0.178

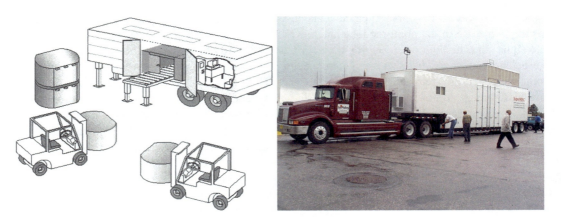

Fig. 19.40 (left) Diagram of the SWB loading configuration; (right) photograph of the mobile measurement system

amplifiers, so the dead time correction is usually negligible for waste samples. The HENC operating parameters is listed in Table 19.11.

The passive calibration slope for the Doubles is 53.8 counts/s g ^{240}Pu, and the triples calibration line has a slope of 5.83 counts/s g ^{240}Pu. The calibration demonstrates that the triples rate can be used down to sub-gram quantities of ^{240}Pu$_{eff}$. In general, the triples rate is used to make cosmic-ray matrix corrections, and the AAS method is used to make the efficiency corrections.

19.4.3.4 SuperHENC

An advanced passive neutron counter was designed, fabricated, and deployed to measure the plutonium content in standard waste boxes (SWBs). The super high-efficiency neutron counter (SuperHENC; see Fig. 19.40 and Ref. [58]) was developed under an agreement between LANL and the Rocky Flats Environmental Technology Site to measure the plutonium content in the waste generated during the facility decommissioning activity. This development occurred in 1998, following HENC drum counter development [54]; the SuperHENC was developed for much larger samples and with advanced performance to make multiplicity counting [55, 59] viable. The primary goal of the design was to produce a mobile assay system for the large SWB containers (1900 L) that has decreased sensitivity to the variable matrix types. The system also includes 200-L drum assay capability. The measurements are based on neutron time-correlation counting of the passive neutron emissions from the ^{240}Pu, and the plutonium isotopic ratios are used to calculate the total plutonium. The high efficiency (40.8%) permits the measurement of the singles, doubles, and triples rates and makes multiplicity counting viable. The normal application of the system makes use of the passive doubles rate for the calibration—together with the AAS method [56]—to correct for the matrix materials in the SWBs. In parallel, the system also makes use of the multiplicity mode analysis to determine the efficiency where the ^{240}Pu mass is directly calculated from the three measured rates (Singles, Doubles, and Triples). For improved delectability limits, advanced methods were incorporated in the SuperHENC to reduce the cosmic-ray neutron background counts. These methods include statistical data filters [60], truncation of high-multiplicity events [61], and a local

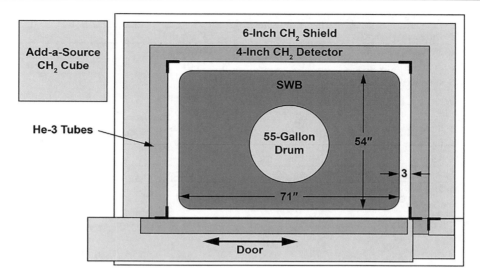

Fig. 19.41 Diagram (top view) of the SuperHENC detector body showing the exterior CH$_2$ shielding and the SWB or drum sample locations. The standard waste box has 7.5 cm of clearance to the walls and roof

neutron coincidence veto (LCV; Ref. [62]) to reduce cosmic-ray origin neutrons. The detector is packaged inside a mobile trailer, as shown in Fig. 19.40, to make the system transportable.

The high detection efficiency and background reduction techniques for SuperHENC provide a detection limit sufficient to segregate wastes at the low-level/transuranic (TRU) criteria of 10–100 nCi/g. The detection system consists of 260 10-atm ^{3}He tubes located in the top, bottom, and four side walls. The ^{3}He detectors are arranged in a dual-row fashion, embedded into an HDPE matrix with optimized spacing and thickness. There are 52 AMPTEK A111 amplifiers in the system, so the dead time is negligible for waste samples. The system is operated in the passive neutron coincidence or time-correlated mode to measure the ^{240}Pu$_{eff}$ spontaneous fission rate. The plutonium isotopic data must be acquired from either acceptable knowledge or direct measurement with an appropriate gamma system. Details of the detector configuration are shown in Fig. 19.41.

Corrections for waste matrix effects on the observed coincidence measure are addressed with the AAS technique. The basis of the AAS method is to measure the matrix perturbation to the counting rate from a small ^{252}Cf source that is extracted from a shield and positioned adjacent to the waste container bottom at six preselected locations. The source is then retracted into the shield for the duration of the passive measurement. The observed AAS perturbation is related to a matrix correction factor through an empirical relationship [56]. This relationship is established through measurements of a set of simulated waste matrices that span the properties expected in actual waste.

In parallel with the AAS result, the SuperHENC gives the plutonium mass derived from the multiplicity counting where the efficiency is measured via the singles (S), doubles (D), and triples (T) rates. This result is independent of the AAS correction factor. The software for SuperHENC makes the decision regarding the best result (AAS versus multiplicity matrix correction) based on the counting rates, the magnitude of the AAS correction, and statistical errors. If the AAS matrix correction is small, the AAS calibration result is preferred because of the good precision using the doubles rate compared with the triples rate. Table 19.12 lists the measurement parameters for the system.

The SuperHENC was calibrated with plutonium standards in an empty SWB. The measurements with loaded SWB containers are corrected back to the empty container calibration using the AAS method or the multiplicity approach. Because negligible multiplication and no neutron self-shielding normally exists over the plutonium mass range of interest (0–350 g), the calibration curve is expected to be a straight line through the origin. Separate calibration lines for the singles, doubles, and triples count rates indicate where the doubles calibration is used for the primary assay result.

Figure 19.42 shows the initial doubles and triples calibrations for both the HENC and SuperHENC for comparison. We see that the SuperHENC has a triples rate that is three times larger than the HENC, which is very beneficial in the multiplicity analysis.

A major source of uncertainty for the SuperHENC is the background subtraction. The neutron backgrounds originate from several sources, including local area source storage (e.g., waste in loaded SWBs); external-area, cosmic-ray neutron sources; cosmic-ray spallation in the detector body; and cosmic-ray spallation neutrons in the sample. The first two of the sources

Table 19.12 Measurement parameters for SuperHENC

Measurement parameter	Value
Neutron efficiency for ^{240}Pu	40.8%
Neutron die-away time	55 µs
Dead time coefficients	
A	0.2416 µs^2
B	0.0242 µs^2
Multiplicity dead time	88.28 ns
Coincidence gate	128 µs
Predelay	1.5 µs
Doubles calibration coefficient	93.02 counts/s·g ^{240}Pu
Multiplication constant (ρ_0)	0.232
Doubles gate fraction	0.750
Triples gate fraction	0.563

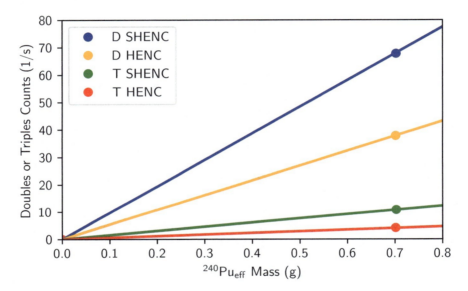

Fig. 19.42 Doubles and triples calibrations for SuperHENC compared with HENC

creates a singles neutron background, and the HDPE shielding on the exterior of the detector essentially eliminates this source. The third source is reduced by the LCV method because many of the coincidence neutrons are detected in a local volume of the detector. The fourth neutron source is significantly reduced by the truncated multiplicity (TM) data reduction in the software. The standard statistical filtering method is also used to eliminate cosmic-ray neutron events with high multiplicity.

The cosmic-ray spallation neutron background in the detector body and in the sample is difficult to shield because the cosmic rays are very penetrating. One meter of overhead concrete provides a factor of 4 reduction in this neutron background. The 150 mm of HDPE shielding on the detector provides less than a 10% background reduction in the coincidence rate. To help reduce the spallation rate, low-Z materials—such as HDPE and aluminum—were used in the detector design. The primary cosmic-ray neutron reduction is obtained from the LCV technique and the TM data analysis. Figure 19.43 shows the SuperHENC background for an empty SWB container for a series of 30 min runs. The doubles rate is shown with and without the LCV circuit in the data collection. Also shown are the data for the LCV plus the TM method.

The minimum mass detectability limit for SuperHENC depends on the location of the system, the type of matrix, and whether doubles or triples rates are used for the assay. At sea level, the detectability limit is a factor of 2–3 lower than at the high elevation (2200 m) at LANL. The detectability limit is determined by the coincidence neutron background generated by cosmic-ray spallation reactions. The detectability limit was defined as three times the standard deviation (3 σ) in the background; repeat measurements of the background were used to predict the minimum detectable mass that gives a signal

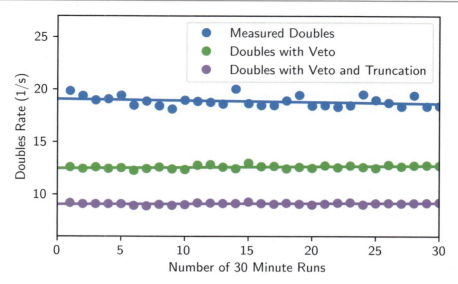

Fig. 19.43 Repeat SWB background measurements (30 min for each point) showing the doubles rate with and without the local neutron coincidence veto and the truncated Doubles (truncated multiplicity)

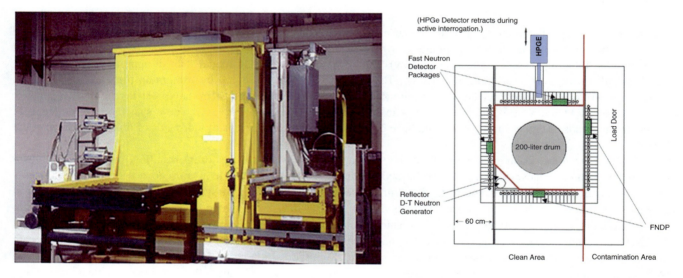

Fig. 19.44 Photograph of an IWAS (left) and a sketch of the horizontal section of the IWAS (right) showing the arrangement of the neutron generator, the ^{3}He proportional tubes, and the retractable HPGe detectors

rate equal to 3 σ of the background scatter. The initial measurements at LANL gave a detectability limit of ~37 mg plutonium for an SWB with a noninterfering matrix. The sensitivity is better than the 100 nCi/g fiducial level for waste handling.

The SuperHENC testing, calibration, and delivery to Rocky Flats Environmental Technology Site were completed in 2000. Additional SuperHENC systems, commercially built by Pajarito Scientific Corporation, have an HRGS spectroscopy system for plutonium isotopic verification built into the back of the mobile trailer.

19.4.3.5 Integrated Waste Assay System

The integrated waste assay system (IWAS; see Refs [63, 64]) was designed by Canberra Industries to quantify plutonium and uranium in 55-gallon drums and 85-gallon overpacks for disposition to the Waste Isolation Pilot Plant (WIPP) in Carlsbad, New Mexico. The IWAS (Fig. 19.44) provides passive and active neutron interrogation and quantitative gamma analysis, allowing rapid characterization of TRU wastes for proper shipment and disposal. The IWAS is based on the HENC design with integrated differential die-away and HRGS sub-systems. The complementary analysis methods have different strengths

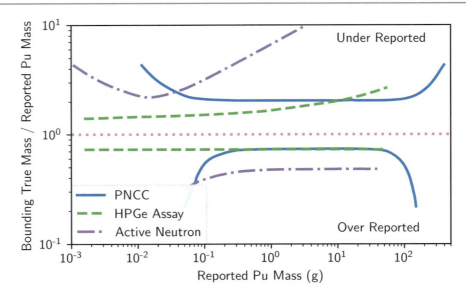

Fig. 19.45 Illustration of the total measurement uncertainty for each of the three integrated waste assay system modalities as a function of plutonium mass for a medium-density waste drum. The curves represent the $\pm 2\sigma$ bounds for the actual mass value relative to the reported plutonium mass. By considering the potential mass values consistent with the assay results, it is possible to better select the appropriate analysis mode to represent the drum contents

and weaknesses (e.g., high-density metallic waste matrices are challenging for the gamma-ray assay but are generally considered benign for the neutron measurements, whereas curium contamination could overwhelm the neutron-counting system but has little impact on the gamma-ray assay). The IWAS employs a multi-modal analysis approach to maximize the likelihood of a successful measurement outcome. This approach is best illustrated by examining the total measurement uncertainty [65] for each of the three IWAS analysis modes (Fig. 19.45). Four such systems have been in continuous operation at the Advanced Mixed Waste Treatment facility in Idaho for approximately 20 years. These systems have collectively performed hundreds of thousands of assays and have been employed to consign tens of thousands of drums to storage at the WIPP.

Combining the three disparate assay methods into a single assay system reduces drum handling, processing time, and personnel exposure. Because all assays are performed by the same system in a single assay sequence, no confusion occurs over item identification number or modification of drum contents between assays. The system is designed to be operated as an automated counting system that can process batches of drums and is fully incorporated in the facility process. Results from each of the individual assay modes are combined automatically by the operating software using a well-defined hierarchical decision matrix (not machine learning) to select the most appropriate analysis result for the container. Multiple parameters are used in the determination of the appropriate result, including drum type, drum net weight, bulk density, moderator loading (AAS correction factor), count rates, reported actinide masses/activities (e.g., ^{235}U and ^{239}Pu), total measurement uncertainty, and process knowledge (e.g., matrix type). Parameters found to be outside preset ranges for the parameter cause warning flags to be issued, and rejection of a given assay mode results in a potential solution for the drum. Integrating multiple analysis modes and analysis into the same system minimizes the likelihood of mischaracterization of the contents and improper dispositioning the drum.

19.4.3.6 Add-a-Source Concept

The AAS concept is used to correct for the effects of non-nuclear matrix material in waste containers. Materials such as concrete, steel, glass, plastic, and paper are often mixed with nuclear material in waste containers such as 55-gallon drums or crates. The matrix materials change the detection efficiency of the emitted neutrons by varying degrees depending on the amount and type of material. This effect is one of the larger contributors to measurement uncertainty in waste measurements. The AAS technique corrects for this effect to minimize the contribution to measurement uncertainty. The AAS correction is applied to the doubles count rate, and then the corrected rate is used in typical coincidence analysis. The AAS technique is used in many waste measurement instruments, including HENC, WCAS, WDAS, and CDAS.

The AAS technique uses a neutron source built into the instrument. The matrix material will have a similar effect on neutrons from the AAS source as it does on neutrons from the nuclear material in the container. The count rate of the AAS source with the assay item present is compared with the known rate of an empty container. If the matrix materials have a small effect on the detection efficiency, the count rate will be about the same, and the correction is minimal. If the matrix materials have a large effect, the AAS count rate will be significantly lower, and a large correction is used. The AAS source material is often ^{252}Cf because it produces fission neutrons with a similar energy to nuclear material. A deuterium-deuterium neutron generator, which has the advantage of being turned on and off and having no radioactive decay, can also be used. The AAS correction can be based on Singles or Doubles and ideally is based on the same value used in the assay (AAS doubles measurement to correct the item's measured doubles rate). The AAS measurement can be performed once at a single location or averaged over multiple measurements at different points around the item container. The larger the container, the more necessary multiple measurements are for an accurate correction. The larger the container, the more room for matrix materials and the stronger the matrix effect will be. In fact, depending on the performance needs of the application, for small enough containers, no correction is needed because the maximum nuclear material that can be present has an acceptably small matrix effect. Also for small containers, the ring ratio can be used as a matrix correction instead (Sect. 19.3). The matrix materials that are possible in the container must be carefully considered to constrain the total measurement uncertainty and to define the calibration requirements. Typically, non-hydrogenous material such as steel, dry dirt, and graphite—in small amounts—have small matrix effects, whereas hydrogenous material such as plastics, water, wet soil, and concrete can have a large effect. The hydrogen in the materials quickly thermalizes neutrons, and they are readily absorbed before escaping the container. The amount of nuclear material present (container size, material mass, packing fraction) has a large effect. Typically, the nuclear material is uniformly distributed inside the drum, and so a volume-averaged correction is appropriate. If the material is localized, then—depending on the detector geometry—the location of nuclear material inside the container can greatly affect the detector efficiency and measurement uncertainty. Because each material affects the detection efficiency differently, the correction is not perfect and only reduces the measurement uncertainty but does not eliminate it.

Add-a-source requires a calibration, which is typically performed by creating mock waste containers with matrix materials that are expected to be in the assay items. The mock containers have sample paths made of metal tubing, which allows either ^{252}Cf sources or plutonium sources to be placed inside the drums safely. The sources are measured at various heights and radii to calculate the volume-averaged effect of the matrix material on the detection efficiency. The AAS measurement is also performed on the drums to determine the effect on the AAS source. The amount of correction needed is plotted against the change in measured AAS strength to create the calibration curve. For unknown samples, the AAS strength can be measured and used to calculate the correction needed. The HENC, AAS approach, and correction performance are shown in Figs. 19.46, 19.47, and 19.48. In Fig. 19.48, the measured doubles values quantify the change in relative doubles efficiency for various drum matrices, which has a range of ~60%. The corrected Doubles demonstrate that the AAS technique is able to correct for the drum matrix to minimize the range in relative efficiency to ~10%. For a technical description of performing an AAS calibration, see Ref. [66].

Fig. 19.46 Photograph of the HENC

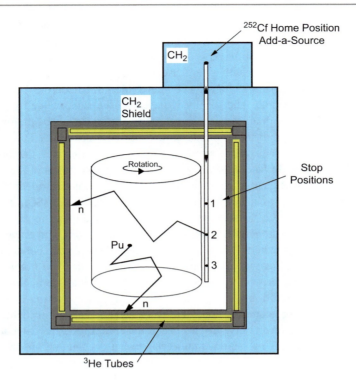

Fig. 19.47 HENC AAS configuration

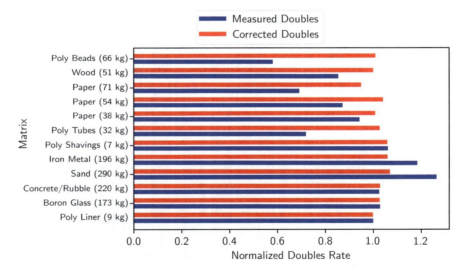

Fig. 19.48 HENC AAS correction performance

19.4.4 Neutron Holdup Measurements

19.4.4.1 Glove Box Assay System

Large neutron slab detectors have been used in the PFPF in Tokai-Mura, Japan, to measure holdup in glove boxes inside this automated MOX fuel fabrication facility [67, 68]. These glove box assay systems were 160 cm high, 100 cm long, and 7.6 cm wide. Each slab contained 20 ^{3}He tubes that were 152 cm in length. The assay for a glove box assay system is based on counting coincident neutrons from spontaneous fission of ^{240}Pu$_{eff}$ to give the total plutonium mass. Monte Carlo calculations were used to design the detector and to study its response before installation. Six slabs were originally installed in pairs on either side of a glove box. The slabs could be moved remotely to measure different locations on a glove box. A standard matrix of measurement positions was assigned for each glove box, and software was written to collect, analyze, and combine all of the measurements. Experience at PFPF has shown a measurement uncertainty of ~5% for neutron assay and 25%–30% for gamma-ray assay.

19.4.4.2 Super Glove Box Assay System

Super glove box assay systems (SBASs; see Ref. [69]), which comprise four detectors, were developed under the agreement between the Joint Nuclear Control Commission (JNC) and the U.S. Department of Energy (LANL). The systems were modified as follows: Two slab detectors routed horizontally were positioned apart on each side of the glove box to cover the glove box horizontally, scanning the glove box vertically from the bottom to the top. The advantage of this scanning procedure is that no horizontal movement of the detectors occurs near the glove box with high radiation dose (especially when the blender is including much MOX inventory), which makes it possible to reduce the operator's radiation exposure. A lifting system that allowed the detectors to be set at designated positions was manufactured. After conducting the profile test for the new system, the SBAS was modified using a californium source and evaluated the performance based on the comparison of the calculation result with the measurement result.

19.4.4.3 Glove Box Unattended Assay and Monitoring System

The glove box unattended assay and monitoring system (GUAM) represents a next-generation concept for holdup monitoring inside the plutonium processing glove boxes. The concept was developed for the Japanese J-MOX fuel fabrication facility in Rokkasho. The system is designed to allow unattended operation, which alleviates the need for intensive operator and inspectorate involvement during attended measurements using the traditional systems discussed in the previous sections (e.g., SBAS). GUAM consists of systems of ^{3}He neutron detectors permanently installed in glove-box shielding. Individual detector layouts are currently under development and will be customized for specific dimensions and material-handling nature of each glove box. Typically, about 16–20 ^{3}He detectors are required per glove box. An example of GUAM implementation is shown in Fig. 19.49. Each ^{3}He detector in the system will be connected to an individual amplifier to enable better monitoring of nuclear material distribution inside the glove box. The data collection will therefore involve many channels (hundreds for a large-scale facility), and list mode or advanced multichannel shift register is foreseen for the data acquisition. GUAM will be used for continuous unattended holdup assay to determine holdup during non-operating periods as well as to establish qualitative nuclear material patterns during operating periods. The efficiency will vary from glove box to glove box but will typically be below 1% (around 0.5%), which will make coincidence counting feasible over long count times (such as overnight). The target total measurement uncertainty is ~8%.

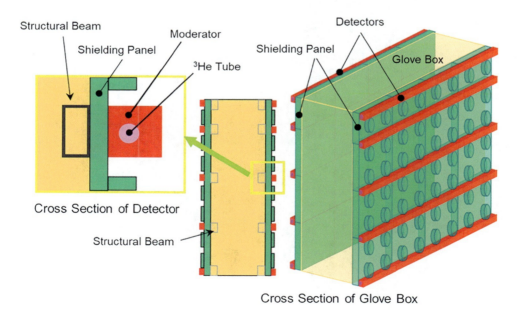

Fig. 19.49 Schematic depiction of an example glove box unattended assay and monitoring system layout on a processing glove box

19.4.5 Bulk Uranium Measurement

The high-efficiency passive counter is a neutron coincidence counter designed to measure the uranium content of large (1–15 kg) items of uranium oxide powder canisters [70, 71]. The units developed for measurements at fuel fabrication plants rely on the detection of coincident neutrons emitted from spontaneous fission events in ^{238}U to determine the ^{238}U mass. The neutron measurement is coupled with a separate gamma-ray measurement to provide an evaluation of the measured mass of ^{235}U. The counters have a ^{252}Cf neutron-detection efficiency on the order of 53%. The counters operate with 148 proportional ^{3}He tubes pressurized to 4 atm and arranged in three rings on each side. The overall uncertainty on the ^{235}U mass is 3.2% [72].

19.4.6 Uranium Fluoride Cylinder Measurements

19.4.6.1 Uranium Fluoride Coincidence Counter

The uranium fluoride coincidence counter (UFCC) is a passive neutron measurement system for determining the ^{235}U mass and enrichment in 5A/5B UF$_6$ cylinders [73]. The system uses neutron self-interrogation to measure the fissile mass in the cylinders. The singles count rate (S) is proportional to the UF$_6$ mass from the (α,n) reaction rate in the fluorine [74]. Also, the doubles-to-singles (D/S) ratio provides a measure of ^{235}U mass that is independent of the isotopic and chemical composition. These two signatures, which are obtained simultaneously, have complementary advantages that can be used together to provide more overall confidence in the ^{235}U mass value than each signature provides when used individually. More specifically, the ^{235}U mass determined with the singles counts has the advantages of good counting statistics and nearly complete insensitivity to fill height and material distribution, but it is sensitive to variations in the ^{234}U content of the UF$_6$. The D/S ratio has the primary advantage of being insensitive to variations in ^{234}U content. If the ^{235}U mass is combined with the total uranium mass from a scale, then the enrichment level in the cylinder can also be calculated.

The UFCC detector has one ring of 12 ^{3}He tubes with polyethylene moderator and a central sample cavity designed to hold a 5A/5B cylinder. The sample cavity has an inside diameter of 14.7 cm (5.82 in.). The ^{3}He tubes have an active length of 91.4 cm (36 in.) and a gas pressure of 4 atm and are made of stainless steel. Located underneath the measurement cavity is an aluminum and polyethylene neutron reflector designed to increase efficiency and flatten the response profile. Figure 19.50 shows the system with the door open to the sample cavity. The junction box is located at the bottom of the counter and houses

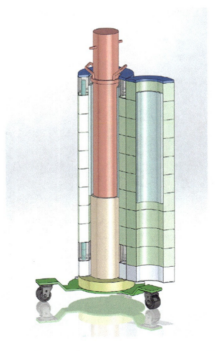

Fig. 19.50 The UFCC with the door open (left), and the MCNP model [73] with a UF$_6$ cylinder in the measurement position (right)

Table 19.13 Detector parameters for the UFCC

Shift register type	JSR-15 or AMSR
Efficiency	17.6%
Die-away time	75 μs
High voltage	1680 V
Dead time coefficients	
A	0.7 μs
B	0.13 μs²
Multiplicity dead time	500 ns
Gate	128 μs
Predelay	2.0 μs
Doubles gate fraction	0.6093
Triples gate fraction	0.3871
ρ_0	N/A

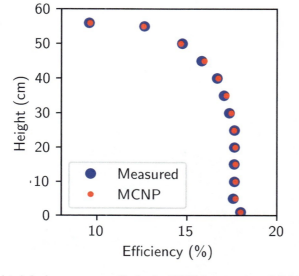
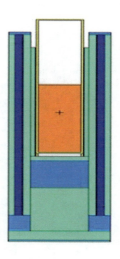

Fig. 19.51 The calculated singles response profile for the UFCC that contains a full 5A/5B UF$_6$ cylinder

six A111 AMPTEK amplifiers (+5 V); each amplifier serves two ^{3}He tubes. The UF$_6$ cylinders can be put in place through the side door or through the top.

The performance parameters for the UFCC are listed in Table 19.13. The detection efficiency was characterized by measuring a ^{252}Cf calibration source centered in the measurement cavity (with no cylinder) and was measured to be 17.7%. The efficiency is increased with the sample inside the cavity because of neutron scattering.

An efficiency profile was measured using a ^{252}Cf source at 13 axial positions inside the sample cavity. Figure 19.51 shows a plot of the measured and simulated response profiles (normalized) with a scale Monte Carlo N-Particle eXtended rendering of the UFCC and a 5A/5B cylinder [73]. The cylinder was modeled with UF$_6$ at the maximum fill level of 24.95 kg, where the UF$_6$ is located at the bottom of the cylinder—the most likely material distribution for UF$_6$ that has been homogenized. Figure 19.51 shows that the UFCC has a flat response profile over the UF$_6$ fill region. This characteristic minimizes uncertainty from positioning, fill height, and material distribution.

For the calibration, the singles and doubles rates were simulated with MCNP over the ^{235}U mass range of 0.1–15 kg; the D/S ratios are shown in Fig. 19.52. The (α,n) source strength cancels in the D/S ratio, and the calibration function is almost linear over the entire maximum filling mass for 5A/5B cylinders.

19.4.6.2 Modification for Uranium Powder Measurements

The UFCC can be modified to measure containers of U$_3$O$_8$ powder or highly enriched uranium (HEU) metal pieces by inserting a 2.54 cm thick polyethlene disc directly under the sample [75]. The UFCC interrogation is then in the active mode,

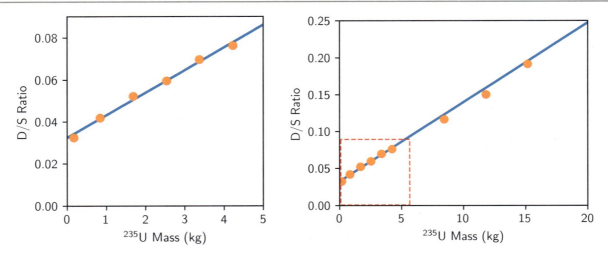

Fig. 19.52 Simulated D/S count rates for a maximum-filled 5B cylinder for enrichment levels ranging from depleted to 90% (right) and depleted to 25% (left; Ref. [74])

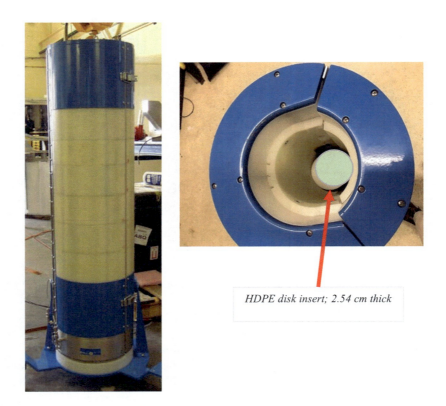

Fig. 19.53 The UFCC configured for active assay of U_3O_8 powder samples

where the ^{252}Cf neutrons are moderated in the polyethlene and induce fission reactions in the ^{235}U. The ^{252}Cf source background is subtarcted from the gross counts, and there is no dependence on the ^{234}U fraction. Figure 19.53 shows the detector and the HDPE insert, which together hold the ^{252}Cf source. The ^{252}Cf source is positioned directly beneath the sample container.

To calibrate the UFCC for U_3O_8 bulk powder samples, the cans were positioned in the UFCC sitting on top of the ^{252}Cf locator disc. The detector parameters settings were as listed in Table 19.13, and the ^{252}Cf source yield was 16,700 n/s.

The U_3O_8 power had enrichments that ranged from NU up to 91% ^{235}U. Figure 19.54 shows the data trend-line fit for samples. All of the measurements were for 450 s, and the small scatter about the calibration curves illustrates the good

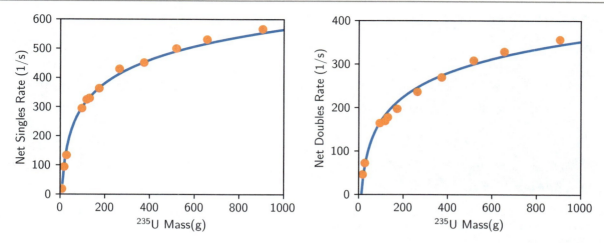

Fig. 19.54 Measured calibration curves for 1000 g of U_3O_8 powder samples that have enrichments in the 0.7%–91% range

precision (1%–5%) for mass values above ~30 g. The D precision benefits from the time correlations between the ^{252}Cf interrogation source and the induced fission reactions in the sample.

Note that the net singles rates continue to increase as the enrichment and ^{235}U mass increase. Part of this increase in the singles rates comes from the ^{234}U (α,n) neutron reactions in the oxide powder that tracks the ^{235}U enrichment.

In summary, the UFCC was designed to measure 5A/B cylinders of UF_6 in the passive self-interrogation mode. The neutrons from the (α,n) reaction in the UF_6 scatter in the polyethylene moderator and induce fission reactions in the ^{235}U. The resulting D rate is proportional to the fissile mass in the sample, and the D/S ratio removes the dependence on the isotopic and chemical properties of the sample.

The purpose of the ^{252}Cf measurements is to extend the UFCC capability to other uranium sample types such as oxides, metals, and liquids that might be found during verification activities. A ^{252}Cf source was introduced to the detector to make the active mode measurements possible because the uranium samples do not have enough neutron emission rate for a passive measurement.

The analysis of the three observables (S, D, and T) led to the conclusion that the singles rate calibration provided the smallest statistical error in the ^{235}U mass value. Also, the singles rate can be easily increased by an order of magnitude by using a stronger ^{252}Cf source (~1E5 n/s) typical of in-field neutron test sources. This method would have the practical benefit of reducing the measurement error to ~1%–2% in 5 min measurements.

The future small modular reactors will require the use of high-assay, low-enrichment uranium, and the UFCC is well suited for the verification of this type of material because the 5A/5B cylinders are used for handling the ~20% enriched uranium.

19.4.7 Neutron Panel and Curtain Detectors for Assay of 30B UF_6 Cylinders

The highly penetrating nature of neutrons makes them an ideal observable for the assay of bulk ^{235}U content within large 30B-type UF_6 storage cylinders. These cylinders have dimensions of 76.2 cm diameter × 207 cm (30 in. diameter × 81.5 in.) and contain up to 2277 kg low-enriched UF_6 (≤5% ^{235}U; see Ref. [76]). Currently, 30B cylinders are verified using the gamma enrichment-meter technique, where a local enrichment assay is conducted based on the 185.7 keV emissions of only a few grams of UF_6. Since the 1970s, passive neutron measurements have been considered as a more direct indicator of bulk characteristics [77, 78], with a series of incremental advances more recently [79–83]. During passive neutron measurements, the correlation between the singles and doubles rates to ^{235}U content are complicated by two systematic factors. Neutron emissions depend on the effect of UF_6 distribution on multiplication and detection efficiency and on the ^{235}U/^{234}U ratio because the primary neutron source in low-enriched UF_6 is the ^{234}U-driven (α,n) reaction on ^{19}F.

On a volumetric scale, 30B cylinders are around two-thirds full of solid UF_6 (5.1 g/cm^3) to avoid rupture during phase changes because the cylinders are typically filled and emptied with gaseous UF_6 and homogenized with UF_6 in the liquid phase (3.6 g/cm^3). UF_6 distributions vary between cylinders and even within a given cylinder over time due to differences in cylinder filling, handling, and thermal cycling. UF_6 distributions have been described as using the "X-factor," where X refers to the percentage of material bound to the walls, and the remaining material fraction is pooled at the bottom of the cylinder [84]. Actual cylinders exhibit more complex variations across all three dimensions [85]. Figure 19.55 shows a recent

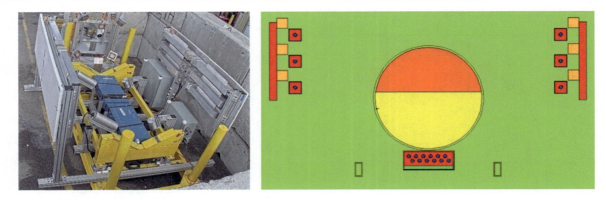

Fig. 19.55 (left) Image of an NDA testbed that shows neutron panels in blue environmental enclosures. At the side in aluminum casings are the neutron curtains, with 12 individual ^{3}He tubes and outer Al-Cd-HDPE shielding. (right) Cross section of geometry that illustrates neutron detector positions relative to the 30B cylinder

Table 19.14 Neutron panel operating parameters

Parameter	Value
Efficiency (simulated 3.5% 30B cylinder)	~0.025
Die-away time	~35 μs
Operating high voltage	1840 V
Gate	64 μs
Predelay	4.50 μs

prototype with an array of ^{3}He detectors positioned around the large 30B cylinder. The operating parameters of the neutron panel are listed in Table 19.14. UF$_6$ distribution variations systematically alter the response of the two primary detectors positioned beneath the cylinder, referred to as the *neutron panels*. These effects are accounted for using the relative distribution, as indicated by the array of 12 individual ^{3}He detectors around the cylinder, referred to as the *neutron curtain*.

Enrichment theory indicates that the ^{235}U/^{234}U ratio is relatively consistent for uranium enriched at a given facility using raw NU feed because the ^{234}U is co-enriched along with ^{235}U—just to a slightly greater extent [86]. Atypical ^{235}U/^{234}U ratios result from HEU downblending (high ^{234}U), use of uranium from reactor returns (low ^{234}U), and use of anthropogenic feed produced by tails stripping (i.e., re-enrichment of depleted uranium [DU], very low ^{234}U). Although raw NU feed is assumed to be the norm, for an enricher with excess separative work units (SWU), the re-enrichment of existing DU inventory could be an economically viable means to generate feed from DU that otherwise requires long-term storage [87]. The analysis described here was prompted by recent 30B measurements where the large ^{235}U/^{234}U ratio variability caused unacceptably high systematic uncertainty for the traditional approach of calibrating Singles and Doubles against ^{235}U mass and enrichment, respectively [85].

The system shown in Fig. 19.55 enables managing effects of both UF$_6$ distribution and ^{235}U/^{234}U ratio. The primary observables are the Singles and Doubles measured by the neutron panel ^{3}He pods directly below the cylinder, each containing 12 10-atm tubes (2.54 cm diameter × 50.8 cm active length), with one signal output per pod. The neutron curtain is an array of 12 4-atm ^{3}He tubes (embedded in rectangular HDPE blocks with cross sections of 7.62 cm × 7.62 cm) distributed around the upper half of the cylinder. The system uses list mode with the ALMM (Sect. 19.2.5) to record each of the 14 pulse trains. This instrument records a set of signals that indicate the relative three-dimensional UF$_6$ distribution based on the singles ratios between the front-to-back and left-to-right neutron curtain detectors as well as the relative vertical distribution based on the singles ratio of the curtain-to-panel detectors.

To remove systematic dependence of ^{235}U/^{234}U ratio from the singles interpretation, the Singles can be calibrated directly against the ^{234}U mass. The improvement seen by applying a set of three linear distribution correction functions in Fig. 19.56 shows that the uncorrected calibration error primarily results from UF$_6$ distribution effects. (For details of experimental results in this section, see Ref. [85]).

Due to the doubles efficiency dependence of $\sim \frac{1}{r^4}$, the neutron panels detect the Doubles from a small, localized volume within the cylinder. The interrogation of this volume is strongly affected by the bulk UF$_6$ distribution and requires a unique analysis approach relative to those typically used for well counters. The approximate point-model sensitivities for the singles and doubles-to-singles ratio may be expressed as $S \propto M_L \cdot \varepsilon \cdot M_{234}$ and $\frac{D}{S} \propto \left(M_L^2 - M_L\right) \cdot \varepsilon$. Taking the ratio of these doubles-

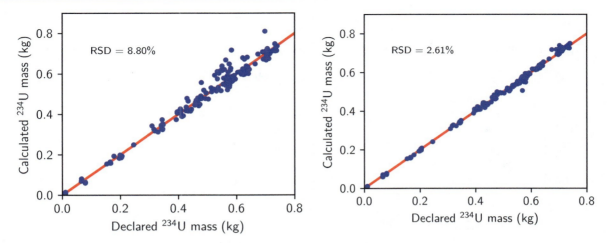

Fig. 19.56 Calibration relationships between neutron panel Singles and declared ^{234}U mass: (left) Background-subtracted only, with added up-down, front-to-back. (right) Left-right distribution corrections to give fully (three-dimensional) UF$_6$ distribution corrected Singles for 221 unique cylinders

to-singles and singles sensitivities reduces the efficiency (ε) and multiplication (M_L) dependences; however, being far from point sources, these relations provide only a heuristic argument for the following empirical relation [85].

$$\frac{D}{S^2} \propto \frac{(M_L - 1)}{M_{234}} \propto \frac{^{235}U}{^{234}U} \cdot \frac{1}{\sqrt{E_{235}}} \tag{19.3}$$

Together, using the Singles to measure ^{234}U mass (M_{234}) and the Doubles-to-(Singles squared) to measure the isotopic ratio, it is possible to directly assay the bulk ^{235}U mass or enrichment (E_{235}) within a 30B UF$_6$ storage cylinder. Because the Doubles-to-(Singles squared) has a slight distribution dependence, it was found to be accounted for using the square root of the three-dimensional correction optimized for the singles-based ^{234}U mass assay.

$$\left(\frac{D}{S^2_{BGsub}}\right)_{\text{Distribution corr}} = \sqrt{(\text{UF}_6 \text{ Distribution corr optimized for } M_{234})} \cdot \left(\frac{D}{S^2_{BGsub}}\right) \tag{19.4}$$

The ^{235}U/^{234}U ratio assay on the left in Fig. 19.57 uses the functional form,

$$\frac{^{235}U}{^{234}U} = a \cdot \ln\left[\sqrt{E_{235}} \cdot \left(\frac{D}{S^2_{BGsub}}\right)_{\text{dist.corr.}}\right] + b \tag{19.5}$$

This approach applies only to LEU cylinders. As in NU and DU, ^{238}U spontaneous fission has a disproportionate contribution to the Doubles, removing the correlation of Doubles to ^{235}U content. These results combine neutron signals with gamma results conducted concurrently using two enhanced LaBr detectors positioned adjacent to the neutron panels in Fig. 19.55. The gamma analysis used the Square Wave Convolute method of calculating the counts under the 186 keV peak [88], resulting in a relative standard deviation of 3.7% for 10 min measurements of 205 cylinders.

The product of these ^{235}U/^{234}U ratio and ^{234}U mass assays yields ^{235}U mass, with high precision as shown on the right of Fig. 19.57. Using the gamma-based enrichment in this method enables a fully independent verification of bulk ^{235}U content. If gamma measurements are unavailable, this algorithm confirms that the neutron signals are consistent with the specified ^{234}U and ^{235}U masses. Assay results improve further if the ^{235}U/^{234}U ratio assay is used to verify the high-precision declared value (determined by the enricher's destructive assay) for use when converting from ^{234}U to ^{235}U mass. For a 10 min assay, this method increases precision by a factor of 2, primarily due to the removal of Doubles statistical uncertainty [85].

The prototype that combines the array of neutron panel and curtain detectors and the associated analysis offer a unique solution for addressing the primary sources of systematic uncertainty for passive neutron measurements of bulk ^{235}U content within the large, low-enriched, type 30B UF$_6$ cylinders.

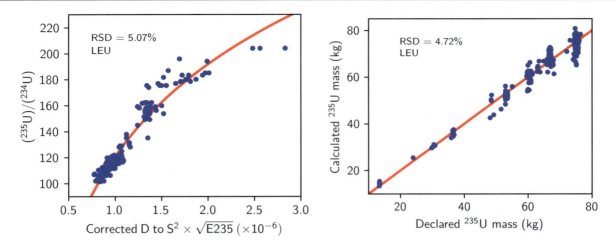

Fig. 19.57 Relationship between the (distribution-corrected D/S^2)·$\sqrt{E_{235}}$ and ^{235}U/^{234}U (left) and results of M$_{235}$ assay calculated as the product of M$_{234}$ assay and ^{235}U/^{234}U assay (right). Both include 205 LEU cylinders and use the gamma enrichment assay for input where required

19.4.8 Fresh Fuel Assemblies

The measurement of linear ^{235}U density in fresh fuel assemblies can be performed by both passive and active techniques. The active techniques are described in Chap. 20. The instruments, in chronological order, are the UNCL, the Euratom Fast Collar (EFC), the fast-neutron passive collar (FNPC), and the passive neutron coincidence collar (PNCC). The first two are active, and the last two are passive; multiple versions of each instrument are available. The UNCL was still in widespread use in 2021, whereas the EFC, FNPC, and PNCC are new instruments designed to address the use of stronger burnable poisons, higher enrichments, and ^{235}U mass loadings. The PNCC, which is used for the measurement of plutonium in fresh mixed-oxide fuel, is described in Sect. 19.4.2.

19.4.8.1 Fast-Neutron Passive Collar

The purpose of the FNPC [89, 90] is to provide a solution to a decades-old safeguards problem with the verification of the fissile concentration in fresh LWR fuel assemblies. The problem is that the burnable poison (e.g., Gd$_2$O$_3$) addition to the fuel rods decreases the active neutron assay for the fuel assemblies [91]. Thus, the FNPC provides a new method for the verification of the ^{235}U linear density (LD) in fresh LEU fuel assemblies that is insensitive to the burnable poison content. The technique makes use of the ^{238}U atoms in the fuel rods to self-interrogate the ^{235}U mass. A benefit of the approach is that the ^{238}U spontaneous fission (SF) neutrons from the fuel rods' induced fission (IF) reactions in the ^{235}U are time-correlated with the SF source neutrons. Thus, the coincidence gate counting rate benefits from both the nu-bar of the ^{238}U SF (2.07) and the ^{235}U IF (2.44) for a fraction of the IF reactions, whereas the ^{238}U SF background has no time-correlation boost. The higher the detection efficiency, the higher the correlated boost because background neutron counts from the SF are being converted to signal doubles counts. This time-correlation in the IF signal increases signal/background ratio that provides a good precision for the net signal from the ^{235}U mass. The FNPC can be configured in either the thermal-neutron mode (no gadolinium or cadmium liners) or in the fast-neutron mode (with the cadmium liners). The fast-neutron energy spectrum makes the technique relatively insensitive to the burnable poison loading, where a cadmium or gadolinium liner on the detector walls is used to prevent thermal-neutron reflection back into the fuel assembly from the detector body.

The other neutron measurement systems [91–93] for the fresh LWR fuel assemblies use AmLi neutron interrogation sources that are no longer commercially available. Thus, the FNPC provides a viable alternative for the verification of the fuel assemblies that does not require external radioactive sources.

The FNPC detector design is shown in Fig. 19.58 (left), where the MCNP code [16] was used for the optimization [94]. There are 28 ^{3}He tubes with 6 bar pressure (model RS-P4-0817-104) and 4 PDT-10A amplifiers inside the PDT junction box [1], which has a compact design 3.5 cm deep, allowing an additional 10 cm of active length to the ^{3}He tubes while keeping the total height the same as the original UNCL. The V notches in the front and back slabs, shown in Fig. 19.58 (left), are to accommodate the size of hexagonal assemblies such as the VVER-1200.

The complete FNPC system—shown in Fig. 19.58 (right), where it is assembled around the LANL mockup 15 × 15 rod PWR fuel assembly. The front slab on the detector is a hinged door that allows fuel assembly insertion through the front of the

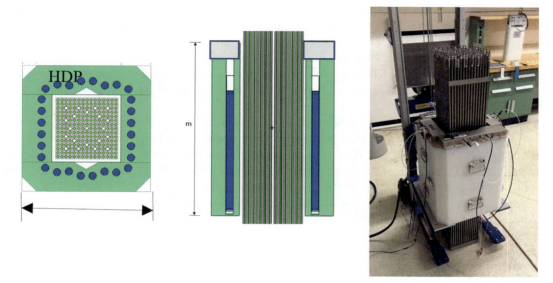

Fig. 19.58 The FNPC detector simulations (left and center) and the full detector positioned around the LANL mockup fuel assembly (right). The inside width of the detector is 23.5 cm

Table 19.15 FPNC performance and operating parameters

Operating parameter	Value
Efficiency (^{252}Cf energy) (no cadmium liners)	0.239
Efficiency (^{252}Cf energy) (with cadmium) liners)	0.212
Die-away time	~70 μs
Operating high voltage	1260 V
Dead time coefficients	
A	0–1.0 μs
B	0.25 μs^2
C	0
Multiplicity dead time	~300 ns
Gate (thermal-neutron mode)	300 μs
Gate (fast-neutron mode)	200 μs
Predelay	2.0 μs
Doubles gate fraction	~0.90
Triples gate fraction	~0.80

collar as well as from above. Table 19.15 lists the system performance parameters, where the efficiency is a function of the cadmium liners and the sample inside the detector. In the fast-neutron cadmium mode, the efficiency for a ^{252}Cf source in the center is 21.2%; the efficiency increases to 23.9% with the cadmium removed. The electronics to support the FPNC are a shift register such as the JSR-15 [5] or the optional advanced list-mode module [9] or PTR-32 [7] and a laptop computer.

Because the passive counting rates are less than 1000 cps, the dead time corrections are negligible, and in practice, the coefficients can be set to zero. The doubles and triples gate fractions are used only for multiplicity analysis and not for the passive calibrations where the gate length is incorporated in the calibration slope.

Calibrations were performed [90] for both fast-neutron and thermal-neutron modes by measuring both with and without the gadolinium/cadmium liners. The LANL 15 × 15 mockup fuel array was used for the measurements with a constant number of rods (204) by varying the number of LEU rods to change the average enrichment. The LEU rods were introduced to give a uniform distribution of the enriched rods mixed with the depleted uranium rods.

The neutron interrogation source strength for the passive measurements is the ^{238}U LD, so the induced counting rates are normalized by the ^{238}U term to provide what is defined as the *gross counts*. The *net counts* are obtained by subtracting the passive neutron background that is determined from the fit of the passive calibration curve for the zero-enrichment intercept. Normalizing to the AmLi source strength has a similar function in the present UNCL systems. Figure 19.59 shows the gross counting rates for the Singles (S), Doubles (D), and Triples (T) in the thermal-neutron mode. The measurement precision for the Doubles and Triples varied from 0.15% to 1.0%. Note that the maximum LD for the measurements was 21.9 g ^{235}U/cm; whereas LD in the range of ~60–65 g ^{235}U/cm can be expected for commercial PWR assemblies. Thus, the net counting rates will significantly increase from the present data for measurements on commercial LD loadings.

Figure 19.60 shows the fast-neutron calibrations with the gadolinium liners for the gross counts for the singles and doubles rates (left) and for the net counts (right). The linear extrapolation was based on MCNP code simulations [94] to the higher-mass loadings. The linear shapes of the three curves is a result of the hard neutron spectrum and the buildup of the ^{234}U (α,n) neutrons as the enrichment increases.

In the thermal-neutron mode, the three observables (S, D, and T) can be used to solve for both the ^{235}U and the effective burnable poison loading; therefore, the operator's burnable poison declaration would not be needed. The statistical error for the ^{235}U mass was below ~2% using this approach, and the errors are quantified in a report about advanced analysis [95].

To summarize, the FNPC can be used to measure the ^{235}U linear density in PWR-type fuel assemblies. The detector is unique from prior systems that are used to verify the ^{235}U content in fuel assemblies in that it does not require the use of an external interrogation source such as AmLi. The self-interrogation is accomplished by using the ^{238}U spontaneous fission neutrons to induce the fission reactions in the ^{235}U that coexists in the same fuel pellets.

The high efficiency of the detector and the low singles counting rates provide both doubles and triples rates with good statistical precision in measurement times of less than 1000 s. The doubles precision for 1000 s is 3.4% in the fast-neutron mode and 1.0% in the thermal-neutron mode.

The availability of the Singles, Doubles, and Triples makes multiplicity counting and advanced analysis possible to determine both the ^{235}U and the burnable poison loadings. The advanced analysis concept for active neutron interrogation has

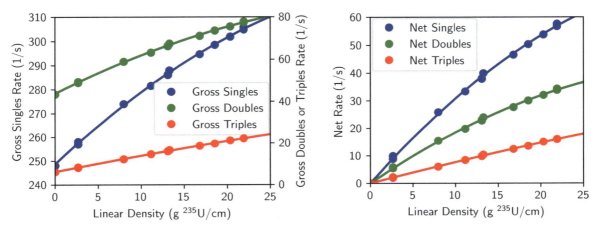

Fig. 19.59 (left) The measured gross counting rates (thermal-neutron mode) for the Singles, Doubles, and Triples as a function of the linear density in the mockup fuel assembly. (right) The net rates for the Singles, Doubles, and Triples after subtracting the background rates

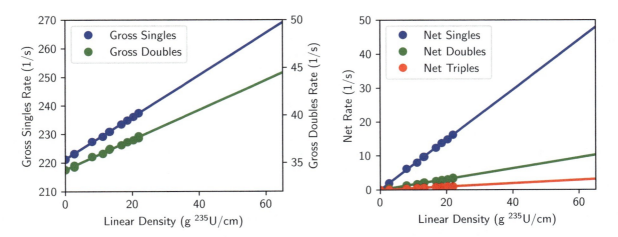

Fig. 19.60 The measured fast-neutron gross counting rates for the Singles and Doubles as a function of the linear density in the mockup fuel assembly, with a linear trend-line fit to the data (left) and the net rates after the background's subtraction (right)

been documented in papers by Root [95, 96]. Future work should include a field test of the passive collar that could easily include both attended and unattended mode data collection in parallel.

19.4.8.2 Passive Neutron Coincidence Collar

The PNCC was developed to measure fresh LWR fuel assemblies in the thermal-neutron mode [97]. The historical measurement problem of the corrections needed for the burnable poison rods was solved by using the singles and doubles rates and/or the doubles and triples rates to solve for both the ^{235}U mass and the gadolinium content separately. This advanced analysis method is described elsewhere [95]. The PNCC has adequate efficiency (24%) to provide statistical precision in 600–1000 s measurement times. The detector hardware and electronics are identical to the FNPC except that the cadmium liners have been removed.

The measurement concept that the ^{238}U spontaneous fission neutrons are used to self-interrogate the ^{235}U mass in the adjacent pellets is the same as for the FNPC. The high efficiency and the time-correlated induced fissions (see Ref. [97]) in the thermal-neutron mode make the Triples precision similar to the Doubles precision; the singles rates are optional for the analysis. The sensitivity to fuel rod removal and substitution with empty rods or steel rods is much better than any of the competing neutron collar systems because the removal of a rod reduces both the source term (^{238}U) as well as the IF from the ^{235}U. If DU rods are used for the substitution, the neutron source term is not reduced but the IF term is still reduced for the doubles and triples rates. The singles rate is sensitive to potential changes in the room neutron background, but the doubles and triples rates are not. The measurement of the singles, doubles, and triples count rates that are time correlated provides a fingerprint for the assembly that would be very difficult to falsify using any combination of false substitute rods.

The PNCC system is shown in Fig. 19.61 (left), and the LANL mock-up PWR fuel assembly is pictured (right). The operating parameters are given in Table 19.16.

The counting rate for the PNCC is less than 1000 cps, so the dead time corrections are negligible, and the coefficients can be defaulted to zero.

To demonstrate the sensitivity of the singles, doubles, and triples rates to the rod removals, a series of measurements was performed using the mockup PWR fuel assembly with the 15 × 15 rod array. The fuel rods were uniformly removed from the

Fig. 19.61 The PNCC during fuel rod exchange (left) and the mockup PWR fuel assembly (right), with the central 30 LEU rods replaced by DU rods

Table 19.16 PNCC detector operating parameters

Operating parameter	Value
Efficiency (^{252}Cf energy) (no cadmium liners)	0.239
Die-away time	~7 μ
Operating high voltage	1260 V
Dead time coefficients	
A	0–1.0 μs
B	0.25 μs^2
C	0
Multiplicity dead time	~300 ns
Gate (thermal-neutron mode)	300 μs
Predelay	2.0 μs
Doubles gate fraction	~0.90
Triples gate fraction	~0.80

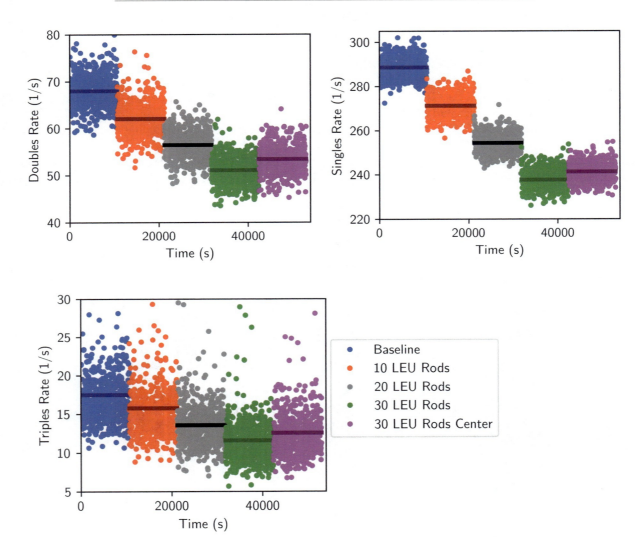

Fig. 19.62 Continuous data collection for the measured singles, doubles, and triples rates (20 s/point) during fuel rod removal steps of 10 LEU rods per step in the mockup PWR fuel assembly

full fuel assembly that contained 204 fuel rods (109 low-enriched uranium [LEU] and 95 DU rods) in three steps of 10 LEU rods per step (5% change). Continuous data were collected in 20 s cycles for approximately 4 hours between steps, as shown in Fig. 19.62 for the singles, doubles, and triples rates. The heavy-shaded bands in the graphs represent the statistical scatter ($\pm 1\sigma$) for a 600 s subinterval of the data. The light-shaded bands represent the $\pm 2\sigma$ scatter for the data.

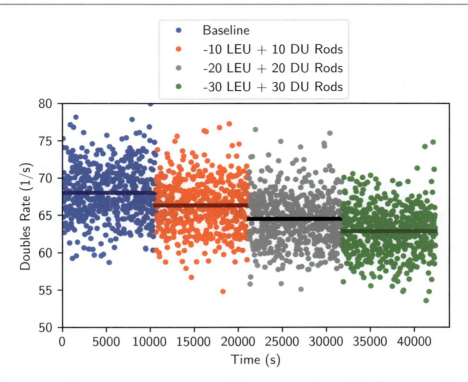

Fig. 19.63 Continuous data collection for the measured doubles rates (20 s/point) during fuel rod substitution steps of 10 LEU rods per step with the DU rod replacements

The singles rates show the most absolute deviation due to the removal of 10 rods; however, the percent deviation is larger for the Doubles (0.85% per rod) and the Triples (1.15% per rod) than for the Singles (0.59% per rod). This result is because the neutron multiplication decrease affects the Doubles and Triples more than the Singles.

Measurements were also performed where DU rods were substituted for the LEU rods. In this case, the ^{238}U spontaneous fission rate has a small increase with the DU substitution, so the net effect from the rod exchange is considerably smaller than for the data shown in Fig. 19.62. Figure 19.63 shows the doubles rates for the same rod exchange pattern as for the prior exchanges. The decrease in the doubles rates from the change are about a factor of 4 smaller than for the steel, lead, and empty rod substitutions; however, the decrease is still clearly visible for a 600 s subinterval of the data for the 10-fuel-rod change.

The sensitivity to rod removal and substitutions in the passive mode is better than for active mode neutron systems because both the neutron source term and the induced fission rates decrease with the rod removals. However, if DU rods are available for the substitution, the perturbation is significantly reduced. The sensitivity was good for all three correlated signatures (S, D, and T), so falsifying the three correlated signals would be very difficult. The doubles rates showed the best sensitivity to the removals where the sensitivity (>3σ change) in 600 s was less than two rods for removals with replacements by steel, lead, or empty rods. However, if DU rods are used for the substitution, the sensitivity was ~9 rods in 600 s. Longer measurement analysis intervals would further improve the sensitivity levels.

The addition of poison rods to the fuel assembly would reduce the doubles and triples rates more than the singles rate, and advanced analysis methods have been developed to simultaneously measure both the ^{235}U mass and the gadolinium content [95]. Future field tests are needed to better understand list-mode data collection and passive operation.

19.4.9 Coincidence with Calorimetry or Gamma Signatures

19.4.9.1 Introduction

One challenge for neutron assay is the measurement of plutonium scrap and residues [98]. These materials tend to have large quantities of light elements, resulting in a particularly large value of α. Typical problematic impurities are beryllium, fluorine, sodium, magnesium, and chlorine. Samples with high beryllium or fluorine content can exhibit values of α as high as 100.

When measuring these items by neutron multiplicity, the limiting factor is the time required to obtain reasonable counting statistics for the triples rate. Even multiplicity counters with efficiencies greater than 50% can take multiple hours to obtain the required statistics. At such long times, calorimetry is often the preferred method for measurement of these items.

To avoid the requirement of the Triples precision, coincidence counting may be used as an alternative. Although the doubles rate has a much better precision, the accuracy of this approach is limited. There are generally three unknown values (^{240}Pu$_{eff}$ mass, alpha, and multiplication) but only two measured values (Singles and Doubles).

19.4.9.2 Known-Alpha by Calorimetry

The known-alpha by calorimetry method was developed to address a particular processing scheme encountered at the Savannah River Site [98, 99]. As part of surveillance and disposal activities, one plutonium-bearing item, the "parent," is often split into multiple "daughter" items. As a result of this process, the parent and the daughters have very similar material characteristics, including impurity content.

This method is a compromise between the long assay times required for calorimetry and the much faster times for coincidence counting. One of the items (either the parent or a daughter) is used as a working standard and is measured on both the calorimeter and the neutron counter. The remaining items are measured by neutron counting only.

For the working standard, the calorimeter value is taken to be the true value for plutonium mass. Because the mass is known, the neutron singles and doubles rates can now be solved for alpha and multiplication. The resulting alpha value is then taken to be the true alpha value for the parent and all daughter items. The multiplication is not assumed to be the same for all items. This assumption is especially true for the parent, which in general, will have a much larger multiplication than the daughters. Once the alpha value is determined, the neutron coincidence rates (Singles and Doubles) for the other items can be used to solve for the ^{240}Pu$_{eff}$ mass and the multiplication. An example of this technique in use is shown in Fig. 19.64. In this case, the parent material was mixed with a dilutant and split into 20 daughters. Because of the addition of the dilutant, the alpha value of the parent did not necessarily match the daughters; however, all of the daughters had the same matrix composition.

One daughter was measured on the calorimeter, and all of the daughters were measured on an AWCC. Because of the low efficiency of the AWCC, the multiplicity results (blue circles in the figure) were extremely imprecise; however, once the known-alpha from the calorimeter was applied, the singles and doubles rates were used to determine the mass with much greater precision (red circles). The error bars in the figure represent the calibration uncertainty (12.9%) of the AWCC, but the average difference between the declared and measured mass values was only 2.9%, suggesting that the random uncertainty is much smaller.

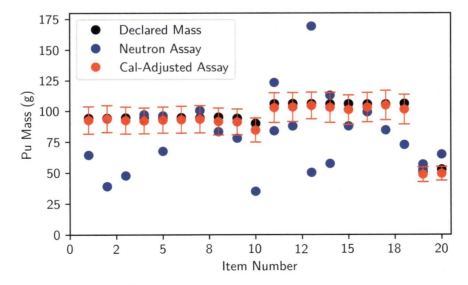

Fig. 19.64 Results from the known-alpha technique using calorimetry

19.4.9.3 Known-Alpha by Gamma Spectroscopy

An alternative to using calorimetry is to rely on the (α,n) reaction gamma rays to estimate the value of alpha for a given sample. Nearly all of the light elements with (α,n) cross sections also emit one or more gamma rays as a result of alpha-induced reactions [100]. In principle, a well-calibrated gamma spectrum could be used to quantify the impurity content of an item and thus predict the alpha value.

In practice, this technique is very imprecise due to multiple factors that affect the (α,n) reaction rates and the resulting gamma rays. The primary challenge is the unknown coupling between the alpha particle and the low-Z target. Chemical form, particle size, mixing, and chemical form can all impact this coupling. Extensive work has been performed to calibrate the measured gamma rays with the true impurity content in the plutonium matrix [101], but in general, the uncertainties remain large.

Qualitatively knowing the identity of light elements can present a significant advantage. The (α,n) neutrons are often emitted with significantly different energy spectra than the spontaneous fission spectrum. As a result, the counter efficiency and induced fission cross sections are different. In general, elements with (α,n) energy higher than fission (beryllium, magnesium) result in a positive bias on multiplicity counting, and those with a lower (α,n) energy (fluorine, chlorine) result in a negative bias [102].

If an impure item contains only a single light element, then in principle, the corresponding gamma-ray emission rate should be proportional to the (α,n) rate and thus correlated with the bias on the neutron counter. In one application, rather than open shipping packages to perform calorimetry on beryllium-bearing items, the 4439 keV gamma ray from the ^{9}Be(α,n)^{12}C reaction was used to predict the bias on a multiplicity counter [103]. This approach was used to verify shipping containers that contained up to 0.75% beryllium by weight.

19.5 Multiplicity Counting

19.5.1 Applications, Advantages, and Disadvantages

Passive neutron multiplicity counting significantly improves measurement accuracy for plutonium metal, oxide, scrap, and residues compared with coincidence counting, as described in Chap. 18. Multiplicity counting has applications in various different areas: improved materials accountability measurements, verification measurements, confirmatory measurements, and excess weapons materials inspections.

Motivated by the need for better accuracy, the goal of neutron multiplicity analysis is to correctly assay in-plant materials without prior knowledge of the sample matrix. The availability of a third measured parameter makes assay possible for many materials, including moist or impure plutonium oxide, oxidized metal, and some categories of scrap and waste. Multiplicity counting can be used for all plutonium samples, but the additional information is beneficial primarily for impure samples. For some material categories, multiplicity may not be helpful because of the limited precision of the triple coincidences. These materials include small plutonium samples, some plutonium-bearing waste, or process residues that are so impure that the high (α,n) reaction rate ruins the precision of the Triples. For pure plutonium metal or oxide, the additional multiplicity information is not needed, and conventional coincidence counting provides better precision and sufficient accuracy. However, if any doubts exist about the plutonium purity, the multiplicity and conventional results can be compared, and the more accurate result can be used.

A useful multiplicity counter should also provide relatively fast assays. At present, a practical goal for assay precision is 1% relative standard deviation (RSD) in 1000 s. The limiting factor here is the poorer precision of the Triples.

The advantages of multiplicity counting are summarized as follows:

- The measurement accuracy for impure plutonium is much greater than for conventional coincidence counting.
- Information on sample self-multiplication and (α,n) reaction rate is obtained.
- Calibration for many material types does not require representative standards.
- Typical measurement time, 1000–2000 s, is short compared with other techniques.
- If a high-efficiency multiplicity counter is used for conventional coincidence counting, one can use very short counting times and obtain somewhat better accuracy.

The disadvantages of multiplicity counting are as follows:

- Multiplicity counters are more costly than conventional coincidence counters.
- Multiplicity counters require somewhat more floor space and height than conventional counters of the same cavity size.
- The measurement time for good precision on triples, typically 1000–2000 s, is longer than the 100–300 s counting time used for most conventional coincidence assays.
- For plutonium samples that do not meet the analysis assumptions, some assay biases still remain. These biases must be removed using correction factors, special calibration procedures, physical standards, or calorimetry on outliers.

19.5.2 Multiplicity Counter Design Goals and Approaches

Thermal-neutron multiplicity counters are high performance, advanced, and expensive. They are designed to minimize the effects of detector-dependent variables such as those summarized earlier in this chapter. In terms of these variables, the goals for multiplicity counter design include the following:

- Maximize the detection efficiency to increase the triple coincidence count rate, which is proportional to the third power of the efficiency, typically 40%–60%.
- Minimize dead time losses by substantially increasing the number of amplifier modules used to read out the ^{3}He tubes. Multiplicity counters use 20 or more circuits compared with 6 in typical coincidence counters. The triples rate is much more sensitive to electronic dead time than the doubles and singles rates.
- Minimize the detector die-away time to decrease accidental coincidences and thereby improve the signal-to-noise ratio for triples.
- Minimize the effects of sample placement—or variable plutonium distribution—by making the radial and axial efficiency profile of the sample cavity as flat as possible.
- Minimize the influence of the emitted neutron energy spectrum on the efficiency.

These goals are achieved by several typical approaches. First, the array of ^{3}He tubes is tightly packed with a small pitch to increase detection efficiency by minimizing losses in polyethylene and maximizing absorption in ^{3}He. This approach also minimizes die-away time by reducing the average distance and time that a thermal neutron travels before reaching ^{3}He. Multiple rows (typically four or five) of ^{3}He tubes are used to minimize the efficiency's dependence on neutron energy. The outer rows will have a higher efficiency for higher-energy neutrons. Multiple rows are also needed to increase the overall detection efficiency and to minimize the fraction of neutrons that escape the detector. Increasing the ^{3}He gas density from 4 atm to as high as 10 atm also increases efficiency and reduces die-away time. The detector and ^{3}He tubes are usually taller, and long graphite plugs are used to fill the cavity ends to minimize vertical efficiency dependence. Finally, a cadmium liner around the sample cavity minimizes thermal-neutron reflection and multiplication, which would otherwise greatly increase the die-away time.

19.5.3 MCNP and Figure-of-Merit Design Calculations

In the design of multiplicity counters, Monte Carlo (often the MCNP code; see Ref. [16]) and figure-of-merit codes are used to supplement what has been learned from past designs of conventional coincidence counters. The codes can be used to study design choices such as tube placement; number, size, and gas pressure of tubes; tube bank layout; placement of different neutron moderator or reflector materials; and the use of cadmium liners. For more details, see Sect. 19.7.5. Figure 19.65 is a schematic used in the Monte Carlo design of the plutonium scrap multiplicity counter pictured in Fig. 19.70.

Because the MCNP code provides an estimate for the efficiency and die-away time, a figure-of-merit (FOM) code can be used to determine the optimum design needed to achieve the desired measurement precision. One FOM code developed for multiplicity-counting analysis [104] determines assay variance from the neutron multiplicity distribution. This distribution is predicted from the detector design parameters obtained from MCNP. The expected values of the sample mass, self-multiplication, and (α,n) reaction rate, and the count time, electronic gate width, and predelay are entered into the code, which then predicts the expected single, double, and triple count rates and determines the assay variance.

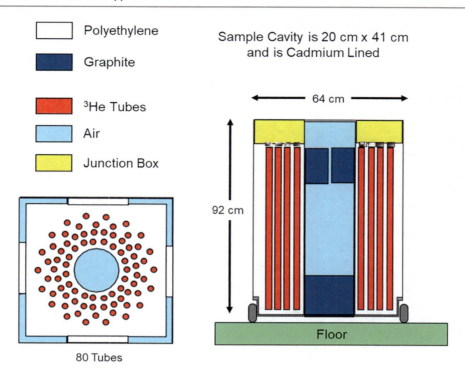

Fig. 19.65 Monte Carlo design schematic for the PSMC

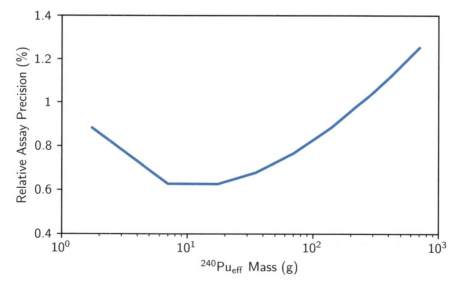

Fig. 19.66 FOM calculation of expected assay precision (RSD) versus ^{240}Pu mass for PuO^2 in a 50% efficient multiplicity counter for 1000 s counts

Once the sample mass and size range have been defined, these calculations can be used to define the target efficiency and die-away time needed to obtain a given assay precision in a given time. Figure 19.66 illustrates the expected assay precision versus ^{240}Pu mass for plutonium oxide samples in a 50% efficient counter for a 1000 s count time. From these results, one can determine whether 50% efficiency is sufficient. For impure samples, the assay precision deteriorates rapidly with increasing (α,n) rates. The FOM calculation is a fast way to estimate the efficiency, die-away time, and count time needed to provide a given assay precision. Then the minimum detector cost to meet the performance requirements can be found.

Table 19.17 Survey of some multiplicity counters in U.S. Department of Energy Facilities [59]

Multiplicity counter name	Location	Application	No. of tube rings	No. of ^{3}He tubes	No. of AMPTEK preamplifiers	Randomizer?	Multiplicity dead time (ns)	Neutron Det. Eff. (%)	Die-Away time (μs)	Cavity size
Five-ring multiplicity counter	LANL	Technology development	5	130	34	Yes	36	53	49	16.5 cm D × 25.4 cm H
Three-ring multiplicity counter	LANL	Technology development	3	60	12	Yes	83	45	63	20 cm D × 30 cm H
Pyrochemical multiplicity counter	LANL	In-plant metals, oxides	4	126	36	No	90	57	47	24 cm D × 38 cm H
Plutonium scrap multiplicity counter	Hanford and Japan	Pu inventory verification	3 ½	80	19	No	121	55	47	20 cm D × 41 cm H
ARIES neutron counter	LANL	Pu metals and residues	3 ½	80	20	Yes	60	55	47	20 cm D × 41 cm H
FB-line multiplicity counter	Savannah River	Metal, oxide inventory	4	113	24	Yes	50	58	50	20 cm D × 41 cm H
30-gallon multiplicity counters	Rocky Flats and Livermore	Inventory verification	3	126	54	Yes	25	42	55	30-gallon drum
High-efficiency neutron counter	LANL	Waste assay	2	113	16	No	171	32	50	55-gallon drum
Epithermal neutron multiplicity counter	LANL	Pu inventory verification	4	121 10 atm	27	Yes	37	65	22	20 cm D × 43 cm H
KAMS neutron multiplicity counter	Savannah River	Receipts verification	3	198 10 atm	54	Yes	19	52	37.3	55-gallon drum
SuperHENC multiplicity counter	Rocky Flats	Standard waste box assay		260 10 atm	32	Yes		40.3		Standard waste box

ARIES advanced recovery and integrated extraction system, *HENC* high-efficiency neutron counter

19.5.4 Existing Multiplicity Counters

There are many different neutron multiplicity counter designs. Table 19.17 lists properties of some neutron multiplicy counters in use for plutonium nondestructive assay.

19.5.4.1 Five-Ring Neutron Multiplicity Counter

The five-ring neutron multiplicity counter (5RMC; Fig. 19.67) is the first thermal-neutron counter designed specifically for multiplicity measurements. Its design is quite old, but as of 2021, it is still used in active research at LANL. It was built with five rings of ^{3}He tubes at 4 atm pressure to ensure a very high neutron-detection efficiency for developing new multiplicity techniques. The original version was built with an aluminum moderator assembly, which was eventually replaced with polyethylene moderator. It was also possible to wrap each ^{3}He tube individually with a removable cadmium liner, and for that reason, the counter was called the *dual-mode multiplicity counter*. With aluminum and cadmium in place, the efficiency was found to be 17%, with an impressively low die-away time of 11.8 μs. Without the cadmium, the efficiency was 53% and die-away was 57 μs, which is readily outperformed by modern multiplicity counters with 10 atm ^{3}He pressure. With

Fig. 19.67 The LANL 5-ring neutron multiplicity counter

polyethylene moderator, the efficiency is 53%, and the die-away time is 49 μs. The 5RMC played a very important role in the development of multiplicity counters because it demonstrated that thermal-neutron multiplicity counters could provide good assays of plutonium samples in reasonable counting times.

19.5.4.2 Three-Ring Multiplicity Counter

The three-ring multiplicity counter (Fig. 19.68) was originally built as an experimental active well coincidence counter. It was converted from an active counter to a passive counter by removing the polyethylene end plugs that held the AmLi sources and replacing them with shorter graphite end plugs. The counter was also upgraded by replacing the 6 original amplifier module boards in the high-voltage junction box with 12 AMPTEK boards and a derandomizer.

This counter has been used for research and development activities, for training classes in multiplicity counting, and for temporary use in IAEA inspections of excess weapons materials at the Hanford facility [105]. However, because this counter is actually a converted conventional coincidence counter, its performance specifications are not as good as a counter specifically designed for multiplicity counting. The neutron-detection efficiency is lower (only 45%), and the spatial and energy response profiles are not as flat. Nevertheless, the counter was able to verify most of the items assayed at Hanford, thereby significantly reducing the number of samples taken for destructive analysis.

19.5.4.3 In-Plant (Pyrochemical) Multiplicity Counter

Based on experience gained with earlier developmental counters, the in-plant or pyrochemical multiplicity counter was designed specifically for in-plant use to optimize the parameters important for multiplicity assay. The counter has two halves so that it could be installed around a glove-box well. Figure 19.69 is the design schematic used in the Monte Carlo calculations to determine the optimum tube spacing—1.59 cm—and the best choice of end plug materials. The result is a very high-performance counter with a single-exponential die-away curve. The individual ring responses are illustrated in Fig. 18.9.

Fig. 19.68 The LANL three-ring multiplicity counter

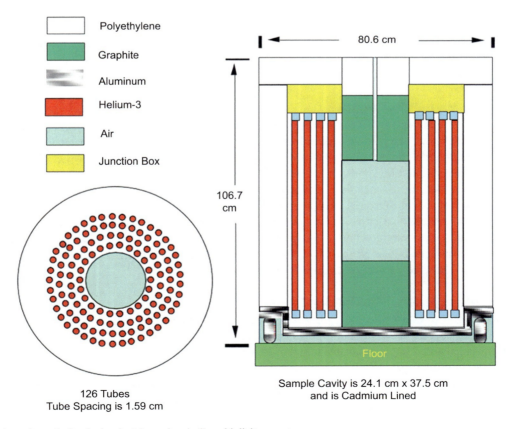

Fig. 19.69 Design schematic for the in-plant (pyrochemical) multiplicity counter

Fig. 19.70 Photo of the PSMC detector shown with the electronics junction box exposed to carry out upgrades to the electronics

The in-plant counter was used in the LANL Plutonium Facility to assay plutonium metal, oxide, and high (α,n) electrorefining salts. The counter was also used at the Livermore Nuclear Materials Facility to assay low- and high-burnup plutonium metal and oxide.

19.5.4.4 Plutonium Scrap Multiplicity Counter

The PSMC is a relatively compact, high-efficiency multiplicity counter for the measurement of impure plutonium and mixed-oxide scrap [106]. The PSMC is shown in Fig. 19.70 and the tube layout in Fig. 19.71. In comparison to the in-plant counter, the PSMC uses far fewer ^{3}He tubes for nearly the same efficiency. This improvement was achieved by reducing the number of ^{3}He tubes in each ring in proportion to the decrease in the neutron flux density in the moderator. Thus, the outermost ring of ^{3}He tubes is only about half filled, and in Table 19.4, the PSMC is described as having 3½ rings of tubes. The PSMC contains 80 4-atm ^{3}He tubes that contain Ar + CH_4 quench gas. The signals from the tubes are processed in 20 amplifiers, which are connected to a derandomizer board. The sample cavity is 41 cm high and 20 cm in diameter. The axial efficiency profile is constant to within ±2% over the height of the cavity, making it easier to place most scrap containers within the flat portion of the efficiency profile. PSMC operating parameters for an instrument located at LANL are summarized in Table 19.18. The parameters listed are specific to this version of PSMC. Some operating parameters of other PSMC instruments—obtained from commercial vendors—could have slightly different values. The table serves as a general performance overview; for operating parameters, users should always refer to the user guide provided with their specific instrument.

The first PSMC instruments were used for inventory verification campaigns in Japan and at Hanford, and additional counters are now commercially available. An inline, active/passive multiplicity counter very similar in design to the PSMC was developed for permanent installation in the LANL Plutonium Facility [107]. This counter, called the *ARIES neutron counter*, has a split body so that it can be installed around a glove-box well.

The PSMC can be applied to impure plutonium and MOX items with mass in the range from a few tens of grams to several kilograms of high-burnup plutonium. The PSMC can also be used for the assay of sub-gram inventory samples that have been

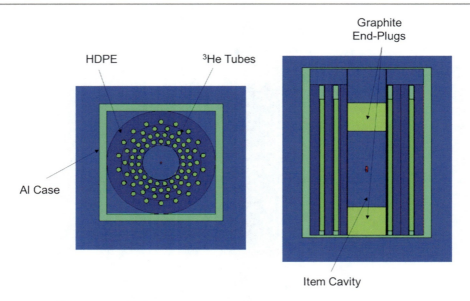

Fig. 19.71 Internal layout of a PSMC from an MCNP model

Table 19.18 Summary of neutron operating parameters for plutonium scrap multiplicity counter

Operating parameter	Value
Efficiency (^{240}Pu energy)	0.55
Operating high voltage	1680 V
Dead time coefficients	
A	0.1515 μs
B	0.0219 μs^2
C	0
Multiplicity dead time	50.0 ns
Gate	64 μs
Predelay	4.50 μs
Doubles gate fraction	0.614
Triples gate fraction	0.391

bagged out of glove boxes. A typical high-burnup MOX sample that contains a few hundred grams of plutonium can be assayed with 1%–2% precision in a 1000 s measurement time.

19.5.4.5 FB-Line Neutron Multiplicity Counter

The FB-Line neutron multiplicity counter (FBLNMC) is designed to measure impure plutonium at the Westinghouse Savannah River Site FB Line Facility [108, 109]. The FBLNMC can be applied to impure samples that range in plutonium mass from a few tens of grams to several kilograms; coincidence counting or multiplicity counting can be used. Monte Carlo calculations helped design the high-efficiency (57%) detector, which has 113 ^{3}He tubes. The axial efficiency profile varies by less than ~2% over the height of the cavity, and the radial efficiency variation over 16 cm is only 1.5% at the midplane of the sample cavity. The energy response profile is identical to that of the pyrochemical multiplicity counter. A derandomizer circuit reduces the dead time by more than a factor of 2–50 ns. The individual ring outputs can be read by auxiliary scalars to diagnose sample anomalies. The ratio of rates in the inner and outer rings can also provide a sensitive indication of the mean energy of the neutrons from a sample and is strongly influenced by the sample matrix or (α,n) neutrons.

19.5.4.6 Thirty-Gallon Multiplicity Counters

The 30-gallon (109-liter) multiplicity counters are an important extrapolation of design concepts to a larger sample volume [110, 111]. The number of ^{3}He tube rings is reduced to three to save cost, and aluminum corner reflectors help maintain a good spatial response. To facilitate loading heavy drums or ATR400 storage containers, a hexagonal design is used; the two front sides form the doors. The mechanical arrangement of the counter and doors is shown in Fig. 19.72. The counter has an

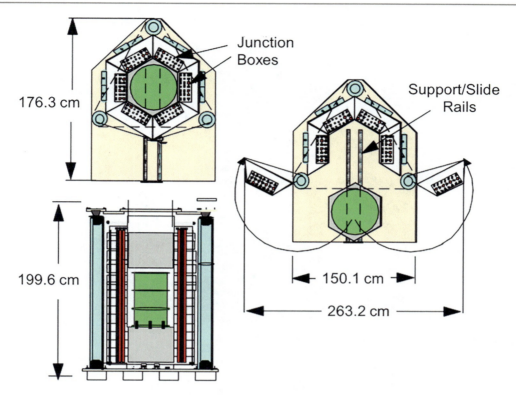

Fig. 19.72 Mechanical schematic for the 30-gallon multiplicity counter

efficiency of 42% and a die-away time of 55 μs, which is sufficient for assay of bulk plutonium in the kilogram range. Fifty-four AMPTEK amplifier modules and a derandomizer circuit are used to obtain an extremely low dead time of 25 ns. One 30-gallon counter was used at Rocky Flats for IAEA inspections of excess weapons materials [112, 113]. Another counter is installed at the Livermore Nuclear Materials Facility, where it is used for inventory verification.

19.5.4.7 High-Efficiency Neutron Counter

The high-efficiency neutron counter (HENC) is a waste-drum counter developed jointly by Canberra Industries and LANL [54]. This counter was designed to be a high-efficiency, low-detectability, passive neutron coincidence counter with multiplicity and segmented AAS matrix correction capability. The design was optimized on the basis of a detectability limit FOM, resulting in a limit of 0.5 mg ^{240}Pu$_{eff}$ by singles counting and 1.7 mg by doubles counting at sea level. An automated drum-handling system opens the assay chamber door, loads drums from the conveyor system, and rotates the drums while they are being assayed. Figure 19.73 is a top view of the HENC.

19.5.4.8 Epithermal Neutron Multiplicity Counter

The thermal-neutron multiplicity counters described above use 4-atm ^{3}He tubes. A new design concept, the ENMC (Fig. 19.74), uses 121 10-atm tubes in closely packed rings with less polyethylene moderator [114]. This configuration enables the ENMC to detect both thermal and epithermal neutrons, resulting in an efficiency of 65% and a die-away time of only 22 μs. The tubes are arranged in four rings surrounding the central well, which is 40 cm deep and 19 cm in diameter (Fig. 19.75). The signals from the 121 ^{3}He tubes are processed in 27 AMPTEK-A111 amplifier modules (with a 190 ns shaping time); the amplifier modules are connected to a derandomizer board. The ENMC's higher efficiency and shorter die-away time lead to dramatic reductions in counting time. At an α value of 1, the ENMC is about 10 times faster than a thermal-neutron multiplicity counter for the same 1% precision. At an α value near 8 (very impure plutonium), the ENMC is about 20 times faster for the same 1% precision. But the practical limit of the ENMC is around α values of 10–12, where the counting time for 1% precision begins to exceed 1 h. For bulk samples of plutonium, ENMC assay times are 5–40 times shorter than prior thermal-neutron multiplicity counters. The largest relative gains are for the most impure items with high (α,n) rates, where such gains reduce counting times from hours to 20 or 30 min. It should also be noted that the ENMC counter

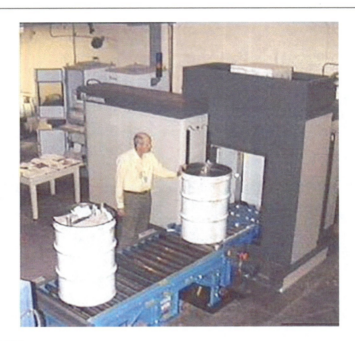

Fig. 19.73 Photograph of the HENC

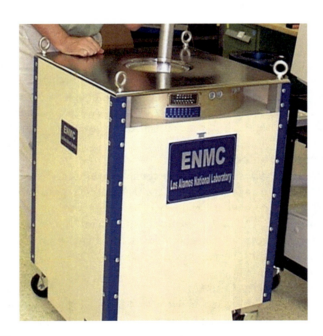

Fig. 19.74 Photograph of the ENMC

design has a relatively flat detection efficiency profile as a function of neutron energy, providing an essentially bias-free assay for α values in the 0.1–0.3 range (impure metal) to the 7–12 range (very impure oxide). Table 19.19 summarizes the operating parameters of the ENMC.

Three essential design features of the ENMC enabled the improvement in die-away time and efficiency:

- The ENMC uses 10-atm ^{3}He detection tubes rather than the more typical 4-atm tubes. The higher pressure ^{3}He captures more of the thermalized and epithermal neutrons. Higher pressure ^{3}He detectors have also been used to upgrade existing neutron counters and have improved detection efficiency and reduced die-away time.

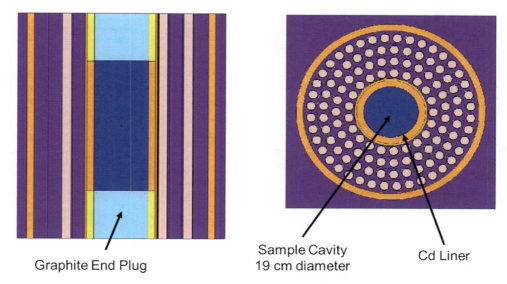

Fig. 19.75 MCNP diagram of the ENMC showing the layout of the ^{3}He tubes in the HDPE detector body

Table 19.19 Summary of neutron operating parameters for the epithermal neutron multiplicity counter [115]

Operating parameter	Value
Efficiency (^{240}Pu energy)	0.65
Operating high voltage	1720 V
Dead time coefficients	
A	0.0954 μs
B	0.0289 μs^2
C	0
Multiplicity dead time	36.8 ns
Gate	24 μs
Predelay	1.50 μs
Doubles gate fraction	0.621
Triples gate fraction	0.404

Table 19.20 Count time comparisons between different counters

Count times (min)					
Detector/reference standard identification	LANL STD-ISO3,6,9,12	LANL STD-11	LANL STD-SRP12-1	Hanford PFP 41–86–03–240	LANL LAO261C10
PSMC	7.3	170	34	1352	27
ENMC	1.6	36	5.6	64	2.7
Time ratio PSMC to ENMC	4.6	4.7	6.1	21	10

- The ENMC uses less polyethylene moderator than conventional designs. With less moderator, some neutrons are detected before they become fully thermalized (hence the name).
- The ENMC uses multiple rings of detectors embedded in polyethylene. Other modern neutron well counters have also used this design, but for the ENMC, the design is particularly important because of the small amount of moderator between the tubes.

The ENMC can be compared with previous multiplicity counters either by comparing the assay time required to achieve a particular precision or by comparing the measured precision for a fixed count time. These two methods are physically equivalent but illustrate the improved performance in different ways. Table 19.20 compares count times (in minutes) and the count time ratios between the ENMC and the PSMC for various reference standards. The standards were chosen to test the

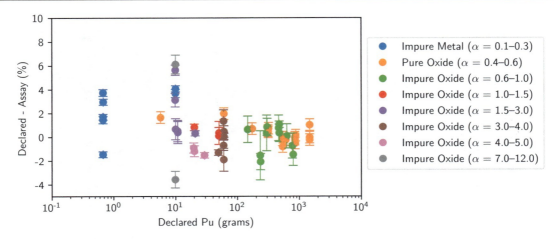

Fig. 19.76 Declared minus ENMC multiplicity assay results versus declared mass; measurement times were variable [115]

limits in precision induced by a high accidentals rate: both high mass and high alpha are represented. The items were counted for a time sufficient to achieve a 1% RSD precision. This comparison clearly shows that with all other parameters equal, the ENMC achieves significantly better multiplicity precision than the PSMC detector (or conversely, a shorter count time for the same precision).

Multiple measurements of reference standards were also made to evaluate the accuracy of the ENMC [115]. These measurements included a broad range of materials, plutonium masses, alpha values, and multiplications. The masses varied from 0.7 g of plutonium to 1451 g of plutonium, alpha in the 0.1–12 range, and multiplication in the 1–1.2 range. The results are plotted in Fig. 19.76 against the known values for these reference standards. The results in Fig. 19.76 show that the ENMC has achieved a high level of operational accuracy over a broad range of test cases. In all cases, a full multiplicity assay was performed.

19.5.4.9 ENMC with Inventory Sample Counter

A variation of the ENMC includes a counter insert called the *inventory sample counter* (INVS). Inserting the INVS counter adds an additional 21 ^{3}He tubes (10 atm) in two rings to the counter system (Fig. 19.77). The signal from the additional 21 tubes is processed in three AMPTEK-A111 amplifiers. A derandomizing buffer is not included in the INVS electronics, and the resulting ENMC/INVS dead time is ~100 ns, which is quite acceptable for the low count rate applications of the ENMC/INVS. The addition of the INVS increases the overall efficiency to 80%, and the die-away time is 19 μs [115]. ENMC/INVS precision was evaluated using a range of standards with masses of 0.7–20 g plutonium and alpha values between 0.2 and 1.5, which demonstrated that precision of 0.2% was achievable for the majority of these standards within a 1-hour measurement time—comparable to calorimetry in a much shorter time (Fig. 19.78).

19.5.4.10 SuperHENC Multiplicity Counter

The largest neutron multiplicity counter built to date also uses epithermal neutron design concepts. The SuperHENC has a cavity large enough to accommodate 1900-L SWBs and also provides AAS matrix corrections using a ^{252}Cf source that is shuffled in and out of the assay cavity [58]. SuperHENC uses 260 10-atm ^{3}He tubes to achieve an efficiency of 40.3%. This counter is mounted in a trailer that is 14.6 m long × 2.6 m wide × 4.1 m high (Fig. 19.79). It was fabricated and used to measure TRU waste containers before shipment to WIPP in Carlsbad, New Mexico [116]. For more detail see section 19.4.3.4.

19.5.4.11 Advanced Inventory Verification Sample System

The advanced inventory verification sample system (AVIS) is an NDA system used to measure small samples, powders, and pellets of bulk MOX. It was designed to support future verification activities at the Japan MOX fuel fabrication facility (J-MOX). The AVIS design has evolved from previously developed conceptual physics and engineering designs for the INVS—a safeguards system for neutron-based NDA of small samples [117–119]—and represents a hybrid between the INVS and the ENMC [114, 115]. The AVIS is an integrated gamma-neutron system. Its split-body design will accommodate installation around a tube under a glove box while keeping a small overall footprint. It was designed to enable NDA measurements with bias defect precision and accuracy to supplement or replace destructive chemical analysis—with cost,

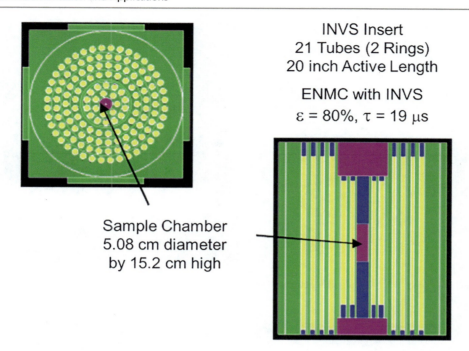

Fig. 19.77 MCNP visualization of the ENMC detector with INVS module inserted

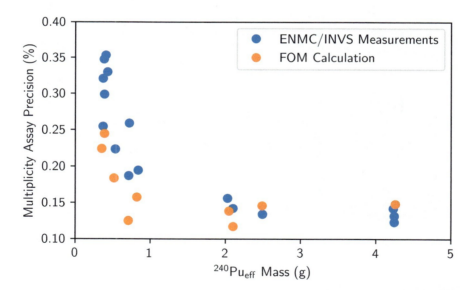

Fig. 19.78 Precision of the ENMC/INVS for small plutonium standards representative of samples collected for onsite laboratories [115]

exposure, and waste-generation savings for the facility. A photo of the AVIS is shown in Fig. 19.80. AVIS operating parameters are summarized in Table 19.21.

The primary technical challenge for AVIS is the high accuracy requirements for the neutron assay [120]). AVIS uses passive neutron coincidence counting from the spontaneous fission of plutonium for the quantification of plutonium in the MOX sample. The primary analysis method used for AVIS is a passive neutron coincidence curve based on calibration standards of well-characterized, representative materials. Additionally, known-alpha and multiplicity analysis will be used concurrently because they provide important diagnostic information necessary to achieve the target measurement accuracy. The expected AVIS measurement accuracy for the neutron measurements is less than 0.5%, 1σ in less than 15 min of measurement time.

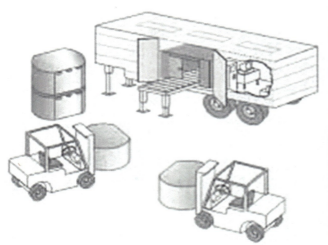

Fig. 19.79 Photograph of the SuperHENC multiplicity counter

Fig. 19.80 Photograph of the AVIS with the ORTEC cryostat to the left and HPGe detector (center) pulled out of the AVIS instrument

Table 19.21 Summary of neutron operating parameters for the AVIS

Operating parameter	Value
Efficiency (^{240}Pu energy)	0.675
Operating high voltage	1740 V
Dead time coefficients	
A	0.2904 µs
B	0.0211 µs^2
C	0
Multiplicity dead time	72.6 ns
Gate	64 µs
Predelay	1.50 µs
Doubles gate fraction	0.7930
Triples gate fraction	0.6225

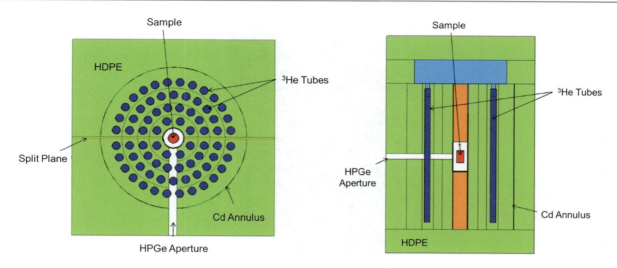

Fig. 19.81 Internal layout of the AVIS from MCNP

AVIS comprises four concentric rings of 74 10-atm ^{3}He tubes (Fig. 19.81) connected to 14 AMPTEK A111 amplifiers. The ^{3}He tubes have an active length of 50.8 cm. The HDPE thickness between the tubes has been increased relative to a standard ENMC and varies slightly between rings to accommodate a whole number of ^{3}He tubes in each ring. A cadmium annulus is positioned approximately 3 cm beyond the outermost ring to absorb room neutrons that have thermalized in the AVIS external shielding to reduce singles background at the facility and make feasible the known-alpha assay methodology. AVIS has a small sample cavity, which is 6 cm in diameter and 12.5 cm high. The sample cavity is lined with cadmium. The planar HPGe detector is positioned with the detector face against the external cadmium ring. A 1 in. hole in the HDPE allows space for the HPGe system. The AVIS will be operated in attended mode using a multiplicity shift register to enable the use of multiplicity analysis. The estimated lower limit of detection at a nuclear facility at sea-level corresponds to 0.09 mg ^{240}Pu$_{eff}$ (0.26 mg plutonium; Ref. [121]).

19.5.4.12 Multiplicity Counter-15

The multiplicity counter-15 (MC-15) is a multiplicity-counting system with 15 ^{3}He tubes [122]. The MC-15 is a joint LANL/Lawrence Livermore National Laboratory/Sandia National Laboratories detector design manufactured by the National Security Campus (formerly known as the Kansas City Plant). The MC-15 shares many similarities (same number of tubes and roughly the same weight) as the previous neutron pod detector system [123]. Some publications refer to the neutron multiplicity array detector as *NoMAD*; the MC-15 and NoMAD are of the same physical design (number and layout of tubes and HDPE). All references for this section are from Refs [20, 122]. unless noted otherwise. Photographs of the MC-15 are shown in Figs. 19.82 and 19.83, and cross-sectional views are shown in Fig. 19.84.

Every ^{3}He tube in the MC-15 has a pressure of 150 psia (10.13 bars) and active dimensions of 2.46 × 38.1 cm (0.97 × 15 in.). The counter's fill gas is a mixture of ^{3}He with 2% CO_2 as a quench gas (in atomic proportion). Table 19.22 lists the specifications of the tubes for the MC-15. The ^{3}He tubes are aligned inside two polyethylene blocks in three rows: a front row of seven tubes, a middle row of six tubes, and a back row of two tubes (shown in Fig. 19.84). The pitch between tubes in a row is 5.08 cm (2.0 in.), and the tubes of the front and middle rows are staggered; the two tubes in the back row line up with the third and fifth tubes in the front row. This design was used to maximize efficiency while minimizing weight and allowing for some spectral information to be obtained (through the use of multiple rows). The system acquires list-mode data (see Sections 19.2.5 and 19.7.2). Two MC-15 detectors can be linked via a cable, allowing for list-mode data to be acquired for all 30 channels.

The overall dimensions of the MC-15 are 38 × 13 × 56 cm (15.0 × 5.1 × 22.0 in.), with a weight of 21 kg (47 lb). The removable canvas bag (shown in black in Fig. 19.82) contains a sheet of cadmium 0.08 cm (0.030 in.) thick on the front side of the detector—facing the source—to minimize the detection of neutrons that have been thermalized by the environment. Each ^{3}He tube is connected to a single amplifier module, which reduces the effects of dead time on the collected data. The model KM-100 amplifier modules were designed at LANL and have been optimized for the ^{3}He tubes in the MC-15 to decrease dead time and increase the maximum counting rate.

Fig. 19.82 Photograph of the MC-15 detector

Fig. 19.83 Measurement setup with two MC-15 detectors

Both measurements and simulations have been performed to characterize the MC-15 detector [20, 124–127]. Figure 19.85 shows the measured efficiency (using a ^{252}Cf source) versus distance. The lines shown are for a fit to calculate the efficiency given a source-to-detector (S/D) and reflector-to-detector (R/D) distance.

MCNP simulations have been used to estimate the response of the MC-15 versus energy. A point source (of varying monoenergetic energies) located 30 cm from the face of the detector was used for all simulations. Figure 19.86 shows the efficiency of each of the 15 tubes versus neutron energy. The circle markers refer to those tubes in Row 1 (the front row of seven tubes), the square markers to those in Row 2 (the middle row of six tubes), and the triangle markers to those in Row 3 (the back row of two tubes). The unfilled circle markers denote the response to ^{252}Cf point source. These same results are shown in Fig. 19.87, but here, the results of each row are summed.

Given the results in Figs. 19.86 and 19.87, it is clear that taking a ratio of counts in a row (Row 1/Row 2, Row 1/Row 3, and Row 2/Row 3) can provide information about the energy of neutrons that reach the detector system. For this reason, these ratios are shown as a function of energy in Fig. 19.88.

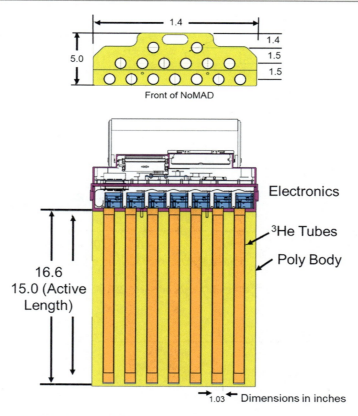

Fig. 19.84 Cross-sectional views of the MC-15 detector [20]

Table 19.22 MC-15 ^{3}He tube specifications [20]

Manufacturer	Reuter-Stokes
Model number	RS-P4-0815-103
Body material	Aluminum 1100
External diameter	1.00 in.
Thickness	1/32 in.
Height (including cladding)	41.6 cm
^{3}He pressure	150 psia
Active length	15.0 in.

The MC-15 has often been used in measurements with the BeRP ball and HEU systems at NCERC. Information on these measurements can be found in Refs. [20, 124, 125, 128, 129]. These measurements include transmission on other materials as well as simulated results. Results from the MC-15 experiments include detailed sensitivity uncertainty analysis (documented in Ref. [20], which is an International Criticality Safety Benchmark Evaluation Project benchmark with the MC-15). These results have been used to aid in validation for computational methods [124] and nuclear data [130]. Examples of data from the MC-15 are given on the discussion of Feynman variance-to-mean technique in Chap. 17, Sect. 17.4.

19.6 Non-Helium-3-Based Counters and Their Applications

19.6.1 Motivations for Alternative Thermal-Neutron Counters

Most of the neutron coincidence counters in current use contain ^{3}He gas tubes because of their high efficiency, reliability, ruggedness, and gamma insensitivity. Tubes that contain BF_3 gas are sometimes used to reduce costs or to operate in higher gamma-ray fields; however, their efficiency is about a factor of 2 less than that of ^{3}He tubes. The main disadvantage of

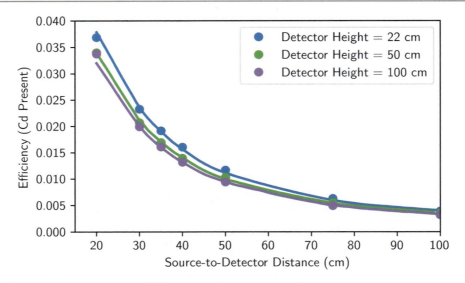

Fig. 19.85 Measured MC-15 efficiency using a bare ^{252}Cf source versus distance [127]

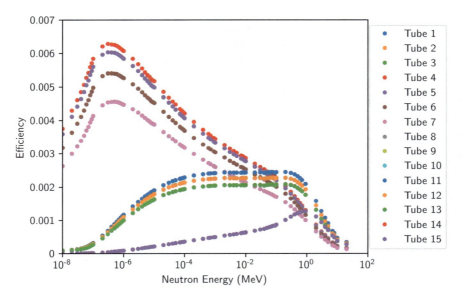

Fig. 19.86 Simulated efficiency as a function of neutron energy is shown for each tube [126]

thermal-neutron-detection (^{3}He, ^{10}B, and ^{6}Li) instruments for coincidence applications is that the neutrons have to slow down to thermal energy via scattering collisions before they are detected in the tubes, and this slowing-down process causes a rather large die-away time (τ) in the detector. As a result, the coincidence gate time (G) in the electronics must be set at a relatively large value (10–100 μs) to detect the time-correlated coincidence neutrons. Ultimately, the large gate length increases the statistical error for high-count-rate applications.

Several historical assay systems based on coincidence counting have used fast-neutron recoil detectors to avoid the die-away-time problem associated with thermal counters. Examples of these detectors are liquid and plastic scintillators and ^{4}He gas recoil counters. The scintillators are sensitive to gamma-ray backgrounds, and the ^{4}He tubes are relatively inefficient. Examples of coincidence systems based on fast plastic scintillators are the random driver, isotopic source assay system, isotopic source assay fissile, and early models of fuel-pin scanners; all have been documented in previous publications [131–133].

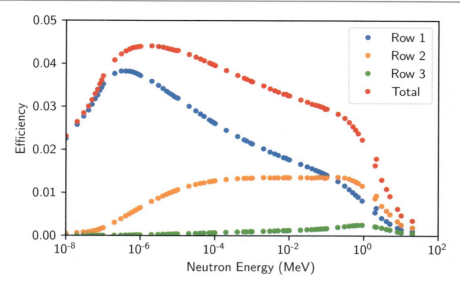

Fig. 19.87 Simulated efficiency as a function of neutron energy is shown for each row, which is summed over the tubes contained by that row [126]

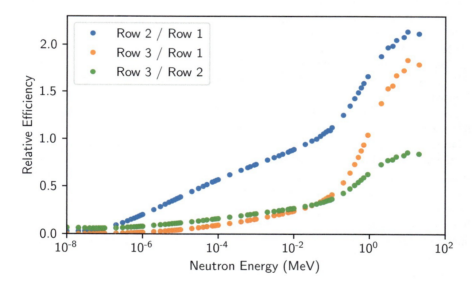

Fig. 19.88 Simulated row ratios show above are the responses of subsets of tubes from each row relative to each other: Row 2 to Row 1 (blue), Row 3 to Row 1 (orange), and Row 3 to Row 2 (green; Ref. [126])

During the early 2000s, the production of ^{3}He gas significantly declined and was anticipated to no longer support the projected demand [134, 135]. The decreased supply and continuing uncertainty of options for future resupply prompted worldwide intense research and development to create and evaluate detection technologies that could provide feasible alternatives to ^{3}He gas [136]. Several alternative detection technologies had been developed and became commercially available. The techniques included scintillation materials with ^{6}Li or ^{10}B compounds [137–140], liquid scintillators for fast-neutron detection [141, 142]), and a variety of designs based on boron-lined proportional counters [143–145]. The latter were among the most extensively considered ^{3}He-alternative technologies with potential for near-term application in the field of nuclear safeguards—largely due to the high thermal-neutron absorption cross section on ^{10}B (3836 barns, 72% of ^{3}He value; reliability and stability associated with the use of well-established proportional detection technology; and compatibility with the existing electronics infrastructure.

Table 19.23 General and technical safeguards ^{3}He replacement requirements used to form technology evaluation criteria

Detector performance parameter requirements	In-plant stability and operational requirements	Cost requirements
Neutron-detection efficiency (ε) Die-away time (τ) Coincidence (FOM) for doubles counting (ε^2/τ) Gamma-ray discrimination Dead time (count-rate capability)	Stability (long term and temperature) Physical size of detector Scalability Sensitivity (radio frequency and microphonics) Safety Reliability and maintenance Component reproducibility Infrastructure Complexity of operation	Affordability Life-cycle costs (cradle to grave)

The technologies proposed for use in nuclear safeguards were presented with a series of specific challenges because neutron measurements for safeguards applications have requirements that are unique to the quantitative assay of SNM. The primary task for neutron measurements in nuclear safeguards and nonproliferation is to determine the mass of the SNM to verify that material has not been diverted. In most cases, the accuracy of the measurements has to be better than 1%–2% to meet IAEA international obligations under agreements such as the Treaty on the Non-Proliferation of Nuclear Weapons. Large-scale nuclear plants, such as MOX fabrication plants, process tons of plutonium per year, and the high accuracy of the measurement systems used to verify the plutonium is critical. The ^{3}He-based neutron NDA systems have been under development, implementation, and continuous improvement for about four decades. The result of this development has resulted in a variety of ^{3}He-based NDA systems that can provide a precision of 0.1% and an accuracy of 0.3% for plutonium inventory sample measurements in actual plant environments. Developing technology to provide a viable ^{3}He alternative capable of achieving similar performance is extremely challenging.

The general and technical ^{3}He replacement requirements for nuclear safeguards applications can be grouped in to three broad categories: detector performance parameter requirements, in-plant stability and operational requirements, and cost requirements. Evaluation of the parameters listed in Table 19.23 is essential for the acceptance of any potential safeguards ^{3}He replacement technology. A detector-level evaluation should address fundamental detector performance parameters and in-plant stability requirements. Operational and cost requirements form an integral part of any system-level evaluation. The latter two requirements are interlinked in the sense that the operational requirements, such as maintenance and the detection system footprint, also need to be factored into the life-cycle costs.

These parameters form the key nuclear safeguards ^{3}He replacement requirements and were developed to provide reference for consistent and reliable evaluation of ^{3}He-alternative technologies [136, 146]. Thermal-neutron-correlated counting requires high neutron-detection efficiency and short die-away time to minimize the contribution of accidental pairs (chance or pileup coincidences) to the doubles rate uncertainty. This effect is reflected in the FOM concept derived in Refs [136, 146]. for thermal-neutron-based detection. Alternative FOMs were proposed to reflect different physics characteristics of various non-helium-3 systems and for fast-neutron detection [147]. The in-plant requirements are essential for reliable, long-term operation and include practical aspects such as scalability yet compact footprint for ease of operator access or low maintenance requirements to minimize long-term burden on the facility and inspectorate.

A series of topical workshops was organized during 2011 through 2014, with participation of international safeguards inspectorates. The workshops resulted in publication of consensus status, observations, needs, and expectations for ^{3}He-alternative technologies for use in nuclear material assay [136, 147, 148]. The last workshop—held at the Joint Research Centre (JRC) in Ispra, Italy—also included practical demonstration of several prototype ^{3}He-alternative coincidence counters [148]. These documents provide a valuable reference and overview of the state-of-the-art technology during this time as well as full-scale instruments that have been developed for safeguards applications. The most feasible examples of such instruments are summarized in the following sections.

19.6.2 Boron-Based Thermal-Neutron Counters

19.6.2.1 High-Dose Neutron Detector

The high-dose neutron detector (HDND) concept was developed to address the need to adapt the nuclear safeguards NDA tools to new challenges associated with developments of advanced nuclear fuel cycle concepts such as pyro-processing or molten-salt reactor applications [149]. Current NDA systems typically employ fission chambers or ^{3}He-based tubes for the measurement of spent fuel material forms. Fission chambers are capable of withstanding the high gamma-ray backgrounds;

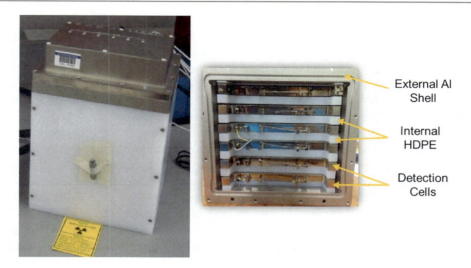

Fig. 19.89 (left) Photograph of the HDND module; (right) internal structure of HDND with six aluminum, sealed, detection cells layered between HDPE slabs

however, they provide very low detection efficiency—on the order of 0.01%. Helium-3-based designs allow for higher detection efficiencies but at the expense of slow signal rise-time characteristics and higher sensitivity to the gamma-ray backgrounds. The HDND was developed based on efforts in the area of ^{3}He-alternative technologies that have yielded a neutron-detection system with performance characteristics similar to systems that use ^{3}He tubes. In addition, the HDND offers features that have the potential to outperform ^{3}He tubes in certain applications, specifically in areas that involve high gamma-ray backgrounds and that require high count-rate capabilities such as spent fuel and pyro-processing materials measurements. The HDND was shown to be capable of tolerating increases in gamma dose rate by almost two orders of magnitude (up to 800 R/hour) with only ~30% reduction in neutron-detection efficiency [150].

The HDND, shown in Fig. 19.89, is based on the same boron-lined proportional counter technology as the high-level neutron counter boron (HLNB; discussed later in this chapter) and uses six narrow (~0.5 cm), boron-lined, parallel-plate, detection cells filled with Ar + CO_2 gas at sub-atmospheric pressure. The detection cells are layered between 1.6 cm thick plates of HDPE. The outer dimensions of HDND are 20 × 20 × 16 cm, with a total height of 25 cm that includes a compact electronics junction box (Fig. 19.89). The pulse rise-time characteristics of HDND correspond to tens of nanoseconds compared with several hundred nanoseconds or even 1–2 microseconds in the case of ^{3}He tubes. The HDND signal-processing electronics include six fast amplifiers, developed by PDT, that exhibit very fast timing characteristics that correspond to pulse pair resolution of 350 ns and a pulse rise time of less than 200 ns. The capabilities of the HDND can be compared with standard PDT-10A characteristics of pulse pair resolution of 700 ns and pulse rise time of 300 ns. Signal from each cell is processed by an individual amplifier, which further reduces system dead time. The individual outputs are recorded in list mode. Additionally, an OR'ed. sum of the six cells is provided, which is compatible with standard shift registers. Access to individual cell signals enables back-to-front ratio measurement in the HDND, which is analogous to the ring-ratio method (Sect. 19.3.4) and is sensitive to average neutron energy associated with different light element impurities present in the measured materials.

For pyro-processing applications, the HDND was developed for measurements of products (U and U/TRU ingots); however, its features—such as broad neutron-detection dynamic range and multi-plate configuration—make it suitable for other areas in the pyro-process, such as monitoring and control of material transfers and moves and confirmatory measurements to establish absence of quantities of plutonium in various output streams. The key HDND capabilities include the following:

- U/TRU ingot assay (using single HDND module or higher-efficiency system with multiple HDND modules to enable correlated counting)
- Confirmatory tool
- Process monitoring

A single HDND module can be used to support all of the listed functionality in singles mode. The main exception is the use of correlated counting for measurement of plutonium content in U/TRU ingots, where multiple HDND units can be combined to achieve sufficient efficiency. Three units have been built and tested to demonstrate this capability [151]. This option, requiring a higher-efficiency system than a single HDND module, would use the doubles-to-singles (D/S) ratio technique, which is sensitive to multiplication and can be used to assay Pu/U content [152]. MCNP reference calibration, confirmed with a representative U/TRU product, would be needed to relate the measured D/S to the plutonium content. The use of HDND for U/TRU ingot measurements based on singles counting would use the curium-ratio method (Chap. 21, Sect. 21.5). A passive singles measurement could be used to assay ^{244}Cm and extract the plutonium content if prior curium/plutonium ratio information is available and if curium tracks plutonium in the process. This option allows for greater flexibility in that the individual HDND module can be placed inside the hot cell and represents a fairly simple technique. It relies on the assumption of constant sample dimensions and negligible low-Z impurities, both of which can be well justified for U/TRU ingots.

The use of HDND as a confirmatory tool would use HDND neutron-counting capability in high-gamma-dose backgrounds and would focus on monitoring the streams with low neutron output to detect potential material diversion, which includes the uranium product or various waste streams such as fission products or metal wastes. In the uranium metal ingots, the enrichment will be low (~1%), and the neutron self-interrogation will be negligible (assuming that the ^{244}Cm was removed in the processing). Thus, the neutron yield will be primarily from the spontaneous fission of the ^{238}U, and the (α,n) neutron yield will be negligible for the metal. Thus, the neutron emission from the ingot will be proportional to the ^{238}U mass; however, the ^{238}U neutron yield is very low (~13.6 n/s kg), and HDND could be used to verify the absence of plutonium and ^{244}Cm in the uranium ingots to detect diversion or off-normal operation. The sensitivity of this approach was evaluated in MCNP and demonstrated using HDND capability to detect a 0.005% weight contamination of plutonium in the uranium ingot with greater than 95% detection probability in a counting time as low as 10 min, assuming optimum HDND operating settings [153].

The use of HDND for process monitoring offers another important capability to monitor nuclear material movement in a facility. HDND monitoring capability was evaluated using ^{252}Cf and ^{137}Cs sources mounted to a moving source platform [151]. The multi-plate internal structure of HDND allows direction-of-motion monitoring by comparing signals from front and back cells, with sideways HDND orientation toward the moving nuclear material. In practical applications, HDND can be positioned to maintain its sideways orientation with respect to key nuclear material transfer paths.

19.6.2.2 MiniHDND

To simplify remote handling, reduce maintenance requirements, and improve capability to measure in difficult-to-access locations, a miniature version of HDND—miniHDND—was developed. This detector introduces a smaller and more versatile neutron-detection instrument that can be placed near process equipment and in confined locations [151]. This development was focused on optimization and update of the HDND hardware for confined areas; it benefits from the key HDND capabilities to expand the HDND process-monitoring capability. The miniHDND was designed with an external enclosure made completely of metal and small footprint to simplify handling. The overall miniHDND dimensions correspond to 6.3 × 6.3 × 29 cm, including an electronics junction box (Fig. 19.90). The miniHDND junction box includes three PDT fast

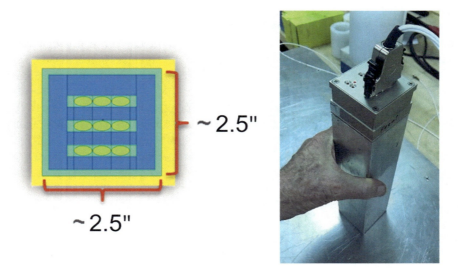

Fig. 19.90 (left) Internal miniHDND layout from the MCNP code; (right) photograph of the miniHDND detector

amplifiers, and it is fully removable to facilitate replacement in case of malfunction or failure. MiniHDND includes three individually sealed detection cells to maintain the direction-of-motion capability as well as capability to extract front-to-back ratio information. The middle cell of the miniHDND is not coated with boron and therefore serves as a gamma monitor.

Due to its compact dimensions and reduced efficiency, the miniHDND is foreseen as a process-monitoring instrument capable of performing simultaneous neutron and gamma measurements in areas such as pyro-processing or molten salt reactor applications. The neutron measurements are performed in singles mode using the front and back boron-lined cells. The gamma measurements are obtained from the central cell without boron coating. The gamma-monitoring capability uses the distinctly different high-voltage curve characteristics of neutron- and gamma-induced signals, where gamma pileup results in a steep increase in count rate with increasing high voltage while the neutron high-voltage characteristics in the plateau region remain relatively unchanged (Figs. 15.8 and 15.10). The miniHDND central cell amplifier can be tuned for optimum gamma-detection performance to extract gamma-induced count rate that can be related to gamma dose rate using a prior calibration.

19.6.2.3 High-Level Neutron Counter Boron

The HLNB was designed using boron-lined technology manufactured by PDT. The underlying boron-lined technology represents a well-established proportional technology that offers a similar level of maturity as ^{3}He-based proportional counters, making it a suitable candidate for development of a robust detection system [143]. The HLNB was developed with design parameters similar to the ^{3}He-based high-level neutron coincidence counter (HLNCC-II; Sect. 19.4.2) and offers similar neutron-detection efficiency characteristics (Table 19.24 and Ref. [154]). The HLNB was developed to mimic the external dimensions of HLNCC-II, with a sample cavity of identical size. The key requirement was achieving performance (count rate capability and plutonium assay uncertainty) similar to HLNCC-II. A side-by-side photograph of HLNB and HLNCC-II is shown in Fig. 19.91 (left).

The HLNB counter comprises six individual boron-lined detector modules. Each detector module contains six individually sealed boron-lined detection cells filled with Ar + CO_2 gas. The individual sealed cells are filled with C_{10} gas (Ar + CO_2 mixture) at sub-atmospheric pressure. The individual detection cells are interleaved with HDPE plates for optimum neutron moderation. The entire system of cells and HDPE is hermetically sealed within an aluminum enclosure. The detectors tightly surround the sample cavity, which measures 17.00 cm (6.69 in.) in diameter and 41.00 cm (16.14 in.) in height. The top and

Table 19.24 Summary of neutron operating parameters for HLNB and HLNCC-II

Operating parameter	HLNB value	HLNCC-II value
Efficiency (^{240}Pu energy)	0.183	0.175
Die-away time	89.5 μs	43.0 μs
Operating high voltage	860 V	1680 V
Dead time coefficients		
A	0.665 μs	0.768 μs
B	0.111 μs^2	0.248 μs^2
Gate	180 μs	64 μs
Predelay	3.0 μs	4.5 μs

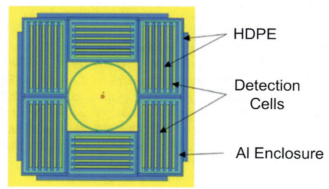

Fig. 19.91 (left) HLNB next to HLNCC-II. The HLNB light-emitting diode display reflects the internal detector layout; (right) MCNP visualization of the HLNB internal structure

Table 19.25 Results of verification measurements in HLNB and HLNCC-II

Analysis method	Declared Pu mass (g)	HLNB				HLNCC-II			
		Assay Pu mass (g)	σ	Declared-assay mass (%)	σ	Assay Pu mass (g)	σ	Declared-assay mass (%)	σ
Known-alpha	2.5	2.51	0.03	−0.48	1.33	2.54	0.01	−1.59	0.43
	15.0	15.06	0.04	−0.58	0.29	14.99	0.02	−0.08	0.15
	50.0	49.79	0.07	0.41	0.13	49.98	0.05	0.03	0.11
Passive calibration	2.5	2.26	0.03	9.26	1.18	2.42	0.03	3.01	1.18
	15.0	15.24	0.11	−1.76	0.72	14.93	0.08	0.27	0.50
	50.0	49.85	0.25	0.29	0.49	50.44	0.21	−0.90	0.41

bottom end plugs consist of an aluminum core surrounded by HDPE. In addition, HDPE slabs that measure 1.27 cm (0.5 in.) thick are added along the external sides of the detectors to boost the detection efficiency through neutron reflection. A view of the HLNB internal layout is shown in Fig. 19.91 (right).

The HLNB signal-processing electronics are housed in a very compact junction box that is only 3.5 cm (1.5 in.) tall. Each of the six detector modules is equipped with 3 amplifiers (each servicing two adjacent detection cells), making a total of 18 amplifiers in the entire system. The HLNB can be used with a standard shift register via a single signal output with all of the amplifiers OR'ed. together. In addition, the HLNB is compatible with list-mode data acquisition and allows a direct readout of all of the 18 amplifiers via an additional multi-pin connector.

To fully evaluate the feasibility of HLNB for verification measurements, a field trial of HLNB was performed at the Plutonium Conversion Development Facility (PCDF) of Japan Atomic Energy Agency [155] using a side-by-side performance comparison with HLNCC-II. The field trial represented a key milestone in the evaluation of HLNB performance and provided an opportunity to evaluate the overall performance of this ^{3}He-alternative system in conditions beyond well-controlled laboratory environment with materials that were representative of realistic deployment conditions. Of key interest was the evaluation of transportability, ease of setup, and functionality in an operational facility with realistic backgrounds, noise interference, and temperature and humidity conditions. The field trial measurements focused on a range of items of operational interest, including medium- to high-mass MOX items. Measured MOX materials included small MOX items with mass corresponding to 2.5–10 g plutonium, medium-mass MOX items of 50 g plutonium and 100 g plutonium, and large-mass MOX items of 200 g plutonium and 1.2 kg plutonium. The large-mass items were used in calibration measurements and demonstrated correlated counting uncertainties comparable to HLNCC-II. Calibration measurements were performed using passive calibration curve as well as known-alpha calibration methods in INCC. The verification results from the field trial are summarized in Table 19.25 and demonstrate good performance of HLNB over the low-to-medium-mass MOX range that is comparable to HLNCC-II. The uncertainties of HLNB results are comparable to HLNCC-II despite the longer die-away time of HLNB. The impact of longer die-away time is typically emphasized with increasing count rate due to increased contribution of accidental coincidences in the longer HLNB gate and therefore is expected to affect predominantly larger-mass items. The results of verification measurements demonstrated that HLNB is capable of satisfying the international target value expected for an HLNCC-II-type counter of 2.1% in 300 s measurement time [155]. The HLNB proved to be a robust technology, with no identified post-transportation issues and operator use similar to HLNCC-II. The reliability and long-term stability of HLNB was demonstrated throughout the 9-month period that the instrument spent at PCDF, performing consistently without the need for any renormalization or modification of calibration parameters.

19.6.2.4 Boron-Coated Straw Counter

A full-scale, fully operational prototype to replace the HLNCC-II was fabricated and tested at the ^{3}He Alternatives for International Safeguards Workshop that took place at the JRC in Ispra, Italy, in October 2014. The prototype, pictured in Fig. 19.92, has overall and cavity dimensions identical to those of the HLNCC-II and weighs 61 kg—only 6 kg heavier than its ^{3}He-based counterpart. It is populated with 804 boron-coated straw (BCS) detectors [156], each 4.4 mm in diameter, and is lined with 2 μm of vapor-deposited $^{10}B_4C$. The BCS detectors are uniformly distributed in the moderator. They are connected together in six groups of 134 detectors. Each readout contains a custom-designed amplifier, fully compatible with standard shift register electronics.

Measurements of performance parameters collected at the JRC, Ispra ^{3}He-alternatives workshop [148] included detection efficiency (ε) and neutron die-away time (τ) and are summarized in Table 19.26. Results demonstrate that the BCS-based counter achieves better performance than the standard ^{3}He-based HLNCC-II, which was also tested at the workshop. The FOM, defined as $ε/\sqrt{τ}$, equals 2.66%/√μs—an improvement over the ^{3}He-based counter—due to the significantly lower

Fig. 19.92 Prototype straw-detector-based neutron coincidence counter

Table 19.26 Performance parameters of the BCS-based coincidence counter

Operating parameter	BCS value	HLNCC-II value
Efficiency (^{252}Cf)	0.136	0.165
Die-away time	26 µs	43 µs
Gate	64 µs	64 µs
Predelay	7.5 µs	4.5 µs

die-away time. The latter is attributed to the more uniform dispersion of neutron absorber throughout the moderator. In this initial benchmark comparison, a full-scale neutron coincidence counter—based on low-cost, boron-coated straw detectors—showed performance similar to the standard ^{3}He-based system in a compact geometry, fully compatible with existing electronics and procedures.

19.7 Additional Concepts

19.7.1 Neutron Imagers

19.7.1.1 Neutron Imaging

The term *neutron imaging* may refer to any one of several diverse imaging techniques, including *transmission* imaging, *emission* imaging, and *stimulated emission* imaging. However, in the present context of passive measurements, it will be used to refer to *emission* imaging. Fast-neutron emission imaging forms an image using the spontaneous neutron emissions from a nuclear material item. A neutron emission "image" may refer to either a single view or a tomographic image reconstructed from multiple views of an object. Although the images produced using emission imaging are typically much less detailed than those produced using transmission imaging, they provide information specifically on the spatial distribution of the neutron-emitting material. Passive imaging measurements are practical only for strong neutron sources, such as curium, plutonium, or uranium that is enriched and in a matrix suitable to produce sufficient (α,n) reactions. Neutron-imaging techniques have the potential to augment conventional neutron counting in scenarios such as quantifying plutonium-bearing materials that are too dense for gamma-ray counting or measuring material within heavy equipment or packaging.

The use of imaging may be desirable for applications where

- determining the location of the neutron-emitting material is essential and cannot be accomplished with sufficient precision using conventional means, perhaps due to the size of detectors or limitations on how they can be positioned with respect to the material;
- the configuration of the neutron-emitting material is itself an important attribute that must be determined or confirmed; or
- it is necessary to quantify a source amid potentially large and varying background from confounding sources, such as material at other points in the process, nearby storage, or other constituents of an assembly.

Fig. 19.93 (left) The MINER detector enclosure; (right) the detector array [160]

Example applications where neutron imaging may be warranted include in-situ characterization of holdup after process upsets or to guide facility decommissioning and decontamination, assessment of unlabeled or mislabeled containerized materials whose form and configuration are not known, or characterization of individual fuel rods in an irradiated fuel assembly.

First-generation imaging instruments reported herein consist primarily of developmental devices or laboratory prototypes that are costly to build and complicated to run, so their use is indicated only for niche applications where the additional complexity of imaging is warranted. Next-generation instruments that are presently under development are substantially reduced in size, making field use of imagers more practical. Note that these examples illustrate core concepts but do not provide a comprehensive discussion of all types of neutron imagers. For instance, this treatment neglects time-encoded imaging, where modulation patterns that correspond to different incident neutron directions are encoded in time [157–159].

19.7.1.2 MINER

The most popular form of fast-neutron imaging uses the kinematics of two sequential neutron scatters from hydrogen to identify the incident direction of each neutron to a cone of response. Imagers that use kinematic reconstruction in this way are referred to as "scatter cameras." The MINER [160] is an exemplar compact and mobile neutron scatter camera developed by Sandia National Laboratories and shown in Fig. 19.93. It followed a laboratory prototype [161] with the intent to achieve similar capability in a package that can be readily deployed. Other examples of deployable instruments exist; for instance, an imaging neutron/gamma spectrometer [162] that operates in a similar manner. A new generation of scatter cameras is presently under development that will substantially decrease the size for a given efficiency [163–165]. This miniaturization is possible largely by replacing photomultiplier tubes (PMTs) with SiPM readout and rack-mount electronics using integrated circuits.

MINER offers modest sensitivity and spatial resolution (appropriate for localization, not true imaging) and excellent gamma-background rejection over a 4π field of view with a moderate channel count, moderate size, and no moving parts.

19.7.1.3 Neutron Coded-Aperture Imager

Coded-aperture imaging is applicable when angular resolution better than $10°$ is desired and sources of interest are known to fall within a limited field of view. For example, this resolution may be required when detailed information regarding the distribution of material in a container is needed rather than the location of its center of mass. In 2013, Oak Ridge National

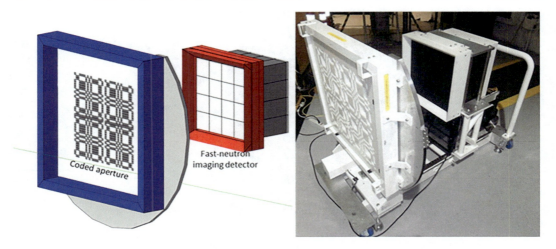

Fig. 19.94 (left) A schematic and (right) a photograph of the P40 NCAI. (Courtesy Oak Ridge National Laboratory)

Table 19.27 Summary of the NCAI

Imager	P40
Mask pattern	Rank 19 MURA
Basis directions	19 × 19
Detector area	44.7 × 44.7 cm^2
Detector pixels	1600 (40 × 40)
Mask-to-detector distance	23 cm to 104 cm
Neutron energies	Fast, thermal
Scintillator(s)	EJ-299-34, EJ-426

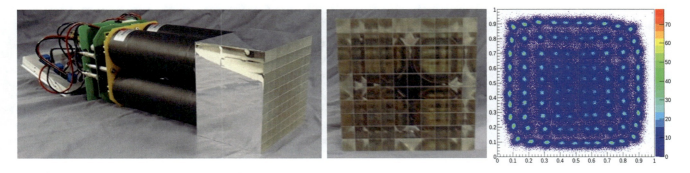

Fig. 19.95 Segmented fast-neutron imaging detector (left and middle) photographs and (right) measured position response that identifies individual pixels

Laboratory developed a prototype neutron coded-aperture imager (NCAI; Fig. 19.94) called *P40* (Ref. [166]; Table 19.27) and a smaller version [167]. Each imager consists of an aperture and a position-sensitive, fast-neutron detector.

The footprint of the P40 imager is 182 cm long × 82 cm wide, and it is 147 cm tall. The mask assembly is designed to perform automated rotation by 90°. A range of masks and mask-to-detector distances can be used. The prototypical capabilities required a large size, but a 2021 compact handheld imager achieved a 6.8-degree angular resolution [168].

The P40 detector is based on developments that occurred during about 2007 to 2013 [169–172]. The active volume of a single detector module, typically referred to as the *pixel block*, consists of a segmented volume of an EJ-299-34 scintillator optically isolated by a 3 M Vikuiti reflector. The scintillator pixels are viewed through a 28 mm thick segmented acrylic light guide by four photomultiplier tubes, shown in Fig. 19.95 (left). In 2014, a ZnS/LiF screen was added to detector modules to enable simultaneous measurement of thermal neutrons. For each interaction in a detector module, signals are captured by waveform digitizers and used to infer the pixel of interaction, particle type, and time of event. The pixel of interaction is

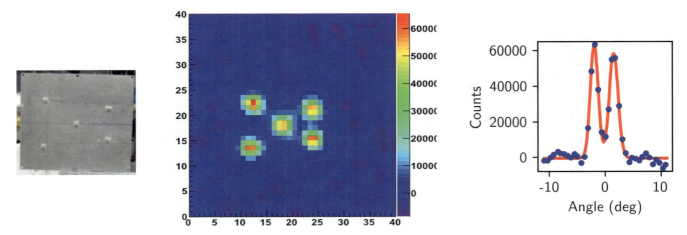

Fig. 19.96 Example high-resolution image of five ^{252}Cf sources: (left) A photograph of the sources, (center) the fast-neutron image, and (right) a vertical slice shown through the image at the point of the closest two sources, spaced 10 cm apart

determined by calculating the position of interaction, then using a lookup table created from calibration data. Example calibration data for the position response of a detector module are shown in Fig. 19.95 (right). Here, the x coordinate is given by the fraction of scintillation light recorded by the PMTs on the right side of the detector, and the y coordinate is given by the fraction of scintillation light recorded by the upper PMTs; the bright locations correspond to individual pixels. The neutron detector panel of the P40 NCAI consists of 16 detector modules that are packed closely together.

Before use, the NCAI imagers require calibration of the pixel-by-pixel position response, energy response, and pulse-shape discrimination response. Due to scattering within the detector, approximately half of neutrons incident on the detector are recorded in the wrong location but with a known spreading kernel.

NCAIs effectively work like neutron pinhole cameras, where the coded aperture is an array of pinholes [173]. Neutrons incident from different parts of the field of view (FOV) are converted to different spatially modulated patterns on a position-sensitive detector. Measurements are performed for mask and antimask configurations. Pixel hit maps are accumulated for fast neutrons, thermal neutrons, and gamma rays. An efficiency correction is applied for each individual pixel. The difference between the mask and antimask hit maps is calculated to eliminate modulation that did not originate from the aperture. Last of all, images are reconstructed from the hit maps by deconvolution. The number of neutrons that originate from a given position in the FOV is determined by adding the neutron counts from all detector pixels that have an unobstructed view of that position and subtracting the neutron counts from all detector pixels that are shielded from that position by the aperture.

An example fast-neutron image of five identical ^{252}Cf neutron sources recorded by the P40 imager is shown in Fig. 19.96. The closest two sources in the image are separated by 10 cm and are clearly resolved. This image shows the potential utility of high-resolution emission imaging for characterizing the configuration of neutron-emitting material. A vertical slice through the image, converted to degrees, shows approximately 2° full width at half maximum angular resolution for this imager configuration.

A second example image in Fig. 19.97 shows the fast-neutron leakage through the target shielding at the Spallation Neutron Source (SNS) at Oak Ridge National Laboratory. This neutron image was recorded by the smaller P24 imager from the mezzanine at the SNS and illustrates the value of localizing neutron backgrounds in a facility.

19.7.1.4 Parallel-Slit Ring Collimator

The parallel-slit ring collimator (PSRC) concept can construct an image with sufficient resolution to identify individual fuel rods within irradiated nuclear fuel assemblies and keep the imager sufficiently compact, efficient, and radiation resistant to be practical. This quantitative NDA capability is desirable to enable new inspection capabilities.

In 2017, work was initiated on the PSRC at Oak Ridge National Laboratory. The PSRC combines measurement of fast-neutron emissions through a movable slit collimator with tomographic reconstruction techniques to produce two-dimensional images of the distribution of neutron activity on a cross-sectional slice through the measurement cavity. The PSRC demonstrated axial fast-neutron emission tomography sufficient to quantify individual fuel rods in spent nuclear fuel assemblies. For instance, rod-by-rod measurement of burnup would enable operators to better plan for storage or disposal than by using burnup codes alone; it could also be used in a similar way to gamma-ray tomography systems to allow

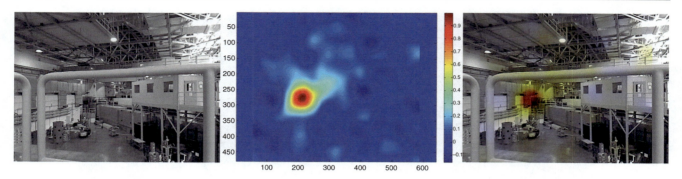

Fig. 19.97 Example (left) photograph, (center) fast-neutron image, and (right) overlay of fast neutrons emanating from the SNS target shielding (images recorded by NCAI P24 imager with camera mounted to it)

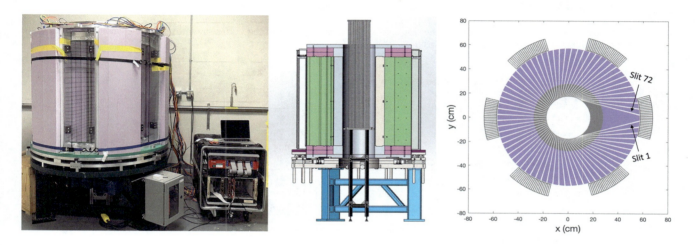

Fig. 19.98 The PSRC imager: (left) A photograph of the prototype, (center) a schematic diagram showing a vertical slice through the center of the inspection cavity, and (right) a simplified horizontal slice illustrating the slit pattern and neutron detectors

safeguards inspectors to verify that a fuel assembly is intact, with no missing or replaced rods [174, 175]. The physics design of the imager is described in Refs [176–180]., the iterative image reconstruction algorithm in Ref. [181], and initial measurements in Ref. [182].

The PSRC consists of an inner inspection cavity, a movable annular collimator with 72 slits, a ring of boron straw neutron detectors, and an outer borated polyethylene neutron shield. The inspection cavity has a diameter of 35.0 cm and a usable height of 100 cm. Although the central cavity passes entirely through the imager, the usable height is defined by the detector active length and the vertical efficiency profile. The large height is needed to integrate the signal from and also inspect a sizable length of fuel rod.

The collimator consists of an inner 10 cm thick steel ring that provides gamma-ray shielding and an outer 29.2 cm thick borated polyethylene ring that effectively modulates neutrons. Each slit in the collimator is 0.3 cm wide at the inner diameter and 0.8 cm wide at the outer diameter and defines a "line of response" along a unique chord through the inspection cavity. Neutron counts correspond primarily to an integral of neutron activity along the chord. A photograph of the imager is shown in Fig. 19.98 (left), along with schematic diagrams that show vertical (center) and horizontal (right) slices through the center of the imager.

Each detector module covers an arc of 30° and has 24 rows of 8 straws; each row of straws is oriented radially and covers 1.25° of the detector ring. The six detector modules are rotated to achieve full coverage. Photographs of a neutron detector module are shown in Fig. 19.99 during assembly (top left), straw rows (bottom left) that are each connected to a preamplifier board, and fully assembled (right). Approximately 0.2% of neutrons emitted from the center of the cavity are detected in the detectors; 13% of the detected neutrons are in the primary line of response, and the remainder penetrate the collimator or scatter down other slits. A summary of PSRC parameters is given in Table 19.28.

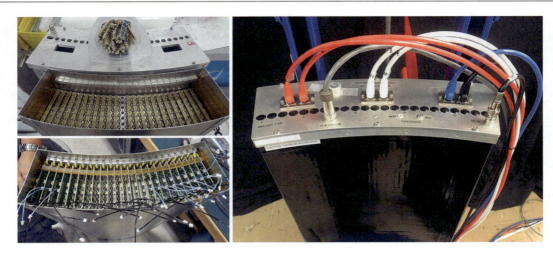

Fig. 19.99 Photographs of a neutron detector module (top left) with the readout end of the boron straw tubes visible, (bottom left) with the preamplifier boards in place, and (right) fully assembled

Table 19.28 Parameters of the PSRC imager

Parameter	Value
Cavity diameter	35.0 cm
Cavity height	100.0 cm
Steel (i.d., o.d.)	35.7 cm, 55.7 cm
Borated polyethylene (i.d., o.d.)	55.7 cm, 114.2 cm
Slit width at collimator (i.d., o.d.)	0.30 cm, 0.80 cm
Detector ring (i.d., o.d.)	115.2 cm, 135.2 cm
System o.d.	145.2 cm
Number of straws	1152
Number of preamplifiers	144
Preamplifiers per detector module	24

i.d. inside diameter, *o.d.* outside diameter

Initial laboratory measurements have been performed using ^{252}Cf neutron sources and a light-water reactor 17 × 17 rod assembly fixture [176]. Where (0,0) is the center slot, measurements were performed for slots (0,0), (0,4), (2,0), (1,3), and (2,1), where pairs of sources were positioned at distances of 1, $\sqrt{2}$, 2, 5, and 4 times the closest possible spacing in the assembly, shown in Fig. 19.100. Data that were reconstructed using the maximum likelihood expectation maximization algorithm for 1000 iterations produced the image of neutron emission rate per pixel; the reconstructed image corresponds to the area within the red box. Careful characterization of the imager is required to extract quantitative values for the neutron emission rates, but resolution sufficient to resolve individual fuel rods is manifest.

19.7.2 List-Mode Data Analysis

List-mode data consist of a list of times that correspond to the arrival time of every pulse, which can be recorded for each amplifier module, giving positional information or, if the amplifier module signals are summed, then just for the detector as a whole. From list-mode data, the traditional gate structure can be applied to determine Singles, Doubles, and Triples that can be used to perform all of the capabilities possible with a shift register. Shift registers are simpler and more common and produce the Singles, Doubles, and Triples as the data "flow" through without storing the individual pulses. List mode requires significant computer storage space for long or high-count-rate measurements, and the processing and analysis of the data are more complicated but more flexible. List mode allows for far more complex data analysis, which is only now beginning to be explored and used.

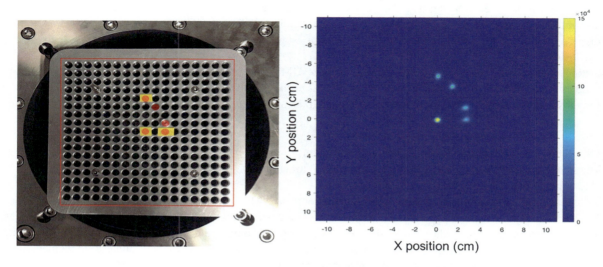

Fig. 19.100 A configuration of (left) neutron source positions and (right) its reconstructed image. The area of the image reconstruction is outlined in red

List-mode data allow the same data to be reprocessed multiple times. The data are stored, so the data could even be reprocessed with techniques developed decades after the measurement was taken. List-mode approaches are continuously being developed. The following are some examples of list-mode data analysis:

- The Singles, Doubles, and Triples could be produced with multiple gate lengths to optimize the best gate length for a specific set of calibration and assay measurements.
- Rossi-α distributions can be produced to calculate the die-away time and gate fraction for a specific measurement, which can be integrated into multiplicity counting.
- The Rossi-α distribution also allows visualization of the reals+accidentals and accidentals gates.
- If list mode is applied at the amplifier module level and position data are available, the source position could be estimated, and the corresponding detector efficiency could be used.
- Machine-learning approaches can be applied because of the much greater amount and resolution of data.
- Data fusion techniques can be applied, integrating the neutron data with, for example, gamma data taken in the same second using the neutron timing information.
- Dead time corrections can be applied with higher resolution and integrated into the pulse train.

19.7.3 Distributed Source-Term Analysis

A new neutron holdup assay method for enrichment facilities—or any facility with a large, distributed volume of material—was recently described [183]. The distributed source-term analysis technique uses Monte Carlo modeling of a centrifuge enrichment cascade hall to derive a calibration curve that relates the average neutron count rate to the mass of uranium holdup. Then, a portable counter similar to the portable handheld neutron counter (PHNC; Sect. 19.3.1) is used to survey the average neutron count rate in the hall. This approach avoids the high attenuation problems of gamma-ray measurement, the difficulties in measuring individual pieces of equipment to obtain the total holdup, and the long measurements required to assay the entire process line.

19.7.4 Unattended Monitoring

Unattended monitoring systems (UMSs) are measurement instrumentation systems that are permanently installed in nuclear facilities to perform quantitative or qualitative NDA measurements [183–185]. These systems operate autonomously in safeguarded nuclear facilities to make, on a 24/7 basis, continuous measurements on relevant nuclear material (NM). UMSs maintain continuity of knowledge (CoK), perform containment and surveillance (C/S) and monitoring of NM, and/or provide qualitative or quantitative measurements—all of which support IAEA inspectors in drawing their safeguards (SG) conclusions.

Operating continuously and without human intervention, UMSs are installed in strategic locations, key measurement points, and/or material balance area boundaries to provide (as needed) CoK, C/S, qualitative measurements, and quantitative measurements. These methods can apply to nuclear material accountancy or near real-time verification of nuclear material flow.

- **C/S applications**—to monitor NM presence and movement
    - Maintain CoK on NM (i.e., spent fuel (SF) transfer to storage)
    - Containment of facility misuse
- **Nuclear material accountancy**—quantitative measurements on NM
    - NDA measurements based on gross gamma/neutron counting or gamma spectrometry (including presence of NM, NM detection through portals, verification of SF assembly burnup, uranium enrichment, and isotopic content), gross neutron or coincidence neutron techniques (i.e., mass of uranium/plutonium material, etc.). Non-nuclear techniques can also be used to determine the thermal power of a research reactor, to measure the mass of NM present in solution (bulk measurement) using pressure sensors correlated to density, and to measure weight, etc.
    - Early detection of NM diversion
    - Early detection of facility misuse

To achieve the above objectives, UMSs are connected to multiple different types of radiation detectors and sensors, which include ^{235}U and ^{238}U fission chambers and ^{3}He tubes for neutron detection; Si-pin diodes, ion chambers (ICs), and doped-optical-fibers coupled to PMTs for gross gamma ray detection; NaI(Tl), CZT, and electrically cooled germanium detectors for spectrometric gamma ray detection. New types of detectors are being considered to detect simultaneously neutron and gamma radiation using pulse-shape discrimination techniques. Non-radiation sensors measure various quantities such as flow (mass of liquid flowing in pipes to determine reactor thermal power, etc.), pressure (density of NM in solution), electrical current, weight, and temperature.

Upon agreement with national authorities, UMSs can be remotely connected to servers at the IAEA headquarters (HQ) to securely accomplish remote data transmission (RDT) of safeguards and/or state-of-health (SoH) data on a regular basis. RDT allows specific inspection regimes or maintenance procedures to be triggered in a timely fashion and early detection of instrumentation issues. The remote connection also can be used to troubleshoot, maintain, and reboot components or to upgrade device firmware and software.

UMS presence in facilities brings many unique advantages for nuclear safeguards because UMSs

- increase the efficiency and effectiveness of safeguards activities;
- are applied to declared NM and facilities and thus contribute to the early detection of misuse of facilities;
- allow for facility processes to be monitored during all stages of their operational cycle up to and through decommissioning;
- reduce the level of intrusiveness to the operation of nuclear facilities;
- place less burden on facility operators to make material available for attended measurements;
- reduce inspectors' radiation exposure;
- maintain CoK on NM, including in difficult-to-access locations;
- allow measurements to be taken 24/7 because they are transparent to facility operations;
- reduce/optimize inspection efforts because UMSs can measure up to 100% of the material compared with sampling plans for attended measurements;
- if remotely connected
    - facilitate data transfers to and availability at HQ for immediate processing and analysis, which allows for near real-time analysis;
    - integration of multiple UMS data sources and online analysis can be performed;
- optimize the cost effectiveness of safeguards activities; and
- play an essential role in case of travel restrictions (e.g., pandemic situation).

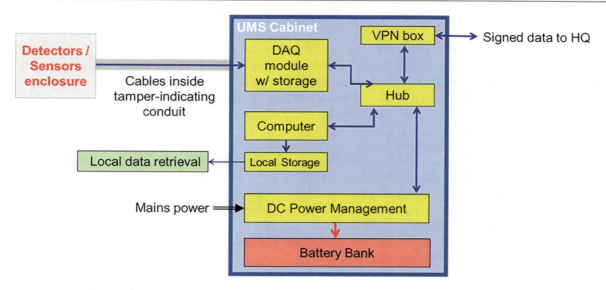

Fig. 19.101 Standard block diagram of a UMS cabinet showing the main modules and data flow (blue arrows)

UMSs are present in almost all stages of the nuclear fuel cycle (or the physical model), from conversion, uranium enrichment, fuel fabrication, nuclear reactors, and research reactors to reprocessing and fuel and waste management.

To address the different challenges confronting IAEA safeguards, a clear taxonomy of systems was defined to increase the visibility and clarify the applicability of the various systems made available to safeguards. To date, seven categories have been introduced—each based on the type of data acquisition (DAQ) module used, which is indirectly related to the type of application. For instance, all NDA systems that involve measurement of uranium/plutonium mass through neutron coincidence methods use a DAQ called the Unattended Data Logger (UDL1) platform. A system of this type is categorized as a UDL-based system (UDBS). The new flagship UDL1 DAQ went into production use in 2021.

Figure 19.101 shows a block diagram of a standard internally DC-powered UMS cabinet. All components are powered through a smart unattended data continuous power (UDCP) management module that charges a bank of backup batteries and includes a watchdog circuit and state-of-health monitoring capabilities. In case of a main power failure, batteries provide power to the UMS for up to several weeks, thus the amount of battery backup required by an application is one of the considerations for cabinet design. The main electronics modules inside the cabinet are DAQs to acquire and store data; a computer for local data storage (from which inspectors can retrieve redundant data in case of emergency or if no remote connection is available), for signing data for remote transmission, and for local access to other components; and a VPN box to establish a secure connection to headquarters for data transmission.

The cabinet has tamper-indicating features and is sealed to prevent unauthorized access. Detectors and sensors installed in sealed enclosures may be located several hundred meters away from the cabinet. To preserve overall system security and data integrity, their cabling connections to the cabinet are run through tamper-indicating protective conduits and junction boxes. A new, active, tamper-detection technique based on time-domain reflectometry is being investigated to supplement existing measures and provide even more robust security.

UMSs are installed in nuclear facilities worldwide—indoors and outdoors—in a wide variety of harsh environmental conditions including low and high temperatures (−20 to +60 °C), high humidity, seaside salt air conditions, high radiation, weather extremes, and sometimes in seismic zones or on moving components.

The highest possible reliability and availability are expected from UMSs to successfully provide uninterrupted, reliable data for drawing safeguards conclusions. Safe operation also must be guaranteed for use in nuclear facilities; therefore,

- components are industrial grade and selected for low power consumption;
- components and systems are certified against international standards (chiefly the International Electrotechnical Commission) for electrical safety and environmental conditions;
- built-in redundancy of critical components (detectors, DAQ, data storage) has been introduced (not indicated in Fig. 19.101) to ensure that no single point of failure can induce a loss of SG data;

- UMS cabinets are equipped with a watchdog circuit that is configured to automatically reboot parts of the system that become unstable;
- remote access allows data transmission, troubleshooting, maintenance, rebooting, and updates;
- in the event of a loss of mains power, limiting DC-power distribution to noncritical components preserves battery life and contributes to system reliability due to increased operational uptime;
- data are securely transferred through a VPN tunnel to ensure their authenticity;
- maintenance cycles have been optimized and correlated with performance monitoring of most individual components.

The enhanced reliability and performance of UMSs during the last decade—through design improvements (e.g., DC power inside the cabinet, standardized subassemblies and build plans, new technologies for components), improved measurement methodologies, upgraded RDT infrastructure, and availability of UMS data during special scenarios in which inspections could not take place—have all led to an increasing use of UMSs (several hundred worldwide at the time of this writing) for both monitoring and NDA applications. By contributing to the early detection of both diversion of NM or misuse of facilities, UMSs improve the efficiency and effectiveness of safeguards activities.

Sophisticated new techniques, technologies, and approaches are under investigation to enhance safeguards, increase technical performance, and increase system reliability.

Examples of UMS systems in Japan are given in Sect. 19.7.4.1. Other examples include the Advanced Thermohydraulic Power Monitor (ATPM; Chap. 24, Sect. 24.8) to measure the thermal power of a research reactor; the Mobile Unit Neutron Detector (MUND; Chap. 19, Sect. 19.3.4.5) to maintain CoK on material transfers; the Nuclear Material Transfer Monitor (NMTM; Chap. 19, Sect. 19.3.4.4), a quantitative assay system to measure the uranium mass in waste packages; the Silo Entry Gate Monitor (SEGM; Chap. 21, Sect. 21.3.5) to verify the direction of motion and number of baskets loaded into a dry storage silo; the Unattended Fork Detector Monitor (UFDM; Chap. 21, Sect. 21.4.2) to verify the burnup and cooling time of spent fuel assemblies; and the VXI Irradiated Fuel Monitor (VIFM, Chap. 21, Sect. 21.5.1) to monitor irradiated CANDU bundle movements within a power station.

19.7.4.1 Examples of Unattended IAEA Safeguards Systems in Japan

Nuclear fuel cycle applications with UMSs in place in Japan range from the reprocessing of spent fuel from LWRs to the production of fuel for fast breeder reactors and MOX for some LWRs. In addition, a high-temperature, gas-cooled reactor, HTTR, is monitored by UMS. The Tokai Reprocessing Plant (TRP) separates plutonium and uranium from the rest of the spent fuel (cladding, fission products, and other wastes) and produces uranium oxide and uranium-plutonium MOX stored in canisters. High-level radioactive waste will be vitrified into canisters for storage at an underground repository in the future. In the Plutonium Fuel Production Facility (PFPF), the nuclear fuel oxides are converted to MOX fuel to be used in the Joyo and Monju fast breeder reactors (FBRs) and some LWRs. The Rokkasho Reprocessing Plant (RRP) is designed as a large-scale commercial plant that produces nuclear fuel oxides to be used at J-MOX, which in turn will produce MOX fuel for LWRs on a commercial scale. A significant number of IAEA safeguards systems are planned, with several already in development.

For the verification of nuclear material in plutonium-handling facilities, coincidence counting systems are a key component used by the IAEA.

Among these, Material Accountancy Glove Box (MAGB) systems are designed to operate continuously and unattended. In typical operation, an IAEA inspector selects the canisters (that usually contain MOX) to be measured, and the operator transfers each canister to the measurement position inside the glove box between two block detectors, each containing multiple ^{3}He tubes. Because the block detectors are located outside the glove box, they are sealed in housings to allow the IAEA to confirm that the detectors have not been moved or subjected to tampering. The detectors are connected to a small, nearby cabinet that contains two redundant DAQ devices that measure in coincidence mode. Once a preset threshold of reals counts per second is reached, a signal triggers an IAEA camera to take pictures of the canister identification number. The MAGB system is designed to measure coincident neutrons and not to determine plutonium isotopic composition. For the latter application, an advanced MAGB system (AMAG) has been developed, which—in addition to the neutron detectors—contains an electromechanically cooled HPGe detector for determining isotopic composition of plutonium via high-resolution gamma spectroscopy. The Rokkasho Reprocessing Plant is outfitted with a larger version of the MAGB system, the Temporary Canister Verification System, whose block detectors contain ^{3}He tubes at different pressures to account for efficiency changes due to geometrical properties.

The Plutonium Canister Assay System and Fresh Assembly Assay System are conceptually similar to the MAGB; in both systems, the material to be measured (canister or fuel assembly) is placed inside collar-type detectors. Similar to the AMAG, the Improved PCAS supplements the collar neutron detector with three electromechanically-cooled HPGe detectors to determine plutonium isotopic composition.

The Vitrified Waste Coincidence Counting (VWCC) system and the VCAS both estimate the mass of actinides inside vitrified waste. The VWCC contains a heavily shielded (~10 cm lead) neutron block detector located a few centimeters from the surface of the vitrified canister to be measured. The canister is rotated around a vertical axis during measurement to compensate for nonuniform actinide distribution. A high (α,n) reaction rate and a generally high neutron source term that results in about 500 kcps pose a challenge for deriving reliable results from coincidence measurements.

The VCAS contains multiple fission chambers at three different heights, each lined with either ^{235}U or ^{238}U. Some of the tubes are shielded with cadmium sheets to allow assessment of canister fill height and neutron energy distribution in order to distinguish between fission and (α,n) neutrons.

By detecting emanating neutrons and gammas, FBR measurement systems are used to detect fuel at key points within the facility: leaving the fresh fuel store; moving through the fuel transfer machine to the reactor core; and being transferred to the ex-vessel storage tank, the canning station, and finally the spent fuel pond. Where applicable, gross counting is used to supplement coincidence counting. Visual examination is no longer possible after the fuel and other items (control rods, material test elements, scrap material after examination) have been transferred into a can, so the Exit Gate Monitor in Joyo contains not only neutron detectors but two CZT detectors to measure the 662 keV ^{137}Cs peak via gamma spectrometry.

The HTTR Door Valve Monitor system uses neutron gross counting and gamma radiation measurements to detect fuel moving across the top of the high-temperature engineering test reactor (HTTR) core.

A completely different type of system had to be implemented after IAEA/SG lost continuity of knowledge at the Fukushima Daiichi LWR following the chain of events triggered by the powerful earthquake in March 2011. Several explosions caused severe damage to the structure of the reactor buildings, thus exposing spent fuel ponds to the outdoor environment. As the operator implemented remote-controlled cranes to clean up the area, SG methods had to be adapted to cover a scenario of equipment being used for the unplanned removal of spent fuel. Unattended monitoring is provided by the Open Air Spent Fuel Monitor, which consists of two outdoor cabinets and four portal neutron detectors that oversee Units 1 through 4 in a harsh environment. Data from the neutron detectors and spectra from two LaBr scintillation detectors are acquired in short time intervals and remotely transmitted for further analysis with the objective of detecting this scenario of unplanned removal of fuel.

19.7.5 Radiation Transport Modeling for Detector Design and Calibration

MCNP [16] is a simulation code widely used in neutron detector design. The MCNP code is most common, but other open-source Monte Carlo codes such as GEANT and openMC are available. The MCNP code is an excellent radiation transport simulation code that accurately simulates almost all relevant physics for neutron assay. The MCNP code allows a single user with a computer to perform accurate, complex, large-scale studies of detector design and assay performance. When done correctly, MCNP can generally be expected to match reality within a few percent.

The MCNP code is not a replacement for actual measurements. The MCNP code is excellent for understanding underlying details and for making initial guiding estimates, but final results should be measurement based whenever possible. Benchmarking MCNP results to measurements is a powerful tool. For example, from MCNP, the efficiency may be found to decrease by 5% (relative) as the source moves from the center to the bottom of the measurement cavity. From measurements, the center efficiency may be known to be 25%, and benchmarking would show the bottom efficiency to be 23.75%. The 23.75% value would be based on a measurement and would more closely match reality than by simulation alone.

The MCNP code can be used to create and optimize detector design aspects such as polyethylene thickness and layout of ^{3}He tubes. This optimization must be done before the materials are purchased, and it can save significant amounts of money. The MCNP code can also be used to model all expected measurement scenarios to characterize expected measurement uncertainties, which is especially valuable when the measurement scenarios would be prohibitively expensive to measure; for example, millions of dollars for a single instrument's calibration. The manufacturing of "off-normal" sources is usually very expensive, and the sources would be valuable only for the calibration. Waste measurements are one example. The various combinations of source and matrix materials would require a large number of sources, and they must be manufactured instead of taken from existing waste to know exactly what is inside. The materials would also be a large amount of waste that would need to be stored or disposed. Performing as many benchmark measurements as possible and then extrapolating changes using the MCNP code enables many measurement applications.

A recent book [186] gives examples of NDA calculations for safeguards.

References

1. PDT., http://www.pdt-inc.com/. Accessed Feb 2023
2. K.D. Ianakiev, M. Iliev, M.T. Swinhoe, Fact Sheet for KM200 Front-end Electronics. Los Alamos National Laboratory report LA-UR-15-24536 (2015)
3. Mirion Products., https://mirionprodstorage.blob.core.windows.net/prod-20220822/cms4_mirion/files/pdf/spec-sheets/jsr-14-neutron-analysis-register.pdf?1557256978. Accessed Mar 2023
4. J.E. Stewart, S.C. Bourret, M.S. Krick, M.R. Sweet, W.C. Harker, A. Gorobets, New shift register circuits for improving precision of neutron coincidence and multiplicity assays, in *Proceedings of the 6th International Meeting on Facilities Operations-Safeguards Interface*, Jackson Hole, Wyoming, September 20–24, 1999
5. Mirion Products, https://www.mirion.com/products/jsr-15-handheld-multiplicity-register-hhmr. Accessed Feb 2023
6. N. Menaa et al., Evaluation of the LANL hand held multiplicity register and Canberra JSR-15, in *Proceedings of the 2007 IEEE Nuclear Science Symposium Conference Record*, Honolulu, HI, October 27–November 3, 2007
7. Pulse Train Recorder PTR-32, https://www.ek-cer.hu/en/pulse-train-recorder-ptr-list-mode-data-acquisition-device-for-neutron-coincidence-counting/
8. D. Henzlova, H.O. Menlove, M.T. Swinhoe, J.B. Marlow, I.P. Martinez, C.D. Rael, Neutron data collection and analysis techniques comparison for safeguards, in *Proceedings of the IAEA Symposium on International Safeguards: Preparing for Future Verification Challenges*, November 1–5, 2010, Vienna, Austria, IAEA-CN-184/178 2010
9. M.R. Newell, R. Rothrock, D. Henzlova, Demonstration of the advanced list mode module, in *Proceedings of the 58th Annual Meeting of the INMM*, Indian Wells, CA, July 16–20, 2017, Los Alamos National Laboratory report LA-UR-17-25577 2017
10. Mirion Products, https://www.mirion.com/products/nda-2000-non-destructive-assay-software. Accessed Feb 2023
11. H. Nordquist, M. Swinhoe, INCC 6.0—LANL's Latest Neutron Coincidence Counting Software. Los Alamos National Laboratory report LA-UR-17-30062 (2017)
12. C. Moss et al., Revised SNAP III Training Manual. Los Alamos National Laboratory report LA-UR-17-30627 (2017), doi:https://doi.org/10.2172/1409806
13. B. Richard, J. Hutchinson, Nickel reflected plutonium metal sphere subcritical measurements, in *International Handbook of Evaluated Criticality Safety Benchmark Experiments*, NEA/NSC/DOC/(95)03/I, FUND-NCERC-PU-HE3-MULT-001 (2016)
14. B. Richard, J. Hutchinson, Tungsten-reflected plutonium-metal-sphere subcritical measurements, in *International Handbook of Evaluated Criticality Safety Benchmark Experiments*, NEA/NSC/DOC/(95)03/I, FUND-NCERC-PU-HE3-MULT-002 (2016
15. D. Reilly, N. Ensslin, H. Smith Jr., S. Kreiner, *Passive Nondestructive Assay of Nuclear Materials* (PANDA) U.S. Nuclear Regulatory Commission (January 1, 1991)
16. The MCNP® Code, https://mcnp.lanl.gov/ Accessed Feb 2023
17. J. K. Mattingly, Polyethylene-Reflected Plutonium Metal Sphere: Subcritical Neutron and Gamma Measurements. Sandia National Laboratories report SAND2009-5804 Rev. 1, https://www.osti.gov/servlets/purl/974870. Accessed Feb 2023, https://doi.org/10.2172/974870 (2009)
18. J. Li, J. Mattingly, SNAP-3 response function and its application. Trans. Am. Nucl. Soc. **108**(1), 491–494 (2013)
19. J. Hutchinson, D. Loaiza, B. Rooney, Correcting for room return neutrons using an empirical relationship. Trans. Am. Nucl. Soc. **96**(1), 453–454 (2007)
20. T. Cutler et al., Copper and polyethylene-reflected plutonium-metal-sphere subcritical measurements, in *International Handbook of Evaluated Criticality Safety Benchmark Experiments*, NEA/NSC/DOC/(95)03/I, FUND-NCERC-PU-HE3-MULT-003 (2019)
21. L.V. East, R.B. Walton, Polyethylene moderated ^{3}He neutron detectors. Nucl. Instrum. Methods **72**(2), 161–166 (1969)
22. J.W. Tape, D.A. Close, R.B. Walton, Total room holdup of plutonium measured with a large-area neutron detector. Nucl. Mater. Manag. **5**(3), 533–539 (1976)
23. D.G. Turner, M.T. Swinhoe, Neutron measurements outside containers of spent fuel for safeguards, in *Proceedings of the 40th Annual INMM Meeting*, Phoenix, AZ, July 1999
24. H. O. Menlove, Manual for the Portable Handheld Neutron Counter (PHNC) for Neutron Survey and the Measurement of Plutonium Samples. Los Alamos National Laboratory report LA-14257-M (2005)
25. D. Parvin, Validation and performance test of the plutonium inventory measurement system (PIMS) at the Rokkasho reprocessing plant (RRP), https://www-pub.iaea.org/mtcd/meetings/PDFplus/2007/cn1073/Papers/4A.3%20Ppr_%20Parvin%20-%20PIMS%20at%20the%20RRP.pdf, Powerpoint presentation https://www-pub.iaea.org/MTCD/Meetings/PDFplus/2007/cn1073/Presentations/4A.3%20Pres_%20Parvin%20-%20PIMS%20at%20the%20RRP.pdf. Accessed Feb 2023
26. IAEA, Safeguards techniques and equipment: 2011 Edition, in *International Nuclear Verification Series No. 1 (Rev. 2)*, International Atomic Energy Agency, Vienna, Austria (2011)
27. T.D. Reilly, M.M. Thorpe, Neutron Assay of 55-Gallon Barrels, in *Nuclear Safeguards Research and Development Program Status Report, May–August 1970*, ed. by G. Robert Keepin. Los Alamos Scientific Laboratory report LA-4523-MS (September 1970), p. 26–29
28. R. Harlan, Uranium and plutonium assay of crated waste by Gamma-ray, singles-neutron, and slow-neutron coincidence counting, in *Measurement Technology for Safeguards and Materials Control*, eds. by T.R Canada, B.S. Carpenter (NBS Special Publication 582, 1980)
29. J.P. Adams, National Low-Level Waste Management Program Radionuclide Report Series, vol. 13: Curium-242, Idaho National Engineering Laboratory (August 1995)
30. P.M. Rinard, H.O. Menlove, Application of Curium Measurements for Safeguarding at Reprocessing Plants Study 1: High-Level Liquid Waste and Study 2: Spent Fuel Assemblies and Leached Hulls. Los Alamos National Laboratory report LA-13134-MS (1996)
31. H.O. Menlove et al., Verification of plutonium content in spent fuel assemblies using neutron self-interrogation, in *Proceedings of the 50th Annual Meeting of the INMM*, Tucson, AZ, July 12–16, 2009

32. L.G. Evans, M.A. Schear, S. Croft, S.J. Tobin, M.T. Swinhoe, H.O. Menlove, Non-destructive assay of spent nuclear fuel using passive neutron albedo reactivity, in *Proceedings of the 51st Annual Meeting of the INMM*, Baltimore, MD, July 11–15, 2010
33. M.T. Swinhoe, H.O. Menlove, D.H. Beddingfield, VCAS (Vitrified Waste Canister Assay System) Monte Carlo Design Study. Los Alamos National Laboratory report LA-UR-03-0281 (2003)
34. M. Zendel, D.L. Donohue, E. Kuhn, S. Deron, T. Bíró, Nuclear safeguards verification measurement techniques, in *Handbook of Nuclear Chemistry*, ed. by A. Vértes, S. Nagy, Z. Klencsár, R.G. Lovas, F. Rösch, (Springer, Boston, 2011). https://doi.org/10.1007/978-1-4419-0720-2_6
35. M.T. Swinhoe, C. Pearsall, T. Pochet, K. Ferstl, Rokkasho Hulls Monitor System Calibration Exercise June 2004. Los Alamos National Laboratory report LA-UR-04-5604 (2004)
36. H.O. Menlove, D.H. Beddingfield, Passive neutron reactivity measurement technique, in *Proceedings of the 38th Annual Meeting of the INMM*, Phoenix, AZ, July 20–24, 1997, Los Alamos National Laboratory report LA-UR-97-2651 (1997)
37. H. Seo et al., Optimization of hybrid-type instrumentation for Pu accountancy of U/TRU ingot in pyroprocessing. Appl. Radiat. Isot. **108**, 16–23 (2015)
38. H. Seo et al., Development of prototype induced-fission-based Pu accountancy instrument for safeguards applications. Appl. Radiat. Isot. **115**, 67–73 (2016)
39. J.B. Franco, H.O. Menlove. G.W. Eccleston, M.C. Miller, A Compact Reactor Gate Discharge Monitor for Spent Fuel. Los Alamos National Laboratory report LA-UR-05-4280 (2005)
40. M.T. Andrews, K.A. Miller, C.D. Rael, M.T Swinhoe, J.B. Marlow, Benchmarking Measurements and MCNP6 Simulations of a Reactor Gate Discharge Monitor. Los Alamos National Laboratory report LA-UR-16-24734 (2016)
41. D.G. Langner, M.S. Krick, D.W. Miller, The use of ring ratios to detect sample differences in passive neutron counting, in *Proceedings of the 33rd Annual Meeting of the INMM*, Orlando, FL, July 19–22, 1992, Los Alamos National Laboratory report LA-UR-92-2327 (1992)
42. H.O. Menlove, C.D. Rael, K.E. Kroncke, K.J. DeAguero et al., Manual for the Epithermal Neutron Multiplicity Detector (ENMC) for Measurement of Impure MOX and Plutonium Samples. Los Alamos National Laboratory report LA-14088 (2004)
43. R. Weinmann-Smith, D.H. Beddingfield, A. Enqvist, M.T. Swinhoe, Variations in AmLi source spectra and their estimation utilizing the 5 ring multiplicity counter. Nucl. Instrum. Methods Phys. Res., Sect. A **856**, 17–25 (2017)
44. R. Weinmann-Smith, S. Croft, M.T. Swinhoe, A. Enqvist, Changes to the ^{252}Cf neutron Spectrum caused by source encapsulation. ESARDA Bulletin **54**, 44–53 (2017)
45. A.P. Belian, L.ReFalo, S. Passerini, A quantitative unattended system to assay uranium in waste, in *Proceedings of the INMM & ESARDA Joint Virtual Annual Meeting*, August 23–26 and August 30–September 1, (2021)
46. W.B. Wilson et al., SOURCES 4C: A code for calculating (α,n), spontaneous fission, and delayed neutron sources and spectra (LA-UR-02-1839) (2002)
47. L. ReFalo et al., Implementation of an integrated, robust and user-friendly electronic device for unattended neutron monitoring and logging, in *Proceedings of the Institute of Nuclear Materials Management Conference* (2012)
48. H.O. Menlove, J.E. Swansen, A high-performance neutron time correlation counter. Nucl. Technol. **71**(2), 497–505 (1985)
49. AMPTEK, https://www.amptek.com/products/charge-sensitive-preamplifiers/preamplifier-discriminators. Accessed Feb 2023
50. H.O. Menlove, Passive/Active Coincidence Collar for Total Plutonium Measurement of MOX Fuel Assemblies. Los Alamos National Laboratory report LA-9288-MS (1982)
51. H.O. Menlove, J. Baca, W. Harker, K.E. Kroncke, M.C. Miller, S. Takahashi, H. Kobayashi, S. Seki, K. Matsuyama, S. Kobayashi, WDAS operation manual including the add-A-source function. Los Alamos National Laboratory report LA-12292-M (1992)
52. K. Frame, Waste Characterization with WDAS and the Add-a-Source Method. Los Alamos National Laboratory report LA-UR-14-24457 (2014)
53. S.Y. Lee et al., Development of an NDA system for high-level waste from the Chernobyl new safe confinement construction site, in *Proceedings of the 51st Annual Meeting of the INMM*, Baltimore, MD, July 11–15, 2010
54. H.O. Menlove, D.H. Beddingfield, M.M. Pickrell, D.R. Davidson, R.D. McElroy, D.B. Brochu, Design of a high efficiency neutron counter for waste drums to provide optimized sensitivity for plutonium assay, in *Proceedings of the 5th Nondestructive Assay and Characterization Conference* (January 14–16, 1997), Los Alamos National Laboratory report LA-UR-96-4585 (1996)
55. M.S. Krick, J.E. Swansen, Neutron multiplicity and multiplication measurements. Nucl. Instr. Meth. **219**(2), 384–393 (1984)
56. H.O. Menlove, Passive neutron waste drum assay with improved accuracy and sensitivity for plutonium using the add-a-source method. J. Nucl. Mater. Manage. **20**(4), 17–26 (1992)
57. Now Mirion Technologies, www.mirion.com/products/wm3100-high-efficiency-passive-neutron-counter-henc. Accessed Mar 2023
58. H.O. Menlove, J.M. Boak, M.L. Collins, M.S. Krick, D.R. Mayo, C.D. Rael, C.M. Brown, J.B. Fanco, D.J. Santi, The SuperHENC mobile passive neutron measurement system for counting plutonium in standard waste boxes, in *Proceedings of the 7th Nondestructive Assay Waste Characterization Conference*, Salt Lake City (May 23–25, 2000), p. 133
59. N. Ensslin, W.C. Harker, M.S. Krick, D.G. Langner, M.M. Pickrell, J.E. Stewart, Application Guide to Neutron Multiplicity Counting. Los Alamos National Laboratory report LA-13422-MS (November 1998)
60. D.H. Beddingfield, H.O. Menlove, Statistical Data Filtration in Neutron Coincidence Counting. Los Alamos National Laboratory report LA-12451-MS (November 1992)
61. M.S. Krick, Thermal neutron multiplicity counting of samples with very low fission rates. Los Alamos National Laboratory Report LA-UR-97-2649, in *Proceedings of the 38th Annual Meeting of the INMM*, Phoenix, AZ, July 19–24, 1997
62. H.O. Menlove, S.C. Bourett, M.S. Krick, Cosmic Ray Neutron Reduction Using Localized Coincidence Neutron Veto Counting. U.S. Patent No. 6,420,712 B1, July 16, 2002
63. R.D. McElroy, Jr., S. Croft, Integrated waste assay system (IWAS) and analysis enhancements, in *Proceedings of the 31st Waste Management Symposium WM'05*, Tucson, AZ (2005)

64. R.D. McElroy Jr., S. Croft, B. Young, L. Bourva, Design and performance of the integrated waste assay system (IWAS), in *Proceedings of the 44th Annual Meeting of the INMM*, Phoenix, AZ, July 13–17, 2003
65. R.D. McElroy Jr., S. Croft, B. Gillespie, B. Young, Total measurement uncertainty analysis for an integrated waste assay system, in *Proceedings of the 44th Annual Meeting of the INMM*, Phoenix, AZ, July 13–17, 2003
66. H.O. Menlove, J. Baca, J.M. Pecos, D.R. Davidson, R.D. McElroy, D.B. Brochu, HENC Performance Evaluation and Plutonium Calibration. Los Alamos National Laboratory report LA-13362-MS (October 1997)
67. M.C. Miller, H.O. Menlove, M. Seya, S. Takahashi, R. Abedin-Zadeh, Holdup counters for the plutonium fuel production facility—PFPF, in *Proceedings of the 31st Annual INMM Meeting*, Los Angeles, CA, July 15–18, 1990, Los Alamos National Laboratory report LA-UR-90-2312 (1990)
68. T.R. Wenz, P.A. Russo, M.C. Miller, H.O. Menlove, S. Takahashi, Y. Yamamoto, I. Aoki, Portable Gamma-Ray Holdup and Attributes Measurements of High- and Variable-Burnup Plutonium. Los Alamos National Laboratory report LA-UR-91-3323 (1991)
69. T. Asano et al., Development of improved hold-up measurement system at plutonium fuel production facility, in *Proceedings of the 38th Annual Meeting of the INMM*, Phoenix, AZ, July 20–24, 1997
70. P. Peerani, R. Jaime, M. Looman, A. Ravazzani, Implementation of a verification procedure for large LEU samples based on passive neutron assay, in *Proceedings of the 25th Annual Symposium on Safeguards and Nuclear Materials Management (ESARDA)*, Stockholm, Sweden, May 13–15, 2003
71. P. Peerani, V. Canadell, J. Garijo, K. Jackson, R. Jaime, M. Looman, A. Ravazzani, P. Schwalbach, M.T. Swinhoe, Development of high-efficiency passive counters (HEPC) for the verification of large LEU samples. Nucl. Instrum. Methods Phys. Res. Sect. A **601**(3), 326–332 (2009)
72. K. Zhao, M. Penkin, C. Norman, S. Balsley, K. Mayer, P. Peerani, C. Pietri, S. Tapodi, Y. Tsutaki, M. Boella, G. Renha Jr., E. Kuhn, *International Target Values 2010 for Measurement Uncertainties in Safeguarding Nuclear Materials*. IAEA STR-368 (International Atomic Energy Agency, Vienna, Austria, November 2010)
73. K. Miller, H. Menlove, J. Marlow, The UF_6 coincidence counter (UFCC) for nondestructive assay of 5a/5b uranium cylinders: Design and benchmark measurements, in *Proceedings of the 55th Annual INMM Meeting*, Atlanta, GA, July 20–24, 2014, Los Alamos National Laboratory report LA-UR-14-24603 (2014)
74. K.A. Miller, M.T. Swinhoe, S. Croft, T. Tamura, S. Aiuchi, A. Kawai, T. Iwamoto, Measured F(α,n) Yield from ^{234}U in Uranium Hexafluoride. Nucl. Sci. Eng. **176**(1), 98–105 (2014)
75. H.O. Menlove, C.D. Rael, UFCC Calibration Measurements for Bulk U_3O_8 Powder Samples. Los Alamos National Laboratory report LA-UR-20-28378 (2020)
76. American National Standards Institute (ANSI), Packaging of uranium hexafluoride for transport, ANSI Report rev. N14.1-2001 of ANSI N14.1-1995, New York NY, 2001
77. R.B. Walton, T.D. Reilly, J.L. Parker, J.H. Menzel, E.D. Marshall, L.W. Fields, Measurements of UF_6 cylinders with portable instruments. Nucl. Technol. **21**(2), 133–148 (1974)
78. R.B. Walton, T.L. Atwell, Portable neutron probe, 'SNAP', in *Nuclear Analysis Research and Development Program Status Report*, January–April 1973," G. Robert Keepin, Ed., Los Alamos Scientific Laboratory Report LA-5291-PR (May 1973), p. 14
79. K.A. Miller, H.O. Menlove, M.T. Swinhoe, J.B. Marlow, A new technique for uranium cylinder assay using passive neutron self-interrogation, in *Proceedings of the IAEA Symposium on International Safeguards: Preparing for Future Verification Challenges*, Vienna, Austria, November 1–5, 2010
80. K.A. Miller, M.T. Swinhoe, H.O. Menlove, J.B. Marlow, Status Report on the Passive Neutron Enrichment Meter (PNEM) for UF_6 Cylinder Assay. Los Alamos National Laboratory report LA-UR-12-21058 (2012)
81. K. Miller, H. Menlove, E. Smith, D. Jordan, C. Orton, P. Schwalbach, J. Morrissey, P. De Baere, T. Visser, R. Veldhof, A study of candidate non-destructive assay methods for unattended UF_6 cylinder verification: Measurement campaign results, in *Proceedings of the 2014 IAEA Symposium on Nuclear Safeguards*, Vienna, Austria, October 20–24, 2014, Conf. No. IAEA-CN-220 (2014)
82. L.E. Smith, K.A. Miller, J. Garner, S. Branney, B. McDonald, J. Webster, M. Zalavadia, L. Todd, J. Kulisek, H.A. Nordquist, N. Deshmukh, S. Stewart, Viability Study for an Unattended UF_6 Cylinder Verification Station: Phase I Final Report. Pacific Northwest National Laboratory Report PNNL-25395, 2016
83. D.P. Broughton, M.T. Swinhoe, Correcting cylinder verification neutron count rates for variations due to UF6 distribution. J. Nucl. Mater. Manage. **47**(4), 27–35 (2019)
84. R. Berndt, E. Franke, P. Mortreau, ^{235}U enrichment or UF_6 mass determination on UF_6 cylinders with non-destructive analysis methods. Nucl. Instrum. Methods Phys. Res., Sect. A **612**(2), 309–319 (2010)
85. D.P. Broughton, *Adaptation of Neutron Safeguards Measurements for the Front-End of the Modern Nuclear Fuel Cycle*, Doctoral Dissertation, University of Texas at Austin, 2021
86. G.S. Solov'ev, A.V. Saprygin, V.V. Komarov, A.I. Izrailevich, ^{234}U content in enriched uranium as a function of the ^{234}U concentration in the initial enrichment. At. Energy **95**(1), 473–475 (2003)
87. Nuclear Energy Agency and International Atomic Energy Agency (NEA/IAEA), *Uranium 2018: Resources, Production and Demand* (OECD Publishing, Paris, FR, 2018)
88. L.E. Smith, J.A. Kulisek, J. Webster, M.A. Zalavadia, R. Guerrero, B.S. McDonald, Refinement of Gamma Spectroscopy Methods for Unattended UF_6 Cylinder Verification, Pacific Northwest National Laboratory report PNNL-26789, Richland, WA (2017)
89. H.O. Menlove, A. Belian, W.H. Geist, C.D. Rael, A new method to measure the U-235 content in fresh LWR fuel assemblies via fast-neutron passive self-interrogation. Nucl. Instrum. Methods Phys. Res., Sect. A **877**, 238–245 (2018)
90. H.O. Menlove, W.H. Geist, M.A. Root, S.E. Sarnoski, Experimental Evaluation of the Fast-Neutron Passive Collar (FNPC) Performance and Comparison with Scintillation Based Collars. Los Alamos National Laboratory report LA-UR-19-32310 (2019)
91. H.O. Menlove, J.E. Stewart, S.Z. Qiao, T.R. Wenz, G.P.D. Verrecchia, Neutron Collar Calibration and Evaluation containing Burnable Neutron Absorbers. Los Alamos National Laboratory report LA-11965-MS (1990)

92. M.T. Swinhoe, H.O. Menlove, P. de Baere, D. Lodi, P. Schwalbach, C.D. Rael, M. Root, A. Tomanin, A. Favalli, A new generation of uranium coincidence fast neutron collars for assay of LWR fresh fuel assemblies. Nucl. Instrum. Methods Phys. Res., Sect. A **1009**, 165453 (2021)
93. J. Beaumont, T.H. Lee, M. Mayorov, C. Tintori, F. Rogo, B. Angelucci, M. Corbo, A fast-neutron coincidence collar using liquid scintillators for fresh fuel verification. J. Radioanal. Nucl. Chem. **314**, 803–812 (2017)
94. W.H. Geist, H.O. Menlove, M.A. Root, Fast Neutron Passive Collar Project Year End Report. Los Alamos National Laboratory report LA-UR-18-29560 (2018)
95. M.A. Root, H.O. Menlove, R.C. Lanza, C.D. Rael, K.A. Miller, J.B. Marlow, J.G. Wendelberger, Using the time-correlated induced fission method to simultaneously measure the 235U content and the burnable poison content in LWR fuel assemblies. Nucl. Technol. **203**(1), 34–47 (2018). https://doi.org/10.1080/00295450.2018.1429112
96. M.A. Root, S. Sarnoski, H.O. Menlove, W.H. Geist, Advanced Analysis to Determine Both the U-235 and Gadolinium Content in Fresh PWR Fuel Assemblies. Los Alamos National Laboratory report LA-UR-19-31922 Version 2 (2019)
97. M.A. Root, S. Sarnoski, H.O. Menlove, W.H. Geist, Verification of Fresh LWR Fuel Assemblies Using the Passive Unattended Neutron Collar. Los Alamos National Laboratory report LA-UR-19-31105 (2019)
98. F.H. DuBose, Calorimeter-Based Adjustment of Multiplicity Determined ^{240}Pu$_{eff}$: Known-α Analysis for the Assay of Plutonium. Savannah River Nuclear Solutions report N-TRT-K-00020, Rev. 0 (2012)
99. R.A. Dewberry, Qualification of the K-Area Neutron Multiplicity Counter for Passive Known Alpha Assays. Savannah River National Laboratory report SRNL-TR-2014-00243, Rev. 0 (2014)
100. H.R. Martin, Reaction Gamma Rays in Plutonium Compounds, Mixtures, and Alloys. RFP-2382, Dow Chemical U.S.A. Rocky Flats Division, Golden, CO (1975)
101. J.E. Narlesky et al., A Calibration to Predict the Concentrations of Impurities in Plutonium Oxide by Prompt Gamma Analysis Revision 2. Los Alamos National Laboratory report LA-UR-15-27057 (2015)
102. R.D. McElroy Jr., Total Measurement Uncertainty in Neutron Coincidence Multiplicity Analysis. Oak Ridge National Laboratory report ORNL/TM-2020/1663 (2020)
103. R.A. Dewberry, KAMS Measurement and Qualification Report for Be-Containing Pu and U/Pu Stabilized Components Received from Hanford/RFETS/LANL/LLNL. Savannah River National Laboratory report SRNS-TR-2009-00264, Rev. 2 (2010)
104. N. Ensslin, M.S. Krick, N. Dytlewski, Assay variance as a figure of merit for neutron multiplicity counters. Nucl. Instrum. Methods A **290**(1), 197–207 (1990)
105. J.E. Stewart et al., Assay of scrap plutonium oxide by thermal neutron multiplicity counting for IAEA verification of excess materials from nuclear weapons production, in *Proceedings of the 37th Annual INMM Meeting*, Naples, FL, July 28–31, 1996, Los Alamos National Laboratory report LA-UR-96-2515 (1996)
106. H.O. Menlove et al., Plutonium Scrap Multiplicity Counter Operation Manual. Los Alamos National Laboratory report LA-12479-M (1992)
107. T.E. Sampson, T.L. Cremers, J.C. Martz, W.R. Dvorzak, An NDA system for automated, inline weapons component dismantlement, in *Proceedings of the 34th Annual Meeting of the INMM*, Scottsdale, AZ, July 18–21, 1993
108. D.G. Langner et al., FB-Line Neutron Multiplicity Counter Operation Manual. Los Alamos National Laboratory report LA-13395-M (1997)
109. D.G. Langner et al., New neutron multiplicity counter for the measurement of impure plutonium metal at westinghouse savannah river site, in *Proceedings of the 39th Annual INMM Meeting*, Naples, FL, July 26–30, 1998, Los Alamos National Laboratory report LA-UR-98-2890 (1998)
110. D.G. Langner, M.S. Krick, K.E. Kroncke, A Large Multiplicity Counter for the Measurement of Bulk Plutonium. Los Alamos National Laboratory report LA-UR-94-2313, Nuclear Materials Management 23 (Proceedings Issue), 474–479 (1994)
111. D.G. Langner et al., Application of Neutron Multiplicity Counting to the Assay of Bulk Plutonium Bearing Materials at RFETS and LLNL. Los Alamos National Laboratory Report LA-UR-95-3320, American Nuclear Society Fifth International Conference on Facility Operations-Safeguards Interface, Jackson Hole, WY, September 1995
112. D.G. Langner et al., The Performance of the 30-Gallon Drum Neutron Multiplicity Counter at Rocky Flats Environmental Technology Site. Los Alamos National Laboratory report LA-UR-96-2569 (1996)
113. D.G. Langner et al., Assay of impure plutonium oxide with the large neutron multiplicity counter for IAEA verification of excess weapons material at the rocky flats environmental technology site," in *Proceedings of the 38th Annual INMM Meeting*, Phoenix, AZ, July 20–24, 1997, Los Alamos National Laboratory report LA-UR-97-2650 (1997)
114. H.O. Menlove, C.D. Rael, K.E. Kroncke, K.J. DeAguero, Manual for the Epithermal Neutron Multiplicity Detector (ENMC) for Measurement of Impure MOX and Plutonium Samples. Los Alamos National Laboratory report LA-14088 (2004)
115. J.E. Stewart, H.O. Menlove, D.R. Mayo, W.H. Geist, L.A. Carillo, G.D. Herrera et al., The Epithermal Neutron Multiplicity Counter Design and Performance Manual: More Rapid Plutonium and Uranium Inventory Verifications by Factors of 5–20. Los Alamos National Laboratory report LA-13743-M (2000)
116. N.M. Abdurrahman, A.P. Simpson, S. Barber, WIPP certification of a new SuperHENC box counter at Hanford. Trans. Am. Nucl. Soc. **93**(1), 819–820 (2005)
117. J.K. Sprinkle, H.O. Menlove, M.C. Miller, P.A. Russo, An Evaluation of the INVS Model IV Neutron Counter. Los Alamos National Laboratory report, LA-12496-MS (1993)
118. D. Davidson et al., A new high-accuracy combined neutron/gamma counter for in-glove box measurements of PuO2 and MOX safeguards samples (OSL—Counter), in *Proceedings of the ESARDA 15th Annual Symposium on Safeguards and Nuclear Material Management*, Rome, Italy, May 11–13, 1993, p. 585–588
119. H.O. Menlove, R. Wellum, M. Ougier, K. Mayer, Performance tests of the high accuracy combined neutron/gamma detector (OSL-counter), in *Proceedings of the ESARDA 15th Annual Symposium on Safeguards and Nuclear Material Management*, Rome, Italy, May 11–13, 1993, p. 363–369
120. IAEA, User Requirements AVIS Version No. 1. IAEA Document Number SG-TE-GNRL-ZZ-1020 (2007)

121. J.B. Marlow, M.T. Swinhoe, H.O. Menlove, C.D. Rael, M.R. Newell, Advanced Verification for Inventory Sample System (AVIS) Manual. Los Alamos National Laboratory report LA-UR-11-07086 (2011)
122. C. Moss et al., MC-15 Users Manual. Los Alamos National Laboratory report LA-UR-18-29563 (2018)
123. B. Richard, J. Hutchinson, Nickel reflected plutonium metal sphere subcritical measurements, in *International Handbook of Evaluated Criticality Safety Benchmark Experiments*, FUND-NCERC-PU-HE3-MULT-001 (2014)
124. J. Arthur et al., Validating the performance of correlated fission multiplicity implementation in radiation transport codes with subcritical neutron multiplication benchmark experiments. Ann. Nucl. Energy **120**, 348–366 (2018)
125. J. Hutchinson et al., Validation of statistical uncertainties in subcritical benchmark measurements: Part II—Measured data. Ann. Nucl. Energy **125**, 342–359 (2019)
126. H. Kistle et al., Simulated energy response in a multiplicity detector. Trans. Am. Nucl. Soc. **123**, 862–865 (2020)
127. M. Nelson et al., Total Efficiency Characterization of the Next Generation Neutron Multiplicity Detector (MC-15). Los Alamos National Laboratory report LA-UR-20-29036 (2020)
128. M. Hua et al., Derivation of the two-exponential probability density function for Rossi-alpha measurements of reflected assemblies and validation for the special case of shielded measurements. Nucl. Sci. Eng. **194**, 56–68 (2020)
129. M. Hua et al., Measurement uncertainty of rossi-alpha neutron experiments. Ann. Nucl. Energy **147**, 107672 (2020)
130. J. Arthur et al., Genetic algorithm for nuclear data evaluation applied to subcritical neutron multiplication inference benchmark experiments. Ann. Nucl. Energy **133**, 853–862 (2019)
131. J.E. Foley, L.R. Cowder, Assay of the Uranium Content of Rover Scrap with the Random Source Interrogation System. Los Alamos Scientific Laboratory report LA-5692-MS (1974)
132. D. Langner, T. Canada, N. Ensslin, T. Atwell, H. Baxman, L. Cowder, L. Speir, T. Van Lyssel, T. Sampson, The CMB-8 Material Balance System. Los Alamos Scientific Laboratory report LA-8194-M (August 1980)
133. T. Gozani, *Active Nondestructive Assay of Nuclear Materials: Principles and Applications*, NUREG/CR-0602 (U.S. Nuclear Regulatory Commission, Washington DC, 1981)
134. R.T. Kouzes, The ^{3}He Supply Problem. Pacific Northwest National Laboratory report PNNL-18388 (2009)
135. D.A. Shea, D. Morgan, The Helium-3 Shortage: Supply, Demand, and Options for Congress. Congressional Research Service report for Congress R41419 (2010)
136. M. Pickrell et al., The IAEA workshop on requirements and potential Technologies for Replacement of ^{3}He detectors in IAEA safeguards applications. J. Nucl. Mater. Manage. **41**(2), 14–29 (2013)
137. K. Soyama, Development and demonstration of a Pu NDA system using ZnS/10B2O3 ceramic scintillator detectors, in *Proceedings of the 55th Annual Meeting of the INMM*, Atlanta, GA, July 20–24, 2014
138. S. Stave, M. Bliss, R. Kouzes, A. Lintereur, S. Robinson, E. Siciliano, L. Wood, LiF/ZnS neutron multiplicity counter. Nucl. Instrum. Methods Phys. Res., Sect. A **784**, 208–212 (2015). https://doi.org/10.1016/j.nima.2015.01.039
139. M. Dallimore, C. Giles, D. Ramsden, G.S. Dermody, The development of a scalable ^{3}He free neutron detection technology and its potential use in nuclear security and physical protection applications, in *Proceedings of the 52nd Annual Meeting of the INMM*, Palm Desert, CA, July 17–21, 2011
140. K.D. Ianakiev, M.P. Hehlen, M.T. Swinhoe, A. Favalli, M. Illiev, T.C. Lin, M.T. Barker, B.L. Bennett, Neutron detector based on particles of 6Li glass scintillator dispersed in organic lightguide matrix. Nucl. Instrum. Methods Phys. Res., Sect. A **784**, 189–193 (2015). https://doi.org/10.1016/j.nima.2014.10.073
141. A.D. Lavietes, R. Plenteda, N. Mascarenhas, L.M. Cronholm, M. Aspinall, M. Joyce, A. Tomanin, P. Peerani, Liquid scintillator-based neutron detector development, in *2012 IEEE Nuclear Science Symposium and Medical Imaging Conference Record (NSS/MIC)*, Anaheim, CA (2012), p. 230–244. doi: https://doi.org/10.1109/NSSMIC.2012.6551100
142. J.L. Dolan, M. Flaska, A. Poitrasson-Riviere, A. Enqvist, P. Peerani, D.L. Chichester, S.A. Pozzi, Plutonium measurements with a fast-neutron multiplicity counter for nuclear safeguards applications. Nucl. Instrum. Methods Phys. Res., Sect. A **763**, 565–574 (2014)
143. D. Henzlova, L.G. Evans, H.O. Menlove, M.T. Swinhoe, J.B. Marlow, Experimental evaluation of a boron-lined parallel plate proportional counter for use in nuclear safeguards coincidence counting. Nucl. Instrum. Methods Phys. Res., Sect. A **697**, 114–121 (2013)
144. J.L. Lacy, A. Athanasiades, L. Sun, C.S. Martin, T.D. Lyons, M.A. Foss, H.B. Haygood, Boron-coated straws as a replacement for ^{3}He-based neutron detectors. Nucl. Instrum. Methods Phys. Res., Sect. A **652**(1), 359–363 (2011)
145. R.T. Kouzes, J.H. Ely, A.T. Lintereur, E.R. Siciliano, Boron-10 ABUNCL Active Testing. Pacific Northwest National Laboratory report PNNL-22567 (2013)
146. L.G. Evans, D. Henzlova, H.O. Menlove, M.T. Swinhoe, S. Croft, J.B. Marlow, R.D. McElroy, Nuclear safeguards ^{3}He replacement requirements. J. Nucl. Mater. Manage. **40**(3), 88–96 (2012)
147. J. Dolan, A. Dougan, D. Peranteau, S. Croft, An international view on ^{3}He alternatives for nuclear safeguards. J. Nucl. Mater. Manage. **43**(3), 17–29 (2015)
148. D. Henzlova, R. Kouzes, R. McElroy, P. Peerani et al., Current Status of Helium-3 Alternative Technologies for Nuclear Safeguards. Los Alamos National Laboratory report LA-UR-15-21201 Rev. 3 (2015)
149. D. Henzlova, H.O. Menlove, High-dose neutron detector development for measuring alternative fuel cycle materials. in *Proceedings of the GLOBAL 2017*, September 24–29, 2017, Seoul (Korea), Paper A-018, Los Alamos National Laboratory report LA-UR-17-23325 (2017)
150. D. Henzlova, H. Menlove, Report on Operational Test of High Dose Neutron Detector. Los Alamos National Laboratory report LA-UR-17-29204 (2017)
151. M. Croce, D. Henzlova, H. Menlove, D. Becker, J. Ullom, Electrochemical Safeguards Measurement Technology Development at LANL. Los Alamos National Laboratory report LA-UR-20-25781. J. Nucl. Mater. Manage. **49**(1), 116–135 (2021)
152. H.O. Menlove, R. Abedin-Zadeh, R. Zhu, The Analyses of Neutron Coincidence Data to Verify Both Spontaneous-Fission and Fissionable Isotopes. Los Alamos National Laboratory report LA-11639-MS (1989)

153. B. Cipiti, M. Browne, M. Reim, The MPACT 2020 milestone: Safeguards and security by design of future nuclear fuel cycle facilities. J. Nucl. Mater. Manage. **49**(1) (2020)
154. D. Henzlova, H.O. Menlove, J.B. Marlow, Design and performance of a ^{3}He-free coincidence counter based on parallel plate boron-lined proportional technology. Nucl. Instrum. Methods Phys. Res., Sect. A **788**, 188–193 (2015)
155. D. Henzlova, H.O. Menlove, M. Tanigawa, Y. Mukai, H. Nakamura, Field test of a full scale ^{3}He-alternative HLNC-type counter: High level neutron counter—boron (HLNB), in *Proceedings of the 39th ESARDA Conference*, Dusseldorf, Germany, May 16–18 2017, Los Alamos National Laboratory report LA-UR-17-22470 (2017)
156. J.L. Lacy, A Athanasiades, C.S. Martin, L. Sun, G.J. Vazquez-Flores, Design and performance of high-efficiency counters based on boron-lined straw detectors, in *Proceedings of the 53rd Annual INMM Meeting*, Orlando, FL, July 15–19, 2012
157. J. Brennan et al., Time-encoded imaging of energetic radiation, in *Proceedings of the SPIE 8852, Hard X-Ray, Gamma-Ray, and Neutron Detector Physics XV*, 885203 (2013). doi: https://doi.org/10.1117/12.2027674
158. J. Brennan et al., Demonstration of two-dimensional time-encoded imaging of fast neutrons. Nucl. Instrum. Methods Phys. Res., Sect. A **802**, 76–81 (2015). https://doi.org/10.1016/j.nima.2015.08.076
159. B.R. Kowash, D.K. Wehe, J.A. Fessler, A rotating modulation imager for locating mid-range point sources. Nucl. Instrum. Methods Phys. Res., Sect. A **602**(2), 477–483 (2009). https://doi.org/10.1016/j.nima.2008.12.233
160. J.E. Goldsmith, M.D. Gerling, J.S. Brennan, A compact neutron scatter camera for field deployment. Rev. Sci. Instrum. **87**, 083307 (2016). https://doi.org/10.1063/1.4961111
161. N. Mascarenhas, J. Brennan, K. Krenz, P. Marleau, S. Mrowka, Results with the neutron scatter camera. IEEE Trans. Nucl. Sci. **56**(3), 1269–1273 (2009). https://doi.org/10.1109/TNS.2009.2016659
162. A.C. Madden et al., An imaging neutron/gamma-ray spectrometer, in *Proceedings of the SPIE 8710, Chemical, Biological, Radiological, Nuclear, and Explosives (CBRNE) Sensing XIV*, 87101L (2013). https://doi.org/10.1117/12.2018075
163. W.M. Steinberger et al., Imaging special nuclear material using a handheld dual particle imager. Sci. Rep. **10**, 1855 (2020). https://doi.org/10.1038/s41598-020-58857-z
164. H. Al Hamrashdi, D. Cheneler, S.D. Monk, A fast and portable imager for neutron and gamma emitting radionuclides. Nucl. Instrum. Methods Phys. Res., Sect. A **953**, 163253 (2020). https://doi.org/10.1016/j.nima.2019.163253
165. M. Monterial, P. Marleau, S.A. Pozzi, Single-view 3D reconstruction of correlated gamma-neutron sources. IEEE Trans. Nucl. Sci. **64**(7), 840–1845 (2017). https://doi.org/10.1109/TNS.2017.2647952
166. P. Hausladen et al., A Deployable Fast-Neutron Coded-Aperture Imager for Quantifying Nuclear Material. Oak Ridge National Laboratory report ORNL/TM-2013/248 (2013). https://doi.org/10.2172/1095660
167. P. Hausladen, J. Newby, F. Liang, M. Blackston, The Deployable Fast-Neutron Coded-Aperture Imager: Demonstration of Locating One or More Sources in Three Dimensions. Oak Ridge National Laboratory report ORNL/TM-2013/446 (2013). https://doi.org/10.2172/1096324
168. J. Boo, M.D. Hammig, M. Jeong, Compact lightweight imager of both gamma rays and neutrons based on a pixelated stilbene scintillator coupled to a silicon photomultiplier array. Sci. Rep. **11**, 3826 (2021)
169. F. Habte, M.A. Blackston, P.A. Hausladen, L. Fabris, Enhancing pixelated fast-neutron block detector performance using a slotted light guide, in *2008 IEEE nuclear science symposium conference record*, Dresden, Germany (2008), p. 3128–3132. doi: https://doi.org/10.1109/NSSMIC.2008.4775016
170. F. Daghighian, R. Sumida, M.E. Phelps, PET imaging: An overview and instrumentation. J. Nucl. Medicine Technol. **18**(1), 5–13 (1990)
171. P.A. Hausladen, M.A. Blackston, R.J. Newby, Position-Sensitive Fast-Neutron Detector Development in Support of Fuel-Cycle R&D MPACT Campaign. Oak Ridge National Laboratory report ORNL/TM-2010/201 (September 2010). https://doi.org/10.2172/1054981
172. R.J. Newby, P.A. Hausladen, M.A. Blackston, J.F. Liang, Performance of Fast-Neutron Imaging Detectors Based on Plastic Scintillator EJ-299-34. Oak Ridge National Laboratory report ORNL/TM-2013/82 (February 2013). https://doi.org/10.2172/1128961
173. S.R. Gottesman, E.E. Fenimore, New family of binary arrays for coded aperture imaging. Appl. Opt. **28**(20), 4344–4352 (1989). https://doi.org/10.1364/AO.28.004344
174. T. Honkamaa et al., A prototype for passive gamma emission tomography, in *IAEA Symposium on International Safeguards: Linking Strategy, Implementation and People*, Vienna, Austria, October 20–24, 2014, CN-220-189, S18-12 (2014), p. 12
175. M. Mayorov et al., Gamma emission tomography for the inspection of spent nuclear fuel, in *2017 IEEE Nuclear Science Symposium and Medical Imaging Conference (NSS/MIC)*, Atlanta, GA, 2017, p. 1–2. doi: https://doi.org/10.1109/NSSMIC.2017.8533017
176. P.A. Hausladen, A.S. Iyengar, L. Fabris, J. Yang, J. Hu, A Design Study of the Parallel-Slit Ring Collimator for Fast Neutron Emission Tomography of Spent Fuel. Oak Ridge National Laboratory report ORNL/SPR- 2018/975 (2018)
177. A.S. Iyengar, The Design of an Imager to Safeguard Spent Fuel Using Passive Fast Neutron Emission Tomography. Doctoral Dissertation, University of Tennessee (2019). https://trace.tennessee.edu/utk_graddiss/5343
178. A. Iyengar, P. Hausladen, J. Yang, L. Fabris, J. Hu, J. Lacy, A. Athanasiades, Detection of fuel pin diversion via fast neutron emission tomography, in *Proceedings of the 39th ESARDA Conference*, Dusseldorf, Germany, May 16–17, 2017
179. P.A. Hausladen, A.S. Iyengar, L. Fabris, J. Yang, J. Hu, A.L. Lousteau, Modeling passive fast-neutron emission tomography of spent nuclear fuel, in *Proceedings of the International Workshop on Numerical Modeling of NDA Instruments and Methods for Nuclear Safeguards*, Luxembourg City, Luxembourg, May 16–18, 2018. ISBN 978-92-79-98443-3, doi:https://doi.org/10.2760/055930, JRC114178
180. A.S. Iyengar, P.A. Hausladen, L. Fabris, J. Yang, J. Hu, Development of a nuclear fuel safeguards verification technology for new facility types: Use of fast neutron emission tomography for spent fuel verification, in *Proceedings of the IAEA Symposium on International Safeguards*, Vienna, Austria, November 5–9, 2018
181. P.A. Hausladen, M.E. Montague, A.S. Iyengar, J. Yang, Tomographic Image Reconstruction for the Parallel-Slit Ring Collimator Fast Neutron Emission Tomography System. Oak Ridge National Laboratory report ORNL/SPR-2020/1510 (2020)
182. P.A. Hausladen, M.E. Montague, L. Fabris, Initial Measurements with the Prototype Parallel-Slit Ring Collimator Fast Neutron Emission Tomography System. Oak Ridge National Laboratory report ORNL/SPR-2021/2208 (2021)

183. D.H. Beddingfield, H.O. Menlove, A New Approach to Hold-Up Measurement in Uranium Enrichment Facilities. Los Alamos National Laboratory report LA-UR-00-2534, *Nucl. Mater. Manage.* **29** (Proceedings Issue) CD-ROM (2000)
184. S.F. Klosterbuer, J.K. Halbig, W.C. Harker, H.O. Menlove, J.A. Painter, J.E. Stewart, Continuous remote unattended monitoring for safeguards data collection systems, in *Proceedings of the IAEA Symposium on International Safeguards*, Vienna, Austria, March 14–18, 1994, IAEA-SM-333/111, Los Alamos Laboratory report LA-UR-94-0256 (1994)
185. J.K. Sprinkle, M. Abhold, M. Browne, S.-T. Hsue, S. F. Klosterbuer, G. E. Bosler, J. K. Halbig, R. F. Parker, D. G. Pelowitz, S. E. Buck, UNARM (Unattended and remote monitoring) overview, in *23rd Annual Meeting, Symposium on Safeguards and Nuclear Material Management*, Bruges, Belgium, May 8–10, 2001, Los Alamos National Laboratory report LA-UR-01-3767 (2001)
186. J. Hendricks, M.T. Swinhoe, A. Favalli, *Monte Carlo N-Particle Simulations for Nuclear Detection and Safeguards: An Examples-Based Guide for Students and Practitioners* (Springer, 2022)

Open Access This chapter is licensed under the terms of the Creative Commons Attribution 4.0 International License (http://creativecommons.org/licenses/by/4.0/), which permits use, sharing, adaptation, distribution and reproduction in any medium or format, as long as you give appropriate credit to the original author(s) and the source, provide a link to the Creative Commons license and indicate if changes were made.

The images or other third party material in this chapter are included in the chapter's Creative Commons license, unless indicated otherwise in a credit line to the material. If material is not included in the chapter's Creative Commons license and your intended use is not permitted by statutory regulation or exceeds the permitted use, you will need to obtain permission directly from the copyright holder.

20. Active Neutron Instrumentation and Applications

M. T. Swinhoe, N. Ensslin, L. G. Evans, W. H. Geist, M. S. Krick, A. L. Lousteau, R. McElroy, M. M. Pickrell, and P. Rinard

20.1 Introduction

As shown in Chaps. 16–19 of this book, passive neutron counting is a relatively straightforward measurement technique; however, it is applicable only to material in which the spontaneous fission and/or (α,n) emission is sufficiently large. In other materials, particularly uranium, similar counting methods can be used by first inducing fission using an external neutron source. Such methods are called *active* methods. The status of such methods was described by Gozani in 1981 [1]. The purpose of this chapter is to provide a description of the current status of such methods, concentrating on those used most commonly for safeguards, security, and safety applications.

20.2 Interrogation Sources

The choice of interrogating source is important to the ultimate performance of any active instrument. Several important factors should be considered.

- Energy spectrum
- Neutron multiplicity
- Stability or half-life
- Commercial availability and cost

The initial energy spectrum of the external source, as modified by the instrument and the item, affects the degree of penetration into the item. If the neutron multiplicity of the external source is greater than one, then a background in the correlated counting rates is created. The induced fission rate, and therefore the counting rates, are proportional to the intensity of the external source. To deduce the fissile mass, it is necessary to know the source intensity. Any variation in source intensity with time needs to be corrected for; otherwise, this variation affects the uncertainty in the measured mass. Finally, any type of external source can be used only if it is available at a reasonable cost. Chapter 13 gives a description of various types of neutron sources. Here we highlight the features that are important for active neutron interrogation (See also Chap. 17, Sect. 17.15.2).

The original version of the chapter has been revised. A correction to this chapter can be found at https://doi.org/10.1007/978-3-031-58277-6_28

Los Alamos National Laboratory strongly supports academic freedom and a researcher's right to publish; as an institution, however, the Laboratory does not endorse the viewpoint of a publication or guarantee its technical correctness.

M. T. Swinhoe (✉) · N. Ensslin · W. H. Geist · M. S. Krick · M. M. Pickrell · P. Rinard
Los Alamos National Laboratory, Los Alamos, NM, USA
e-mail: swinhoe@lanl.gov; wgeist@lanl.gov; mpickrell@lanl.gov

L. G. Evans · A. L. Lousteau · R. McElroy
Oak Ridge National Laboratory, Oak Ridge, TN, USA
e-mail: evanslg@ornl.gov; lousteaula@ornl.gov; mcelroyrd@ornl.gov

© The Author(s) 2024, Corrected Publication 2024
W. H. Geist et al. (eds.), *Nondestructive Assay of Nuclear Materials for Safeguards and Security*,
https://doi.org/10.1007/978-3-031-58277-6_20

20.2.1 AmLi

AmLi interrogation sources have many good characteristics as active neutron sources. The energies of the emitted neutrons are largely below the fission threshold of ^{238}U, reducing the contribution from that isotope. Neutrons from AmLi sources are emitted singly and do not directly contribute to correlated counting rate backgrounds because they are (α,n) emitters. The long half-life makes AmLi sources very suitable for the determination of detector stability because of its predictable intensity; however, the neutron yield per curie is relatively low, and so the activity of the ^{241}Am material is very high. This case results in high-intensity, 60 keV gamma, which is usually shielded with a tungsten capsule. The biggest disadvantage of AmLi sources is that the number of current suppliers of this type of source is extremely limited.

AmLi sources have been used in the active well coincidence counter (AWCC; Sect. 20.5.2), the uranium neutron collar (UNCL; Sect. 20.5.3), the Euratom fast collars (Sect. 20.5.4), and the fast-neutron coincidence collar (Sect. 20.5.5).

20.2.2 Californium-252

Californium-252 is a very useful as a reference source for detector functionality checking. The specific fission rate of ^{252}Cf is very high, and so the mass required to make typical sources is less than a microgram; however, the neutron spectrum does extend well beyond the ^{238}U fission threshold, which leads to interference when ^{235}U is being measured in low-enriched items. Additionally, the high neutron multiplicity causes backgrounds in correlated counting systems that have to be carefully subtracted. Californium-252 sources have been used successfully as external interrogation sources (Chap. 17, Sects. 17.15.6 and 20.5.2). Californium sources typically contain a few percent of ^{250}Cf when they are new; ^{250}Cf has a longer half-life than ^{252}Cf, and so the fraction of neutrons from ^{250}Cf gradually increases. This condition makes the calculation of the intensity of older californium sources difficult.

Californium-252 sources have also been used in an instrument to measure the plutonium content of fast breeder reactor fuel pins [2].

20.2.3 AmBe, PuBe

These (α,n) sources were once commonly used to check detectors. Because beryllium is a more prolific (α,n) emitter than lithium, the sources can be made more intense. AmBe and AmBe sources possess some of the advantages of AmLi (no correlated neutrons), but the neutron energy spectrum extends well beyond the fission threshold of ^{238}U.

20.2.4 SbBe

SbBe photoneutron sources produce neutrons at ~23 keV, which is a convenient neutron energy to penetrate items and well below the fission threshold of ^{238}U. The antimony has to be activated in a reactor to create ^{124}Sb, which has a half-life of only 60 days, making the practical in-field use a scheduling problem.

An instrument based on an SbBe source was described in Ref. [3], and SbBe sources have been successfully used in safeguards in the Phonid device [4].

20.2.5 Curium-244

Curium-244 sources are similar to ^{252}Cf sources but are not widespread. The specific spontaneous fission rate of ^{244}Cm is much smaller than that of ^{252}Cf, and the mass of a useful source is in the range of grams. The advantages of ^{244}Cm over ^{252}Cf are the longer half-life and the lack of a competing longer-lived isotope to interfere with the neutron production rate. The sources are not readily commercially available.

20.2.6 Neutron Generators

Both deuterium-tritium (DT) and deuterium-deuterium (DD) neutron generators (Chap. 13, Sect. 13.8) have been widely available for decades. They have been used for waste measurements and have recently been applied to safeguards measurements in the AWCC (Sect. 20.5.2), the UNCL [5], and the fast-neutron coincidence collar (FNCL; Ref. [6]). The neutron energy in both types of tube is high and causes significant fission in ^{238}U. The tubes can be steady state or pulsed. In both cases, the measurement and monitoring of the tube intensity is not trivial. Small variations in intensity can lead to difficulties with measuring correlated counting rates and increase random uncertainties.

20.3 Design Principles

The design principles of active interrogation incorporate many of the same aspects as passive systems (Chap. 19, Sect. 19.1), such as detection efficiency, die-away time, and consideration of background counting rates. In addition, in active systems, the optimization of the coupling between the interrogation neutron source and the item being measured is an important design feature that allows one to maximize a uniform fission rate over the full volume of the item while minimizing the direct contribution of the interrogating neutrons to the counting rates in the detector.

20.4 Active Singles Counting

As we have seen in Chap. 19, Sect. 19.3, singles counting is a straightforward measurement of the neutron count rate. In active neutron methods, the opportunity exists to measure the item with and without the external source present, which provides a good method of subtracting neutron background—an advantage for singles counting in active compared with passive methods. This section provides details of singles counting applications for active neutron methods.

20.4.1 Shuffler

A shuffler has a movable ^{252}Cf source that induces fission in fissile nuclei (^{235}U, ^{239}Pu) and is then moved away from the sample being assayed while a neutron counter measures delayed neutrons from fission. Shufflers are most often used to measure ^{235}U. The virtue of the shuffler technique is that the delayed neutron signal is proportional to the ^{252}Cf strength, yet the background is very low because the source is shielded when delayed neutrons are counted. Shufflers were first conceived in 1969 as a method to assay uranium nondestructively. About 20 shufflers have been built in the United States, and several others have been built in France and the United Kingdom. A comprehensive description of shuffler operation is given in Refs. [7, 8].

Many different designs of shufflers exist—each tailored to the specific item they are designed to measure. In general, shufflers achieve good precision measurement of fissile material by using high-intensity ^{252}Cf sources (e.g., 10^9 n/s). Such sources need extensive shielding to reduce the neutron background in the interrogation chamber during the delayed neutron-counting period and to have acceptable dose to personnel.

20.4.2 Differential Die-Away Technique

The differential die-away technique uses a pulsed fast-neutron source in a chamber that has a long die-away time for thermal neutrons. The fast-neutron population dies away quickly, leaving a longer-lived thermal-neutron population. A fast-neutron detector will see only the fast die-away of the fast neutrons unless fast neutrons are produced by thermal-neutron-induced fission of fissile material in the chamber. The method is extremely sensitive, being able to detect 1 mg of ^{235}U in 200 L drums in the ideal case. Because the interrogation source is thermal neutrons, the method is highly affected by self-shielding. A comprehensive description of the development of the technique is given in Ref. [9]. This technique has been applied in combination with passive techniques (Chap. 19, Sect. 19.4.3) and also for spent fuel (Chap. 21).

20.5 Active Coincidence Counting

As in passive methods, coincidence counting provides a signature (primarily Doubles) that is indicative of fission, which is particularly useful when the external source does not cause correlated counts in the instrument. The following section describes active instruments that measure the doubles count rate from induced fission.

20.5.1 Active Well Coincidence Counter

20.5.1.1 AmLi-Driven Active Well Coincidence Counter

The ^{235}U mass assay of bulk uranium items, such as oxide canisters, fuel pellets, and fuel assemblies, is not achievable by traditional gamma-ray assay techniques due to the limited penetration of the item by the characteristic ^{235}U gamma rays. Instead, fast-neutron interrogation methods, such as active neutron coincidence counting, must be used. For international safeguards applications, a commonly used active neutron system is the AWCC, a ^{3}He proportional tube-based neutron coincidence well counter in which AmLi neutron sources located in the detector end plugs provide a source of fast interrogating neutrons. The AmLi neutrons induce fission in the fissile materials contained within the item of interest. The fissile mass of the item is determined from the observed neutron coincidence rate using an empirically derived calibration curve.

Figure 20.1 illustrates the design of the AWCC [10]. The appearance is very similar to that of a passive coincidence counter except for the two (~5 × 10^{4} n/s) AmLi neutron sources mounted above and below the assay chamber. Two rings of ^{3}He tubes give high efficiency for counting coincidence events from induced fissions. The AmLi sources produce no coincident neutrons but do cause many accidental coincidences that dominate the assay error (see Chap. 17, Sect. 17.15.5); thus the polyethylene

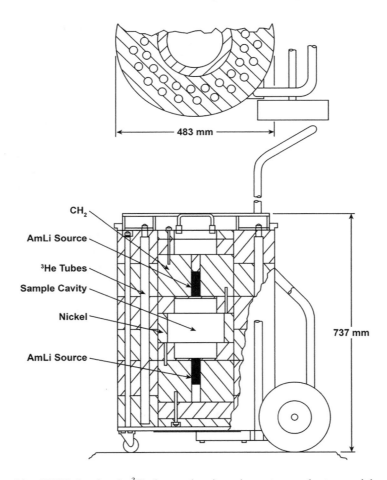

Fig. 20.1 Schematic diagram of the AWCC showing the ^{3}He detector locations, the neutron moderators, and the cadmium-lined sample cavity

Table 20.1 Performance characteristics of the AWCC

Characteristic	Thermal mode	Fast mode
Detection efficiency		26%
Die-away time		50 μs
Range	0–100 g ^{235}U	100–20,000 g ^{235}U
Low-enrichment U_3O_8	11 counts/s − g ^{235}U	0.18 counts/s − g ^{235}U
High-enrichment metal	NA	0.08 counts/s − g ^{235}U
Absolute precision for large items (1000 s)	0.3 g ^{235}U	18 g ^{235}U
Sensitivity limit[a] for small items (1000 s)	1 g ^{235}U	24 g ^{235}U

[a]Defined as net coincidence signal equal to 3σ of background in 1000 s counting times

moderator and cadmium sleeves are designed for most efficient counting of the induced fission neutrons but inefficient counting of the (α,n) neutrons from the AmLi interrogation source.

The nickel reflector on the interrogation cavity wall gives a more penetrating neutron irradiation and a slightly better statistical precision than would be obtained without it. With the nickel in place, the maximum item diameter is 17 cm. For larger items, the nickel can be removed to give a cavity diameter of 22 cm. The end plugs have polyethylene disks that serve as spacers; the disks can be removed to increase the cavity chamber height. Removal of the disks on the top and bottom plugs allows the cavity to accommodate an item that is 35 cm tall.

A cadmium sleeve on the outside of the detector reduces the background rate from low-energy neutrons in the room. A cadmium sleeve in the detector well removes thermal neutrons from the interrogation flux and improves the shielding between the ^{3}He detectors and the AmLi source. With this cadmium sleeve in place, the AWCC is said to be configured in the "fast mode." The neutron spectrum is relatively high energy, and the counter is suitable for assaying large quantities of ^{235}U. With the cadmium sleeve removed, the AWCC is in the "thermal mode." The neutron spectrum is moderated to a relatively low energy, and the sensitivity of the counter is greatly enhanced, but the penetrability of the interrogation neutrons is very low. In the thermal mode, the counter is suitable for assaying small or low-enriched uranium items.

Table 20.1 summarizes the performance characteristics of the AWCC for both the fast and thermal modes of operation. The absolute assay precision is nearly independent of the mass being assayed (see Chap. 17, Sect. 17.15.5). In general, the AWCC is best suited for high-mass, highly enriched uranium items and should not be used for low-^{235}U-mass samples except for well-defined items in the thermal mode. The AWCC can also be used for the passive assay of plutonium by removing the AmLi sources.

The AWCC is portable, lightweight, stable, and insensitive to gamma-ray backgrounds (up to ~5 mSv/h). This last feature makes it applicable to ^{233}U-Th fuel-cycle materials, which generally have very high gamma-ray backgrounds from the decay of ^{232}U. Variants have been deployed using internal gamma-ray shielding to extend the applicability of the AWCC to surface exposure rates from the item up to 5 Sv/h [11].

The AWCC has been evaluated for several measurement problems that are of interest to inspectors, including

1. highly-enriched-uranium (93% ^{235}U) metal buttons weighing approximately 1–4 kg, which are input materials to fabrication facilities;
2. cans of uranium-aluminum scrap generated during manufacture of fuel elements;
3. cans of uranium-oxide powder;
4. mixtures of uranium oxide and graphite;
5. uranium-aluminum ingots and fuel pins; and
6. materials testing reactor (MTR) fuel elements.

Typical calibration curves are shown in Figs. 20.2 and 20.3 for cases (1), (3), and (4). All of the calibration curves show the effects of neutron absorption within the uranium, and Fig. 20.2 also shows the opposing effect of self-multiplication within the metal.

Field tests with MTR fuel elements [13] have shown that it is possible to obtain ~1% accuracy in assay times of 400 s. The advantage of the AWCC over the traditional gamma-ray assay for MTR fuel elements is that the AWCC has no problems with different plate geometries and lower ^{235}U enrichments. For applications to MTR-type fuel elements and plates, the AWCC is reconfigured as shown in Fig. 20.4 [14]. The two AmLi sources are positioned in the interior of the polyethylene insert that holds the MTR elements. Figure 20.5 shows the calibration curve for typical MTR fuel plates and elements.

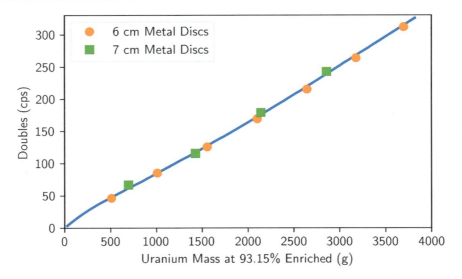

Fig. 20.2 AWCC response as a function of HEU uranium mass for 6 and 7 cm metal discs stacked together to obtain the total masses shown [12]

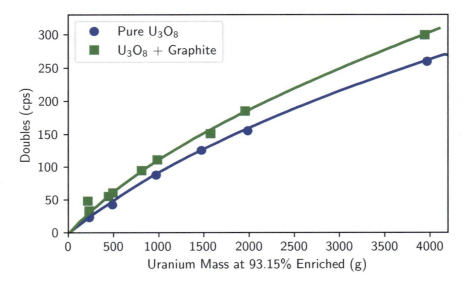

Fig. 20.3 AWCC response as a function of uranium mass for highly enriched uranium oxide powder and mixtures of uranium oxide and graphite [12]

20.5.1.2 DD Neutron-Generator-Driven AWCC

Limited availability of AmLi in recent years prompted investigation of alternative neutron interrogation sources. This section discusses the use of the DD neutron generator as a replacement to the AmLi source [15]. The DD/AWCC replaces the two AmLi sources with a single neutron generator installed in the lower end plug of the AWCC (Fig. 20.6). The DD generator affords only a single source of interrogating neutrons, leading to a more pronounced vertical response profile such that the induced fission rate decreases as a function of height. Unlike the AmLi source, the DD generator may be switched on and off, simplifying passive/active coincidence measurements and background corrections. The higher neutron energy (2.2 MeV) of the DD generator in comparison with the AmLi source affords greater penetration into the item but also increases the sensitivity to ^{238}U so that the DD/AWCC measurement requires prior knowledge of the ^{235}U enrichment. The ^{238}U sensitivity is roughly three times greater compared with AmLi interrogation.

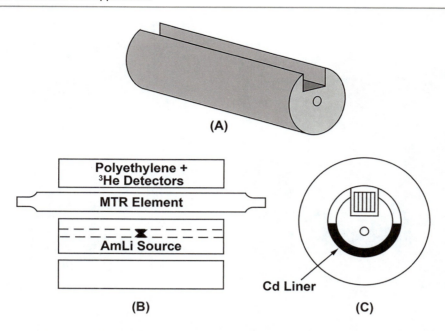

Fig. 20.4 Horizontal configuration of the AWCC with a polyethylene insert used for the assay of MTR fuel plates and elements

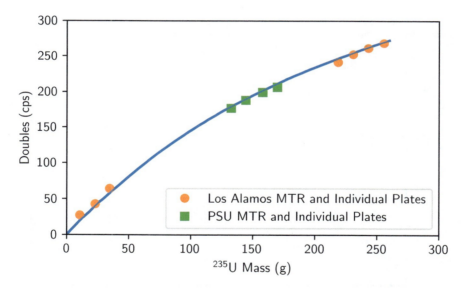

Fig. 20.5 Calibration curve for MTR fuel plates and elements measured in the AWCC. Two elements were measured—one from Los Alamos National Laboratory and the other from Pennsylvania State University (PSU)

With the neutron generator operating in the steady state or continuous mode, the mass dependent coincidence response is similar to that of the AmLi (Fig. 20.7), although the increased sensitivity to ^{238}U requires the introduction of a ^{235}U effective mass (similar in concept to the ^{240}Pu effective mass in passive coincidence counting). The ^{235}U$_{eff}$ mass is given as

$$m_{235\text{eff}} = m_{235} + 0.043 \cdot m_{238.} \tag{20.1}$$

The ability to switch the neutron generator on and off allows alternative analyses to be performed by measuring the delayed neutron signal. Operated in a slow-pulse mode (e.g., generator on for 10 s and off for 10 s), the DD/AWCC behaves in the same manner as the ^{252}Cf shuffler (Sect. 20.4.1) but without the complication of the source storage module or mechanical

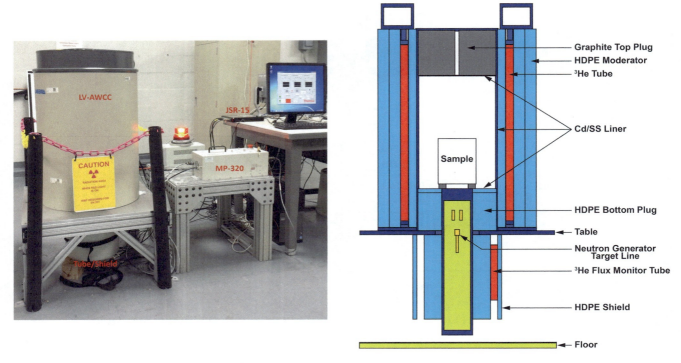

Fig. 20.6 (left) Photograph of the modified LV-AWCC with the neutron generator tube installed in the bottom end plug. (right) Illustration of the arrangement of the DD neutron generator tube within the LV-AWCC assay cavity [15]

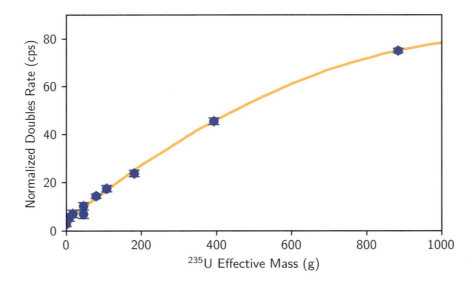

Fig. 20.7 Measured DD/AWCC doubles rate as a function of ^{235}U effective mass over enrichments in the 0.2%–93% range (generator yield 2.3×10^5 n/s; Ref. [15])

source transfer mechanism. The delayed neutrons are counted following each irradiation period, and the cycle is repeated until the desired measurement precision is achieved.

Operated in a fast-pulse mode—typically 250 Hz with a 5% duty cycle—the time dependence of the neutron count rate (Fig. 20.8) is reminiscent of the response from the differential die-away systems (Sect. 20.4.2). Due to the assay chamber size and timing properties of the AWCC, the signal-to-noise ratio in the die-away region is too poor to extract any meaningful mass information. Instead, the delayed neutron signal is counted during the second half of the irradiation cycle. This fast-pulse method results in a greater fraction of detectable delayed neutrons (50%) than is achievable in either the slow-pulse mode or

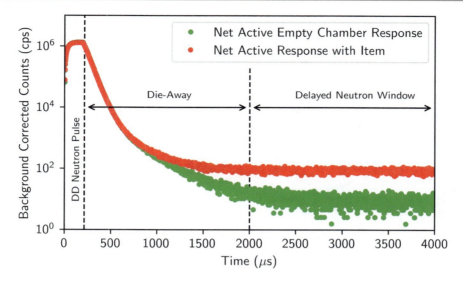

Fig. 20.8 Plot of the count rate as a function of time for the DD/AWCC with the MP320 operating at a 250 Hz repetition rate with the multichannel scaler sweep synched to the generator pulse. The plot from the 20 min count represents the sum of 300,000 generator pulses and 2 μs per channel [16]

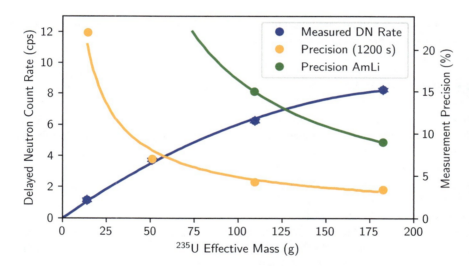

Fig. 20.9 Plot of the measured delayed neutron rates and the observed measurement precision for the DD/AWCC operated in pulsed, fast active interrogation mode at 2.3×10^5 n/s

the traditional ^{252}Cf shuffler. Additionally, because the typical pulsed neutron generator will produce the same average neutrons per second independent of the pulse duty cycle, the apparent source strength of the pulsed neutron generator is much larger than the comparable ^{252}Cf shuffler source, which is exposed only part of the time. For example, a 2×10^5 n/s pulsed generator that operated with 5% duty cycle would be equivalent to a 4×10^7 ^{252}Cf source.

Figure 20.9 shows the delayed neutron (DN) count rate response as a function of mass for the fast-pulsed DD neutron interrogation along with the measurement precision, for a total 1200 s measurement time. The measurement precision for the equivalent measurement with two AmLi sources is shown for comparison. Figure 20.10 provides a comparison of the expected measurement precision as a function of mass for the AWCC operated as a traditional AmLi-driven coincidence counter, as a steady-state DD neutron-generator-driven coincidence counter, and as a pulsed DD neutron-generator-driven delayed neutron counter. The ^{238}U contributes to the delayed neutron count rates and must be accounted for—in this case, $m_{235\text{eff}} = m_{235} + 0.083 \cdot m_{238}$. The performance presented in Figs. 20.9 and 20.10 were obtained with a ThermoFisher MP320 DD neutron generator operating at a yield of 2.3×10^5 n/s. Unlike the coincidence measurement, the measurement precision

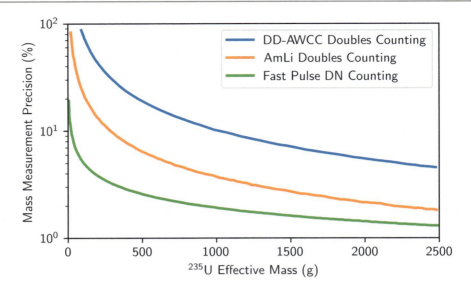

Fig. 20.10 Plot of the expected measurement precision of the AWCC measurement for the AWCC operated as a traditional AmLi-driven coincidence counter, as a steady-state DD neutron-generator-driven coincidence counter, and as a pulsed DD neutron-generator-driven delayed neutron counter

can be improved by increasing the interrogating source strength. The measurement precision achievable in the delayed counting mode will improve with increasing generator yield.

20.5.2 AWCC Operation with ^{252}Cf

The principle of time-correlated induced fission (TCIF) was described in Chap. 17, Sect. 17.15.2. The AWCC has been adapted to use ^{252}Cf sources as a possible replacement for AmLi for the measurement of bulk uranium [17, 18]. The measurement functions similarly: calibration curves are produced from known sources, and then the measured doubles rate of an unknown source is used to determine the ^{235}U mass. Any departure from the conditions of the calibration measurements will result in a less accurate assay. The use of ^{252}Cf causes a stronger sensitivity to any differences; for example, spectrum modifications that change the amount of fission in ^{238}U.

Any variable that would change the Doubles from the ^{252}Cf source must be tightly controlled. The doubles rate from the ^{252}Cf source can be much greater than that of the ^{235}U Doubles, so a small change in ^{252}Cf neutron-detection efficiency can result in a large error in the determination of ^{235}U mass. In an example measurement, the ^{252}Cf doubles background was measured to be 18,000, which is 63% of the 28,400 Doubles measured with a 181 g ^{235}U source. Variables that should be controlled include the radial and axial position of the uranium item, the shape of the uranium container and uranium within, the orientation of both the ^{252}Cf and uranium sources, and the presence of any nonnuclear material; for example, stands or the thickness of the uranium sample container.

Other variables should be controlled or included in the calibration. The uranium itself scatters the ^{252}Cf neutrons that affect the doubles background even without considering induced fission. The measurement performs best when the uranium volume in the calibration and assay items is held constant and the enrichment is changed or when the uranium volume changes consistently with the change in mass and the enrichment is held constant. Although it may be tempting to subtract the contribution from ^{238}U, it is prohibitively difficult due to the complex coupling of Doubles produced from ^{252}Cf, ^{235}U, and ^{238}U; induced fission between the three isotopes; and changes in ^{252}Cf scattering due to ^{238}U. Instead of attempting to break the measured Doubles into its constituent components, the analysis should simply use a calibration curve, which automatically accounts for all contributions to the Doubles. The ^{252}Cf decay should be accounted for due to its short half-life.

A thermal calibration curve created with the uranium standards is shown in Fig. 20.11. The statistical uncertainties are better than 0.5%. A simple mass verification with the 102 g ^{235}U item resulted in a 4.9% difference in declared and assay mass.

TCIF performs best in thermal mode because the probability of inducing fission is greater, so the fraction of Doubles from ^{235}U compared with ^{252}Cf is higher. Still, fast mode does allow the measurement of a higher-mass range. A calibration curve is shown in Fig. 20.12.

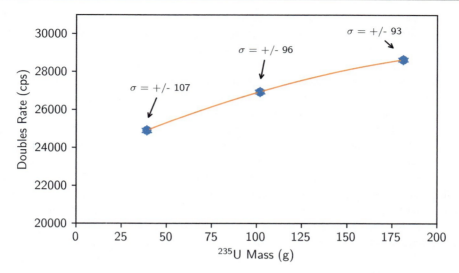

Fig. 20.11 Doubles rate versus ^{235}U mass calibration curve with 1σ statistical uncertainties. Both the calibration curve and the uncertainties were determined using INCC software [18]

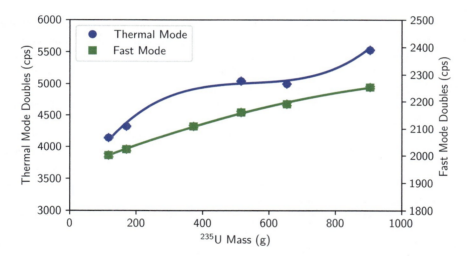

Fig. 20.12 Fast- and thermal-mode AWCC time-correlated induced fission calibration curves [18]

20.5.3 Neutron Collar for Fresh Fuel Assembly Measurements

Active neutron coincidence counting is used for fresh nuclear fuel assembly measurements in low-enriched uranium fuel fabrication facilities worldwide. For safeguards purposes, of interest is to measure full fresh nuclear fuel assemblies because they constitute the output product from the fuel fabrication facility and the input to the nuclear reactors. Enriched uranium is often transferred from one installation or country to another in the form of fuel assemblies. The UNCL [19] is the primary nondestructive assay (NDA) system used by international safeguards inspectorates to perform verification measurements of uranium content in light-water reactor fresh fuel assemblies.

The UNCL is based on the same physics principles as the AWCC. The standard assay approach uses an active neutron interrogation technique [20] for measurement of the fissile ^{235}U content in fresh nuclear fuel assemblies. Coincidence counting is performed in both active and passive modes. In the active mode, the method employs an AmLi neutron source—housed in polyethylene—to induce fission reactions in the fissile component of the fuel assembly. This method enables coincidence counting of the resulting fission neutrons emitted from the fuel assembly. Coincidence counting distinguishes detected fission neutrons from the undesired neutron counts from the ^{241}AmLi interrogation source and room background. The method uses the time signatures for these reactions that originate from the neutron arrival times in the detector. Prompt fission neutrons are emitted almost simultaneously from fission events in time-correlated bursts and are thus detected close together in time. Neutrons emitted from the AmLi interrogation source are detected randomly in time. An active coincidence neutron-counting rate can be derived and used to quantify ^{235}U mass. The AmLi is selected because the emitted neutron flux lies below the fission threshold of ^{238}U.

The passive signal from a fresh nuclear fuel assembly is dominated by neutrons emitted from the spontaneous fission of ^{238}U; therefore, in principle, the ^{238}U content can be verified separately by a passive measurement with the ^{241}AmLi source removed. When no interrogation source is present, the passive neutron coincidence rate gives a measure of the fertile ^{238}U component of the fuel assembly through the detection of spontaneous fission neutrons. When the interrogation source is added, the increase in the total coincidence rate gives a measure of ^{235}U. The net signal of interest is obtained by the subtraction of the passive coincidence rate from the active coincidence rate. This fresh fuel assay technique is applied to fissile content determination in boiling-water reactor (BWR), pressurized-water reactor (PWR), and the water-water energetic reactor (WWER) or the Russian-designed, water-cooled, water-moderated reactors (VVER) fresh fuel assemblies.

The UNCL assay system design was developed at Los Alamos National Laboratory (LANL) in 1981 [19]. The UNCL was succeeded by the UNCL-II design, which was developed in 1989 [20] and is now commercially available. Two separate UNCL-II configurations were developed to accommodate different detector body sizes for the assay of PWR and BWR fuel assemblies. The UNCL-II has higher detection efficiency than the original UNCL design and features an additional lift-out door that contains the ^{241}AmLi source [20].

The UNCL-II comprises three banks of ^{3}He gas-filled proportional tubes and uses one ^{241}AmLi source embedded in a high-density polyethylene (HDPE) body. The ^{3}He neutron detector tubes are 2.54 cm in diameter and 33 cm long (active length), with 20 and 16 tubes in the PWR and BWR configurations, respectively, compared with 18 tubes of the same dimensions in the original UNCL unit. The HDPE body performs three basic functions in the system: general mechanical support, interrogation source neutron moderation to thermalize the interrogating neutron source flux, and slowing down of induced fission neutrons before their detection in the ^{3}He tubes to increase the probability of detection. For inspection applications, it is desirable to make the system portable. The weight of the UNCL-II detector system is 48 kg and 39 kg for the PWR and BWR configurations, respectively, compared with 38 kg for the original UNCL unit.

The neutron-detection response of the collar is uniform throughout a fuel assembly in the absence of burnable poisons. The interrogation neutron flux decreases with distance from the ^{241}AmLi source; however, the detection efficiency increases with distance from the source as fission neutrons are moderated within the assembly [21] and fast-neutron multiplication is increased in the central region of the assembly. The UNCL-II detector body can be used both with and without cadmium liners. With the cadmium liners in place, the collar operates in the fast mode, and epithermal neutrons dominate the interrogating neutron flux. Without the cadmium liners, the collar operates in the standard thermal mode, and thermal neutrons can return from the polyethylene and interrogate the fuel assembly.

The UNCL-II assay system is shown in Fig. 20.13. For applications, the collar is positioned on a cart that is moved next to a fuel assembly. For acquiring neutron counts, the assay system is used with data acquisition electronics, such as a multiplicity shift register connected to a local computer to derive the ^{235}U mass (per unit length) from the acquired data using the IAEA neutron coincidence counting software package INCC [22].

The original tests and evaluations of the UNCL were performed at both PWR [23] and BWR [24] fuel fabrication facilities. These measurements have since been updated. Menlove et al. [20] is currently the main reference manual for neutron collar operation and calibration.

The neutron collar measures the coincidence rate, Doubles, which is used to derive the ^{235}U mass per unit length of the fuel assembly (g.cm^{-1}), also known as the linear mass density. The linear mass density is proportional to the uranium enrichment for a given type of fuel assembly. The sample region is only a small fraction of the total length of the fuel assembly, centered in the midplane of the detector body; therefore, the uranium linear mass density is multiplied by the operator-declared active length of the fuel assembly (cm), which is verified separately with a measurement using a handheld gamma detector, for example, to calculate the total uranium mass (g).

An absolute calibration curve and calibration cross-reference method were established [20] to facilitate the use of different neutron collars among facilities for routine safeguards inspection measurements. The measured response of each individual neutron collar is cross-referenced to an absolute calibration curve to derive the uranium linear mass density [20]. The shape of this absolute calibration curve was established for a single UNCL-II neutron collar, and then all other neutron collars are cross-calibrated to the reference detector.

The absolute calibration curve was established by measuring the reals rate response of a single UNCL-II neutron collar unit to several standard BWR and PWR fuel assemblies that cover a wide range of uranium enrichments and dimensions, as well as burnable poison rod loadings. These absolute calibration measurements were performed using the LANL BWR and PWR

Fig. 20.13 Uranium neutron coincidence collar UNCL-II design (BWR size)

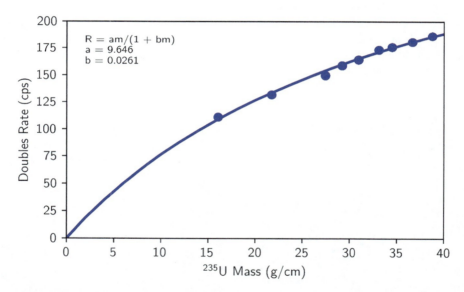

Fig. 20.14 PWR reference calibration curve (no cadmium) using UNCL-II [20]

reference assemblies to establish the shape of the calibration function both with and without cadmium liners. A corresponding collar operational and calibration procedure [20] has been developed to implement the calibration at different nuclear facilities.

The cross-reference calibration method holds under the assumption that all neutron collars have the same behavior for a given fuel type and differ only in the absolute magnitude of the doubles response, D, depending on their source and detection efficiency [20]. Therefore, the absolute calibration curve for a reference UNCL-II may be used for all neutron collars if correction factors are applied to account for deviations from the reference conditions due to the individual collar characteristics and calibration conditions [25].

The method of correcting the measured doubles rate for AmLi source strength, efficiency, burnable poison, heavy metal linear density, and other factors is described in Chap. 17, Sect. 17.15.4. The calibration curve in Fig. 20.14 corresponds to the

PWR reference curve for the thermal mode (no cadmium) UNCL-II [20]. Other modes of operation and fresh fuel type have similarly shaped calibration curves.

The neutron collar absolute calibration method and results are described in Ref. [21]. An alternative description of the calibration correction factors is available in Ref. [24]. The original work [21] recognizes that achieving this kind of standardization and universal calibration is complex because fresh nuclear fuel assemblies include many different enrichments, pin configurations, fuel masses, and burnable poison loadings. Peerani et al. [25] have assessed uncertainties in the calculated response introduced by the parameterization of the correction factors. Since the establishment of the absolute calibration method, several studies have been performed to extend the calibration correction factors for different fuel parameters, including for increased burnable poison loadings in modern fuel designs. Moreover, studies have been performed to determine how to extend the measured calibration approach by using modeling and simulation to derive accurate collar models for calibration [26, 27].

Recent updates to the standard UNCL assay method include the evaluation of AmLi source replacement options [15, 28, 29] and the use of list-mode data acquisition to study neutron collar spatial response [30].

20.5.4 Euratom Fast Collar for PWR and BWR Fresh Fuel Assemblies

The Euratom fast collar (denoted EFCP or EFCB for the PWR or BWR versions, respectively) is designed to address modern reactor fuel assemblies that have higher burnable poison loadings to enable higher burnups [31]. Models of the instruments are shown in Fig. 20.15. The EFCP is cadmium lined and operates in fast mode but was designed with higher efficiency to reduce the measurement time and enable high throughput. The higher efficiency is created by two rows of 10 atm pressure, 1 in. diameter ^{3}He tubes—33 for the PWR case and 27 for the BWR case. An AmLi source is still used to induce fission in the ^{235}U. The effect of the poison rods is greatly reduced. Twelve poison rods of 5.2% gadolinium in the UNCL thermal mode affect the Doubles by 20% but in the EFCP cause only a 3% effect. Measurement time is also greatly improved. A 20 min measurement with the UNCL in fast mode resulted in a 7% ^{235}U mass uncertainty, whereas the same measurement with an EFCB resulted in 2.6% uncertainty. The improved design reduces count time to ~30 min from ~1 hour and reduces the measured mass uncertainty from ~4% to 2% [32].

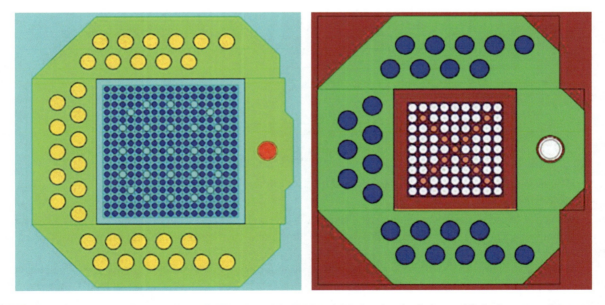

Fig. 20.15 Horizontal cross section through the EFCP (left) and the EFCB (right) showing the fuel assembly in the center of the measurement cavity. Polyethylene is shown as green in both diagrams. The ^{3}He detectors have a 1 in. diameter (yellow in the top diagram and blue in the bottom). The AmLi source is located on the right side of the polyethylene panel in each case

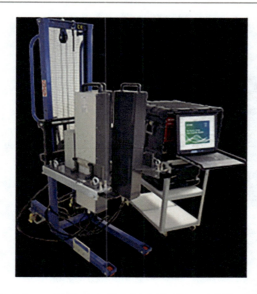 

Fig. 20.16 (left) Fast-neutron coincidence collar alongside data acquisition without active interrogation block; (right) computer drawing showing detector positioning

20.5.5 Fast-Neutron Coincidence Collar

The prevalence of legacy ^{3}He-based neutron detectors is waning somewhat while replacement instruments implement non-helium-3 technologies across most applications, including safeguards. The FNCL, shown in Fig. 20.16, is a replacement candidate developed by the International Atomic Energy Agency [33] for the uranium neutron coincidence collar (Sect. 20.5.2) for the NDA of fresh nuclear fuel assemblies that contain burnable poisons. Both instruments rely on neutron sources—particularly AmLi due to its soft spectrum and lack of intrinsic multiplicity—for active interrogation of fresh fuel assemblies that possess a negligible intrinsic neutron source term. The use of coincidence counting suppresses the contributions from the neutron background and interrogation source to count neutrons from induced fissions of ^{235}U. A passive measurement—without interrogation sources—is performed to account for the contributions of ^{238}U spontaneous fission. The result of the assay is the neutron doubles rate and the linear mass density (g/cm) of ^{235}U.

The FNCL consists of three panels, each containing four isolated detection volumes of EJ-309 (a liquid plastic scintillator) for 12 total detectors. Plastic scintillators rely on the pulse shape to discriminate between gamma and fast-neutron events. Each cube-shaped detector is approximately 10 cm, with its own photomultiplier tube. The side of each detector that faces the measurement cavity is shielded by lead (10 mm) to reduce gamma interference and by a cadmium liner (1 mm) to prevent neutrons moderated by the FNCL from inducing fission chains. The space between detector elements is filled with polyethylene to reduce neutron crosstalk by slowing down neutrons below the fast detection threshold.

Measurements of three types of fuel assemblies (PWR, BWR, and VVER) are supported by unique interrogation source blocks, as shown in Fig. 20.17. Each block holds two AmLi sources (that are easily extracted between passive and active acquisitions) embedded in HDPE. The interior face is also lined with cadmium to reduce reflection from the assembly and to also maintain a greater-than-thermal-neutron interrogation energy to induce fissions deeper within the assembly instead of the pins adjacent to the source.

A fresh fuel assay comprises normalization measurements that are performed under the technician account such that inspectors cannot alter the normalization; the assay measurements are performed by the inspector account and can receive only the active normalization from the technician. The normalization requires the measurement—in sequence—of a background, a gamma-energy calibration (typically ^{137}Cs), a pulse-shape discrimination calibration (typically using the neutron interrogation sources or ^{252}Cf), an initial source measurement (using interrogation sources), and finally the AmLi normalization. Once all normalization steps are complete, the normalization is applied, and fuel assembly assays can commence in the inspector account.

The FNCL offers some advantages and disadvantages compared with the ^{3}He-based UNCL. Several advantages include a fast detector response that practically eliminates the accidentals rate (by using a 60 ns coincidence gate compared with the 64 μs coincidence gate of the UNCL), direct detection of fast neutrons from fission (no need to moderate for thermal capture),

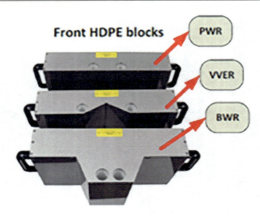

Fig. 20.17 AmLi active interrogation source holders for several fuel assembly designs

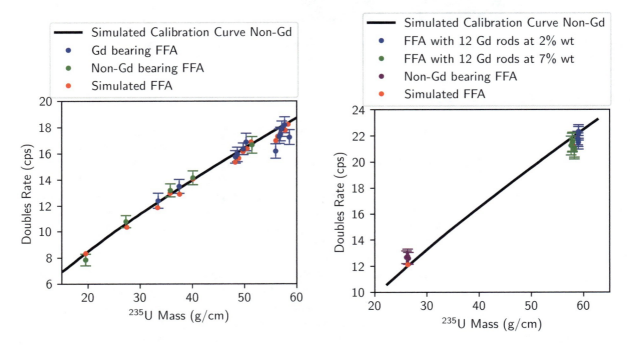

Fig. 20.18 Assays of fresh fuel assemblies at KEPCO Nuclear Fuel, Republic of Korea (left), and at the Indústrias Nucleares do Brasil in Brazil (right), with 3σ random uncertainty shown

and full waveform capture for reanalysis. Some drawbacks include imperfect pulse-shape discrimination, sensitivity to gammas, possibility of crosstalk, and complexity of the normalization procedure.

Field tests of the FNCL were conducted in Brazil at the Indústrias Nucleares do Brasil [34] and the Republic of Korea at the KEPCO Nuclear Fuel [35]—both fuel fabrication plants—for a variety of enrichments and burnable poisons of 39 fresh fuel assemblies. MCNPX-PoliMi [36, 37] models, coupled with SimPLis (Ref. [38]; a post processor for liquid scintillators), produced synthetic list-mode data to obtain the ^{235}U mass calibration for the FNCL. A typical event rate of 4×10^5 per second was present for all active measurements, using total (two) AmLi activities of 129 GBq and 84 GBq at each site. The assay measurements from both sites are shown in Fig. 20.18 and display good agreement between measured and known ^{235}U linear densities across a large range.

20.5.6 Boron-Based Collar

A neutron collar [39] has been designed that is based on similar technology as the high-level neutron counter boron detector and the high-dose neutron detector (Chap. 19, Sect. 19.6.2). The modules are constructed with polyethylene layers between each boron cell to provide neutron thermalization. One module has been experimentally benchmarked with simulations using the MCNP code, including a notch that allows use with WWER fuel (as in the case of the FNCL, Sect. 20.5.4). The use of MCNP calculations, extrapolation to a full instrument shows a figure of merit that is almost three times higher than the UNCL-I.

20.6 Active Multiplicity Counting

20.6.1 Introduction

Most active multiplicity measurements to date have been carried out with conventional AWCCs (Sect. 20.5.1.1) that contain AmLi sources to induce fission events in the uranium. The AWCC measurement cavity is typically configured to have a detection efficiency of ~27%–30% and a neutron die-away of ~52 μs. The only hardware change is the use of a multiplicity shift register to collect neutron singles, doubles, and triples coincidence events. The AmLi sources do not produce a very high count rate, so a variety of existing multiplicity shift registers (including those without derandomizer circuits) or list-mode modules can be used for this application; however, the AmLi neutron sources produce a large number of accidental coincidence events that limit the assay precision, as described in more detail in Chap. 18, Sect. 18.3.

20.6.2 Measurement Software

Active multiplicity analysis (Chap. 18, Sect. 18.3.1) requires the use of a software package to acquire and analyze data from the multiplicity shift register. Measurement control options, quality control tests, and calibration and least-squares fitting options are also needed in the software. These tasks can be accomplished with the INCC software [22] or several commercial software packages that use the same analysis algorithms. These codes include data collection and analysis algorithms for passive coincidence counting by several different algorithms, passive multiplicity counting, active coincidence counting, and active minus passive coincidence counting. At this time, the active multiplicity option in the INCC code determines the neutron multiplication from the triples/doubles ratio but does not determine the uranium mass. The calculation of sample coupling and ^{235}U mass requires manual calculations.

20.6.3 Active Epithermal Neutron Multiplicity Counter

The epithermal neutron multiplicity counter (ENMC) has also been used for active multiplicity measurements, made possible by the design and construction of a new set of active end plugs [40]. Several different design ideas were studied, including varying the end plug material, recessing the interrogation neutron sources, and shielding the interrogation sources with reflector materials. The optimum solution was to construct the end plugs from polyethylene without recessing or shielding the interrogation sources. Because the ENMC uses more—and higher-pressure—^{3}He tubes than the AWCC, the MCNP-calculated efficiency for the active mode was near 55%, with a die-away time around 19 μs. This large reduction in the die-away time relative to the AWCC reduces the number of accidental coincidences from the AmLi interrogation neutron source. Running the ENMC in active mode will reduce the assay time for uranium samples by a factor of approximately 10 over the standard AWCC.

20.6.4 Future Detectors

In the future, neutron counters that are optimized specifically for active multiplicity counting could become available. Two design goals exist for reducing the accidental coincidence background from the AmLi sources. The first goal is to design a counter that minimizes the interrogation neutron-detection efficiency and maximizes the induced fission neutron-detection efficiency, which will result in less-random interrogation neutrons that accidentally fall within the coincidence gate. The second goal is to build a counter with as small a die-away time as possible, which will allow for a reduction in the coincidence gate width that reduces the accidental coincidence rate proportionally. Both goals may be realized by the development of a liquid-scintillator-based neutron counter [41, 42]. This counter could have a good efficiency of ~25% and a die-away time and coincidence gates as short as 30 ns. This result would yield another order of magnitude reduction in assay time for highly enriched uranium (HEU) and allow much smaller samples (~100 g or more) of HEU to be measured by active multiplicity counting.

References

1. T. Gozani, *Active Nondestructive Assay of Nuclear Materials: Principles and Applications*, NUREG/CR-0602 (U.S. Nuclear Regulatory Commission, Washington DC, 1981)
2. H.O. Menlove, R.A. Forster, J.L. Parker, D.B. Smith, Californium-252 assay system for FBR-type fuel pins. Nucl. Technol. **20**(2), 124–133 (1973)
3. H.O. Menlove, R.A. Forster, D.L. Matthews, A photoneutron antimony-124-beryllium system for fissile materials assay. Nucl. Technol. **19**(3), 181–187 (1973)
4. J.K. Sprinkle, R. Bardelli, L. Becker, L. Lezzoli, P. Rochez, P. Schillebeeckx, U. Weng, The measurement capabilities of Phonid 3b, in *Proceedings of the 13th Annual ESARDA Symposium on Safeguards and Nuclear Material Management*, Avignon, France May 14–16, 1991, p. 453–457
5. R.D. McElroy Jr., S.L. Cleveland, The DD Neutron Generator as an Alternative to Am(Li) Isotopic Neutron Source in the Uranium Neutron Coincidence Collar. ORNL/TM-2017/736, Oak Ridge National Laboratory report, Oak Ridge, TN (2017). https://doi.org/10.2172/1427636
6. R.D. McElroy, S.E. O'Brien, A.L. Lousteau, Initial characterization of a DD neutron generator driven fast neutron coincidence collar, in *Proceedings of the INMM & ESARDA Virtual Annual Meeting*, August 23–26 and August 30–September 1, 2021
7. P.M. Rinard, Shufflers, Chapter 9 in *Passive Nondestructive Assay of Nuclear Materials—2007 Addendum* https://www.lanl.gov/org/ddste/aldgs/sst-training/technical-references.php. Accessed Mar 2023
8. P.M. Rinard, Application Guide to Shufflers. Los Alamos National Laboratory report LA-13819-MS (September 2001)
9. J.T. Caldwell, R.D. Hastings, G.C. Herrera, W.E. Kunz, E.R. Shunk, The Los Alamos Second-Generation System for Passive and Active Neutron Assays of Drum-Size Containers. Los Alamos National Laboratory report LA-10774-MS (1986)
10. H.O. Menlove, Description and Operation Manual for the Active Well Coincidence Counter. Los Alamos Scientific Laboratory report LA-7823-M (1979)
11. R.D. McElroy, Jr., A. Lousteau, Characterization of the High Activity Active Well Coincidence Counter for Highly Enriched Uranium Residues. Oak Ridge National Laboratory technical report ORNL/TM-2018/1065, Oak Ridge, TN (November 2018)
12. H.O. Menlove, N. Ensslin, T.E. Sampson, Experimental Comparison of the Active Well Coincidence Counter with the Random Driver. Los Alamos Scientific Laboratory report LA-7882-MS (1979)
13. M.S. Krick, P.M. Rinard, Field Tests and Evaluations of the IAEA Active Well Coincidence Counter. Los Alamos National Laboratory report LA-9608-MS (December 1982)
14. R. Sher, Active Neutron Coincidence Counting for the Assay of MTR Fuel Elements. Los Alamos National Laboratory report LA-9665-MS (February 1983)
15. R.D. McElroy, S.L. Cleveland, The D-D Neutron Generator as an Alternative to Am(Li) Isotopic Neutron Source in the Active Well Coincidence Counter. Oak Ridge National Laboratory Technical report ORNL/TM-2017/57, Oak Ridge, TN, March (2017)
16. R.D. McElroy, Jr., S.L. Cleveland, The D-D neutron generator as an alternative to the Am(Li) isotopic neutron source in active neutron coincidence counting for international nuclear safeguards, in *Proceedings of the 59th Annual Meeting of the INMM*, Baltimore, MD, July 22–26, 2018
17. H.O. Menlove, C.D. Rael, J.B. Marlow, The Optimization and Calibration of the AWCC Using ^{252}Cf Interrogation and the Comparison with an AmLi Neutron Source, in *Proceedings of the 39th ESARDA Conference*, Düsseldorf, Germany, May 16–18, 2017, p. 303–312
18. R. Weinmann-Smith, V.E. Lucas, L. Crabtree, Bulk U-235 Assay in the AWCC with Cf-252 Interrogation. Los Alamos National Laboratory report LA-UR-22-23910 (2022)
19. H.O. Menlove, Description and Performance Characteristics for the Neutron Coincidence Collar for the Verification of Reactor Fuel Assemblies. Los Alamos National Laboratory report LA-8939-MS (August 1981)
20. H.O. Menlove, J.E. Stewart, S.Z. Qiao, T.R. Wenz, G.P.D. Verrecchia, Neutron Collar Calibration and Evaluation for Assay of LWR Fuel Assemblies Containing Burnable Neutron Absorbers. Los Alamos National Laboratory report LA-11965-MS (November 1990)
21. W.H. Geist, Introduction to Active Neutron Coincidence Counting Using the Collar. Fuel Fabrication Facility Workshop, February 16–20, 2008, Los Alamos National Laboratory report LA-UR-09-00154
22. W. Geist, J. Longo, M. Krick, W. Harker, INCC Software Users Manual. Los Alamos National Laboratory report LA-UR-01-6761 (2009)
23. C. Beets, Optimization of NDA Measurements in Field Conditions for Safeguards Purposes. Centre D'Etude de L'Energie Nucleaire Third Progress Report BLG553, Contract RB/2274 (January 1982)
24. H.O. Menlove, A. Keddar, Field Test and Evaluation of the IAEA Coincidence Collar for the Measurement of Unirradiated BWR Fuel Assemblies. Los Alamos National Laboratory report LA-9375-MS (December 1982)
25. P. Peerani, J. Tanaka, Calibration of neutron collars for fresh fuel element verification, in *Proceedings of the 27th ESARDA Symposium on Safeguards and Nuclear Material Management*, London (UK), May 10–12, 2005

26. A. Favalli, S. Croft, M.T. Swinhoe, Perturbation and Burnable Poison Rod Corrections for BWR Uranium Neutron Collar, in *Proceedings of the 33rd ESARDA Symposium on Safeguards and Nuclear Material Management*, Budapest, Hungary, May 16–20, 2011. Los Alamos National Laboratory report LA-UR-11-02837
27. S.Y. Lee, D.H. Beddingfield, MCNPX-Based Determination of UNCC Correction Factors (Benchmark Problems for MCNPX Simulation). Los Alamos National Laboratory report LA-UR-09-03573 (2009)
28. H.O. Menlove, S.H. Menlove, C.D. Rael, The development of a new, neutron, time correlated, interrogation method for measurement of ^{235}U content in LWR fuel assemblies. Nucl. Instrum. Methods Phys. Res. Sect. A **701**, 72–79 (2013)
29. M.A. Root, H.O. Menlove, R.C. Lanza, C.D. Rael, K.A. Miller, J.B. Marlow, J.G. Wendelberger, Using the time-correlated induced fission method to simultaneously measure the ^{235}U content and the burnable poison content in LWR fuel assemblies. Nucl. Technol. **203**(1), 34–47 (2018). https://doi.org/10.1080/00295450.2018.1429112
30. L.G. Evans, A.D. Nicholson, C.L. Britton Jr., K.J. Dayman, M.N. Ericson, A.S. Moore, Development of a Neutron List Mode Collar (LMCL) and a List Mode Response Matrix Analysis Concept. Oak Ridge National Laboratory report ORNL/TM-2021/2012 (June 2021)
31. L.G. Evans, M.T. Swinhoe, H.O. Menlove, P. Schwalbach, P. De Baere, M.C. Browne, A new fast collar for safeguards inspection measurements of fresh low enriched uranium fuel assemblies containing burnable poison rods. Nucl. Instrum. Methods Phys. Res. Sect. A **729**, 740–746 (2013)
32. M.T. Swinhoe, H.O. Menlove, P. De Baere, D. Lodi, P. Schwalbach, C.D. Rael, M. Root, A. Tomanin, A. Favalli, A new generation of uranium coincidence fast neutron collars for assay of LWR fresh fuel assemblies. Nucl. Instrum. Methods Phys. Res. Sect. A **1009**, 165453 (2021)
33. UM6834—SD7750D Fast Neutron Collar (FNCL) Data Acquisition System 21 User Manual rev. 2 https://www.caen.it/download/?filter=VeryFuel
34. J.S. Beaumont, T.H. Lee, M. Mayorov, et al., A fast-neutron coincidence collar using liquid scintillators for fresh fuel verification. J. Radioanal. Nucl. Chem. **314**, 803–812 (2017). https://doi.org/10.1007/s10967-017-5412-x
35. J. Beaumont, T. Lee, M. Mayorov, J. Jeon, A. Bonino, M. Grund, M. Dutra, G. Renha, S. Ahn, K. Kim, J. Park, G. Yang, Field testing of a fast-neutron coincidence collar for fresh uranium fuel verification. Nucl. Instrum. Methods Phys. Res. Sect. A **962**, 163682 (2020)
36. E. Padovani, S.A. Pozzi, S.D. Clarke, E.C. Miller, MCNPX-PoliMi User's Manual, C00791 MNYCP, Vol. 1, Radiation Safety Information Computational Center, Oak Ridge National Laboratory (2012)
37. S.A. Pozzi, E. Padovani, M. Marseguerra, MCNP-PoliMi: A Monte-Carlo code for correlation measurements. Nucl. Instrum. Methods Phys. Res. Sect. A **513**(3), 550–558 (2003)
38. A. Tomanin, P. Peerani, G. Maenhout, The SimPLiS Code: A Simulation Post-Processor for Liquid Scintillators. Universiteit Gent report (2013). http://hdl.handle.net/1854/LU-4426267
39. W.H. Geist, D.C. Henzlova, H.O. Menlove, Boron-Lined Parallel-Plate Collar. Los Alamos National Laboratory report LA-UR-16-27428 (2016)
40. W.H. Geist, J.E. Stewart, H.O. Menlove, N. Ensslin, C. Shonrock, Development of an Active Epithermal Neutron Multiplicity Counter (ENMC), in *Proceedings of the 41st Annual Meeting of the INMM*, New Orleans, LA, July 16–20, 2000, Los Alamos National Laboratory report LA-UR-00-2721 (2000)
41. K.C. Frame, W.H. Geist, J.P. Lestone, A.P. Belian, K.D. Ianakiev, Characterizing the detector response and testing the performance of a new liquid scintillator counter for neutron multiplicity measurements of enriched uranium, in *Proceedings of the 44th Annual Meeting of the INMM*, Phoenix, AZ, July 13–17, 2003, Los Alamos National Laboratory report LA-UR-03-4899 (2003)
42. K.C. Frame, W.A. Clay, W.H. Geist, J.P. Lestone, A.P. Belian, K.D. Ianakiev, Neutron and Gamma Pulse shape discrimination in a liquid scintillator counter for neutron multiplicity measurements of enriched uranium, in *Proceedings of the 45th Annual Meeting of the INMM*, Orlando, FL, July 18–22, 2004, Los Alamos National Laboratory report LA-UR-04-4868 (2004)

Open Access This chapter is licensed under the terms of the Creative Commons Attribution 4.0 International License (http://creativecommons.org/licenses/by/4.0/), which permits use, sharing, adaptation, distribution and reproduction in any medium or format, as long as you give appropriate credit to the original author(s) and the source, provide a link to the Creative Commons license and indicate if changes were made.

The images or other third party material in this chapter are included in the chapter's Creative Commons license, unless indicated otherwise in a credit line to the material. If material is not included in the chapter's Creative Commons license and your intended use is not permitted by statutory regulation or exceeds the permitted use, you will need to obtain permission directly from the copyright holder.

21 Spent Fuel Measurements

A. C. Kaplan-Trahan, A. P. Belian, M. Croce, D. C. Henzlova,
P. Jansson, G. Long, G. E. McMath, J. R. Phillips, E. Rapisarda,
M. A. Root, and H. R. Trellue

21.1 Introduction

Spent nuclear fuel is one of the most complex nuclear materials on Earth. Typical power reactor assemblies contain thousands of fission product isotopes and emit more than 10^8 neutrons and 10^{15} gamma rays per second. Some of the assemblies are cooled for a year or two and then reinserted into the reactor to optimize the burnup (Sect. 21.2.2) of the residual fissile material; others, having reached their maximum design burnup, are placed in the storage pond at the reactor site until they are sent to a long-term storage facility or to a waste disposal site. This sort of irregular history can further complicate spent fuel accountancy by rendering simulations unreliable. Safeguarding such complex nuclear material requires innovative and robust nondestructive assay (NDA) technology. The ideal NDA technique for spent fuel would be one that does not rely on a calibration curve or knowledge of operating history, is inexpensive, and measures the entire assembly at once instead of a limited region without touching the fuel [1]. Though the perfect system does not exist, a variety of promising techniques and systems for measuring the fissile content or burnup credit and verifying operator declarations of spent fuel assemblies have been in development for several decades. Techniques vary from simple, poolside configurations with a single detector to complex, underwater systems with hundreds of advanced detectors. Regardless of the approach, assay of research and power reactor spent fuel is a critical safeguards capability.

This chapter describes the physical characteristics of reactor fuel (Sect. 21.1), gamma-ray and neutron detection-based measurement techniques and systems (Sects. 21.3 and 21.4), and measurement techniques and systems based on other signatures, including Cerenkov radiation, X-rays, and heat (Sect. 21.5).

Los Alamos National Laboratory strongly supports academic freedom and a researcher's right to publish; as an institution, however, the Laboratory does not endorse the viewpoint of a publication or guarantee its technical correctness.

A. C. Kaplan-Trahan (✉) · A. P. Belian · M. Croce · D. C. Henzlova · G. Long · G. E. McMath · J. R. Phillips · M. A. Root · H. R. Trellue
Los Alamos National Laboratory, Los Alamos, NM, USA
e-mail: atrahan@lanl.gov; anthony.belian@srnl.doe.gov; mpcroce@lanl.gov; henzlova@lanl.gov; gracel@lanl.gov; gem@lanl.gov; margaret@lanl.gov; trellue@lanl.gov

P. Jansson
Uppsala University, Stockholm, Sweden

E. Rapisarda
International Atomic Energy Agency, Vienna, Austria
e-mail: E.Rapisarda@iaea.org

21.2 Characteristics of Reactor Fuel

21.2.1 Physical Description

As of 2020, 443 nuclear power reactors and 220 research reactors are operating around the world [2–4]. The most common power reactor types are pressurized-water reactors (PWRs) and Russian-designed, water-cooled, water-moderated reactors (VVERs [a type of PWR]); followed by boiling-water reactors (BWRs) and pressurized heavy-water reactors (PHWRs). The fuel and core parameters of standard PWRs, VVERs, BWRs, and PHWRs are compared in Table 21.1 [5, 6]. Two commonly found, pool-type research reactors are materials testing reactors (MTRs) and training, research, isotopes, general atomics reactors (TRIGAs), parameters of which are given in Table 21.2 [7].

Reactor fuel initially consists of fissile material (^{235}U or ^{239}Pu) plus fertile material (^{238}U). These materials are usually present in ceramic oxide or carbide forms in power reactors because of their improved corrosion characteristics, relative ease of fabrication, and better radiation-damage characteristics. In research reactors, uranium-aluminum alloys are typically formed into plates or pins and clad in pure aluminum.

Table 21.1 UO$_2$ fuel and core parameters for power reactors

	PWR	VVER	BWR	PHWR
Fuel parameters				
Cladding diameter (cm)	0.95	0.91	1.23	1.31–1.53
Cladding thickness (cm)	0.06	0.14	0.08	0.038
^{235}U initial enrichment	2.1%/2.6%/3.1%	1.6–4.9%	2.8% average	Natural, 0.7%
Pellet diameter (cm)	0.82	0.78	1.04	1.22–1.44
Pellet column height (cm)	1.5	373	1.04	59.79
Assembly diameter (cm)		15–23.6		
Assembly height (cm)		373		
Assembly array	14 × 14 17 × 17	Hexagonal	7 × 7 10 × 10	Concentric circles
Fuel pins/assembly	201 or 264	126–312	63	19, 28, or 37
Core parameters				
Number of fuel assemblies	193	163–300	740	3672 or 4704
Design fuel burnup	47 GWd/tU	55 GWd/tU	44 GWd/tU	7 GWd/tU
Refueling cycle	1/3 fuel/year	12–18 mos	¼ fuel/year	2 year

Table 21.2 Fuel and core parameters for research reactors

	MTR	TRIGA
Fuel parameters		
Cladding material	Aluminum alloy	Type 304 stainless steel
Cladding diameter (cm)	7.10 × 0.1275	1.27; 3.81
Cladding thickness (cm)	0.057; 0.039	0.041; 0.051
Fuel material	U$_3$O$_8$; U$_3$Si$_2$-Al	UzrHxEr
^{235}U initial enrichment	20–95%	8.5–19.7
Pellet diameter (cm)	6.24 × 0.05	3.8
Pellet height (cm)	59.79	38.1
Assembly array	Plates	Rod clusters
Fuel pins/assembly	19–23 plates/element	1
Core parameters		
Number of fuel assemblies	10–24	60–100
Reactor power	<25 MW	
Design fuel depletion	50%	30%
Refueling cycle		>10 years

21.2.2 Burnup and Depletion

Two common terms for fuel irradiation—*burnup* and *depletion*—are often used to describe the extent to which power reactor and research reactor fuel have been burned. *Burnup* is defined as the integrated energy released by fission of the heavy nuclides initially present in the fuel. *Depletion* is defined as the percentage of ^{235}U atoms consumed in the reactor. Burnup has units of megawatt or gigawatt days (thermal output of the reactor) per metric ton (1000 kg) of initial uranium: MWd/tU or GWd/tU. Depletion is typically expressed as a percentage of ^{235}U removed from the fuel. The term *burnup credit* is also used in the context of spent fuel verification and refers to the reactivity loss due to fuel irradiation when performing criticality safety analyses, which without burnup credit, usually assumes conservatively that the fissile material content corresponds to the initial enrichment.

Table 21.1 includes average design values for reactor fuel burnup. Actual values can range up to 60 GWd/tU for PWRs and VVERs, 45 GWd/tU for BWRs, and 15 GWd/tU for PHWRs. High burnup power reactor fuel can contain significant amounts of neutron-absorbing isotopes (poisons) such as ^{135}Xe or ^{149}Sm. These isotopes remove neutrons from the system, shortening the effective die-away and lowering the passive neutron signal. The relatively low power and flux obtained in research reactors result in a comparatively negligible buildup of neutron absorbers.

21.2.3 Initial Enrichment and Cooling Time

Both initial enrichment and cooling time affect the isotopic composition of spent fuel, even for assemblies with the same burnup or depletion. The ratio of ^{235}U to ^{238}U in fuel before irradiation determines the initial enrichment of the assembly. The rate of buildup of the principal neutron-emitting transuranic isotopes is relatively insensitive to the initial fuel density and power levels; however, the initial ^{235}U enrichment and the fuel irradiation history can significantly influence the rate of buildup. Lower-enriched fuel has less fissile material per unit volume and therefore requires a higher neutron fluence to achieve the same burnup as a higher-enriched fuel. The higher neutron fluence results in more ^{242}Cm and ^{244}Cm buildup, and a correspondingly higher neutron emission rate. Higher initial enrichment also results in more ^{137}Cs buildup and subsequently higher gamma-ray emissions.

The amount of time that elapses following removal of an assembly from a reactor core is called the *cooling time*. The gamma-ray signal emitted from spent fuel is highly dependent on the cooling time because of the relatively short half-lives of several predominant fission products (see Table 21.4). For short cooling times, gamma-ray activity is dominated by the buildup and decay of short-lived fission products, such as ^{95}Nb and ^{95}Zr. For long cooling times, ^{137}Cs dominates the gamma-ray activity.

21.2.4 Assembly Multiplication

Multiplication of a spent fuel assembly is a key characteristic that quantitatively captures the interplay between fission chains and neutron absorbers. Though multiplication itself is not a declared or therefore verifiable parameter, it is used as a stepping stone to determine the more-challenging declared parameters such as initial enrichment and cooling time. The concept of item multiplication is discussed in more detail in Chap. 14. Net multiplication (M_N), or the neutrons produced per source neutron, is a commonly used form of multiplication to describe spent fuel assemblies. Another commonly used form of multiplication is the leakage multiplication (M_L)—or total neutron profit per source neutron times the probability of escape divided by the probability of non-fission interactions. Equation 21.1 shows the relationship between M_N and M_L:

$$M_L = \frac{p_L \times M_N}{1-p}, \tag{21.1}$$

where
p_L = probability of leakage and
p = probability of fission.

21.2.5 Fission Product Yields

In spent fuel, the primary sources of energy are the fissioning of ^{235}U, ^{239}Pu, and ^{241}Pu, with some contribution from the fast fissioning of ^{238}U. Table 21.3 [8] lists the relative number of fissions for each of these four isotopes as a function of fuel burnup for one typical case. The table shows that the plutonium isotopes begin to contribute a significant part of the total after only a few GWd/tU exposure, and above 38 GWd/tU exposure the plutonium isotopes become the dominant source of fissions.

Each fission results in the formation of two medium-mass fission products. The primary difference between the fission product formation from ^{235}U and ^{239}Pu is the slight shift to higher atomic mass number of the ^{239}Pu curve with respect to the ^{235}U curve, as shown in Fig. 21.1. The shift can be seen in the relative fission yields of ^{106}Ru from ^{235}U and ^{239}Pu. Fissions in ^{239}Pu yield 11 times as much ^{106}Ru as fissions in ^{235}U; thus a measurement of the gamma-ray output of ^{106}Ru can determine the relative proportion of ^{239}Pu fissions to total fissions. In contrast, an isotope such as ^{137}Cs has nearly identical fission yields from both ^{235}U and ^{239}Pu. The ^{137}Cs gamma-ray output can be used to determine the total number of fissions.

Most of the fission products are initially rich in neutrons and undergo beta decay to approach stability. In addition to emitting beta particles, the fission products also emit gamma rays, which result in measurable signatures. Table 21.4 [9] lists the most dominant isotopes, along with their gamma rays and half-lives. The fission yields are the number of nuclei of each isotope (in percent) produced on average per thermal-neutron fission in ^{235}U or ^{239}Pu. In addition to the fission product gamma rays, gamma rays from the activation of fuel cladding and structural materials such as ^{54}Mn, ^{58}Co, and ^{60}Co are also present; they are included in Table 21.4.

Table 21.3 Percentage of total fissions as a function of fuel exposure for PWR fuel with an initial enrichment of 4.2% ^{235}U

Exposure (GWd/tU)	^{235}U	^{238}U	^{239}Pu	^{241}Pu
0.03	94.4	5.6	0.01	0.00
1.0	93.8	5.7	0.56	0.00
5.7	86.7	5.9	7.3	0.05
13.7	74.5	6.3	18.3	0.8
21.8	62.2	6.9	28.0	2.8
29.8	51.5	7.4	35.5	5.3
37.9	42.0	7.9	41.7	8.1
45.9	33.2	8.5	46.9	10.9
54.0	25.4	8.9	51.4	13.5
59.4	20.8	9.2	54.0	15.1

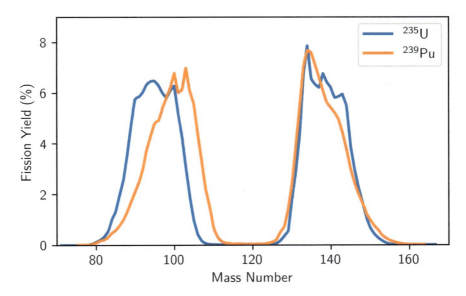

Fig. 21.1 Mass distribution of fission products for the thermal fission of ^{235}U and ^{239}Pu

Table 21.4 Isotopes measurable by gamma rays in a typical irradiated fuel assembly [9]

Fission product isotopes	Half-life	Fission yield in ^{235}U (%)	Fission yield in ^{239}Pu (%)	Gamma-ray energy (keV)	Branching ratio (%)
^{95}Zr	64.0 days	6.50	4.89	724.2	43.1
				756.7	54.6
^{95}Nb	35.0 days	6.50	4.89	765.8	99.8
^{103}Ru	39.4 days	3.04	6.95	497.1	86.4
				610.3	5.4
^{106}Ru-Rh	366.4 days	0.40	4.28	622.2	9.8
				1050.5	1.6
^{134}Cs	2.06 years	1.27×10^{-5a}	9.89×10^{-4a}	604.7	97.6
				795.8	85.4
				801.8	8.7
				1167.9	1.8
				1365.1	3
^{137}Cs	30.17 years	6.22	6.69	661.6	85.1
^{140}Ba-La	12.7 days			1596	
^{144}Ce-Pr	284.5 days	5.48	3.74	696.5	1.3
				1489.2	0.3
				2185.6	0.7
^{154}Eu	8.5 years	2.69×10^{-6a}	9.22×10^{-5a}	996.3	10.3
				1004.8	17.4
				1274.4	35.5
Activation products					
^{54}Mn	312.2 days			834.8	100.0
^{58}Co	70.3 days			811.1	99.0
^{60}Co	5.27 years			1173.2	100.0
				1332.5	100.0

^{a}Europium-154 and ^{134}Cs values are given only for direct production of the isotope from fission. Actually, each of these isotopes is produced primarily through neutron absorption. For PWR fuel material irradiated to 25 GWd/tU, the "accumulated fission yields" of ^{154}Eu and ^{134}Cs were calculated as 0.15 and 0.46%, respectively

The gamma rays from fission products and the neutrons from transuranic nuclides completely mask the gamma rays and neutrons from the uranium and plutonium isotopes in the fuel. Even though the atom densities for ^{137}Cs, ^{235}U, and ^{239}Pu are comparable at the 46.8 GWd/tU burnup level, the number of ^{137}Cs gamma rays being emitted is 4.5×10^7 and 2.7×10^7 times the number of principal ^{235}U and ^{239}Pu gamma rays, respectively. The ^{137}Cs isotope is just one of about 10 dominant gamma-emitting fission products.

21.2.6 Transuranic Nuclides

Uranium present in a neutron flux can also capture neutrons and build up transuranic nuclides, as illustrated in Fig. 21.2 [8]. Many of these nuclides produce neutrons through spontaneous fission or (α,n) reactions; the spontaneous fission and (α,n) neutron production rates for the primary neutron sources—^{238}Pu, ^{240}Pu, ^{242}Pu, ^{241}Am, ^{242}Cm, and ^{244}Cm—are included in Tables 13.1 and 13.3 (Chap. 13). Therefore, the passive neutron signal is dominated by the transuranic nuclides, with the neutron yield of the curium isotopes exceeding that of plutonium by two orders of magnitude. In the vast majority of spent power reactor fuel assemblies, spontaneous fission of ^{244}Cm is the primary source of passive neutrons. In research reactors, power is often too low to build up significant amounts of transuranic nuclides, and ^{240}Pu therefore often emerges as the primary passive neutron emitter via spontaneous fission.

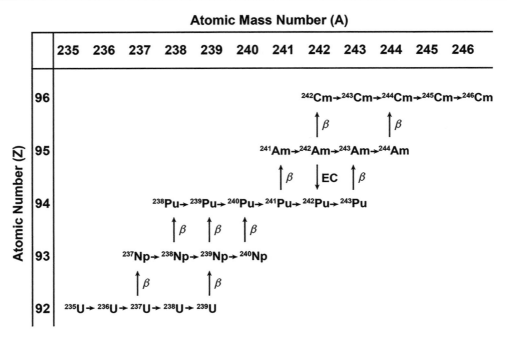

Fig. 21.2 Principal neutron capture reactions (horizontal arrows) and beta decay reactions (vertical arrows) leading to the formation of transuranic nuclides in irradiated fuel. Both stable and metastable states exist for ^{240}Np, ^{242}Am, and ^{244}Am

21.3 Gamma-Ray-Based Measurement Techniques

Gamma rays emitted from fission products within spent fuel provide insight into the verifiable parameters of the fuel including (foremost) burnup and depletion. Additionally, gamma-ray signatures can be used to determine cooling time or initial enrichment to improve the accuracy of the burnup determination. Total gamma-ray activity, gamma-ray spectroscopy, and tomography are techniques used in the characterization and measurement of spent fuel, and a variety of measurement systems and detectors use these techniques.

21.3.1 Total Gamma-Ray Activity

The total gamma-ray activity of irradiated fuel is the sum of the activities from each fission product and transuranic element, with each activity given by an equation like Eq. 21.2. Most of the gamma-ray activity comes from a few important fission products, namely ^{134}Cs, ^{137}Cs, ^{154}Eu, ^{144}Pr, ^{95}Zr, and ^{95}Nb. For cooling times greater than 1 year, the total gamma-ray activity is roughly proportional to burnup.

A rapid way to determine the consistency of operator-declared values for burnup and cooling time has been developed using total gamma-ray activity [10, 11]. The total activity is divided by the declared burnup and is plotted as a function of the declared cooling times. The result is a relationship of the form aT^b, where a and b are scaling parameters, and T is the cooling time. An example of this relationship for PWR fuel assemblies is given in Fig. 21.3. The shape of the curve is due primarily to the half-lives of ^{134}Cs and ^{137}Cs.

Total gamma-ray activity should be measured when a fuel assembly is raised above the storage rack to avoid interference from ^{60}Co, ^{58}Co, and ^{54}Mn activation products in the structural material. Under these conditions, the consistency of operator-declared values for cooling time and burnup has been verified to an accuracy of ~10% using a standard total gamma detector such as an ion chamber. Fuel assemblies with unusual irradiation histories have been identified easily. An ion chamber designed to be placed against the side of a fuel assembly has been incorporated into the fork detector described in Sect. 21.4.2.

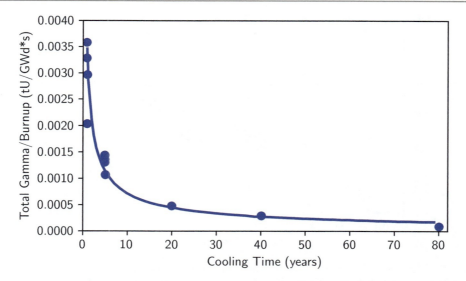

Fig. 21.3 Total detected gamma-ray activity divided by burnup as a function of cooling time for simulated PWR fuel assemblies of various initial enrichments. The fitted curves illustrate how the passive gamma activity can be used to verify the consistency of operator-declared values for burnup and cooling time

21.3.2 Gamma-Ray Spectroscopy

More detailed information on fuel burnup, initial enrichment, and cooling time can be obtained by high-resolution gamma-ray spectroscopy. The concentration of individual fission products can be determined by measuring single fission product gamma-ray activities; these isotopic concentrations can be related to fuel burnup. Some gamma-ray intensity ratios can also be used to calculate fission product isotopic ratios that can be related to burnup. The measured gamma-ray activity of a fission product is proportional to the number of fissions that occurred during irradiation if the following conditions are met.

- The fission product has equal fission yields for the major uranium and plutonium fissioning nuclides.
- The neutron capture cross section of the fission product is low enough so that the observed fission product concentration is due only to heavy element fission and not to secondary neutron capture reactions.
- The fission product half-life is long compared with the fuel irradiation time, so the quantity of fission product present is approximately proportional to the number of fissions.
- The fission product gamma rays are relatively high energy (500 keV or more) to escape from the fuel pin.

If these conditions are met, the measured gamma-ray activity I(counts/s) from the fission product is proportional to the number N of fission product nuclei formed during irradiation:

$$I = \varepsilon k S N \lambda e^{-\lambda T}, \tag{21.2}$$

where
ε = absolute detector efficiency
k = branching ratio
S = attenuation correction
λ = fission product decay rate
T = cooling time.

After solving Eq. 21.2 for N, the fuel burnup can be calculated from the equation

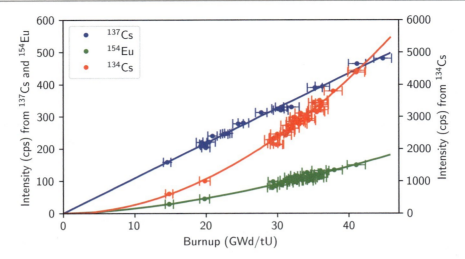

Fig. 21.4 Intensity of ^{137}Cs, ^{154}Eu, and ^{134}Cs gamma rays as a function of assembly burnup. Gamma-ray intensties are normalized for cooling time. The ^{137}Cs relationship is best fit by a linear function, whereas the ^{154}Eu and ^{134}Cs relations are best fit by power functions. (Figure reproduced from Ref. [12])

$$at\ \%\ burnup = 100 \times \frac{\frac{N}{Y}}{U}, \quad (21.3)$$

where

Y = effective fission product yield
U = number of initial uranium atoms.

Cesium-137 is the most widely accepted indicator of fuel burnup because its neutron absorption cross sections are negligible, its yields from both ^{235}U and ^{239}Pu are approximately the same, and its 30-year half-life makes a correction for reactor power history unnecessary [11]. The absolute ^{137}Cs activity is linearly proportional to burnup and can determine burnup to an accuracy of 1–4% for individual fuel rods [12]. The normalized intensities of ^{137}Cs gamma rays, as well as ^{134}Cs and ^{154}Eu, are shown as a function of burnup in Fig. 21.4. For a 1 cm diameter oxide fuel pin, 39% of the 662 keV ^{137}Cs activity is absorbed within the pin. This strong self-absorption implies that, in practice, passive gamma-ray measurements of fuel assemblies are limited to the outer rows. For short cooling times, the 757/766 keV gamma rays from ^{95}Nb/^{95}Zr often prove the most useful for verifying spent fuel burnup [13].

The burnup of irradiated fuel can also be determined from the ratios of some fission product isotopes. The isotopic ratios can be determined from gamma-ray activity ratios using equations like Eq. 21.2 The two most commonly used isotopic ratios are ^{134}Cs/^{137}Cs and ^{154}Eu/^{137}Cs, though higher resolution measurements, such as ^{106}Ru/^{137}Cs and ^{144}Ce/^{137}Cs, enable further ratio measurement [14, 15]. Gamma-ray ratios as a function of burnup are shown in Fig. 21.5.

Cesium-134 is produced by neutron capture on the fission product ^{133}Cs; therefore, its production requires two neutron interactions. The first is the neutron that causes fission of the uranium or plutonium, and the second is the ^{133}Cs (n,γ) reaction. Because these interactions are the primary source of ^{134}Cs, the concentration of ^{134}Cs within the fuel is approximately proportional to the square of the integrated flux, as seen in Fig. 21.4. By dividing the concentration of ^{134}Cs by the concentration of ^{137}Cs, which is directly proportional to the integrated flux, the ratio becomes approximately proportional to the burnup (total flux; Refs [16–18]). The ^{154}Eu/(^{137}Cs)^3 isotopic ratio also has a fairly linear dependence on exposure [9, 19]. Inclusion of ^{106}Ru and ^{144}Ce improves the accuracy of initial enrichment, burnup, and cooling time assessments. More measured parameters enable solutions of the inverse problem even when operating history is inconsistent.

21.3.3 Microcalorimetry

As discussed in the previous sections, verifying fissile material content and irradiation history of spent nuclear fuel is challenging for gamma-ray spectroscopy due to the presence of intense fission product activity. Associated Compton-

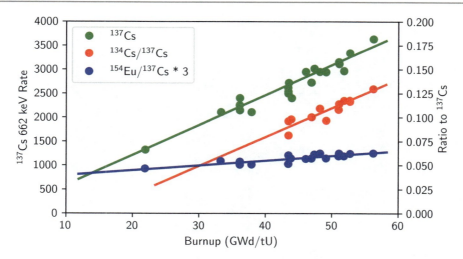

Fig. 21.5 Gamma-ray ratios as a function of burnup for standard PWR and BWR assemblies [15]

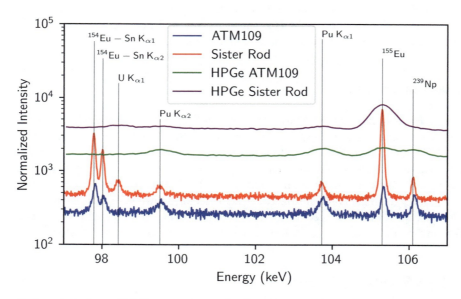

Fig. 21.6 Comparison of microcalorimeter and HPGe spectra for two high-burnup spent fuel samples

scattering background typically obscures low-energy peaks that result from actinides. Ultra-high-resolution microcalorimeter spectrometers (Chap. 9, Sect. 9.4.4) offer a way to improve nondestructive gamma-ray isotopic analysis. Improved energy resolution compared with high-purity germanium (HPGe) detectors can reduce the effect of large Compton-scattering background from high-energy fission product gamma rays and more accurately quantify actinide signatures in the presence of neighboring peaks.

Example spectra from high-burnup spent fuel samples are shown in Fig. 21.6. ATM109 is BWR fuel from the Quad Cities I reactor. This material had an initial enrichment of approximately 3 wt%, estimated burnup of 67–70 GWd/MTU, and a discharge date of September 1992 [20]. The sister rod material is PWR fuel from the North Anna reactor; the material has an initial enrichment of 3.59–4.55 wt%, estimated burnup of 50–58 GWd/MTU, and a discharge date between 1989 and 2011 [21]. Several actinide signatures are visible and can be quantified relative to fission product peaks. Neptunium-239 (106.1 keV) and plutonium X-rays (99.5 and 103.7 keV) result from the decay chain of ^{243}Am, a long-lived actinide that is produced by the irradiation process. The uranium X-ray at 98.4 keV is understood to have a significant component from self-induced X-ray fluorescence [22]. These peaks are resolved in the microcalorimeter data from both samples, whereas the peaks are much less clear in the HPGe data. Systematic uncertainty has the potential to be significant in analysis of HPGe data due to

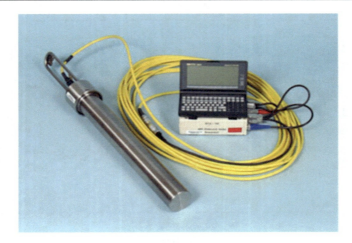

Fig. 21.7 Irradiated fuel attribute tester from Ref. [13]

peak overlaps. The improved resolution of microcalorimetry nearly eliminates this type of uncertainty contribution and therefore allows peaks in this region to be confidently used for quantitative analysis. Microcalorimetry also provides a significant reduction in the number of counts required to achieve a given statistical precision, and simulations suggest that this advantage is especially pronounced for fuel with a shorter cooling time [23].

21.3.4 Attribute Measurements

The International Atomic Energy Agency (IAEA) uses the spent fuel attribute tester (SFAT) and the irradiated fuel attribute tester (IRAT) to verify spent fuel using gamma spectroscopy [13]. The SFAT system uses a shielded, low-resolution detector (typically CdZnTe, or CZT) to measure attributes of spent fuel in a storage rack through an air/water collimator tube. The SFAT provides a quantitative verification of the presence of spent fuel using the ^{137}Cs gamma rays and identifies activation products such as ^{60}Co as well. The maximum dose rate obtained with a 5 cm diameter, 6 m long collimator from 40 GWd/tU exposure fuel with a 2-year cooling time is ~10 mR/h. The IRAT detector, shown in Fig. 21.7, comprises a small CZT detector and may be suspended in a spent fuel pool specifically to differentiate between irradiated non-fuel and irradiated fuel items. IRAT measures fuel from the side as opposed to the top and therefore requires fuel assembly movement from the storage rack.

21.3.5 Silo Entry Gate Monitor

The silo entry gate monitor (SEGM) system is used during spent fuel dry storage campaigns to monitor the loading of spent fuel assemblies into multi-basket storage silos. The directionality of the SEGM system means that removals of fuel baskets from the silo would be detected. To monitor the insertion of a basket into the silo, two detectors are located inside the silo's verification (sealing) tube, parallel to the path of the basket insertion within the silo and physically separated by a fixed distance (typically 0.5 m). The detectors are referred to as the *upper* and *lower* detectors. Figure 21.8 shows a scheme of the SEGM IRD100 detectors inside a stand-alone silo sealing tube with IRD100 detectors.

The purpose of the detectors' physical separation is to confirm the directionality of movement. For example, during basket loading, a peak will appear first on the upper detector and subsequently on the lower detector, thus providing the ability to confirm that the basket was inserted into the silo. Conversely, if a peak appeared first on the lower detector and subsequently on the upper detector, this case would indicate removal of a basket from the silo.

The primary considerations for selection and fabrication of detector assemblies are
- the ability to be physically placed and mounted into the sealing and/or verification tube(s);
- the position of the detectors within the tube, which should be such that a signal is detectable for all of the baskets, i.e., not masked by background counts from the baskets already loaded and not shielded by the silo walls; and
- the time response of the detectors and their relative positions to ensure that the assembly can detect all baskets with a sufficient time separation between the detected responses to verify insertion and confirm non-removal.

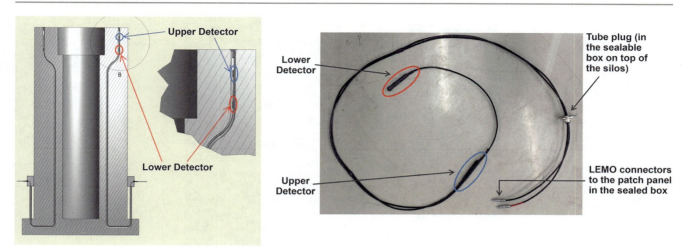

Fig. 21.8 Sealing tube on a stand-alone silo with the SEGM detectors mapped to a close-up view (left) and an IRD100 detector pair for a MACSTOR verification tube assembly (right)

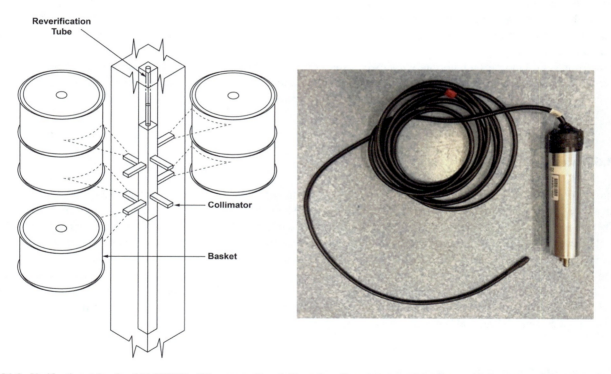

Fig. 21.9 Verification tube for MACSTOR 400 center silos (left) and a fiber detector with photomultiplier tube (right). Two detector photomultiplier tube assemblies are combined before the detectors are inserted into the verification tube

Any access points to verification or sealing tubes that contain detectors must be sealed to ensure that the detector assembly is not subject to tampering.

Fiber detectors were developed for a different type of silo—the MACSTOR 400—in which four silos share one common verification tube with collimators facing each of the 40 basket locations (10 basket locations for each of the four silos), as shown in Fig. 21.9. The detector system consists of two scintillating fiber detectors grouped together within a plastic tube for easy use. The lower detector has a 100 cm active length, and the upper detector has a 45 cm active length. With this configuration, a single detector pair is capable of aligning with all four upper and lower collimators.

Fig. 21.10 SMOPY system positioned for an underwater measurement

When the detectors align with the basket position, a step function is produced in the data instead of a simple peak. In this situation, the basket does not move away from one or both of the detectors during loading; therefore, the detectors will continue to see the gamma radiation emitted by the basket under an almost constant solid angle and, consequently, at an almost constant rate.

Data from SEGM systems are normally reviewed using Integrated Review and Analysis Package (IRAP) software, which features different automatic peak-recognition algorithms for a variety of applications and can correlate the peaks seen in the detectors to operator declarations for the purposes of verification. Step function detection could be used as a further verification that the silo is full.

21.3.6 Safeguards MOX Python

The utility of gamma spectroscopy for spent fuel characterization can be taken a step further by incorporating an additional signal such as passive neutron, as was done with the Safeguards MOX Python (SMOPY) detector system shown in Fig. 21.10. SMOPY is a transportable system capable of discriminating mixed-oxide (MOX) versus low-enriched uranium (LEU) fuel, fully characterizing LEU fuel through verification of burnup, cooling time, and plutonium content and performing partial defect testing on LEU fuels [24]. The SMOPY system comprises a cylindrical measurement head located within a carrier that fits the spent fuel rack on the bottom and the operator's fuel-moving tool on top. The measurement head typically contains CZT detectors or fission chambers. The portable electronics cabinet for data processing and analysis is located poolside. Accuracy of burnup measurements with SMOPY is on the order of 5% for LEU fuel, and the system can readily and reliably differentiate between LEU and MOX assemblies using the neutron emission rate.

21.3.7 Passive Gamma Emission Tomography

Gamma emission tomography can be used to verify the declarations of spent fuel by creating an axial image of emission locations to detect pin-level diversion. The passive gamma emission tomography (PGET) system, developed in collaboration with multiple member-state support programs, performs three simultaneous measurements at one axial position on the fuel assembly: gross neutron counting, medium-resolution gamma spectroscopy, and two-dimensional emission tomography. Neutron data can be used for burnup verification (Sect. 21.4.1), and the spectroscopy data can be used as a qualitative check on the cooling time (Sect. 21.2.3) or to verify non-fuel items. To perform a measurement, the PGET is lowered into the spent fuel pond to a depth of 10–13 m and placed either over an empty position in the fuel rack or on a separate stand, as shown in Fig. 21.11 [25]. The fuel-handling machine is then used to retrieve an assembly from a position in the fuel rack, lower it through the central hole in the PGET torus, and hold the fuel stationary for the duration of the measurement (on the order of 5 min). The fuel is then returned to its position in the rack, and another assembly can be retrieved.

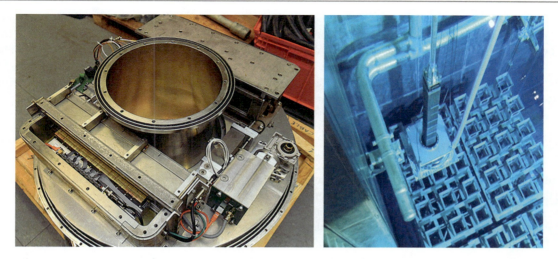

Fig. 21.11 Passive gamma emission tomography instrument (with water-tight cover removed) and performing spent PWR fuel measurement underwater. Image courtesy of the IAEA [25]

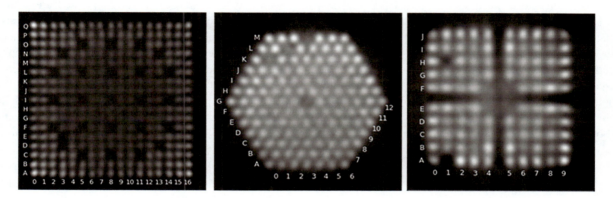

Fig. 21.12 Example reconstructions from a PWR 17 × 17 assembly (left), a VVER-440 (center), and a BWR spent fuel assembly. Pixel intensity in the images correlates with gamma activity; brighter pixel regions indicate higher activity. Locations of removed pins and locations in the grid without pins (water or instrument channels) can be seen in each image [25, 26]

The PGET measurements are made by a set of 182 discrete CZT detectors set behind 10 cm thick tungsten collimators and two ^{10}B neutron detectors. During a measurement, the detector assembly is rotated within the stationary torus through 360° to measure the emissions at one axial position of the assembly. Gamma data from all detectors are collected into four broad energy bins to form the sinograms that are subsequently reconstructed into cross-sectional images that represent the relative activity of fission products at that axial position. The main fission-product signatures are from ^{137}Cs, ^{154}Eu (for cooling times up to about 20 years), and ^{144}Ce-Pr (for cooling times on the order of a few years). Figure 21.12 shows example images from three fuel assemblies with missing pins. The PGET is the only instrument capable of verifying the declared number of pins in sealed pin containers commonly used to store leaking pins. Drawbacks of the instrument are that it is tedious to the operator because it requires significant fuel movements, and it is complicated to install. Future challenges for spent fuel tomography would be to measure the relative pin-to-pin burnup, enabling detection of pin diversion and replacement between fuel cycles.

21.4 Neutron-Based Measurement Techniques

Neutron measurements of spent fuel are relatively easy to make and provide a uniform signal, measuring neutrons from across the entire radial area of the fuel assembly. The detection equipment is simple and easy to operate, and a preliminary analysis of the data can typically be performed at the facility. A neutron detector or multiple detectors are placed adjacent to the fuel assembly, and the signal is analyzed using a multichannel analyzer. Gamma-ray-insensitive detectors, such as ^{235}U fission chambers, are useful for irradiated fuel measurements because the ratio of gamma rays to neutrons emitted from spent fuel can be as high as 10^{10} to 1 [27]. Helium-3 detectors that have sufficient gamma-ray shielding to prevent interference from pileup may also be used for irradiated fuel measurements. Although neutrons can penetrate the fuel assembly more readily than gamma rays, their attenuation in water is more severe. The gamma-ray signal decreases by a factor of 10 in roughly 36 cm, whereas the neutron signal decreases by a factor of 10 in ~10 cm. Thus it is important to position the neutron detector close to the fuel assembly and to be extremely consistent in positioning for repeated measurements.

21.4.1 Total Neutron Activity

The total neutron output of irradiated fuel can serve as an indicator of burnup and has both advantages and disadvantages relative to the gamma-ray output. The neutron signal comes only from the fuel, not from the cladding materials. Attenuation of the neutron signal within the fuel assembly is less than attenuation of the gamma-ray signal; in fact, induced fissions within the assembly result in nearly equal response from both interior and exterior fuel pins. The neutron measurements can be made soon after the fuel is discharged from the reactor, but the gamma-ray signal is still dominated by the decay of short-lived isotopes that reflect recent reactor power levels. A disadvantage of the neutron signal is that the quantity of primary neutron-emitting isotopes is only indirectly correlated to exposure. Also, neutron detectors can be sensitive to gamma rays, although the fission chambers used for measurements today are almost completely insensitive to gamma rays.

The five principal neutron sources in a PWR fuel assembly with a typical exposure of 31.5 GWd/tU are plotted in Fig. 21.13 as a function of cooling time. For other fuel assemblies or different burnup levels, ^{241}Am or ^{246}Cm may also be significant neutron sources; however, ^{244}Cm and ^{242}Cm are usually the two dominant neutron-emitting isotopes. Because of the short half-lives of these two isotopes (18.1 and 0.45 years, respectively), the cooling time of the fuel must be known to correctly interpret the total neutron output. Operator-declared values for the cooling time can be verified by measuring the total gamma-ray activity, as described in Sect. 21.3.2.

Figure 21.14 shows the calculated relationship between the total neutron output and burnup for 5-year-cooled 5% initial enrichment PWR fuel assemblies.

Above 10 GWd/tU exposure, the relationship can be approximated by following the empirical power function, shown in Eq. 21.4:

$$neutron\ rate = \alpha(burnup)^\beta. \tag{21.4}$$

Equation 21.4 has been demonstrated for a wide variety of light-water reactor (LWR) fuels [28, 29] and makes it possible to determine burnup from the observed total neutron output. The value of β is usually between 3.0 and 4.0.

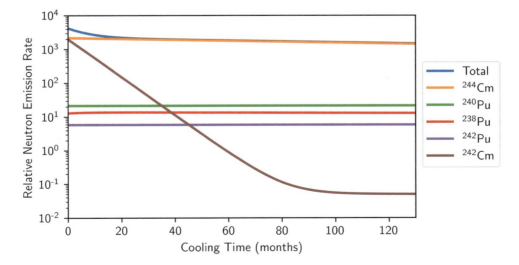

Fig. 21.13 The five principal neutron sources in a PWR fuel assembly with an exposure of 31.5 GWd/tU. At long cooling times, the decay of ^{242}Cm follows the decay of its parent, ^{242}Am, a metastable state of ^{242}Am with a half-life of 153 years

21 Spent Fuel Measurements

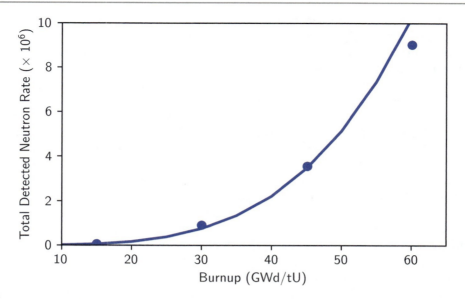

Fig. 21.14 Calculated dependence of total neutron output on burnup for simulated PWR fuel

Fig. 21.15 (left) The UFDM (Photo courtesy of ANTECH; Ref. [32]) and (right) the mini UFDM

Noncontinuous irradiation histories can also have a significant effect on the neutron emission rate. The effect of noncontinuous irradiation history is short term, due primarily to the increased buildup of ^{242}Cm following any period of cooling time. During any non-irradiation period, ^{241}Pu ($t_{1/2}$ = 14.29 year) decays to form ^{241}Am ($t_{1/2}$ = 432.6 year). When this material is reinserted into a high-neutron flux, large amounts of ^{242}Cm form through neutron capture in ^{241}Am. If the fuel were measured after a longer cooling time (~2 years), most of the ^{242}Cm would have decayed, and the measured neutron emission rates would be more consistent with the rates obtained from fuel material that had undergone continuous irradiation.

21.4.2 Unattended Fork Detector Monitor

The unattended fork detector monitor (UFDM) is used during cask-loading campaigns to verify the spent fuel assemblies in the pool before loading them into casks or baskets for long-term or interim dry/wet storage. Designed to be connected to the fork detector, the system measures the total neutron and gamma radiation emitted from a spent fuel assembly [30, 31].

The UFDM is a watertight polyethylene detector head, shaped like a square U, whose two arms contain identical sets of detectors for measuring opposite sides of the fuel assembly simultaneously, as shown in Fig. 21.15. Each arm contains an ionization chamber that operates in the current mode to measure gamma-ray emission and two fission chambers that operate in the pulse mode to measure neutron emission. In the original design, each arm included one bare and one cadmium-encased fission chamber; however, the current generation of fork detectors features only bare fission chambers. Another fork detector design modification replaces the fission chambers with ^{10}B-lined tubes wrapped in heavy tungsten shielding to absorb gamma radiation and minimize gamma pileup in the neutron signal. The fork detector is available with three different apertures: one for PWR, one for BWR, and one for VVER spent fuel assemblies.

The UFDM measurement system (Fig. 21.15) is designed to be sufficiently portable to hand-carry on a plane. The components for data collection and unattended operation are housed inside a Rimowa Pilot case with dimensions of 53 cm × 45 cm × 27 cm and a total weight (case plus components) of 18 kg. All components are direct-current powered. Like all unattended systems, the UFDM ensures redundant data collection, battery backup, and tamper-indication. In particular, the pilot case is sealable, and the detector cables are protected inside tamper-indicating conduit. The backup battery ensures continuous data acquisition for up to 3 days even in the event of a mains power failure. If remote data transmission is to be implemented, the system has space to accommodate a virtual private network box (optional).

Review of the data collected by the UFDM system is performed using IRAP, the standard data analysis and review platform used by IAEA and Euratom [33]. The touchscreen display allows inspectors to perform onsite data evaluation to confirm that the expected number of assemblies was loaded into the cask and to qualitatively verify that the measured neutron and gamma count rates correspond to the operator's declaration.

A more thorough analysis for partial defect verification is done at IAEA Headquarters, whereby the measurements of the UFDM system are compared with the results of ORIGEN calculations using the operator's declarations. By integrating the ORELLA interface module, it is possible to perform ORIGEN depletion analysis directly from IRAP. Measurement results from 57 VVER-1000 irradiated spent fuel assemblies are shown in Fig. 21.16. The histograms show the percentage difference between the calculated and the measured count rates. The calculated neutron and gamma count rates are corrected by an experimental calibration coefficient that considers the response of the electronics units. Typically, standard deviations of the difference distributions shown in Fig. 21.16 are in the range of 5–10%.

21.4.3 Passive Neutron Albedo Reactivity

The ratio of the total neutron-detection rate in high and low multiplying regions, also known as the *passive neutron albedo reactivity* (PNAR) ratio, is proportional to the assembly multiplication as demonstrated with simulated data in Fig. 21.17 [34, 35]. To implement the PNAR method in spent fuel assay, a neutron-absorbing material such as cadmium is introduced to the system to absorb neutrons and lower the assembly multiplication. A second total neutron measurement is then performed with the neutron-absorbing material removed, and the ratio is calculated. Alternatively, PNAR can be implemented by

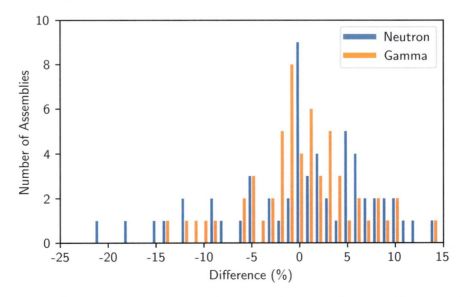

Fig. 21.16 Distribution of the difference between calculated and measured rates of neutron and gamma for 57 VVER-1000 assemblies

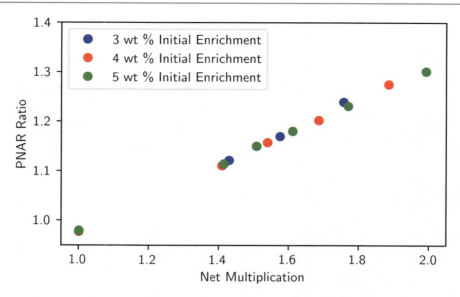

Fig. 21.17 PNAR ratio as a function of net multiplication for simulated BWR assemblies from Ref. [35]

including both cadmium-wrapped and bare detectors in the measurement system and using the ratio of the two detector configurations. Correct implementation of the PNAR technique is essential; if the assembly is repositioned between the two measurements, for example, with and without cadmium, uncertainty introduced from the repositioning of the assembly could prove to be larger than the PNAR effect, particularly if the measurement system is asymmetric [36].

21.4.4 Active Neutron Measurements

Differential die-away analysis (DDAA) is an active measurement technique that uses a pulsed interrogation source—typically a deuterium-tritium neutron generator—to measure fission neutron lifetimes in a system. The source is positioned near the irradiated fuel, where it produces an induced fission signal proportional to the amount of fissile material. Typical neutron source strengths must be on the order of 10^8 to 10^9 n/s to induce a fission signal that is comparable in size to the passive neutron yield in power reactor fuel. The regular pulse frequency of the generator allows the induced fission neutrons to be binned according to their detection times relative to the pulse trigger. The resulting data can be used to get time-window-specific signals, which include the following:

- Neutron generator signal. In the very early time window after the pulse trigger, the neutrons detected will be proportional to the neutron-generator output (uninfluenced by the fuel) and can be used for normalization.
- Active fuel signal. In the early- to mid-range time windows, both the neutrons detected and the die-away time of those neutrons are a direct measure of the net multiplication of the fuel and indicative of other fuel parameters, such as initial enrichment and fissile content.
- Passive fuel signal. In the very late time window after the induced fission signal has dissipated, the neutrons detected are from the passive signal from the fuel required for normalization and are proportional to fuel burnup. To use this feature, the pulse period of the generator must be large enough to allow for the induced fission signal to dissipate; otherwise, a separate passive measurement would have to be taken.

The DDAA technique has the benefit of being highly sensitive and providing a wide range of information for both spent and fresh fuel measurements in a much shorter time frame than comparable techniques [37–40]. DDAA uses the singles rate only and therefore does not need the large efficiency or long measurement times that a coincidence counter requires. Additionally, this technique's unique ability to take both active and passive data in a single measurement allows for shorter measurement times compared with a californium-driven interrogation system. DDAA's largest drawback is its technical complexity. The requirement for a trigger time for the interrogating neutrons necessitates an expensive technical component (a neutron generator) that is not required for a californium-driven active interrogation system or a passive system.

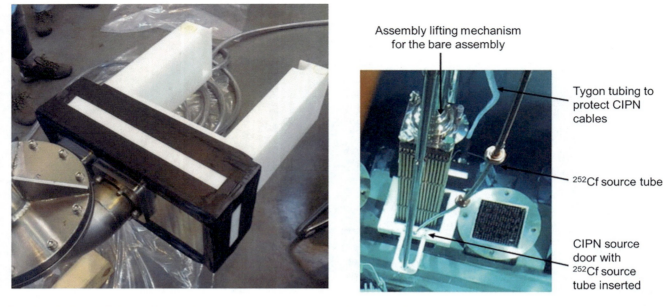

Fig. 21.18 (left) Californium interrogation prompt neutron instrument for characterizing power reactor fuel; (right) view of the californium interrogation prompt neutron instrument from the source-door side during spent fuel measurements

The californium interrogation prompt neutron (CIPN) instrument is another example of an NDA system that uses active neutron measurements [41]. As the name implies, CIPN uses a ^{252}Cf source to induce fission in power reactor fuel and measures the subsequent neutrons with three fission chambers. The system also contains two ionization chambers for simultaneous total gamma-ray measurement; thus the system resembles an active version of the familiar fork detector. The CIPN instrument is shown in Fig. 21.18. Due to its similarity to the fork detector, CIPN represents a simple instrument based on well-established technologies capable of providing additional information through active neutron measurements. The major distinction with respect to the UFDM is in handling experience due to the use of a removable source door. The source door is used for reproducible positioning of ^{252}Cf source and can be removed to position the instrument around the spent fuel assembly. An example of underwater positioning of CIPN around a spent fuel assembly is shown in Fig. 21.18 (right).

CIPN was tested on a range of spent fuel assemblies with burnup between 17 and 38 GWd/tU, 2–3% enrichment, and about a 30-year cooling time [42]. During this spent fuel measurement demonstration, typical measurement times for passive and active measurements corresponded to <5 min, and the overall measurement times (including CIPN positioning and ^{252}Cf source handling) corresponded to 15–20 min. CIPN requires two separate measurements—an active and a passive measurement—to provide the net active signal used for multiplication measurement. Typical required neutron source strength is $\sim 10^8$ n/s, which can be obtained from a $\sim$ 50 μg ^{252}Cf source. The use of a ^{252}Cf source for the active measurement requires that the source be adequately shielded during passive measurements. Additionally, because ^{252}Cf has a half-life of 2.65 years, replacement sources must be purchased periodically, adding to the cost of maintaining the instrument. To mitigate these concerns, CIPN can be modified, replacing the ^{252}Cf source with a neutron generator.

21.4.5 Passive Differential Die-Away

The characteristic neutron die-away from irradiated fuel can also be measured passively with a technique called *differential die-away self-interrogation* (DDSI). The DDSI method uses the fact that the spontaneous fissions and (α,n) reactions within irradiated fuel emit neutrons that will thermalize between fuel pins and can induce fission events or chains. With a ^{3}He-based detector system and list-mode data acquisition, neutrons from the spontaneous and induced fission events are detected, producing a Rossi-α distribution, which is a histogram of times between neutron detections (see Chap. 17, Sect. 17.2.3). In traditional active differential die-away measurements, *time zero* is defined as the time at which neutrons are emitted from the generator, and neutron die-away is calculated using emission as a starting point. In contrast, in DDSI, each detected neutron represents its own time zero. Neutron die-away is therefore quantified via coincidence counting. Although this technique has

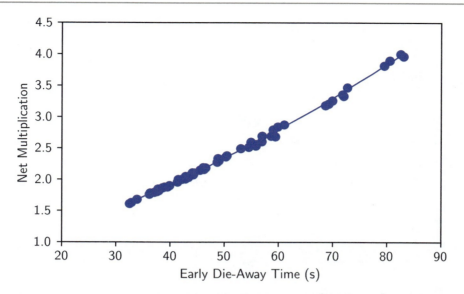

Fig. 21.19 Net multiplication as a function of early die-away time for 56 simulated PWR fuel assemblies with varying initial enrichments, burnups, and cooling times

the benefit of avoiding the need for a neutron generator or other interrogating source, it has the added complication of requiring a high-efficiency system and list-mode data collection to produce Rossi-α distributions [43].

The Rossi-α distribution as measured with DDSI is well characterized by a double exponential function, with the fast exponential component reflecting the die-away time of the detector system itself and the slow component reflecting the die-away time of the neutrons in the fuel assembly. A single exponential fit to the early time region of the Rossi-α distribution detected from irradiated fuel has been shown to be nearly linearly proportional to assembly net multiplication [44]. The exponential term, called the *early die-way time*, is a robust parameter for measuring multiplication regardless of operating condition, initial enrichment, cooling time, or burnup [43]. Net multiplication as a function of early die-away time is shown in Fig. 21.19 for 56 different simulated PWR fuel assemblies [45].

21.4.6 Neutron Coincidence Counting

Neutron coincidence counting—or detection of multiple neutrons emitted from the same fission event or chain—can be used in combination with active interrogation to directly measure fuel depletion. Interrogating a spent fuel assembly continuously with neutrons will induce fission in the remaining fissile isotopes, and the detected coincidence rate is therefore proportional to the depletion. Historically, coincidence counting of spent power reactor fuel has been considered impractical because of the high rate of accidentals and low efficiency of typical neutron-counting systems. Newly designed systems with relatively high efficiency, such as DDSI (discussed in Sect. 21.4.2), have made coincidence counting of power reactor fuel possible. Coincidence counting of spent fuel also has the advantage of excluding the background neutron signal from (α,n) reactions.

The advanced experimental fuel counter (AEFC) is an NDA instrument that measures research reactor spent fuel via active neutron coincidence counting [46]. Six ^{3}He tubes form a half-circle around a fuel funnel into which spent fuel is inserted in the underwater system. On the other side of the fuel funnel, a ^{252}Cf source produces interrogating neutrons. Use of a spontaneous fission active interrogation source lowers the doubles rate uncertainty as a result of the time-correlated induced fission (TCIF) effect (see Chap. 17, Sect. 17.15.2). TCIF occurs when an interrogating source emits multiple neutrons simultaneously, effectively multiplying the correlated neutrons in the system. All neutrons from the fission source itself—as well as any induced fission events or chains—are correlated by definition, and if detected within the same time gate, are classified as doubles events. The ratio of Doubles to accidentals increases because of TCIF, which therefore lowers the doubles rate uncertainty.

The AEFC measured a series of MTR assemblies and determined the residual ^{235}U mass (i.e., depletion) with a root mean square error of approximately 6 g [47]. The calibration curve that relates the net active doubles rate determined with active neutron coincidence counting to residual ^{235}U mass takes the form of a second order polynomial forced through the origin.

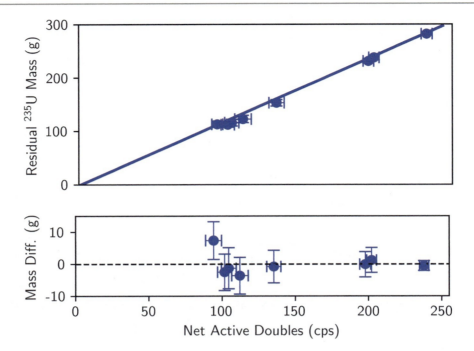

Fig. 21.20 Calibration curve that relates residual ^{235}U mass with net active doubles rate for measured MTR fuel assemblies [47]

The shape of the curve is the result of competing effects of self-shielding and multiplication within the fuel and is therefore geometry and density dependent. MTR measurements with the AEFC were used to construct a calibration curve, given in Fig. 21.20.

21.5 Other Spent Fuel Measurement Techniques and Systems

21.5.1 VXI Integrated Fuel Monitor (VIFM)

Safeguards measures on CANDU reactors rely on neutron and gamma ray detectors operated in unattended mode, which are placed in strategic locations to follow the movement of irradiated fuel bundles during on-load refueling and their transfer to the spent fuel pond. In particular, the discharge of irradiated fuel bundles from the reactor fuel channel is monitored by the core discharge monitor (CDM) located in the calandria area, while the movement of the irradiated bundles to the spent fuel pond is monitored by the bundle counter (BC). The CDM provides a tally of the number of irradiated fuel bundles discharged from the fuel channel and stored in the fueling machine, while the BC provides both counting of the bundles being transferred to the spent fuel pond and confirmation of the declared direction of bundle movement. Optionally, so-called Yes/No monitors (Y/N) are located beside accesses to the spent fuel duct in order to monitor possible removal of bundles. In the spent fuel pond, the loading of baskets with spent fuel bundles is monitored by a special version of the BC, the underwater bundle counter (UWBC). Taken together, CDM, BC and Y/N detectors form the VXI Integrated Fuel Monitor (VIFM).

In Fig. 21.21, the usual locations of the CDM, BC and Y detector assemblies can be seen on a typical CANDU-600 layout.

21.5.1.1 Core Discharge Monitor

In CANDU reactors, core discharge monitors (CDMs) are used to assess the number of irradiated spent fuel bundles that leave the reactor core. The detector assembly consists of neutron and gamma ray sensors, usually present in pairs within the same detector assembly to provide redundancy and improve the reliability of unattended operation. A high-density polyethylene block encased in cadmium is used as the moderator for the neutron detectors. The cadmium absorbs previously moderated neutrons reflected by the wall of the reactor vault, this reduces the background count rate which minimizes false positives.

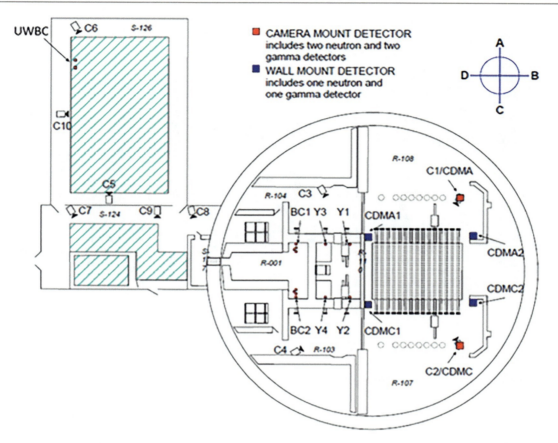

Fig. 21.21 Typical CANDU-600-type station layout. The usual locations of core discharge monitor (CDM), bundle counters (BC and UWBC) and Yes/No monitor (Y) detector assemblies are highlighted. Face A and Face C are the sides of the reactor where the fueling machines attach to the fuel channels

The neutron detectors consist of gas-filled fission chambers with a deposit of ^{235}U. The fission chambers used are Reuter Stokes model RS-P6 with a sensitivity of 0.14 cps/nv (thermal neutrons). A radiation-hardened preamplifier is connected to the fission chamber, which generates the necessary high voltage and shapes and amplifies the output signal.

The gamma ray detectors consist of silicon PIN diode detectors. These solid-state detectors are Bot Engineering model IRD30A [48], which feature an integrated preamplifier and can measure gamma ray fields with a sensitivity of approximately 0.3 cps/mR/h. Optionally, an ion chamber can be added to the assembly as a second gamma ray detector.

A drawing of the CDM detector assembly is shown in Fig. 21.22.

CDMs come in two different geometries:

- Wall-mounted CDMs, where the detector assembly is mounted on the wall of the reactor vault on the side of the calandria. Typically, two assemblies are mounted on each face of the reactor where the fueling machines attach to the fuel channels (indicated as face A and C).
- Camera-mounted CDMs, where the detector assembly is mounted on the floor of the reactor vault in front of the calandria. A single assembly each is mounted on the A and C faces of the reactor, where the fueling machines attach.

Note: A facility would have either camera- or wall-mounted CDM detectors, not both (as shown in Fig. 21.21).

In on-load reactors, neutron radiation is used for the purpose of counting the discharged bundles. An overview of the CDM response characteristics is given in Ref. [49]. The peaks that appear in the neutron trace, as well as their shape, are correlated to the movement of irradiated fuel outside the reactor core. It should be noted that the neutron detectors are sensitive to the neutrons emitted by the irradiated fuel only once it emerges from the reactor's radiological shield. CDMs mostly detect delayed and photoneutrons. The main component of the measured count rate for the fuel bundles comes from delayed neutrons, which are emitted in the irradiated fuel bundle via beta decay of certain fission fragments. Photoneutrons are

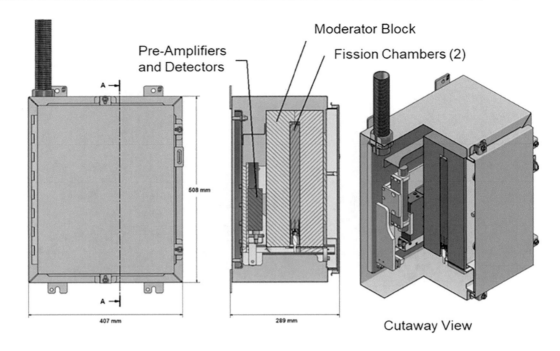

Fig. 21.22 Drawing of the CDM detector assembly

produced by the gamma radiation emitted from the fuel bundle impinging on the heavy water present in the fuel channel and fueling machine.

Because of their higher sensitivity to the rotation of the fueling machine magazine, gamma ray detectors are used primarily as a secondary data source when interpreting non-standard events that appear in the neutron trace.

The CDM signature strongly depends on the design of the fueling method used in the reactor. There are two basic schemes for CANDU reactors: the fuel separator method and the fuel latch method. In both fueling schemes, fresh bundles are pushed from one end of a fuel channel while irradiated bundles are removed from the other end. Both fresh and irradiated fuel bundles are stored in the fueling machine magazines in pairs.

In the fuel latch method, the movement of fuel bundles along the hydraulic flow of the coolant is restrained by means of downstream latches inside the pressure tube. Those fuel latches hold the bundles in place also during the fueling operations, allowing only two irradiated bundles at a time to exit the reactor's radiological shield before being collected into the fueling machine magazine. In this case, the CDM presents a cleaner signature, where the peaks directly correspond to the irradiated bundle pairs removed from the fuel channel and loaded in the fueling machine magazine. Figure 21.23 shows an example of the CDM signature for an 8-bundle push at a reactor using the fuel latch method.

By contrast, the CDM signature for reactors using the fuel separator method shows a more complex structure, mainly because the fuel is held in place by the shield plug at the end of the fuel channel, so that the entire fuel column moves while fresh bundles are pushed into the reactor. In this fueling method, some fuel bundles exit the radiological shield and sit for some time in the fueling machine snout before they are collected in the fueling machine magazine. As can be seen in Fig. 21.24, for a typical 8-bundle push, this mechanism leads to a 6-peak signature in the neutron trace of the CDM (different from the four peaks seen with the fuel latch method), followed by other structures related to the rotation of the fueling machine magazine as it puts the shield plug back into the end of the fuel channel.

The rising and falling slopes are the fundamental characteristics of the signature through which they can be identified and sorted. For fuel bundles, the count rates build up as the delayed neutrons emitted by the irradiated material are detected as the bundle emerges from the radiological shield. In this case, the rising slope is a direct indication of the speed at which bundles are moved. For fueling machine operations that are not bundle pair discharges, the change in count rate is due mainly to photo neutrons, and can lead to dramatically different rising slopes. For example, a rotation of the fueling machine magazine that brings the bundles closer to the CDM detector will induce a sudden, much steeper increase in the neutron count, while signals

21 Spent Fuel Measurements

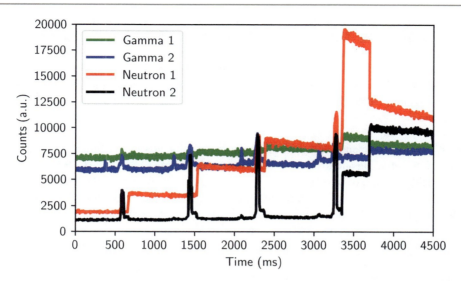

Fig. 21.23 Typical 8-bundle discharge of a CANDU reactor that uses the latch method. The neutron traces from the two fission chambers are plotted in red and black, while the two gamma traces from the Si-PIN detectors are plotted in green and blue

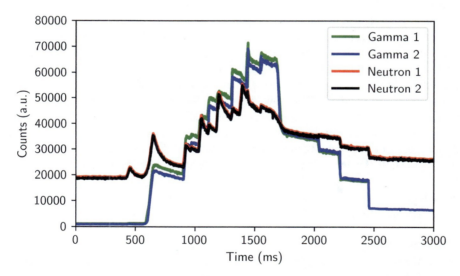

Fig. 21.24 Typical 8-bundle discharge of a CANDU reactor that uses the fuel-separator method. The neutron traces from the two fission chambers are plotted in red and black, while the two gamma traces from the Si-PIN detectors are plotted in green and blue

from fuel bundles sitting in the fuel channel end fitting for some time during downstream fueling machine preparation will present a more gradual increase in neutron count rate.

For fuel bundles, the exponential behavior of the falling count rate is indicative of the expected decay of the delayed neutrons. Any deviation from this behavior is an indication that the neutron count rate has dropped due to extra shielding of the neutron source—for example, by the re-insertion of the shield plug or a rotation of the fueling machine magazine that moves the bundles away from the CDM detector—and therefore, that the signal is not due to a bundle discharge.

From these considerations, the CDM neutron signature shown in Fig. 21.24 can be interpreted as follows:

- The first two peaks are due to one bundle and then three bundles, respectively, pushed outside the radiological shield and waiting in the fuel channel end fitting for the downstream fueling machine to allow them into the magazine;
- The third to sixth peaks correspond to bundle pairs being loaded into the fueling machine. The second small peak appearing on the tail of the first peak is due to the rotation of the fueling machine magazine;

- The subsequent structures show sudden changes in neutron count rate typical of the rotation of the fueling machine magazine, with the last drop in counts due to the re-insertion of the shield plug into the channel.

Various algorithms were written in the past years for bundle-detection in core discharge monitors. The functionality is mostly based on fixed thresholds and detection of peak shapes starting from the considerations described in Sect. 2.1. However, deviations from the usual patterns often result in misdetection of the peaks related to the bundle pairs loaded into the fueling machine. In addition, the software has to be parameterized to fit each reactor type or installation and needs regular readjustment.

In the effort to improve the performance of the CDM signature detection algorithm, comprehensive research was done [50] to create a model of CANDU refueling and to analyze the effects of different events on the CDM signatures. Machine learning approaches based on a Hidden Markov model are discussed in Ref. [51]. Recently, the IAEA investigated new machine learning algorithms and methods to increase bundle detection accuracy in the CDM for CANDUs using the fuel separator method. The aim was to require as little knowledge of the specific instrument behavior as possible, so that the algorithm can be reused in different facilities and reactor types [49]. A decision-tree based solution was tested, and showed an accuracy of 84–95% in identifying peaks coming from the bundle pairs loaded into the fueling machine magazine. This result confirmed the potential of applying a novel machine-learning-based algorithm to CDM signature recognition. Other methodologies, such as recurrent neural networks, should be evaluated before adopting a novel algorithm for safeguards evaluations.

21.5.1.2 Bundle Counter

In CANDU reactors, BCs are used to assess the number of irradiated spent fuel bundles that are transferred to the spent fuel pond, and to detect any bundle movement to and from the spent fuel pond. The sensor assembly consists of collimated gamma ray detectors to increase the spatial accuracy of the measurement. The collimator is designed to limit the field of view of the detector, so that the detector only sees the radiation field when the irradiated material passes below the collimator (Fig. 21.25, right). The detector housings are made from stainless steel and the detector cable is inside a flexible stainless-steel guide sleeve, which forms a watertight envelope with the detector housing (Fig. 21.25, left).

The gamma ray detectors consist of silicon PIN diode detectors. The solid-state gamma ray detectors are Bot Engineering model IRD30AP [52], which have the same specifications as the IRD30 [48] used for the CDM, but with a motion sensing circuit and additional electronics for detector operation verification. The IRD30AP provides a constant state-of-health signal, as well as a real-time indication of any detector movement, which might indicate tampering. Applications requiring higher detector sensitivity use the IRD30HS detector [53], a variant of the IRD30AP which incorporates a larger area silicon detector to provide a tenfold increase in detector sensitivity.

A minimum of two detector locations is required to determine direction of movement. Bundle counter detector assemblies come in either 2-point or 4-point configurations—detailed in the following sections.

21.5.1.2.1 2-Point Bundle Counter

In a 2-point configuration, the two measurement positions (points) for the bundle counter detector assemblies are chosen in the area of the spent fuel port, where the bundles are being pushed along a track towards the spent fuel pond. The bundles move

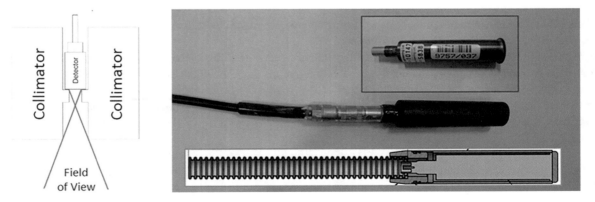

Fig. 21.25 Left: Sectional view of collimator and detector. Right: IRD30 detector (inset photo, top) and wrapped in waterproofing (optional, used in applications where detector might be exposed to water). Detector is placed inside housing (inset drawing, bottom)

horizontally past the detectors. Each position has two gamma ray detectors to provide redundancy and improve the reliability of the unattended operation. Figure 21.26 shows a 2-point bundle counter configuration scheme.

Due to the collimator, the counts show a sharp change when the bundles pass below the detectors at the measurement points and a plateau while the bundle is within each detector's field of view. As the distance between the two measurement points is approximately equal to the length of a single fuel bundle, both detectors cannot observe the same bundle simultaneously. The overlap of signals from the two measurement points allows the type of transfer (single or double bundle) to be determined. A typical 2-point BC signature for a double-bundle movement is shown in Fig. 21.27. The small decrease in counts between the two plateaus corresponds to the gap between the two bundles.

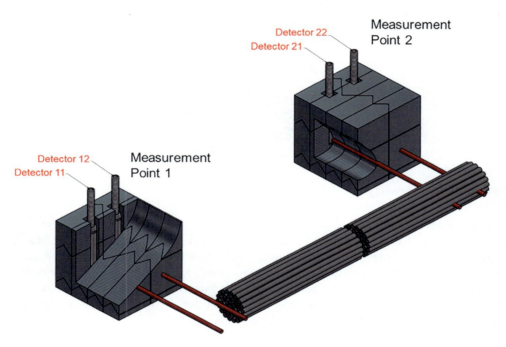

Fig. 21.26 A 2-point configuration bundle counter (drawing not to scale, as the distance between the detectors is usually one fuel bundle in length). The collimated field of view of the detectors is highlighted in red

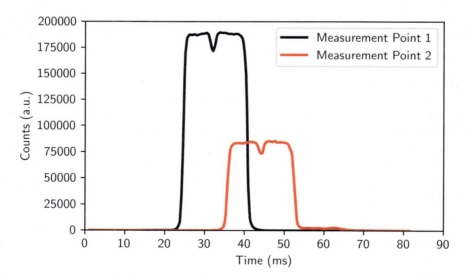

Fig. 21.27 Typical 2-point BC signature showing the fuel passing in and out of the field of view of the two measurement points. Note the overlap in the signals, indicating that a double-bundle movement has been detected

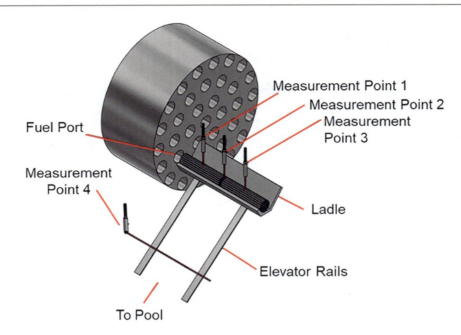

Fig. 21.28 A 4-point bundle counter configuration

21.5.1.2.2 4-Point Bundle Counter

In a 4-point configuration, the four selected measurement positions (points) for the bundle counter detector assemblies are after the spent fuel port where the bundles are pushed onto a ladle, which is then lowered into the spent fuel pond via an elevator. The bundles pass the first three measurement points horizontally and the fourth point during their descent. Normally, each measurement point has only one gamma detector, as basic directionality information can be determined with any combination of two measurement points, thus obviating the need for detector redundancy as in the 2-point configuration. Figure 21.28 shows a scheme of the 4-point bundle counter configuration.

Due to the collimator, the counts show a sharp change when the bundles pass below the detectors at the measurement points and a plateau while the bundle is within each detector's field of view. Measurement points 1 and 3 are separated by a distance greater than the bundle length, therefore these two detectors cannot observe the same bundle simultaneously. The overlap of signals from these two measurement points allows the type of transfer (single or double bundle) to be determined. The signal from the fourth measurement point monitors the actual movement toward the spent fuel pond, thus serving as an extra control on the total time of the transfer. A typical 4-point bundle counter signature for a double-bundle movement is shown in Fig. 21.29. The signal in the fourth measurement point is shorter because the detector is looking at the ends of the bundles, rather than the sides.

21.5.1.2.3 Underwater Bundle Counter

The VIFM detector suite includes also an underwater implementation of the bundle counter (UWBC) to count fuel bundles being placed in baskets in the spent fuel pond in preparation for transfer to dry storage. The UWBC ensures continuity of knowledge for spent fuel bundles as they enter the waste management part of the fuel cycle.

The UWBC uses the same gamma ray detectors as the normal bundle counter, but with a special watertight housing for underwater operations. The UWBC always has a 2-point design, and the detector assembly includes a collimator to narrow the detector's field of view. Typically, each measurement position contains two detectors for redundancy purposes.

The gamma ray detectors are installed in the part of the fuel pond where basket loading takes place (Fig. 21.30). Since they are paired (two gamma ray detectors each at two measurement points), UWBCs provide directional confirmation of bundle movement. Figure 21.31 shows a typical underwater bundle counter signature. As bundles pass the detectors in a vertical orientation, the duration of increased counts is normally short in time. In addition, due to the geometry of the spent fuel pond and the basket loading station, the two measurement points are normally close to each other, resulting in a narrow peak separation. For this reason, both the detectors and data acquisition for the underwater bundle counter must operate at a high sampling rate.

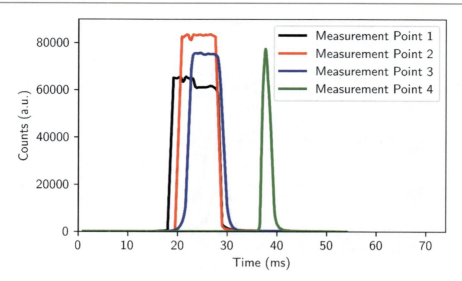

Fig. 21.29 Typical 4-point BC signature showing the fuel passing in and out of the field of view of the four measurement points. The overlap in the signals of point 1 (black) and point 3 (blue) indicates that a double-bundle movement has been detected

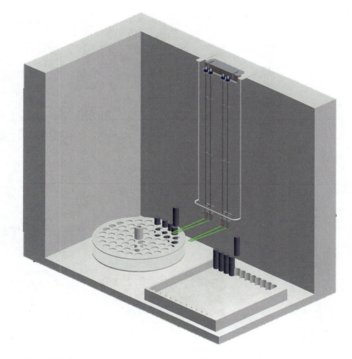

Fig. 21.30 Layout of the underwater bundle counter (UWBC) in a spent fuel pond to count the bundles being loaded into baskets. The collimated field of view of the detectors is highlighted in green

21.5.1.3 Yes/No Monitor

In CANDU reactors, in order to ensure that no fissile material has been diverted, Yes/No monitors (Y/N) are used to monitor locations where radiation should be absent. The detector assembly consists of silicon PIN diode gamma ray detectors installed in a tamper-indicating enclosure that can be inserted into a penetration or mounted on a wall (Fig. 21.32). The solid-state gamma ray detectors are Bot Engineering model IRD30AP [52], the same as those used for bundle counters. Typically, each detector assembly contains two detectors for redundancy purposes.

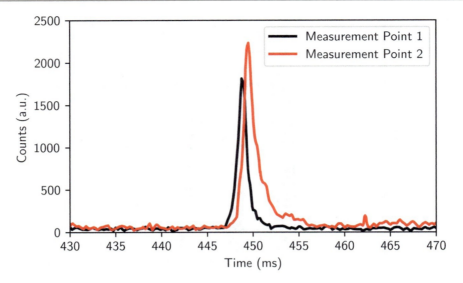

Fig. 21.31 Typical signature of the underwater bundle counter showing the fuel passing in and out of the field of view of the two measurement points

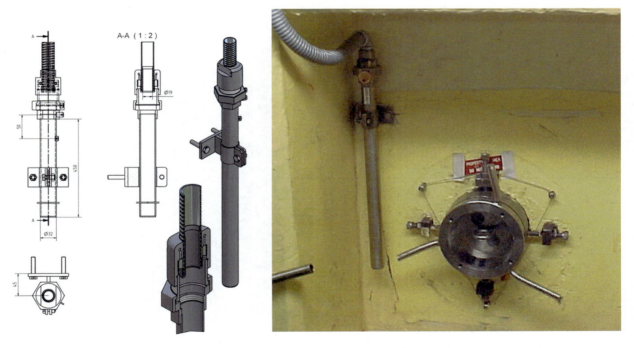

Fig. 21.32 Design of the Yes/No monitor detector assembly (left) and a picture of an installed detector (right)

The Yes/No monitor detector assembly is typically installed at access ports that are not part of the normal discharge path, such as near the fresh fuel loading ports, exit ports or emergency gates. As only one measurement point is foreseen, the Y/N monitor does not provide information on the direction of movement, but it is used to confirm absence of movement.

21.5.2 Cerenkov Viewing Devices

Electromagnetic Cerenkov radiation is emitted whenever a charged particle passes through a medium with a velocity that exceeds the phase velocity of light in that medium. In water, the phase velocity of light is ~75% of the value in vacuum. Any electron that passes through water and has at least 0.26 MeV of kinetic energy will exceed this velocity and emit Cerenkov radiation. Irradiated fuel assemblies are a prolific source of beta particles, gamma rays, and neutrons. All three types of emissions can produce Cerenkov light, directly or indirectly.

The most significant production of Cerenkov light is from high-energy fission-product gamma rays that interact with the fuel cladding or storage water. This interaction produces electrons and positrons by Compton scattering and other effects such as pair production. Calculations of the number of Cerenkov photons generated from these beta particles indicate that Cerenkov light production in the visible wavelength range of 4000–7000 Å is negligible for gamma rays with energies below 0.5 MeV but rises steeply with higher energy. A 2 MeV gamma ray produces 500 times more Cerenkov light than a 0.6 MeV gamma ray.

Cerenkov viewing devices (CVDs) are used by international inspectorates to nondestructively inspect and verify spent nuclear fuel by measuring the Cerenkov light [54]. Several versions of CVDs are in regular use by inspectorates, two of which are discussed in the following sections: the improved Cerenkov viewing device (ICVD; Ref. [54]) and the digital Cerenkov viewing device (DCVD; Ref. [55]), both shown in Fig. 21.33.

The ICVD is a handheld device that filters ambient visible light and intensifies light in the near-ultraviolet (UV) range (i.e., Cerenkov light). Due to its compact size and ease of use, it is the instrument most commonly used by the IAEA to verify spent LWR fuel assemblies in fuel ponds. The ICVD is used to verify the presence or absence of spent nuclear fuel. It can verify whether spent fuel, void, or some other object is present in a particular location. The results of the analysis are compared with the operator declaration, though it is important to note that the ICVD does not record any quantitative data.

The DCVD has been used since 2004 as an alternative to the ICVD to nondestructively inspect and verify spent nuclear fuel, especially fuel with low burnup and/or long cooling times [56]. The Cerenkov glow can be mapped, allowing the predicted Cerenkov glow to be compared with the measured glow to determine the presence or lack thereof of spent fuel pins within the assembly, as shown in Fig. 21.34. The DCVD is shown in use in Fig. 21.35.

The DCVD is capable of verifying a broad range of spent fuel burnups and cooling times. Burnups as low as 1.1 MWd/kgU have been measured [57], and cooling times can range from fuel that has been freshly discharged to fuel that has been cooled for up to 40 years [58]. Additionally, the DCVD is able to not only qualitatively map the UV emissions from spent fuel assemblies but can also quantify the amount of Cerenkov light emitted, which allows the detection of partial defects—in addition to gross defects—in spent fuel assemblies. The UV-light sensitivity of the DCVD comes from a charge-coupled device chip [56]. It also includes a motorized zoom lens. When used to make a measurement, the DCVD is mounted onto the fuel-handling machine railing. To reduce background and noise effects on the measurement, three to five measurements are taken for each assembly.

The DCVD is non-intrusive, and its measurements are relatively quick because it does not require the fuel assemblies to be moved to make a measurement [55]. However, it provides no fission product information and has limited sensitivity, meaning that only relatively large diversions can be detected [57]. The more recent version of the DCVD, the enhanced DCVD

Fig. 21.33 Photos of the digital Cerenkov viewing device (left) and the improved Cerenkov viewing device (right)

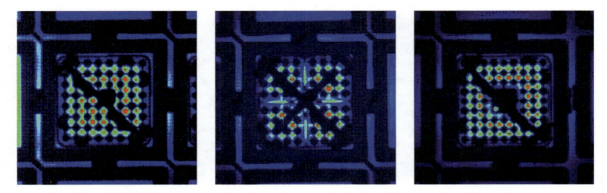

Fig. 21.34 Digital Cerenkov viewing device Cerenkov light image that shows dark points where pins are missing, and light points where pins are present [56]

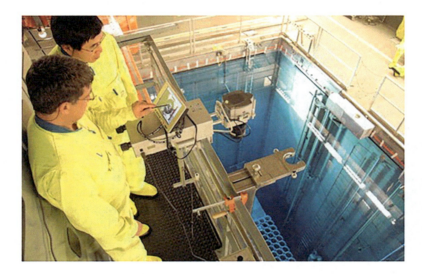

Fig. 21.35 An early version of the digital Cerenkov viewing device attached to a spent fuel viewing platform. (Image courtesy of Channel Systems Inc.)

(DCVD-e), incorporates sophisticated software to highlight possible missing pins in real time, which makes the inspector's job easier and less time consuming [56]. The most recent version of the Cerenkov viewing device approved for use by the IAEA is the next-generation Cerenkov viewing device [59].

21.5.3 Curium/Plutonium Ratio

The elemental curium content in spent fuel grows as a power function of burnup, whereas the plutonium in-growth is fairly linear [60]. Therefore, the ratio of these elements also follows a power law and, if known, can be used to determine the fuel burnup. Table 21.5 provides calculated curium-to-plutonium ratios for PWR fuel in an accountability tank at different burnup levels. X-ray spectroscopy with a sensitive detector system such as micro-calorimetry (discussed in Sect. 21.3.5) would provide the elemental composition information necessary to complete curium-to-plutonium ratio analysis.

21.5.4 Calorimetry

Specialized underwater calorimeters capable of housing entire spent power reactor fuel assemblies may be used to quantify the decay heat emitted from spent fuel. Radioactive decays within spent fuel emit characteristic amounts of heat. The total amount

Table 21.5 Atom densities and ratio of curium/plutonium for three different levels of burnup

Burnup (GWd/tU)	Atom density (atoms/tU)		Cm/Pu
	Cm	Pu	
15	1.57×10^{21}	1.36×10^{25}	1.16×10^{-4}
30	4.26×10^{22}	2.00×10^{25}	2.13×10^{-3}
45	2.99×10^{23}	2.39×10^{25}	1.27×10^{-2}

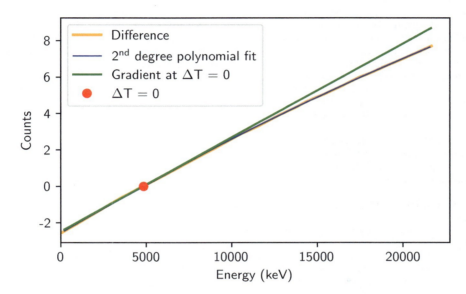

Fig. 21.36 Temperature difference between the inside of a spent fuel calorimeter and the spent fuel water surrounding the instrument. The difference in temperature is used to quantify assembly burnup

of heat emitted from a fuel assembly as a function of time is therefore related to the amount of decay occurring within the assembly, and the relationship is quantified via a calibration curve.

Once a fuel assembly has been placed in the calorimeter and the system sealed, a measurement flow pump circulates water within the calorimeter, creating an overpressure environment. At this point, the flow pump is stopped, and a circulation pump circulates the water within the calorimeter as the water temperature is constantly monitored. The measurement stops after a predetermined temperature increase is obtained, and the amount of time required to reach this desired temperature increase corresponds to the decay heat. A correction must also be made for the small amount of gamma radiation and heat that escapes the calorimeter. Obtaining a single measurement point for calibration takes approximately 5 h [61]. Figure 21.36 shows the temperature evolution within a spent fuel calorimeter [62].

References

1. A. Lebrun, G. Bignan, Nondestructive assay of nuclear low-enriched uranium spent fuels for burnup credit application. Nucl. Technol. **135**(3), 216–229 (2001)
2. IAEA, *Power Reactor Information System* (International Atomic Energy Agency, Vienna, 2020)
3. IAEA, *Research Reactor Database* (International Atomic Energy Agency, 2020). https://nucleus.iaea.org/RRDB/RR/ReactorSearch.aspx. Accessed Mar 2023
4. J. Hu, I.C. Gauld, J.L. Peterson, S.M. Bowman, U.S. Commercial Spent Nuclear Fuel Assembly Characteristics: 1968–2013, U.S. Nuclear Regulatory Commission (2016). https://doi.org/10.2172/1330516. Accessed Mar 2023
5. J.J. Duderstadt, L.J. Hamilton, *Nuclear Reactor Analysis* (Wiley, New York, 1976)
6. A.V. Nero, *A Guidebook to Nuclear Reactors* (University of California Press, Berkeley, 1979)
7. IAEA, *Research Reactors: Purpose and Future* (International Atomic Energy Agency, Vienna, 2016)
8. G.E. Bosler, J.R. Phillips, W.B. Wilson, R.J. LaBauve, T.R. England, Production of Actinide Isotopes in Simulated PWR Fuel and Their Influence on Inherent Neutron Emission. Los Alamos National Laboratory report LA-9343 (1982)

9. D. Cobb, J. Phillips, G. Bosler, G. Eccleston, J. Halbig, C. Hatcher, S.-T. Hsue, Nondestructive Verification and Assay Systems for Spent Fuels. Los Alamos National Laboratory report LA-09041 (1982)
10. P.M. Rinard, A Spent-Fuel Cooling Curve for Safeguards Applications of Gross-Gamma Measurements. Los Alamos National Laboratory report LA-09757-MS (1983)
11. J.R. Phillips, G.E. Bosler, J.K. Halbig, S.F. Klosterbuer, H.O. Menlove, Nondestructive Verification with Minimal Movement of Irradiated Light-Water Reactor Fuel Assemblies. Los Alamos National Laboratory report LA-09438-MS (1982)
12. P. Jansson, *Studies of Nuclear Fuel by Means of Nuclear Spectroscopic Methods*. Doctoral Dissertation, Acta Universitatis Upsaliensis, Uppsala, Sweden (2002)
13. IAEA, *Safeguards Techniques and Equipment: 2011 Edition* (International Atomic Energy Agency, Vienna), p. 2011
14. D. Vo, A. Favalli, B. Grogan, P. Jansson, H. Liljenfeldt, V. Mozin, P. Schwalbach, A. Sjoland, S. Tobin, H. Trellue, S. Vaccaro, Passive gamma analysis of the boiling-water-reactor assemblies. Nucl. Instrum. Methods Phys. Res. Sect. A **830**(1), 325–337 (2016)
15. A. Favalli, D. Vo, B. Grogan, P. Jansson, H. Liljenfeldt, V. Mozin, P. Schwalbach, A. Sjoland, S.J. Tobin, H.R. Trellue, S. Vaccaro, Determining initial enrichment, burnup, and cooling time of pressurized-water-reactor spent fuel assemblies by analyzing passive gamma spectra measured at the Clab interim-fuel storage facility in Sweden. Nucl. Instrum. Methods Phys. Res. Sect. A **820**(1), 102–111 (2016)
16. D.M. Lee, J.R. Phillips, J.K. Halbig, S. Hsue, L.O. Lindquist, E.M. Ortega, J.C. Caine, J. Swansen, K. Kaieda, E. Dermendjiev, New developments in nondestructive measurement and verification of irradiated LWR fuels, in *Advisory Group Meeting on the Nondestructive Analysis of Irradiated Power Reactor Fuel* (International Atomic Energy Agency, Vienna, 1979). https://archive.org/stream/measurementtechn582cana/measurementtechn582cana_djvu.txt
17. J.R. Phillips, K. Barnes, T.R. Bement, Correlation of the cesium-134/cesium-137 ratio to fast reactor burnup. Nucl. Technol. **46**(1), 21–29 (1979)
18. D.K. Min, H.J. Park, K.J. Park, S.G. Ro, H.S. Park, *Determination of Burnup, Cooling Time and Initial Enrichment of PWR Spent Fuel by Use of Gamma-Ray Activity Ratios* (International Atomic Energy Agency, Vienna, 1998)
19. J.R. Phillips, J.K. Halbig, D.M. Lee, S.E. Beach, T.R. Bement, E. Dermendjiev, C.R. Hatcher, K. Kaieda, E.G. Medina, *Application of Nondestructive Gamma-Ray and Neutron Techniques for the Safeguarding of Irradiated Fuel Materials* (Los Alamos Scientific Laboratory, Los Alamos, 1980)
20. PNNL, FY 2009 Progress: Process Monitoring Technology Demonstration at PNNL. Pacific Northwest National Laboratory (2009). DOI: https://doi.org/10.2172/985022
21. R. Montgomery, J.M. Scaglione, B. Bevard, Post-irradiation examinations of high burnup PWR rods, in *WM2018 Conference Proceedings*, 2018
22. A.S. Hoover, C.R. Rudy, S.J. Tobin, W.S. Charlton, A. Stafford, D. Strohmeyer, S. Saavedra, Measurement of plutonium in spent nuclear fuel by self-induced X-ray fluorescence, in *Proceedings of the 50th Annual INMM Meeting*, Tucson, AZ, July 12–16, 2009
23. M. Croce, D. Henzlova, H. Menlove, D. Becker, J. Ullom, Electrochemical safeguards measurement technology development at LANL. J. Nucl. Mater. Manage. **49**(1), 116 (2020)
24. A. Lebrun, M. Merelli, J.-L. Szabo, M. Huver, R. Arlt, J. Arenas-Carrasco, *SMOPY a New NDA Tool for Safeguards of LEU and MOX Spent Fuel* (International Atomic Energy Agency, Vienna, 2003)
25. T. White, M. Mayorov, A. Lebrun, P. Peura, T. Honkamaa, J. Dahlberg, J. Keubler, V. Ivanov, A. Turunen, Application of passive gamma emission tomography (PGET) for the verification of spent nuclear fuel, in *Proceedings of the 59th Annual Meeting of the INMM*, Baltimore, MD, July 22–26, 2018
26. T. White, M. Mayorov, N. Deshmukh, E. Miller, L.E. Smith, J. Dahlberg, T. Honkamaa, SPECT reconstruction and analysis for the inspection of spent nuclear fuel, *Proceedings of the IEEE Nuclear Science Symposium and Medical Imaging Conference* (NSS/MIC), Atlanta, GA, October 21–28, 2017
27. A.C. Kaplan, M. Flaska, A. Enqvist, J.L. Dolan, S.A. Pozzi, EJ-309 pulse shape discrimination performance with a high gamma-ray-to-neutron ratio and low threshold. Nucl. Instrum. Methods Phys. Res. Sect. A **729**(1), 463–468 (2013)
28. S.-T. Hsue, J. Stewart, K. Kaieda, J. Halbig, J. Phillips, D. Lee, C. Hatcher, Passive Neutron Assay of Irradiated Nuclear Fuels. Los Alamos Scientific Laboratory report LA-07645-MS (1979)
29. J.R. Phillips, G.E. Bosler, J.K. Halbig, S.F. Klosterbuer, D. Lee, H.O. Menlove, Neutron Measurement Techniques for the Nondestructive Analysis of Irradiated Fuel Assemblies. Los Alamos National Laboratory report LA-09002-MS (1981)
30. P.M. Rinard, G.E. Bosler, Safeguarding LWR Spent Fuel with the Fork Detector. Los Alamos National Laboratory report LA-11096-MS (1987)
31. S. Vaccaro, I.C. Gauld, J. Hu, P. De Baere, J. Peterson, P. Schwalbach, A. Smejkal, A. Tomanin, A. Sjoland, S. Tobin, D. Wiarda, Advancing the Fork detector for quantitative spent nuclear fuel verification. Nucl. Instrum. Methods Phys. Res. Sect. A **888**(1), 202–217 (2018)
32. ANTECH, B2102 Series Fork Detectors Fact Sheet, https://www.antech-inc.com/product/fork-detector-iaea/. Accessed Aug 2023
33. A. Smejkal, R. Linnebach, A. Angelo, J. Regula, J. Longo, S. Bertl, IRAP: A new system for integrated analysis and visualization of multi-source safeguards data: Challenges and techniques, in *IAEA Symposium on International Safeguards: Building Future Safeguards Capabilities*. Book of Abstracts, 2018
34. H.O. Menlove, D.H. Beddingfield, Passive Neutron Reactivity Measurement Technique. Los Alamos National Laboratory report LA-UR-97-2651 (1997)
35. S.J. Tobin, P. Peura, C. Belanger-Champagne, M. Moring, P. Dendooven, T. Honkamaa, Measuring spent fuel assembly multiplication in borated water with a passive neutron albedo reactivity instrument. Nucl. Instrum. Methods Phys. Res. Sect. A **897**, 32–37 (2018)
36. A.C. Trahan, G.E. McMath, P.M. Mendoza, H.R. Trellue, U. Backstrom, A. Sjoland, Preliminary results from the spent nuclear fuel assembly field trials with the differential die-away self-interrogation instrument, in *Proceedings of the 59th Annual Meeting of the INMM*, Baltimore, MD, July 22–26, 2018
37. A.V. Goodsell, V. Henzl, M.T. Swinhoe, C. Rael, D. Desimone, W.S. Charlton, Comparison of fresh fuel experimental measurements to MCNPX results using the differential die-away instrument for nuclear safeguards applications. ESARDA Bull. **53**, 2–12 (2015)
38. G.E. McMath, M.T. Swinhoe, A.C. Trahan, H.R. Trellue, A. Tumulak, Fresh fuel measurements utilizing active interrogation differential die-away (DDA) instrument, in *Proceedings of the 58th Annual Meeting of the INMM*, Indian Wells, CA, July 16–20, 2017

39. G.E. McMath, V. Henzl, P.M. Mendoza, C.D. Rael, M.A. Root, C.J. Thomson, A.C. Trahan, H.R. Trellue, A. Sjoland, U. Backstrom, Initial results of nondestructive assay of commercial spent nuclear fuel with differential die-away (DDA) instrument, in *Proceedings of the 61st Annual Meeting of the INMM*, International Virtual Experience, July 12–16, 2020
40. C. Thompson, G. McMath, U. Backstrom, W.S. Charlton, V. Henzl, P. Mendoza, C. Rael, M. Root, A. Sjoland, A. Trahan, H.R. Trellue, Improved evaluation of safeguards parameters from spent fuel measurements with the Differential Die-Away (DDA) instrument. Nucl. Instrum. Methods Phys. Res. Sect. A **1029**, 166462 (2022)
41. D. Henzlova, H.O. Menlove, C.D. Rael, H.R. Trellue, S.J. Tobin, S.H. Park, J.-M. Oh, S.-K. Lee, S.-K. Ahn, I.-C. Kwon, H.-D. Kim, Californium interrogation prompt neutron (CIPN) instrument for non-destructive assay of spent nuclear fuel—design concept and experimental demonstration. Nucl. Instrum. Methods Phys. Res. Sect. A **806**, 43–54 (2016)
42. D. Henzlova, H.O. Menlove, H.R. Trellue, C.D. Rael, R.A. Weldon, T. Burr, J.S. Hendricks, J. Hu, S.J. Tobin, Field Trial of Californium Interrogation Prompt Neutron (CIPN) Technique Developed for the Next Generation Safeguards Initiative Spent Fuel Research Effort. Los Alamos National Laboratory report LA-UR-14-27932 (2014)
43. A.C. Trahan, G.E. McMath, P.M. Mendoza, H.R. Trellue, U. Backstrom, L. Poder Balkestahl, S. Grape, V. Henzl, D. Leyba, M. Root, A. Sjoland, Results of the Swedish spent fuel measurement field trials with the differential die-away self-interrogation instrument. Nucl. Instrum. Methods Phys. Res. Sect. A **955**, 163329 (2020)
44. A.C. Kaplan, V. Henzl, H.O. Menlove, M.T. Swinhoe, A.P. Belian, M. Flaska, S.A. Pozzi, Determination of spent nuclear fuel assembly multiplication with the differential die-away self-interrogation instrument. Nucl. Instrum. Methods Phys. Res. Sect. A **757**(1), 20–27 (2014)
45. R.A. Weldon, M.L. Fensin, H.R. Trellue, Total neutron emission generation and characterization for a next generation safeguards initiative spent fuel library. Prog. Nucl. Energy **80**, 45–73 (2015)
46. H.O. Menlove, M.T. Swinhoe, J.B. Marlow, C.D. Rael, Advanced Experimental Fuel Counter (AEFC) Calibration and Operation Manual. Los Alamos National Laboratory report LA-14359-M (2008)
47. I. Levi, A.C. Trahan, K. Ben-Meir, O. Ozeri, A. Krakovich, A. Pesach, O. Rivin, C.D. Rael, M.T. Swinhoe, H.O. Menlove, N. Hazenshprung, J.B. Marlow, I. Neder, Nondestructive measurements of residual ^{235}U mass of Israeli research reactor-1 fuel using the advanced experimental fuel counter. Nucl. Instrum. Methods Phys. Res. Sect. A **964**, 163797 (2020)
48. RL-VI-IRD30—Solid State Gamma Detector. https://www.bot.engineering/product-category/18
49. A. Alessandrello, F. Mingrone, S. Bertl, Core discharge monitor machine learning algorithm for CANDUs with fuel separator method, in *Proceedings of the INMM & ESARDA Joint Virtual Annual Meeting*, August 23–26 & August 30–September 1, 2021. https://resources.inmm.org/annual-meeting-proceedings/core-discharge-monitor-machine-learning-algorithm-candus-fuel-separator
50. J. Budzinski, H. Boeck, Report on the analysis of CDM data. Atomic Institute of the Austrian Universities report number AIAU--23302 (April 2003)
51. J. Budzinski, Machine learning techniques for the verification of refueling activities in CANDU-type nuclear power plants (NPPs) with direct applications in nuclear safeguards, Bibliographic information available from INIS: http://inis.iaea.org/search/search.aspx?orig_q=RN:38037550; available from Univ. Wien Bibliothek, Dr. Karl Lueger-Ring 1, 1010 Wien (AT) (2006)
52. RL-VI-IRD30AP—Tamper Indication and Authentication AddßOn for IRD30 Detectors. https://www.bot.engineering/product-category/18
53. RL-IRD30HS High Sensitivity Dose Tolerant Gamma Detector. https://www.bot.engineering/product-category/18
54. IAEA, Safeguards Techniques and Equipment: 2011 Edition. International Nuclear Verification Series No. 1 (Rev 2), International Atomic Energy Agency (2011)
55. C. Orton, D. Parise, S. Lackner, S. Jung, D. Belemsaga, L.T. Carbonell, F. Muelhauser, G. Bernasconi, A. Lebrun, The international atomic energy agency's experience verifying spent fuel using the digital Cerenkov viewing device, in *Proceedings of the 58th Annual Meeting of the INMM*, Indian Wells, CA, July 16–20, 2017
56. "DCVD" Channel Systems. www.channelsystems.ca/products/digital-cerenkov-viewing-device-dcvd/dcvd. Accessed Mar 2023
57. E. Branger, *Enhancing the Performance of the Digital Cerenkov Viewing Device*. Ph.D. thesis, Uppsala University, Sweden (2018)
58. J. Chen, A.F. Gerwing, P.D. Lewis, M. Larsson, K. Jansson, B. Lindberg, E. Sundkvist, U. Meijer, M. Thorsell, M. Ohlsson, *Long-Cooled Spent Fuel Verification Using a Digital Cherenkov Viewing Device*. IAEA-SM-367/14/07 (International Atomic Energy Agency Symposium on International Safeguards, Vienna, 2001)
59. "Development and Implementation Support Programme for Nuclear Verification 2022–2023," International Atomic Energy Agency Safeguards Report STR-400 (January 2022)
60. P.M. Rinard, H.O. Menlove, Applications of Curium Measurements for Safeguarding at Reprocessing Plants. Los Alamos National Laboratory report LA-UR-95-2788 (1995)
61. F. Sturek, L. Agrenius, *Measurements of decay heat in spent nuclear fuel at the Swedish interim storage facility, Clab* (Svensk Karnbranslehantering AB, Stockholm, 2006)
62. P. Jansson, M. Bengtsson, U. Bäckström, K. Svensson, M. Lycksell, A. Sjöland, Data from calorimetric decay heat measurements of five used PWR 17 × 17 nuclear fuel assemblies. Data Brief **28**, 104917 (2020)

Open Access This chapter is licensed under the terms of the Creative Commons Attribution 4.0 International License (http://creativecommons.org/licenses/by/4.0/), which permits use, sharing, adaptation, distribution and reproduction in any medium or format, as long as you give appropriate credit to the original author(s) and the source, provide a link to the Creative Commons license and indicate if changes were made.

The images or other third party material in this chapter are included in the chapter's Creative Commons license, unless indicated otherwise in a credit line to the material. If material is not included in the chapter's Creative Commons license and your intended use is not permitted by statutory regulation or exceeds the permitted use, you will need to obtain permission directly from the copyright holder.

22 Perimeter Radiation Monitors

M. G. Paff and J. Toevs

22.1 Introduction

Radiation portal monitors (RPMs) have uses in a wide variety of application spaces. Certain facilities, such as scrap metal processing plants, use RPMs to screen all incoming feed material to protect their facility, workers, and product from scrap metal contaminated with radiation sources that can include orphan industrial or medical radiation sources. In the nuclear security arena, RPMs are frequently used to screen people, vehicles, and cargo containers as they enter or exit a country via a land border crossing, an international airport, or a seaport. At nuclear facilities, perimeter radiation monitors are located at the periphery of nuclear-material and radioactive-contamination control areas to detect accidental or covert removal of radioactive materials. Depending on the application and the design basis threat (DBT), different types and configurations of perimeter monitors—such as contamination monitors, diversion monitors, or material out of regulatory control (MORC) monitors—may be used. Contamination monitors detect contamination on the surface of a person or an object where the radiation comes from an extended area viewed without intervening absorbers. Diversion monitors must be able to detect small, possibly shielded quantities of nuclear material, e.g., various forms of uranium and plutonium. In this case, the small source size and the possible presence of absorbers reduce the radiation intensity. In other applications, any MORC, such as orphan industrial sources and contaminated scrap metal, may be of interest. This chapter discusses various applications of perimeter monitors and gives primary emphasis to RPMs used to screen for MORC.

Contamination is of concern in at least two applications: contaminated goods that cross international borders (e.g., goods contaminated due to nuclear power accidents) and contamination of people or equipment that exits a nuclear facility (i.e., concern regarding spreading contamination beyond site boundaries). The need to detect contamination predated security concerns. When the need to detect nuclear and radiological material arose, handheld contamination monitors were available; however, security personnel had to interpret an analog meter display to use this type of instrument, and their attention was distracted from the security search. Automatic portal monitors [1] and handheld monitors [2] were developed to eliminate the distraction. Modern monitors provide visual and/or audible alarm signals that allow the operator to devote full attention to the security search. For fixed monitors, these alarms can be at the lane and/or at a central alarm station depending on local concept of operations and security protocols. At nuclear facilities, the responsibility of employers to furnish top-grade, contamination-monitoring equipment to employees has fostered development of automatic, high-sensitivity contamination monitors [3]. These, as well as modern MORC monitors, are designed for high sensitivity, dependability, and easy maintenance.

Diversion monitors meet Department of Energy [4] and Nuclear Regulatory Commission [5] requirements to search each person, package, or vehicle that leaves a nuclear material access area. Contamination monitors meet radiation safety standards for monitoring persons who are leaving a radioactive contamination area. RPMs for countering smuggling of nuclear and radiological materials are evaluated against a variety of programmatic and international standards, including the ANSI N42

Los Alamos National Laboratory strongly supports academic freedom and a researcher's right to publish; as an institution, however, the Laboratory does not endorse the viewpoint of a publication or guarantee its technical correctness.

M. G. Paff (✉) · J. Toevs
Los Alamos National Laboratory, Los Alamos, NM, USA
e-mail: marc.paff@culmen.com; toevs@lanl.gov

Fig. 22.1 RPMs with large plastic scintillators and ^{3}He proportional counters used to monitor international passengers who arrive at an airport. (Photo courtesy of National Nuclear Security Administration NA-213 Nuclear Smuggling Detection and Deterrence)

Fig. 22.2 RPM with large plastic scintillators and ^{3}He proportional counters used to monitor vehicles entering a country. (Photo courtesy of National Nuclear Security Administration NA-213 Nuclear Smuggling Detection and Deterrence)

series [6]. In all cases, visual or manual searches may be ineffective, but radiation monitors sense radiation emitted by the materials and can conduct unobtrusive, sensitive, and efficient searches. The monitors provide timely notice of contamination, diversion, or smuggling before the controlled material can leave an access area or cross an administrative line.

Examples of MORC monitors are the portals (shown in Fig. 22.1, with their detectors positioned beside passing pedestrians and, in Fig. 22.2, for screening vehicles at an international border crossing) and the handheld survey meter (shown in Fig. 22.3 being manually screened over a vehicle). The versatile handheld monitors have many applications—even to contamination monitoring—but their effectiveness depends on the operator's thoroughness in scanning. In contrast, portal monitors are fully automatic and reduce operator error.

Thousands of RPMs are operated by customs and/or border protection agencies at any place where goods or people cross international borders, such as vehicle/pedestrian border crossings, container seaports, ferry terminals, and international airports. Portal monitors can be designed for specific conveyances such as people, trucks, trains, or suitcases, and their settings and performance can be optimized depending on a nation's or organization's DBT, such as a specific quantity of

Fig. 22.3 Monitoring with handheld instruments is highly effective when the operator is well trained and motivated. (Photo courtesy of National Nuclear Security Administration NA-213 Nuclear Smuggling Detection and Deterrence)

Fig. 22.4 Example of a walk-in contamination monitor. (Photo courtesy of ThermoFisher Scientific)

nuclear material or a gamma fluence from any MORC. Some standards [6–9] set suggested baseline criteria for such systems, but actual requirements and needs will be unique due to differing DBT.

New contamination monitor designs locate detectors inside an enclosure (Fig. 22.4), where an individual being monitored must stand near radiation detectors for an extended period. The longer monitoring period and the proximity of the occupant and the detectors improve detection sensitivity.

22.2 Background Radiation Effects

Radiation monitors are influenced by background radiation and the variation of its intensity with time. The intensity of the background radiation influences the effectiveness of monitoring. Alarm thresholds must be set well above background intensity to avoid alarms from counting statistics (one cause of false alarms). The required threshold setting becomes

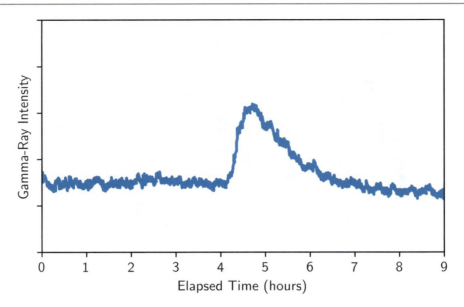

Fig. 22.5 A background intensity record showing a road bed monitor count-rate increase during brief, intense precipitation

proportionately higher as the background intensity increases, causing the monitoring sensitivity to decrease. In modern RPMs, these thresholds are continuously updated based on measured background rates between occupancies. In an occupied monitor, rapid variations in background intensity—which can be caused by natural background radiation processes, movement of radioactive materials, or radiation-producing machinery (such as non-intrusive inspection equipment at airports and border crossings)—may be mistaken for nuclear material signals and cause another type of false alarm. An example of a natural background radiation process that leads to rapid intensity variation is the decay of ^{226}Ra in soil. Its gaseous daughter, ^{222}Rn, can escape the soil to decay in the atmosphere. These daughter products, which are themselves radioactive, may attach to dust particles that form condensation points for raindrops. When the raindrops fall to the ground, they temporarily increase background intensity (Fig. 22.5). An artificial background radiation process with rapid intensity variation occurs because RPMs are often collocated with non-intrusive inspection equipment, such as X-ray scanners for hand luggage at airports and larger X-ray scanners at ports of entry for detecting taxable or illegal non-declared goods (such as tobacco products and narcotics). If not properly accounted for in portal monitor design and site layout, radiation from X-ray scanners may trigger false alarms in RPMs when occupied.

22.3 Characteristics of Diversion and Contamination Signals

22.3.1 Radiation Sources

As described in Chap. 2 and Chap. 13, nuclear materials can be detected by their spontaneous radiation emissions. These radiation emissions—alpha, beta, gamma ray, X-ray, and neutron—each have a different ability to penetrate materials. Alpha radiation is not very penetrating and is easily stopped by several centimeters of air. Except when contamination detectors almost touch the emitter, alpha radiation contributes no signal to a radiation monitor. More penetrating forms of radiation that easily pass through air can be detected at a distance; however, the shielding provided by detector cabinets and nuclear-material packaging may exclude all but gamma-ray and neutron signals. One important aspect separates the signals available to contamination and MORC monitors: contamination usually lies on a surface where its radiation is readily detected, whereas MORC monitors must sense penetrating radiation from material that is, in some cases, intentionally shielded. Therefore, contamination monitors often detect many forms of radiation, but MORC monitors detect primarily gamma rays and neutrons. The nuclear-material diversion, smuggling, and RPMs discussed in the remainder of this chapter mainly detect gamma rays, but some have neutron sensitivity depending on the application and portal monitor model. Similarly, the detectable radiation signal will depend strongly on specific site characteristics, such as the conveyance (pedestrian, luggage, vehicle, and cargo container), the average conveyance speed, the RPM pillar spacing, and local radiation background conditions.

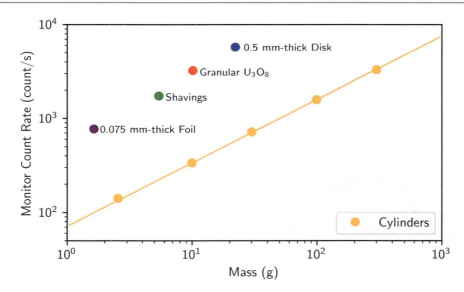

Fig. 22.6 The form of HEU influences the self-absorption of gamma rays. The emitted intensity varies with the surface area; therefore, the intensity varies as the 2/3 power of the mass

Internal absorption of source radiation also could significantly alter detection signals. For example, nuclear materials shield their own gamma radiation; the extent of self-absorption depends on the physical form of the material. Figure 22.6 illustrates self-absorption in different shapes and sizes of highly enriched uranium (HEU). Thin uranium materials—such as powders and foils—emit most of their radiation, whereas more compact shapes—such as spheres and cylinders—absorb most of it. The cylinders in Fig. 22.6 emit in proportion to their surface area, which increases as the two-thirds power of the mass, giving rise to the straight line in the plot.

22.3.2 Time-Varying Signals

Diversion or contamination signals are usually present in a monitor for only a short time interval. Unless the occupant is stationary, signals from nuclear material will vary during the monitoring period as the occupant moves toward and away from the detectors. Figure 22.7 illustrates the net signal in a vehicle monitor as a cargo container carrying various nuclear and radiological materials passes through. The time integral of the variable signal is ~60% of that for a stationary occupant. Good monitor design ensures that the monitoring period matches the intense part of the signal as closely as possible. Techniques for obtaining this optimum situation are discussed in Sect. 22.4.

Another important feature of the radiation signature is whether it approximates more a point-like or a distributed source (Fig. 22.7). Many modern RPMs provide a temporal count profile as the conveyance passes through the RPM. A source of concern—such as nuclear material or an orphan radiological source—is more likely to be point-like and produce a narrower count spike in the temporal profile. Naturally radioactive bulk commodities, such as a truckload of fertilizer, generally would exhibit a much broader elevated count profile as the conveyance passes through the portal monitor. However, exceptions exist, such as large shipments of contaminated scrap metal, which can produce a profile similar to a large shipment of naturally radioactive bulk cargo. Similarly, a patient after a recent nuclear medicine procedure might produce a narrower count profile yet not constitute a security threat. However, depending on the speed of transit and the sensitivity of the detectors, the statistical variation between counting increments could be so large as to make distinguishing between bulk and point sources impossible.

A complementary effect that diminishes signals in a monitor is the reduction in background intensity caused by an occupant. Ambient background radiation from the monitor's surroundings can be partly absorbed by the person or vehicle that occupies the monitor. The reduction in intensity may be negligible for pedestrians but is much greater for motor vehicles and even more so for fully loaded intermodal shipping containers. Figure 22.8 illustrates the reduction caused by the presence of a truck in a vehicle monitor. The reduction ranges from 10% to 45% for different-sized vehicles and loadings, i.e., small vehicles to trucks pulling 40 ft. intermodal shipping containers loaded to their maximum carrying capacity. Because the

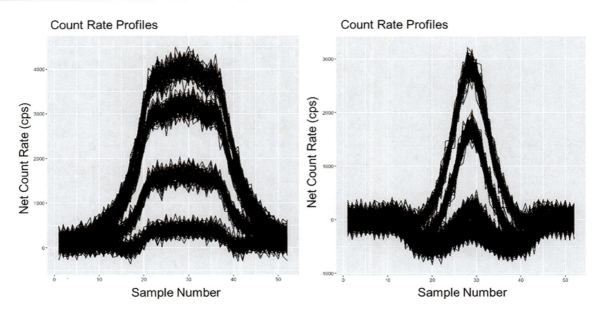

Fig. 22.7 Signal intensity profile results from a vehicle RPM that measures distributed MORC in scrap metal (left) and a point-like source (right) in a 40 ft. cargo container passing through an RPM

Fig. 22.8 Five gamma response profiles of an HEU source traversing a vehicle RPM in a 40 ft. cargo container. Background is suppressed as the truck approaches the RPM, but gamma count rate spikes when HEU is directly between the two detection pillars

monitor's gross counting alarm threshold is typically based on background measured before the beginning of an occupancy, a much larger signal could be required to alarm an occupied monitor than one that is unoccupied. Background suppression effects are generally considered when evaluating the minimum detectable quantity of a threat material for a specific lane based on expected conveyance type. Therefore, for optimized sensitivity, conveyance-specific lanes—such as dedicated lanes for pedestrians, trucks, and personal vehicles at land border crossings—are preferable.

22.4 Signal Analysis

22.4.1 Detecting Radiation Signals

Many RPMs rely on large inexpensive gamma detectors, such as polyvinyltoluene, and employ gross-counting alarm algorithms. Such detectors allow for high efficiency but provide only low resolution, thus complicating any attempts to identify the specific radionuclide that is causing an elevated gamma radiation reading. Spectroscopic portals that use higher resolution detector materials also exist but are less widely adopted mainly due to their higher cost and higher technical complexity. Gross-counting radiation monitors use signal analysis to decide whether a measurement indicates a background signal alone or a background signal plus an additional radiation signal. Unfortunately, statistical variations in background and monitoring measurements preclude a simple comparison. Although the expected background may be determined from a long, precise measurement, each monitoring measurement is necessarily short and imprecise. If background measurements have an expected count rate (B), individual measurements will range many standard deviations higher and lower than B, so 1 standard deviation in this instance is the square root of the count B. Comparisons of monitoring measurements must allow statistical variations of several times the square root of B to exclude false alarms. Each monitoring measurement is usually compared with an alarm threshold equivalent to that given by Eq. 22.1. An alarm is sounded when the measurement equals or exceeds the alarm threshold M,

$$M = B + N\sqrt{B}, \tag{22.1}$$

where N = alarm increment (number of standard deviations). Alarms are real when they result from real signals and are false alarms when they result from statistical variation or background changes.

Such methodology can be applied to portal monitors in a variety of ways. Many RPM designs consist of two pillars, each with multiple radiation detectors. Alarm decisions could be made for the entire summed output of all gamma detectors or for each individual pillar or each individual detector. Trade-offs exist between the different options, such as increased positional sensitivity versus increased false alarm rate probability. For example, when an RPM pillar is located too close to an X-ray-producing machine without proper mitigation strategies, algorithm adjustments, such as using only the total portal summed algorithm and not also the pillar vertically summed algorithm, may be sufficient to alleviate X-ray interference alarm issues.

Nuisance or innocent alarms are another type of benign alarm caused by real radiation, but the source of the radiation is not one of concern to the operator. For example, someone entering or exiting a nuclear facility might set off a diversion monitor alarm due to significant gamma emissions stemming from a recent nuclear medicine procedure. Or an RPM at a border crossing might be triggered by a cargo container with a sufficient loading of naturally occurring radioactive material (NORM), such as various ceramics, construction materials, and fertilizers, as opposed to the smuggled nuclear material or radioactive MORC that is of concern to the portal monitor operators.

For neutron alarm decision making, a sequential probability ratio test is recommended because Gaussian statistics cannot correctly be applied to the much lower neutron background and source count rates relative to the much higher common gamma background count rates.

22.4.2 Digital Signal Analysis

Alarm decisions can be made by digital circuits and microprocessors. For example, Eq. 22.1 can be implemented by comparing the result of a 0.4 s monitoring measurement contained in a digital register with a stored alarm threshold derived from an earlier background measurement. The stored alarm threshold might have been derived from an earlier 20 s background measurement divided by 50 to obtain B, plus an added multiple of the square root of B. In this case, the comparison is a numerical one, and no calibration is required. This single-interval method does have a shortcoming: it is not continuous, so the measurement interval might not match the most intense part of the signal (Fig. 22.7); however, digital logic methods are easily changed to overcome such shortcomings. The improvements described in the following sections include the moving-average method, the stepwise method, and the sequential hypothesis test.

A digital method that performs well in free-passage monitors uses a moving average of monitoring measurements. Short measurement periods (for example, 0.2 s) are used, and the counts from four or more measurement periods are summed and compared with the alarm threshold. After the first group of four or more periods, each new measurement is added, and the oldest measurement is subtracted from the sum. Every new sum is compared with the alarm threshold; measurements then continue unless an alarm occurs or the monitor is no longer occupied. Monitoring is continuous, and many decisions are made; therefore, the alarm threshold must be higher than for the single-interval test (described in the preceding paragraph) to achieve the same statistical-alarm probability. The moving-average method obtains greater sensitivity because it measures the most intense part of the signal.

22.4.3 Long-Term Monitoring (Facility Safeguards)

Long-term monitoring is a technique that achieves high sensitivity through repeated measurements applied in conjunction with but independent of other standard techniques [10]. The method can detect repeated instances of contamination or diversion of nuclear material in quantities too small to detect in normal monitoring. One application of the method sums the net monitoring results for pedestrians who enter an area and compares this sum with the sum for pedestrians who leave the area. Any difference between the two could signify contamination or diversion of nuclear material.

The long-term monitoring method calculates the net signal during occupancy by subtracting from each measurement an average background determined before and after the measurement. Although individual measurements are imprecise, the average net signal for hundreds of passages is quite precise. In fact, this method provides the most precise measurement of the average background radiation attenuation by monitor occupants.

In addition to being able to average monitoring results for large populations, the method can require identification of each occupant so that data for each individual can be recorded. Then analysis of long-term averages of the incoming and outgoing measurements for an individual can identify cases of repeated contamination or diversion of small quantities of nuclear material. For cases where each outgoing passage involves contamination or diversion and each incoming passage does not, the long-term monitoring method is 10 times more sensitive than other methods.

22.5 Radiation Detectors

Perimeter monitors use a different type of radiation detector depending on whether they are designed to detect contamination or diverted nuclear material. Gas proportional counters are most appropriate for detecting the radiation from contamination, and scintillators are most appropriate for detecting the penetrating radiation from diverted material. The general properties of gas proportional counters and inorganic scintillators are discussed in Chaps. 4 and 15; organic scintillators, which are widely applied to perimeter diversion monitoring, are discussed in this section, along with gas-flow proportional counters for perimeter contamination monitoring. These inexpensive, large-area detectors are well adapted to the requirements of perimeter monitoring.

When discussing perimeter radiation monitors, it is important to distinguish between primary and secondary screening instrumentation. In primary screening, RPMs typically use large-volume detectors in a gross-counting mode discussed previously to detect anomalous elevations in radiation relative to local natural background; however, some primary screening RPMs will use other detector materials, such as NaI(Tl) or high-purity germanium (HPGe), to add spectroscopic capabilities in primary screening, i.e., the ability not only to detect conveyances emitting elevated levels of radiation but also to identify the source of radiation. In secondary screening, primary alarms are further investigated and adjudicated using spectroscopic handheld radiation detectors with sufficiently high energy resolution and sufficiently reliable radionuclide identification algorithms to distinguish nonthreatening sources of radiation (medical isotopes, NORM) from materials (such as MORC) that need to be detained for further analysis by the responsible reach-back or "expert support" organization.

22.5.1 Plastic Scintillators

Plastic scintillation detectors are solid organic scintillators that contain fluorescent compounds dissolved in a solidified polymer solute [11]. These materials have low density and low atomic number, so they lack strong photoelectric absorption. They detect gamma rays by detecting Compton recoil electrons, and they detect neutrons by detecting recoil protons. These detectors do not display full-energy peaks; they display a continuous spectrum from the Compton edge down to zero energy. Although organic scintillators are poor energy spectrometers and have low intrinsic detection efficiency, they make excellent large-area, low-cost radiation counters. Their low cost results from the use of inexpensive materials and simple packaging; NaI

(Tl) crystals, on the other hand, are expensive to grow and to protect from moisture and thermal shock. Nevertheless, plastic scintillators do not inherently have to be used in only a gross counting mode. Many vendors apply a variety of techniques to plastic scintillators to provide information beyond only gross counting. Examples include the use of energy-windowing techniques to perform classification analysis, such as distinguishing the higher-energy gamma-ray emissions of NORM from lower gamma-energy-emitting materials more likely to be threat or industrial sources. Going beyond classification techniques, various reconstruction and Bayesian techniques also exist to provide some level of radionuclide identification capability in plastic and other organic scintillator materials. Throughput, statistics, and the DBT will inform what types of technologies are acceptable at various locations.

The large size of plastic scintillators gives them good total detection efficiency even though their intrinsic efficiency is low. Total efficiency is the product of a detector's intrinsic efficiency and the fraction of emitted photons that strike the detector. The latter factor depends on the size of the detector. The large detector size also provides more uniform monitoring than would an array of small detectors.

Plastic scintillation detectors do have some shortcomings. They produce only ~10% as much light as NaI(Tl) detectors, and their large size makes uniform light collection difficult. Uniform light collection is important to minimize the spread in pulse heights that result from detection of radiation in different parts of a detector. Reference [12] describes methods for obtaining total internal reflection of scintillation light and for making a large detector's response homogeneous. Fogging and degradation have also been observed in plastic scintillators that are exposed to many years of extreme temperature oscillations in harsh outdoor environments. Moisture from humidity can get trapped in the plastic scintillator, causing fogging, and therefore efficiency loss over time. Encapsulation techniques for plastic scintillators show promise in minimizing these degradation effects.

Low photomultiplier noise is important in organic scintillators because the Compton pulse-height spectrum extends down to zero pulse height. Noise sets a practical limit to the pulse amplitude that can be detected; this bias level limits the intrinsic detection efficiency. The bias level influences detection efficiency over a broad range of incident gamma-ray energy, as illustrated in Fig. 22.9. Bialkali photocathodes can operate near the 0.045 V bias level at room temperature. For portal monitors that use a single energy window, such as for nuclear material detection, low photomultiplier noise is of less concern.

22.5.2 Gas-Flow Detectors (Facility Safeguards)

An inexpensive form of gas proportional counter is the very large area gas-flow proportional counter. Very thin detector windows (100 µg/cm^2) transmit the low-energy radiation emitted by surface contamination into the detector interior, which is a thin, large-surface-area cavity. An argon-methane mixture slowly flows through the cavity and then is burned or recirculated

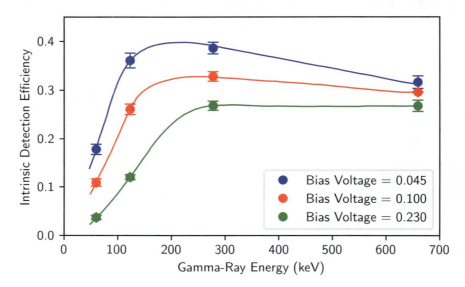

Fig. 22.9 Low-bias voltage is essential to good intrinsic detection efficiency in organic scintillators. The data illustrated here were taken with a detector that yielded 2 V pulse height for incident 662 keV gamma rays

with a small quantity of new gas. Argon is the counting gas, and methane lowers the operating voltage and quenches discharges between counter electrodes. Discharges caused by contaminants in the counting gas or by secondary emission from metallic counter parts cause electronic noise. The flat-slab geometry has a non-uniform electric field and gain, so the instrument serves as a counter rather than an energy spectrometer. Although the very large gas-flow proportional counter is a noisy detector, its good low-energy response and low cost make it attractive for contamination monitoring where measurements can be repeated freely without significant penalty.

22.6 Perimeter Monitor Components

The perimeter radiation monitor previously shown in Fig. 22.1 monitors pedestrians, and those shown in Fig. 22.10 monitor motor vehicles. Each monitor has similar components (Fig. 22.11). The detectors sense radiation and transmit information to the monitor's control unit, which provides power, signal conditioning, and signal analysis. The control unit usually has an occupancy sensor to determine when to measure background and indicator lamps and sounders to announce alarms.

Fig. 22.10 RPMs at the Port of Algeciras, Spain. (Courtesy of the Spanish Agencia Estatal de Administración Tributaria)

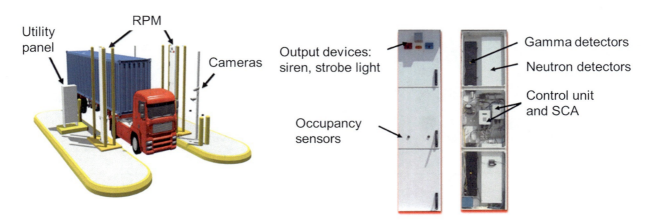

Fig. 22.11 Basic components of a perimeter radiation monitor. The monitor must detect radiation, sense the presence of an occupant, make decisions, and announce the result. (Photo courtesy of National Nuclear Security Administration NA-213 Nuclear Smuggling Detection and Deterrence)

22.6.1 Components and Their Functions

The components and functions of a radiation monitor are described as follows:

- *Detectors* sense radiation from a particular region of space, usually the region between two or more detectors.
- *Signal conditioning electronics* transform the detected radiation charge pulses into voltage pulses that can be transmitted to another device for analysis.
- *Single-channel analyzers* (SCAs) or *multichannel analyzers* (MCAs) select the pulses in a desired energy region. The output is a standard logic pulse. For spectroscopic portal monitors, an MCA is required to create the energy spectrum that is fed into the radionuclide identification algorithm.
- *Control units* count the SCA logic pulses and use the result to derive alarm levels or monitoring measurements. They also test background measurements against high- and low-background thresholds to detect malfunction, display each new background result, and compare monitoring measurements with the alarm threshold (Sect. 22.4). Control units use the occupancy sensor to determine when to measure background and when to monitor, and they assist with monitor calibration.
- *Occupancy sensors* sense the presence of a person or vehicle and, if important, the direction of travel.
- *Output devices* communicate monitoring results by visual signals (flashing lights) and audible signals (chirps).
- *Speed sensors* capture the transit speed of the conveyance, which is valuable when analyzing an occupancy that triggered an RPM alarm.
- *Cameras* and video systems are integrated with many RPMs at border crossings and seaports to collect records of license plates or other distinguishing features for any occupancy.
- *Central alarm stations* are located at sites that have many RPMs, such as an airport or a seaport. All portal monitors are linked via a central alarm station, where operators can monitor all portal monitors across a large site and track and adjudicate alarms and faults as they arise.
- *Power supplies* convert line power to the direct-current voltages needed to operate the detectors and electronics.

Some of these devices and their functions are discussed in more detail in the following sections.

22.6.2 Signal Electronics

Noise is present in any detection system, and some of it can be eliminated by combining two voltage-level discriminators to form an SCA. An SCA acceptance window that is limited to a particular band of energies can optimize the performance of a radiation monitor. For example, because the intense part of the gamma-ray spectrum of HEU lies in a narrow energy region, an acceptance window limited to that gamma-ray-energy region gives the best detection sensitivity for uranium, even when poor spectrometers (such as organic scintillators) are used. A well-chosen upper-level discriminator will also significantly reduce the signal contribution from higher-energy gamma-ray emissions of various common NORM sources, as well as those from natural background, which improves signal-to-noise ratio for the portal.

Table 22.1 gives an example of how uranium detection sensitivity varies with the size of the acceptance window. The figure of merit S divided by the square root of B relates the net signal S in a particular window to the standard deviation of the background in the same window. The greater the figure of merit, the easier it is to detect a uranium source; and the lower is the minimum detectable quantity. For the values shown in Table 22.1, source detection was improved by ~50% with an optimum SCA window. For single energy window systems intended for both HEU and plutonium detection, a compromise must be made in choosing the energy window to find an optimum window to accommodate detection of both of these nuclear materials.

Table 22.1 Figures of merit and detected mass for three SCA windows[a]

SCA window (V)	Energy window (keV)	$S/\sqrt{B}$	^{235}U mass detected (g)
0.3–0.85	70–215	7.87	10
0.21–1.5	46–385	6.93	12.2
0.3–7.0	70–1735	6.0	15.2

[a]Source, spherical masses; background, 21 μR/h

Scalers count the SCA logic pulses during a measurement period. At the end of each counting interval, the scaler transfers its sum to the decision logic. When the monitor is unoccupied, many such sums are averaged to obtain a precise background value. During monitoring, each sum is compared with the alarm threshold.

Some modern RPM designs include MCA logic to perform spectroscopic or energy windowing analysis on occupancies in addition to the gross counting analysis that an SCA provides.

22.6.3 Power Supplies

High voltage for detectors is provided by a regulated electronic circuit that maintains an essentially constant output voltage. To use a single power supply for multiple scintillation detectors, each photomultiplier voltage-divider circuit is provided with a series potentiometer to adjust its gain.

Monitors that use nuclear instrument module (NIM) electronic modules for amplifier, SCA, and high-voltage power supplies use low-voltage power from the NIM bin. Where microprocessor electronics are used, low-voltage power supplies can operate from trickle-charged batteries. This feature makes the monitor's controller insensitive to short-term power failure. Without backup power, a monitor must restart after each power loss—with some operating delay.

In case of long-term power failure, some type of backup line power should be provided for the entire monitor. This requirement is often met by facility backup power; if not, it can be provided temporarily by commercial power units. A 12 V battery in the RPM is also often used for short-term backup power. In other cases, hand monitoring suffices as a backup during a power outage.

22.6.4 Diagnostic Tests

Simple diagnostic tests can identify faults in radiation monitors as soon as they occur. The tests may be performed by separate modules or incorporated in the program of a microprocessor control unit. Background tests simply compare the measured background with high and low thresholds. A malfunctioning monitor may have a high or low background because of an inoperative or noisy detector. Inadvertent shielding of the detector or storage of radioactive material near the monitor will also be detected by a background test. To detect such anomalies as they occur, each new background value is usually checked and, if necessary, flagged by an audible or visual alarm.

More complex diagnostic techniques examine the monitor's counting statistics to determine if the counts originate from radiation detection or noise. Reference [13] describes a long-term analysis method that can diagnose noise even in the presence of sources or varying background intensity.

Variance analysis is suitable for short- or long-term analysis and is also used for detector calibration. This technique calculates the mean and variance of a group of counts. If these quantities are nearly identical, the variance analysis test quickly establishes that the detectors are operating properly. Noise can be detected in a single measurement set, and minor noise problems that may influence nuisance-alarm frequency can be detected by averaging the results for many sets.

Much information on proper functioning of RPM equipment can also be gleaned by analyzing large batches of historical data for a portal, e.g., daily files that contain constantly updated count rates for each gamma and neutron detector. Looking at data over a long period of time can pinpoint potential maintenance issues with individual detectors, such as improper alignment of a gamma detector, which results in strong gamma background oscillations with daily temperature cycles.

22.7 Monitor Calibration

Improper calibration is a common cause of problems such as frequent nuisance alarms and lack of sensitivity. Calibration involves adjusting the detector gains so that they all provide the same response to a calibration source and then adjusting the SCA to respond to radiation in the desired energy region. Gas counters require little calibration, but scintillation detectors must be calibrated periodically.

Calibrating a scintillation detector begins by setting the photomultiplier high voltage to a chosen value, typically 1000 V, and continues by adjusting each detector's gain potentiometer to obtain the same pulse-height response for a test source (for example, a 5 µCi ^{137}Cs source due to its single energy emission and easily recognizable Compton edge). The source is placed in the same relative position next to each detector, and the pulse-height distribution is observed at the amplifier output using a pocket MCA. Next, the amplifier gain is adjusted to give the desired Compton edge location for the detector response to ^{137}Cs. The desired Compton edge location will have been determined through optimization studies for the response to materials of interest relative to a fixed and optimized energy window used for the gross counting algorithm.

22.7.1 Single-Channel Analyzer Calibration

Both upper- and lower-level discriminators must be adjusted to form the SCA window. The upper-level discriminator can first be set to a desired value by using an oscilloscope or an MCA. For monitoring different materials, material-specific energy window upper levels should be established.

The lower-level discriminator can be set in the same fashion to 60 keV; however, lower settings that are still above the noise could improve performance. One way to set lower values is to adjust the discriminator while making source-in and source-out intensity measurements until a maximum value of the figure of merit $S/\sqrt{B}$ is achieved (Sect. 22.6.2). This slow procedure can be replaced by a variance analysis technique for much quicker results. The discriminator is decreased to the point where the variance analyzer indicates only noise, and then it is raised slightly to the point where noise is no longer indicated.

Alternatively, the SCA window defined by the lower- and upper-level discriminators may have fixed values, and the detector gains therefore must be periodically adjusted to maximize sensitivity to a desired source in the fixed energy window.

22.7.2 Periodic Calibration Checks

For facility monitoring, a daily test is important to determine whether the monitor is functioning properly. If a low-intensity source (1 μCi of ^{133}Ba as a weapons-grade Pu surrogate, for example) is used for the daily test, both operation and calibration are verified. A more thorough test with nuclear material is performed on a quarterly basis. Additional information on monitor calibration is available in Ref. [10]. For portal monitors at border crossings, calibrations could be performed on an annual basis or during corrective maintenance, such as replacing a detector or an electronic component that could affect the portal's sensitivity. Gamma and neutron detector efficiencies should be periodically monitored because minimum detectable quantity calculations will assume some minimum detector efficiency. Failing detectors would require replacement to maintain desired RPM sensitivity levels to materials of interest.

22.8 Monitor Evaluation Methods

Laboratory evaluation can verify a monitor's ability to detect radioactive material reliably and can reveal shortcomings in design. Summaries of evaluations have been published for pedestrian nuclear material monitors [14] and for contamination monitors [3]. These evaluations were performed with monitors that were operated for long periods without recalibration while their statistical alarm frequency and detection sensitivity were determined.

Statistical alarm frequency and sensitivity are interdependent, and determining one has little meaning without determining the other. Statistical alarm testing requires recording alarms in a constant background environment over a long enough period to observe 10^5 or more decisions. A timing switch is used to operate the monitor periodically, and the background is updated between monitoring periods. The statistical alarm probability is obtained by dividing the observed number of alarms by the total number of monitoring tests performed. The statistical alarm probability per passage of an occupant is then the product of statistical alarms per test and the average number of tests per passage. This type of testing ignores background reduction by an occupant—a factor that may overestimate the statistical alarm frequency in normal operation.

Monitor sensitivity can be determined by observing the probability for a monitor to detect the passage of nuclear material or contamination test sources. The background intensity and the method of passing the test source through the monitor must be regulated [15], as well as other factors that affect performance. Because some spatial variation always occurs in detector efficiency, testing should be done in the least sensitive part of the monitor; for example, at shoe level for a pedestrian monitor. The test source should be carried through the monitor by different individuals who are walking in their usual manner. For a general discussion of monitor testing, see Ref. [16].

22.9 Examples of Perimeter Monitors

22.9.1 Handheld Perimeter Monitors

Handheld contamination monitors usually measure the dose rate for a single type of radiation, although some multipurpose instruments use filters to sense different types of radiation. Contamination monitors are simple, inexpensive, analog or digital devices that are operated sporadically and are usually powered by batteries. An example of a handheld contamination monitor is shown in Fig. 22.12. The sensitivity of handheld contamination monitors varies a great deal; most sense radiation intensities above 0.1 µR/h, although NaI(Tl) monitors can operate at the natural background intensity of a few µR/h.

Handheld radioisotope identification devices, shown in Fig. 22.13, have scintillation or semi-conductor detectors and battery-operated electronics. The instruments usually have rechargeable batteries and are operated to screen pedestrians and vehicles in secondary inspections after a portal monitor has registered elevated radiation levels. Each detector sounds an audible signal when it senses a significantly increased radiation intensity and will provide radionuclide identification in a matter of minutes or less. Besides their use as perimeter monitors, these highly sensitive gamma-ray-detecting instruments can

Fig. 22.12 Example of a digital handheld contamination monitor. (Photo courtesy of Nucleonix Systems)

Fig. 22.13 Handheld radioisotope identification devices from different vendors being evaluated for performance. (Photo courtesy of National Nuclear Security Administration NA-213 Nuclear Smuggling Detection and Deterrence)

be used as area radiation monitors or as survey monitors for salvaged equipment. They sense radiation intensities of a few μR/h and can detect ~0.5 μCi of ^{137}Cs in a rapid but careful search [17]. They can detect a few grams of HEU or a fraction of a gram of low-burnup plutonium under worst-case conditions (25 μR/h background intensity and maximum self-absorption in the nuclear material). Better performance will always be obtained under routine circumstances. Frequent statistical alarms—one or two per minute—are easily tolerated in these instruments because alarms in a specific area locate the radioactive material. Occasional alarms that are not repeated in the same area do not detract from monitoring effectiveness because they verify that the instrument is operating.

At seaports and border crossings, a variety of handheld spectroscopic radiation detectors (Fig. 22.13) are used in secondary inspections to identify the radiation source that caused a primary alarm. A personal radiation detector or survey meter may be used initially to locate hot spots, especially when performing a secondary inspection on a large conveyance, such as a 40 ft. cargo container. Two-minute measurements at specific locations and at hot spots with the spectroscopic handheld detectors generally suffice to positively identify the source of radiation. For example, fertilizer contains significant amounts of the radioactive isotope ^{40}K, and at a seaport, a cargo container filled with bags of fertilizer could set off a gamma alarm in a primary RPM. Measurements with a handheld HPGe or NaI(Tl) would confirm the presence of ^{40}K, and the shipping manifest would confirm the presence of fertilizer. However, the 2 min measurements are required to find any threat sources that might be present in addition to the NORM. If no threat source is found, the container would be cleared to continue.

At an airport, a passenger could have received ^{99m}Tc for a routine medical procedure in recent days and could set off a gamma alarm at a pedestrian portal monitor. Airport security personnel would identify the medical isotope in a secondary inspection and question the passenger about any recent medical procedures. At a border crossing, a truck could set off a gamma alarm in the portal monitor. The border police would perform a secondary inspection and identify ^{60}Co in the back of the truck. This is neither a medical isotope in a person nor NORM expected to be found in certain bulk cargo materials. The truck driver claims to be transporting scrap metal. The truck is detained, and on closer inspection, an orphan ^{60}Co source is found amongst the scrap metal.

22.9.2 Automatic Pedestrian Monitors

Automatic pedestrian contamination monitors for use with pedestrians are commercially available as traditional walk-through portals with gas-flow proportional counters that detect quantities below 1 μCi of ^{137}Cs and as high-sensitivity wait-in monitors that detect quantities below 100 nCi of fission or activation products. Figure 22.3 illustrates a portal that achieves high sensitivity by requiring pedestrians to place their body surfaces against the proportional counters. The proximity between body surface and detector and an extended monitoring period both help to achieve high sensitivity.

Pedestrian RPMs (Fig. 22.1) can detect less than 10 g of HEU and less than 0.3 g of low-burnup plutonium under worst-case operating circumstances. The typical statistical alarm frequency is 1 per 4000 passages.

22.9.3 Automatic Vehicle Monitors

Automatic contamination monitors for use with vehicles are rare because the interior surfaces of a vehicle usually must be monitored closely. An exception is the Los Alamos National Laboratory roadbed monitor that has a detector positioned below the vehicles to sense activated accelerator target material that may be transported from a facility. This monitor alarms at about twice background intensity. It provided the first evidence of contaminated Mexican steel introduced into the United States in 1983.

Vehicle RPMs for detecting MORC and smuggled nuclear and radiological materials are often simple, drive-through portals like the one shown in Fig. 22.14. Except for detector spacing, vehicle portals are similar to pedestrian portals. Moving-vehicle portals detect intensity increase of ~15% above background; for worst-case conditions, they can detect less than 10 g of low-burnup plutonium with less than 1 statistical alarm per 4000 passages.

22.9.4 Other Portal Monitor Examples

Many other RPM applications exist, and only a few others are mentioned here:

Fig. 22.14 The nuclear material portal monitor measures vehicles that pass slowly (8 km/h) through the detector columns for elevated emissions of gamma and neutron radiation. (Photo courtesy of National Nuclear Security Administration NA-213 Nuclear Smuggling Detection and Deterrence)

- *Mobile detection systems*: Radiation detection systems can be mounted in vehicles, such as dedicated systems permanently mounted in larger vans or smaller removable systems that fit in the trunk or rear cargo area of any vehicle. Such systems can be used as stationary temporary checkpoints. Alternatively, they can be used in a mobile search mode, though the constantly fluctuating gamma background—especially in urban environments—can make this type of use challenging.
- *Train portal monitors*: Due to the size and speed of trains, portal monitors designed for screening trains generally require larger total detector volume, and performing a secondary inspection on a long cargo train can be operationally challenging.
- *Conveyor portal monitors*: Airports frequently use miniaturized portal monitors to screen checked luggage for nuclear and radiological contraband.
- *Relocatable portal monitors*: A variety of relocatable portal monitors can be quickly set up and taken down to monitor traffic at special events or accommodate sudden dramatic changes in flows of people, such as during humanitarian crises.
- *Spectroscopic portal monitors*: During the primary screening occupancy, these monitors categorize or identify sources of radiation by using spectroscopic information from gamma-ray detectors, such as NaI or HPGe. However, such systems come at a higher price relative to gross-counting RPMs and may require more complex calibration and maintenance procedures. Another option, energy windowing, uses relative count rate in different energy regions of the spectrum to distinguish higher-energy contributions from NORM relative to lower-energy contributions from threat sources, industrial sources, and medical isotopes. Therefore, energy windowing generally can provide only categorization and not specific radionuclide information.

References

1. EG&G Inc., Santa Barbara Division (R. W. Hardy, R. B. Knowlen, C. W. Sandifer, and W. C. Plake), U.S. Patent No. 3,670,164 (1972)
2. W.E. Kunz, *Portable Monitor for Special Nuclear Materials*, Los Alamos Scientific Laboratory report LASL-77-18 (1977)
3. M. Littleton, *High sensitivity portal monitors—A review*, Institute of Nuclear Power Operations report 82-001-EPN-01 (1982)
4. *Physical Protection of Special Nuclear Material*, U.S. Department of Energy Order 5632.2 (1979)
5. *Control of Personnel Access to Protected Areas, Vital Areas, and Material ACCESS areas*, U.S. Atomic Energy Commission Regulatory Guide 5.7 (1973)
6. *American National Standard for evaluation and performance of radiation detection portal monitors for use in homeland security*, ANSI N42.35–2016 (American National Standards Institute, 2016)

7. C. Blessinger, J. Livesay, *Methods for Calculation of the Minimum Detectable Quantity of Radiation Portal Monitors*, ORNL/SPR-2020/1508 (2020)
8. *Detection of Radioactive Materials at Borders*, IAEA-TECDOC-1312 (IAEA, WCO, EUROPOL, and INTERPOL joint report, 2002)
9. *Radiation Protection Instrumentation – Installed Radiation Portal Monitors (RPMs) for the Detection of Illicit Trafficking of Radioactive and Nuclear Materials*, IEC 62244:2019 (International Electrotechnical Commission, 2019)
10. C.N. Henry, J.C. Pratt, A new containment and surveillance portal monitor data analysis method. ESARDA **10**, 126–131 (1979)
11. C.R. Hurlbut, *Plastic Scintillators: A Survey*, in *Trans. American Nuclear Society* winter meeting, San Francisco, CA, November 10–15, 1985 (available from Bicron Corp., Newbury, OH)
12. P.E. Fehlau, G.S. Brunson, Coping with plastic scintillators in nuclear safeguards. IEEE Trans. Nucl. Sci. **30**, 158 (1983)
13. E. Appel, M. Giannini, A. Serra, A new method of self-diagnosis for pulse measuring systems. Nucl. Instrum. Methods **192**, 341 (1981)
14. P.E. Fehlau, *An applications Guide to Pedestrian SNM Monitors*, Los Alamos National Laboratory report LA-10633-MS (1986)
15. P.E. Fehlau, T.E. Sampson, C.N. Henry, J.M. Bieri, W.H. Chambers, *On-Site Inspection Procedures for SNM Doorway Monitors*, Los Alamos Scientific Laboratory report LA-7646 (NUREG/CR-0598) (1979)
16. P.E. Fehlau, Standard Evaluation Techniques for Containment and Surveillance Radiation Monitors, *ESARDA* 15; also available as Los Alamos National Laboratory report LA-UR-82-1106 (1982)
17. P.E. Fehlau, *Hand-Held Search Monitor for Special Nuclear Materials User's Manual*, Los Alamos National Laboratory report LALP-84-015 (1984)

Open Access This chapter is licensed under the terms of the Creative Commons Attribution 4.0 International License (http://creativecommons.org/licenses/by/4.0/), which permits use, sharing, adaptation, distribution and reproduction in any medium or format, as long as you give appropriate credit to the original author(s) and the source, provide a link to the Creative Commons license and indicate if changes were made.

The images or other third party material in this chapter are included in the chapter's Creative Commons license, unless indicated otherwise in a credit line to the material. If material is not included in the chapter's Creative Commons license and your intended use is not permitted by statutory regulation or exceeds the permitted use, you will need to obtain permission directly from the copyright holder.

23 Principles of Calorimetric Assay

M. P. Croce, D. S. Bracken, R. N. Likes, C. R. Rudy, and P. A. Santi

23.1 Introduction

Calorimetry is the quantitative measurement of heat or thermal power. In contrast to adiabatic calorimeters that measure the total heat energy released in a reaction (for example, to determine the energy content of food), heat-flow calorimeters are used in nuclear material assay to measure the thermal power from nuclear decay in an item. *Calorimetric assay* refers to the combination of the thermal power measurement and knowledge of the isotopic composition to determine the mass of each nuclide in an item. It can be implemented in a completely nondestructive manner using high-resolution gamma-ray spectroscopy isotopic analysis, or the nondestructive calorimetry measurement can be combined with precision destructive isotopic analysis.

Calorimetric assay is most extensively used for materials with high plutonium concentrations, where it is an especially important technique for nuclear materials accountability and shipper-receiver confirmatory measurements. It is well-suited to the assay of plutonium, americium, and tritium that have relatively high specific activity from short-lived nuclides. Uranium (with the possible exception of ^{233}U materials) consists of longer-lived isotopes that are difficult to measure with calorimetry. The method is routinely applied to a wide range of materials, including metals, alloys, oxides, fluorides, mixed oxides, waste, and scrap. When applied to concentrated homogeneous, plutonium-bearing materials, calorimetry is comparable in accuracy to mass measurement and chemical analysis, with precisions approaching 0.1%. For high-density scrap, calorimetry plus gamma-ray spectroscopy can approach a precision and accuracy of 1% if the scrap has homogeneous isotopic composition and up to 1.6% accuracy for inhomogeneous salts [1–3].

The calorimetry result is generally independent of item geometry, composition, impurities, and packaging. Calorimeter instrumentation is relatively simple and can be optimized for a wide variety of items and measurement applications. In general, the technique has better uncertainty but is less rapid and less portable than other nondestructive assay (NDA) techniques described in this book. Calorimetry can often provide accurate reference measurements for improving the calibration of other assay techniques such as neutron coincidence counting (Chap. 17 and Ref. [4]).

This chapter discusses the calorimetric assay method, principles of heat production in nuclear materials, specific power determination, calorimeter types, methods of operation, and basic sources of assay uncertainty. Chapter 24, Calorimetric Assay Instruments, describes existing calorimeters of various types, including small calorimeters for laboratory use and bulk assay calorimeters for in-plant applications.

Los Alamos National Laboratory strongly supports academic freedom and a researcher's right to publish; as an institution, however, the Laboratory does not endorse the viewpoint of a publication or guarantee its technical correctness.

M. P. Croce (✉) · D. S. Bracken · C. R. Rudy · P. A. Santi
Los Alamos National Laboratory, Los Alamos, NM, USA
e-mail: mpcroce@lanl.gov; psanti@lanl.gov

R. N. Likes
Fort Lewis College, Durango, CO, USA

23.2 Calorimetric Assay Method

Calorimetric assay requires two types of information: the measured thermal power W (watts) of an item and the material's effective specific power P_{eff} (watts/gram). The mass m (grams) of nuclear material within an item can then be determined as follows:

$$m = \frac{W}{P_{eff}}, \quad (23.1)$$

where W is the result of a calorimeter measurement. P_{eff} is often calculated from the isotopic composition of the material that has been measured by either gamma-ray spectroscopy or destructive analysis. If isotopic fractions R_i are known, the mass m_i of each isotope in an item can be determined by

$$m_i = \frac{R_i W}{P_{eff}} \quad (23.2)$$

Understanding calorimetric assay is therefore about understanding how P_{eff} and W are determined. The following sections expand on these concepts.

23.3 Heat Production in Nuclear Materials

Since Rutherford and Barnes reported in 1903 on experiments that established a direct connection between heat emission of radium and its radioactivity, calorimetry has been known to provide a sensitive measurement of radioactive decay [5]. Calorimeters used in nuclear material assay operate on the principle that almost all of the energy associated with the decay of radioactive material in an item is absorbed within the form of heat within the calorimeter. Although the decay of all radioactive materials generates heat, calorimetry is typically used only for assay of plutonium, americium, and tritium because of their relatively short half-lives and thus high specific activities that generate heat at a high enough rate to be measured accurately. Figure 23.1 illustrates the typical applicability of calorimetry for a range of material types organized by specific power output.

Calorimetry is a good match for assay of nuclear materials that have relatively high specific thermal power output, such as plutonium, americium, and tritium, which are routinely quantified with calorimetric assay. For reference, the nuclear decay parameters for the plutonium isotopes, [241]Am, and tritium (including the specific thermal power output for these materials) are presented in Table 23.1.

Alpha and beta decay are the most important nuclear decay modes. Alpha decay is the dominant decay mode of the plutonium and americium isotopes. The range of alpha particles in various materials relative to their range in air is given by the Bragg-Kleeman rule (Eq. 13.2, Chap. 13). The range of 5–6 MeV alpha particles is on the order of 5 μm in common materials; therefore, virtually all of the energy released by alpha decay will remain within the item as heat.

Plutonium-241 and tritium are important beta-decaying nuclides. The determination of energy deposition in the sample is more complicated with beta decay or electron capture than with alpha decay due to the energy of the escaping neutrino or antineutrino. Neutrinos are extremely penetrating regarding matter so that a portion of the beta emission disintegration energy is not locally deposited. The average energy deposited locally in an absorber in the form of heat can be approximated as roughly one-third of the maximum disintegration energy that accompanies the beta decay of a particular radionuclide. The range of a 20 keV beta particle is on the order of microns in common materials, and its energy will remain within the item as heat.

Other nuclear reactions such as fission or (alpha, n) reactions are negligible contributors to the thermal power produced by most nuclear materials. The energy lost through neutron emission is many orders of magnitude less than the total disintegration energy. Gamma-ray escape also corresponds to a very small fraction of the total decay energy for materials typically measured by calorimetry. The upper limit on lost energy due to escaping gamma-rays from plutonium isotopes and [241]Am was calculated in Ref. [3] and is presented in Table 23.2. In addition to the heat loss due to escaping gamma rays, Table 23.2 also presents the relative heat contribution due to spontaneous fission energy based on branching ratios and an average reaction Q value of 200 MeV. The maximum energy loss was calculated assuming that 20% of the spontaneous fission energy is completely lost from the calorimeter due to the loss of neutrons and photons.

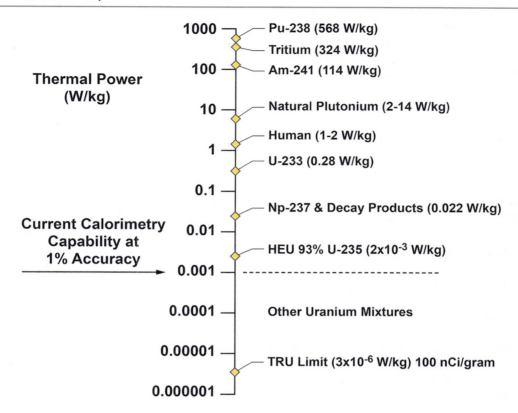

Fig. 23.1 Relative thermal power of actinides [6]

Table 23.1 Nuclear parameters of commonly assayed nuclides [7]

Isotope	Dominant decay mode[a]	Specific power (P_i) (mW/g)	% Standard deviation	$T^{1/2}$ (y)[b]	% Standard deviation
^{238}Pu	α	567.57	0.05	87.74	0.05
^{239}Pu	α	1.9288	0.02	24,119	0.11
^{240}Pu	α	7.0824	0.03	6564	0.17
^{241}Pu	β	3.412	0.06	14.348	0.15
^{242}Pu	α	0.1159	0.22	376,300	0.24
^{241}Am	α	114.2	0.37	433.6	0.32
^{3}H	β	324	0.14	12.3232	0.017

[a] For all of the nuclides listed, the dominant decay mode has a branching ratio of >99.99%
[b] Half-lives provided are the values used with the Specific Power values

Table 23.2 Maximum energy loss per decay due to gamma-rays or spontaneous fission [3]

Isotope	P_i (mW/g)	Photon energy fraction (%)[a]	Spontaneous fission branching ratio (%)	Spontaneous fission energy fraction (%)	Max energy loss (%)[b]
^{238}Pu	567.57	0.031	1.8e−07	6.6e−06	0.031
^{239}Pu	1.9288	0.0013	3.0e−10	1.1e−08	0.0013
^{240}Pu	7.0824	0.00054	5.8e−06	2.2e−04	0.00058
^{241}Pu	3.412	0.025	2.4e−14	8.7e−10	0.025
^{242}Pu	0.1159	0.028	5.5e−04	2.2e−02	0.032
^{241}Am	114.2	0.509[c]	4.3e−10	1.5e−08	0.51

[a] The energy loss due to gamma escape will be significantly less than total energy
[b] Maximum fractional energy due to escaping neutrons and gamma rays
[c] The 59.54 keV gamma ray that is often effectively captured provides 0.39% of total energy

Additional sources of heat production could be significant for materials with high fission product content, such spent fuels or spent fuel separation process solutions. Chemical reactions, such as radiolysis in high-activity aqueous solutions, could be a significant factor in the measured thermal power. Possible errors from such sources can be estimated by using gamma-ray spectroscopy to quantify and correct for fission products, by observing any time-dependent power emission resulting from short-lived nuclides, or by comparing calorimetry with other techniques [8].

23.4 Specific Power Determination

23.4.1 Definition of Specific Power

Each radionuclide produces a specific power, P_i (W/g), that depends on its half-life, energy released per decay, and molar mass. Table 23.1 summarizes recommended specific power values for important nuclides [7]. Values of P_i were directly measured for ^{239}Pu, ^{240}Pu, ^{241}Pu, and ^{242}Pu using isotopically pure samples. For ^{238}Pu and ^{241}Am, values were calculated from nuclear data. Note that the specific power of ^{238}Pu is much greater (~250 times) than that of ^{239}Pu because of the shorter alpha half-life of ^{238}Pu. Similarly, the specific power of ^{241}Am is also considerably larger than that of ^{239}Pu. Tritium undergoes a low-energy beta decay but has a very short half-life, and therefore a relatively high specific power.

Specific power values can be calculated from nuclear data as follows: Each radionuclide decays at a constant mean rate proportional to its half-life ($T_{1/2}$, in years). For each alpha decay, it is a good approximation that the total nuclear decay energy (Q, in MeV) is converted to heat. For beta decays, a significant and variable fraction of the decay energy is transferred to an escaping neutrino or antineutrino, and an average deposited energy should be considered as Q. The ratio m/A, where m is the mass of the radionuclide in grams and A is its atomic mass in grams per mole, is proportional to the number of atoms of the radionuclide. These factors can be combined with a unit conversion factor to give the specific power P_i of a radionuclide in watts per gram [9]:

$$P_i = \frac{2119.3 \cdot Q_i}{T^i_{\frac{1}{2}} A_i} \tag{23.3}$$

23.4.2 Effective Specific Power

The effective specific power P_{eff} is used to consider thermal power contributions from multiple radionuclides in an item, as is the case for most plutonium materials. P_{eff} can be computed as a sum over n individual specific powers weighted by individual isotopic fractions R_i:

$$P_{\mathit{eff}} = \sum_{i=1}^{n} R_i P_i \tag{23.4}$$

If only one radionuclide is known to be present in the item (e.g., tritium), the equation simplifies to $P_{\mathit{eff}} = P_i$. The effective specific power is dominated by nuclides with a high specific power, high mass fraction, or both. ^{238}Pu and ^{241}Am are particularly important for most plutonium materials due to their high specific powers, whereas ^{239}Pu contributes significantly due to its high mass fraction. The isotopic composition of plutonium materials changes significantly over time due to radioactive decay, and therefore, the effective specific power does as well. ^{241}Am content increases with time because ^{241}Am is produced by the decay of ^{241}Pu, and thus the power from a plutonium sample also increases as a function of time. The impact of radioactive decay on the effective specific power of a high burnup plutonium item is illustrated in Table 23.3.

Uncertainty in calorimetric assay is often limited by uncertainty in the effective specific power of the material being measured. P_{eff} is most often determined computationally using Eq. 23.4 or from plutonium isotopic analysis software, such as FRAM or MGA (discussed in Chap. 10). This method is appropriate when accurate isotopic composition information is available for the material, typically from nondestructive gamma-ray spectroscopy (Chap. 9) or destructive analysis [10, 11]. Decay correction of isotopic composition and P_{eff} is typically done using software but can also be done manually [12]. P_{eff} can also be determined empirically by measuring the thermal power output of a well-known mass of a specific

Table 23.3 Calculation of effective specific power for a plutonium item on two different dates, illustrating the impact of radioactive decay on the effective specific power of an item

Date	Isotope	R_i	P_i (mW/g)	$R_i \times P_i$ (mW/g)	Power (%)
06/20/1986	^{238}Pu	0.0120	567.57	6.811	58.34
	^{239}Pu	0.6253	1.9288	1.206	10.33
	^{240}Pu	0.2541	7.0824	1.780	15.25
	^{241}Pu	0.0668	3.412	0.228	1.95
	^{242}Pu	0.0418	0.1159	0.005	0.05
	^{241}Am	0.0144	114.2	1.644	14.08
SUM				11.674	100.00
6/20/2021	^{238}Pu	0.0096	567.57	5.449	32.58
	^{239}Pu	0.6637	1.9288	1.280	7.65
	^{240}Pu	0.2690	7.0824	1.905	11.39
	^{241}Pu	0.0131	3.412	0.044	0.26
	^{242}Pu	0.0446	0.1159	0.005	0.04
	^{241}Am	0.0704	114.2	8.040	48.08
SUM				16.723	100.00

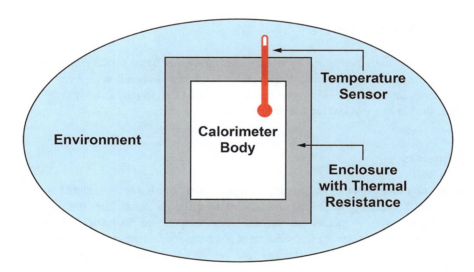

Fig. 23.2 A calorimeter consists of a chamber isolated from its environment by a thermal resistance. The temperature difference between the chamber and the external environment provides a measure of thermal power produced in the chamber. (Figure from Ref. [3])

material at a specific time. Chemical analysis can be applied if needed to determine the fraction of nuclear material in the sample. The need for a decay correction could limit the applicability of an empirical value of P_{eff}.

23.5 Calorimeters for Nondestructive Assay

A calorimeter consists of a chamber isolated from the environment by a thermal resistance (Fig. 23.2). The difference between the temperature of the environment and the chamber is then a measure of the thermal power produced in the chamber, according to Eq. 23.5.

$$\frac{dQ}{dt} = k(T_{cal} - T_{env}), \tag{23.5}$$

where
Q = heat energy
t = time

k = thermal conductivity between calorimeter chamber and environment
T_{cal} = internal calorimeter chamber temperature
T_{env} = external environment temperature.

Generally, the temperature of the environment is held constant using a water bath or other high-heat-capacity medium. Thermal conductivity k is set by the design of the calorimeter and determines its applicable range of measured power. Adiabatic calorimeters that measure total reaction energy operate in the limit of $k = 0$. Calorimeters designed for the assay of nuclear materials are heat-flow calorimeters with $k > 0$ because the spontaneous decay of radionuclides produces a nearly constant thermal power during the measurement. Instruments designed for measuring high-power items have high k values to limit the maximum temperature of the chamber. Instruments designed for low-power items have low k values to maximize sensitivity.

When an item that produces heat at a constant rate is placed in the calorimeter chamber, the system will reach an equilibrium condition in which the rate of heat transfer to the environment is equal to the rate of heat production in the sample. The rate of heat transfer is directly proportional to the temperature difference between the chamber and the environment, as described by Eq. 23.5. It is important that the value of k remain constant because the measurement of the rate of heat transfer is essentially a precision temperature measurement. The calibration function for a given calorimeter depends on k.

23.6 Types of Heat-Flow Calorimeters

Although calorimetric assay is usually performed by a heat-flow calorimeter based on the concept of Fig. 23.2, no universal calorimeter design is suitable for all applications. Each system is custom designed with many specifications in mind, as described in Chap. 24. Depending on the specific requirements, the final design may employ features of one or more of the calorimeter designs described in this section: twin, over/under, and gradient.

23.6.1 Twin Calorimeters

Twin calorimeters employ two chambers that are identical except that only one holds the sample being measured and the other remains an empty reference. Subtracting the reference chamber temperature signal from the sample chamber temperature signal is an extremely effective way to cancel out environmental temperature fluctuations or other noise sources. Twin calorimeters typically have an increased level of accuracy and precision over single-cell calorimeters.

An example of a twin calorimeter is the isothermal twin-bridge calorimeter developed at Mound Laboratory, used extensively for in-plant assays [13]. Figure 23.3 shows a cross section of this calorimeter type. The design consists of two identical thermal elements—sample side and reference side—each separated from the outer wall by an air gap or solid gap that serves as the thermal resistance. Each thermal element has heater wires wrapped around its sample chamber for calibration purposes. Axial heat loss from the sample chamber is minimized using plastic end caps with Styrofoam baffles and metal plates to provide a thermal short from the sample to the calorimeter walls. The calorimeters were referred to as twin-bridge calorimeters because two resistance thermometers that consist of thin, high-purity nickel wire are wound over the entire length of each thermal element and connected in a Wheatstone bridge circuit [13]. Section 23.6.4 provides additional details on both Wheatstone bridge sensors and solid-state sensors that are commonly used within heat-flow calorimeters.

To establish a constant reference environment for the measurement, twin-bridge calorimeters were typically immersed in a water bath whose temperature is held constant to about 1 millidegree Celsius. More recently, twin calorimeters and single-cell calorimeters are being placed within an isothermal air bath to maintain a constant reference temperature (e.g., the calorimeter shown in Ref. [14]). The use of two identical thermal elements in a twin calorimeter design yields a difference signal that is 10–100 times more stable than the reference environment. When a sample is placed in the chamber, the temperature rises until heat losses through the nickel windings, thermal gap, and outer jacket to the water bath equal the heat generated by the sample. When this equilibrium is attained, the temperature difference measured by the bridge potential is proportional to the amount of heat being generated. Precise heat-flow calorimeters of this type are designed and constructed so that the thermal paths between the sample chamber and the environment remain constant. Additional care is taken to keep the temperature of the environment constant and to minimize the heat distribution error associated with the location of the sample in the chamber.

23 Principles of Calorimetric Assay

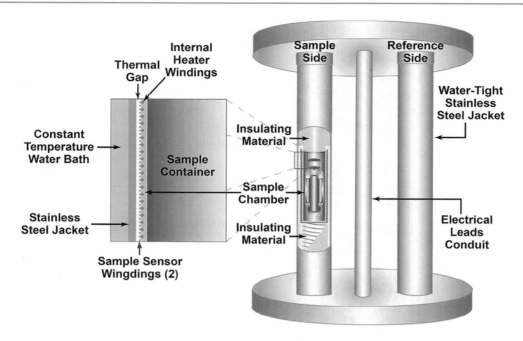

Fig. 23.3 Cross section of a typical isothermal, twin-bridge calorimeter with two identical thermal cells—one for the sample and one for reference. (Figure from Ref. [3])

Twin calorimeters typically provide the best precision, accuracy, sensitivity, and long-term stability for power measurements because of the internal cancellation of thermal effects; however, the two thermal elements and the environmental bath result in an instrument that is larger and requires more floor space than other designs. Twin-bridge calorimeters have been designed for sample size diameters in the 1–30.5 cm range, with an upper limit dictated only by nuclear criticality safety considerations and a lower limit in the 0.1–0.2 g ^{239}Pu/L range [13]. An application of a twin-bridge calorimeter is described in Chap. 24, Sect. 24.4.1.

23.6.2 Over/Under Calorimeters

In the over/under design, the sample chamber is mounted above a somewhat shortened reference chamber [3], as illustrated in Fig. 23.4. Both chambers share a single isothermal water jacket that is smaller than the environmental bath used by the twin calorimeter. The over/under design requires less floor space but is taller. The calorimeter shown in Fig. 23.4 can be mounted on a mobile cart along with its circulation bath.

The over/under calorimeter has size and cost advantages over the twin design and provides cancellation of some thermal effects; however, the accuracy is not as good as the full twin calorimeter.

23.6.3 Gradient Calorimeter

The gradient calorimeter (Fig. 23.5) consists of a series of concentric cylinders with the sample chamber cylinder inside the reference chamber cylinder [3]. The physical gap between the reference chamber cylinder and the sample chamber cylinder determines the thermal resistance of the calorimeter and thus its sensitivity. Outside the reference chamber is a jacket that provides a uniform thermal heat sink. The jacket is in contact with a thin, circulating water bath. This design leads to small size and low material and fabrication costs. The gradient design is well suited for transportable instruments, glove-box installations, or large samples where a full twin design might be too bulky. The gradient calorimeter is usually operated in the constant temperature power-replacement method described in Sect. 23.7.2. A Mound gradient bridge calorimeter design is described in Chap. 24, Sect. 24.4.2.

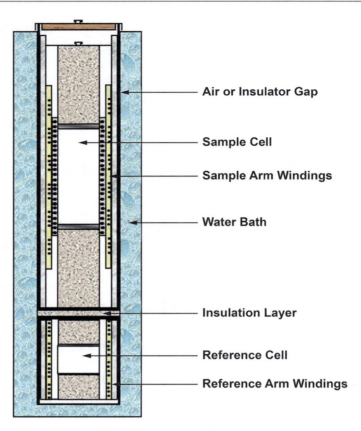

Fig. 23.4 A transportable calorimeter of the over/under design, with the sample cell above the reference cell and a thin water jacket. (Figure from Ref. [3])

23.6.4 Temperature Sensors

Two types of temperature sensors are commonly employed in calorimeters: resistance and thermoelectric. Both are applicable to the calorimeter designs described in the previous section. Resistance sensors are based on materials such as nickel that have a significant change in their electrical resistivity with temperature. For instance, high-purity nickel wire has an electrical resistance that changes linearly with temperature, with a sensitivity of 0.6% change in resistance per °C change in temperature. Resistance sensors are commonly implemented as a nickel wire helically wound interleaved (bifilar winding) around the calorimeter chamber(s) in a Wheatstone bridge configuration or as thermistors. This design has been proven over many decades, and many calorimeters of this type are in use. In twin-bridge calorimeters, the sample chamber and the reference chamber each have two nickel-wire windings. The four windings are connected as a transposed-arm, Wheatstone bridge and are supplied with a constant current, as illustrated in Fig. 23.6 [15]. Placement of a heat-producing item within the sample side of the calorimeter would cause an increase in temperature in the nickel-wire windings on that chamber and subsequently increase the electrical resistance of the sample arms of the Wheatstone bridge circuit relative to the electrical resistance of the reference arms of the circuit. By applying the constant electrical current across the Wheatstone bridge (shown as I in Fig. 23.6), the resistance change between the sample and reference sides of the bridge causes an imbalance in the bridge, leading to a change in the voltage that is measured across the bridge (shown as V in Fig. 23.6). The voltage signal from the Wheatstone bridge, sometimes called *bridge potential* or *BP*, is proportional to the temperature difference between the sample and reference chambers. Although Wheatstone bridge calorimeters can perform high-precision thermal power measurements, the Wheatstone bridge circuit is sensitive to any changes in the resistance of the nickel wire. Therefore, these calorimeters are sensitive to environmental factors, such as physical stress or movement, that could strain the wire and subsequently change the wire's resistance.

A modern alternative to using Wheatstone bridge thermal sensors within a twin calorimeter are solid-state sensors, which use thermoelectric sensors as heat-flow sensors [16–18]. Thermoelectric sensors use junctions of dissimilar materials that

23 Principles of Calorimetric Assay

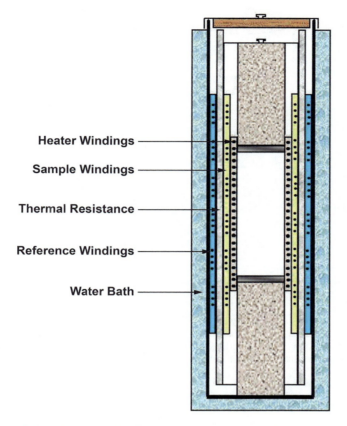

Fig. 23.5 Cross section of a gradient bridge calorimeter design. The environmental heat sink, which would be small or only an air chamber, is not shown. (Figure from Ref. [3])

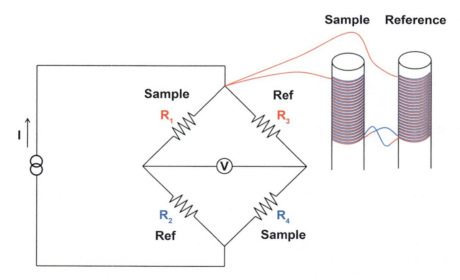

Fig. 23.6 A Wheatstone bridge circuit is a very sensitive and accurate configuration to measure the temperature difference between the sample and reference parts of a calorimeter based on resistance sensors. (Figure from Ref. [15])

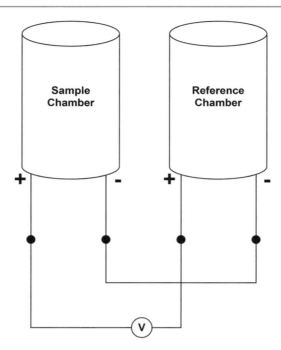

Fig. 23.7 Electrical schematic of twin thermoelectric calorimeter. The differential voltage, V, is analogous to the Wheatstone bridge potential, BP. Individual voltmeters can optionally be connected to the test points, indicated by black dots, to verify the operation of each chamber sensor

produce a voltage via the Seebeck effect in response to temperature differences across the junction. A thermocouple is a simple implementation of a thermoelectric sensor, though calorimeters typically use many stacked junctions to increase the measured voltage. In contrast to resistance sensors, thermoelectric sensors can be monitored by a simple connection to a voltmeter so that no constant current source is required. The differential voltage between sample and reference sensors is monitored with a single voltmeter, as shown in Fig. 23.7. Thermoelectric sensors tend to be more stable with respect to mechanical strain, which makes them a good choice for calorimeters that are frequently relocated. These practical advantages often lead to the selection of thermoelectric sensors for new calorimeter systems.

Calorimeters based on both types of sensors typically have a non-zero baseline value (V_0) that results from small differences in the sample and references sensors. V_0 is defined as the equilibrium signal voltage with no intentional thermal power present in the calorimeter chambers. Repeated measurements of the baseline value provide a way to monitor and correct for sensor drift over time.

23.7 Methods of Operation for Heat-Flow Calorimeters

23.7.1 Passive Mode

The passive mode is often used when high precision and accuracy are desired, a relatively long measurement time is acceptable, and heat standards are available for calibration or bias correction. In this mode, an item is placed into the sample chamber of the calorimeter and is passively allowed to reach thermal equilibrium. When the calorimeter signal is determined to reach equilibrium (Fig. 23.8), a calibration function is applied to the sensor voltage to determine the measured power.

The accuracy of the passive mode depends on the accuracy of the calibration function, which is determined through repeated measurements of electrical or radioactive heat standards that span the power range of items being measured. For most calorimeters, the calibration function is well-described by a quadratic function:

$$V = V_0 + S_1 W + S_2 W^2, \qquad (23.6)$$

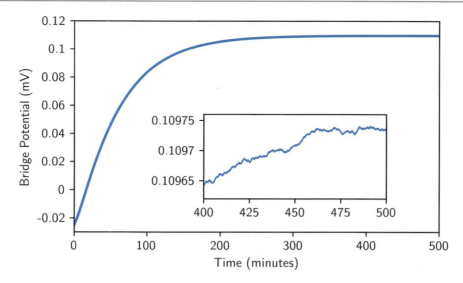

Fig. 23.8 Data from a passive mode calorimeter show the signal as it approaches an equilibrium value while the temperature of the system stabilizes after placing an item in the calorimeter chamber. The inset shows the data from 400 to 500 min on an expanded vertical scale

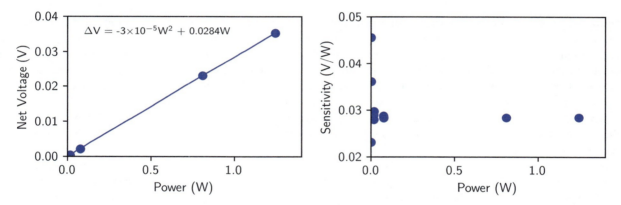

Fig. 23.9 (left) Calibration data for a twin-bridge, water-bath calorimeter operating in passive mode. (right) Differential sensitivity as a function of standard power for the data points shown in the calibration curve [3]

where
V = equilibrium calorimeter signal (volts)
V_0 = baseline value with no power in calorimeter (volts)
W = thermal power (watts)
S_1 = baseline sensitivity (volts/watt)
S_2 = differential sensitivity (volts/watt^2)

S_1 and S_2 can be determined by a least-squares fit to a data set that consists of measured V versus certified W values, as in Fig. 23.9 [3]. Typically the function is very nearly linear with $S_1 >> S_2$. The quadratic function can be solved for the usual case of $S_2 < 0$, as follows, to determine the measured power W of an item:

$$W = \frac{-S_1}{2S_2} - \sqrt{\left(\frac{-S_1}{2S_2}\right)^2 - \left(\frac{V_0 - V}{S_2}\right)} \tag{23.7}$$

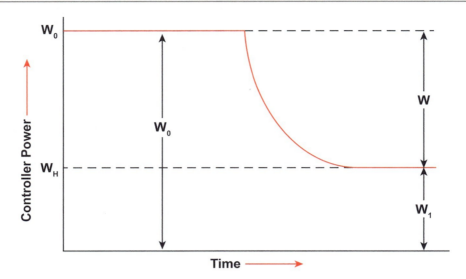

Fig. 23.10 Illustration of internal heater power during the measurement of an item using a servo-controlled calorimeter

23.7.2 Constant Temperature Power-Replacement Mode

The power-replacement mode (historically known as the *servo-control mode*) is often used to reduce measurement time relative to passive mode at the possible expense of precision and accuracy. In this mode of operation, the sample chamber is held at constant temperature by a servo-control mechanism. Heater power is used to maintain the sample chamber at a higher temperature than the bath. When a sample is placed in the calorimeter, to hold the chamber at constant temperature, the servo controller reduces the heater power by the amount of the sample's power. Thus, the thermal power from the sample replaces the original power being provided by the heater. This mode, also called the *isothermal mode*, may be used with any calorimeter bridge design. Use of the power-replacement mode typically reduces assay time by one-third to one-half because the time required to reach equilibrium depends principally on the thermal resistance and heat capacity of the sample and less on the calorimeter body.

Before samples are assayed by the constant temperature mode, the baseline power level is set 10–20% above the estimated wattage of any calibration standards or actual samples. A baseline power run is made to establish the empty chamber equilibrium power level W_0, also known as *basepower*. This run takes less time than a sample assay because no additional heat capacity is in the chamber.

When a radioactive sample is placed in the calorimeter, the heater power drops as the servo controller tries to maintain a constant temperature in the chamber (Fig. 23.10). The new equilibrium power level is W_1, and the sample power W is given by:

$$W = W_0 - W_1 \tag{23.8}$$

The constant temperature servo-control method is one of the fastest methods of calorimeter operation, especially when used with preconditioning of the sample in a separate temperature-controlled environment. For the same calorimeter operated in passive or servo mode, servo-mode operation tends to provide relatively lower precision and accuracy due to additional noise introduced by the servo-mode controller and heater.

23.7.3 Alternate Methods

Calorimeters can be operated with alternate methods that combine features of passive and servo control. For example, the replacement method [13] is based on measuring an item in passive mode and recording the equilibrium signal value, then removing the item and applying electrical heater power to exactly match that signal value. The item power is therefore equal to the heater power.

23.7.4 Heat Standards for Calorimetric Assay

The two types of heat standards commonly used are electrical and radioactive. Regardless of the type used, traceability to primary standards is essential [10]. Electrical heat standards have the advantage of being readily available and capable of providing adjustable thermal power. They can be integrated into the calorimeter chamber or designed to be inserted in the sample location. However, the effect of wires connected to the electrical heater must be considered when designing or using electrical heat standards. Accurate electrical calibration requires that the electrical heater produce the identical change in calorimeter output as an equivalent amount of power from a sample [10], but the electrical connections to the heater are both thermally conductive and a source of electrical heat due to their finite resistance. The problem of power dissipation in the wires leading to the heater can be minimized in the twin-design calorimeter by running the wires through both the sample and reference chambers so that the effects cancel. Radioactive heat standards have the advantage of more directly simulating thermal power from assayed nuclear material items. Radioactive heat standards, such as ^{238}Pu heat standards, can be certified as traceable to national electrical standards by measuring the ^{238}Pu heat standard in a twin-cell calorimeter operating in a differential mode, which compares the heat output of the radioactive heat standard against the measured heat from a certified electrical heater. If available, calibration with radioactive heat standards, such as certified ^{238}Pu items, is recommended.

23.8 Assay-Time Considerations

The time necessary to complete a calorimetric assay of an item is dependent on a variety of factors that influence the time necessary for the system to attain equilibrium conditions. The most important factors are (1) the heat capacity and thermal conductivity of the item and calorimeter chamber and (2) the difference between initial and equilibrium temperatures. Small calorimeters that measure low-heat-capacity, thermally conductive items—such as ^{238}Pu heat sources for radioisotope thermal generators—can passively reach equilibrium in less than 1 h. Large calorimeters that measure waste drums filled with insulating materials can take anywhere from tens of hours to several days to reach thermal equilibrium; however, most calorimeter measurements for nuclear material assay may be expected to each passive equilibrium in 10 h or less.

Table 23.4 (from Ref. [3]) gives some examples of how assay times for a twin-cell, water-bath calorimeter can be affected by the packaging and materials that are present within the item. For the measured data listed in the table, each item was pre-equilibrated to a temperature of 24 °C before being placed into a twin-bridge, water-bath calorimeter operating in passive mode that was kept at an internal initial temperature of 25 °C. The container holding the matrix had a volume of ~3 L. Two different measurements were performed: one with the container holding only the matrix and one with the container holding the matrix and a 0.8 W ^{238}Pu heat source. The columns marked "Time" in the table indicate the time required for the calorimeter to come to thermal equilibrium. The container was used for all of the results presented in Table 23.4.

As illustrated by the results in Table 23.4, one of the major factors that can affect the assay time is the amount of physical mass associated with the item being measured—with the larger-mass items taking a longer amount of time to reach thermal equilibrium than smaller-mass items. Another factor is the thermal properties, such as thermal conductivity and specific heat

Table 23.4 Item measurement time dependence on matrix material

Matrix type	No heat source		0.8-W ^{238}Pu heat source	
	Mass (kg)	Time (h)[a]	Mass (kg)	Time (h)[a]
Air	0.668	4.8	0.766	5.0
Poly beads	1.722	25.0	1.723	18.7
Al foil (1)	0.094	6.8	0.094	5.0
Al foil (2)	0.286	6.0	0.287	5.8
Copper shot	15.820	25.3	15.824	21.5
Salt	3.102	15.0	3.358	15.0
Al bars/foil	3.636	17.0	3.636	15.0
Sand	4.580	15.0	4.580	13.8
Steel shot	13.782	27.0	13.782	30.0
Lead shot	20.738	12.5	20.739	12.5
Poly beads	1.728	20.0	–	–
Sand	4.670	16.5	–	–

[a]Time required for the calorimeter to come to thermal equilibrium

capacity, of the item and item packaging. Other factors that can affect the assay time include the difference between the initial temperature of the item relative to the final equilibrium temperature of the item in the calorimeter, the type of calorimeter being used, and the operation mode of the calorimeter (passive or servo; Ref. [3]).

23.8.1 Calorimeter Design and Operating Method

The physical dimensions and thermal properties of a given calorimeter affect both the equilibrium time and the sensitivity. Decreasing the thermal resistance of a calorimeter will reduce the time required to come to equilibrium, but it will also reduce the sensitivity. Because reduced sensitivity will lead to reduced accuracy, the choice of calorimeter design must reflect a trade-off between assay time and assay accuracy for the sample range. The composition and dimensions of the gap between the sample and the outer calorimeter wall are important factors.

The constant temperature servo-control mode of operation is effective in reducing assay time. When the calorimeter body is maintained in an equilibrium state, the effect of its time constant is minimized, and the assay time is principally dependent on the thermal time constant of the sample. Assay time can be reduced further via preconditioning. *Preconditioning* refers to "pre-equilibration" of an item by storing it at nearly the same temperature as its equilibrium temperature inside the calorimeter. In principle, this technique can be used with passive operation but is best suited to servo-control mode because the calorimeter temperature will be regulated at a constant value regardless of item power.

Several advantages and disadvantages are associated with preconditioning. The primary advantage is increased throughput. The preconditioning bath can be designed to accommodate more than one sample, and it can be used overnight to prepare samples for assay the next day. The disadvantages are that additional floor space, electronics, and operator action are required. Also, the preconditioning bath temperature must be carefully maintained near the internal temperature of the calorimeter. Any change in the calorimeter's baseline power (basepower) level will require adjustment of the preconditioning bath temperature.

23.8.2 Equilibrium Prediction

Assay time can be reduced by mathematically fitting calorimeter data to predict the equilibrium result before the system actually reaches thermal equilibrium. The technique has been shown to yield good results with time savings of approximately 30–50% on a variety of calorimeters [19–25]; however, additional uncertainty terms may be introduced, and prediction techniques should be validated for specific measurement applications.

Prediction is based on the principle that the calorimeter's response function can be approximated by a sum of exponential terms:

$$f(t) = A + \sum_i B_i e^{-t/C_i} \tag{23.9}$$

The number of terms is empirically determined as the minimum required for a good fit to the calorimeter data as determined from the residuals (for example, as in Fig. 23.11 adapted from Ref. [21]). A single exponential term may be appropriate if the system is dominated by one thermal time constant. Values of A, B_i, and C_i are determined from the fitting process for each data set. Equation 23.9 is based on the fact that a calorimeter consists of different regions, each having different values for thermal resistance and heat capacity. The differing thermal properties of each region determine the time required for each region to come to equilibrium with its surroundings. The time required for the entire calorimeter to reach equilibrium is therefore a sum of several exponential response functions. Samples that are poorly packaged—with high thermal resistance or high heat capacity—will be assayed with less time savings.

Other approaches that have been explored to speed up calorimetry measurements include the usage of numerical analysis techniques such as the Aitken transformation and spectral analysis using Fourier transformation to extrapolate the equilibrium value based on an analysis of the initial approach to equilibrium [26]. Although Ref. [26] is limited in scope as a feasibility study to determine the accuracy and precision in extrapolating the equilibrium value using only the first hour of calorimetry data, the results appear promising that the techniques could be adapted to reduce the measurement time associated with calorimetry.

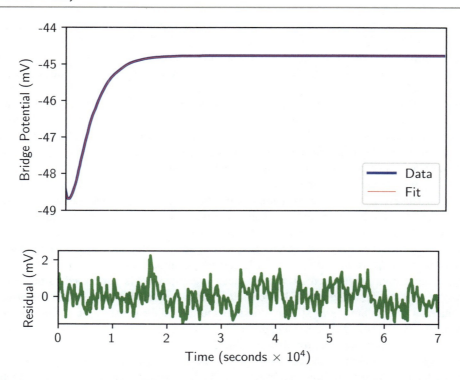

Fig. 23.11 Prediction based on a sum of exponential terms can often provide an excellent description of the data, as shown by the residuals plot (Figure adapted from Ref. [21])

Table 23.5 Sources of error in calorimetric assay [8, 27, 28]

Error source	Approximate magnitude (%)
Power measurement	
Calorimeter imprecision	<0.1
Heat distribution error	≤0.1
Calibration error	<0.1
Interference error	<0.1
Effective specific power determination	
Empirical method	
Sample power	<0.2
Sample plutonium content	<0.2
Computational method	
Gamma-ray spectroscopy	0.5
Mass spectroscopy	0.1

23.9 Sources of Uncertainty

As described in Sect. 23.2, determination of plutonium content is a two-step process: determination of calorimetric power and determination of effective specific power either by empirical or computational methods. The sources of error in this process are summarized in Table 23.5 [8, 27, 28].

Sources of error in the power determination include

- imprecision of the calorimeter system—the variance of the system response as a result of variations in room temperature, bath temperature, humidity, sample weight, sample loading and unloading stresses, and so forth;

- heat distribution error—the variance of the system response resulting from spatial distribution of the sample in the sample chamber;
- calibration or bias correction error—the variance of the system response to calibration method and standards; and
- interference error—heat production from interfering processes such as fission product reactions or chemical reactions.

Estimations of the impacts of systematic uncertainty on the overall uncertainty in the power determination have been made in Ref. [28], with an expanded discussion on the impacts of calibration on the overall uncertainty presented in Ref. [29].

Sources of error in the determination of P_{eff} depend on the method used for this determination: empirical or computational. When isotopic measurements of the item can be made—by using either destructive assay methods [30] or gamma-ray spectroscopy—it is most appropriate to use the computational method for determining P_{eff} using Eq. 23.4. The error associated with P_{eff} using the computational method depends on which technique is being used to determine the isotopic composition. As noted in Table 9.3, the typical precision that can be achieved when determining P_{eff} based on a gamma-ray measurement using a high-purity germanium detector can range between 0.1% and 2% depending on the transmission of gamma rays from the item. Use of the computational method with gamma-ray spectroscopy for isotopic analysis is the most common technique for calorimetric assay [3]. When destructive assay techniques are used to determine the isotopic composition, the uncertainty on P_{eff} is normally 0.1% for isotopically homogenous items [3].

To determine P_{eff} using the empirical method, a calorimeter measurement is performed to determine the total power produced by a sample of the material being assayed. A chemical analysis of the material is then performed to determine the amount of nuclear material within the item [3]. The possible sources of error using the empirical method are errors in determination of sample power and errors in determination of the content of the plutonium sample. Approximate size of the error associated with the empirical method is provided in Table 23.5.

Once P_{eff} has been determined by either the computational or the empirical method, the total assay uncertainty may be estimated by combining the error in the calorimeter power determination with the uncertainty in the effective specific power:

$$\frac{\sigma(m_{Pu})}{m_{Pu}} = \sqrt{\left(\frac{\sigma_W}{W}\right)^2 + \left(\frac{\sigma_{P_{eff}}}{P_{eff}}\right)^2} \tag{23.10}$$

In the case of using the computational method with gamma-ray spectroscopy for determining P_{eff}—the most common technique used in conjunction with calorimetry—the total uncertainty associated with the measured plutonium mass will be primarily dependent on the uncertainty in P_{eff} rather than the uncertainty associated with the measured thermal power from the calorimeter [6].

23.10 Measurement Performance

Calorimetric assay can measure plutonium-bearing items with a low bias and a high precision, and the results are often comparable to destructive analysis methods. For impure or heterogeneous items, calorimetric assay can have lower uncertainties than destructive assay techniques due to the sampling error associated with destructive assay [31].

The precision and accuracy of calorimetric assay has been consistently demonstrated within the U.S. Department of Energy (DOE) Calorimetry Exchange Program (CALEX). The CALEX Program distributed identical PuO_2 items that contained 400 g of plutonium with 5.86% ^{240}Pu by weight. The program tabulated the results from the facility's measurements yearly. Each facility collects data in a manner suitable for its own operations. The plutonium content and isotopic composition reference values of the mother lot of PuO_2 material used for these standards were measured by coulometry and mass spectrometry/alpha counting by four analytical laboratories. More general information on the CALEX program can be found in Ref. [32].

Calorimeter biases for 23 calorimeters at five DOE facilities are presented in Fig. 23.12. The dashed vertical lines separate the data submitted by each laboratory. These data were collected for the CALEX program over a 15-month period starting in October 1993. All measurements have a bias of less than ±0.8%. The average bias is 1.0004, with an average standard deviation of ±0.0002. The error expected on a single measurement would have a relative standard deviation (RSD) of 0.3%. The power of the CALEX standard during this period was ~1 W.

The results of multiple calorimetric assay measurements of the CALEX standards by three facilities are shown in Table 23.6 [33]. The calorimetry and gamma-ray measurements used to determine plutonium mass variabilities and biases

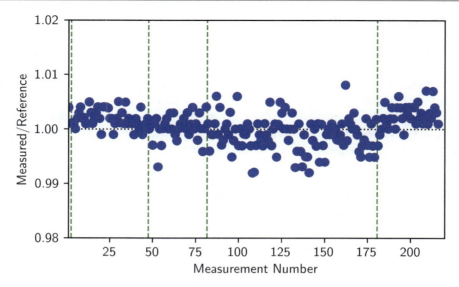

Fig. 23.12 Calorimeter measurement biases for heat measurements of the CALEX standards. Measurements were taken over a 15-month period by 5 DOE laboratories using 23 different calorimeters. The dashed vertical lines separate the data submitted by each laboratory. (Figure from Ref. [3])

Table 23.6 Calorimetry/gamma-ray assay measurement of CALEX standards[a]

Facility	Within-facility variability, g	Within-facility variability, % RSD	Bias, g	Bias, % RSD
A[b]	1.5	0.38	0.03	0.01
B[b]	1.5	0.38	−0.40	−0.10
C[c]	1.4	0.36	0.04	0.01

[a] All masses are shown in grams of plutonium decayed to a common date
[b] Measurements were made using multiple water-bath, twin-bridge calorimeters
[c] Measurements were made using an air-bath calorimeter

Table 23.7 CALEX precision and bias data for thermal power and P_{eff}

	Power		P_{eff}	
Facility	% Bias[a]	% RSD[b]	% Bias[a]	% RSD[b]
A	0.11	0.61	−0.23	0.10
B	0.08	0.22	0.07	0.26
C	−0.01	0.17	0.13	0.21
D	−0.08	0.30	0.02	0.20
E	0.17	0.21	−0.18	0.48

[a] Measurement Bias = 100 × [Measured–Accepted]/Accepted]
[b] Relative Standard Deviation is based on repeated measurement of the same item

reported in Table 23.6 were taken over a 1-year period. The within-facility variability and the bias of the calorimetric assay were calculated from results reported by each facility decayed to a common date. Each facility used different gamma-ray analysis codes for the isotopic measurements. For some, the reported values are the averages of measurements of the standard item with different calorimeters.

A breakdown of the CALEX data taken at five different DOE facilities in terms of the precision and bias of their respective power and P_{eff} measurements is presented in Table 23.7. The data were collected over an 8-year period—from 1990 to 1998. Not all facilities reported results each year; therefore, the averages contained data from a maximum of 8 years and a minimum of 5 years. Presented in Table 23.7 are the average percent measurement bias and percent RSD from repeated measurements for P_{eff} and item power. The percent bias and percent RSD are comparable for power and P_{eff} measurements on this item [3].

23.11 Calorimetric Assay Advantages

Calorimetric assay offers several distinct advantages over other NDA techniques and chemical analysis [3]:

- The calorimeter heat measurement is independent of material and matrix type; self-attenuation does not bias results.
- No physical standards representative of the materials being assayed are required.
- The thermal power measurement is traceable to the U.S. or other national measurement systems through electrical standards used to directly calibrate the calorimeters or to calibrate secondary ^{238}Pu heat standards.
- Calorimetric assay can be used to prepare secondary standards for neutron and gamma-ray assay systems [34, 35].
- The heat from the entire item volume is measured. As a result, the response of a well-designed calorimeter is independent of the location of the heat sources inside the measurement chamber.
- Calorimetry is very precise and nearly bias free. Biases can be quantitatively determined during instrument calibration.
- Calorimetry measurements are typically much longer than other NDA techniques with typical measurement times between 1 and 8 h.

References

1. J.G. Fleissner, M.W. Hume, *Comparison of Destructive and Nondestructive Assay of Heterogeneous Salt Residues*. RFP-3876 (1986)
2. V.L. Longmire, T.L. Cremers, W.A. Sedlack, S.M. Long, Isotopic ratios and effective power determined by gamma-ray spectroscopy vs mass spectroscopy for molten salt extraction residues, in *Nuclear Materials Management. XXXI Proceedings Issue* (1990)
3. D.S. Bracken, R.S. Biddle, L.A. Carrillo, P.A. Hypes, C.R. Rudy, C.M. Schneider, M.K. Smith, Application Guide to Safeguards Calorimetry. Los Alamos National Laboratory Report LA-13867-M (2002)
4. W. Strohm, S. Fiarman, R. Perry, A demonstration of the in-field use of calorimetric assay for JAEA inspection purposes. Nucl. Mater. Manage. **XIV**(3), 182 (1985)
5. E. Rutherford, H. Barnes, Heating effect of the radium emanation. Nature **68**, 622 (1903). https://doi.org/10.1038/068622a0
6. D.S. Bracken, C.R. Rudy, Principles and Application of Calorimetric Assay. PANDA Addendum, Los Alamos National Laboratory report LA-UR-07-5226 (2007)
7. Standard Test Method for Nondestructive Assay of Plutonium, Tritium, and ^{241}Am by Calorimetric Assay, ASTM Standard C 1458 (American Society of Testing and Materials, 2016)
8. Calibration Techniques for the Calorimetric Assay of Plutonium-Bearing Solids Applied to Nuclear Materials Control, ANSI N15.22–1975 (American National Standards Institute, Inc., New York, 1975) and 1986 revision
9. W. Rodenburg, An Evaluation of the Use of Calorimetry for Shipper-Receiver Measurements of Plutonium. Mound Laboratory report MLM-2518, NUREG/CR-0014 (1978)
10. Methods for Chemical, Mass Spectrometric, and Spectrochemical Analysis of Nuclear-Grade Plutonium Dioxide Powders and Pellets, ANSI N104–1973 (American National Standards Institute, Inc., New York, 1973)
11. Methods for Chemical, Mass Spectrometric, Spectrochemical, Nuclear, and Radiochemical Analysis of Nuclear-Grade Plutonium Metal, ANSI N572-1974 (American National Standards Institute, Inc., New York, 1974)
12. D. Reilly, N. Ensslin, H. Smith Jr., S. Kreiner, *Passive Nondestructive Assay of Nuclear Materials* (PANDA) (U.S. Nuclear Regulatory Commission, January 1, 1991)
13. F. O'Hara, J. Nutter, W. Rodenburg, M. Dinsmore, Calorimetry for Safeguards Purposes. Mound Laboratory report MLM-1798 (1972)
14. G. Jossens, C. Mathonat, G. Etherington, Large volume and sensitive calorimeter for nuclear applications: Latest developments for greater accuracy and shorter measurement times, in *Proceedings of the 52nd Annual Meeting of the INMM*, Palm Desert, CA, July 17–21, 2011
15. P. Hypes, P. Santi, W. Geist, Chapter 20: Analytical techniques in nuclear safeguards, in *Handbook of Radioactivity Analysis*, ed. by Michael L'Annunziata, 4th ed (Academic, 2020)
16. D.S. Bracken, P. Hypes, Solid-state calorimeter, in *Proceedings of the 41st Annual Meeting of the INMM*, New Orleans, LA, July 16–20, 2000
17. D.S. Bracken, R. Biddle, C. Rudy, Performance evaluation of a commercially available heat flow calorimeter and applicability assessment for safeguarding special nuclear materials, in *Proceedings of the 39th Annual Meeting of the INMM*, Naples, FL, July 26–30, 1998
18. D.S. Bracken, R. Biddle, R. Cech, Design and performance of a vacuum bottle solid-state calorimeter, in *Proceedings of the 38th Annual Meeting of the INMM*, Phoenix, AZ, July 20–24, 1997
19. C.L. Fellers, P.W. Seabaugh, Real-time prediction of calorimetric equilibrium. Nucl. Instrum. Methods **163**, 499 (1979)
20. R.A. Hamilton, Evaluation of the Mound Facility Calorimeter Equilibrium Prediction Program. Rockwell Hanford report RHO-SA-114 (1979)
21. M.K. Smith, D.S. Bracken, Calorimeter prediction based on multiple exponentials. Nucl. Instrum. Methods Phys. Res. Sect. A **484**, 668–679 (2002). https://doi.org/10.1016/S0168-9002(01)02054-X
22. R.B. Perry, S. Fiarman, Recent developments in fast calorimetry, in *Proceedings of the 29th Annual Meeting of the INMM*, Las Vegas, NM, June 26–29, 1988
23. C.L. Fellers, P.W. Seabaugh, Real-time prediction of calorimeter equilibrium. Nucl. Instrum. Methods **163**, 499 (1979)
24. R.L. Mayer, Application of prediction of equilibrium to servo-controlled calorimetry measurements, in *Proceedings of the 28th Annual Meeting of the INMM*, Newport Beach, CA, July 12–15, 1987

25. J.R. Wetzel, T.E. Sampson, T.L. Cremers, Calorimeter in the Aries recovery system, in *Proceedings of the 38th Annual Meeting of the INMM*, Phoenix, AZ, July 20–24, 1997
26. D.K. Hauck, P.A. Santi, Prediction algorithms for performing calorimetry measurements in about one hour, in *Proceedings of the 52nd Annual Meeting of the INMM*, Palm Desert, CA, July 17–21, 2011
27. W. Rodenburg, Some Examples of the Estimation of Error for Calorimetric Assay of Plutonium-Bearing Solids. Mound Laboratory report MLM-2407, NUREG-0229 (1977)
28. M.K. Smith, D. Bracken, C. Rudy, P. Santi, An analysis of the systematic components of calorimetry uncertainty, in *Proceedings of the 46th Annual Meeting of the INMM*, Phoenix, AZ, July 10–14, 2005
29. D.K. Hauck, S. Croft, D.S. Bracken, Expressing precision and bias of calorimetry, in *Proceedings of the 51st Annual Meeting of the INMM*, Baltimore, MD, July 11–15, 2010
30. Standard Test Method for Chemical, Mass Spectrometric, and Spectrochemical Analysis of Nuclear-Grade Plutonium Dioxide Powders and Pellets, ASTM Standard C 697–16 (2016)
31. T.L. Welsh, L.P. McRae, C.H. Delegard, A.M. Liebetrau, W.C. Johnson, M.S. Krick, J.E. Stewart, W. Theis, R.J. Lemaire, J. Xiao, Comparison of NDA and DA measurements techniques for excess Pu powders at the Hanford site: Operator and IAEA experience, in *Proceedings of the 36th Annual Meeting of the INMM*, Palm Desert, CA, July 9–12, 1995
32. B. Srinivasan, K.J. Mathew, U.I. Narayanan, W.F. Guthrie, T.E. Sampson, Plutonium accountability measurements by calorimetry and gamma spectrometry: Evaluation of measurement uncertainties and method performance. J. Radioanal. Nucl. Chem. **282**, 963–970 (2009)
33. Calorimetry Exchange Program, Quarterly/Annual Data Report, Calendar Year 1998, prepared by M. Irene Spaletto and David T. Baran, NBL-353, May 1999
34. In-Field Calibration of Neutron Correlation Counter via Calorimetry and High Count Rate Gamma-Ray Isotopic Abundance Measurements, IAEA-SM-293/126, Nuclear Safeguards Technology 1986 2, 239–249 (1987)
35. J.P. Lestone, T.H. Prettyman, J.D. Chavez, Performance of the skid-mounted tomographic gamma scanner for assays of plutonium residues at RFETS, in *Proceedings of the 41st Annual INMM Meeting*, New Orleans, LA, July 16–20, 2000

Open Access This chapter is licensed under the terms of the Creative Commons Attribution 4.0 International License (http://creativecommons.org/licenses/by/4.0/), which permits use, sharing, adaptation, distribution and reproduction in any medium or format, as long as you give appropriate credit to the original author(s) and the source, provide a link to the Creative Commons license and indicate if changes were made.

The images or other third party material in this chapter are included in the chapter's Creative Commons license, unless indicated otherwise in a credit line to the material. If material is not included in the chapter's Creative Commons license and your intended use is not permitted by statutory regulation or exceeds the permitted use, you will need to obtain permission directly from the copyright holder.

24. Calorimetric Assay Instruments

M. P. Croce, D. S. Bracken, X. Brochard, R. N. Likes, C. R. Rudy, and P. A. Santi

24.1 Introduction

Calorimeters for different measurement applications vary widely in design due to factors such as measurement power range, physical size of assayed items, facility requirements, precision and accuracy requirements, and measurement time limitations. This chapter discusses considerations for specific calorimeter systems and describes some instruments currently in use throughout the nuclear fuel cycle for the nondestructive assay (NDA) of nuclear materials.

The heat-flow calorimeter architecture in common use for assay of nuclear materials is based on a design described by Watson and Henderson in 1928 for the measurement of radium and its decay products [1]. The twin resistance bridge design developed in 1947 at the former Mound Laboratory proved to be a successful concept upon which many subsequent instruments were built. Research laboratories, universities, and industry have continued the development of calorimetric assay instrumentation for a wide range of applications, including analysis of small laboratory samples, nuclear material accounting, and spent fuel characterization.

24.1.1 Calorimetric Assay System Components

A typical system designed for the calorimetric assay of nuclear materials consists of the following components:

- A precisely machined heat-flow calorimeter body (An electric heater circuit is often built into the body of the calorimeter either for calibration or power-replacement mode operation.)
- Either a Wheatstone bridge circuit or solid-state thermoelectric sensors for the precise measurement of temperature differences between the sample chamber and the environmental heat sink (The electric circuitry includes digital voltmeters and, in the case of Wheatstone bridge sensors, a constant current source.)
- A computer to control calorimeter operation, data collection and storage, and data analysis

- Electrical or radioactive heat source calibration standards (Electrical measurements are used to determine temperature differences and to control electrical heater circuits for sample power duplication. Alternatively, radioactive heat source standards may be available to the system for calibration, bias correction, and power determination.)

24.1.2 Calorimeter Design Considerations

In the design of a calorimetric assay instrument, many important factors dictate the appearance of the final product. It is not possible to design a single, universal calorimeter that is applicable to all measurement situations. This section lists some of the factors that influence calorimeter design.

- *Sample size.* The physical size of the sample dictates the dimensions of the sample chamber; tight, thermal coupling of the sample to the calorimeter is essential to minimize assay time. Calorimeters have been made for items that range in size from small analytical laboratory samples to large waste drums.
- *Sample power.* In general, high-power samples require low-sensitivity calorimeters with low thermal resistance, and low-power samples require sensitive, small-sized calorimeters with high thermal resistance.
- *Calibration.* Calorimeter design is a function of the calibration method to be used—radioactive heat standards or electrical standards.
- *Construction materials.* The heat capacities and thermal conductivities of the materials used in construction influence performance.
- *Throughput.* The available time per assay influences the choice of calorimeter type and method of operation and the number of calorimeters needed.
- *Accuracy.* The desired assay accuracy must be weighed against the needed throughput and available space in choosing the type of calorimeter and the method of operation.
- *Facility environment.* Calorimeter design is affected by the working environment and the available floor space. An inline production facility environment requires different design considerations than a laboratory setting.

24.2 Small Calorimeters for Laboratory Samples

Small calorimeters can be used in analytical chemistry laboratories to provide NDA of small nuclear material samples. They can be used to determine effective specific power by the empirical method: a small sample of plutonium is assayed nondestructively by calorimetry and then destructively by chemistry. Also, small calorimeters provide a means for evaluating sampling errors.

24.2.1 Mound Analytical Calorimeter

The Mound analytical calorimeter was developed specifically for use in an analytical chemistry laboratory [2]. It is a compact instrument (Fig. 24.1) designed for sample sizes up to 5 cm^3 (~10 g of plutonium). The calorimeter includes an automatic sample loader with a pre-equilibrium position and is designed to fit inside a standard 3 ft. glove box. For samples that produce 10 mW of power, the precision and accuracy are better than 0.1%. Performance data for the analytical calorimeter are summarized in Table 24.1 [2, 3].

A wide variety of plutonium-bearing materials—including metal, oxide, and mixed oxide—have been measured in the analytical calorimeter, dissolved, and analyzed for plutonium content. Comparisons of the results demonstrated that calorimetric assay is both accurate and precise.

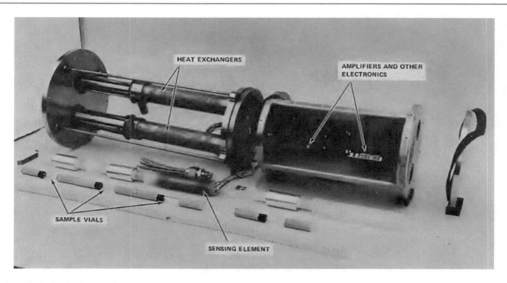

Fig. 24.1 Mound analytical calorimeter for measurement of small laboratory samples [2]

Table 24.1 Performance data for the Mound analytical calorimeter [2, 3][a]

Accuracy	0.02 mW
Precision	0.02 mW
Range	1–23 mW
Measurement sensitivity	1 mW in a 5 cm^3 container
Assay time	30–100 min

24.2.2 LANL Small-Sample Calorimeter

The Los Alamos National Laboratory (LANL) small-sample calorimeter was designed for the assay of items that could fit within a measurement chamber of 4 cm in diameter by 7.6 cm in height in the form of pellets and powders [4]. The Small Sample Calorimeter 3 (SSC3), shown in Fig. 24.2, consists of a matched pair of cylindrical thermopile heat-flow sensors that were manufactured by International Thermal Instruments Company.[1] The calorimeter is placed inside a small water bath. The unique feature of this system is its portability, which makes it useful for onsite measurements in the field. Because the thermopile heat-flow sensors are insensitive to vibration or mechanical strain, the system can be moved from one location to another without concerns of having to wait for the sensors to settle as is typical for calorimeters that use a Wheatstone bridge sensor design.

Performance data for the SSC3 are given in Table 24.2 [4]. The calorimeter has a precision that ranges from ~0.04% at 1 W to 1.6% at 1 mW [4] based on measurements of ^{238}Pu heat standards, electrical heat standards, and the measurement of a well-characterized sub-gram plutonium item. The high precision is due in part to the use of solid-state thermopiles in the construction of the calorimeter.

24.3 Transportable Calorimeters

Some calorimeters have been designed to be transportable for use at different nuclear fuel cycle facilities or within different physical locations within the same facility. This portability provides auditors and nuclear material inspectors the ability to verify inventory items independently and allows the facility to potentially reduce the number of calorimeters needed to meet their measurement needs.

---

[1] International Thermal Instruments Co., P.O. Box 309, Del Mar, CA 92014, http://www.iticompany.com

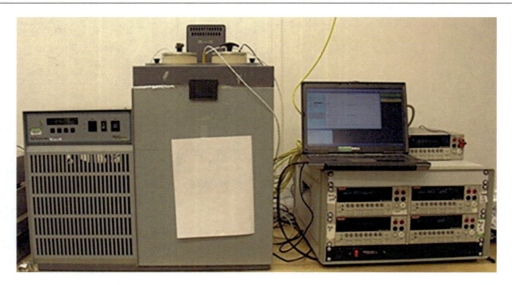

Fig. 24.2 (left) LANL small-sample calorimeter and (right) its data acquisition system. (Photograph taken from Ref. [4])

Table 24.2 Performance data for the LANL SSC3 [4]

Accuracy	0.05 mW
Precision	0.02–0.4 mW
Range	1 mW to 1.1 Watts
Measurement sensitivity	0.08 mW
Assay time	3–5 h

One of the key design considerations in enabling calorimeters to be transportable is the reduction of the physical size and weight associated with the calorimeter. A significant contributor to the size and weight of a calorimetry system is the external medium that is used to maintain a constant temperature around the calorimeter. For calorimetry systems that are intended to work within a water-based environment, reducing the amount of water associated with the calorimeter system can improve the transportability of the system as a whole.

The Mound transportable calorimeter described in Ref. [5] is an example of a calorimeter that was originally designed and built by Mound Laboratories for use with a large, static water bath. It is similar in design to the larger in-plant, twin-bridge systems described in Sect. 24.4.1 except that the calorimeter was placed in a 55-gallon steel drum, which reduced the volume of water needed from 200 gallons. Although the reduction in the volume of the water bath could, in theory, increase the effects of changes in room temperature on the precision of the calorimeter, the bath control system that consists of a solid-state heat exchanger, along with a circulatory pump and digital controller, was robust enough to maintain the water temperature to within 0.001 °C [5].

Figure 24.3 shows the calorimeter (within the steel drum) and its associated electronics housed in an aluminum cart (left). The system was designed so that it could be moved by one person. The sample chamber can accommodate samples up to 12.5 cm in diameter and 25 cm in height. The data acquisition system reads the bridge potential and monitors the temperature within the water reservoir.

The capabilities of this calorimeter were demonstrated in Ref. [5] by performing measurements on highly enriched uranium (HEU) samples that ranged in mass from 989 g to 3954 g, which corresponds to thermal powers between 1 mW and 7.9 mW. After the calorimeter was calibrated using ^{238}Pu heat standards that ranged from 0.0007–0.022 W [5], the calorimeter was able to measure HEU items with a precision of 12–18% relative standard deviation (RSD) for the lower-mass items and up to 0.92% RSD for the higher-mass item.

Fig. 24.3 Mound transportable calorimeter system showing electronics (left) and twin-bridge calorimeter (center; photograph taken from Ref. [5])

24.4 Calorimeters for Nuclear Material Accounting

24.4.1 Mound Twin-Bridge Production Calorimeter

The twin-bridge production calorimeter (Fig. 24.4) is designed for the assay of plutonium metal, oxide, and high-density scrap and waste [6]. The calorimeter is constructed for use in facility environments. Sample and reference cells are contained in a controlled, constant-temperature environmental water bath and are ~12 cm in diameter to accept standard cans in which plutonium materials are packaged. Calorimeters that can accommodate gallon-size cans of scrap and waste are also produced in this design. The instrument shown in Fig. 24.4 has two twin-bridge calorimeters inside the water bath.

Each twin-bridge calorimeter can include a servo controller for constant temperature operation, a sample pre-equilibrium chamber, or provisions for the use of end-point prediction techniques. These features can reduce assay time and increase throughput. All electrical measurements are made with bridge circuits or digital voltmeters and are fed directly into the computer data acquisition system. Standard resistors and voltage sources traceable to the National Institute of Standards and Technology (NIST) are used to ensure accurate measurements. Plutonium-238 heat standards are used for calibration.

Performance data for the twin-bridge calorimeter for low-burnup plutonium with an effective specific power of 2.3 mW/g are summarized in Table 24.3 [6, 7]. Similar performance data have been reported for high-burnup plutonium with an effective specific power of ~14.5 mW/g [8].

24.4.2 ARIES Gradient Calorimeter

One of the main practical considerations when selecting a calorimetry system for use within a nuclear material accounting program is how the system will physically exist with other equipment and processes within the facility. In the case of the LANL Advanced Recovery and Integrated Extraction System (ARIES), which dismantles the cores of nuclear weapons and converts the plutonium into oxide, the integration of a calorimetry system with the other NDA instrumentation needed to achieve both the required accuracy for material control and accountability and the capacity needed for process control required the calorimeter to be compact, with a minimal overall height [9]. This configuration was achieved through the selection of a gradient-bridge-type calorimeter, as seen in Fig. 24.5. The figure shows the calorimeter installed within the rigid framework that houses the other NDA instruments associated with the ARIES project, along with a robot that lifts the item at the staging area, loads the item into the various measurement stations, and then returns the item back to the staging area once the assay has

 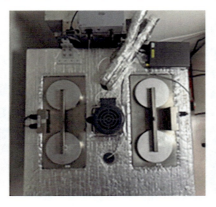 

Fig. 24.4 Mound twin-bridge production calorimeter. The water bath contains two twin-bridge calorimeters

Table 24.3 Performance data for the Mound twin-bridge production calorimeter [6, 7]

Accuracy	0.1–0.2%
Precision	0.1–0.2%
Range	0.23–5.8 W
Assay time	8–12 h

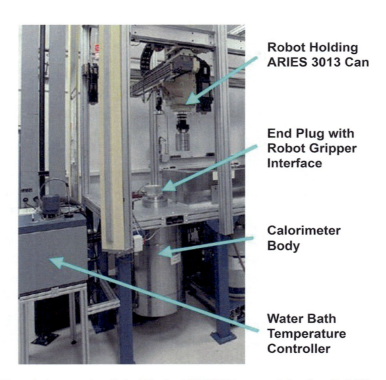

Fig. 24.5 ARIES gradient bridge calorimeter as installed within the ARIES NDA system (taken from Ref. [9])

been completed [9]. The gradient bridge calorimeter that is used with the ARIES NDA system can measure the full-size DOE standard 3013 can (12.5 cm diameter 25 cm tall; Ref. [10]). The calorimeter is equipped with internal heater windings that allow the calorimeter to operate in either servo or passive mode.

The design of the ARIES calorimeter body is similar to the gradient bridge design shown in Fig. 23.5 and described in Chap. 23, Sect. 23.6.3. The thermal baffles above and below the sample chamber are made of Styrofoam. Sample weight is transmitted to the support structure by a central post so that the accuracy of the bridge potential readings is not affected by

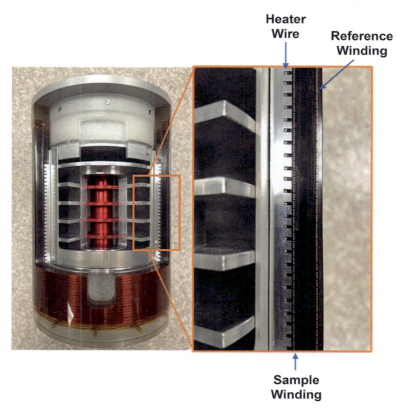

Fig. 24.6 (left) Cutaway model of a gradient bridge calorimeter with a calorimeter can inserted. (right) The heater, sample, and reference windings can be seen in this expanded view of the calorimeter body

Fig. 24.7 ARIES gradient-bridge calorimeter

strain in the windings. To help visualize the inner construction of a gradient calorimeter, Fig. 24.6 presents a cutaway of a gradient calorimeter to show both the sample and reference windings within the calorimeter, as well as the location of the heater windings that allow the system to operate in servo mode.

The ARIES calorimeter uses a temperature-controlled heat sink that consists of a water jacket through which water continuously flows from the temperature-controlled reservoir shown on the left of Fig. 24.7 [10]. The water jacket surrounds

Table 24.4 Performance of the ARIES gradient bridge calorimeter [10]

Accuracy	0.02%
Precision	0.08%
Range	1–12 W
Assay time	5–9 h

the calorimeter body. The lines used to bring water to and from the calorimeter can be better seen on the left side of the calorimeter. Using a temperature-controlled heat sink instead of a large water bath further reduces the footprint associated with the calorimeter and minimizes the amount of water that is present within this portion of the facility.

The performance of the ARIES calorimeter was determined and monitored using a combination of ^{238}Pu heat standards and the DOE Calorimetry Exchange (CALEX) standard—which is a PuO_2 item that contains 400 g of plutonium with 5.86% ^{240}Pu by weight—and NDA standards that were specifically created for the ARIES program for use with all of the NDA instrumentation [9]. Throughout the history of the ARIES program, the calorimeter has been operated in both passive and servo modes, with little difference in performance observed between the two operating modes. The performance of the ARIES calorimeter is listed in Table 24.4.

24.5 LANL-Mound Heat Standard Certification Calorimeter

Certification of ^{238}Pu heat standards at LANL is done with a calorimeter system developed by LANL using Mound calorimeter components. The system uses a combination of the replacement and differential methods to certify plutonium heat standards to NIST-traceable electronics. To meet this certification, the calorimeter needs to have the capability of heating both the sample and reference chamber with programmable heaters. To achieve NIST traceability, the meters used to measure the bridge potential and heater outputs need a current NIST certification. By performing multiple measurement sequences, this method produces high-fidelity results.

The ability to apply heat to both the sample and reference chambers provides this method with high-fidelity results and shorter runtimes. By heating both chambers during the baseline measurement before the sample is loaded, the system is establishing a unique baseline at the estimated wattage (W_{Est}) of the unknown sample. The wattage estimate is set within 1% of the predicted decay of the standard. The measured bridge potential of the baseline measurements incorporates systematic errors that would otherwise be omitted. Another advantage of supplying heat to the reference chamber during the measurement of a standard is that the bridge potential is being driven to zero. By keeping the bridge potential near zero, this method is using very small variances in its calculations and therefore increases the accuracy of the measurement. Another advantage to this method is that the sample chamber will be preheated at the completion of the baseline measurements. By preheating the sample chamber before the sample is loaded and driving the bridge potential to zero, measurement times are greatly reduced.

The following is a description of a full sequence consisting of three total measurements:

1. The estimated wattage (W_{Est}) is applied to both the sample and the reference chambers of the calorimeter and is allowed to come to thermal equilibrium while the sample chamber is empty. Power applied is continuously monitored for accuracy and stored as wattage actual (W_{a1}). Once thermal equilibrium is achieved, the baseline bridge potential value is stored as (BP_{b1}). Then the standard is loaded into the sample chamber of the calorimeter.
2. After the sample is loaded, the wattage estimate (W_{Est}) is applied to the reference chamber only, and the calorimeter is allowed to come to thermal equilibrium. Once thermal equilibrium is achieved, the bridge potential value is stored as (BP_S). The standard is then removed from the calorimeter.
3. The estimated wattage (W_{Est}) is applied to both the sample and the reference chambers of the calorimeter and allowed to come to thermal equilibrium while the sample chamber is empty. Power applied is continuously monitored for accuracy and stored as wattage actual (W_{a2}). Once thermal equilibrium is achieved, the baseline bridge potential value is stored as (BP_{b2}).

Then, BP_{b1} and BP_{b2} are averaged to BP_b:

$$\frac{BP_{b1} + BP_{b2}}{2} = BP_b. \tag{24.1}$$

The power W_s for the unknown standard for one sequence is

Fig. 24.8 The LANL-Mound calorimeter system used for certification of ^{238}Pu heat standards includes three sizes of calorimeters in a common water bath with an automated sample loader mechanism

$$W_s = W_{Est} + \frac{BP_s - BP_b}{S}, \qquad (24.2)$$

where S is the sensitivity of the calorimeter.

Also, if the system is not capable of supplying an exact power to the heaters, W_{a1} and W_{a2} can be averaged together to substitute for

$$\frac{W_{a1} + W_{a2}}{2} = W_{Est}. \qquad (24.3)$$

To certify a standard, multiple sequences are needed to ensure accuracy. The typical number of sequences is seven. Note that when running a subsequent sequence, the final baseline bridge potential BP_{bx} from the previous sequence is used as the first baseline bridge potential in the following sequence. Once all sequences are complete, all of the calculated sample wattages are averaged together for the final reported wattage.

The system, pictured in Fig. 24.8, includes three sizes of twin calorimeters to accommodate the range of ^{238}Pu heat standards. The largest calorimeter can accept items overpacked in 1-quart SAVY containers. An automated sample loader allows for continuous operation of the many individual measurements needed to certify a standard.

24.6 Large-Volume Calorimeter for Waste Drum Assay

The large-volume calorimeter (LVC; Ref. [11]) is capable of measuring the power output from a standard 208 L drum 60 cm in diameter. With special positioning considerations, cylindrical items of up to 66 cm in diameter and up to 100 cm long can be measured in the LVC. The LVC uses thermopile heat-flow sensors. The footprint of the calorimeter is 104 cm wide × 157 cm deep and 196 cm high in the closed position. The space for a standard electronics rack is also necessary.

The 208 L drums are lifted and placed onto the LVC pedestal using a drum handler. The pedestal is exposed by lifting the entire LVC shell and sensors. A photograph of the calorimeter in the open position is presented in Fig. 24.9 (left). The pedestal is a circular insulating plug of extruded polystyrene that prevents item heat leakage from the bottom of the calorimeter. The

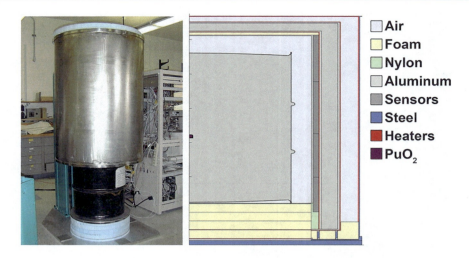

Fig. 24.9 The LVC in the open position with a 208 liter drum (black cylinder) visible within the calorimeter. (left). Radial cross sectional diagram of the LVC. (right)

calorimeter consists of three concentric cylinders closed on the top and open on the bottom for the insertion of the 208 L drums and pedestal.

The LVC uses two conductive temperature zones heated by silicone-rubber-encapsulated wire surface heaters to provide a constant reference temperature to the cold side of the thermopile heat-flow sensors. Temperature control is achieved through servo-controlled feedback loops for each heater. The temperature feedback signal is obtained from each heater through a four-wire resistance readout of a thermistor. The LVC does not use any water or other significant neutron-moderating or -reflecting materials for temperature control. The LVC does not have the ability to actively cool.

To maintain a relatively small overall size, the LVC does not use any compensating chamber to reduce thermal noise in the reference temperature. Drift of the reference temperature is the largest source of noise in the system.

24.7 Spent Fuel Assembly Calorimeters

The General Electric irradiated fuel storage facility near Morris, Illinois, has developed an in-basin calorimeter for underwater measurements of the heat generated by irradiated fuel assemblies [12]. The calorimeter is similar in size and shape to a water boil-off calorimeter developed earlier by Pacific Northwest Laboratory for above-water, hot-cell measurements [13]. The General Electric calorimeter is operated in the unloading pit of the fuel storage basin at a depth of ~40 ft. Although the calorimeter was developed to provide heat-generation information for planning future irradiated fuel storage needs, it was also possible to correlate the measured heat with fuel burnup.

The in-basin calorimeter (Fig. 24.10) is 4.6 m long and consists of two concentric steel pipes [12]. The 41 cm diameter inner pipe forms the sample chamber, which can be fitted with inserts to support either boiling-water-reactor (BWR) or pressurized-water-reactor (PWR) fuel assemblies. The annular space between the two pipes contains 6 cm of urethane insulation to reduce heat transfer from the calorimeter to the basin water. Temperature measurements inside the sample chamber and outside the calorimeter are made with platinum resistance temperature detectors. A recirculation pump maintains a homogeneous water temperature inside the sample chamber. Gamma radiation monitors are installed on the calorimeter to measure radiation heat losses and axial fuel assembly burnup profiles.

The calorimeter is usually operated with a constant-temperature environment. After a fuel assembly is loaded into the calorimeter, a water-tight head is bolted on to seal the sample chamber [12]. The rise in internal water temperature is monitored over a 5 h period. Water temperature outside the calorimeter is usually stable to 0.1 °F if water circulation is provided in the basin. The rate of change of internal water temperature (typically 2 °F/h) is proportional to the thermal output of the fuel assembly. The calorimeter is calibrated with a 4 m long pipe wrapped with electrical heater tape. All measurements are corrected for heater lead power losses, heat capacity variations between calibration and actual fuel, and gamma radiation heat losses.

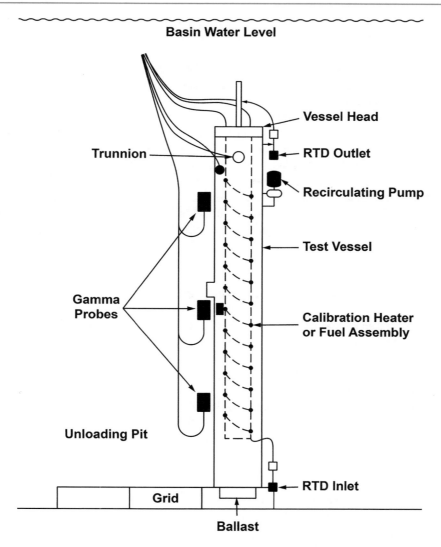

Fig. 24.10 General Electric-Morris Operations in-basin calorimeter for measurement of irradiated fuel assemblies in the unloading pit of the fuel storage basin. RTD = platinum resistance temperature detector. (Figure adapted from Ref. [12])

Reference [12] reports a series of 24 measurements of 14 PWR fuel assemblies with operator-declared burnups of 26–40 GWd/tU and cooling times of 4–8 years. The in-basin calorimeter measured thermal powers of 360–940 W with a precision of 1%. The measured power was compared with that calculated from reactor fuel burnup codes, and the agreement varied from 15% (if the codes assumed constant irradiation histories) to 1% (if the codes used the actual irradiation histories). For assemblies with the same cooling times, measured power was proportional to burnup to within ~3%.

A similar calorimeter was developed by Swedish Nuclear Fuel and Waste Management Company for use at the Swedish central interim storage facility for spent nuclear fuel (Clab) to determine the decay heat from assemblies before their encapsulation and final disposal [14]. The calorimeter was developed to evaluate three different calorimetric measurement techniques [14].

- *Temperature Increase Method.* This technique is similar to what was described in Ref. [12]—with the rate of change in the internal water temperature proportional to the thermal output of the assembly—except that the water within the calorimeter is cooled down below the pond temperature before starting the measurement. In addition, instead of measuring the rise in temperature over a certain period of time, the measurement is concluded based on a specified amount of change in temperature within the water of the calorimeter.
- *Recirculation method.* In this case, water is circulated from the pool through the calorimeter. When the system reaches thermal equilibrium, the temperatures of both the water outlet and inlet are measured, with the difference between the two temperatures being proportional to the thermal output of the assembly.

- *Equilibrium method.* In this traditional heat-flow calorimetry method, the calorimeter is isolated from the environment, and the temperature inside the calorimeter is measured when the system reaches thermal equilibrium.

Based on the results of their measurements, the equilibrium method was ruled out as a viable option due to each measurement taking at least 1 week per fuel assembly [14]. Because the recirculation method required more attention from the operator than the temperature increase method, for practical purposes, the temperature increase method was chosen, which allowed for one assembly to be measured per day. Due to the location of the calorimeter within one of the fuel-handling pods at Clab, the pool temperature was not kept constant and depended on both the total decay heat contained within the handling pool as well as the temperature of the sea water outside the facility used for cooling [15]. Based on more than 100 total calorimetry measurements of spent fuel assemblies (which included both BWR and PWR assemblies), the results have a standard deviation between 0.9% and 4.2% depending on the fuel type and power, with better results being observed with higher-power assemblies [14]. An effort to improve the performance of the calorimeter applied equilibrium predication algorithms to measurements that were performed using the equilibrium method [15]. Although this effort yielded a measurable reduction in assay time for measurements performed using the more precise equilibrium method, the reduced assay times were still too long to be practical [15].

24.8 Future Application: Advanced Thermo-Hydraulic Power Monitor

The basic principle of the advanced thermo-hydraulic power monitor (ATPM) is that the total energy produced in the reactor can be calculated based on the reactor coolant return temperature, the reactor coolant supply temperature, and the coolant volumetric flow rate to the reactor core.

This calculation can be used to determine if substantial amounts of fissile material might have been generated in the reactor or to confirm the declared operation of the reactor.

The total energy produced in the reactor is determined by the following formula:

$$E_{Produced} = \Delta T \times \varphi(Coolant) \times \rho_{Water} \times c_{Water}, \quad (24.4)$$

where

$E_{Produced}$: the total energy produced in the reactor (W)
ΔT: the temperature difference between the coolant supply temperature and the return temperature (°C) measured at the beginning and the end of the cooling loop, respectively
$\varphi(Coolant)$: the volumetric flow rate through the reactor core (L/s)
ρ_{Water}: the density of water (0.9965 kg/L)
c_{Water}: the heat capacity of the water (4181.3 J/kg/°C)

The measurement principle is as follows:

1. Two transducers are mounted onto the pipe, which is completely filled with the fluid (Fig. 24.11).
2. The ultrasonic signals are emitted alternately by one transducer and received by the other. This process can be seen in Fig. 24.11, where the darker arrows from the transducer on the left propagate to the transducer on the right, and conversely, the lighter arrows from the transducer on the right propagate to the transducer on the left.
3. The physical quantities are determined from the transit times of the ultrasonic signals [16].

As the fluid flows where the ultrasound signals propagate, the transit time of the ultrasonic signal in the direction of flow is shorter than the transit time against the flow direction. The transit time difference Δt (Fig. 24.12) is proportional to the average coolant velocity along the propagation path of the ultrasonic signals. A flow profile correction is then performed to obtain the area average-flow velocity, which is proportional to the volumetric flow rate. The integrated microprocessors control the entire measuring cycle. The received ultrasonic signals are checked for measurement usability and evaluated for their reliability. Noise signals are eliminated.

The ATPM is a complete system that consists of ultrasonic transducers and temperature sensors mounted on the primary cooling loop elements, a cabling system for bringing the detector signals from the measurement locations to an electronics

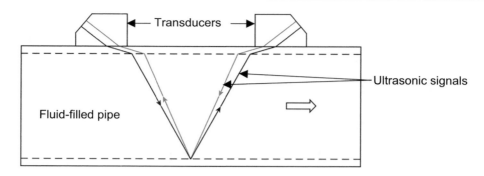

Fig. 24.11 Path of the ultrasonic signals in the flowing fluid

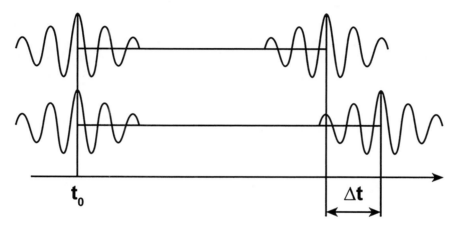

Fig. 24.12 Illustration of the measurement concept of transit time difference (Δt)

cabinet in a low-radiation-level area, electronics for data gathering and retrieval, and software for data management and analysis. The system requires AC (alternating current) power only at the electronics enclosure and has a substantial battery-backup capability. The electronics enclosure is a tamper-indicating 19 in. industrial cabinet, which is fixed in place and can be located up to 300 m from the measurement location.

The sensor enclosures are mounted on two separate sections of the non-insulated, single-walled pipe of the cooling loop and are protected by sealed metal covers. One set of temperature probes is mounted on the hot coolant supply of the cooling loop and the ultrasonic transducers, and the second set of temperature probes is mounted on the cold coolant that returns to the reactor. Initial calibration of the system is based on operator-declared information.

ATPM systems can be single or dual channel to monitor one or two cooling loops, respectively. In the standard configuration, two pair of ultrasonic transducers are connected by means of coaxial cables to redundant data acquisition units inside the collect cabinet. A general system layout is shown in Fig. 24.13. Data from the data acquisition are saved to files on a collect computer inside the cabinet, copied to removable media, and analyzed to monitor the power of the nuclear research reactor. Data can be collected and analyzed by inspectors during servicing visits to the facility or through remote monitoring from International Atomic Energy Agency Headquarters, where available.

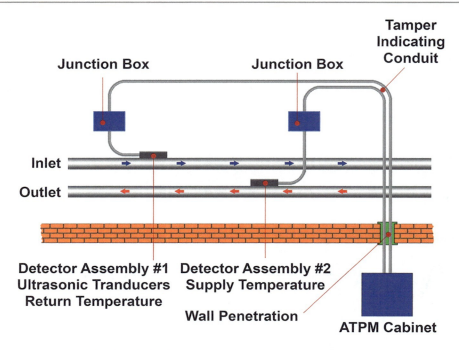

Fig. 24.13 Diagram of a typical advanced thermo-hydraulic power monitor system configuration

References

1. S.W. Watson, M.C. Henderson, The heating effects of thorium and radium products. Proc. R. Soc. A: Math. Phys. Eng. Sci. **118**(779), 318–334 (1928)
2. C. Fellers, W. Rodenburg, J. Birden, M. Duff, J. Wetzel, Instrumentation Development for the Enhanced Utilization of Calorimetry for Nuclear Material Assay, in *Proceeding of the American Nuclear Society Topical Conference on Measurement Technology for Safeguards and Materials Control*, Kiawah Island, South Carolina, November 26–28, 1979, NBS Special Publication 582 (1980), p. 192
3. W. Rodenburg, An Evaluation of the Use of Calorimetry for Shipper-Receiver Measurements of Plutonium, Mound Laboratory report MLM-2518, NUREG/CR-0014 (1978)
4. P. Santi, K. Perry, Calibration and Characterization of the Small Sample Calorimeter, Los Alamos National Laboratory report LA-UR-12-24,053 (2012)
5. C. Rudy, D.S. Bracken, P. Staples, L. Carrillo, R. Cech, M. Craft, J. McDaniel, D. Fultz, Transportable calorimeter measurements of highly enriched uranium, in *Proceeding of the 38th Annual Meeting of the INMM*, Phoenix, AZ, July 20–24, 1997
6. F. O'Hara, J. Nutter, W. Rodenburg, M. Dinsmore, Calorimetry for Safeguards Purposes, Mound Laboratory report MLM-1798 (1972)
7. W. Rodenburg, Some Examples of the Estimation of Error for Calorimetric Assay of Plutonium-Bearing Solids, Mound Laboratory report MLM-2407, NUREG-0229 (1977)
8. W. Strohm, W. Rodenburg, R. Carchon, Demonstration of the Calorimetric Assay of Large Mass, High Burn-Up PuO_2 Samples, *Nuclear Materials Management XIII (Proceedings Issue)* 269 (1984)
9. T.E. Sampson, T.L. Cremers, The ARIES Nondestructive Assay (NDA) system, in *Proceeding of the 42nd Annual Meeting of the INMM*, Indian Wells, CA, July 15–19, 2001
10. T.E. Sampson, T.L. Cremers, T.R. Wenz, W.J. Hansen, N.L. Scheer, T.A. Kelley, W.C. Harker, D.R. Mayo, D.S. Bracken, G.D. Herrera, B.A. Gullien, The ARIES Nondestructive Assay System: Description and Process Demonstration Results, Los Alamos National Laboratory report LA-14143 (2004)
11. D.S. Bracken, Performance testing of a large volume calorimeter, in *Proceeding of the 45th Annual Meeting of the INMM*, (Orlando, FL, July 18–22, 2004)
12. B. Judson, J. Doman, K. Eger, Y. Lee, In-Plant Test Measurements for Spent-Fuel Storage at Morris Operation. General Electric Company report, NEDG-24922-3 (1982)
13. J. Creer, J. Shupe, Jr., Development of a Water Boil-Off Spent-Fuel Calorimeter System, Pacific Northwest Laboratory report PNL-3434 (1981)

14. F. Sturek, L. Agrenius, O. Osifo, Measurements of decay heat in spent nuclear fuel at the Swedish interim storage facility, Clab, Svensk Kärnbränslehantering AB, SKB-report R-05-62, ISSN 1402–309 (2006)
15. H. Liljenfeldt, P.A. Santi, Applying fast calorimetry analysis on spent fuel calorimeter – 15620, in *Proceeding of the Waste Management Conference*, Phoenix, AZ, March 15–19, 2015
16. L.C. Lynnworth, Y. Liu, Ultrasonic flowmeters: Half-century progress report 1955–2005. *Ultrasonics* **44**(Supplement), e1371–e1378 (2006). https://doi.org/10.1016/j.ultras.2006.05.046

Open Access This chapter is licensed under the terms of the Creative Commons Attribution 4.0 International License (http://creativecommons.org/licenses/by/4.0/), which permits use, sharing, adaptation, distribution and reproduction in any medium or format, as long as you give appropriate credit to the original author(s) and the source, provide a link to the Creative Commons license and indicate if changes were made.

The images or other third party material in this chapter are included in the chapter's Creative Commons license, unless indicated otherwise in a credit line to the material. If material is not included in the chapter's Creative Commons license and your intended use is not permitted by statutory regulation or exceeds the permitted use, you will need to obtain permission directly from the copyright holder.

Basics of Uncertainty

25

J. Stinnett, P. A. Santi, and M. T. Swinhoe

Rigorous discussion of uncertainties is the subject of hundreds of textbooks, but a basic understanding can be built within one chapter. This chapter aims to describe the vocabulary of counting statistics, the fundamentals of uncertainty models, and the basics of error propagation applied to typical problems and provides brief discussions on particular issues for nondestructive assay (NDA) applications. Significantly more detail can be found on any of these topics in the references, particularly in Bevington's "Data Reduction and Error Analysis for the Physical Sciences" [1].

25.1 Introduction to Uncertainty

In a world without uncertainty, measurements could exactly yield the "true value" of the target variable; an enrichment meter would report 3.750000000 wt% ^{235}U, and that would exactly reflect the true enrichment of the item. In reality, measurements are subject to both *systematic errors* and *random errors*. These errors cause the measured values to differ from the true value. Systematic errors, or *biases*, cause measurement results to differ from the true values in reproducible ways and are generally due to poor measurement design, a failure to follow measurement procedures, or mistakes in calculations. For instance, for the infinite-thickness uranium enrichment measurement technique discussed Chap. 8, Sect. 8.3, if the calibration is done incorrectly (e.g., the container thickness of the calibration measurement is entered as 1.0 mm of steel instead of 1.5 mm): All of the subsequent measurements will be biased by a multiplicative factor of $e^{-0.05\mu\rho}$ (approximately 0.94, meaning all of the results will be biased by 6%). Even if the measurement is repeated many times, the average result will be significantly different from the true value. In general, measurement plans are designed to eliminate systematic errors.

Random errors are the random fluctuations in measurements that arise from statistical chance. Random errors can be minimized but can never be eliminated entirely. For example, suppose the activity of a radiation source were to be measured with a hypothetical 100% efficient detector, meaning that every single decay would be observed. For all nuclear decay processes, the decay rate is the inverse of the average time between decays. Because radioactive decay is a random process (specifically, a Poisson process), suppose the time between the first and second decays is 1.95 μs. If that is all of the data we have, we might conclude that the activity of the source is 0.51 MBq. But suppose the next decays give the sequence 1.95, 0.27, 0.63, 2.99, 0.21, 0.15, 0.28, 1.85, 1.43, and 0.52 μs. Based on this expanded data, the average time between decays is 1.03 μs, which works out to 1.08 MBq of radioactive material. Intuitively, the answer based on more data is likely the better answer, but to answer how it is better, uncertainty analysis is needed.

Because random and possibly systematic errors affect our measurements, not only will measured values differ from true values, but measured values from repeated or different measurements will differ from each other. Understanding these differences is one of the motivations for performing the uncertainty analyses described in this chapter. For example, suppose a

---

Los Alamos National Laboratory strongly supports academic freedom and a researcher's right to publish; as an institution, however, the Laboratory does not endorse the viewpoint of a publication or guarantee its technical correctness.

J. Stinnett (✉) · P. A. Santi · M. T. Swinhoe
Los Alamos National Laboratory, Los Alamos, NM, USA
e-mail: stinnett@lanl.gov; psanti@lanl.gov; swinhoe@lanl.gov

certified reference material that contains 1.00 kg of effective ^{240}Pu mass is measured in the High-Level Neutron Coincidence Counter (see Chap. 19, Sect. 19.4.2), which gives a measurement result of 1.20 kg. Does the discrepancy of 0.20 kg indicate a problem with the measurement, or is it consistent with uncertainty arising from random errors? Without more information and without performing uncertainty analysis, the answer is not clear.

In general, the term *uncertainty* is not always used consistently. The Guide to the Expression of Uncertainty in Measurement (GUM) uses uncertainty to refer to both the general concept of doubt in a measurement result as well as various quantitative measures of the concept [2]. The traditional approach to uncertainty in safeguards has used different terminology and apparently different concepts than the GUM approach. However, a group of experts has published a paper that compares the two approaches, showing how they can be reconciled [3]. This paper also gives a useful description of uncertainties in safeguards and how they are used.

The terms *accuracy* and *precision* are frequently used in discussions of uncertainty but not always correctly. Accuracy is a measure of how close a measurement is to the true value, whereas precision is a measure of the reproducibility of a result. In other words, an accurate measurement has little to no effect from biases, whereas a precise measurement has a small random error.

Finally, random variables are said to be "independent" if one does not affect the probability of the other. For example, if one flips a fair coin multiple times, the outcome of a flip does not affect the others, so each outcome is independent of the others. This concept of independence is related to but not the same as correlation. Correlation describes how much the deviation of one random event from the expected value affects the deviation of another random event from its expected value.

Although these concepts sound similar, consider two random variables Y_1 and Y_2:

- Y_1 is uniformly distributed on $[-1,1]$, meaning it can take any value in this range with equal probability. The expected (or average) value of Y_1 is zero.
- $Y_2 = Y_1^2$

These two random variables are clearly not independent but, perhaps surprisingly, they are uncorrelated. For any given value of Y_2, there are still two possible values for Y_1 with equal probability, and the expected (average) value of Y_1 given a value of Y_2 is still zero. Because the expected value of Y_2 is the same given any value of Y_1, these two variables are uncorrelated.

25.2 Uncertainty Models

Different mechanisms for errors can impact measurement results in significantly different ways. Consider a mechanical mass scale. If the scale is improperly tared, all of the measurements would be biased by a fixed amount. On the other hand, if the scale is calibrated incorrectly, any measurement could be biased by a fixed but unknown percentage. The former is an "additive error," whereas the latter is a "multiplicative error," and both are examples of systematic errors.

In the additive model, the observed value Y_i is the sum of an unknown systematic bias a, the true value μ_i, and a random error term ϵ_i.

$$Y_i = a + \mu_i + \epsilon_i. \tag{25.1}$$

In the multiplicative model, the systematic effect a is instead assumed to be multiplicative

$$Y_i = (1+a) \cdot \mu_i \cdot e^{\epsilon_i}. \tag{25.2}$$

However, with a simple transformation, this model can be reverted to the additive model

$$ln(Y_i) = ln(1+a) + ln(\mu_i) + \epsilon_i. \tag{25.3}$$

25.3 Probabilities, Means, and Standard Deviations

Consider the outcome of any measurement. This measurement y_i is an observation of the true value y, but due to random errors, there will be a non-zero discrepancy. Instead of performing a single measurement, suppose we instead took a series of measurements, $y_1, y_2, \ldots y_n$. What information can this series convey that a single measurement might not? Let us explore the outcome of a series of 25 independent mass measurements of the same certified 1.00 kg item.

The effects of measurement uncertainty can be clearly seen in Fig. 25.1. Although the true mass is not changing, the measurement results are randomly varying.

More information can be extracted using two common statistics: the sample mean and the sample standard deviation.

$$\text{Sample mean} = \bar{y} = \frac{1}{n} \sum_{i=1}^{n} y_i \qquad (25.4)$$

$$\text{Sample standard deviation} = s = \sqrt{\frac{1}{n-1} \sum_{i=1}^{n} (y_i - \bar{y})^2} \qquad (25.5)$$

These statistics can be used for better estimates of the true value from the measurement result. In this case, a better estimate for the mass is 1.01 ± 0.09 kg. These statistics will be discussed in more depth in Sect. 25.3.1 on the Central Limit Theorem. They also convey some information about the sample distribution of the data, which can be more easily visualized with the histogram in Fig. 25.2.

The sample distribution is only an approximation of the "true" or "parent" distribution. How well the sample distribution approximates the true distribution depends on the number of samples or measurements that have been taken. If an infinite number of samples could be taken, then the sample distribution would exactly match the true (or parent) distribution.

For many NDA measurements, this parent distribution is the Gaussian distribution, also called the *normal* or *bell curve distribution*. The probability density function (pdf) of a Gaussian-distributed random variable x with mean μ and standard deviation σ is given by

$$f(x) = \frac{1}{\sqrt{2\pi\sigma^2}} \exp\left(-\frac{(x-\mu)^2}{2\sigma^2}\right). \qquad (25.6)$$

A common shorthand notation for a Gaussian-distributed random variable X with mean μ and variance σ^2 is

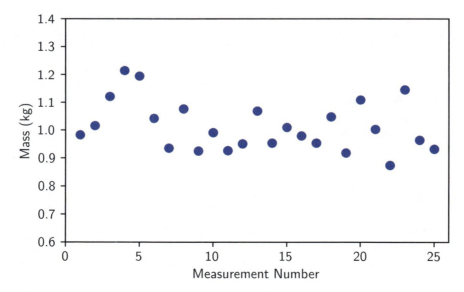

Fig. 25.1 Results from 25 independent measurements of the same certified 1.00 kg item. This sequence has a mean 1.01 kg and sample standard deviation of 0.09 kg

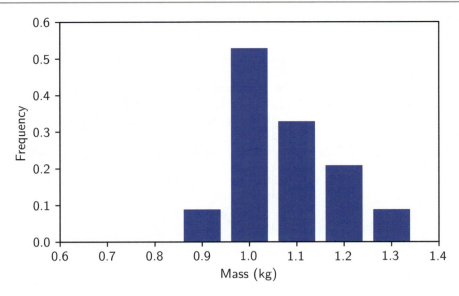

Fig. 25.2 Histogram of 25 measurements of a 1 kg item

$$X \sim N(\mu, \sigma^2). \tag{25.7}$$

The probability density function describes the relative likelihood of obtaining a sample x_i. A probability density function $f(x)$ has a few important properties. First, for any observation x_i of an observation from the parent distribution f(x), the probability of x_i being in the interval [a,b] is given by

$$Probability(a \leq x_i \leq b) = \int_a^b f(x)dx. \tag{25.8}$$

Second, a probability density function must be "normalized," meaning that the total area under the curve is equal to 1.

$$\int_{-\infty}^{\infty} f(x)dx = 1. \tag{25.9}$$

Applying a Gaussian fit to the data gives the mean and sample standard deviation, which are estimates of these quantities for the parent distribution. This fit is shown in Fig. 25.3.

Intuitively, one might expect the shape of the distribution to change as the number of data points increases, with the shape becoming more fixed as the number of data points approaches infinity. Although it is not possible to perform an infinite number of measurements, suppose these mass measurements were conducted 10,000 times. The distribution of these data, along with the Gaussian fit, is shown in Fig. 25.4.

With a very large number of measurements, the sample distribution very closely resembles the parent distribution, though some random variation can still be observed in Fig. 25.4.

25.3.1 Law of Large Numbers and Central Limit Theorem

Previously, the sample mean and sample standard deviation were defined. It is important to understand that these are not the same as the mean and standard deviation of the parent distribution. The sample mean and sample standard deviation will approach the true mean and standard deviation as the number of measurements goes to infinity; this statement is the *Law of Large Numbers*.

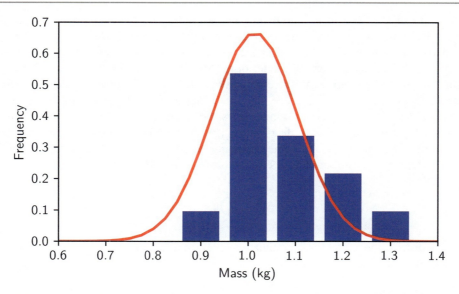

Fig. 25.3 Histogram of 25 measurements of a 1 kg item. A Gaussian fit is shown, with mean 1.01 kg and standard deviation 0.09 kg

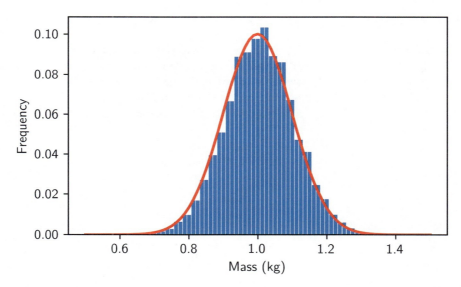

Fig. 25.4 Histogram of 10,000 measurements of a 1 kg item. A Gaussian fit is shown, with mean 0.999 kg and standard deviation 0.010 kg

$$Mean = \mu = \lim_{n \to \infty} \frac{1}{n} \sum_{i=1}^{n} y_i \qquad (25.10)$$

$$Standard\ deviation = \sigma = \lim_{n \to \infty} \sqrt{\frac{1}{n} \sum_{i=1}^{n} (y_i - \mu)^2} \qquad (25.11)$$

Note that the standard deviation is the square root of the variance (noted as σ^2). One common error is using the standard deviation instead of the sample standard deviation when evaluating a sequence of measurement data. Note that the sample standard deviation (Eq. 25.5) has a factor of *(n−1)* instead of *n* because the sample mean $\bar{y}$ was determined from the observed data. Correctly applying the *(n−1)* factor produces an unbiased estimate of the standard deviation of the underlying parent distribution, which is explained in more detail in Chapter 6 of Ref. [4].

In general, the objective of a measurement is to obtain the best possible measurement of the true value μ, though with practical limitations (such as number of replicate measurements, equipment and method used, measurement time, etc.), as well as to have some understanding of how good of a measurement that is. An important result in probability theory is the Central Limit Theorem, which establishes that the sum of a series of random variables tends to a Gaussian distribution regardless of the parent distribution of the random variables. There are multiple forms to this theorem, but the most useful one is this: consider a sequence of measurements $y_1, y_2, \ldots y_n$ that are independent and identically distributed, where the parent distribution has finite variance σ^2. Then, as the number of samples approaches infinity, sample mean $\bar{y}_n$ approaches a Gaussian random variable regardless of the parent distribution. The distribution of these sample means, $\bar{y}_n$, is referred to as the *sampling distribution*, which is centered about the true population mean and has variance scaled by sample size. The square root of the variance of the sampling distribution is referred to as the *standard error*.

$$\bar{y}_n \to N\left(\mu, \frac{\sigma^2}{n}\right) \tag{25.12}$$

In addition, the sampling distribution allows for the estimation of how much $\bar{y}_n$ is changing as the number of samples increases since the standard error decreases with sample size n.

25.3.2 More on Standard Deviations

The standard deviation is broadly used when discussing measurement precision, alarm thresholds, acceptance criteria, and more.

For example, a measurement result can be reported with an accompanying uncertainty, e.g., 1.0 ± 0.1 kg. The uncertainty reported here is generally equal to the standard deviation, though the convention for some professions is to report 2σ uncertainties; it is important to understand how these uncertainties are being quoted. Consider some Gaussian random variable y with mean $\mu = 100$ and a standard deviation $\sigma = 10$, as illustrated in Fig. 25.5.

As indicated in Fig. 25.5, the probability that a measurement of a Gaussian random variable will yield a result within the range of $\mu \pm 1\sigma$, $\mu \pm 2\sigma$, or $\mu \pm 3\sigma$ are as follows:

$$P(\mu - \sigma < y < \mu + \sigma) \approx 68.3\%$$

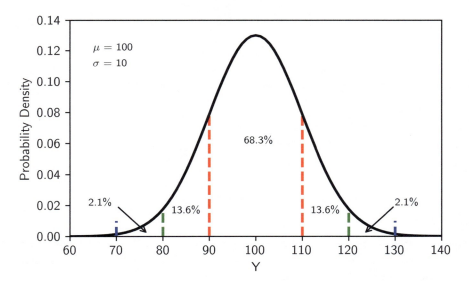

Fig. 25.5 Illustration of a Gaussian (or normal) distribution with a mean $\mu = 100$ and a standard deviation $\sigma = 10$. The region between the red lines indicates the portion of the curve covered by $\pm 1\sigma$ from the mean, whereas the region between the green lines represents the area covered when the limits are expanded to $\pm 2\sigma$, and the region between the blue lines represents the area when the limits are expanded to $\pm 3\sigma$

$$P(\mu - 2\sigma < y < \mu + 2\sigma) \approx 95.4\%$$

$$P(\mu - 3\sigma < y < \mu + 3\sigma) \approx 99.7\% \tag{25.13}$$

This property of the normal distribution is known as the empirical rule. Because σ is a measure of dispersion, quoting precisions this way conveys a confidence interval in the result. In the example report of 1.0 ± 0.1 kg, assuming it is a 2-standard-deviation report, there is a 95.4% chance that the true mass is in the range of 0.8–1.2 kg. This result means that there is a 4.6% chance that the mass is outside of this range. It should be noted the empirical rule does not extend to all other distributions.

Alarm thresholds are very commonly set such that a result that is 3σ or 5σ above the average will trigger an alarm.

$$(y > \mu + 3\sigma) \approx 0.13\%$$

$$(y > \mu + 5\sigma) \approx 3 \times 10^{-5}\% \tag{25.14}$$

Choosing an alarm threshold is a balance between the detection probability and the false alarm probability. With a 3σ threshold, on average, one in every 741 measurements will cause a false alarm. If this average is based on a count rate as with many radiation portal monitors, which might update every second, this average would cause several alarms per hour. On the other hand, if it is an infrequently performed measurement, a lower threshold may be acceptable.

25.3.3 The Poisson Distribution

Previously, the probability density function and the Gaussian distribution were defined, but another kind of distribution occurs very frequently with NDA. Whereas the probability density function allowed the random variable to take a continuous range of values, some probability functions describe discrete random variables, like the number of counts in a channel. These kinds of functions are called *probability mass functions* or *pmfs*. For a probability mass function $f(k)$

$$Probability(X = k) = f(k) \tag{25.15}$$

Like the probability density functions, any probability mass function $f(k)$ must also be normalized

$$\sum_k f(k) = 1. \tag{25.16}$$

The most frequently encountered pmf in NDA is the Poisson distribution. The Poisson distribution is a discrete, non-negative distribution that describes the probability of a given number of events that occur over a given time interval. In NDA applications, this concept applies broadly to count data, such as the number of counts in a channel or region of a gamma-ray spectrum or the number of neutrons detected during a measurement.

$$f(k; \lambda) = P(X = k) = \frac{\lambda^k e^{-\lambda}}{k!} \; for \; k = 0, 1, 2\ldots \tag{25.17}$$

This distribution has several interesting properties. The average number of events expected, λ, is also equal to the variance

$$\lambda = \sigma^2. \tag{25.18}$$

For measurements that involve counting discrete events, such as counting the number of neutrons detected over a period of time or counting the number of events within a specific channel of a gamma-ray spectrum, the average number of events expected is the number of counts in the detector or in the channel (λ = N, where N is the number of counts). Therefore, from Eq. 25.18, we can see that the standard deviation in a measurement where events is counted is equal to the square root of the number of counts:

$$\sigma = \sqrt{\lambda} = \sqrt{N}. \qquad (25.19)$$

If two or more different Poisson random variables are summed together, the resulting sum is also a Poisson random variable. Therefore, when adding together the number of counts from different channels that lie within a region of interest in gamma-ray spectroscopy, the resulting sum (often referred to as the *Integral*) is also a Poisson random variable that has an uncertainty equal to the square root of sum of counts.

Suppose a radiation source was measured that, on average, produced 12 cps on a detector and that a sequence of 1 s measurements was performed. These data will be Poisson distributed with a rate parameter of 12. Figure 25.6 shows the first 60 s of this hypothetical simulation.

As before, it is easier to glean information by looking at a histogram of all of the data rather than a sequence chart (Fig. 25.7).

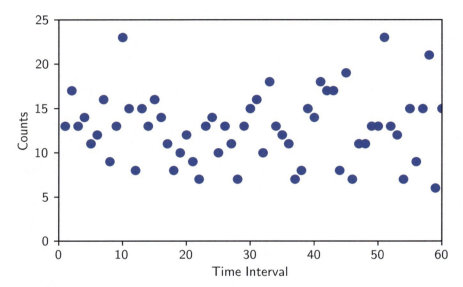

Fig. 25.6 First 60 s of count data

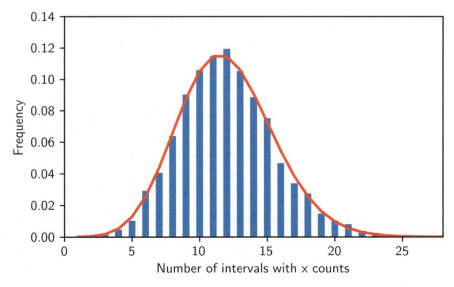

Fig. 25.7 Histogram that summarizes 3600 independent 1 s measurements. A Poisson fit was computed using the sample mean (12.00) as the rate parameter and is displayed in red

Notice that unlike the Gaussian distribution, the Poisson distribution is "skewed," meaning that the distribution is not symmetric about the mean value. It is more probable for a random sample to be below the mean than above the mean.

25.4 Bottom-Up Uncertainty and Error Propagation

Often the quantities of interest are not directly measured. Perhaps the activity of a source needs to be computed from a single measurement, or the total nuclear material mass in a vault needs to be computed from separate measurements of each source. It is necessary to correctly account for all sources of uncertainty in the final result. In this section, the "bottom-up" uncertainty approach is discussed. In this approach, one starts with uncertainties in each of the parameters (the "bottom" of the process) and the effects are propagated "up" to the quantity of interest. This approach is also sometimes called a *first principles* approach because the measurement uncertainty is derived from a model of the process.

25.4.1 Error Propagation

To compute the uncertainty in a computed value $f(y_1, y_2, \ldots y_n)$ from the uncertainties in each of the variables, the "error propagation equation" is

$$\sigma_f^2 = \sum_{i=1}^{n} \sigma_{y_i}^2 \left(\frac{\delta f}{\delta y_i}\right)^2 + 2 \sum_{i=1}^{n-1} \sum_{j=i+1}^{n} \sigma_{y_1 y_2}^2 \left(\frac{\delta f}{\delta y_i}\right) \left(\frac{\delta f}{\delta y_j}\right). \tag{25.20}$$

When the input variables $y_1, y_2, \ldots y_n$ are independent, this result simplifies to

$$\sigma_f^2 = \sum_{i=1}^{n} \sigma_{y_i}^2 \left(\frac{\delta f}{\delta y_i}\right)^2. \tag{25.21}$$

A derivation of this equation, as well as further explanation, can be found in Chap. 3 of Ref. [1].

This approach is widely used in NDA applications, where a desired value is computed from measured values and nuclear data. For example, consider a measurement of total plutonium mass from a passive neutron assay (giving the ^{240}Pu effective mass m_{240}) and isotopics derived from high-resolution gamma spectroscopy (giving the ^{240}Pu effective fraction Pu_{240eff}).

$$m_{Pu} = \frac{m_{240}}{^{240}Pu_{eff}}. \tag{25.22}$$

Assume that the uncertainties in each of the values on the right are known. Then applying error propagation to this gives

$$\sigma_{m_{Pu}}^2 = \sigma_{m_{240}}^2 * \left(\frac{\delta m_{Pu}}{\delta m_{240}}\right)^2 + \sigma_{^{240}Pu_{eff}Pu_{240eff}}^2 \cdot \left(\frac{\delta m_{Pu}}{\delta^{240}Pu_{eff}}\right)^2 \tag{25.23}$$

$$\sigma_{m_{Pu}}^2 = \sigma_{m_{240}}^2 \cdot \left(\frac{1}{^{240}Pu_{eff}}\right)^2 + \sigma_{^{240}Pu_{eff}}^2 \cdot \left(\frac{m_{240}}{^{240}Pu_{eff}^2}\right)^2 \tag{25.24}$$

This equation is often rearranged to give

$$\frac{\sigma_{m_{Pu}}^2}{\left(\frac{m_{240}}{^{240}Pu_{eff}}\right)^2} = \frac{\sigma_{m_{240}}^2}{(m_{240})^2} + \frac{\sigma_{P^{240}Pu_{eff}}^2}{\left(^{240}Pu_{eff}\right)^2} \tag{25.25}$$

Table 25.1 Common cases for error propagation with uncorrelated random variables (x and y) and c being a known constant with zero uncertainty

Function	Absolute uncertainty	Relative uncertainty
$f(x) = x + c$	$\sigma_f = \sigma_x$	$\dfrac{\sigma_f}{f} = \dfrac{\sigma_x}{x+c}$
$f(x) = c \cdot x$	$\sigma_f = c \cdot \sigma_x$	$\dfrac{\sigma_f}{f} = \dfrac{\sigma_x}{x}$
$f(x, y) = c \cdot x + y$	$\sigma_f = \sqrt{c^2 \cdot \sigma_x^2 + \sigma_y^2}$	$\dfrac{\sigma_f}{f} = \dfrac{\sqrt{c^2 \cdot \sigma_x^2 + \sigma_y^2}}{c \cdot x + y}$
$f(x, y) = c \cdot x \cdot y$	$\sigma_f = \sqrt{(c \cdot y \cdot \sigma_x)^2 + (c \cdot x \cdot \sigma_y)^2}$	$\dfrac{\sigma_f}{f} = \sqrt{\left(\dfrac{\sigma_x}{x}\right)^2 + \left(\dfrac{\sigma_y}{y}\right)^2}$
$f(x, y) = \dfrac{x}{y}$	$\sigma_f = \dfrac{x}{y} \cdot \sqrt{\left(\dfrac{\sigma_x}{x}\right)^2 + \left(\dfrac{\sigma_y}{y}\right)^2}$	$\dfrac{\sigma_f}{f} = \sqrt{\left(\dfrac{\sigma_x}{x}\right)^2 + \left(\dfrac{\sigma_y}{y}\right)^2}$

$$\frac{\sigma_{m_{Pu}}}{m_{Pu}} = \sqrt{\left(\frac{\sigma_{m_{240}}}{m_{240}}\right)^2 + \left(\frac{\sigma_{240_{Pu_{eff}}}}{240_{Pu_{eff}}}\right)^2} \tag{25.26}$$

Note that a value $\frac{\sigma_x}{x}$ is generally called the *relative uncertainty*. Thus, we can see that in this case, the relative uncertainties add in quadrature. Suppose the relative uncertainty from the neutron measurement was 10.0%, and the relative uncertainty from the gamma isotopics measurement was 2.0%. Then the relative uncertainty in the total plutonium mass is 10.2%.

If a function can be analytically defined, then these computations can be carried out. Table 25.1 shows results for some common cases. As seen in Table 25.1, in some cases it is more convenient to work with the relative uncertainty in propagating error than the absolute uncertainty.

Equation 8.10 (Chap. 8, Sect. 8.3.2) for the determination of the net 186 keV counting rate using a two-region-of-interest method is an example of the third entry in Table 25.1.

$$R = C1 - f\,C2, \tag{25.27}$$

with R as $f(x, y)$, C1 as y, C2 as x, and c equal to f. The uncertainty on R, σ_R, using the relationship in the table, is given by

$$\sigma_R = \sqrt{[f^2 \cdot \sigma_{C2}^2] + \sigma_{C1}^2}. \tag{25.28}$$

It must be emphasized that the expression for the uncertainty of a quantity as the square root of the counts (Eq. 25.19) applies only when the quantity is the recorded number of counts. In this example, the quantities C1 and C2 are not counts but counting rates, which require two measured quantities—the number of counts collected (N1 and N2, respectively) within the measurement time (t_m). In the case of C1,

$$C1 = \frac{N1}{t_m}. \tag{25.29}$$

Because modern computers can measure time very accurately, the uncertainty on t_m is considered as inconsequential and thus t_m is treated as a constant. Using the second entry in Table 25.1 as well as Eq. 25.19 for the uncertainty in *N1*, we get

$$\sigma_{C1} = \frac{\sigma_{N1}}{t_m} = \frac{\sqrt{N1}}{t_m}. \tag{25.30}$$

Or using Eq. 25.29,

$$\sigma_{C1} = \sqrt{\frac{C1}{t_m}}. \tag{25.31}$$

Substituting back into Eq. 25.28,

$$\sigma_R = \sqrt{\left[f^2 \cdot \left(\frac{C2}{t_m}\right)\right] + \left(\frac{C1}{t_m}\right)}. \qquad (25.32)$$

It should be noted that this analytical approach to computing an uncertainty is not always feasible or possible. The function (s) may be too complex to make manually working out all derivatives practical; or rather than a single function, the quantity of interest might be the output of a complex computer code. In these complex cases, the uncertainty can be computed using numerical approximations for the derivatives.

Alternatively, this type of bottom-up uncertainty can be found using Monte Carlo simulation. In this approach, input parameters are randomly sampled according to their respective or combined probability distributions, and the distribution of the output quantity of interest can be studied. This method eliminates the need for computing derivatives and can be applied even when the function of interest is very complex. It can also be used to study the probability distribution of the function of interest. This approach is explored further in Ref. [5] and in Chap. 5 of Ref. [1].

25.4.2 Calibration

Many nondestructive measurement results depend on a calibration curve of some kind. A calibration curve is usually established by making measurements of items of well-known "reference" material, whose masses and isotopic compositions have been established by destructive analysis. In a simple case such as the passive neutron calibration curve method (Chap. 17, Sect. 17.9) for the determination of the ^{240}Pu$_{eff}$ mass of an item, the calibration curve is shown in Fig. 17.26. A curve is fitted to the measured counting rates as a function of mass. Fitting different curves to calibration data is discussed in Ref. [1]. There are some important points to note. The masses of the reference materials must cover the expected range of the masses of the unknown items. It is unwise to use the calibration curve outside of the mass limits of the reference materials used. In addition, the reference material must be similar to the material types of the unknown items. The necessary similarity depends on the measurement technique being used. For example, for the neutron passive calibration curve method (Chap. 17, Sect. 17.9), the neutron multiplication of the unknown must be similar to that of the reference standards, which requires that the unknowns have similar shapes and density to the reference material. The chosen form of the calibration curve needs to represent the expected shape of the reference measurement; otherwise large error will result. In the case of the known alpha method (Sect. 17.11), the expected shape is a straight line. For the passive neutron calibration curve method, the expected shape is quadratic. For active coincidence counting (Sect. 17.15.4), the curve may be cubic.

The uncertainties on the fitting parameters are calculated from the deviation of the reference measurements from the chosen form of the calibration curve, including the uncertainties on the mass and counting rates. These uncertainties are then included in the uncertainty propagation on the final result. Detailed discussions on the calculation of the uncertainties are contained in Refs. [6–13].

The traditional method used in fitting a calibration curve is to use the item mass as the independent variable and the counting rate as the dependent variable. This method gives a curve in which it is easy to visualize effects such as neutron multiplication or self-shielding; however, the extraction of the result requires the inverse solution of the calibration equation, which in some cases may be complex. The alternative method is to use the counting rate as the independent variable and the item mass as the dependent variable. In this case, the result can be obtained directly from the counting rate, but the shape of the curve is less intuitively obvious. The uncertainties that result from both methods are usually similar.

25.4.3 Total Measurement Uncertainty

A key safeguards question is, "Does the result of a measurement 'verify' the operator's declaration?" Stated another way, "Is the difference between measurement result and the declared value consistent within the expected uncertainty?" To answer this question, it is necessary to establish the uncertainty of the measurement instrument. This uncertainty is called the *total measurement uncertainty* (TMU) and is obtained in a bottom-up way from the contributions of all known factors that contribute to a measurement. These factors include both random and systematic contributions. Table 25.2 shows an example of the factors that are considered for a hypothetical instrument. The main systematic uncertainties arise from random uncertainties of the measurements of the items that are used to make the calibration. The main random uncertainty arises from the measurement of the unknown item itself. Typically, all of these uncertainties are combined in quadrature to make a

Table 25.2 Contributions to total measurement uncertainty of hypothetical instrument

Source	Uncertainty estimate
Calibration contribution:	
Counting statistics (systematic for calibration) (2 h)	1.4%
Pu mass via weight (systematic)	0.1%
Pu isotopics (via isotope dilution mass spectrometry)	0.2%
Moisture variability (destructive analysis values)	0.25%
Position reproducibility of item (estimated from loading system)	0.2%
Electronics drift (measured value)	0.05%
Total systematic uncertainty	1.45%
Assay contribution:	
Counting statistics (random) (1 h)	2.0%
Pu isotopics (via isotope dilution mass spectrometry)	0.2%
Moisture variability (from Monte Carlo)	0.5%
Position reproducibility of item (estimated from loading system)	0.2%
Fill height variability (from Monte Carlo using estimated density range)	0.51%
Electronics drift (measured value)	0.05%
Room background variability (cross talk)	0.2%
Total random	2.15%
TMU	2.6%
Note: Uncertainty terms are combined as follows: $\sigma = \sqrt{\sigma_1^2 + \sigma_2^2 + \ldots + \sigma_i^2}$	

value that can be used in the assessment of the difference between measured and declared values. The factors contributing to the TMU can be different for each combination of instrument and material type and are typically established by the instrument designer during the testing and calibration of the instrument. They often include Monte Carlo estimates of expected changes to the item, such as density, moisture content and position.

A recent evaluation of total measurement uncertainty for neutron coincidence and multiplicity measurement is given in Ref. [14].

25.5 Top-Down Uncertainty

In the "top-down" or "empirical" approach, the measurement uncertainty is estimated from a set of many measurements, typically quality control or verification dataset. This approach is sensitive to systematic errors, and by evaluating measurements against known standards, biases can be detected, quantified, and in many cases eliminated.

Consider the example of mass measurement data in Sect. 25.2. After 25 measurements, a mean value of 1.01 kg and a sample standard deviation of 0.09 kg were computed. In a future measurement of a similarly sized item, it has been demonstrated that for this system, a typical random uncertainty is approximately 0.09 kg.

Ideally, top-down and bottom-up approaches should yield the same uncertainties. This comparison can be used to determine if other sources of uncertainty are present that are not being sufficiently accounted for. Compared against bottom-up approaches, top-down approaches can yield larger values for the uncertainties [15]. This result may be due to incomplete or inaccurate models of a measurement that produce lower uncertainties in bottom-up approaches. The difference in uncertainty between top-down and bottom-up approaches is often called the *dark uncertainty* [16, 17].

A bottom-up approach is best suited for optimizing the performance of a measurement system by determining how the uncertainties associated with each aspect of the measurement is contributing to the overall systematic and random uncertainty of the measurement result. In contrast, the top-down approach is best suited for monitoring and ensuring that a measurement system is performing as needed or as expected to meet the performance goals established for that measurement system. Figure 25.8 provides an example of applying a top-down approach to data taken by a measurement system on a reference standard with well-known properties by plotting the difference between the measured and reference values as a function of time when the measurements were performed. In this illustration, the measurement system collects data during specific intervals or measurement campaigns. By treating the data as one continuous data set, the overall bias and precision of the system can be determined. Upon closer inspection, each measurement campaign has a specific short-term systematic bias associated with it, as indicated by the fact that the average value for the data within each measurement campaign (noted by the solid line passing

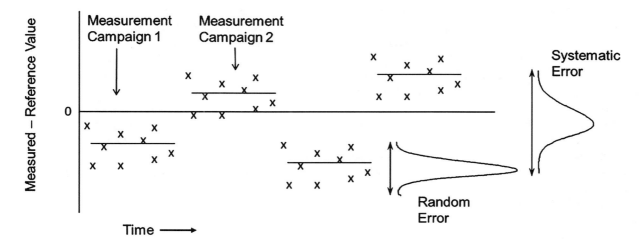

Fig. 25.8 An illustration of how a top-down analysis is performed. In this figure, the difference between measured and reference values is shown as a function of time. In this case, short-term systematic error is evident for each measurement campaign by the solid line passing through each data set, indicating the average value achieved during that specific campaign

through the data) lies off the horizontal axis that corresponds to the measured value that equals the reference value. The spread in the data associated with each measurement campaign indicates the random error or precision associated with the measurement system. Short-term systematic uncertainties can be encountered in situations where the data being analyzed include measurements from different physical detectors that are temporarily deployed within a facility to perform the same NDA measurement on a specific set of items.

Based on the nature of the notional data shown in Fig. 25.8, an additive model could be selected to perform a top-down analysis of the data. Unlike the additive model shown in Eq. 25.1, the model that would be needed to analyze these data would need to include a term to account for the short-term systematic error that is present within the data.

$$Y_i = a + a_i + \mu_i + \epsilon_i, \tag{25.33}$$

where a_i is the unknown short-term systematic bias.

A practical example of a top-down approach to error analysis can be seen in Chap. 23, Sect. 23.10, where the results of the Calorimetry Exchange Program (CALEX) were discussed (Fig. 23.12). Here, replicate measurements were performed by five laboratories on a set of certified PuO_2 reference standards that were produced from the same batch of material and were shown to have identical properties. Because the true values for these materials were known from more precise and accurate destructive analysis techniques, both the measurement uncertainty and the bias can be determined in a top-down approach. As noted in Chap. 23, Sect. 23.10, an analysis of the combined data set produced an average bias of 1.0004 with a standard deviation of 0.0002, which provides indications of the bias and precision associated with performing a calorimetry measurement on a PuO_2 item that produces approximately 1 W of thermal power.

Upon closer inspection of the data presented in Fig. 23.12, it is clear that different facilities have different measurement precisions and biases, with each facility's data separated by the dashed lines associated with measuring the CALEX standard. Table 23.7 provides the top-down analysis results for five U.S. Department of Energy facilities that participated in the CALEX program from 1990 to 1998. The results in Table 23.7 provide the precision and bias data for both the calorimetry measurement of the thermal power as well as the gamma-ray isotopic measurement of the effective specific power (P_{eff}). For many of the measurement systems shown in Table 23.7, the bias in the measurement systems is smaller than the precision of the instrument expressed as a percent relative standard deviation (% RSD).

$$\%RSD = 100 \cdot \frac{\sigma}{\bar{x}}. \tag{25.34}$$

In instances where the measured bias of a system is smaller than the precision of the system, it is effectively considered bias-free because the true value has been shown to lie within 1σ of the measured value.

A bottom-up uncertainty estimate might account for or identify root causes of this variation between the results from the different facilities, but it may not be capable of accounting for all of them. Some sources of random uncertainty are difficult or impossible to quantify, such as effects from how precisely a particular operator places a container within the calorimeter; however, the impact of these random errors on the overall measurement is accounted for in the top-down approach. Lastly, because the true masses were known for these certified reference materials, any bias in the measurement can be detected and quantified. This method could be used to apply a correction during the analysis of the items or to find and eliminate the source of the bias.

More sophisticated top-down analysis techniques are used in situations where measurements of reference standards are not possible. Such techniques are used by organizations such as the International Atomic Energy Agency (IAEA), which uses NDA techniques to verify that a facility's nuclear material inventory has not been diverted from their intended peaceful use. In this case, paired operator inspector verification measurements are assessed by using one-item-at-a-time testing to detect significant differences [18]. The statistic parameter used is known as the *operator-inspector difference* (OID), which is the relative difference between the operator's declared value for the item being verified and the inspector's measurement of the item.

$$OID = \frac{(Operator - Inspector)}{Operator} \cdot 100\% \qquad (25.35)$$

Complications associated with the top-down uncertainty quantification of OIDs from a given measurement system include uncertainties associated with both the operator's declarations and the inspector's measurements, accounting for both short-term systematic error that could exist within an inspection period and overall systematic error and lack of replicate measurements performed on the same item. Details of the uncertainty quantification analysis of similar data can be found in Ref. [17].

25.6 Measurement Control

Once uncertainties have been established for a measurement system by using either a top-down or bottom-up approach, a measurement control program can be instituted to ensure that the system continues to perform as expected or required. An effective measurement control program will incorporate periodic measurements of either reference or working standards in and amongst item measurements. The data collected from the measurements of standards can then be presented on a control chart that overlays the $\pm 2\sigma$ and $\pm 3\sigma$ uncertainties for the measurement system onto a plot of measured values of the standard as a function of time. An example of a control chart is shown in Fig. 25.9. If the measurement system being monitored was

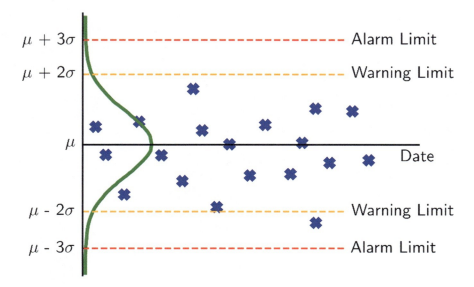

Fig. 25.9 An example of a control chart

operating properly, one would expect that 95% of the measurement control data will fall within the $\pm 2\sigma$ limits, whereas 99.7% of the data should lie within the $\pm 3\sigma$ limits. Indications of potential problems with a measurement system under measurement control include having one data point lie above the $\pm 3\sigma$ alarm limit, having two of three data points lie outside the $\pm 2\sigma$ warning limit, and observing trends in the data, such as consecutive points all above or all below the true value or a series of consecutive points with upward or downward trends. Additional information on statistical control limits can be found in Section 6.3.4 of Ref. [19].

25.7 Guide to the Expression of Uncertainty in Measurement

In 2008, the Joint Committee for Guides in Metrology (JCGM), consisting of seven organizations including the International Organization for Standardization (ISO), the International Union of Pure and Applied Chemistry (IUPAC), and the International Union of Pure and Applied Physics (IUPAP) published the "Guide to the Expression of Uncertainty in Measurement" (the GUM) to establish "general rules for evaluating and expressing uncertainty in measurement that can be followed at various levels of accuracy and in many fields" [2]. The methodology for evaluating and expressing uncertainty that is described in the GUM is consistent with a bottom-up approach where the measurement process is modelled through mathematical relationships between the value of a particular quantity that is the objective of a measurement (referred to as the *measurand*) and its input components. Uncertainty in the measurand is determined using standard error propagation methods from the standard uncertainties of the input quantities.

The GUM method identifies two specific categories of standard uncertainty that can be assigned to an input component. Type A uncertainties are evaluated using statistical analysis of a series of observations that use many of the principles discussed in Sects. 25.3 and 25.4. Unlike Type A uncertainties, which are derived from the underlying data itself, Type B uncertainties are determined using an assumed probability distribution for an input component based on the additional available information regarding that measurement. In essence, whereas a Type A uncertainty will be determined based on the data generated by the measurement itself, Type B uncertainties are derived based on ancillary information that is available regarding the measurement, such as external calibration reports, manufacture specifications, published data, and standard guides and methods [2].

25.8 Use of International Target Values and Standards for Uncertainty Estimation

To assist organizations and safeguards practitioners in evaluating the reliability and effectiveness of both NDA and destructive assay measurement techniques, the concept of international target values (ITV) was proposed in June 1991 by a consultants group meeting convened by the IAEA. ITVs are "values of uncertainties, expressed as relative random and systematic error standard deviations" that should be achievable in performing an NDA measurement on a single item under typical conditions that are experienced in either a typical nuclear fuel cycle laboratory or in a safeguards inspection [20]. ITVs are established based primarily on top-down analysis of variance applied to measured data from a given set of NDA measurements performed on a specific type of nuclear material and reviewed by subject matter experts. Details of the methodology used to determine ITVs can be found in Section 3 of Ref. [21]. In cases where an NDA measurement technique is being applied for the first time to a process or within a facility, the ITVs can provide a good initial estimate regarding what accuracy and precision can be achieved with that technique. Current ITV values are available at an IAEA website [22]. Although ITVs represent the uncertainties that can be achieved with the current state of practice for a given NDA measurement technique, it should be noted that they do not represent regulatory requirements that may exist for those measurements [20].

In addition to the ITVs produced by the IAEA, another source of information on expected uncertainties for certain NDA techniques can be found within standards that have been written for these techniques by organizations such as The American Society for Testing and Materials(ASTM). Within these standards, the typical sources of bias for a given NDA measurement are discussed, as well as factors that can impact the precision of measurements. These standards often provide estimates for expected precision and accuracy of the measurement technique based on previously published data. Examples of standards that include discussions of uncertainty can be found in Refs. [23–25].

References

1. P.R. Bevington, D.K. Robinson, *Data Reduction and Error Analysis for the Physical Sciences* (McGraw-Hill, 2003)
2. Guide to the Expression of Uncertainty in Measurement (GUM), JCGM 100:2008
3. O. Alique, Y. Aregbe, R. Bencardino, R. Binner, T. Burr, J.A. Chapman, S. Croft, A. Fellerman, T. Krieger, K. Martin, P. Mason, C. Norman, T. Prohaska, D. Trivedi, S. Walsh, D. Wegrzynek, B. Wright, J. Wüster, Statistical error model-based and GUM-based analysis of measurement uncertainties in nuclear safeguards—a reconciliation. ESARDA Bull. **64**(1), 10–29 (2022). https://doi.org/10.3011/ESARDA.IJNSNP.2022.2
4. L. Wasserman, *All of Statistics: A Concise Course in Statistical Inference* (Springer, New York, 2004)
5. C.E. Papadopoulos, H. Yeung, Uncertainty estimation and Monte Carlo simulation method. Flow Meas. Instrum. **12**(4), 291–298 (2001)
6. E. Bonner, T. Burr, T. Krieger, K. Martin, C. Norman, Comprehensive uncertainty quantification in nuclear safeguards. Sci. Technol. Nucl. Install. 2017, Article ID 2679243, 16 pp. (2017). https://doi.org/10.1155/2017/2679243
7. T. Burr, T. Krieger, C. Norman, Approximate Bayesian computation applied to metrology for nuclear safeguards. J. Phys. Conf. Ser. **1141**, 012127 (2018)
8. S. Croft, T. Burr, A. Favalli, A. Nicholson, Analysis of calibration data for the uranium active neutron coincidence counting collar with attention to errors in the measured neutron coincidence rate. Nucl. Instrum. Methods Phys. Res., Sect. A **811**, 70–75 (2016)
9. T. Burr, S. Croft, T. Krieger, K. Martin, C. Norman, S. Walsh, Uncertainty quantification for radiation measurements: bottom-up error variance estimation using calibration information. Appl. Radiat. Isot. **108**, 49–57 (2016)
10. T. Burr, S. Croft, D. Dale, A. Favalli, B. Weaver, B. Williams, Emerging applications of bottom-up uncertainty quantification in nondestructive assay. ESARDA Bull **53**, 54–61 (2015)
11. T. Burr, S. Croft, K. Jarman, Uncertainty quantification in application of the enrichment meter principle for nondestructive assay of special nuclear material. J. Sens. 2015, Article ID 267462, 10 pp. (2015). https://doi.org/10.1155/2015/267462
12. P. Kang, C. Koo, H. Roh, Reversed inverse regression for the univariate linear calibration and its statistical properties derived using a new methodology. Int. J. Metrol. Qual. Eng. **8**(28) 10 pp (2017)
13. S. Croft, T. Burr, Calibration of nondestructive assay instruments: an application of linear regression and propagation of variance. Appl. Math. **5**(5), 785–798 (2014). https://doi.org/10.4236/am.2014.5507
14. R.D. McElroy Jr., Total Measurement Uncertainty in Neutron Coincidence Multiplicity Analysis, Oak Ridge National Laboratory report ORNL/TM-2020/1663 (November 2020)
15. T. Burr, S. Croft, A. Favalli, T. Krieger, B. Weaver, Bottom-up and top-down uncertainty quantification for measurements. Chemom. Intell. Lab. Syst. **211**, 104224 (2021)
16. M. Thompson, S.L.R. Ellison, Dark uncertainty. Accredit. Qual. Assur. **16**, 483–487 (2011)
17. T. Burr, T. Krieger, C. Norman, Approximate Bayesian computation applied to nuclear safeguards metrology. ESARDA Bull **57**, 50–59 (2018)
18. E. Bonner, T. Burr, T. Guzzardo, T. Krieger, C. Norman, K. Zhao, D.H. Beddingfield, T. Lee, M. Laughter, W. Geist, Ensuring the effectiveness of safeguards through comprehensive uncertainty quantification. J. Nucl. Mater. Manage. **44**(2), 53–61 (2016)
19. *DOE Standard Nuclear Materials Control and Accountability*, DOE-STD-1194-2011 (U.S. Department of Energy, Washington, D.C., 2011). https://www.standards.doe.gov/standards-documents/1100/1194-astd-2011-cn3-2013/@@images/file
20. C. Norman, K. Krzysztoszek, E. Bonner, K. Zhao, M.T. Serrano Reina, International target values-looking forward and lessons learned, in *Proceeding of the INMM & ESARDA Virtual Annual Meeting*, August 23–26 and August 30–September 1, 2021
21. K. Zhao, M. Penkin, C. Norman, S. Balsley, K. Mayer, P. Peerani, C. Pietri, S. Tapodi, Y. Tsutaki, M. Boella, G. Renha Jr., E. Kuhn, International Target Values 2010 for Measurement Uncertainties in Safeguarding Nuclear Materials, IAEA STR-368 (International Atomic Energy Agency, Vienna, Austria, November 2010)
22. International Target Values for Measurement Uncertainties in Safeguarding Nuclear Materials, Vienna, Austria, IAEA STR-368,Rev. 1.1 (September 2022). https://nucleus.iaea.org/sites/connect/ITVpublic/ITV%20tables/Forms/Index%20of%20ITV%20Tables.aspx. Accessed Mar 2023
23. Standard Test Method for Determination of Plutonium Isotopic Composition by Gamma-Ray Spectrometry, ASTM Standard Test Method C1030–10 (American Society for Testing and Materials, Philadelphia, PA, 2018)
24. Standard Test Method for Nondestructive Assay of Plutonium in Scrap and Waste by Passive Neutron Coincidence Counting, ASTM Standard Test Method C1207-10 (American Society for Testing and Materials, Philadelphia, PA, 2010)
25. Standard Test Method for Nondestructive Assay of Plutonium, Tritium, and ^{241}Am by Calorimetric Assay, ASTM Standard Test Method C1458-16 (American Society for Testing and Materials, Philadelphia, PA, 2016)

Open Access This chapter is licensed under the terms of the Creative Commons Attribution 4.0 International License (http://creativecommons.org/licenses/by/4.0/), which permits use, sharing, adaptation, distribution and reproduction in any medium or format, as long as you give appropriate credit to the original author(s) and the source, provide a link to the Creative Commons license and indicate if changes were made.

The images or other third party material in this chapter are included in the chapter's Creative Commons license, unless indicated otherwise in a credit line to the material. If material is not included in the chapter's Creative Commons license and your intended use is not permitted by statutory regulation or exceeds the permitted use, you will need to obtain permission directly from the copyright holder.

26 Nuclear Material Accounting and Control Measurements

G. E. McMath, A. L. Lousteau, and S. Smith

26.1 Overview

Nuclear material accounting and control (NMAC), a vital element in a strong nuclear security and safeguards regime, is the collective set of activities that help to deter and detect unauthorized removal of nuclear material by maintaining and reporting accurate and timely inventories of all nuclear material present. NMAC is used for both nuclear safeguards and nuclear security, though the implementation may differ given the different objectives of the two. The purpose of nuclear safeguards is to detect and deter a State's clandestine nuclear weapons program. The purpose of nuclear security is to prevent and detect attempts to steal nuclear material by a non-State actor. NMAC protects against both of these very different threats/adversaries. Although NMAC is the name most commonly used internationally, domestic programs often use material control and accountability (MC&A) or nuclear material control and accounting (NMC&A). Unless referring to a specific program, this chapter will use the term *NMAC*. Even though the legal obligations, requirements, and implementation of an NMAC program differ among countries, facilities, and agencies, the basic NMAC principles are the same. This chapter focuses on the role of nondestructive assay, general guidance, and best practices related to measurements and measurement control as they relate to an effective NMAC program.

NMAC is one of nine primary security elements that comprise a robust nuclear security regime as recommended by the Office of International Nuclear Security under the United States (U.S.) Department of Energy (DOE). The nine security elements are as follows:

- Nuclear Material Accounting and Control
- Physical Security
- Transport Security
- Cyber Security
- Insider Threat Mitigation
- Sabotage Mitigation
- Response
- Drills and Exercise Program
- National Level Infrastructure

Los Alamos National Laboratory strongly supports academic freedom and a researcher's right to publish; as an institution, however, the Laboratory does not endorse the viewpoint of a publication or guarantee its technical correctness.

G. E. McMath (✉)
Los Alamos National Laboratory, Los Alamos, NM, USA
e-mail: gem@lanl.gov

A. L. Lousteau · S. Smith
Oak Ridge National Laboratory, Oak Ridge, TN, USA
e-mail: lousteaula@ornl.gov; smithsk@ornl.gov

Many of these security elements are interdependent and must work in a complementary manner to ensure defense in depth. For example, NMAC relies on both physical security and insider threat mitigation to achieve its goal of providing information on, control of, and assurance of the presence of nuclear material. This goal is achieved through integrated systems of inventory, tracking, access control, and the timely detection of loss or diversion of nuclear material [1].

To achieve the stated goals of an NMAC program, eight key elements have been identified as essential to success:

- NMAC Program Management
- Nuclear Material Accounting Records
- Physical Inventory
- Measurements and Quality Control
- Nuclear Material Control
- Nuclear Material Movements
- Detection, Investigation, and Resolution of Irregularities
- Assessment and Performance Testing

For more information on each of these elements, refer to the International Atomic Energy Agency (IAEA) Nuclear Security Series No. 25-G Implementing Guide on *Use of Nuclear Material Accounting and Control for Nuclear Security Purposes at Facilities* [1] or the relevant domestic guidance available.

In the U.S., MC&A requirements are outlined in DOE Order 474.2, *Nuclear Material Control and Accountability* [2]. In this case, the eight key elements listed above are essentially collapsed down to five elements. The key elements necessary for a successful MC&A program as identified in the DOE Order are

- Program Management
- Material Control
- Measurement and Measurement Control
- Accounting
- Physical Inventory

Outside of the DOE, the U.S. Nuclear Regulatory Commission governs NMC&A through requirements provided in Title 10 Code of Federal Regulations, Part 74, where elements of the program are delineated by the strategic significance of the material. Different programs and implementations of NMAC may be distinct; however, the principles and good practices are universal.

26.2 NMAC Measurement System Goals

The surety of the presence of nuclear material typically requires two things: appropriate and accurate measurement of the material and continuity of knowledge/chain of custody of the material. Measurements provide accountability, and control ensures continuity of knowledge.

In both domestic and international nuclear safeguards and security, accountability of material provided by NMAC system measurements ultimately focuses on closing the material balance. To evaluate material balance, the combined results of all measurement systems are used to identify non-zero differences in the nuclear material active inventory. These non-zero differences are referred to domestically as *inventory differences* (IDs) and internationally as *material unaccounted for* (MUF). Ideally, all non-zero differences should be able to be explained by the combined measurement uncertainty. Cases where the measurement uncertainty cannot explain the discrepancy warrant further investigation into whether a theft, diversion, or operational issue has resulted in a loss or gain of nuclear material.

Material Balance Equation:

$$\text{ID (or MUF)} = \text{Beginning Inventory} + \text{Additions} - \text{Removals} - \text{Ending Inventory} \tag{26.1}$$

An NMAC system's performance requirements are determined by the type and amount of nuclear material that is being handled in a particular facility. These guidelines, laid out in regulatory documents as well as the international standards of accountancy, set a standard for the allowable amount of uncertainty across all NMAC measurement systems in a facility. These measurement uncertainty values are essential to closing a material balance. This chapter will focus on the measurement aspect and the role of NDA in an NMAC program.

26.3 NMAC Measurements

NMAC measurements are performed at key measurement points (KMPs) as part of established material balance areas (MBAs). The measurements are split into two types: destructive analysis (DA) and nondestructive analysis (NDA). The six fundamental types of NDA measurements for NMAC are as follows (Figs. 26.1, 26.2, 26.3, 26.4, 26.5, and 26.6):

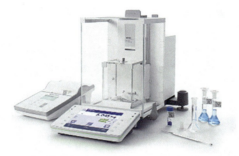

Fig. 26.1 Weight—Provides gross weight of combined material (nuclear and other)

Fig. 26.2 Volumetric—Provides volume of combined material (nuclear and other)

Fig. 26.3 Gamma ray—Provides isotopic information of nuclear material (can provide mass with assumptions)

Fig. 26.4 Neutron—Provides mass of nuclear material (requires additional information)

Fig. 26.5 Calorimetric—Provides mass of nuclear material (requires additional information)

Fig. 26.6 Calculation—Provides mass and isotopics when direct measurement is not possible

The first two measurement types (weight and volumetric) are fairly straightforward and are not nuclear specific; however, they can be prone to complications and errors. Mass measurements can be performed with a scale or balance and can provide gross weights of all material. Volumetric is similar but is used for liquids or gases and can require calibrated tank-level sensors. Both types can give bounding information; however, because neither provides knowledge of the constituents of what is measured, they are limited in the information they can provide on their own. Combined with chemical composition and isotopic information (from DA or other means), these measurement systems can provide accountability data and make up most measurement systems used for NMAC. The next three types of measurements (gamma ray, neutron and calorimetric) have their own chapters in this book and are essential in providing the accurate data necessary to have a successful NMAC program. Lastly, calculations are used when direct measurements are not possible, which is often the case with spent nuclear fuel (see Spent Fuel Chap. 21) and occasionally holdup measurements, where a technically justified calculation is used in place of an impractical measurement scenario. Regardless of the measurement, the standard measurement process is consistent:

1. Determine the measurement objective
2. Collect information regarding the item, either from previous measurements, design drawings, or process knowledge
3. Select a suitable measurement technique/system
4. Qualify the measurement system
5. Calibrate the measurement system
6. Perform measurements
7. Report measurement results with associated uncertainty

Best practices related to this measurement process are described in the following sections.

26.3.1 Measurement Objectives

When making any NDA measurement—and specifically for NMAC—it is crucial to determine the objective of the measurement. All measurement objectives should be clearly expressed in NMAC plans and procedures. The two high-level measurement purposes for NMAC measurements are *confirmatory* and *accountability*.[1] A confirmatory measurement can be qualitative and generally verifies the presence and state of material that is already known through the measurement/verification of two attributes. These measurements are valid only when an item is in control, i.e., no loss in continuity of knowledge, which is typically achieved through the use of seals or tamper-indicating devices (TIDs), used to assure that the item has not been altered since its last accountability measurement. Confirmatory measurements do not alter accounting values. An accountability measurement is quantitative and establishes the inventory values for the accounting records. For example, if a confirmation is needed of nuclear material in a container that has not been accessed since its accounting measurement (i.e., in a tamper-indicating container), a two-factor check of the TID integrity and a gross mass measurement may be all that are needed. To fully characterize the container, a neutron or calorimetry measurement, along with gamma techniques to acquire the isotopic information, could be used to determine the nuclear material mass.

26.3.2 Measurement Selection and Qualification

Once the objective of the measurement is determined, the measurement criteria must be identified to select an appropriate measurement technique. For example, the selection of measurement technique and instrument must consider the associated uncertainty, the measurement point in the process, and its impact on the measurement objective or material balance closure if used for physical inventory.

One useful reference for target values is the *International Target Values for Measurement Uncertainties in Safeguarding Nuclear Materials*, IAEA Report STR-368, Rev. 1.1, published September 2022 [3]. However, it should be noted that uncertainties given in this reference could have been determined using measurement configurations that differ from those in use at a given site. In those instances, it may be more reasonable to consult method-specific ASTM standards for uncertainty guidance (DOE-STD-1194-2019, Sect. 6.3.3, p. 54).

Additional questions that should be addressed to ensure that a measurement technique is suitable include the following:

- What are the minimum measurement requirements?
- What type of material is being measured?

---

[1] The term *verification measurement* is sometimes used interchangeably with accountancy or confirmatory measurement, but it is a domestic term used when verifying the accounting system's values.

- What radionuclides could be present?
- What is the mass range expected to be measured?
- What is the matrix of the material, and is it uniform?
- What is the geometry of the items to be measured?
- What precision and accuracy are required?
- What background conditions exist in the measurement area?
- Do building or facility constraints exist?
- How will the measurement system be implemented?
- What measurement controls will be necessary?
- What is the throughput expectation?
- What is the budget?

The weight of each of these questions depends on the application and must be considered with respect to the measurement need.

Once a measurement system is selected, it should be qualified for use. In the U.S., accountability measurement systems are required to be qualified and approved before use for accountability purposes, and a measurement control program is required to ensure the quality of measurements. The DOE Order does not dictate how a program meets these requirements; however, the IAEA provides the following documentation suggestions for ensuring quality control of a measurement program [1]:

- Assurance that personnel are qualified to perform measurements
- Maintenance and recertification of measurement equipment
- Control of standards
- Maintenance and re-certification of standards
- Calibration of equipment (frequency and method)
- Equipment recalibration and recertification
- Verification of measurement equipment performance
- Actions to be taken in case of equipment failures
- Measures to ensure that the measurement equipment, standards, and methods used are appropriate for the material being measured
- Complete documentation of all measurement results, including the results of measurements of standards
- Control charts used to monitor measurement of standards
- Measures to be taken when a measurement system appears to be out of control

Only upon approval from the competent authority shall a given measurement system be used for accountability purposes.

26.3.3 Measurement Calibrations

A measurement calibration provides a mathematical relationship that correlates a measurement system response with a characteristic of the measured item (e.g., a measured doubles rate with an effective mass of an item). Different measurement techniques could have different calibration requirements. Calibration data should always bound expected measurement conditions because the detector response cannot be assumed outside of the calibration range.

When performing a calibration, traceable physical standards should be used whenever possible. These standards should be representative of the materials to be measured, and the materials should be measured under the same measurement conditions as the accountability measurements to reduce measurement uncertainty. Once the calibration is established, a series of validation measurements should also be performed using different standards. This process helps ensure that the system is calibrated and accurately reflects the response of the measurement system. It is important to understand the calibration requirements for each system, and these requirements should be well documented.

26.3.4 Measurement Control

To assure the quality of all accountability measurements, a robust measurement control program is also required. A measurement control program ensures that a measurement system is performing as expected and with sufficient quality to meet the required measurement objectives. Measurement controls also ensure that over time, the procedures, equipment, and analysis used for both accounting and confirmatory measurements are accurate and fit for purpose. Measurement control is primarily achieved by performing control measurements (standardized, repeatable measurements) on a regular basis that are applicable to the use cases of the NMAC program. Control measurements are performed using one or several well-characterized standards or materials.

During the system qualification process, a series of control measurements (generally 30+ measurements) are performed to establish a mean value, warning limits (commonly $\pm 2\sigma$), and alarm limits (commonly $\pm 3\sigma$) for the system. These control limits are used to determine whether a given system is in control and may be used for accountability measurements. The rules used to determine whether a system is in or out of control may vary across facilities and instruments but should be clearly defined in NMAC documentation. Some common conditions that suggest a system is out of control include

- measurement results fall outside of the alarm limits,
- two out of three measurement results fall outside of the warning limits, and
- seven measurement results fall on one side of the mean value.

For a normally distributed measurement, these three conditions give roughly equivalent probabilities of false alarm (0.27%, 0.25%, and 0.39%, respectively), meaning that ~3 out of 1000 measurements will cause an alarm purely due to statistics and not instrument failure. Many other conditions can be added; however, care should be taken to keep the false alarm probability roughly the same across all conditions.

Based on their needs and applications, each NMAC program should evaluate their measurement control charts, limits, and rules for out of control. Guidance and numerous standards, such as DOE Standard 1194-2019 [4], ANSI ASQC B1-B3 [5], and the NIST/SEMATECH Engineering Statistics Handbook [6] are available. The critical function of a good measurement control program is that it will quickly catch an instrument that is starting to go or is out of control. If a control measurement violates the control limits, the NMAC program must have a documented methodology for addressing those irregularities, performing an investigation, and bringing the system back into control.

The pre-defined frequency of these control measurements depends on the type of measurement system and the materials measured. For example, control measurements are generally performed for a scale at the beginning (and sometimes end) of each day that the scale is used for accountability purposes, whereas control measurements for a calorimeter may be performed only weekly. This process provides confidence in the accountability measurements performed between successful control measurements. If a control measurement fails, the accountability measurements since the last successful control measurement are questionable. The results of these control measurements are recorded and tracked over time to warn of or alert the program to any gradual changes in the system. For example, if a detector was losing efficiency slowly over a long period of time, it would be extremely difficult to identify unless long-term trend analyses are performed. Trend analyses should be performed at least annually and can be focused on unnatural patterns such as stratification, mixture, systematic trends, or repetition. Control limits may be adjusted periodically when measurement parameters that influence the system response change. Parameters that can necessitate control limit adjustment include changes to the physical system or measurement geometry, changes in operating procedures, calibration changes, or instrument repairs.

26.3.5 Measurement Procedures

To ensure that measurements are performed correctly and to control or reduce measurement uncertainty introduced by an operator, detailed procedures and training are required. Measurement procedures should be documented, controlled, and approved before use. They should be clear, concise, and easy to follow. Measurement procedures should include the following at a minimum:

- Purpose of measurement
- System setup, if applicable
- Prerequisites and training requirements

- Calibration/validation procedure
- Measurement control requirements
- Measurement steps
- Method for recording data (both control measurement and accountability measurement)

A formalized training program also helps ensure that measurements are conducted only by authorized personnel who understand the necessary control, calibration, validation, and/or accountability measurement processes and can identify anomalous conditions.

26.4 Measurement Reporting

NMAC measurements must be well documented, and reports must include all relevant information including the measurement results and the associated uncertainty. Measurement and uncertainty analyses should be documented in a way that can be peer reviewed. It is important to note that determination of the total measurement uncertainty is not trivial. All major contributing factors beyond simple counting statistics must be considered. The measurement uncertainty is extremely important in distinguishing between an inventory difference (loss or gain of material) and a statistical variation.

It is considered a best practice to have all measurement and analysis results peer reviewed to ensure that errors have not been made. Complete and correct documentation provides confidence that measurements are reliably performed and allows for investigation if the results show irregularities/anomalies. The following information should be included with any measurement:

- Unique identification number of item or process being measured
- Unique identification numbers of the personnel who perform the measurement
- Unique identification number of the measurement equipment/system
- Time, date, and place of the measurement
- Calibration standards used, if applicable
- Control measurement results, if applicable
- Measurement result and associated uncertainty
- Isotopic information, if applicable
- Signatures or identification number scans of measurement operators and reviewers

26.5 Summary

NDA measurements are an integral part of an effective NMAC program. To meet the NMAC goals of providing information on, control of, and assurance of the presence of nuclear material, a robust measurement and measurement control program is needed. The foundation of a strong program is accurate and complete documentation. Regardless of the question—What instrument was used? Who performed the calibration? Who analyzed the measurement results? Who changed or approved the procedure?—all answers should be easily traced. In addition to providing confidence in the measured values, good documentation provides the information necessary to investigate and reconcile any irregularities or inventory differences quickly and accurately. A strong NMAC program not only has the ability to quickly detect these irregularities/anomalies but also to determine what happened and why it happened. Then the program staff can implement corrective actions so the problem does not reoccur. Accurate, complete, and persisting records make this outcome possible.

References

1. Use of Nuclear Material Accounting and Control for Nuclear Security Purposes at Facilities, IAEA Nuclear Security Series No. 25-G (International Atomic Energy Agency, Vienna, Austria, 2015)
2. Nuclear Material Control and Accountability, DOE O 474.2 (U.S. Department of Energy, Office of Health, Safety and Security (2016). https://www.directives.doe.gov/directives-documents/400-series/0474.2-BOrder. Accessed Mar 2023
3. International Target Values for Measurement Uncertainties in Safeguarding Nuclear Materials, IAEA STR-368, Rev. 1.1 (International Atomic Energy Agency, Vienna, Austria, September 2022). https://nucleus.iaea.org/sites/connect/ITVpublic/Resources/International%20Target%20Values%20for%20Measurement%20Uncertainties%20in%20Safeguarding%20Nuclear%20Materials.pdf. Accessed Aug 2023
4. Nuclear Materials Control and Accountability, DOE-STD-1194-2019 (U.S. Department of Energy, Office of Environment, Health, Safety and Security, 2019)
5. Guide for Quality Control Charts, Control Chart Method of Analyzing Data, Control Chart Method of Controlling Quality During Production, ANSI/ASQC B1-B3-1996 (American National Standards Institute, 1996)
6. *NIST/SEMATECH e-Handbook of Statistical Methods* (2013). http://www.itl.nist.gov/div898/handbook/. Accessed Mar 2023

Open Access This chapter is licensed under the terms of the Creative Commons Attribution 4.0 International License (http://creativecommons.org/licenses/by/4.0/), which permits use, sharing, adaptation, distribution and reproduction in any medium or format, as long as you give appropriate credit to the original author(s) and the source, provide a link to the Creative Commons license and indicate if changes were made.

The images or other third party material in this chapter are included in the chapter's Creative Commons license, unless indicated otherwise in a credit line to the material. If material is not included in the chapter's Creative Commons license and your intended use is not permitted by statutory regulation or exceeds the permitted use, you will need to obtain permission directly from the copyright holder.

Nuclear Data for Nondestructive Assay

P. A. Santi and T. D. Reilly

Nondestructive assay (NDA) measurement techniques rely on nuclear data to accurately convert the detection of gamma rays, neutrons, and heat into quantities of interest in nuclear safeguards and nuclear security. The specific nuclear data used in these techniques have been compiled within this chapter, which consists of both tables of data and an index of tables and figures presented in this book that contain relevant or useful nuclear data (Tables 27.1, 27.2, 27.3, 27.4, 27.5, 27.6, 27.7, 27.8, 27.9, 27.10, 27.11, 27.12, 27.13, 27.14, 27.15, 27.16, 27.17, and 27.18). Where possible, the data presented in these tables have been updated to reflect the most current consensus. The data presented reflect what is either currently used by various measurement techniques or data prescribed by various standards that may differ slightly from the most recently established consensus values for these data.

Table 27.1 Principal nondestructive assay gamma-ray signatures

Isotope	Energy[a] (keV)	Activity[a] (γ/g − s)	Mean free path[b] (mm) High-Z, ρ	Low-Z, ρ
^{234}U	120.9	8.06×10^4	1.0	69.9
^{235}U	143.8	8.76×10^3	0.4	74.4
	185.7	4.56×10^4	0.7	81.1
^{238}U	766.4[c]	3.94×10^1	9.8	141.6
	1001.0[c]	1.05×10^2	13.1	160.5
^{238}Pu	152.7	5.89×10^6	0.5	76
	766.4	1.39×10^5	9.8	141.6
^{239}Pu	129.3	1.45×10^5	0.3	71.6
	413.7	3.37×10^4	4.0	108.6
^{240}Pu	45.2	3.75×10^6	0.1	33.1
	160.3	3.38×10^4	0.4	77.3
	642.5	1.09×10^3	7.9	130.8
^{241}Pu	148.6	7.12×10^6	0.4	75.3
	208.0[d]	1.99×10^7	0.9	84.3
^{241}Am	59.5	4.55×10^{10}	0.2	47
	662.4	4.62×10^5	8.2	132.6

[a]Data for energy and activity are extracted from *NuDat 2.8*, http://www.nndc.bnl.gov/nudat2/chartNuc.jsp
[b]The mean free path is the absorber thickness that reduces the gamma-ray intensity to 1/e = 0.37. The mean free path in uranium or plutonium oxide ($\rho = 10$ g/cm^3) is given for the high-density, high-atomic-number case (high-Z, ρ). The mean free path in aluminum oxide ($\rho = 1$ g/cm^3) is given for the low-density, low-atomic-number case (low-Z, ρ). Attenuation data are from J. H. Hubbell, "Photon Cross Sections, Attenuation Coefficients, and Energy Absorption Coefficients from 10 keV to 100 GeV," U.S. Department of Commerce, National Bureau of Standards report NSRDS-NBS 29 (August 1969)
[c]From the ^{238}U daughter ^{234m}Pa; equilibrium assumed
[d]From the ^{241}Pu daughter ^{237}U; equilibrium assumed

Los Alamos National Laboratory strongly supports academic freedom and a researcher's right to publish; as an institution, however, the Laboratory does not endorse the viewpoint of a publication or guarantee its technical correctness.

P. A. Santi (✉) · T. D. Reilly
Los Alamos National Laboratory, Los Alamos, NM, USA
e-mail: psanti@lanl.gov

© The Author(s) 2024
W. H. Geist et al. (eds.), *Nondestructive Assay of Nuclear Materials for Safeguards and Security*,
https://doi.org/10.1007/978-3-031-58277-6_27

Table 27.2 Common nuclides used for energy calibration of gamma-ray nondestructive assay systems

Nuclide	Half-life	Energy (keV)[a]	Remarks
^{241}Am	432.6 year	59.54	Many others but weaker by factors of 10^4 or greater
^{137}Cs	30.08 year	661.66	Monoenergetic source + Ba X-rays
^{133}Ba	10.55 year	81.0, 276.40, 302.85, 356.01, 383.85	Several others but much weaker
^{60}Co	5.27 year	1173.23, 1332.49	
^{22}Na	2.6 year	511.0	Annihilation radiation
		1274.54	
^{109}Cd	1.26 year	88.03	Monoenergetic source
^{54}Mn	312 days	834.8	Monoenergetic source
^{65}Zn	244 days	511.0	Annihilation radiation
		1115.5	
^{57}Co	272 days	122.06, 136.47	Two other higher-energy lines but much weaker
^{75}Se	120 days	121.12, 136.00, 264.65, 279.54, 400.66	Several others but much weaker
^{152}Eu	13.52 year	121.78, 344.3, 778.9, 867.4, 964.1, 1085.8, 1112.1, 1408.0	Several others but weaker
^{55}Fe	2.74 year	5.9, 6.5	Mn K X-rays for low-energy calibration

[a]Energies are from the *NuDat 2.8* Database, http://www.nndc.bnl.gov/nudat2/chartNuc.jsp

Table 27.3 Selected gamma-rays from natural background

Nuclide	Parent series	Energy (keV)[a]	Remarks
^{212}Pb	^{232}Th	238.6	Also present in sources containing ^{228}Th (e.g., ^{232}U)
^{214}Pb	^{238}U	295.2	Also present in sources containing ^{222}Rn (e.g., ^{226}Ra)
^{214}Pb	^{238}U	351.9	Also present in sources containing ^{222}Rn (e.g., ^{226}Ra)
^{208}Tl	^{232}Th	583.2	Also present in sources containing ^{228}Th (e.g., ^{232}U)
^{214}Bi	^{238}U	609.3	Also present in sources containing ^{222}Rn (e.g., ^{226}Ra)
^{228}Ac	^{232}Th	911.2	Can distinguish ^{228}Th sources from ^{232}Th
^{214}Bi	^{238}U	1120.3	Also present in sources containing ^{222}Rn (e.g., ^{226}Ra)
^{40}K	^{40}K	1460.8	Often the most intense peak in the background
^{214}Bi	^{238}U	1764.5	Also present in sources containing ^{222}Rn (e.g., ^{226}Ra)
^{208}Tl	^{232}Th	2614.5	Also present in sources containing ^{228}Th (e.g., ^{232}U)

[a]*NuDat 2.8*, https://www.nndc.bnl.gov/nudat2/chartNuc.jsp

Table 27.4 Major K and L X-rays of uranium and plutonium

X-ray	Transition (Final-initial)	Uranium			Plutonium		
		Energy (keV)[a]	Intensity (%)[a,b]	Width (eV)[c]	Energy (keV)[a]	Intensity (%)[a,b]	Width (eV)[d]
$K_{\alpha 2}$	K—L_2	94.65	62.5	104.3	99.52	63.2	111
$K_{\alpha 1}$	K—L_3	98.44	100	103.0	103.74	100	114
$K_{\beta 3}$	K—M_2	110.42	11.5	107.3	116.24	11.6	116
$K_{\beta 1}$	K—M_3	111.30	22.6	105.0	117.23	22.8	110
$K_{\beta 2}$	K—$N_{2,3}$	114.3, 114.56	8.7		120.44, 120.7	8.9	
$L_{\alpha 2}$	L_3—M_4	13.44	11.4[e]	12.4	14.08	11.4[e]	
$L_{\alpha 1}$	L_3—M_5	13.62	100	12.4	14.28	100	
L_ℓ	L_3—M_1	11.62	16.7		12.12	7.2	
$L_{\beta 2}$	L_3—N_5	16.43	21	13.3	17.26	21.3	
$L_{\beta 4}$	L_1—M_2	16.58	100[e]	27.5	17.56	100[e]	
$L_{\beta 1}$	L_2—M_4	17.22	100[e]	13.5	18.29	100[e]	
$L_{\beta 3}$	L_1—M_3	17.45	87	23.7	18.54	82	
$L_{\gamma 1}$	L_2—N_4	20.17	22.6	15.7	21.42	23	

[a]G. Zschornack, *Handbook of X-Ray Data* (Springer-Verlag, Berlin, Heidelberg, 2007)
[b]Intensities relative to $K_{\alpha 1}$, $L_{\alpha 1}$, and $L_{\beta 4}$ shown in percent
[c]*CRC Handbook of Chemistry and Physics*, 103rd edition (Internet Version 2022), John R. Rumble, ed., CRC Press/Taylor and Francis, Boca Raton, FL
[d]G. Barreau, H. G. Börner, T. von Egidy, R. W. Hoff, "Precision Measurements of X-Ray Energies, Natural Widths, and Intensities in the Actinide Region," *Zeitschrift für Physik A Atoms and Nuclei* 308, 209 (1982)
[e]Note for L lines, vacancies in L_1, L_2, and L_3 are not necessarily proportional. Different intensity ratios between shells will be observed depending on excitation conditions

Table 27.5 Decay characteristics of uranium isotopes and their progenies[a]

Isotope	Half-life (years)	Activity (dis/s − g)	Specific power (mW/g isotope)
^{232}U	68.9 ± 0.4	8.274×10^{11}	717.6 ± 4.2
^{233}U	$(1.592 \pm 0.002) \times 10^5$	3.565×10^8	0.28039 ± 0.00036
^{234}U	$(2.455 \pm 0.006) \times 10^5$	2.302×10^8	0.17925 ± 0.00044
^{235}U	$(7.038 \pm 0.005) \times 10^8$	7.996×10^4	$(6.0959 \pm 0.0044) \times 10^{-5}$
^{236}U	$(2.342 \pm 0.004) \times 10^7$	2.393×10^6	$(1.7532 \pm 0.0030) \times 10^{-3}$
^{238}U	$(4.468 \pm 0.003) \times 10^9$	1.244×10^4	$(1.0220 \pm 0.0053) \times 10^{-5}$
^{228}Th	1.9125 ± 0.0009	3.033×10^{13}	$(1.6222 \pm 0.0008) \times 10^5$

[a] *NuDat 3* Database, http://www.nndc.bnl.gov/nudat3/

Table 27.6 Major gamma-rays from uranium isotopes[a]

Isotope	Gamma-ray energy (keV)	Branching ratio (%)	Specific intensity (gamma/s − g of isotope)
^{232}U	57.8	2.00×10^{-1}	1.65×10^9
	129.1	6.82×10^{-2}	5.64×10^8
	270.2	3.16×10^{-3}	2.61×10^7
	327.9	2.83×10^{-3}	2.34×10^7
^{233}U	54.7	1.68×10^{-2}	5.99×10^4
	97.1	2.03×10^{-2}	7.24×10^4
	119.0	3.63×10^{-3}	1.29×10^4
	120.8	2.82×10^{-3}	1.01×10^4
	146.3	6.50×10^{-3}	2.32×10^4
	164.5	6.26×10^{-3}	2.23×10^4
	245.3	3.57×10^{-3}	1.27×10^4
	291.3	5.25×10^{-3}	1.87×10^4
	317.2	7.37×10^{-3}	2.63×10^4
^{234}U	53.2	1.23×10^{-1}	2.83×10^5
	120.9	3.50×10^{-2}	8.06×10^4
^{235}U	143.8	1.10×10^1	8.76×10^3
	163.4	5.08×10^0	4.06×10^3
	185.7	5.70×10^1	4.56×10^4
	202.1	1.08×10^0	8.64×10^2
	205.3	5.02×10^0	4.01×10^3
^{236}U	49.5	7.80×10^{-2}	1.87×10^3
	112.8	1.90×10^{-2}	4.55×10^2
^{238}U	63.3	3.70×10^0	4.60×10^2
In equilibrium with ^{234m}Pa	92.4	2.13×10^0	2.65×10^2
	92.8	2.10×10^0	2.61×10^2
	258.2	7.64×10^{-2}	9.50×10^0
	742.8	1.07×10^{-1}	1.33×10^1
	766.4	3.17×10^{-1}	3.94×10^1
	786.3	5.44×10^{-2}	6.77×10^0
	1001.0	8.42×10^{-1}	1.05×10^2
^{228}Th	238.6	4.35×10^1	5.41×10^3
In equilibrium with ^{208}Tl	583.2	3.06×10^1	3.81×10^3
	2614.5	3.59×10^1	4.46×10^4

[a] NuDat 3, http://www.nndc.bnl.gov/nudat3/chartNuc.jsp

Table 27.7 Decay characteristics of nuclides commonly assayed by calorimetric assay[a]

Isotope	Dominant decay mode[b]	Specific power (P_i) (mW/g)	% Std. Dev.	$T^{1/2}$ (y)[2]	% Std. Dev.
^{238}Pu	α	567.57	0.05	87.74	0.05
^{239}Pu	α	1.9288	0.02	24,119	0.11
^{240}Pu	α	7.0824	0.03	6564	0.17
^{241}Pu	β	3.412	0.06	14.348	0.15
^{242}Pu	α	0.1159	0.22	376,300	0.24
^{241}Am	α	114.2	0.37	433.6	0.32
^{3}H	β	324	0.14	12.3232	0.017

[a]"Standard Test Method for Nondestructive Assay of Plutonium, Tritium, and ^{241}Am by Calorimetric Assay," American Society of Testing and Materials Standard C1458 (2016)
[b]For all the nuclides listed, the dominant decay mode has a branching ratio > 99.99%

Table 27.8 Gamma rays of ^{238}Pu[a]

	Energy (keV)	Intensity (%)	Error
	43.498	3.92×10^{-2}	8×10^{-4}
U K$_{\alpha2}$	94.654	5.79×10^{-5}	1.5×10^{-6}
U K$_{\alpha1}$	98.434	9.25×10^{-5}	2.3×10^{-6}
	99.853	7.29×10^{-3}	8×10^{-5}
K$_{\beta3}$	110.421	1.16×10^{-5}	3×10^{-7}
K$_{\beta1}$	111.298	2.20×10^{-5}	6×10^{-7}
	152.720	9.29×10^{-4}	7×10^{-6}
	200.97	3.9×10^{-6}	2×10^{-7}
	705.9	5.3×10^{-8}	2×10^{-7}
	708.4	4.1×10^{-7}	7×10^{-8}
	742.81	5.2×10^{-6}	2×10^{-7}
	766.39	2.2×10^{-5}	2×10^{-6}
	786.3	3.2×10^{-6}	3×10^{-7}
	804.4	1.2×10^{-7}	5×10^{-8}
	808.3	7.9×10^{-7}	2×10^{-8}
	851.7	1.24×10^{-6}	1.5×10^{-7}

[a]E. Browne and J. K. Tuli, *NuDat 3* Database, *Nuclear Data Sheets* 108, 681 (2007)

Table 27.9 Gamma rays of ^{239}Pu[a]

	Energy (keV)	Intensity (%)	Error	Energy (keV)	Intensity (%)	Error
	30.040	2.17×10^{-4}	6×10^{-6}	243.38	2.53×10^{-5}	5×10^{-7}
	38.661	1.04×10^{-2}	1.3×10^{-4}	244.92	5.1×10^{-6}	5×10^{-7}
	41.930	1.46×10^{-4}	1.5×10^{-5}	248.95	7.2×10^{-6}	7×10^{-7}
	46.210	7.21×10^{-5}	1.1×10^{-6}	255.38	8.00×10^{-5}	1×10^{-6}
	46.680	4.65×10^{-5}	2.5×10^{-6}	263.95	2.65×10^{-5}	1×10^{-6}
	51.624	2.72×10^{-2}	2.2×10^{-4}	265.7	1.6×10^{-6}	3×10^{-7}
	54.039	1.94×10^{-4}	2.5×10^{-6}	281.2	2.1×10^{-6}	3×10^{-7}
	56.828	1.15×10^{-3}	1.3×10^{-5}	285.3	1.9×10^{-6}	4×10^{-7}
	65.708	5.2×10^{-5}	3×10^{-6}	297.46	4.98×10^{-5}	8×10^{-7}
	67.674	1.52×10^{-4}	2.3×10^{-6}	302.87	5.1×10^{-6}	4×10^{-7}
	77.592	3.8×10^{-4}	5×10^{-5}	307.85	5.5×10^{-6}	4×10^{-7}
	78.430	1.54×10^{-4}	2.2×10^{-6}	311.78	2.58×10^{-5}	7×10^{-7}
	89.640	2.7×10^{-5}	2×10^{-6}	316.41	1.32×10^{-5}	4×10^{-7}
U K$_{\alpha2}$	94.654	3.5×10^{-3}	3×10^{-4}	319.68	4.8×10^{-6}	5×10^{-7}
	96.140	3.79×10^{-5}	1.8×10^{-6}	320.862	5.42×10^{-5}	7×10^{-7}
U K$_{\alpha1}$	98.434	5.6×10^{-3}	5×10^{-4}	323.84	5.39×10^{-5}	7×10^{-7}
	98.780	1.47×10^{-3}	7×10^{-5}	332.845	4.94×10^{-4}	3×10^{-6}
	103.060	2.16×10^{-4}	5×10^{-6}	336.113	1.12×10^{-4}	2×10^{-6}
K$_{\beta3}$	110.421	7.0×10^{-4}	7×10^{-5}	341.506	6.62×10^{-5}	1.4×10^{-6}
K$_{\beta1}$	111.298	1.33×10^{-3}	1.3×10^{-4}	345.013	5.56×10^{-4}	5×10^{-6}

(*continued*)

Table 27.9 (continued)

	Energy (keV)	Intensity (%)	Error	Energy (keV)	Intensity (%)	Error
$K_{\beta 2}$	114.445	5.2×10^{-4}	5×10^{-5}	354	7×10^{-7}	3×10^{-7}
	115.380	4.6×10^{-4}	5×10^{-5}	361.89	1.22×10^{-5}	6×10^{-7}
	116.260	5.67×10^{-4}	1.1×10^{-5}	367.073	8.9×10^{-5}	2×10^{-6}
	119.70	2.1×10^{-5}	1×10^{-5}	368.554	8.8×10^{-5}	2×10^{-6}
	123.62	2.37×10^{-5}	9×10^{-7}	375.054	1.554×10^{-3}	9×10^{-6}
	124.51	6.81×10^{-5}	1.8×10^{-4}	380.191	3.05×10^{-4}	6×10^{-6}
	125.21	5.63×10^{-5}	1.5×10^{-6}	382.75	2.59×10^{-4}	5×10^{-6}
	129.296	6.31×10^{-3}	4×10^{-5}	392.53	2.05×10^{-4}	2×10^{-5}
	141.657	3.2×10^{-5}	7×10^{-7}	393.14	3.5×10^{-4}	3×10^{-5}
	143.35	1.73×10^{-5}	7×10^{-7}	399.53	5.9×10^{-6}	3×10^{-7}
	144.201	2.83×10^{-4}	6×10^{-6}	406.8	2.5×10^{-6}	5×10^{-7}
	146.094	1.19×10^{-4}	3×10^{-6}	411.2	7×10^{-6}	3×10^{-7}
	158.1	1.0×10^{-6}	1×10^{-7}	413.713	1.466×10^{-3}	1.1×10^{-5}
	160.19	6.2×10^{-6}	1.2×10^{-6}	422.598	1.22×10^{-4}	2×10^{-6}
	161.45	1.23×10^{-4}	2×10^{-6}	426.68	2.33×10^{-5}	6×10^{-7}
	167.81	2.9×10^{-6}	7×10^{-7}	428.4	1.0×10^{-6}	1×10^{-7}
	171.393	1.1×10^{-4}	2×10^{-6}	430.08	4.3×10^{-6}	1.3×10^{-7}
	173.7	3.1×10^{-6}	8×10^{-7}	445.72	8.8×10^{-6}	6×10^{-7}
	179.22	6.6×10^{-5}	1×10^{-6}	451.481	1.894×10^{-4}	1.6×10^{-6}
	188.23	1.09×10^{-5}	1.1×10^{-6}	457.61	1.49×10^{-6}	2×10^{-8}
	189.36	8.3×10^{-5}	1×10^{-6}	461.25	2.27×10^{-6}	2×10^{-8}
	195.68	1.07×10^{-4}	1×10^{-6}	463.9	2.8×10^{-7}	3×10^{-8}
	203.55	5.69×10^{-4}	3×10^{-6}	473.9	5×10^{-8}	3×10^{-8}
	225.42	1.51×10^{-5}	5×10^{-7}	481.66	4.6×10^{-6}	2×10^{-7}
	237.77	1.44×10^{-5}	6×10^{-7}	487.06	2.65×10^{-7}	2.1×10^{-8}
	242.08	7.3×10^{-6}	5×10^{-7}	493.08	8.7×10^{-7}	3×10^{-8}
	526.4	5.7×10^{-8}	1.9×10^{-8}	703.68	3.95×10^{-6}	2×10^{-8}
	550.5	4.2×10^{-7}	3×10^{-8}	714.71	7.9×10^{-8}	8×10^{-9}
	579.4	8.6×10^{-8}	1.7×10^{-8}	718	2.8×10^{-6}	2×10^{-7}
	582.89	6.15×10^{-7}	1.8×10^{-8}	720.3	2.9×10^{-8}	3×10^{-9}
	586.3	1.53×10^{-7}	1.5×10^{-8}	727.9	1.24×10^{-7}	6×10^{-9}
	596	3.9×10^{-8}	2×10^{-8}	736.5	3×10^{-8}	1×10^{-8}
	597.99	1.67×10^{-6}	5×10^{-8}	747.4	8.1×10^{-8}	1.6×10^{-8}
	599.6	2×10^{-7}	2×10^{-8}	756.4	2.8×10^{-6}	5×10^{-7}
	606.9	1.2×10^{-7}	1.2×10^{-8}	766.47	1.3×10^{-7}	2×10^{-8}
	612.83	9.5×10^{-7}	5×10^{-8}	792.9	2×10^{-8}	4×10^{-9}
	617.1	1.34×10^{-6}	7×10^{-8}	805.9	2.7×10^{-8}	4×10^{-9}
	618.28	2.04×10^{-6}	6×10^{-8}	808.4	1.21×10^{-7}	6×10^{-9}
	619.21	1.21×10^{-6}	8×10^{-8}	813.7	4.5×10^{-8}	5×10^{-9}
	624.78	4.37×10^{-7}	2×10^{-8}	816	2.4×10^{-8}	4×10^{-9}
	633.15	2.53×10^{-6}	3×10^{-8}	821.3	5×10^{-8}	1.1×10^{-8}
	639.99	8.7×10^{-6}	2×10^{-7}	832.5	2.96×10^{-8}	2.3×10^{-9}
	645.94	1.52×10^{-5}	3×10^{-7}	840.4	4.8×10^{-8}	5×10^{-9}
	649.32	7.1×10^{-7}	5×10^{-8}	843.78	1.34×10^{-7}	7×10^{-9}
	652.05	6.6×10^{-6}	2×10^{-7}	879.2	3.6×10^{-8}	4×10^{-9}
	654.88	2.25×10^{-6}	3×10^{-8}	891	7.5×10^{-8}	8×10^{-9}
	658.86	9.7×10^{-6}	2×10^{-7}	918.7	8×10^{-9}	3×10^{-9}
	664.58	1.66×10^{-6}	3×10^{-8}	940.3	5×10^{-8}	5×10^{-9}
	668.2	3.9×10^{-8}	1.3×10^{-8}	955.6	3.1×10^{-8}	3×10^{-9}
	674.05	5.15×10^{-7}	1.6×10^{-8}	957.6	3.2×10^{-8}	3×10^{-9}
	690.81	9×10^{-7}	2.5×10^{-7}	979.7	2.8×10^{-8}	5×10^{-9}
	693.2	2×10^{-8}	2×10^{-9}	986.9	2.1×10^{-8}	4×10^{-9}
	697.8	7.4×10^{-8}	1.5×10^{-8}	992.7	2.7×10^{-8}	4×10^{-9}
	701.1	5.12×10^{-7}	1.6×10^{-8}	1005.7	1.8×10^{-8}	3×10^{-9}

[a]E. Browne and J. K. Tuli, *NuDat 3* Database, *Nuclear Data Sheets* 122, 205 (2014)

Table 27.10 Gamma rays of ^{240}Pu[a]

	Energy (keV)	Intensity (%)	Error
	45.244	4.47×10^{-2}	9×10^{-4}
U K$_{\alpha 2}$	94.654	2.54×10^{-5}	7×10^{-7}
U K$_{\alpha 1}$	98.434	4.05×10^{-5}	1.1×10^{-6}
	104.234	7.14×10^{-3}	9×10^{-5}
K$_{\beta 3}$	110.421	5.08×10^{-6}	1.3×10^{-7}
K$_{\beta 1}$	111.298	9.6×10^{-6}	3×10^{-7}
K$_{\beta 2}$	114.445	3.73×10^{-6}	1.0×10^{-7}
	160.308	4.02×10^{-4}	4×10^{-6}
	212.46	2.9×10^{-5}	3×10^{-6}
	642.35	1.3×10^{-5}	1×10^{-6}
	687.57	3.5×10^{-6}	2×10^{-7}

[a]E. Browne and J. K. Tuli, *NuDat 3* Database, *Nuclear Data Sheets* 107, 2649 (2006)

Table 27.11 Gamma rays of ^{241}Pu[a]

	Energy (keV)	Intensity (%)	Error
	44.2	4.165×10^{-6}	
	44.86	8.33×10^{-7}	
	56.32	2.499×10^{-6}	
	56.76	9.8×10^{-7}	
	71.6	2.87×10^{-6}	1×10^{-7}
	77.1	2.06×10^{-5}	5×10^{-7}
U K$_{\alpha 2}$	94.654	3.12×10^{-4}	1.1×10^{-5}
U K$_{\alpha 1}$	98.434	4.99×10^{-4}	1.7×10^{-5}
	103.68	1.012×10^{-4}	6×10^{-7}
K$_{\beta 3}$	110.421	6.25×10^{-5}	2.1×10^{-6}
K$_{\beta 1}$	111.298	1.18×10^{-4}	4×10^{-6}
	114	8.1×10^{-5}	7×10^{-6}
K$_{\beta 2}$	114.445	4.59×10^{-5}	1.6×10^{-6}
	121.2	6.86×10^{-7}	
	148.567	1.86×10^{-4}	1.4×10^{-6}
	159.955	6.54×10^{-6}	8×10^{-8}

[a]M. S. Basunia, *NuDat 3* Database, *Nuclear Data Sheets* 107, 2323 (2006)

Table 27.12 Gamma rays of ^{237}U[a]

	Energy (keV)	Intensity (%)	Error
	26.3446	2.43	0.06
	33.196	0.13	5×10^{-3}
	43.42	2.4×10^{-2}	2×10^{-3}
	51.01	0.34	1×10^{-4}
	59.5409	34.5	0.8
	64.83	1.282	0.017
K$_{\alpha 2}$	97.069	15.4	0.4
K$_{\alpha 1}$	101.059	24.5	0.6
K$_{\beta 3}$	113.303	3.07	0.07
K$_{\beta 1}$	114.234	5.81	0.14
K$_{\beta 2}$	117.463	2.27	0.05
	164.61	1.86	0.03
	208.005	21.2	0.3
	221.80	2.12×10^{-2}	7×10^{-4}
	234.4	2.05×10^{-2}	7×10^{-4}
	267.54	0.712	0.01

(*continued*)

Table 27.12 (continued)

	Energy (keV)	Intensity (%)	Error
	332.35	1.2	0.16
	335.37	9.51×10^{-2}	2.2×10^{-3}
	337.7	8.9×10^{-3}	5×10^{-4}
	368.62	3.92×10^{-2}	1.7×10^{-3}
	370.94	0.1073	1.7×10^{-3}

[a] M. S. Basunia, *NuDat 3* Database, *Nuclear Data Sheets* 107, 2323 (2006)

Table 27.13 Gamma rays of ^{241}Am[a]

	Energy (keV)	Intensity (%)	Error	Energy (keV)	Intensity (%)	Error
	26.3446	2.27	0.12	383.81	2.82×10^{-5}	8×10^{-7}
	33.196	0.126	3×10^{-3}	406.35	1.45×10^{-6}	2.2×10^{-7}
	59.5409	35.9	0.4	419.33	2.87×10^{-5}	8×10^{-7}
$K_{\alpha 2}$	97.069	1.14×10^{-3}	4×10^{-5}	426.47	2.46×10^{-5}	7×10^{-7}
	98.97	2.03×10^{-2}	4×10^{-4}	452.6	2.4×10^{-6}	3×10^{-7}
$K_{\alpha 1}$	101.059	1.81×10^{-3}	6×10^{-5}	454.66	9.7×10^{-6}	4×10^{-7}
	102.98	1.95×10^{-2}	4×10^{-4}	459.68	3.6×10^{-6}	3×10^{-6}
$K_{\beta 3}$	113.303	2.27×10^{-4}	7×10^{-6}	468.12	2.88×10^{-6}	2.1×10^{-7}
$K_{\beta 1}$	114.234	4.3×10^{-4}	1.4×10^{-5}	512.5	1.15×10^{-6}	2.3×10^{-7}
$K_{\beta 2}$	117.463	1.68×10^{-4}	5×10^{-6}	514	2.6×10^{-6}	3×10^{-7}
	123.052	1×10^{-3}	3×10^{-5}	522.06	1×10^{-6}	3×10^{-7}
	125.30	4.08×10^{-3}	1×10^{-4}	573.94	1.25×10^{-6}	1.9×10^{-7}
	139.44	5.3×10^{-6}	1.1×10^{-6}	586.59	1.31×10^{-6}	2×10^{-7}
	146.55	4.61×10^{-4}	1.2×10^{-5}	590.28	2.86×10^{-6}	2.1×10^{-7}
	150.04	7.4×10^{-5}	2.2×10^{-6}	597.48	7.4×10^{-6}	3×10^{-7}
	159.26	1.4×10^{-6}	5×10^{-7}	619.01	5.94×10^{-5}	8×10^{-7}
	164.69	6.67×10^{-5}	2.5×10^{-6}	627.18	5.6×10^{-7}	1.7×10^{-7}
	165.81	2.32×10^{-5}	1.1×10^{-6}	632.93	1.26×10^{-6}	1.9×10^{-7}
	169.56	1.73×10^{-4}	4×10^{-6}	641.47	7.1×10^{-6}	3×10^{-7}
	175.07	1.82×10^{-5}	1×10^{-6}	653.02	3.77×10^{-5}	1.2×10^{-6}
	191.96	2.16×10^{-5}	1×10^{-6}	662.4	3.64×10^{-4}	9×10^{-6}
	204.06	2.9×10^{-6}	1.9×10^{-7}	669.83	3.8×10^{-7}	1.2×10^{-7}
	208.01	7.91×10^{-4}	1.9×10^{-5}	676.03	6.4×10^{-7}	1.3×10^{-7}
	221.46	4.24×10^{-5}	1.1×10^{-6}	680.10	3.13×10^{-6}	1.7×10^{-7}
	232.81	4.6×10^{-6}	3×10^{-7}	688.72	3.25×10^{-5}	9×10^{-7}
	234.33	7×10^{-7}	3×10^{-7}	696.6	5.34×10^{-6}	2.1×10^{-7}
	246.73	2.4×10^{-6}	3×10^{-7}	722.01	1.96×10^{-4}	4×10^{-6}
	260.8	1.21×10^{-6}	1.9×10^{-7}	729.72	1.33×10^{-6}	1.4×10^{-7}
	264.89	9×10^{-6}	4×10^{-7}	737.34	8×10^{-6}	3×10^{-7}
	267.58	2.63×10^{-5}	8×10^{-7}	755.9	7.6×10^{-6}	3×10^{-7}
	275.77	6.6×10^{-6}	4×10^{-7}	767	5×10^{-6}	1.9×10^{-7}
	291.3	3.1×10^{-6}	3×10^{-7}	770.57	4.74×10^{-6}	2.2×10^{-7}
	292.77	1.42×10^{-5}	5×10^{-7}	772.4	2.66×10^{-6}	1.5×10^{-7}
	304.21	1.01×10^{-6}	2.1×10^{-7}	801.94	1.36×10^{-6}	1.4×10^{-7}
	322.52	1.52×10^{-4}	4×10^{-6}	828.5	2.4×10^{-7}	6×10^{-8}
	332.35	1.49×10^{-4}	3×10^{-6}	860.7	8×10^{-8}	3×10^{-8}
	335.37	4.96×10^{-4}	1.1×10^{-5}	862.7	5.3×10^{-7}	6×10^{-8}
	337.7	4.29×10^{-6}	2.3×10^{-7}	887.3	2.2×10^{-7}	5×10^{-8}
	358.25	1.2×10^{-6}	2.4×10^{-7}	921.5	1.9×10^{-7}	4×10^{-8}
	368.65	2.17×10^{-4}	5×10^{-6}			
	370.94	5.23×10^{-5}	1.3×10^{-6}			
	376.65	1.38×10^{-4}	3×10^{-6}			

[a] M. S. Basunia, *NuDat 3* Database, *Nuclear Data Sheets* 107, 3323 (2006)

Table 27.14 Isotopes measurable by gamma rays in typical irradiated fuel assembly

Fission product isotopes	Half-life	Fission yield in ^{235}U (%)	Fission yield in ^{239}Pu (%)	Gamma-ray energy (keV)	Branching ratio (%)
^{95}Zr	64.0 days	6.50	4.89	724.2	43.1
				756.7	54.6
^{95}Nb	35.0 days	6.50	4.89	765.8	99.8
^{103}Ru	39.4 days	3.04	6.95	497.1	86.4
				610.3	5.4
^{106}Ru − Rh	366.4 days	0.40	4.28	622.2	9.8
				1050.5	1.6
^{134}Cs	2.06 years	1.27×10^{-5} [a]	9.89×10^{-4} [a]	604.7	97.6
				795.8	85.4
				801.8	8.7
				1167.9	1.8
				1365.1	3
^{137}Cs	30.17 years	6.22	6.69	661.6	85.1
^{140}Ba − La	12.7 days			1596	
^{144}Ce − Pr	284.5 days	5.48	3.74	696.5	1.3
				1489.2	0.3
				2185.6	0.7
^{154}Eu	8.5 years	2.69×10^{-6} [a]	9.22×10^{-5} [a]	996.3	10.3
				1004.8	17.4
				1274.4	35.5
Activation products					
^{54}Mn	312.2 days			834.8	100.0
^{58}Co	70.3 days			811.1	99.0
^{60}Co	5.27 years			1173.2	100.0
				1332.5	100.0

[a]Europium-154 and ^{134}Cs values are given only for direct production of the isotope from fission. Actually, each of these isotopes is produced primarily through neutron absorption. For pressurized-water reactor fuel material irradiated to 25 GWd/tU, the "accumulated fission yields" of ^{154}Eu and ^{134}Cs were calculated as 0.15% and 0.46%, respectively
[b]D. Cobb, J. Phillips, G. Bosler, G. Eccleston, J. Halbig, C. Hatcher, and S.-T. Hsue, "Nondestructive Verification and Assay Systems for Spent Fuels," Los Alamos National Laboratory, 1982

Table 27.15 Spontaneous fission neutron yields

Isotope A	Protons Z	Neutrons N	Half-Life[a]	Fission Yield[b] (n/s − g)	Spontaneous multiplicity ν[b,c,d]	Thermal multiplicity ν^c
^{232}Th	90	142	1.40×10^{10} year	1.01×10^{-7}	2.13	1.9
^{232}U	92	140	68.9 year	1.43×10^{-2}	1.71	3.13
^{233}U	92	141	1.59×10^5 year	3.7×10^{-4}	1.76	2.4
^{234}U	92	142	2.45×10^5 year	6.82×10^{-3}	1.81	2.4
^{235}U	92	143	7.04×10^8 year	1.05×10^{-5}	1.86	2.41
^{236}U	92	144	2.34×10^7 year	4.24×10^{-3}	1.89	2.2
^{238}U	92	146	4.47×10^9 year	1.34×10^{-2}	1.98	2.3
^{237}Np	93	144	2.14×10^6 year	1.14×10^{-4}	2.05	2.70
^{238}Pu	94	144	87.74 year	2.56×10^{3e}	2.19	2.9
^{239}Pu	94	145	2.41×10^4 year	1.49×10^{-2}	2.16	2.88
^{240}Pu	94	146	6.56×10^3 year	1.04×10^3	2.154	2.8
^{241}Pu	94	147	14.4 year	2.06×10^{-3}	2.25	2.8
^{242}Pu	94	148	3.75×10^5 year	1.74×10^{3e}	2.149	2.81
^{241}Am	95	146	432.7 year	1.14	2.5	3.09
^{242}Cm	96	146	163 day	1.98×10^7	2.54	3.44
^{244}Cm	96	148	18.1 year	1.11×10^7	2.71	3.46
^{249}Bk	97	152	320 day	1.0×10^5	3.40	3.7
^{252}Cf	98	154	2.645 year	2.29×10^{12}	3.757	4.06

[a]N. E. Holden and D. C. Hoffman, "Spontaneous Fission Half-Lives for Ground-State Nuclide," *Pure Appl. Chem.* 72 (8), 1525–1562 (2000)
[b]R. T. Perry and W. B. Wilson, "Neutron Production from (a,n) Reactions and Spontaneous Fission in ThO$_2$, UO$_2$ and (U$_x$PU)O$_2$ Fuels," Los Alamos National Laboratory report LA-8869-MS (June 1981)
[c]*Evaluated Nuclear Data File ENDF/B-V* (available from and maintained by the National Nuclear Data Center at Brookhaven National Laboratory)
[d]P. Santi and M. Miller, "Reevaluation of Prompt Neutron Emission Multiplicity Distributions for Spontaneous Fission," *Nucl. Sci. Eng.* 160, 190 (2008)
[e]S. Croft and A. Favalli, "Review and Evaluation of the Spontaneous Fission Half-Lives of ^{238}Pu, ^{240}Pu, and ^{242}Pu and the Corresponding Specific Fission Rates," *Nuclear Data Sheets* 175, 269 (2021)

Table 27.16 (α,n) neutron yields

Isotope	Half-Life[a,b]	Alpha Yield[a,b] (α/s·g)	Average alpha energy (MeV)[a,b]	(α,n) yield in oxide (n/s·g)[c]	(α,n) yield in UF$_6$/PuF$_4$ (n/s·g)[d,e]
^{232}Th	1.40×10^{10} year	4×10^3	4.00	2.2×10^{-5}	
^{232}U	68.9 yr[f]	8.0×10^{11}	5.30	1.49×10^4	2.6×10^6
^{233}U	1.59×10^5 year	3.57×10^8	4.82	4.8	7.0×10^2
^{234}U	2.45×10^5 year	2.3×10^8	4.76	3.0	5.8×10^2
^{235}U	7.04×10^8 year	8.0×10^4	4.40	7.1×10^{-4}	0.122
^{236}U	2.34×10^7 year	2.4×10^6	4.48	2.4×10^{-2}	3.96
^{238}U	4.47×10^9 year	1.2×10^4	4.19	8.3×10^{-5}	0.014
^{237}Np	2.14×10^6 year	2.6×10^7	4.77	3.4×10^{-1}	
^{238}Pu	87.74 year	6.3×10^{11}	5.49	1.34×10^4	2.2×10^6
^{239}Pu	2.41×10^4 year	2.3×10^9	5.15	3.81×10^1	5.6×10^3
^{240}Pu	6.56×10^3 year	8.4×10^9	5.15	1.41×10^2	2.1×10^4
^{241}Pu	14.3 year[f]	9.3×10^7	4.89	1.3	1.7×10^2
^{242}Pu	3.75×10^5 year[f]	1.5×10^8	4.90	2.0	2.7×10^2
^{241}Am	432.6 year[f]	1.3×10^{11}	5.48	2.69×10^3	4.4×10^5
^{242}Cm	163 day	1.2×10^{14}	6.10	3.76×10^6	
^{244}Cm	18.1 year	3.0×10^{12}	5.80	7.73×10^4	
^{249}Bk	320 day	8.8×10^8	5.40	1.8×10^1	
^{252}Cf	2.645 year[f]	1.9×10^{13}	6.11	6.0×10^5	

[a]C. M. Lederer and V. S. Shirley, Eds., *Table of Isotopes*, 7th ed. (John Wiley & Sons, Inc., New York, 1978)
[b]N. E. Holden and D. C. Hoffman, "Spontaneous Fission Half-Lives for Ground-State Nuclide," *Pure Appl. Chem.* 72 (8), 1525–1562 (2000)
[c]R. T. Perry and W. B. Wilson, "Neutron Production from (α,n) Reactions and Spontaneous Fission in ThO$_2$, UO$_2$, and (U$_x$PU)O$_2$ Fuels," Los Alamos National Laboratory report LA-8869-MS (June 1981)
[d]UF$_6$: T. E. Sampson, "Neutron Yields from Uranium Isotopes in Uranium Hexafluoride," *Nucl. Sci. Eng.* 54, 470 (1974); W. B. Wilson, J. E. Stewart, and R. T. Perry, "Neutron Production in UF$_6$ from the Decay of Uranium Nuclides," *Trans. Am. Nucl. Soc.* 38, 176 (1981)
[e]PuF4: W. B. Wilson, Los Alamos National Laboratory memorandum T-2-M-1432 to N. Ensslin (1983)
[f]*NuDat 3.0*, https://www.nndc.bnl.gov/nudat3/

Table 27.17 Thick-target yields from (α,n) reactions

Element[a]	Neutron yield per 10^6 alphas of energy 4.7 MeV (^{234}U)	Neutron yield per 10^6 alphas of energy 5.2 MeV (avg. Pu)	References	Average neutron energy for 5.2 MeV Alphas[b]
Li	0.16 ± 0.04	1.13 ± 0.25	c	0.3
Be	44 ± 4	65 ± 5	d	4.2
B	12.4 ± 0.6	17.5 ± 0.4	b, c, e	2.9
C	0.051 ± 0.002	0.078 ± 0.004	b, c, d	4.4
O	0.040 ± 0.001	0.059 ± 0.002	b, c, d	1.9
F	3.1 ± 0.3	5.9 ± 0.6	b, c, e	1.2
Na	0.5 ± 0.5	$1.1 + 0.5$	f	
Mg	0.42 ± 0.03	0.89 ± 0.02	b, c, d	2.7
Al	0.13 ± 0.01	0.41 ± 0.01	b, c, d	1.0
Si	0.028 ± 0.002	0.076 ± 0.003	b, c, d	1.2
Cl	0.01 ± 0.01	0.07 ± 0.04	f	

[a]Natural Isotopic Composition
[b]G. J. H. Jacobs and H. Liskien, "Energy Spectra of Neutrons Produced by Alpha Particles in Thick Targets of Light Elements," *Ann. Nucl. Energy* 10, 541 (1983)
[c]J. K. Bair and J. Gomez del Campo, "Neutron Yields from Alpha-Particle Bombardment," *Nucl. Sci. Eng.* 71, 18 (1979)
[d]D. West and A. C. Sherwood, "Measurements of Thick-Target (α,n) Yields from Light Elements," *Ann. Nucl. Energy* 9, 551 (1982)
[e]W. B. Wilson and R. T. Perry, "Thick-Target Neutron Yields in Boron and Fluorine," Los Alamos National Laboratory memorandum T-2-M-1835 to N. Ensslin (1987)
[f]J. Roberts, "Neutron Yields of Several Light Elements Bombarded with Polonium Alpha Particles," U.S. Atomic Energy Commission report MDDC-731 (1947)

Table 27.18 Nuclear data tables and figures in PANDA

X-ray and Gamma-ray data tables and figures	Table/figure
Major K X-rays of uranium and plutonium	Table 2.1
Major NDA gamma-ray signatures	Table 2.2
Differential cross section for Compton scattering as a function of scattered electron energy and incident photon energy (graph)	Fig. 3.9
Energy of the Compton edge and backscatter versus the energy of the incident gamma ray (graph)	Fig. 3.11
Common nuclides used for energy calibration	Table 7.1
Selected gamma-ray energies from natural background	Table 7.2
Intense gamma radiation from uranium nuclides	Table 8.1
Decay characteristics of uranium isotopes and their progenies	Table 8.4
Decay characteristics of isotopes used in plutonium isotopic measurements	Table 9.1
Americium-241 contribution to ^{237}U gamma-ray peaks (graph)	Fig. 9.2
Useful plutonium gamma rays in various energy regions	Table 9.2
Energies and relative intensities of the major K and L X-rays of uranium and plutonium	Table 12.1
Isotopes measurable by gamma rays in typical irradiated fuel assembly	Table 21.4
Gamma-ray attenuation data tables and figures	Table/figure
Linear attenuation coefficient of NaI (graph)	Fig. 3.3
Photoelectric mass attenuation coefficient of lead (graph)	Fig. 3.6
Mass attenuation coefficient of various elements (graph)	Fig. 3.15
Attenuation properties for lead and tungsten	Table 3.2
Total mass attenuation coefficients of various element (graph)	Fig. 6.1
Far-field correction factors for slab, cylinder, and sphere as a function of transmission	Table 6.1
Mean free paths and infinite thicknesses for 186 keV photons in uranium compounds	Table 8.2
Material composition correction factors for uranium compounds	Table 8.3
Energy dependence of the photon mass attenuation coefficients for H, N, O, U, and Pu (graph)	Fig. 11.3
Mass attenuation coefficient vs energy for uranium and plutonium (graph)	Fig. 12.6
Mean free path of 400, 100, and 20 keV photons in water and 50 g/L uranium solution (graph)	Fig. 12.7
The 122 keV mean free path vs uranium concentration (uranyl nitrate in 4-M nitric acid) (graph)	Fig. 12.13
Gamma-ray spectra	Table/figure
High-resolution gamma-ray spectrum of highly enriched uranium (93% ^{235}U)	Fig. 2.10
Gamma-ray spectrum of highly enriched uranium (0–250 keV)	Fig. 2.11
Gamma-ray spectrum of depleted uranium	Fig. 2.12
Gamma-ray spectrum of plutonium with 14% ^{240}Pu and 1.2% ^{241}Am	Fig. 2.13
Gamma-ray spectrum of low-burnup plutonium with approximately 6% ^{240}Pu	Fig. 2.14
Gamma-ray spectrum of uranium ore	Fig. 2.16
Gamma-ray spectra from natural, 17% enriched, and 90% enriched uranium measured with a 25% efficiency coaxial HPGe detector	Fig. 8.1
Gamma-ray spectrum of 4.46% enriched uranium item in the 60–240 keV region measured with a planar HPGe detector with resolution 520 eV at 122 keV	Fig. 8.8
Gamma-ray spectrum of a 20.1% ^{235}U item measured with a coaxial HPGe detector with resolution of 1.0 keV at 186 keV	Fig. 8.10
Gamma-ray spectrum of low-burnup plutonium (93.82% ^{239}Pu)	Fig. 9.3
Gamma-ray spectrum of high-burnup plutonium (82.49% ^{239}Pu)	Fig. 9.4
Gamma-ray spectra in 30–60 keV region from freshly separated solutions of plutonium	Fig. 9.5
Gamma-ray spectra of PuO$_2$ in 60–120 keV region	Fig. 9.6
Gamma-ray spectra of PuO$_2$ in 120–500 keV region	Fig. 9.7
Gamma-ray spectra of PuO$_2$ in 500–800 keV region	Fig. 9.8
Neutron data tables and figures	Table/figure
Spontaneous fission neutron yields	Table 13.1
Measured prompt fission multiplicity distributions	Table 13.2
(α,n) reaction neutron yields	Table 13.3
(α,n) Q-values, threshold energies, and coulomb barriers	Table 13.4
Thick target yields from (α,n) reactions	Table 13.5
Mass stopping forces for commonly encountered elements	Table 13.6
Characteristics of ^{252}Cf	Table 13.9

(*continued*)

Table 27.18 (continued)

Neutron data tables and figures	Table/figure
Characteristics of some isotopic (α,n) sources	Table 13.10
Reactions for producing neutrons	Table 13.11
Average number of collisions required to reduce a neutron's energy from 2 MeV to 0.025 eV by elastic scattering	Table 14.1
Total neutron cross section of ^{239}Pu (graph)	Fig. 14.3
Low-energy total neutron cross section of boron (graph)	Fig. 14.4
Low-energy total neutron cross-section of cadmium (graph)	Fig. 14.5
Fission cross section for some important fissile (^{235}U, ^{239}Pu) and fertile (^{238}U, ^{240}Pu) isotopes (graph)	Fig. 14.6
Nuclear data for ^{nat}UO$_2$	Table 14.2
Neutron cross section of common materials	Table 14.3
^{3}He(n,p), ^{10}B(n,α) and ^{6}Li(n,α) cross sections as a function of incident neutron energy (graph)	Fig. 15.6
Primary neutron production rates from spontaneous fission and (α,n) in PuO$_2$ and PuF$_4$ for four plutonium isotopic compositions	Table 16.1
Primary neutron production rates in uranium metal, UO$_2$, UO$_2$F$_2$, and UF$_6$ for four uranium isotopic compositions (depleted uranium, natural uranium, and low-enriched uranium)	Table 16.2
Primary neutron production rates in U metal, UO$_2$, UO$_2$F$_2$, and UF$_6$ for four uranium isotopic compositions (highly enriched uranium)	Table 16.3
Specific neutron production rates from (α,n) reactions in PuO$_2$ versus added moisture for 10% ^{240}Pu, 16% ^{240}Pu, and 25% ^{240}Pu (graph)	Fig. 16.3
Specific neutron production rates from (α,n) reactions in PuO$_2$ versus fluorine contamination for 16% ^{240}Pu (graph)	Fig. 16.5
Cosmic ray neutron energy spectrum (graph)	Fig. 16.26
Common curium isotope properties	Table 19.8
The five principal neutron sources in a pressurized-water reactor fuel assembly with an exposure of 31.5 GWd/tU (graph)	Fig. 21.13
Data tables and graphs relevant for calorimetric assay	Table/figure
Nuclear parameters of commonly assayed nuclides	Table 23.1
Maximum energy loss per decay due to gamma-rays or spontaneous fission	Table 23.2

Open Access This chapter is licensed under the terms of the Creative Commons Attribution 4.0 International License (http://creativecommons.org/licenses/by/4.0/), which permits use, sharing, adaptation, distribution and reproduction in any medium or format, as long as you give appropriate credit to the original author(s) and the source, provide a link to the Creative Commons license and indicate if changes were made.

The images or other third party material in this chapter are included in the chapter's Creative Commons license, unless indicated otherwise in a credit line to the material. If material is not included in the chapter's Creative Commons license and your intended use is not permitted by statutory regulation or exceeds the permitted use, you will need to obtain permission directly from the copyright holder.

Correction to: Active Neutron Instrumentation and Applications

M. T. Swinhoe, N. Ensslin, L. G. Evans, W. H. Geist, M. S. Krick, A. L. Lousteau, R. McElroy, M. M. Pickrell, and P. Rinard

Correction to:
Chapter 20 in: W. H. Geist et al. (eds.), *Nondestructive Assay of Nuclear Materials for Safeguards and Security,*
https://doi.org/10.1007/978-3-031-58277-6_20

The original version of this chapter "Active Neutron Instrumentation and Applications" was inadvertently published without the following author's name "L. G. Evans". The author's name has been updated in the chapter.

Open Access This chapter is licensed under the terms of the Creative Commons Attribution 4.0 International License (http://creativecommons.org/licenses/by/4.0/), which permits use, sharing, adaptation, distribution and reproduction in any medium or format, as long as you give appropriate credit to the original author(s) and the source, provide a link to the Creative Commons license and indicate if changes were made.

The images or other third party material in this chapter are included in the chapter's Creative Commons license, unless indicated otherwise in a credit line to the material. If material is not included in the chapter's Creative Commons license and your intended use is not permitted by statutory regulation or exceeds the permitted use, you will need to obtain permission directly from the copyright holder.

The updated version of this chapter can be found at
https://doi.org/10.1007/978-3-031-58277-6_20

Index

A

Absorption edge
    densitometry, 241–249
    discontinuity, 241
    energy, 142, 241
Absorption efficiency, 55
Accidental coincidence rate, 433, 602
Activation products, 220, 609, 610, 614, 653, 726
Active well coincidence counter (AWCC), 377, 379, 432, 433, 453, 469, 470, 472, 474, 507, 542, 586–595, 601
Add-a-source, 460, 494, 519, 520, 526–527
Adiabatic calorimeter, 657, 662
Alpha decay
    energy, half-lives, yields, 10
    heat production, 660
    particle range, 9
(α, n) reaction
    Coulomb barrier, 294–296
    gamma rays, 294, 300
    neutron sources, 302, 567, 586
    neutron yields, 294
    n spectrum, 369
    Q value, 294, 296, 298
    thick target yield, 295, 297, 298
    thin target yield, 369
    threshold energy, 296, 297
Alpha ratio, 297, 362, 421
AmBe, neutron source, 586
AmBe, neutron spectrum, 300
AmLi, neutron source, 356, 412, 415, 471, 472, 586, 588, 601
AmLi, neutron spectrum, 300, 356, 431
^{241}Am-^{237}U peaks, 169, 172
Analog-to-digital converter (ADC), 59, 71–73, 117, 349
Atomic mass number (A), 9, 11, 19, 143, 608
Atomic number (Z), 9–12, 14, 15, 28, 30, 32, 36, 38–40, 46, 50, 83, 135, 237, 247, 272, 275, 277, 278, 280, 288–290, 295, 299, 337, 646
Attenuation coefficients
    compound materials, 30
    curves, 38
    linear, 27, 28, 40, 82, 91, 312
    mass, 17, 29–30, 38, 237, 240–242, 246, 275, 277
    power law dependence, 246
Attenuation correction factor
    approximate forms, 97–99
    Compton-scattering-based, 284
    far-field assay, 88
    holdup measurement, 224–225
    intensity ratio, 416
    internal standard, 284
    interpolation and extrapolation, 97–99
    numerical computation, 91–99
    precision, 99
    segmented gamma scan, 222–223
    transmission (γ-ray), 27, 40, 88, 248
    XRF, 165, 174, 275
Attenuation, fundamental law, 27–29, 84
Attribute measurement, 3, 185–189
Auger electron, 10, 14, 272
Autocorrelation, 389, 390, 396–397, 409

B

Background radiation
    cosmic rays, 21
    ^{40}K, 22
    natural radioactivity, 21, 22
Backscatter peak, 35, 36, 52, 53, 257, 276
Barn, 308, 561
Baseline restoration (BLR), 68, 70, 412
Beta decay, 8, 9, 11–12, 20, 170, 278, 279, 293, 294, 608, 658, 660
BF$_3$ neutron detector
    gamma-ray sensitivity, 329, 338
    neutron capture cross section, 352
    pulse-height spectrum, 107, 334
Binding energy, electron, 10, 30, 31
^{10}B neutron detector, 325, 329, 333–337
Branching intensity, gamma ray, 9
Bremsstrahlung, 15, 16, 31, 248, 250, 251, 254, 255, 277–279
Burnup
    calorimeter measurement, 181, 658, 668, 672, 673
    definition, 607
    gamma-ray assay, 20, 611
    neutron assay, 557
Burnup indicator
    ^{137}Cs, 608
    ^{134}Cs/^{137}Cs, 612
    ^{154}Eu/^{137}Cs, 612
    fission product ratio, 612
    total gamma-ray activity, 610–611, 618
    total neutron output, 423, 618, 619

C

Calibration standards, 83–85, 154, 155, 165, 174, 225, 227–229, 253, 256, 432, 469, 473, 668, 678, 716
Calorimeter
    adiabatic, 657, 662
    air chamber, 665
    analytical calorimeter, 678–679
    ARIES Gradient, 681–684
    assay error sources, 671
    assay time, 668–670

Calorimeter (*cont.*)
  bulk calorimeter, 657
  components, 677–678, 684
  design, 662–665, 668, 670, 677–682
  electrical calibration, 668
  equilibrium time, 669
  gradient bridge, 665, 681–684
  heat flow, 657, 662–669, 677, 688
  heat source calibration, 678
  irradiated fuel, 686, 687, 726
  isothermal, 662, 668
  large-volume, 685–686
  mound transportable, 680, 681
  mound twin-bridge production, 681
  over/under bridge, 663
  sensitivity, 662–664, 670, 685
  twin-bridge, 662–664, 667, 669, 673, 680–682
Calorimeter operation
  differential method, 684
  end-point prediction, 681
  isotopic assay, 657
  replacement method, 668, 683
  sample preconditioning, 668, 670
  servo-control, 668, 670
CdTe detector, 46, 50
Cerenkov radiation, 605, 633
^{252}Cf
  prompt gamma-ray spectrum, 293
  prompt neutron spectrum, 292, 302, 622
Compton background
  single ROI subtraction, 111
  step function, 110, 112
  straight-line subtraction, 109–112
  two-standard subtraction, 111
Compton edge, 34–36, 52, 189, 344, 646, 650, 651, 728
Compton scattering, 28–30, 32–36, 38, 50–52, 55, 61, 82, 108, 110, 113, 116, 156, 284, 331, 342, 613, 633, 728
Compton suppression79, 389
Cosmic rays
  background, 21, 214, 301, 383, 493
  neutrons, 301, 383, 521, 522, 524
Criticality, 5, 84, 186, 225, 319, 355, 374, 418, 423, 517, 521, 559, 607, 663
Cross section
  ^{10}B, 339
  barn, 308
  definition, 308
  ^{3}He, 330, 338, 339, 353, 356, 381, 515, 534, 557, 598
  ^{1}H and ^{4}He elastic scattering, 339
  ^{6}Li neutron capture, 334, 352, 354
  macroscopic, 307, 312–318, 515
  microscopic, 307, 312, 313, 318, 515
  table of neutron, 311
Curium, neutrons, 503, 504

D

Data throughput/resolution, 66, 70, 122–123, 350
Deadtime correction, neutron
  coincidence counter, 413, 427, 442, 452, 485, 544
  empirical correction, 412, 466
  shift register, 408, 412–418, 422, 452, 453, 465, 467
  updating and nonupdating, 395–396, 414, 415
Deadtime/pileup corrections, gamma
  pulser based, 124–126
  pulser-peak precision, 125

reference-source, 126–128
Delayed gamma rays, 20, 186, 294, 304
Delayed neutrons
  detectability limit, 501
  energy spectrum, 291, 292, 368
Densitometer, K-egde
  Allied General Nut. Services, 250, 251, 259, 262
  Japan, 250
  Karlsruhe, 250, 251
  Los Alamos, 250, 258
  Oak Ridge Y-12, 259
  performance, 250
  portable K-edge, 261, 262
  Savannah River plant, 250, 261
Densitometer, edge, LIII
  Los Alamos, 250, 251, 264–266
  New Brunswick Lab, 250, 251, 264, 265
  performance, 262, 265
  Savannah River Lab, 251, 264, 265
Densitometry
  absorption-edge, 241–251
  characteristic concentration, 239, 243, 244, 250
  matrix effects, 246–248
  measurement precision, 239–241, 243–244, 250
  measurement sensitivity, 244–246
  sample cell thickness, 244–246, 262, 263
  single energy, 237, 249
  two energy, 240, 241
  x-ray generator, 248, 250, 257, 264, 266
  XRF comparison, 237, 251–252
Detectability limit, 460, 501, 502, 524, 525, 551
Detector design, neutron
  collimation, 221, 276
  ^{3}He tube arrangement, 355, 381–382
  moderator thickness, 316, 381–383
Detector efficiency, gamma-ray
  full-energy peak, 54, 55, 134–135, 178
  geometric efficiency, 55
  intrinsic, 55, 56
  relative, 50, 53
Detector, fast n, ^{4}He and CH_4, 339–340
Detector, gamma-ray
  gas-filled, 43–44
  scintillation, 43–45, 49, 51, 53, 54
  selection, 43, 47
  solid state, 43–49, 53, 54, 61
Detector, ^{3}He and BF_3, 333–338, 359
Detector, neutron
  ^{10}B lined, 329, 333, 338–339
  die-away time, 392, 394, 396, 410, 411, 441, 447, 450, 471, 475, 479, 480, 522, 534, 544, 545, 551, 560, 562, 566, 567, 587
  efficiency table, 332
  fission chamber, 325, 329, 333, 338, 341–342, 446, 447, 503, 574, 617–618, 625
  gamma-ray sensitivity, 329, 331–333, 337–339, 342, 352, 354, 502
  gas-filled thermal-n, 333
  gas mixture, 330, 331, 335
  loaded scintillator, 352
  neutron interaction probability, 332, 333, 336, 342–344
  operating voltage, 329, 335, 339
  scintillators, 325, 332, 333, 339, 342–344, 346, 347, 349, 351–355
Detector resolution, gamma
  Fano factor, 54
  full width half maximum (FWHM), 53, 60, 106
  measurement, 61, 104, 106–108, 200, 201, 248
  theoretical, 54, 107, 138

Die-away time
  measurement, 394, 397, 451–453, 468, 475, 478, 480, 484, 515, 521, 524, 530, 566, 572, 573, 587
Differential die-away counter, 587, 592

E

Effective Z, 86
Elastic scattering, neutron
  energy loss, 376
  ^{1}H and ^{4}He cross section, 339
Electron
  binding energy, 10, 14, 31–33, 36, 272
  capture reaction, 23
Electron volt (eV), 7
Energy calibration
  internal, 102
  linear, 102, 104, 105
Energy spectrum
  (α, n) reaction, 294–300, 658
  ^{252}Cf prompt gamma rays, 293, 302
  ^{252}Cf prompt neutrons, 291, 292, 302
  delayed fission neutrons, 293, 294
  neutron measurement, 291, 292, 383
  spontaneous-fission n, 291, 292, 294, 300
Enrichment meter, 108, 152, 200–202, 282, 533, 693

F

Far-field assay, 88
Fertile isotopes, definition
  fission cross sections, 291, 312
Feynman variance technique, 397–408
Filters
  gamma ray, 39–40
  Pu isotopic assay, 195
Fission chamber
  pulse-height spectrum, 334, 341
  spent fuel measurement, 562, 617–619, 622, 624, 626
Fission product
  activity ratio, 612
  gamma rays, 17–21, 38, 40, 254, 608–612, 633, 726
  mass distribution, 608
  solution assay, 259
  yields, 608–609
Fission reaction
  cross sections, 186, 291, 310–313, 316, 317, 341, 430, 431, 460, 543, 729
  fragments, 20, 290–294, 310, 325, 341
  induced, 359, 389, 390, 532, 536
  spontaneous, 187, 289, 359, 364, 365, 368, 369, 609
Fork detector, 576, 610, 619–620, 622

G

Gamma rays
  delayed, 20, 186, 293, 304
  fission product, 19–21, 38, 40, 254, 608, 609, 611–612, 633, 726
  (γ,n) reactions, 300–301
  heat production, 660
  prompt, 20, 24, 293, 304
  reaction cross section, 333, 356
  from (α,n) reactions, 294–300, 361, 727
  shielding, 27, 40, 81, 302, 320, 331, 571, 589, 618
  signatures, 17, 23, 37, 610, 719, 728
  spent fuel measurement, 331, 337, 610–617
Gamma-ray spectrum
  Compton edge, 34–36, 52, 189, 344, 646, 728
  escape peaks, 18, 31, 37–38, 191, 221
  full-energy interact rate, 79, 123, 134, 142, 143, 155
  full-energy peak, 31, 34, 36, 37, 51–56, 59, 61, 65, 68, 72–74, 108, 115, 220
  full width half maximum, 44, 47, 60, 106, 107, 195
  plutonium, 39, 47, 53–55, 169–182
  single-channel analyzer, 107, 649
  spent fuel, 610–612
  thorium, 17, 22
  uranium, 17, 19, 22, 150, 151, 154, 158–160, 162, 164
  uranium ore, 22, 159, 728
Gamma sensitivity
  neutron capture cross section, 611
  plateau curve, 335
  pulse-height spectrum, 65, 71, 107, 108, 252
Gas proportional counter
  BF$_3$, 333–338, 359
  ^{3}He, 317, 329, 333–338, 359, 378, 381, 442, 500, 525, 588, 640
  He and CH$_4$, 339–340
Gaussian function, 105–108, 113, 115–117
Ge detector
  geometry, 91
  hyperpure, 46
  Li-drifted, 46
  resolution, 46, 106, 195, 196
Geiger-Mueller (GM) detector, 44

H

Half life
  alpha decay, 10
  definition, 9
  spontaneous fission, 291, 302
  total, 290, 291, 294, 295, 302
Heat measurement, 657, 673, 674, 686
Heat production, 657–660, 662, 672
^{3}He neutron detector
  gamma sensitivity, 336, 337
  neutron capture cross section, 352
  plateau curve, 335
  pulse-height spectrum, 334, 335, 337
High level neutron counter (HLNC)
  detection efficiency, 565, 566
  efficiency profile, 550
  HLNCC-II, 515–518
High-voltage bias supply, 61–62
Holdup
  causes and mechanisms, 224–233
  magnitude, 232
  statistical modeling, 225, 233
Holdup measurement
  attenuation correction, 229–232
  calibration, 227–229
  generalized geometry, 225–228, 232, 233
  glove box assay system, 445, 528–529
  radiation signatures, 529
  slab neutron detector, 499–500
  SNAP-II, 495
  typical accuracy, 186, 233

I

Induced fission multiplicity, 422, 425, 428, 441, 447, 456, 460
Inelastic-scattering, 23, 252, 309, 310, 321, 376

Internal conversion, 10–12, 14–16, 129, 160
International target values (ITVs), 251, 286, 457, 484, 566, 707, 713
Interval distribution, 346, 392–395, 445
Intrinsic efficiency, 54–56, 86, 132, 135, 139, 194, 204, 254, 339, 341, 356, 379, 497, 647
Inventory sample counter (INVS), 487, 554
Inverse-square law
    sample rotation, 130–131
Ionization chambers, 43, 44, 328, 620, 622
Irradiated fuel
    active assay, 293
    burnup, 611, 686
    burnup codes, 570
    calorimeter, 613, 614
    Cerenkov, 633–634
    ^{137}Cs, 20, 612
    ^{134}Cs/^{137}Cs, 612
    ^{154}Eu/^{137}Cs, 612
    exposure, 608, 612
    fission chamber, 618
    fission product yields, 607–609
    fork detector, 619–620
    gamma-ray assay, 609, 611
    gamma-ray spectra, 22, 39, 609
    neutron assay, 617–618
    neutron capture reactions, 610
    neutron production, 289, 294
    physical attributes, 186, 617
    TLD measurement, 355
    US fuel assembly inventory, 568, 609, 633, 686, 688, 726, 728
Isotopic composition, 1, 4, 8, 17, 55, 149, 158–160, 169–182, 185, 194–220, 254, 261, 289, 293, 296, 298, 305, 359–364, 366–369, 389, 390, 406, 421, 424, 427, 430, 441, 451, 472, 483, 514, 518–521, 576, 577, 607, 657, 658, 660, 672, 703, 727, 729

L
Leached hull assay, 40
Least-squares fit
    linear, 208
List mode, 359, 395, 397, 418, 445–449, 452, 453, 465, 467, 468, 470, 487, 488, 491–494, 499, 529, 534, 537, 541, 557, 563, 566, 572–573, 598, 600, 601, 622, 623

M
Mean free path
    gamma ray, 23, 29, 152, 280, 719
    neutron, 316, 328
Microcalorimetry, 181, 612–614
Moderating power and ratio, 318
Monte Carlo calculations
    moderator design, 381–383
    photon transport, 91
    sample multiplication, 426
Multichannel analyzers (MCAs), 50, 59, 60, 62, 65, 68, 71–75, 78–80, 101, 102, 107, 118, 155, 156, 195, 232, 250, 254, 261, 265, 284, 490, 512, 617, 649–651
Multiplication
    correction factors, 422, 425, 428, 471, 472
    K_{eff} factor, 319, 371, 374
    leakage, 360, 370–376, 386, 403–408, 423–426, 442, 449, 463, 464, 475, 607
    sample self-, 412, 422–425, 428, 429, 451, 470, 543
Multiplicity
    active, 439, 440, 469–472, 474, 475, 477, 601–602
    counting, 289, 293, 334, 342, 383, 439–442, 445, 447, 453–462, 464–480, 484, 485, 487, 492, 494, 503, 507, 520, 522, 523, 538, 543–559, 573, 601–602
    point model, 447, 449, 454, 455, 458, 462–465, 467, 475
    prompt n, 304, 403
    two-energy point model, 447, 462, 464
    weighted point model, 447, 455, 458, 462–465

N
NaI(Tl) detector
    linear attenuation coefficient, 29, 728
    resolution, 45, 60, 66, 232
Near-field assay
    numerical computation, 91
    ^{239}Pu in solution, 105
Neutron coincidence circuit
    accidental rate, 411, 432
    auto-and cross-correlation, 396–397
    die-away time, 390, 392, 394, 411, 420, 425
    gate length, 396, 408, 411
    nonupdating/updating deadtime, 395–396
    reduced-variance logic, 480
    shift register, 408–418
    updating one-shot, 395, 412
    variable deadtime counter, 394, 396
Neutron coincidence counters
    active well, 356, 377, 419, 432, 453, 469, 470, 474, 507, 547, 586, 588–594
    Euratom Fast Collar, 536, 586, 598–599
    fast neutron coincidence collar, 586, 587, 599–600
    fast neutron passive collar, 434, 536
    55-gal dmm, 502
    fuel-pin tray, 560
    high level, 336, 427, 428, 515–518
    high level neutron counter boron, 563–567, 601
    inventory sample, 487, 554
    passive neutron coincidence collar, 486, 519, 536, 539–541
    solution, 601
    uranium collar, 432–434, 518, 596, 597
Neutron cross section
    ^{10}B, 325, 334, 339, 560
    cadmium, 310
    common materials, 315, 729
    energy dependence, 310–312
    fission, 291, 302, 313, 316, 317, 341, 430, 431, 460, 543
    ^{3}He and ^{6}Li, 352
    ^{239}Pu, 310, 313
    ^{235}U, 313, 341, 431
Neutron energy-velocity relation, 308
Neutron multiplicity, 292, 293, 302, 304, 347, 374, 390, 403, 411, 439–480, 487, 503, 542–544, 546, 547, 550, 551, 553, 554, 557, 585, 586, 601
Neutron multiplicity counter
    bias correction, 455–456, 470
    calibration, 442, 454–456
    characterization, 451–454
    epithermal neutron multiplicity counter, 445, 452, 460, 465, 551–554
    five ring neutron multiplicity counter, 459, 546–547
    high efficiency neutron counter, 446, 457, 487, 492, 521–522, 543, 549, 551
    in-plant multiplicity counter, 460
    multiplicity counter-15, 557–559
    performance, 456–462
    plutonium scrap multiplicity counter, 460

superHENC multiplicity counter, 546, 554
thirty gallon multiplicity counter, 550–551
three ring neutron multiplicity counter, 547–548
Neutron production rate
  PuO_2 plus fluorine, 24, 303, 367, 369
  PuO_2 plus moisture, 366, 369
  spent fuel, 159, 331, 610
  uranium and plutonium, 360, 366
  ^{234}U thin target, 369
Neutron pulse train, 390–395, 418, 442, 445, 466
Neutron reactions
  absorption, 19, 310, 360, 429, 471, 485, 502, 589, 612
  (α, n) yield, 294, 295, 297, 362, 440, 726
  delayed neutrons from fission, 20, 219, 290, 293, 376, 587
  energy leakage spectrum, 319, 374–379
  energy losses, 376
  inelastic scattering, 23
  mean free path, 316
  notation, 309
  prompt neutrons from fission, 291, 293, 376, 397
  reaction rate, 298, 316
  scattering, 23, 301, 383, 429, 531
  spontaneous fission yield, 291, 363
Neutron shielding, 320–321
Neutron singles counters
  box counter, 516
  ^{252}Cf hydrogen analyzer, 302
  high-dose neutron detector, 486, 503, 521–522, 601
  long counter, 246, 529
  ^{238}Pu heat source counter, 215
  shuffler, 219, 587
  slab detector, 499, 508
  SNAP assay probe, 221–222, 495–499
Neutron sources
  AmBe and AmLi, 412, 415, 586, 588, 601
  (α, n), 428, 567
  energy and dose, 195, 302, 321
  spontaneous fission, 321, 356
(n,2n) and (n,n′) reactions, 300
Nuclide identification
  manual methods, 190–192
  spectral analysis, 193–194, 220
  template methods, 192–193

P

Pair production, 28–30, 36–38, 54, 633
Passive neutron collar, 518
Peak area determination
  complex fit, tailing functions, 108, 115
  multiples, known shape, 113–114
  peak fitting, 111, 115–117, 156
  region of interest sums, 111–112
  simple Gaussian fit, 112–113
Peak position determination
  five channel method, 108
  graphical, 108, 221
Peak width determination
  analytical interpolation, 108
  graphical, 221
  second-moment method, 108
Perimeter monitor
  alarm threshold, 641, 644, 645, 649, 650, 698
  automatic vehicle monitor, 653
  calibration, 651
  contamination, 639–641, 646, 647, 651–653
  diagnostic tests, 650
  electronics, 649, 651
  hand-held, 652–653
  long-term monitoring, 646
  moving-average method, 645
  nut-material diversion, 642
  pedestrian, 640–644, 646, 648, 653
  performance, 640, 651
  portal, 640, 642, 643, 645, 647, 649, 650, 652
  sequential hypothesis test, 645
  statistical alarm test, 651
  stepwise method, 645
Photoelectric effect, 30–32, 45, 237, 330
Photomultiplier tubes (PMTs), 44, 45, 61, 62, 74, 343, 348–350, 353, 354, 370, 568–570, 574, 599, 615
Pileup rejection, 61, 69–70, 73, 118, 119, 122, 123
Plutonium
  gamma-ray spectrum, 17–23, 53, 61, 102, 171–173, 220, 728
  neutron production, 297, 359–378, 484, 485, 503, 729
  production reaction, 23
  specific power, 171, 181, 182, 194, 197, 657, 660–661, 671, 672, 678, 681
Plutonium isotopic assay
  high americium content, 39, 128, 159, 170, 210, 211, 214, 220, 248, 262, 302, 637, 658, 728
  Lawrence Livermore Lab, 204
  Los Alamos, 179, 367
  mass ratio, 171
  response function analysis, 197, 204, 205
  Rockwell Hanford, 487, 546, 547
  Tokai-Mura, Japan, 260, 528
Poisson statistics, 119, 125, 475, 477
Pole-zero compensation, 67–68, 327
Preamplifier, 45, 46, 54, 59, 61–68, 70, 72–79, 118, 122–124, 127, 266, 327, 328, 412, 416–418, 442, 445, 446, 452, 454, 465, 467, 487–489, 508, 512, 517, 546, 571, 572, 625
Prompt γ and n spectrum and multiplicity, 291–294, 302, 728
^{242}Pu correlation, 179–180, 207, 210, 215
Pu decay characteristics
  neutron counter, 169
  ^{238}Pu heat source, 215, 669
  standards, 451
^{239}Pu effective mass, 430
^{240}Pu effective mass
  neutron coincidence, 213, 694
  neutron totals, 303, 362, 369
^{242}Pu gamma rays, 170, 172–175, 204, 207, 210, 215–217
Pu, isotopic assay spectral regions, 169, 171–178
Pulse-shape discrimination (PSD), 45, 332, 342, 346–348, 351–354, 356, 370, 574, 599, 600

Q

Q-value, 10, 294–296, 298, 328, 728

R

Radiation damage, Ge detector, 47
Radiation dose, Appendix B
  neutron sources, 321
  shielding calculations, 321, 322
Radioactivity in soil, 355
Random driver, 560

Rate-related loss corrections (gamma-ray)
  ADC deadtime, 117, 122
  data throughput, 122–123
  electronic correction, 115
  Poisson statistics, 119–120, 125
  pulse pileup, 117, 119
  pulser-based, 124–126
  reference-source based, 126–128
Reaction rate, neutron, 316, 441, 442, 510, 544, 577
Reactor fuel characteristics, 605–609
Receipts, Assay Monitor, 508
Region of interest (ROI) selection, 108
Relative efficiency
  curve, 136, 137, 139–145, 161, 162, 178, 207–209, 215, 216, 218, 254
Rossi-alpha distribution, 331

S
Scintillation detectors
  ^{10}B, Gd, and ^{6}Li loaded, 352, 355, 560
  gamma ray, 53, 342–347, 352–354
  light output, 351
  NaI(Tl), 53, 151, 646
  plastic/liquid, 339, 344, 351, 355
  ZnS(Ag), 352
Secular equilibrium, 159, 162, 169, 205, 214, 218
Segmented gamma scanner, 222–224
Shalev spectrometer, 356
Shielding, gamma ray
  neutron, 321, 331, 571, 618
Shift register circuit
  AMPTEK Module, 266, 279, 331, 334, 417, 442, 487, 488, 515, 517, 521, 523, 531, 546, 547, 551, 554, 557
  dead-time correction, 412–418
  multiplication bias correction, 455–456, 461, 470
  multiplicity shift register, 410
Si(Li) detector, 47, 490
Signal-to-noise ratio, 62, 65–67, 70, 119, 203, 517, 544, 592, 649
Slab neutron detector, 382, 499–500, 508, 528
SNAP-II Assay Probe
  cylinder measurements, 495
  holdup measurements, 495
Specific power, 159, 170, 171, 181, 182, 194, 197, 657–658, 671, 672, 678, 681, 705, 721, 722
Spectrum stabilizer, 73–74
Spontaneous fission
  fragment mass distribution, 289
  half lives, 291, 302, 390, 726
  neutron multiplicity and yield, 292, 293
  neutron sources, 302, 303
  neutron spectrum, 339
Sum peaks, 69, 122, 129, 191, 195, 199

T
Thermal neutrons, 45, 186, 292, 293, 308–310, 312, 316, 320, 321, 331–338, 341, 346, 352–354, 378, 381, 391, 396, 422, 425, 429, 431–433, 442, 445, 447, 474, 485, 503, 504, 507, 514, 515, 536–540, 544, 546, 547, 551, 559–567, 569, 570, 587, 589, 596, 625
Thermoluminescent dosimeter
  holdup measurements, 355
  spent fuel assay, 620
  spent fuel measurements, 355
Thorium, gamma-ray spectrum, 17

U
Uranium
  atom and weight fraction, 365
  compounds, infinite thickness, 152, 157, 162, 202, 693
  gamma-ray spectrum, 155, 160, 161
  natural isotopic abundance, 204–205
  neutron production rate, 363, 364, 366, 729
  ^{234}U origin, 149
  uranium ore, spectrum, 22, 728
Uranium enrichment assay
  enrichment meter equation, 282
  gas-phase monitor (UF$_6$), 164
  infinite thickness, 152–157, 693
  60–240 keV region, 161, 205, 728
  120–1010 keV region, 161–162, 197, 205, 206
  peak ratio, 157, 162
  relative efficiency curve, 161, 162
  UF$_6$ slab neutron detector, 164
  visual estimation, 162–164
  wall correction, 165–167

V
Vehicle monitor, 643, 653

W
Waste, low-level
  detectability limit, 501
  fluorescence yield, 14, 272
  55-gal drum assay, 223, 486, 487, 501, 502, 519
  generator, 5
  line shape, 116, 174
  measurement, 5, 501
  100 nCi/g activity limit, 219
  nomenclature, 14
  production, 501
  segmented gamma scanning, 222
  tomographic gamma scanning, 222
  U and Pu, energy and intensity, 212

X
X-ray fluorescence assay
  attenuation correction, 165, 271, 281, 283–284
  beta-particle-induced, 284
  excitation sources, 165, 274, 276, 279, 284
  measurement geometry, 276
  reprocessing plant solutions, 251, 284
  sensitivity, 254, 284

Z
Z (atomic number), 9–11, 14, 28, 30, 32, 36–40, 84, 272, 278, 290, 295, 337, 646

Printed in the United States
by Baker & Taylor Publisher Services